D0152609

Decimal Equivalents of Common Fractions

4ths	8ths	16ths	32nds	64ths	To 4 Places	To 3 Places	To 2 Places	4ths	8ths	16ths	32nds	64ths	To 4 Places	Places	Places
				1/64	.0156	.016	.02					33/64	.5156	.516	.52
			1/32		.0312	.031	.03				17/32		.5312	.531	.53
				3/64	.0469	.047	.05					35/64	.5469	.547	.55
		1/16			.0625	.062	.06			9/16			.5625	.562	.56
				5/64	.0781	.078	.08					37/64	.5781	.578	.58
			3/32		.0938	.094	.09				19/32		.5938	.594	.59
				7/64	.1094	.109	.11					39/64	.6094	.609	.61
	1/8				.1250	.125	.12		5/8				.6250	.625	.62
				9/64	.1406	.141	.14					41/64	.6406	.641	.64
			5/32		.1562	.156	.16				21/32		.6562	.656	.66
				11/64	.1719	.172	.17					43/64	.6719	.672	.67
		3/16			.1875	.188	.19			11/16			.6875	.688	.69
				13/64	.2031	.203	.20					45/64	.7031	.703	.70
			7/32		.2188	.219	.22				23/32		.7188	.719	.72
				15/64	.2344	.234	.23					47/64	.7344	.734	.73
1/4					.2500	.250	.25	3/4					.7500	.750	.75
				17/64	.2656	.266	.27					49/64	.7656	.766	.77
			9/32		.2812	.281	.28				25/32		.7812	.781	.78
				19/64	.2969	.297	.30					51/64	.7969	.797	.80
		5/16			.3125	.312	.31			13/16			.8125	.812	.81
				21/64	.3281	.328	.33					53/64	.8281	.828	.83
			11/32		.3438	.344	.34				27/32		.8438	.844	.84
				23/64	.3594	.359	.36					55/64	.8594	.859	.86
	3/8				.3750	.375	.38		7/8				.8750	.875	.88
				25/64	.3906	.391	.39					57/64	.8906	.891	.89
			13/32		.4062	.406	.41				29/32		.9062	.906	.91
				27/64	.4219	.422	.42					59/64	.9219	.922	.92
		7/16			.4375	.438	.44			15/16			.9375	.938	.94
				29/64	.4531	.453	.45					61/64	.9531	.953	.95
			15/32		.4688	.469	.47				31/32		.9688	.969	.97
				31/64	.4844	.484	.48					63/64	.9844	.984	.98
					.5000	.500	.50						1.0000	1.000	1.00

Online Services

Delmar Online
To access a wide variety of Delmar products and services on the World Wide Web,
point your browser to:

http://www.delmar.com
or email: info@delmar.com

thomson.com
To access International Thomson Publishing's
home site for information on more than 34 publishers
and 20,000 products, point your browser to:

http://www.thomson.com
or email: findit@kiosk.thomson.com

A service of I(T)P®

Construction Methods, Materials, and Techniques

Construction Methods, Materials, and Techniques

William P. Spence

Delmar Publishers

An International Thomson Publishing Company ITP®

Albany • Bonn • Boston • Cincinnati • Detroit • London • Madrid
Melbourne • Mexico City • New York • Pacific Grove • Paris • San Francisco
Singapore • Tokyo • Toronto • Washington

Notice to the Reader

Publisher does not warrant or guarantee any of the products described herein or perform any independent analysis in connection with any of the product information contained herein. Publisher does not assume, and expressly disclaims, any obligation to obtain and include information other than that provided to it by the manufacturer.

The reader is expressly warned to consider and adopt all safety precautions that might be indicated by the activities herein and to avoid all potential hazards. By following the instructions contained herein, the reader willingly assumes all risks in connection with such instructions.

The publisher makes no representation or warranties of any kind, including but not limited to, the warranties of fitness for particular purpose or merchantability, nor are any such representations implied with respect to the material set forth herein, and the publisher takes no responsibility with respect to such material. The publisher shall not be liable for any special, consequential, or exemplary damages resulting, in whole or part, from the readers' use of, or reliance upon, this material.

Cover photo copyright © 1998 PhotoDisc, Inc.
Cover Design: Lachina Publishing Services

Delmar Staff
Publisher: Alar Elken
Acquisitions Editor: John Anderson
Production Coordinator: Larry Main
Art and Design Coordinator: Nicole Reamer
Editorial Assistant: John Fisher

COPYRIGHT © 1998
By Delmar Publishers
a division of International Thomson Publishing Inc.

The ITP log is a trademark under license.

Printed in the United States of America

For more information, contact:

Delmar Publishers
3 Columbia Circle, Box 15015
Albany, New York 12212-5015

International Thomson Publishing Europe
Berkshire House 168-173
High Holborn
London, WC1V 7AA
England

Thomas Nelson Australia
102 Dodds Street
South Melbourne, 3205
Victoria, Australia

Nelson Canada
1120 Birchmont Road
Scarborough, Ontario
Canada, M1K 5G4

International Thomson Editores
Campos Eliscos 385, Piso 7
Col Polanco
11560 Mexico D F Mexico

International Thomson Publishing GmbH
Konigswinterer Strasse 418
53227 Bonn
Germany

International Thomson Publishing Asia
221 Henderson Road
#05-10 Henderson Building
Singapore 0315

International Thomson Publishing—Japan
Hirakawacho Kyowa Building, 3F
2-2-1 Hirakawacho
Chiyoda-ku, Tokyo 102
Japan

All rights reserved. No part of this work covered by the copyright hereon may be reproduced or used in any form or by any means—graphic, electronic, or mechanical, including photocopying, recording, taping, or information storage and retrieval systems—without the written permission of the publisher.

4 5 6 7 8 9 10 XXX 03 02 01 00

Library of Congress Cataloging-in-Publication Data

Spence, William Perkins, 1925-
    Construction methods, materials, and techniques / William P.
Spence.
        p.   cm.
    Includes index.
    ISBN 0-314-20537-3
    1. Building.   2. Building materials.   I. Title.
TH145.S827   1998   87-20499
624--dc21   CIP

Contents

v

Preface

This book provides a detailed look at the major construction methods and building systems and the vast array of materials and products provided by the manufacturers supplying the construction industry.

The first part of the book gives a brief overview of aspects of the construction industry and of the content of this book, as well as information about some of the industry's major professional and technical organizations. The use of metrics in construction is explained, supported by additional details in Appendix C.

Early in the book is a discussion of the properties of materials. This discussion relates to all of the materials in the remainder of the book. The architect, engineer, and material manufacturer all know the importance of this information, and readers cannot understand technical bulletins, reports, and manufacturers' technical publications without a firm grasp of the properties presented.

The remainder of the book is organized following the sixteen divisions of the MasterFormat developed by the Construction Specifications Institute and Construction Specifications Canada. Division 1 addresses general requirements.

Division 2 pertains to the building site, analysis of various types of soils, and commonly used foundation systems. The study of soils is critical to the ultimate design of the foundation.

Division 3 presents a detailed study of the manufacture, types, characteristics, and properties of concrete. Consideration of the use of admixtures, proportions, water, mixing, and placing are included. Drawings and photos are used extensively to illustrate cast-in-place and precast concrete structural systems.

Division 4 includes detailed information on mortar, the key to satisfactory masonry construction. The materials and techniques involved with clay brick and tile, concrete masonry, and stone construction are explained in detail and generously illustrated.

Ferrous and nonferrous metals are presented in Division 5. Their characteristics, mechanical properties, and practical applications are discussed. Steel frame construction systems are illustrated.

One of the largest divisions in the book is Division 6, which covers the vast array of wood and plastic materials. Their properties, characteristics, and recommended applications are explained. Several chapters detail wood structural framing systems and the methods and materials of light wood-frame construction. The remainder of this division is used to present information on paper and pulp products and plastic materials useful in building construction.

Insulating, waterproofing, and sealing buildings against the weather are covered in Division 7. Walls, ceilings, and floors need to be properly insulated and sealed against moisture penetration. Various bonding agents, sealers, and sealants available are covered. Bituminous materials are another form of waterproofing that also serve as surfaces for parking areas and roads. The roofing systems for residential and commercial building conclude this division.

The types, styles, methods of operation, and materials used for doors and windows is extensive. Many of the products available are illustrated in Division 8, as well as factory stock and custom-made storefronts. Many of these use some form of glass, so the types, properties, and uses for the various glass products available are discussed. Finally, the exterior of the commercial building is covered with some form of cladding system. An entire chapter is devoted to discussing and illustrating cladding systems.

Finishing the interior of the building involves the most diverse range of products in any area of construction. Division 9 includes interior finishes; decorative and protective coatings; gypsum, lime, and plaster materials and products; acoustical finishes and materials; and all types of finish flooring. The construction and finish of interior walls, partitions, and ceilings in residential and commercial buildings is included.

Division 10 covers some of the specialty products, such as visual displays, screens, grills, service walls, and identifying signs. Other products included as specialty items are shelving, partitions, fire protection devices, telephone enclosures, and toilet and bath accessories.

Division 11 includes items specified as equipment. Examples include ecclesiastical equipment, library equipment, and vending equipment.

Furnishings form the final touch to a completed building, and Division 12 classifies these into seven major groups. Artwork, window treatments, and rugs and mats offer a variety of types and styles. Casework and furniture for use in commercial buildings form the largest group.

A most interesting assembly of special construction features is found in Division 13. This division discusses and illustrates a diverse offering, including air-supported structures, prefabricated and pre-engineered assemblies, sound and seismic control devices, nuclear reactors, and radiation shielding components.

Division 14 discusses and illustrates the variety of conveying systems available, such as conveyors, elevators, escalators, moving walks, and material-handling systems.

Division 15, one of the most complex divisions, covers the mechanical systems used in residential, commercial, and industrial buildings. Extensive information is presented on fire protection systems, plumbing systems, and heating, refrigeration, air-conditioning, and ventilation. The range of systems and products is great, and improvements occur regularly.

The book concludes with Division 16, Electrical Systems. It carries the discussion from the generation and transmission of electrical power to the service entrances in residential and commercial buildings. Internal power distribution systems are illustrated, and extensive information on lighting is available. Equipment for controlling and operating the electrical system, as well as equipment used for communication, such as alarm, television, public address, and other communication systems, is presented.

At the end of each chapter are materials providing the reader with a means of reviewing what has been read and reinforcing the learning experience. Definitions of key terms are listed. Sources of additional information are provided to enable the reader to explore areas in greater depth.

A master glossary at the end of the book provides an additional means of locating definitions. Appendix B contains a listing of organizations that contribute much to the improvement and quality of construction. Many have an extensive offering of technical manuals and the reader is encouraged to contact them for a catalog. Detailed metric information is given in Appendix C. This is essential, because construction is moving into a metric design metric material future.

DISCLAIMER

The information presented in this book was secured from a wide range of manufacturers, professional and trade associations, government agencies, and architectural and engineering consultants. In some cases generalized or generic examples are used. Every effort was made to provide accurate presentations. However, the author and publisher assume no liability for the accuracy of applications shown. It is essential that appropriate architectural and engineering staff be consulted and specific information about products be obtained directly from the manufacturer.

ACKNOWLEDGMENTS

A major factor in the organization, writing, and illustrating of this book was the help given by hundreds of manufacturer representatives. Representatives of many of the professional and technical organizations supporting the construction industry also made important contributions. A special note of appreciation is due the consultants located at universities across the country for their assistance in reviewing the manuscript and the illustrations.

Finally, I dedicate this book to my wife, Bettye Margaret Spence, for her steady encouragement, editorial assistance, and the many hours she spent at her computer preparing the manuscript.

INTRODUCTION TO CONSTRUCTION METHODS, MATERIALS, AND TECHNIQUES

Architects, engineers, construction managers, general contractors, specialty contractors, and manufacturers and suppliers of construction materials should have an overview of the various aspects of the construction industry.

Part I provides a general discussion of the industry, the organizations involved, and properties of materials used in construction.

Chapter 1 gives an overview of the various segments of the industry and their relationship to the total picture.

Chapter 2 focuses on the people, organizations, and specifications, particularly those provided in the Construction Specifications Institute MasterFormat.

Chapter 3 presents information on the properties of materials. ▲

Building Structures and Systems: An Overview

This chapter will help you to:

1. Develop an understanding of the scope of activities involved in building construction.

2. Get an overview of the construction methods, materials, and techniques included in this book.

3. Develop an appreciation for the range of engineering, technical, and architectural expertise involved in construction projects.

A building is the relationship of its many parts. It is the result of the complex interdependent aspects of meeting a predetermined need, the design process, the application of current technology to materials and construction methods, and the actual construction processes. A building is the result of the technology that both restricts and permits the expansion of the design possibilities. Through the years the technology of building construction has changed rapidly, and this continues with constant innovations in materials and methods.

The following discussion gives a brief introduction to the various technical factors involved in designing and constructing a building. In addition, this chapter gives an overview of the sixteen divisions in this book that cover the methods, materials, and techniques of construction.

SPACE

The design of a building involves the utilization of space and all the factors that go with it. This includes cost, physical requirements (such as a series of desks or room for mechanical equipment), psychological requirments that reflect on the attitudes and behavior of those using the space, and the need for those in the building to carry out the activities expected to occur in the building. In addition factors involving the use of materials, energy efficiency and economic utilization of space are considered (Fig. 1.1).

As space considerations are studied and the internal design of individual and group spaces is determined, the relationship between spaces and activities and reasonable distances of travel becomes important. Traffic flow within the building and to and from outdoors come into focus (Fig. 1.2). Obviously, considerations of safety and comfort come into play and the accessibility for the handicapped must be considered. Services such as stairs, elevators, communications between spaces, restroom facilities, security, and special areas such as employee lounges and lunchroom become part of the task. Color, texture, and decoration within the spaces affect the psychological reactions of the occupants. Over all of this are the specifications imposed by building codes, national standards, and guidelines recommended for satisfactory space planning.

Figure 1.1 The design of a building includes a study of the proper utilization of space.

Figure 1.2 Decisions pertaining to traffic flow are an important part of the design of a building.

THE BUILDING STRUCTURE

As the space requirements are being determined, the structure comes into consideration. Decisions about the type of structural frame are based on many factors, such as cost, availability, load-carrying capability, appearance, ease of erection, usable life, and spans. Structural engineers have great flexibility in designing the structural frame to enclose the proposed spacial needs (Fig. 1.3).

A building and its component parts are subjected to various loads. The following discussion gives a very general explanation of the factors affecting structure.

Foundation and Walls

The material used to construct the walls and foundation must resist compressive and horizontal forces. Compressive forces are due to the weight of floors, building contents, and roof. Horizontal forces are due to wind, subsurface water, and soil (Fig. 1.4). The walls and foundation must rest on a base that will resist the vertical forces imposed on it by the wall. This is accomplished by some form of footing. For very large buildings and for heavy loads pilings are used (Fig. 1.5). The bearing surface of the footing or piling resting on the soil

Figure 1.3 Engineering the structural frame to permit the building to accommodate the desired activities is the work of the structural engineer.

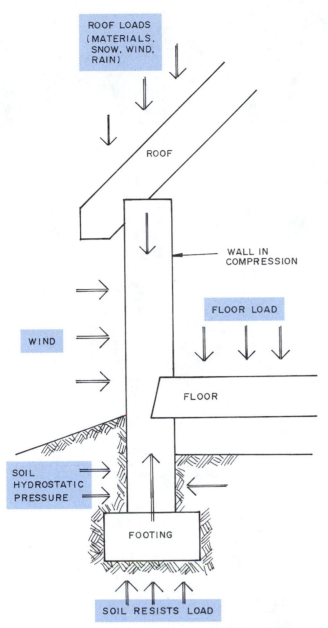

Figure 1.4 The building structure is subject to various loads and forces.

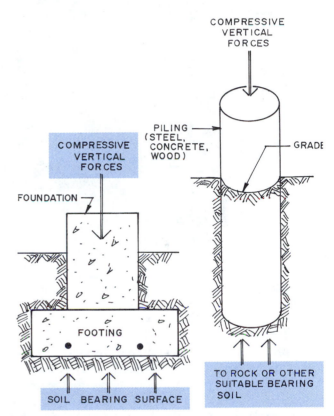

Figure 1.5 The building is supported by some type of footing or piling.

must be large enough to resist the downward forces from the vertical load. A properly designed footing is essential to stability. Failure of walls due to compression is unusual because the materials used generally equal or exceed these forces. When they do fail, it is most often because the wall is not stable and overturns or buckles (Fig. 1.6). If a load is centered on a properly designed wall, the wall remains stable. If the forces generated by a load become off-center, the wall could begin to tilt and possibly buckle. For example, if the footing begins to settle on one side, the load becomes off-center and the wall tends to buckle. Under these conditions one side of

the wall has increased tension forces tending to pull it apart while the other has increased compressive forces that may cause it to crush the material. This is typical of what may happen when a high wind strikes a wall, forcing it to tilt.

The stability of walls and the stability of the building are increased by the interconnection of walls. The rectangle formed by connecting walls in Fig. 1.7 is much more stable than a single wall standing alone. The stability of walls can be increased by adding pilasters or by stepping the wall as shown in Fig. 1.8. You can find detailed information about foundations in Chapter 6.

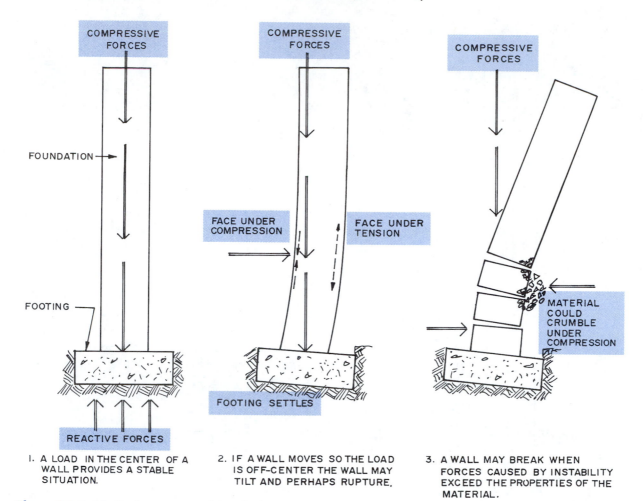

Figure 1.6 Stable footings are essential to the integrity of the foundation.

Columns

A column is a slender vertical support used to carry the weight of materials placed above it. It is subject to the vertical compressive forces and horizontal forces as discussed for walls (Fig. 1.9). Sometimes a series of columns replaces a wall, thus opening up a large area free of dividing walls. Columns are topped with a horizontal structural member upon which the area above the column is built. The spacing between columns is regulated by their capacity to carry the imposed load and the capacity of the horizontal member to carry required loads between columns (Fig. 1.10). Columns can be masonry, steel, wood, or concrete (Fig. 1.11). Study the material in Chapters 8, 9, 17, and 21.

Arches

An arch is a curved structural construction that spans an opening between two points. Block arches (brick and stone) carry only the in-plane forces that cause the bricks or stones to compress tightly together, forming a strong unit. They are usually made from wedge-shaped blocks called voussoirs. Rigid arches are made from a continuous curved, rigid material such as concrete, steel, or laminated wood (Fig. 1.12).

Rigid Frames

A rigid frame is a structural framework in which the wall columns and roof beams are rigidly connected. They do not use hinged joints. Rigid frames are commonly used in metal building framing (Fig. 1.13). Sometimes rigid frames are referred to as arches because they react in a similar manner. Additional information can be found in Chapters 17 and 21.

Beams

A beam is a structural member used to carry transverse loads, such as those imposed by a joist, girder, rafter, or purlin. They typically are supported by the foundation, piers, or columns (Fig. 1.14). Beams can be fabricated from concrete, wood, or steel. They are subject to the forces discussed in Chapter 3. These include bending stress and horizontal and vertical shear. The stiffness (modulus of elasticity) is also a critical factor.

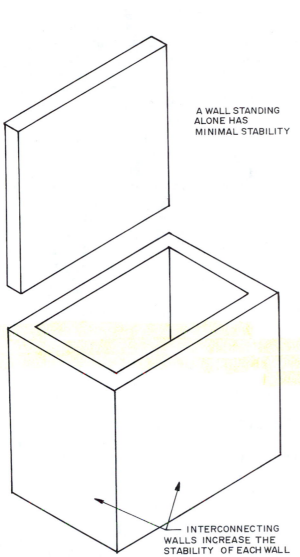

Figure 1.7 The interconnection of walls increases their stability.

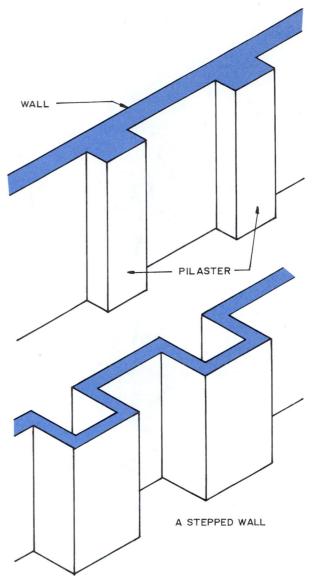

Figure 1.8 Using pilasters or stepping the wall increases the wall's stability.

Trusses

A truss is a structural unit composed of assembled members that utilize a triangular arrangement to produce a rigid framework. They are made from wood and steel and commonly are used to form the structure for floors and roofs. They span long distances, reducing the need for intermediate support (Fig. 1.15).

Structural Frames

Various chapters of this book discuss the commonly used structural frames in detail. The following discussion gives a thumbnail sketch of these.

The most common structural frame used for residential work in the United States is wood frame. Although it can be assembled in various ways using solid wood and remanufactured wood products, the illustration in Fig. 1.16 is typical of the on-site platform framing in wide use. There is an increasing use of factory-assembled housing units that are transported to the job in large assemblies. The wood frame is typically enclosed with brick, wood, hardboard, plywood, vinyl, or aluminum siding (Fig. 1.17).

Metal light frame construction is finding increasing use in residential construction. It is widely used in commercial construction. The increasing difficulty to secure quality wood framing in large quantities will possibly increase the acceptance of metal light frame construction (Fig. 1.18).

Heavy timber frames (often called post-and-beam) use solid wood or laminated wood columns and beams. The roof is framed with widely spaced beams or some

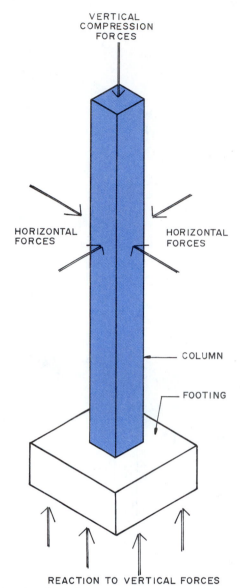

Figure 1.9 A column is subjected to vertical and horizontal forces.

buildings. The exterior wall encloses the structure and supports the floor and the roof (Fig. 1.22).

Various pre-engineered manufactured steel frames are also used for small commercial buildings. One type uses a rigid frame (Fig. 1.13) that provides support for the exterior wall and the roof. Other types use steel columns and various types of trusses for joists or beams. Multistory steel framed buildings use steel girders for framing the floors and roof, which are supported by steel columns (Fig. 1.23).

Cast-in-place concrete frames can take various forms, one of which is shown in Fig. 1.24. The structural frame is formed, reinforcing steel is placed in the form, and the concrete is poured. After the concrete hardens, the forms are stripped. Sometimes the floor and roof deck are cast at the same time as the beams and joists.

Precast concrete structural members are cast in a plant under carefully controlled conditions. After they have cured, they are shipped to the construction site, where they are lifted into place with a crane. One example of an assembled precast concrete frame is shown in Fig. 1.25. Another widely used precast technique is tilt-up construction. Structural wall panels are cast in a horizontal position on the floor of the building. They are then tilted into position with a crane (Fig. 1.26).

An unusual structure used to enclose large areas is a tension fabric structure. The fabric, which is often a polyester substrate covered with a polyvinyl-chloride outer coating, is supported by steel cables, posts, rigid arches, and beams (Fig. 1.27). It can be designed in many shapes and forms. Study Chapters 8, 9, 14, 17, 20, and 21 to learn about all of these major structural systems.

THE ENCLOSING WALLS

As shown in the previous section on structures, some enclosures also serve structural purposes. Many others are non–load bearing and serve to protect the interior from the weather and provide an attractive exterior. Materials used for enclosures include clay, glass, plastic, concrete, stone, metal, and wood products. In addition, the enclosing walls must provide for access into the building and for the admission of natural light and ventilation.

Clay products include brick and tile (Fig. 1.28). These are strong and highly weather resistant. They are poor when considering energy efficiency.

Concrete products include stucco and various precast concrete wall panels, which may be load or non–load bearing. Stone wall panels are used extensively to finish exterior walls (Fig. 1.29).

Glass products are used in windows and curtain-wall spandrel panels. Glass spandrel panels are made from heat-strengthened glass with a layer of weather-resistant ceramic color fused to the exterior surface.

form of truss. Floors and roof decks are thick tongue-and-groove wood planks. A typical frame is shown in Fig. 1.19.

Large rigid structural arches are curved members supported at the sides or ends. Typical structural arch shapes are shown in Fig. 1.20. These frequently are laminated wood members and span distances exceeding 200 ft. A structural arch supported at each end by flexible anchorage is identified as a two-hinge arch. An arch that has flexible anchorage at both ends and at the top center is identified as a three-hinge arch (Fig. 1.21). A hinge joint is a joint that permits action similar to a hinge in which there is no appreciable separation of the adjacent members.

Solid masonry structures find some use in residential construction and extensive use in small commercial

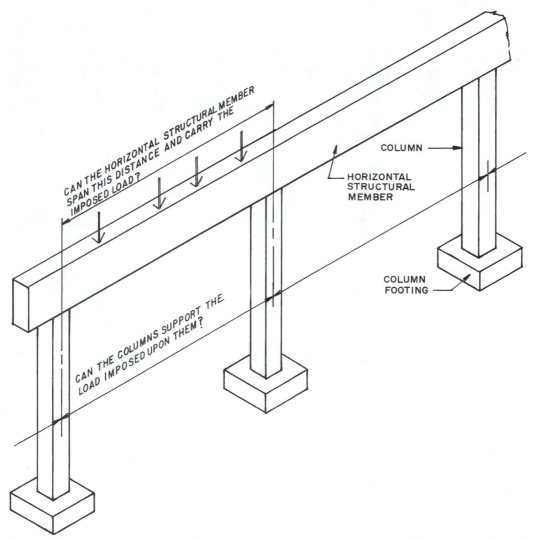

CAN THE HORIZONTAL STRUCTURAL MEMBER SPAN THIS DISTANCE AND CARRY THE IMPOSED LOAD?

COLUMN

HORIZONTAL STRUCTURAL MEMBER

CAN THE COLUMNS SUPPORT THE LOAD IMPOSED UPON THEM?

COLUMN FOOTING

Figure 1.10 The size and spacing of the columns depend on the column specifications and the load to be placed on them.

These strong, weather-resistant panels resist checking or crazing (Fig. 1.30). Other glass exterior wall panels are transparent and extend from the floor to the ceiling (Fig. 1.31). Plastic wall panels are used in the same manner as glass but are much lighter than glass.

Metal curtain-wall panels are available with aluminum alloy, galvanized, or aluminized steel spandrels. They are finished with a variety of coatings (Fig. 1.32). Stone-faced curtain-wall panels are also available, which give the finished stone exterior appearance but are much lighter than solid stone facings.

Wood enclosure products include solid wood siding, plywood, and hardboard. These products are available in a variety of designs and textures. They are finished by staining or painting. Enclosure materials are discussed in detail in Chapters 11, 13, 14, 19, 22, 24, 29, 30, and 31.

THERMAL FACTORS

A variety of things affect the thermal efficiency of a building. One major consideration is the climate. If the building is in a location where the stresses of cold are more important than heat stresses in the summer, the designer will have to adapt the building to this situation. The designer must consider the orientation of the building, the use of glass, prevailing winter winds, and other factors. For example, in cold climates winter winds must be blocked or their effect on the building reduced in some way. In hot climates provision for avoiding summer heat gain is more important.

The materials used also are major factors in thermal efficiency. Materials factors that can reduce heat gain and heat loss include energy-efficient glazing, insulation, sheathing, siding, type and color of roofing, and other factors (Fig. 1.33). Since the exterior walls

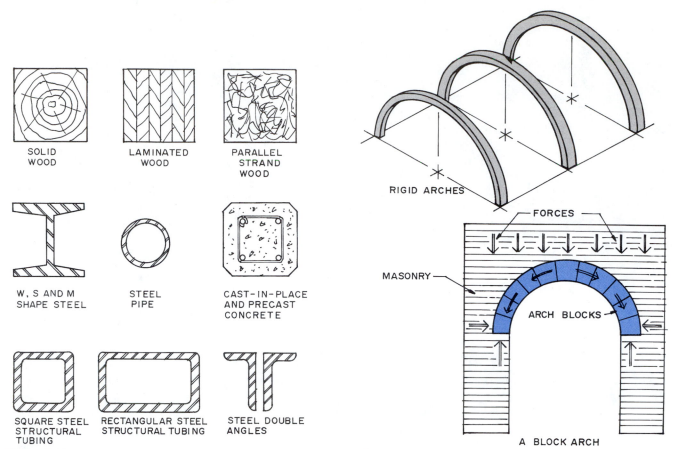

SOLID
WOOD

LAMINATED
WOOD

PARALLEL
STRAND
WOOD

W, S AND M
SHAPE STEEL

STEEL
PIPE

CAST-IN-PLACE
AND PRECAST
CONCRETE

SQUARE STEEL
STRUCTURAL
TUBING

RECTANGULAR STEEL
STRUCTURAL TUBING

STEEL DOUBLE
ANGLES

Figure 1.11 Commonly used types of columns.

RIGID ARCHES

FORCES

MASONRY

ARCH BLOCKS

A BLOCK ARCH

Figure 1.12 Arches may be rigid or block types.

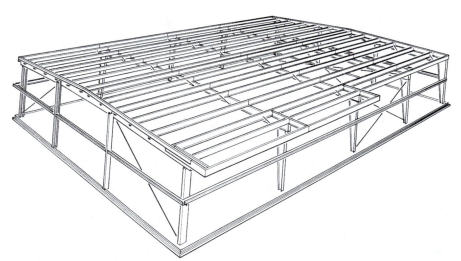

Figure 1.13 This metal frame uses rigid frames to span the building and support the roof load. (*Artwork provided by and reflecting the products, components, and systems of Butler Manufacturing Co., P.O. Box 917, Kansas City, Mo. 64141.*)

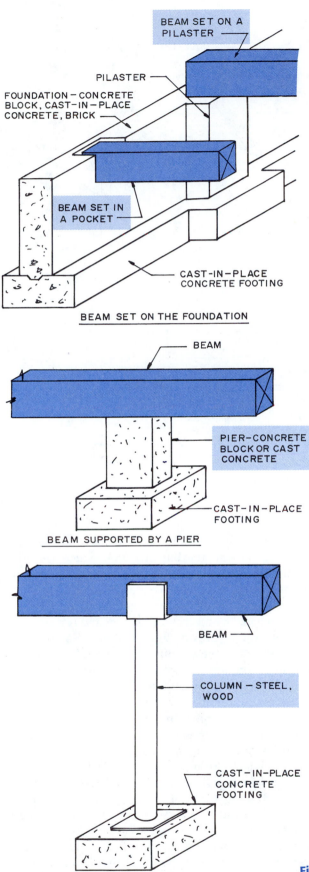

BEAM SET ON A PILASTER

PILASTER

FOUNDATION – CONCRETE BLOCK, CAST–IN–PLACE CONCRETE, BRICK

BEAM SET IN A POCKET

CAST–IN–PLACE CONCRETE FOOTING

BEAM SET ON THE FOUNDATION

BEAM

PIER–CONCRETE BLOCK OR CAST CONCRETE

CAST–IN–PLACE FOOTING

BEAM SUPPORTED BY A PIER

BEAM

COLUMN – STEEL, WOOD

CAST–IN–PLACE CONCRETE FOOTING

BEAM SET ON A COLUMN

are an assembly of material, the control of heat loss and gain can be regulated by their design. Some examples are in Fig. 1.34. Design of the floor and ceiling assemblies is considered in the same manner. Reducing air infiltration is also a part of the basic design. Selecting quality doors and windows greatly reduces air infiltration, as does sealing all leakage points in the exterior wall.

As thermal factors are considered, it is necessary to consider possible problems due to high humidity inside the building. Activities in a building generate considerable moisture in the air. Dehumidifiers on furnaces can help lower the humidity, as can various mechanical ventilation systems. For example, exhaust fans in the kitchen of a residence or the restroom of a commercial building can efficiently exhaust excess moisture to the outside. If a building has tight construction and air infiltration has been reduced to a minimum, special efforts must be made to remove excess humidity and improve the quality of the air. Water vapor is kept from penetrating the exterior wall from the inside of the building by installing a plastic vapor barrier on the inside of the wall behind the finished interior wall material. High humidity is an important consideration when designing roofs. Roofs need a vapor barrier on the inside of the ceiling to keep water vapor off the bottom surface of the roof sheathing. This also requires adequate ventilation between the ceiling and the bottom of the roof (Fig. 1.35). Chapters relating to thermal efficiency include Chapters 25, 29, 30, and 46.

BUILDING SYSTEMS

A building has several systems that enable it to serve its design function. These include heating, air-conditioning, ventilation, water, sewage disposal, electrical, fire protection, sound and signal, and mechanical transportation.

Heating, Air-Conditioning, and Ventilation

Heating, air-conditioning, and ventilation systems maintain a comfortable indoor climate by controlling air temperature and humidity; removing odors, dust, and pollen; and replacing stale air with fresh air from outside the building. Without these systems many buildings could not be used. Actually, most buildings now in use rely almost entirely on these systems to regulate the indoor climate rather than on opening windows to change the temperature or freshen the interior air.

Heating systems may be fueled by coal, oil, gas, electricity, or solar energy. Gas, electric, and solar systems are the most environmentally friendly. Air-conditioning is

Figure 1.14 Beams are supported by the foundation, piers, or columns.

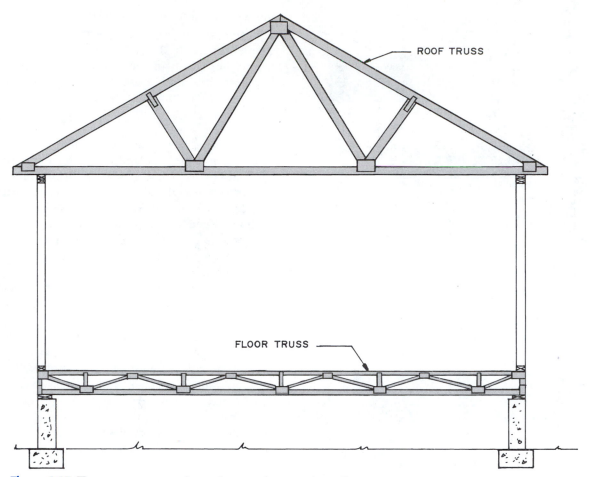

ROOF TRUSS

FLOOR TRUSS

Figure 1.15 Trusses are structural members used to span long distances.

accomplished by compressors or chilled water that use the same ducts or pipes used for heating. Residential heating systems generally are hot air systems that use electric blowers to move heated air through ducts to the spaces within the building (Fig. 1.36). A hot water system uses pipes to move heated water to radiators (Fig. 1.37).

Larger buildings are usually heated by gas- or oil-fired hot water systems. The fuel heats water in a boiler, and the water is pumped to radiators. Fresh air is often brought into the building through grills on the exterior wall located behind the radiator. The fans on the radiator move the heated air about the space and pull a small amount of fresh outside air into the building. In some localities the outside air is not very fresh. A piping diagram for a multistory building with hot water heat is in Fig. 1.38. Large systems like this have a water chiller that circulates cold water through the pipes to the radiators to provide for cooling the air.

Air Handling and Ventilation

Tied in with heating and air-conditioning in large buildings is some form of air-handling and ventilation system.

The principal purpose of a ventilation system is to introduce fresh air into the building and reduce the odors caused by the activities within the structure. The amount of ventilation required depends on the items contributing to odors, such as body odor, clothing, furniture, dust, and odors produced by mechanical devices and restrooms.

Ventilation is provided by blowers that force air through ducts to the conditioned spaces. The air is filtered and conditioned by some type of air-handling unit. The air-handling unit cools, heats, dehumidifies, and filters the air and provides the required ventilation. Air-handling equipment in a large building is placed in an air-handling equipment room. The air-handling unit introduces outside air into the building and exhausts stale interior air. Some systems have a heat exchanger that transfers heat in the air to be exhausted to the cooler incoming air, thus saving the heat. A typical unit is diagrammed in Fig. 1.39. Some ventilation systems are designed to serve a single area, such as a commercial kitchen or a welding area in a factory, and are independent from the system serving other less noxious areas, such as offices. Study Chapter 46 for additional information.

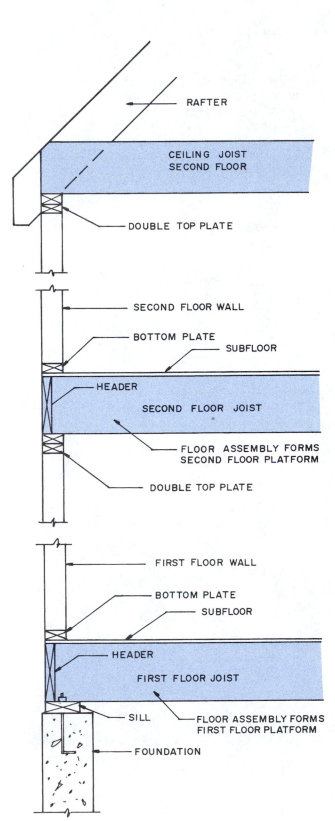

Figure 1.17 A wood siding used in light frame construction. *(Courtesy Georgia-Pacific)*

Figure 1.18 This apartment building is built with lightweight steel framing. *(Courtesy Marino Industries Corp.)*

Figure 1.16 Platform framing is used for light frame construction.

ITEM OF INTEREST

CONSTRUCTION SAFETY—FALLS

Worker injuries due to falls on construction jobs are a major type of accident and lead to lost time, medical expenses, contractor expenses, and increased insurance costs. In the United States, the Occupational Health and Safety Act (OSHA) strictly regulates safety procedures and devices used to protect workers in various situations. Although the regulations change over the years, contractors are responsible for records such as a written "Fall Protection Plan" and accident records. They also must provide fall protection training for each employee and use approved fall prevention systems. The general contractor is responsible not only for the safety of its employees, but also for that of subcontractors on the site. Access to the site by visitors is strictly controlled. Manufacturers of fall protection equipment have developed fall protection systems that meet OSHA and ANSI (American National Standards Institute) requirements and that provide construction workers with the required margin of safety. Contractors and equipment manufacturers must keep in constant contact with the regulatory agencies because requirements change over time and the new requirements are expected to be observed.

Figure A Miller® Equipment's new Trailing Rope Grab brings a new level of convenience, protection, and productivity to vertical lifeline applications, enabling a worker to quickly climb with both hands free. Unlike other devices, which require the worker to pause during climbing to make manual adjustments, the Trailing Rope Grab automatically moves up and down the lifeline. (*Courtesy Miller® Equipment*)

Figure B This wall-form hook assembly is designed to work with a full-body harness. The hooking snap can be placed with one hand and is self-locking. When the hook is in place, the worker can use both hands for his work. The hook is made from heat-treated and zinc-plated forged alloy steel and fits most wedge bolt slots or holes on metal forms used for cast-in-place concrete. Although very strong the hook only weighs 16 oz. (453 g). (*Courtesy DBI/SALA*)

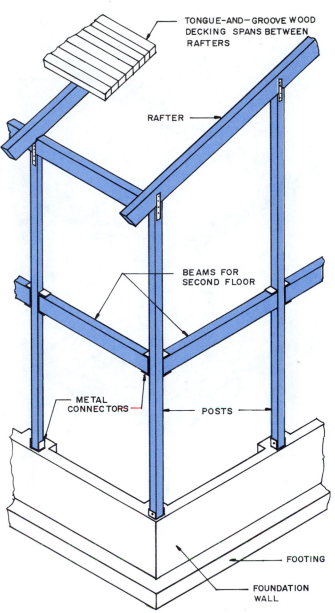

Figure 1.19 Heavy timber frames use metal connectors to join the posts and beams.

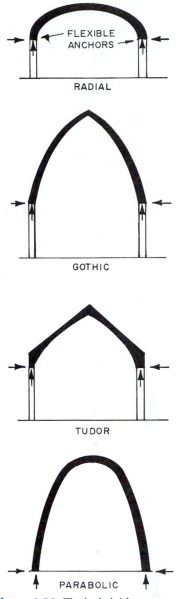

Figure 1.20 Typical rigid structural arches. These are two-hinge arches.

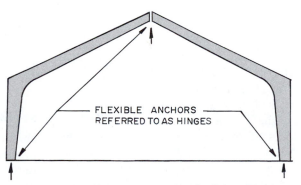

Figure 1.21 A three-hinge arch has flexible anchorages at each end and at the top center.

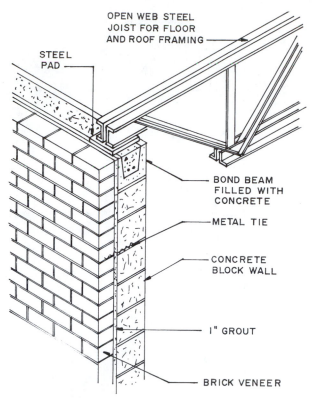

Figure 1.22 A typical solid masonry load-bearing wall.

Water Systems

Potable water is distributed within a building by pipes sized to produce the amount of flow required. Separate systems are required for cold water and hot water. The source of water for both may be from a water main owned by a public or private water company or from privately owned wells. As the building is planned, provisions are made to provide ways to install the pipe both horizontally on each floor and vertically between floors. A piping diagram for a small two-story residence is in Fig. 1.40. Figure 1.41 shows the major lines of a distribution system that provides water to the various rooms in a multistory building through branch lines on each floor. You can find detailed information on plumbing systems in Chapter 45.

Gas Systems

In some areas natural gas is a major source of energy. Propane, a gas occurring in petroleum and natural gas, is also used as a fuel. These are distributed within a building by a network of pipes. The natural gas is piped to the building in a distribution system similar to that for water. It enters the building using a pipe that takes off from the main, which usually is located on the front of the site. The gas moves through a pressure governor to control the gas pressure as it enters the building, then through a meter to measure the amount used, and finally on to a system of pipes that distributes it within the building. Since gas is flammable, explosive, and causes asphyxiation, its installation and use are closely regulated by building codes.

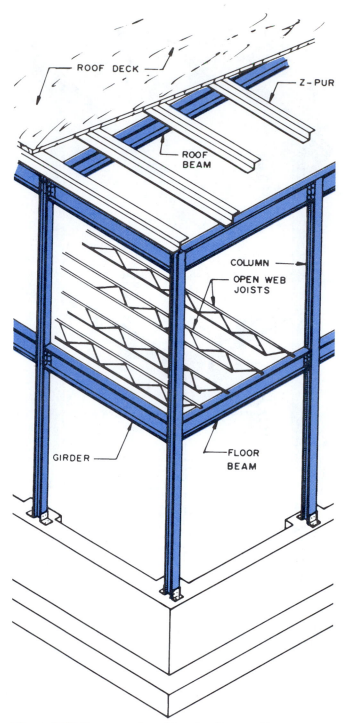

Figure 1.23 One type of steel framing for a multistory building.

Sewage Disposal Systems

A sewage disposal system removes fluid waste and organic matter to protect the occupants of the building from health problems that can be caused by exposure to these materials. The system is composed of a network of large-diameter pipes that must carry the waste horizontally on each floor to vertical soil stacks, which carry it to the house drain below the building. The house drain ties into the house sewer, which connects to the public

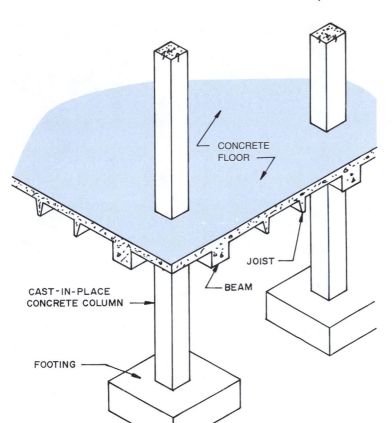

Figure 1.24 A monolithically cast concrete floor, joist, and beam structural system.

Figure 1.26 Precast concrete wall sections permit rapid construction. *(Courtesy Precast/Prestressed Concrete Institute)*

Figure 1.25 A building with a precast concrete structural frame. *(Courtesy Precast/Prestressed Concrete Institute)*

Figure 1.27 A tension fabric structure. *(Photo by Robert Reck, Courtesy Birdair, Inc.)*

Figure 1.28 Clay brick is a widely used exterior wall construction material.

Figure 1.30 This building uses glass spandrel panels to enclose the structure. *(Courtesy PPG Industries, Inc.)*

Figure 1.29 Stone veneers were used on the U.S. Capitol. *(Courtesy Bybee Stone Co.)*

Figure 1.31 Transparent glass panels are used for exterior walls and curved glass for door enclosures. *(Courtesy Laminated Glass Corporation)*

Figure 1.32 Curtain-wall panels are available in aluminum and steel. *(Courtesy Robertson: A United Dominion Company)*

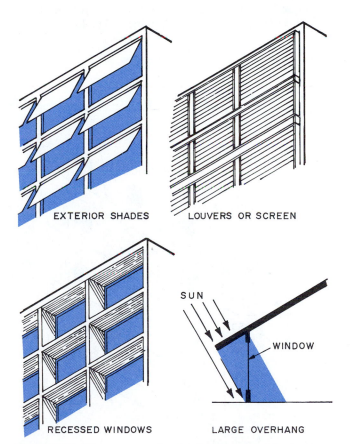

Figure 1.33 Ways to use solar shading to control heat gain.

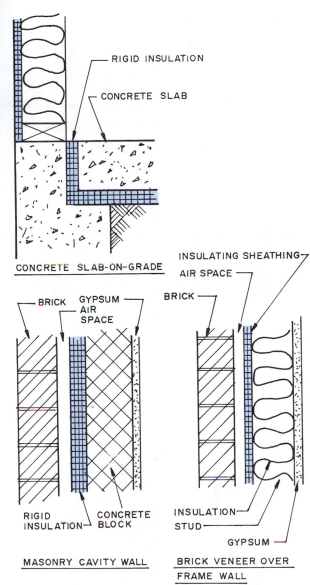

Figure 1.34 A few of the ways heat loss and gain can be controlled by the design of the walls and floor.

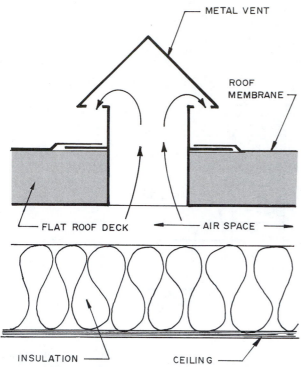

Figure 1.35 One way to vent the air space below a flat roof so moisture in the air is vented to the outside.

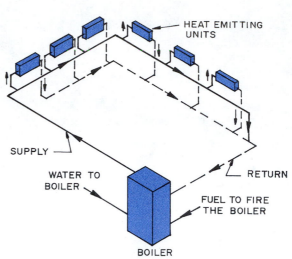

Figure 1.37 A hot water heating system as used in residences and small one-story commercial buildings.

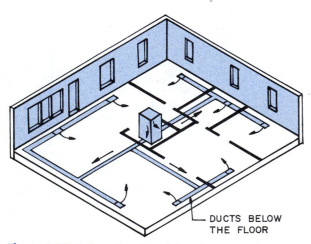

Figure 1.36 A downflow hot air heating system has ducts below the floor or in the concrete slab floor.

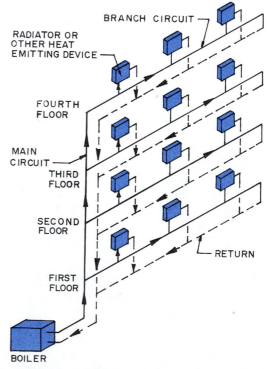

Figure 1.38 A two-pipe hot water system in a multistory building.

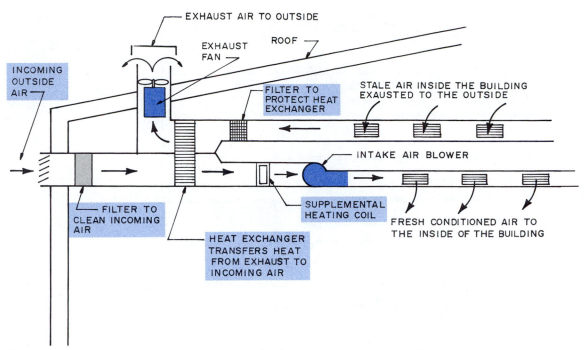

Figure 1.39 This air-handling system has a heat exchanger to salvage the heat from the exhausted air.

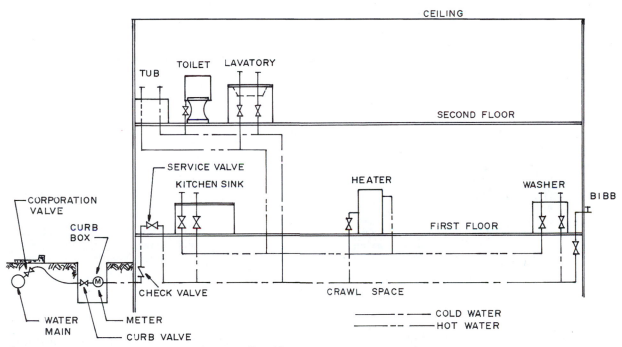

Figure 1.40 A water piping diagram for a small residence.

sewer or, for residential and some small commercial buildings, to some form of septic tank (Fig. 1.42). A diagram for a typical waste disposal system in a commercial building is shown in Fig. 1.43. In small buildings, such as a residence or small commercial building, the system can be run below the ground-level floor and between the floor joists on floors at the second-floor level.

Large multistory commercial buildings have to provide space on the floor plan for running the plumbing. A plumbing wall is shown in Fig. 1.44. This space runs up several floors through the building and is an important factor in developing the floor plan. Chapter 45 explains the types of waste systems in common use.

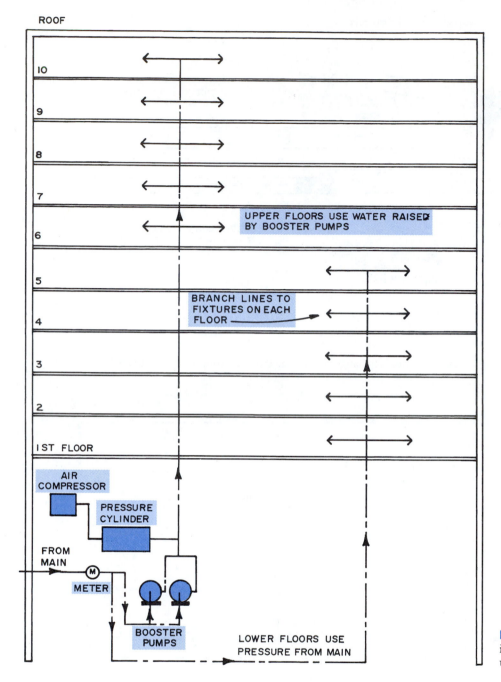

ROOF

10

9

8

7

UPPER FLOORS USE WATER RAISED BY BOOSTER PUMPS

6

5

BRANCH LINES TO FIXTURES ON EACH FLOOR

4

3

2

1ST FLOOR

AIR COMPRESSOR

PRESSURE CYLINDER

FROM MAIN

METER

BOOSTER PUMPS

LOWER FLOORS USE PRESSURE FROM MAIN

Figure 1.41 A simplified diagram illustrating one possible water distribution system for a multistory building.

Electrical Systems

The heating, air-conditioning, and lighting systems within a building require electrical power to function. In addition, there are other heavy demands for electrical power by the activities that are to take place within the building. Estimating the amount of electrical service required and planning the actual layout require a knowledge of the design solutions of all the factors involving a need for electricity.

The source of electrical power is from the lines of a public or privately owned electrical utility. Residences and small commercial buildings are supplied from these lines by transformers located on a pole or on a concrete pad at the corner of the building site (Fig. 1.45). Large buildings require additional large transformers that may be installed inside a building or on concrete pads outside the building. These are subject to strict building regulations. Oil-filled inside transformers are located in a fireproof transformer vault. Dry type transformers do not need to be in a vault and can be placed in the electrical room.

Provision must be made for the location of switchgear within the building. Switchgear includes disconnect

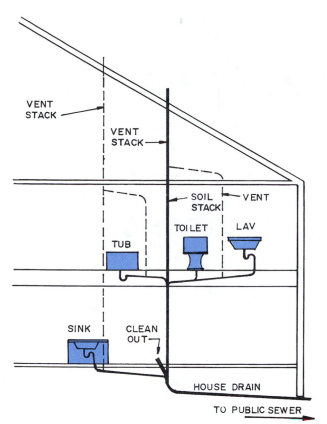

Figure 1.42 The basic elements of a residential waste disposal system.

switches, secondary switches, circuit breakers, and fuses. The electrical room is often in the basement. It must be well ventilated because dry transformers are often located here and the heat generated must be exhausted from the room (Fig. 1.46). Switchgear may be located outdoors and is housed in metal, waterproof cabinets designed for this purpose.

Circuits are run from the electrical room to electric panels or closets located in various parts of the building. Individual circuits are run from these panels, providing service to lighting circuits, convenience outlets, and special electrical equipment. A typical example is in Fig. 1.47. More details are in Chapter 47.

Fire Protection Systems

Fire protection systems are required to provide protection of the structure and its contents against damage by fire and to protect the occupants from injury or death. The design of the structure is the first step in protection against fire. The selection of materials and traffic patterns within the structure are critical. In addition, consideration of how to prevent the movement of a fire within a building is mandatory. For example, heat and air-conditioning ducts require dampers to prevent the flow of smoke and fire through the building. Provisions for venting smoke and gases through the roof must be made (Fig. 1.48). Openings between stories must have automatic closing devices if a fire occurs. Divisions within a floor should be compartmentalized into fire-containing areas with fire walls and automatic doors to

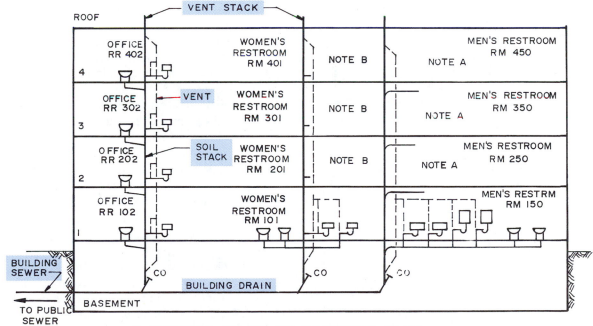

A. RESTROOMS 250, 350, AND 450 IDENTICAL TO 150.
B. RESTROOMS 201, 301, AND 401 IDENTICAL TO 101.

Figure 1.43 A simplified diagram of a waste disposal system for a multistory building.

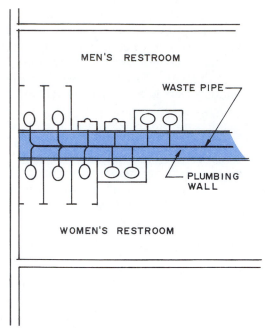

Figure 1.44 A plumbing wall provides space to run the plumbing and other services vertically through the building.

Figure 1.45 A slab-mounted transformer used to supply electricity to several small buildings, such as single-family residences.

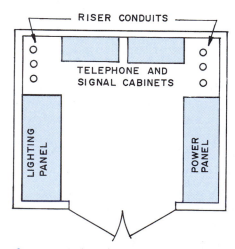

Figure 1.46 A typical electrical room.

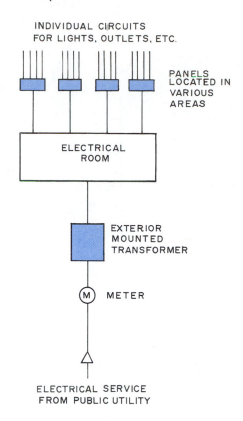

Figure 1.47 This diagram shows the flow of electricity into and within a large building.

close them off. Fire detection and alarm systems, automatic sprinkler systems, and standpipe and hose systems within the building are required.

Fire detection and alarm systems for residential construction are often simple smoke detectors or more sophisticated light-beam smoke detectors. Another type is activated by the presence of a flame or heat. These sound an alarm to alert the occupants.

Commercial fire detection and alarm systems generally have ionization smoke detectors, heat detectors, and manual stations that are activated by a person pressing the alarm button. These systems can also activate a sprinkler system and open smoke control doors. The system can be designed to operate in a specific area of the building and to indicate on a control panel which area in the building has a fire.

Automatic sprinkler systems are installed near the ceiling and consist of a network of pipes that provide water to outlet pipes connected to sprinkler heads (Fig. 1.49). The branch line is large enough to supply water to a series of sprinkler heads. The sprinkler heads open automatically when the air temperature reaches a predetermined temperature, which is generally from 135 to 160°F. Considerable additional information is in Chapter 44.

Figure 1.48 Roof vents are required to permit the exhaust of smoke during a fire. *(Courtesy the Bilco Company)*

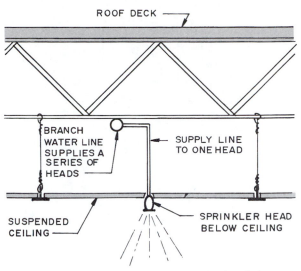

Figure 1.49 Water is supplied to sprinkler heads by large-diameter branch lines.

Signal and Sound Systems

Typical signal systems include doorbells, public television, intercoms, music, fire alarms, security systems, telephone, closed-circuit television, clocks, and time-activated signals. Extensive signal systems are computer controlled and become an important part of the activities within a building. As the building is planned, space must be provided on various levels for installing equipment and plans must be made for running the cables within floors and between floors.

Typical distribution systems include the use of conduit, boxes, and various duct and raceway systems. Some are in or below the floor. A large building may require a room just for installation of the telephone terminals. Some equipment can be installed in closet-size spaces. A major system, such as a telephone system, may require space on each floor for a telephone equipment closet from which all circuits on that floor can be run (Fig. 1.50). Other communications equipment terminals may be in this room if codes permit.

Closed-circuit television is widely used for security purposes in apartments, industrial, and commercial buildings. Some central location for personnel to monitor the system is necessary. Closed-circuit television is also used to monitor activities in areas where it is too dangerous for humans to work, such as in certain parts of a nuclear generation plant.

Mechanical Transportation Systems

Systems to transport people and materials within a building include elevators, escalators, and moving sidewalks and ramps.

Elevators provide vertical transportation. Passenger elevators are designed for commercial use, institutional purposes (such as hospitals), and residential use. The location of elevators within the building and the number needed for the occupancy of the building are carefully considered by the architect. Elevators should have easy access, provide a minimum of waiting time, operate automatically, and have the required safety devices. Freight elevators are designed to carry the expected loads, to load easily, and to move the required amount of materials per hour. If a freight elevator is also to carry passengers, it must meet the requirements for passenger service. Freight elevators may be the traction type or hydraulic. Light-duty freight elevators carry loads of up to 2,500 pounds. General purpose freight elevators are designed to carry loads of 10,000 pounds or more.

Escalators are moving stairways (Fig. 1.51). Since they are always available, they do not cause the backup that occurs when people are waiting for an elevator. Even if escalators are not operating, they can be mounted in the same manner as a fixed stair. The location is an important part of the architectural design. In some buildings escalators start from the lobby and move to the next floor. In merchandising facilities they become part of the traffic flow pattern of the store. They can be staggered within the building, providing greater flexibility in routing traffic. Elevators require a vertical shaft, which limits them to the same location on each floor.

Moving walks transport people and materials horizontally, and moving ramps provide horizontal transportation and some vertical rise. Generally, moving walks do not have a rise or fall of more than 5 degrees and moving ramps have code limitations on the amount of rise, typically 15 degrees maximum (Fig. 1.52). Moving ramps can run between floors in a building if the space is available to allow for the long distances required. Review Chapter 43 for more information on conveying systems.

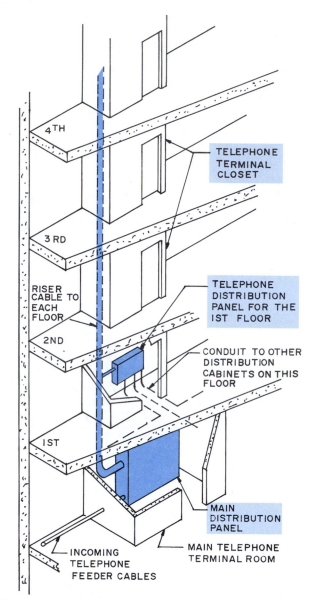

Figure 1.50 A typical riser arrangement for a telephone installation.

Figure 1.51 Escalators move people from one floor to another. *(Courtesy Otis Elevator Company)*

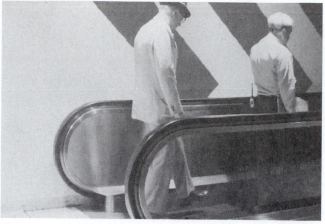

Figure 1.52 Moving walks are used to assist people to move long distances and up inclines as found in large airports. *(Courtesy Otis Elevator Company)*

REVIEW QUESTIONS

1. What is the purpose served by foundations?
2. What are the common types of arches?
3. What types of structural frames are used for residential and commercial buildings?
4. What is a tension fabric structure?
5. What types of enclosing walls are in common use?
6. What are the various types of building systems required to enable a building to serve the occupants?
7. What type of fire protection devices are available?

KEY TERMS

This chapter provides a brief overview of the information included in this book. The terms used in this chapter are defined and discussed in the specific chapters covering the material and in the master glossary at the end of the book.

SUGGESTED ACTIVITY

Make a tour of your locality and record where various types of construction are underway. Use this information to make more detailed visits as you study the chapters in this book. Keep a log of the type of construction activity underway and the date the particular activity was begun.

Construction: A Dynamic Industry

This chapter will help you to:

1. Understand the major divisions within the construction industry and the activities that occur in each.

2. Be aware of the basic construction materials and typical applications for each.

3. Realize the place of environmental considerations raised when construction projects are proposed.

4. Know the purposes served by zoning ordinances and building codes, and codes in use in the United States and Canada.

5. Develop an appreciation for the place of professional and trade associations and their contribution to the construction industry.

6. Understand the Construction Specifications Institute MasterFormat and its sixteen divisions.

7. Use metric units in designing and reading construction drawings.

The construction industry is one of the major industries in the United States. Construction, with all of its related activities, is vital to the economic health of the country, and the construction industry reacts to and causes reactions in the nation's economy.

The Associated General Contractors of America, Inc., has described the construction industry as follows:

> The construction industry is a brawny, hearty giant stretching to embrace all kinds of construction activity, from the erection of towering skyscrapers, construction of an interstate highway, or the establishment of a massive dam on a wilderness river to major maintenance and alterations. (Fig. 2.1).

Construction involves the coming together of skilled workers, architects, engineers, machines, and a vast array of materials to follow a carefully drawn plan. Supporting these is a broad range of industries that manufacture the materials and components designated for the project. These can include the manufacture of sand, beams, doors, windows, siding, roofing, finish materials, clay products, cement and cast concrete products, asphalt, glass, and thousands of other products. Many manufacturing endeavors involve the shaping, forming, cutting, and assembling of various materials into a useful product. Each area has a tremendous choice of materials and requires extensive technical knowledge. Consider the electrical and mechanical products available and the technical knowledge required to design, manufacture, specify, and install them.

The people involved in construction need to be well educated and constantly kept up-to-date as new materials and methods of construction are developed. They need to be involved in their professional organizations, attend conferences, and read their professional journals. In this way they become aware of new methods and materials and the proper use of each.

THE CONSTRUCTION INDUSTRY

The construction industry can be divided into three major areas—*building construction, heavy construction,* and *electrical, mechanical, and special trades.* Building

Figure 2.1 The construction industry is a dynamic, challenging field offering a wide range of job opportunities. *(Courtesy the Associated General Contractors of America)*

Figure 2.2 This commercial building represents the building construction aspects of the vast construction industry. *(Courtesy Precast/Prestressed Concrete Institute)*

construction includes residential, commercial, and industrial construction (Fig. 2.2). Heavy construction includes projects such as highways, bridges, canals, dams, subways, tunnels, piping systems (for example, for natural gas), water control construction, and construction of electrical power and communications networks (Figs. 2.3 through 2.6).

CONSTRUCTION CONTRACTORS

The actual construction activity is performed by construction contractors. They are generally divided into general contractors and specialty contractors. *General contractors* assume the responsibility for the construction of an entire project at a specific cost and by a specified date. They are responsible for all the work performed even though their employees do not do the work. A general contractor signs contracts with specialty contractors who perform the required work within their technical areas. These contractors are often called subcontractors.

Specialty contractors do the work required in a limited area, such as all the electrical, plumbing, roofing, bricklaying, carpentry, concrete, or other work needed. They are independent contractors employed by the general contractor to perform specific work at a specified cost.

Figure 2.3 This bridge over the Ohio River is one of the projects built by contractors working in heavy construction. *(Courtesy West Virginia Highway Department)*

They bring their employees, supervisors, and tools to the job. Although they work directly for their supervisors, the general contractor has job superintendents on the site to oversee the work of all specialty contractors (Fig. 2.7).

Figure 2.4 Dams and underground projects are part of the heavy construction field. *(Courtesy Tennessee Valley Authority)*

Figure 2.6 The construction of electrical power and communications networks involves working at great heights. *(Courtesy Tennessee Valley Authority)*

Figure 2.5 Water control projects present unusual challenges and dangers. *(Courtesy U.S. Army Corps of Engineers)*

CONSTRUCTION MATERIALS

The production of materials used in construction is a vast industry requiring highly trained managers and technical workers. These materials include about every type that is available. They include naturally occurring materials, such as wood and stone. They include materials made by processing natural materials into usable form, such as steel, aluminum, and plastics. They include those formed by assembling a combination of ingredients to produce a usable product, such as coatings.

These materials are remanufactured to produce other construction-oriented products. Steel is manufactured into beams, columns, doors, nails, screws, and hundreds

of other products. Aluminum is used to make windows, siding, roofing, gutters, and other products, especially those exposed to the elements (Fig. 2.8). Wood is manufactured into lumber, plywood, composition products, doors, windows, siding, cabinets, flooring, and other products (Fig. 2.9). Plastic is finding an ever-increasing use in construction and is widely used in piping, electrical products, siding, windows, doors, and other products. So when you think of materials, think of the processes used to obtain them and then the processes used to make them into usable construction products.

It is important to select materials that perform as expected. This can include such considerations as strength, weathering characteristics, cost, construction methods, and appearance. Industrial standards have been established to govern the manufacture of most materials and to specify their proper use. Building codes also regulate the use of materials. The persons specifying materials must be knowledgeable about both industrial and government standards.

Key people in the selection of materials are the architects and engineers employed to design the project. They consider the aesthetic value of the material as well as its structural properties. The desires of the owner are considered and recommendations are made to the owner by

Figure 2.8 Aluminum is widely used for products exposed to the elements. (*Courtesy the Aluminum Association*)

Figure 2.7 This electrical worker is employed by the subcontractor responsible for installing all electrical components. (*Courtesy International Brotherhood of Electrical Workers*)

Figure 2.9 Wood-framed windows may be clad with plastic or aluminum. (*Courtesy Andersen Windows, Inc.*)

these experienced designers. These are the people who must know the materials that are available and their properties, advantages, and disadvantages. They must also keep up-to-date on new materials and building code changes.

Specific information about materials to be used on a construction project is one of the major aspects of the job. As the architect and consulting engineers develop the design, the selection and specification of materials is of major importance. These choices relate not only to appearance, but also to the cost, durability, and strength requirements for each situation. Materials are specified by symbols on the working drawings. However, since it is not possible to give detailed information on drawings, specifications are written. **Specifications** are produced as a written document organized around the various trade areas, such as electrical or concrete. They give detailed information about the materials to be used, the quality of workmanship, methods of installation, and the scope of the work. The specifications plus the working drawings must provide a complete description of the project. They become part of the legal contract between the owner and the contractor.

When a contractor bids on a job, the information about materials on the working drawings and in the specifications is what is used to determine costs. The ultimate cost depends on the material decisions made and how accurately they are specified. Poorly written material specifications could permit a contractor to substitute a less costly material or one that is unsatisfactory to the owner, even though it meets the material specifications as written.

Architects and engineers must have an extensive knowledge of materials, their costs, advantages, disadvantages, properties, and availability. They must constantly learn about new materials coming on the market and assess their reliability. This is accomplished by reading professional journals, attending conferences and conventions, and visiting with representatives of materials manufacturers and suppliers. An extensive reference library is essential to successful selection and specification of materials. One important resource is the Sweet's Catalog File published by McGraw-Hill Information Systems Company.

CONSTRUCTION AND THE ENVIRONMENT

Before a large construction project can be started, an environmental impact study must be made to determine the effects it will have on the environment of the surrounding area. This impact study is required by the U.S. government as a result of the National Environmental Policy Act (NEPA). The study determines how the proposed project will affect the natural features of the area, such as forests, rivers, and wetlands. There are also social impacts, such as building a paper mill near a pleasant residential community or even building it many miles away. The odor will have an impact for many miles around. Harvesting trees to produce the needed wood pulp will also have an impact.

USE OF RESOURCES

The construction industry is a major consumer of many different types of materials. Clays are mined for brick and tile, and trees are harvested to produce lumber. Portland cement is manufactured from mined limestone, silica, alumina, and iron. Many plastics are produced using oil as the major material. These and other examples show that the materials in the construction industry are produced from a dwindling source of supply—our natural resources.

The average residence uses about 10,000 board feet of lumber that must be large in size and, therefore, cut from mature, large-diameter trees, which are a very small percentage of the timber currently available. Although trees are a renewable resource, it takes many years to produce a new large-diameter tree (Fig. 2.10). Most of our natural resources are not renewable. When they are exhausted, they cannot be replaced.

Environment-Friendly Materials

Much is being done to better utilize our natural resources and to produce useful products using materials that for years have been thrown away. For example, major progress has been made in the wood industry. Structural members made by bonding thin wood pieces or wood chips with waterproof adhesives use waste products to produce products equal to or better than those made from solid wood (Fig. 2.11).

Recycling materials is another way being used to conserve the natural resources. Wall panels are made

Figure 2.10 The U.S. Forest Service oversees vast areas of timber on government-owned property. *(Courtesy U.S. Forest Service, U.S. Department of Agriculture)*

Figure 2.11 This wood beam is manufactured using wood strands and waterproof adhesive to form a parallel strand lumber beam. *(Courtesy Trus Joist MacMillan (Trade name Parallam® PSL))*

using recycled newsprint and gypsum, carpet pads are made from recycled rubber, tile is made from recycled auto windshield glass and lightbulb glass, heat energy is saved by heat recovery units on ventilation systems, shelving is made with a recycled paper core, and nails are made from industrial remelts. A great deal of aluminum is recycled and used with newly mined ores to produce useful products.

Some designers are giving much more attention to specifying environment-friendly materials. A number of associations have been formed to promote the use of recycled and energy-efficient materials. Some of these are included in Appendix B.

Manufacturers of construction materials make a point of indicating which of their products have recycled material in them and the energy-efficiency ratings achieved.

Some have materials catalogs devoted to recycled products (Fig. 2.12). One such product is Nature Guard insulation, which is made from recycled newspaper (Fig. 2.13). It is nontoxic, nonflammable, and does not cause the itchy skin common with other insulations. Another product is a fiberboard sheet material that contains about 20 percent wastepaper pulp and 40 percent chips from waste wood. It is asphalt impregnated and used as wall sheathing and roof insulation board (Fig. 2.14).

Considerable savings of materials can be accomplished by remodeling a building rather than destroying it, throwing the bricks, lumber, etc., in the local landfill, and building a new structure. Remodeling can also contribute to preserving our natural resources by utilizing recycled and energy-efficient materials. It is important to use materials we have traditionally thrown away and to use more efficiently the materials we produce from our primary natural resources. Architects should specify the expanded use of alternative materials, and builders should be willing and able to use them. Owners of buildings under design and construction should be informed about the value of alternative materials and should be encouraged to accept them.

One major change that is likely to occur in the years immediately ahead is the use of concrete instead of wood for all parts of typical residences and small commercial buildings. Since wood is in diminishing supply and the materials to make concrete exist abundantly all over the world, the floors, walls, and roofs will more often be cast-in-place and precast concrete units. Concrete lasts much longer than wood, resists hurricanes and earthquakes, and can be crushed and recycled as aggregate for new concrete products. Steel framing is also finding increasing use in residential and small commercial construction.

CONSTRUCTION METHODS

Architectural working drawings show in detail the assembly methods required by the architect and engineer. Specifications also detail specific methods to be used.

Figure 2.12 This publication describes plastic products made utilizing recycled materials. *(Courtesy American Plastics Council)*

Figure 2.13 This insulation is made from recycled paper. *(Courtesy Louisiana Pacific)*

Figure 2.14 This fiberboard sheathing is made using recycled paper and waste wood chips. *(Courtesy Huebert Fiberboard, Inc.)*

When something is not specified, it is up to the knowledge and skill of the employees of the subcontractor and the general contractor to use approved methods of construction. Sometimes it is necessary to get back to the architect or engineer for additional information. The methods of installing, assembling, joining, or otherwise utilizing the materials specified are just as important as the specification of the materials. Quality materials improperly installed do not produce a satisfactory job. Industrial and trade associations do a lot of work exploring the best ways to assemble materials, and they disseminate this information widely. Building codes also specify acceptable construction methods. The general contractor should select subcontractors who will use the very latest and most effective methods.

ZONING AND CODES

A study of the materials of construction involves not only the technical aspects related to the material but also the influences on the material and its use by the local zoning ordinances, national building codes, and various standards established by trade and professional organizations and governmental agencies (Fig. 2.15). In addition, the gradual conversion to the metric system must be considered.

Zoning Ordinances

Zoning ordinances are local regulations that control the use of land and buildings, height and bulk of buildings, density of population, building coverage of the lot in relation to open space, setbacks, and ancillary facilities, such as parking. They are for the protection of the property owner. They attempt to group buildings that house similar activities together. For example, in a sec-

Figure 2.15 The construction industry relies on an extensive array of codes, standards, and other technical publications for proper design and construction practices.

tion of a city zoned for single-family dwellings, it would be inappropriate to build a manufacturing plant.

Building Codes

Building codes regulate the design and construction of buildings to provide *minimum standards* to safeguard life or limb, health, property, and public welfare. They control the design, construction, use, and occupancy of buildings and the location and maintenance of the buildings within their jurisdiction. In some cases, equipment is also regulated. Building codes apply to the construction, alteration, moving, repair, and demolition of buildings or other structures within their jurisdiction.

In most cases, municipalities use one of the regional model building codes. Special local requirements can be required in addition to those in the model code.

There are several major regional model codes. The International Conference of Building Officials (ICBO) publishes the *Uniform Building Code (UBC)*. It is revised regularly, and a revised code is published every three years. The conference also publishes an extensive number of related codes and textbooks, including the *Uniform Mechanical Code, Uniform Plumbing Code, Uniform Fire Code*, and the *One and Two Family Dwelling Code*.

The Uniform Building Code is widely used in the area west of the Mississippi River. It is also used in various parts of Illinois, Louisiana, Michigan, and Indiana. It is used overseas from the U.S. Virgin Islands to Alaska, in Guam and other U.S. territories in the Pacific, and in the state of Hawaii. It is also used as the basis for codes in Saudi Arabia, Japan, and South Korea.

The BOCA National Building Codes are prepared by the Building Officials and Code Administrators International, Inc. BOCA publishes a variety of other codes including *The Basic Housing Code*. BOCA National Codes are used in the eastern and midwestern United States. They are in use from Kentucky and Virginia up to the Canadian border and west, including Illinois and Wisconsin. In addition, they are used in North Dakota, Nebraska, and Kansas, and in parts of Oklahoma, Missouri, and Texas.

The Southern Building Code Congress International, Inc. (SBCCI) promulgates and holds the copyright to the *Standard Building Codes*. This code uses performance-based requirements that encourage the use of innovative building designs, materials, and construction systems while at the same time recognizing the merits of the more traditional materials and systems. SBCCI publishes a series of other codes, including the *Standard Plumbing Code, Standard Mechanical Code, Standard Gas Code*, and the *Standard Fire Prevention Code*. They also publish a magazine and a variety of other publications. The Standard Building Codes are widely used in the southeastern United States as far west as Texas and Oklahoma and on the eastern seaboard as far north as

Delaware. A number of countries overseas also utilize the codes.

The Council of American Building Officials (CABO) was created by the three nationally recognized model code organizations, BOCA, ICBO, and the SBCCI, to establish a communications channel between building officials and congressional, federal, and industry organizations.

There are other codes produced by various associations and institutes. For example, the National Fire Protection Association publishes the *National Fire Codes* and the American Concrete Institute publishes the *ACI Building Code Requirements for Reinforced Concrete.*

Another major publication is the multivolume *ASTM Standards in Building Codes.* When you use these volumes you satisfy code requirements of many of the U.S. and Canadian codes and have standards referenced by MASTERSPEC®, SPECSystem™, SPECTEXT®, Army Corps of Engineers, and the Naval Facilities Engineering Command (NAVFAC). MASTERSPEC and SPECSystem are publications of the American Institute of Architects. SPECTEXT is reviewed and updated by the Construction Specifications Institute.

Canadian Building Codes

The major building codes in Canada include Canadian Housing Code, National Building Code of Canada, National Fire Code of Canada, Canadian Plumbing Code, and the Canadian Farm Building Code. The National Building Code of Canada is issued by the Associate Committee on the National Building Code, National Research Council of Canada.

TRADE ASSOCIATIONS

Other organizations are devoted to the advancement of knowledge about materials and methods used in construction. Some of these have developed codes that influence the content of the building codes and establish industry-wide standards. Trade associations are a major factor in materials development. A **trade association** is an organization whose membership comprises manufacturers or businesses involved in the production or supply of materials and services in a particular area. An example is the American Institute of Steel Construction.

The goal of a trade association is to promote the interests of its membership. This includes dissemination of information on the proper use of the materials, products, and services of the members. Some trade associations support research into the proper use and improvement of construction materials and better construction methods. As they do this, they develop material *specifications* and *performance procedures* for the trade area. Often these become accepted nationally and become national standards. Some trade associa-

tions are involved in programs of *certification.* Products made according to their specifications and standards are identified with a seal, label, or stamp. An example is the certification stamp on plywood issued by APA - The Engineered Wood Association (Fig. 2.16).

Following are brief descriptions of some of the associations involved with materials used in the construction industry.

The *American Concrete Institute (ACI)* is a technical and educational organization dedicated to improving the design, construction, manufacture, and maintenance of concrete products and reinforced concrete structures. ACI disseminates its information through conventions, seminars, publications, and volunteer committee work. The ACI series of technical publications is extensive, and one, *The ACI Manual of Concrete Practice,* is a major reference for the industry. ACI publishes several magazines.

The *Portland Cement Association (PCA)* is a nonprofit organization dedicated to improving and extending the uses of portland cement and concrete. It is supported by voluntary financial contributions by member companies. PCA serves as a clearinghouse for concrete design practices and construction methods. It has an extensive service program that offers publications, motion pictures, and slides depicting up-to-the-minute developments. It offers many educational programs and is involved in developing computer programs for construction-related activities. PCA provides research and development-engineering services that include troubleshooting field problems, engineering investigations, development services, and assistance with the manufacture of portland cement.

The *American Institute of Steel Construction (AISC)* is a nonprofit association serving the fabricated structural steel industry. Its objectives are to improve and advance the use of fabricated structural steel through research and engineering studies to develop the most efficient and economical design of structures. The institute publishes manuals, textbooks, specifications, and technical

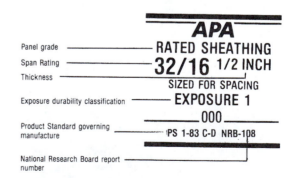

Figure 2.16 This stamp on plywood products indicates that APA - The Engineered Wood Association has certified that it meets their specifications. *(Courtesy APA - The Engineered Wood Association)*

books. The most widely used publication is the *Manual of Steel Construction* (Fig. 2.17). AISC publishes current technical information in two quarterly publications, *Engineering Journal* and *Modern Steel Construction*.

The *Copper Development Association, Inc.*, disseminates technical information pertaining to the use of copper, bronze, and brass products in building construction.

The *Aluminum Association* is the primary source for statistics, standards, and information on aluminum and the aluminum industry. Members of the association produce almost all domestic primary aluminum ingots and most of the semifabricated mill products. The association works to increase public awareness of aluminum, helps companies with operational problems, develops voluntary standards and technical data, collects marketing and statistical data, and works with product development and educational programs. It offers a variety of publications and audio-visual presentations. A most significant publication is the *Aluminum Association Construction Manual*.

The *Gypsum Association* has as its primary function the promotion of gypsum products. It serves as an information center by gathering and disseminating information pertaining to gypsum and gypsum products. Representatives of member companies contribute to the program of the association. Activities include mining and manufacturing, review of specifications incorporated into building codes, and research in such areas as fire testing, structural design, sound testing, and safety.

The *Asphalt Institute* works to advance the most efficient use of bituminous concrete through research, engineering, and educational activities. It assists in extending and developing the asphalt business. Basically it is an engineering organization with its engineering staff located in offices throughout the United States. The institute is active in educational activities and offers a wide range of technical publications and quarterly bulletins.

A number of organizations represent various aspects of the *wood industry*. Some of these are:

The *Southern Pine Inspection Bureau* establishes grades and inspects softwood lumber in the southeastern part of the United States.

The *West Coast Lumber Inspection Bureau* establishes grades and inspects Douglas fir, hemlock, cedar, and other softwood lumber in California and the western regions of Washington and Oregon.

The *Redwood Inspection Service* establishes grades and inspects redwood lumber.

The *Western Wood Products Association* is an association of wood product manufacturers in twelve western states, and it grades pines, firs, spruces, and cedars.

The *Northwestern Lumber Manufacturers Association* establishes grades and inspects northern and eastern pines, cedars, balsam fir, spruces, and other softwoods in New England, New York, New Jersey, and Pennsylvania.

The *National Hardwood Lumber Association* establishes grades and inspects hardwood lumber such as walnut, oak, birch, and maple.

APA - The Engineered Wood Association establishes grades and supervises the manufacture of softwood plywood. It offers an extensive variety of technical publications (Fig. 2.18).

The *American Institute of Timber Construction* performs research, testing, and structural analysis for solid and laminated timber construction.

The *National Forest Products Association* has many publications used by builders and designers.

The *Southern Forest Products Association* has technical publications related to the use of southern pine in building construction (Fig. 2.19).

The *California Redwood Association* has publications related to the technical and aesthetic uses of redwood.

The *Cedar Shake and Shingle Bureau* has technical publications related to the use of cedar shakes and shingles.

The *National Particleboard Association* represents most of the manufacturers of particleboard and medium density fiberboard. They have many technical publications.

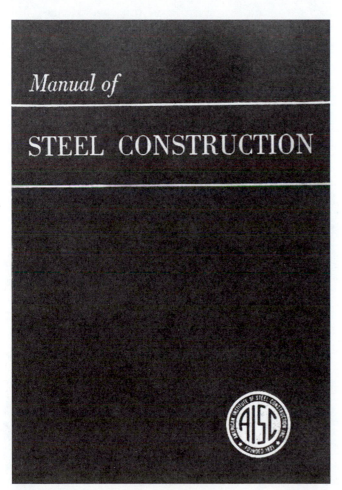

Figure 2.17 The *Manual of Steel Construction* is a publication of the American Institute of Steel Construction. *(Courtesy American Institute of Steel Construction)*

DATA FILE

AMERICAN PLYWOOD ASSOCIATION

Nailed Structural-Use Panel & Lumber Beams

When roof load or span requirements are too great to allow use of commonly available dimension lumber or timbers, a box beam constructed of lumber and APA trademarked structural panels can solve the problem. It offers an inexpensive alternative to steel or glued laminated wood beams.

Nailed structural-use panel and lumber beams have many other advantages, too. Among them are

- Stiffness and strength
- Light weight
- No shrinkage, warping, or twisting
- Ease of fabrication
- Materials availability
- Speedy, easy installation
- Easily insulated where necessary

Parts of a box beam are shown in the diagram. The lumber flanges carry most of the bending stresses and the structural panel webs transmit the shear stresses. Vertical stiffeners between the flanges act as shear splices at panel butt joints, distribute concentrated loads and end reactions, and resist web buckling. The fasteners transfer the stresses between the lumber and panel parts.

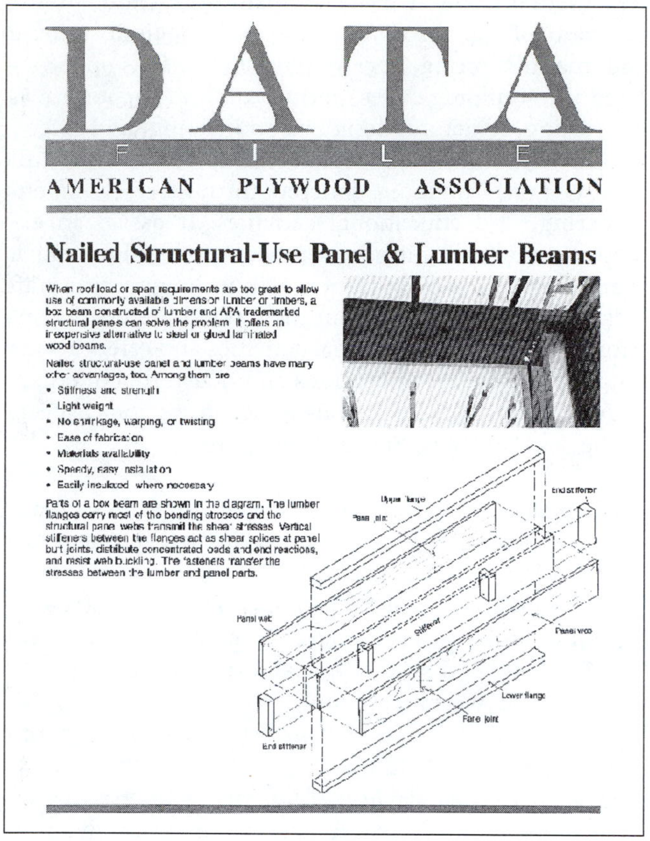

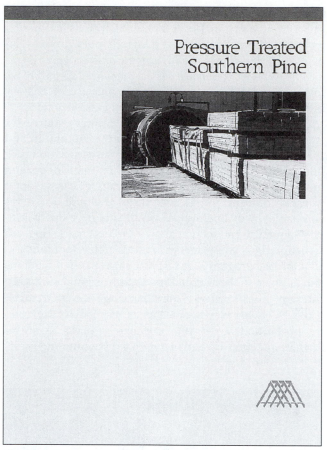

Figure 2.18 This is one of the many technical publications of APA - The Engineered Wood Association. *(Courtesy APA - The Engineered Wood Association)*

Figure 2.19 A technical publication of the Southern Forest Products Association. *(Courtesy Southern Forest Products Association)*

Addresses for some of these associations are in Appendix B.

OTHER ORGANIZATIONS

In addition to trade associations, many professional, private, and governmental organizations make major contributions to the safe and efficient use of construction materials and methods. Descriptions of several of these organizations follow.

Department of Housing and Urban Development

The purpose of the Department of Housing and Urban Development (HUD) is to promote the development of housing in the nation's communities. An important function is the stimulation of housing production through mortgage insurance and other federal subsidies. Included are improving the quality of housing built and establishing more efficient land-planning standards (Fig. 2.20).

National Institute of Building Sciences

The National Institute of Building Sciences is chartered by the U.S. Congress as an authoritative national source of knowledge and advice on matters of building science and technology and as a leadership forum for the purpose of bringing together the building community to serve the public interest by

- ▲ Promoting a more rational building regulatory environment.
- ▲ Identifying and facilitating the development and use of new technology and processes.
- ▲ Disseminating recognized technical information.
- ▲ Facilitating building science and research.
- ▲ Convening knowledgeable and affected interests to seek solutions to building community problems and needs.

General Services Administration

The General Services Administration (GSA) is an agency of the federal government responsible for the

Product Standards are developed by manufacturers, distributors, and users. The program is administered by the National Institute of Standards and Technology. The purpose of Voluntary Product Standards is to establish nationally recognized requirements for products and to provide all concerned interests with a base for common understanding of the characteristics of the product. When manufacturers make references to standards in connection with their product in advertising, on labels, or any other way, they become legally liable for provisions of the standard.

Product Standards are identified by code numbers, beginning with the letters "PS" followed by an identification number. The last two digits indicate the year the standard was issued or revised. The standards relating to construction in materials include PS 1-83, Construction and Industrial Plywood, PS 2-92, Performance Standard for Wood-Based Structural-Use Panels, and PS 20-7, American Softwood Lumber Standard.

Standard Reference Materials Program

The U.S. Department of Commerce offers the Standard Reference Materials Program (SRMP) through the National Institute of Standards and Technology (NIST) located in Gaithersburg, Maryland. The NIST includes the work of the former National Bureau of Standards and has assumed additional responsibilities in the area of material standards.

The Standard Reference Materials Program has available more than one thousand different materials that are certified for their chemical composition, chemical properties, or physical properties. Other materials are also available. Each bears a name and identifying number. Standard Reference Materials (SRMs) are well-characterized materials produced in quantity to improve measurement science. They are certified for chemical and physical properties and issued with a NIST certificate that contains the characterization of the material and its intended use. Standard Reference Materials are prepared and used for three main purposes:

1. To develop accurate methods of analysis
2. To calibrate measurement systems
3. To ensure the long-term adequacy and integrity of measurement quality assurance programs.

SRM reports by a code number and name such as SRM 1881 Portland Cement. This particular sample is for x-ray spectroscopic and chemical analysis. The sample received is in three sealed vials, each containing 5 g of material.

NIST also issues Research Materials (RM) that are not certified. These materials have one or more properties that are sufficiently well established to be used for

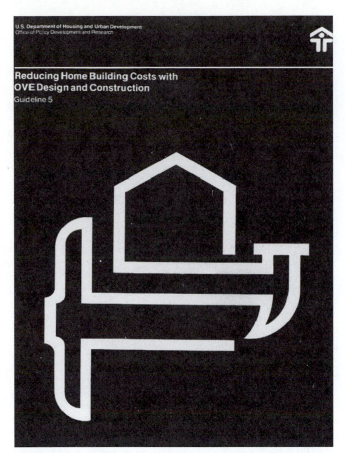

Figure 2.20 A guideline publication of the U.S. Department of Housing and Urban Development.

products purchased by the government. It develops federal and interim specifications that standardize the sizes and quality of materials and products purchased by government agencies.

Department of Commerce

The Department of Commerce, through the National Bureau of Standards, provides for the development of Product Standards (PS) and standard materials samples through its Standard Reference Materials (SRM) service.

Voluntary Product Standards

Manufacturers have long realized the importance of standardizing sizes, quality of materials, terminology, and other factors relating to their products. For example, standardized sizes and grades of construction and industrial plywood benefit both the manufacturer and contractor. This leads to efficient, economical growth within an industry.

The federal government encourages and assists in this process by issuing Voluntary Product Standards.

ITEM OF INTEREST

TYPES OF CONSTRUCTION

Building codes require that every building be classified into one of the types of construction specified by the building code. The classifications consider fire resistance of materials and construction and provide specific fire resistance ratings for the structural elements in hours for each type of construction. The architect and engineer must be thoroughly familiar with these requirements to be able to produce an approved structure.

Following are the general specifications for the types of construction as detailed in the Standard Building Code published by the Southern Building Code Congress International, Inc., Birmingham, Alabama.

Types of Construction

Type I is construction in which the structural members, including exterior walls, interior bearing walls, columns, beams, girders, trusses, arches, floors, and roofs, are of noncombustible materials and are protected so as to have fire resistance not less than that specified for the structural elements as specified in this code.

Type II is construction in which the structural members, including exterior walls, interior bearing walls, columns, beams, girders, trusses, arches, floors, and roofs, are of noncombustible materials and are protected so as to have fire resistance not less than that specified for the structural elements as specified in this code.

Type III is construction in which fire resistance is attained by the sizes of heavy timber members (sawn or glued laminated) being not less than indicated in this section, or by providing fire resistance not less than one hour where materials other than wood of heavy timber sizes are used; by the avoidance of concealed spaces under floors and roofs; by the use of approved fastenings, construction details, and adhesives for structural members; and by providing the required degree of fire resistance in exterior and interior walls.

Type IV is construction in which the structural members, including exterior walls, interior bearing walls, columns, beams, girders, trusses, arches, floors, and roofs, are of noncombustible materials. Type IV construction may be protected or unprotected. Fire resistance requirements for structural elements of Type IV construction shall be as specified in this code. For provisions governing combustibles in concealed spaces, see this code.

Type V is construction in which the exterior bearing and nonbearing walls are of noncombustible material and have fire resistance not less than that specified in Table 600; bearing portions of interior walls are of material permitted in Table 600 and have fire resistance not less than that specified in Table 600; and beams, girders, trusses, arches, floors, roofs, and interior framing are wholly or partly of wood or other approved materials and have fire resistance not less than that specified in this code. Type V construction may be either protected or unprotected. Fire resistance requirements for structural elements of Type V construction shall be as specified in this code.

Type VI is construction in which the exterior bearing and nonbearing walls and partitions, beams, girders, trusses, arches, floors, and roofs and their supports are wholly or partly of wood or other approved materials. Type VI construction may be either protected or unprotected. Fire resistance requirements for structural elements of Type VI construction shall be as specified in this code.

Mixed types of construction are those in which two or more types of construction not separated by fire walls occur in the same building. The area of the entire building shall not exceed the least area permitted based on occupancy for the types of construction used in the building.

---

the calibration of an apparatus, the assessment of a measurement method, or for assigning values to materials.

The SRM Standard Reference Materials catalog is available from the U.S. Government Printing Office.

Underwriters Laboratories Inc.

Underwriters Laboratories Inc. (UL) is a nonprofit organization that establishes, maintains, and operates laboratories for the examination and testing of devices, systems, and materials to determine their relation to hazards to life and property. Its objectives are

1. To determine the relation of various materials, devices, products, equipment, constructions, methods, and systems to hazards appurtenant thereto or to the use thereof, affecting life and property. This is accomplished by scientific investigation, study, experiments, and tests.

2. To ascertain, define, and publish standards, classifications, and specifications for materials, devices, products, equipment, constructions, methods, and systems affecting such hazards.

3. To disseminate information tending to reduce and prevent bodily injury, loss of life, and property from such hazards.

Underwriters Laboratories engineering functions are divided into six departments:

▲ Burglary Protection and Signaling
▲ Casualty and Chemical Hazards
▲ Electrical
▲ Fire Protection
▲ Heating, Air-Conditioning, and Refrigeration
▲ Marine

UL publishes yearly directories of manufacturers who have demonstrated an ability to produce products meeting UL requirements. They publish an extensive list of standards that have been accepted as "UL Standards for Safety." Products manufactured to these standards are permitted to carry the UL symbol indicating approval (Fig. 2.21).

UL standards are designated by the initials UL followed by an identifying number. For example, UL943 covers ground fault circuit interrupters.

Figure 2.21 Manufacturers whose products meet the Safety Standards of Underwriters Laboratories Inc. (UL) may be authorized by UL, subject to testing and evaluation, to use this UL mark on those products. *(Reproduced with the permission of Underwriters Laboratories Inc.)*

American Society of Testing and Materials

The American Society for Testing and Materials (ASTM) is a nonprofit corporation formed for the development of standards on characteristics and performance of materials, products, systems, and services and for the promotion of related knowledge. The standards include test methods, definitions, recommended practices, classifications, and specifications.

ASTM standards are developed by those who are producers, users, consumers, and government and academic representatives who have expertise in the specific areas under study and who voluntarily choose to work within the ASTM system. Standards developed by the society's committees are published annually in a 48-volume publication, *Book of ASTM Standards.* Section 4 contains five volumes that encompass the standards for the construction industry. In addition, ASTM has a yearly *Publications Catalog,* which is a comprehensive listing of all ASTM publications.

ASTM specifications are designated by the initials ASTM followed by an identification number and the year of last revision. One example is *High-Strength Bolts for Structural Steel Joints,* ASTM A325-79.

ASTM also publishes *Standards Adjuncts,* which includes supporting material such as reference objects, charts, tables, and drawings used in conjunction with ASTM test methods. They are identified by a single letter of the alphabet followed by a code number and the word adjunct. An example is C856 adjunct, *Petrographic Examination of Hardened Concrete.*

ASTM also publishes *Special Technical Publications,* which are books devoted to specific technical, scientific, and standardization topics. They are identified by the letters STP followed by a code number. An example is STP 691 *Durability of Building Materials and Components.*

American National Standards Institute

The American National Standards Institute (ANSI) is the coordinating organization for the national standards system in the United States. It is a federation of approximately nine hundred companies and two hundred trade, technical, professional, labor, and consumer organizations.

ANSI serves the nation in three ways:

1. *It coordinates the voluntary development of national standards.* Through cooperative efforts of its member organizations, ANSI's councils, boards, and committees coordinate the efforts of the hundreds of organizations in the United States that develop standards (Fig. 2.22).

ANSI helps identify what standards are needed and establishes timetables for their completion. It

Figure 2.22 This American National Standard was developed by the Hardwood Plywood & Veneer Association. *(Courtesy Hardwood Plywood & Veneer Association)*

arranges for competent organizations to develop them or, if a standard-developing organization does not exist, it forms a group of persons who have the necessary competencies to develop the standard.

2. *It establishes national consensus standards.* ANSI provides and administers the only recognized system in the United States for establishing standards, regardless of the originating source, as American National (Consensus) Standards. The standards approval procedures ensure that all concerned national interests (trades, labor, manufacturing, consumers, etc.) have had an opportunity to participate in a standard's development or to comment on its provisions. They further ensure that the standard has achieved general recognition and acceptance for use. The approval process

includes the right to appeal actions at several levels during the review.

3. *It provides for the effective representation of U.S. interests in international standardization carried out by such nontreaty organizations as the International Organization for Standardization (ISO) and the International Electrotechnical Commission (IEC).* ANSI is the official representative of the United States to the international community pertaining to standards development. It manages, coordinates, finances, and administratively supports effective participation in ISO and IEC. It helps govern ISO through membership on its Council Executive Committee and Planning Committee.

There are now more than 10,000 American National Standards in virtually every field and discipline. Standards deal with dimensions, ratings, terminology and symbols, test methods, and performance and safety specifications for materials, equipment, components, and products in the following fields: construction, electrical and electronics, heating, air-conditioning and refrigeration, information systems, medical devices, mechanical, nuclear, physical distribution, piping and processing, photography and motion pictures, textiles, and welding.

An American National Standard is designated by code numbers beginning with the letters ANSI followed by letters indicating the organization that formulated the standard, such as ACI (American Concrete Institute). These letters are followed by an identification number and the date the standard was issued or released. A typical example is ANSI/ACI 517-70, *Practice for Curing Concrete.*

International Organization for Standardization (ISO)

The International Organization for Standardization (ISO) is the world body that coordinates the production of engineering and product standards for worldwide use. Each industrial nation in the world has a national standards group. In the United States, it is the American National Standards Institute (ANSI) and in Canada it is the Canadian Standards Association. Most nations are represented on ISO through their national standards organizations. Since most of the world is metric, the United States needs to continue to work closely with ISO and accept its metric standards.

National Association of Home Builders of the United States

The National Association of Home Builders (NAHB) is a federation of approximately eight hundred state and local builders associations in the United States. It has a wide range of committees and councils. It established and staffs the National Housing Center in Washington, D.C. It publishes a number of newsletters and magazines and manages a National Housing Center Library. It issues reports pertaining to economic research and analysis and offers seminars, workshops, and conferences.

NAHB Research Foundation

The NAHB Research Foundation is a subsidiary of the National Association of Home Builders. It serves as a center of technical research for the residential and light frame construction industries. It has four divisions: (1) The Laboratory Services Division performs tests on materials and components under contract from their manufacturers. It could be a part of a product certification program. (2) The Industrial Engineering Division designs and performs cost and economic studies of alternative construction techniques and materials. (3) The Building Systems Division provides technical studies of whole functional systems in small buildings. (4) The Special Studies Division initiates projects in new areas of interest to the residential construction industry.

Canadian Organizations

The following are a few of the Canadian organizations active in various aspects of the construction industry.

The Canada Mortgage and Housing Corporation (CMHC) is an agency of the federal government and is responsible for administering the National Housing Act. This legislation is designed to improve the housing and living conditions in Canada. Among its responsibilities is conducting research into the social, economic, and technical aspects of housing and related fields. It publishes the results of these activities.

The Canadian Housing Information Centre (CHIC) is a part of the Canadian Mortgage and Housing Corporation. It provides thousands of publications related to all aspects of housing, building, and community development.

The Canadian Wood Council (CWC) is a national federation of forest products associations. It is involved in developing and promoting technical product information to assist members of the specifying and regulatory community. It has an extensive listing of technical manuals and books.

THE MASTERFORMAT OF THE CONSTRUCTION SPECIFICATIONS INSTITUTE

MasterFormat™ is a system of numbers and descriptive titles used to organize the extensive array of construction information into a standard order, which facilitates the retrieval of information and serves as a means for communicating within the construction industry. Since the design and completion of construction projects involves individuals in many technical fields, the ability to communicate by having a standard system for identifying and referring to construction information is essential. The MasterFormat system was developed by the Construction Specifications Institute and Construction Specifications Canada. It is used by U.S. and Canadian construction companies and material suppliers, the U.S. Department of Defense, the McGraw-Hill Information Systems as a basis for their Sweet's Catalog Files, and the R.S. Means Company construction cost data publications.

The numbers and titles in MasterFormat are divided into sixteen basic groupings called divisions. Each division has a title and identifying number. Within each division specifications are identified in numbered sections, each of which covers a particular part of the work of the division. Each section is identified by a five-digit number. The first two digits are the same as the division number. The divisions and sections are shown in Table 2.1.

Each section is divided into four levels of detail, identified as level one, level two, level three, and level four. Level two titles cover the broad categories and are the major section titles. The hyphenated three-digit numbers are the level three titles. They limit the area covered by the level two numbers and titles. Level four titles are very limited and are used for specific elements of work. They appear below the level three titles and are not numbered (Table 2.2).

Table 2.1 CSI MasterFormat Contents

MasterFormat™
Level Two Numbers and Titles
Introductory Information
00001   Project Title Page
00005   Certifications Page
00007   Seals Page
00010   Table of Contents
00015   List of Drawings
00020   List of Schedules
Bidding Requirements
00100   Bid Solicitation
00200   Instructions to Bidders
00300   Information Available to Bidders
00400   Bid Forms and Supplements
00490   Bidding Addenda
Contracting Requirements
00500   Agreement
00600   Bonds and Certificates
00700   General Conditions
00800   Supplementary Conditions
00900   Addenda and Modifications
Facilities and Spaces
         Facilities and Spaces
Systems and Assemblies
         Systems and Assemblies

Division 1
General Requirements
01100   Summary
01200   Price and Payment Procedures
01300   Administrative Requirements
01400   Quality Requirements
01500   Temporary Facilities and Controls
01600   Product Requirements
01700   Execution Requirements
01800   Facility Operation
01900   Facility Decommissioning

Division 2
Site Construction
02050   Basic Site Materials and Methods
02100   Site Remediation
02200   Site Preparation
02300   Earthwork
02400   Tunneling, Boring, and Jacking
02450   Foundation and Load-Bearing Elements
02500   Utility Services
02600   Drainage and Containment
02700   Bases, Ballasts, Pavements, and
         Appurtenances
02800   Site Improvements and Amenities
02900   Planting
02950   Site Restoration and Rehabilitation

Division 3
Concrete
03050   Basic Concrete Materials and
         Methods
03100   Concrete Forms and Accessories
03200   Concrete Reinforcement
03300   Cast-in-Place Concrete
03400   Precast Concrete
03500   Cementitious Decks and
         Underlayment
03600   Grouts
03700   Mass Concrete
03900   Concrete Restoration and Cleaning

Division 4
Masonry
04050   Basic Masonry Materials and Methods
04200   Masonry Units
04400   Stone
04500   Refractories
04600   Corrosion-Resistant Masonry
04700   Simulated Masonry
04800   Masonry Assemblies
04900   Masonry Restoration and Cleaning

Division 5
Metals
05050   Basic Metal Materials and Methods
05100   Structural Metal Framing
05200   Metal Joists
05300   Metal Deck
05400   Cold-Formed Metal Framing
05500   Metal Fabrications
05600   Hydraulic Fabrications
05650   Railroad Track and Accessories
05700   Ornamental Metal
05800   Expansion Control
05900   Metal Restoration and Cleaning

Division 6
Wood and Plastics
06050   Basic Wood and Plastic Materials and
         Methods
06100   Rough Carpentry
06200   Finish Carpentry
06400   Architectural Woodwork
06500   Structural Plastics
06600   Plastic Fabrications
06900   Wood and Plastic Restoration and
         Cleaning

Division 7
Thermal and Moisture Protection
07050   Basic Thermal and Moisture Protection
         Materials and Methods

Table 2.1 *Continued*

07100	Dampproofing and Waterproofing	10520	Fire Protection Specialties
07200	Thermal Protection	10530	Protective Covers
07300	Shingles, Roof Tiles, and Roof Coverings	10550	Postal Specialties
		10600	Partitions
07400	Roofing and Siding Panels	10670	Storage Shelving
07500	Membrane Roofing	10700	Exterior Protection
07600	Flashing and Sheet Metal	10750	Telephone Specialties
07700	Roof Specialists and Accessories	10800	Toilet, Bath, and Laundry Accessories
07800	Fire and Smoke Protection	10880	Scales
07900	Joint Sealers	10900	Wardrobe and Closet Specialties

Division 8
Doors and Windows

Division 11
Equipment

08050	Basic Door and Window Materials and Methods	11010	Maintenance Equipment
		11020	Security and Vault Equipment
08100	Metal Doors and Frames	11030	Teller and Service Equipment
08200	Wood and Plastic Doors	11040	Ecclesiastical Equipment
08300	Specialty Doors	11050	Library Equipment
08400	Entrances and Storefronts	11060	Theater and Stage Equipment
08500	Windows	11070	Instrumental Equipment
08600	Skylights	11080	Registration Equipment
08700	Hardware	11090	Checkroom Equipment
08800	Glazing	11100	Mercantile Equipment
08900	Glazed Curtain Wall	11110	Commercial Laundry and Dry Cleaning Equipment

Division 9
Finishes

09050	Basic Finish Materials and Methods	11120	Vending Equipment
09100	Metal Support Assemblies	11130	Audio-Visual Equipment
09200	Plaster and Gypsum Board	11140	Vehicle Service Equipment
09300	Tile	11150	Parking Control Equipment
09400	Terrazzo	11160	Loading Dock Equipment
09500	Ceilings	11170	Solid Waste Handling Equipment
09600	Flooring	11190	Detention Equipment
09700	Wall Finishes	11200	Water Supply and Treatment Equipment
09800	Acoustical Treatment	11280	Hydraulic Gates and Valves
09900	Paints and Coatings	11300	Fluid Waste Treatment and Disposal Equipment

Division 10
Specialties

10100	Visual Display Boards	11400	Food Service Equipment
10150	Compartments and Cubicles	11450	Residential Equipment
10200	Louvers and Vents	11460	Unit Kitchens
10240	Grilles and Screens	11470	Darkroom Equipment
10250	Service Walls	11480	Athletic, Recreational, and Therapeutic Equipment
10260	Wall and Corner Guards		
10270	Access Flooring	11500	Industrial and Process Equipment
10290	Pest Control	11600	Laboratory Equipment
10300	Fireplaces and Stoves	11650	Planetarium Equipment
10340	Manufactured Exterior Specialties	11660	Observatory Equipment
10350	Flagpoles	11680	Office Equipment
10400	Identification Devices	11700	Medical Equipment
10450	Pedestrian Control Devices	11780	Mortuary Equipment
		11850	Navigation Equipment
10500	Lockers	11870	Agricultural Equipment
		11900	Exhibit Equipment

(continued on next page)

Table 2.1 *Continued*

Division 12		13800	Building Automation and Control
Furnishings		13850	Detection and Alarm
12050	Fabrics	13900	Fire Suppression
12100	Art		
12300	Manufactured Casework	Division 14	
12400	Furnishings and Accessories	Conveying Systems	
12500	Furniture	14100	Dumbwaiters
12600	Multiple Seating	14200	Elevators
12700	Systems Furniture	14300	Escalators and Moving Walks
12800	Interior Plants and Planters	14400	Lifts
12900	Furnishings Restoration and Repair	14500	Material Handling
		14600	Hoists and Cranes
Division 13		14700	Turntables
Special Construction		14800	Scaffolding
13010	Air-Supported Structures	14900	Transportation
13020	Building Modules		
13030	Special Purpose Rooms	Division 15	
13080	Sound, Vibration, and Seismic	Mechanical	
	Control	15050	Basic Mechanical Materials and
13090	Radiation Protection		Methods
13100	Lightning Protection	15100	Building Services Piping
13110	Cathodic Protection	15200	Process Piping
13120	Pre-Engineered Structures	15300	Fire Protection Piping
13150	Swimming Pools	15400	Plumbing Fixtures and Equipment
13160	Aquariums	15500	Heat-Generation Equipment
13165	Aquatic Park Facilities	15600	Refrigeration Equipment
13170	Tubs and Pools	15700	Heating, Ventilating, and Air
13175	Ice Rinks		Conditioning Equipment
13185	Kennels and Animal Shelters	15800	Air Distribution
13190	Site-Constructed Incinerators	15900	HVAC Instrumentation and Controls
13200	Storage Tanks	15950	Testing, Adjusting, and Balancing
13220	Filter Underdrains and Media		
13230	Digester Covers and Appurtenances	Division 16	
13240	Oxygenation Systems	Electrical	
13260	Sludge Conditioning Systems	16050	Basic Electrical Materials and Methods
13280	Hazardous Material Remediation	16100	Wiring Methods
13400	Measurement and Control	16200	Electrical Power
	Instrumentation	16300	Transmission and Distribution
13500	Recording Instrumentation	16400	Low-Voltage Distribution
13550	Transportation Control Instrumentation	16500	Lighting
13600	Solar and Wind Energy Equipment	16700	Communications
13700	Security Access and Surveillance	16800	Sound and Video

The numbers and Titles use in this book are from *MasterFormat* (1995 edition). *MasterFormat* is published by the Construction Specifications Institute (CSI) and Construction Specifications Canada (CSC), and is reproduced with permission from CSI, 1996.

Table 2.2 Levels of the CSI MasterFormat Classification System

Level Two	09200	Plaster and Gypsum Board
Level Three	−205	Furring and Lathing
Level Four		Gypsum Lath
		Lead-lined Lath
		Metal Lath
		Plaster Accessories
		Veneer Plaster Base
	−210	Gypsum Plaster
		Acoustical Gypsum Plaster
		Fireproofing GypsumPlaster
		Gypsum Veneer Plaster
	−220	Portland Cement Plaster
		Adobe Finish
		Portland Cement Veneer Plaster

Table 2.3 SI Metric Base Units

Quantity	Name of Unit	Symbol
Length	meter	m
Mass	kilogram	kg
Time	second	s
Electric current	ampere	A
Thermodynamic temperature	kelvin	K
Amount of substance	mole	mol
Luminous intensity	candela	cd

METRICS IN CONSTRUCTION

The United States is in the process of converting to the International System of Units or, as it is referred to, the **SI metric system.** As this occurs, it will produce changes in the standard sizes of materials and other quantities used, such as length, volume, and weight. Some industries in the United States already design and produce products that are totally metric. Others are still in the process of change, and metric units have not reached the workplace. The construction industry is one of these. As we await decisions pertaining to the metric sizes of construction materials, it is important that those in the construction industry become familiar with metric units. It is essential for any who work outside the United States.

SI Metric Base Units

The seven base units of measurement in the SI metric system are length, mass, electric current, time, quantity of substance, thermodynamic temperature, and luminous intensity. These are indicated by standard symbols as shown in Table 2.3. Notice that some are capital letters and some are lowercase. It is important to show the correct case of the symbol.

Following is a brief discussion of each of these units.

Length. The basic unit of length is the meter. Subunits are decimeter, centimeter, and millimeter. A decimeter is one-tenth of a meter, a centimeter is one-hundredth of a meter, and a millimeter is one-thousandth of a meter.
Mass. Basically this is considered as weight. Very light weights are given in grams (g) and heavier weights in kilograms (1,000 grams) (kg).

Electric current. Amperes (A) are used to measure electric current. An ampere is the unit of the rate of flow of electric current.
Time. Time is measured in seconds in the same manner as the customary system.
Substance. The quantity of substance is the mole. It is the amount of atoms in carbon 12 and is used mostly in chemistry.
Temperature. The metric unit for temperature is the kelvin (K).
Luminous intensity. The metric unit of luminous intensity is the candela (cd).

Prefixes

A prefixes is a word added to the beginning of a word that alters the word's meaning. Metric prefixes are added to SI metric units to change the multiples of the unit. For example, the prefix kilo means 1,000, so a kilogram is 1,000 grams. The prefix milli means 1/1,000 of the unit. For example, a millimeter is one-thousandth of a meter. It takes 1,000 millimeters to equal a meter. The commonly used prefixes are in Table 2.4.

Derived SI Units

Derived SI units are combinations of base or other metric units generated by multiplying or dividing. For example, area is derived by multiplying the length and width of a rectangle (in meters), producing a result in square meters (m^2). This is a unit derived from the base unit, meter. SI derived units with special names that are used in engineering calculations are in Table 2.5. Following are some SI units derived from SI Base units that are commonly used in construction work.

Volume. Small volumes are specified in cubic millimeters (mm^3) or cubic centimeters (cm^3) and larger volumes

Table 2.4 Metric Prefixes

Prefix	Symbol	Means to Multiply by:	or by:
Mega	M	1 000 000	10^6
Kilo	k	1 000	10^3
Hecto	h	100	10^2
Deca	da	10	10
Deci	d	0.1	10^{-1}
Centi	c	0.01	10^{-2}
Milli	m	0.001	10^{-3}
Micro	μ	0.000 01	10^{-6}
Nano	n	0.000 000 001	10^{-9}

Table 2.5 SI Derived Units with Special Names

Quantity	Name	Symbol	Expression
Frequency	hertz	Hz	$Hz = s^{-1}$
Force	newton	N	$N = kg \cdot m/s^2$
Pressure, stress	pascal	Pa	$Pa = N/m^2$
Energy, work, quantity of heat	joule	J	$J = N \cdot m$
Power, radiant flux	watt	W	$W = J/s$
Electric charge, quantity	coulomb	C	$C = A \cdot s$
Electric potential	volt	V	$V = W/A$ or J/C
Capacitance	farad	F	$F = C/V$
Electric resistance	ohm	Ω	$\Omega = V/A$
Electric conductance	siemens	S	$S = A/V$ or Ω^{-1}
Magnetic flux	weber	Wb	$Wb = V \cdot s$
Magnetic flux density	tesla	T	$T = Wb/m^2$
Inductance	henry	H	$H = Wb/A$
Luminous flux	lumen	lm	$lm = cd \cdot sr$
Illuminance	lux	lx	$lx = lm/m^2$

(for example, for concrete) in cubic meters (m^3). A cubic meter is the volume of a cube the edges of which are one meter. The liter (L) is the measurement for liquid volume.

Area. The basic unit for area is the square meter (m^2), while smaller areas are specified in square centimeters (cm^2) or square millimeters (mm^2). A square meter is the area of a square with edges of one meter. The hectare (ha) is the metric measurement used in surveying.

Pressure. The metric unit for pressure (replacing pounds per square inch) is kilopascals (kPa) and megapascals (MPa) for stress.

Heavy loads. The metric ton (t) is used to denote large loads, such as those used in excavation.

Plane and solid angles. The radian (rad) and steradian (sr) denote plane and solid angles used in lighting work and some engineering calculations. The units degree (°), minute ('), and second (") are used in surveying.

Conversion Factors

Metric projects are designed in metric units and conversion to English units should not be necessary. There may be times when a project designed in English units must be converted to metric units. Tables 2.6 and 2.7 list frequently required conversion factors.

U.S. Metric Policy

The Metric Conversion Act of 1975 as amended by the Omnibus Trade and Competitiveness Act of 1988 established the modern metric system (Système International d'Unités or SI) as the preferred system of measurement in the United States. It requires that, to

Table 2.6 Metric-to-English Conversion Table

	From	To	Multiply by:
Length	mm	in.	0.039
	m	ft.	3.280
	km	mi.	0.621
Area	mm^2	$in.^2$	0.002
	m^2	$ft.^2$	10.763
	km^2	$mi.^2$	0.386
Volume	m^3	$ft.^3$	35.314
	L	gallon	0.219
Speed	m/s	mi./hr.	2.236
Flow	m^3/s	gal./day	19.005
	L/s	$ft.^3/min.$	2.119
Mass	kg	lbs.	2.204
	g	oz.	0.035
Density	kg/m^2	$lbs./ft.^2$	0.204
	kg/m^3	$lbs./ft.^3$	0.062
Force	N	lbf	0.224
	kN	kip	0.224
Luminance	lx	ft.-candle	0.093
Energy and power	kJ	Btu	0.948
	W	Btu/hr.	3.412
	$W/(m^2 \cdot °C)$	$Btu/ ft.^2/hr./°F$	0.176
	W	HP	0.001

the extent feasible, the metric system be used in all federal procurement, grants, and business-related activities by September 30, 1992. The U.S. Department of Commerce established the Interagency Council on Metric Policy (ICMP). Its subcommittee, the Metrication Operating Committee, established a Construction Subcommittee to facilitate the metrication of all federal construction. The Construction Subcommittee established the goal of instituting the use of metric design in all new federal facilities by January 1994.

Table 2.7 English-to-Metric Conversion Table

	From	To	Multiply By:
Length	in.	mm	25.4
	ft.	m	0.305
	miles	km	1.609
Area	in.^{2}	mm^2	645.16
	ft.^{2}	m^2	0.092
	mi.^{2}	km^2	2.590
Volume	in.^{3}	mm^3	16 387.1
	yds.^{3}	m^3	0.765
	quarts	L	1.137
Speed	mi./hr.	m/s	0.447
	gal./day	m^3/d	0.005
Flow	ft.^{3}/min.	L/s	0.472
Mass	lbs.	kg	0.453
	oz.	g	28.350
	lbs/in.^{2}	kg/m^2	703.07
Density	lbs./ft.^{2}	kg/m^2	4.882
	lbs./ft.^{3}	kg/m^3	16.018
Force	lbf	N	4.448
	kip	kN	4.448
Luminance	ft.-candle	lx	10.764
Energy and power	Btu	kJ	1.055
	Btu/h	W	0.293
	Btu/ft.^{2}/hr./ $^\circ$F	W/m$^{2.\circ}$C)	5.678
	HP	kW	0.746

KEY TERMS

building code Regulations that specify the minimum standards pertaining to the design and construction of buildings to safeguard life or limb, health, property, and public welfare.

metric system An international system of units, referred to as SI in all languages based on decimal arithmetic.

specifications A written contract document giving descriptions of a technical nature of materials, equipment systems, standards, and workmanship.

trade association An organization of manufacturers and businesses involved in the production and supply of materials and services to various areas of the construction industry.

zoning ordinance A local municipal ordinance that controls the use of land and building, the height and bulk of buildings, density of population, building coverage of the lot, setbacks, ancillary facilities to implement a master plan.

REVIEW QUESTIONS

1. What are the three major divisions of the construction industry?
2. What is the difference between a general contractor and a specialty contractor?
3. What is a specification document?
4. How does a construction contractor know how the various parts of a building should be assembled?
5. What is the purpose of zoning ordinances?
6. What purposes do building codes serve?
7. What are the major regional building codes in the United States and Canada?
8. What purposes are served by trade associations?
9. What are the purposes of the National Institute of Building Sciences?
10. Why are product standards important to the construction industry?
11. What purposes are served by the Underwriters Laboratories, Inc.?
12. How does the work of the American Society for Testing and Materials affect the construction industry?
13. What is the purpose of the MasterFormat published by the Construction Specifications Institute?
14. What are the metric units used for length, mass, and luminous intensity?

SUGGESTED ACTIVITIES

1. Convert the following customary units to metric units: 2″ × 4″, 15′-6″, 100 pounds, 72°F

2. Contact one of the trade organizations and request information concerning their activities and a catalog of publications. Share with others in the class. This could be used to build a permanent file of information.

3. Try to get your school to secure a set of Sweet's General Building and Renovation Catalog File. The file is also available on compact discs. If sets cannot be purchased, contact local architects and general contractors, and try to secure one of their older sets for use in the classroom.

4. Ask a general contractor and a specialty contractor to address the class to explain the operation of their companies and how they cooperate to bid on and construct a project.

5. Ask the local building official (inspector) to address the class and review local zoning ordinances and give a brief review of the building code used.

ADDITIONAL INFORMATION

American National Metric Council
1735 N. Lynn St., Suite 950
Arlington, VA 22209-2022

National Institute of Science and Technology
U.S. Department of Commerce
Bldg. 101, Rm 813
Gaithersburg, MD 20899

U.S. Metric Association
10245 Andasol Ave.
Northridge, CA 91325

Metric (SI) in Everyday Science and Engineering by Stan
   Jakuba. Available from
   Society of Automotive Engineers, Inc.,
   400 Commonwealth Drive
   Warrendale, PA 15096-0001.

3

Properties of Materials

This chapter will help you to:

1. Understand the requirements to be considered when selecting construction materials.

2. Know the properties of materials that must be considered when selecting and installing construction products.

3. Discuss the technical aspects of mechanical, thermal, electrical, and chemical properties.

The properties of materials are determined by their structure. The basic properties of materials are known from testing and research. Using computers, architects and engineers rapidly process data on materials, enabling them to make maximum use of the materials currently available to them.

New materials are developed each year, and the architect, contractor, and engineer must be informed about their properties so they can use these materials in the most effective and safe manner. For a material to be suitable for a particular application it must have predictable behavior. If the properties of the material are known, the designer can predict its behavior and check it against that specified by codes and standards. The material must measure up to these to be an acceptable choice.

The following requirements are considered when selecting materials:

▲ *Integrity of shape.* The product must retain its shape within specified tolerances when under a design load. Excessive shape change can cause a failure.

▲ *Strength.* The material must carry the design loads and keep deflection and deformation within design limits. *Deflection* is the distance a loaded member bends from its unloaded condition. *Deformation* is a change from the original shape due to a load on the member. The material for structural parts of a building must resist the expected tension, compression, shear, and torsion.

▲ *Structural integrity.* The material must resist structural changes due to temperature and other environmental conditions that might cause it to fail. For example, the internal structure of most plastics changes under elevated temperatures, causing the material to fail under a load.

▲ *Useful life.* The material must remain in a serviceable condition for a predictable period of time. In some cases this includes the time it was in storage before it was used. For example, some types of roofing materials have a guaranteed useful life of twenty-five years.

▲ *Specified functions.* If a product has to perform a special function, such as one that requires optical, magnetic, or thermal qualities, the selection of materials becomes very limited.

49

▲ *Cost*. The selection of materials is influenced by their cost.

Decisions also must be made concerning other considerations, such as *appearance, availability, maintenance,* and *ease of installation.*

MATERIALS GROUPS AND PROPERTIES

The basic construction materials can be grouped into three types: metals, ceramics, and organics.

Metals are extracted from ores and refined. Due to the addition of a variety of ingredients, metals have a wide variation in properties. Most metals are ductile and strong, and they are especially useful when tensile forces are expected to occur.

Ceramic materials are extracted from the soil. When they are refined they produce a variety of products. Typical ceramic materials include sand, limestone, glass, brick, cement, gypsum, mortar, and mineral wool insulation. They are hard, rigid, brittle, and heavy. They are especially useful when compression forces are expected to occur.

Organic materials include wood, bitumens, and many synthetic materials based on a chemical compound containing carbon. Examples include wood and paper products, asphalts, rubber, and plastics. These materials vary a great deal in their properties and this influences how they can be best used in construction.

The four properties of materials that most influence decisions in construction are *mechanical, thermal, electrical,* and *chemical.*

MECHANICAL PROPERTIES

Mechanical properties are a measure of a material's ability to resist mechanical forces or stresses. They are related to the response of a material to static (continuous) and dynamic (intermittent) loads. Mechanical properties are ascertained by standard tests to determine the reaction of materials to various conditions.

Stresses

There are three possible types of stresses—tension, compression, and shear. **Tensile stresses** are created when forces pull on a member and tend to increase its length. **Compressive stresses** push on a member and tend to shorten it. **Shear stresses** produce forces that work in opposite directions parallel with the plane of the force or that cause a rotating motion such as that exerted by a wrench (Fig. 3.1).

The amount of stress a member can withstand is determined by making standardized tests as specified by the American Society for Testing and Materials (ASTM). These tests determine how the material be-

haves under a load. The load causes a stress-strain condition. Stress and strain occur together when a material is placed under load.

Stress and Strain

Stress is the intensity of internally distributed forces that resist a change in the form of a body. It is measured in pounds per square inch (psi), meganewtons per square meter (MN/m^2) or megapascals (MPa). One pascal (pa) equals $1 \ N/m^2$.

Stress is equal to the *load* divided by the *area* upon which it is acting. Normal stress is indicated by the Greek letter sigma, σ.

$$\sigma \ (\text{Stress}) = \frac{\text{Load}}{\text{Area}} = \text{psi, MN/m}^2, \text{ or MPa}$$

This applies to materials that have the force applied parallel with the axis of the part. It is important to know if a load is to be distributed over an area or concentrated on a small point. For example, if a 100 lb. load is distributed over a column 2 in. square the stress is 25 psi. If it was concentrated on the end of a rod with a cross-sectional area of 1 in.^{2} the stress would be 100 psi (Fig. 3.2).

Normal **strain** is the change per unit length in a linear dimension of a body that accompanies a change in stress. Strain is measured in inches of change per inch of the original dimension (in./in.) or in millimeters per meter (mm/m) or given as a percent. Strain is indicated by the Greek letter epsilon, ε.

Strain is equal to the change in the original dimension divided by the original dimension. If strain is to be stated as a percent of the dimensional change, the result is multiplied by 100.

$$\epsilon \ (\text{Strain}) = \frac{\text{Change in the}}{\text{Original dimension}} = \text{in./in. or mm/m}$$

$$= \frac{0.0010''}{120} = 0.000008 \ \text{in./in.}$$

In Fig. 3.2 the deformation of the column was 0.0010″ and the column was 120″ long. The strain was 0.000008 in./in.

Stress-Strain Diagrams

The behavior of a material under load can be represented on a stress-strain diagram (Fig. 3.3). A stress-strain diagram can be given for compression, tension, or shear. The data used to lay out the diagram are obtained from testing standard specimens of the material in a universal testing machine. (Information on testing machines can be found in Chapter 15.)

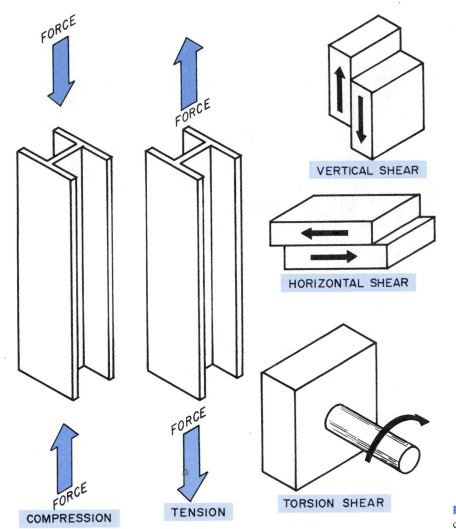

VERTICAL SHEAR

HORIZONTAL SHEAR

TORSION SHEAR

COMPRESSION

TENSION

Figure 3.1 Stresses are created by tension, compression, and shear forces.

The strain values are laid out on the horizontal axis (abscissa) and the stress values are on the vertical axis (ordinate). An increase in stress values will produce a proportionate increase in the strain, up to the elastic limit. This produces a straight line on the diagram from 0 to the elastic limit stress value.

As the stress-strain data are located on the diagram, the properties of the samples tested that can be noted are

Proportional limit
Elastic limit
Yield point
Yield strength
Ultimate or tensile strength
Breaking strength
Ductility
Elongation
Reduction of area
Modulus of elasticity

Proportional limit is the point up to which stress is proportional to the strain. Up to this limit the material behaves elastically. It has no permanent deformation. In other words, the deflection caused by the load is present as long as the load is applied. When the load is removed the material returns to its original position. This behavior is called **elastic deformation.** For example, the wind blows on a tall flag pole causing it to bend slightly. When the wind stops the pole returns to its original upright position.

The **elastic limit** is the highest stress that can be imposed on a material without permanent deformation. *Permanent deformation* is a permanent, irreversible deforming (changing shape) of the material and is called **plastic deformation.**

The *yield point* is the highest stress a material will bear and still retain its original shape.

The **yield strength** is the amount of stress required to produce a small, specific amount of permanent deformation. When a material is stressed to or beyond the

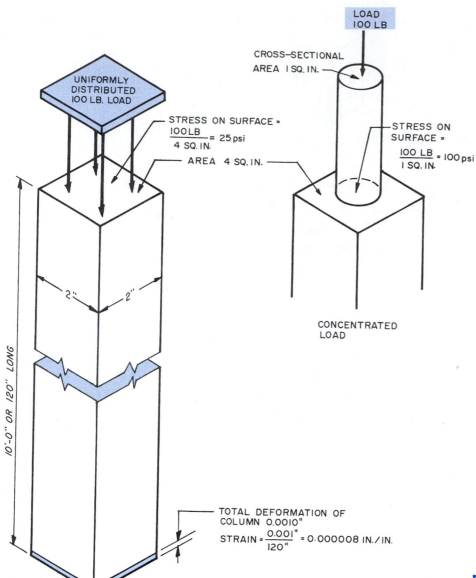

Figure 3.2 Columns are subjected to stress and strain forces.

yield strength, permanent deformation will occur. A material stressed in this range will not return to its original shape. Not all materials exhibit a yield point.

Ultimate or **tensile strength** is the maximum tensile stress a material can sustain up to the point of failure.

Breaking strength is the point at which the material actually fails.

Figure 3.4 shows the test results for a tensile test of a specimen of a mild steel as used in construction. This specimen has a yield strength of about 37,500 psi, an ultimate tensile strength of 66,000 psi, and a breaking strength of about 51,500 psi.

Ductility refers to the ability of a material to be deformed plastically without actually breaking or fractur-

ing. As tensile tests are conducted, ductility is determined as a percent of *elongation* and percent of *reduction of area* of the specimen. For example, copper is a very ductile material. As it is tested under tension, it stretches for some distance without breaking, indicating that it is ductile. As the copper specimen stretches it gets smaller in cross-sectional area.

Elongation is a measure in percent of the amount in inches the specimen stretches divided by the original length in inches multiplied by one hundred.

Reduction of area is a measure in percent of the difference between the final cross-sectional area and the original cross-sectional area, multiplied by one hundred.

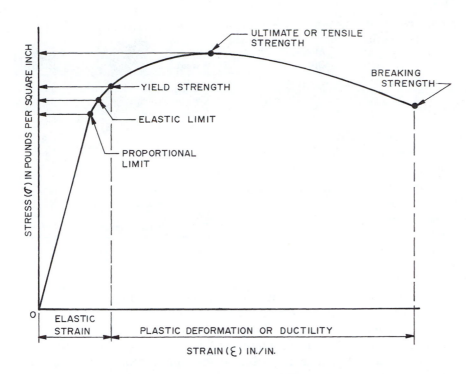

Figure 3.3 A typical stress-strain diagram. Note that since this material sustained considerable strain between the ultimate strength and before it fractured, it is a ductile material.

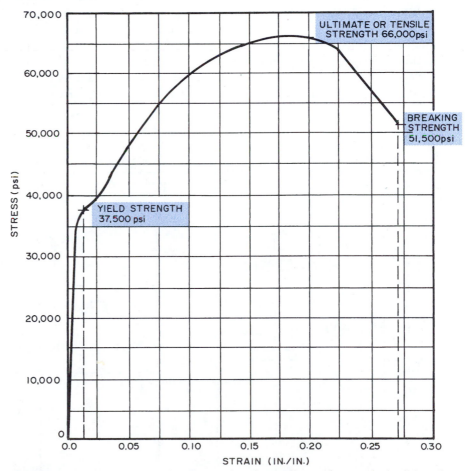

Figure 3.4 A diagram illustrating the results of a tensile test for a typical mild steel specimen.

Modulus of elasticity (E) is a proportional constant between stress and strain. For stresses within the elastic range, E equals stress divided by strain.

$$E = \frac{Stress}{Strain}$$

It is indicated by symbol E and is often referred to as the *E-value*. The modulus of elasticity is basically a measure of stiffness or rigidity. For beams, E-values are an indication of their resistance to deflection. The modulus of elasticity (E) defines the stiffness of a material, governs deflections, and influences buckling behavior. Table 3.1 lists E-values for selected construction materials. Remember that there is no one value for a particular kind of material because of the many variations available. For example, there are many types of plastics, each having its own stress-strain behavior. Within materials, such as steel, there are variations in the alloying ingredients used, so there are many types of steel. Steel has high E-values and is used where loads must be carried with minimal deflection, such as in beams. Plastics having low E-values are not useful for this purpose. The E-value of wood is higher than that of plastics but lower than that of concrete. Wood is used to carry light structural loads over short distances. One example is its use as residential joists and rafters. A comparison of the relative stiffness (E-value) of selected construction materials is in Fig. 3.5.

Ceramic materials have low E-values and do not deform plastically before they fracture. Therefore, they usually do not have a yield strength specified. Portland

Table 3.1 E-Values for Selected Construction Materials[a]

Material Group	Material	10^6 psi
Metals	Aluminum alloys	25
	Cast iron, malleable	25
	Copper alloy	13–17
	Steel, carbon	29
Ceramic Materials	Brick	1.8–2.2
	Concrete	1.0–5.0
	Glass	9–11
	Stone	10–15
Organic Materials	Particleboard	0.9–2.0
	Plastics, molded	0.1–1.9
	Plastics, reinforced	0.1–2.2
	Wood, 2″, visually graded	0.9–1.6

[a]These are approximate. Values will vary with specific samples.

cement concrete has a very low yield point when tested in tension. Concrete and other ceramic materials, such as brick and tile, are very brittle and are generally used to carry compressive loads but not tensile loads. In situations in which concrete is under tension, steel reinforcing is added to the member. Since there is no transition from elastic to plastic for ceramic materials, the yield strength, ultimate strength, and breaking strength are almost the same.

Plastics also have a low E-value. Therefore they have a very limited elastic region. A small elastic stress produces a large deflection. Plastic materials are not as satisfactory for load-carrying situations as are other materials. Reinforced plastics (with glass fiber reinforcing)

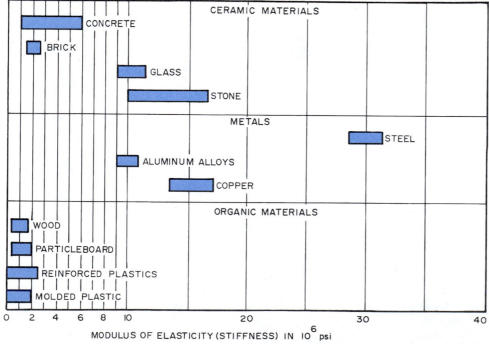

Figure 3.5 A comparison of the modulus of elasticity for selected construction materials.

have higher E-values than wood but are not currently used for structural purposes. Plastics in general have a rather low modulus of elasticity.

A study of these figures shows that metals, especially steel, are the strongest, have a good elastic range, and are ductile. Ceramic materials are good in compression, poor in tension, have no elastic range, and are, therefore, brittle. Molecular materials have low mechanical properties but are useful in construction applications in which these are not a major factor.

In addition to the mechanical performance indicated by the stress-strain data, other mechanical properties sometimes must be known when a material is being selected. These are compressive strength, hardness, fatigue strength, damping capacity, and impact strength.

Compressive Strength

Compressive strength is a measure of the ability of a material to resist forces that are tending to decrease its length. A load on the foundation wall or a column is an example. It is measured in pounds per square inch (psi) or kilograms per square millimeter (kg/mm^2). Refer to Fig. 3.1. Ceramic materials have high compressive strength and are often used when compressive forces are present. A stress-strain diagram may be given for compression.

Hardness

Hardness is a measure indicating the ability of a material to resist indentation or surface scratching. It is the result of several properties of a material, such as elasticity, ductility, brittleness, and toughness. Hardness is related to the tensile strength of the material. Generally, a material that has a high hardness rating also will have a high tensile strength. This is not true for all materials; concrete is one example.

Hardness tests are made according to ASTM test procedures. There are a number of hardness testing machines on the market. Basically, the machine forces a hardened steel ball into the material. The impression made by the ball is measured and is used to calculate the hardness number. Materials with high yield strength and E-values generally are hard and have a wear-resistant surface.

Fatigue Strength

Resistance of a material to a *cyclic load,* one that varies in direction and/or magnitude, is called **fatigue strength.** This is illustrated by bending a wire back and forth until it breaks. Most materials are lower in fatigue strength than they are in tensile strength. A failure due to fatigue occurs slowly. The product that fails due to fatigue offers some useful life before failure. This is one factor considered when the useful life of a product is established. Fatigue strength in buildings and other

structures, such as bridges and elevators, is commonly caused by vibrations and rotational motion produced by machinery. For example, fatigue on the structure of a steel building due to machinery can be reduced by mounting the machinery on vibration-absorbing pads made for that purpose. Many failures occur within products, such as electrical switches, pumps, and other mechanical equipment.

Damping Capacity

Damping capacity is the ability of a material to absorb vibrational energy. A material that has a good damping capacity also is a good sound-absorbing material.

Impact Strength

Impact strength is the ability of a material to resist a very rapidly applied load, such as the strike of a hammer. It is an indication of the toughness of the material. A material with a high impact strength will absorb the energy of impact without fracturing. Impact strength is affected by strength and ductility. Metals tend to be strong and ductile and have high impact strength. Ceramics are strong in compression but are brittle (lack ductility) and break under impact. Plastics, in general, are very ductile but low in strength and do not absorb impact very well, although there are exceptions to this.

THERMAL PROPERTIES

Thermal properties pertain to the behavior of a material when subjected to a change in temperature. When a material has its thermal energy level changed by heating or cooling, it may undergo a physical change, such as melting, vaporizing, or solidifying. It will also change in size and temperature.

Changes in temperature do not occur uniformly throughout a material. This depends on the transfer of heat energy and the rate at which the material absorbs heat. These factors—melting temperature, thermal conductivity, and thermal expansion—are of great significance when selecting construction materials.

Melting Temperature

The **melting temperature** is that point at which a material turns from a solid to liquid form. The melting temperature of materials varies a great deal (Table 3.2). In general, materials with high melting temperatures, such as ceramics, perform better at high temperatures. They also tend to retain their mechanical properties over a wider range of temperatures. The melting point of metals is below ceramics, so metals do not perform as well under service conditions having high temperatures. This is why steel structural members must be protected with a fire-resistant material. Although steel will not burn, it

Table 3.2 Melting Temperatures of Selected Materials[a]

Material Group	Material	Melting Temperatures °F[b]	°C
Metals	Aluminum alloys	2700	1494
	Copper alloys	1800–2000	990–1102
	Stainless steels	2700	1494
	Steels and low alloys	2700	1494
Ceramic Materials	Brick	2000–5000	1102–2782
	Concrete	3000	1662
	Plate glass	1500	822
Organic Materials	Acrylics (PMMA)	200–300	94–150
	ABS plastics	200–300	94–150
	Polyethylene (PE)	200	94
	Vinyl (PV)	200–300	94–150
	Wood	Chars at 400	206

[a]These are approximate melting temperatures. Actual temperatures will vary depending on the composition of the material.
[b]To convert °F to °C, subtract 32 from °F and multiply by 0.56.

does lose strength as its temperature rises. Molecular materials perform the poorest in conditions of high temperature. Service temperatures of commonly used construction materials are in Fig. 3.6.

Thermal Conductivity and Thermal Conductance

Thermal conductivity (k) is the ability of a material to transfer heat from an area of high temperature to an area having a lower temperature. For example, when you grind a piece of metal, heat is generated. Pretty soon the end you are holding gets so hot you cannot hold it. Heat from the grinding operation has been conducted to the other end of the metal part.

Thermal conductivity is a measure of the number of Btu's (British thermal units) that can pass through one square foot of a material one inch thick when the temperature on one surface is 74°F and is 75°F on the other. It is expressed as Btu/in./ft.²/hr./°F. The number of Btu's conducted through this piece of material from one side to the other in one hour is its thermal conductivity. Expressed in metric units, thermal conductivity is $W(m^2 \cdot °C)$ or watts per square meter per degree Celsius.

Metals tend to have the highest thermal conductivity. The molecular structure of metal is such that it transmits heat easily. Ceramics have much lower conductivities, and the organic materials are the lowest (Table 3.3). Materials with low thermal conductivity have excellent insulation qualities.

Thermal conductance (C), or the coefficient of heat transfer, is used with products for which the one-inch thickness used with the conductivity values is not a useful measure. Conductance is the number of Btu's per hour that will pass through one square foot of a material at a stated thickness in one hour when one face is 74°F and the other is 75°F. Examples for selected materials are in Table 3.4.

Table 3.3 Thermal Conductivity (k) of Selected Materials[a]

Material	Btu/in./hr./ft.²/°F[b] (English units)	W(m²·°C) (metric units)
Polyurethane foam	0.11–0.14	0.016–0.020
Polystyrene foam	0.17–0.28	0.024–0.040
Softwoods	0.80	0.12
Most other plastics	1–2	0.14–0.28
Plate glass	5	0.72
Hardwoods	1.10–0.16	0.16–0.02
Brick	9	1.3
Concrete	12	1.7
Steel, stainless	100	14
Steel, carbon	300–500	43–70
Aluminum alloys	800–1600	115–230
Copper alloys	500–2500	70–360

[a]These values are approximate. Actual values will vary depending on the composition of the material.
[b]To convert from English units to metric multiply the English units by 0.144.

Table 3.4 Thermal Conductance (C) of Selected Materials[a]

Material	Btu/in./hr./ft.²/°F[b] (English units)	W(m²·°C) (metric units)
Insulation board, 1/2"	0.75	4.62
8" hollow concrete block	0.90	5.11
Float and plate glass, 1/4"	1.13	6.42
Gypsum wallboard, 1/2"	2.20	12.50
Metal lath and 3/4" plaster	7.70	43.70

[a]Values are approximate. Actual values will depend on the composition of the material.
[b]To convert from imperial units to metric units, multiply the imperial units by 5.68.

Thermal Expansion

Most materials expand when heated. The amount of expansion differs depending on the material. The rate of expansion is expressed by the **coefficient of thermal expansion.** The units indicate the amount of change in

ITEM OF INTEREST

ON-THE-JOB DEATHS

In 1994 the Bureau of Labor Statistics reported that there were 6,588 occupation-related deaths, which was a 9 percent increase from the year before. Construction managers should note that 10 percent of these were by falls, of which 5 percent were in the construction industry. One-fifth of these deaths were from falling from or through roofs.

Transportation accidents were the largest single category, accounting for 42 percent of the deaths. Since construction projects involve the use of all sorts of vehicles, the construction manager should be certain to establish safety procedures for traffic on the site.

Contact with objects and equipment, such as moving or falling machinery and building materials, caused 15 percent of the deaths.

Exposure to harmful substances and environments, including contact with electrical current, caused 10 percent of the deaths.

This death rate averages eighteen work-related deaths per day. Of these 85 percent died the day of the accident and 97 percent were dead within thirty days from the date of the accident.

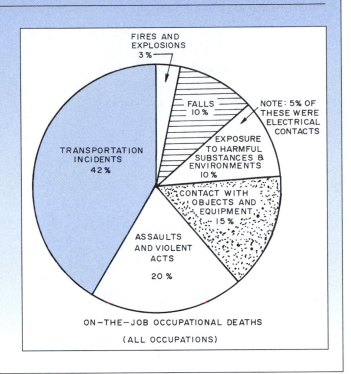

ON-THE-JOB OCCUPATIONAL DEATHS
(ALL OCCUPATIONS)

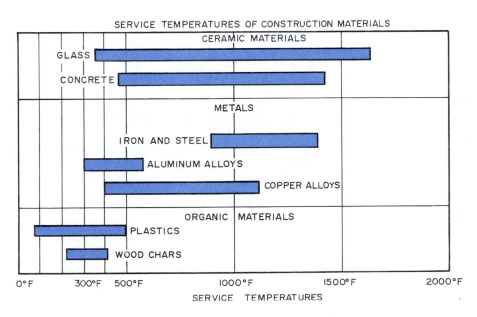

Figure 3.6 Service temperatures of construction materials.

inches per inch of length per °F temperature change in millionths of an inch: 10^{-6} (in./in./°F). This can indicate the amount of expansion or contraction that occurs as temperatures change. For example, the coefficient of thermal expansion for aluminum is 15×10^{-6} inches per inch of length per °F or 0.000015 inch per inch of length. In metric terms, thermal expansion is expressed as $W(m^2 \cdot °C)$ or watts per square meter per degree centigrade.

Plastics tend to have the highest coefficient of thermal expansion, metals are much lower than plastics, and ceramics are slightly lower than metal (Table 3.5). Notice that the coefficients of thermal expansion tend to be inverse to the melting temperature. Therefore, the higher the thermal expansion the lower the melting temperature.

Table 3.5 Coefficients of Thermal Expansion of Selected Materials[a]

Material	10^{-6}/in./in./°F[b] (English units)	$W(m^2 \cdot °C)$ (metric units)
Brick	3.1	5.6
Glass, float and plate	4.4	7.9
Reinforced concrete	6.2	11.2
Gypsum plaster	7.0	12.6
Gypsum wallboard	9.0	16.2
Cast iron	6.0	10.8
Steel, mild	6.7	12.0
Steel, high strength	6.9	12.4
Copper	9.3	16.7
Brass	10.4	18.7
Aluminum	12.8	23.0
Vinyls	30–100	54–180
Acrylics (PMMA)	35–50	63–90
Plastic (ABS)	45–50	81–90
Epoxies	50–60	90–108
Softwood (width)	1.7	3.1
Hardwood (width)	2.5	4.5

[a]These values are approximate. Actual values will depend upon the composition of the material.
[b]To convert imperial units to metric, multiply imperial units by 1.8.

When materials in a project change size, stresses develop. If provision is not made to permit materials to expand and contract, buckling and fractures will occur.

The amount of expansion in some materials is great. Steel, for example, has an expansion coefficient of 0.0000065 in./in./°F. If a steel beam 100 feet long has a temperature change from 70°F (23°C) to 100°F (40°C), the beam will expand 0.25 in. (6 mm) in length. If this beam happened to be aluminum, it would gain 0.50 in. (12 mm) in length.

ELECTRICAL PROPERTIES

Electrical conductivity (G) is a measure of the ability of a material to conduct electric current. Conductivity is measured in mhos. The symbol for conductivity is G.

Electrical resistance (R) is the ability of a material to oppose the flow of electrical current, thus causing a loss of power. The symbol for electrical resistance is R. Resistance is measured in ohms. Conductivity and resistance are reciprocals of each other:

$$\text{Conductivity} = \frac{1}{\text{Resistance}}$$

In general, metals have very high values of conductivity and low resistance. Ceramics and organic materials have low conductive values and high resistance, and they are considered *electrical insulators*, such as the protective plastic cover on electric wire. Ceramic and glass insulators are used on power transmission lines. They are described by their *resistance* to the flow of electricity. Therefore, materials that have high resistance have low conductivity and those having high conductivity have low resistance. Notice that materials that have a high thermal conductivity also have high electrical conductivity. This is because the free electrons in these materials allow heat and electrical current to be conducted easily.

Electrical conductivity is indicated in units of mhos per ft. of conductor length. A mho is equal to the conductance of a conductor in which a potential difference of one volt maintains a current of one ampere (Table 3.6). The best electrical conductors, in order of conductance, are silver, copper, and gold. Copper is widely used because it is an excellent conductor and less costly than silver. Aluminum is also a good conductor but not as good as those just mentioned. It is used for many electrical applications because it is less costly than copper. Following is a comparison, assuming silver has a conductivity rating of 100.

Silver	100
Copper	96
Aluminum	92
Iron	17

Table 3.6 Electrical Conductivity (G) of Selected Materials[a]

Material	Electrical Conductivity (G) Mhos (Γ)
Copper alloys	2,000,000.
Aluminum alloys	1,000,000.
Steel, carbon	166,666.
Steel, stainless	33,333.
Brick	0.0000002
Concrete	0.0000002
Vinyl	0.00000000001
Plate glass	0.0000000000001
Plastic (ABS)	0.000000000000001

[a]These values are approximate. Actual values for a particular piece of material depend on the composition of the material, its length, its cross-sectional area, and temperature.

Electrical resistance depends on the composition of the material, its length, its cross-sectional area, and the temperature. Electrical resistance (R) decreases as the cross-sectional area increases. Resistance increases as the length of a conductor increases. This is why a long extension cord made from a small diameter wire often gets very hot. A comparison of the electrical resistivity of selected metals, assuming that copper has a rating of 1.0, follows:

Copper	1.00
Aluminum	1.59
Brass	4.40
Iron	6.67
Steel	8.62

The resistances of selected materials are listed in Table 3.7.

Table 3.7 Electrical Resistance of Selected Materials[a]

Material	Electrical resistance (R) Ohms (Ω)
Copper alloys	.0000005
Aluminum alloys	.000001
Steel, carbon	.000006
Steel, stainless	.00003
Brick	4,000,000.
Concrete	4,000,000.
Vinyl	100,000,000,000.
Plate glass	10,000,000,000,000.
Plastic (ABS)	1,000,000,000,000,000.

[a]These values are approximate. Actual values of a particular piece of material depend on the composition of the material, its length, its cross-sectional area, and temperature.

CHEMICAL PROPERTIES

Construction materials are chemically degraded by the environment in which they are placed. This degradation is called *corrosion*. Corrosion is due to *chemical or electrochemical action*. The most common form of chemical corrosion is oxidation. **Oxidation** is the reaction of a material with the oxygen in the atmosphere. Iron rusting is an example. Other corrosive chemicals are present in the air, water, and soil to which materials are exposed. Some of these are sulfate ions in groundwater, soils, and seawater; sulfur dioxide gas in the air; and alkali in soils. Electrochemical action is due to a minute electrical current that flows between materials.

Nonmetallic materials are much more resistant to deterioration than most metals. In many cases they are used as protective coatings on metals and are not affected by electrochemical corrosion. Under some conditions nonmetals do experience deterioration. For example, concrete is damaged chemically by seawater and groundwater. If the water reaches the reinforcing steel in the concrete, it begins to rust, accelerating the deterioration. The durability of concrete depends on the chemical composition of the aggregate, the cement used, and the way it is mixed and placed.

Some plastics swell and crack when subjected to certain organic solvents. The solvents have a chemical reaction with one or more of the compounds making up the plastic. Other plastics deteriorate when exposed to oxygen and sunlight. Some types of foamed, rigid insulation deteriorate this way. Even glass can be susceptible to chemical attack under certain conditions.

Metals usually corrode due to electrochemical action. This action is similar to the way a wet-cell automobile battery operates. Corrosion occurs when minute amounts of electricity that occur naturally in the atmosphere or soil flow from one metal called an *anode* to a dissimilar metal called a *cathode* through a current-carrying medium (moisture) called the electrolyte. The anode deteriorates and the cathode remains unaffected by the electrochemical action. The electrolyte is usually a water or gas, such as carbon dioxide or sulfur dioxide (Fig. 3.7). The metal that is destroyed (corrodes) is one that is high on the galvanic table. Metals are ranked by their tendency to be anodic or cathodic. In Table 3.8,

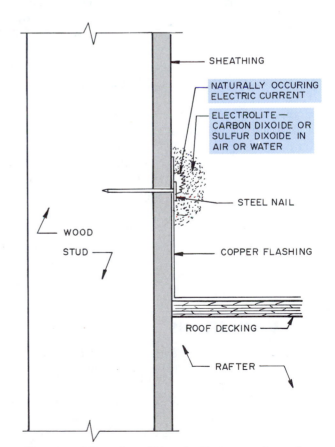

Figure 3.7 Natural conditions produce corrosive action when dissimilar metals are in contact with each other.

Table 3.8 Galvanic Series of Selected Metals and Alloys

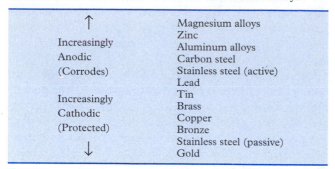

↑	Magnesium alloys
	Zinc
Increasingly	Aluminum alloys
Anodic	Carbon steel
(Corrodes)	Stainless steel (active)
	Lead
Increasingly	Tin
Cathodic	Brass
(Protected)	Copper
	Bronze
↓	Stainless steel (passive)
	Gold

selected metals are ranked from those most anodic to those most cathodic. When a material near the top of the list, such as steel, is placed in contact with a material near the bottom of the list, such as copper, and an electrolyte is present, corrosion will occur. The steel (anode) will be sacrificed and the copper (cathode) will be protected. Metals that are widely separated on the galvanic table should not be in contact with each other. Those close together, such as cadmium-plated steel and aluminum, have practically no difference in electrolytic potential and, therefore, almost no galvanic action.

Corrosion can be prevented by breaking up this electrical circuit by removing one of the elements. Most often this is done by painting or covering the metal surfaces with a material that will not conduct electricity. This not only breaks the electrical circuit but also protects the metal from other corrosive materials in the air or soil. It is also possible to use the same type of metal in all parts. For example, aluminum gutters can be hung with aluminum nails rather than copper nails. Or one part of the installation could be insulated from the other with material such as plastic.

Sometimes a metal, such as steel, is coated with another metal, such as zinc. The zinc, which is high on the galvanic series, is purposely sacrificed to save the steel.

Aluminum and stainless steel are anodic and will corrode. However, they do develop some natural corrosion resistance due to exposure to oxygen. The oxygen forms a transparent oxide film that protects the metal.

REVIEW QUESTIONS

1. What are the requirements of material to be considered when selecting it for a specific application?
2. What are the major material groups?
3. What four properties of materials are most important when selecting construction materials?
4. How does a tensile stress differ from a compressive stress?
5. What is the difference between stress and strain?
6. What properties can be noted on a stress-strain diagram?
7. Give an example in which fatigue strength is an important consideration when designing a building.
8. What is indicated by a measure of impact strength?
9. How does the application of heat to a material influence its behavior?
10. What are the major thermal factors to consider when selecting a material?
11. Which of the three major material groups have the lowest thermal conductivity?
12. What is meant by electrical conductivity and electrical resistance?
13. What is the relationship between electrical conductivity and resistance?
14. What is the most common form of chemical degradation?
15. What causes materials to be damaged by oxidation?
16. Which materials are more resistant to chemical degradation?
17. What does the galvanic series table reveal about metals?

KEY TERMS

breaking strength The point at which a material actually begins to fail.

coefficient of thermal expansion The change in the size of a material per unit of dimension per degree change in the temperature (in./in.°F).

compressive strength A measure of the ability of a material to resist forces that tend to decrease its length.

compressive stresses Stresses created when forces push on a member and tend to shorten it.

damping capacity A measure of the ability of a material to absorb vibrational energy.

ductility The ability of a material to be stretched or deformed without fracturing.

elastic deformation The ability of a material to return to its original position after a load has been removed.

elastic limit The highest stress that can be imposed on a material without permanent deformation.

electrical conductivity A measure of the ability of a material to conduct electric current.

electrical resistance The ability of a material to oppose the flow of electricity.

fatigue strength A measure of the ability of a material to carry a load when subjected to repeated loading and unloading stresses.

hardness A measure of the ability of a material to resist indentation or surface scratching.

impact strength A measure of the ability of a material to resist mechanical shock.

melting temperature The temperature at which a material turns from a solid to a liquid.

modulus of elasticity The proportional constant between stress and strain. The modulus of elasticity is the stress divided by the strain. It is a measure indicating stiffness.

oxidation A reaction between a material and oxygen in the atmosphere.

plastic deformation An irreversible deforming of a material.

proportional limit The upper limit at which stress is proportional to strain.

shear stress The result of forces acting parallel to an area but in opposite directions, causing one portion of the material to "slide" past another.

strain The change per unit length in a linear dimension of a body due to applied stress.

stress The intensity of internally distributed forces that resist a change in the form of a body. These include tensile, compressive, and shear stresses.

tensile strength The maximum tensile stress a material can sustain at the point of failure.

tensile stresses Stresses created when forces pull on a member and tend to increase its length.

thermal conductivity A measure of the ability of a material to transfer heat by conduction.

thermal properties The behavior of a material when subjected to a change in temperature.

ultimate strength The maximum amount of tension, compression, or shear that a material can sustain without failure.

yield strength The stress required to produce a small amount of permanent deformation.

SUGGESTED ACTIVITIES

1. Prepare samples of various materials and test for compression and tensile strength. Keep an accurate record of what each sample contains and the conditions of its preparation. For example, prepare concrete samples, keeping a record of the ingredients, curing time, and how it was contained during curing. Vary certain ingredients in various samples. Test each for compression strength and compare the results to try to verify how the differences in mixture, curing time, and methods of curing influenced the strength. Similar tests can be made on other materials, such as various species of wood and different types of metals.

2. Run some experiments on thermal conduction by applying heat to identical size samples of various materials and recording the time it took to feel heat at the end opposite the end exposed to heat. Although this is rather crude, it will illustrate the property of thermal conduction.

3. Most schools have material testing equipment or even an entire testing laboratory. Have the instructor in this area help the class devise some more accurate, scientifically based experiments and help the class run these in the laboratory. For example, with the proper equipment the class could possibly test a sample, such as a metal, and record the elastic limit, yield point, and breaking strength.

4. If a hardness tester is available, run hardness tests on various materials.

5. Try to set up a device that will subject a material, such as a metal, to constant rapid flexing and see which material has the best fatigue strength.

6. Can you figure some way to try to find which materials have the higher coefficients of thermal expansion?

7. Look in a physics book and find out how to set up an electrolysis experiment. Try it using various metals as anodes and cathodes to demonstrate the principle behind galvanic action that occurs between building materials.

ADDITIONAL INFORMATION

James A. Jacobs and Thomas F. Kilduff, *Engineering Materials Technology,* Prentice Hall, Upper Saddle River, N.J.

William O. Fellers, *Materials Science, Testing, and Properties for Technicians,* Prentice Hall, Upper Saddle River, N.J.

PART II

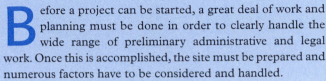

GENERAL REQUIREMENTS AND SITE CONSTRUCTION

Before a project can be started, a great deal of work and planning must be done in order to clearly handle the wide range of preliminary administrative and legal work. Once this is accomplished, the site must be prepared and numerous factors have to be considered and handled.

Part II includes materials covering Division 1 and Division 2 of the Construction Specifications Institute MasterFormat.

Chapter 4 covers the basic site materials and methods involved with preparing to start construction, handling matters that occur during construction, and completing the final work on the site per the contract drawings and documents.

Chapter 5 discusses types of soils and how they may influence decisions on design and final contours, roads, parking, and other features.

Chapter 6 presents a detailed description of foundations, ranging from those typically used on residential and small commercial buildings to the choices available to the engineer as multistory buildings are designed. ▲

D I V I S I O N 1

GENERAL REQUIREMENTS
CSI MASTERFORMAT™

The items specified in CSI Division 1 are not covered in this book. A brief description of these items can be found in Appendix A.

Courtesy H.H. Robertson, A United Dominion Company

D I V I S I O N 2

SITE CONSTRUCTION
CSI MASTERFORMAT™

Courtesy GOMACO Corporation, Ida Grove, Iowa

The Building Site

This chapter will help you to:

1. Understand the preliminary work that must be taken on the site before construction can begin.

2. Be aware of the problems that may occur during excavation and techniques to control these.

3. Be familiar with paving materials and design.

4. Realize the dangers from seismic activities and the types of forces developed.

Site work planning requires the services of building architects, landscape architects, engineers, and specialists. The on-site work includes a wide range of activities—subsurface investigation; demolition; preparing the site for construction; dewatering excavations; shoring and underpinning; earthwork; tunneling; constructing piles and caissons; paving; installing various types of piping, water distribution, sewage, and drainage systems; constructing ponds and reservoirs; constructing site improvements, such as fences and walks; and landscaping (Fig. 4.1).

Building codes have sections devoted to site work and demolition. Details are shown on site plans developed by the architect with the assistance of engineers specializing in the activities required. Additional information is contained in reports of subsurface investigations, including a geotechnical report and soil test boring data. (See Chapter 5 for details.)

SITE PLANS

The **site plan** shows the results of the land survey, the existing contours of the land, and the desired finished contours. Locations of the buildings, roads, wells, and other features are shown. Locations of utilities (gas, electric service, public water and sewer lines) are indicated. Setbacks and easements must be located. The location and size of parking areas, vehicle access routes, streets, curbs, and other details such as fences, walks, and retaining walls are shown. A typical site plan is shown in Fig. 4.2.

ENVIRONMENTAL CONSIDERATIONS

As a part of the preliminary work, the project must address environmental considerations by making an impact study as required by the National Environmental Policy Act. This study determines how the project and site work will affect the natural features in the surrounding area. When the plan is found to be sound, site work can proceed. However, the general contractor must be constantly vigilant to maintain conditions friendly to the environment. For example, soil runoff into a nearby stream or lake must be curtailed. Discharge of toxic waste

Figure 4.1 Site preparation involves the activities of a number of construction trades.

materials must be prohibited. The existence of flood zones must be acknowledged and decisions must be made concerning how they impact on the site.

PREPARING THE SITE

On some jobs preparing the site requires the demolition of structures. In addition to following demolition regulations, the waste material must be disposed of in an approved manner. Preparing the site also includes removing trees and brush in areas to be occupied by the structure, parking areas, roads and driveways; removing existing improvements above and below grade; stripping and saving existing soil; and removing and replanting existing trees. The site plan and landscape plan give the information needed to accomplish these.

Demolition

The required site work frequently involves demolishing or partially demolishing an existing structure. Demolition work is subject to control by the local building codes. Safety practices must be followed, and waste material must be safely and properly disposed of as regulated. No demolition is to occur until the plans and schedule are approved by the local building official. This also involves the removing and capping existing utilities and removing existing hazardous materials.

SUBSURFACE INVESTIGATION

A key factor in the design of a structure is the determination of the characteristics of the soil. This **site investigation** requires field tests made by a soils engineer and tests made on soil samples in a soil testing laboratory or by field test procedures. The soil report is used by the archi-

tect and design engineers to determine what must be done as the building is designed and constructed. This can influence footing design, retaining walls, paving, dewatering requirements, and other factors. The soil tests are performed on subsurface test borings, and the results are detailed on a soil test report. The report includes a detailed geologic description of the site, subsurface conditions, and recommendations based on the findings. Recommendations may pertain to site grading, foundation support, paving design, retaining walls, and other aspects of the project to be built.

Soil investigations for shallow depths (around 8 ft. or 2.4 m) can be made by digging an excavation and taking soil samples. Investigations can be made using a *penetration method,* which involves driving a test rod into the soil and recording the number of blows it takes with a standard driving hammer to penetrate the soil a predetermined distance. A typical report is shown in Fig. 4.3.

Soil investigations are also made by *drilling test holes* using a hollow-stem auger and wash boring techniques. The auger is mounted on the end of a drill rod that is rotated, causing the auger to penetrate the soil (Fig. 4.4). Water run into the pipes under pressure washes away waste material and lubricates the drilling process. The location, depth, and number of borings vary with site conditions and are based on the judgment of the soil engineer. The locations of boreholes are shown on a drawing of the site (Fig. 4.5). The height of the water table can sometimes be noted in the borehole. The findings are reported on some type of drawing as shown in Fig. 4.6.

It should be noted that there are times when unexpected site conditions crop up after contracts have been signed and site work begins. For example, as a site was being excavated for a four-story masonry building massive foundations from a previous unknown building were found. Who pays to blast them to pieces and haul them away? A thorough subsurface investigation will help prevent this. However, contracts should be written to cover this unexpected occurrence.

EARTHWORK

Earthwork includes grading, excavating, backfilling, compacting, laying base courses, stabilizing the soil, setting up slope protection, establishing erosion control, carrying out soil treatment, and building earth dams.

Grading

Rough **grading** involves adjusting the level of the ground to facilitate the excavation and construction of the building. Usually the top soil is removed and stockpiled for use during finish grading. Finish grading occurs after the building is complete. It brings the site elevations up to the levels shown on the site or grading plan.

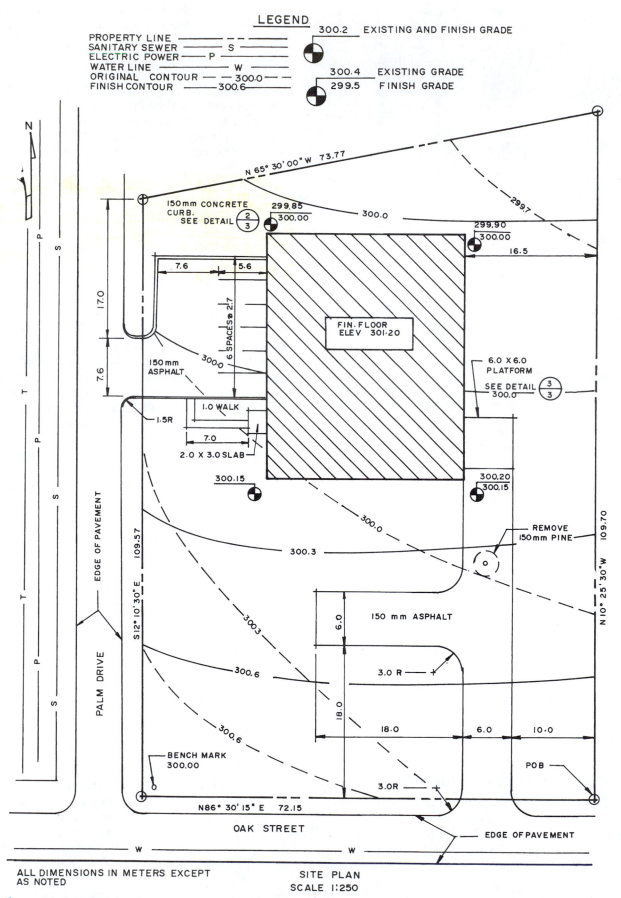

Figure 4.2 A typical site plan.

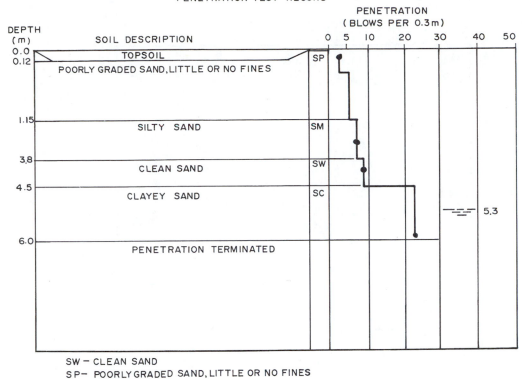

PENETRATION TEST RECORD

Figure 4.3 The results of a typical soil investigation using the penetration method.

Figure 4.4 Deep test holes are made with a drilling rig.

Excavation

Excavations are required for footings, foundations, and other aspects of a project, such as underground utilities. A wide variety of equipment is available. This includes power shovels, backhoes, draglines, clamshells and cranes, trenching machines, wheel-mounted belt loaders, and tractors with a bulldozer.

The excavation usually begins with the removal of the topsoil, which is piled out of the way of the construction area. If the footings are shallow they are usually dug with a backhoe. In cold climates they will have to be several feet deep to get below the frost line. Another consideration is the need to excavate to the depth at which the soil tests have shown the soil to have the required load-bearing capacity. A building with a basement will have a large excavation and a considerable amount of soil must be removed and stockpiled in a **spoil bank.**

Sometimes excess soil must be removed from the site. Multistory buildings require extensive excavations, which in some cases go several stories into the ground to get to bedrock or other adequate supporting soil. If rock must be removed, the cost of excavation increases rapidly. Weak and thinly layered rock can be loosened with power shovels, tractor-mounted rippers, backhoes, and pneumatic hammers. Other rock formations can be fractured by explosives. The fractured rock can then be removed from the excavation by a front-end loader.

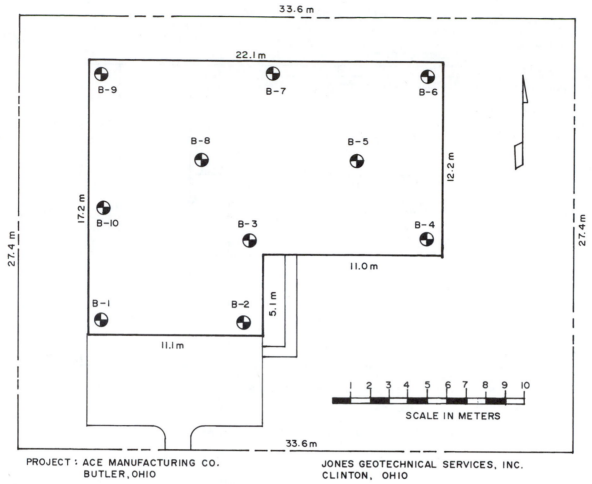

Figure 4.5 A site plan locating the boreholes.

Like soil, rock must be stockpiled or removed from the site.

Deep excavations present an ever-present danger of collapse of the sides. The depth and type of soil influence what must be done. If the excavation is not too deep and the site is large enough, the side of the excavation can be sloped. The angle of the sloped surface will vary with the type of soil. This angle, called the **angle of repose,** will range from 34° for cohesionless soil to 90° for stable rock. Maximum allowable slopes are specified by OSHA regulations. Deeper excavations can use a bench system as shown in Fig. 4.7. The particles in cohesionless soils tend to not stick together, which can result in a slope failure unless the angle of repose is small. Cohesive soil grains tend to bond, providing good shear strength. If these solutions to preventing cave-in are not acceptable, the excavation sides must be shored with sheeting. This may be wood or metal and must be braced according to the pressures to be against it. It is also necessary to install sheeting when excavating trenches for utilities.

Sheeting

If the sides of the excavation cannot be sloped to provide protection from slides, some form of **sheeting** must be used. These include sheet piling, lagging, and slurry walls.

Sheet piling may be wood, aluminum, steel, or precast concrete, placed vertically (Figs. 4.8, 4.9, and 4.10). Sometimes steel and concrete sheet pilings are left in place after construction is finished. Wood pilings are usually removed. Generally, sheet pilings are driven into the earth before excavation begins (Fig. 4.11). Another sheeting technique uses vertical steel piles called soldier piles that are driven into the soil. Horizontal wood lagging is then placed between them (Fig. 4.12).

Since the sheeting is subject to soil and subsurface water pressures, some form of bracing is needed as the excavation gets deeper. In shallow excavations the sheeting may be driven deep enough below the bottom of the excavation so that bracing is not required. Vertical sheeting has horizontal beams called wales placed at intervals (Fig. 4.8). Narrow excavations can use *cross bracing,* also

LOG OF SUBSURFACE PROFILE

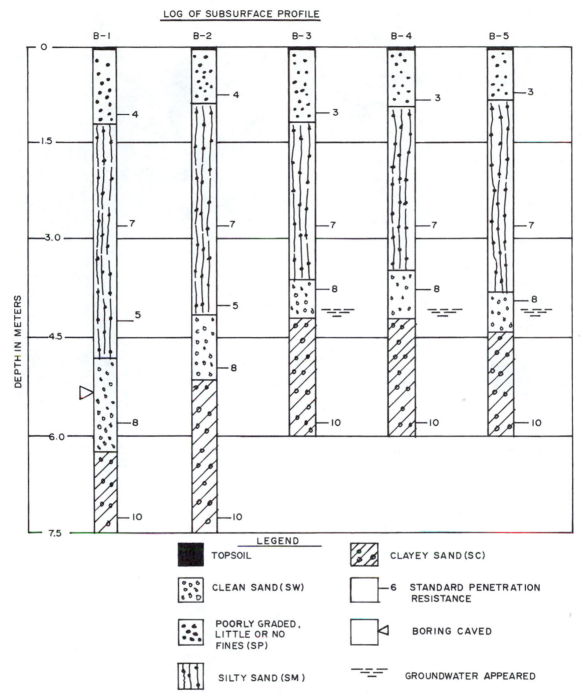

Figure 4.6 A typical report on the findings from test borings at predetermined locations on the site.

shown in Fig. 4.8. The bracing may be tightened with screws like a jack, or it may be hydraulically operated. Wider excavations require the use of rakers or cross-lot bracing as shown in Fig. 4.13. Cross-lot bracing used on excavations wider than trenches requires vertical steel posts to be driven within the excavation to support the horizontal cross braces. Rakers are used on wide excavations where cross bracing would be impractical. The rakers are set on an angle and transfer the forces to a

footing set in the bottom of the excavation. Both types block the open space in the excavation and interfere with any work that occurs within the excavation.

Some subsoils permit the use of tiebacks. Tiebacks are steel cables or tenons that are inserted into holes drilled through the sheeting and into the rock or subsoil. The drilled hole is filled with concrete grout. When this has set, the cables are tightened with hydraulic jacks and fastened to the wales (Fig. 4.14). This provides an excavation free

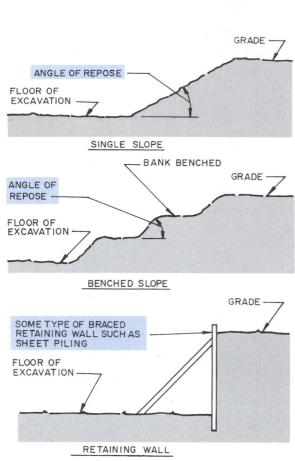

Figure 4.7 Ways to contain the sides of excavations to prevent them from collapsing into the excavation.

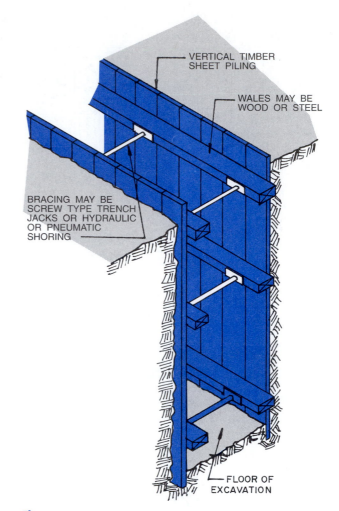

Figure 4.8 Vertical wood sheet piling can be used to support the sides of an excavation.

of barriers. Screw tieback anchors are another type in use (Fig. 4.15). They are installed with the same rotary drilling equipment used in grouted anchor construction. The screw anchor is rapidly installed and does not require the drilling of an anchor hole or the injection of grout. They are especially useful in loose sandy and clay soils where a drilled hole may collapse. Screw anchors can be withdrawn and reused.

Another sheeting technique uses a slurry wall. A **slurry wall** serves as sheeting that protects the excavated area, and it becomes a part of the permanent foundation. It may be cast in place or built from precast concrete units. The excavation is dug with a narrow clamshell bucket, which establishes the width of the wall. The type of soil present is important to the success of this method. To help prevent collapse of the excavation walls the excavation is filled with a **slurry** composed of bentonite clay and water, which stabilizes the wall. The clamshell bucket moves through the slurry to continue digging. After the required depth is reached, a welded cage of steel reinforcing is lowered into

the cavity. Concrete is poured from the bottom of the excavation, filling it to the top. The slurry rises above the concrete and is pumped off (Fig. 4.16). Once the entire wall is poured and has reached sufficient strength, the soil can be excavated from one side. Steel tiebacks are set in holes drilled through the wall and into the subsoil as the excavation deepens. (See Fig. 4.14.)

Another technique is to insert precast concrete wall sections into the slurry instead of site casting the concrete wall. The surface of the precast sections that are on the side to be excavated are coated to keep the slurry from bonding to them. A tongue-and-groove joint or a rubber gasket seals the sections. After the slurry remaining in the excavation has hardened, the soil can be dug away. The slurry on the back side of the wall remains and aids in waterproofing. The slurry on the excavated side falls free, leaving the surface of the precast units exposed.

The sides of an excavation can also be stabilized by using **soil anchors.** This involves drilling holes into the

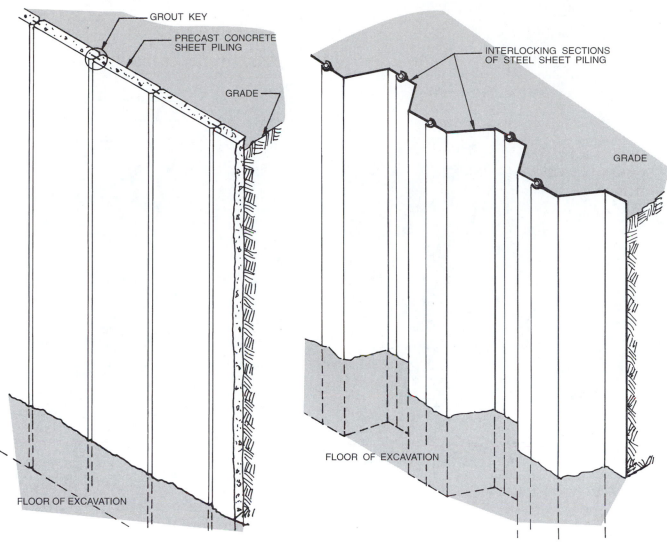

Figure 4.9 Precast concrete sheet piling installed vertically.

Figure 4.10 Metal sheet piling has interlocking connections between the sheets.

soil embankment, removing the drill, and filling the hole with grout. Then the soil nail, which is usually a section of steel reinforcing bar, is pushed into the grouted hole. A centering device is attached to the nail so it remains in the center of the grouted hole. This is much like forming a cast pile that is made on an angle. (Additional information is in Chapter 6.)

Cofferdams and Caissons

Cofferdams are temporary watertight enclosures built in an area of water-bearing soil or directly in the water. They hold out the water from the area inside so that construction can proceed in a dry, stable environment. Water is pumped from within the cofferdam, and pumps are maintained during construction to remove any leakage that may occur. Cofferdams typically are built using sheet piling, soldier beams with lagging, or as a double-

wall structure. A typical double-wall cofferdam is shown in Fig. 4.17. The cofferdam extends through the areas of permeable water-bearing soil formations, through any impervious rock formations that have low bearing capacity, and on to solid bedrock, which is used to support the foundation.

A **caisson** is a watertight shell in which construction work is carried on below water level. Caissons may be open or pneumatic. The top of the open caisson is exposed to the weather, and work is performed under normal atmospheric pressure (Fig. 4.18). Pneumatic caissons are air- and watertight. They are open on the bottom so soil excavation can be accomplished. The caisson is kept filled with air under pressure to keep water from entering as soil is excavated at the bottom. Workers must go through decompression when they leave to avoid suffering from the bends, a painful and sometimes fatal disorder. Pneumatic caissons are not

Figure 4.11 On shallow excavations steel sheet pilings are driven before the excavation begins.

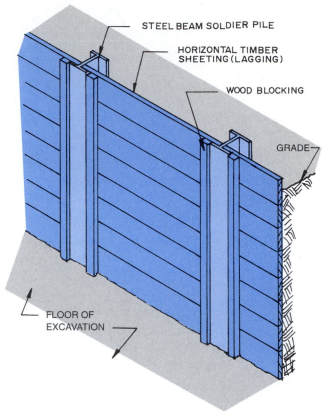

STEEL BEAM SOLDIER PILE

HORIZONTAL TIMBER SHEETING (LAGGING)

WOOD BLOCKING

GRADE

FLOOR OF EXCAVATION

Figure 4.12 This form of sheeting used steel soldier piles and horizontal timber lagging.

used a great deal (Fig. 4.19). The structure forming the caisson is frequently left in place and is filled with concrete, forming a caisson pile.

DEWATERING TECHNIQUES

Dewatering is the process used to lower the subsurface water on a site so that the excavation remains dry and stable. It is usually started before excavation is started and continues as the excavation proceeds. Subsurface water is removed using a system of pumps, pipes, and well points. Basically, a well point is a perforated unit placed at the bottom of a pipe driven into the soil (Fig. 4.20). A series of well points are driven in and around the area to be excavated. They are connected to horizontal pipes above the ground called headers. The headers are connected to centrifugal pumps. The water is drawn up into the header pipes and ejected onto the side of the site where it can be drained away from the excavation (Fig. 4.21). More than one series of well point, header, pump assembly is usually required, especially as the excavation is dug deeper (Fig. 4.22).

Water can also be kept out of an excavation by constructing a watertight wall (Fig. 4.23). This could be a slurry wall or one constructed with sheet piling, as discussed under the Cofferdams heading. The wall must resist the hydrostatic pressure of the subsurface water and the pressure generated by the soil itself. It must reach a depth where it penetrates a stratum of impermeable material so a watertight seal can be achieved. Pumps should be kept available to remove any water that may seep into the excavation.

UNDERPINNING TECHNIQUES

Excavation work involves the protection of buildings next to the excavated area. Occasionally the excavation for a new building is so close to an existing building that it becomes necessary to support the existing building during construction of the new one. This **underpinning** requires considerable engineering work, and special considerations are needed for each situation.

It is important to take soil samples and have the test results evaluated by trained engineers. The soil should be

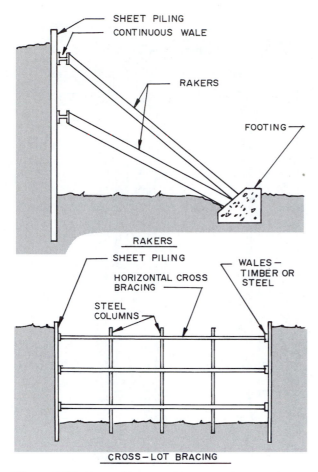

Figure 4.13 Sheet piling on wide excavations can be braced with rakers and/or cross-lot bracing.

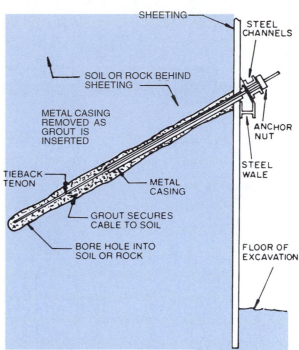

1. BORE A HOLE IN SOIL OR ROCK WITH A ROTARY DRILL. INSERT A METAL CASING TO KEEP THE SOIL FROM FILLING THE HOLE.

2. INSERT STEEL TIEBACK TENONS INTO THE HOLE. FILL THE HOLE WITH GROUT AS THE STEEL CASING IS REMOVED. LET THE GROUT SET.

3. WHEN THE GROUT HAS SET POST−TENSION THE TENON WITH A HYDRAULIC JACK AND ANCHOR TO THE WALE.

Figure 4.14 Sheeting can be supported by tiebacks anchored in stable earth or rock.

evaluated to the depths required to ascertain if the new footings will be bearing upon a suitable stratum. Two frequently used techniques are to use trenches dug below the foundation of the existing building or to use needles.

The *trench technique* uses trenches dug at intervals beneath the foundation. The building is supported by the remaining undisturbed soil. The required underpinning is installed in the trenched area and additional trenches are dug through the unexcavated area. This is repeated until the existing building has adequate underpinning (Fig. 4.24).

Needles are heavy wood timbers or steel beams that are run horizontally through the wall of the building and are supported on each end. The spacing between needles in a wall is determined by an engineer. In Fig. 4.25 the building is supported with needles run through the foundation, then the area under the foundation is excavated and a new footing and foundation are built. Notice the sheeting used under the building to prevent cave-in of the excavation wall.

There are various ways to underpin a foundation. It may have a new footing and foundation or a series of

Figure 4.15 Screw anchors are used as tiebacks to support retaining walls on excavations. (*Courtesy A.B. Chance Company*)

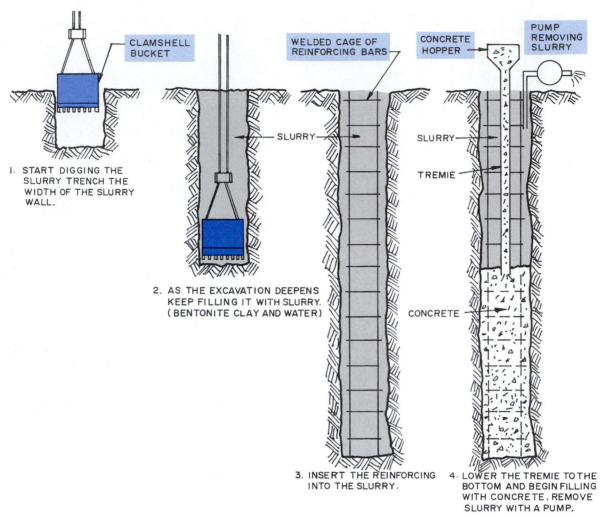

CLAMSHELL BUCKET

1. START DIGGING THE SLURRY TRENCH THE WIDTH OF THE SLURRY WALL.

2. AS THE EXCAVATION DEEPENS KEEP FILLING IT WITH SLURRY. (BENTONITE CLAY AND WATER)

WELDED CAGE OF REINFORCING BARS

SLURRY

3. INSERT THE REINFORCING INTO THE SLURRY.

CONCRETE HOPPER

PUMP REMOVING SLURRY

SLURRY

TREMIE

CONCRETE

4. LOWER THE TREMIE TO THE BOTTOM AND BEGIN FILLING WITH CONCRETE. REMOVE SLURRY WITH A PUMP.

Figure 4.16 A typical procedure for constructing a slurry wall.

piles on each side (Fig. 4.26). The exact design must be prepared by an experienced engineer.

PAVING

The most commonly used paving materials for roads and parking areas are asphaltic concrete or concrete. Paving occurs after the subbase and base are in place. A typical situation with a flexible pavement, such as asphaltic concrete, is shown in Fig. 4.27. The lowest strata is the *natural subgrade* which is consolidated by compaction. The *subbase* course is laid on top of the natural subgrade. It is a natural soil that is of a higher quality than the natural subgrade. It provides additional support as required to distribute the loads imposed on the finished surface. The *base course* is laid over the subbase and is the surface upon which the *finished wearing surface* is laid. It is a granular base of high quality, such as crushed stone, gravel, slag, or some combination of these. Sometimes sand is also mixed with these. It is of-

ten treated with a bitumen to bind the materials. The subbase and base courses are extended beyond the edge of the next layer to provide a cone dispersement of the loads.

Figure 4.28 shows a section through the base material for a typical rigid pavement, such as concrete. In this design the concrete slab has to be thick, reinforced, and strong enough to resist the imposed loads. The base material helps carry the loads, but the concrete slab actually must carry most of the weight. The loads are transmitted over the entire width of the slab, which reduces the forces on the base materials.

Asphalt Materials

Asphalt is a dark brown to black cementitious material in semisolid or solid form made up of bitumens found in deposits of natural asphalt. Asphalt is also refined from petroleum. Following are the types of asphalts that find some use in paving.

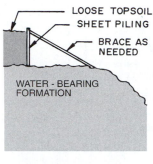

1. SHEET PILING IS USED TO HOLD THE LOOSE TOPSOIL FORMATION.

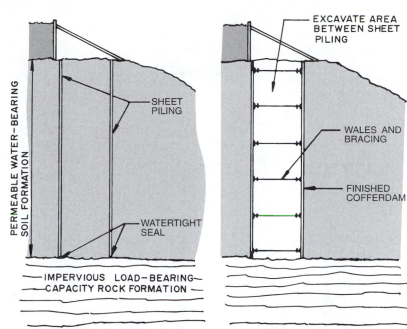

2. SHEET PILINGS ARE DRIVEN THROUGH THE PERMEABLE WATER-BEARING SOIL FORMATION TO IMPERVIOUS ROCK. THE PILING FORMS A WATER-TIGHT SEAL WITH THE ROCK.

3. EXCAVATE THE AREA BETWEEN THE SHEET PILING AND BRACE WITH WALES AND CROSS BRACING AS THE EXCAVATION PROCEEDS.

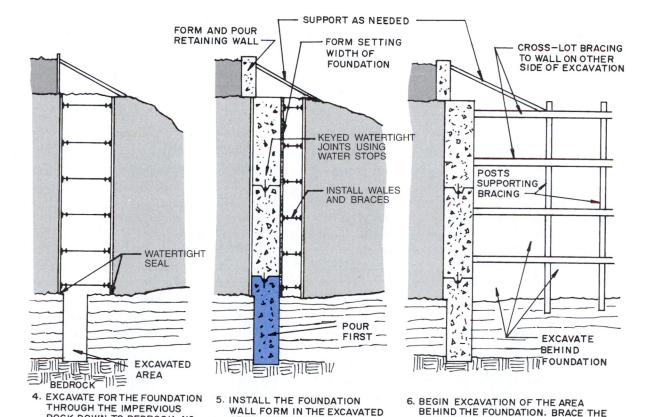

4. EXCAVATE FOR THE FOUNDATION THROUGH THE IMPERVIOUS ROCK DOWN TO BEDROCK. NO SHEET PILING NEEDED IN THIS AREA.

5. INSTALL THE FOUNDATION WALL FORM IN THE EXCAVATED AREA. INSTALL REINFORCING AND FILL WALL CAVITY WITH CONCRETE. POUR IN SECTIONS WITH WATER STOPS BETWEEN THE SECTIONS.

6. BEGIN EXCAVATION OF THE AREA BEHIND THE FOUNDATION. BRACE THE FOUNDATION AS THE EXCAVATION PROCEEDS. WHEN COMPLETELY EXCAVATED THE WALL CAN BE SUPPORTED WITH CROSS-LOT BRACING.

Figure 4.17 A cofferdam can be constructed to provide a watertight area in which to construct a footing and foundation.

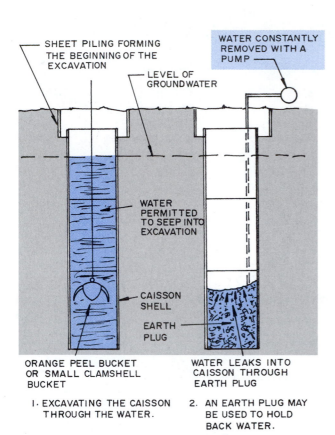

Figure 4.18 Typical open-air caissons.

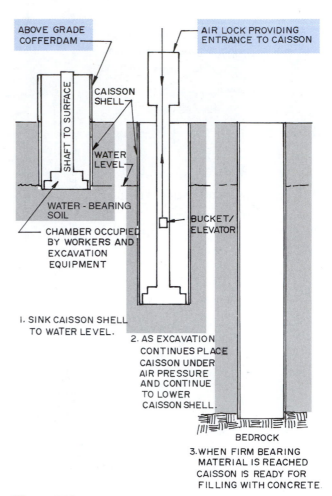

Figure 4.19 Pneumatic caissons operate under elevated air pressure.

Cutback asphalt is a broad classification of residual asphalt materials left after petroleum has been processed to produce gasoline, kerosene, diesel oil, and lubricating oil. The *residual asphalt* is blended with various solvents, producing cutback asphalt. There are three classifications, rapid curing (RC), medium curing (MC), and slow curing (SC). They are mixed with granular soil to stabilize a roadbed before paving or to serve as a binder for a finished road surface for light traffic.

The *emulsion* group of asphalts consists of emulsified asphalt. An emulsion consists of droplets of one liquid dispersed within another liquid with which the first cannot mix. The emulsion is made by adding heated, fluid asphaltic cement into water to which an emulsifying agent, such as soap or bentonite clay, has been added. A stabilization agent, protein, is added to prevent the particles from blending together within the mix. The suspended particles of asphalt blend with the aggregate or soil particles as the water drains away and evaporates.

Emulsified asphalts are either anionic or cationic, depending on the emulsifying agents used. Anionic emulsions are those in which the asphalt globules in the mix are negatively charged. Cationic emulsions have asphalt globules that are positively charged. Each comes in three grades. They are used for the same purposes as cutback asphalts.

Asphalt Paving

Four types of asphalt paving are in common use. They are made with either cutback or emulsified asphalt.

Asphalt concrete consists of asphalt cement and graded aggregates carefully proportioned and mixed in an asphalt plant at controlled temperatures (Fig. 4.29). The mix is transported to the site, spread by an asphalt paving machine, and rolled while it is still hot (Fig. 4.30).

Cold-laid asphalt is made in the same way as asphalt concrete but the asphalt liquid is cold.

Asphalt macadam is laid using a penetration method. A coarse aggregate is laid over the base, compacted to a smooth surface, and sprayed with an asphalt emulsion or hot asphalt cement. This is covered with fine aggregates and rolled, forcing the fine aggregate into the voids between the larger aggregate.

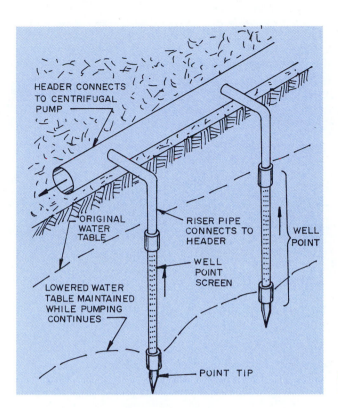

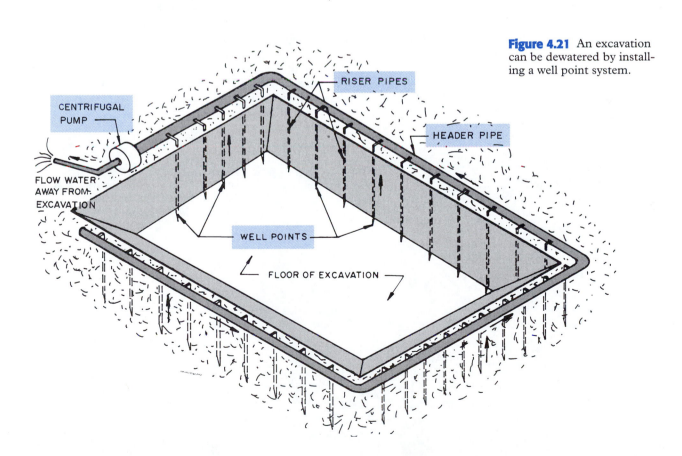

Figure 4.20 Well points are driven into the water-bearing soil.

Figure 4.21 An excavation can be dewatered by installing a well point system.

I T E M   O F   I N T E R E S T

BLASTING

Many parts of the country have considerable rock formations extending to the surface of the earth. Before a foundation, utility trenches, or other subsurface work can be dug, it often is necessary to blast out the rock in the way of the excavations. This requires the services of a qualified blaster. A well-prepared blaster can do the job safely and at the least cost. Select a blaster who has a record of successful projects, is licensed as required by the state, and has adequate insurance coverage.

Drilling and blasting is expensive and a careful analysis of the site must be made when bidding the job. For example, a site may have rock ledges, but the building location may be clear of them. However, when the time comes to run utility lines or, even worse, to install a septic system, a hidden rock ledge is found. After the house has been built, it is generally too late to blast, so the rock is removed with airhammers. The septic tank will require a large excavation and possibly not work because the laterals would also be in rock. When rock is known to be in an area, drilling into the soil to check for rock is essential to determine whether blasting is needed before construction is started.

The actual situation varies of course but special care must be exercised if there are structures nearby. This may require more drilled holes and smaller charges or shooting fewer holes at one time. Blast before any construction is begun. Use heavy blasting mats over the charged area to reduce the throwing of material. If you know rock is below the surface of the soil drill through the soil into the rock. There is no need to go to the expense of excavating the soil down to the rock. Another way to save money is

Figure A These are the items used for blasting rock. They include a stick of dynamite, shotgun blasting cord with a blasting cap on the end, and packing. The blasting cord is a plastic tube filled with explosive power. When ignited it sends a ball of fire down the inside of the tube to the blasting cap, which ignites the dynamite.

to move the location of the building, if you can, to miss the rock ledge and still get it on the lot within the setbacks. Underground utility line locations sometimes also can be moved to miss a rock ledge. Another possibility if rock is hit several feet below the surface is to build on it and raise the house higher above the surface of the earth. The extra height can be covered by fill dirt, which is cheaper than blasting.

Figure 4.22 This excavation is being kept dry with a well point system as the floor is covered with aggregate moved in with a boom conveyor. *(Courtesy ROTEC Industries)*

....

I T E M   O F   I N T E R E S T

BLASTING

Continued

Special caution and technical knowledge are required if it is necessary to blast close to an existing foundation or other below-surface construction, such as the casing for a well. Concrete does not resist lateral forces as well as it does compressive forces. Before this is done the insurance companies involved need to be notified. They will generally make an on-site visit and possibly take photos or videos of the foundation and building to record its condition before the blasting occurs.

Figure C The blast blows debris over a wide area unless the hole is covered with a blasting mat.

Figure B The blaster inserts the blasting cap into the dynamite and lowers it into the hole by the blasting cord. The hole is filled with ammonium nitrate, an inexpensive explosive.

Photos courtesy Pat Paquette, Operations and Sales manager, Dyno New England, Inc. Photos by Carolyn Bates.

Old asphalt surfaces can be renewed by applying a surface treatment consisting of a 1 in. (25.4 mm) *asphalt sealer.* The treatment may consist of several layers of a liquid asphalt covered with a mineral aggregate or a single layer of cold or hot asphalt mix that is plant mixed and rolled after it is machine laid (Fig. 4.31). Other sealers used include a coal-tar-latex emulsion and vinyl and epoxy resin sealers. These have greater tensile strength than regular coal-tar sealers, and some are available in colors.

Additional information on bitumens is in Chapter 26.

Concrete Paving

Roads, driveways, and parking lots are often paved with concrete. The subgrade, subbase, and base course preparation is much like that described for asphalt paving. Side forms are placed along the edges of the paving to contain the plastic concrete and are used as rails upon which the concrete placing spreader rides (Fig. 4.32). Some placing spreaders are designed to ride inside the side forms, and others ride outside the forms.

The pavers may be form or slip-form types. The *form type* paver rides on the metal side forms, places the concrete, strikes it off, and consolidates it (Fig. 4.33). *Slip-form* pavers place a slab without the side forms. The concrete is of a consistency that it develops sufficient strength to be self-supporting as it leaves the paver. The slip-form paver spreads, consolidates, and finishes the concrete slab (Fig. 4.34). Some types of pavers can be used for form or slip-form placement.

Information on concrete is in Chapter 7.

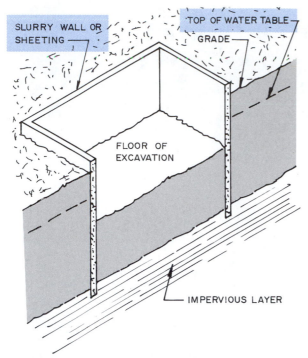

Figure 4.23 Some type of watertight wall can be constructed to hold back subsurface water so construction can proceed.

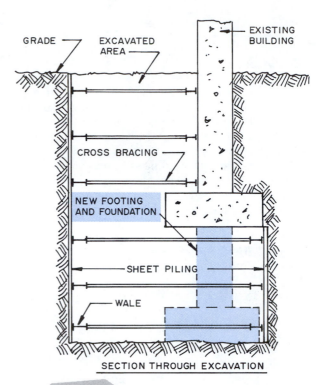

SECTION THROUGH EXCAVATION

SITE WORK ACTIVITIES

Site work includes many other activities, such as erosion control, dam construction (Fig. 4.35), and tunneling (Fig. 4.36). Tunneling alone involves excavating, ventilation, lining the tunnel, grouting, and support systems. Foundation work is a big part of site work and includes spread footings, piles, and caissons (see Chapter 6). In marine areas dredging, underwater work, and constructing seawalls, jetties, and docks are required. Utilities require site work for water, sewer, gas, oil, and steam distribution systems (Fig. 4.37). Ponds, reservoirs, and sewage lagoons are part of the site work plan. Finally, there are extensive site improvements to be done, such as walks, fences, irrigation systems, and landscaping.

SEISMIC CONSIDERATIONS

Seismic forces are destructive forces caused by earthquakes. Although accurate prediction of earthquakes is difficult, certain sections of the country are known to have subsurface conditions that make them possible earthquake zones, as shown in Fig. 4.38. These **seismic areas** have special building code requirements regulating the design and construction of buildings and other structures (Fig. 4.39).

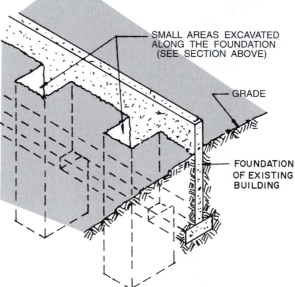

Figure 4.24 An existing building can be supported by using the trench underpinning technique if the remaining soil can carry the weight.

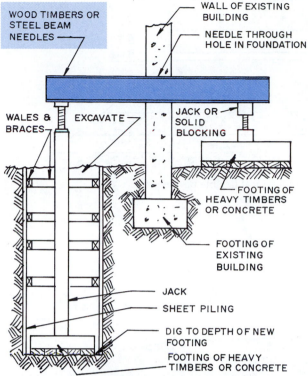

WOOD TIMBERS OR STEEL BEAM NEEDLES

WALL OF EXISTING BUILDING

NEEDLE THROUGH HOLE IN FOUNDATION

WALES & BRACES

EXCAVATE

JACK OR SOLID BLOCKING

FOOTING OF HEAVY TIMBERS OR CONCRETE

FOOTING OF EXISTING BUILDING

JACK

SHEET PILING

DIG TO DEPTH OF NEW FOOTING

FOOTING OF HEAVY TIMBERS OR CONCRETE

1. INSTALL NEEDLES THROUGH THE FOUNDATION AND RAISE TO CARRY THE WEIGHT OF THE BUILDING.

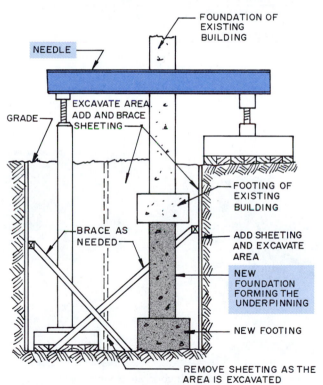

FOUNDATION OF EXISTING BUILDING

NEEDLE

GRADE

EXCAVATE AREA. ADD AND BRACE SHEETING

FOOTING OF EXISTING BUILDING

BRACE AS NEEDED

ADD SHEETING AND EXCAVATE AREA

NEW FOUNDATION FORMING THE UNDERPINNING

NEW FOOTING

REMOVE SHEETING AS THE AREA IS EXCAVATED

2. EXCAVATE THE AREA BELOW THE BUILDING FOOTING AND INSTALL THE UNDERPINNING.

Figure 4.25 A typical underpinning installation using needles to support the building.

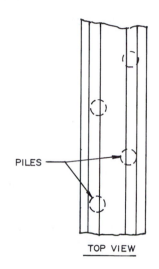

PILES

TOP VIEW

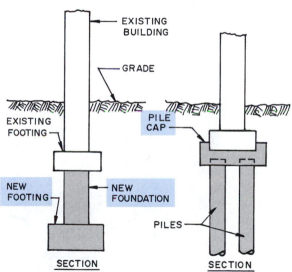

EXISTING BUILDING

GRADE

EXISTING FOOTING

NEW FOOTING

PILE CAP

NEW FOUNDATION

PILES

SECTION

SECTION

Figure 4.26 Two types of underpinning.

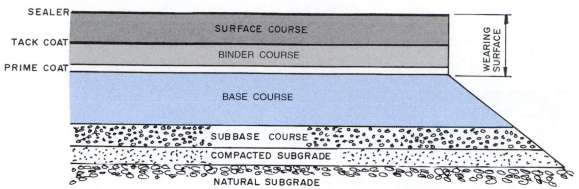

Figure 4.27 The base courses and wearing surface courses for a typical flexible pavement.

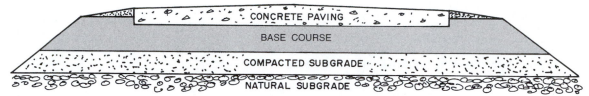

Figure 4.28 The base courses and concrete paving for a typical rigid pavement.

Figure 4.29 An asphalt plant. *(Courtesy The Asphalt Institute)*

Figure 4.30 Laying asphalt concrete paving. *(Courtesy The Asphalt Institute)*

Figure 4.31 Asphalt concrete paving is rolled after it is machine laid. *(Courtesy The Asphalt Institute)*

Figure 4.32 Laying concrete paving. *(Courtesy Bid-Well Division of CMI Corporation)*

Figure 4.33 The concrete paving is leveled, consolidated, and finished. *(Courtesy Bid-Well Division of CMI Corporation)*

Figure 4.34 The concrete curb and gutter are slip-formed in a continuous process. *(Courtesy GOMACO Corporation)*

ROCK RIPRAP USED TO CONTROL EROSION

Figure 4.35 Site work includes the construction of dams and the installation of riprap. *(Courtesy U.S. Army Corps of Engineers)*

Figure 4.36 Tunneling requires extensive shoring and control of subsurface water. *(Courtesy U.S. Army Corps of Engineers)*

Figure 4.37 Preliminary site work includes installing a wide range of utilities.

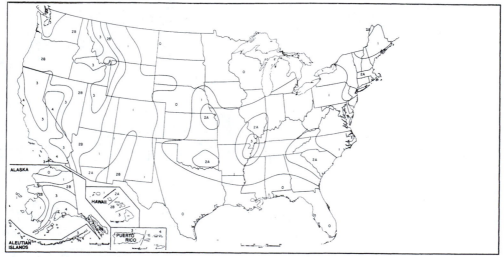

Courtesy International Conference of Building Officials

SEISMIC ZONE MAP OF THE UNITED STATES

ZONE 0 — No damage.
ZONE 1 — Minor damage. Distant earthquakes may cause damage to structures with fundamental periods greater than 1.0 seconds. corresponds to intensities V and VI of the Modified Mercalli Intensity Scale.
ZONE 2 — Moderate damage. Corresponds to intensity VII of the Modified Mercalli Intensity Scale.
ZONE 3 — Major damage. Corresponds to intensity VIII and higher on the Modified Mercalli Intensity Scale.

Figure 4.38 Seismic zone map of the United States. *(Courtesy International Conference of Building Officials)*

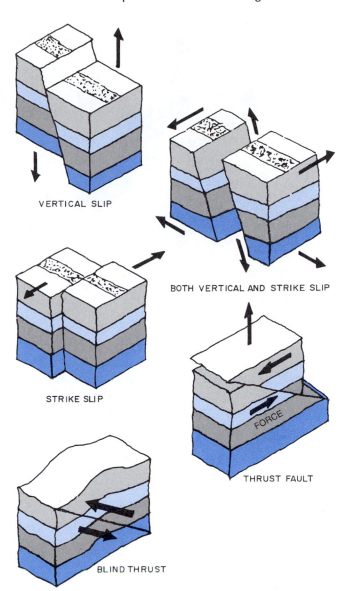

Figure 4.39 The exterior walls and the top stories collapsed when an earthquake occurred near this unreinforced masonry building. *(Courtesy U.S. Department of the Interior, Geological Survey)*

Figure 4.40 Typical types of faults. *(Courtesy U.S. Department of the Interior, Geological Survey)*

An earthquake is the oscillatory movement of the earth's surface that follows a release of energy in the earth's crust. When subjected to deep-seated forces, the crust may first bend, but when the stress exceeds the strength of the rocks, they break and "snap" to a new position. This produces vibrations called seismic waves. The waves travel along the surface and through the earth. Some of these vibrations have a high enough frequency to be audible. Others are of low frequency.

An earthquake is caused by the slippage of the earth along a fault plane. A *fault* is a fracture in the earth's crust along which two blocks of crust have slipped with respect to each other. Common movements include one crustal block moving horizontally in one direction while the block facing it moves horizontally in the other direction. This is referred to as a *strike slip*. In other cases one block may move up vertically while the abutting block moves downward. This is referred to as a *vertical slip*. In some cases the movement along the fault may have both vertical and horizontal movement. A *thrust fault* results when sections of rock press together, forcing one side up over the other. A *blind thrust* raises the surface into folded hills without breaking the surface (Fig. 4.40).

Frequently these slippages of strata occur deep in the earth and cause no surface rupture but do cause surface vibration and shaking. These fault slippages generate ground motions that radiate vertically and horizontally in all directions (Fig. 4.41). Earthquakes frequently produce ruptures of the surface as well as a rolling, waving motion of the ground surface. The ground shakes as this occurs, and the shaking is what causes the most damage to buildings and other structures rather than the more visible and dramatic ground rupture.

The magnitude of an earthquake is measured by the *Richter scale*, which is based on the maximum single movement recorded on a seismograph. The measurement most widely used by seismologists is the *moment*

Figure 4.41 Fault slippages generate motions in all directions. *(Courtesy U.S. Department of Interior, Geological Survey)*

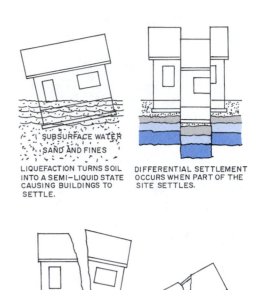

LIQUEFACTION TURNS SOIL INTO A SEMI–LIQUID STATE CAUSING BUILDINGS TO SETTLE.

DIFFERENTIAL SETTLEMENT OCCURS WHEN PART OF THE SITE SETTLES.

RUPTURE WILL CAUSE THE BREAKUP OF THE STRUCTURE.

GROUNDSHAKES MOTION WILL CAUSE BUILDINGS TO BREAK INTO RUBBLE.

Figure 4.42 The effects of an earthquake can cause a variety of destruction, resulting in building failure. *(Courtesy U.S. Department of Interior, Geological Survey)*

magnitude. It is based on the size of the fault on which the earthquake occurs and the amount of earth slippage. The larger the fault and the larger the actual slip the higher the moment magnitude of the earthquake.

No buildings should be located over known active geologic faults. The types of destruction of buildings during an earthquake include ground rupture, ground shaking, liquefaction, and differential settling or some combination of these. Illustrations of these are in Fig. 4.42. *Liquefaction* occurs when the motions of the earthquake transform the soil into a semiliquefied state that is much like quicksand. It occurs in areas with loose sands and silts that have high water tables (Fig. 4.43).

Other reactions triggered by earthquakes include tidal waves, seiche, and landslides. *Tidal waves* occur when the earthquake occurs in the ocean. *Seiche* is a sloshing wave movement occurring in lakes and enclosed bodies of water, such as bays. It can generate waves that will go over the top of a dam. A *landslide* is the failure of a slope, permitting rocks, earth, and trees to flow down the slope.

The design of footings for buildings in seismic areas is specified by local codes. The design varies with the soil classification. Spread footings and pier foundations are connected with ties capable of resisting specified forces in tension and compression. Piles are designed to resist the maximum imposed curvatures from seismic forces. Piles and pile caps are interconnected by ties. Connections between structural steel piles and the pile cap must be designed for a specified tensile force. Concrete-filled steel and tube piles must have the specified minimum reinforcement, and longitudinal reinforcement is also required. Cast-in-place piles also must have the minimum reinforcement specified by the code,

Figure 4.43 An earthquake moved more than 1.5 meters of silt over a portion of this forest. *(Courtesy U.S. Department of Interior, Geological Survey)*

and the reinforcing must be sized and placed according to the code.

Masonry construction is usually required to meet the requirements for seismic design set forth by the American Concrete Institute, American Society of Civil Engineers, and the Masonry Council. Seismic requirements for reinforced concrete are set forth by the American Concrete Institute plus any additional requirements specified by local codes. Structural steel seismic requirements are usually those set forth by the American Institute of Steel Construction, Inc., in their publication *AISC Seismic Provisions for Structural Steel Buildings,* plus other requirements specified by the local code. Seismic requirements for wood and timber construction follow the recommendations of the National Forest and Paper Association and additional local code specifications.

REVIEW QUESTIONS

1. What information is shown on the site plan?
2. What types of activities might be required to prepare the site for construction?
3. What information is gained by making a subsurface investigation?
4. What type of work might be required under the earthwork section of a construction contract?
5. What are the various ways the sides of an excavation can be prepared to protect workers from caving earth?
6. What purpose does dewatering serve?
7. When would underpinning be necessary?
8. What are the layers of material used to pave a parking area with asphaltic concrete?
9. What are the sources of asphalt?
10. What are the four types of asphalt paving in common use?
11. What is the difference in laying a concrete paving between form type pavers and slip-form pavers?
12. Describe what happens in the earth when an earthquake occurs.
13. What organizations provide information about designing structural systems for buildings located in an area of possible seismic activity?

KEY TERMS

angle of repose The angle of the sloped surface of the sides of an excavation.

asphalt Dark brown to black hydrocarbon solids or semisolids having constituents that are bitumens in nature that gradually liquefy when heated.

caisson A watertight structure within which work can be carried out below the surface of water.

cofferdam A temporary watertight enclosure around an area of water-bearing soil or around an area of water from which the water is pumped, allowing construction to take place in the water-free area.

dewatering Pumping subsurface water from an excavation to maintain dry and stable working conditions.

grading Adjusting the level of the ground on the site.

seismic area A geographic area where earthquake activity may occur.

sheeting Wood, metal, or concrete members used to hold up the face of an excavation.

site investigation An investigation and testing of the surface and subsoil of the site to record information needed to design the foundation and structure.

site plan A drawing of a construction site showing location of the building, contours of the land, and other features.

slurry A liquid mixture of water, bentonite clay, or portland cement.

slurry wall A wall built with a slurry, used to hold up the sides of an area to be excavated.

soil anchors Metal shafts grouted into holes drilled into the sides of an excavation to stabilize it.

spoil bank An area where soil from the excavation is stored.

underpinning Placing a new foundation below an existing foundation.

SUGGESTED ACTIVITIES

1. Secure site plans for recent projects from a local architect or contractor. Prepare a detailed list of the things shown on the plan. See if they had any environmental problems with each site; if so, record what these were and the actions taken.

2. Visit local building sites where excavation is underway or just completed. Prepare a written report of the things you saw. Did you notice any especially good practices or any unsafe conditions?

3. Prepare sketches showing various ways the sides of excavations can be stabilized.

4. Examine the local building codes and list the requirements for site control of erosion, tree removal, and spoil banks.

5. If your area is in an identified seismic area, cite the building code requirements specified because of possible seismic activity.

ADDITIONAL INFORMATION

Ambrose, J., *Building Construction: Site and Below-Grade Systems,* Van Nostrand Reinhold, New York, 1993.

Colley, B. C., *Practical Manual of Site Development,* Mc-Graw-Hill, New York, 1985.

DeChiara, J., and Koppelman, L. E., *Time-Saver Standards for Site Planning,* McGraw-Hill, New York, 1984.

Sitt, F. A., *Architect's Detail Library,* Van Nostrand Reinhold, New York, 1990.

5

Soils

This chapter will help you to:

1. Learn the major types of soils.

2. Know how soils are classified.

3. Be familiar with some of the commonly used soil tests.

4. Understand how factors related to soils can affect design decisions.

5. Be aware of the various ways to stabilize soils.

Soils are the end result of mechanical and chemical weathering of rock. Naturally occurring abrasive and mechanical forces wear down large rock masses into smaller particles due to heat and gravity. This produces gravels, sands, and fine silts. Mechanically weathered soil particles are three-dimensional in shape.

The less stable minerals in rocks produce very small, two-dimensional flakelike particles in crystalline form. These are produced by chemical action in clayey soils.

A knowledge of soils, their characteristics, and their properties is essential for those who design foundations. Building codes specify maximum design loads for various types of soils. Soil investigations on the site are required to produce the information needed.

TYPES OF SOILS

Soils are classified by the sizes of particles and their physical properties. The following five types of soils are those typically found. However, often soil is a mixture of these.

▲ **Gravel** is a hard rock material with particles larger than ¼ in. (6.4 mm) in diameter but smaller than 3 in. (76 mm).
▲ **Sand** is fine rock particles smaller than ¼ in. (6.4 mm) in diameter to 0.002 in. (0.05 mm).
▲ **Silt** is fine sand with particles smaller than 0.002 in. (0.05 mm) and larger than 0.00008 in. (0.002 mm).
▲ **Clay** is a very cohesive material with microscopic particles (less than 0.00008 in. or 0.002 mm).
▲ **Organic matter** is partly decomposed vegetable matter.

Rock particles larger than 3 in. (76 mm) are called cobbles or boulders and are not classified as soil.

SOIL CLASSIFICATIONS

The commonly used soil classification systems are the Unified Soil Classification System (USCS) and the system of the American Association of State Highway and Transportation Officials (AASHTO).

Important to understanding soil classification are the liquid limit, plastic limit, plasticity index, and the shrinkage limit of the soil. These states of soil consistency are indicated in terms of water content (Fig. 5.1). The **liquid limit (LL)** of a soil is the water content expressed as a percentage of the dry weight at which the soil will start to flow when tested by the shaking test. The **plastic limit (PL)** of a soil is the percent moisture content at which the soil begins to crumble when it is rolled into a thread ⅛ in. (3 mm) in diameter. The **plasticity index (PI)** is the difference between the liquid limit and the plastic limit. This indicates the range in moisture content over which the soil would remain in a plastic condition. The **shrinkage limit (SL)** is the water content at which the soil volume is at its minimum.

Test methods, practices, and guides for classifying soils and preparing test specimens are detailed in the American Society of Testing and Materials publication *ASTM Standards on Soil Stabilization with Admixtures.*

Unified Soil Classification System

The *Unified Soil Classification System* was developed by the U.S. Army Corps of Engineers and is used to classify soils for use in roads, embankments, and foundations. Soils are classified according to the percentage of grain-size particles of each of the established soil grain sizes in the soil distribution index, the plasticity index, the liquid limit, and the organic-matter content. Soils are grouped in fifteen classes. Eight of these are coarse-grained soils that are identified by the letter symbols GW, GP, GM, GC, SW, SP, SM, and SC. Six classes are fine grained and are identified as ML, CL, OL, MH, CH, and OH. The one class of highly organic soils is identified as Pt. Soils on the borderline between two classes are given a dual classification, such as GP-GW. Descriptive details are in Table 5.1.

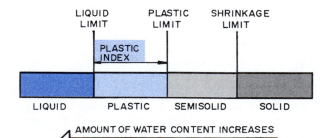

Figure 5.1 When the water content in a soil increases, the soil becomes more fluid.

American Association of State Highway and Transportation Officials System

The *AASHTO soil classification system* classifies soils according to those properties that affect their use in highway construction and maintenance. In this system the mineral soil is classified in one of seven basic groups, ranging from A-1 through A-7 on the basis of grain-size distribution, liquid limit, and plasticity index. Soils in the A-1 through A-3 groups are sands and gravels. They have a low content of fines (very small particles). Soils in the A-4 through A-7 groups are silts and clays and are therefore fine-grained soils. A special grade, A-8, is reserved for highly organic soils.

When laboratory tests are made on soil samples, the A-1, A-2, and A-7 groups can be further classified into groupings such as A-1-a, A-1-b, A-2-4, A-2-5, A-2-7, A-7-5, and A-7-6. Details are shown in Table 5.2.

FIELD CLASSIFICATION OF SOIL

Sometimes it is not possible to get soil samples tested in a laboratory, so field tests are made. The soil particles are separated by size by screening them through various sieves. Frequently used sieve sizes are shown in Table 5.3. Following are generalized descriptions of field tests made by a soils engineer. Detailed information can be obtained from the U.S. Department of Interior, Bureau of Reclamation.

Table 5.4 shows the division of soil into various fractions based on the size of the soil particles. Although this is important when considering the properties of a soil, other properties, such as the shape of the particles (angular or sharp distinct edges versus subangular or rounded edges), influence the ability of a soil to interlock particles.

Testing for Soil Coarseness

All soil particles larger than 3 in. in diameter are removed from the sample. The remaining particles are separated through a No. 200 sieve, which permits the smallest particles that can be seen by the naked eye to pass through. If more than 50 percent of the sample by weight does not pass through the sieve, the sample is a *coarse-grained soil*. The coarse sample is now divided into particles larger and smaller than ¼ in. (6 mm). If more than 50 percent of this sample by weight is larger than ¼ in. (6 mm), the sample is classified as *gravel*. If more than 50 percent by weight is smaller than ¼ in. (6 mm), the sample is classified as *sand*.

Dry Strength Test

A field test for dry strength, which is an indication of plasticity, can be done by wetting a soil sample until it

Table 5.1 Classification of Soils Using the Unified Classification System

Type	Letter Symbol	Description	Rating as Subgrade Material	Rating as Surfacing Material
Gravel and gravelly soils	GW	Well-graded gravel; gravel-sand mixture; little or no fines	Excellent	Good
	GP	Poorly graded gravel; gravel-sand mixture; little or no fines	Good	Poor
	GM	Gravel with silt; gravel-sand-silt mixtures	Good	Fair
	GC	Clayey gravels; gravelly sands; little or no fines	Good	Excellent
Sand and sandy soils	SW	Well-graded sands; gravelly sands; little or no fines	Good	Good
	SP	Poorly graded sands; gravelly sands; little or no fines	Fair	Poor
	SM	Silty sands; sand-silt mixtures	Fair	Fair
	SC	Clayey sands; sand-clay mixtures	Fair	Excellent
	ML	Inorganic salts; fine sands; rock flour; silty and clayey fine sands with slight plasticity	Fair	Poor
Silts and clays with liquid limit greater than 50[a]	CL	Inorganic clays of low to medium plasticity; gravelly clays; silty clays; lean clays	Fair	Fair
	OL	Organic silts of low plasticity	Poor	Poor
Silts and clays with liquid limit less than 50[a]	MH	Inorganic silts; micaceous or diatomaceous fine sandy or silty soils; elastic silts	Poor	Poor
	CH	Inorganic clays of high plasticity	Very poor	Poor
	OH	Organic clays of medium to high plasticity; organic silts	Very poor	Poor
Highly organic soils	Pt	Peat and other highly organic soils	Unsuited for subgrade material	Unsuited for surfacing

[a] The liquid limit is the water content, expressed as a percentage of the weight of the oven-dried soil, at the boundary between the liquid and plastic states of the soil.
Courtesy U.S. Department of Interior, Bureau of Reclamation.

Table 5.2 The AASHTO System of Soil Classification

Typical Material	A-1 Sand and Gravel A-1-a	A-1-b	A-2 Gravel or Silty or Clayey Sand A-2-4	A-2-5	A-2-6	A-2-7	A-3 Fine Sand	A-4 Silt	A-5 Silt	A-6 Clay	A-7 Clay
No. 10 sieve	50% max.										
No. 40 sieve	30% max.	50% max.					51% min.				
No. 200 sieve	15% max.	25% max.	35% max.	35% max.	35% max.	35% max.	10% max.	36% min.	36% min.	36% min.	36% min.
Fraction passing No. 40 sieve											
Liquid limit	6% max.	6% max.	40% max.	41% min.	40% max.	41% min.	—	40% max.	41% min.	40% max.	41% min.
Plasticity index	—	—	10% max.	10% max.	11% min.	11% min.	—	10% max.	10% max.	11% min.	11% min.

From *Standard Specifications for Transportation Materials and Methods of Sampling and Testing, Fifteenth Edition,* Copyright 1990 by the American Association of State Highway and Transportation Officials, Washington, D.C. Used by Permission.

Table 5.3 Some Frequently Used Standard Sieves

U.S. Standard Sieve Sizes	Opening Size Inch	Millimeter
3 in.	3	76.2
1½ in.	1.50	38.1
¾ in.	0.75	19.0
No. 4	0.186	4.76
No. 10	0.078	2.00
No. 40	0.017	0.425
No. 100	0.006	0.150
No. 200	0.003	0.075

Table 5.4 Soil Size Fractions

Constituent	U.S. Standard Sieve No.
Cobbles	Above 3 in.
Gravel	
Coarse	3–¾ in.
Fine	¾ in.–No. 4
Sand	
Coarse	No. 4–No. 10
Medium	No. 10–No. 40
Fine	No. 40–No. 200
Silts and clays	Below No. 200

is about the consistency of a stiff putty. Mold it into a ball about 1 in. (25 mm) in diameter. After it has dried, hold it between the thumb and forefinger of both hands and squeeze. If it does not break, the soil is highly plastic and is characteristic of clays. If the sample breaks but it is difficult to cause the sections to powder when rubbed between the fingers, it has a medium plasticity. If the sample easily breaks down into a powder, it has low plasticity.

Toughness Test

The toughness test ascertains the consistency of the soil sample near the plastic limit. All soil particles larger than the opening in a No. 40 (about ⅟₆₄ in. or 0.4 mm) sieve are removed. A soil specimen of about a ½ in. (12.5 mm) cube is molded to the consistency of putty. Water may be added if necessary to get this consistency. If the sample is too sticky the soil must be spread out in a thin layer to allow some moisture to evaporate. When the proper consistency is achieved, roll the soil between your hands into a thread about ⅛ in. (3 mm) in diameter. The thread is folded and rerolled repeatedly until it finally loses its plasticity and crumbles. When it crumbles, it has reached its plastic limit. The crumbled pieces are lumped together and kneaded until the lump again crumbles. The tougher the thread near the plastic limit and the stiffer the lump when it crumbles the more colloidal clay there is in the soil. Weakness of the thread at the plastic limit and easy crumbling of the lump indicate either inorganic clay with low plasticity or organic clays.

Shaking Test

The test for identifying the character of fines in a soil is the *shaking test*. After removing particles larger than No. 40 (about ⅟₆₄ in. or 0.4 mm) sieve size, form a soil sample into a ball about ¾ in. (19 mm) in diameter. Add enough water to make the sample soft but not sticky.

Place the ball in your open palm and shake horizontally, striking it against the other hand. A positive reaction is the appearance of a glossy surface caused by water on the surface. Squeeze the sample in your hand and

the ball stiffens and the water and gloss disappear. The ball will finally crack or crumble. *Very fine sands* give the quickest reaction. Inorganic silts give a moderately quick reaction, and clays give no reaction.

SOIL VOLUME CHARACTERISTICS

When soil is excavated, hauled, placed, and compacted its volume changes. When soil is excavated and moved it increases in volume. This is called *swell*. The grains loosen and air fills the voids between them. The amount of swell varies in different types of soil. It is measured as a percentage of the original volume.

When the soil is placed and compacted, the air is forced out between the grains and the soil's volume decreases. This is called *shrinkage*. The amount of shrinkage varies with the type of soil and is measured as a percentage of the original volume.

The volume of one cubic yard of soil **in situ** (undisturbed soil) is referred to as **bank measure.** When disturbed it will produce more than one cubic yard and when compacted produce less than one cubic yard of soil. The weight and volume changes for typical types of soil are shown in Table 5.5.

Soil volume also changes due to a change in the moisture content. Clays are especially susceptible to changes in moisture. As moisture content increases, the soil expands, sometimes to the point of lifting a footing or slab. If the moisture decreases, the soil will shrink, sometimes causing the footing or slab to settle (Fig. 5.2).

SOIL WATER

The two major types of water present in soils are capillary water and gravitational water. *Capillary water* is retained in the microscopic pores of the individual sand particles and enables them to bind together. *Gravitational water* is water flowing within the soil in the voids between particles. It can be removed by pumping.

The *water table* or **groundwater** *level* is the level of the water below the surface. Since it is free-flowing, the water below the soil surface can be varied by pumping water from wells, by a dry or rainy season, and by installing drain pipes to remove it from the construction area.

Table 5.5 Typical Weights, Swell, and Shrinkage for Selected Types of Soils

	Loose		Weight[a] Bank Measure		Compacted			
	lb./yd.³	kg/m³	lb./yd.³	kg/m³	lb./yd.³	kg/m³	Swell	Shrinkage
Clay, natural	2300	1300	2900	1720	3750	2220	38%	20%
Common earth, dry	2100	1245	2600	1540	3250	1925	24%	10%
Sand with gravel, dry	2900	1720	3200	1900	3650	2160	12%	12%
Sand, dry	2400	1400	2700	1600	2665	1580	15%	12%

[a]Weights, swell, and shrinkage of actual samples may be larger or smaller.

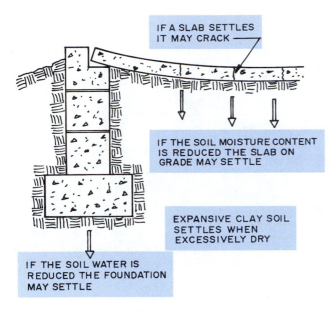

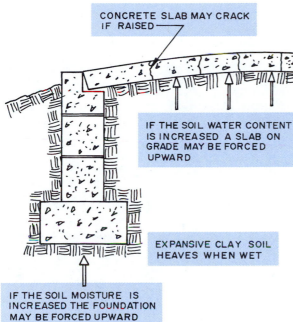

Figure 5.2 Expansive clay soils heave when wet and settle when excessively dry.

The moisture content of soil is an important factor when compacting it. The density (pcf or kg/m^3) varies with the moisture content. The moisture in proper amounts enables the soil to be compacted to its maximum density. This is called the *optimum moisture content* and is determined by laboratory tests of soil samples.

SOIL-BEARING CAPACITIES

The acceptable load-bearing capacities of the various types of soils are specified by the local building code. In Table 5.6 are the maximum allowable load-bearing val-

Table 5.6 Presumptive Load-Bearing Values of Foundation Materials

Class of Material	Load-Bearing Pressure lb./ft.^{2}	kg/m^2
Crystalline bedrock	12,000	58 560
Sedimentary rock	6,000	29 280
Sandy gravel or gravel	5,000	24 400
Sand, silty sand, clayey sand, silty gravel, and clayey gravel	3,000	14 640
Clay, sandy clay, silty clay, and clayey silt	2,000	9760

From *The BOCA National Building Code/1993*
Courtesy Building Officials and Code Administrators International, Inc.
Copyright 1993. All rights reserved.

ues for supporting soils under spread footings placed at or near the surface of the site. In many cases it is necessary to test soil samples to determine the maximum allowable pressure. These surface values must be adjusted for deep footings and load-bearing strata under piles as specified by the building code.

If the soil in the area of the structure is uniform in composition and the footing design is adequate, the building will remain stable. Even if it settles a little over time the settlement will be uniform. If, however, the soil composition varies and is not compensated for in the footing design, the building could begin to settle unevenly. Actions of others could also affect existing structures. For example, extensive pumping of subsurface water will cause soil to settle because of a lowering of the water table.

SOIL ALTERATIONS

After the soil tests have been made and the soil's properties determined, the soil can be modified to improve its characteristics. Commonly used processes to alter the properties include compaction, stabilization, and dewatering.

Compaction

Compaction of soil refers to increasing its density by mechanically forcing the soil particles closer together. This expels the air in the voids between the particles. Soil density can also be increased by *consolidation,* which involves removing the water in the voids between soil particles, permitting the particles to come closer together. Compaction produces immediate results. Consolidation requires much longer to increase the soil density.

The amount of compaction that can be obtained depends on the physical and chemical properties of the soil, its moisture content, the compaction method used, and the thickness of the soil layer being compacted. As compaction occurs the air and water between the soil particles are forced out of the soil. Excess water must be drained away. The soil engineer will use the results of the soil tests to decide the best way to achieve the required compaction density.

There are many ways compaction can occur. These include a kneading action, vibration, static weight, explosives, and impact. Pneumatic-tired rollers are multi-tired units that effect compaction of the soil by providing a kneading action. The rows of tires are staggered to give complete coverage, and some units have wheels mounted to give a wobbly effect that increases the kneading action. The weight of such units can be varied by adding ballast. They are effective on most soils but least effective on sands and gravels.

Tamping foot rollers, such as the sheep's-foot roller in Fig. 5.3, use static weight to compact the soil. The tamping foot rollers are available in a range of foot sizes and shapes. As they pass over the soil the feet sink into it, compacting it below the surface. With repeated passes the feet do not penetrate as deep and eventually walk on the top surface. Tamping rollers are most effective on cohesive soils.

Another type of static-weight compactor has smooth steel wheels or drums. It is best when used to compact granular bases, asphalt bases, and asphalt pavements. It is not effective on cohesive soils because it tends to compact the surface, forming a crust over a loose, non-compacted subsurface (Fig. 5.4).

A variety of compaction devices use vibration. Small hand-operated vibratory compactors are used on compacting areas where it may be difficult to get larger equipment, such as preparing the base for a concrete floor or sidewalk. Larger vibratory compactors include tamping foot rollers and smooth drum rollers. The vibratory action provides compaction in addition to static weight.

Impact rammers are used to provide compaction in tight areas. The vertical movement of the rammer provides considerable compaction. Units are also available for mounting on the end of a backhoe boom.

Loose, saturated, granular soil can be compacted by subjecting it to a sudden shock and vibration. This causes the soil particles to fall into a denser pattern, displacing the water between the particles. The weight of the particles forces the water to flow from the soil. Explosives are typically used for this purpose. The spacing, depths, and sizes of the explosive charges are determined by experienced soil engineers.

Another way to increase the density of cohesionless soils is to use vibrocompaction. This uses a vibratory probe, which is a large-diameter tube with a vibrating device mounted on one end. It is lifted by a crane and driven into the soil by the weight of the tube and the vibrations. Figure 5.5 shows this process using a patented compacting probe. The probe is inserted in the soil by a powerful vibrator. The frequency and duration of vibration can be monitored to achieve max-

Figure 5.3 This compactor has sheep's-foot rollers that provide a tamping action. The front-mounted bulldozer is used to rough-grade the surface. (*Courtesy Caterpillar, Inc.*)

Figure 5.4 Smooth-wheeled rollers are used to compact granular materials and asphalt pavements.

imum compaction for the prevailing soil conditions. The degree of compaction depends on the soil type, spacing of probes, frequency, and the duration of vibration.

Another compaction probe is shown in Fig. 5.6. This shows the vibroflotation process. The probe has a powerful cylindrical vibrator that penetrates the soil by vibration assisted by compressed air or water jets. When the probe reaches the desired depth the space created by the vibroflot is filled with gravel by the simultaneous introduction of the material and withdrawal of the probe. The vibration during withdrawal ensures the efficient compaction of the added material and the neighboring soil (Fig. 5.7).

ITEM OF INTEREST

SUBCONTRACTOR AGREEMENTS

Seldom does a general contractor have the personnel to do all of the tasks required to construct a building. Typically, many areas, such as electrical and mechanical work, are handled by subcontractors employed by the general contractor. It is important that the general contractor have a detailed written agreement indicating what the subcontractor is to do and what parts the general contractor will handle. Following are some suggestions as to what should be in the contract.

The first consideration is to develop a schedule that includes the starting and finishing dates. It should include who is to pay for expenses caused by any delays. It could state what type of delay would constitute a breach of contract, thus permitting the general contractor to employ a new subcontractor.

Responsibility for obtaining required permits and notifying inspectors when the job has progressed to that point should be noted. Some statement pertaining to failure to do the job to the required quality expected is necessary.

Of great importance is a clearly specified payment schedule. This includes the costs of materials provided, labor, and the terms of payment.

Who is to supply the materials is also specified. Typically, a plumber supplies the pipe but the general contractor may prefer to purchase the fixtures. Drywall installers typically prefer to do the work but expect the general contractor to have the materials available when they appear on the job.

If changes are to be made, the agreement should specify that the general contractor reviews these changes with the owner. The subcontractor should not proceed with changes unless they are approved by the general contractor.

All trades are subject to meeting safety standards, and even though the general contractor may have the overall responsibility, the contract should indicate these safety standards must be met by all subcontractors.

The contract should indicate that the subcontractor is responsible for having all the required licenses for his or her trade and for having a comprehensive insurance policy with premiums paid. Although legal differences exist across the country, it is advantageous to have the subcontractor agree to indemnify the general contractor if the subcontractor fails to pay workers compensation or injury claims and the general contractor is held responsible for paying them.

The general contractor should have a written contract with the subcontractors. *(Photo courtesy U.S. Gypsum)*

The subcontractor should provide a written warranty for the work done so the owner can contact the subcontractor if something fails and needs repair.

A big problem on all construction jobs is cleanup. The contract should specify daily and final cleanup responsibilities and should be in great detail.

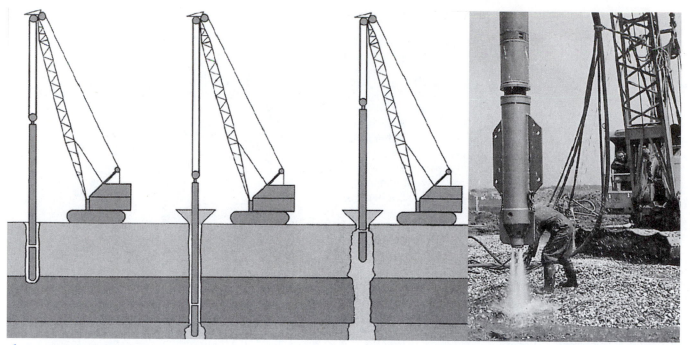

Figure 5.5 Vibrocompaction consolidates the soil by inserting a patented compaction probe into the soil and using a powerful vibrator to drive the probe into the soil. (*Courtesy Franki, International*)

Soil Stabilization with Admixtures

Test methods and specifications on soil stabilization with admixtures are detailed in the publication *ASTM Standards on Soil Stabilization with Admixtures.* Following are explanations of some of these.

Blending Soils

Soils can be blended by bringing in soil and mixing it with the original with a motor grader or disc. Power shovels or deep-cutting belt loaders can be used to cut deeply through the soil, mixing that below the surface with that on top.

Lime-Soil Stabilization

Clays and silty clay soils can be stabilized by mixing lime, which produces a chemical reaction. Clays expand when they are wet, causing damage to concrete surfaces over them. The concrete slabs heave up and crack due to the expansion of the clay. The lime reduces the amount of expansion and forms a moisture barrier that protects the expansive clay from subsurface water. In some cases it can remove the need to excavate unsatisfactory soil and replace it.

Usually slaked or hydrated lime or quicklime is spread over the base material and blended into the base with a pulverizer-type machine. Water may be added if conditions warrant. The blending is generally 12 to 18

in. (300 to 450 mm) deep and compacted to the final thickness. The compacted layer will continue to gain strength for many months.

Asphalt-Soil Stabilization

Asphalts blended with granular soils produce a durable, stable soil and may be used as a finished surface for low-traffic roads or a stabilized base course for a higher quality pavement. Some soils require the addition of fines with the asphalt to fill the spaces between the soil particles.

Cement-Soil Stabilization

Predominantly granular soils that have minute amounts of clay particles can be stabilized by blending with portland cement. In some cases fly ash is used to replace some of the portland cement. The portland cement is uniformly spread over the surface and blended with the soil using a pulverizer-type machine. After this has been accomplished to the specified depth, the surface is fine-graded and compacted. Water can be sprinkled on the surface during blending if the soil moisture content is low. Since portland cement sets up rather rapidly, compaction immediately after blending is required. Initial compaction can be with tamping or pneumatic rollers, followed by a steel smooth-wheel roller.

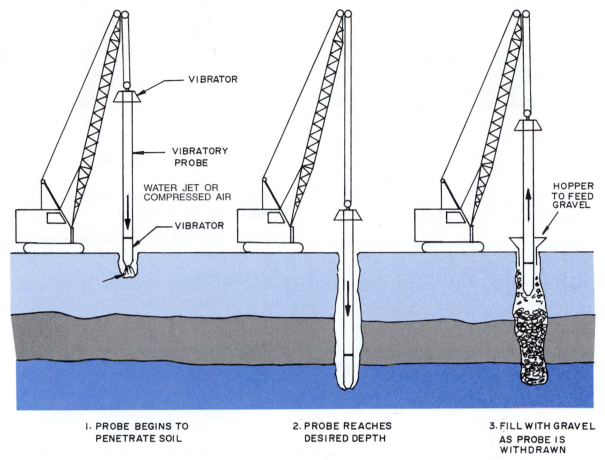

VIBRATOR

VIBRATORY
PROBE

WATER JET OR
COMPRESSED AIR

VIBRATOR

HOPPER
TO FEED
GRAVEL

I. PROBE BEGINS TO
PENETRATE SOIL

2. PROBE REACHES
DESIRED DEPTH

3. FILL WITH GRAVEL
AS PROBE IS
WITHDRAWN

Figure 5.6 The vibroflotation compaction method inserts a vibrating probe into the soil. Penetration is assisted by jets of compressed air or water from the bottom of the probe.

Figure 5.7 A typical vibrator unit mounted on the end of a probe. (*Courtesy Franki, International*)

Salt-Soil Stabilization

Coarse crushed rock salt can be used to stabilize well-graded clay or loamy soil having some limestone fines. The salt can be blended in the soil in dry form or mixed with water to form a brine. The salt and water form a bond with soil particles, creating a stable soil.

The base soil is pulverized to the specified depth, the salt added, and the mixture blended. This layer is then compacted, fine-graded, and watered and finished with a steel smooth roller. It can take up to several weeks before the soil is cured.

REVIEW QUESTIONS

1. How are soils classified?
2. What are the classes of soils?
3. How is the plasticity index of a soil determined?
4. What soil classification system is used by the U.S. Army Corps of Engineers?

5. What association has a soil classification system for use on highway construction?
6. Describe the field test for the dry strength of a soil.
7. What happens to the volume of a given amount of soil after it has been excavated?
8. What types of water are identified in soils?
9. Why are the load-bearing capacities of soil important?
10. In what ways can the properties of the soil be modified to improve its characteristics?
11. What admixtures are used to stabilize soil?

KEY TERMS

bank measure The volume of soil in situ in cubic yards.

clay A very cohesive material with microscopic particles—less than 0.00008 in. (0.002 mm).

compaction Compressing the soil to increase its density.

gravel Hard rock material in particles larger than ¼ in. (6.4 mm) in diameter but smaller than 3 in. (76 mm).

groundwater Water existing below the surface of the earth which passes through the subsoil.

in situ Undisturbed soil.

liquid limit Related to soils, it is the water content expressed as a percentage of dry weight at which the soil will start to flow when tested by the shaking test.

organic matter Partly decomposed vegetable matter.

plastic limit Related to soils, it is the percent moisture content at which the soil begins to crumble when it is rolled into a thread ⅛ in. (3 mm) in diameter.

plasticity index The difference between the liquid limit and the plastic limit.

sand Fine rock particles smaller than ¼ in. (6.4 mm) in diameter to 0.002 in. (0.05 mm).

shrinkage limit Related to soils, it is the water content at which the soil volume is at its minimum.

silt Fine sand with particles smaller than 0.002 in. (0.05 mm) and larger than 0.00008 in. (0.002 mm).

SUGGESTED ACTIVITIES

1. Have a representative of the state highway division visit the class and discuss local soils and their classifications.

2. If there is a laboratory near that does soil testing, try to arrange a visit.

3. If you can secure sieves of the various sizes used for soil testing, run some tests on local samples for coarseness.

4. Run dry strength and toughness field tests on soil samples from various sites.

5. Pack soil samples in a container of known volume, such as a quart or gallon. Ram each sample a predetermined number of times. Then dump it in a loose condition into other containers and measure the increase in volume. Compute the percentage of swell for each sample. Although this is a crude test, it will possibly give some interesting results if the soil samples are chosen from over a wide geographic area.

6. Visit a road construction site, observe the grading and compaction process, and report on the compaction and soil stabilization techniques being used.

ADDITIONAL INFORMATION

Merritt, F. S., and Ricketts, J. T., *Building Design and Construction Handbook*, McGraw-Hill, New York, 1994.

Head, K. H., *Soil Technicians Handbook*, John Wiley and Sons, New York, 1989.

Venkatramaiah, C., *Geotechnical Engineering*, John Wiley and Sons, New York, 1991.

American Association of State Highway and Tranportation Officials System
444 Capital St., NW.
Washington, DC 20001

6

Foundations

This chapter will help you to:

1. Understand the factors to be considered when designing foundations.
2. Identify the various types of foundations.
3. Discuss the features and construction of each type of foundation.

The foundation of a structure transmits the weight of the structure to the earth or rock at or below ground level. Residential and small commercial buildings consist of two major parts—a superstructure, which is the portion above grade, and the foundation, mostly below grade. Large commercial buildings often have a third part, the substructure, which is habitable space below grade (Fig. 6.1). In this case, the foundation is typically caisson piles that penetrate the soil well below the substructure.

FOUNDATION DESIGN

The design of the foundation begins as soon as the basic architectural concepts are developed. Since there are so many factors to consider, the actual foundation may be in a period of flux during the preliminary design stages. Codes require an investigation of subsoil conditions. These data may dictate the choice of foundation used. The size of the building, loads to be carried, conditions on the site, climate, type of structural system to be used, and other such factors are considered by the foundation engineer.

The architect is involved with factors such as the shape and size of the building, orientation, space utilization, and material selection. The structural engineer designs the structural aspects of the building. The foundation engineer must work with both as foundation designs are developed. The final building is the result of the cooperative effort of these three engineering staffs (Fig. 6.2).

Foundation Loads

Foundations are designed to carry the loads imposed by the structure and any additional loads produced by occupancy. These loads are divided into dead loads and live loads. **Dead loads** are those, such as the weight of materials used to construct the building, that act continuously on the foundation. **Live loads** are those that are not constant, such as the weight of furniture, people, wind, or snow. Loads may be **vertical** or **lateral** (horizontal), **static** (fixed) or **dynamic** (moving), **concentrated** (on one spot) or **uniform** (spread evenly over a surface) (Fig. 6.3).

Figure 6.1 These high-rise buildings have foundations extending many stories into the earth.

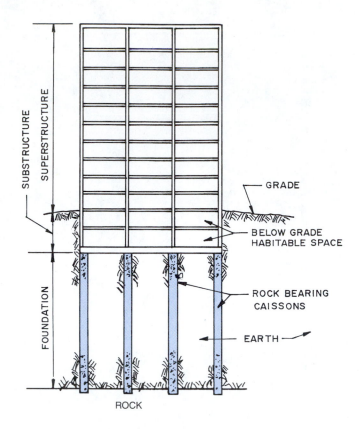

MULTISTORY BUILDING WITH
CAISSON FOUNDATION

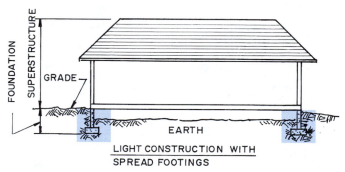

Figure 6.2 Multistory construction extends to rock or other satisfactory bearing surface. Light construction uses footings in the earth just below the frost line.

Foundation Types

Foundations are classified into three major types—spread, pile, and caisson pile foundations.

Spread foundations transfer the loads to the soil through the footings at the bottom of a foundation wall or column. **Pile foundations** use long wood, concrete, or steel piles driven into the earth. They get their load-carrying ability by friction on the sides of the pile or from the end resting on load-bearing strata. **Caisson pile foundations** are formed by drilling holes in the earth and filling them with concrete. They get their load-carrying ability in the same manner as piles (Fig. 6.4).

SPREAD FOUNDATIONS

Spread foundations use spread footings to distribute the building loads over a large enough area of soil to provide adequate bearing capacity. They are usually near the surface of the site and are considered shallow footings. Following are the most commonly used spread footings.

The most common spread footing is a *continuous wall footing* (strip footing). Columns and piers are placed on **independent footings** that are usually square or rectangular. Both types generally are reinforced with steel reinforcing bars (Fig. 6.5). Spread footings must be placed on undisturbed soil. If the excavation for the

footing is dug too deep it must be filled with concrete. In some cases compaction is allowed but only if it is carefully engineered to provide a known load-bearing capacity. Wall footings on sloped sites are usually stepped using horizontal steps (Fig. 6.6). The maximum slope is usually specified by local codes. *Grade beam foundations* support the building on a grade beam that rests on a series of piles or piers. A **grade beam** is a reinforced concrete beam serving as the foundation wall (Fig. 6.7).

A widely used shallow foundation is a *slab-on-grade* as shown in Fig. 6.8. There are many designs for this type of footing. It is used in warm climates where the surface

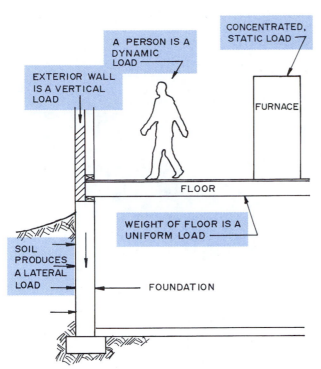

Figure 6.3 Loads on the foundation may be vertical, lateral, static, dynamic, concentrated, or uniform.

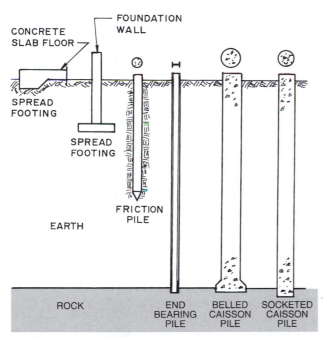

Figure 6.4 Spread, pile, and caisson foundations.

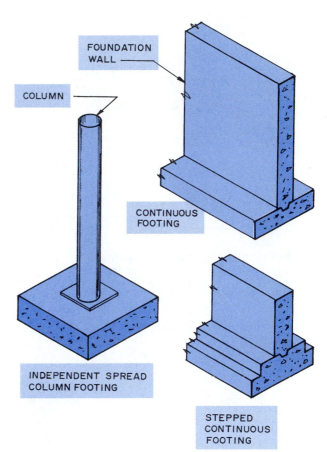

Figure 6.5 Spread footings are used under foundation walls, piers, and columns.

soil has the necessary bearing capacity. The slab can be thickened within the building to support interior load-bearing partitions. Other designs permit the spread footing to be set below the frost line.

Raft or **mat foundations** are reinforced concrete slabs several feet thick covering the entire area of the foundation (Fig. 6.9). This slab spreads the weight of the building over the entire area below the building and reduces the load per square foot on the soil. This is especially useful when the soil characteristics vary across the area. To avoid construction joints, the entire slab must be poured continuously. Another raft design uses a T-beam construction that has a solid slab poured and two-directional beams poured above or below the slab (Fig. 6.10). When the beam is inverted, the slab becomes the basement floor. The beams' cavities are dug in the soil, which must be able to stand without caving in as the beams and slab are poured. When the beams are on top of the slab, the spaces between them must be filled before pouring a concrete floor.

Steel grillage footings use steel beams to reinforce the footing. The base layer is sized to the area needed to support the load. The upper layer of beams is joined at right angles to the lower level, and the entire thing is encased in concrete (Fig. 6.11).

Combined footings are used to support walls and columns near the property line, in which case it is not possible to allow the footing to cross the line onto the neighboring property. The combined footing ties the outside row of columns to the next row within the building. If the outer row of columns were to be placed on the edge of individual footings the loads would not be symmetrical

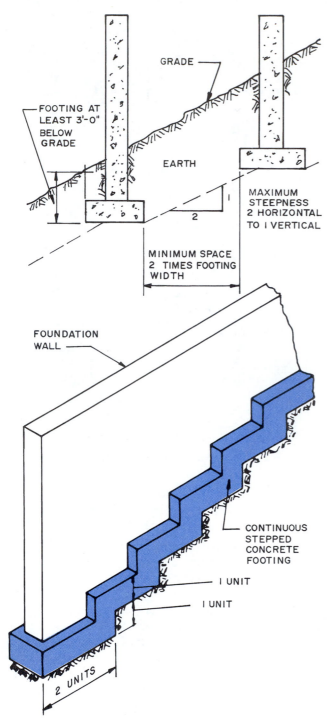

FOUNDATION WALL

CONTINUOUS STEPPED CONCRETE FOOTING

I UNIT

I UNIT

2 UNITS

Figure 6.6 Typical stepped footings.

FOOTING AT LEAST 3'-0" BELOW GRADE

GRADE

EARTH

MAXIMUM STEEPNESS 2 HORIZONTAL TO I VERTICAL

MINIMUM SPACE 2 TIMES FOOTING WIDTH

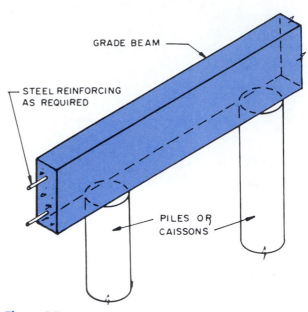

GRADE BEAM

STEEL REINFORCING AS REQUIRED

PILES OR CAISSONS

Figure 6.7 Grade beams cast on piles or piers form the foundation.

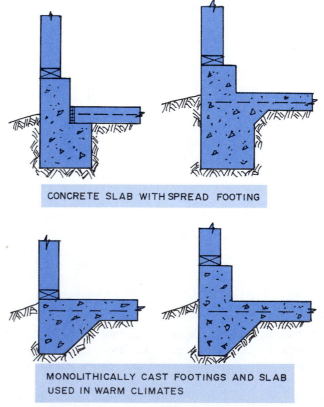

CONCRETE SLAB WITH SPREAD FOOTING

MONOLITHICALLY CAST FOOTINGS AND SLAB USED IN WARM CLIMATES

Figure 6.8 These are monolithically cast footings and slabs used in warm climates.

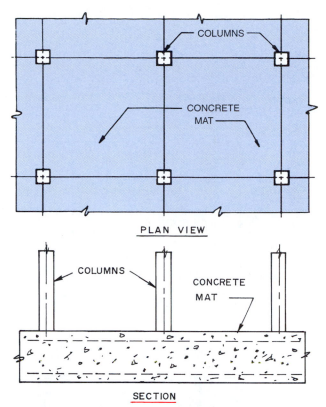

PLAN VIEW

SECTION

Figure 6.9 This is a solid poured concrete mat foundation supporting several structural columns.

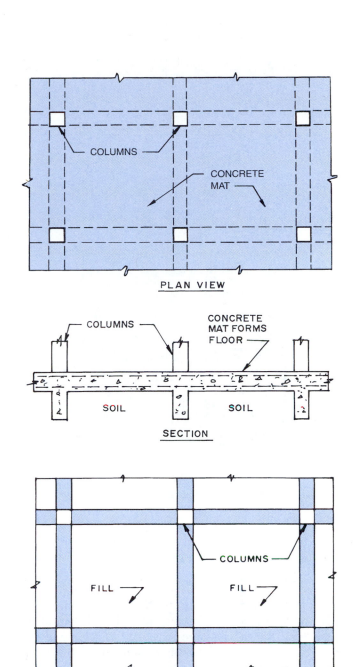

PLAN VIEW

SECTION

Figure 6.10 Two types of T-beam mat constructions.

and the footings would tend to settle unevenly or rotate (Fig. 6.12).

Cantilever footings are a form of combined footing that permits erection of wall columns near the edge of the footing. A beam rests on independent footings and cantilevers over the outer footing to support the column (Fig. 6.13). The footings may be reinforced concrete or steel grillage.

Continuous footings are reinforced concrete members extending continuously under several columns. They are especially useful if the soil varies in the area because they reduce uneven settlement between the columns. They can be flat or T-footings as shown in Fig. 6.14.

Tieback Anchors

Tieback anchors are used to support sheeting and foundations. When used with sheeting they reduce the need for braces running into the excavation, as discussed in Chapter 4. In addition, they are used to anchor foundations and slurry walls, also discussed in Chapter 4.

Two types of tieback anchors are grouted anchors and helical anchors. *Grouted anchors* are installed by drilling through the foundation or slurry wall into the soil and rock to the specified depth (Fig. 6.15). The steel tiebacks

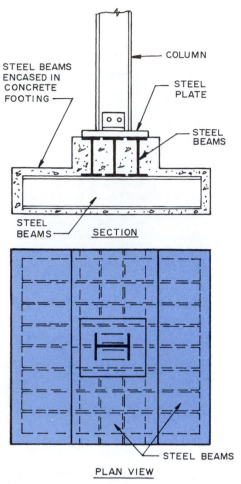

Figure 6.11 A steel grillage column footing.

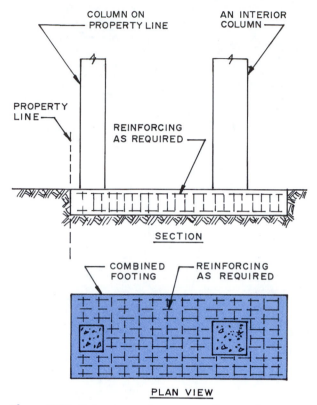

Figure 6.12 Columns on the property line can be supported with a combined footing.

are inserted into the holes and the anchor zone area is filled with pressure-injected grout. After the grout has hardened, the tiebacks are tensioned and secured to a steel chuck that maintains the tension (Fig. 6.16). A typical construction detail is shown in Fig. 6.17.

Helical tiebacks are used for anchoring foundations in light construction into the soil behind the wall. The helical anchor is drilled into the soil. The helical fins pull it into the soil and serve as the anchor (Fig. 6.18). A typical detail is in Fig. 6.19.

PILE FOUNDATIONS

Pile foundations use long wood, steel, and concrete **piles** driven into the earth with a pile hammer. This delivers a series of blows to the top of the pile, driving it into the earth. The commonly used driven piles are shown in Fig. 6.20. These include timber, steel H-piles, pipe, precast concrete, and cast-in-place concrete with steel shells.

Timber piles are tree trunks trimmed and stripped of bark and treated with preservatives. They are driven with the small end down and may have a pointed metal shoe on this tip. The minimum tip is 6 in. (152 mm) in diameter. The top or butt end, minimum diameter 12 in. (305 mm), has a metal pile ring to protect it during driving. These work best in soils relatively free of rock. Parts below the water level will last for years, but the part above the groundwater level may decay (Fig. 6.21).

Wood piles that reach rock or other load-bearing strata transmit the load through the end and are called *end-bearing piles*. Those that get their load-bearing capacity from friction between the sides of the pile and the earth are called *friction piles* (Fig. 6.22).

H-piles are HP structural steel shapes and are especially good for driving into soil with rocks or thin rock strata. Since they have a small surface area, they cannot rely solely on friction but must be driven to end bearing on rock. They are available in widths from 8 to 14 in. (200 to 355 mm) and can be driven to depths of 150 to

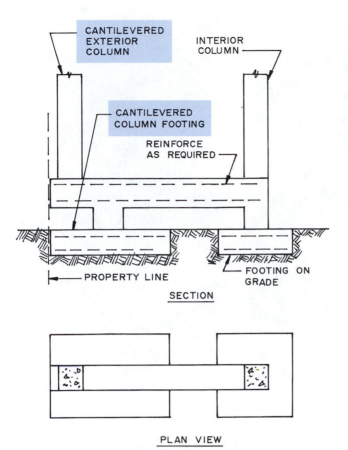

Figure 6.13 Columns on the property line can be supported with a cantilevered footing.

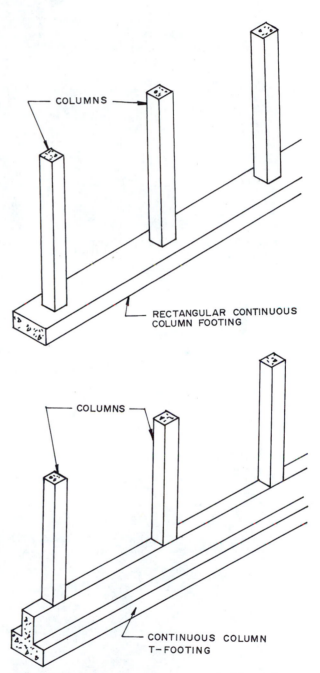

Figure 6.14 Columns can be supported with continuous footings.

200 ft. (45 to 61 m). Long lengths are developed by driving shorter sections and welding on another section.

Pipe piles are heavy-gauge steel pipe that may be driven with an open end. Small-diameter piles generally have the end closed. As open-end piles are driven, the earth inside is removed to decrease resistance. The soil can be removed by inserting a pipe into it and blowing it out. Water may be mixed with the soil to help loosen it for removal. Concrete is placed inside the pipe, starting the fill from the bottom. If the pipe fills with water a concrete seal is inserted at the bottom and the water pumped out. Pipe diameters may be any practical size, with 8 to 18 in. (203 to 457 mm) frequently used, and wall thicknesses from ¼ to ½ in. (6 to 12 mm) are common (Fig. 6.23).

Precast and pretensioned concrete piles are made in a variety of ways. They may be cylindrical, square, or octagonal and are available tapered or uniform in size their entire length. Although most are solid in cross section, some cylindrical types are hollow. The piles are reinforced with steel to withstand the installation stresses as well as the building loads. Piles driven in clay soils have a blunt point;

Figure 6.15 The drilling rig is boring holes through a slurry wall for a tieback anchor. *(Courtesy Franki Canada Limited)*

Figure 6.16 This shows a multiple tenon tieback being grouted under pressure. *(Courtesy Franki Canada Limited)*

I T E M　O F　I N T E R E S T

EARTHQUAKE DESIGN

Major earthquakes around the world have shown the inadequacy of many of the current construction designs. Attempts are being made to design structures that stretch and bend in moderate earthquakes but return to their original shapes. Girders and columns designed for a major earthquake will be expected to absorb energy by permanently deforming but not failing. Earthquakes have caused extensive damage to welded joints in steel-framed buildings, and research is underway to find ways to counteract that damage. Masonry structures will have to be built with steel reinforcing rods grouted into the cavities. Concrete blocks and bricks fall apart if unreinforced. There will be constant changes in foundation design. For example it has been found that reinforced concrete columns that have failed in an earthquake survive if they are enclosed in a steel jacket.

Building codes in many parts of the country have no seismic provisions. However, this is gradually changing as authorities begin to realize that earthquakes can occur in many places not currently specified as seismic areas.

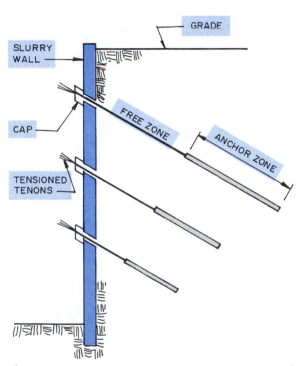

Figure 6.17 A typical detail showing the placement of tieback anchors.

Figure 6.18 Helical tiebacks are screwed into the soil. (*Courtesy A.B. Chance Co.*)

a tapered point is used in sand and gravel. Precast concrete piles can be lengthened by welding together the reinforcing bars where the two sections meet. Some are installed by jetting with water. A pipe is cast in the center of the pile and high-pressure water is forced through it, blasting the soil as the pile penetrates it (Fig. 6.24).

Cast-in-place piles can be cased with a metal shell, or they can be uncased and shell-less. One type of pile uses a tapered steel shell driven with an internal steel mandrel. A mandrel is a steel core inserted into a hollow pile to reinforce the pile shell while it is being driven into the earth. When the required depth has been reached, the mandrel is removed and the shell is filled with concrete (Fig. 6.25). The shells are in 8 ft. lengths, and sections are added to increase the length as the pile is driven.

A *shell-less uncased pile* is formed by driving a steel shell with an interior driving core, removing the core, and filling the shell with concrete as the steel shell is removed from the earth (Fig. 6.26). The Franki pressure-injected footing construction procedure is shown in Fig. 6.27. It uses a steel casing into which zero-slump concrete is rammed to produce a pedestal footing.

A variety of *composite piles* are used when it reduces cost or has other advantages. For example, a wood pile (low in cost) could be driven into the area below ground water and a concrete pile could be placed on top. Several examples are in Fig. 6.28.

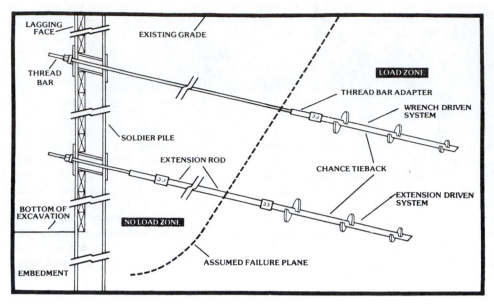

Figure 6.19 A typical installation detail of helical tieback anchors. *(Courtesy A.B. Chance Co.)*

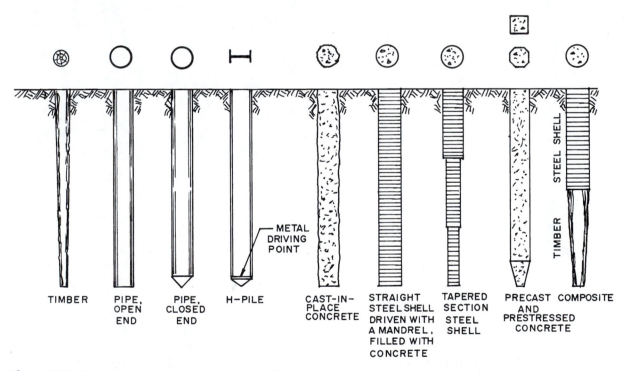

Figure 6.20 Some of the commonly used driven piles.

Figure 6.21 A pile hammer drives these wood piles to stable soil. *(Courtesy Canadian Wood Council)*

Figure 6.23 These pipe piles are ready to be driven by the pile drivers in the background. *(Courtesy Franki Canada Limited)*

Figure 6.22 After the piles are driven they are cut to length. *(Courtesy Canadian Wood Council)*

The foundation drawings will show how the piles are to be placed. Usually they are in clusters with a cast-in-place reinforced concrete pile cap. The cap distributes the load equally on the piles below the cap (Fig. 6.29). Entire walls can be supported by a series of piles, as shown in Fig. 6.30. The footings can be supported by piles, or grade beams can be run over the pile caps to provide a foundation for construction of the wall. Mat foundations can also be supported on piles.

Individual pile caps and caissons are interconnected by ties that carry tension and compression loads as specified by the building code.

Pile Hammers

Piles are driven into the earth with **pile hammers.** The commonly used types include drop, single-acting steam, double-acting steam, diesel, vibratory, and hydraulic.

Drop hammers use a heavy weight lifted up vertical rails called leads. When the weight reaches the top of the rail, it is released and falls onto the pile, driving it into the earth. Repeated cycles drive the pile to the required depth. A drop hammer can weigh from 500 to 3,000 lb. (226 to 1360 kg), and it can be dropped from a height of 5 to 20 ft. (1.5 to 6.0 m), depending on the pile and soil conditions. It makes four to six blows per minute (Fig. 6.31).

Single-acting steam hammers are mounted on top of the pile. The hammer (called a ram) is lifted by the steam or compressed air. When the ram reaches the top of the stroke it is released and free-falls onto the top of the pile. The steam raises the ram and the cycle repeats. It makes forty or more blows per minute (Fig. 6.32).

Double-acting steam hammers operate in the same manner as single-acting hammers except on the downstroke the steam pressure is applied to the backside of a piston, which increases the energy per blow.

Diesel pile hammers are self-contained driving units that include a vertical cylinder, a ram, an anvil, and the parts (fuel pump, injectors, fuel tank, etc.) that make up a diesel power unit. The ram is placed on top of the pile and released. As the ram falls, the fuel pump injects diesel fuel into the combustion chamber. As the ram

Figure 6.24 Precast concrete hollow cylinder piles. *(Courtesy Precast/Prestressed Concrete Institute)*

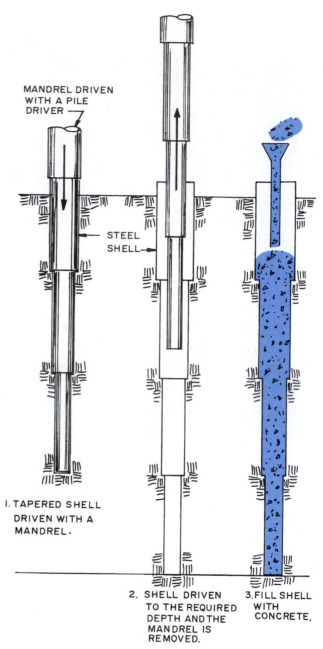

MANDREL DRIVEN WITH A PILE DRIVER

STEEL SHELL

1. TAPERED SHELL DRIVEN WITH A MANDREL.

2. SHELL DRIVEN TO THE REQUIRED DEPTH AND THE MANDREL IS REMOVED.

3. FILL SHELL WITH CONCRETE.

Figure 6.25 Shells for cast-in-place concrete piles can be driven with a mandrel and filled with concrete after the required depth has been reached and the mandrel removed.

continues to fall, the fuel is compressed and the heat generated causes an explosion. This drives the anvil hard against the pile and causes the ram to rise, thus starting a new cycle.

Vibratory pile hammers are held by a crane and mounted on the end of the pile. A motor drives eccentric weights in opposing directions, producing vibrations transmitted to the pile. This agitates the soil, reducing the friction between the pile and the soil and allowing the pile to penetrate the soil. Vibratory pile hammers are especially effective for driving into water-saturated noncohesive soils. They have difficulty in dry sand and will not penetrate tight cohesive soils (Fig. 6.33). They are effective in driving sheet piling.

Hydraulic pile hammers operate much the same as described for steam hammers, but hydraulic fluid is used to move the ram.

CAISSON PILE FOUNDATIONS

A *caisson* is a watertight structure used to provide a dry working area for building foundations or structures below water. A *caisson pile* is a pile cast-in-place in the earth by driving a tube into the soil, emptying the earth from

inside the tube, and filling it with concrete and required reinforcing steel (Fig. 6.34).

Caisson piles are end-bearing units drilled using large-diameter augers. A steel casing is lowered into the hole as drilling continues. The soil must be able to stand without collapsing when the casing is removed. Caisson piles are bored through the weaker soil until they strike rock or other acceptable bearing soil. When rock is not reached, the bottom is belled using a belling bucket on the end of the drill to provide a larger end-bearing surface (Fig. 6.35). Shaft diameters vary from

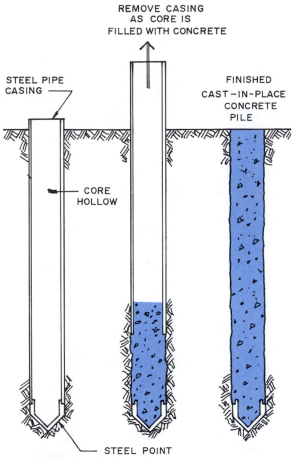

Figure 6.26 A typical cast-in-place concrete pile when the steel shell is removed.

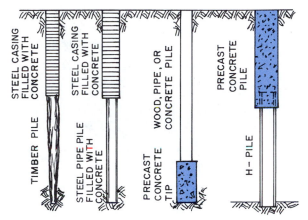

Figure 6.28 Typical composite piles.

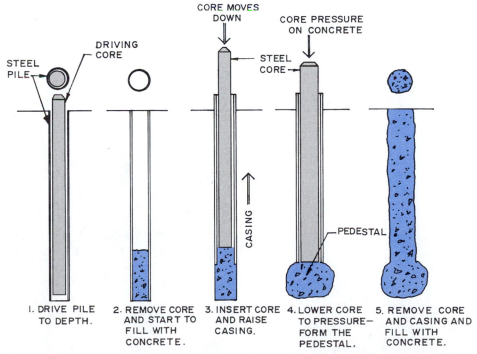

Figure 6.27 The Franki pressure-injected process produces a pile with a pedestal footing.

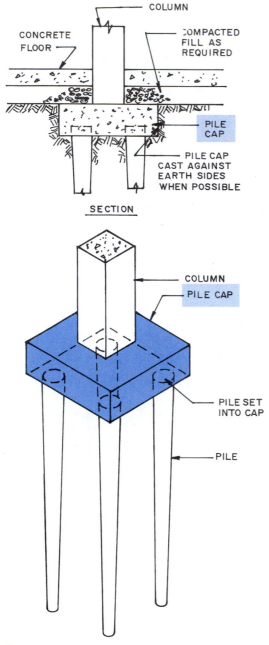

Figure 6.29 A typical pile cap supporting a column.

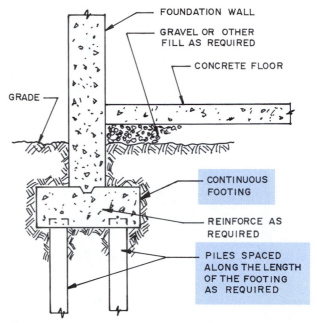

Figure 6.30 Footings can be supported on a series of pilings.

Figure 6.31 Piles are driven by pile hammer. *(Courtesy Franki Canada Limited)*

Figure 6.33 Vibratory pile hammers have a vibrating unit mounted on the end of the pile. (*Courtesy Franki Canada Limited*)

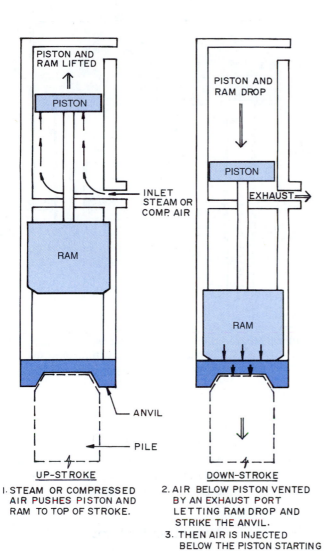

PISTON AND RAM LIFTED

PISTON

RAM

INLET STEAM OR COMP. AIR

ANVIL

PILE

UP-STROKE

1. STEAM OR COMPRESSED AIR PUSHES PISTON AND RAM TO TOP OF STROKE.

PISTON AND RAM DROP

PISTON

EXHAUST

RAM

DOWN-STROKE

2. AIR BELOW PISTON VENTED BY AN EXHAUST PORT LETTING RAM DROP AND STRIKE THE ANVIL.

3. THEN AIR IS INJECTED BELOW THE PISTON STARTING ANOTHER CYCLE.

Figure 6.32 This is a generalized example showing how a single-acting steam hammer operates.

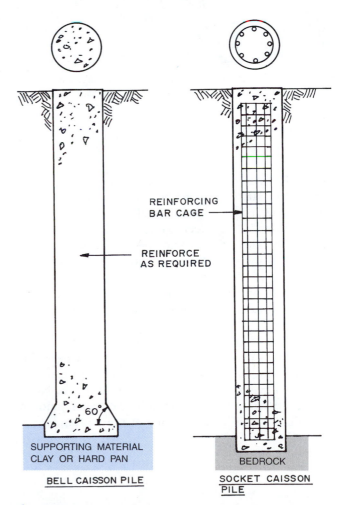

REINFORCING BAR CAGE

REINFORCE AS REQUIRED

60°

SUPPORTING MATERIAL CLAY OR HARD PAN

BELL CAISSON PILE

BEDROCK

SOCKET CAISSON PILE

Figure 6.34 Caisson piles may rest on bedrock or other approved bearing material as specified by local codes.

Figure 6.35 The bottom of a caisson pile is belled with a belling bucket. (*Courtesy Franki Canada Limited*)

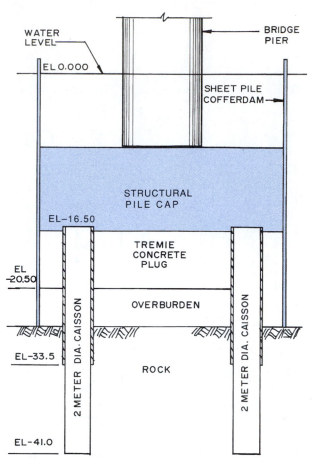

Figure 6.36 A section through a bridge pier constructed on caisson piles. The piles support a structural pile cap carrying the bridge pier. (*Courtesy Franki Canada Limited*)

18 in. (460 mm) to 5 or 6 ft. (1.5 to 1.8 m) or more. The bottoms of large-diameter borings are inspected by a person lowered into them. If all is as expected, the required reinforcing steel is lowered into the hole and the hole is filled with concrete. As it is filled, the steel casing is usually removed, although steel casings are sometimes left in place.

The size and the reinforcing depend on the local conditions. Local codes specify allowable soil-bearing capacity.

An example of the use of caissons built under difficult conditions is illustrated by the section drawing in Fig. 6.36. This shows the construction of a bridge pier to be supported by six 2 m–wide (6 ft. 6 in.) caissons. Notice the construction of a sheet pile cofferdam and the watertight tremie concrete plug at the bottom. In this

case the caissons were 23 m (75 ft. 5 in.) deep with 7 m (72 ft. 11½ in.) rock sockets. The tops of the caissons were 16 m below the mean water line. The work was performed from floating platforms, as shown in Fig. 6.37.

Helical Pier® Foundation Systems

Helical Pier foundations use helical (screw) steel piers installed at intervals between the footing forms. They are installed by screwing them into the soil (Fig. 6.38). The torque required to drive each pier correlates to the bearing capacity of the soil below and is used to ascertain the depth the anchor must be placed to support the required structural load. It can also be used to underpin settling foundations of existing buildings by using a special support bracket (Fig. 6.39).

CAISSONS
SHEET PILE COFFERDAM

Figure 6.37 The floating work station was used to install a sheet cofferdam and six 2 m–diameter caissons that formed the support for the bridge pier detached in Fig. 6.36. *(Courtesy Franki Canada Limited)*

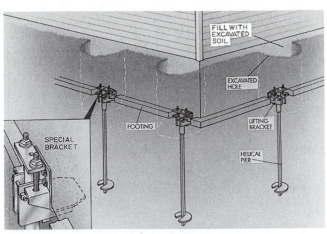

Figure 6.39 Helical Piers® can be used to underpin the foundation of existing buildings. *(Courtesy A.B. Chance Co., Helical Pier Foundation Systems is a registered trademark of the A.B. Chance Co.)*

Figure 6.38 The Helical Pier® Foundation System uses steel helical metal piers drilled into the soil before the footing is poured. *(Courtesy A.B. Chance Co., Helical Pier Foundation Systems is a registered trademark of the A.B. Chance Co.)*

REVIEW QUESTIONS

1. What loads act upon a foundation?
2. What are the major types of foundations?
3. What types of piles are available?
4. What types of equipment are used to drive piles?
5. How are caisson piles installed?

KEY TERMS

caisson pile foundation A foundation formed by drilling holes in the earth and filling them with concrete.

cast-in-place piles Concrete piles cast in a hollow metal shell driven into the earth or an uncased drilled hole.

concentrated load Any load that acts on a very small area of a structure.

dead load A permanent load that provides steady pressure on the building structures, such as roofing materials.

dynamic load Any nonstatic load.

grade beam A reinforced structural member supporting the exterior wall of a structure and bearing directly upon columns or piers.

independent footing A footing that supports a single structural element, such as a column.

lateral load A load moving in a horizontal direction, such as a wind load.

live load Nonpermanent moving or movable external load on a structure, such as furniture.

mat foundation A large reinforced concrete slab on the earth covering the entire area of the building foundation.

pile A wood, steel, or concrete column, usually driven into the soil, to be used to carry a vertical load.

pile foundation A foundation that uses long wood, concrete, or steel piles driven into the earth.

pile hammer A machine for delivering blows to the top of a pile, driving it into the earth.

spread foundation A foundation that distributes the load over a large area.

static load Any load that does not change in magnitude or position with time.

tieback anchors Steel anchors grouted into holes drilled in the excavation wall to hold the sheeting, thus reducing the number of braces required.

uniform load Any load that is spread evenly over a large area.

vertical load A load acting in a direction perpendicular to the plane of the horizon

SUGGESTED ACTIVITIES

1. Secure foundation drawings from local architects and general contractors and list the details shown, such as the type of foundation specified, reinforcing, and other details. See if a log of test borings was made and identify the characteristics of the soils as the bore hole goes deeper.

2. Visit construction sites and observe the excavation, forms, piers, reinforcing, and other materials used to construct the foundation. Report these with a written statement. What type of footing was used?

3. If any buildings in your area are being built on piles try to visit the site and take photos to illustrate the process from locating the piles to establishing the finished elevation of the tops of the piles. What type of piles were being used?

4. When visiting in-process foundation construction, report on any actions taken to keep the excavation dry.

5. As you visit construction sites observe situations you think are not safe and prepare remarks so you can lead a class discussion on safety. (While on the site do not charge the supervisor or workers with allowing unsafe practices. Simply observe and remember.)

ADDITIONAL INFORMATION

Ambrose, J. E., *Building Construction: Site and Below-Grade Systems,* Van Nostrand Reinhold, New York, 1993.

Ambrose, J., *Simplified Design of Building Foundations,* John Wiley and Sons, New York, 1988.

Merritt, F. E., and Ricketts, J. T., *Building Design and Construction Handbook,* McGraw-Hill, New York, 1994.

CONCRETE AND MASONRY

Concrete and masonry are one of the most widely used construction materials. They are readily available in most parts of the world and can use a wide variety of naturally occurring materials. Concrete can be made by simple hand-batching methods or in large quantities in a computer-controlled batching plant. It can be used for a wide range of applications and serve successfully for many years.

Masonry construction has been used since early times, and historical records and architectural excavations have produced evidence of masonry manufacture and use. Masonry is simple to use and can be made from a variety of materials.

Chapters 7, 8, and 9 explain the factors involved in the manufacture of various concretes and the recommended placing and curing processes. Methods for forming, reinforcing, and placing cast-in-place concrete are discussed, as well as the range of activities involved in forming and placing precast concrete units.

Chapters 10 through 14 include information on the types of masonry units, sizes, technical data, and a wide range of applications. Methods for using masonry units for building construction are illustrated. ▲

Courtesy Precast/Prestressed Concrete Institute

DIVISION 3

CONCRETE
CSI MASTERFORMAT™

Courtesy Precast/Prestressed Concrete Institute

7

Concrete

This chapter will help you to:

1. Identify the various types of portland cement and cite the purposes for which they can be used.

2. Be familiar with the properties of portland cement and how they affect concrete.

3. Discuss the place of water in concrete and its desired properties.

4. Discuss the types of aggregate used in concrete and their desirable characteristics.

5. Understand the purposes of various concrete admixtures and how they effect the concrete.

6. Understand what is required to produce concrete with properties for various applications.

7. Be aware of the commonly used concrete tests and what they reveal about the mix.

Concrete is a major material of construction used all over the world. The raw materials needed are available in most parts of the world. It does not require complex or expensive equipment to make concrete. In many parts of the world wood resources are meager, and what is available is too expensive for use in construction as is done in the United States. Concrete is an inexpensive construction material, but it has no form of its own or tensile strength. Forms must be built to shape it into useful form, and reinforcing steel must be added to provide tensile strength.

Concrete is a solid, hard material produced by combining in proper proportions portland cement, coarse and fine aggregates (sand and stone), and water. The chemical reaction between the cement and water produces heat and a hardening of the mass.

PORTLAND CEMENT

Portland cement is a fine, pulverized material consisting of compounds of lime, iron, silica, and alumina. The exact composition of types of portland cement varies, but the composition of Type 1, Normal, is representative (Table 7.1). The manufacture of portland cement requires the combination of these elements in proper proportion under carefully controlled conditions. The basic process is shown in Fig. 7.1.

The lime in portland cement is commonly derived from limestone, marble, marl, or seashells. Iron, silica, and alumina are obtained from mining clays containing these elements. In some areas iron furnace slag and flue dust are used, as are sand, chalk, and bauxite. These ingredients are crushed in the primary crusher and sent to a vibrating screen. Pieces falling through the screen go to storage. Those that are too big go to a secondary crusher and on to storage. After this, either the wet or dry process is used.

The *dry process* grinds the raw material to a powder and blends it in mixing silos. From here it goes to the kiln. In the *wet process*, the ground materials are mixed with water to form a slurry, blended, and moved to a kiln. The *kiln* is a rotating cylinder operating at 2600–3000°F (1600–1780°C). It burns the materials into a clinker. The clinker is cooled, a small amount of

Table 7.1 Oxide Composition of Type 10 (ASTM Type I) or Normal Portland Cement

Oxide Ingredient	Range, %
Lime, Ca_O	60–66
Silica, SiO_2	19–25
Alumina, Al_2O_3	3–8
Iron, Fe_2O_3	1–5
Magnesia, MgO	0–5
Sulfur trioxide, SO_3	1–3

Courtesy Portland Cement Association

Table 7.2 Types of Portland Cement

Canadian Designation	United States Designation	Type
10	I	Normal
20	II	Moderate
30	III	High early strength
40	IV	Low heat of hydration
50	V	Sulfate resisting
Air-Entrained Types		
10A	IA	Normal air-entrained
20A	IIA	Moderate air-entrained
30A	IIIA	High early strength air-entrained

Courtesy Portland Cement Association

gypsum is added, and then it is ground into a powder so fine it would pass through a sieve with 40,000 openings per square inch. The gypsum is added to retard the **curing** process.

Most portland cement is shipped in bulk in railroad cars or large trailers designed especially for this material. Bulk cement is sold by the ton (2000 lb. or 907.2 kg), and smaller quantities are bagged with a volume of one cubic foot (0.028 m^3) weighing 94 pounds (42 kg).

ASTM-Designated Types of Portland Cement

Cement is an easily molded temporary paste when mixed with water. After a short time it hardens or sets to a rigid mass. The setting and hardening is called **hydration.** It is the result of a chemical reaction between the portland cement and water. Hardening is *not* simply a process of evaporation or drying.

Portland cements are made following ASTM Designation C150, which provides eight types of cement. These are Type I, Normal; Type IA, Normal, Air-entrained; Type II, Moderate; Type IIA, Moderate, Air-entrained; Type III, High Early Strength; Type IIIA, High Early Strength, Air-entrained; Type IV, Low Heat of Hydration; and Type V, Sulfate Resisting (Table 7.2). In Canada these are specified as types 10, 20, 30, 40, and 50.

Type I: Normal

Type I is a general-purpose portland cement used whenever the special properties of the other types are not necessary, such as sulfate attack from soil or water or objectionable temperature rise in the concrete due to hydration. It is used for purposes such as pavements, sidewalks, reinforced concrete structural members, bridges, tanks, water pipes, and masonry building units.

Type II: Moderate

Type II is used when protection against moderate sulfate attack, found in some soils and groundwater, is necessary. It also generates less heat by hydrating at a slower rate than Type I. This makes Type II useful in structures

that have larger masses of concrete, such as abutments, piers, and large retaining walls.

Type III: High Early Strength

Type III portland cement provides high strength faster than Type I or II. It is used when forms are to be removed as soon as possible or to get strength faster in cold weather, thus reducing the curing period when special protection from freezing is necessary. High early strength normally reduces the curing time required to one week or less.

Type IV: Low Heat of Hydration

Type IV is used when it is necessary to reduce the rate and amount of heat generated by hydration. It develops strength slower than Type I. Its major use is for structures having large concrete masses, such as dams and nuclear plants.

Type V: Sulfate Resisting

Type V portland cement is used only in concrete that is exposed to severe sulfate conditions. This most frequently occurs in soils and groundwater in areas where sulfate concentrations are high.

White Portland Cement

White portland cement is a true portland cement, but during the manufacturing process the components are controlled so the finished products will be white. It is used mainly for architectural purposes, such as precast curtain-wall and facing panels, terrazzo floors, stucco, finish-coat plaster, and tile grout.

Types IA, IIA, IIIA: Air-Entrained

There are three types of **air-entrained portland cement,** IA, IIA, and IIIA. These are the same as Types I, II, and III except that small quantities of air-entraining materials are interground with the clinker and gypsum.

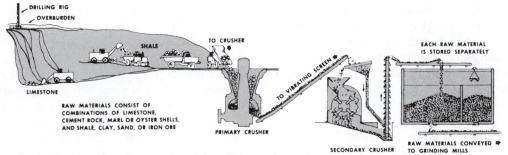

1. Stone is first reduced to 5-in. size, then to ¾ in., and stored.

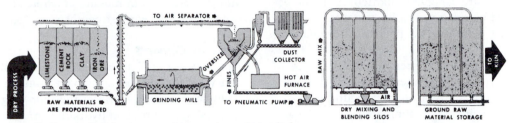

2. Raw materials are ground to powder and blended.

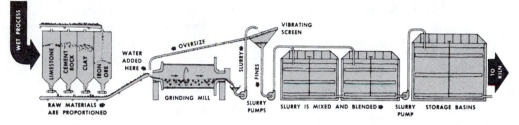

2. Raw materials are ground, mixed with water to form slurry, and blended.

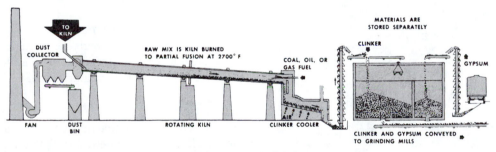

3. Burning changes raw mix chemically into cement clinker.

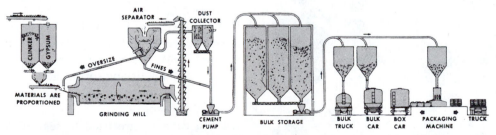

4. Clinker with gypsum is ground into portland cement and shipped.

Figure 7.1 The steps in the manufacture of portland cement. *(Courtesy Portland Cement Association)*

Air-entraining entrains millions of microscopic air bubbles into the concrete (Fig. 7.2). This greatly improves the durability of concrete that is exposed to moisture and the freezing and thawing cycles in the winter. It also increases the concrete's resistance to surface scaling, which is often caused by salts used to remove ice on the surface. Air-entraining improves the workability of concrete and reduces segregation of the aggregate, bleeding, and the amount of water needed.

Blended Hydraulic Cements

Hydraulic cements are capable of hardening under water. There are two blended hydraulic cements, portland blast-furnace-slag cements and portland pozzolan cements.

ASTM C595 sets requirements for two types of portland blast-furnace-slag cements, IS and IS-A (air-entrained). Blast-furnace slag is interground with the portland cement clinker or separately ground and blended with portland cement. Blast-furnace slag content of these cements is 25 to 65 percent of the weight of the cement.

The four types of portland pozzolan cements are IP, IP-A, P, and P-A. The "A" indicates air-entraining. They are manufactured by intergrinding portland cement clinker with a suitable pozzolan. **Pozzolan** is a siliceous or siliceous and aluminous material that chemically reacts with calcium hydroxide to form compounds possessing cementitious properties. Pozzolans may be manufactured or natural. Natural pozzolans include pumicite, volcanic ash, and volcanic tuff. Tuff is a porous rock formed by the consolidation of volcanic ashes and dust. Processed natural pozzolans include clay and shale that are burned or calcined in a kiln, then crushed and ground. All four types can be used for general concrete construction. Types P and P-A are used in massive concrete structures that do not require high early strengths.

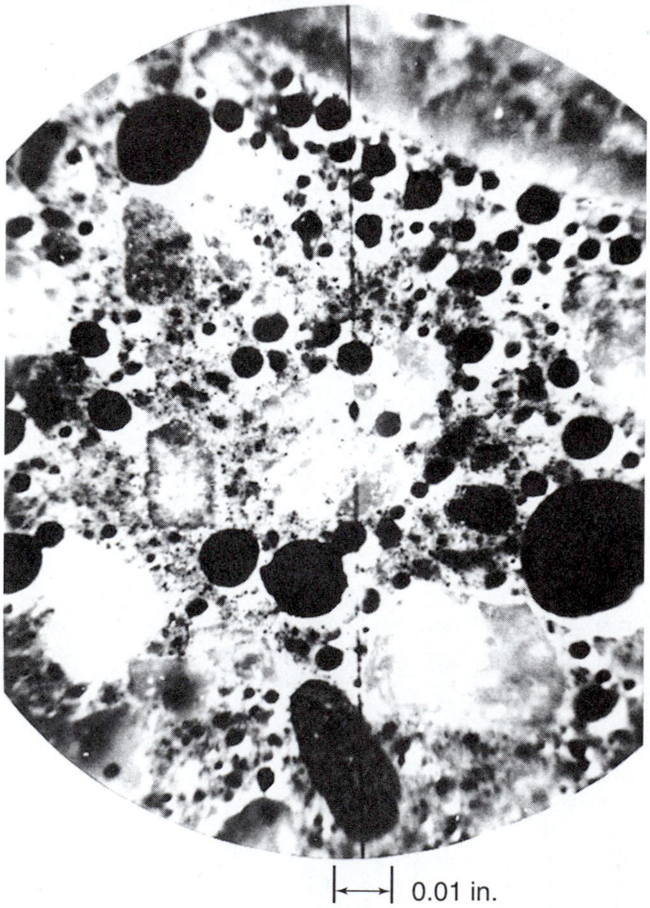

Masonry Cements

Masonry cements are used for the manufacture of masonry mortars. They are covered under ASTM, C91. They may contain some of the following: portland cement, slag cement, and hydraulic lime. In addition, they may contain one or more of the following: hydrated lime, limestone, chalk, calcareous shell, talc, slag, or clay. Mortar cements must have excellent workability, plasticity, and water retention properties.

Other Cements

There are other types of cements that are not covered by ASTM specifications. These include some special types of portland cement.

Waterproof Portland Cement

Waterproof portland cement has a small amount of calcium stearate or aluminum stearate added to the portland cement clinker during grinding.

Plastic Cements

Plastic cements have plasticizing agents added to Type I or Type II portland cement during the milling operation. They are used in stucco and plaster.

Expansive Cement

Expansive cement is a hydraulic cement that expands during the early hardening period. The three types available are K, M, and S. They can be used to compensate for the effects of drying shrinkage.

Oil-Well Cement

Oil-well cements are used to seal oil wells and are a blend of portland cement and hydraulic cement. They must be slow setting and withstand high pressures and temperatures.

Regulated Set Cement

Regulated set cement is a hydraulic cement that can be compounded to have a set time from one to two minutes to as long as fifty or sixty minutes.

|← →| 0.01 in.

Figure 7.2 A polished section of air-entrained concrete as seen through a microscope. *(Courtesy Portland Cement Association)*

Properties of Portland Cement

The specifications for portland cement place limits on its chemical composition and physical properties. A knowledge of these properties is necessary to evaluate the results of tests on the cement itself and on concrete made with the cement (Fig. 7.3).

Fineness

The rate of hydration increases as the cement increases in fineness, which accelerates the strength development of the concrete. The effects of greater fineness on strength are most apparent during the first five to seven days. The compound composition and fineness of portland cements are in Table 7.3.

Soundness

Soundness is the ability of a hardened paste to retain its volume after setting. The major problem is a delayed destructive expansion after the paste has hardened. Excessive amounts of hard-burned free lime or magnesia are the main cause. Cement can be tested following ASTM standards to determine soundness.

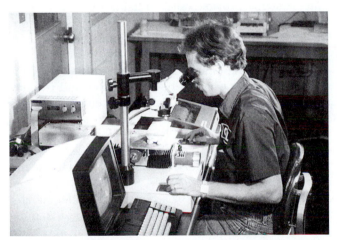

Figure 7.3 Test results on portland cement are made in a cement laboratory. *(Courtesy Portland Cement Association)*

Table 7.3 Potential Compound Composition and Fineness of Cements

Type of Portland Cement	Compound Composition, Percentage				Wagner Fineness, m^2/g
	C_3S	C_2S	C_3A	C_4AF	
Type I	55	19	10	7	0.18
Type II	51	24	6	11	0.18
Type III	56	19	10	7	0.26
Type IV	28	49	4	12	0.19
Type V	38	43	4	9	0.19
White	33	46	15	2	

Courtesy Portland Cement Association

Consistency

Consistency is the relative mobility of a fresh mixture, in other words its ability to flow. Cement paste tests for consistency are made using the tests specified by ASTM C191.

Setting Time

Setting time tests are made to determine if the cement paste undergoes setting and hardening during the first few hours. The initial set must not occur before the concrete finishing operations can be completed. The final set should occur without unnecessary delay. Setting times of cement pastes and concrete do not correlate directly because of water loss, the surface on which the concrete has been poured, and temperature differences.

False Set

False set results in the loss of plasticity without the development of much heat shortly after the concrete has been mixed. False set causes no difficulty if the concrete is given additional water. Chemical admixtures can be added to delay the occurrence of false set.

Compressive Strength

The compressive strength of cement is a major physical property. It is found by testing 2 in. (50 mm) cubes of mortar as specified by ASTM C150. Compressive strength is influenced by the type of cement (its compound composition and fineness). Cement compressive strengths cannot be used to determine concrete compressive strengths because of variances in concrete mixtures. Actual concrete samples must be tested to find the compressive strength of concrete. Compressive strengths of cements are shown in Table 7.4.

Heat of Hydration

Heat of hydration is the heat generated when the cement and water chemically react. The *amount* of heat generated depends on the chemical composition of the cement. The *rate* of heat generated is affected by the fineness of the cement, the chemical composition, and the temperature during hydration. Structures with a large concrete mass may experience a significant rise in temperature unless the heat can be rapidly dissipated. If not controlled, it may create undesirable stresses. In cold weather the excessive heat may be beneficial because it helps to maintain favorable curing temperatures. The approximate amounts of heat generated are shown in Table 7.5.

Loss of Ignition

Loss of ignition is a test to find out how much weight a sample of portland cement will lose when heated to 1652–1832°F (900–1000°C). A loss of ignition may be caused by improper and prolonged storage or by adulteration during transfer and transport.

Table 7.4 Relative Compressive Strength Requirements as Affected by Type of Cement[a]

Type of Portland Cement	Compressive Strength, Minimum, Percentage of Strength of Type I at 7 days			
	1 day	3 days	7 days	28 days
I	—	64	100	143[c]
IA	—	52	80	114[c]
II[b]	—	54	89	143[c]
IIA[b]	—	43	71	114[c]
III	64	125	—	—
IIIA	52	100	—	—
IV	—	—	36	89
V	—	43	79	107
IS	—	64	100	125
IS-A	—	52	80	100
IS(MS)	—	36	64	125
IS-A(MS)	—	27	50	100
IP	—	64	100	125
IP-A	—	52	80	100
P	—	—	54	107
P-A	—	—	45	89

[a]ASTM Designations: C150-77 and C595-77.
[b]A strength reduction of one-fifth to one-third is allowed if the optional heat of hydration or the chemical limit on the sum of the C_3S and C_3A is specified. See ASTM specification.
[c]Optional specification requirement.
Note: When suffixes MH or LH are added to any of last eight cement types listed, the strength requirement is 80% of the values shown.
Courtesy Portland Cement Association

Table 7.5 Approximate Relative Amounts of Heat Generated during Hydration[a]

Portland Cement Type	Percent of Hydration, Relative to Type 1, Normal
Type 1, Normal	100
Type 2, Moderate	80 to 85
Type 3, High early strength	up to 150
Type 4, Low heat of hydration	40 to 50
Type 5, Sulfate resisting	60 to 75

[a]First seven days.
Courtesy Portland Cement Association

Specific Gravity

The specific gravity of portland cement is about 3.15. Portland blast-furnace-slag and portland pozzolan cements may be as low as 2.90. The specific gravity is not an indication of the quality of a cement. It is used when calculating mix designs.

Consistency

Consistency is an indication of the ability of a fresh mix to flow. Mortars are mixed and tested for flow, and the findings are used to regulate the water content. The consistency of concrete is measured by a slump test.

Storing Portland Cement

Portland cement is moisture sensitive, so it must be protected from dampness. Sacked material must be stored on pallets, whether it is in a warehouse with a concrete floor or temporarily on the site in the open. Warehouse storage must be watertight, and bags must not touch the exterior walls. Bags should be packed closely together to reduce airflow and should be covered with plastic or tarpaulins if they are to be stored for long periods.

Cement in bags tends to pack if it is stored a long time. This can be corrected by rolling the bags on the floor before opening them. If lumps remain, the cement should be tested before using it on jobs where properties are important.

Bulk cement is stored in watertight bins or silos. Dry low-pressure aeration or vibration should be used to make the cement flow better. When cement is loaded into bins or silos it swells, so a unit will store only about 80 percent of its rated capacity.

WATER

Water used in making concrete should be clear and free of sulfates, acids, alkalis, and humus. Most municipal water systems provide water suitable for use. Potable water from wells also is usually acceptable. Water from lakes, ponds, or rivers should be carefully checked for its suitability before using it.

Water of questionable quality can be used for concrete if test mortar cubes have seven-day and twenty-eight-day strengths equal to 90 percent of samples made with drinkable water. These tests should be made following ASTM C109 specifications. ASTM C191 specifies how to test the samples to see if impurities in the water adversely shorten or lengthen the setting time.

Alkali carbonate and bicarbonate, chloride, sulfate, carbonates of calcium and magnesium, iron salts, inorganic salts, acid waters, and alkaline waters all have their effect on the concrete. Also sugar, silt, suspended particles of clay or fine rock, oils in suspension, and algae in water can make mixing water unsuitable.

Seawater containing up to 35,000 ppm (parts per million) of dissolved salts is generally suitable as mixing water for unreinforced concrete. Concrete made with seawater has a higher early strength than normal concrete and usually has lower strength after twenty-eight days. Seawater may be used for reinforced concrete, but the reinforcement must be protected from corrosion and the concrete must contain entrained air. It is also necessary to check for sea salt on aggregates that were exposed to salt water. Not more than 1 percent sea salt on aggregates is acceptable.

The Environmental Protection Agency and some state agencies prohibit discharging untreated wash water used in reclaiming sand and gravel from returned concrete or mixer washout operations into rivers, streams, and lakes.

AGGREGATES

Approximately 60 percent to 80 percent of concrete is made up of the **aggregates.** The cost of concrete and its properties are directly related to the aggregates. Natural aggregates are composed of rocks and minerals. *Minerals* are naturally occurring inorganic substances that have distinctive physical properties and a composition that can be expressed by a chemical formula. *Rocks* are usually composed of several minerals. For example, limestone is basically calcite (a mineral) with small amounts of quartz, feldspar, and clay (all minerals). The crushing and weathering of rock produces stone, gravel, sand, silt, and clay. Rocks and minerals commonly found in aggregates are shown in Table 7.6.

Aggregates must conform to ASTM specifications. They must be clean, hard, strong, free of absorbed chemicals, and free of coatings of clay, humus, and other fine materials. Aggregates containing some shale, shaly

rocks, soft or porous rocks, and some types of chert are not to be used because they do not weather well and cause popouts on the exposed concrete surface.

Normal weight concrete weighs about 135 to 160 lb. per ft.^{3} (2150 to 2550 kg/m^3). It uses sand, gravel, crushed stone, and air-cooled blast-furnace-slag aggregates. It is specified in ASTM C33. *Structural lightweight concrete* weighs about 85 to 115 lb. per ft.^{3} (1350 to 1850 kg/m^3) and uses expanded shale, clay, slate, and slag aggregates. It is specified in ASTM C330. *Insulating concrete* weighs about 15 to 90 lb. per ft.^{3} (250 to 1450 kg/m^3) and uses pumic, scoria, perlite, vermiculite, and diatomite aggregates. It is specified in ASTM C332. *Heavyweight concrete* weighs 175 to 400 lb. per ft.^{3} (2800 to 6400 kg/m^3) and uses barite, limonite, magnetite, ilmenite, hematite, iron, and steel slugs as aggregate. It is specified in ASTM C637.

Characteristics of Aggregates

The major characteristics for aggregates, listed in Table 7.7, are described below.

Resistance to Abrasion and Skidding
Abrasion resistance is important when the aggregate is to be used in an area that is subject to heavy abrasive use, such as a factory floor. Aggregate for abrasion resistance is tested following ASTM C131 standards. To give skid resistance, the siliceous particle content of the fine aggregate should be 25 percent or more.

Resistance to Freezing and Thawing
The freeze-thaw properties of an aggregate are important in concrete to be exposed to a wide range of temperatures. Important considerations are the porosity, absorption, permeability, and pore structure of the aggregate. Suitable aggregates may be chosen from those used in the past that have given good results. Unknown materials should be tested using ASTM tests.

Compressive Strength
The strength of aggregates under compression is an important factor to consider when choosing the material to be used. This is tested when hardened concrete samples are given standard compression tests.

Shape and Texture of Particles
The shape and texture of aggregate particles influence the properties of fresh concrete more than cured concrete. Rough-textured, angular, elongated particles require more water than do smooth, rounded aggregates. Therefore, angular particles require more cement to maintain the required water-cement ratio.

The cement tends to bond better to angular particles than to smooth particles. Therefore, this must be

Table 7.6 Rock and Mineral Constituents in Aggregates

Minerals	Igneous Rocks	Metamorphic Rocks
Silica	Granite	Marble
Quartz	Syenite	Metaquartzite
Opal	Diorite	Slate
Chalcedony	Gabbro	Phyllite
Tridymite	Peridotite	Schist
Cristobalite	Pegmatite	Amphibolite
Silicates	Volcanic Glass	Hornfels
Feldspars	Obsidian	Gneiss
Ferromagnesian	Pumice	Serpentinite
Hornblende	Tuff	
Augite	Scoria	
Clay	Perlite	
Illites	Pitchstone	
Kaolins	Felsite	
Chlorites	Basalt	
Montmorillonites		
Mica		
Zeolite	**Sedimentary Rocks**	
Carbonate	Conglomerate	
Calcite	Sandstone	
Dolomite	Quartzite	
Sulfate	Graywacke	
Gypsum	Subgraywacke	
Anhydrite	Arkose	
Iron Sulfide	Claystone, Siltstone,	
Pyrite	Argillite and Shale	
Marcasite	Carbonates	
Pyrrhotite	Limestone	
Iron Oxide	Dolomite	
Magnetite	Marl	
Hematite	Chalk	
Goethite	Chert	
Ilmenite		
Limonite		

For brief descriptions, see "Standard Descriptive Nomenclature of Constituents of Natural Mineral Aggregates" (ASTM C294). Courtesy Portland Cement Association

Table 7.7 Characteristics and Tests of Aggregates

Characteristic	Significance	Test Designation	Requirement or Item Reported
Resistance to abrasion	Index of aggregate quality; wear resistance of floors, pavements	ASTM C131 ASTM C295 ASTM C535	Maximum percentage of weight loss
Resistance to freezing and thawing	Surface scaling, roughness, loss of section, and unsightliness	ASTM C295 ASTM C666 ASTM C682	Maximum number of cycles or period of frost immunity; durability factor
Resistance to disintegration by sulfates	Soundness against weathering action	ASTM C88	Weight loss, particles exhibiting distress
Particle shape and surface texture	Workability of fresh concrete	ASTM C295 ASTM D3398	Maximum percentage of flat and elongated pieces
Grading	Workability of fresh concrete; economy	ASTM C117 ASTM C136	Minimum and maximum percentage passing standard sieves
Bulk unit weight or density	Mix design calculations; classification	ASTM C29	Compact weight and loose weight
Specific gravity	Mix design calculations	ASTM C127, fine aggregate ASTM C128, coarse aggregate ASTM C29, slag	—
Absorption and surface moisture	Control of concrete quality	ASTM C70 ASTM C127 ASTM C128 ASTM C566	—
Compressive and flexural strength	Acceptability of fine aggregate failing other tests	ASTM C39 ASTM C78	Strength to exceed 95% of strength achieved with purified sand
Definitions of constituents	Clear understanding and communication	ASTM C125 ASTM C294	—

Courtesy Portland Cement Association

considered when flexural strength or high compressive strength is specified.

Specific Gravity

Specific gravity is a measure of the relative density of an aggregate. It is a ratio of its weight to the weight of an equal volume of water. It is used to determine the absolute volume occupied by the aggregate. Most natural aggregates have specific gravities between 2.4 and 2.9.

Absorption and Surface Moisture

The absorption and surface moisture of an aggregate are determined so the net water content of the concrete can be controlled and the correct **batch** weights determined.

Moisture conditions are designated in four categories:

1. Oven dry. Completely dry and fully absorbent.
2. Air dry. Dry on the surface but with some interior moisture.
3. Saturated surface dry. Neither absorbing water nor contributing water to the mix. The surface is dry but the voids and interior of the aggregate are fully saturated.
4. Damp. Contains an excess of surface moisture.

Surface moisture can increase the *bulk* (volume) of sand and fine sands bulk more than coarse sands. Since moisture in sands delivered on a job can vary, it is necessary to weigh the aggregate rather than using volume measures when portioning concrete ingredients.

Test for Moisture The test to determine the amount of moisture on fine aggregates is made according to ASTM C70, Surface Moisture in Fine Aggregate, which depends on the displacement of water by a known weight of moist aggregate. The density of the aggregate must be known to use this method.

Another method, ASTM C566, Total Moisture Content of Aggregate by Drying, involves weighing a sample, heating it until it is dry, and reweighing it. The difference in weights is used to calculate the moisture content.

Chemical Stability

Chemically stable aggregates do not react chemically with cement, which could cause harmful reactions. Some aggregates contain minerals that do react with alkalies in cement, causing abnormal expansion and cracking in the concrete. Field records of aggregates provide evidence of their chemical stability. If this is unknown, a laboratory test based on ASTM specifications should be made.

Harmful Materials

Included with aggregates may be a number of harmful materials, such as silt, organic materials, coal, and soft rock particles. Aggregate specifications limit the amount of these materials that may be present. A series of ASTM tests is used to identify these harmful materials in aggregate samples.

Grading Aggregates

Aggregates are graded into standardized sizes by passing them through a sieve as specified by ASTM C136 and CSA A23.2.2. The sieves used for fine aggregates are in seven sizes and there are ten sizes for coarse aggregates (Table 7.8).

The limits used are specified for the percentage of material passing each sieve. Figure 7.4 shows the grading limits for fine aggregates. Notice that almost all of the particles of fine sand pass through the No. 4 sieve, but only 10 to 20 percent pass through the No. 100 sieve. The coarse sand graph shows that about 98 percent pass through the No. 4 sieve and only about 3 percent pass through the No. 100 sieve. Each product is therefore an accumulation of the particles of various percentages passing through the various sieves.

When concrete mixes are designed, limits are specified for the percentage of material passing each sieve. This, plus specifying the maximum aggregate size, af-

Table 7.8 Fine and Coarse Sieve Sizes

Fine Sieve Sizes
³⁄₈ in. (9.5 mm)
No. 4 (4.75 mm)
No. 8 (2.36 mm)
No. 16 (1.18 mm)
No. 30 (600 μm)
No. 50 (300 μm)
No. 100 (150 μm)

Coarse Sieve Sizes	
Size Number	Nominal Size (sieves with square openings)
1	3½ to 1½ in. (90 to 37.5 mm)
2	2½ to 1½ in. (63 to 37.5 mm)
357	2 in. to No. 4 (50 to 4.75 mm)
467	1½ in. to No. 4 (37.5 to 4.75 mm)
57	1 in. to No. 4 (25.0 to 4.75 mm)
67	¾ in. to No. 4 (19.0 to 4.75 mm)
7	½ in. to No. 4 (12.5 to 4.75 mm)
8	³⁄₈ in. to No. 8 (9.5 to 2.36 mm)
3	2 to 1 in. (50.0 to 25.0 mm)
4	1½ to ¾ in. (37.5 to 19.0 mm)

Courtesy Portland Cement Association

fects the relative proportions of aggregate, as well as the cement and water requirements, workability, economy, porosity, shrinkage, and durability of concrete. The best mixes are those that do not have a large deficiency or excess of any size. Very fine sands are uneconomical. Very coarse sands produce a harsh, unworkable mix. A wide range of particle sizes is necessary to fill the voids

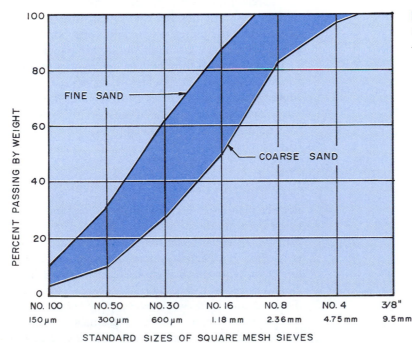

Figure 7.4 Grading limits for fine aggregates. *(Courtesy Portland Cement Association)*

between aggregates (Fig. 7.5). As the range of sizes fills the voids, there is less volume of voids, which decreases the cement needed.

Figure 7.6 shows a set of sieves and a shaker used to grade sands. Each layer has a sieve with a fine mesh. The sand remaining on each sieve can be weighed to get the percent of the sample for each sieve. Following is an example of a typical test using 1000 g of sand.

Sieve	Weight of Sand Retained in Grams	Percent Retained	Percent Cumulative
4 (5 mm)	4	4	5.0
8 (2.5 mm)	115	11.5	15.5
16 (1.25 mm)	210	21.0	36.5
30 (630 μm)*	262	26.2	62.7
50 (315 μm)	201	20.1	82.8
100 (160 μm)	172	17.2	100.0
		Total	302.5

* micromillimeter

The *fineness modulus* of the sand sample can be found by adding the cumulative percent retained figures and dividing by 100. In the previous illustration, this total was 302.5, which gives a fineness modulus of 3.02. A range of 2.30 to 2.60 are fine sands, 2.61 to 2.90 a medium sand, and 2.91 to 3.10 a coarse sand. The sand in the example was coarse.

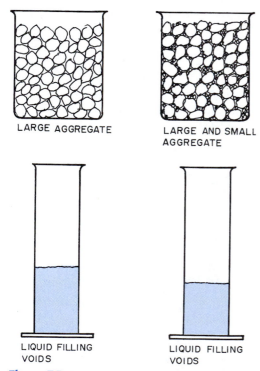

Figure 7.5 The voids between the large sizes of aggregates are filled with finer ranges of aggregates, reducing the amount of portland cement needed.

Figure 7.6 A set of sieves and a shaker used to grade sands. (*Courtesy ELE International, Inc.*)

Structural Lightweight Aggregates

Lightweight aggregates are used to produce lightweight structural concrete. Lightweight concrete has a density ranging from 90 to 115 lb./ft.^{3} (1440 to 1850 kg/m^3), depending on the aggregate. Shale, slate, clay, and slag are generally used. Pelletized clay or crushed shale or slate is placed in a rotary kiln and heated to about 1800°F (1000°C). Gases in the material cause it to expand, forming small air cells in the pieces. Another process, *sintering*, involves crushing the aggregate, mixing it with a small amount of finely ground coal or coke, and spreading it over a grate, at which time the coke is ignited. Blowers increase the temperature of the burning, causing the material to expand and bond into large pieces. These are crushed forming the smaller aggregate.

Expanded blast furnace slag aggregate is produced by treating the slag with steam or water while it is in a molten state, which causes it to expand and bond into large particles. These are then crushed to the desired size.

Sintered fly ash is produced by pelletizing fly ash and then sintering it.

Insulating Lightweight Aggregates

Perlite, diatomite, vermiculite, pumice, and scoria are commonly used as insulating lightweight aggregate. *Perlite* is siliceous volcanic rock that contains moisture. It is crushed and heated, causing it to expand and form a honeycomb structure. It is used as loose fill insulation and plaster and concrete aggregate. *Diatomite* is a diatomaceous earth, which is composed of silicified skeletons of microscopic one-celled animals. *Vermiculite* is a hydrated magnesium-aluminum-iron silicate that occurs in thin layers having moisture between them. When vermiculite is crushed and heated the layers expand forming dead air cells. *Pumice* is a type of volcanic rock that is porous and light in weight. *Scoria* is a type of volcanic slag; scoria also is used to describe a form of blast-furnace slag. It is a cellular material and is crushed to form aggregate of various sizes.

Heavyweight Aggregates

Heavyweight aggregates are used to produce heavyweight concretes that have densities up to 400 lb./ft.^{3} (6400 kg/m^3). These are used for radiation shielding and other applications in which great weight is desired.

Frequently used heavyweight aggregates include ferrophosphorus, barite, goethite, hematite, ilmenite, limonite, magnetite, and steel shot and punchings.

Testing Aggregates for Impurities

Aggregates must be free of clay, silt, and organic impurities. If present, these could influence the composition of the cement paste and, therefore, the strength and setting of the concrete.

Organic impurities in fine aggregate are determined by ASTM C40, Organic Impurities in Sands for Concrete. This test involves putting 4½ oz. (130 g) of fine aggregate into a 12 oz. (340 g) bottle. Next, 2½ oz. (70 g) of a 3 percent solution of sodium hydroxide are added, and the bottle is shaken. After the bottle is allowed to stand for twenty-four hours, the color of the solution is examined. If it is a very light tan the amount of organic material is not serious. If it is darker than this it should not be used without washing and retesting (Fig. 7.7). If there is a doubt, the aggregate should be used to make mortar samples that are tested for strength using ASTM C87.

Excess clay and silt in an aggregate can cause an increase in shrinkage and affect durability, and they tend to separate from the other aggregate particles. Aggregate specifications usually limit clay and silt passing through a No. 200 (75 μm) sieve to 2 or 3 percent for sand and less than 1 percent in coarse aggregate. Tests are made based on ASTM C117, Sieve in Mineral Aggregates by Washing. Testing for clay lumps is made following ASTM C142, Clay Lumps and Friable Particles in Aggregates.

Handling and Storing Aggregates

Aggregates should be stored in layers of uniform thickness. In stockpiles, each truck load should be discharged tight against the previous load. Aggregate is removed from stockpiles with a front-end loader. If removing it from a tall, conical pile, the loader should move back and forth across the face of the pile so as to reblend sizes. Washed aggregates must have sufficient

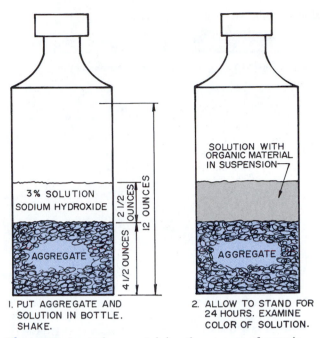

Figure 7.7 A test for ascertaining the amount of organic impurities in fine aggregate.

time in storage to drain to a uniform moisture content. Fine aggregates should not be dropped from a bucket or conveyor because the wind tends to blow away the very fine particles. Grades should be kept from intermingling by building dividers or storing in completely separate locations. Above all, aggregates must be stored and handled so they are not contaminated by unwanted substances, such as earth and leaves.

ADMIXTURES

Admixtures are ingredients added to concrete, other than portland cement, aggregates, and water. They may be added before or during mixing. Admixtures change the properties of concrete, so they should be used sparingly and only on the advice of a concrete specialist.

Concrete should be workable, finishable, strong, durable, watertight, and wear resistant. Whenever possible, these properties should be obtained by careful selection of suitable types of aggregate, portland cement, and the water-cement ratio. If this is not possible, or special circumstances such as freezing weather exist, admixtures will be of benefit. A summary of admixtures and their uses is in Table 7.9.

Admixtures are of the following types:

Air-entraining agents
Retarders
Water reducers

Table 7.9 Admixtures by Classification

Desired Effect	Type of Admixture	Material
Improve durability	Air entraining (ASTM C260)	Salts of wood resins Some synthetic detergents Salts of sulfonated lignin Salts of petroleum acids Salts of proteinaceous material Fatty and resinous acids and their salts Alkylbenzene sulfonates
Reduce water required for given consistency	Water reducer (ASTM C494, Type A)	Lignosulfonates Hydroxylated carboxylic acids (Also tend to retard set so accelerator is added)
Retard setting time	Retarder (ASTM C494, Type B)	Lignin Borax Sugars Tartaric acid and salts
Accelerate setting and early strength development	Accelerator (ASTM C494, Type C)	Calcium chloride (ASTM D98) Triethanolamine
Reduce water and retard set	Water reducer and retarder (ASTM C494, Type D)	(See water reducer, Type A, above)
Reduce water and accelerate set	Water reducer and accelerator (ASTM C494, Type E)	(See water reducer, Type A, above. More accelerator is added)
Improve workability and plasticity	Pozzolan (ASTM C618)	Natural pozzolans (Class N) Fly ash (Class F and G) Other materials (Class S)
Cause expansion on setting	Gas former	Aluminum powder Resin soap and vegetable or animal glue Saponin Hydrolyzed protein
Decrease permeability	Dampproofing and waterproofing agents	Stearate of calcium, aluminum, ammonium, or butyl Petroleum greases or oils Soluble chlorides
Improve pumpability	Pumping aids	Pozzolans Organic polymers
Decrease air content	Air detrainer	Tributyl phosphate
High flow	Superplasticizers	Sulfonated melamine formaldehyde condensates Sulfonated naphthalene formaldehyde condensates

Courtesy Portland Cement Association

Accelerators
Pozzolons
Workability agents
Superplasticizers
Dampproofing and permeability-reducing agents
Bonding agents
Concrete coloring agents
Hardeners
Grouting agents
Gas-forming agents

Air-Entraining Agents

Air-entraining admixtures are used to entrain microscopic air bubbles in concrete. The entrained air bubbles are distributed uniformly throughout the cement paste. Entrainment can be produced by using air-entrained portland cement or by adding an air-entraining admixture to the concrete as it is being mixed. The entrained air constitutes 2 to 8 percent of the volume of the concrete. Air-entrained portland cement has the air-entraining material ground into it as it is being manufactured. If it is added to the concrete mix, it can be added before or during the mixing process. Numerous commercial air-entraining admixtures are manufactured from a variety of materials. Some ingredients used in air-entraining admixtures include polyethylene oxide polymers, some fats and oils, sulfonated compounds, and detergents. Air-entraining admixtures are specified by ASTM C226.

Entrained air bubbles improve the *durability* of the concrete, which increases the resistance to damage due to the freeze-thaw cycle and to damage from de-icers, which cause *scaling*. Air-entraining gives improved *workability* during placement and superior *watertightness*. It also improves resistance to *sulfate attack* from soil water and seawater. An important feature is that properly proportioned air-entrained concrete requires less water per cubic yard than non-air-entrained concrete of the same slump. This improves the **water-cement ratio** and is responsible for some of the advantages mentioned earlier.

Air-entrained concrete is used in cold climates where concrete, such as paving and architectural concrete, is exposed to the freeze-thaw cycle. It is also effective for concrete exposed to the soil and water, where sulfate attack is possible.

Retarders

Retarding admixtures are used to retard the setting time of the cement paste in the concrete. They often are used to retard set time in hot weather, because hydration is accelerated by the heat. This uses up some of the water needed to give the plasticity required for placement and finishing. Without a retarder in hot weather, more water is required to get the desired slump, which produces lower-strength concrete. Retarding admixtures tend to reduce the water required, which produces a better water-cement ratio and increases the ultimate strength.

Retarders also help when it is necessary to pour large amounts of concrete or where placement is difficult. For example, if concrete must be pumped a considerable distance, retarders will enable it to be moved, placed, and finished before too much set occurs. They are sometimes used in concrete mix trucks that have to travel an unusually long distance to a job. They also reduce the temperature rise caused by the heat of hydration in large concrete masses.

A variety of chemicals are used as retarders. There is a reduction in early strength during the first one to three days. Retarders can also cause shrinkage, which may cause cracking. Before retarders are used, tests should be made with job materials and conditions.

Often set time can be slowed by means other than the use of retarders. Higher air temperatures during hydration often are the cause of an increased rate of hardening. Often all that is needed to retard set time in hot weather is to cool the mixing water or aggregates or both. If ice is placed in the mixing water it should be completely melted before the water is used. Set time can also be reduced by shading the concrete so it is not directly exposed to the rays of the sun.

Water Reducers

Water-reducing admixtures reduce the amount of water needed to produce a concrete of a given consistency. They can also be used to increase the amount of slump without requiring additional water. This means a lower water-cement ratio can be used. The lower the water-cement ratio the greater the strength of the concrete.

Lignin solfonic acids and metallic salts are commonly used. Some water-reducing admixtures reduce set time and sometimes cause increased drying shrinkage. Some introduce entrained air.

Accelerators

An **accelerator admixture** speeds up the strength development of concrete. Strength development can also be accelerated by using Type III high-early-strength portland cement, by increasing the amount of cement to lower the water-cement ratio, or by curing at higher temperatures. Accelerators are used in cold weather to develop strength faster in order to offset freeze damage.

Calcium chloride is a frequently used accelerator. It should be added to the concrete mix in solution form and not added in dry form. The amount added varies depending on conditions and the desired set time, but it never exceeds 2 percent by weight of the cement. Commercially available calcium chloride ($CaCl_2$) is available in regular flake and concentrated flake types. Regular flake contains at least 77 percent $CaCl_2$ and concentrated flake has 94 percent $CaCl_2$. This must be known so the correct amount of weight can be calculated.

Experience has shown that using 2 percent or less of calcium chloride has no significant corrosive effect on steel reinforcing materials if the concrete is of high quality. Calcium chloride is not recommended for use in (1) prestressed concrete, because of the possibility of corrosion; (2) in concrete containing embedded aluminum, such as electrical conduit; (3) in concrete subject to alkali-aggregate reaction or soils or water containing sulfates; (4) in nuclear shielding concrete; (5) in floor slabs to receive dry-shake metallic finishes; and (6) in hot weather.

Some commercial accelerators are made from chemicals other than calcium chloride. Some are in a nonhygroscopic powder form and others are concentrated liquids. Since they are free of chlorides, they are useful in concrete in which steel is embedded. They produce a high early set, reducing normal set time from three hours to one hour, and the concrete develops higher-than-normal strength during the initial three-day curing period.

Pozzolans

A pozzolan is a siliceous or siliceous and aluminous material that, when finely ground and with the presence of moisture, chemically reacts with calcium hydroxide at ordinary temperatures to form compounds possessing cementitious properties. (See the earlier discussion under hydraulic cements for more information.)

Pozzolan materials are sometimes added to concrete to help reduce internal temperatures. This is especially helpful when pouring large masses of concrete. Some pozzolans are used to reduce or eliminate potential expansion of concrete from alkali-reactive aggregates. Some pozzolans improve resistance to sulfate attack.

Pozzolans can replace 10 to 35 percent of the cement. They substantially reduce the twenty-eight-day strength of concrete and require that continuous wet curing and favorable temperatures be continued for a longer period than for normal concrete.

Pozzolans affect concrete in many different ways. The concrete should be tested using the specific cement and aggregates to see if it is suitable.

Workability Agents

If fresh concrete is harsh because of improper aggregate grading or incorrect mix proportions, workability agents can be added to improve workability. **Workability** is a term used to describe the ease with which concrete can be placed and consolidated. If concrete is to have a troweled finish, workability is important. Improved workability may be needed if the concrete has to be placed in forms containing considerable reinforcing or if it must be pumped.

Workability can be improved by increasing the cement content. The addition of entrained air is the best workability agent. Some organic materials, such as alginates and cellulose derivatives, will increase slump. Finely divided materials can be used as admixtures to improve the workability of mixes deficient in aggregates passing through the No. 50 and No. 100 sieves. When added to mixes not deficient in fines, additional water is usually required. This may reduce strength, increase shrinkage, and adversely affect other properties of concrete. Some finely divided materials used are pozzolan and others that are inert chemically. Fly ash and pozzolans used as workability agents must meet ASTM specifications. These also tend to reduce early strength and entrained air, so additional air-entraining admixture must be used. In any case, as with any admixture, the effects of workability agents on the concrete must be considered before deciding to use them.

Superplasticizers

When cement and water mix, the wet cement particles form into small clumps that inhibit the proper mixing of the cement and water. This reduces workability and inhibits hydration. *Superplasticizers* are admixtures that coat the cement particles, causing them to break away from the lumps and disperse in the water. Superplasticizers give each cement particle a negative charge, which causes them to repel each other, thus providing a more thorough dispersement.

Superplasticizers can be used to

▲ Reduce the water and cement used at a constant water-cement ratio, giving a concrete the same strength as a normal mix but reducing the amount of cement used.

▲ Produce normal concrete at normal water-cement ratios that is so workable it can be placed with little or

no vibration or compaction and not have excessive bleed or segregation.

▲ Produce a concrete of higher strength by reducing the water content required while maintaining the normal cement content.

Some superplasticizers are formulated to give slightly accelerated early strength and high early strength. Others are used with water-reducing admixtures. These tend to retard and lower early strength but increase ultimate strength. It is also recommended that superplasticizers be added directly to the mix in the concrete mixer truck to enable the slump to be maintained during long hauls or delivery delays.

Commonly used superplasticizers include a higher molecular weight condensed sulfonate naphthalene formaldehyde, sulfonated melamine formaldehyde, and modified ligninsulfonates, all of which are in liquid form.

Superplasticizers have a short duration—thirty to sixty minutes—before the concrete has a rapid loss in workability. Specific information on the various products should be obtained from the manufacturer.

Permeability-Reducing and Dampproofing Agents

Sound, dense concrete that has a water-cement ratio of 0.50 by weight and that is properly placed and cured will be watertight. Concretes that have low cement contents, a deficiency in fines, or a high water-cement ratio can have the permeability reduced by using *permeability-reducing agents* such as certain soaps, stearates, and petroleum products.

Permeability is a measure of the amount of water that passes through channels running between the outer faces of the concrete. The permeability-reducing agent reduces the flow of water through these channels. They should generally not be used in well-proportioned mixes because they will increase the amount of mixing water required and increase rather than decrease permeability.

It should be noted that other things can be done to reduce permeability that may reduce the need for a permeability-reducing agent. Concrete that is properly placed, compacted, and cured will have reduced permeability. Air-entraining admixtures also reduce permeability by increasing the plasticity and by reducing the amount of water needed and bleeding. In effect, air-entraining is an excellent permeability-reducing admixture. Superplasticizers disperse cement particles throughout the mix, which helps reduce permeability.

Dampproofing admixtures are used to reduce moisture that is transferred by capillary action. This occurs when one side of the concrete is exposed to moisture and the other to the air, as in a slab on the ground. The surface exposed to the air tends to dry. Capillary action occurs as moisture flows to a dry surface. The effectiveness of dampproofing admixtures varies, and manufacturers' test data need to be studied.

There are other ways to dampproof concrete. Various types of film can be applied to the damp side. A typical example is putting a plastic membrane on the ground below a concrete slab floor. Various coatings can be applied, such as asphalt, sodium silicate, and metallic-aggregate mixed with portland cement. Above-grade concrete and masonry can be dampproofed by applying a coat of silicone sealer. This also protects against freeze-thaw cycles, efflorescence, weathering, and staining.

Bonding Agents

Often fresh concrete must be placed over a concrete surface that has already set. It is important to get a good bond between them. When fresh concrete is poured over hardened concrete, the fresh concrete shrinks, breaking its bond to the hardened concrete. The hardened surface must be prepared so the fresh concrete will firmly bond to the aggregate in the hardened surface. The condition of the surface is of great importance. It must be dry, clean (free of dirt, dust, grease, paint, etc.), and at the proper temperature.

Bonding admixtures can be added to portland cement mixtures or applied to the surface of old concrete to increase the bond strength. These admixtures are usually water emulsions of certain organic materials, such as a liquid acrylic polymer, which may be added to the portland cement with or without mixing water. The pores of the concrete absorb the water and the resin unites into a mass and bonds the two layers of concrete.

Bonding is also accomplished by exposing the surface of the aggregate in the hardened concrete, applying a cement paste slurry to the hardened surface, and immediately pouring the new layer of concrete over it. Another type of bonding agent that is not an admixture is a two-compound moisture-insensitive epoxy adhesive applied to the hardened concrete and immediately covered by the fresh concrete. It also forms a waterproof membrane.

Concrete Coloring Agents

Concrete can be colored by mixing pure, finely ground *mineral oxides* with the dry portland cement. Thorough mixing is necessary to produce a uniform color. Mixing time may sometimes be longer than normal. Oxides added to normal portland cement are usually limited to earthy colors and pastels because of the cost and graying effect of the cement. White portland cement produces clearer, brighter colors and is preferred. Some color agents are compounded to give water-reducing and set-controlling properties.

Concrete can also be colored by exposing the aggregate. Colored aggregates are spread on the surface of the freshly cast concrete and floated in place. At the proper degree of set, the unhydrated paste is washed away.

Color can be placed in the surface of concrete before it sets. One way is to use natural mineral oxides of cobalt, chromium, or iron or ochres and umbers ground to fine powder. These are mixed with portland cement into a topping mix that is troweled over the surface of the concrete.

Another method uses synthetic oxides mixed with a fine silica sand. This is spread over the unset but floated concrete surface. This mixture is then floated and troweled into the uncured surface and cured in the normal way. Another method is to use the colors mixed with a metallic aggregate and dry portland cement. This mixture is applied as a dry shake to the surface, floated, troweled, and cured. Some dry shakes also give additional hardness to the surface.

Another method is to simply paint the surface after it has been completely cured and neutralized. Neutralization can occur naturally by aging and weathering or can be accomplished with a neutralizing agent, such as zinc sulfate.

Hardeners

When a concrete surface is to be subjected to heavy wear, such as a factory or warehouse floor, its life can be extended by using a liquid chemical or dry powder hardener. One form of *chemical hardener* is a colorless, nontoxic, nonflammable liquid containing magnesium and zinc fluosilicates and a wetting agent. The wetting agent reduces the surface tension of the liquid hardener, which makes it easier for it to enter the pores of the concrete. The hardening agent produces a chemical reaction with the free lime and calcium carbonates in the portland cement. This reaction densifies the surface, making it more wear resistant and impenetrable to many liquids and chemicals. This liquid hardener is applied to the surface of the cured concrete. Several coats are recommended. Similar hardeners are made using baume sodium silicate that reacts with the lime to form an insoluble crystal within the pores of the concrete. Other similar products are available from a variety of suppliers.

Concrete surfaces are also hardened using *dry powder hardeners*. One product uses quartz silica aggregates and alkali-fast inorganic oxides, which color the concrete. These two are mixed with portland cement and plasticizing agents, giving a dry shake that is applied to the freshly poured concrete. This provides a high-strength surface with color provided by the oxides. Another type uses a finely ground iron aggregate and, if desired, inorganic oxides for color. These are mixed with portland cement and plasticizing agents and applied as a dry shake on the surface of the freshly laid concrete.

Grouting Agents

Portland cement grouts are widely used for many purposes, such as stabilizing foundations, filling cracks in concrete walls, filling joints, grouting tenons and anchor bolts, and other such applications. Grout properties can be altered by using the various admixtures discussed in this chapter.

Gas-Forming Agents

Gas-forming agents are added to concrete or grout to cause a slight expansion in it before it hardens. Aluminum powder is one of several gas-forming agents in use. It reacts with the hydroxides in hydrating cement and produces small hydrogen gas bubbles. This also helps eliminate voids caused by settlement of the concrete or grout.

A typical application is to expand the grout under a machine base or column where it is difficult to reach to be certain the area is fully grouted. After concrete or grout has hardened, gas-forming agents will not overcome shrinkage. When they are used in large amounts gas-forming agents produce lightweight cellular concrete.

BASICS OF CONCRETE

Concrete is made up of two parts, aggregates and a paste. The paste is a portland cement and water mixture with some entrapped air. The paste coats the particles of aggregate and hardens and bonds them together. The paste hardens due to a chemical reaction between the portland cement and water. This process is called *hydration*. The aggregates used range from fine to coarse, with various degrees of size within each.

Figure 7.8 shows the range in proportions of materials used in typical designs for rich and lean mixes. Notice that the volume of cement varies from 7 percent to 15 percent and water from 14 percent to 21 percent. A small percentage is air, and the rest, the major volume, is aggregate. This means that the selection of aggregates is vital for the production of quality concrete. It is essential that a continuous graduation of particle size be maintained.

Figure 7.9 shows a cross section cut through a sample of hardened concrete. The paste must completely fill the spaces between the particles and coat each particle.

Water-Cement Ratio

If all conditions are held constant the quality of the hardened concrete is determined by the water-cement ratio. The water-cement ratio is the ratio by weight between the water and cement used to make the paste. For example a mix that uses 0.62 lb. of water per one pound of cement has a water cement ratio of $\frac{0.62}{1.00}$ or 0.62. In metric measure this would be 620 g of water per kilogram of cement, for a metric water cement ratio of 620.

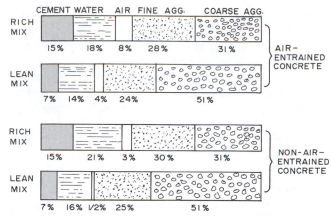

Figure 7.8 The range in proportions of materials used in concrete by volume. (*Courtesy Portland Cement Association*)

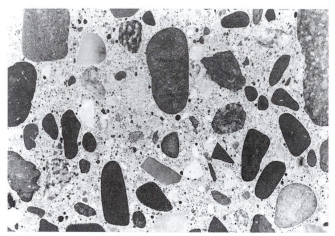

Figure 7.9 A cross section of hardened concrete showing how the cement-and-water paste coats each particle of aggregate and fills the voids between the particles. (*Courtesy Portland Cement Association*)

Stated another way, it is 62 percent of the weight of the cement.

The water-cement ratio should be the lowest value required to meet the design considerations. If too much water is used, giving a high water-cement ratio, the paste is thin and will be porous and weak when hardened. The effect of the water-cement ratio on the strength of concrete is illustrated in Fig. 7.10. Only a very small amount of water is needed for hydration to occur. A water-cement ratio of 0.31 will produce hydration, but usually more water is added so the concrete is workable and can be properly placed. The lower the water-cement ratio the stronger the concrete (Table 7.10).

Concrete exposed to the elements must have the following for durability:

▲ Air entrainment
▲ Low water-cement ratio
▲ Quality cement and aggregate
▲ Proper curing
▲ Proper construction practices

The required water-cement ratios for various conditions are shown in Table 7.11. These data do not give consideration to relative strength. When durability is not the major consideration, the water-cement ratio is selected based on compressive strength. This often requires that tests be made using the actual job materials. If flexural strength is the basis for concrete design, tests are made to find the relationship between the water-cement ratio and flexural strength.

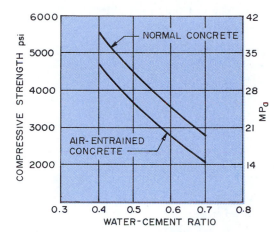

.Figure 7.10 This shows the effect of the water-cement ratio on the strength of concrete. *(Courtesy Portland Cement Association)*

Table 7.10 Maximum Permissible Water-Cement Ratios for Concrete When Strength Data from Trial Batches or Field Experience Are Not Available

Specified Compressive Strength F'_c, psi[a]	Maximum Absolute Permissible Water-Cement Ratio, by Weight	
	Non-air-entrained concrete	Air-entrained concrete
2500	0.67	0.54
3000	0.58	0.46
3500	0.51	0.40
4000	0.44	0.35
4500	0.38	[b]
5000	[b]	[b]

[a]28-day strength. With most materials, the water-cement ratios shown will provide average strengths greater than required.
[b]For strengths above 4500 psi (non-air-entrained concrete) and 4000 psi (air-entrained concrete) proportions *should be established* by the trial batch method.
1000 psi ≈ 7 MPa.
Courtesy Portland Cement Association

Table 7.11 Maximum Water-Cement Ratios for Various Exposure Conditions

Exposure Condition	Normal-Weight Concrete, Absolute Water-Cement Ratio by Weight
Concrete protected from exposure to freezing and thawing or application of deicer chemicals	Select water-cement ratio on basis of strength, work ability, and finishing needs
Watertight concrete	
In fresh water	0.50
In seawater	0.45
Frost-resistant concrete	
Thin sections; any section with less than 2-in. cover over reinforcement and any concrete exposed to deicing salts	0.50
All other structures	0.45
Exposure to sulfates	
Moderate	0.50
Severe	0.45
Placing concrete under water	Not less than 650 lb. of cement per cubic yard (386 kg/m^3)
Floors on grade	Select water-cement ratio for strength, plus minimum cement requirements

Courtesy Portland Cement Association

Minimum Cement Content

In addition to the specification of the water-cement ratio, the minimum cement content is also given. This ensures that the concrete will have good finishability, good wear-resistance, and good appearance (Table 7.12).

Aggregates

Aggregates must be properly graded and of proper quality. *Grading* pertains to the particle size and the distribution of particles. Properly graded aggregate produces the most economical concrete because it allows the use of the maximum size coarse aggregate. This reduces the amount of water and cement required. The reduction in cement required reduces the cost.

Table 7.12 Minimum Cement Requirements for Concrete Used in Flatwork

Maximum Size of Aggregate, In.	Cement, lb. per Cubic Yard
1½	470
1	520
¾	540
½	590
⅜	610

1 in. ≈ 25 mm
100 lb./yd.^{3} ≈ 60 kg/m^3
Courtesy Portland Cement Association

ITEM OF INTEREST

CONCRETE ON THE MOON

A portland cement–based concrete experiment was performed during a flight of the space shuttle. It was designed by researchers at Master Builders, Inc., Cleveland, Ohio, and students at the University of Alabama, Huntsville. It involved the mixing and curing of concrete in microgravity. The findings will be useful when future bases will be established on the moon. The concrete was cured for seven days in microgravity, and the samples were tested and compared with similar earth cast and cured concrete. This revealed the effect of gravitational differences in the cured concrete.

The mixer developed for this experiment contained a 12 in. long, 4 in. diameter Plexiglas cylinder that held the dry components and served as the mixing chamber. Another device was developed to hold the required water and a wa-ter-reducing admixture. It was programmed to inject the water into the chamber, after which a blade on top of the assembly stirred the mixture. The cylinder was removed and the cured sample recovered after the shuttle landed.

They found that water and cement particles would combine in microgravity, but it produced a concrete that had no air voids, so it was stronger than a similar sample mixed on earth. The space concrete did not have the capillary channels created when water rises to the top (bleed) as it does on concrete poured on the earth.

It appears that concrete could possibly be made on the moon and that materials mined there could be mixed with portland cement. Concrete might be the most important material for building lunar structures.

The concrete mixing apparatus sealed and ready for installation in the space shuttle. *(Courtesy Master Builders, Inc.)*

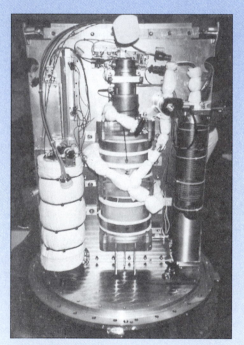

This view of the cargo bay of the Space Shuttle Endeavor 115 nautical miles above Earth shows a number of the experiments in place, including the making of concrete in space experiment. *(Courtesy National Aeronautics and Space Administration)*

The concrete mixing chamber and related equipment used on the flight of the space shuttle to produce and cure concrete samples in space for testing upon return to Earth. *(Courtesy Master Builders, Inc.)*

The maximum size of aggregate used depends on the size and shape of the members being formed and the amount of reinforcing steel. The maximum size aggregate acceptable can be no more than *one-fifth* the narrowest dimension between the sides of the forms or *three-fourths* the clear space between steel, such as reinforcing bars, ducts, conduit, or bundles of bars. Aggregate in unreinforced slabs on the ground should not exceed *one-third* the slab thickness. For leaner concrete mixes a finer grade of sand is used to improve workability. Richer mixes use a coarse grade of sand for greater economy.

Entrained Air

Entrained air should be used in all concrete exposed to freezing and thawing cycles. It should be used in all concrete paving regardless of the temperatures. The required percentage of entrained air is shown in Table 7.13. The amount of entrained air required decreases as the maximum size of the aggregate increases. Entrained air reduces the amount of water required, producing a lower water-cement ratio.

Slump

Slump is a measure of the consistency of concrete. *Consistency* is the ability of fresh concrete to flow. This is measured by the slump test explained later in this chapter. **Slump** is the decrease in height of a molded mass of fresh concrete that occurs immediately after it is removed from a standard metal slump cone. The higher the slump (the more the sample lowers), the wetter the concrete mixture. A measure of slump can only be used to compare mixes of identical design. Slump is usually specified in the concrete specifications. Recommended slumps are shown in Table 7.14. These slumps are for concrete **consolidated** by mechanical vibration.

CONCRETE DESIGN

Normal-Weight Concrete

The design of normal-weight concrete first depends on the required strength and durability. Consideration of workability, plasticity, and cost are factored. The goal is to produce a concrete that meets these requirements.

As discussed earlier, the *water-cement ratio* is a basic premise used when designing normal-weight concrete. Based on research and experience, water-cement ratios for various applications can be recommended as shown in Table 7.11.

Another way mix design is accomplished is by using the *absolute volume* of material amounts. The absolute volume of a loose material, such as aggregate, is the total volume, including the particles and air spaces between them. This includes the absolute volume of the cement, aggregate, water, and trapped air.

$$\text{Absolute volume} = \frac{\text{weight of dry material}}{\text{specific gravity} \times \text{unit weight of water}}$$

For example, the absolute volume of 100 pounds of aggregate having a specific gravity of 2.5 would be

$$\text{Absolute volume} = \frac{100 \text{ lb.}}{2.5 \times 62.5 \text{ lb. (one cubic foot of water)}}$$
$$= 0.64 \text{ ft.}^3$$

Publications of the Portland Cement Association give detailed instructions on concrete design by water-cement ratio and absolute volume.

Lightweight Insulating Concrete

The design of lightweight insulating concrete depends on the aggregate used and the desired compressive strength. The amount of water required varies greatly,

Table 7.13 Recommended Average Total Air-Content Percentage for Level of Exposure

| | Nominal Maximum Sizes of Aggregates | | | | | |
Exposure	⅜ in. (10 mm)	½ in. (13 mm)	¾ in. (19 mm)	1 in. (25 mm)	1½ in. (40 mm)	2 in. (50 mm)
Mild	4.5	4.0	3.5	3.0	2.5	2.0
Moderate	6.0	5.5	5.0	4.5	4.5	4.0
Extreme	7.5	6.0	6.0	6.0	5.5	5.0

1 in. ≈ 25 mm
Courtesy Portland Cement Association

Table 7.14 Recommended Slumps for Various Types of Construction

Concrete Construction	Slump, In. Maximum[a]	Minimum
Reinforced foundation walls and footings	3	1
Plain footings, caissons, and substructure walls	3	1
Beams and reinforced walls	4	1
Building columns	4	1
Pavements and slabs	3	1
Mass concrete	2	1

[a]Slumps shown are for consolidation by mechanical vibration. May be increased 1 in. for consolidation by hand methods, such as rodding and spading.
Courtesy Portland Cement Association

depending on the circumstances. An air-entraining agent is recommended for some mixes. Detailed information is available from the Portland Cement Association.

Lightweight Structural Concrete

Lightweight structural concrete can be designed to produce structural members that are 25 to 35 percent lighter than the same member made using normal-weight concrete and have no loss in strength. Since the aggregate is cellular and weights are different from normal aggregate, its design is usually derived from testing trial batches, experience, and other reliable test data.

CONCRETE TESTS

Hardened, cured concrete and freshly mixed concrete are tested to make certain they meet the specifications written for the concrete. This is especially important when working in various parts of the country because aggregates and water differ and the mix must be adjusted accordingly.

Tests with Fresh Concrete

When testing fresh concrete, it is essential to take samples that are representative of the batch. Samples must be taken and handled following the specifications in ASTM C172. Except for slump and air-content tests, the sample must be at least one cubic foot (0.030 m³). The sample must be used within fifteen minutes after it has been taken from the batch and be protected from sources of rapid evaporation during the test. Samples taken at the very first and last of a batch are not representative.

Slump Test

Each load of transit-mixed concrete has a certificate listing its ingredients and their proportions. An on-site slump test is made to see if it has the required consis-

tency. A mix with a high slump may be too wet, and one with a low slump may be too stiff.

The slump test is made following ASTM C143 specifications (Fig. 7.11). A standard *slump cone* is 8 in. (200 mm) in diameter at the bottom and 12 in. (305 mm) high. The cone is placed on a flat surface and is held in place by standing on the foot pieces. It is filled about 1/3 full and rodded twenty-five times with a 5/8 in. (16 mm) diameter, 24 in. (600 mm) long rod with a rounded tip. The second 1/3 is poured and rodded as above, being certain the rod penetrates the surface of the layer below. After the top layer has been rodded, the excess concrete is struck off, leveling the top surface, and the mold is carefully lifted. The amount the concrete will slump is measured from the top of the cone. For example, if the top of the concrete is 4 in. below the top of the cone, the slump is 4 in.

Another method for testing slump is the *ball penetration test* as specified by ASTM C360 (Fig. 7.12). The depth to which a 30 lb. (13.6 kg), 6 in. (150 mm) diameter hemisphere will sink into the fresh concrete is measured. When calibrated for a particular set of materials, the results can be related to the slump. The concrete is placed in a container at least 18 in. (450 mm) square and at least 8 in. (200 mm) deep.

Unit Weight Test

The unit weight test involves weighing a properly consolidated specimen in a calibrated container following ASTM C138 standards. It can determine the quantity of concrete produced per batch and also indications of air content.

Air Content Test

Methods for measuring air content include the pressure method (ASTM C231), the volumetric method (ASTM C173), and the gravimetric method (ASTM C138).

The *pressure method* requires the sample be placed in a pressure air meter and subjected to an applied pressure. The air content can be read directly. When lightweight aggregates are used, this method also compresses the air in them; therefore, tests by the pressure method are not recommended for concrete with lightweight aggregates (Fig. 7.13).

The *volumetric method* measures air content by agitating a known volume of concrete in an excess of water. This method is suitable for concrete containing all types of aggregate.

The *gravimetric method* uses the same test as used for the unit weight test of concrete. The actual unit weight of the sample is subtracted from the theoretical unit weight as determined from the absolute volumes of the ingredients, assuming no air is present. The mix proportions and specific gravities of the ingredients must be known. The difference in weight is given as a percentage as is the air content. This method requires laboratory control, so it is not suitable for on-site use.

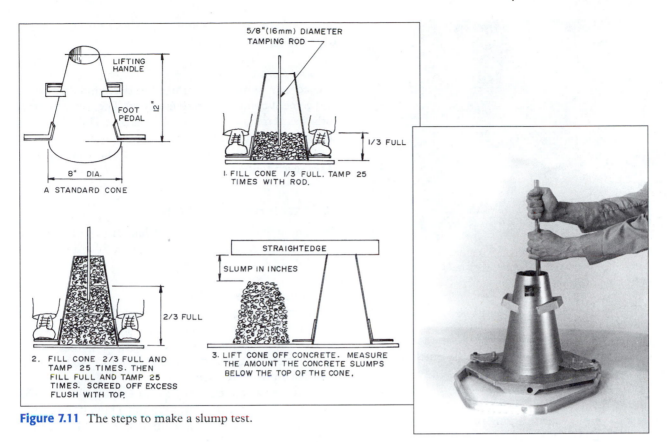

Figure 7.11 The steps to make a slump test.

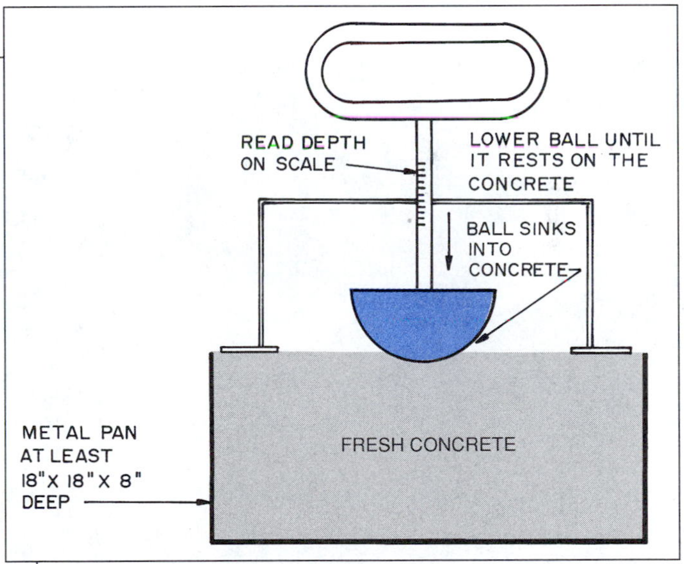

Figure 7.12 The ball penetration test is used to test slump. *(Courtesy Portland Cement Association)*

Figure 7.13 This is a pressure-type air meter used to ascertain air content of concrete. *(Courtesy Portland Cement Association)*

A quick check for air content can be made using a pocket-size air indicator. This is only for a quick on-site test and is not suitable for a standard ASTM test.

Temperature Test

Sometimes concrete specifications place a limit on the temperature of the fresh concrete because it influences the properties. There is no standard method for measuring temperature. Armored thermometers are placed into the fresh concrete sample and left until the reading becomes stable (Fig. 7.14).

Cement Content Test

The cement content test is used to determine the water and cement content of fresh concrete. The water-cement ratio has a major influence on strength. Therefore, this test gives an estimate of the strength potential without waiting for samples to harden and cure, which usually takes seven to twenty-eight days.

Tests with Hardened Concrete

Specimens for strength tests of hardened concrete are made and cured according to ASTM C31 (in the field) and ASTM C192 (in the laboratory).

Compressive Strength Test

The compressive strength test is made following ASTM C31 specifications. It is one of the most frequently required tests. Specifications indicate the required curing times before tests are made, but seven to twenty-eight days are commonly used.

Concrete specimens are cast in cylinders 6 in. (152 mm) in diameter and 12 in. (305 mm) high if the coarse aggregate is 2 in. (50 mm) or smaller (Fig. 7.15). Larger cylinders are specified for aggregate above 2 in. (50 mm). The cylinder is filled in three equal layers. Each layer is rodded twenty-five times with a ⅝ in. (16 mm)

THERMOMETER

Figure 7.14 This armored thermometer is placed in the fresh concrete to take its temperature. *(Courtesy Portland Cement Association)*

Figure 7.15 Test cylinders being cast on the job site. Notice the cones of concrete on the left from a slump test. *(Courtesy Portland Cement Association)*

diameter rod with a round end. After the last layer is rodded and the mold is struck off full, the ends are covered with a glass or metal plate to prevent evaporation (Fig. 7.16). After twenty-four hours the hardened specimen is removed from the mold and placed in a curing location. Some specimens are stored under controlled conditions in a test laboratory, while others may be stored on the site and cured under actual field conditions.

After the specified curing period has passed, the ends of the samples are capped with a commercially available

capping material, per ASTM C617 (Fig. 7.17). The capped specimen is placed in a compression-testing machine and loaded until it fractures. Cylindrical specimens are tested according to ASTM C39 (Fig. 7.18).

Flexural Strength Test

The flexural strength test is used to determine the flexural or bending strength of the concrete. The concrete sample is formed in a mold in the shape of a beam. Samples with aggregates up to 2 in. (50 mm) should have a minimum cross section of 6 × 6 in. (150 × 150 mm). Large-aggregate samples should have a minimum cross section dimension of three times the maximum size of the aggregate. The span of the test beam should be three times the depth of the beam plus two additional inches. A 6 × 6 in. (150 × 150 mm) beam would be 20 in. (508 mm) long.

The mold is filled in two layers with one rodded stroke for every 2 in.^{2} (13 cm^2) of area. The top is struck flush with the mold and the sample is cured with controlled temperature and moisture (Fig. 7.19).

The cured specimen is tested as shown in Fig. 7.20. It is supported on each end, and pressure is applied to the midpoint until the specimen breaks. The ultimate flexural strength is read on a dial in pounds per square inch (kilopascals).

Abrasion Test

The abrasion test is used to ascertain the resistance to wear of hardened concrete samples. A hardening admixture or surface coating is used with the sample concrete mix. The test is made on a machine that rolls steel balls under pressure in a circular motion on the surface of the specimen. The specimen is weighed before and after the test. The loss in weight determines the ability to resist abrasion (Fig. 7.21).

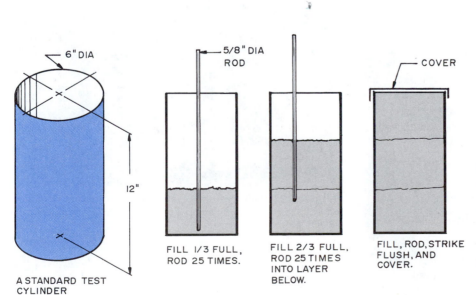

6" DIA

5/8" DIA ROD

COVER

12"

A STANDARD TEST CYLINDER

FILL 1/3 FULL, ROD 25 TIMES.

FILL 2/3 FULL, ROD 25 TIMES INTO LAYER BELOW.

FILL, ROD, STRIKE FLUSH, AND COVER.

Figure 7.16 Concrete samples are cast in standard test cylinders and cured.

Figure 7.17 The cured test cylinders are capped in preparation for the compression test. *(Courtesy Portland Cement Association)*

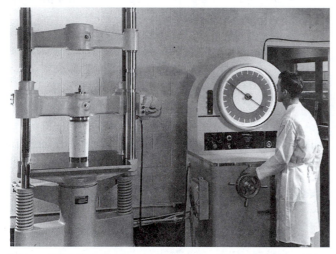

Figure 7.18 Standard, cured concrete cylinders are tested in a compression testing machine. *(Courtesy Portland Cement Association)*

Figure 7.19 A test beam is cast in a mold and cured. *(Courtesy Portland Cement Association)*

Figure 7.20 The cured test beam is loaded to the breaking point to find the ultimate flexural strength. *(Courtesy Portland Cement Association)*

Figure 7.21 This abrasion test machine determines the ability of a cured concrete sample to resist wear. *(Courtesy Portland Cement Association)*

Freeze-Thaw Test

The cured specimens are placed in a freeze-thaw tester, which is a cabinet much like a freezer. It is run through a series of freeze-thaw cycles. The loss between the original weight and final weight of the specimen is used to determine which samples withstand the freeze-thaw cycle best (Fig. 7.22).

Accelerated Curing Tests

Accelerated curing tests are used when it is desired to get acceptance of structural concrete without the usual twenty-eight-day curing period. ASTM C684 has three methods for making accelerated strength tests.

Nondestructive Tests

These nondestructive tests are used to evaluate the strength and durability of hardened concrete. Commonly used tests are rebound, penetration, and pull-out, and dynamic or vibration tests.

Rebound Tests

Rebound tests are made with a Schmidt rebound hammer (Fig. 7.23). It measures the distance of rebound of a spring-loaded plunger after it has struck the concrete surface. The reading is related to the compressive strength of concrete.

Penetration Tests

The penetration method uses a Windsor probe, which is a power-activated gun that drives a hardened alloy probe into the concrete. The exposed length of the probe is measured and related by a calibration table to the com-

Figure 7.23 A rebound hammer can be used to ascertain the compressive strength of cured concrete. *(Courtesy Portland Cement Association.)*

pressive strength of the concrete. This leaves a small indentation in the concrete surface.

Pull-Out Tests

The pull-out test requires a steel rod with an enlarged end be cast in the concrete. A device used to pull the rod from the concrete measures the force required. This gives the shear strength of the concrete. It has the disadvantage of damaging the surface of the concrete.

Dynamic or Vibration Tests

A dynamic or vibration test uses the principle that the velocity of sound in a solid can be measured by either recording the time it takes short impulses of vibrations to pass through a sample or determining the resonant frequency of a specimen. High velocities indicate a very good concrete while very low velocities indicate a poor concrete.

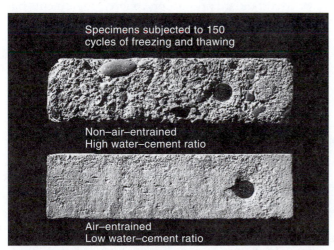

Specimens subjected to 150 cycles of freezing and thawing

Non–air–entrained
High water–cement ratio

Air–entrained
Low water–cement ratio

Figure 7.22 These are samples of test specimens subjected to 150 freezing and thawing cycles. Notice how the air-entrained sample is highly resistant to damage from repeated freezing and thawing. *(Courtesy Portland Cement Association)*

REVIEW QUESTIONS

1. What are the ingredients in concrete?
2. How does concrete harden?
3. What ingredients normally make up portland cement?
4. What are the designations for the different types of portland cement?
5. What is meant by air-entraining?
6. How does the addition of air-entraining materials to portland cement improve the concrete?
7. What ingredients may be used in masonry cements?
8. How does the fineness of the cement affect the concrete?
9. Explain how portland cement should be stored.

10. What natural aggregates are used in making concrete?
11. What aggregates should be avoided when making concrete?
12. What aggregates are used in insulating concrete?
13. What grains of sand are best for making concrete?
14. What are the commonly used admixtures?
15. What is the recommended water-cement ratio?
16. What five factors influence the quality of finished concrete exposed to the elements?
17. What can be done in hot weather to help retard rapid setting times?

KEY TERMS

accelerator An admixture used to speed up the setting of concrete.

admixture A material other than portland cement, aggregate, and water that is added to concrete to alter its properties.

aggregate An inert mineral, such as gravel or crushed stone, mixed with portland cement and sand to form concrete.

air-entrained concrete Concrete with an admixture added that produces millions of microscopic air bubbles in the concrete.

batch The amount of concrete mixed at one time.

concrete A solid, hard material produced by combining portland cement, aggregates, sand, water and sometimes admixtures.

consolidation The process of compacting freshly placed concrete in forms.

curing Protecting concrete after placing so that proper hydration occurs.

hydration A chemical reaction between portland cement and water that produces heat and hardening of the concrete.

portland cement A fine, pulverized material composed of compounds of lime, iron, silica, and alumina that has the capability of binding together dissimilar materials.

pozzolan A siliceous or siliceous and aluminous material blended with portland cement that chemically reacts with calcium hydroxide to form compounds possessing cementitious properties.

retarders An admixture used to slow the setting of concrete.

slump A measure of the consistency of concrete.

water-cement ratio A ratio by weight of water to portland cement used to make the paste for concrete.

workability The ease or difficulty with which concrete can be placed and worked into its final location.

SUGGESTED ACTIVITIES

1. Visit a local concrete batch plant. Ask the supervisor to show the process of producing concrete and to explain instructions given to drivers of the delivery trucks.

2. If the concrete plant has a test lab, see if you can observe some of the tests made for the contractor.

3. As covered in Chapter 3, mix samples of concrete using various types of cement, varying the proportions of the ingredients and using different aggregates. Cure the samples and make compression tests. Compare the results and explain the differences that occur.

4. Prepare identical concrete samples and cure these the same number of days but under different temperatures, such as at room temperature, above 100°F, and below freezing (put in freezer). Test for compression strength. Examine the broken samples and report any differences in the appearance of the surfaces.

5. Prepare identical concrete samples, cure them under ideal conditions, and then expose them to various conditions to see how they survive. For example, immerse in salt water, fresh water, gasoline, oil, and other materials to which concrete is often exposed. Report your findings in a written paper.

6. On construction sites visited, observe how the aggregates were stored. Report the good and poor practices observed.

7. Prepare a list of admixtures and tell what purposes they serve.

8. Mix standard proportion concrete ingredients but vary the amount of water used. Measure the slump of each sample.

ADDITIONAL INFORMATION

ACI Building Code Requirements for Structural Concrete and Commentary, American Concrete Institute, Detroit, Mich., 1993.

Waddell, J.J., and Dobrowolski, J.A., *Concrete Construction Handbook,* McGraw-Hill, New York, 1994.

Other resources are

Technical publications from the Portland Cement Association, 5420 Old Orchard Road, Skokie, Ill. 60077-1083.

8

Cast-In-Place Concrete

This chapter will help you to:

1. Understand the processes for preparing, transporting, handling, and placing concrete.

2. Discuss the finishes used on concrete surfaces.

3. Tell how concrete should be cured.

4. Cite various types of formwork used for cast-in-place concrete.

5. Identify and describe all types of concrete reinforcing materials.

6. Describe cast-in-place on-grade concrete slabs and the types of joints commonly used.

7. Explain how cast-in-place concrete walls, beams, and columns are formed, reinforced, and poured.

8. Prepare sketches illustrating the various types of monolithically cast slab and beam floors and roofs.

9. Describe the procedure for casting and erecting tilt-up concrete walls.

10. Explain briefly what is meant by lift-slab construction.

Cast-in-place concrete is produced by setting wood, metal, molded plastic, or wood-fiber forms in place; placing reinforcing material in the forms; and pouring the concrete over the reinforcing, filling the form. Cast-in-place concrete members must be designed by a professional engineer. The examples in this chapter are for illustration purposes only and are not intended to be used as design solutions.

An engineer can design cast-in-place concrete members in a wide range of sizes and shapes, with a variety of surface textures and colors. Although some concrete structural members can be precast and shipped to the site (see Chapter 9), some parts of a structure generally are cast on the site. These include spread footings, foundation caissons, pilings, piers, flat on-grade slabs, and any members too large to be precast and moved to the site. Some designs have irregular-shaped features that are difficult to precast and transport, so the pieces are cast-in-place. For example, construction of a dam requires huge intricate forms and massive amounts of concrete all site-built and cast-in-place (Fig. 8.1).

Cast-in-place concrete structural members usually are heavier than steel, wood, or precast concrete members. This increases the load on the foundation and is a major design consideration. Also, prefabricated steel and wood members and precast concrete members can be erected rapidly and in weather not suitable for pouring cast-in-place concrete. There are continuing developments to make cast-in-place concrete faster and easier. A wide range of forms is available as is equipment such as concrete pumps and power finishing machines needed to speed up the process. Cast-in-place concrete is a widely used and effective construction material.

BUILDING CODES

Reinforced concrete structural members and prestressed concrete must be designed and constructed according to the provisions in the building code. This includes provisions to resist seismic forces if they are a factor in the area. Codes specify how to bend the reinforcement, what the surface conditions must be, how to place the concrete in the forms, and what the coverage of the reinforcing with concrete must be. For example, concrete that is cast against the earth and that will remain on the earth requires

Figure 8.1 This dam is being built with site-built forms and cast-in-place concrete. Notice the concrete bucket used to fill the form. *(Courtesy Tennessee Valley Authority)*

3 in. (76.2 mm) of concrete cover over the reinforcement. Concrete walls, joists, and slabs not exposed to ground or earth typically require ¾ to 1½ in. (19 to 38 mm) concrete over the reinforcing, depending on the diameter of the reinforcing bars. Codes also include specifications for placing concrete in corrosive environments, for thicknesses over reinforcing for fire protection, for resistance to frost action, and for vertical and lateral loads.

PREPARING THE SITE

Concrete cast on the site involves much planning. Among the things to be considered are access to the site, preparation of temporary roads, and sources of suitable water and aggregates. Areas must be cleared and prepared for stockpiling materials and protecting them from the weather and contamination and for setting up the batching plant or other source of concrete. An overall safety plan for the site must be developed, including plans for scaffold and form construction.

There are various ways to produce, deliver, finish, and cure concrete. Those used depend on the circumstances under which the job is operating. The following sections discuss the major activities in preparing and placing concrete for cast-in-place construction.

PREPARING CONCRETE

Batching

Concrete is usually prepared in batches. A **batch** is the amount of concrete mixed at one time. The quantities of each dry ingredient are usually weighed. Water and admixtures are specified by either weight or volume. The use of volume measurements is discouraged because they are inaccurate. Moisture in the aggregate changes

the weight, but this is not accounted for in volume measures. Aggregates, especially sand, tend to fluff when handled, so the actual volume of sand can vary from batch to batch. When concrete is produced by a continuous mixer, volumetric measure is used. The job specifications usually establish the percentage of accuracy allowed when measuring ingredients. These typically range from 1 percent to 2 percent, so accurate weighing facilities are essential.

Mixing

Concrete is mixed until it is uniform in appearance. All ingredients must be evenly distributed. If an increased amount of concrete is needed, an additional mixer should be used rather than overloading or speeding up those in operation. It is important to follow the manufacturer's recommendations and to keep the mixing blades clean. Bent or worn blades should be replaced.

Stationary Mixing

On a large job the concrete is often mixed on the job using a stationary mixer (Fig. 8.2). This may be a tilting or

Figure 8.2 Concrete from a stationary mixer is being loaded into a transit truck for movement to the place on the site where it is needed. *(Courtesy Portland Cement Association)*

nontilting type and may be manual, semiautomatic, or automatically controlled. Some mixers have data for various mix designs stored on computer tape. Generally, the batch is mixed one minute for the first cubic yard and an additional fifteen seconds for each additional $\frac{1}{2}$ yd.^{3} (0.35 m^3) or fraction thereof. The mixing time is measured from the time when all ingredients are placed in the mixer. All water must be added before one-fourth of the mixing time has elapsed.

About 10 percent of the mixing water is placed in the drum before the dry ingredients are added. Then water is added uniformly with the dry ingredients, saving 10 percent to be added after all dry ingredients are in the drum.

Ready-Mix Concrete

Ready-mix concrete may be fully mixed in a central mixing plant and delivered to the site in a truck mixer that operates at agitating speed or in-a special nonagitating truck. The concrete may be partially mixed in a central mixer and completed in a truck mixer as it is moved to the site, or the dry ingredients may be placed in a transit truck mixer, water added, and the entire mixing done by the truck. When the entire mix is made in the truck mixer, seventy to one hundred rotations of the drum at the rotating speed specified by the manufacturer is enough to produce a uniform mix. All revolutions after one hundred should be at a slower agitating speed so as not to overmix the batch. Concrete must be delivered and discharged within 1½ hours or before the drum has revolved three hundred times after the introduction of water.

Remixing

Fresh concrete in the drum tends to stiffen even before the concrete has hydrated to initial set. It can be used if remixing will restore sufficient plasticity so it can be compacted in the forms. Under special conditions a small amount of water can be added, but it must not exceed the allowable water-cement ratio, designated slump, and allowable drum revolutions, and it must be remixed at least half the minimum required mixing time or number of revolutions.

TRANSPORTING, HANDLING, AND PLACING CONCRETE

Before the fresh concrete arrives on the job, preparations for moving it to its point of placement must be complete. Delays in placing the concrete can cause a loss of plasticity. Also, the method of moving the concrete must not result in the segregation of concrete materials. **Segregation** is the tendency of the coarse aggregate to separate from the sand-cement mortar. In some cases the heavy aggregate settles to the bottom and the sand-cement mortar rises to the top. This produces unsatisfactory final results.

There are many ways to move and place concrete. As discussed earlier, truck agitators, truck mixers, and nonagitating trucks are used most frequently to bring concrete to the site. Concrete is also mixed on the site with a stationary mixer. Concrete is moved about the site to the point of placement with cranes using concrete buckets and with barrows and buggies, chutes, belt conveyors, pneumatic guns, and concrete pumps. Some of these are shown in Fig. 8.3.

Moving Concrete

If transporting units are to be filled from a hopper, the concrete should pour straight into them, as shown in Fig. 8.4. Concrete coming from a conveyor should be directed straight down by using deflector plates and a down pipe (Fig. 8.5). If the concrete is moved with a chute, the chute should also be placed perpendicular to the receiving unit through a down pipe to prevent segregation of the mix (Fig. 8.6).

For many jobs, such as pouring a garage floor or residential basement walls, *wheelbarrows* or *manually pushed buggies* provide an adequate flow of material. This requires construction of on-ground runways from 2 in. (50 mm) lumber. If the pour is above grade, ramps and scaffolding must be built (Fig. 8.7). To speed up the flow of concrete, powered buggies can be used.

Often a slab or foundation can be poured directly from the *transit-mix* or *ready-mix concrete truck*. If the truck can back up close enough and the pour is not too high above grade, much of the pour can be directly from the truck (Fig. 8.8). In some cases wheelbarrows or buggies might be used for part of the pour, while the rest is poured directly from the truck.

Multistory slabs, beams, and columns are poured from *buckets* lifted by cranes or from concrete pumps. The buckets are filled on the ground, lifted over the point of pour, and opened to deposit the concrete as shown in Fig. 8.9. Buckets are available in 2 yd.^{3} (1.5 m^3) capacities and may be round or square. They are opened by hand and pour from the bottom.

Concrete pumps are heavy-duty piston pumps that force concrete through a pipe ranging from 6 to 8 in. (152 to 203 mm) in diameter (see Fig. 8.3). They can place concrete over long distances, ranging up to about 100 ft. (30.5 m) vertically and 800 ft. (244 m) horizontally. Pumps can also place concrete below grade as required for foundations of multistory buildings, whose foundations must be several stories below grade. The maximum aggregate size for an 8 in. (203 mm) pipeline is 3 in. (75 mm), for a 7 in. (178 mm) pipeline is 2½ in. (64 mm), and for a 6 in. (152 mm) line is 2 in. (50 mm). The pump requires an uninterrupted flow of plastic concrete so that the concrete does not begin to set and produce unwanted joints in the slab, beams, or column. On the end of the discharge line is a choke that controls the flow of concrete.

Tower crane with concrete bucket

Power buggy

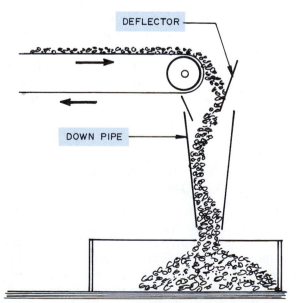

Concrete pump

Figure 8.3 Some of the ways concrete is moved for placement. *(Courtesy Portland Cement Association)*

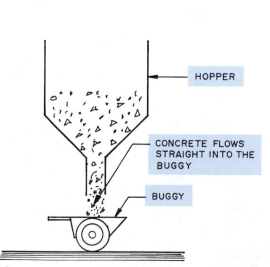

Figure 8.4 Hoppers should discharge straight into the transporting unit.

Figure 8.5 Concrete transported with a conveyor is directed straight down with a deflector and a down pipe.

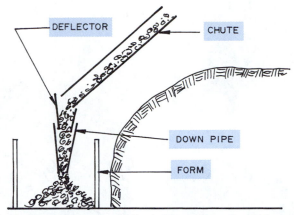

Figure 8.6 Concrete delivered by a chute should be directed with a down pipe.

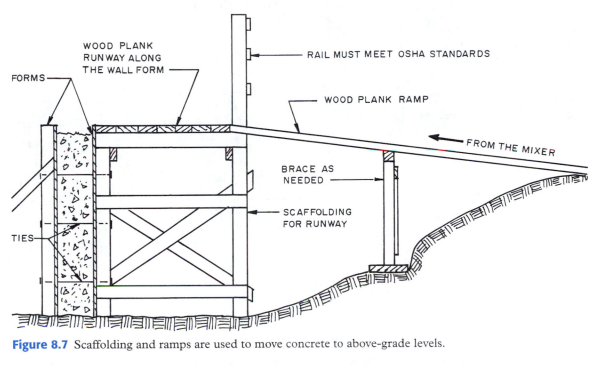

Figure 8.7 Scaffolding and ramps are used to move concrete to above-grade levels.

Figure 8.8 Many jobs permit the concrete to be poured directly from the transit truck onto the area of placement. (*Courtesy Portland Cement Association*)

Figure 8.9 A pour being made using a concrete bucket.

Figure 8.10 Concrete can be placed with a telescopic conveyor. This permits a long range of placement. Notice the down flow discharge tube. *(Courtesy ROTEC Industries)*

Figure 8.11 Concrete is being consolidated with a mechanical vibrator. *(Courtesy Portland Cement Association)*

The pump and pipeline are thoroughly flushed after each use.

Concrete is also moved with a telescopic conveyor, shown in Fig. 8.10. Notice the concrete truck discharging onto a belt that raises the mix to the telescopic conveyor.

Placing Concrete

Concrete is placed after the base for on-grade pours is ready or the forms for walls, columns, and beams are erected and reinforcing is in place. Concrete should be placed continuously as near as possible to its final location. In slab construction, work is started along one end and each batch is discharged against the one previously placed. If the concrete is to be thick, as in a foundation wall, it should be placed in layers 6 to 20 in. (150 to 500 mm) thick for reinforced members and 15 to 20 in. (400 to 500 mm) thick for mass work. Each layer should be consolidated before a second layer is placed on it. It is necessary to work fast so the first layer is still plastic when the next layer is placed on it.

Consolidation is the process of compacting the freshly placed concrete to the forms and around the reinforcing steel to remove air pockets and pockets of stone. Consolidation can be done by hand by pushing a rod into the concrete or with a mechanical vibrator (Fig. 8.11). The vibrator should be lowered vertically

into the concrete. It should not be used to move concrete along in a form. Rodding or vibrating should extend through the layer being consolidated and about 6 in. (150 mm) into the layer below. It is important not to overconsolidate. Too much consolidation tends to force the heavy aggregate to the bottom and the lighter cement paste to the top. The correct amount is judged by the person doing the work.

Placing Concrete in Cold Weather

Concrete placed in cold weather gains strength slowly. Fresh concrete must be protected from freezing. The critical period after which concrete is not seriously damaged by several freezing cycles depends on the ingredients, conditions of mixing, placing, curing, and long-term drying. The concrete designer must consider the heat of hydration, use of special cements and admixtures, and the temperature of the concrete, which is influenced by heating the aggregates and water. A formula is used to determine the temperature of fresh concrete. The final temperature of the combined ingredients should be well below 100°F (38°C) with 60°F to 80°F (15°C to 27°C) most frequently used. The temperatures of all batches should be about the same. If the overall temperature of the concrete exceeds 100°F (38°C) the concrete may flash set. When water is heated, most of the cement is not added to the mix until most of the water and aggregates have been mixed in the drum.

After placement, the concrete must be protected from freezing. The length of time depends on the concrete mix and local conditions. Common methods include insulated blankets and air heaters.

Placing Concrete in Hot Weather

In addition to maintaining low concrete temperature by cooling the aggregates and water, precautions must be taken to maintain a low temperature while placing the concrete. This could involve shading and painting white the mixers, chutes, hoppers, pump lines, and other concrete handling equipment. Forms can be cooled with water, and the subgrade can be moistened before the concrete is placed. The concrete must be transported from the mixing station to the point of placement as quickly as possible. In hot weather the mix should be in place 45 to 60 min. after mixing.

Pneumatic Placement

Concrete can be placed by pneumatically forcing a dry mixture of sand, aggregate, and cement through a hose and mixing it with water at a nozzle. This is referred to as pneumatically placed concrete or *shotcrete*. It is used to form thin sections in difficult locations and to cover large areas. It is used to place concrete in free-form shapes, such as coating a dome, for applying protective coatings, and for repairing concrete surfaces. Typical applications include forming swimming pools, covering rock outcroppings along highways to prevent rock falls, and providing underground support, such as tunnel and coal mine shaft linings. The surface can be covered with metal lath and the concrete sprayed over it. This can be done without the construction of expensive forms.

FINISHING CONCRETE

After the concrete has been placed and consolidated, it is screeded (Fig. 8.12). **Screeding** (also called strike-off) involves removing the excess concrete with a **screed** to bring it flush with the tops of the forms. Then the surface is *bullfloated* and *darbied*. This is done immediately after strike-off to lower high spots, fill low spots, and embed large aggregate that may be on the surface. A bullfloat has a long handle connected to a float (Fig. 8.13). A **darby** has a shorter handle and is used for shorter distances. This work must be done before any bleed water appears on the surface. **Bleed** refers to the water that

rises to the surface very soon after the concrete is placed on the forms.

When the bleed water sheen has evaporated the surface is ready for final finishing. *Any finishing operation performed on the surface of the concrete while bleed water is present will cause it to scale and dust.* Final finishing includes one or more of the following: edging, jointing, floating, troweling, and brooming.

Edging rounds off the edge of the slab to prevent chipping. **Jointing** forms control joints in the slab. A groove ¼ the thickness of the slab is formed across the slab at intervals specified by the architect. Its purpose is to provide a weak spot where the slab can crack when stresses exceed the strength of the concrete. Control joints can be formed in the wet concrete, sawed after the concrete has hardened, or formed by inserting plastic or hardwood strips in the concrete (Fig. 8.14). Other joints used are isolation and construction. *Isolation joints* provide a space between a slab and a wall, allowing each to move without disturbing the

Figure 8.12 After the concrete has been placed it is screeded. *(Courtesy Portland Cement Association)*

Figure 8.13 A bullfloat is used to lower high spots and fill in low spots. *(Courtesy Portland Cement Association)*

other. *Construction joints* are formed where one pour is to end and a joining one will meet it (Fig. 8.15).

After the concrete is edged and jointed it can be floated. This is done with a wood or metal handheld **float** or a finishing machine with float blades (Fig. 8.16). *Floating* embeds aggregate slightly below the surface, removes imperfections, compacts the mortar at the surface for final finishing, and keeps the surface open, allowing excess moisture to escape. Marks left by edging and jointing are removed by floating. If they are wanted for decorative purposes they need to be rerun after floating.

The final finish might be accomplished by troweling, brooming, or forming a pattern or texture in the surface. **Troweling** produces a hard, dense, smooth surface. The trowel is a steel-bladed handheld tool. *Brooming* involves roughing the surface with a steel-wire or coarse-fiber broom. It provides a slip-resistant surface. Patterns can be formed in the surface by placing divider strips in the concrete. For example, it can be divided to look like flagstones. Aggregate can be embedded in the surface and excess paste washed away, exposing the aggregate.

CURING CONCRETE

As explained in an earlier chapter, adding water to portland cement produces a chemical reaction called **hydration.** This reaction produces a hard cement paste that bonds the aggregate into a solid mass. Hydration continues for an indefinite period at a decreasing rate as long as water is in the mix and the temperature is favor-

able (73°F (23°C) is recommended). Therefore, concrete should be protected, so moisture remains in the mix during the early hardening period and the temperature is maintained. Concrete that is protected and kept moist for seven days has about twice the compressive strength of unprotected concrete (exposed to air with no attempt to keep moisture in the mix). The curing process is essential to producing concrete members with the expected compressive strength.

The length of the curing period depends on the type of cement, the design of the mix, the specified strength required, the size of the member poured, and weather conditions.

Following are some of the curing methods used.

Forms may be left in place and the exposed concrete surfaces kept moist.

Curing compounds are sprayed on the surface to retard the evaporation of moisture. After the forms have been removed the exposed concrete can be sprayed with curing compound.

The exposed concrete should be covered with some type of *sheet material* to hold moisture in the concrete. Among those in use are waterproof curing paper and plastic film. Tar paper can be used, but it often stains the concrete.

Wet covering materials, such as burlap or moisture-retaining fabrics, can be placed over the concrete. They should be kept wet over the entire curing period.

Continuously *sprinkling* the surface is an excellent way of curing. The sprinkling must be done so that the concrete is always wet.

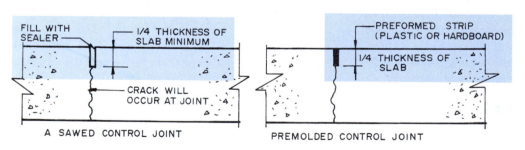

Figure 8.14 Control joints provide a place for the slab to crack without being visible or running at angles across the slab.

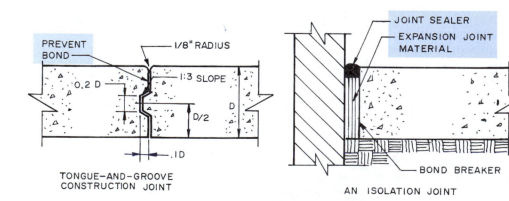

Figure 8.15 Construction joints occur where two pours will meet. Isolation joints separate a pour from a wall, column, or other abutting form.

Figure 8.16 The slab can be floated with a power finishing machine. *(Courtesy Portland Cement Association)*

Figure 8.17 Foundation walls can be formed rapidly using standard metal forms. *(Courtesy Precise Forms, Inc.)*

FORMWORK

A variety of standard metal forms (steel and aluminum) are available in a range of sizes, along with assembly and bracing systems (Fig. 8.17). Plywood is used for carpenter-built forms. Although these are used for general wall construction, they are the main material for constructing more intricate and complex forms. An excellent material for carpenter-built forms is exterior high-density overlay plywood, although other plywoods made with waterproof glue are used. Waferboard and solid wood are also used for form construction (Fig. 8.18), and a number of molded plastic and waxed cardboard forms are available.

The formwork sometimes must support the working deck in addition to the reinforcing steel, concrete placing and finishing equipment, the concrete, and the construction workers. It must be designed and erected to carry these vertical loads as well as to resist lateral forces. The resident engineer or building inspector usually verifies that the form construction is safe (Fig. 8.19).

The finish produced by forms ranges from the untreated surface left by the forms to one produced by *form liners* (Fig. 8.20). Form liners are molded plastic sheets that have been molded from actual concrete, masonry, or wood patterns. They are bonded to plywood sheets and are secured inside the form, producing a textured surface (Fig. 8.21).

Other forms are used to construct cast-in-place structural members. Columns are usually round, square, or rectangular. Round columns are formed with circular forms, such as the molded fiberglass column form shown in Fig. 8.22. These particular forms are available in diameters from 12 to 48 in. (304.8 to 1219.2 mm) and in lengths to 20 ft. (6 m). The manufacturer of column forms can supply data on maximum allowable lateral pressure, which on the forms pictured is 2,250 psf (108000 Pascals or 11009 Kg/m^2).

Figure 8.18 These wall and column forms are made of plywood panels and lumber frames and walls. *(Courtesy APA - The Engineered Wood Association)*

Custom forms are offered by suppliers of concrete forms. One example is shown in Fig. 8.23. This rectangular column has heavily rounded corners, as has the spandrel beam that is connected to the column. These forms were custom-made so the aesthetic goals of the architect could be achieved. Examples of the forms used are in Fig. 8.24.

An example of one-sided formwork is shown in Fig. 8.25. This form uses vertical trusslike structural members against which the form sheets are placed. The bracing is set on footings and is positioned with a number of screw-adjusted feet. A close look at the bracing being secured is shown in Fig. 8.26.

Circular formwork is shown in Fig. 8.27. The horizontal wales are curved metal units, and the form is supported with round steel braces bolted to the concrete floor.

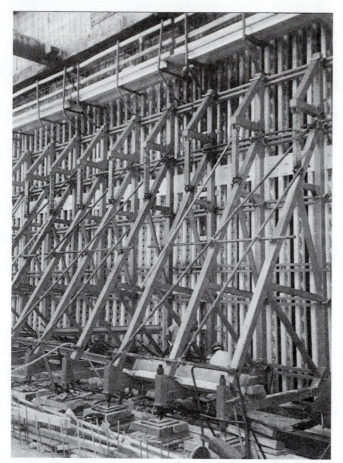

Figure 8.19 This tubular steel bracing supports a working deck, the form, steel, and concrete for pours of 20 ft. high and greater. *(Courtesy PERI Formwork Systems, Inc.)*

Figure 8.20 Form liners, moldable plastic sheets with a textured surface, are secured to the inside of the form. *(Courtesy Symons Corporation)*

Sculptured surface on a cast concrete wall surface.

Figure 8.21 Form liners produce a wide range of surface textures on poured concrete. *(Courtesy Symons Corporation)*

Surface texture on a cast concrete column.

Surface texture on a cast concrete wall panel.

Figure 8.22 This fiberglass column form has the reinforcing steel in place. The workers are installing a metal ring to which the bracing will be connected. *(Courtesy Molded Fiberglass Concrete Forms Co.)*

Figure 8.23 A rectangular column and spandrel beam cast-in-place using custom-built molded fiberglass forms. *(Courtesy Molded Fiberglass Concrete Forms Co.)*

An example of climbing formwork is shown in Fig. 8.28. After the section is poured and has reached sufficient strength, the form is raised to the next level and positioned, reinforcing is set in place, and the pour is repeated. A side view in Fig. 8.29 shows the work platform and the formwork being lifted to the next level.

Large projects use a variety of forms. Figure 8.30 shows the forms being installed for part of the construction of a dam. Straight wall forms, horizontal slabs, and forms for slanting concrete surfaces are seen. Form construction and installation is a major part of a large project such as this.

There are a number of ways to construct cast-in-place concrete floors and roofs. Figure 8.31 shows pans set for pouring a waffle floor. Between the pans are the reinforcing bars needed in the ribs. The reinforcing for the floor is placed over the pans. The large areas without pans are column heads cast over each column below the

floor. These are discussed in greater detail later in this chapter.

Flying formwork is built in large sections that are moved up by a crane to the next floor and reused (Fig. 8.32). The formwork is supported by metal trusses, providing a strong, rigid unit. This reduces costs by eliminating some of the labor required to strip a form and rebuild it on the next floor.

Another type of form used for residential and light commercial foundation construction is an insulated stay-in-place form (Fig. 8.33). The form is made from expanded polystyrene, and the units are notched to facilitate stacking them to form the cavity for the cast-in-place concrete wall. The formwork is nonstructural, and although it remains in place after the wall is poured, the reinforced concrete is designed to carry the loads. The form provides insulation and acoustical values to the wall. It can be used to build walls above and below grade (Fig. 8.34).

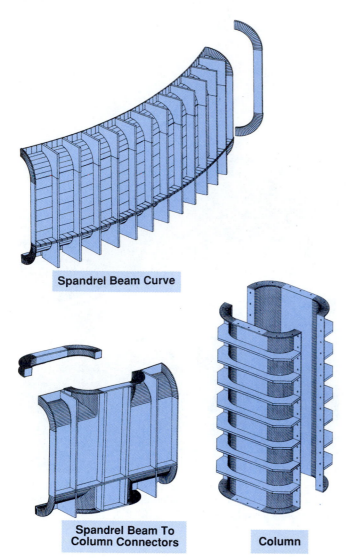

Spandrel Beam Curve

Spandrel Beam To Column Connectors

Column

Figure 8.24 Custom-built molded fiberglass forms. *(Courtesy Molded Fiberglass Concrete Forms Co.)*

Figure 8.26 These women are adjusting the tubular steel braces supporting the one-sided concrete formwork. *(Courtesy PERI Formwork Systems, Inc.)*

Figure 8.25 This one-sided form is 22 ft. (6.7 m) high and is supported with heavy tubular steel bracing. *(Courtesy PERI Formwork Systems, Inc.)*

Figure 8.27 A plywood skin is used on this circular formwork. Notice the curved wales and braces bolted into the concrete floor. *(Courtesy PERI Formwork Systems, Inc.)*

Figure 8.28 The climbing formwork is raised to the next level and used to pour the next exterior wall panel. *(Courtesy PERI Formwork Systems, Inc.)*

Figure 8.29 This side view of climbing formwork shows the ends of panels that have been cast and the form and work deck being raised for the next pour. *(Courtesy PERI Formwork Systems, Inc.)*

Figure 8.30 Notice the variety of concrete forms required on this dam construction project. *(Courtesy PERI Formwork Systems, Inc.)*

Figure 8.31 These pans with steel reinforcing in place are typical of those used to form cast-in-place concrete floors and roofs on multistory buildings. *(Courtesy U.S. Army Corps of Engineers)*

Figure 8.32 This unit of flying formwork has molded fiberglass pans supported by a steel framework. It is being set in place by a crane to be used to pour the floor directly below it. *(Courtesy Molded Fiberglass Concrete Forms Co.)*

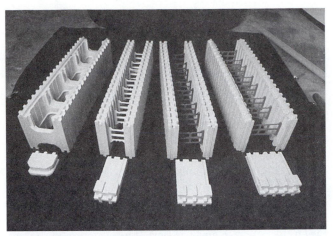

Figure 8.33 These expanded polystyrene concrete forms remain in place after the foundation is cured. *(Courtesy American ConForm Industries.)*

Figure 8.34 The polystyrene forms are spaced with plastic inserts and braced with wood posts and wales. *(Courtesy American ConForm Industries)*

CONCRETE REINFORCING MATERIALS

Concrete has no useful tensile strength, so steel reinforcing is added. Steel has high tensile strength. Concrete and steel have about the same coefficient of thermal expansion, concrete bonds to steel, and steel is not corroded by concrete, so they work together to provide a satisfactory structural system.

The two commonly used steel reinforcing materials are reinforcing bars with associated hooks and stands and welded wire reinforcement. A wide range of fibers also find use as concrete reinforcing.

Steel Reinforcing Bars

Reinforcing bars (also called *rebars*) are hot rolled steel rods that may be plain (smooth) or deformed. The deformed type has surface ridges, which gives better bonding to the concrete. The smooth type is used for special applications. The bars are available in 60 ft. (18.3 m) lengths. The inch-size bars are available in eleven standard diameters. The metric bars are made in eight diameters (Table 8.1). The diameters are identified by the bar size designation. Inch-size bar designations represent $1/8$ in. of bar diameter. For example, a No. 4 bar is $4/8$ or $1/2$ in. in diameter. Metric designations give the diameter in millimeters. A No. 20 M bar has a diameter of 20 mm (actually 19.5 mm as shown in Table 8.1).

Table 8.1 Steel Reinforcing Bar Sizes

ASTM Inch-Size Steel Reinforcing Bars			
Bar Size Designation	Weight in Pounds Per Foot	Nominal Dimensions Diameter in Inches	Cross-Sectional Area in Square Inches
#3	0.376	0.375	0.11
#4	0.668	0.500	0.20
#5	1.043	0.625	0.31
#6	1.502	0.750	0.44
#7	2.044	0.875	0.60
#8	2.670	1.000	0.79
#9	3.400	1.128	1.00
#10	4.303	1.270	1.27
#11	5.313	1.410	1.56
#14	7.650	1.693	2.25
#18	13.60	2.257	4.00

ASTM Metric-Size Steel Reinforcing Bars			
Bar Size Designation	Mass (kg/m)	Nominal Dimensions Diameter (mm)	Area (mm²)
#10M	0.785	11.3	100
#15M	1.570	16.0	200
#20M	2.355	19.5	300
#25M	3.925	25.2	500
#30M	5.495	29.9	700
#35M	7.850	35.7	1000
#45M	11.775	43.7	1500
#55M	19.625	56.4	2500

Copyright American Society for Testing and Materials. Reprinted with permission.

Reinforcing bars are manufactured to ASTM standards A615, A616, A617, and A706. They are made in grades 40, 50, 60, and 75. These refer to the minimum yield strength of the steel—40,000, 50,000, 60,000, and 75,000 psi (276, 345, 414, 517 MPa) (Table 8.2). In Canada metric reinforcing bars are available in grades 300 (43,600 psi or 300 MPa), 350 (50,800 psi or 350 MPa), and 400 (58,000 psi or 400 MPa). Grade 400W is a weldable bar.

Reinforcing bars are made from three types of steel—rail steel, axle steel, and billet steel. The bars are available galvanized or coated with epoxy to prevent corrosion.

Examples of the markings stamped into the rebar are in Fig. 8.35. Shown are the type of steel, bar size, grade mark, and identification of the mill making the bar.

The higher strength bars are used where space is tight and smaller diameter higher strength bars will provide the strength needed. Concrete columns and beams are members for which high-strength rebar is often used.

Reinforcing steel for concrete structural members is specified on the structural drawings. It is generally the deformed type, although smooth, plain reinforcement bars are used for some applications. The engineering drawings give the size, location, and bending information (Fig. 8.36). The steel fabricator cuts the bars to length and makes the required bends. The bends are made as specified by the engineer on the engineering drawings. Information pertaining to the design of bends can be obtained from the Concrete Reinforcing Steel Institute.

The types of bends are standardized and identified by number. A few of these are shown in Fig. 8.37. Some of the standard hook bends are shown in Fig. 8.38. Reinforcing bars must be bent cold unless specific approval is given by the design engineer. The formed bars are wired together into bundles and tagged. The tag has the name of the fabricator, the address of the job, fabrica-tion data, and the mark that locates their place on the structural drawing. The bent bars are usually delivered to the site by truck. They should be unloaded and stored as near as possible to the place where they are to be used. Care should be taken so the steel for the roof and upper floors is not stored on top of that to be used on the lower floors.

Some preassembly of reinforcing bars is necessary before they are placed in the forms. Examples are column spirals, column ties, and footing bars (Fig. 8.39). The steel in beams usually involves a set of bottom bars and stirrups. The bottom bars resist tension forces that exist in the bottom of the beam. The stress is dissipated from the bars into the concrete through the bond between the concrete and the bars. The concrete in the top is under compression, and concrete is able to resist compression forces. Codes specify the amount of concrete needed to protect the steel during a fire (Fig. 8.40). The ends of the bottom bars are bent into hooks that help dissipate tension forces at the end of the beam. Tension forces remaining in the bearing ends of the beam are resisted by stirrups. Most applications use U-stirrups, but in some cases closed stirrup ties are required (Fig. 8.41).

The bottom bars in the beam are raised above the bottom of the form with one of the several types of bar supports. These could be bolsters or chairs. Bars in the top of the slab are supported with high chairs. Bar supports are available in wire, precast concrete, cementitious fiber-reinforced, and all-plastic types (Fig. 8.42). Reinforcement that rests directly on the ground is supported by bar supports made from concrete, which become completely sealed from the earth by the concrete. Steel wire chairs tend to provide a passage for moisture and rust to reach the bottom reinforcing bars. Stirrups are tied to the bottom bar with wire and the top (hook) is supported with a small-diameter bar (Fig. 8.43).

Table 8.2 Grades, Strengths, and Types of Reinforcing Steel

Type of Steel	Steel Type Symbol	Yield Strength (psi)	Tensile Strength (psi)	Yield Strength (MN/m²)	Tensile Strength (MN/m²)
ASTM A615					
Billet steel					
Grade 40	S	40,000	70,000	276	483
Grade 60		60,000	90,000	414	621
Grade 75		75,000	100,000	517	690
ASTM A616					
Rail steel					
Grade 50	R	50,000	80,000	345	552
Grade 60		60,000	90,000	414	621
ASTM A617					
Axle steel					
Grade 40	A	40,000	70,000	276	483
Grade 60		60,000	90,000	414	621

Courtesy Concrete Reinforcing Steel Institute

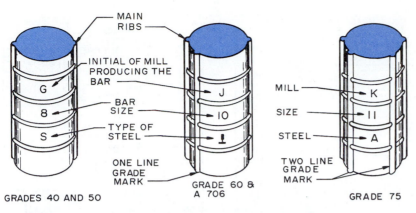

LINE SYSTEM TO INDICATE GRADE MARKS

FIELD IDENTIFICATION
SYMBOLS FOR STEEL TYPE

S – BILLET STEEL (ASTM A 615)
↓ – RAIL STEEL (ASTM A 616)
A – AXLE STEEL (ASTM A 617)
W – LOW ALLOY STEEL (ASTM A 706)

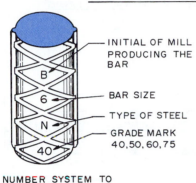

NUMBER SYSTEM TO
INDICATE GRADE MARKS

Figure 8.35 Markings stamped into reinforcing bars tell the bar size, type of steel, and grade mark and identify the mill that made them. *(Courtesy Concrete Reinforcing Steel Institute)*

If a member is formed from several pours, the reinforcing bars are extended beyond the end of the pour and into the next pour. The bars in the second pour overlap those from the first pour a distance specified by codes. Figure 8.44 shows two possible connections. Connections between reinforcing bars in columns require that they be spliced end-to-end and secured with a mechanical splicing device or by welding (Fig. 8.45). Connections should be made as specified by the structural engineer.

Welded Wire Reinforcement

Welded wire reinforcement (WWR)—sometimes called welded wire fabric (WWF)—is an assembly of steel reinforcing wires made from rod that is worked either cold-drawn or cold-rolled or both. The transverse wires are electrically resistance-welded to each of the longitudinal wires to form square or rectangular grids (Fig. 8.46). The material is available in rolls and sheets (Table 8.3). The wires may be *plain* (W) or *deformed* (D). The plain welded wire bonds to concrete by the positive mechanical anchorage at each welded intersection. The deformed wire uses the deformations in the surface of the wire in addition to the welded intersections for bonding and anchoring.

Welded wire reinforcement is widely used to reinforce concrete structures. The smaller diameter wires provide more uniform stress distribution and crack control. The selection of wire sizes by the engineer provides the needed cross-sectional area of the reinforcing steel.

In addition to uncoated wire, two coatings are available. One is a hot-dipped galvanized coating specified by ASTM A641 or A153. It is usually applied to the wire before it is welded. The other is an epoxy coating specified by ASTM A884. This is applied after the sheets have been welded. Specifications for plain and deformed welded wire are provided in publications of the American Society for Testing and Materials and the Canadian Standards Association (Table 8.4).

The terms used when discussing WWR are building fabric, pipe fabric, and structural welded wire reinforcement. *Building fabric* includes products with wire sized up to W4.0. Wire spacings are generally 4 × 4 in. and 6 × 6 in. (101 × 101 mm and 152 × 152 mm). Sheets and rolls are available in building fabric styles. *Pipe fabric* is formed into cylindrical pipe forms (circular, elliptical, and arch types). Wire sizes go up to W12 generally with spacings of 2 × 6 in. (50 × 152 mm), 2 × 8 in. (50 × 203 mm), and 3 × 6 in. (76 × 152 mm) in the standard styles. Pipe fabric is usually in roll form. *Structural welded fabric* reinforcement wire sizes include anything over D4 or W4. They have a variety of wire spacings, from 3 in. to 18 in. (76 to 457 mm) in both directions. Generally, structural welded wire is furnished in sheet or mat form.

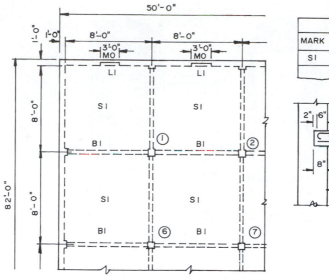

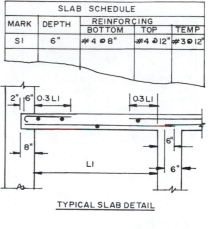

PARTIAL FLOOR FRAMING PLAN

TYPICAL SLAB DETAIL

SLAB SCHEDULE				
MARK	DEPTH	REINFORCING		
		BOTTOM	TOP	TEMP
S1	6"	#4 @ 8"	#4 @ 12"	#3 @ 12"

BEAM AND GIRDER SCHEDULE							
MARK	SIZE		BOTTOM		TOP	STIRRUPS	
	W	D	"A" BARS	"B" BARS		NO.– SIZE	SPACING FROM FACE OF SUPPORT
G1	12"	30"	3 #8	3 #8	2 #9 NON-CONTINUOUS ENDS 3 #8 AT COLUMNS #8 IN TOP LAYER	20 #3	1 @ 2", 3 @ 6", 2 @ 9", 2 @ 10", 2 @ 12"
B1	10"	20"	3 #8	3 #6	3 #10 AT COLUMNS	10 #3	1 @ 2", 2 @ 12", 2 @ 18"
L1	8"	12"	2 #6	—	—	NONE	

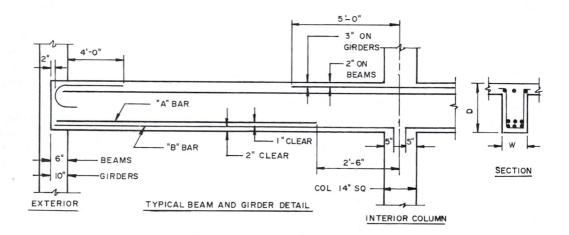

EXTERIOR

TYPICAL BEAM AND GIRDER DETAIL

INTERIOR COLUMN

SECTION

BENDING DETAILS							
MARK	SIZE	LENGTH	TYPE	A	B	C	D
B1	8	8'- 6"	2	8	8'-0"		
S1	4	4'- 6"	1	6	4'-0"		

TYPE 1 TYPE 2 TYPE 3

Figure 8.36 Examples of typical details for reinforcing cast-in-place concrete structural members in a building.

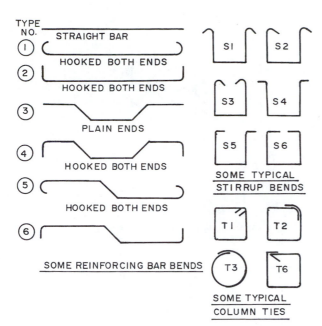

Figure 8.37 Some of the standard bends used with steel reinforcing bars. *(Courtesy American Concrete Institute)*

Welded wire reinforcement is specified by listing the longitudinal wire spacing, transverse wire spacing, longitudinal wire size, and the transverse wire size. An example for WWR style is $12 \times 12 - W12 \times W5$. This means

Longitudinal wire spacing 12 in. apart.
Transverse wire spacing 12 in. apart.
Longitudinal wire type and area-W means plain wire, 12 means area of 12 in.^{2}/ft.
Transverse wire type and area-W means plain wire, 5 means area of .05 in.^{2}/ft.

Longitudinal wire spacings are available in 2, 3, 4, 6, 8, 10, 12, and 16 in. In some cases spacing to 24 in. is available. Soft metric conversions are 51, 76, 102, 152, 203, 254, 305, and 406 mm. Transverse wire spacings are 3, 4, 6, 8, 12, and 16 in., soft converted to 76, 102, 152, 203, 305, and 406 mm. The designations for common styles of welded wire reinforcement are detailed in Table 8.5. Notice most sizes are available in both W and D type wire.

Metric styles and spacings of welded wire reinforcement are soft metricated as shown in Table 8.5. Structural WWR and building fabric styles decimal sizes are rounded to whole numbers. Wire sizes for pipe fabric carry the decimal sizes to one-tenth decimal place and are not rounded to whole numbers.

Following are examples of metric specifications:

Structural WWR—305×305-MD71 $\times$ MD71. This is equivalent to inch size 12×12-D11 $\times$ D11.
Building fabric—152×152-MW19 $\times$ MW19. This is equivalent to inch size 6×6-W2.9 $\times$ W2.9.

Pipe fabric—76×152-MW12.9 $\times$ MW12.9. This is equivalent to inch size 3×6-W2.0 $\times$ W2.0.

The spacings between wires are in millimeters (mm) and wire sizes are in square millimeters (mm^2). Metric plain wire is MW and deformed wire is MD. When conversion to hard metric design is accomplished in the future, the probable metric sizes will be based on 5 and 10 mm^2/m (square millimeters per meter) increments as shown in Table 8.6. Most wire manufacturers can furnish wire sizes in 1 mm^2 increments (0.001 in.^{2}) for exact area requirements.

Detailed information relating to the manufacture, specifications, properties, design information, and building code requirements are available from the Wire Reinforcement Institute, Inc. The American Concrete Institute publication ACI 318, Building Code Requirements for Reinforced Concrete, contains design data information on welded wire reinforcement.

Welded wire sheets are shipped in bundles in quantities varying with the size and weight of the sheets. Typical bundles weigh 500 to 5000 lb. (227 to 2268 kg). They are bound together with steel strapping. The bundles should never be lifted off the truck by the steel strapping. Lifting eyes can be specified when bundles are to be lifted with cranes.

Bundles are usually delivered to the site by truck. In some cases they are placed on wood dunnage so a forklift can remove them from the truck. Some bundles that are too long or too heavy for a forklift must be lifted with a crane. In these cases the lifting eyes are used or appropriate slings must be provided.

The engineer specifies the amount of reinforcement required and the correct position for it within the slab or wall. The sheets must be placed on supports or, in the case of walls, firm support spacers are used to maintain their position as the concrete is placed. The supports are usually concrete, steel, or plastic chairs, as discussed earlier. The amount of splice for sheets is also specified by the engineer. ACI 318, Chapter 12, is generally used to design splices. Slab-on-grade splices can generally be less than a structural splice, since the steel reinforcement is used primarily for crack control.

Fiber Reinforcement

In addition to steel reinforcing bars and welded wire fabric, a number of fibers are used to reinforce concrete. They are mixed with the concrete as it is prepared in the mixer. A number of manufacturers produce these products, and research will probably increase their effectiveness and use in the future. Under some conditions they may reduce the amount of reinforcing bars required, and they may replace welded wire fabric in some installations. Fibers also are used along with rebar and welded wire to produce more desirable properties in concrete.

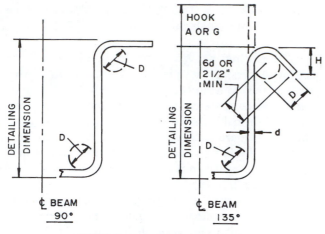

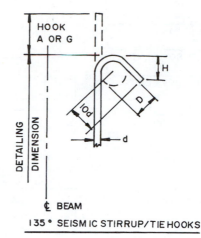

STIRRUP AND TIE HOOKS

Bar Size	D, In.	90-deg hook A or g	135-deg hook A or G	H, approx
#3	1½	4	4	2½
#4	2	4½	4½	3
#5	2½	6	5½	3¾
#6	4½	1-0	7¾	4½
#7	5¼	1-2	9	5¼
#8	6	1-4	10¼	6

135 deg seismic stirrup/tie hook dimensions, in.*

Bar Size	D, In.	135-deg hook A or G	H, approx
#3	1½	5	3
#4	2	6½	4½
#5	2½	8	5½
#6	4½	10¾	6½
#7	5¼	1-0½	7¾
#8	6	1-2¼	9

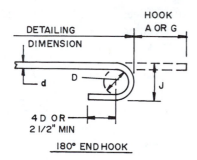

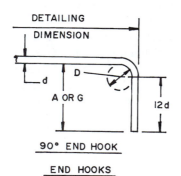

END HOOKS

Bar size	Finished bend diameter D, in.	180-deg hooks A or G, in.	J, in.	90-deg hooks A or G, in.
#3	2¼	5	3	6
#4	3	6	4	8
#5	3¾	7	5	10
#6	4½	8	6	1-0
#7	5¼	10	7	1-2
#8	6	11	8	1-4
#9	9½	1-3	11¾	1-7
#10	10¾	1-5	1-1¼	1-10
#11	12	1-7	1-2¾	2-0
#14	18¼	2-3	1-9¾	2-7
#18	24	3-0	2-4½	3-5

Figure 8.38 Some of the standard hook bends used with steel reinforcing bars. (*Courtesy Concrete Reinforcing Steel Institute*)

I T E M   O F   I N T E R E S T

STRIP-APPLIED WATERSTOPS

Waterstops in concrete construction are critical in construction, contraction, and expansion joints as well as for securing the perimeter of any penetrations in the concrete wall (see figure). Strip-applied waterstops are typically 1 × ¾ in. (25 × 19 mm), although some are larger. They are made from a variety of materials, which have varying characteristics and therefore different applications and installation procedures. Types commonly available include bituminous strips, bentonite-based strips, hydrophilic rubber strips and vinylester gaskets, and ethylene vinyl acetate strips. The concrete worker must know the proper way to install each type, including the method of bonding each to the concrete. The engineer designing the concrete structure must choose the product best suited for the conditions that will exist. When in doubt, the manufacturer should be consulted.

Bituminous strips are typically some blend of refined bituminous hydrocarbon resins and plasticizing compounds reinforced with inert mineral fillers. They resist fresh and salt water and acids but not oil or oil by-products. They are bonded with an asphalt primer.

Bentonite-based strips are a blend of sodium bentonite clay and butyl rubber or sodium bentonite clay with various binders and fabrics. Various grades are available that resist exposure to salt water, fresh water, and certain chemicals. The concrete worker can bond these to the concrete with a water-based latex adhesive. Some also prefer to use a few nails. One type has an adhesive back that is simply pressed onto the concrete surface.

Hydrophilic rubber strips are made from various types of rubber that have been modified to make them hydrophilic. "Hydrophilic" refers to the product having a strong affinity for water. These strips swell when exposed to water and shrink to normal size when dry. They can resist fresh and salt water and various chemicals. Epoxy or polyurethane sealants are used to bond them to the concrete.

Hydrophilic vinylester gaskets are made from a vinylester that is hydrophilic. Like hydrophilic rubber strips, they expand when wet and shrink when dry. They are available in several compositions, one for fresh water, an-

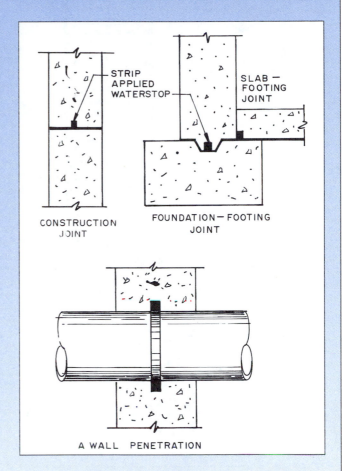

CONSTRUCTION JOINT

FOUNDATION—FOOTING JOINT

STRIP APPLIED WATERSTOP

SLAB—FOOTING JOINT

A WALL PENETRATION

other for salt water, and a third for selected chemicals. The concrete worker can secure these to the concrete with a xylene-based glue, nails, or screws.

Ethylene vinyl acetate strips are a closed-cell cross-linked foam produced from ethylene vinyl acetate. It resists exposure to oil-based products, fresh and salt water, and selected chemicals. The concrete worker bonds these strips to the concrete with a trowel-applied structural grade epoxy adhesive.

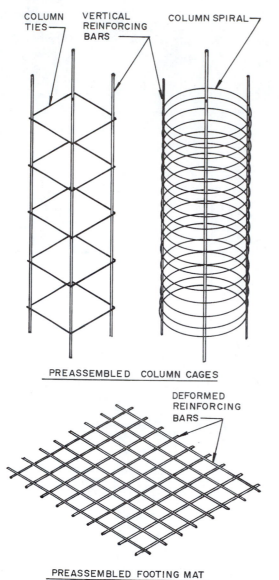

PREASSEMBLED COLUMN CAGES

PREASSEMBLED FOOTING MAT

Figure 8.39 Some reinforcing is preassembled and placed in the form.

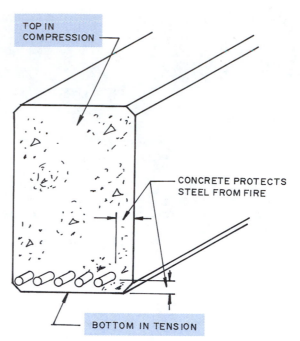

Figure 8.40 Codes specify the amount of concrete coverage required over reinforcing to protect it during a fire.

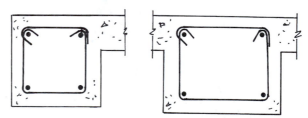

Figure 8.41 A typical two-piece closed stirrup.

The types of fibers available include glass fibers, polymeric (polypropylene, polyethylene, polyester, acrylic, and aramid), steel, asbestos, carbon, and natural fibers (wood, sisal, coconut, bamboo, jute, okwara, and elephant grass). Each of these fibers has different characteristics, and consultation with a concrete specialist should occur before using them. The design of the mix, strength, fatigue resistance, durability, shrinkage control, and other factors need to be considered. The Portland Cement Association has extensive material on fiber reinforced concrete.

In addition to being used in batch-mixed concrete, fiber reinforcement is also used in pneumatically placed concrete (shotcrete). The fiber is added with the dry concrete mixture and fed through a hose to a nozzle where water is injected. The mix is then sprayed onto the desired surface. This is referred to as SFRC (sprayed fiber reinforced concrete).

Another application of fiber reinforcement is its use in slurry infiltrated fiber concrete (SIFCON). Fibers are preplaced in a form or mold, and fine-grained cement slurry is pumped or poured in the bed of fibers. Air space is infiltrated between the fibers while they conform to the shape of the mold. The mold can be vibrated to assist in air infiltration. This technique is used in precast and cast-in-place concrete in projects such as precast slabs and pavement overlays.

TYPICAL TYPES AND SIZES OF WIRE BAR SUPPORTS

SYMBOL	BAR SUPPORT ILLUSTRATION	BAR SUPPORT ILLUSTRATION PLASTIC CAPPED OR DIPPED	TYPE OF SUPPORT	TYPICAL SIZES
SB	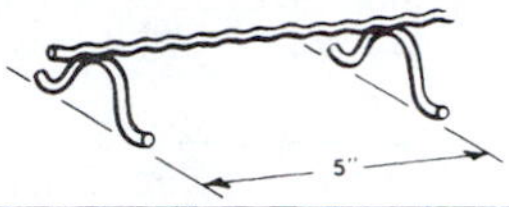 5"	CAPPED 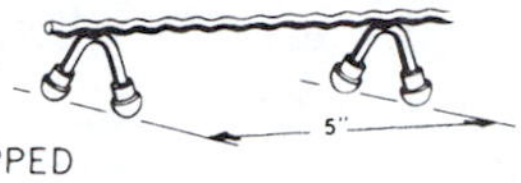 5"	Slab Bolster	¾, 1, 1½, and 2 inch heights in 5 ft. and 10 ft. lengths
SBU*	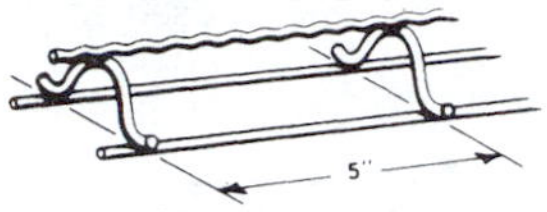 5"		Slab Bolster Upper	Same as SB
BB	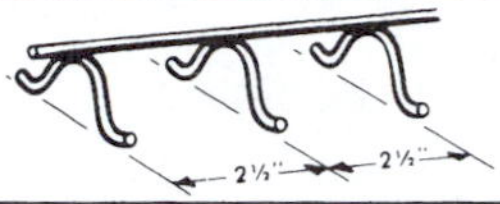 2½" 2½"	CAPPED 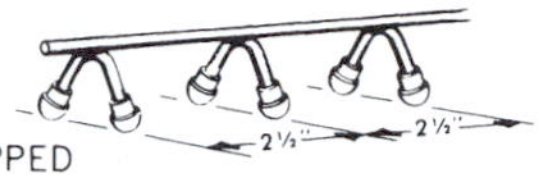 2½" 2½"	Beam Bolster	1, 1½, 2, over 2" to 5" heights in increments of ¼" in lengths of 5 ft.
BBU*	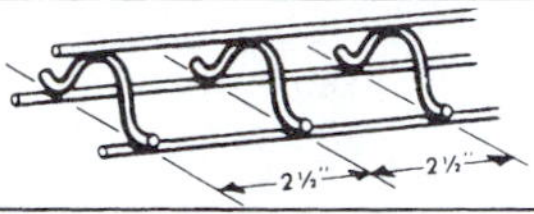 2½" 2½"		Beam Bolster Upper	Same as BB
BC		DIPPED 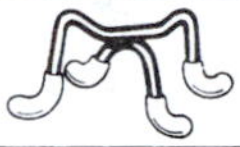	Individual Bar Chair	¾, 1, 1½, and 1¾" heights
JC	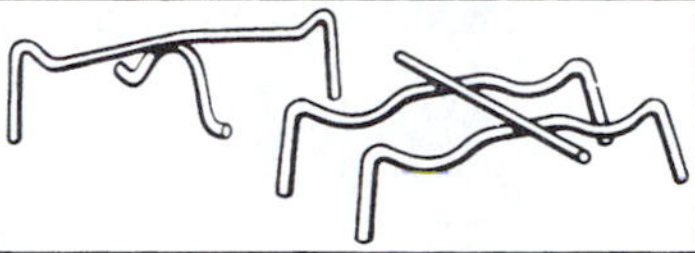	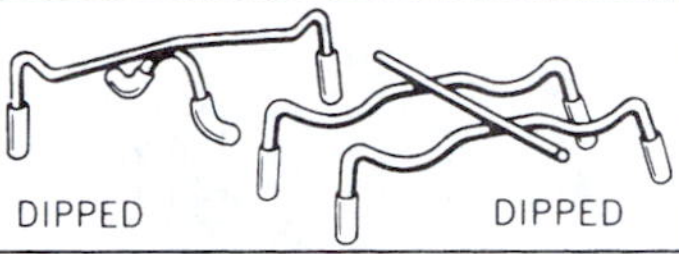 DIPPED DIPPED	Joist Chair	4, 5, and 6 inch widths and ¾, 1 and 1½ inch heights
HC	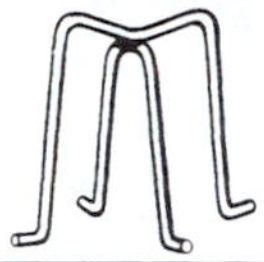	CAPPED 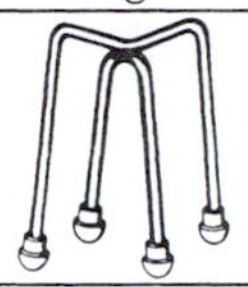	Individual High Chair	2 to 15 inch heights in increments of ¼ inch
HCM*	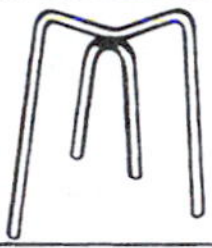		High Chair for Metal Deck	2 to 15 inch heights in increments of ¼ in.
CHC	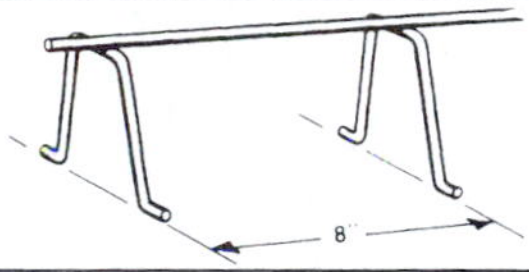 8"	CAPPED 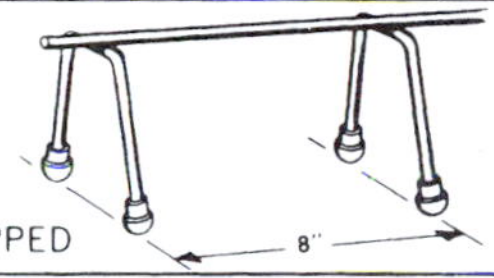 8"	Continuous High Chair	Same as HC in 5 foot and 10 foot lengths
CHCU*	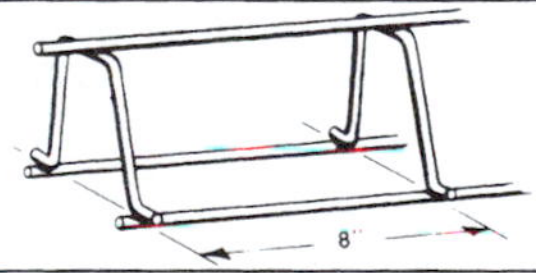 8"		Continuous High Chair Upper	Same as CHC
CHCM*	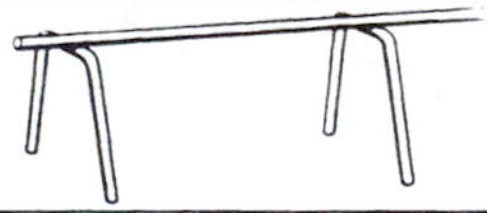		Continuous High Chair for Metal Deck	Up to 5 inch heights in increments of ¼ in.
JCU**	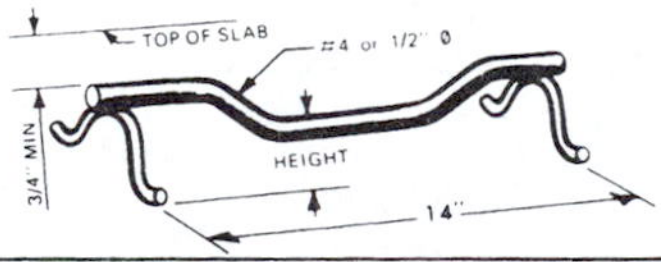 TOP OF SLAB #4 or 1/2" Ø HEIGHT 3/4 MIN 14"	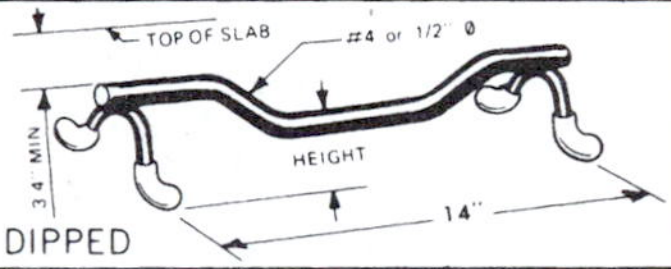 TOP OF SLAB #4 or 1/2" Ø HEIGHT 3/4 MIN 14" DIPPED	Joist Chair Upper	14" Span Heights −1" thru +3½" vary in ¼" increments
CS	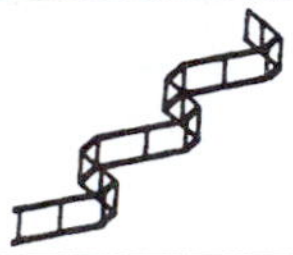		Continuous Support	1½" to 12" in increments of ¼" in lengths of 6'-8"

*Usually available in Class 3 only, except on special order.
**Usually available in Class 3 only, with upturned or end bearing legs.

Figure 8.42 Types and sizes of reinforcing bar supports. *(Courtesy Concrete Reinforcing Steel Institute)*

TYPICAL TYPES AND SIZES OF PRECAST CONCRETE BAR SUPPORTS

SYMBOL	BAR SUPPORT ILLUSTRATION	TYPE OF SUPPORT	TYPICAL SIZES	DESCRIPTION
PB	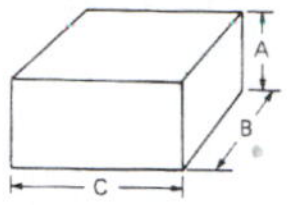	Plain Block	A—¾" to 6" B—2" to 6" C—2" to 48"	Used when placing rebar on grade. When "C" dimension exceeds 16" a piece of rebar should be cast inside block.
WB	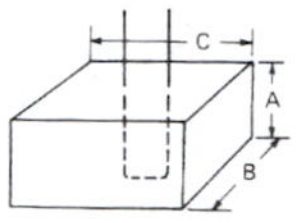	Wired Block	A—¾" to 4" B—2" to 3" C—2" to 3"	Commonly 16 ga. tie wire is usually cast in block, commonly used against vertical forms or in positions necessary to secure the block by tying to the rebar.
TWB	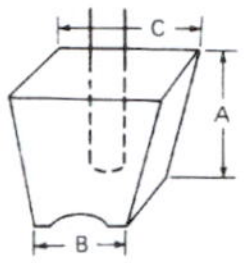	Tapered Wired Block	A—¾" to 3" B—¾" to 2½" C—1¼" to 3"	Commonly 16 ga. tie wire is usually cast in block, commonly used where minimal form contact is desired.
CB	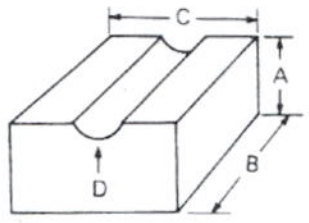	Combination Block	A—2" to 4" B—2" to 4" C—2" to 4" D—fits #3 to #5 bar	Commonly used on horizontal work.
DB	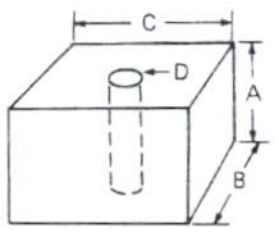	Dowel Block	A—3" B—3" to 5" C—3" to 5" D—hole to accommodate a #4 bar	Used to support top mat from dowel placed in hole. Block can also be used to support bottom mat.

TABLE IV—TYPICAL TYPE AND SIZES OF CEMENTITIOUS FIBER-REINFORCED BAR SUPPORTS

SYMBOL	BAR SUPPORT ILLUSTRATION	TYPE OF SUPPORT	TYPICAL SIZES	DESCRIPTION
SB	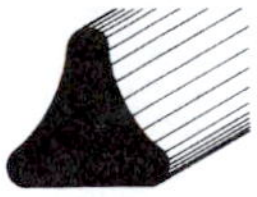	Slab Bolster	Heights, ¾" to 3" Lengths, 3'-0", 1'-0", 3" and 2"	Used to support lower slab reinforcing steel. Also available in circular shape for architectural work and square shape for heavy reinforcement.
BB		Beam Bolster	Same sizes as above.	Supports lower beam rebar and/or heavy slab rebar.
BC	 Single Cover 	Bar Chair	Heights, ¾" to 2"	Used for supporting reinforcing steel in slab and deck construction. Available with or without tie wire in three types—single, double and triple cover.
HC	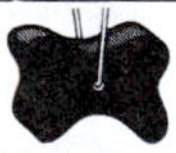 Double Cover Triple Cover	High Chairs	Heights, ¾" to 3"	Used for supporting upper slab rebar directly or for supporting lower slab rebar. All are available with or without tie wire.
CS	 Double Clips  Single Clip	Clip-on Spacers	Concrete cover, ¾" to 2"	Available with either double or single clips. Used to provide the necessary cover between vertical rebar and the form work.
VS	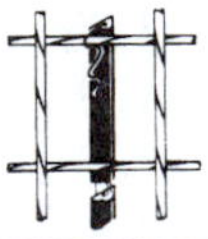	Vertical Spacers	Concrete cover, ¾" to 3"	Available with hooks for attachment to rebar. Used to provide cover between vertical rebar and form work. Triangular in shape and 1'-0" in length.

Figure 8.42 *Continued*

TYPICAL TYPES AND SIZES OF ALL-PLASTIC BAR SUPPORTS

SYMBOL	BAR SUPPORT ILLUSTRATION	TYPE OF SUPPORT	TYPICAL SIZES	DESCRIPTION
BS	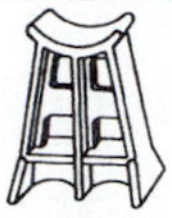	Bottom Spacer	Heights, ¾" to 6"	Generally for horizontal work. Not recommended for ground or exposed aggregate finish.
BS-CL		Bottom Spacer, Clamp On	Heights, ¾" to 2"	Generally for horizontal work, provides bar clamping action. Not recommended for ground or exposed aggregate finish.
HC	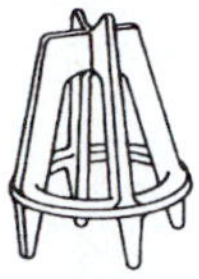	High Chair	Heights, ¾" to 5"	For use on slabs or panels.
HC-V		High Chair, Variable	Heights, 2½" to 6¼"	For horizontal and vertical work. Provides for different heights.
WS		Wheel Spacer	Concrete cover, ⅜" to 3"	Generally for vertical work. Bar clamping action and minimum contact with forms. Applicable for column reinforcing steel.

Figure 8.42 *Continued*

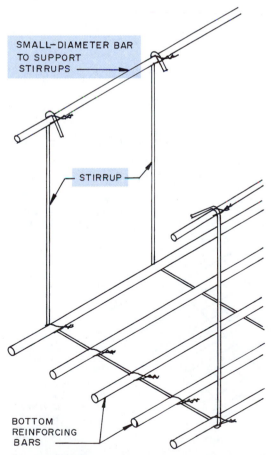

Figure 8.43 Stirrups are supported with small-diameter reinforcing bars.

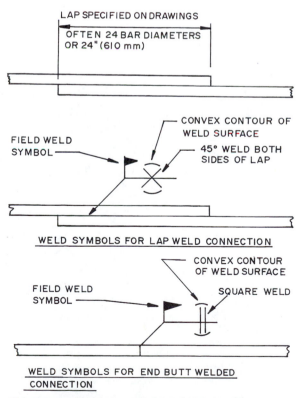

Figure 8.44 Welds for splicing reinforcing bars are specified on the drawings.

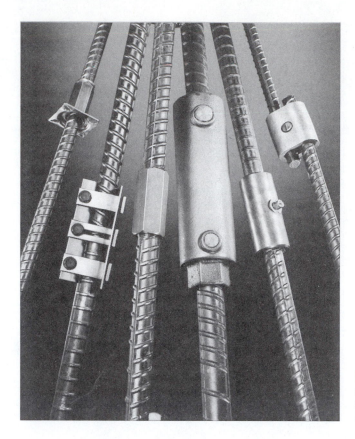

Figure 8.45 Mechanical reinforcing bar splicing devices. *(Courtesy ERICO)*

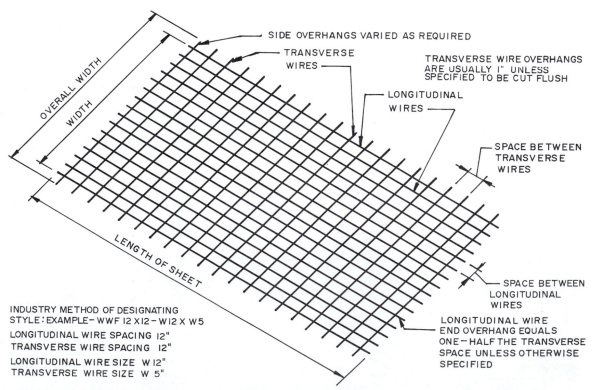

SIDE OVERHANGS VARIED AS REQUIRED

TRANSVERSE WIRES

TRANSVERSE WIRE OVERHANGS ARE USUALLY 1" UNLESS SPECIFIED TO BE CUT FLUSH

OVERALL WIDTH

WIDTH

LONGITUDINAL WIRES

SPACE BETWEEN TRANSVERSE WIRES

LENGTH OF SHEET

SPACE BETWEEN LONGITUDINAL WIRES

LONGITUDINAL WIRE END OVERHANG EQUALS ONE-HALF THE TRANSVERSE SPACE UNLESS OTHERWISE SPECIFIED

INDUSTRY METHOD OF DESIGNATING
STYLE: EXAMPLE- WWF 12 X12- W12 X W5

LONGITUDINAL WIRE SPACING 12"
TRANSVERSE WIRE SPACING 12"

LONGITUDINAL WIRE SIZE W 12"
TRANSVERSE WIRE SIZE W 5"

Figure 8.46 Identification of the parts of a sheet of welded wire fabric. *(Courtesy American Society for Testing and Materials. Reprinted with permission.)*

Table 8.3 Common Sheet Sizes for Plain and Deformed Welded Wire Fabric

	Customary in. ft.	Metric mm m		Customary in. ft.	Metric mm m
U.S. (except west coast)	96 × 12.5	2438 × 3.8	Canada	48 × 8	1219 × 2.4
	96 × 15	2438 × 4.6		96 × 12	2438 × 3.7
	96 × 20	2438 × 6.1		96 × 14	2438 × 4.3
U.S. (west coast)	84 × 20	2134 × 6.1		96 × 16	2438 × 4.9
	84 × 25	2134 × 7.6		96 × 20	2438 × 6.1

Courtesy Wire Reinforcement Institute

Table 8.4 Specifications Covering Welded Wire Fabric

U.S. Specification	Canadian Standard	Title[a]
ASTM A82	CSA G30.3	Cold-Drawn Plain Steel Wire for Concrete Reinforcement
ASTM A185	CSA G30.5	Welded Plain Steel Wire Fabric for Concrete Reinforcement
ASTM A496	CSA G30.14	Deformed Steel Wire for Concrete Reinforcement
ASTM A497	CSA G30.15	Welded Deformed Steel Wire Fabric for Concrete Reinforcement

[a]The titles of the American Society for Testing and Materials specifications and the Canadian Standards Association are identical.

Table 8.5 Common Styles of Welded Wire Fabric

Yield Strength (min.) (fy in psi)	Style Designation (W=Plain, D=Deformed)	Steel Area (in.²/1'-0") Longit.	Trans.	Metric[a] Style Designation
65,000 (W only)	4 × 4-W1.4 × W1.4	.042	.042	102 × 102 MW 9.1 × MW 9.1
	4 × 4-W2.0 × W2.0	.060	.060	102 × 102 MW 13.3 × MW 13.3
	6 × 6-W1.4 × W1.4	.028	.028	152 × 152 MW 9.1 × MW 9.1
	6 × 6-W2.0 × W2.0	.040	.040	152 × 152 MW 13.3 × MW 13.3
	4 × 4-W2.9 × W2.9	.087	.087	102 × 102 MW 18.7 × MW 18.7
	6 × 6-W2.9 × W2.9	.058	.058	152 × 152 MW 18.7 × MW 18.7
70,000 (W & D)	4 × 4-W/D 4 × W/D 4	.120	.120	102 × 102 MW 25.8 × MW 25.8
	6 × 6-W/D 4 × W/D 4	.080	.080	152 × 152 MW 25.8 × MW 25.8
	6 × 6-W/D 4.7 × W/D 4.7	.094	.094	152 × 152 MW 30.3 × MW 30.3
	12 × 12-W/D 9.4 × W/D 9.4	.094	.094	304 × 304 MW 60.6 × MW 60.6
72,500 (W & D)	6 × 6-W/D 8.1 × W/D 8.1	.162	.162	152 × 152 MW 52.3 × MW 52.3
	6 × 6-W/D 8.3 × W/D 8.3	.166	.166	152 × 152 MW 53.5 × MW 53.5
	12 × 12-W/D 9.1 × W/D 9.1	.091	.091	304 × 304 MW 58.7 × MW 58.7
	12 × 12-W/D 16.6 × W/D 16.6	.166	.166	304 × 304 MW 107.1 × MW 107.1
75,000 (W & D)	6 × 6-W/D 7.8 × W/D 7.8	.156	.156	152 × 152 MW 100.6 × MW 100.6
	6 × 6-W/D 8 × W/D 8	.160	.160	152 × 152 MW 51.6 × MW 51.6
	12 × 12-W/D 8.8 × W/D 8.8	.088	.088	304 × 304 MW 56.8 × MW 56.8
	12 × 12-W/D 16 × W/D 16	.160	.160	304 × 304 MW 103.2 × MW 103.2
80,000 (W & D)	6 × 6-W/D 7.4 × W/D 7.4	.148	.148	152 × 152 MW 95.5 × MW 95.5
	6 × 6-W/D 7.5 × W/D 7.5	.150	.150	152 × 152 MW 48.4 × MW 48.4
	12 × 12-W/D 8.3 × W/D 8.3	.083	.083	304 × 304 MW 53.5 × MW 53.5

[a]These are soft conversions from the inch sizes.
Courtesy Wire Reinforcement Institute

Table 8.6 Metric Wire Area, Diameter, and Mass

Size[a] (MW = Plain) (mm²)	Metric Units[a]		
	Area (mm²)	Diameter (mm)	Mass (kg/m)
MW200	200	16.0	1.57
MW130	130	12.9	1.02
MW120	120	12.4	.941
MW100	100	11.3	.784
MW90	90	10.7	.706
MW80	80	10.1	.627
MW70	70	9.4	.549
MW65	65	9.1	.510
MW60	60	8.7	.470
MW55	55	8.4	.431
MW50	50	8.0	.392
MW45	45	7.6	.353
MW40	40	7.1	.314
MW35	35	6.7	.274
MW30	30	6.2	.235
MW26	26	5.7	.204
MW25	25	5.6	.196
MW20	20	5.0	.157
MW19	19	4.9	.149
MW15	15	4.4	.118
MW13	13	4.1	.102
MW10	10	3.6	.078
MW9	9	3.4	.071

[a]True metric sizes are based on 5 and 10 square (mm²) increments per meter (m). Metric wire sizes can be specified in 1 mm² increments.
Available in MW and MD wire.
Courtesy Wire Reinforcement Institute

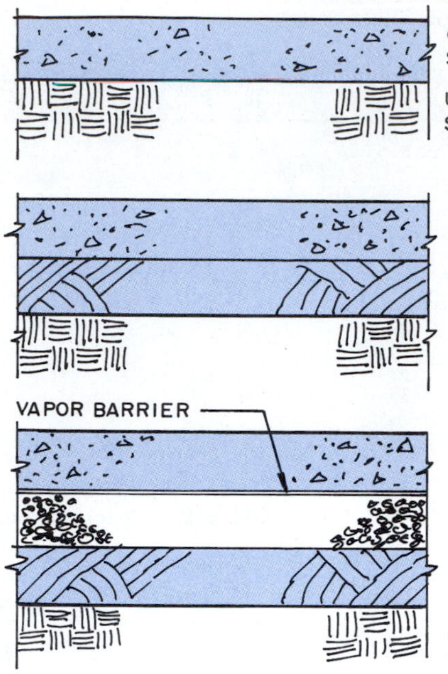

Figure 8.47 Methods for constructing on-grade concrete slabs.

CAST-IN-PLACE ON-GRADE CONCRETE SLABS

Casting On-Grade Slabs

On-grade concrete slabs require preparation of the slab bed. This varies with the type of soil, but often a base of compacted gravel is required over the soil. Some soils only require compacting before pouring the slab. To control moisture penetration a plastic sheet can be laid over the base. Some place 2 or 3 in. (50 to 75 mm) of compacted sand over the plastic sheet (Fig. 8.47). Finally, the reinforcing is placed over this base and held the required distance above the surface with concrete bricks or bar supports.

The common types of slabs include unreinforced, lightly reinforced, and structurally reinforced. *Unreinforced* slabs rely entirely upon the earth for support against tension forces. If subjected to excess loads, they could crack. As the slab cures, the surface may also crack due to shrinkage. *Lightly reinforced* slabs also depend on the earth for total support. However, they are reinforced with welded wire fabric or one of the fiber reinforcements. In some cases both are used. This helps hold together surface cracks that occur during curing. *Structurally reinforced* slabs contain steel reinforcing bars and often welded wire fabric and/or fiber reinforcing. The design of the reinforcing and thickness of the slab varies with the loads to be carried. Some designs depend on the soil for support, but the reinforcing helps control tensile stress. Other designs have sufficient reinforcing so that the slab can extend from one support, such as a foundation, to another without depending on the earth for support. These slabs are designed by a professional engineer. Selected examples are shown in Fig. 8.48.

Joints are necessary when building concrete slabs on grade to help control cracking, reduce the size of the pour, and separate the slab from surfaces where it should not bond. *Control joints* are used to provide a weakened place in the slab where it can crack (if necessary), and the crack will be be relatively hidden. This prevents cracks from running across slabs at all angles. Control joints are spaced 15 to 20 ft. (4.6 to 6.1 m) apart. They are formed by sawing into the slab after it has begun to harden or by placing molded strips into the concrete before it hardens.

Construction joints are used to separate a large area to be poured into smaller more manageable areas. They also help prevent cracking because each joint serves as a control joint, allowing for expansion and contraction within a large slab.

Isolation joints are used to keep a slab from bonding to some abutting part of the building. This permits each part to move independently. Typical isolation joints are ⅛ to ¼ in. thick (3 to 6 mm) asphalt impregnated fiber or molded plastic strips (Fig. 8.49).

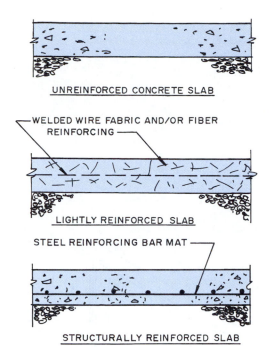

Figure 8.48 Various degrees of reinforcing used on concrete slabs.

Reinforcing Cast-In-Place Concrete Walls

Reinforced concrete cast-in-place walls may rest on a continuous concrete footing and be below grade, such as a basement wall, or extend above grade, forming the exterior or interior walls of a building. When the footing is poured, metal dowels are inserted that project above the top of the footing. Some footings have a key cast into them that forms a tie at the bottom of the wall (Fig. 8.50). One side of the wall form is set on the footing and braced. The vertical bars are wired to the dowels. The top of each vertical bar is wired to a horizontal bar, which maintains the spacing. Now the other horizontal bars are located and wired to the vertical bars, forming a grid. The bars usually are wired together every second or third bar. This forms a reinforcing wall mat. The mat is secured to the top of the form so that it remains in a vertical position during pouring (Fig. 8.51). Sometimes the mat is formed flat on the ground and then lifted and placed on the footing with a crane. If the design calls for a second mat, the mat closest to the form is completed first, then the second is built in the same manner.

If the wall is to be topped with a concrete slab floor or roof, the working drawings will show the required reinforcing for the connection. Typically, vertical reinforcing extends above the top of the wall and is bent to be cast into the slab (Fig. 8.52). Corners between walls also have to be tied together. Usually some form of a hook or elbow bar is used, as shown in Fig. 8.53. After all steel is

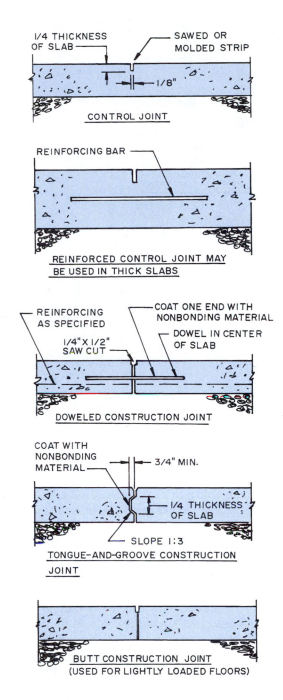
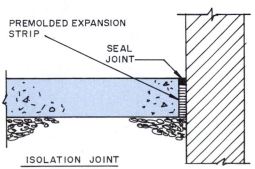

Figure 8.49 Commonly used construction joints.

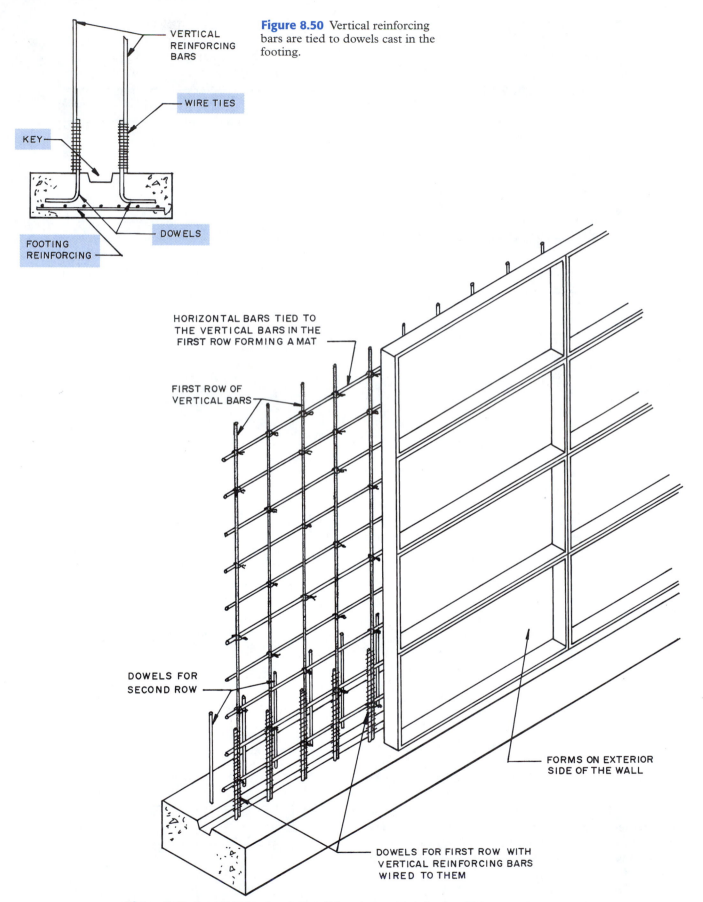

VERTICAL
REINFORCING
BARS

WIRE TIES

KEY

DOWELS

FOOTING
REINFORCING

Figure 8.50 Vertical reinforcing bars are tied to dowels cast in the footing.

HORIZONTAL BARS TIED TO
THE VERTICAL BARS IN THE
FIRST ROW FORMING A MAT

FIRST ROW OF
VERTICAL BARS

DOWELS FOR
SECOND ROW

FORMS ON EXTERIOR
SIDE OF THE WALL

DOWELS FOR FIRST ROW WITH
VERTICAL REINFORCING BARS
WIRED TO THEM

Figure 8.51 An example of typical wall forming and reinforcing. This example shows the first vertical mat in place.

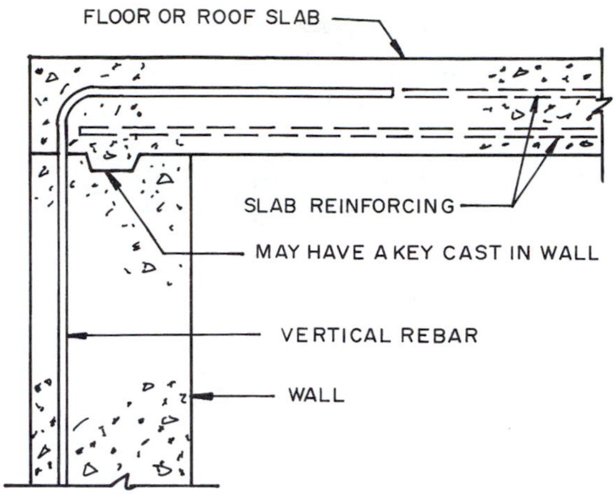

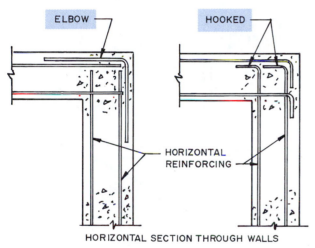

Figure 8.52 A typical wall-to-slab connection.

Figure 8.54 A straight wall form with reinforcing in place. The form supports a working deck, and the wall is being poured using a concrete pump. *(Courtesy PERI Formwork Systems, Inc.)*

Figure 8.53 A typical corner reinforcement for cast-in-place walls.

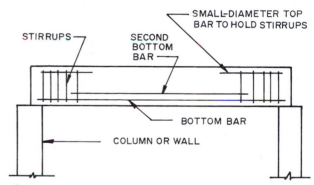

Figure 8.55 Typical reinforcing for a single-span cast-in-place beam.

in place, it should be rechecked before the other side of the form is set in place and braced. The wall is now ready to be poured (Fig. 8.54).

Reinforcing Cast-In-Place Beams

A simple single-span cast-in-place beam rests on end supports. One possible arrangement for placing the re-inforcing is shown in Fig. 8.55. The stirrups are held in place with a small-diameter top bar or the top member of a truss bar.

Continuous beam casting is common in structures with a cast-in-place concrete structural system. The beam ties together the structure from column to column. The bottom of the beam at midspan is in maximum tension, and this force dissipates toward the ends of the beam. The stirrups at the end transfer these tension forces to the concrete. Over the column, the top

of the beam is subject to tension forces due to bending, so appropriate top bars are placed over each column (Fig. 8.56).

Reinforcing Cast-In-Place Columns

Cast-in-place concrete columns are typically round, square, or rectangular (Fig. 8.57). They are reinforced with vertical bars, which help carry compressive loads and resist tension forces due to lateral loads, such as the load caused by a high wind. The vertical bars may be arranged in a square or rectangular pattern or a circular pattern. Either pattern can be used for square and circular columns.

Tied columns have vertical bars tied together with small-diameter smooth steel bars placed horizontally and wired to the vertical bars (Fig. 8.58). These ties hold the vertical bars in position during the pour and help

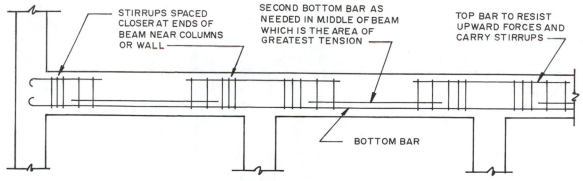

Figure 8.56 Typical reinforcing for a cast-in-place continuous beam.

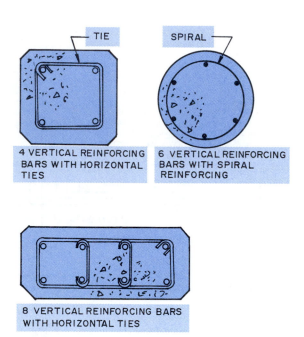

Figure 8.57 Cast-in-place concrete columns are commonly round, square, or rectangular.

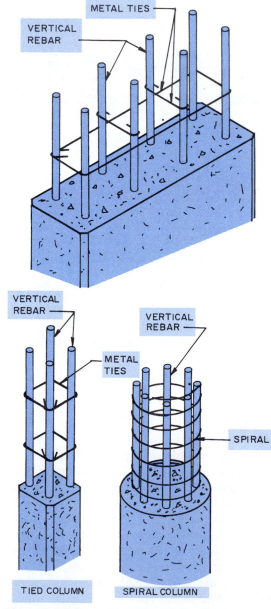

Figure 8.58 Typical column construction.

them resist outward buckling when under load. *Spiral* columns act in the same manner as tied columns except they also restrain the concrete inside them. This enables the column to bend or bow under load rather than cracking.

The size of vertical bars, their spacing, and the size and spacing of ties or spirals are indicated on the working drawings by the structural engineer. After the reinforcing is in place in the column form, the concrete is poured (Fig. 8.59).

Figure 8.59 These columns are about to be poured from a concrete bucket.

CAST-IN-PLACE CONCRETE FRAMING SYSTEMS

A typical cast-in-place concrete framing system utilizes cast-in-place columns, one-way or two-way concrete slabs, and cast-in-place joists, beams, and girders. The exterior can be finished in a number of ways, including precast concrete spandrel panels. The flat slab concrete floor and roof slabs are among the simplest to form, reinforce, and pour. A generic example is shown in Fig. 8.60.

Several types of reinforced concrete framing systems are used with cast-in-place concrete construction. In some cases the beams and girders are cast with the floor or roof slab. The two commonly used reinforced concrete floor and roof systems are the flat slab and the flat plate. The *flat slab* is a concrete slab reinforced in two or more directions and supported by columns with dropped panels and capitals, which enlarge the columns at the top, or by beams or joists. When two-way reinforcement is used in the slab, it is called a two-way flat slab. A *flat plate* floor or roof is much like the flat slab. Its reinforcing runs in two directions, but it is supported by columns that are not enlarged where they meet the slab.

General Construction Procedures

After the foundation is in place, the load-bearing walls and columns are poured (Fig. 8.61). When they have cured enough to carry the floor or roof load, the forms for the slab are built. Generally, the beams are formed and poured with the floor, forming a monolithic unit (Fig. 8.62).

The forms are supported with temporary joists and beams of wood or metal, which are supported by temporary shores. A *shore* is a column whose length can be adjusted. The entire temporary form support system must be designed by an engineer so it can carry the required loads.

Sharp corners on concrete members are hard to cast and tend to break away if struck, leaving a ragged edge. Therefore, wood or plastic inserts are placed in the form to produce a beveled or rounded corner (Fig. 8.63). The interior surfaces of the forms are coated with a form-release

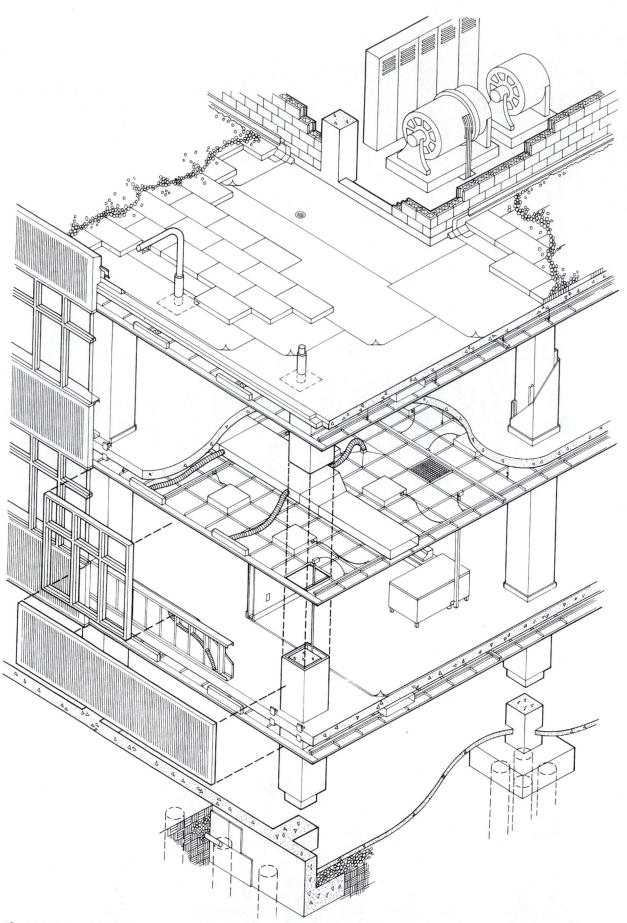

Figure 8.60 A typical flat plate cast-in-place concrete structure. (*Reproduced with permission from* The Building Systems Integration Handbook, *Richard Rush, ed., Butterworth-Heinmann Publisher, Stoneham, Mass., 1986.*)

Figure 8.61 Pouring a column formed with a molded plastic column form. Notice the bracing attached to the metal ring around the column. *(Courtesy Molded Fiberglass Concrete Forms Co.)*

compound, which prevents the concrete from bonding to the form and aids in removing the form with minimum damage to the concrete member and the form. Forms are expensive and must be preserved for reuse. Finally, the reinforcing steel is placed as specified by the engineer's placing drawings.

Before the pour begins, the entire assembly must be inspected and approved. As mentioned earlier, usually the beams, girders, joists, and floor or roof slab are poured monolithically. The exposed surface of the slab is finished as specified. This is the same procedure as described earlier for on-grade slabs. If additional floors are to be built above the one poured, there will be reinforcing protruding through the slab to be joined to the reinforcing in the columns on the next floor or hooked to lay into the roof slab. After the structure has reached sufficient strength, the formwork for the next floor is constructed. The formwork for the lower floor is stripped, and the beams and slabs are reshored with temporary posts to provide added support.

One-Way Flat Slab Floor and Roof Construction

The one-way flat slab systems in common use include a one-way solid slab, one-way flat slab with beams and girders, and one-way flat slab with joists and beams.

One-Way Solid Slab Construction

One-way solid slab construction has the slab supported on two sides by beams or load-bearing walls, as shown in Fig. 8.64. The main beams are on one axis, and the slab spans the distance between them. The reinforcing bars or prestressed tendons are placed perpendicular to the supporting walls or beams.

One-Way Flat Slab with Beams and Girders

Another one-way slab construction is shown in Fig. 8.65. It utilizes a one-way flat slab with beams and girders. The construction follows the same procedures described for one-way solid slab construction, but this slab is supported by cast-in-place concrete beams that are supported by cast-in-place girders. The girders rest on columns. The slab spans the distance between beams. Usually the girders, beams, and slab are cast monolithically.

One-Way Flat Slab with Joists and Beams

Another widely used system is a one-way flat slab with joists and beams (Fig. 8.66). The joists span the beams and support the slab. The beams are supported by the columns. This is often used when the span between beams is not large. As the span increases, the thickness of the slab increases, which adds to the total weight.

The beams, joists, and slab are cast monolithically. The joists are formed with molded fiberglass or metal domes that rest on the temporary framing and shoring (Fig. 8.67). When the forms are removed, the exposed ceiling appears as shown in Fig. 8.68. This can be treated in many ways. For example, it can be painted, left natural, sprayed with acoustical plaster, or covered with a suspended ceiling.

Two-Way Flat Slab Floor and Roof Construction

As mentioned earlier, the two-way flat slab has reinforcing running in two (perpendicular) or more directions. The systems in general use are the two-way flat slab, two-way solid slab, two-way flat plate, and the two-way joist (or waffle) slab. This construction enables a building structure to be designed in nearly square bays.

Two-Way Flat Slab Construction

Two-way flat slab construction utilizes a flat slab supported by thickening the slab at each column with a drop panel or a drop panel and a capital (Fig. 8.69).

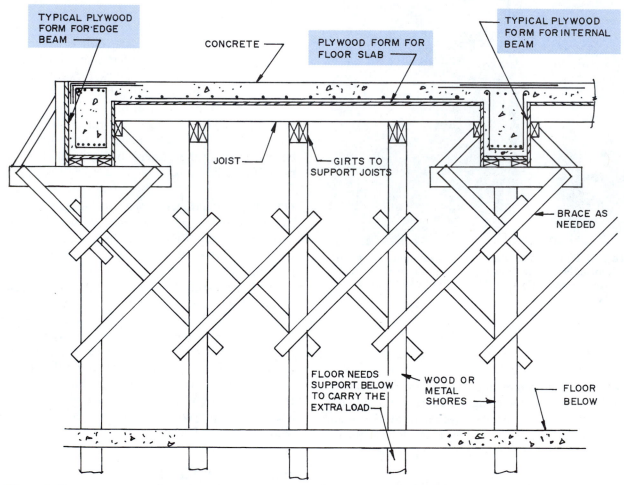

Figure 8.62 A generalized example of forms used to cast a flat slab and beams monolithically. Other types of support are available. The actual size and design of the supporting structure must be prepared by an engineer.

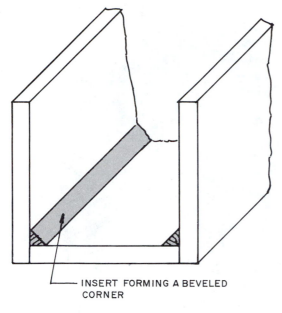

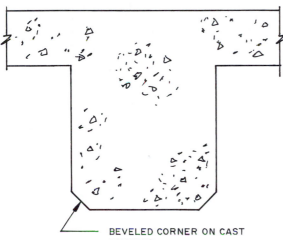

Figure 8.63 Inserts are placed in the corners of the forms to produce beveled corners.

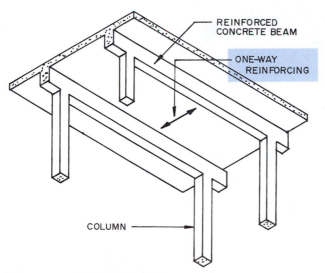

Figure 8.64 One-way solid slab with beam construction has the main beams on one axis. The slab spans the distance between beams.

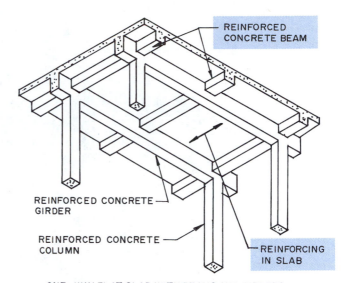

Figure 8.65 A one-way flat slab with beams and girders has a slab that spans the beams, which are supported by the girders.

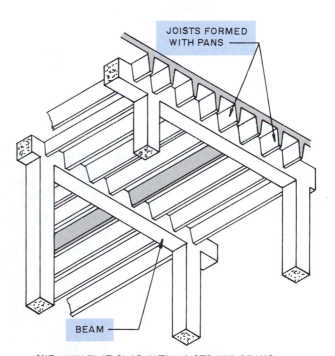

Figure 8.66 A one-way flat slab with joists and beams has joists that span the distance between the beams.

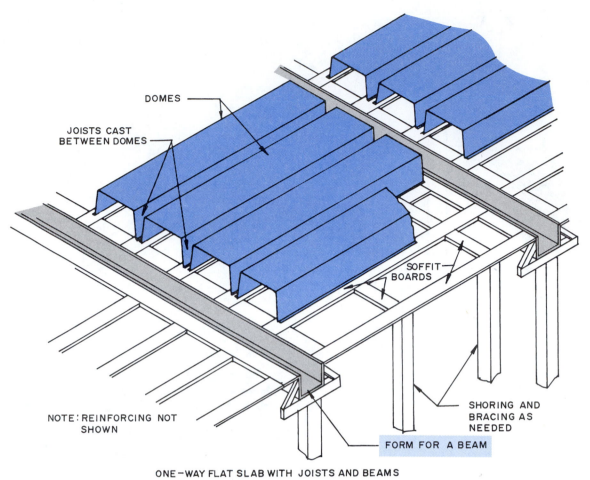

DOMES

JOISTS CAST
BETWEEN DOMES

SOFFIT
BOARDS

NOTE: REINFORCING NOT
SHOWN

SHORING AND
BRACING AS
NEEDED

FORM FOR A BEAM

ONE-WAY FLAT SLAB WITH JOISTS AND BEAMS

Figure 8.67 The beams, joists, and flat slab are cast monolithically using domes to form the joists and the base for the floor.

Figure 8.68 The ceiling formed with a cast-in-place flat slab with joists and beam construction. *(Courtesy Molded Fiberglass Concrete Forms Co.)*

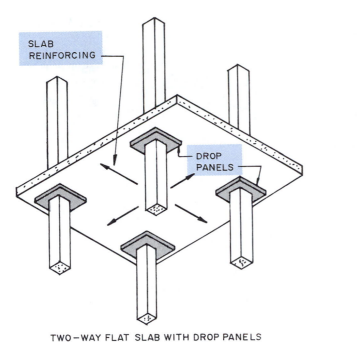

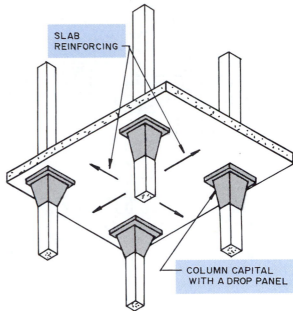

Figure 8.69 Two-way flat slab construction using columns with drop panels, and drop panels with a column capital.

The slab reinforcing runs in two directions. The increased thickness at each column helps resist the shear forces at the top of the column. This system is useful for buildings that will have heavy loads. In buildings designed for lighter loads, the thickness of the slab above the column can be reduced. A simplified partial plan view and sections are in Fig. 8.70.

The slab reinforcing is divided into column strips and middle strips. Each is about one-half span. The column strips carry the bend forces developed in the area over the column, while the middle strips reinforce the slab in the area between columns. The slab reinforcement is shown on the engineering drawings. The drop panel usually has no additional reinforcing.

Two-Way Solid Slab

The two-way solid slab has the flat slab supported by beams that run between columns. The slab and beams are cast monolithically (Fig. 8.71). This construction can be used for buildings designed to carry heavy loads. The slab is reinforced in two directions.

Two-Way Flat Plate

The two-way flat plate uses the same general construction as the flat slab, but it has no drop panel or capital on the top of each column (Fig. 8.72). A simplified partial plan view of a two-way flat plate is in Fig. 8.73. Notice the sections showing typical reinforcing of the column strip and middle strip, each of which is about one-half the span. The design is very similar to the flat slab.

Two-Way Joist (Waffle) Flat Slab

The two-way joist framing system consists of a series of concrete joists cast at right angles to each other to form a grid. The slab spans the distances between the joists and they are poured monolithically (Fig. 8.74). The system is designed around a series of columns, usually equally spaced. The joists are formed by domes placed on a wood deck that is supported by shoring. Areas over the columns are left open so a solid concrete slab area called a column head can be poured (Fig. 8.75).

A simplified plan view and sections showing the layout of the joists and columns and typical details for the column strip and middle strip are in Fig. 8.76. After the pour has reached sufficient strength, the supporting structure is removed and the domes dropped out (Fig. 8.77). The exposed ceiling may be finished in several ways or covered with a suspended ceiling (Fig. 8.78).

Tilt-Up Wall Construction

Tilt-up concrete construction is a form of site-cast concrete. Wall panels are cast in a horizontal position near the building or, most frequently, on the concrete on-grade floor slab. A bond-breaking agent is placed on the slab so the wall panel will not adhere to it. The surfaces of the panel can be finished in many ways—textured (from a form liner), washed to expose embedded aggregate, troweled smooth, or painted.

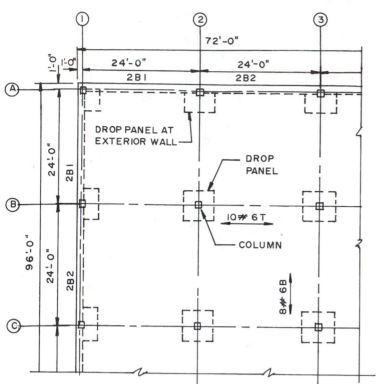

TYPICAL FRAMING PLAN SHOWING COLUMNS AND DROP PANELS

NOTE: DIMENSIONS ON THESE DRAWINGS ARE FOR ILLUSTRATION PURPOSES ONLY.

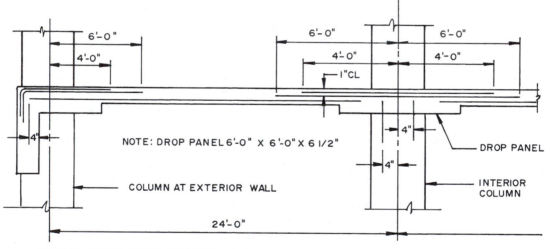

NOTE: DROP PANEL 6'-0" X 6'-0" X 6 1/2"

COLUMN AT EXTERIOR WALL

DROP PANEL

INTERIOR COLUMN

TYPICAL DETAIL FOR DROP PANEL AND FLAT SLAB CONSTRUCTION

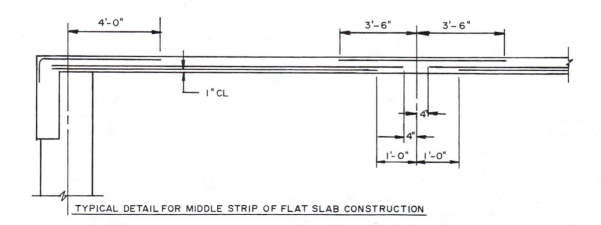

TYPICAL DETAIL FOR MIDDLE STRIP OF FLAT SLAB CONSTRUCTION

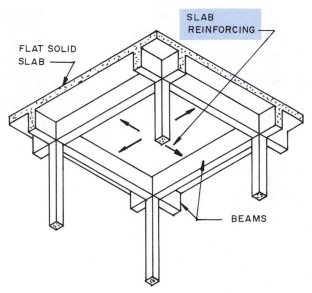

SLAB REINFORCING

FLAT SOLID SLAB

BEAMS

TWO—WAY SOLID SLAB AND BEAM

Figure 8.71 A two-way solid slab and beam is supported by beams running between the columns.

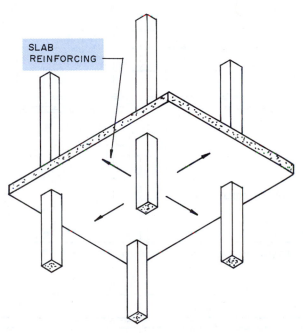

SLAB REINFORCING

TWO-WAY FLAT PLATE

Figure 8.72 This two-way flat plate does not have drop panels or column capitals.

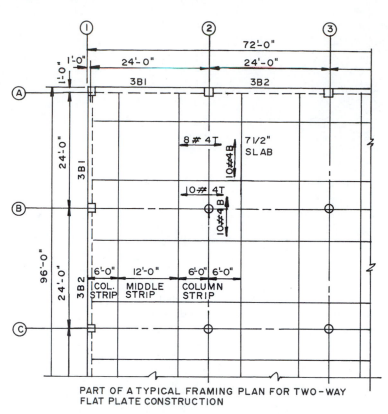

Figure 8.73 A partial drawing showing typical details for two-way flat plate construction.

PART OF A TYPICAL FRAMING PLAN FOR TWO-WAY FLAT PLATE CONSTRUCTION

NOTE: DIMENSIONS ARE FOR ILLUSTRATION PURPOSES ONLY.

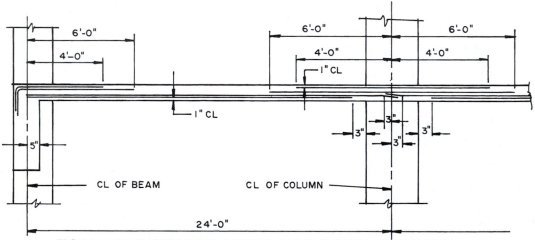

TYPICAL COLUMN STRIP DETAIL FOR TWO-WAY FLAT PLATE CONSTRUCTION

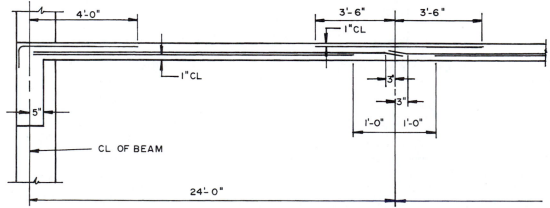

TYPICAL MIDDLE STRIP DETAIL FOR TWO-WAY FLAT PLATE CONSTRUCTION

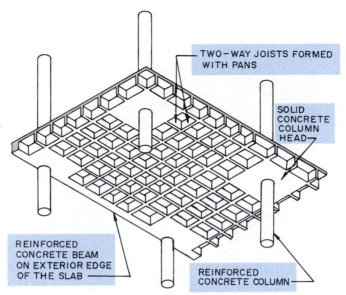

TWO-WAY JOISTS FORMED WITH PANS

SOLID CONCRETE COLUMN HEAD

REINFORCED CONCRETE BEAM ON EXTERIOR EDGE OF THE SLAB

REINFORCED CONCRETE COLUMN

TWO-WAY JOISTS WITHOUT BEAMS (WAFFLE FLAT PLATE CONSTRUCTION)

Figure 8.74 The two-way joist framing system has cast-in-place joists at right angles to each other, forming a wafflelike ceiling.

Figure 8.75 These fiberglass domes are set on a wood platform supported by shoring. (*Courtesy Molded Fiberglass Concrete Forms Co.*)

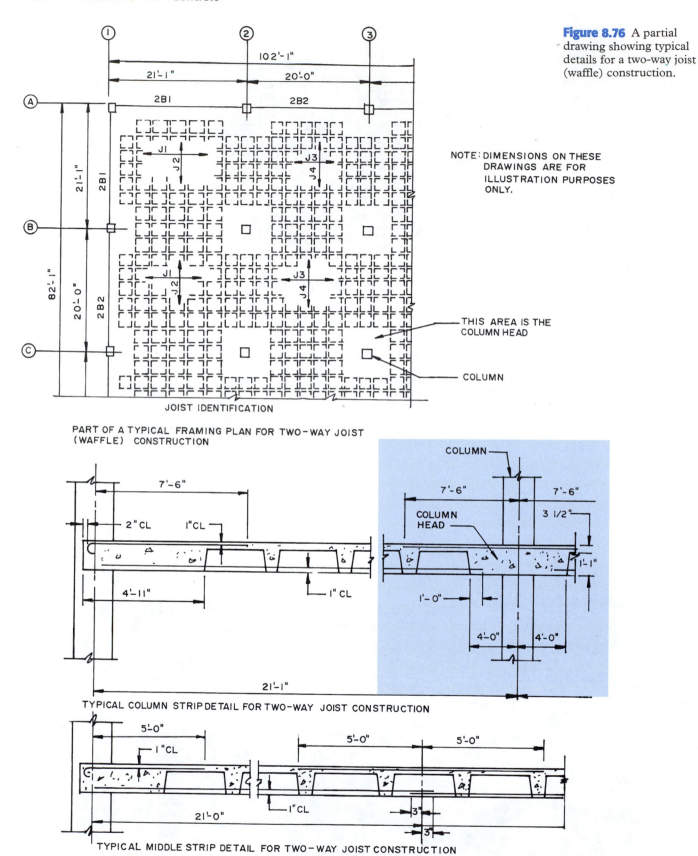

Figure 8.76 A partial drawing showing typical details for a two-way joist (waffle) construction.

NOTE: DIMENSIONS ON THESE DRAWINGS ARE FOR ILLUSTRATION PURPOSES ONLY.

THIS AREA IS THE COLUMN HEAD

COLUMN

JOIST IDENTIFICATION

PART OF A TYPICAL FRAMING PLAN FOR TWO-WAY JOIST (WAFFLE) CONSTRUCTION

TYPICAL COLUMN STRIP DETAIL FOR TWO-WAY JOIST CONSTRUCTION

TYPICAL MIDDLE STRIP DETAIL FOR TWO-WAY JOIST CONSTRUCTION

Figure 8.77 These molded fiberglass domes are being removed after the temporary supports have been removed. *(Courtesy Molded Fiberglass Concrete Forms Co.)*

Figure 8.79 This site-cast concrete wall is being tilted up into a vertical position. *(Courtesy Portland Cement Association)*

Figure 8.78 The interior concrete ceiling after the molded fiberglass forms have been removed. *(Courtesy Molded Fiberglass Concrete Forms Co.)*

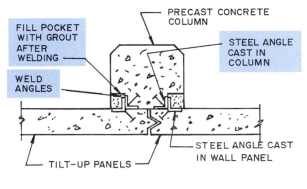

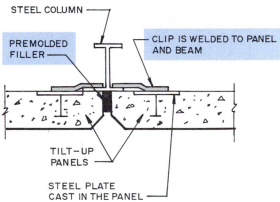

Figure 8.80 Some of the ways tilt-up wall panels can be connected to columns.

After the wall panel has developed sufficient strength, it is lifted into position with a crane and secured to the footing (Fig. 8.79). The panels are temporarily braced until all wall and roof structural members are in place. The panels are secured to columns, which may be precast concrete, steel, or concrete that is cast-in-place after the walls are up (Fig. 8.80).

Panels generally range in thickness from about 5½ to 7½ in. (140 to 190 mm) and may be several stories high. They may contain openings for windows and doors. Key to success is the selection and placement of steel reinforcing, which not only must serve when the wall is erect and carrying a roof or floor load but also must withstand the bending loads generated when the panel is tilted and lifted into place. Pick up hangers must be carefully located in the panel so the crane can attach to and lift the wall. The design of the panels must be made by an experienced engineer.

Generally, the wall panels are load bearing and will support floors and roofs. Typical decking includes hollow-core units, single or double tees, and metal open-web joists. Common designs are shown in Fig. 8.81.

Panels have ¾ in. (18 mm) chamfers around the outside edges. Sharp corners are hard to cast and break off easily when hit. The wall panels usually are not connected to each other but have a ¾ in. (18 mm) space sealed with backer rods and a weatherproof sealant (Fig. 8.82). A backer rod is a flexible, compressible rod made of foam plastic. It is placed in the joint to limit the depth to which a sealant can enter the joint. The wall panels are

tied together by the roof framing and floor slab. Each panel acts as an independent member. A typical wall-floor connection is made by leaving out a section of the floor along the wall when the on-grade slab is cast. After the wall is up this section is cast, tying the wall to the floor (Fig. 8.83). Corner joints may be butted or butt a column (Fig. 8.84).

The wall panels may be supported by continuous footings, isolated footings, or caissons. They are aligned with dowels or metal angles. The void between the bottom of the panel and the top of the footing is filled with grout.

A generic example of a complete assembly using open-web joists is shown in Fig. 8.85. Notice that on the floor is the form used to cast the wall panel with a window opening. The bar joists rest on the top of the wall panels. The roof consists of metal decking covered with rigid insulation sheets and a built-up finished roof covering.

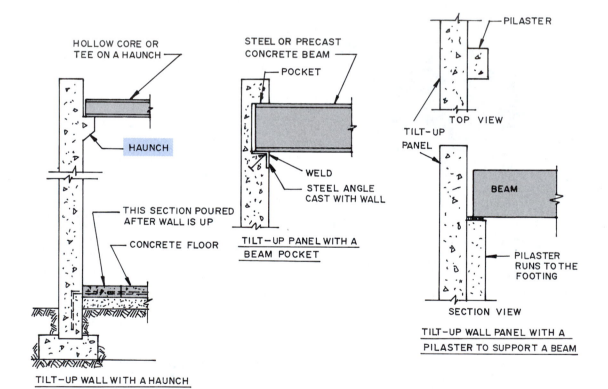

Figure 8.81 These load-bearing tilt-up wall panels are designed to carry floor and roof loads.

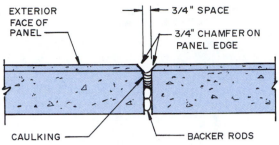

EXTERIOR FACE OF PANEL — 3/4" SPACE — 3/4" CHAMFER ON PANEL EDGE — CAULKING — BACKER RODS

Figure 8.82 Butting tilt-up wall panels can be sealed with backer rods and a weatherproof caulking.

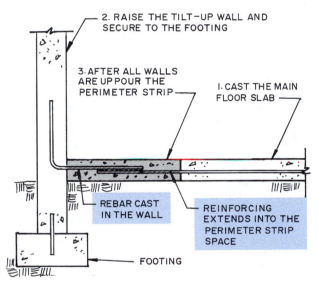

Figure 8.83 One way to tie the wall to the floor slab.

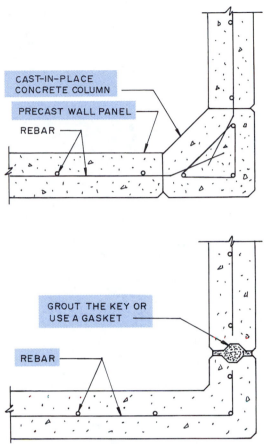

Figure 8.84 Typical ways to form corners on tilt-up construction.

Lift-Slab Construction

Lift-slab construction involves casting a reinforced concrete slab for each floor of the building and the roof one on top of the other. The ground floor on-grade slab is cast first. It is coated with a bond release compound that prevents the next slab that is cast on top of it from bonding to it. Slabs are cast for each floor and for the roof, if it is to be a concrete slab. Each is coated with a bond release compound. The slabs can be reinforced concrete flat slabs or prestressed. They are usually two-way flat slabs.

The slabs are lifted by hydraulic jacks mounted on the tops of the columns. They slide up to the desired elevation and are secured to each column by welding,

bolting, or a series of pins (Fig. 8.86). This is repeated for each floor.

Slip-forming is used to cast-in-place tall structures such as grain elevators and stairwells. It involves constructing the formwork, which is then slowly lifted upward by jacks supported by vertical steel rods. The forms are filled with concrete and reinforcing bars as the form is raised. The concrete gains sufficient strength to carry the weight of the newly poured concrete as the form slowly moves upward. Keys to success are a proper concrete mix and the equipment to lift it to high levels as the structure increases in height (Fig. 8.87).

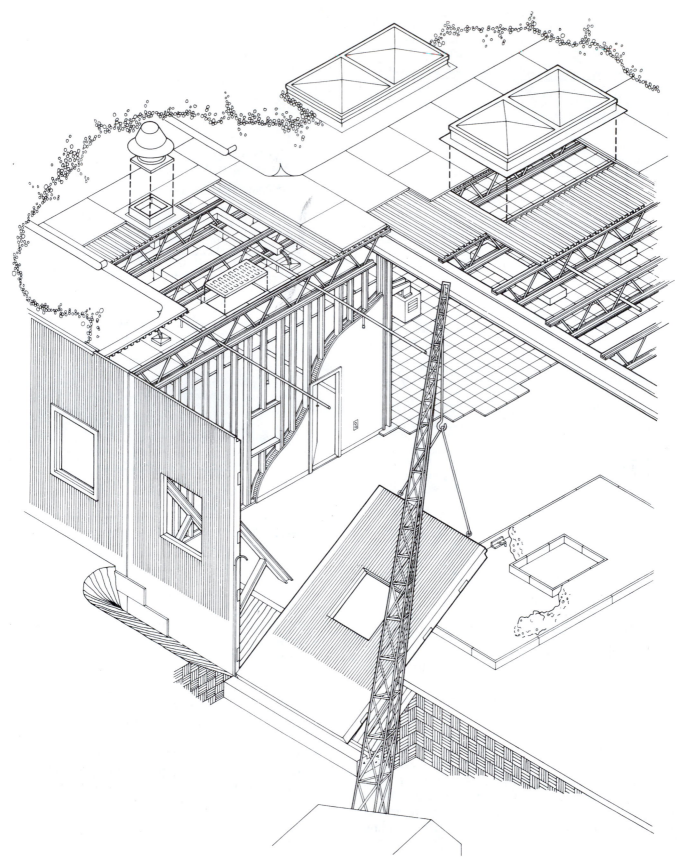

Figure 8.85 A generic example of tilt-up construction. *(Reproduced with permission from* The Building Systems Integration Handbook, *Richard Rush, ed., Butterworth-Heinmann Publishers, Newton, Mass., 1986.)*

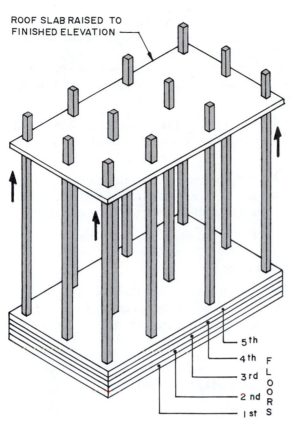

Figure 8.86 Lift-slab construction casts the floor and roof slabs on grade and lifts them to their required elevation.

Figure 8.87 Tower cranes are used to lift concrete to high levels. (*Courtesy Portland Cement Company*)

REVIEW QUESTIONS

1. How long can ready-mix concrete remain in the mixer and still be used?
2. What devices are used on the site to move concrete to the point of placement?
3. How is concrete consolidated?
4. What may happen if the temperature of the concrete mix exceeds 100°F (38°C)?
5. Why are freshly poured concrete slabs bullfloated and darbied?
6. What is the difference between an isolation joint and a construction joint?
7. What is meant by hydration?
8. What are the curing methods in common use?
9. What is the purpose of form liners?
10. How is the tensile strength of concrete beams increased?
11. What two styles of reinforcing bars are used?
12. What is the diameter in inches of a No. 8 bar?
13. What purpose do stirrups serve when installing reinforcing bars in a form?
14. How can you decide how many inches of concrete are needed to give fire protection to reinforcing bars?
15. What styles of wire are used to make welded wire reinforcement?
16. What types of coatings are used on welded wire reinforcement?
17. What is meant by the following specification: WWR − 6 × 6 − W8.1 × W8.1?
18. What types of fibers are used in the production of fiber reinforced concrete?
19. What is the difference between lightly reinforced and structurally reinforced concrete slabs?
20. What is the purpose of using dropped panels and capitals when casting flat concrete slabs?
21. What is meant by tilt-up construction?
22. What is unique about the slip-forming technique?

KEY TERMS

batch   An amount of concrete mixed at one time.

bleed   Water that rises to the surface of concrete shortly after it has been placed in the form.

cast-in-place concrete   Concrete members formed and poured on the building site.

concrete pump   A pump that moves concrete through hoses to the place it is to be placed.

consolidation   The process of compacting freshly placed concrete in the forms and around reinforcing steel to remove air pockets and pockets of stone.

darby   A handheld straightedge several feet long used to level the surface of freshly poured concrete before it is floated.

float   A flat hand tool used to smooth the surface of freshly poured concrete after it has been leveled with a darby.

hydration   A chemical reaction between the water and portland cement that bonds the aggregate into a solid mass.

jointing   Forming control joints in a slab.

ready-mix   Concrete mixed in a central plant and delivered to the site by truck.

screed   A tool used to strike off the surface of concrete so it is flush with the top of the form.

screeding   The process of striking off the surface of freshly poured concrete with a screed so it is flush with the surface of the form.

segregation   The tendency of coarse aggregate to separate from the other ingredients so the mixture is no longer homogeneous.

troweling   Producing a final smooth finish on freshly poured concrete with a steel bladed tool after it has been floated.

SUGGESTED ACTIVITIES

1. Visit a construction site where a cast-in-place concrete structural frame is being erected. Prepare a report giving full details about what you saw. For example, what types of forms and reinforcing were used? How were they pouring the concrete, and was it site-mixed or from a batch plant? How often are concrete samples taken from the delivery truck and on-site mixed batches? Who does the testing? What type of structural system was used?

2. Review the local building codes and list briefly the regulations relating to cast-in-place concrete structural systems.

3. Observe the pouring of floor and roof slabs and report on the use of various joints observed, the surface finish, and the protection used during the curing phase.

4. Collect small samples of concrete reinforcing materials. Prepare a display and lead a discussion on reinforcing techniques.

5. If any tilt-up wall construction is being done in your area, try to get some photos and prepare a sketch of a panel with the reinforcing and panel-lifting devices.

ADDITIONAL INFORMATION

Newman, M., *Structural Details for Concrete Construction*, McGraw-Hill, New York, 1994.

Ramsey, C.G., and Sleeper, H.R., Hoke, J.R., eds., *Architectural Graphic Standards*, John Wiley and Sons, New York, 1991.

Other resources include

ACI Building Code Requirements for Structural Concrete and Commentary, American Concrete Institute, Detroit, Mich.

Numerous publications from the Portland Cement Association, 5420 Old Orchard Road, Skokie, Ill. 60077-1083.

Publications of the Wire Reinforcement Institute, 2911 Gleneagle Drive, Findlay, Ohio 45840.

Technical publications from the American Concrete Institute, P.O. Box 19150, Detroit, Mich. 48219-0150.

Technical publications from the Concrete Reinforcing Steel Institute, 933 North Plum Grove Road, Schaumburg, Ill. 60173.

Precast Concrete

This chapter will help you to:

1. Describe and identify the major groups of precast concrete units and the members within each.

2. Explain how precast units are manufactured.

3. Cite the advantages and limitations of using precast concrete structural units.

4. Explain the differences between prestressed and nonprestressed precast concrete units.

5. Explain the differences between pretensioned and posttensioned structural concrete units.

6. Describe the various types of precast concrete slab units and their applications.

7. Be aware of the types and typical sizes of precast concrete columns, beams, girders, and wall panels.

8. Discuss the procedure for erecting precast concrete units and the types of connections used.

Structural precast concrete units are used to form the structural system of a building. These include columns, beams, girders, floor and roof slabs, and exterior and interior wall panels (Fig. 9.1). They are cast under factory-controlled conditions and moved to the job site.

Precast concrete units can be classified into two major groups—*precast structural concrete units* and *precast architectural units.* Typical precast structural units include floor and roof slabs, beams, girders, columns, and wall panels. Precast architectural concrete units include wall panels and other features. These may or may not be structural.

Precast units are cast in *beds,* which are permanent forms made of metal, wood, or fiberglass. The quality of the surface of the form produces the finished surface of the precast unit. Both pretensioned and posttensioned units can be produced (Fig. 9.2).

ADVANTAGES OF PRECAST CONCRETE UNITS

Precast construction is widely used because it affords a number of advantages when compared with **cast-in-place** concrete. These advantages include the following:

1. Casting takes place in computer-controlled facilities where an experienced crew produces units under close supervision.
2. The control of materials, mixing, and placing the concrete is rigid, producing higher quality concrete. Samples of the concrete are taken regularly and tested per ASTM standards. Typically, 5000 psi (35 MN/m^3) concrete is used.
3. Units are finished and cured under carefully controlled conditions. The concrete is mechanically vibrated in the form and cured. To speed up production, the units are covered with insulation blankets and are steam cured (autoclaved). When Type III, high early strength, portland cement is used, the units can be removed from the form after twenty-four hours (Fig. 9.3).
4. Forms are on the ground, making the placing of steel reinforcing, casting, and curing faster and easier.

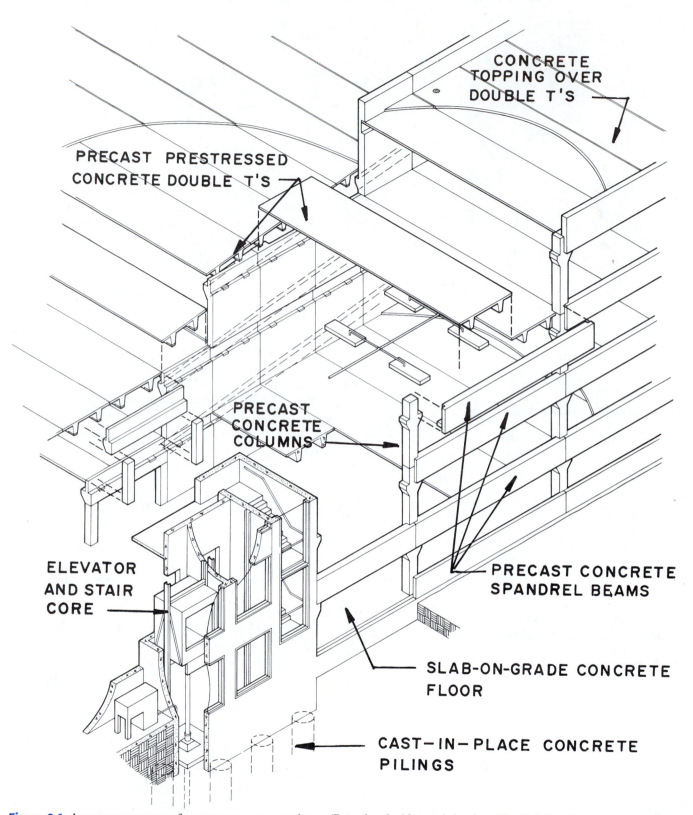

CONCRETE
TOPPING OVER
DOUBLE T'S

PRECAST PRESTRESSED
CONCRETE DOUBLE T'S

PRECAST
CONCRETE
COLUMNS

ELEVATOR
AND STAIR
CORE

PRECAST CONCRETE
SPANDREL BEAMS

SLAB-ON-GRADE CONCRETE
FLOOR

CAST-IN-PLACE CONCRETE
PILINGS

Figure 9.1 A structure system of precast concrete members. *(Reproduced with permission from* The Building Systems Integration Handbook, *Richard Rush, ed., Butterworth-Heinmann Publishers, Newton, Mass., 1986.)*

Figure 9.2 A cast concrete unit being removed from the casting bed. (*Courtesy Portland Cement Association*)

Figure 9.3 Precast units are removed from the casting bed and stored until they have sufficient strength to be moved to the site. (*Courtesy Portland Cement Association*)

5. The forms are used over and over, reducing these costs.

6. Generally, precast units can be smaller and lighter than cast-in-place units for the same span and loads.

7. Members can be cast and stored until needed, speeding up the erection process. They are erected with a crane, much the same as structural steel.

8. Inclement weather does not slow down precast construction as easily as it does cast-in-place jobs.

After the units have achieved adequate strength, they are delivered to the job site, usually by truck. A limiting factor is that their size is determined by what can be moved by truck or railroad. Highway regulations limit load sizes and weights. Extra-large precast units can be cast by moving the forms to the job and casting them on the site.

BUILDING CODES

Building codes have extensive specifications pertaining to the design and installation of precast concrete members. In addition to engineering design data, the codes refer to areas such as reinforcements, connections, lifting devices, fabrication, shop drawings, tendon anchorage, and grout. The codes also specify the requirements for meeting fire codes.

NONPRESTRESSED AND PRESTRESSED PRECAST UNITS

Nonprestressed units are cast in molds in a plant, then cured and shipped to the job site. They are reinforced in the same manner as cast-in-place concrete and are not under tension forces (Fig. 9.4). Some large or unusual beams, girders, and many columns and lintels are cast this way.

PRESTRESSED PRECAST CONCRETE

Prestressed concrete units have stresses introduced before they are placed under a load. There are two types: pretensioned and posttensioned.

Pretensioned Units

Pretensioning is used to produce units that are standardized in size or when enough identical units are needed to make it economically possible to cast them. They are cast in forms, some as long as 800 ft. (250 m) or more. High tensile strength steel reinforcing strands are stretched from one end of the form to the other. One end is anchored to a large fixed abutment. The other end passes through the abutment on the other end and has hydraulic stressing equipment fastened to it. The desired design stress is applied to the cable by pulling it in tension. When the design tension is reached, the cable is clamped to the abutment.

After the strand is stressed, the concrete is placed in the form. As it hardens it bonds to the strands. After twenty-four hours samples of the concrete are taken and tested. If they have attained the required strength, the strands are cut from the abutments. The resulting forces produced by the strands prestress the concrete. The units are lifted from the bed and moved to storage, from which they are shipped to job sites. The forms are now ready for laying up another pretensioned member.

A comparison of nonprestressed (Fig. 9.5) and prestressed (Fig. 9.6) concrete shows that the nonpre-

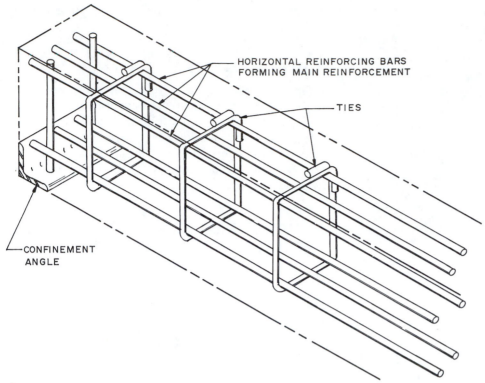

Figure 9.4 A typical reinforcing cage for a precast nonprestressed concrete beam.

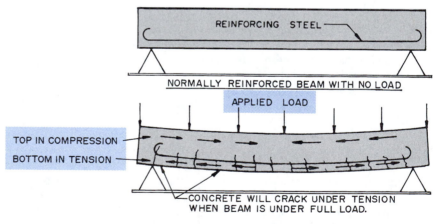

Figure 9.5 A cast-in-place concrete beam will deflect under load and possibly crack on the bottom because concrete cannot resist the tension forces.

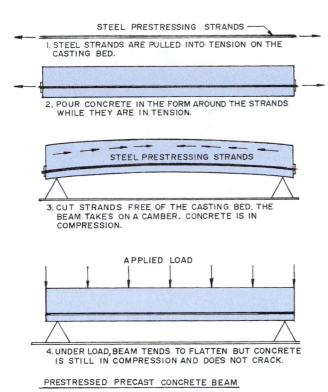

1. STEEL STRANDS ARE PULLED INTO TENSION ON THE CASTING BED.

2. POUR CONCRETE IN THE FORM AROUND THE STRANDS WHILE THEY ARE IN TENSION.

STEEL PRESTRESSING STRANDS

3. CUT STRANDS FREE OF THE CASTING BED. THE BEAM TAKES ON A CAMBER. CONCRETE IS IN COMPRESSION.

APPLIED LOAD

4. UNDER LOAD, BEAM TENDS TO FLATTEN BUT CONCRETE IS STILL IN COMPRESSION AND DOES NOT CRACK.

PRESTRESSED PRECAST CONCRETE BEAM

Figure 9.6 A prestressed concrete beam has a slight camber, which places the concrete in compression. When under load the concrete is still in compression.

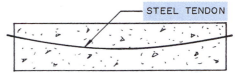

1. CONCRETE IS POURED OVER A STEEL TENDON THAT IS GREASED OR IN A TUBE PREVENTING IT FROM BONDING TO THE CONCRETE. THE TENDON IS DRAPED.

2. THE CONCRETE IS ALLOWED TO CURE. THEN THE TENDONS ARE TENSIONED AND ANCHORED TO THE ENDS OF THE BEAM PRODUCING SOME CAMBER.

Figure 9.7 Posttensioned precast concrete members develop camber by applying tension after the member is cast and cured.

stressed unit has no **camber** (arch) and deflects slightly under load, but the pretensioned member has some camber due to the tension in the strands. When under load, a pretensioned member tends to move toward a level position. The stressed strands produce high compressive stresses in the lower part of the unit and a tensile stress in the upper part. As the unit is loaded these stresses are reduced.

Posttensioned Units

Posttensioning involves applying stresses to the concrete unit after it has been cast and hardened. It is used primarily with cast-in-place concrete but finds some use with precast concrete units.

Posttensioning involves casting the unit using normal reinforcement and, in addition, running several posttensioning strands through the unit in such a way that the concrete does not bond to them as it sets. The strands may be greased or placed in thin-walled metal tubes to keep the concrete from bonding with them (Fig. 9.7). After the concrete has hardened, the posttensioning strands are anchored firmly on one end and a hydraulic jack is used on the other to apply tension to the strand. Once the desired tension is reached this end is also anchored (Fig. 9.8). The steel strands may be left unbonded. If they are in steel

Figure 9.8 This concrete beam is being posttensioned with a hydraulic jack tensioning the cables in each tendon. The steel protruding on the top of the beam will be embedded in the concrete deck to be poured on top of the beam. (*Courtesy Portland Cement Association*)

tubes they can be bonded by filling the space between the strands and the tube with high-pressure grout. Figure 9.9 shows the structural framing and tenons in tubes on a posttensioned concrete structure.

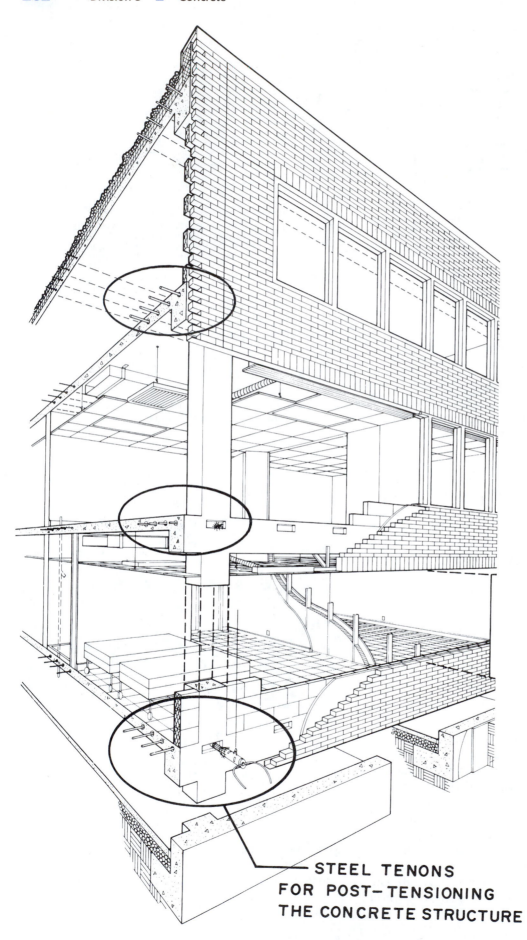

STEEL TENONS
FOR POST-TENSIONING
THE CONCRETE STRUCTURE

Figure 9.9 A posttensioned concrete structure. (*Reproduced with permission from* The Building Systems Integration Handbook, *Richard Rush, ed., Butterworth-Heinmann Publishers, Newton, Mass., 1986.*)

PRECAST CONCRETE SLABS

The major types of slabs used for floor and roof construction are solid flat slabs, tongue-and-groove planks, channel slabs, hollow core, double tees, and single tees (Fig. 9.10). Some are prestressed and others precast with untensioned reinforcing.

Solid slabs and *channel slabs* are used for short spans and minimum slab depth. Spans of 25 ft. (7.6 m) with thicknesses of 2 to 8 in. (50 to 200 mm) and widths of 8 to 12 ft. (2.4 to 3.7 m) are common. Slabs made thicker than this to span longer distances become very heavy and are not economical.

Hollow core slabs are used for spans ranging up to 40 ft. (12 m) in length and are 6 to 12 in. (150 to 300 mm) thick. The hollow core, created by removing concrete from an area of the slab that has little influence on strength, lightens the member. A topping of 2 in. (50 mm) thick concrete is often applied after installation to increase structural performance and fire rating.

Double tees and *single tees* are used to span the longest distances. Single tees come in widths of 6, 8, 10, and 12 ft. (1.8, 2.4, 3.0, and 3.7 m) and depths of 16 to 48 in. (406 to 1219 mm). Double tees come in widths of 4, 8, 10, and 12 ft. (1.2, 2.4, 3.0, and 3.7 m) and depths from 10 to 40 in. (254 to 1016 mm). Standard lengths range to over 100 ft. (30.5 m). Lengths over 60 ft. (18.3 m) present special transportation problems.

Tees are cast with a rough or smooth surface on top. The rough surface is used when a concrete topping is to be applied over them. This topping is usually 2 in. (50 mm) thick. Structural continuity across the units can be provided by placing steel reinforcing bars in the topping.

Normal weight and lightweight concrete are used to cast these units. Lightweight concrete is more expensive than normal weight, but it reduces the weight on the structure. Since double tees do not require temporary support to prevent tipping, they are easier and more economical to erect than single tees.

Precast Concrete Columns

Precast columns are usually combined with precast concrete beams, forming a post-and-beam structural framework. Most precast concrete columns are reinforced conventionally. If they are prestressed, this is mainly to reduce stresses on the column during transportation and handling.

Columns may be rectangular or square. Square sizes range from 10 to 24 in. (254 to 619 mm). Rectangular columns vary depending on the design. Columns up to 60 ft. (18 m) in length can be transported. Those above that size need special arrangements. Some columns have **haunches** on two or three sides (Fig. 9.11). Concrete used may range from 5,000 to 11,000 psi (48 to 76 MPa).

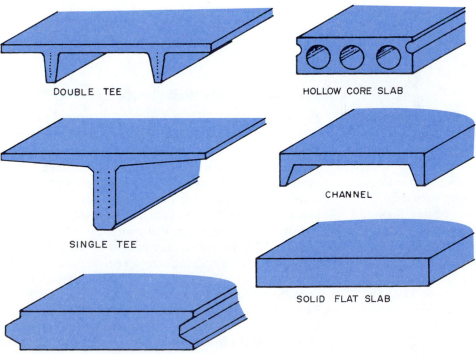

Figure 9.10 Precast concrete units used for floors and roof decks.

DOUBLE TEE

HOLLOW CORE SLAB

SINGLE TEE

CHANNEL

SOLID FLAT SLAB

TONGUE-AND-GROOVE PLANK

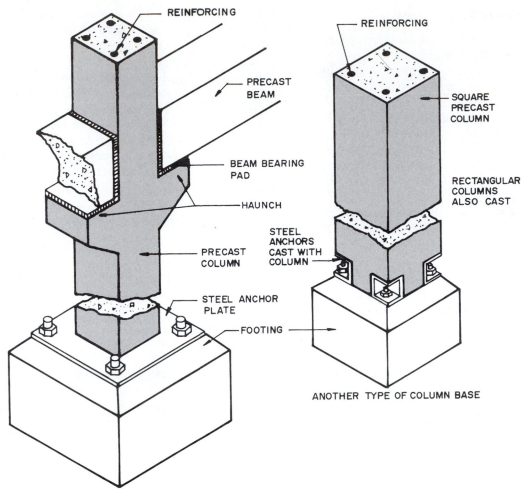

Figure 9.11 Typical precast concrete columns.

Beams and Girders

Precast prestressed concrete beams and girders can be used in any building in which precast construction is desired. They are made in several standard shapes, and special designs can be produced (Fig. 9.12). The L-shaped and inverted tee beams provide a bearing surface for precast floor units, such as single and double tee units. This allows lower overall height than does resting floor and roof units on top of rectangular beams.

Rectangular beams range in depth from 18 to 48 in. (457 to 1219 mm) and are 12 to 36 in. (305 to 914 mm) wide. Inverted tee and L-beams range in depth from 18 to 60 in. (457 to 1524 mm) and in width from 12 to 30 in. (305 to 762 mm). Beam ledges are usually 6 in. (152 mm) wide and 12 in. (305 mm) deep. Beams are available in increments of 2 or 4 in. (50 or 100 mm).

Precast Concrete Wall Panels

Precast concrete wall panels may be prestressed or conventionally reinforced. Their design varies considerably. For example, they may be load bearing or non–load bearing, they may have a flat, ribbed, or other surface configuration, and they may be hollow, solid, or of a sandwich construction. They may be used in connection with a precast concrete framing system, cast-in-place concrete, or a steel framing system. Flat panels can be two stories high, and ribbed panels can be cast up to four stories high. Panels with openings are not prestressed.

Solid panels are typically $3^1/_2$ to 10 in. (89 to 254 mm) thick. Sandwich and hollow core panels vary from $5^1/_2$ to 12 in. (140 to 305 mm) thick. Ribbed panels are made in thicknesses from 12 to 24 in. (305 to 610 mm). Most panels are designed with an 8 ft. (2.4 in.) width (Fig. 9.13).

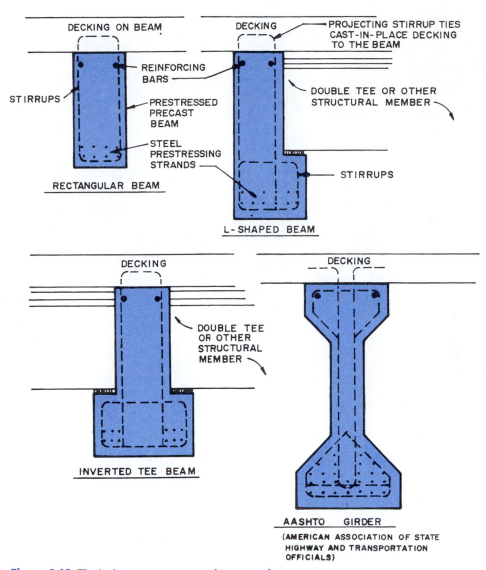

Figure 9.12 Typical precast prestressed concrete beams.

Multistory load-bearing panels can be cast with a haunch or other type of support for carrying floor and roof slabs (Fig. 9.14).

CONNECTING PRECAST UNITS

Types of Connections

The types of connections include bolt connections, welded connections, posttensioned connections, and doweled connections.

Bolt connections speed up erection and allow final alignments to be made after the crane has placed the member. Generally ½, ¾, and 1 in. (12, 18, and 25 mm) bolts with

national coarse threads are used. Steel washers may be required under certain conditions. Bolts should be tightened to the recommended torque.

Welded connections are strong and easy to make on the site. They should be made following the details on the erection drawings, which include the size, type, length of the weld, type of electrode, and required temperatures. Welded connections are made before the unit is released by the hoisting device. This tends to tie up the hoisting device for longer periods than if bolted connections were used. The amount of heat generated may cause the concrete to crack and the metal to become distorted. Large or long continuous welds should be used sparingly, and the welding temperature should be carefully controlled.

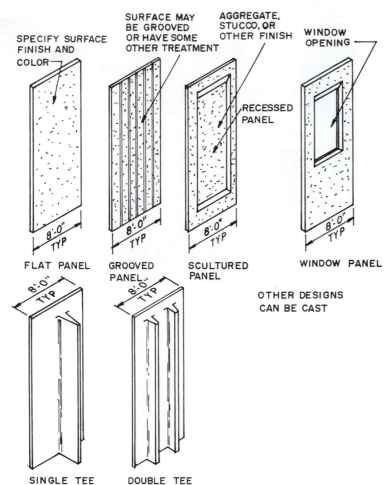

SPECIFY SURFACE FINISH AND COLOR

SURFACE MAY BE GROOVED OR HAVE SOME OTHER TREATMENT

AGGREGATE, STUCCO, OR OTHER FINISH

WINDOW OPENING

RECESSED PANEL

8'-0" TYP

FLAT PANEL

GROOVED PANEL

SCULTURED PANEL

WINDOW PANEL

OTHER DESIGNS CAN BE CAST

SINGLE TEE

DOUBLE TEE

Figure 9.13 A wide variety of precast panels can be manufactured, including ones with openings for doors and windows.

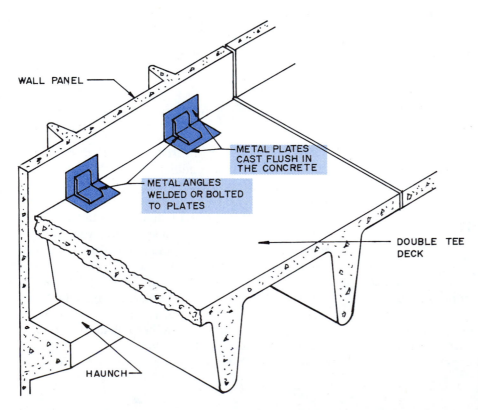

WALL PANEL

METAL PLATES CAST FLUSH IN THE CONCRETE

METAL ANGLES WELDED OR BOLTED TO PLATES

DOUBLE TEE DECK

HAUNCH

Figure 9.14 Multistory precast concrete panels can be cast with a haunch to carry floor and roof decking.

Posttensioned connections are made using either bonded or unbonded tendons. Bonded tendons are installed in holes cast in the member and are bonded to the member after tensioning by filling the hole with grout. Unbonded tendons are not grouted in place but are coated with an organic coating to prevent corrosion.

Doweled connections use reinforcing bars grouted into dowel holes in the precast member. The strength of the connection depends on the dowel diameter, the depth it is in the member, and the bond developed by the grout. A number of dowel systems are manufactured that use a sleeve with the dowel. When used it is important to follow the manufacturer's directions (Fig. 9.15).

Grout, Mortar, and Drypack

Grouts and mortars are used to transfer loads between members. They are also used to fill noncritical voids. Load-bearing grouts should be nonshrinking, and grouts manufactured for this specific purpose should be used (Fig. 9.16).

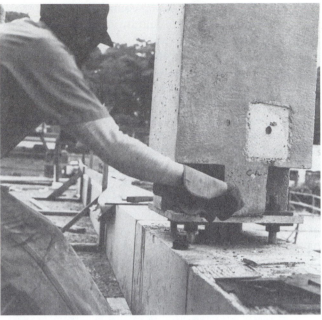

1. The column is set on the anchor bolts and plumbed.

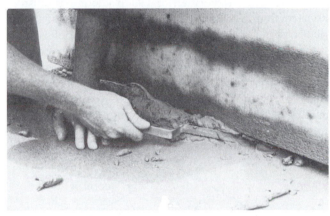

2. Grout is forced below the column after it is set and plumbed.

Figure 9.16 Nonshrinking load-bearing grout is placed below precast columns and precast walls. (*Courtesy Precast/Prestressed Concrete Institute*)

Drypack describes a method of placing grout. The dry grout material has only enough water added to produce a stiff granular mix, which is packed into an opening.

Connection Details

The following detail drawings are typical of connections that are used, but they do not represent the only possible designs and are for general illustration purposes only. They are not intended to be copied and used as design solutions. All member sizes, shapes, reinforcing, and connections must be designed by a professional engineer to meet the requirements of each specific situation.

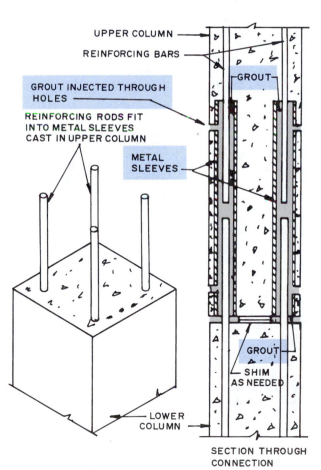

Figure 9.15 An example of a doweled connection made using a metal sleeve.

A typical *column-to-column* connection is shown in Figs. 9.17 and 9.15. The column can be shimmed as needed to get it plumb.

Column-to-foundation connections are shown in Figs. 9.18 and 9.19. In Fig. 9.18 anchor bolts are cast in the foundation. The column base plate rests on leveling nuts, which are used to get the column plumb. The area below the base plate is filled with grout. In Fig. 9.19 the anchoring nuts are located outside the column. The installation is basically the same as the previous example.

Typical *beam-to-column* connections include the use of dowels or haunches. Figure 9.20 shows the use of metal dowels that are cast with the column and fit into holes cast in the beam. Figure 9.21 shows precast columns with haunches designed to support the beam. Typically, the beam is connected to the column by welding an angle to metal anchor plates that were cast into the beam and column.

Connections of *precast walls to a foundation* are shown in Fig. 9.22. This typically uses metal plates that are welded or an anchor rod that is posttensioned. One way to connect a precast concrete interior wall to a floor in multistory construction is in Fig. 9.23. The wall has a metal anchor plate cast in it with a pocket above. An anchor bolt is in the wall below and is secured to the wall above with a nut.

There are a number of ways to make floor and roof panel connections. The installation of *hollow core deck units* to a beam is shown in Fig. 9.24. Reinforcing rod is cast in concrete between butting units. Additional rods may be placed in the cores on either side of the joint.

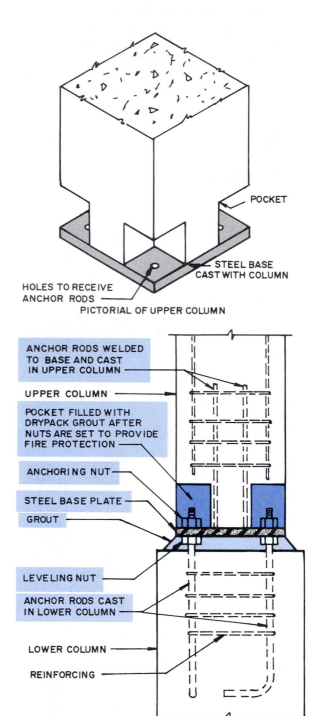

Figure 9.18 This column is anchored to the foundation with bolts located within the base of the column.

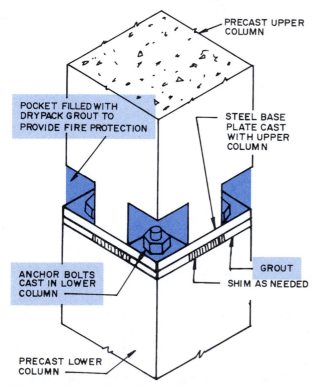

Figure 9.17 A typical column-to-column connection.

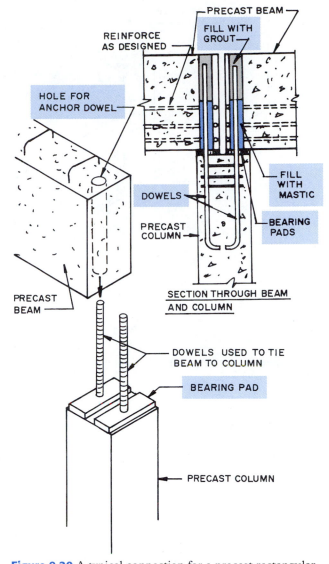

Figure 9.20 A typical connection for a precast rectangular beam to a precast column.

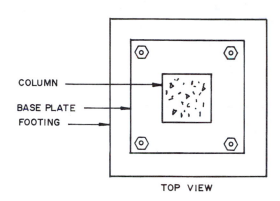

TOP VIEW

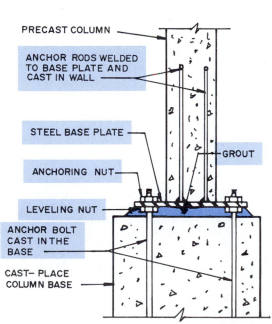

Figure 9.19 This column is anchored to the foundation with bolts located outside the base of the column.

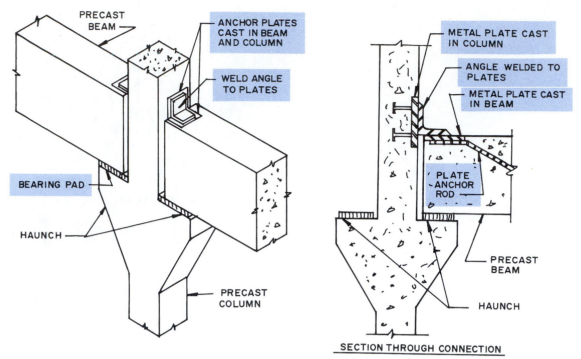

Figure 9.21 Precast rectangular beams are supported on column haunches.

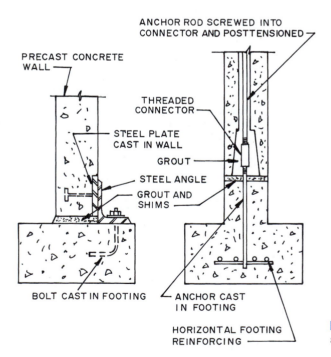

Figure 9.22 Two ways to connect a precast wall to the foundation.

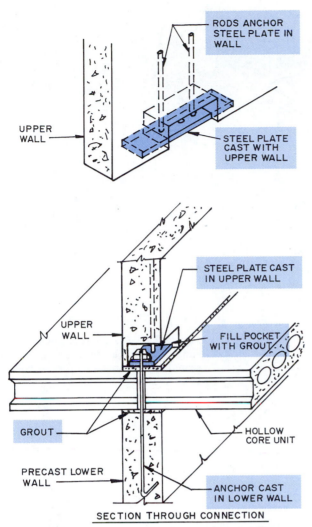

Figure 9.23 shows the section through the connection with labels: RODS ANCHOR STEEL PLATE IN WALL, UPPER WALL, STEEL PLATE CAST WITH UPPER WALL, STEEL PLATE CAST IN UPPER WALL, UPPER WALL, FILL POCKET WITH GROUT, GROUT, PRECAST LOWER WALL, HOLLOW CORE UNIT, ANCHOR CAST IN LOWER WALL.

SECTION THROUGH CONNECTION

Figure 9.23 Anchor bolts can be used to connect precast interior walls when hollow core decking is used.

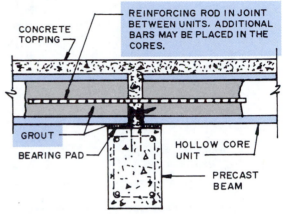

Figure 9.24 A typical detail for installing hollow core units on a precast rectangular beam.

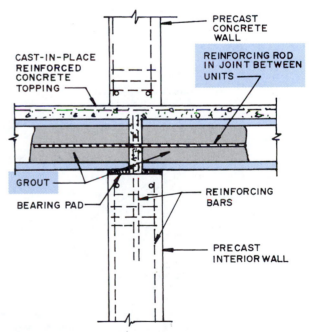

Figure 9.25 A typical detail for supporting hollow core units on a precast interior wall.

Figure 9.25 shows a detail for connecting hollow core units to an interior precast concrete wall.

One method for installing *single and double tees* is to use a precast L-shaped beam that is supported on column haunches (Fig. 9.26). The L-beam is connected to the column by welding metal plates that were cast into each unit. The tee member is welded to the L-beam. Another technique is to support inverted tee beams on the precast column haunch and place the precast tee decking units as shown in Fig. 9.27. A section through this construction in Fig. 9.28 shows the details.

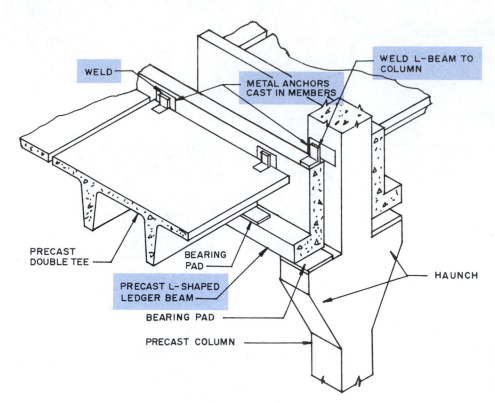

WELD

WELD L—BEAM TO
COLUMN

METAL ANCHORS
CAST IN MEMBERS

PRECAST
DOUBLE TEE

BEARING
PAD

PRECAST L—SHAPED
LEDGER BEAM

BEARING PAD

PRECAST COLUMN

HAUNCH

Figure 9.26 Double tees can be supported on a ledger beam that is seated on a haunch of a precast concrete column.

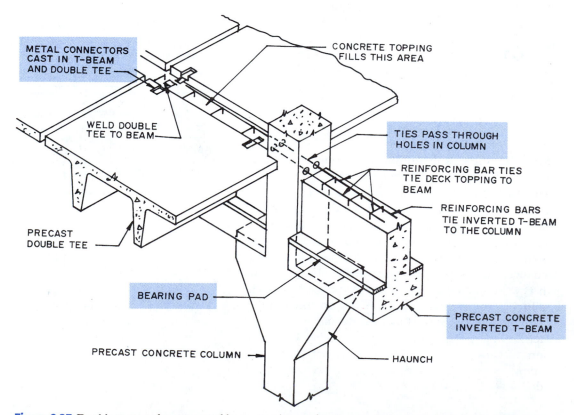

METAL CONNECTORS
CAST IN T-BEAM
AND DOUBLE TEE

CONCRETE TOPPING
FILLS THIS AREA

WELD DOUBLE
TEE TO BEAM

TIES PASS THROUGH
HOLES IN COLUMN

REINFORCING BAR TIES
TIE DECK TOPPING TO
BEAM

REINFORCING BARS
TIE INVERTED T-BEAM
TO THE COLUMN

PRECAST
DOUBLE TEE

BEARING PAD

PRECAST CONCRETE
INVERTED T-BEAM

PRECAST CONCRETE COLUMN

HAUNCH

Figure 9.27 Double tees can be supported by precast inverted tee beams that rest on the haunches of precast concrete columns.

Chapter 9 ▲ Precast Concrete 213

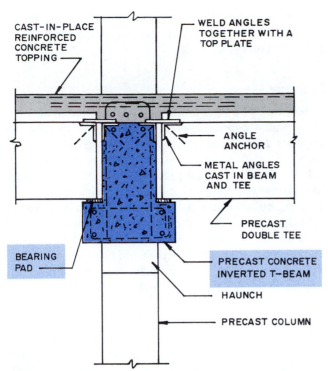

Figure 9.28 A section through a precast inverted tee beam supporting double tees.

ERECTING PRECAST CONCRETE

Planning

After the structural design work has been completed and reviewed, the plans go to the company that will cast the units. Here they plan the work using standard molds when possible and building special molds as needed.

A major factor is the sequencing of the manufacture and delivery of the members. This needs to be detailed before the job starts. A production and delivery schedule must be planned so the units can be delivered to the site when they are needed. Included is time for the units to cure and gain sufficient strength to permit them to be moved to the job site.

Consideration of the erection procedures also begins before the units are cast. The design engineer must discuss the use of lifting hardware and connections with the erector. The manufacturer may have specific details for the design and manufacture of the units, which the engineer should observe. Precast units sometimes have lifting devices cast in them. Their location must be worked out by the designer so the unit can be lifted without damage and be in the position needed for connection.

Precast units are erected and connected as shown on the erection drawings. The erector must comply with the specified dimensional tolerances, hardware sizes, weld lengths, and torque on bolts. Connections should be planned so workers can make them from a stable work platform. Planning should include provision for activities such as welding, posttensioning, and grouting. The erector should install temporary bracing during construction. Final adjustment and alignment of precast structural units can then be made without the need for a crane to support them.

The erection procedure should include plans to maintain the stability and equilibrium of the structure during construction as well as after the structure is completed. For example, when an L-shaped or inverted tee ledger beam is subjected to eccentric loading, the beam is subject to torsion and may tilt slightly on the column below (Fig. 9.29). The connections should be designed to resist this motion.

Transporting Units to the Site

Deliveries of precast members are usually made by truck (Fig. 9.30). This requires the shipper to do some careful planning. The distance to the job, the type, size and weight of the member, the type of vehicle required, the weather, road conditions, and other such factors must be considered. At the site, proper equipment for unloading must be available. Finally, the members must be carefully stored until needed to avoid structural damage.

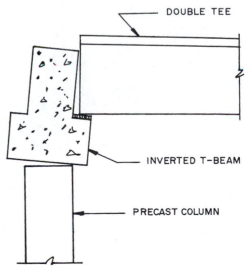

Figure 9.29 If a beam is loaded eccentrically during construction, torsion forces may cause it to tilt.

At the site the units are placed so they can be reached easily when needed and near the point where they are to be used. The erection contractor will have made arrangements to have the needed hoisting equipment available to unload the members when they arrive and to lift them to the point of erection.

Rigging

Rigging refers to the equipment, cables, slings, and other hoisting equipment used to lift the precast units into place (Fig. 9.31). This is the work of specialists.

Prestressed members must never be lifted by their center only, and they should not be lifted or stored upside down. They must be kept in the upright position in which they will be installed. If lifting devices are not in the member, it can usually be safely lifted by slings placed near each end.

Placing Precast Units

The foundations and footings are cast-in-place and carefully located. Tolerances are rather tight, and assembly will halt if tolerances have not been carefully maintained.

The structure may be built using precast columns and beams. The columns are lifted and placed with a crane, then plumbed and grouted with drypack grout.

Columns are joined by beams (Fig. 9.32) and the floor decking is set in place. Floor decking often is hollow core units (Fig. 9.33).

Framing Plan

An example of a floor **framing plan** for a small building using hollow core units is in Fig. 9.34. The specific way these are drawn varies but the information shown is consistent. Single or double tees also are often used for floor decking. Double tees are preferred to single tees because they do not require bracing during erection (Fig. 9.35).

Some buildings use precast structural walls instead of columns. These are erected on foundations and braced until members abutting them are in place (Fig. 9.36).

If precast exterior wall panels are used to enclose the building, they are lifted in place and secured to the structural frame (Fig. 9.37).

Figure 9.30 Most precast units are transported to the site by truck and set in place with a crane. (*Courtesy Precast/Prestressed Concrete Institute*)

Figure 9.31 Precast units are lifted into place with a crane and proper cables, slings, and other hoisting devices. (*Courtesy Precast/Prestressed Concrete Institute*)

Figure 9.32 Precast prestressed beams are connected to the columns. (*Courtesy Precast/Prestressed Concrete Institute*)

Figure 9.33 This precast hollow core floor decking is secured to the beams. (*Courtesy Precast/Prestressed Concrete Institute*)

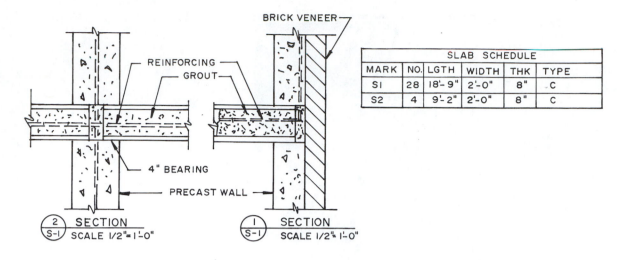

SLAB SCHEDULE					
MARK	NO.	LGTH	WIDTH	THK	TYPE
S1	28	18'-9"	2'-0"	8"	C
S2	4	9'-2"	2'-0"	8"	C

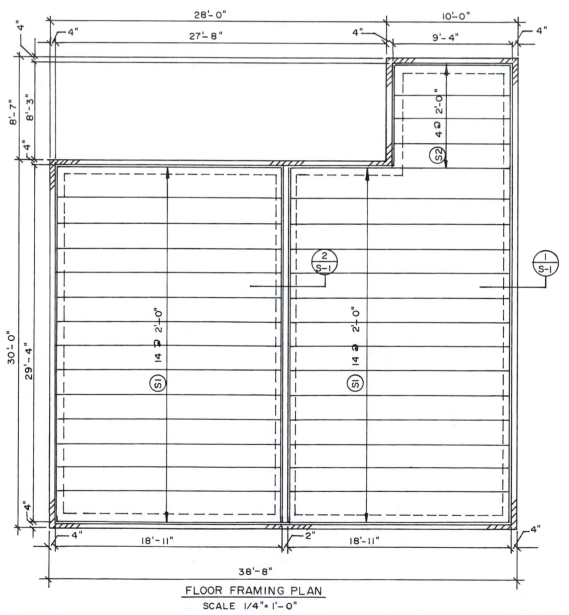

FLOOR FRAMING PLAN
SCALE 1/4"= 1'-0"

Figure 9.34 A framing plan for a building with hollow core floor units. Included are drawings showing construction details and a schedule giving details about the units required.

PRECAST TEE UNITS

PRECAST INTERIOR WALL

PRECAST BEAM

PRECAST COLUMN

Figure 9.35 The underside of a deck built using precast double tees. (*Courtesy Precast/Prestressed Concrete Institute*)

Figure 9.36 Precast interior and exterior walls are connected to a foundation or beam. (*Courtesy Precast/Prestressed Concrete Institute*)

Figure 9.37 Precast exterior wall panels are connected to the structural frame. (*Courtesy Precast/Prestressed Concrete Institute*)

REVIEW QUESTIONS

1. Give some examples of precast concrete structural units.
2. How do nonprestressed structural units differ from pretensioned units?
3. How much camber does a nonprestressed beam develop?
4. What are the commonly used types of precast prestressed floor and roof slabs?
5. Which type of precast pretensioned slabs span the greatest distances?
6. What is the main value in prestressing precast concrete columns?
7. What is the difference between posttensioning and pretensioning?

KEY TERMS

camber An arching in a structural member due to tension applied by the steel reinforcing.

cast-in-place Members cast in forms that are erected on the job site in the locations where members are needed.

double tees T-shaped precast floor and roof units that span long distances unsupported.

drypack A stiff granular grout.

framing plan A drawing showing the location of structural members.

grout A mortar of a consistency that permits it to be placed into joints, spaces, and voids.

haunch A projection used to support a member as a beam.

nonprestressed units Structural units cast in molds in a factory that have reinforcing members not under tension as they are being cast.

posttensioned Tension is applied to steel reinforcing after the concrete has been poured and allowed to harden.

prestressed units Structural cast in molds in a factory that have stresses introduced as they are being cast or after casting.

pretensioned Tension is applied to steel reinforcing before the concrete is poured in the mold.

rigging Hoisting equipment used to lift materials above grade.

SUGGESTED ACTIVITIES

1. If there is a precasting plant within reasonable driving range it is well worth the effort to make a visit. Ask the supervisor to explain and show the process from beginning to end. Take photos and prepare an exhibit detailing the process.

2. Explain how a precast plant prepares the concrete and how the quality control tests are taken.

3. See if you can get the structural plans for a precast concrete structure from a local architect or contractor. Prepare a report citing the size and spacing of columns, beams, and girders. What type of floor and roof decking were specified, and what are the sizes of each? Is there anything else on the plan that you would like to report?

4. Make a model with wood pieces representing precast concrete beams, girders, columns, wall panels, and decking.

5. Examine the local building code and briefly cite the specific areas related to precast concrete structures.

6. If you are fortunate enough to be able to visit the site of a precast concrete structure being erected, take photos and notes describing what you saw. For example, what types of connectors were used? How were members lifted into place? What safety devices were the workers wearing?

ADDITIONAL INFORMATION

Newman, M., *Structural Details for Concrete Construction,* McGraw-Hill, New York, 1994.

Other resources include

Design and construction manuals from the Prestressed/Precast Concrete Institute, 175 W. Jackson Blvd., Chicago, Ill. 60604.

Numerous publications from the Portland Cement Association, 5420 Old Orchard Road, Skokie, Ill. 60077-1083.

Publications from the organizations listed in the "Professional and Technical Organizations" section in Appendix.

D I V I S I O N 4

MASONRY
CSI MASTER FORMAT™

Courtsey Italian Government Travel Office

Mortars for Masonry Walls

This chapter will help you to:

1. Describe the various materials used to produce mortar.

2. Cite the various ways mortar can be obtained for use on the construction site.

3. Describe the ingredients used to manufacture masonry cements.

4. Identify and cite the uses of the types of standard mortars.

5. Discuss the desirable properties of plastic mortar.

6. Discuss the properties of quality hardened mortar.

7. Be aware of the frequently used mortar tests.

8. Specify how to cure masonry mortar laid in cold weather.

Mortar is the bonding agent used to join masonry units into an integral structure. In addition to bonding the units together, it also must (1) seal the spaces between the units so they are not penetrated by air or moisture, (2) tie steel reinforcement, ties, and anchor bolts into the walls, (3) provide a design of lines of color and shadows, and (4) allow for the adjustment of the slight variations that occur in masonry units.

MORTAR COMPOSITION

Mortar is composed of cementitious materials, clean carefully graded mortar sand, clean water, and sometimes a coloring agent or an admixture. The actual composition of mortars varies according to the intended use. For general masonry applications, mortars may contain portland cement or masonry cement, sand, hydrated lime or lime putty, and water. Mortars for special applications, such as prefabricated masonry units that must be moved into place, require additives that increase compressive and tensile strength and have greater bonding capabilities.

Cementitious Materials

Cementitious materials provide the bonding ingredient in mortars. They are made according to ASTM (American Society for Testing and Materials) and CSA (Canadian Standards Association) specifications as shown in the following listing:

Masonry cement—ASTM C91
    (Types M, S, or N), CSA A8 (Types S or N)
Portland cement—ASTM C150
    (Types I, IA, II, IIA, III, or IIIA), CSA A5 (10, 20, 30, or 50)
Blended hydraulic cement—ASTM C595
    (Types IS, IS-A, IP, IP-A, I(PM), or I(PM)-A) (Pozzolan modified portland cement), CSA A362 (Type 10S)
Hydrated lime for masonry purposes—ASTM C207
    (Types S, SA, N, or NA)
Quicklime for structural uses (for lime putty)—ASTM C5

You can find detailed information on various cements in Chapter 7. Most mortars are now made with mortar cement. Masonry cement mortars are made by combining masonry cement, clean carefully graded sand, and enough clean water to produce a plastic, workable mix.

Masonry Cements

Masonry cements are hydraulic cements used in mortars for masonry construction. They produce a mortar that has greater plasticity and water retention than if portland cement were used. Typically they may include portland cement, slag cement, blast-furnace slag cement, portland-pozzolan cement, natural cement, and hydraulic lime. Manufacturers may include chalk, clay, talc, calcareous shell, or limestone to add the properties desired.

Masonry cements meet the specifications found in ASTM C91. This report specifies three types of *masonry cement*—Type N, Type S, and Type M. They are used to produce *mortars* Types N, O, S, and M as specified in ASTM C270.

Type N masonry cement is used in Type N and O mortars. It can be blended with portland cements for use in Type S and M mortars.

Other Mortar Ingredients

Lime is used to help stabilize the volume of the mortar by controlling shrinking and expansion.

Masonry sand may be natural or a manufactured product that meets the requirements of ASTM C144, The Standard Specification for Aggregate for Masonry Mortar. In Canada the standard is CSA A82.56, Aggregate for Masonry Mortar. Since sand makes up the major portion of a mortar, the quality of the finished mortar depends heavily on clean, quality sand. The gradation requirements for sand specified in ASTM C144 are shown in Table 10.1.

Table 10.1 Aggregate Gradation for Masonry Mortar

Sieve Size No.	Gradation Specified, Percent Passing ASTM C144[a]	
	Natural Sand	Manufactured Sand
4	100	100
8	95 to 100	95 to 100
16	70 to 100	70 to 100
30	40 to 75	40 to 75
50	10 to 35	20 to 40
100	2 to 15	10 to 25
200	—	0 to 10

[a]Additional requirements: Not more than 50% shall be retained between any two sieve sizes, nor more than 25% between No. 50 and No. 100 sieve sizes. Where an aggregate fails to meet the gradation limit specified, it may be used if the mortar will comply with the property specification of ASTM C270 (Table 2).
Courtesy American Society for Testing and Materials

Water used to produce mortar must be clean and free of acids, alkalies, and other organic materials. Water containing soluble salts will cause the mortar to effloresce. **Efflorescence** appears on the finished masonry wall as a white powdery substance that discolors the mortar and masonry nits. The water used should be potable.

Mortar can be *colored* by adding various organic pigments. White mortar is made with white masonry cement or white portland cement, lime, and white sand. Colored mortars are made with white masonry cement and pigments, which are typically some type of mineral oxide compound such as iron, chromium, cobalt, or manganese oxides. Carbon black is used to produce a dark gray or black mortar.

Admixtures may sometimes be used to alter the properties of the mortar. However, the possibility of their creating unforeseen problems is great. Admixtures usually are not used unless laboratory tests have been made to verify the changes they will have on the mortar. For example, certain admixtures may affect the hydration process and hardening properties. Consultants such as those at the Portland Cement Association can provide helpful information.

In Table 10.2 you can see some of the commonly available admixtures (modifiers) and their benefits and possible problems they may cause. You can find more information on admixtures in Chapter 7.

TYPES OF MORTAR

Four types of mortar are specified for use in the United States. These are M, S, N, and O and are specified by ASTM C270. The Canadian Standards Association Standard A179 specifies two types, S and N.

Mortar types are identified by either proportion or property specifications, but only by one of these, not both. In Table 10.3 you can find the specifications for mortar *proportion* indicated by the combinations of portland cement or blended cement, masonry cement, hydrated lime or lime putty, and the aggregate. Blended cement is a combination of portland cement and masonry cement. Mortar types, classified under *property specifications* are based on the compressive strength, water retention, and air content (Table 10.4). This data is developed by testing 2 in. (50.8 mm) cubes of hardened mortar to find the compressive strength. The compression test is discussed later in this chapter.

Type M mortar is a high-strength mortar with a compressive strength of 2500 psi (17 MPa). It has better durability than the other types. Type M is recommended for masonry below grade and in contact with earth and for conditions of severe frost action. It is also used for reinforced masonry.

Type S mortar is a medium-high-strength mortar with a compressive strength of 1800 psi (12.5 MPa). It is often permitted in wall construction instead of Type M

Table 10.2 Admixtures—Benefits and Concerns

Admixture	Primary Benefits	Possible Concerns
Air-entraining	Freeze-thaw durability, workability	Effect on compressive and bond strengths
Bonding	Wall tensile (and flexural) bond strength	Reduced workability, bond strength regression upon wetting, corrosive properties
Plasticizer	Workability, economy	Effect on hardened physical properties under field conditions
Set accelerator	Early strength development	Effectiveness at cold temperatures, corrosive properties, effect on efflorescence potential of masonry
Set retarder	Workability retention	Effect on strength development, effect on efflorescence potential of masonry
Water reducer	Strength, workability	Effect on strength development under field conditions with absorptive units
Water repellent	Weather resistance	Effectiveness over time
Pozzolanic	Increase density and strength	Effect on plastic and hardened physical properties under field conditions
Color	Esthetic versatility	Effect on physical properties, color stability over time

Courtesy Portland Cement Association

Table 10.3 Proportion Specifications for Mortar

	United States-ASTM C270					
	Parts by Volume					
Mortar type	Portland cement or blended cement	Masonry cement type M	S	N	Hydrated lime or lime putty	Aggregate[a]
M	1	—	—	1	—	4½ to 6
	—	1	—	—	—	2¼ to 3
	1	—	—	—	¼	2¹³⁄₁₆ to 3¾
S	½	—	—	1	—	3⅜ to 4½
	—	—	1	—	—	2¼ to 3
	1	—	—	—	Over ¼ to ½	
N	—	—	—	1	—	2¼ to 3
	1	—	—	—	Over ½ to 1¼	
O	—	—	—	1	—	2¼ to 3
	1	—	—	—	Over 1¼ to 2½	

	Canada—CSA A179M				
	Parts by Volume				
Mortar type	Portland cement	Masonry cement type S	N	Hydrated lime or lime putty	Aggregate[a]
S	—	—	1	—	2¼ to 3
	½	—	1	—	3½ to 4½
	1	—	—	½	3½ to 4½
N	—	—	1	—	2¼ to 3
	1	—	—	1	4½ to 6

[a]The total aggregate shall be equal to not less than 2¼ and not more than 3 times the sum of the volumes of the cement and lime used.
Notes: 1. Under both ASTM C270, Standard Specification for Mortar for Unit Masonry, and CSA A179, Mortar and Grout for Unit Masonry, aggregate is measured in a damp, loose condition and 1 cu ft. of masonry sand by damp, loose volume is considered equal to 80 lb. of dry sand (in SI units 1 cu m of damp, loose sand is considered equal to 1280 kg of dry sand).
2. Mortar should not contain more than one air-entraining material.
Courtesy Portland Cement Association and the Canadian Standards Association. Canadian material presented with the permission of the Canadian Standards Association; material is reproduced from CSA Standard A179-94 (Mortar and Grout for Unit Masonry), which is copyrighted by CSA, 178 Rexdale Blvd., Etobicoke, Ontario M9W 1R3. Although use of this material has been authorized, CSA shall not be responsible for the manner in which the information is presented, nor for any interpretations thereof. This material may not be updated to reflect amendments made to the original content. For up-to-date information, contact CSA.

Table 10.4 Property Specifications for Laboratory-Prepared Mortar[a]

		United States		
Mortar specification	Mortar type	Minimum 28-day Compressive Strength, psi	Minimum water retention, %	Maximum air content, %[b]
ASTM C270	M	2500	75	12[c]
	S	1800	75	12[c]
	N	750	75	14[c]
	O	350	75	14[c]

		Canada		
		Minimum Compressive Strength, MPa		Minimum water retention, %
Mortar specification	Mortar type	7-day[d]	28-day	
CSA A179	S	7.5	12.5	70
	N	3	5	70

[a]The total aggregate shall be equal to not less than 2¼ and not more than 3½ times the sum of the volumes of the cement and lime used.
[b]Cement-lime mortar only (except where noted).
[c]When structural reinforcement is incorporated in cement-lime or masonry cement mortar, the maximum air content shall be 12% or 18%, respectively.
[d]If the mortar fails to meet the 7-day requirement but meets the 28-day requirement, it shall be acceptable.
Courtesy Portland Cement Association and the Canadian Standards Association. Canadian material presented with the permission of the Canadian Standards Association; material is reproduced from CSA Standard A179-94 (Mortar and Grout for Unit Masonry), which is copyrighted by CSA, 178 Rexdale Blvd., Etobicoke, Ontario M9W 1R3. Although use of this material has been authorized, CSA shall not be responsible for the manner in which the information is presented, nor for any interpretations thereof. This material may not be updated to reflect amendments made to the original content. For up-to-date information, contact CSA.

because it has almost the same allowable strength and can be used above or below grade. Type S has better workability and more water retention than Type M. It is used for reinforced and unreinforced masonry and has high tensile bond strength.

Type N mortar is a medium-strength mortar with a compressive strength of 750 psi (5 MPa). It is used for above-grade general construction involving exposed masonry load-bearing walls for which requirements for compressive strength and lateral strength are not high.

Type O mortar is a medium-strength mortar with a compressive strength of 350 psi (2.5 MPa). It is used for general interior purposes, such as non-load-bearing walls for which compressive strength does not exceed 100 lb. per sq.in. It must not be in contact with the soil or exposed to saturated freezing conditions.

A guide for selecting masonry mortars for various uses above and below grade is in Table 10.5.

Sources of Mortar

Mortars are available ready-mixed, dry-batch, and on-site mixed.

Ready-mixed mortar is mixed at a central batch plant and transported to the job site, where a slump test is made. If required, additional water is added to get the desired slump. This mortar contains a retarding, set-controlling admixture that keeps the mortar workable and in a plastic condition for more than twenty-four hours. The mix is delivered by a ready-mix truck and is stored on the site in large tubs (Fig. 10.1). Ready-mixed mortar must meet the requirements of ASTM C1142, Specifications for Ready-Mixed Mortar for Unit Masonry.

Dry-batch mortar has the required ingredients blended and packaged in bags or delivered to the site in a sealed truck. It is stored at the site in a sealed hopper. When mortar is needed, the dry ingredients are moved to an on-site mixer and the required water is added. Dry-batching provides control over the proportions of the various ingredients in the mortar. Premixed dry mortar ingredients are also available packaged in water-proof bags.

On-site mixed mortar can be made by storing the dry ingredients in a silo-type mixer in which an auger-type mixing device receives the correct proportions of dry ingredients and blends them together. The dry ingredients are stored in separate chambers, from which they are fed to the mixer. The water is then injected from a pressurized water source, such as a city water main, and the mixer produces the finished mortar. The process is

Table 10.5 Guide for the Selection of Masonry Mortars (United States)[a]

Location	Building segment	Mortar type Recommended	Alternative
Exterior, above grade	Load-bearing walls	N	S or M
	Non-load-bearing walls	O[b]	N or S
	Parapet walls	N	S
Exterior, at or below grade	Foundation walls, retaining walls, manholes, sewers, pavements, walks, and patios	S[c]	M or N[c]
Interior	Load-bearing walls	N	S or M
	Non-load-bearing partitions	O	N

[a]Adapted from ASTM C270. This table does not provide for specialized mortar uses, such as chimney, reinforced masonry, and acid-resistant mortars.
[b]Type O mortar is recommended for use where the masonry is unlikely to be frozen when saturated or unlikely to be subjected to high winds or other significant lateral loads. Type N or S mortar should be used in other cases.
[c]Masonry exposed to weather in a nominally horizontal surface is extremely vulnerable to weathering. Mortar for such masonry should be selected with due caution.
Courtesy American Society for Testing and Materials

Figure 10.1 Ready-mixed mortar may be batched at a central location and delivered to the job site ready to use. It is stored in mortar containers (tubs) until it is needed. *(Courtesy Portland Cement Association)*

usually computer controlled. The mortar is discharged into tubs or wheelbarrows for distribution on the site. This produces a very accurately proportioned mortar (Fig. 10.2).

For small jobs the ingredients typically are delivered to the site as separate items. The mortar cement is in bags, and the sand is in bulk. The sand must be protected from moisture and dirt. The mortar is often proportioned by the worker putting so many shovelfuls of sand into a small power-operated mixer and adding the required number of bags of mortar cement (Fig. 10.3). Water may be added by a certain number of bucketfuls. The gasoline-powered mixer produces the mortar, which usually is moved to the masons by wheelbarrow. Obviously, this is often done on small jobs, and the mix is based on experience. The proportions of the ingredients can vary considerably, and

Figure 10.2 This is a storage bin/batcher in which the dry mortar ingredients (masonry cement, portland cement, lime, and sand) are stored in separate compartments. A computer batches the dry ingredients by weight and dry blends them prior to their being sent to the site or to a mixing silo where water is added. *(Courtesy Portland Cement Association)*

Figure 10.3 Small quantities of on-site mixed mortar should be batched in a portable mixer. *(Courtesy Portland Cement Association)*

Figure 10.4 Mortar must be workable, cling to vertical surfaces, extrude from joints, and not drop off. *(Courtesy Portland Cement Association)*

this can often be seen after the mortar has cured and the joints over the wall do not have a uniform shade. To be most effective the proportions should be carefully measured and the power mixer run from three to five minutes. Undermixing produces a nonuniform mortar, and overmixing may reduce the strength of the mortar.

PROPERTIES OF PLASTIC MORTAR

The properties that affect mortar in a plastic condition are workability and water retention.

Workability

Workability is a term used to describe the condition of a mortar that will spread easily, cling to vertical surfaces, extrude easily from joints but not drop off, and permit easy positioning of masonry units (Fig. 10.4). It is a condition of the mortar that is influenced by other properties such as consistency, water retention, setting time, weight, adhesion, and penetrability. Masons judge workability by observing how the mortar slides or adheres to their trowel.

Water Retention

Mortar must have good **water retention,** which means it resists rapid loss of the mixing water to the air or to an absorptive masonry unit. Rapid loss of mix water causes the mortar to stiffen, making it difficult to get a good bond or a watertight joint. Mortar that has good water retentivity remains workable, enabling the mason to properly place the masonry units. Low-absorption masonry units may float when placed on a mortar with too

much water retentivity. This causes the mortar joint to bleed.

Water retentivity is increased by entrained air, very fine aggregate, or cementitious materials. Water retention is measured by a flow test.

Mortar Flow Test

The water retention limit of mortar is measured by initial flow and flow after suction tests made in a laboratory as described in ASTM C91.

The *initial flow test* is made by placing a truncated cone of mortar with a 4 in. (100 mm) diameter on a metal flow table. The table is mechanically dropped ½ in. (12 mm) twenty-five times in fifteen seconds. The mortar will flow into an enlarged circular shape. The diameter of this shape is measured and compared with the original diameter. Allowable *initial flow* should be in a range of 100 percent to 115 percent (Fig. 10.5).

The *flow after suction test* is used to determine flow after loss of water to an absorbent masonry unit. The mortar sample is placed for one minute in a vacuum device that removes some water. The mortar is then tested as described above and the flow measured. The *flow after suction* should be in a range of 70 percent to 75 percent.

PROPERTIES OF HARDENED MORTAR

The properties essential to quality hardened mortar are bond strength, durability, compressive strength, low volume change, appearance, and rate of hardening.

Bond Strength

Bond strength refers to the *degree of contact* between the mortar and the masonry units and also to the **tensile**

Figure 10.5 The mortar flow test is used to measure the water retention limit. *(Courtesy Portland Cement Association)*

Figure 10.6 The tensile bond strength test measures the extent of the bond between the mortar and the masonry unit. *(Courtesy Portland Cement Association)*

bond strength available for resisting forces that tend to pull the masonry units apart.

The degree of contact between masonry units is essential to watertight joints and tensile bond strength. Good bond strength requires good, workable, water-retentive mortar, good workmanship, full joints, and masonry units with a medium rate of absorption (**suction rate**).

Tensile bond strength is necessary to withstand forces such as wind, structural movement, expansion of clay masonry units, shrinkage of mortar or concrete masonry units, and temperature changes. Tensile bond strength is tested by bonding together samples of the concrete masonry units with the mortar, curing the mortar, and pulling them apart on a tensile testing machine (Fig. 10.6).

The major factors affecting the bond strength are (1) the characteristics (strength) of the masonry units, (2) quality of the mortar, (3) workmanship of the mason, and (4) curing conditions.

Bond is high on *textured surfaces* and low on smooth surfaces. *Suction rates* of masonry units influence bond. Concrete masonry units tend to retain moisture after they are cured and have relatively low suction rates. Some clay bricks have very high suction rates, and unless they are wetted before using they will pull water from the mortar, resulting in a poor bond. After the bricks are wetted, the surface should be permitted to dry before the bricks are used.

Mortar flow influences tensile bond strength. As the water content increases, the bond strength increases. This indicates that it is wise to use the highest water content possible and yet retain a workable mortar. As water content increases, mortar compressive strength decreases. Bond strength takes precedence over compressive strength.

Good *workmanship* requires a minimum of time elapse between spreading the mortar and placing the masonry unit. Some water in the mortar will evaporate and some will be sucked away by the masonry unit on which it is placed, leaving insufficient water to form a good bond on the next masonry unit. After placing a unit on the mortar and getting its initial alignment, it should not be moved, tapped, or slid in any way. This would break the initial bond, and it cannot be reestablished. The mortar must be replaced if this happens.

Good curing conditions require the maximum amount of water possible be in the mortar, because it is needed for hydration. The laid units should be covered with plastic to retain moisture while curing. Under severe dry conditions it may be necessary to keep the wall wet with a fine mist spray for several days. It must also be protected from freezing by using insulating blankets.

Durability

Mortar must have the *durability* to withstand the forces of weathering. Frost or freezing will not damage mortar joints unless they leak and are water soaked. High compressive strength mortar usually has good durability.

Air-entrained mortar also provides protection against the freeze-thaw cycle.

Compressive Strength

The **compressive strength** of mortar depends mainly on the type and quantity of cementitious material used in the mortar. It increases as cement content increases and decreases as air-entrainment, lime, or water content increases. The compressive strength of mortar is found by testing standard cured 2 in. (50.8 mm) square cubes in a laboratory compression testing machine following ASTM C270 standards (Fig. 10.7). Mortar can be tested in the field using ASTM C780 standards.

The compressive strength of a wall depends not only on the mortar but also on the masonry unit, design of the structure, workmanship, and curing.

Low Volume Change

Mortars have low volume change. The actual shrinkage during curing of a mortar joint is minuscule.

Appearance

The mortar joints in a wall should have a uniform color or shade. Each batch of mortar should have exactly the same proportions of each ingredient. The time of tooling also causes variances in the shade of the joint. If the mortar is tooled when it is fairly hard, a darker shade will occur than if it is tooled when fairly soft. White mortar cement should be tooled with a glass or plastic joint tool because a metal tool will darken the joint.

Any pigments added to color the mortar must be carefully measured. It is best to premix the color with enough mortar cement to do the entire job rather than one batch at a time.

Joint color and shade also are affected by atmospheric conditions, admixtures, and the moisture content of the masonry units.

Rate of Hardening

The rate of hardening of a mortar due to hydration is the speed at which it develops the strength to resist an applied load. Mortar that hardens very slowly can delay construction because it will not support many brick courses above it. Mortar that has a very high rate of hardening can be difficult for the mason to use. The rate of hardening must be consistent from batch to batch, allowing sufficient time to place the masonry units and to tool the joints but permitting construction to continue at a normal pace.

Loss of water also can influence the stiffening of mortar. This is especially a concern in hot weather. In high temperatures masons frequently lay fewer masonry units on shorter beds before stopping to tool the joints.

COLORED MORTAR

Architects frequently specify the color of mortar. White mortar is made using white masonry cement or white portland cement, lime, and sand. Colored mortars are made using white masonry cement or white portland cement and get color by using color pigments, colored sand, or colored masonry cements. The final color achieved is the result of blending these ingredients. Trial batches are made and cured until the desired color is developed.

Color pigments are a form of mineral oxide such as iron, manganese, chromium, and cobalt oxides. Carbon black is used to produce a dark gray to black mortar.

Following are recommended pigments.

Color	Pigment	Maximum Amount Used*
Gray to black	Carbon	3
Green	Chromium oxide	10
Blue	Cobalt oxide	10
Reds, yellows, browns, blacks	Iron oxide	10

*Maximum amount of pigment as a percent of the weight of the portland cement.

MORTAR IN COLD WEATHER

When mortar is placed and cured the temperature of the mortar should be kept in the range of 60°F to 80°F (15.7°C to 26.9°C). The water in the mortar is needed for **hydration,** a chemical reaction between the masonry cement and the water. This leads to the hardening of the mortar, which will be slowed or stopped if the

Figure 10.7 The cube compressive strength test measures the compressive strength of the mortar. (*Courtesy Portland Cement Association*)

cement paste in the mortar drops below 40°F (4.5°C). The cement paste is the mixture formed by water and the mortar cement.

Mortar can be laid when the air temperature is above 40°F (4.5°C) using normal procedures. When the air temperature is below 40°F (4.5°C), the mortar water will usually have to be heated. However, the temperature of the *mortar* should never go above 120°F (50°C) because higher temperatures will cause the mortar to set

up too fast and the mortar will have a loss of compressive and bond strength. If the temperature of the air is falling as the mortar is laid the minimum *mortar temperature* should be 70°F (21°C). You can find the recommended mortar temperatures for various air temperatures in Tables 10.6 and Table 10.7.

The finished laid masonry should be covered when work is stopped. When air temperatures are above 40°F (4.5°C) the walls should be covered with plastic sheets

Table 10.6 Recommendations for All-Weather Masonry Construction[a]

| Air temperature, °F | Construction Requirements | |
	Heating of materials	Protection
Above 100 or above 90 with wind velocity greater than 8 mph	Limit open mortar beds to no longer than 4 ft. and set units within one minute of spreading mortar. Store materials in cool or shaded area.	Protect wall from rapid evaporation by covering, fogging, damp curing, or other means.
Above 40	Normal masonry procedures. No heating required.	Cover walls with plastic or canvas at end of work day to prevent water entering masonry.
Below 40	Heat mixing water. Maintain mortar temperatures between 40°F and 120°F until placed.	Cover walls and materials to prevent wetting and freezing. Covers should be plastic or canvas.
Below 32	In addition to the above, heat the sand. Frozen sand and frozen wet masonry units must be thawed.	With wind velocities over 15 mph, provide windbreaks during the work day and cover walls and materials at the end of the work day to prevent wetting and freezing. Maintain masonry above 32°F using auxiliary heat or insulated blankets for 24 hours after laying units.
Below 20	In addition to the above, dry masonry units must be heated to 20°F.	Provide enclosure and supply sufficient heat to maintain masonry enclosure above 32°F for 24 hours after laying units.

[a]Adapted from recommendations of the International Masonry Industry All Weather Council and requirements of ACT530 1/ASCE 6/TMS 602. (References 4 & 13).
Courtesy American Concrete Institute

Table 10.7 Canadian Protection Requirements[a]

Mean daily air temperature, °C	Protection
0 to 4	Masonry shall be protected from rain or snow for 24 hours.
−4 to 0	Masonry shall be completely covered for 24 hours.
−7 to −4	Masonry shall be completely covered with insulating blankets for 24 hours.
−7 to below	The masonry temperature shall be maintained above 0°C for 24 hours by enclosure and supplementary heat.

[a]The amount of insulation required to properly cure masonry in cold weather shall be determined on the basis of the expected air temperature and wind velocity (wind-chill factor) and the size and shape of the structure. This material is presented with the permission of the Canadian Standards Association; material is reproduced from CSA Standard A179-94 (Mortar and Grout for Unit Masonry), which is copyrighted by CSA, 178 Rexdale Blvd., Etobicoke, Ontario M9W 1R3. Although use of this material has been authorized, CSA shall not be responsible for the manner in which the information is presented, nor for any interpretations thereof. This material may not be updated to reflect amendments made to the original content. For up-to-date information, contact CSA.

to protect them from rain or light snow. Below 40°F (4.5°C) the walls should be covered with plastic or canvas to prevent their freezing or becoming wet. Some use insulated plastic or canvas-covered blankets. When air temperatures fall below 32°F (0°C) the wall must be covered with insulated blankets and a source of heat must be provided to keep the mortar from freezing for at least twenty-four hours. You can find protection recommendations in Tables 10.6 and 10.7.

SURFACE BONDING MORTARS

A variety of surface bonding mortars are available from various manufacturers. These are applied by trowel, brush, or spray to any masonry surface, and they present a new surface to which plaster, stucco, concrete, and cement-based paints will adhere.

Another form of surface bonding is the application of a cement mortar that is reinforced with glass fibers to both the surfaces of concrete block walls that were laid up without mortar. This bonds them into a solid wall and provides a waterproof coating. Tests indicate it is as strong in bending flexure as walls laid with conventional mortar joints. If the surface between the blocks is not flat and smooth, vertical compressive strength is reduced.

REVIEW QUESTIONS

1. What are the basic materials used to produce mortar?
2. What is meant by the workability of a mortar?
3. Why are mortar water-retention properties important?
4. What steps are performed to make a mortar flow test?
5. What does a flow after suction test reveal?
6. What is meant by the tensile bond strength of mortar?
7. What type of masonry surfaces produce the best mortar bond?
8. How does the water content in mortar influence bond and compressive strengths?
9. How does air-entrainment influence the compressive strength of mortar?
10. How does the volume change in mortar as it cures affect the mortar joint?
11. What things can cause a change in the shade of the mortar joint?
12. What are the types of mortar used in the United States and Canada?
13. Which type of mortar has the highest strength and where is it used?
14. Where would you use the lowest strength mortar?
15. Which type of mortar is used for general construction of exposed masonry load-bearing walls?
16. What materials can be used to produce colored mortar?
17. How can you keep mortar waiting to be used from having excessive loss of moisture?
18. What should you do with mortar that has become stiff due to hydration while it was waiting to be used?
19. What is the lowest temperature in which mortar can be laid before protection from freezing is required?
20. What is the temperature range within which the mortar should be kept?
21. When laying masonry units in below freezing weather how long must the mortar be protected from freezing?

KEY TERMS

admixtures Materials other than cement, water, lime, or aggregates added to a mortar being mixed to modify its properties.

cementitious materials Materials that have cementing properties.

compressive strength The maximum compressive stress a material can withstand.

efflorescence A white powdery substance that sometimes leaches from the mortar after it has hardened, causing a white stain on the masonry units.

hydration A chemical reaction between the cementitious materials and water.

mortar A plastic mixture of cementitious materials, water, and a fine aggregate, such as sand.

mortar flow A measure of the consistency of freshly mixed mortar related to the diameter of a molded truncated cone specimen after the sample has been vibrated a specified number of times.

suction rate The weight of water absorbed when a brick is partially immersed in water for one minute expressed in grams per minute or ounces per minute.

tensile bond strength The ability of a mortar to resist forces tending to pull the masonry units apart.

water retention The property of a mortar that prevents the rapid loss of water by absorption into the masonry units.

workability The property of freshly mixed mortar that determines the ease and homogeneity with which it can be spread and finished.

SUGGESTED ACTIVITIES

1. Prepare a number of mortar samples using different variations of ingredients, including portland cement mortar and masonry cement mortar. Cure each under identical conditions and test their compressive strength.

2. Get a supply of clay masonry and concrete masonry units. Mix mortars of different consistencies and use them to lay up a small wall. Observe and record the ease or difficulty in using the mortar and maintaining the desired size mortar joint. After the walls have cured an equal time, try to separate the masonry units. Do some seem to have greater bonding strength?

3. Prepare several samples of mortar, varying the water content slightly. Then conduct an initial mortar flow test and report your results.

4. Prepare several samples of mortar using various ingredients. After they have cured, make compression tests and record your results.

ADDITIONAL INFORMATION

Numerous publications from the Portland Cement Association, 5420 Old Orchard Rd., Skokie, Ill. 60077-1083.

Publications of the Canadian Standards Association, 178 Rexdale Blvd., Etobicoke, Ontario, Canada, M9W 1R3.

11

Clay Brick and Tile

This chapter will help you to:

1. Describe how clay bricks are made.

2. Identify the various clay masonry products and explain where they are typically used.

3. Be aware of the grades, types, and classes of clay masonry products.

4. Use information about the properties of clay masonry products to select materials for various applications.

Clay brick is made from surface or deep mined clays that have the necessary plasticity when mixed with water to permit molding to the desired shape. The clay also must have the tensile strength to hold this shape while in a plastic condition and contain clay particles that will fuse together when subjected to high temperatures. Properly manufactured brick is fire resistant. It is made in small units, providing the designer the opportunity to use a variety of designs and patterns.

CLAYS

Clay is found in three forms: surface clay, shale, and fireclay.

Surface clays are found near the surface of the earth and are strip-mined. Most bricks are made from surface mined clays. **Shales** are clays that have been subjected to high pressures, causing them to be relatively hard. **Fireclays** are found at deeper levels than the first two and have more uniform physical and chemical properties. Fireclays can withstand higher temperatures and are used where these will occur, such as in the lining of a fireplace firebox.

Clays contain a variety of materials, but they are predominantly silica and alumina with smaller amounts of metallic oxides and other ingredients. Clays are divided into two classes, calcareous and noncalcareous. **Calcareous clays** have about 15 percent calcium carbonate and have a yellow color when burned. **Noncalcareous clays** contain silicate of alumina, feldspar, and iron oxide. The iron oxide causes them to take a buff, red, or salmon color when burned. The color varies depending on the amount of iron oxide.

MANUFACTURING CLAY BRICKS

The manufacture of clay bricks requires seven steps: (1) winning and storage, (2) processing the raw materials, (3) forming the bricks, (4) drying, (5) glazing (if required), (6) burning and cooling, and (7) drawing and storage of the finished bricks.

Winning and Storage

Winning is a term used to describe the mining of the clay. Most bricks are made from surface mined clays dug from open pits. Fireclays are obtained from underground mines. The clays are moved by truck or rail to the plant where they are crushed and moved to storage piles.

Processing the Raw Materials

Clays are removed from the various storage piles and blended to produce the desired chemical composition and physical properties. The blended clays are then moved to crushers where stones are removed and the clay lumps are reduced to a maximum of about 2 in. (50 mm) in diameter.

This material is then moved by conveyor to grinders. Here it is ground to a fine powder and passed over vibrating screens. Material too large to go through the screen is sent back to the grinder. The fine material is placed in storage.

Forming the Bricks

The three major methods used to form bricks are the soft-mud process, the stiff-mud process, and the dry-press process.

The **stiff-mud process** is most widely used. It is a high-production process in which clay that contains 12 to 15 percent moisture is passed through a de-airing machine (a vacuum), which removes air pockets. The clay is then forced by an auger through a die, producing a continuous column of clay of the desired size and shape (Fig. 11.1). As the clay leaves the die, the desired surface texture is applied. Surface texture is applied by attachments that scratch, brush, roll, or in some way leave markings on the face of the brick. Common textures include smooth, matt, rugs, barks, stippled, sand-molded, water-struck, and sand-struck (Fig. 11.2).

The column then passes through a cutter which cuts the bricks to size. They are then moved by conveyor to the inspection area. Imperfect bricks are removed and sent back to be reprocessed. The good bricks are placed on drier cars for transfer to a drier kiln.

The **soft-mud process** is used for making clay bricks from clays having too much natural water to permit the use of the stiff-mud process. The bricks are formed in molds lubricated with water or sand. Brick formed in water-lubricated molds are called *water-struck* while those made in sand-lubricated molds are called *sand-struck*. Water-struck bricks have a relatively smooth surface. Sand-struck bricks have a matte textured surface.

The **dry-press process** is used with clays having 10 percent or less moisture. The mix is formed into bricks in steel molds under high pressure.

Drying

The moisture content of green bricks (unfired, newly formed bricks) varies depending on the clay and the process used. Once formed, the bricks are placed in a low temperature drier kiln for one to two days. The temperature and humidity are carefully controlled to prevent rapid shrinkage and possible cracking. From here they are glazed, if required, or go directly to high-temperature kilns.

Glazing

Some bricks have a ceramic glaze applied to one or more surfaces. This is applied after the brick has been dried. **Glaze** is a sprayed coating of mineral ingredients that melts and fuses to the brick when subjected to the required temperature. The glaze forms a smooth, glasslike coating and is available in a wide range of colors.

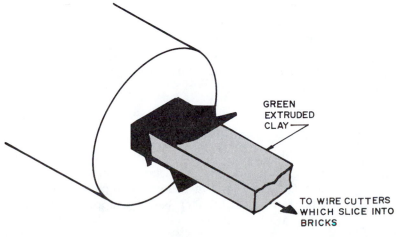

Figure 11.1 Clay bricks are cut from an extruded column of green (freshly compounded) clay.

GREEN
EXTRUDED
CLAY

TO WIRE CUTTERS
WHICH SLICE INTO
BRICKS

Antique smooth   Bark   Matt

Sandmold   Smooth   Water-struck

Figure 11.2 A few of the brick textures available.

Burning and Cooling

Burning involves raising the temperature of the dried bricks to a predetermined level. The two types of kilns in use are a periodic kiln and a tunnel kiln. The *periodic kiln* is filled with bricks stacked so air can circulate between them. The temperature in the kiln is raised, held, and lowered. The *tunnel kiln* is a long, narrow building through which the bricks are moved on cars. When the bricks reach the other end of the kiln they have been fired and cooled. The burning process for both methods takes from 40 to 150 hours, depending on desired end results.

The burning process involves several stages: water-smoking, dehydration, oxidation, vitrification, flashing, and cooling.

Water-smoking removes free water by evaporation and requires temperatures up to about 400°F (204°C). *Dehydration* removes additional moisture and requires temperatures ranging from 300 to 1800°F (150 to 980°C). *Oxidation* temperatures range from 1000 to 1800°F (540 to 980°C) and *vitrification* from 1600 to 2400°F (870 to 1315°C). These last processes transform the clay into a solid, ceramic material. *Flashing*, if required, follows at this point. It is accomplished by adjusting the fire to reduce the atmosphere in the kiln (insufficient oxygen to support combustion). This produces a variation in the colors and color shading of the bricks.

As bricks are burned they have considerable *shrinkage*. Allowance for this is made when they are cut to size or molded. The higher the burning temperatures the more the shrinkage and the darker the color of the brick. Therefore, dark bricks are usually slightly smaller than light colored bricks. Some size variation is always possible, and some distortion caused by the burning process is normal.

After the bricks have been burned and flashed as required, the cooling period begins. This takes forty-eight to seventy-two hours. The rate of cooling affects the color of the brick and controls cracking and checking.

The color of brick is related to the chemical composition of the clay or shale used and the temperature during the burn. The iron in clay turns red in an oxidizing fire and purple in a reducing fire. As mentioned earlier, the higher the temperature of the burn the darker the color.

Drawing and Storage

After the cooling stage is complete, the bricks are removed from the kiln, sorted, graded, and moved to storage. Often they are stacked on wood pallets for loading by a forklift. Each pallet load is wrapped in plastic to keep the bricks dry.

STRUCTURAL CLAY MASONRY UNITS

Structural clay masonry units are classified as solid masonry and hollow masonry.

Solid Masonry

Bricks are classified as **solid masonry** if they have cores whose area does not exceed 25 percent of the gross cross-sectional area of the brick. The cores (holes) help the drying and burning of the unit and reduce its weight.

Bricks are made in a wide range of types and sizes. A check with a brick supplier is sometimes necessary to verify the size of the unit to be supplied. The commonly manufactured bricks are shown in Fig. 11.3. It shows both modular and nonmodular units.

Modular bricks are those whose *actual size* plus a mortar joint can be assembled on a standard unit or module. The modular unit for inch-size bricks is 4 in. The actual size plus the thickness of a mortar joint ($\frac{1}{2}$ in.) is the *nominal size*. The actual and nominal sizes for modular inch bricks are shown in Table 11.1. Brick sizes

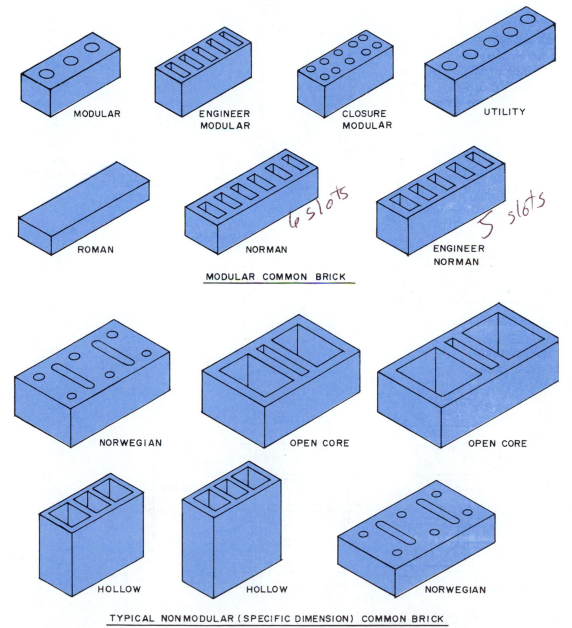

Figure 11.3 Some of the more commonly used types of modular and nonmodular brick.

are specified by three dimensions, *width, thickness,* and *length* given in that order. Figure 11.4 shows the application of the modular size of a standard inch modular brick. Using the 4 in. module, three courses of brick produces an 8 in. module. The actual sizes of nonmodular inch bricks are in Table 11.2.

The *actual size* of modular metric brick is the size of the manufactured product. The *modular size* is the size stated when specifying the brick. For example, when ordering an 89 × 57 × 190 mm actual size modular metric brick, 90 × 57 × 190 mm is specified. The *nominal size* is the modular size plus the 10 mm mortar joint. Modular metric brick sizes are in Table 11.3 and nonmodular sizes are in Table 11.4.

A standard metric modular brick can be laid on a 200 mm multiple, as shown in Fig. 11.5.

Table 11.1 Modular Inch-Size Common Brick Sizes

Unit Name		Actual Dimensions[a]	Nominal Dimensions[b]	Joint Thickness	Modular Courses
Modular	w	3½	4		
	h	2¼	2⅔	½	3C = 8″
	l	7½	8		
Engineer modular	w	3½	4		
	h	2¾	3⅕	½	5C = 16″
	l	7½	8		
Closure modular	w	3½	4		
	h	3½	4	½	1C = 4″
	l	7½	8		
Roman	w	3½	4		
	h	1⅝	2	½	2C = 4″
	l	11½	12		
Norman	w	3½	4		
	h	2¼	2⅔	½	3C = 8″
	l	11½	12		
Engineer norman	w	3½	4		
	h	2¾	3⅕	½	5C = 16″
	l	11½	12		
Utility	w	3½	4		
	h	3½	4	½	1C = 4″
	l	11½	12		

[a]Actual size unit as manufactured.
[b]Specified unit size plus intended joint size.
Courtesy Brick Institute of America

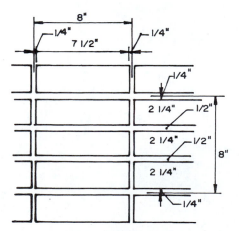

STANDARD MODULAR BRICKS SIZED
FOR 1/2" MORTAR JOINT

    NOMINAL   4"x   2 2/3" x 8"

    ACTUAL   3 1/2"x 2 1/4" x 7 1/2"

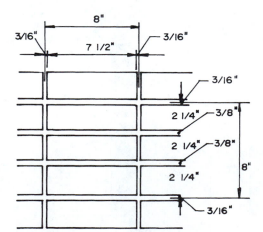

STANDARD MODULAR BRICKS SIZED
FOR 3/8" MORTAR JOINT

    NOMINAL   4" x   2 2/3" x 8"

    ACTUAL   3 5/8"x 2 1/4" x 7 1/2"

Figure 11.4 Modular bricks are sized for an 8 in. module with ½ or ⅜ in. mortar joints.

Hollow Masonry

Hollow clay masonry is a clay masonry unit whose net cross-sectional area in the plane of the bearing surface is not less than 60 percent of the gross cross-sectional area of that face. A hollow brick may have a cored area from 25 to 40 percent of the gross cross-sectional area of the bearing surface.

Structural clay facing tiles are hollow clay units whose cores exceed 25 percent of the gross cross-sectional area (Fig. 11.6). They are used in load-bearing and non-load-bearing walls. They are available with a glazed or an unglazed finished surface. A smooth, colored glaze is the most frequently used surface, but they also are available in matte, speckled, and mottled finishes. Surfaces with a

Table 11.2 Specified Inch-Size Nonmodular Common Brick Sizes

Unit Name		Actual Dimension[a]
Standard	w	$3^5/_8$
	h	$2^1/_4$
	l	8
Engineer standard	w	$3^5/_8$
	h	$2^{13}/_{16}$
	l	8
King	w	3
	h	$2^3/_4$
	l	$9^5/_8$
Queen	w	3
	h	$2^3/_4$
	l	8

[a]Anticipated manufactured dimension
Courtesy Brick Institute of America

Table 11.3 Modular Metric Common Brick Sizes[a]

Unit Name		Actual Dimensions (comparable inch size)	mm[b]	Metric Modular Size (mm)[c]	Metric Nominal Size (mm)[d]	Vertical Coursing
Modular	w	$3^1/_2$	89	90	100	
	h	$2^1/_4$	57	57	67	3:200 mm
	l	$7^1/_2$	190	190	200	
Engineer modular	w	$3^1/_2$	89	90	100	
	h	$2^3/_4$	70	70	80	5:400 mm
	l	$7^1/_2$	190	190	200	
Closure modular	w	$3^1/_2$	89	90	100	
	h	$3^1/_2$	89	90	100	2:200 mm
	l	$7^1/_2$	190	190	200	
Roman	w	$3^1/_2$	89	90	100	
	h	$1^5/_8$	41	40	50	4:200 mm
	l	$11^1/_2$	292	290	300	
Norman	w	$3^1/_2$	89	90	100	
	h	$2^1/_4$	57	57	67	3:200 mm
	l	$11^1/_2$	292	290	300	
Engineer norman	w	$3^1/_2$	89	90	100	
	h	$2^3/_4$	70	70	80	5:400 mm
	l	$11^1/_2$	292	290	300	
Utility	w	$3^1/_2$	89	90	100	
	h	$3^1/_2$	89	90	100	2:200 mm
	l	$11^1/_2$	292	290	300	

[a]Metric sizes are hard conversions.
[b]Actual size unit as manufactured.
[c]Size used when specifying metric brick.
[d]Modular size plus 10 mm mortar joint.
Courtesy Brick Institute of America

Table 11.4 Specified Metric Nonmodular Common Brick Sizes[a]

Unit Name	Actual Dimensions (comparable inch size)		mm[b]	Metric Modular Size (mm)[c]	Vertical Coursing
Standard	w	3⅝	92	90	
	h	2¼	57	57	3:200 mm
	l	8	203	190	
Engineer standard	w	3⅝	92	90	
	h	2¹³/₁₆	71	70	5:400 mm
	l	8	203	190	
King	w	3	76	?	
	h	2¾	70	70	5:400 mm
	l	9⅝	245	?	
Queen	w	3	76	?	
	h	2¾	70	70	5:400 mm
	l	8	203	190	
Standard closure	w	3½	89	90	
	h	3⅝	92	90	2:200 mm
	l	8	203	190	

[a]Metric sizes are hard conversions.
[b]Actual size unit is manufactured.
[c]Size used when specifying metric brick.
Courtesy Brick Institute of America

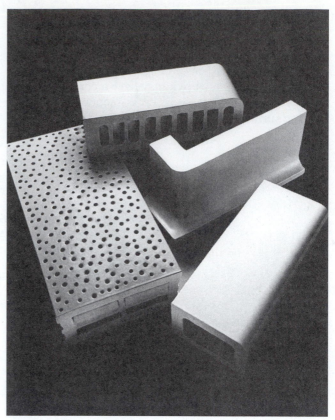

Figure 11.6 A wide variety of structural clay facing tile is available. *(Courtesy Stark Ceramics, Inc.)*

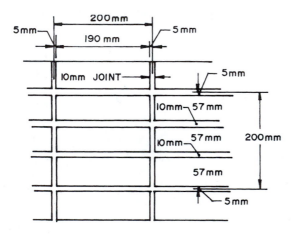

ACTUAL SIZE METRIC MODULAR BRICK
89 X 57 X 190 mm
MORTAR JOINT 10 mm

Figure 11.5 Standard metric modular brick are laid on a 200 mm multiple.

rough texture or covered with small-diameter holes of varying sizes provide acoustical treatment. They meet or exceed ASTM requirements for imperviousness, resistance to fading and crazing, zero flame spread, zero toxic fumes, and resistance to scratching. They can be made to provide for radiation protection and can be installed over wood and metal stud walls and any type of masonry or concrete wall (Fig. 11.7). Selected sizes of structural clay facing tile are in Fig. 11.8.

MATERIAL SPECIFICATIONS FOR CLAY MASONRY

The specifications for clay masonry units have been developed by the American Society for Testing and Materials. When specifying clay masonry units the appropriate ASTM specifications should be included. The ASTM specifications for brick, hollow brick, and structural clay tile are based on the weathering index.

Weathering Index

The **weathering index** reflects the ability of clay masonry units to resist the effects of weathering. It is the product of the number of **freezing cycle days** and the annual *winter rainfall*.

A freezing cycle day is a day when the temperature of the air rises above or falls below 32°F (0°C). Winter rainfall is measured in inches between the first and last killing frosts in the fall and spring.

The weathering index ranges from 0 to more than 500. Regions rated higher than 500 are *severe weathering regions,* those between 50 and 500 are *moderate regions,* and areas below 50 are *negligible regions* (Fig. 11.9).

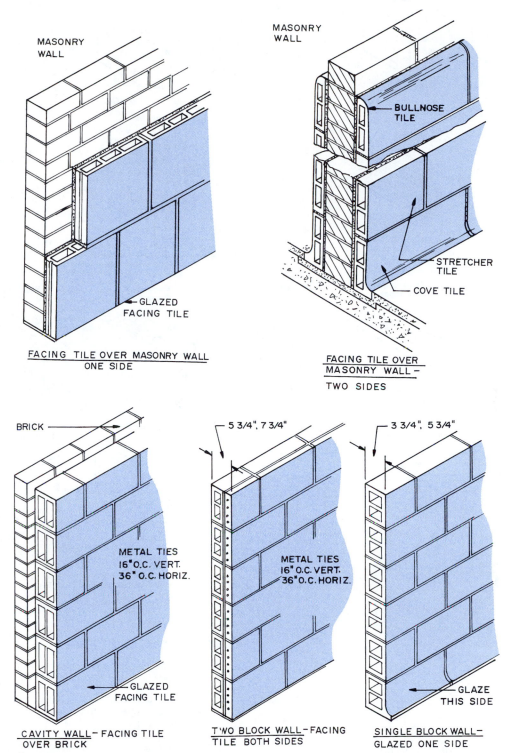

Figure 11.7 Typical installation details for structural clay facing tile.

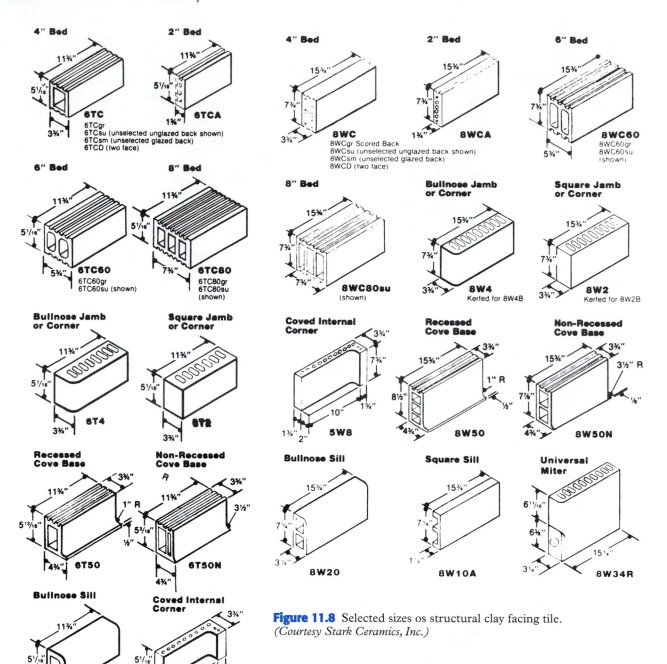

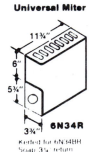

Write for complete shapes, sizes and specifications sheets. Request 6T or 8W series.

Figure 11.8 Selected sizes os structural clay facing tile. *(Courtesy Stark Ceramics, Inc.)*

Grades, Types, and Classes

Building (common) brick is available in three grades, SW, MW, and NW, as specified by ASTM C62 (Table 11.5).

Facing brick is available in two grades, SW and MW, and three types, FBX, FBS, and FBA, as specified by ASTM C216 (Table 11.6).

Hollow brick is available in two grades, SW and MW, and four types, HBS, HBX, HBA, and HBB, as specified by ASTM C652 (Table 11.7).

Load-bearing structural clay tile is available in two grades, LBX and LB, as specified by ASTM C34 (Table 11.8). Load-bearing tile will carry building loads in addition to its own weight.

Non-load-bearing structural clay tile is available in one grade, NB, as specified by ASTM C56 (Table 11.9). Non-load-bearing tile carries only its own weight.

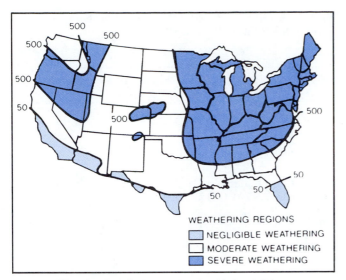

Weathering Indexes in the United States

Figure 11.9 The weathering index is an indication of the ability of clay masonry units to resist the effects of the weather. *(Courtesy Brick Institute of America)*

Table 11.5 Grades and Uses of Solid Building Brick (ASTM C62)

Grade (based on weathering index)	Use
SW Severe weathering	For wet locations below grade where bricks may be frozen, such as in foundations
MW Moderate weathering	For vertical masonry surfaces exposed to the weather in relatively dry conditions where freezing can occur
NW Negligible weathering	For use as backup or interior masonry where no freezing occurs

Courtesy American Society for Testing and Materials

Table 11.6 Grades, Types, and Uses of Solid Facing Brick (ASTM C216)

Grade (based on weathering index)	Use
SW Severe weathering	Masonry in wet locations, in contact with the ground, and subject to freezing
MW Moderate weathering	Exterior walls and other exposed masonry above grade where freezing can occur
Type (based on appearance of the finished wall)	**Use**
FBX	High degree of physical perfection, minimum variation in color, minimum variation in size
FBS	Wider color range and size variations than permitted in Type FBX
FBA	Nonuniform in size, color, and texture

Courtesy American Society for Testing and Materials

Unglazed structural facing tile is available in two grades, FTX and FTS, as specified by ASTM C212. This is based on face shell thickness and factors affecting the appearance of the finished wall (Table 11.10).

Ceramic glazed facing brick and structural clay facing tile is available in two grades, S and SS, and two types, I and II, as specified by ASTM C126 (Table 11.11).

Pedestrian and light traffic paving brick is available in three classes, SX, MX, and NX, and three types, I, II, and III, as specified by ASTM C902 (Table 11.12). This is used on patios, walkways, floors, and driveways.

PROPERTIES OF CLAY BRICK AND TILE

Finished clay units vary considerably in their physical properties. The properties of the raw materials used and the effects of production processed greatly influence these properties. The major properties are compressive strength, durability, absorption, color, and texture.

Compressive Strength

The compressive strength depends on the clay, how the units are made, and the temperature and length of the burn. In general, plastic clays used in the stiff-mud process produce units with higher compressive strengths. Higher burn temperatures produce higher compressive strengths in almost any clay or process used. Bricks vary in compressive strength from 1500 psi (10.35 MN/m^2) to more than 5000 psi (34.50 MN/m^2).

It should be noted that the compressive strength in a wall of brick depends not only on the compressive strength of the brick but also on that of the mortar. For example, a brick may have a compressive strength of 4000 psi, but when it is combined with a mortar the allowable compressive strength will be limited by the compressive strength of the mortar used.

Durability

Durability refers to the ability of a clay masonry unit to resist damage due to freezing and thawing cycles while it is subjected to moisture. Durability is a result of fusion of the clays during burning. High burn temperatures tend to produce a harder, more durable unit.

Absorption

Clay units absorb a certain amount of water. This property affects the bond strength of the mortar to the brick.

Table 11.7 Grades, Types, and Uses of Hollow Brick (ASTM C652)

Grade (based on weathering index)	Use
SW Severe weathering	High degree of resistance to disintegration by weathering when brick may be permeated with water and frozen
MW Moderate weathering	Moderate degree of resistance to frost action where brick is not likely to be permeated with water when it is exposed to freezing temperatures

Type (based on appearance of the finished wall)	Use
HBS	Visible interior and exterior walls where graded variations in color and size than specified for HBX are acceptable
HBX	Visible interior and exterior walls where a small variation in color and size are acceptable
HBA	Nonuniform in size, color, and texture
HBB	Color and texture are not a consideration Size variation greater than specified for HBX is acceptable

Courtesy American Society for Testing and Materials

Table 11.8 Grades and Uses of Structural Clay Load-Bearing Tile (ASTM C34)

Grade (based on weathering index)	Use
LBX	Tile exposed to the weather and as a base for the applications of stucco
LB	Tile not exposed to frost or earth May be used in exposed masonry if covered with 3 in. or more of other masonry

Courtesy American Society for Testing and Materials

Table 11.9 Grade and Use of Structural Clay Non-Load-Bearing Tile (ASTM C56)

Grade	Use
NB	Non-load-bearing walls, partitions, fireproofing, and furring

Courtesy American Society for Testing and Materials

Table 11.10 Grades and Uses of Unglazed Structural Clay Facing Tile (ASTM C212)

Grade	Use
FTX	Exposed masonry with minimum variation in color and dimensions, smooth face, mechanically perfect
FTS	Smooth or rough textured, with moderate absorption and variation in dimensions, medium color range, and minor surface finish defects

Courtesy American Society for Testing and Materials

The properties of the clays, the process used to make the brick, and the burn temperature affect absorption.

In general, plastic clays and high-temperature burns produce clay units that have lower absorption. Units made with the stiff-mud process usually have lower absorption than those made with other processes. The rate at which a clay unit absorbs moisture is called the **suction rate.** Suction refers to the tendency of a brick to take up moisture in pores and small openings in the surface by capillary action. It does not refer to moisture penetrating the brick itself but just surface water. Suction affects the bond strength of the mortar to the brick. The strongest bond is achieved when the unit has a suction rate that does not exceed 0.7 ounces (20 grams) of water per minute. If a brick has a greater suction rate than this, it should be wetted and the surface allowed to dry before the brick is laid.

The initial rate of absorption (suction) can be determined in the laboratory using ASTM 67 testing procedures. The masonry unit is immersed in $1/8$ in. (3 mm) of water for one minute. It is removed and weighed. The gain in weight from its dry condition is an indication of the *initial rate of absorption* (IRA) and is measured in grams per minute per 30 in.^{2} The IRA values range from 1 to 50 or more.

Color

The color of clay masonry units depends on the clays used, the burning temperature, and the method of controlling color during the burn. Burned clays vary widely in color, from creams and buffs through reds and purples. Iron oxide, commonly found in clay, is the dominant oxide that influences color. It produces a red unit in an oxidizing fire and becomes purple if burned in a reduced atmosphere (called flashing). Underburning produces a salmon colored unit that is softer, has a lower compressive strength, and a higher absorption rate than red bricks. Bricks that are overburned are called clinkers and tend toward dark red to black when made from clays high in iron oxide. Buff clays are used to produce bricks ranging from yellow-brown to dark tan.

Since the composition of clays varies, the color of the units produced has some variation within each unit. The application of surface coatings, such as glazes, enables

Table 11.11 Grades, Types, and Uses of Ceramic Glazed Structural Clay Facing Tile (ASTM C126)

Grade	Use
S Select	For masonry with narrow mortar joints (¼″)
SS Select sized or ground edge	For masonry where the face dimension variation is small
Type	**Use**
I Single-face units	Where only one finished face is to be exposed
II Two-faced units	Where two opposite finished faces are to be exposed

Courtesy American Society for Testing and Materials

Table 11.12 Classes, Types, and Uses of Pedestrian and Light Traffic Paving Block (ASTM C902)

Class (based on weathering index)	Use
SX	Where brick may be frozen when saturated with water
MX	Where resistance to freezing is not necessary
NX	Interior use when an effective sealer or water-resistant surface coating will be applied
Type (based on traffic)	**Uses of the brick**
I	Where brick will be exposed to extensive abrasion, as in driveways
II	Where brick will be exposed to intermediate traffic, as on floors in stores
III	Where bricks will be exposed to low traffic, as in residences

Courtesy American Society for Testing and Materials

units to be produced with almost any color desired. Chemicals can be introduced into the clay that produce a range of colors when they vaporize.

Texture

The surface texture of a finished clay masonry unit is produced by the surface of the die or mold used to form the unit or by attachments that cut, scratch, roll, or in some other way alter the surface as the clay unit leaves the die. Some of the standard textures are smooth, matt (with horizontal or vertical markings), barks, rugs, sand-molded, stippled, water-struck, and sand-struck (Fig. 11.2).

Heat Transmission

The ability of clay brick walls to transmit or resist the transmission of heat directly influences the surface temperature of interior walls. In most applications resistance to heat transmission is very important. In others, such as the storage and transmission of solar heat, the ability to transmit heat is important. Selected examples of heat transmission (called the U value) and heat resistance (called the R value) are in Table 11.13. These figures vary for different types of brick because of the variances in the clay used and the density of the units.

Sound Transmission

A major factor when designing walls is the ability of the material used to block the transmission of sound through the wall. Walls can reduce sound transmission by absorbing or reflecting sound or by reducing the diaphragm action of the materials.

Clay products absorb very little sound because they have a high density and a hard surface. They do reflect sound, which helps reduce transmission through the wall, but this may cause problems within the room due to the reflection of the sound back and forth around the room. Diaphragm action occurs when sound waves cause a material to vibrate. Clay masonry has good resistance to diaphragm action because of its weight and stiffness.

The transmission of sound is given in decibels. The reduction in sound transmission afforded by several brick and tile units are in Table 11.14.

Fire Resistance

The fire resistance of a material is an indication of its ability to prevent materials behind it from igniting. Fire resistance is stated in hours and is usually called the *fire rating*. These ratings are developed by testing the units under actual fire conditions set up in a test laboratory. Most floors, walls, and ceilings are made up of several materials so fire ratings of these assemblies are also

Table 11.13 Coefficients of Heat Transmission and Resistance of Solid and Cavity Clay Brick[a]

Brick	Transmission (U)	Resistance (R)
4″ (100 mm) solid brick	2.27	0.44
6″ (150 mm) solid brick	1.52	0.66
8″ (200 mm) solid brick	1.14	0.88
12″ (300 mm) solid brick	0.76	1.32
10″ (250 mm) cavity brick	0.34	2.96

[a]Values reflect an average rating. Get actual values from the manufacturer of the masonry units to be used.

known. Table 11.15 lists the fire ratings for selected clay masonry units.

CERAMIC TILE

Ceramic tile includes wall tile, mosaic tile, quarry tile, and paver tile. Wall tiles are $\frac{5}{16}$ in. thick, mosaic tiles $\frac{1}{4}$ in. thick, quarry tiles, $\frac{3}{8}$, $\frac{1}{2}$, and $\frac{3}{4}$ in. thick, and paver tiles $\frac{1}{4}$, $\frac{3}{8}$, and $\frac{1}{2}$ in. thick. They are made in a variety of trim types and flat tiles. Refer to manufacturers' catalogs for information about the extensive designs available.

Ceramic Tile Grade Marking and Certification

There are three grades of ceramic tile. *Standard grade* is the best and meets all of the specifications. *Second grade* meets all specifications, but it has facial defects noticeable from a distance of 10 ft. *Decorative thin wall tile grade* meets all specifications except breaking strength.

Ceramic wall tile, mosaic tile, quarry tile, paver tile, and special-purpose tile are shipped in sealed cartons that have the grade indicated by a label glued to the carton. The color of the label also indicates the grade. Standard grade labels are blue and second grade yellow. The label of decorative thin wall tile is orange. Examples of the labels are in Fig. 11.10.

Types of Ceramic Tile

Ceramic wall tiles are fired clay tiles that are usually glazed and widely used on interior surfaces, such as walls, floors, showers, and countertops. They are avail-

able in a variety of shapes and sizes (Fig. 11.11). They are not expected to withstand excessive impact or be subject to freezing and thawing. High gloss and matt finishes are commonly used and the surface may be smooth or textured. A wide range of colors is available, including solid, multishaded, and speckled variations. Much tile is imported from foreign manufacturers. Distributors' catalogs should be consulted for sizes available. These tiles are often supplied mounted on a backing sheet so a number of tiles, properly spaced, can be applied to the wall at one time (Fig. 11.12).

Ceramic mosaic tile may be either natural clay or porcelain in composition. The tiles are smaller than wall tiles and usually square or hexagonal in shape. Standard sizes are in Fig. 11.11. These individual tiles are assembled on a backing material forming sheets from 9 × 9 in. (225 × 225 mm) to 12 × 24 in. (300 × 600 mm). The entire sheet is embedded in the adhesive on the wall or floor as if it were one big tile (Fig. 11.13).

Ceramic tile can be installed over wood or metal studs or furring or any masonry or concrete wall. Some construction details are in Fig. 11.14.

To get metric sizes the industry has soft-converted inch sizes to the nearest millimeter. For examples a 4¼ × 4¼ in. tile is 108 × 108 mm. Likewise, a $\frac{5}{16}$ in. thick tile is 8 mm thick.

Quarry tiles are hard, burnt unglazed or glazed clay tiles that are red, brown, or buff in color, depending on the clay. They may be extruded or dry-pressed. They are made in a variety of shapes as shown in Fig. 11.11. Quarry tiles generally are from ½ to ¾ in. (12 to 20 mm) thick. A number of trim tiles are made for use with quarry tiles. These include straight base, cover base,

Table 11.14 Sound Transmission Class Ratings of Selected Clay Masonry Units

Brick	STC Class
4″ (100 mm) brick wall	45
6″ (150 mm) SCR brick wall	51
8″ (200 mm) solid face brick wall	52
10″ (250 mm) reinforced brick wall	59
12″ (300 mm) solid brick wall	59
Clay Tile	**STC Class**
4″ (100 mm) structural clay tile wall	39
8″ (200 mm) structural clay tile wall	45

Table 11.15 Fire Resistance of Selected Brick and Tile Walls

Solid Brick	Fire Rating in Hours
4″ (100 mm) brick	1
6″ (150 mm) brick	3
8″ (200 mm) brick	4
Hollow Core Brick	
8″ (200 mm) brick wall	2
8″ (200 mm) brick wall plastered both sides	4
Structural Facing Tile	
4″ (100 mm) tile wall plastered one side	1
6″ (150 mm) tile wall	2
8″ (200 mm) tile wall plastered both sides	4

MANUFACTURER'S NAME
AND ADDRESS

GRADE SEAL

STANDARD GRADE

THIS TILE COMPLIES WITH
RECOMMENDED STANDARD SPECIFICATIONS
FOR CERAMIC TILE ANSI A137.1-1988

MANUFACTURER'S NAME
AND ADDRESS

GRADE SEAL

SECONDS

THIS TILE COMPLIES WITH
RECOMMENDED STANDARD SPECIFICATIONS
FOR CERAMIC TILE ANSI A137.1-1988

MANUFACTURER'S NAME
AND ADDRESS

GRADE SEAL

DECORATIVE THIN WALL TILE

THIS TILE COMPLIES WITH
RECOMMENDED STANDARD SPECIFICATIONS
FOR CERAMIC TILE ANSI A137.1-1988

Figure 11.10 Grade labels for ceramic wall, mosaic, quarry, paver, and special-purpose tile. *(Courtesy Tile Council of America)*

double bullnose, and internal and external corner base units.

Paver tile is the same as quarry tile but thinner, ranging from ³⁄₈ to ⁵⁄₈ in. (9.5 to 16 mm). Paver tiles are used in areas where the loads and traffic are less.

OTHER CLAY TILE PRODUCTS

Clay tile is a vitrified clay product, such as clay pipe, used for drainage and sewer systems.

Roof tile is burnt clay used for finished roofing. The tiles are made in flat, plain shingles, single-lap tiles, and interlocking tiles. They are made in various styles, reflecting the countries where they have been used for centuries. Among these are Spanish, French, Roman, and English. Dimensions vary, so manufacturers should be consulted for design data (Fig. 11.15).

Architectural Terra-Cotta and Ceramic Veneer

Architectural terra-cotta is a hard fired clay that has been used for hundreds of years for decorative purposes. Modern terra-cotta products are extruded and molded or pressed to shape. Most of these products are custom-made to the designs and specifications of the architect.

A machine-made unit having either a flat or ribbed back and a flat face is called *ceramic veneer.* Molded or pressed units are called *architectural terra-cotta.*

Ceramic veneer may be an earth red if unglazed but may have a transparent ceramic glaze, a nonlustrous glaze with a satin or matte finish or a ceramic color glaze in either a solid color or a mottled blend of colors. A polychrome finish is used, which means that two or more colors are applied to separate areas and each color is burned separately. The surface may be smooth, scored, combed, or roughened.

Ceramic veneer may be adhesion or anchor type (Fig. 11.16). The adhesion type is available in sizes up to 600 in.^{2} (0.385 m^2). Widths can be up to 24 in. (600 mm), with panel lengths up to 36 in. (900 mm) and up to 1⁵⁄₈ in. (40 mm) thick. Ceramic veneer tiles are bonded to the wall surface by setting them in a coat of mortar applied to the wall.

The anchor-type panels are larger and thicker, approaching 2¹⁄₂ in. (62 mm). They are secured to the wall with metal ties, and the open area in the ribs is filled with grout.

Architectural terra-cotta is also a red earth color when unglazed. It is available with a smooth, plain surface that may be unglazed, or it may have a transparent glaze, a nonlustrous glaze that gives a satin or matte finish, a ceramic color glaze, or a polychrome finish. Sculptural and decorative reproductions of classic architectural ornamentation can be made from terra-cotta and are generally handmade in special molds. A typical example of a terra-cotta molding is in Fig. 11.17.

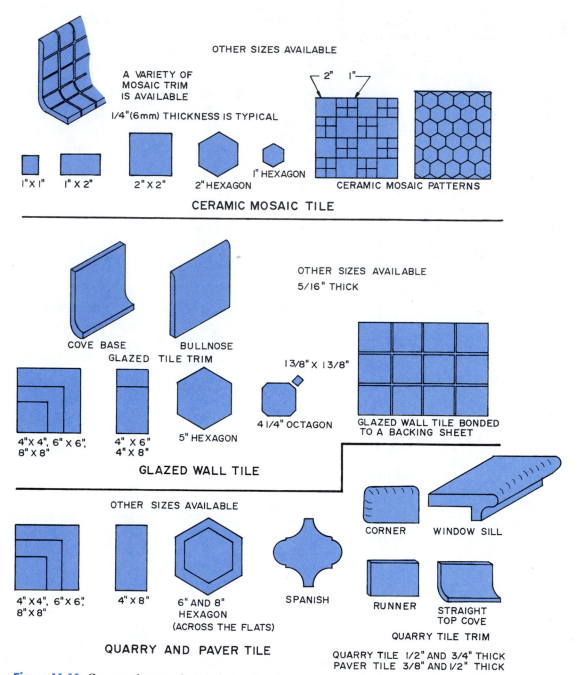

Figure 11.11 Commonly manufactured glazed wall, ceramic mosaic, quarry, and paver tile.

Figure 11.12 Glazed wall tiles are assembled with a backing sheet permitting the installation of a large area at one time. *(Courtesy American Olean Tile Company)*

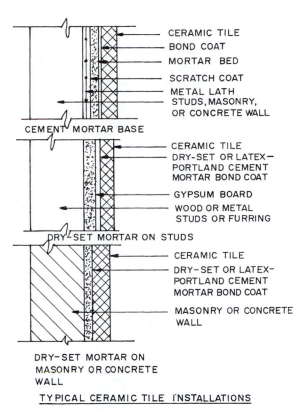

CEMENT MORTAR BASE
- CERAMIC TILE
- BOND COAT
- MORTAR BED
- SCRATCH COAT
- METAL LATH
- STUDS, MASONRY, OR CONCRETE WALL

DRY-SET MORTAR ON STUDS
- CERAMIC TILE
- DRY-SET OR LATEX–PORTLAND CEMENT MORTAR BOND COAT
- GYPSUM BOARD
- WOOD OR METAL STUDS OR FURRING

DRY-SET MORTAR ON MASONRY OR CONCRETE WALL
- CERAMIC TILE
- DRY-SET OR LATEX–PORTLAND CEMENT MORTAR BOND COAT
- MASONRY OR CONCRETE WALL

TYPICAL CERAMIC TILE INSTALLATIONS

Figure 11.14 Ceramic tile installation details.

Figure 11.13 Ceramic mosaic tiles are commonly used on floors. *(Courtesy American Olean Tile Company)*

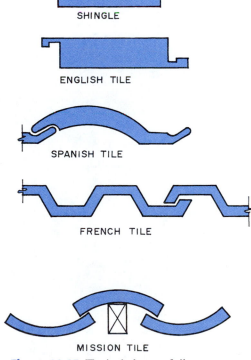

SHINGLE

ENGLISH TILE

SPANISH TILE

FRENCH TILE

MISSION TILE

Figure 11.15 Typical clay roof tile patterns.

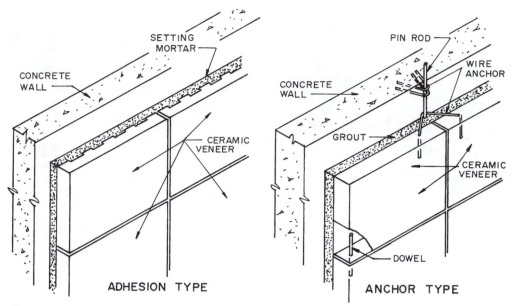

Figure 11.16 Ceramic veneer may be adhesion or anchor type.

Figure 11.17 These terra-cotta units form a decorative finish to this arched opening.

REVIEW QUESTIONS

1. What are the forms in which clay is found?
2. How do fireclays differ from surface clays?
3. How do the colors of calcareous and noncalcareous clays differ?
4. What methods are used to form bricks?
5. What are some of the common surface textures used on face bricks?
6. Why are the temperature and humidity carefully controlled when drying bricks?
7. How long does the burning process take?
8. What stages do the brick go through in the burning process?
9. How does the burning temperature influence the size and color of the bricks?
10. What are the classifications of structural clay masonry units?
11. How can you identify solid masonry clay units?
12. What is the module used when making standard bricks?
13. What is the actual size of a standard clay brick?
14. What is the allowable hollow cored area for a hollow brick?
15. What two factors are considered when establishing the weathering index?
16. What are the grades of common building brick?
17. What is the range of compressive strength for common bricks?
18. What unit of suction provides conditions for the best mortar bond?

KEY TERMS

architectural terra-cotta Clay masonry units made with a textured or sculptured face.

burning Curing bricks by placing them in a kiln and subjecting them to a high temperature.

calcareous clays Clays containing at least 15 percent calcium carbonate.

dry-press process Process used to make bricks when the clay contains 10 percent or less moisture.

fireclays Deep mined clays that withstand heat.

freezing cycle day A day when the temperature of the air rises above or falls below 32°F (0°C).

glaze A ceramic coating fused to the surface of bricks at high temperatures, forming a glasslike coating.

hollow clay masonry A unit whose core area is 25 to 40 percent of the gross cross-sectional area of the unit.

noncalcareous clays Clays containing silicate of alumina, feldspar, and iron oxide.

shales Clays that have been subjected to high pressures, causing them to become relatively hard.

soft-mud process Process used to make bricks when the clay contains moisture in excess of 15 percent.

solid masonry A unit whose core does not exceed 15 percent of the cross-sectional area of the unit.

stiff-mud process Process used to make bricks from clay that has 12 to 15 percent moisture.

suction rate The rate at which clay masonry units absorb moisture.

surface clays Clays obtained by open pit mining.

weathering index A value that reflects the ability of clay masonry units to resist the effects of weathering.

winning A term used to describe the mining of clay.

SUGGESTED ACTIVITIES

1. Collect samples of clay masonry products for use in the classroom. Label each, giving as much information about its properties as you can collect.

2. Design and run some tests on samples of clay masonry units, such as compression tests, hardness tests, and moisture absorption tests.

ADDITIONAL INFORMATION

Hornbostel, C., *Construction Materials: Types, Uses, and Applications*, John Wiley and Sons, New York, 1991.

Merritt, F.S., and Ricketts, J.T., *Building Design and Construction Handbook*, McGraw-Hill, New York, 1994.

Ramsey, C.G., Sleeper, H.R., and Hoke, J.R., eds. *Architectural Graphic Standards*, John Wiley and Sons, New York, 1991.

Additional resources are

Publications from the Brick Institute of America, 11490 Commerce Park Drive, Reston, Va. 22091-1525.

C H A P T E R 12

Concrete Masonry

This chapter will help you to:

1. Understand how concrete masonry units are manufactured.

2. Use information about the physical properties of concrete masonry units when making material selection decisions.

3. Recognize the many types of concrete masonry units and be able to choose those suitable for various applications.

oncrete masonry units are manufactured in a wide range of standard sizes and custom-designed architectural units. They are one of the most widely used construction materials, and they find use in structural and nonstructural applications. For example, they are used for foundations, piers, columns, pilasters, and for other structural purposes. They also are used for non-load-bearing walls, for fire protection of steel, for sidewalks, drives, and patios, and as a backing for other materials, such as clay brick.

MANUFACTURE OF CONCRETE MASONRY UNITS

Concrete masonry units are made of a relatively dry mix of portland cement, aggregates, water, and in some cases admixtures. The dry materials are carefully weighed and moved to a mixer. The mixer adds the required water and mixes the batch for a predetermined time. The mixed batch is discharged into the hopper of the block machine. Here it is fed into molds and consolidated by pressure and vibration. The freshly molded blocks are called *green blocks*. They are moved from the block machine on steel pallets to a curing rack.

The curing rack full of green units is moved to a low-pressure steam kiln or an autoclave for hardening. In a low-pressure steam **kiln** the green units are allowed to attain an initial set before steam is introduced. This takes one to three hours, and the temperature is kept at 70°F to 100°F (22°C to 38°C). Then the steam is introduced, which provides heat and moisture. The temperature is gradually raised to 150°F to 180°F (66°C to 82°C), depending on the composition of the concrete. This condition is maintained for ten to twenty hours until the units reach the required strength. After the blocks are removed, they are stored in a protected condition and attain almost full strength in two to four more days.

The **autoclave** uses high-pressure steam. The molded units are placed in the autoclave and allowed to set for two to five hours. Then they are gradually heated with saturated steam under a pressure of 150 psi (1035 kPa). This takes two to three hours. Once the maximum temperature, 350°F (178°C), and pressure are reached the units soak for five to ten hours. Finally, the pressure is gradually released over a thirty-minute period and the units go to storage. They can be used twenty-four hours after they are removed from the autoclave. These units have greater stabilization against volume changes caused by moisture than the low-pressure units (Fig. 12.1).

PHYSICAL PROPERTIES OF CONCRETE MASONRY UNITS

The physical properties of concrete, discussed in Chapter 7, apply to concrete masonry units. The properties of the units are determined by cement paste and aggregates. The differences that occur when compared to concrete for general use are due to different mix compositions, methods of consolidation, and curing processes. Concrete masonry units are generally made with less cement per cubic yard and a lower water-cement ratio. The aggregate is finer, with ³⁄₈ in. (10 mm) being the largest size. Concrete masonry units have a large volume of void spaces between aggregate particles, while normal concrete should have no voids.

The properties of importance when considering the use of concrete masonry units are weight, compressive strength, water absorption, and the coefficient of thermal expansion (Table 12.1). The required properties are established by national building codes and ASTM standards.

The *weight* of a concrete masonry unit varies with the design of the block and the mix used to make it. It is necessary to know weights so the dead loads of the structure can be calculated.

Compressive strength data give a means of determining a unit's ability to carry loads and withstand structural stresses. The strength varies depending on the wetness of the mix. The wetter mixes give the highest strength but cause difficulties in making the units. The manufacturer develops units and tests them so the user will know the compressive strength of the products available. The compressive strengths in Table 12.1 are based on the gross bearing area of the block, including the core spaces. Compressive strength of the net area (actual surface, excluding the cores) is 1.8 times the values shown.

Using the graph in Fig. 12.2, design compressive data can be obtained if the compressive strength of the concrete masonry unit (based on actual area) and the type of mortar are known.

Tensile strength is about 5 percent to 10 percent of compressive strength. Flexural strength is about 15 percent to 20 percent of compressive strength, and the modulus of elasticity ranges from 300 to 1200 times the compressive strength.

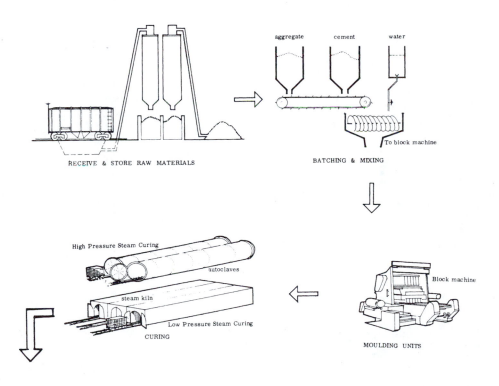

Figure 12.1 Manufacturing process used to produce concrete masonry units. (*Courtesy National Concrete Masonry Association*)

Table 12.1 Properties of Concrete Blocks with Various Aggregate

Customary Units			
Aggregate	**Compressive Strength in psi (gross area)**	**Coefficient of Thermal Expansion per °F × 10⁻⁶**	**Water Absorption (lb./ft.³ of concrete)**
Sand and gravel	1200–1800	5.0	7–12
Limestone	1100–1800	5.0	8–12
Air-cooled slag	1100–1500	4.6	9–13
Expanded shale	1000–1500	4.5	12–15
Cinders	700–1000	4.5	12–18
Expanded slag	700–1200	4.0	12–18
Pumice	700–900	4.0	13–20
Scoria	700–1200	4.0	12–18
Metric Units			
Aggregate	**Compressive Strength in kg/cm² (gross area)**	**Coefficient of Thermal Expansion mm/mm/°C × 10⁻⁶**	**Water Absorption (kg/m³)**
Sand and gravel	84.4–127	9.0	128–190
Limestone	77.3–127	9.0	128–190
Slag	77.3–105.5	8.3	144–208
Expanded shale	70.3–105.5	8.1	192–240
Cinder	49.2–70.3	4.5	192–288
Expanded slag	49.2–84.4	7.2	192–288
Pumice	49.2–63.3	7.2	208–320
Scoria	49.2–84.4	7.2	192–288

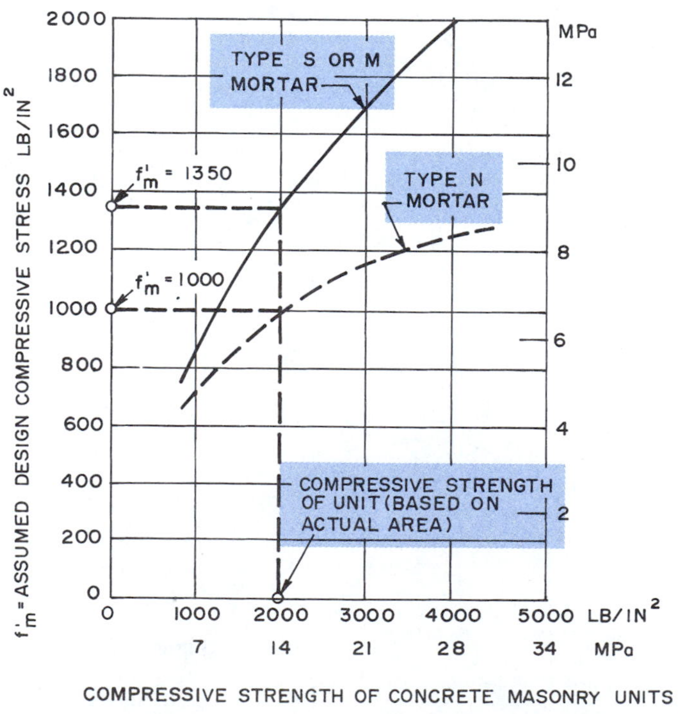

Figure 12.2 The compressive strength of a wall made from concrete masonry units depends on the type of mortar and the strength of the concrete masonry units. *(From "Masonry Structural Design for Buildings," Department of Army and the Air Force.)*

Water absorption varies with the *density* of the concrete masonry unit. Absorption is a measure of the pounds of water per cubic foot of concrete. Units made with a dense aggregate have much lower absorption rates than those made with lightweight aggregate. High water absorption is not acceptable for many applications. Units having a large number of *interconnected pores and voids* have a high absorption rate (also called suction rate). Units with a high suction rate and high absorption by aggregate have a high permeability to water, air, and sound and are more likely to be damaged by freezing. Units with unconnected pores, such as those found in lightweight aggregate and in the air-entrained cement paste, provide for sound absorption, reduce thermal conductivity, and minimize permeability to water.

Units used for exterior walls that are not to be painted should be low absorption units. Painting and other waterproof coatings greatly reduce water absorption.

It should be remembered that concrete masonry units should be in an air-dry condition (in equilibrium with the surrounding air). If they are used before this occurs, they will shrink slightly as moisture is lost. This causes tensile and shearing stresses, which may cause cracking.

The *coefficient of thermal expansion* is used to calculate the amount of expansion that can be expected as temperatures change. Concrete masonry units expand when heated and contract when cooled. These changes are reversible, and the unit returns to its original size after the temperature returns to the point at which the change started. Since the aggregate makes up about 80 percent of the unit, the coefficient of expansion of the aggregate is a major factor influencing expansion and contraction of the concrete masonry unit.

Other properties that are a result of those just discussed include insulation value, coefficient of heat transmission, sound-absorbing properties, and fire resistance.

The *insulation value* of units made with porous, lightweight aggregate is better than those using denser material. The units with lightweight aggregate also have a lower *coefficient of heat transmission*. Heat transmission, U, is stated in British thermal units, Btu, per hour per square foot per degree Fahrenheit for each degree difference in temperature between the air on the cool and warm sides of a wall. Concrete masonry walls are poor insulators and good heat transmitters. Therefore they require the addition of insulation materials to be energy efficient (Table 12.2).

Concrete masonry units resist the *transmission of sound*. Hollow block made with lightweight aggregate is recommended. The addition of a plaster interior or exterior finish increases this property. Concrete masonry units that have an open, porous surface absorb sound better than denser units with smooth surfaces. Painting the surface fills these pores and reduces sound-absorb-

Table 12.2 Coefficients of Heat Transmission for Selected Concrete Masonry Units

Hollow Masonry			
Size	Aggregate	Btu/hr./ft.²/°F	Watts/m²/°C
8″	Limestone	0.53	3.0
8″	Sand and gravel	0.53	3.0
8″	Cinders	0.37	2.1
8″	Expanded clay, slag, or shale	0.31	1.8

ing properties. Acoustical concrete block units are manufactured that combine a sound-deadening liner panel with a concrete unit with open slots on the face.

Building codes are very strict on required *fire resistance ratings* for various parts of a building. Products such as concrete masonry units must be carefully tested before they are given a fire resistance rating. The rating is the number of hours the material can be exposed to a flame before it fails. The rating varies depending on the aggregate. Plaster on the concrete unit is an effective way of increasing its fire resistance. A method for estimating the fire resistance rating of concrete masonry is in Table 12.3.

TYPES OF CONCRETE MASONRY UNITS

Concrete masonry units, often called concrete blocks, are made in a wide range of sizes and types. Many are made based on a special design prepared by an architect. They can be divided into three classifications: (1) concrete brick (solid units), (2) concrete block (hollow and solid units), and (3) special units.

Modular Sizes

Concrete masonry units are made in modular sizes and are specified by their nominal or modular size. They are made on a 4 in. module. The metric modular unit is 100 mm.

The nominal or modular size is the theoretical size without allowance for a mortar joint. The actual size is $\frac{3}{8}$ in. less than the nominal size, so the actual unit plus a $\frac{3}{8}$ in. mortar joint gives the modular size (Fig. 12.3). Concrete units are also related to standard bricks. For example, as shown in Fig. 12.4, the unit is two bricks wide, three brick courses high, and two bricks long. This allows for the mortar joints between the standard modular bricks. A standard modular brick is $4 \times 2\frac{2}{3} \times 8$ in. nominal.

True metric modular concrete blocks modular size is $200 \times 200 \times 400$ mm. Actual modular size is $190 \times 190 \times 390$ mm ($7\frac{1}{2} \times 7\frac{1}{2} \times 15\frac{3}{8}$ in.). American metric

Table 12.3 Estimated Fire Rating in Hours for Concrete Masonry

Type of Aggregate	Minimum Equivalent Thickness in Inches for Fire Ratings in Hours			
	1 hour	2 hours	3 hours	4 hours
Pumice or expanded slag	2.2*	3.2	4.0	4.7
Expanded shale, clay, or slate	2.6	3.6	4.2	5.1
Limestone, cinders, or unexpanded slag	2.7	4.0	5.0	5.9
Calcareous gravel	2.8	4.2	5.3	6.2
Siliceous gravel	3.0	4.5	5.7	6.7

How to Figure Equivalent Thickness for Cored Blocks

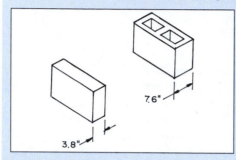

Equivalent thickness is the solid thickness that would be obtained if the same amount of concrete contained in a hollow unit were re-cast without core holes.

Calculating Estimated Fire Resistance. Example: A 7.6 in. (actual size) hollow masonry wall is constructed of expanded slag units reported to be 55%* solid. What is the estimated fire resistance of the wall? (modular units)

Eq Th = 0.55 × 7.6 in. = 4.2 in. From table: 3 hr fire resistance requires 4.00 in. Use 3 hr est. resistance.

	Minimum Equivalent Thickness In Millimeters for Fire Ratings in Hours			
	1 hour	2 hours	3 hours	4 hours
Pumice or expanded slag	55.9	76.2	101.6	119.4
Expanded shale, clay, or slate	66.0	91.4	106.7	129.5
Limestone, cinders, or unexpanded slag	68.6	101.6	127.0	149.9
Calcareous gravel	71.1	106.7	134.6	157.5
Siliceous gravel	76.2	144.8	114.8	170.2

How to Figure Equivalent Thickness for Cored Blocks

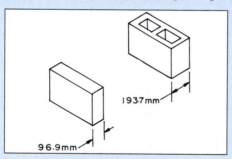

Equivalent thickness is the solid thickness that would be obtained if the same amount of concrete contained in a hollow unit were re-cast without core holes.

Calculating Estimated Fire Resistance. Example: A 193.7 mm (actual size) hollow masonry wall is constructed of expanded slag units reported to be 55%* solid. What is the estimated fire resistance of the wall? (modular units)

Eq Th = 0.55 × 193.68 mm = 106.43 mm. From table: 3 hr fire resistance requires 101.6 mm. Use 3 hr est. resistance.

From "Fire Safety with Concrete Masonry," 35C, with permission, National Concrete Masonry Association.

modular blocks soft-converted from inch sizes are (actual size) 194 × 194 × 397 mm (7⅝ × 7⅝ × 15⅝ in.), which is quite similar to the true metric block. The mortar bed is 10 mm. See Figs. 12.3 and 12.4.

Concrete Brick

Concrete bricks are made either solid or with a depressed area called a frog. The frog reduces the weight. They are laid with a ⅜ in. mortar joint. A standard modular brick is 4 × 2⅔ × 8 in. nominal. Other sizes are made by some block manufacturers.

Slump Brick and Block

Slump brick and block are made with a concrete mix that permits the unit to slump a little when it is removed from the mold. This produces units having irregular faces and some differences in height and surface texture. They give an unusual appearance when laid into a wall.

Concrete Block

A wide variety of types of concrete blocks are available, many designed for a special use. Some of these are shown in Fig. 12.5. They are produced in three major

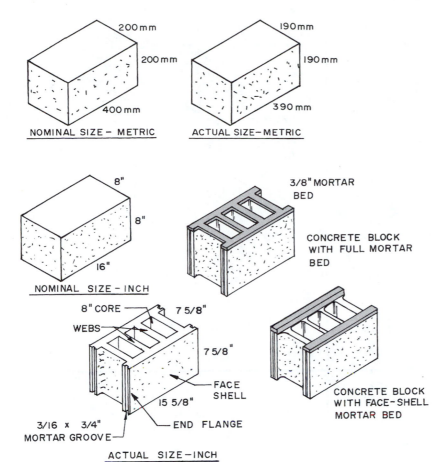

Figure 12.3 Standard inch and metric concrete blocks.

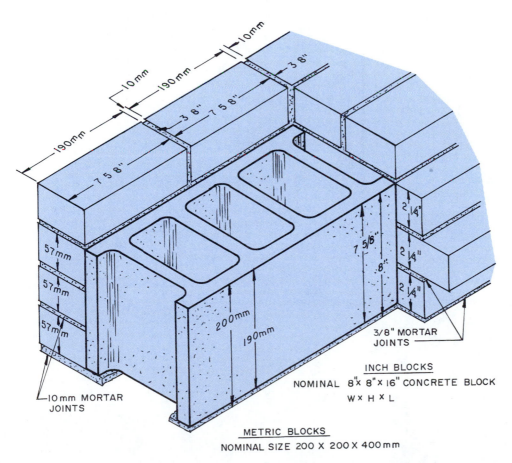

Figure 12.4 The sizes of standard concrete blocks and bricks are related.

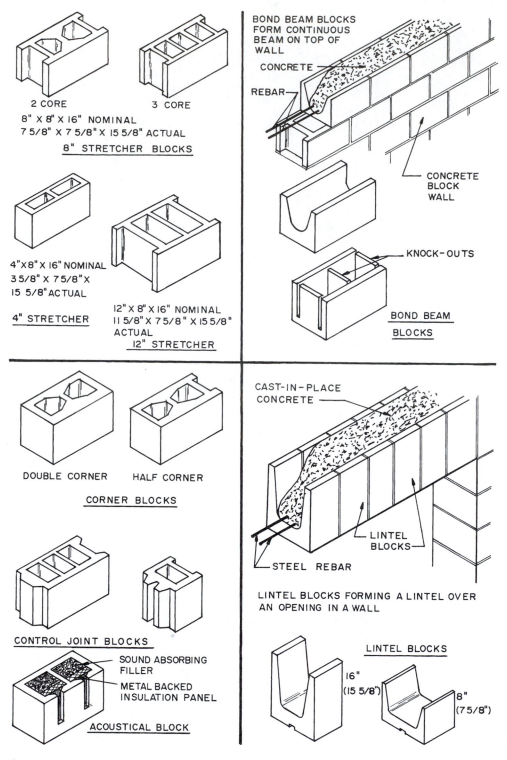

Figure 12.5 Shapes of commonly available concrete masonry units.

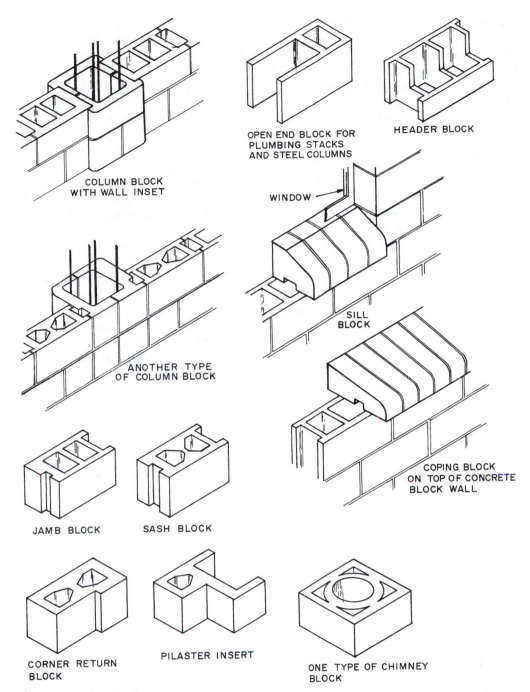

COLUMN BLOCK
WITH WALL INSET

OPEN END BLOCK FOR
PLUMBING STACKS
AND STEEL COLUMNS

HEADER BLOCK

WINDOW

ANOTHER TYPE
OF COLUMN BLOCK

SILL
BLOCK

JAMB BLOCK

SASH BLOCK

COPING BLOCK
ON TOP OF CONCRETE
BLOCK WALL

CORNER RETURN
BLOCK

PILASTER INSERT

ONE TYPE OF CHIMNEY
BLOCK

Figure 12.5 *Continued*

groups: (1) solid load-bearing, (2) hollow load-bearing, and (3) non-load-bearing units. A *solid load-bearing unit* is one whose cross-sectional area in every plane parallel to the bearing surface is not less than 75 percent of the gross cross-sectional area measured in the same plane. A *hollow concrete block* is one whose cross-sectional area in every plane parallel to the bearing surface is less than 75 percent of the gross cross-sectional area measured in the same plane.

Many units are made with two rather than three cores. The two-core unit has the advantage of having an increase of the face shell thickness and the center web. This increases strength, reduces cracking due to shrinkage during curing, produces a lighter unit, reduces heat concentration in the wall, and permits the cores to line up vertically so plumbing and electrical runs can be made inside the wall.

The *stretcher unit* is most commonly used in foundation and wall construction. The most common widths are 4, 6, 8, 10, and 12 in. with 8 in. height and 16 in. length. Half blocks are also available.

Corner blocks are used to make an exposed corner and for piers and pilasters. They are made with one or both ends flush.

A *corner return block* permits a full-block face to appear on one wall with a recess to help turn the corner.

Header blocks have a recess that holds the header unit in a masonry bonded wall.

Pilaster blocks are used when the design calls for pilasters to reinforce a wall. They also provide a bearing surface for beams. Some are made in halves and others are a single unit. They may have rebar and concrete in the core if greater strength is necessary.

Control joint blocks are used when vertical shear-type control joints occur in a wall. This relieves stresses in long masonry walls.

Bond beam blocks are filled with concrete and reinforcing bar to form a continuous reinforced concrete beam along the top of a wall. They also are used to span above lintels and below sills.

Sash blocks and *jamb blocks* are used where windows are to be installed. One style is used for wood windows and the other for metal windows.

A *joist block* permits a floor joist to rest on the wall. The end is covered with the concrete wing, so from the exterior it appears as a normal block wall.

Open-end units are used to provide a vertical cavity for running plumbing or electrical conduit or to enclose a steel beam.

Lintel blocks are used to form a steel-reinforced concrete lintel over an opening in a wall, such as a window opening. They can also be used to form bond beams. More commonly, precast concrete lintels are used above and below openings in concrete block walls.

Sill and *coping blocks* are used to provide a finished cap. Sill blocks are placed below the window to direct water away from the wall. Coping blocks are placed on top of the exterior wall to seal out moisture.

Partition blocks are available in 4 and 6 in. thickness and are used for non-load-bearing walls.

Chimney blocks are used to quickly form a concrete surround around a fireclay flue lining.

Solid top blocks are used where a solid, coreless surface is needed, such as on the top of a foundation.

A number of types of blocks are used to form floors. One type, called a *soffit block,* is shown in Fig. 12.6. Notice that it has reinforcing bars and a poured concrete deck on top of it.

Split-face blocks are concrete units that are split in half lengthwise after they have hardened. This produces a rough, irregular surface that is to be the exposed face. They may be hollow or solid units. Color variations are produced by using aggregates of various colors and by putting mineral colors in the mix.

Faced blocks have the exposed surface covered with a ceramic glaze or a plastic overlay, or they are polished by grinding the surface smooth.

Decorative blocks are made with many different face designs (Fig. 12.7). Typically they are pierced or recessed to produce an unusual wall surface. A common use is for a decorative privacy wall. Common sizes are 8 × 16 in. and 12 × 12 in. Special designs can be made by the block manufacturer to meet the requirements of the architect.

Special units are custom-made blocks designed by an architect to face a project for which a special surface treatment is desired. The design possibilities are unlimited. Block manufacturers can produce almost any design, provided a mold can be made to produce the unit.

Decorative walls can be produced by using standard concrete masonry units and varying the bond pattern (Fig. 12.8).

MATERIAL SPECIFICATIONS

The American Society for Testing and Materials has specifications for four types of concrete masonry units. These are used by block manufacturers to produce quality blocks (Table 12.4).

Grades

Load-bearing concrete masonry units are available in two grades, S and N (Table 12.5). Grade S units are restricted to above-grade applications, and if they are used in exterior walls they must have a protective coating. Grade N units can be used above and below grade and may be exposed to the weather or moisture penetration.

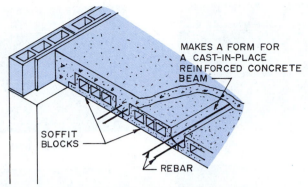

Figure 12.6 Soffit blocks are used to form floors.

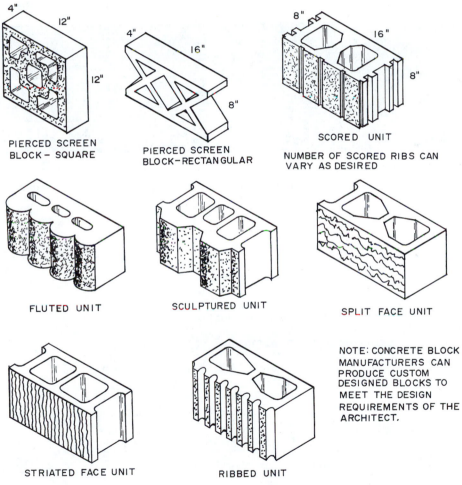

Figure 12.7 Selected examples of decorative concrete masonry units.

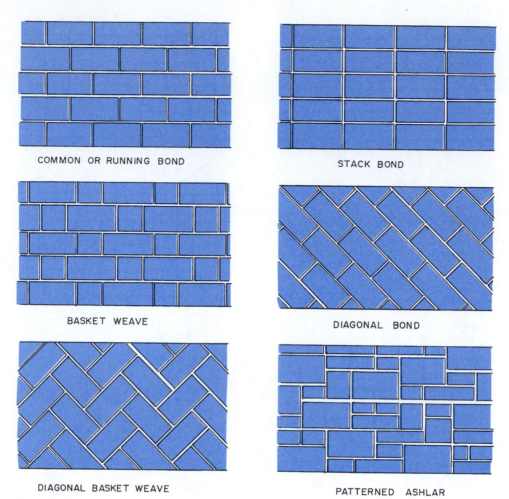

COMMON OR RUNNING BOND

STACK BOND

BASKET WEAVE

DIAGONAL BOND

DIAGONAL BASKET WEAVE

PATTERNED ASHLAR

Figure 12.8 Wall patterns that can be produced using standard concrete masonry units.

Table 12.4 ASTM[a] Standards for Concrete Masonry Units

Standard	Type of Unit
ASTM C55	Concrete building brick, solid veneer, and split block
ASTM C90	Hollow load-bearing concrete masonry units
ASTM C129	Hollow non-load-bearing concrete masonry units
ASTM C145	Solid[b] load-bearing concrete masonry units

[a]American Society for Testing and Materials.
[b]Units with 75% or more net area.

Table 12.5 Grades, Types, Weights, and Uses of Concrete Masonry Units (ASTM C90)

Grade	Use
N	General use above and below grade
S	Above grade only in walls not exposed to weather or with weather-protective coating if exposed to the weather

Type	Use
I	Moisture controlled during manufacture
II	Non-moisture-controlled units

Weight	Use
Normal weight	Manufactured using concrete weighing more than 125 pcf (2000 kg/m^3)
Medium weight	Manufactured using concrete weighing between 105 and 125 pcf (1680 to 2000 kg/m^3)
Lightweight	Manufactured using concrete weighing 105 pcf (1680 kg/m^3) or less

Table 12.6 Grades and Uses of Solid Load-Bearing Concrete Block, ASTM C145 and Hollow Load-Bearing Concrete Block, ASTM C90

Grade	Use
N-1, N-2	Exterior walls above or below grade that may be exposed to moisture and weather, and for interior walls and backup
S-1, S-2	Above-grade walls not exposed to weather or exterior walls if covered with weather-protective coatings

Table 12.7 Grades and Uses of Concrete Building Brick ASTM C55

Grade	Use
N-1, N-2	On interior and exterior walls where high strength and resistance to moisture and frost action are required
S-1, S-2	Where moderate strength and resistance to moisture and frost action are required

When used below grade, protective coatings are recommended and often are required by building codes.

Types

Two types of concrete masonry units are specified by ASTM C90, Type I and Type II. Type I units are manufactured with specific limits on moisture content. Type II has no moisture control limits. Moisture limits on block manufacturing have the advantage of minimizing shrinkage of the units after they have been laid. Shrinkage could cause cracking of the wall. Uses of the various grades and types are in Tables 12.6 and 12.7.

Weights

ASTM C90 establishes three weights of concrete masonry units—normal, medium, and lightweight. Normal weight units are made from concrete weighing more than 125 pcf (2000 kg/m^3), medium weight uses concrete weighing 105 to 125 pcf (1680 to 2000 kg/m^3), and lightweight uses concrete 105 pcf (1680 kg/m^3) or less.

REVIEW QUESTIONS

1. What are the two ways concrete masonry units are cured?
2. What advantages are there to autoclaved concrete masonry units?
3. What two things influence the properties of concrete masonry units?
4. How do the tensile strength and flexural strength of concrete masonry units compare with the compressive strength?
5. What affects the water absorption of concrete masonry units?
6. How is the insulation value of concrete masonry units improved?
7. What are the three major classifications of concrete masonry units?
8. What is the design module used when sizing concrete masonry units?
9. What are the grades established for load-bearing concrete masonry units?

KEY TERMS

autoclave A high-pressure steam room that rapidly cures green concrete units.

concrete masonry Factory manufactured concrete units, such as concrete brick and block.

kiln A low-pressure steam room in which green concrete units are cured.

special units Concrete masonry units that are designed and made for a special use.

SUGGESTED ACTIVITIES

1. If you are near a plant that produces concrete masonry units, arrange for a visit. Before the visit prepare a list of specific things to look for and ask questions about, such as the actual proportions of the concrete used, curing procedures, and tests used to ensure quality control of the products produced.

2. Secure a supply of concrete blocks and lay up a short wall and turn a corner. Vary the amount of water and mortar cement in several batches of mortar and prepare a written report citing what this did to the finished wall before and after curing.

ADDITIONAL INFORMATION

Ramsey, C.G., Sleeper, H.R., and Hoke, J.R., eds., *Architectural Graphic Standards,* John Wiley and Sons, New York, 1991.

Other resources are

Various publications available from the Portland Cement Association, 5420 Old Orchard Rd., Skokie, Ill. 60077-1083.

Various publications available from the National Concrete Masonry Association, P.O. Box 781, Herndon, Va. 22070.

CHAPTER 13

Stone

This chapter will help you to:

1. Be able to discuss the characteristics and uses of various stones used on buildings.

2. Identify the various types of commercially available stone.

3. Understand the processes used to quarry and work stone to make it useful for building construction.

4. Select stone for a project based on the requirements of the job.

For centuries stone was used as a material to build structural load-bearing walls, but now it is used mainly as a veneer or facing material. This greatly reduces the weight of the building yet permits the designer to use the beauty of stone as a finish material (Fig. 13.1).

Rock is a solid mineral material, occurring in individual pieces or large masses, such as a hill. **Stone** is rock that is selected or shaped to size for building purposes.

BASIC CLASSIFICATIONS OF ROCK

Rock is divided into three basic categories depending on its origin: igneous, sedimentary, and metamorphic.

Igneous Rock

Igneous rock is formed by change of a molten material, usually deep in the earth, to a solid. Commonly used forms include granite, serpentine, and basalt.

Granite

Granite is an igneous rock having crystals or grains of visible size. It consists mainly of quartz, feldspar, mica, and other colored minerals. Colors include black, gray, red, pink, brown, buff, and green. It is hard, strong, nonporous, and durable. Granite is one of the most permanent building stones. It can be used under severe weather conditions and in contact with the ground. Granite stones are finished with a range of textures, including a highly polished surface. Granite is used for windowsills, cornices, columns, floors, and wall veneers.

Serpentine

Serpentine is an igneous rock named after its major ingredient, serpentine. It ranges from an olive green to a greenish black. It has a fine grain and is dense. Since some types deteriorate due to weathering, its major uses are on interiors. It can be cut into thin sections, $\frac{7}{8}$ to $1\frac{1}{4}$ in. (22 to 32 mm), so it is used as paneling, windowsills, stools, stair treads and risers, and landings.

Stone on the interior of the African Pavilion in the Smithsonian Institution, Washington, D.C.

Stone used on the exterior of a building.

Figure 13.1 Stone is used as a finish material on the interior and exterior of buildings. (*Courtesy Bybee Stone Company, Inc., and Cold Spring Granite Co.*)

Basalt

Basalt is an igneous rock that ranges in color from gray to black. It is fine grained and is used mainly for paving stones and retaining walls.

Sedimentary Rock

Sedimentary rock is formed of materials (sediments) deposited on the bottom of bodies of water or on the surface of the earth. Major types include sandstone, shale, and limestone.

Sandstone

Sandstone is a sedimentary rock composed of sand-sized grains cemented together by naturally occurring mineral materials such as silica, iron oxide, and clay. Quartz grains predominate in the sandstone used for building construction. The two most familiar forms are brownstone, used mainly in wall construction, and bluestone, used for paving and wall copings. Colors include gray, brown, light brown, buff, russet, red, copper, and purple.

Since sandstone's hardness and durability depend on the cementing material, there is a wide range in weight and porosity. Sandstone is used for panels and can have a variety of surface finishes, such as chipped, hammered, and rubbed faces.

Shale

Shale is a sedimentary rock derived from clays and silts. It is weak along planes and is in thin laminations. It is not suitable as a concrete aggregate. Shale that is high in limestone is ground into small particles used in making cement, bricks, and tiles. It is basically gray in color but is found ranging from black to red, yellow, and blue.

Limestone

Limestone is a sedimentary rock composed mainly of calcite and dolomite. There are three types of limestone. *Oolitic* is a calcite-cemented calcareous stone formed from shells. It is very uniform in composition and structure. *Dolomitic* limestone consists mainly of magnesium carbonate. It has a greater compressive strength than oolitic. *Crystalline* limestone consists mainly of calcium carbonate crystals. It has high tensile and compressive strengths.

Limestone is used for building stones and is available as dimension (cut), ashlar, and rubble stones. It is used for paneling, veneer, window stools and windowsills, flagstone, mantels, copings, and facings (Fig. 13.2). It is

Figure 13.2 Stone is a popular material for interior use, such as this fireplace. (*Courtesy Mr. and Mrs. Roy Register*)

also crushed to form crushed-stone aggregate and burnt to produce lime.

Metamorphic Rock

Metamorphic rock is either igneous or sedimentary rock that has been altered in appearance, density, and crystalline structure by high temperature and/or high pressure. Major types used in construction include marble, quartzite, shist, and slate.

Marble

Marble is a metamorphic rock made up largely of calcite or dolomite that has been recrystallized. There are a number of types of marble. The colors vary from white through gray and black. Other colors found include red, violet, pink, yellow, and green. The presence of oxides of iron, silica, graphite, carbonaceous matter, and mica produce these color variations.

Marble is used as wall panels and column facings as well as window stools, windowsills, and floors. Some types are used on the exterior of buildings, and others are limited to interior applications. The surface can be ground to a fine, polished condition.

Quartzite

Quartzite is a metamorphic rock that is often confused with granite. It is a variety of sandstone composed mainly of granular quartz that is cemented by silica, which produces a coarse, crystalline appearance. It has high tensile and crushing strengths. Quartzite is used for building stone, gravel, and aggregate in concrete. Its colors can be brown, buff, tan, ivory, red, or gray.

Shist

Shist is a metamorphic rock generally made up of silica with smaller amounts of iron oxide and magnesium oxide. The color depends on the mineral makeup, but blue, green, brown, gold, white, gray, and red are common. It is commonly available in rubble veneer and flagstone and is used for interior and exterior wall facing, patios, and walks.

Slate

Slate is a hard, brittle metamorphic rock consisting mainly of clays and shales. The major ingredients are silicon dioxide, aluminum oxide, iron oxide, potassium oxide, magnesium oxide, and sometimes titanium, calcium, and sulfur. Slate is found in parallel layers, which enables it to be cut into thin sheets.

Slate is produced in three textures: sand-rubbed, honed, and natural cleft. It is cut into three types: roof tiles, random flagging, and dimension slate (cut to size) (Fig. 13.3). It is commonly used for interior and exterior wall facing, flooring, flagstones, countertops, coping, and windowsills and window stools.

Figure 13.3 Slate is a beautiful roofing material. (*Courtesy Evergreen Slate Co., Inc.*)

TYPES AND USES OF STONE

Commercially, stone is used in several types: rubble stone, rough stone, monumental stone, dimension stone, flagstone, broken and crushed stone, and stone powder and dust.

Rubble stone consists of irregular fragments from a quarry that have one good face. The pieces are irregular in shape and sized usually in pieces 12 in. (300 mm) by 24 in. (600 mm). They are cut and fitted by a mason.

Rough building stone, sometimes called fieldstone, occurs in naturally found rock masses. The stones generally are used in the shapes as found (Fig. 13.4).

Monumental stone is used for monuments, gravestones, and similar purposes.

Dimension stone, also referred to as cut stone, is cut to size at a stone mill and shipped to the site. The surface may be rough, as occurs when it is split, or polished. It is used as veneer on interior and exterior walls, floors, copings, stair treads, and other similar uses. *Ashlar stone,* a form of dimension stone, is a cut rectangular stone that is smaller than other dimension stone and is generally rectangular with square corners and faces.

Flagstone is thin, rather flat stone from ½ to 4 in. (12 to 100 mm) in thickness. Flagstones laid over a concrete base are usually ¾ to 1 in. (18 to 25 mm) thick. If laid over a sand or loam base, pieces 1¼ to 1½ in. (31 to 37 mm) thick are required. The surface may be left rough or polished. Random flagstones are the natural shape of the pieces with minor shaping. Trimmed flagstones are random-shaped pieces with several edges sawed straight. Trimmed rectangular flagstones are pieces with four sides sawed, forming square and rectangular pieces. Typical patterns for laying flagstones are in Fig. 13.5.

Broken and crushed stone includes irregular shapes and crushed pieces of stone of one type of rock that are graded for hardness and size and used as aggregate in concrete for surfacing roads and driveways and as aggregate for surfacing fiberglass asphalt shingles and built-up roofing.

Stone powder and *stone dust* are used as fill in paints and asphalt paving surfaces.

Figure 13.4 Fieldstone is laid in random design, forming a textured, attractive wall.

QUARRYING AND PRODUCING BUILDING STONES

A quarry is an excavation from which stone used for building is taken by blasting or cutting (Fig. 13.6). Broken stone is produced by blasting the rock. The larger pieces can be rebroken or cut into smaller units for use as an exterior finish material. The rest is crushed and sorted into various sizes for use as an aggregate.

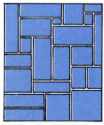
RANDOM RECTANGULAR

EUROPEAN

SEMI–RANDOM RECTANGULAR

IRREGULAR

Figure 13.5 Some of the commonly used patterns for laying flagstone.

Most stone used in building construction is dimensional stone produced by cutting large blocks in the quarry. These are often cut by a channeling machine, which makes a cut one to three inches wide. Some machines use a rotating chisellike cutter, and others use a wire that runs over pulleys and moves a quartz sand cutting agent over the stone. This produces a saw-type cut in the stone.

Large blocks are removed from the quarry to a mill, where they are cut to the sizes and thicknesses needed (Fig. 13.7). The architectural drawings specify the shape and size of each stone. Holes are drilled in each block as indicated for lifting and anchoring it in place (Fig. 13.8).

Figure 13.6 A typical quarry showing a cut shelf of rock. The rock-sawing equipment is on top of the shelf. (*Courtesy Bybee Stone Company, Inc.*)

Figure 13.7 Large blocks of rock are moved from the quarry to the mill to be cut into desired shapes and sizes. (*Courtesy Bybee Stone Company, Inc.*)

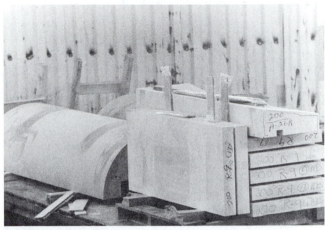

Figure 13.8 Finished pieces of stone cut to the size and shape indicated on the architectural drawings. Notice the markings on the ends indicating the location of each on the building. (*Courtesy Bybee Stone Company, Inc.*)

CHOOSING A STONE

When selecting the stone to be used in a building, the architect must consider the following: (1) cost, (2) strength, (3) durability, (4) hardness, (5) grain and color, and (6) texture and porosity.

The *cost* requirement is influenced by the ease with which the stone can be shaped. Soft stones have lower production costs. Another cost factor is accessibility. Cost increases as the distance the cut stone must be transported increases. Some stone is easier to quarry, so quarrying costs are another factor that must be considered.

In general, the commonly used stone has sufficient *strength* for the purposes for which it is to be used. However, this must be a design consideration when choosing stone. If the stone is exposed to the weather it must have sufficient *durability* to withstand the freeze-thaw cycles and erosive conditions. *Hardness* is very important for stones that are to be used on floors, steps, patios, and other areas exposed to traffic. *Grain and color* are considered when the appearance of the stone in its finished position is decided. Colors vary widely and must be specified. The *texture* of the stone has a great influence on the finished appearance. Fine-grained stones can have a smoother, polished surface, while coarse-grained stones present a more open face. The *porosity* pertains to the ability of the stone to resist penetration by moisture. Porous rock tends to permit some of the minerals to dissolve and stain the exposed face. It also is not durable and will be damaged by the freeze-thaw cycle (Fig. 13.9).

Figure 13.9 Stone selected for carving is carefully chosen for proper grain, color, durability, hardness, and porosity. (*Courtesy Bybee Stone Company, Inc.*)

REVIEW QUESTIONS

1. What is the main use for stone in buildings built today?
2. What are the three basic categories of rock?
3. How is igneous rock formed?
4. What colors of granite rock are commonly available?
5. In what part of a building is serpentine rock used?
6. How is sedimentary rock formed?
7. What materials bind together the sand-sized grains forming sandstone?
8. Shale that has a high limestone content is used to make what masonry materials?
9. What are the three types of limestone?
10. What are the main differences in the three types of limestone?
11. What are the major types of metamorphic rock used in building construction?
12. In what colors is marble found?
13. Where is marble commonly used in building construction?
14. Where is shist used in construction?
15. In what textures is slate produced?
16. What type of machine is used in a quarry to cut rock into manageable sizes?
17. What factors must a designer consider when choosing stone for a building?

KEY TERMS

igneous rock Rock formed deep in the earth by a change of molten material to a solid.

metamorphic rock Igneous or sedimentary rock that has been altered by high temperature or pressure.

rock A solid mineral material found naturally in large masses.

sedimentary rock Rock formed from sediments deposited on the bottoms of bodies of water and on the surface of the earth.

stone A rock selected or processed by shaping to size for building or other use.

SUGGESTED ACTIVITIES

1. Collect samples of various types of stone typically used in building construction. Write a report citing the characteristics of each type. Your building materials supplier can help with samples.

2. Test selected stone samples for compressive strength. Note variances between samples of the same stone and between the different types of stone. Report why you think these variances may have occurred.

ADDITIONAL INFORMATION

Hornbostel, C., *Construction Materials: Types, Uses, and Applications,* John Wiley and Sons, New York, 1991.
    Other sources listed in the "Professional and Technical Organizations" section of Appendix B.
    Publications from the Building Stone Institute, P.O. Box 507, Purdys, New York 10578.

Masonry Construction

This chapter will help you to:

1. Select the type of masonry bearing wall best suited for a particular situation.

2. Understand the purpose of expansion and control joints in masonry construction.

3. Recognize the types of brick and concrete masonry wall constructions and the properties associated with them.

4. Explain the advantages and disadvantages of the various mortar joints.

5. Be aware of the process used to lay brick and concrete masonry units.

6. Evaluate the quality of stone wall construction.

7. Recognize proper construction of structural clay tile walls.

The engineer designing load-bearing masonry buildings must consider magnitude and direction of all forces acting on the building. This includes things such as live loads, dead loads, and lateral loads and forces due to temperature change, impact, or unequal settlement of the foundation.

Masonry wall construction can be of clay brick, concrete masonry units, stone, or structural tile. Load-bearing walls carry floor and roof loads on the exterior and interior of a building (Fig. 14.1). Non-load-bearing masonry walls are used on the exterior to protect a building from the elements and for interior partitions. Masonry units provide acoustical control and fireproof construction, and they are highly resistant to damage. Masonry walls are easy to design and construct and often produce a more economical building than other methods. They can be used to construct one-story low-rise buildings without lateral bracing. Walls above certain heights require lateral bracing. Masonry is a heavy material, and this limits its use in high-rise buildings. This weight requires extensive footings and produces compressive stresses and can cause possible buckling. Lateral forces are controlled by various design features, such as building internally reinforced masonry walls.

MASONRY BEARING WALLS

A masonry load-bearing wall may be unreinforced or reinforced, solid masonry or cavity type. It may be a composite wall built using several different types of units, such as clay brick over concrete blocks.

Reinforced masonry is used when the flexural, compressive, and shear stresses exceed those permitted for partially reinforced or unreinforced masonry. The amount of steel reinforcing is required by building codes.

Cavity wall construction is generally used on exterior walls because they can be built to control moisture penetration and can be insulated. Solid masonry walls do not have these advantages and are therefore generally used for interior partitions.

Composite masonry walls have an exterior veneer of a quality masonry unit, such as brick, tile, or stone, and the hidden interior portion of the wall is built from a

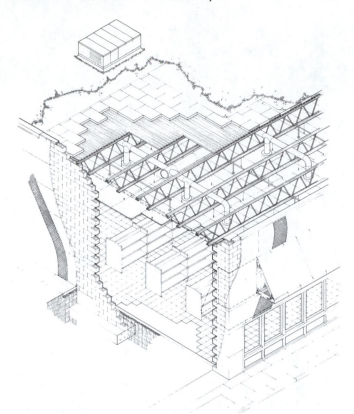

Figure 14.1 Masonry is used to provide structural support as well as finished interior and exterior walls. This building has masonry load-bearing walls and a bar joist roof. (*Reproduced with permission from* The Building Systems Integration Handbook, *Richard Rush, ed., Butterworth-Heinmann Publishers, Newton, Mass., 1986.*)

cheaper unit, such as concrete block. The engineer designing the wall must consider how differences in thermal expansion, moisture absorption, and load-bearing capabilities affect the wall.

MASONRY ARCHES

Masonry **arches** are a form of curved construction used to span an opening and usually consist of wedge-shaped blocks called voussoirs. When rectangular blocks are used, the mortar between them assumes a wedge shape. Arches vary in shape from a flat horizontal arch through semicircular and semielliptical forms to pointed shapes.

A masonry arch may be constructed using soldier courses, two or three rowlock courses, or alternating rowlock and soldier courses (Fig. 14.2). The block arch uses the compressive strength of the brick or stone to span the opening and carry the load. The horizontal thrust generated is resisted by the masonry mass of the wall against which it butts. Two equal arches close together provide opposite but equal thrust, thus resisting the horizontal forces (Fig. 14.3).

The technical terms used to describe the parts of block arches are in Fig. 14.4. Block arches include

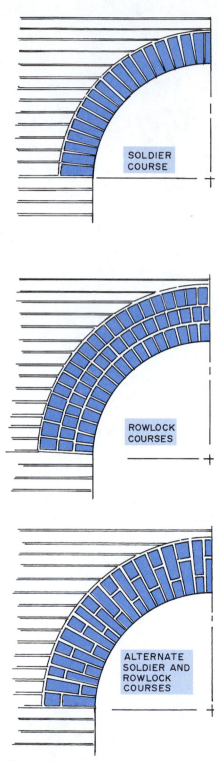

Figure 14.2 Typical masonry arches.

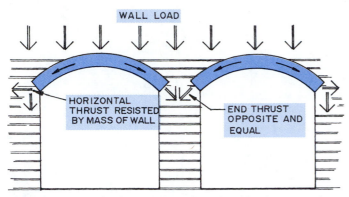

Figure 14.3 The horizontal thrust is resisted by the mass of the masonry wall.

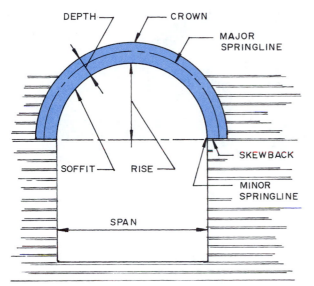

Figure 14.4 Terms used to identify the parts of a block arch.

Roman, Gothic, elliptical, segmental, tudor, parabolic, and jack (Fig. 14.5).

CONTROL AND EXPANSION JOINTS

Within masonry walls, movements occur that produce stresses that may cause cracking. These stresses can be created by expansion and contraction due to changes in temperature or moisture content, structural movements due to settling of the foundation, and concentration of stresses at openings in the wall. Masonry walls often use units of different materials. These expand and contract at different rates and can create stresses. Examples include brick veneer over concrete block and the use of metal lintels over openings. The locations of control and expansion joints must be carefully determined as part of the wall design.

Expansion joints provide a space between adjacent parts of masonry construction and permit a limited

amount of movement. The actual spacing varies with the type of masonry unit, size of the area, and the reinforcing used. Vertical expansion joints are placed near the corners or where the walls change direction. Horizontal joints are placed above masonry walls that butt structural frames or the bottom of floor or roof structures (Fig. 14.6).

Control joints provide tension relief between parts of a masonry wall that may change from their original dimensions. One major contributor is the fact that masonry walls tend to contract as they dry after they are laid. The tensile strength of the wall tries to resist these stresses and the wall cracks if the strength of the wall is exceeded. This cracking is controlled by using carefully placed control joints and adequate steel reinforcing in the wall.

Control joints are continuous vertical joints built into the masonry wall as the units are laid. The locations of control joints must be carefully determined as part of the wall design. Generally they are located where there are changes in wall height or wall thickness and at openings, wall intersections, returns in U, T, and L shaped buildings, and junctions between walls and columns (Fig. 14.7).

Various types of control joint construction are shown in Fig. 14.8. Control joints must be carefully sealed with a specified elastic joint sealer. Regular inspection and maintenance of the seal is required to keep the wall watertight. Control joints are generally not used below grade because they could develop leaks. Also, the temperatures below grade are more constant, so expansion and contraction are usually less.

BRICK MASONRY CONSTRUCTION

Some of the commonly used brick masonry wall constructions are shown in Fig. 14.9. The fire resistance of these walls varies because the clays used vary in different parts of the country. The interior space in cavity walls tends to collect moisture, which can be controlled by providing for the drainage of this moisture. This construction uses flashing at the base and weep holes located just above the bottom of the flashing (Fig. 14.10). Moisture that enters the cavity moves to the bottom of the wall and is directed out of the building through weep holes.

Weep holes are openings left in the mortar joint. They can be formed by pieces of plastic or metal pipe laid in the joint, by omitting mortar in places along the joint or by placing oiled rope in the joint and pulling it out after the mortar begins to set. Weep holes are often spaced 18 to 24 in. (457 to 508 mm) apart (Fig. 14.11).

The flashing is copper or plastic sheet material. Aluminum flashing is not used because it has an unfavorable chemical reaction with the mortar. As the masonry units are being laid the masons lower a wood strip

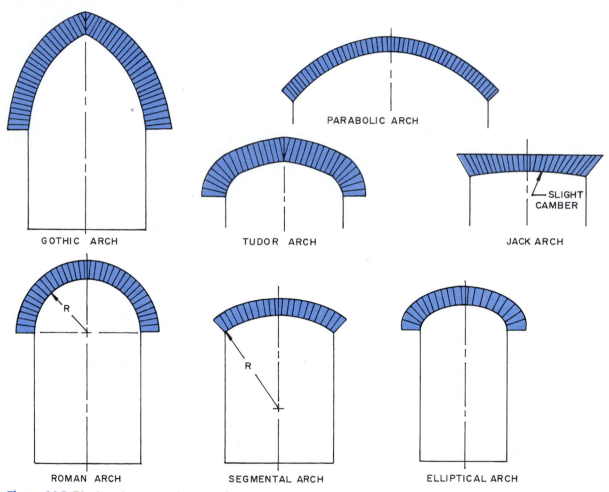

Figure 14.5 Block arches may take many forms.

with wires into the cavity below the area where bricks are being laid. The strip catches mortar accidentally dropped in the cavity and is raised up and cleared. This keeps mortar from filling the area above the flashing and closing the weep holes. Moisture penetration of brick walls can also be controlled by parging the width between the two courses of brick with mortar or grout.

Some examples of brick bearing walls supporting floor and roof joists and decking are in Figs. 14.12 through 14.16. Examples of roof construction are in Fig. 14.17.

Details for typical brick curtain wall construction are in Fig. 14.18. The curtain wall is tied to columns with metal reinforcing. A *curtain wall* is an exterior nonbearing wall between columns not supported by the beams or girders of the structural frame.

Openings in brick masonry walls are spanned with a lintel. These are typically steel angles, but precast concrete or concrete masonry **lintel blocks** are also used. In some cases reinforced brick lintels can be used. The choice and design depend on the span of the opening and the load to be carried by the lintel (Fig. 14.19). Precast concrete and concrete masonry lintels generally carry the loads imposed on the wall and steel lintels carry the face brick. Some classic architectural styles use exposed concrete or stone lintels.

Brick masonry *columns* are typically designed to be built without requiring the bricks to be cut. The interior core has reinforcing bars and is filled with concrete (Fig. 14.20). **Pilasters** are built the same as columns but are an integral part of the load-bearing wall.

Brick masonry walls use various types of metal reinforcing (Fig. 14.21). The mortar and metal ties work together to provide the needed structural bond. A solid wall two bricks wide can use header bricks as a cross tie or metal ties. The header bricks also provide an interesting bond pattern. A cavity wall uses metal ties set in the mortar. **Brick veneer** over a concrete block or sheathed wood frame wall uses metal ties and straps. Brick walls veneered over tile units can use header bricks or metal ties.

Grouted reinforced brick masonry walls use vertical and horizontal reinforcing bars in the wall cavity as shown in Fig. 14.22. The amount and placement of the reinforcing are engineering decisions. The reinforcing increases the strength of the wall to resist lateral loads and reduce buckling. A typical layout for a wall with openings is in Fig. 14.23.

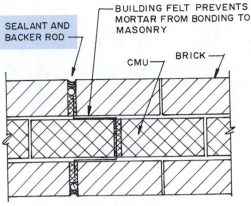
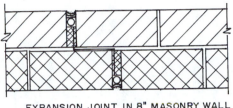
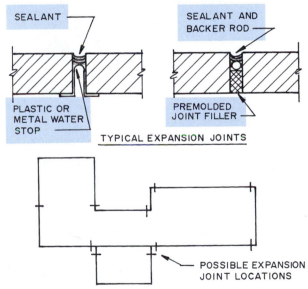

Figure 14.6 Expansion joints used in masonry construction.

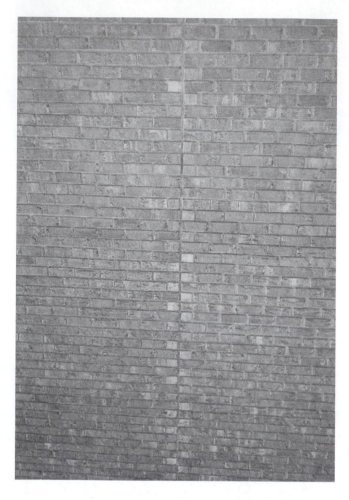

Figure 14.7 Commonly used locations for control joints in masonry construction.

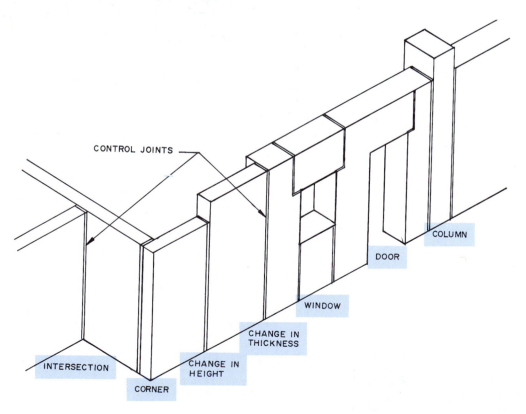

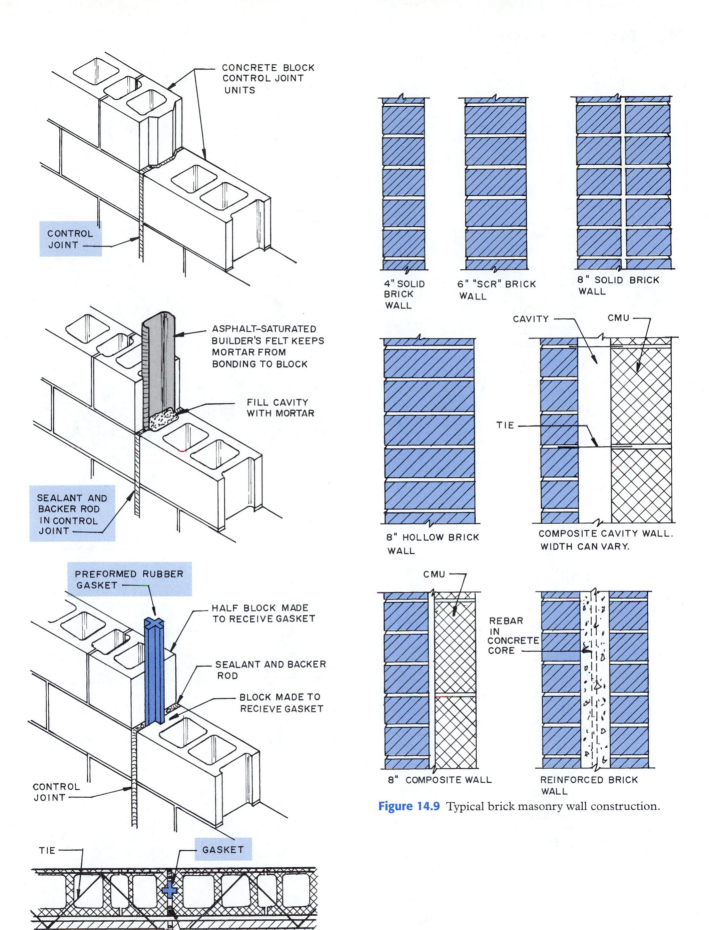

Figure 14.8 Typical control joint construction.

CONCRETE BLOCK
CONTROL JOINT
UNITS

CONTROL
JOINT

ASPHALT–SATURATED
BUILDER'S FELT KEEPS
MORTAR FROM
BONDING TO BLOCK

FILL CAVITY
WITH MORTAR

SEALANT AND
BACKER ROD
IN CONTROL
JOINT

PREFORMED RUBBER
GASKET

HALF BLOCK MADE
TO RECEIVE GASKET

SEALANT AND BACKER
ROD

BLOCK MADE TO
RECIEVE GASKET

CONTROL
JOINT

TIE

GASKET

4" BRICK VENEER

CONTROL JOINT

4" SOLID
BRICK
WALL

6" "SCR" BRICK
WALL

8" SOLID BRICK
WALL

8" HOLLOW BRICK
WALL

CAVITY

CMU

TIE

COMPOSITE CAVITY WALL.
WIDTH CAN VARY.

CMU

8" COMPOSITE WALL

REBAR
IN
CONCRETE
CORE

REINFORCED BRICK
WALL

Figure 14.9 Typical brick masonry wall construction.

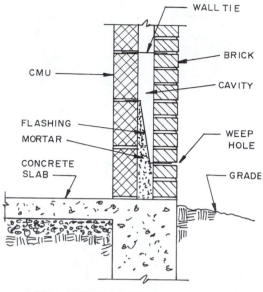

A WALL WITH FLASHING AND WEEP HOLES

- WALL TIE
- BRICK
- CAVITY
- CMU
- WEEP HOLE
- FLASHING
- MORTAR
- CONCRETE SLAB
- GRADE

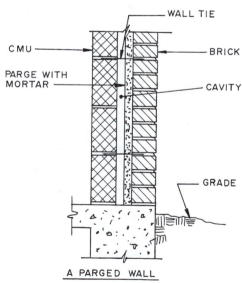

A PARGED WALL

- WALL TIE
- CMU
- BRICK
- PARGE WITH MORTAR
- CAVITY
- GRADE

Figure 14.10 Two techniques used to control moisture inside a brick cavity wall.

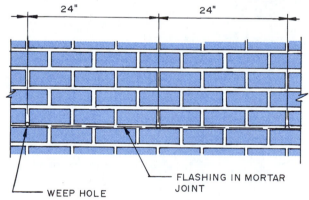

- 24"
- 24"
- WEEP HOLE
- FLASHING IN MORTAR JOINT

Figure 14.11 A typical location pattern for weep holes.

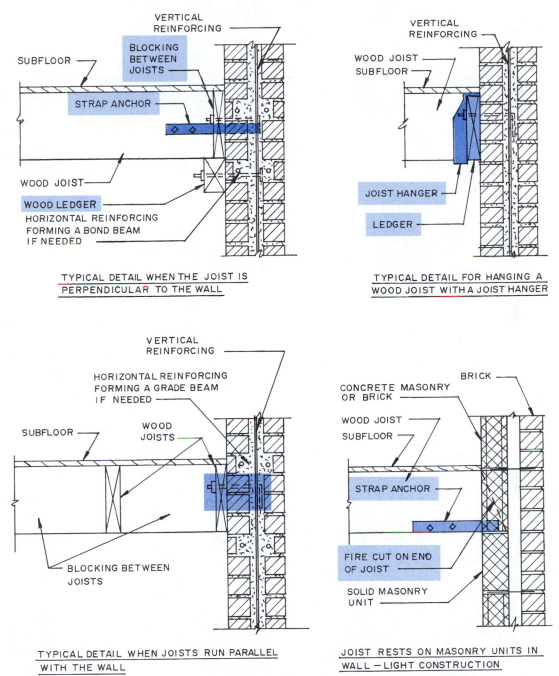

Figure 14.12 Commonly used ways to frame wood joists and beams to brick masonry walls.

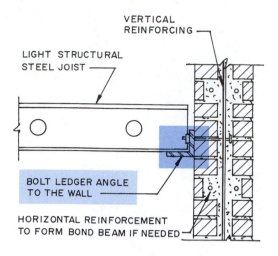

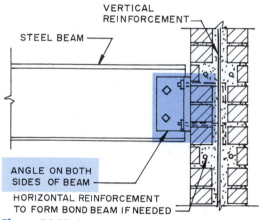

Figure 14.13 Several ways to frame steel joists or beams to brick masonry walls.

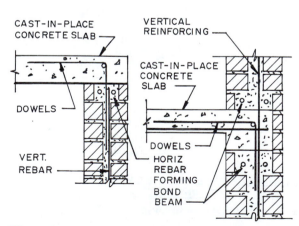

Figure 14.14 Two ways to support cast-in-place concrete decks with brick masonry walls.

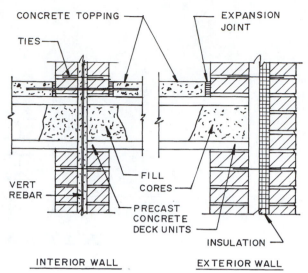

Figure 14.15 Details for framing precast concrete floor and roof units on brick masonry walls.

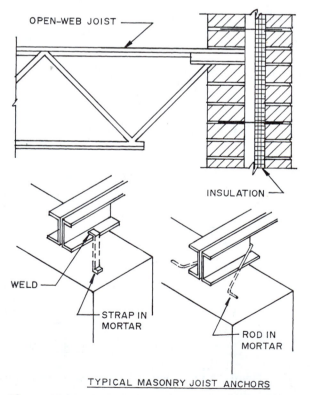

Figure 14.16 Open-web joists are anchored to masonry walls with metal straps set in the mortar joint.

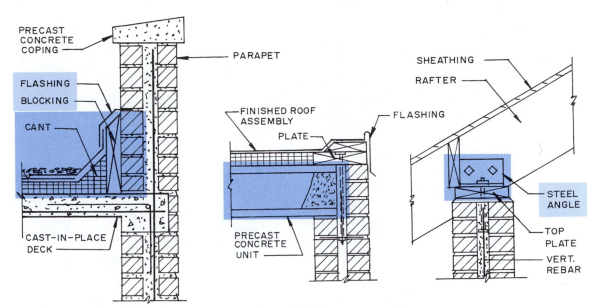

Figure 14.17 Examples showing roof construction with brick masonry walls.

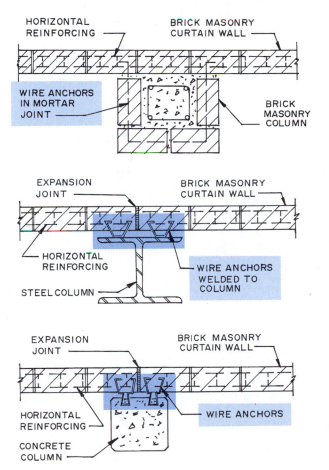

Figure 14.18 Typical brick masonry curtain wall construction.

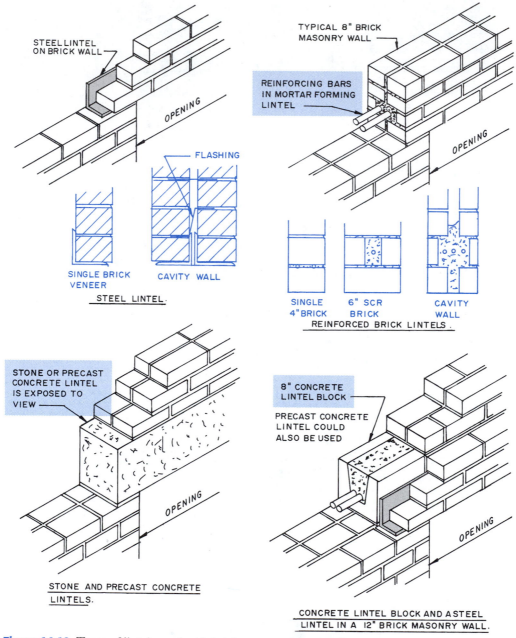

Figure 14.19 Types of lintels used with brick masonry construction.

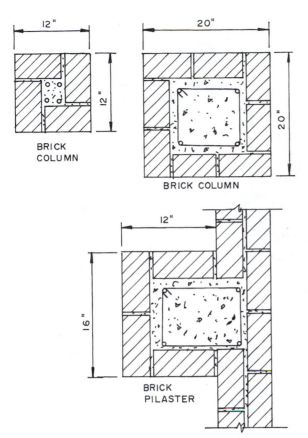

Figure 14.20 Reinforced brick masonry pilaster and column construction.

Insulation

Solid masonry walls are poor insulators. They conduct heat and cold. To improve the energy efficiency of a masonry wall insulation can be placed on the exterior face, within the wall, or on the interior surface. Rigid insulation consists of rigid foam plastic sheets adhered to the masonry with layers of acrylic polymer stucco material reinforced with fiberglass mesh applied over it. This forms an attractive hard exterior finish. Since the masonry is completely hidden, less costly units may be used (Fig. 14.24).

Cavity walls permit the insulation to be placed inside the wall. If rigid insulation is used, it is bonded to the inside face of the masonry units forming the inside wall (Fig. 14.25). Another way to insulate interior surfaces is to construct a stud wall over it and insulate using batt insulation (Fig. 14.26).

The cores of concrete block walls can be filled with a granular insulation. The webs permit considerable energy loss, so rigid insulation should be applied to the exterior or interior surface.

BUILDING A BRICK MASONRY WALL

Part of the design process is to specify the brick bond pattern and the type of mortar joint required. Bricks are also laid in several directions, providing some structural and decorative properties. The terms used to describe brick in various positions are shown in Fig. 14.27.

Brick Bond Patterns

Brick bond patterns commonly used for standard and oversize bricks are shown in Fig. 14.28. Notice the running bond consists entirely of stretchers. The English bond alternates courses of headers and stretchers. The common bond has a header course every sixth course. When the longer, narrower Norman and Roman bricks are used, the bond patterns in Fig. 14.29 are used.

Types of Mortar Joints

Mortar joints can take several forms. The appearance of the finished wall depends on the type of joint. Joints in common use are in Fig. 14.30. The mortar joint is finished by troweling or tooling. *Troweling* involves striking off the excess mortar using the mason's **trowel.** This does not produce the most watertight joint. The weathered, concave, and V-joints resist leakage the best. They are called *tooled joints.* Tooled joints are the result of compressing the mortar into the joint and against the face of adjacent units. Ruled, flush, and flush and rodded joints are not as watertight. Extruded, beaded, struck, and raked joints are most likely to eventually let moisture penetrate the wall.

Information on the types of mortar is in Chapter 10.

Laying Bricks

When possible, masonry walls are sized to permit the use of bricks without cutting (Fig. 14.31). A ⅜ to ½ in. (9.5 to 13 mm) joint is commonly used. Each unit length then is the length of the brick plus the mortar joint. (Brick sizes are in Chapter 11.)

The basic process for laying a brick masonry wall is shown in Fig. 14.32. The outer edge of the wall is located on the footing by snapping a chalk line. Then the first course is laid without mortar to see if a brick must be cut to close the wall. A brick is laid and leveled at each end of the wall. Working from each end toward the center, the first course is laid in mortar, making certain it is straight and level. Then several courses are laid up on each corner. A chalk line is run for the second course, and it is laid in mortar. After several courses are laid up, again the bricks are raised at each corner and then filled in toward the center.

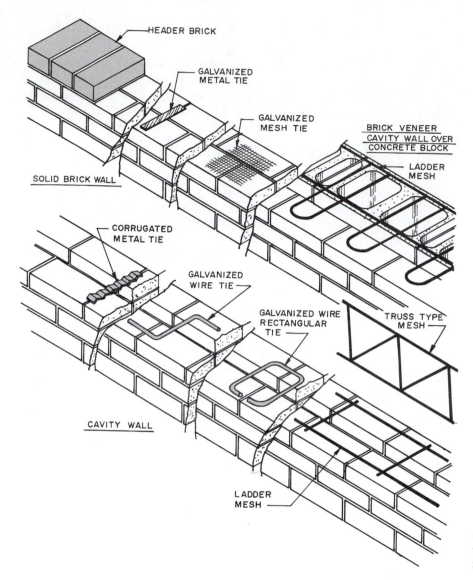

Figure 14.21 Brick masonry walls use various types of metal wall ties.

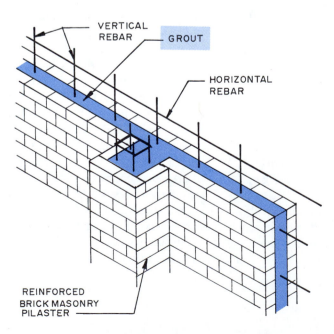

Figure 14.22 A grout-filled reinforced brick masonry wall with horizontal and vertical rebars.

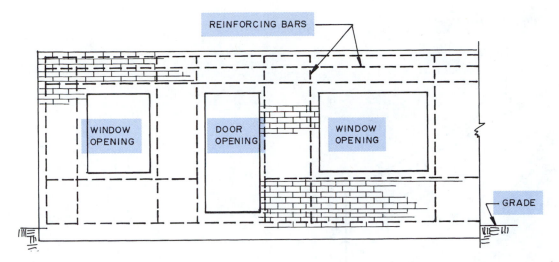

Figure 14.23 Typical reinforcing in brick masonry load-bearing walls that have openings.

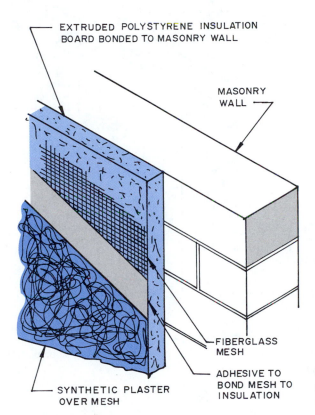

Figure 14.24 Rigid insulation may be applied to the exterior of masonry walls.

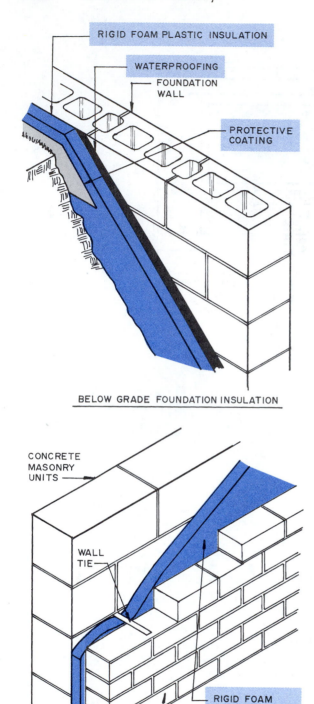

RIGID FOAM PLASTIC INSULATION

WATERPROOFING

FOUNDATION WALL

PROTECTIVE COATING

BELOW GRADE FOUNDATION INSULATION

CONCRETE MASONRY UNITS

WALL TIE

RIGID FOAM PLASTIC SHEET INSULATION

BRICK

AIR SPACE

EXTERIOR WALL INSULATION

Figure 14.25 Rigid plastic foam insulation is used on all types of masonry construction.

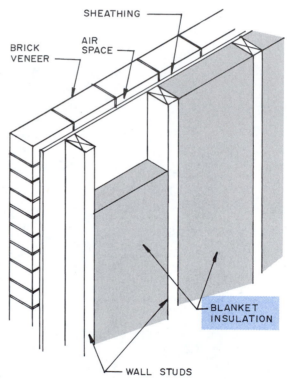

SHEATHING

BRICK VENEER

AIR SPACE

BLANKET INSULATION

WALL STUDS

Figure 14.26 This brick veneer is applied over a frame wall insulated with batt-type insulation.

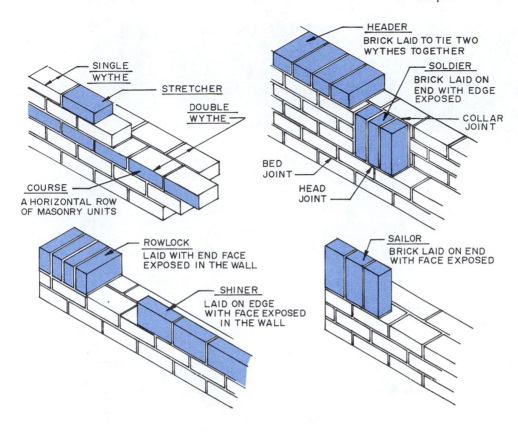

Figure 14.27 Terms used to describe brick in various positions in a wall.

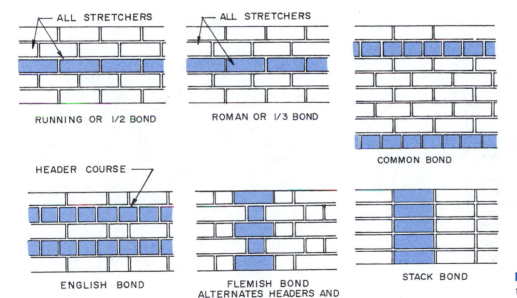

Figure 14.28 Brick bond patterns used with standard and oversize brick masonry units.

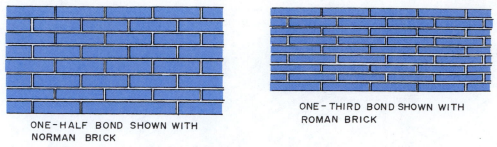

ONE-HALF BOND SHOWN WITH
NORMAN BRICK

ONE-THIRD BOND SHOWN WITH
ROMAN BRICK

Figure 14.29 Bond patterns used with Norman and Roman brick masonry units.

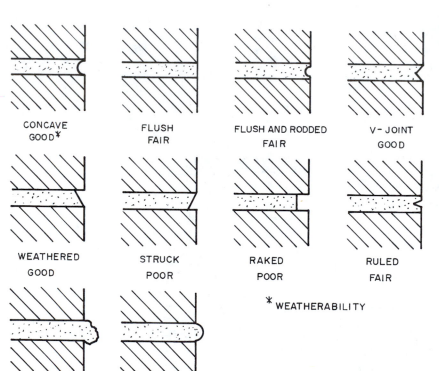

CONCAVE
GOOD *

FLUSH
FAIR

FLUSH AND RODDED
FAIR

V-JOINT
GOOD

WEATHERED
GOOD

STRUCK
POOR

RAKED
POOR

RULED
FAIR

EXTRUDED
POOR

BEAD
POOR

* WEATHERABILITY

Figure 14.30 Commonly used mortar joints.

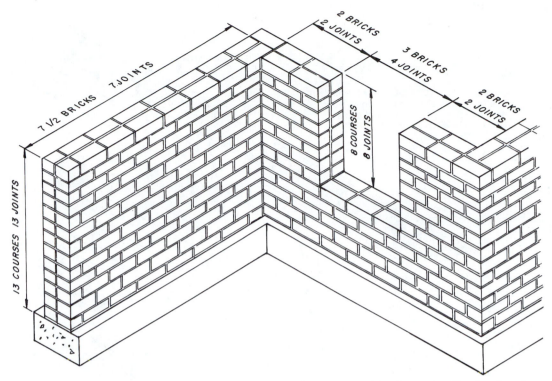

Figure 14.31 When designing a structure, the lengths of walls should be determined by the size of the masonry units used. This reduces the need to cut the units to odd lengths to complete the wall.

A **story pole** is used to check the height of each course as it is laid. A story pole has the height of each course marked on it.

In Fig. 14.33 are the basic steps generally followed to lay brick. The mason must work rapidly yet carefully in placing the bricks and maintaining a uniform mortar line. Because bricks are laid from each corner to the center of the wall, the closure is made by placing the final brick in the opening that is left. Again, this requires careful work by the mason so closure can be made without cutting a brick. The basic steps for closure are in Fig. 14.34.

After the mortar has set sufficiently the joints are tooled. The tooling forms the joints as shown in Fig. 14.30. Various shaped tooling tools are used to shape the joints. After tooling, the wall is lightly brushed with a fiber brush to remove loose mortar particles. Generally some weeks after the wall has cured it is washed with muriatic acid (HCl) and thoroughly rinsed with clean water. This removes any remaining mortar stains on the face of the brick.

CONCRETE MASONRY CONSTRUCTION

Some of the frequently used concrete masonry wall constructions are shown in Fig. 14.35. Since the composition of materials used to make **concrete masonry** varies, their compression strength, fire resistance, coefficient of thermal expansion, and other properties must be known so the designer can specify the type of construction required.

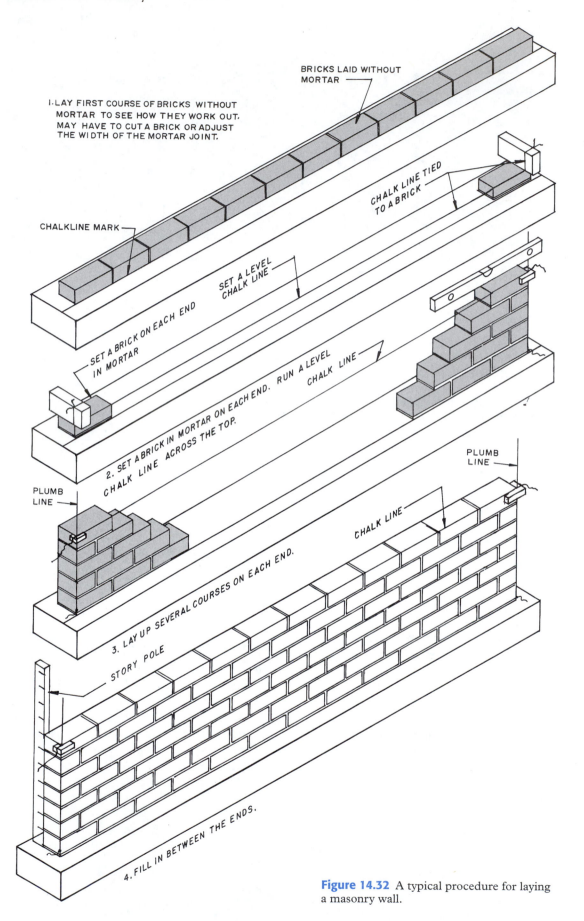

BRICKS LAID WITHOUT MORTAR

1. LAY FIRST COURSE OF BRICKS WITHOUT MORTAR TO SEE HOW THEY WORK OUT. MAY HAVE TO CUT A BRICK OR ADJUST THE WIDTH OF THE MORTAR JOINT.

CHALK LINE TIED TO A BRICK

CHALKLINE MARK

SET A LEVEL CHALK LINE

SET A BRICK ON EACH END IN MORTAR

2. SET A BRICK IN MORTAR ON EACH END. RUN A LEVEL CHALK LINE ACROSS THE TOP.

CHALK LINE

PLUMB LINE

PLUMB LINE

CHALK LINE

3. LAY UP SEVERAL COURSES ON EACH END.

STORY POLE

4. FILL IN BETWEEN THE ENDS.

Figure 14.32 A typical procedure for laying a masonry wall.

1. Lay a trowel of mortar on the top of the bricks.

2. Tilt the trowel and lay a windrow of mortar along the row of bricks.

3. Spread the mortar to cover four or five bricks.

4. Cut off the mortar projecting over the edge of the bricks.

5. Make a shallow furrow in the line of mortar.

6. The brick is positioned to receive the mortar for the head joint.

Figure 14.33 The basic steps for laying bricks. (*From* Concrete and Masonry, *U.S. Department of the Army*) *(continued on next page)*

7. Lay mortar on the end of the brick.

Figure 14.33 *Continued*

8. Place the brick on the mortar bed and press against the adjacent brick to close the head joint. Check it for levelness and correct height. Strike off surplus mortar that is squeezed out of the joints.

1. The bed of mortar is in place.

2. Place mortar on both ends of the closure brick.

3. Lower the brick in place being careful not to dislodge the mortar on the ends. tap into place and cut off excess mortar.

Figure 14.34 Procedure for closure of a row of bricks. (*From* Concrete and Masonry, *U.S. Department of the Army*)

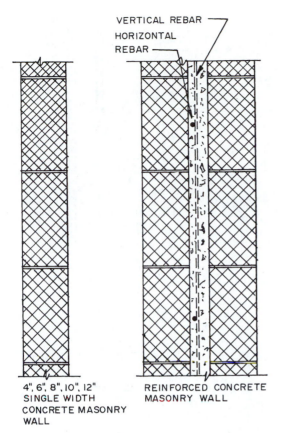

VERTICAL REBAR
HORIZONTAL REBAR

4", 6", 8", 10", 12"
SINGLE WIDTH
CONCRETE MASONRY
WALL

REINFORCED CONCRETE
MASONRY WALL

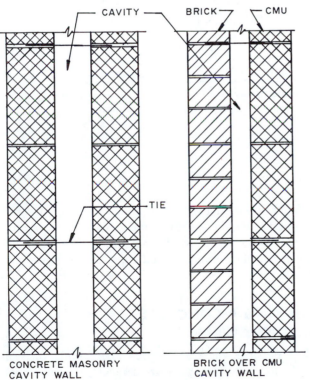

CAVITY   BRICK   CMU

TIE

CONCRETE MASONRY
CAVITY WALL

BRICK OVER CMU
CAVITY WALL

Figure 14.35 Concrete masonry wall constructions.

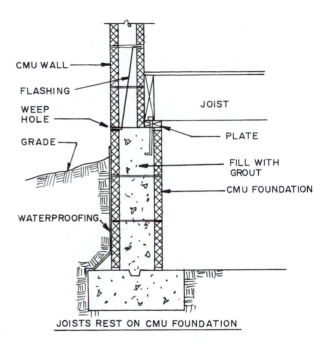

CMU WALL

FLASHING

WEEP HOLE

GRADE

WATERPROOFING

JOIST

PLATE

FILL WITH GROUT

CMU FOUNDATION

JOISTS REST ON CMU FOUNDATION

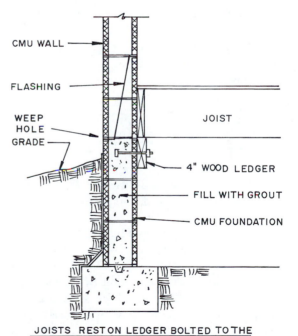

CMU WALL

FLASHING

WEEP HOLE
GRADE

JOIST

4" WOOD LEDGER

FILL WITH GROUT

CMU FOUNDATION

JOISTS REST ON LEDGER BOLTED TO THE
CMU FOUNDATION

Figure 14.36 Concrete masonry units are widely used for foundation construction.

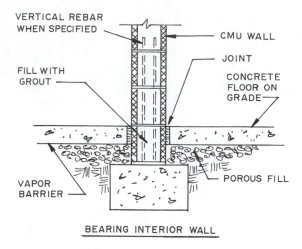

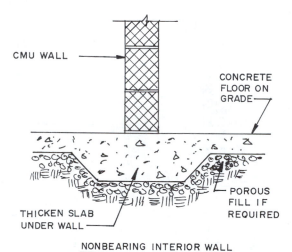

Figure 14.37 Interior concrete masonry unit walls may be load-bearing or non–load bearing.

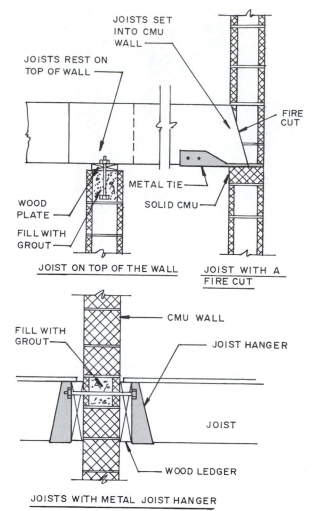

Figure 14.38 Ways to support floor joists using concrete masonry foundations and walls.

Construction details for load-bearing concrete masonry foundations are in Fig. 14.36. Interior concrete masonry walls may be load bearing or non–load bearing as shown in Fig. 14.37. Floor joists may rest on top of a concrete masonry wall, be set into it, or be supported by ledgers bolted through the wall (Fig. 14.38). Wood joists must bear at least 4 in. on the foundation. Some examples of roof construction are in Fig. 14.39. Openings are spanned using steel, precast concrete, or concrete lintel blocks as shown in Fig. 14.40.

Concrete masonry columns are designed to be built using standard-size blocks to reduce the need to cut units to special lengths. The interior core is filled with concrete and specified vertical reinforcing. Customized concrete masonry column units are available in some areas (Fig. 14.41). Pilasters are built the same as columns but are a part of the foundation. Custom-cast concrete masonry pilaster units are also available (Fig. 14.42).

Pilasters are used to stabilize long concrete masonry walls and provide a larger bearing surface for beams and girders resting on the wall. They are spaced along the wall at specified intervals and may have a reinforced concrete core.

BUILDING A CONCRETE MASONRY WALL

When possible, the designer sets the foundation wall lengths to use the standard-size concrete masonry units. This includes full blocks (16 in.) and half blocks (8 in.). A ⅜ in. or ½ in. mortar joint is commonly used. A typical example is in Fig. 14.43. The process for laying block is basically the same as for brick. First, the blocks are spaced along the footing without mortar to establish spacing and to see if some blocks may need to be cut to nonstandard size (Fig. 14.44).

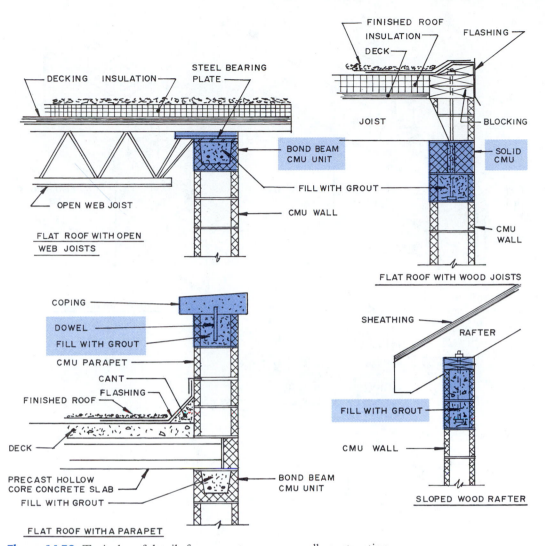

Figure 14.39 Typical roof details for concrete masonry wall construction.

The basic steps for laying up a concrete masonry wall are shown in Fig. 14.45. The mason begins by laying a full mortar bed on the footing and laying up each corner several courses high. The height of the courses is checked with a story pole. A chalk line is run the length of the wall to establish the level height of the first course, and it is laid on a full mortar bed. Courses above it are usually laid without placing mortar on the webs. Mortar is placed on the face shells, and each course is laid from the corners to the center of the wall. As the blocks meet, the closure must be made as shown in Fig. 14.46. The walls are checked for levelness and plumb with long mason's levels. After the mortar has set properly the joints are tooled to the desired contour, and the mortar crumbs are brushed off with a soft brush.

STONE MASONRY CONSTRUCTION

Stone laid in mortar may be rubble or ashlar masonry. **Rubble** is stone found naturally, so it is in irregular shapes and sizes. **Ashlar** masonry uses units cut into squared shapes. Stone may be laid in a random or coursed pattern. The coursed pattern maintains horizontal lines, and the random pattern courses fall in many directions (Fig. 14.47). One form of rubble stone masonry, ledgerock, is a rock formed naturally into thin layers of varying thicknesses.

When stone is laid, the grain in each piece should run in a horizontal direction. The stone is stronger in this position and tends to better resist weathering. Rubble stone is difficult to lay because the mason must select

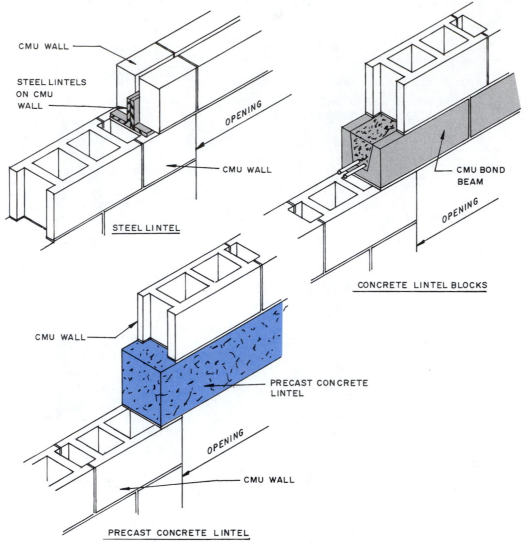

Figure 14.40 Lintels used in concrete masonry construction.

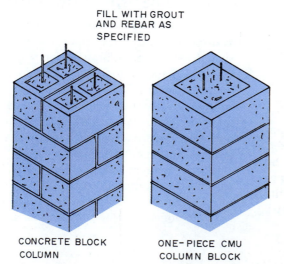

Figure 14.41 Concrete masonry units can be used to construct columns. Several types of special column units are available.

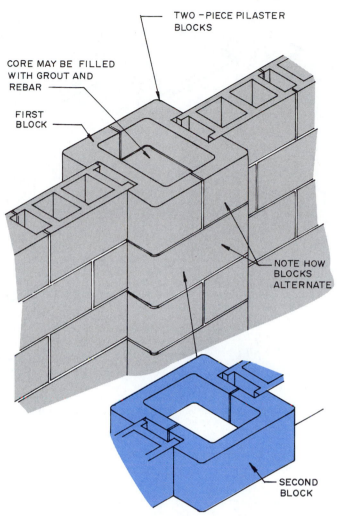

TWO-PIECE PILASTER
BLOCKS

CORE MAY BE FILLED
WITH GROUT AND
REBAR

FIRST
BLOCK

NOTE HOW
BLOCKS
ALTERNATE

SECOND
BLOCK

Figure 14.42 Pilasters can be built using concrete masonry pilaster units.

pieces to fit the open space and work in with the shapes of adjoining stones. Sometimes it is necessary to trim the stone with a hammer. The mortar joints are irregular and often quite large. The squared ashlar stones are easier to lay than rubble, but it is still necessary for the mason to select stones that will fit together as needed. Since the stones are squared, the mortar joint can be of relatively uniform width over the entire wall. Joints in ashlar masonry are usually ⅜ to ¾ in. Large squared blocks are called *dimension stone*. They usually require a hoist to set them in place. Mortar joints in ashlar masonry are usually raked after setting and allowed to set thoroughly. Then the raked space is filled with mortar (called pointing) and tooled to the specified shape. Great care is taken to avoid getting mortar on the face of the stone. The face is cleaned with a mild soap and soft brush and flushed clean with water.

Stone masonry is usually a veneer over a backup wall. Concrete masonry units are frequently used. Some form of metal tie is used to bond the stone veneer to the backup wall (Fig. 14.48), and much the same as is done with brick masonry construction. Anchors should be chromium-nickel, stainless steel, or a zinc alloy. Building codes often ban the use of galvanized steel. Copper, brass, and bronze ties may cause some staining. Construction details for several types of stone masonry veneered wall construction are shown in Fig. 14.49.

Some stone is cut into large thin panels and these are scored to accept some form of metal tie to secure them in place to the backup wall. The manufacturers of various types of stone panels have engineered systems for securing the panels in place. A number of connection methods are used. The first veneer panel on the base can be set on the foundation or other supporting masonry, or the panel can overhang the supporting structure and be secured with metal connectors as shown in Fig. 14.50. As the panels are placed up the wall, some form of metal connector is used to secure it to the backup wall.

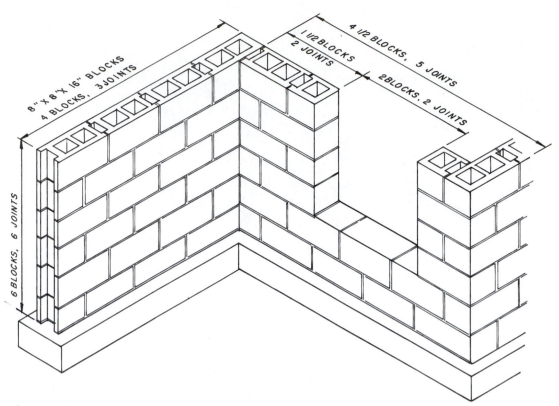

Figure 14.43 Concrete masonry walls should be designed to use standard-size units.

Figure 14.44 Concrete masonry units are cut with a saw that has a water-cooled diamond saw blade. *(Courtesy Portland Cement Association)*

1. Lay the first course on a full mortar bed. This shows the bed for the mortar blocks. Notice the blocks on th right were laid without mortar to establish spacing.

2. The blocks for the first corner are being laid.

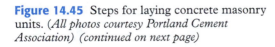

3. The corner blocks are laid up several courses. Notice that mortar has not been place on the webs.

Figure 14.45 Steps for laying concrete masonry units. (*All photos courtesy Portland Cement Association*) (*continued on next page*)

4. The height of each course is checked with a story pole.

5. Blocks are laid up on each corner and worked toward the center of the wall.

6. The first course has been laid across the footing and the corners have been raised several courses. The mason is laying the second course.

7. A level is used to check the wall for plumb.

8. A level is used to check the wall for levelness.

9. The mortar is tooled to the desired contour when it has properly set.

10. Mortar crumbs are removed with a soft brush after the tooling is finished.

Figure 14.45 *Continued*

1. The closure opening has mortar buttered on all sides.

2. The closure block has mortar on the ends and is slid into the opening. Excess mortar is cut away.

Figure 14.46 Procedure for closure of a course of concrete blocks. *(Courtesy Portland Cement Association)*

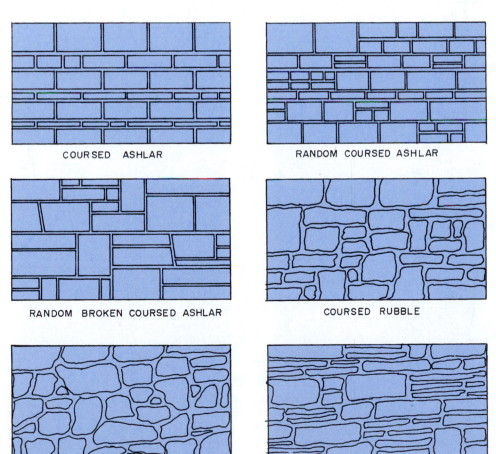

COURSED ASHLAR

RANDOM COURSED ASHLAR

RANDOM BROKEN COURSED ASHLAR

COURSED RUBBLE

RANDOM RUBBLE

UNCOURSED LEDGEROCK

Figure 14.47 Patterns for laying ashlar and rubble stone.

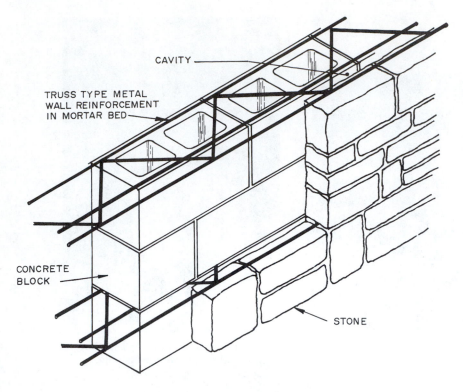

CAVITY

TRUSS TYPE METAL
WALL REINFORCEMENT
IN MORTAR BED

CONCRETE
BLOCK

STONE

Figure 14.48 Stone masonry is usually laid over a backup wall and connected to it with metal ties. This example is a cavity wall.

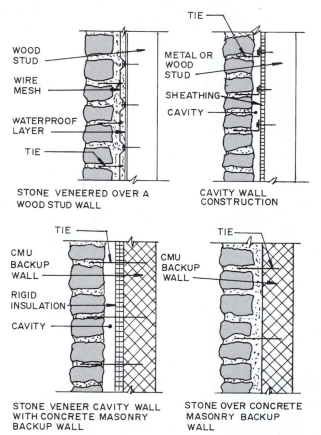

WOOD STUD

WIRE MESH

WATERPROOF LAYER

TIE

STONE VENEERED OVER A
WOOD STUD WALL

TIE

METAL OR
WOOD
STUD

SHEATHING

CAVITY

CAVITY WALL
CONSTRUCTION

TIE

CMU
BACKUP
WALL

RIGID
INSULATION

CAVITY

STONE VENEER CAVITY WALL
WITH CONCRETE MASONRY
BACKUP WALL

TIE

CMU
BACKUP
WALL

STONE OVER CONCRETE
MASONRY BACKUP
WALL

Figure 14.49 Some typical stone veneer wall constructions.

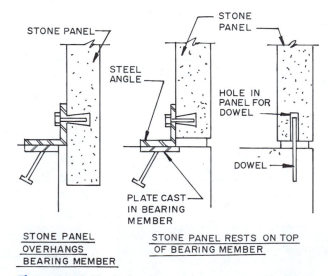

STONE PANEL

STONE
PANEL

STEEL
ANGLE

HOLE IN
PANEL FOR
DOWEL

DOWEL

PLATE CAST
IN BEARING
MEMBER

STONE PANEL
OVERHANGS
BEARING MEMBER

STONE PANEL RESTS ON TOP
OF BEARING MEMBER

Figure 14.50 Vertical stone panels may rest on the bearing member or overlap it.

Connections to masonry walls and structural steel frames are shown in Fig. 14.51. Connections to a coping are also shown in this figure. Columns often are cased with stone veneer. Typically, metal connectors tie the stone veneer to the column (Fig. 14.52). Preassembled column covers are available and are bonded with high-strength epoxy adhesives.

Details for external corners are shown in Fig. 14.53. These details are specified by the architect in connection with advice from the stone masonry supplier. As discussed with brick and concrete masonry, provision must be made for expansion of the stone veneer panels and for vertical and horizontal control joints.

Stone veneered walls usually use stone windowsills (Fig. 14.54) and various types of decorative moldings (Fig. 14.55), balusters (Fig. 14.56), and modillions and consoles. A **modillion** is a scroll supporting the **corona** (overhanging member of the cornice) under a cornice. A **console** is a decorative vertical scrolled bracket projecting from the wall to support a cornice, door, or window. *Balusters* are vertical members in a stair rail used to support the handrail. These are shown in Fig. 14.57.

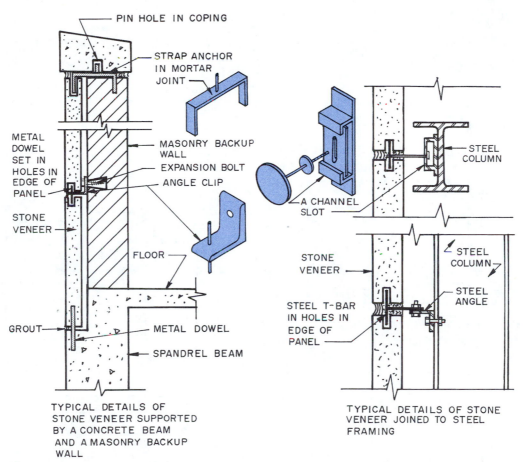

Figure 14.51 Typical details for installing vertical stone panels over masonry, concrete, and steel framing.

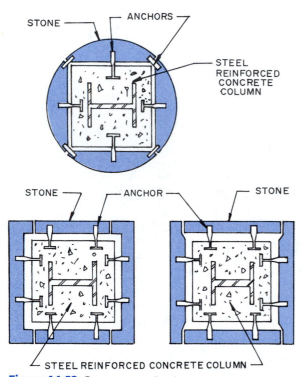

Figure 14.52 Stone-encased columns.

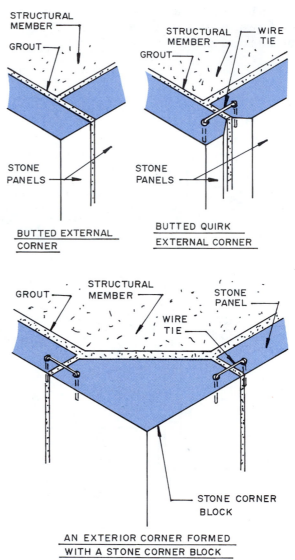

Figure 14.53 Methods for corner construction when installing stone wall panels.

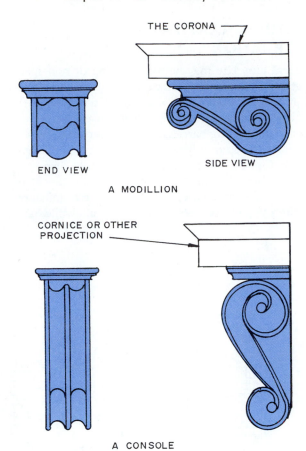

THE CORONA

END VIEW   SIDE VIEW

A MODILLION

CORNICE OR OTHER PROJECTION

A CONSOLE

Figure 14.56 Typical modillion and console applications.

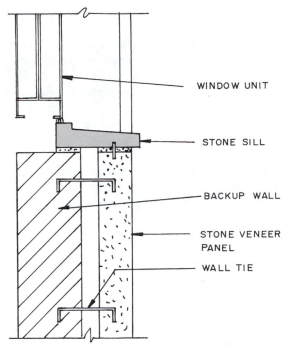

WINDOW UNIT

STONE SILL

BACKUP WALL

STONE VENEER PANEL

WALL TIE

Figure 14.54 Typical stone sill details. Actual design varies with wall construction and the windows used.

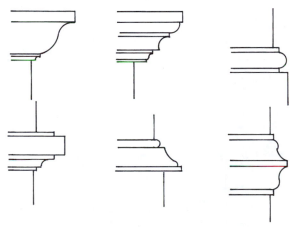

Figure 14.55 Examples of moldings and trim that can be produced in various types of stone.

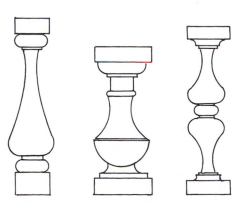

Figure 14.57 Stone balusters can be made to suit the designs of the architect.

STRUCTURAL CLAY TILE CONSTRUCTION

Structural clay tile may be load bearing or non–load bearing. Load-bearing units may form an unfaced wall, or they may be used as a backup wall and faced with another material. Non-load-bearing tile walls are used as interior partitions, fireproofing barriers, and masonry screens.

Examples of commonly used structural tile construction are in Fig. 14.58. It should be noted that there are many possible variations depending upon the expected end result, loads, and exposed face. Some tile units are utility block used to construct a backup wall, while oth-

ers have glazed exposed faces available in a variety of colors and textures.

The cavity walls use flashing and weep holes as described for brick construction. They may be insulated with rigid plastic foam insulation. Floor joists and rafters also are installed much like those on brick and concrete masonry. Typical examples are in Fig. 14.59. Tile can be used as a veneer over a backup wall of a different material, such as concrete masonry units or stud walls (Fig. 14.60).

Openings in structural clay tile may be spanned using tile lintel units as shown in Fig. 14.61.

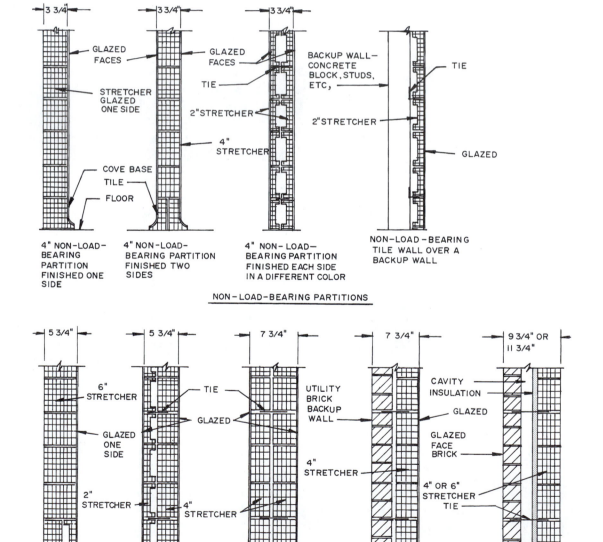

Figure 14.58 Typical wall and partition construction using structural tile.

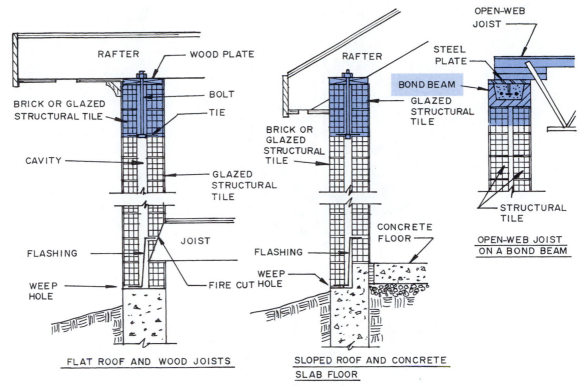

Figure 14.59 Examples of floor and roof construction used with structural tile.

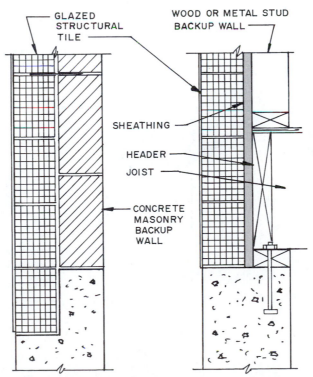

Figure 14.60 Structural tile can be installed over backup walls.

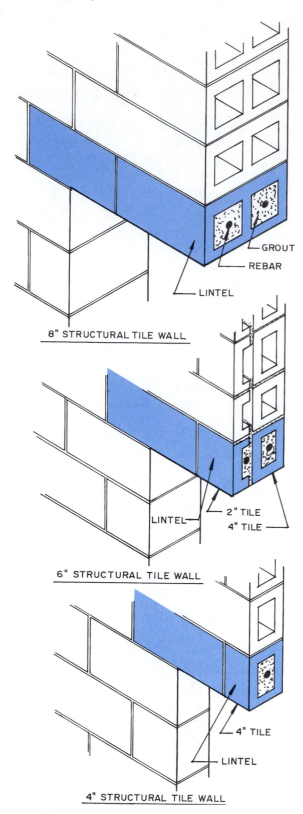

8" STRUCTURAL TILE WALL

GROUT

REBAR

LINTEL

6" STRUCTURAL TILE WALL

LINTEL

2" TILE

4" TILE

4" STRUCTURAL TILE WALL

4" TILE

LINTEL

Figure 14.61 Lintels can be constructed with structural tile.

BUILDING A STRUCTURAL CLAY TILE MASONRY WALL

As with brick and concrete masonry, structural clay tile walls should be designed to use standard size units to avoid having to cut them. The face sizes of a standard stretcher block are 5 1/16 in. high by 11 3/4 in. long (5 1/2 in. nominal) and 7 3/4 in. high by 15 3/4 in. long (8 × 16 in. nominal). The designer must have knowledge of these sizes and other special units available.

High-strength mortars are used with structural tile. The local codes and information from the tile manufacturer should be used when specifying the mortar to be used. The construction requires the use of metal ties as discussed for brick and concrete masonry. Expansion joints must be planned.

Corners can be built in several ways. Typical internal and external corner constructions are shown in Fig. 14.62. Sills on openings can be constructed using horizontal bull-nose units as shown in Fig. 14.63.

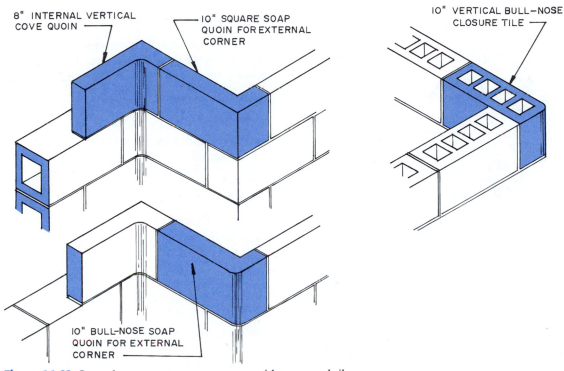

Figure 14.62 Several ways to construct corners with structural tile.

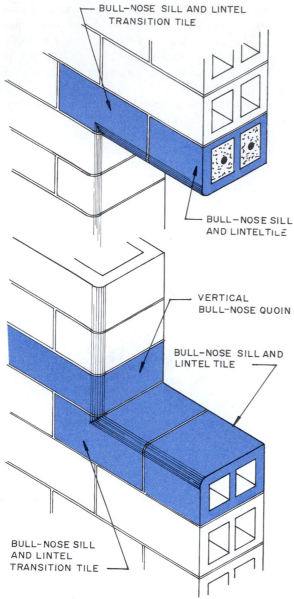

BULL-NOSE SILL AND LINTEL TRANSITION TILE

BULL-NOSE SILL AND LINTEL TILE

VERTICAL BULL-NOSE QUOIN

BULL-NOSE SILL AND LINTEL TILE

BULL-NOSE SILL AND LINTEL TRANSITION TILE

Figure 14.63 One way to frame openings in structural tile walls.

REVIEW QUESTIONS

1. What are the commonly used types of masonry load-bearing walls?
2. Why are control and expansion joints required in masonry walls?
3. Why are weep holes used?
4. What can be done to reduce moisture penetration through masonry walls?
5. What types of metal flashing can be used with masonry wall construction?
6. How are masonry units supported over openings in the wall?
7. How can the heat and cold conduction of masonry walls be reduced?
8. What are the commonly used brick patterns used in wall construction?
9. What are the advantages and disadvantages of the various types of mortar joints?
10. What size mortar joint is commonly used?
11. What is the difference between rubble and ashlar stone used for wall construction?
12. What types of structural clay tile are used for wall construction?

KEY TERMS

arch A curved construction spanning an opening and supported at the sides or ends.

ashlar stone Stone cut into rectangular shapes.

brick A solid or hollow masonry unit made from clay or shale, molded into a rectangular shape, and fired in a kiln to produce a hard, strong unit.

brick bond The arrangement of masonry units when laid to form a wall.

brick veneer A facing of brick laid against a wall but not structurally bonded to it.

cavity wall A masonry wall consisting of outer and inner width separated by a continuous air space.

concrete masonry unit A solid masonry unit, rectangular in shape, made from portland cement and suitable aggregates.

console A decorative stone scrolled bracket projecting from a wall to support a cornice, door, or window without damaging the structure.

control joint A groove formed in concrete and masonry structures to regulate the location and amount of cracking due to dimensional changes in various parts of the structure.

corona The overhanging vertical member of a cornice.

expansion joint A gap between adjacent parts of a structure that permits their relative movement, such as movement caused by temperature changes.

lintel block A precast concrete structural member used to span an opening in a wall and carry the weight of the masonry units above it.

modillion A stone scroll supporting a corona.

pilaster A column structurally attached to a masonry or concrete wall.

reinforced masonry Masonry units assembled with steel reinforcement rods embedded in the assembly so that the two materials act together to resist forces.

rubble Rough stones in their natural irregular shapes and sizes.

story pole A rod divided into equal parts with each part equal to the height of one course of masonry units.

structural clay tile Rectangular masonry units made from clay and fired to harden them for use in load-bearing and non-load-bearing walls and partitions.

trowel A hand tool with a flat, broad steel blade used to apply, shape, and spread mortar.

weep hole A small opening left in a wall from which accumulated water can drain to the exterior.

SUGGESTED ACTIVITIES

1. Check the local building code and make a record of the requirements pertaining to the construction of masonry load-bearing walls.

2. Using standard 8 × 8 × 16 in. concrete blocks, design the side and rear walls of a building so it will be approximately 10 ft. wide and 25 ft. long. Size the walls so it is not necessary to cut any of the concrete blocks. If the wall is to be about 10 ft. high, how many blocks will it take to lay the back and one side?

3. Secure a quantity of clay bricks and mortar mix. Prepare a story pole using ½ in. mortar joints. Lay up a length of wall about 6 ft. long and 4 ft. high. Try making several types of finished mortar joints.

4. Collect samples of all of the masonry units you can find. Most clay brick manufacturers have sample boards with thin sections of the bricks they manufacture glued onto a strong backing.

5. Pick an exterior wall of a brick building and measure the width and height. Calculate the number of standard and oversize bricks in the wall. Get the local cost of bricks and figure the cost of the masonry units in this wall. Refer to Chapter 11 for sizes of masonry units.

ADDITIONAL INFORMATION

Hornbostel, C., *Construction Materials,* John Wiley and Sons, New York, 1991.

Ramsey, C. G., Sleeper, H. R., and Hoke, J. R., eds. *Architectural Graphic Standards,* John Wiley and Sons, New York, 1991.

Stitt, F.A., *Architect's Detail Library,* Van Nostrand Reinhold, New York, 1990.

Other resources are

Numerous publications from the Brick Institute of America, Reston, Va.

Numerous publications from the National Concrete Masonry Association, Herndon, Va.

PART IV

METALS, WOOD, AND PLASTICS

Metals were not widely used in building construction until relatively recent times. Discoveries leading to processes for recovering them from the earth and for improving their properties led to their current widely used applications.

Chapter 15 includes technical information about the types of ferrous metals and their alloying elements, properties, and uses in the construction industry. Chapter 16 gives detailed technical information about the wide range of nonferrous metals, their properties, and applications. Chapter 17 illustrates the details for structural steel framing.

Wood is a renewable resource, and as such it is a unique building material. It has properties that make it useful for structural purposes as well as a decorative material. In addition, wood is widely used for other items, such as cabinets and furniture. Chapters 18 and 19 describe the various species of wood used in construction, as well as the many uses of wood and products produced with wood as the basic material. Chapters 20 and 21 illustrate the complexities of wood light frame construction and heavy timber construction.

Chapter 22 describes in detail the processes used to finish the interiors and exteriors of light frame construction. This includes products such as wood, gypsum, tile, asphalt, metals, and plastics.

Chapter 23 provides a description of various papers and paper products used in construction. These important products find extensive use but are often overlooked.

Chapter 24 gives extensive details about plastic—the newest construction material. A wide variety of plastics are used in construction, but their properties vary greatly. This chapter cites the proper applications for this relatively new material. ▲

DIVISION 5

METALS

CSI MASTERFORMAT™

Photo courtesy H.H. Robertson

Ferrous Metals

This chapter will help you to:

1. Explain the processes for mining and processing iron ore and for producing pig iron and steel.

2. Develop a knowledge of the properties of ferrous metals that you can use when making material selection decisions.

3. Use with confidence the various steel identification systems and the Unified Numbering System for Metals and Alloys

Ferrous metals are those in which the chief ingredient is the chemical element iron (ferrum). Iron (chemical symbol Fe) is found in the earth's crust in large quantities mixed with other minerals. To be useful iron must be extracted from the mined ore, have impurities removed and ingredients added to alter its properties, and then be formed into useful products.

Ferrous metal products are widely used in the construction industry. They are a major construction material, and architects, engineers, and contractors should be familiar with the various types, their properties, and the proper applications for each. Although a ferrous metal product may fail, it generally is not because it is a poor material but because the choice of the type of ferrous metal for a particular application was incorrect. When the properties of a ferrous metal are known, its performance can be accurately determined during the engineering design process.

HISTORY

Ferrous metals had little use in construction until the late eighteenth century, when cast iron found limited use as a structural material. By the early nineteenth century wrought iron and cast iron were finding limited use for structural purposes, but they were not used extensively because cast iron was brittle and the production of wrought iron was limited and expensive.

A breakthrough occurred in the 1850s when the Bessemer process for removing impurities from molten iron was developed. Among the steps in the purification process was the reduction of the carbon content that makes iron brittle. The Bessemer process involved blowing an air stream into the vessel holding the molten iron. This enabled steel, a greatly improved material, to be produced rather quickly and in large quantities, thus reducing the cost. Steel was also a material that was easy to form into useful products because of its ductile and malleable characteristics. In the late 1860s the open hearth steelmaking process was developed in Europe. This provided another source of steel, one of the most valuable and widely used construction materials. Other processes used to make steel include electric processes and the basic oxygen process.

IRON

Iron is found in large quantities in the earth's crust. Pure iron, free from impurities and other elements, is ductile and soft but not strong enough for structural purposes. It has high magnetic properties, oxidizes (rusts) easily, and does not resist attack by acids and some chemicals. For commercial purposes iron must have **alloying elements** added to improve its characteristics. Iron can be hardened by heating and then cooling rapidly and can be made more workable by **annealing**—heating it and allowing it to cool slowly.

The typical commercial iron product is called **pig iron.** It contains 3 to 5 percent carbon and traces of other elements such as manganese, sulfur, silicon, and phosphorus. Pig iron is the base material used to produce various types of iron and steel.

Iron is used for a wide range of purposes in construction. For example, iron particles may be used as the abrasive for sandblasting, and iron is the main ingredient in cast iron, steel, stainless steel, and iron alloys. It finds some use as an aggregate for specialized concretes and as the basis for some color pigments.

Mining and Processing Iron Ore

Iron is found in rock, gravel, sand, clay, and mud. It is mined in open pit and underground mines. Common iron-bearing minerals are pyrite (FeS_2), siderite ($FeCO_3$), and hematite (Fe_2O_3). These contain up to 70 percent iron. Jasper and taconite rock contain 20 to 30 percent iron.

The surface ores are blasted, loaded onto trucks, trains, and ships, and moved to blast furnaces. If the iron content of the ore is less than about 50 percent it is too costly to ship any distance. These lower-content ores are processed at the site by a **beneficiation** process that removes some of the unwanted elements, leaving an ore with a higher iron content. Beneficiation involves grinding the ore to remove the unwanted elements and then increasing the size of the ground particles by a process called agglomeration. **Agglomeration** involves pelletizing the ore particles into ball-like pellets, which are easier to ship than the finely ground ore. The iron ore dust that is produced is recycled by sintering. **Sintering** involves fusing the iron ore dust with coke and fluxes into a clinker that is high in iron content.

Ores that have a very high iron content are made into pellets and briquettes containing more than 90 percent iron. These pellets are so pure they are not used in producing pig iron but go directly to the steelmaking process. The process of producing these pure pellets is called *direct reduction.*

Producing Iron from Iron Ore

Iron ore is converted into pig iron in a *blast furnace* by smelting and reduction. **Smelting** is a process in which the ore is heated, permitting the iron to be separated from impurities with which it may be chemically or physically mixed. **Reduction** is a process that separates the iron from oxygen with which it is chemically mixed.

The Blast Furnace

The blast furnace separates the iron from the waste materials and sinters the ore and flue dust. A very large blast furnace is shown in Fig. 15.1. It operates twenty-four hours a day and is computer controlled and fed by a continuous conveyor.

Coal, oil, and natural gas are the commonly used fuels in a blast furnace. Most operate on *coke,* which is produced from coal. The coke not only provides heat to melt the ore but also helps separate the iron from its oxides.

A **flux** is a mineral added to the molten ore in the blast furnace. The flux combines with impurities and forms a **slag** that floats on top of the molten material. Limestone and dolomite are typical basic fluxes. Sand, gravel, and quartz are acid fluxes.

The heart of the blast furnace is a tall, cylindrical shaft 150 to 200 ft. (45.8 to 61 m) tall and about 30 ft. (9.2 m) in diameter at the base. It is usually smaller in diameter at the top. It is lined with a refractory brick made of a material such as magnesia that will withstand temperatures approaching 3000°F (1662°C) (Fig. 15.2).

Figure 15.1 This very large blast furnace is capable of producing 8,000 tons of molten iron per day. *(Courtesy Bethlehem Steel Corporation)*

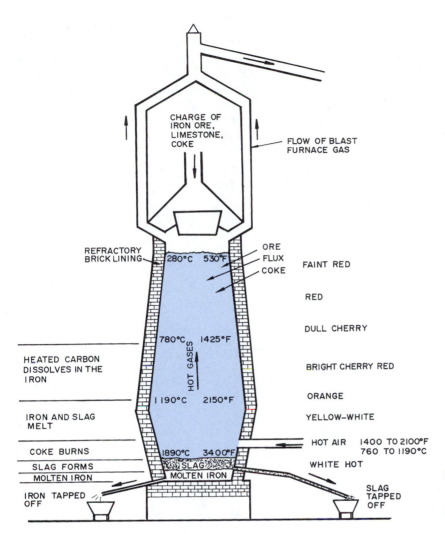

CHARGE OF IRON ORE, LIMESTONE, COKE

FLOW OF BLAST FURNACE GAS

REFRACTORY BRICK LINING — 280°C 530°F

ORE
FLUX
COKE

FAINT RED

RED

DULL CHERRY

780°C 1425°F

HOT GASES

BRIGHT CHERRY RED

HEATED CARBON DISSOLVES IN THE IRON

1190°C 2150°F

ORANGE

YELLOW–WHITE

IRON AND SLAG MELT

COKE BURNS

1890°C 3400°F

HOT AIR 1400 TO 2100°F
760 TO 1190°C

WHITE HOT

SLAG FORMS

MOLTEN IRON

SLAG

MOLTEN IRON

IRON TAPPED OFF

SLAG TAPPED OFF

Figure 15.2 A blast furnace receives the charge at the top and hot air at the bottom causing combustion, which frees the iron from the oxide-forming pig iron.

A charge of ore, hot coke (fuel), and limestone (flux) are loaded at the top. A blast of hot air is injected at the bottom. As the air works its way up through the charge, the oxygen in the hot air combines with the hot coke. This causes combustion, which produces the heat necessary to melt the ore. As the gases from combustion pass through the heated ore, a chemical reaction occurs that frees the iron from its oxide. The gases are removed from the top of the furnace and pass through cleaners, which retain the dust. The cleaned heated air is passed through brick heaters that heat the new incoming air. A blast furnace must be operated continuously.

The molten pig iron settles to the bottom of the furnace and is tapped off every few hours. At this point it is impure and brittle. It contains about 4 to 5 percent carbon. This high carbon content is what makes pig iron brittle and not as useful as steel. It has no ductility and cannot be rolled into useful products, such as beams or studs. Pig iron must have additional processing in a steelmaking furnace to reduce the carbon content. The impurities, flux, and oxygen combine to form a slag on top of the molten iron. The slag is tapped off every few hours. Slag is used as an aggregate for concrete, loose fill, and road surfacing.

The molten pig iron is moved to furnaces called mixers. Here it is mixed to equalize the chemical composition and kept liquid until it is to be used. The molten iron sometimes goes into a casting machine in which it is cast into ingots (**pigs**) and is allowed to cool. In other operations it is moved directly to basic oxygen steelmaking furnaces. In this case, energy is saved because the iron is already in molten form. You can see the total process in Fig. 15.3.

The pig iron ingots are usually made to specifications based on their end use. The elements that are carefully controlled include carbon, sulfur, silicon, phosphorus, and manganese.

Cast Irons

Iron containing almost no carbon is identified as a *wrought iron*. It is a mixture of low-carbon iron and a large amount of slag. Wrought iron is soft, tough, and ductile (easily worked). *Ingot iron* is a very low-carbon iron that has no slag and is also tough, ductile, and soft.

Cast irons have carbon contents above 1.7 percent and include white, gray, and malleable types. *White cast irons* have a low silicon content and are cooled rapidly. They are hard and brittle and have few applications for construction uses.

Gray cast irons are produced by increasing the silicon content and cooling the molten metal slowly. They are tougher and softer than white cast iron and are gray in color. They may have other elements, such as nickel, copper, and chromium, added to them. Gray cast irons are widely used for all types of castings, such as sewer pipe, ornamental railings, and decorative lampposts.

Malleable cast iron is produced by reheating white cast iron, holding the required temperature for a long time, and then cooling it slowly. It is not as brittle as white and gray cast iron because it has a lower carbon content, and it has greater ductility. Malleable cast iron is used for hardware and other cast items requiring **toughness** and resistance to breakage.

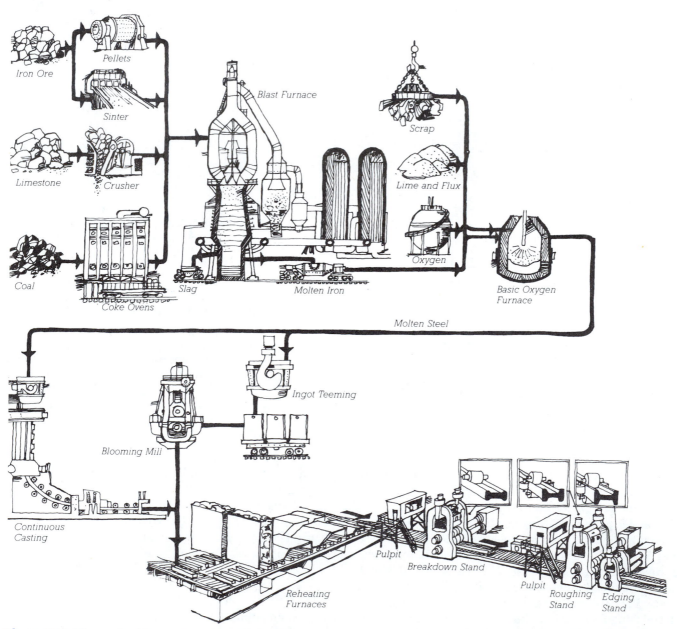

Figure 15.3 The steelmaking process, from preparing the ore through processing into steel to producing steel beams. (*From "Steelmaking Flowline" by permission of the American Iron and Steel Institute*)

STEELMAKING

The production of steel involves (1) controlling impurities in the pig iron, (2) adding alloying elements as needed to get the required properties, and (3) lowering the carbon content of the pig iron. This involves combining molten pig iron, scrap metal, and fluxes in a steelmaking furnace from which the molten steel is cast into ingots or run through a strand-casting machine. Steel is produced using the basic oxygen process or with an electric furnace.

Basic Oxygen Process

This furnace has a large pear-shaped vessel lined with refractory material, as you can see in Fig. 15.4. The charge, consisting of molten pig iron, metal, scraps, and fluxes, is added at the top. A jet of high-purity oxygen is shot into the vessel through a water-cooled lance. The heat of the molten pig iron is great enough to start a chemical reaction between the oxygen and the carbon and other impurities. This **oxidation** produces the heat necessary to melt the charge without additional heat. The slag and molten steel are tapped off. A modern basic oxygen furnace can be seen in Fig. 15.5. It operates behind huge pollution control doors that remain closed during the steelmaking process. This enables dust and gases generated during the process to be collected by the pollution control system. A pour of the molten steel can be seen in Fig. 15.6.

Electric Steelmaking Processes

High-grade steels, such as stainless, tool, heat-resisting, and alloy, are generally produced in electric furnaces. These furnaces can develop the high temperatures needed to produce the conditions required to produce these metals. These furnaces use arc radiation and electric resistance to current flow to produce the high temperatures needed. High-purity oxygen is injected when needed, which oxidizes impurities. Since the process can be carefully controlled, there is less loss of the alloying elements. The furnaces are charged with iron, scrap metal, fluxes, and alloying elements.

The two types of electric furnaces are the electric arc furnace and the induction furnace. The electric arc furnace can produce steels in large quantities. The induction furnace is used to produce smaller quantities of special grades of steel that require the use of expensive alloying elements.

The Electric Arc Furnace

The electric arc furnace has a circular steel shell somewhat like a large cooking pot. Several cylindrical electrodes project into the furnace from the top. A high-voltage electric current is passed through the electrodes, causing an arc to pass between them. This generates the heat needed to melt the charge. The furnace is tilted and the steel is poured off under the slag (Fig. 15.7). Figure 15.8 shows a furnace in operation.

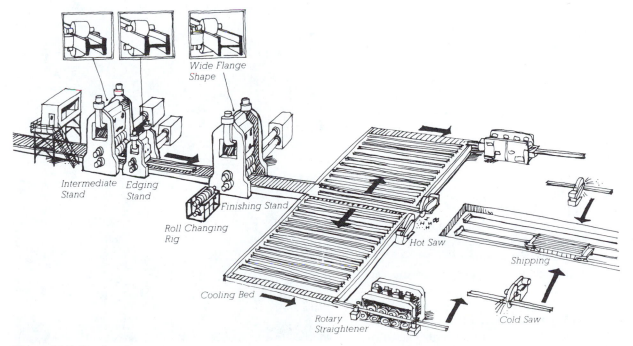

Figure 15.3 *Continued*

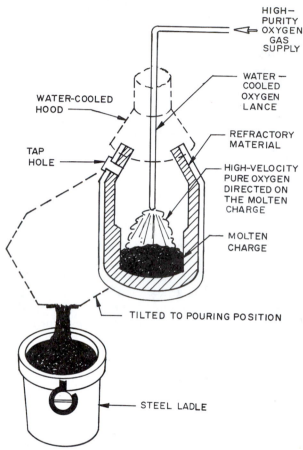

Figure 15.5 This is a basic oxygen steelmaking furnace. Notice its size compared to the man in the picture. It is operating behind closed pollution control doors, which allow dust and gases to be collected by the pollution control system. (*Courtesy Bethlehem Steel Corporation*)

Figure 15.4 The basic oxygen process uses a jet of high-purity oxygen plus the heat of the molten pig iron to start the process.

The Induction Furnace

This furnace has a cylindrical vessel made of magnesia and insulated with refractory materials. Outside of this are windings of copper tubing through which high-voltage alternating current is passed. This creates an induction current that is resisted by the charge. This resistance generates the heat needed for melting the charge. The vessel is tilted and the steel is poured off below the slag.

Casting the Steel

When the steel is ready to be poured from the furnace, it is either cast into *ingot molds* or run through a **strand-casting machine.**

Figure 15.6 An internal view of a basic oxygen furnace during a pour. (*Courtesy Bethlehem Steel Corporation*)

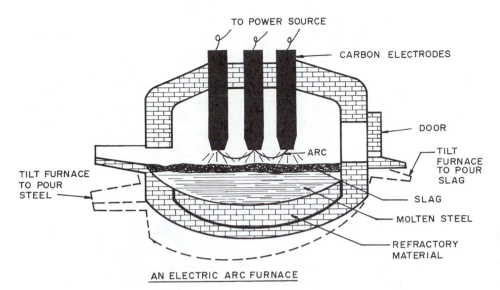

Figure 15.7 The electric arc furnace passes high-voltage electric current between electrodes, producing an arc that melts the charge.

Figure 15.8 An electric steelmaking furnace. (*Courtesy Bethlehem Steel Corporation*)

Ingot molds can be several feet in diameter and six to eight feet high. They are coated inside to prevent the steel from damaging the surface. After an ingot has begun to harden it is removed from the mold and taken to a soaking pit. Here the inside, which is still molten, solidifies.

Strand-casting machines produce a continuous ribbon of steel that begins to harden as it passes through a series of rollers. When it has hardened, it reaches a horizontal conveyor and is cut to the required lengths. This process takes about thirty minutes (Fig. 15.9). Notice the parts of the operation shown in Figs. 15.10 through Fig. 15.13.

MANUFACTURING STEEL PRODUCTS

Steel products are manufactured by rolling, extruding, cold-drawing, forging, and casting. Items produced by casting are referred to as *cast steel products* and all of the others are called *wrought products*. The most frequently used products are produced by hot-rolling and cold finishing.

Cast steel products are made by pouring molten steel into sand molds. The molten steel goes through a process to alter its properties to suit the purpose served by the finished product.

Most wrought products are produced by *hot-rolling* and *cold-rolling*. The ingots produced in the steel mill are used to produce slabs, blooms, and billets, which can be seen in Fig. 15.14. When steel is produced in the form of slabs, blooms, or billets it is called *semifinished*. When semifinished steel is fabricated into a product that is used to produce a finished product or fabricated directly to produce a finished product, it is *finished* steel.

Slabs are large rectangular semifinished steel pieces that are made in a number of sizes. Their width is typically more than twice their thickness. The thickness is usually about 10 in. (254 mm) or less. A *bloom* is an oblong of semifinished steel with a square cross section that is usually larger than 6 in. (152 mm) square. A continuous bloom caster is in Fig. 15.15. Notice the size of the machine and the working environment. *Billets* are also oblongs of semifinished steel, but they are smaller in cross section and longer than blooms.

Figure 15.14 also lists typical finished products that use these semifinished steel products. Notice that plates and skelp (skelp is used to produce pipe and tubing) are rolled from slabs. Finished steel plates resting on cooling beds can be seen in Fig. 15.16. The plate stock in Fig. 15.17 is entering a four-high finishing stand where it will be rolled to the final desired length and thickness by being passed back and forth through the finishing stand rollers.

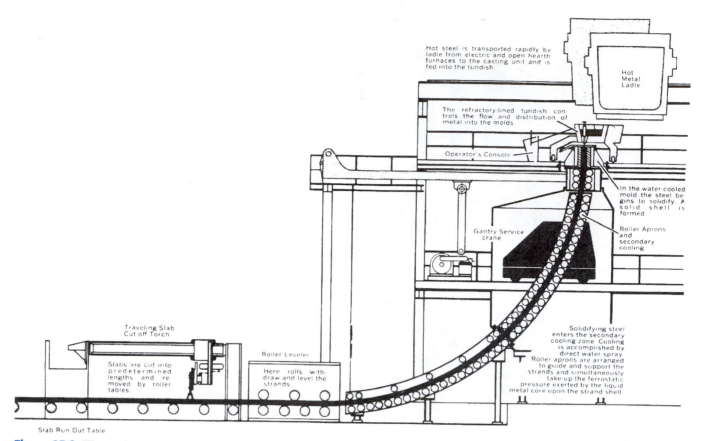

Hot steel is transported rapidly by ladle from electric and open hearth furnaces to the casting unit and is fed into the tundish.

Hot Metal Ladle

The refractory-lined tundish controls the flow and distribution of metal into the molds.

Operator's Console

In the water-cooled mold the steel begins to solidify. A solid shell is formed.

Gantry Service Crane

Roller Aprons and secondary cooling

Traveling Slab Cut-off Torch

Slabs are cut into predetermined lengths and removed by roller tables

Roller Leveler

Here rolls withdraw and level the strands.

Solidifying steel enters the secondary cooling zone. Cooling is accomplished by direct water spray. Roller aprons are arranged to guide and support the strands and simultaneously take up the ferrostatic pressure exerted by the liquid metal core upon the strand shell.

Slab Run Out Table

Figure 15.9 The molten steel can be cast into a continuous slab by a strand-casting machine. *(Courtesy American Iron and Steel Institute)*

Figure 15.10 This is the control room of a multiline continuous steel-casting machine. Notice the use of remote-controlled video cameras plus the glass windows that allow the operator to oversee the operation. *(Courtesy Bethlehem Steel Corporation)*

Figure 15.11 This is the steel strand produced by the continuous casting machine. The slab is cut into predetermined lengths by an automatic torch machine. *(Courtesy Bethlehem Steel Corporation)*

Figure 15.12 This view shows the multiline continuous caster moving the slab after it has been cut to length. *(Courtesy Bethlehem Steel Corporation)*

Figure 15.13 The slabs produced on the continuous slab casting machine are loaded on railroad cars for shipment to a slab yard for inspection and conditioning prior to delivery to the appropriate mill for rolling into finished steel products. *(Courtesy Bethlehem Steel Corporation)*

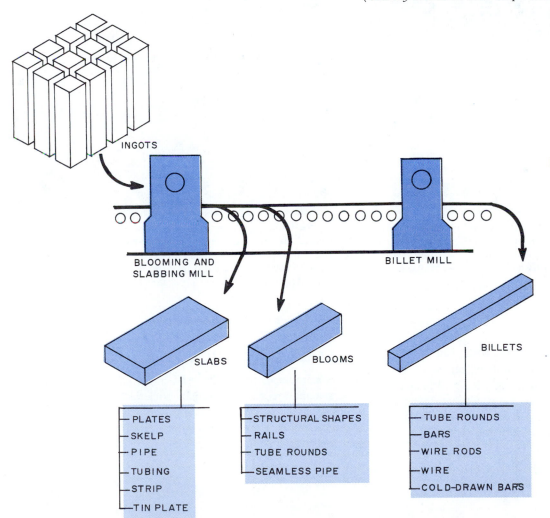

Figure 15.14 The steps in producing ingots into semifinished steel products, which are then used to produce finished steel products.

Figure 15.15 This is a highly sophisticated continuous bloom, computer-controlled casting machine that is fed by an electric furnace steelmaking operation and has a rated annual capacity of 1.3 million tons. Notice the working conditions and protective clothing worn by those directly involved with the operation. (*Courtesy Bethlehem Steel Corporation*)

Figure 15.17 The plate stock is run through a four-high finishing machine until it reaches the desired length and thickness. This takes many passes back and forth through the computer-controlled machine. (*Courtesy Bethlehem Steel Corporation*)

Figure 15.16 These perfectly formed steel plates are resting on a cooling bed. They remain here until their temperature drops below 300°F (150°C), after which they are moved to the marking area. (*Courtesy Bethlehem Steel Corporation*)

Figure 15.18 These structural steel beams have been roll-formed hot and are being transferred to a "walking" cooling bed before they are cut to the desired lengths. Notice the beam in the center is just out of the machine and is very hot. (*Courtesy Bethlehem Steel Corporation*)

All types of structural shapes, rods, rails, strip, and seamless pipe are manufactured using blooms. Finished structural beams are shown in Fig. 15.18. One beam has just left the hot-rolling process and will be moved to the cooling beds on the side. Billets are used to produce solid steel bars and rounds, wire rods and wire, seamless pipe, and cold-drawn bars.

STEEL IDENTIFICATION SYSTEMS

A number of nationally used metal and alloy numbering systems are administered by societies, trade associations, and individual users and producers of metals and alloys. Among these are numbering systems by the American Society for Testing and Materials (ASTM), SAE/Aerospace

Materials Specifications (AMS), American Welding Society (AWS), American Iron and Steel Institute (AISI), Society of Automotive Engineers (SAE), Federal Specifications, Military Specifications (MIL), and the American Society of Mechanical Engineers (ASME). Following are brief descriptions of several of these systems.

The American Iron and Steel Institute (AISI) and the Society of Automotive Engineers (SAE) series of identifying numbers for carbon and alloy steels uses four-digit numbers and in some cases five digits. The first two digits indicate the type of steel (carbon or alloy) and the last two digits indicate the carbon content. For example, in AISI 1030 the 10 indicates a carbon content of about 0.30 (actual range is 0.28 to 0.34). An AISI 4012 indicates a molybdenum steel alloy with 0.15 to 0.25 molybdenum. It also has 0.09–0.14 carbon, 0.75–1.00 manganese, 0.035 phosphorus maximum, 0.040 sulfur maximum, and 0.15–0.35 silicon.

The American Society for Testing and Materials (ASTM) sets standards for steel designations using an arbitrary number to indicate chemical compositions and to specify minimums for strength and ductility. These specifications regulate specific chemical elements that directly affect the fabrication and erection of the steel. This makes it possible for various proprietary steels of different chemical compositions to conform to ASTM performance standards. Structural steels used in construction are designated by ASTM standards (Fig. 15.19). Typical steels used for construction purposes are identified in Table 15.1. Detailed information on these and other systems is available from the sponsoring organization.

One system, the Unified Numbering System for Metals and Alloys (UNS), was developed jointly by the Society of Automotive Engineers, Inc., and the American So-

ciety for Testing and Materials. It provides a means for describing the composition of several thousand metals designations, it cross-references those of the organizations mentioned above.

The Unified Numbering System for Metals and Alloys

The Unified Numbering System (UNS) provides the uniformity required for indexing, record keeping, data storage and retrieval, and cross-referencing. The UNS is not a specification but is used as an identifier of a metal or alloy for which controlling limits have been established in specifications published elsewhere.

The UNS has eighteen series of designations for metals and alloys, seventeen of which are active. Each UNS designation has a single-letter prefix followed by five digits. The letter is used to identify the family of metals, such as S for stainless steels and A for aluminum. The prefixes are listed in Table 15.2.

The significance of the digits can vary with the UNS series. Therefore, their meaning for each series can serve a different purpose. Effort has been made to relate UNS digits to those established by related trade organizations. In the published UNS manual, *Metals and Alloys in the Unified Numbering System*, the UNS numbers are cross-indexed with the systems of other organizations whose documents describe materials that are the same as or similar to those covered by the UNS numbers. These are identified by the letters for each organization shown in Table 15.3.

The base elements in the specification of composition of the metal are identified by standard chemical symbols for the elements. Some of these are shown in Appendix E.

Figure 15.19 These general purpose structural steel members are most likely to be ASTM A36 (Unified Number K02600) carbon steel having a yield point of 36 ksi.

Table 15.1 Identification Number, Type, and Yield Point of Selected Structural Steels

Unified Number	ASTM Number	Type	Yield Point, ksi[a]
K02600	A36	Carbon steel	36
K11510	A242	High-strength low-alloy, corrosion resistance 5 to 8 times carbon steel	42–50
K11630	A514	High-yield-strength, quenched and tempered alloy steel	90–100
K02703	A529	Carbon steel	42
K02303	A572	High-strength low-alloy, niobium-vanadium, structural quality, corrosion resistance four times carbon steel	42–65
K11430	A588	High-strength low-alloy, corrosion resistance four times carbon steel	42–50
K01803	A633	Normalized high-strength low-alloy steel	42–60
K01600	A678	Quenched and tempered steel	50–75
K12043	A852	High-strength quenched and tempered alloy steel	70

[a]Kips per square inch (ksi). One ksi equals 1000 lb. per square inch.

Table 15.2 The Unified Numbering System (UNS) for Identifying Metals and Alloys

Series Designation	Family of Metals
Axxxxx	Aluminum and aluminum alloys
Cxxxxx	Copper and copper alloys
Exxxxx	Rare earth and similar metals and alloys
Fxxxxx	Cast irons
Gxxxxx	AISI and SAE carbon and alloy steels
Hxxxxx	AISI and SAE H-steels
Jxxxxx	Cast steels (except tool steels)
Kxxxxx	Miscellaneous steels and ferrous alloys
Lxxxxx	Low melting metals and alloys
Mxxxxx	Miscellaneous nonferrous metals and alloys
Nxxxxx	Nickel and nickel alloys
Pxxxxx	Precious metals and alloys
Rxxxxx	Reactive and refractory metals and alloys
Sxxxxx	Heat and corrosion-resistant steels (including stainless), valve steels, and iron-base "superalloys"
Txxxxx	Tool steels, wrought and cast
Wxxxxx	Welding filler metals
Zxxxxx	Zinc and zinc alloys

Courtesy Society of Automotive Engineers

Table 15.3 Numbering Systems Cross-Referenced to UNS

Cross-Reference Prefix	Specifying Organization
AA	(Aluminum Association) numbers
ACI	(Steel Founders Society of America) numbers
AISI	(American Iron and Steel Institute) including SAE (Society of Automotive Engineers) numbers (carbon and low-alloy steels)
AMS	(SAE/Aerospace Materials Specification) numbers
ASME	(American Society of Mechanical Engineers) numbers
ASTM	(American Society for Testing and Materials) numbers
AWS	(American Welding Society) numbers Federal Specification Numbers
MIL	(Military Specification) numbers
SAE	(Society of Automotive Engineers) "J" numbers

Courtesy Society of Automotive Engineers

STEEL AND STEEL ALLOYS

Steel is a term generally applied to plain carbon steels that are alloys of iron and carbon with a carbon content less than 2 percent. Other steel products include stainless and alloy steels.

Plain Carbon Steels

Plain carbon steels have iron as the major element (more than 95 percent), but they also contain impurities such as sulfur, nitrogen, and oxygen. Other elements may be present as residual impurities or as alloys added to change the properties of the steel. These can include phosphorus, nickel, aluminum, copper, silicon, and manganese.

The properties of carbon steel are varied not only by the elements that are added to alter their chemical composition but also by the type of mechanical and heat treatment used in producing the steel. For example, during production, carbon steel could be hot- or cold-rolled, cast, cooled slowly, or cooled rapidly. All of these influence the final properties.

Control of the *carbon content* is the major factor in establishing the properties of carbon steel. As the carbon content increases, the strength and hardness increase and ductility decreases. The differences in the amount of carbon among the types of steel is very small, usually in terms of hundredths of a percent. The percentage of carbon for several types of carbon steel are shown in Table 15.4.

Carbon steels contain other elements, such as manganese, phosphorus, and sulfur. The amount varies with the chemical design of the metal. For example AISI/SAE 1008 carbon steel contains 0.10 percent maximum carbon, 0.30 to 0.50 percent manganese, 0.040 percent maximum phosphorus, and 0.050 percent maximum sulfur.

Carbon steels are used for many products used in construction, including structural shapes, bars, sheet and strip products, plate, pipe, tubing, wire, nails, rivets, and screws. Carbon steel is also used to produce cast products, which are typically from a medium-grade carbon steel. The castings are often heat treated to relieve internal strain developed during the casting process.

Alloy Steels

An alloy steel contains one or more alloying elements other than carbon (such as chromium, nickel, molybdenum) that have been added in amounts exceeding a specified minimum to produce properties not obtained in carbon steels. These elements give particular physical, mechanical, and chemical properties to the steel. Stainless steels, specialty steels, and tool steels are not considered alloy steels although they do contain alloying elements.

Standard alloy steels generally have elements of carbon, manganese, and silicon occurring naturally. Alloy steel

that has one additional alloying element is identified as a *single alloy steel*. Adding two alloying elements produces a *double* or *binary alloy steel* and three elements a *triple* or *ternary steel*. The major alloying elements and the property changes they produce can be found in Table 15.5.

As can be seen from reviewing Table 15.5, alloying elements are used to improve properties such as hardness, performance of the material at high and low temperatures, strength, ductility, workability, wear resistance, electromagnetic properties, and electrical resistance or conductivity. The architect and structural engineer are especially involved with those alloys having increased strength, wear resistance, and resistance to **corrosion,** expansion, contraction, and ductility.

There are many alloy steels specified by the various organizations mentioned earlier. The uses range widely and

Table 15.4 Carbon Content of Carbon Steels

Type of Steel	Percent Carbon Content	Characteristics
Extra soft grade	0.05–0.15	Ductile, soft, tough Used for wire, rivets, pipe, sheets
Mild structural grade	0.15–0.25	Ductile, strong Used for boilers, bridges, buildings
Medium grade	0.25–0.35	Harder than mild structural Used for machinery and general structural purposes
Medium hard grade	0.35–0.65	Harder than medium grade Resists abrasion and wear
Spring grade	0.85–1.05	Strong and hard Used to make springs
Tool steel	1.05–1.20	Hardest and strongest of carbon steel<None>

Table 15.5 Properties Imparted to Steel Alloys by Alloying Elements

Element	Properties
Aluminum	A deoxidizer used to control the grain size within the structure of the steel, promotes surface hardening
Boron	Increases the depth of hardness
Carbon	Increases hardness, strength High percentages contribute to brittleness
Chromium	Increases hardness, corrosion, and wear resistance
Cobalt	Hardens or strengthens the ferrite, resists softening at high temperatures Used in high-speed tool steels
Copper	Increases resistance to atmospheric corrosion and increases yield strength
Manganese	Increases strength and resistance to wear and abrasion
Molybdenum	Increases corrosion resistance, raises tensile strength and elastic limit, reduces creep, improves impact resistance
Nickel	Increases elastic limit and internal strength Increases strength and toughness in heat treated steel In some steels increases hardness, fatigue, and corrosion resistance
Niobium	Retards softening during tempering operations, increases resistance to creep at high temperatures, increases ductility and impact strength.
Phosphorus	Increases corrosion resistance and strength
Silicon	Increases hardenability, strength, and magnetic permeability in low-alloy steels
Sulfur	Improves machining properties Especially useful in mild steels
Titanium	Prevents intergranular corrosion of stainless steels, is a deoxidizer, increases strength in low-carbon steels
Tungsten	Used in tool steels to promote hardness In stainless steels helps maintain strength at high temperatures
Vanadium	Improves resistance to thermal fatigue and shock
Zirconium	Inhibits grain growth and is a deoxidizer

include such types as alloy steel electrode, alloy steel welding wire, alloy steel, high-strength low-alloy steel, and others. Some of these have direct applications to products used in construction. Examples include high silicon content, which improves magnetic permeability making it useful in transformers, motors, and generators. Nickel improves toughness, so nickel alloys are used on the cutting tools used in rock-drilling machines and air hammers. Since alloy steels are more expensive than carbon steels, they typically are not used for structural members unless special requirements exist, such as functioning in areas with high temperatures or very high strengths. They are also used in the manufacture of power tools and heavy construction equipment. Some common uses for carbon and alloy steels are shown in Table 15.6.

Types of Structural Steel

Of major importance to the architect and engineer is the steel specified for structural purposes. They have low to medium carbon content. The American Institute of Steel Construction (AISC) publication "Code of Standard Practice for Steel Buildings and Bridges" includes those things that must be specified in construction documents, including columns, beams, trusses, bearing plates, and various fastening devices and connectors. The AISC publication "Specification for Structural Steel Buildings" details information on structural steels for use in building construction.

Table 15.6 Construction Uses for Selected Carbon and Alloy Steels

UNS Designation	ASTM Designation	Construction Use
K02600	A36	Structural steel members
K20504	A53	Welded and seamless pipe
K11510	A242	Structural members
K02303	A572	Structural members
K11430	A588	Structural members
K02703	A529	Structural members
K01803	A633	Structural members
K03000	A500	Cold-formed, welded, and seamless tubing
K10600	A678	Quenched and tempered steel plate
K03000	A501	Hot-formed, welded, and seamless tubing
K02706	A325	High-strength bolts
K03900	A490	High-strength alloy steel bolts
	A307	Machine bolts and nuts
	A328	Sheet piling
KJ0300	A27	Cast steel items
J31575	A148	High-strength cast steel items
	615	Billet steel bars for reinforcing concrete
	616	Rail-steel bars for reinforcing concrete
	606, 607	Sheet steel

Courtesy Society of Automotive Engineers

Structural steels fall into four major classifications:

1. Carbon steel (ASTM A36, A529, UNS K02600)
2. Heat-treated construction alloy steel (ASTM A514, UNS K11630)
3. Heat-treated high-strength carbon steel (ASTM A633, A678, A852, UNS K01803, K01600, K12043)
4. High-strength low-alloy steel (ASTM A242, A572, A588 UNS K11510, K02303, K11430)

Carbon steels must meet maximum content requirements for manganese and silicon. Copper requirements have minimum and maximum specifications. There are no other minimums specified for other alloying elements.

Heat-treated construction alloy steels have more stringent alloying element specifications than carbon steel. They produce the strongest general-use structural steel. The structural members of the bridge in Fig. 15.20 could be of this type.

Heat-treated high-strength carbon steels are brought to the desired strength and toughness by heat treating. Heat treating refers to the process of heating and cooling metals to produce changes in the physical and mechanical properties.

High-strength low-alloy steels are a group of steels to which alloying elements have been added to produce improved mechanical properties and greater resistance to atmospheric corrosion. The carbon range is typically from 0.12 to 0.22 percent. The percent of alloying agents varies with the manufacturer of the steel. These are available from manufacturers and listed under specific trade names. ASTM has specifications to cover all the trade name steels. One type that is of special interest is known by the general name **weather-**

Figure 15.20 These bridge structural members may have been made from heat-treated construction alloy steel.

ing steels. These are designed for architectural exterior applications where the exposure to the atmosphere causes the steel to form a natural, rust-colored, self-healing oxide coating. It is never painted but left with the oxide coating protecting and coloring the exterior material.

High-strength low-alloy steels can be worked by many metal processing operations, including hot- and cold-forming, punching, shearing, gas cutting, and welding. In construction, alloy steels find use in high-strength bolts, cables used in prestressed concrete, sheet and strip material, plates, various bars and structural shapes, wire, tubing and pipe.

Stainless and Heat-Resisting Steels

Stainless steels have outstanding corrosion and oxidation resistance at a wide range of temperatures. *Heat-resisting steels* maintain their basic mechanical and physical properties when subjected to high temperatures. Typically, standard stainless steels have 10 to 25 percent chromium.

The chromium alloying element gives the stainless steel its corrosion resistance qualities and produces a thin, hard, invisible film over the surface, which inhibits corrosion. Nickel and manganese increase strength and toughness and make the material easier to fabricate.

Other alloying elements that may be used to produce desired characteristics include zirconium, titanium, sulfur, silicon, selenium, phosphorus, molybdenum, and niobium. Refer to Table 15.5 for additional information on how each of these alters the characteristics of the product.

The chemical compositions of several types of stainless steel frequently used in construction are shown in Table 15.7. Notice the small ranges allowed within each alloying element in each type of stainless steel.

Classifying Stainless and Heat-Resisting Steels

Stainless and heat-resisting steels are classified into three major groups based on their chemical composition and their reaction when heat treated: ferritic steels, martensitic steels, and austenitic steels (Table 15.8).

Table 15.7 Chemical Compositions of Stainless Steels Used in Construction

| UNS Designation | AISI/SAE Designation | Chemical Composition (percent) | | | | | | |
		Fe	Cr	Ni	Mn	C	Mo	Si
S20100	201[a]	67.51–73.51	16.0–18.0	3.5–5.5	5.5–7.5	0.15 max.		1.00 max.
S20200	202[a]	63.51–70.01	17.0–19.0	4.0–6.0	7.5–10.0	0.15 max.		1.00 max.
S30100	301[b]	70.84–74.84	16.0–18.0	6.0–8.0	2.0 max.	0.15 max.		1.00 max.
S30200	302[b]	67.84–70.84	17.0–19.0	8.0–10.0	2.0 max.	0.15 max.		1.00 max.
S30400	304[b]	66.28–70.77	18.0–20.0	8.0–10.5	2.0 max.	0.15 max.		1.00 max.
S31600	316[b]	61.83–68.83	16.0–18.0	10.0–14.0	2.0 max.	0.10 max.	2.0–3.0	1.00 max.
S43000	430[b]	79.81–81.05	14.0–18.0		1.0 max.	0.12 max.	0.75–1.25	1.00 max.

[a]Contains maximum 0.06% phosphorus, 0.03% sulfur, 0.25% nitrogen
[b]Contains maximum 0.04% phosphorus, 0.03% sulfur.
Courtesy American Iron and Steel Institute

Table 15.8 AISI/SAE and UNS Designations and Characteristics of Heat-Resisting and Stainless Steels

Grain Structure	Chief Alloying Elements	Series	AISI/SAE Designation	UNS Designation	Characteristics	Steel Type
Martensitic (ferromagnetic and hardenable by heat treatment)	Chromium 4%–6%	500	502	S50200	Retain their mechanical properties at high temperatures	Heat-Resisting steels
	Chromium 11.50–13.50%	400	410	S41000	Moderate corrosion resistance, high strength and hardness	
Ferritic (ferromagnetic and nonhardenable	Chromium 16%–18%	400	430	S43000	Very good corrosion resistance, particularly at high temperatures	
Austenitic (nonmagnetic and hardenable by cold working)	Chromium 17%–19% Nickel 8%–10% Manganese 2.0% Max.	300	302	S30200	Excellent corrosion resistance, high strength and ductility Suitable for many fabrication techniques	Stainless steels
	Chromium 17%–19% Nickel 4%–6% Manganese 7.5%–10%	200	202	S20200	Excellent corrosion resistance at high temperatures, high strength and toughness	

Courtesy American Iron and Steel Institute

Ferritic stainless steels have a chromium content of about 16 to 18 percent and a low carbon content of 0.12 maximum. These are nonhardenable steels, which means they cannot be hardened by heat treating. Ferritic steels have excellent corrosion resistance. Although they have especially good corrosion resistance at high temperatures, they do lose strength under these conditions. They are used when high temperatures exist and corrosion resistance is important and when a low coefficient of thermal expansion is helpful.

Martensitic stainless steels have a chromium content of 11.50 to 13.50 percent and a carbon content of 0.15 percent maximum. They can be hardened by heat treating and can be worked hot, forged, or formed cold. They are typically used when hardness, strength, and abrasion resistance are critical, such as in a steam turbine.

Austenitic stainless steels have a chromium content of about 17 to 19 percent and a nickel alloying element of 8 to 10 percent. They have a maximum carbon content of 0.15 percent. They are nonmagnetic and harden when worked cold. At high temperatures they have high strength and are very tough and corrosion resistant. They have a high coefficient of thermal expansion, which must be taken into consideration by the engineer specifying the use of this material. Of all of the stainless steels they have the least resistance to corrosion by attack of sulfur gases. Typical uses include curtain wall panels, railings, and door and window finished surfaces. Some types are used in the manufacture of food preparation equipment and various appliances in hospitals.

Designations for Stainless and Heat-Resisting Steels
The American Iron and Steel Institute (AISI) designations for heat-resisting and stainless steels are based on a three-digit numbering system ranging from 200 to 500 as shown in Tables 15.7 and 15.8. The first digit indicates the group, and the last two digits indicate the type within the group. Some types have the prefix TP, which refers to tubular grades.

Stainless Steel Uses in Construction
Typical applications for stainless steels frequently specified by architects and engineers are shown in Table 15.9. Of the ferritic stainless steels, Type 430 is most commonly used in construction for such applications as column covers, trim, grills, and gutters. Of the austenitic stainless steels, Types 301, 302, and 304 are used for applications such as storefronts, doors, windows, and railings. Types 201 and 202 are used for the same applications as 301 and 302. Other uses include drinking fountains, kitchen sinks, flatware, and cooking utensils (Fig. 15.21).

STEEL PRODUCTS
Steel of various types is used extensively in construction, ranging from something as small as a nail to huge beams and columns. Following are a few of the products in common use.

Structural Steel Products
Rolled structural steel shapes are used extensively in construction. They carry the live and dead loads imposed on the structure. The commonly used shapes are in Fig. 15.22. These include S (standard beam) shapes, W (wide flange) shapes, WWF (welded wide flange) shapes, M (miscellaneous) shapes, H-sections, HP (bearing pile) shapes, C (channel) shapes, ST (structural T) shapes cut from M, W, and WWF shapes, L (angle) shapes with equal and unequal legs, HSS (hollow structural section), square and rectangular shapes, pipe, bar (flat, round, and square), and plate (Fig. 15.23). These are specified on architectural drawings as shown in Table 15.10. The S beam has narrow flanges that taper at a 17 percent slope. The W beam has wide flanges that have very little slope. The WWF shapes are made by welding the flanges to the web, but the S and W beams are rolled to shape from a single piece of steel.

Table 15.9 Stainless Steels Used in Construction

UNS Designation	AISI Designation	Typical Applications
S30100	301	Gutters, trim, flashing, household and industrial appliances
S30200, S30400	302, 304	Storefronts, curtain walls, doors, windows, railings, household and industrial appliances
S31600	316	All exterior marine uses, on seacoast building exteriors, in chemical, petroleum, and paper manufacturing facilities
S43000	430	Exterior applications in areas exposed to salt water atmosphere, column covers, trim, grills, gutters
S20100	201	Gutters, trim, flashing, household and industrial appliances
S20200	202	Storefronts, curtain walls, doors, windows, railings, household and industrial appliances

Figure 15.21 Drinking fountains use stainless steel for all parts exposed to moisture. *(Courtesy EBCO Manufacturing Co.)*

Open-web steel joists are a widely used steel product. They are lightweight and produced by welding structural steel shapes, such as angles and bars, into a Warren truss unit. They are manufactured in short-span, long-span, and deep long-span series. The short-span joists are manufactured to span clear openings from 8 ft. (2.4 m) to 60 ft. (18.3 m) and in depths from 8 in. (203 mm) to 30 in. (762 mm). The long-span joist series is a heavier joist and is made to span clear openings from 25 ft. (7.6 m) to 96 ft. (29.2 m) and depths from 18 in. (457 mm) to 48 in. (1219 mm). The deep long-span series spans a clear opening from 89 ft. (27 m) to 144 ft. (43.9 m) and in depths from 52 in. (13 mm) to 72 in. (1829 mm). They are used to support floor and roof loads. The openings in the web permits the passage of plumbing, heating ducts, and electrical runs (Fig. 15.24).

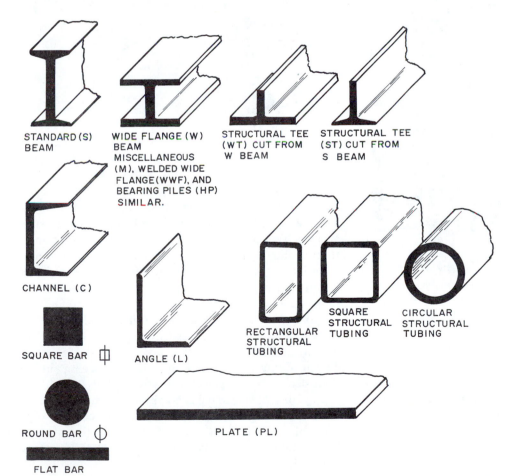

STANDARD (S) BEAM

WIDE FLANGE (W) BEAM MISCELLANEOUS (M), WELDED WIDE FLANGE (WWF), AND BEARING PILES (HP) SIMILAR.

STRUCTURAL TEE (WT) CUT FROM W BEAM

STRUCTURAL TEE (ST) CUT FROM S BEAM

CHANNEL (C)

SQUARE BAR

ANGLE (L)

RECTANGULAR STRUCTURAL TUBING

SQUARE STRUCTURAL TUBING

CIRCULAR STRUCTURAL TUBING

ROUND BAR

PLATE (PL)

FLAT BAR

Figure 15.22 Commonly available structural steel shapes.

Table 15.10 Symbols and Abbreviations for Structural Steel Members

Structural Shape	Symbol	Order of Presenting Data	Sample of Abbreviated Note
Square bar	▢	Bar, size, symbol, length	Bar 1 ▢ 5′–9″
Round bar	φ	Bar, size, symbol, length	Bar ⅝ φ 7′–8″
Plate	PL	Symbol, thickness, weight	PL ½ × 24
Angle, equal legs	L	Symbol, leg 1, leg 2, thickness, length	L 2 × 2 × ¼ × 9′ – 6″
Angle, unequal legs	L	Symbol, long leg, short leg, thickness, length	L 4 × 3 × ¼ × 6′ – 2″
Channel	C	Symbol, depth, weight, length	C 6 × 10.5 × 12′ – 7″
American Standard beam	S	Symbol, depth, weight, length	S 10 × 35.0 × 13′ – 6″
Wide flange beam	W	Symbol, depth, weight, length	W 16 × 64 × 20′ – 2″
Structural tee	ST, WT	Symbol, depth, weight	ST 10 × 50
Pipe	Name of pipe	Name, diameter, strength	Pipe 3 X-strong
Miscellaneous	M	Symbol, depth, weight, length	M 14 × 17.2 × 19′–0″
Bearing pile	HP	Symbol, depth, weight, length	HP 14 × 117 × 36′–0″
Welded wide flange	WWF	Symbol, depth, weight, length	WWF 14 × 84 × 15′–6″

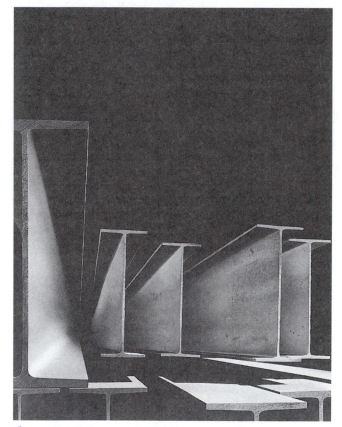

Figure 15.23 Structural steel H beams are available in a wide range of sizes. *(Courtesy Bethlehem Steel Corporation)*

Figure 15.24 These open-web joists have corrugated steel decking on top. *(Courtesy Vulcraft Division of Nucor Corporation)*

Sheet Steel Products

Many products are made by rolling them to shape from flat steel sheets. The thickness of steel sheets is given by gauge numbers as shown in Table 15.11. Common among these are roofing, siding, decking, and light gauge steel framing systems. Steel decking can be seen in Fig. 15.24.

Many styles of steel roof and siding systems are available. A variety of coatings are used to protect the steel surface. Typical coatings include galvanizing, zinc-aluminum alloy with a paint over it, siliconized polyester in a variety of colors, fluoropolymer paint finish, and weathering copper coating. A few of the available wall panel configurations are shown in Fig. 15.25.

Corrugated *steel floor decking* is used with structural steel framing. This decking usually spans 6 to 15 ft. (1.8 to 4.6 m) and has a site-cast concrete topping (See Fig. 15.10). Cellular floor decking provides openings in the floor slab for running electrical systems (Fig. 15.26).

Steel roof decking may have a site-cast lightweight concrete or gypsum topping, or it may be covered with some form of insulation board and finished roofing system, such as tar and gravel. Examples of decking products are in Fig. 15.27.

Table 15.11 Standard Thicknesses for Some Basic Steel Sheets

Minimum Thickness	
Equivalent Inches	Millimeters
0.3937	10.0
0.3543	9.0
0.3150	8.0
0.2756	7.0
0.2362	6.0
0.2165	5.5
0.1969	5.0
0.1890	4.8
0.1772	4.5
0.1654	4.2
0.1575	4.0
0.1496	3.8
0.1378	3.5
0.1260	3.2
0.1181	3.0
0.1102	2.8
0.0984	2.5
0.0866	2.2
0.0787	2.0
0.0709	1.8
0.0630	1.6
0.0551	1.4
0.0472	1.2
0.0433	1.1
0.0394	1.0
0.0354	0.90
0.0315	0.80
0.0276	0.70
0.0256	0.65
0.0236	0.60
0.0217	0.55
0.0197	0.50
0.0177	0.45
0.0157	0.40
0.0138	0.35

Light gauge steel framing systems use studs, joists, channels, and runners to frame wall and floor systems. Metal studs are used in the same manner as wood studs. They have metal runners for top and bottom plates. Metal joists are welded to the top runner and have a perimeter channel for a header. The floor and roof deck can be metal or plywood (Fig. 15.28). Additional information can be found in Chapter 20.

Expanded steel mesh is made by slitting metal sheets and stretching them to form diamond-shaped openings (Fig. 15.29). It is used for many purposes, including gratings, decks, partitions, and as a base for the application of trawled stucco and plaster. Expanded metal mesh is usually designated by the size of the width of the mesh opening and the gauge of the steel sheet. For example a ¾-16 expanded metal mesh has diamond openings ¾ in. (18 mm) wide on 16-gauge material. The length of the diamond is approximately twice the width, or 1½ in. (38 mm) in this example.

Metal lath is another sheet steel product (Fig. 15.30). It is a form of expanded steel that has a flat or ribbed mesh ⅜ in. (9 mm) in height. It is used as a base for plaster. Expanded metal corner beads are available.

Other Products

Welded wire fabric is used to reinforce concrete slabs. Several commonly used sizes are in Table 15.12. More information is given in Chapter 8. Inch sizes of welded wire fabric are designated by two numbers and two letter-number combinations. The first two numbers, such as 6 × 6, gives the spacing of the wires in inches. The first number is the spacing of the longitudinal wires. The second number gives the spacing of the transverse wires in inches. The first letter-number combination gives the type and size of the longitudinal wire and the second for the transverse wire. The W indicates a smooth wire and the number following it gives the cross-sectional area of

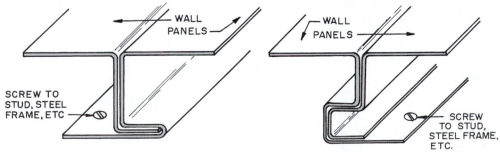

TYPICAL STEEL EXTERIOR WALL PANELS

SCREW TO STUD, STEEL FRAME, ETC

WALL PANELS

WALL PANELS

SCREW TO STUD, STEEL FRAME, ETC.

Figure 15.25 Metal siding is available in a variety of configurations and colors.

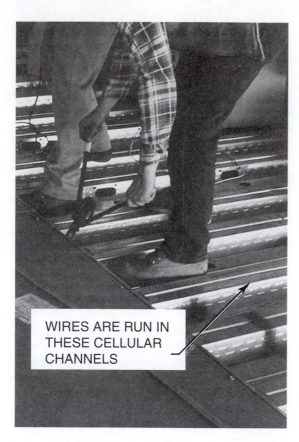

WIRES ARE RUN IN THESE CELLULAR CHANNELS

Figure 15.26 Cellular steel floor decking is used in multistory construction. *(Courtesy H.H. Robertson, A United Dominion Company)*

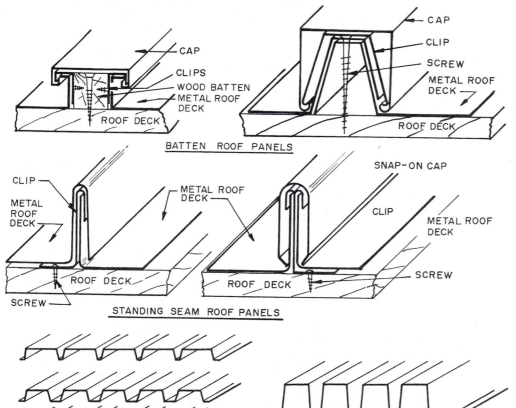

CAP
CLIPS
WOOD BATTEN
METAL ROOF DECK
ROOF DECK

CAP
CLIP
SCREW
METAL ROOF DECK
ROOF DECK

BATTEN ROOF PANELS

CLIP
METAL ROOF DECK
METAL ROOF DECK
ROOF DECK
SCREW

SNAP-ON CAP
CLIP
METAL ROOF DECK
ROOF DECK
SCREW

STANDING SEAM ROOF PANELS

TYPICAL METAL DECK PROFILES

Figure 15.27 Several of the many types of steel roof decking available.

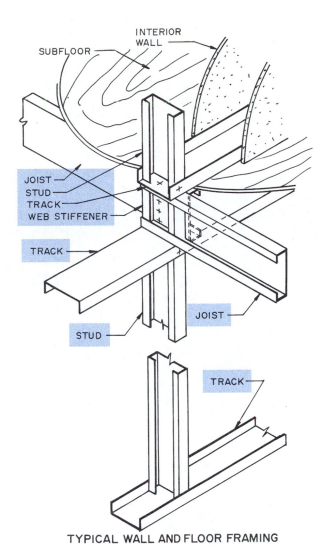

TYPICAL WALL AND FLOOR FRAMING

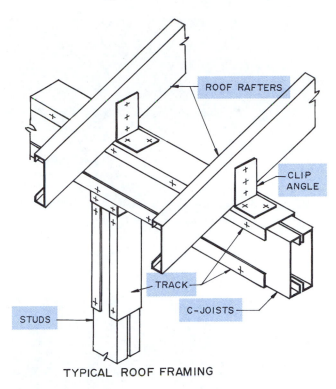

TYPICAL ROOF FRAMING

Figure 15.28 Construction details of lightweight steel framing. *(Courtesy Marino Industries Corporation)*

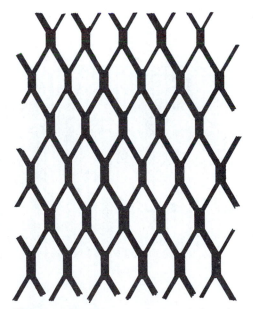

Figure 15.29 One type of expanded steel mesh.

the wire (Fig. 15.31). When the W is replaced by a D, it indicates a deformed wire was used. The area is given in hundredths of an inch per foot. For example, a W8.0 wire is a smooth wire with a cross-sectional area of 0.08 in.^{2}.

Longitudinal wires are spaced 2, 3, 4, 6, 8, and 12 in. Transverse wires are spaced 4, 6, 8, and 12 in.

Metric welded wire fabric is specified in the same manner as inch sizes but the sizes are in millimeters. Typical sizes are also shown in Table 15.12. The first two numbers are the wire spacing in millimeters. The last two numbers indicate the type of wire and give the cross-sectional area in square millimeters (mm^2).

Reinforcing bars are placed in concrete members to improve the tensile strength of concrete (Fig. 15.32). They are made to ASTM requirements for yield and tensile strength, cold bend, and elongation. Three grades are available: structural, intermediate, and hard. The structural has the lowest yield point and tensile strength, and the hard has the highest. Round bars are most commonly used, but it is available in squares. The

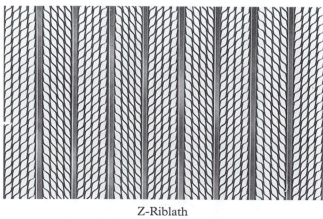

Junior diamond mesh lath

Paper backed metal lath

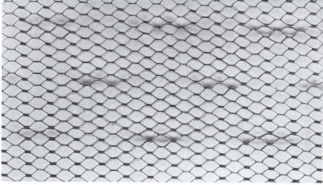

3/8 in. Riblath

Self-furring diamond mesh lath

Z-Riblath

Figure 15.30 Examples of metal lath and related products used for plaster walls. *(Courtesy USG Corporation)*

bar may be smooth or deformed (Fig. 15.33). The deformed bar has surface projections that provide a better bond to the concrete. They are available in sizes 3 through 18. The number indicates the diameter in eighths of an inch. For example a No. 5 bar has a diameter of ⅝ in. Metric sizes are given by numbers as shown in Table 15.13. A variety of accessories are used to hold reinforcing bars in place until the concrete is poured. These are made from steel wire (Fig. 15.34). Detailed information is available in Chapter 8.

Many fasteners used in construction are made from steel. Commonly used fasteners include bolts, nails, rivets, and screws. *Standard steel bolts* are available in a wide range of sizes and head types (Fig. 15.35). *High-strength steel bolts* are used where tensile strength is important, such as in structural steel framing. Additional details are in Chapter 17. Commonly used *nails* are shown in Fig. 15.36. Some nails are specified in length by inches and others by the term penny (d). The penny lengths are shown in Table 15.14.

Wood screws most commonly used have flat, round, or oval heads. Their lengths are specified in inches and their diameters by wire gauge numbers. A variety of slots are used in the heads (Fig. 15.37).

Table 15.12 Selected Sizes of Welded Wire Fabric (Inch and Metric)

Inch Sizes		
Wire Style Designation (W or D)	**Steel Area (in.²/ft.)**	
	Longitudinal Wire	**Transverse Wire**
4 × 4–W1.4 × W1.4	0.028	0.028
4 × 4–W2.0 × W2.0	0.060	0.060
4 × 4–W2.9 × W2.9	0.087	0.087
6 × 6–W1.4 × W1.4	0.028	0.028
6 × 6–W2.0 × W2.0	0.040	0.040
6 × 6–W2.9 × W2.9	0.058	0.058

Metric Sizes		
Wire Style Designation (MW or MD)	**Steel Area (mm²/m)**	
	Longitudinal Wire	**Transverse Wire**
102 × 102–MW9 × MW9	88.9	88.9
102 × 102–MW14 × MW14	127.0	127.0
102 × 102–MW19 × MW19	184.2	184.2
152 × 152–MW9 × MW9	59.3	59.3
152 × 152–MW14 × MW14	84.7	84.7
152 × 152–MW19 × MW19	122.8	122.8

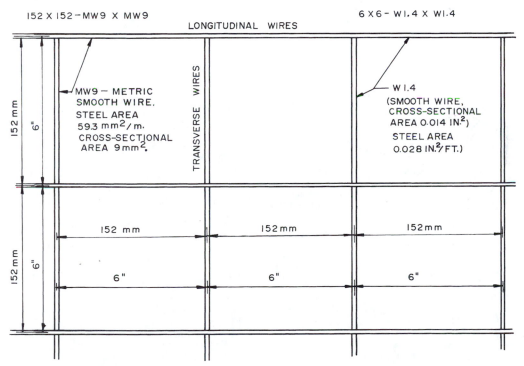

Figure 15.31 Welded wire fabric is specified by giving the wire spacing, the style of wire, and the cross-sectional area.

Figure 15.32 This tunnel is to be cast-in-place concrete and is heavily reinforced with steel reinforcing bars. *(Courtesy U.S. Army Corp of Engineers)*

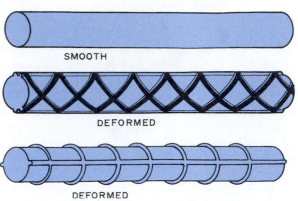

Figure 15.33 Typical types of reinforcing bar.

Table 15.13 Metric Reinforcing Bar Sizes and Diameters

Metric Bar Number	Diameter (mm)
10	11.3
15	16.0
20	19.5
25	25.2
30	29.9
35	35.7
45	43.7
55	56.4

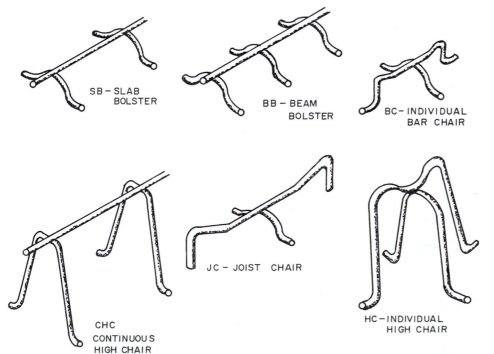

SB – SLAB BOLSTER

BB – BEAM BOLSTER

BC – INDIVIDUAL BAR CHAIR

JC – JOIST CHAIR

CHC CONTINUOUS HIGH CHAIR

HC – INDIVIDUAL HIGH CHAIR

Figure 15.34 Bolsters and chairs used to support reinforcing bars in beams and slabs. Individual chairs support one or possibly two bars. Bolsters are continuous and support many bars.

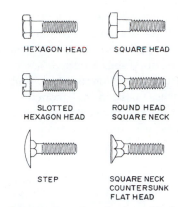

Figure 15.35 Standard steel bolts.

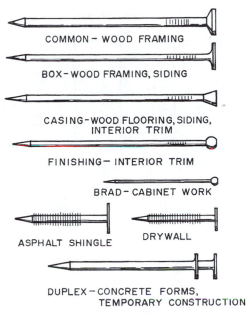

Figure 15.36 Commonly used nails.

Table 15.14 Lengths and Diameters of Common, Box, Casing, and Finishing Nails

Size	Length (in.)	American Steel Wire Gauge Number		
		Common	Box and Casing	Finishing
2d	1	15	15½	16½
3d	1¼	14	14½	15½
4d	1½	12½	14	15
5d	1¾	12½	14	15
6d	2	11½	12½	13
7d	2¼	11½	12½	12½
8d	2½	10¼	11½	12½
9d	2¾	10¼	11½	12½
10d	3	9	10½	11½
12d	3¼	9	10½	11½
16d	3½	8	10	11
20d	4	6	9	10
30d	4½	5	9	
40d	5	4	8	

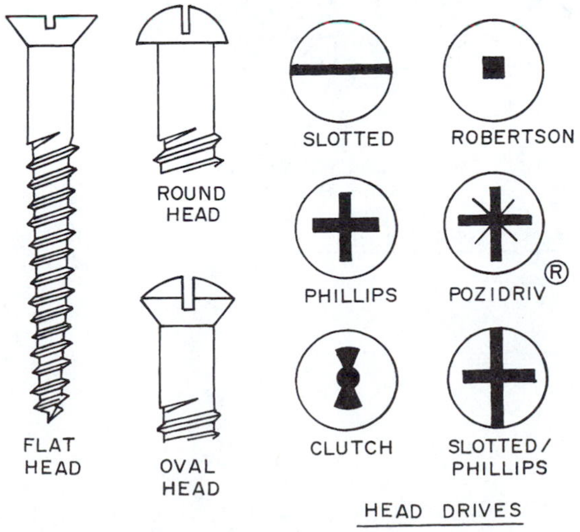

FLAT
HEAD

ROUND
HEAD

OVAL
HEAD

SLOTTED

PHILLIPS

CLUTCH

ROBERTSON

POZIDRIV ®

SLOTTED/
PHILLIPS

HEAD DRIVES

WOOD SCREWS		
DIA.	DECI. EQUIV.	LENGTH (INCHES)
0	.060	¼–⅜
1	.073	¼–½
2	.086	¼–¾
3	.099	¼–1
4	.112	¼–1½
5	.125	⅜–1½
6	.138	⅜–2½
7	.151	⅜–2½
8	.164	⅜–3
9	.177	½–3
10	.190	½–3½
11	.203	⅝–3½
12	.216	⅝–4
14	.242	¾–5
16	.268	1–5
18	.294	1¼–5
20	.320	1½–5
24	.372	3–5

Figure 15.37 Common head types used on wood screws.

Rivets are permanent fasteners used to join two things, such as two or more sheets of steel or a beam to a column. Standard rivets are available in a variety of diameters and head shapes (Fig. 15.38). These are used when the back side of the material through which the rivet protrudes is accessible so a head can be formed on it. Some rivets have heads formed while the steel is cold. Others require the rivet to be heated before forming the back head.

Blind rivets are used when the joint is accessible from only one side. After the rivet is inserted, a head is formed by a drive pin, a pull stem, or a chemical expansion in the protruding end (Fig. 15.39).

The constructor will encounter hundreds of other fastening devices. These include concrete anchors, self-drilling fasteners, set screws, washers, eye bolts, U-bolts, and many types of hooks. Numerous other steel prod-

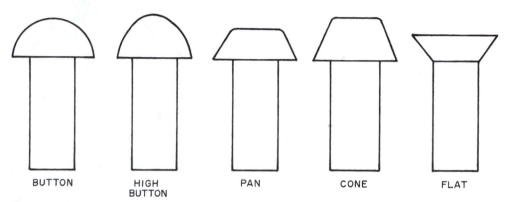

BUTTON   HIGH   PAN   CONE   FLAT
              BUTTON

Figure 15.38 Common types of rivets.

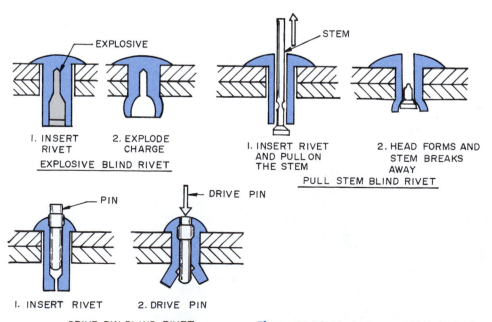

EXPLOSIVE

1. INSERT   2. EXPLODE
   RIVET   CHARGE

EXPLOSIVE BLIND RIVET

STEM

1. INSERT RIVET   2. HEAD FORMS AND
   AND PULL ON   STEM BREAKS
   THE STEM   AWAY

PULL STEM BLIND RIVET

PIN   DRIVE PIN

1. INSERT RIVET   2. DRIVE PIN

DRIVE PIN BLIND RIVET

Figure 15.39 Typical types of blind rivets.

ucts are used in building construction, such as locks, hinges, gutters, and flashing.

TESTING METALS

Engineers and construction managers must know the capabilities of metals to be used. These are determined by making standardized tests following the procedures of the American Society for Testing and Materials (ASTM). Some of the more frequently used tests include tests of hardness, tensile strength, and fatigue.

Hardness testing determines the resistance of a metal to penetration. A Tinius Olsen Air-O-Brinell metal hardness tester is shown in Fig. 15.40. To make a test the operator adjusts the air regulator until the desired Brinell load in kilograms is indicated. The piece to be tested is placed on the anvil and the operator pulls the plunger valve. The plunger, which has a hardened ball on the bottom, impacts the metal and forms an impression in the surface. The plunger valve is depressed and the plunger and ball retract from the metal. The impression diameter is measured using optical instruments to provide an automatic and precise reading. The

computer makes it possible for the operator to customize test parameters and test report formats. Test data can be recalled by individual test number, lot identifier, heat number, date, time, or other specific data.

Tensile testing determines the mechanical properties of a metal when it is subjected to a force that tends to pull it apart. The tensile test determines the ultimate strength, yield point, yield strength, and factors related to ductility. The Tinius Olsen Universal Testing Machine is shown in Fig. 15.41. It is available in capacities up to 400,000 lb. (2000 kN). The test data is recorded by a four-channel digital indicating system. The console contains the servo controls and the digital indicating system. The sample to be tested is gripped with pinion-type wedge grips in the crossheads. Once the sample is secure in the grips, the lower crosshead moves down, placing the material being tested under tension. Data are recorded and the stress-strain curve values are displayed. At the end of the test the report is printed by the computer, stored, and transmitted to a host computer. The DS-50/2 data system includes a computer, color monitor, high-speed graphics printer, and required testing software.

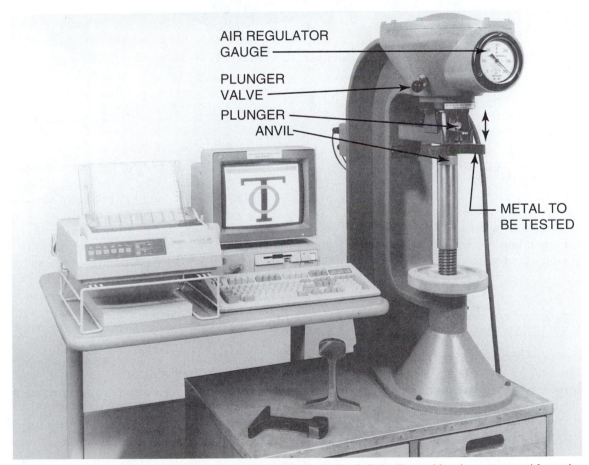

Figure 15.40 This is the Tinius Olsen Model DS/AOB No. 2 Air-O-Brinell metal hardness tester with semi-automatic and automatic optical diameter measurement system and a computer system providing various printouts. *(Photo Courtesy Tinius Olsen Testing Machine Company, Willow Grove, Pa.)*

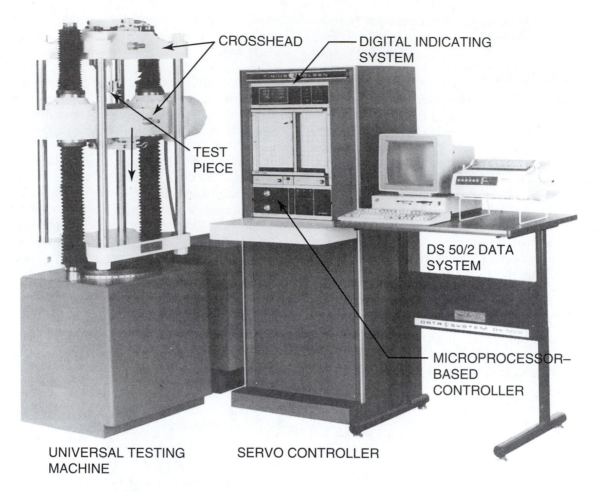

CROSSHEAD

DIGITAL INDICATING SYSTEM

TEST PIECE

DS 50/2 DATA SYSTEM

MICROPROCESSOR–BASED CONTROLLER

UNIVERSAL TESTING MACHINE

SERVO CONTROLLER

Figure 15.41 This is a Tinius Olsen 120,000 lb. (600 kN) capacity servo-controlled Super "L" Universal Testing Machine with DS-50/2 data acquisition and control system. *(Photo Courtesy Tinius Olsen Testing Machine Company, Willow Grove, Pa.)*

Fatigue testing determines the stress level a metal can withstand without failure when subjected to an infinitely large number of repeated alternating stresses. Typical tests include subjecting a specimen, such as a beam, to rotating motion while under a bending movement. Another test is a bending test in which the metal is bent back and forth but not rotated.

REVIEW QUESTIONS

1. What is the key element that influences the properties of iron?
2. What process is used to remove unwanted elements from the freshly mined iron ore?
3. What fuels and fluxes are commonly used in a blast furnace?
4. What uses are made of the slag removed from the blast furnace?
5. What are the commonly produced cast irons?
6. What are the main properties of white cast iron?
7. Why is malleable cast iron less brittle than white cast iron?
8. How does the alloying element nickel affect the properties of steel?
9. If carbon in steel is below 0.90 percent, how does this affect the properties of the steel?
10. How do the alloying elements copper and cobalt influence the properties of steel?
11. What are the three basic steelmaking processes?
12. How does carbon steel differ from alloy steel?
13. What are the two outstanding properties of stainless steel?
14. Explain the purpose of each part of the AISI steel designation code.

15. Why does stainless steel resist corrosion?
16. What are the three groups of stainless and heat-resisting steels?
17. Identify the following structural steel shapes: W, S, HP, ST, HSS, C, M.
18. How is welded wire fabric used in building construction?
19. What tests are used on metal to determine its hardness?
20. What does a tensile test show?
21. What does a fatigue test show?

KEY TERMS

agglomeration A process that bonds ground iron ore particles into pellets to facilitate handling.

alloying elements Elements added to change the properties of a material, such as chromium, nickel, or molybdenum added to steel.

annealing Heating a metal to a high temperature, followed by controlled cooling to relieve internal stresses.

beneficiation A process of grinding and concentration that removes unwanted elements from iron ore before the ore is used to produce steel.

cast iron A hard, brittle metal made of iron that contains a high percentage of carbon.

corrosion The deterioration of metal by chemical or electrochemical reaction due to exposure to the atmosphere, chemicals, or other agents.

ferrous metals Metals with iron as the chief ingredient.

flux A mineral added to molten iron that causes impurities to separate into a layer of molten slag on top of the iron.

malleability The property of a metal that permits it to be formed mechanically, as by rolling or forging, without fracturing.

oxidation The chemical reaction of a substance, such as metal, with oxygen.

pig An ingot of cast iron.

pig iron A high carbon content iron produced by the blast furnace and used to produce cast iron and steel.

reduction A process in which iron is separated from the oxygen with which it is chemically mixed by smelting the ore in a blast furnace.

sintering A process that fuses iron ore dust with coke and fluxes into a clinker.

slag A molten mass composed of fluxes and impurities removed from iron ore in the furnace.

smelting A process in which iron ore is heated, separating the iron from the impurities.

strand-casting machine A machine that casts molten steel into a continuous strand of metal that hardens and is cut into required lengths.

toughness A measure of the ability of a material to absorb energy from a blow or shock without breaking.

weathering steel A form of high-strength low-alloy steel that resists corrosion by naturally forming an oxide layer that protects it from additional corrosion.

SUGGESTED ACTIVITIES

1. Collect representative samples of ferrous metals and prepare a sheet for each one, identifying the type and its properties.

2. Make a tour of your campus or some other area and make a list of all the applications of ferrous metals you can find. Take photos if possible.

3. Test samples of ferrous metals for tensile strength and hardness and keep a record of your findings for each sample. Compare your findings with those of others who tested the same materials.

4. Find products made from ferrous metals produced by various manufacturing processes, such as casting and extruding.

ADDITIONAL INFORMATION

Merritt, F.S., and Rickett, J.T., *Building Design and Construction Handbook*, McGraw-Hill, New York, 1994.
Zahner, L.W., *Architectural Metals: A Guide to Selection, Specifications, and Performance*, John Wiley and Sons, New York, 1994.
Other resources are
Publications on wire reinforcing for concrete from the Wire Reinforcement Institute, 2911 Gleneagle Drive, Findlay, OH 45840-2908.

16

Nonferrous Metals

This chapter will help you to:

1. Understand the effects of galvanic corrosion and how to design to eliminate it.

2. Select nonferrous metals for a wide range of applications.

3. Know the properties of nonferrous metals and how these will influence the performance of the material.

Nonferrous metals are those containing no or very little iron. In other words, all metals other than iron and steel are nonferrous. Some of those commonly found in construction projects include aluminum, copper, lead, tin, and zinc. Nickel is used mainly as an alloying element. Titanium, widely used in aircraft, aerospace, and military applications, may be used as a construction material in the future.

GALVANIC CORROSION

Both ferrous and nonferrous metals are subject to **galvanic corrosion.** In the presence of an electrolyte (moisture in the atmosphere), dissimilar metals in contact will corrode more rapidly than similar metals in contact. For example, an aluminum gutter secured with copper nails produces galvanic action at the point of contact. The presence of moisture in the atmosphere sets up an electrolytic action that causes the aluminum to corrode. The metal that is higher on the table of electrolytic corrosion potential will be sacrificed (Table 16.1). For example, when aluminum is plated with zinc to protect it from possible corrosion due to electrolytic action, the zinc corrodes, protecting the aluminum. Steel is also zinc coated and is called **galvanized.** When it is not possible to avoid contact between dissimilar metals, they can be given a coat of unleaded paint or separated with a plastic or other nonconducting material. A joint could be caulked or otherwise sealed to keep out moisture, thus eliminating the electrolyte.

ALUMINUM

Aluminum (chemical symbol Al) is a versatile material that is used widely in the construction industry. The construction industry is one of the largest consumers of aluminum. Aluminum is light in weight, having a specific gravity of only 2.7 times that of water and approximately one-third that of steel. Pure aluminum melts at 1220°F (665°C), which is considerably lower than that of other structural metals. It also is relatively weak as far as mechanical properties are concerned. Aluminum elastically deforms about three times more than steel under comparable loading. It can be strengthened by alloying, cold-working, or **strain hardening.** Aluminum

Table 16.1 The Galvanic Series

Electrolytic Potential	Metals and Alloys
High potential (anode +)	Magnesium
	Magnesium alloys
	Aluminum (pure and several cast and wrought alloys)
	Zinc
	Cadmium
	Aluminum (wrought 2024 and 356.0 cast)
	Iron or steel
Electric current flows from positive (+) to negative (−)	Cast iron
	Stainless steel
	Lead
	Tin
	Nickel (active)
	Brass
	Copper
	Bronze
Low potential (cathode −)	Chromium stainless steel
	Silver
	Titanium
	Platinum
	Gold

alloys do not lose ductility or become brittle at cryogenic (low) temperatures.

Aluminum is a good conductor of electricity. Compared to copper wire of the same diameter, aluminum's conductivity is roughly 65 percent of that of copper.

HISTORY

People probably were using alum, one of the aluminum compounds found in nature, as early as 500 B.C. These naturally occurring compounds were used as astringents (a substance that contracts the tissues of the body), to fix dyes in cloth, and in the tanning of animal skins.

The Romans used a natural potassium sulfate they called alumen. This was purified into crystalline alum around 1200 A.D., and by the 1500s alum was produced from the clay in which it occurred.

The actual production of aluminum as we know it today occurred much more recently. Aluminum is the most recently developed and used metal in construction. Various attempts were made by German, French, and Danish chemists and scientists to isolate aluminum. About 1850 one was successful in isolating a small quantity of the metal by decomposing anhydrous aluminum chloride with potassium.

In 1855 an aluminum ingot was exhibited at the Paris Exposition. It was a rarity worth more per pound than gold. Emperor Napoleon III commissioned Henri Sainte-Claire Deville to find a way to produce large amounts of aluminum to be used for army equipment. He successfully produced a few tons but not the huge commercial quantities required.

In 1886 an American, Charles Martin Hall, and a Frenchman, Paul Louis-Toussaint Héroult independently developed the production method used today, in which the alumina is dissolved in molten cryolite and is decomposed electrolytically.

Since that time, industry has begun to produce large amounts of aluminum, with the first to appear in construction products around 1925. Applications have expanded since then to include hundreds of internal and external uses.

Mining Aluminum

Aluminum is found in most rocks and clays, but concentrations of the aluminum oxide content should be about 45 percent to be economically extracted. Aluminum ores are called **bauxites.** Most ores are mined by open pit mining. The ore is crushed, washed, screened, ground, and dried.

Refining Bauxite Ore

Aluminum is refined using a two-step process. The first, the *Bayer process,* produces a very pure alumina (Al_2O_3). Alumina is an oxide of aluminum in crystal form. The second step is to reduce the alumina to a metallic aluminum, which at this point is about 99 percent pure. This process is referred to as the *Hall-Héroult process.* It is named after the two men who developed the electrolytic method of aluminum production.

The Bayer Process

The ground dried bauxite is mixed in a digester with soda ash, crushed lime, sodium hydroxide, and hot water. Live steam and mechanical agitators stir the mixture (Fig. 16.1). The mixture is then pumped to digester tanks where, under high pressure and the injection of steam, the mixture is churned. The chemical reaction forms sodium aluminate and the insoluble impurities form a waste material. The mixture flows through a pressure-reducing tank into a settling tank where the waste is removed. The sodium aluminate solution passes through a filter to a cooling tower into the precipitator where aluminum hydrate is added. The mixture is agitated with compressed air, and cooling continues allowing the sodium aluminate to precipitate as aluminum hydrate. This is pumped into filter tanks that separate the aluminum hydrate from the solution. This is calcined in a rotary kiln operating at about 2000°F

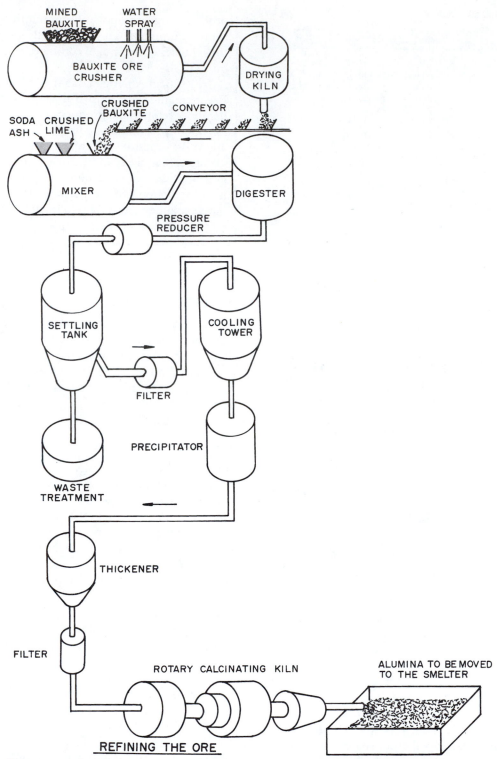

Figure 16.1 The first step in producing aluminum, the Bayer process, is to get the alumina from the bauxite ore.

(1100°C), which produces the alumina used to produce the metallic aluminum in the second step, the Hall-Héroult electrolytic process.

The Hall-Héroult Electrolytic Process

Aluminum is produced from the oxide **alumina** by a reduction process (Fig. 16.2). **Reduction** is an electrolytic process that uses a carbon-lined vessel containing molten cryolite and alumina. An electric current is passed through this liquid using large carbon anodes that are suspended in it. When the current is passed through the liquid, molten aluminum separates and settles to the bottom of the vessel where it is siphoned off. Pure aluminum may be cast in molds for use in products requiring its properties. It may also be sent to a holding furnace where alloying elements are added to alter its physical properties.

Aluminum Alloys

When aluminum is produced it is between 99.5 and 99.9 percent pure. In this form it is relatively soft and ductile and has a tensile strength of around 7000 psi (48 258 kPa). For many products, aluminum must have alloying elements added to alter its physical properties.

Aluminum alloys are in two classifications, wrought alloys and casting alloys. **Wrought alloys** are those that are mechanically worked by processes such as forging, drawing, extruding, or rolling to form sheet material. **Cast alloys** are those used to produce a product for which the molten metal is cast into the shape of the finished product, such as a grille, in a sand mold or permanent mold.

Wrought Aluminum Alloy Classifications

The major alloying elements added to aluminum are manganese, copper, magnesium, silicon, and zinc. A wide variety of aluminum alloys are identified by a code system developed by the Aluminum Association, Inc. Each wrought aluminum alloy is specified by a four-digit code number, as shown in Table 16.2.

The *first digit* specifies the alloy series and indicates the major alloying element. The *second digit* in the 1xxx series represents a modification of impurity limits and in the 2xxx through 9xxx series represents a modification of the alloy. The *last two digits* in the 1xxx series indicate an aluminum purity above 99 percent. For example, a 1050 specifies an aluminum containing 0.50 percent more aluminum than the minimum 99 percent. In the 2xxx through 9xxx series the last two digits are arbitrary numbers identifying the alloy in the series. Examine Tables 16.3 and 16.4 for a breakdown of these classification systems.

Cast Aluminum Alloy Classifications

The cast alloy is specified by a three-digit number followed by a decimal (Table 16.5). The *first digit* identifies the alloy series and the *second and third digits* the specific alloy or purity. The decimal indicates whether the alloy composition is for final casting (.0) or for ingot (.1 or .2).

Table 16.2 Wrought Aluminum and Aluminum Alloy Designation System

Aluminum 99.00% Minimum 1xxx	
Aluminum Alloys Grouped by Major Alloying Elements	
Copper	2xxx
Manganese	3xxx
Silicon	4xxx
Magnesium	5xxx
Magnesium and silicon	6xxx
Zinc	7xxx
Other elements	8xxx
Unused series	9xxx

Courtesy the Aluminum Association

Table 16.3 Breakdown of 1XXX Aluminum Designations

1 X XX		
Indicates 99% pure commercial aluminum	0 indicates no special impurity control	Indicates pure aluminum content
	1–9 indicates special specific impurity controls	above 99% in hundredths of a percent

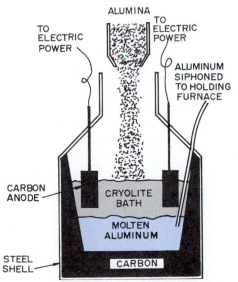

Figure 16.2 The second step in producing aluminum is the reduction of the alumina using the Hall-Héroult process.

Table 16.4 Breakdown of 2XXX–9XXX Aluminum Designations

X X XX		
1 commercially pure	0 indicates original alloy developed	in 1xxx series indicates impurity limits
2–9 indicates major alloying element as shown in Table 16.2	2–9 indicate modifications of the original alloy	2XXX through 9XXX series—arbitrary numbers identifying the alloy in the series

Table 16.5 Cast Aluminum and Aluminum Alloy Designation System

Aluminum 99.00% minimum	1xx.x
Aluminum Alloys Grouped by Major Alloying Elements	
Copper	2xx.x
Silicon, with added copper and/or magnesium	3xx.x
Silicon	4xx.x
Magnesium	5xx.x.
Zinc	7xx.x.
Tin	8xx.x
Other element	9xx.x
Unused series	6xx.x

Courtesy the Aluminum Association

Unified Numbering System Designations

The Unified Numbering System (UNS) is described in detail in Chapter 15. It uses a five-digit number with a letter prefix to classify and identify various types of aluminum and aluminum alloys that are specified by other organizations, such as the Aluminum Association (AA), American Society for Testing and Materials (ASTM), Society of Automotive Engineers/Aerospace Materials Specifications (AMS), Military Specifications (MS), and Federal Specifications (FS). The UNS pulls together those materials with like specifications from these organizations and gives them a unified number. In Table 16.6 are a few selected examples to show you how this system pulls these together.

Temper Designations

Temper is the degree of hardness and strength imparted to a metal by some process such as by **heat treating** or cold-working. Tempering refers to the process of bringing a metal to the proper degree of hardness and elasticity so it can be used for the purpose intended.

Wrought aluminum alloys fall into two classes, heat-treatable and non-heat-treatable.

Heat-treatable alloys are those whose strength characteristics are improved by heat treating. The strength is also improved by adding alloying elements such as copper, zinc, silicon, and magnesium.

Table 16.6 Selected Examples of Aluminum Alloy UNS Designations

UNS Designations	Material	Numbers Used by Various Organizations
A02400	Aluminum foundry alloy, casting	AA 240.0, AMS 4227
A91035	Wrought aluminum alloy, non-heat-treatable	AA 1035
A92011	Wrought aluminum alloy, heat-treatable	AA 2011, ASTM B210, FS QQ-A-22513, SAE J454

Courtesy Society of Automotive Engineers

Non-heat-treatable alloys have alloying elements added that do not cause an increase in strength when heat treated. The strength of non-heat-treatable alloys depends on alloying elements such as iron, magnesium, manganese, and silicon. These alloys are strengthened by cold-rolling or strain hardening.

Some temper specifications apply only to cast aluminum, and others apply only to wrought aluminum. Alloys specified as F, H, or O can be hardened by cold-working and may or may not be non-heat-treatable. Heat-treatable aluminum alloys use the T and W specifications.

The specification of an aluminum alloy requires the designation of temper or the metallurgical condition. The following temper designation system for aluminum alloys was developed by the Aluminum Association, Inc. The **temper designation** consists of letters and numbers that are placed after the alloy number and separated from it by a dash.

Temper designations include:

F   (as fabricated): No special control over thermal or work-hardening conditions is used.
O   (annealed): Wrought products have been heated to effect recrystallation, which produces the lowest strength. Cast products are annealed to improve stability and ductility.
H   (strain-hardened): Wrought products are strain-hardened through cold-working. The H is followed by one or two digits. See Table 16.7.
W   (solution heat-treated): The alloy is heated to about 1000°F (542°C) and then quenched.
T   (thermally treated): The product has been heat treated and then strain hardened. The T is followed by one or two digits. See Table 16.8.

Some applications for aluminum alloys are shown in Table 16.9. Table 16.10 gives some detailed uses of aluminum in construction.

Table 16.7 Subdivision of H Temper: Strain Hardened

First Digit Indicates Specific Treatment:

H1—Strain hardened only
H2—Strain hardened and partially annealed
H3—Strain hardened and stabilized
H4—Strain hardened and lacquered or painted

Digit (0–8) Indicates the Degree of Strain Hardening as Identified by a Minimum Value of the Ultimate Tensile Strength[a]:

0—Annealed

2—Tempers whose ultimate tensile strength is midway between 0 and 4.

4—Tempers whose ultimate tensile strength is midway between 0 and 8.

6—Tempers whose ultimate tensile strength is midway between 4 and 8.

8—Tempers whose ultimate tensile strength exceeds that of 8 by 2 ksi or more.

1, 3, 5, 7—Tempers whose ultimate tensile strength falls between those defined above.

[a]The tensile strength for each number is specified by Aluminum Association tables.
Courtesy the Aluminum Association

Table 16.8 Subdivisions of T Temper: Thermally Treated

First Digit Indicates Specific Sequence of Treatments:

T1—Naturally aged after cooling from an elevated temperature shaping process
T2—Cold-worked after cooling from an elevated temperature shaping process and then naturally aged
T3—Solution heat treated, cold-worked, and naturally aged
T4—Solution heat treated and naturally aged
T5—Artificially aged after cooling from an elevated temperature shaping process
T6—Solution heat treated and artificially aged
T7—Solution heat treated and stabilized (overaged)
T8—Solution heat treated, cold-worked, and artificially aged
T9—Solution heat treated, artificially aged, and cold-worked
T10—Cold-worked after cooling from an elevated temperature shaping process and then artificially aged

Second Digit Indicates Variation in Basic Treatment:

Examples:
T42 or T62—Heat treated to temper by user

Additional Digits Indicate Stress Relief:

Examples:
TX51—Stress relieved by stretching after solution heat treating
TX52—Stress relieved by compressing after solution heat treating or cooling

Courtesy the Aluminum Association

Table 16.9 Typical Applications of Aluminum Alloys

Typical Applications of Non-Heat-Treatable Aluminum Alloys

AA Alloy Series	AA Typical Alloys	UNS Designation	Typical Applications
1xxx	1350	A91350	Electrical conductors
	1060	A91060	Chemical equipment, tank cars
	1100	A91100	Sheet metal work, cooking utensils, decorative
3xxx	3003	A93003	Sheet metal work, chemical equipment, storage tanks
4xxx	4043	A4043	Welding electrodes
	4343	A94343	Brazing alloy
5xxx	5005	A95005	Decorative and automotive trim, architectural and anodized, sheet metal work, appliances
	5050	A95050	
	5454	A95454	
	5456	A95456	
5xxx (3% Mg)	5083	A95083	Marine, welded structures, storage tanks, pressure vessels, armor plate, cryogenics
	5086	A95086	
	5454	A95454	
	5456	A95456	

Typical Applications of Heat-Treatable Aluminum Alloys

AA Alloy Series	AA Typical Alloys	UNS Designation	Typical Applications
2xxx (Al-Cu)	2011	A92011	Screw machine products
	2219	A92219	Structural, high temperature
2xxx (Al-Cu-Mg)	2014	A92014	Aircraft structures and engines, truck frames and wheels
	2024	A92024	
	2618	A92618	
4xxx	4032	A94032	Pistons
6xxx	6061	A96061	Marine, truck frames and bodies, structures, architectural furniture
	6063	A96063	
7xxx (Al-Zn-Mg)	7004	A97004	Structural, cryogenic, missile
	7005	A97005	
7xxx (Al-Zn-Mg-Cu)	7001	A97001	High-strength structural and aircraft
	7075	A97075	
	7178	A97178	

Courtesy the Aluminum Association

Aluminum Castings

The conditions involved with the production of aluminum castings influence their physical properties. The metallurgist has to consider not only the characteristics of the alloy and the use to which the product will be put but also how it will be cast. In general, aluminum cast in permanent (metal) molds is stronger than sand-cast products. The cost of producing metal molds is high, but that can be justified when a large number of products will be cast because the molds can be reused. A sand casting mold produces only one casting.

Heat-treated alloys typically are stronger and have greater ductility than non-heat-treatable alloys. After casting, heat-treatable alloy castings are first subjected to a very high temperature (but below melting point of the alloy), then quenched in cold or hot oil or water and left to age at room temperature. This can help control the characteristics of the alloy.

Aluminum Finishes

As mentioned earlier, when aluminum is exposed to oxygen in the atmosphere it forms a natural protective **oxide layer**, so under ordinary circumstances no protective coating need be applied. However, various finishes are applied to aluminum products to improve their appearance and usefulness. The Aluminum Association specifies three categories of finishes: mechanical, chemical, and coatings. Aluminum can also have an **anodized** finish or be left natural. The flow chart in Fig. 16.3 illustrates these and shows their relationships.

The Aluminum Association classifies finishes with a letter and a two-digit number. Mechanical finishes are designated by M and chemical finishes by a C. Coatings are designated by letters denoting a particular type and anodic coatings use the letter A. You can see the specifications and designations for these in Tables 16.11, 16.12, and 16.13.

Natural Finishes

Natural finishes may be controlled or uncontrolled. *Uncontrolled finishes* are produced on the surfaces of wrought and cast products by the condition of the surfaces of the rollers, molds, or extruding dies used to form them. Hot-rolled products have a brighter surface. *Controlled finishes* are produced by varying the smoothness of the rollers or mold surfaces. The sheet

Table 16.10 Specific Construction Product Applications

UNS Designation	AA Alloy Number	Use
A93003	3003	Flashing, ducts, garage doors, curtain wall panels, grilles, louvers, siding, shingles, termite shields
A91235	1235	Vapor barriers, insulation
A04430	B443.0	Cast products, such as hardware, curtain wall castings, architectural letters
A96063	6063	Extruded products, such as curtain wall, door, and window frames; grilles, mullions; railings; thresholds

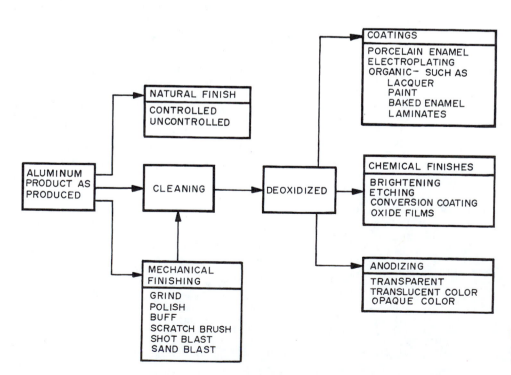

Figure 16.3 The finishing processes that can be used on various aluminum products.

Table 16.11 Mechanical Finishes on Aluminum

Type of Finish	Designation[a]	Description	Examples of Methods of Finishing[b]
As fabricated	M10	Unspecified	
	M11	Specular as fabricated	
	M12	Nonspecular as fabricated	
	M1X	Other	To be specified.
Buffed	M20	Unspecified	
	M21	Smooth specular	Polished with grits coarser than 320. Final polishing with a 320 grit using peripheral wheel speed of 30 m/s (6,000 feet per min.). Polishing followed by buffing, using tripoli based buffing compound and peripheral wheel speed of 36 to 41 m/s (7,000 to 8,000 feet per min.).
	M22	Specular	Buffed with tripoli compound using peripheral wheel speed 36 to 41 m/s (7,000 to 8,000 feet per min.).
	M2X	Other	To be specified.
Directional textured	M30	Unspecified	
	M31	Fine satin	Wheel or belt polished with aluminum oxide grit of 320 to 400 size; peripheral wheel speed 30 m/s (6,000 feet per min.).
	M32	Medium satin	Wheel or belt polished with aluminum oxide grit of 180 to 220 size; peripheral wheel speed 30 m/s (6,000 feet per min.).
	M33	Coarse satin	Wheel or belt polished with aluminum oxide grit of 80 to 100 size; peripheral wheel speed 30 m/s (6,000 feet per min.).
	M34	Hand rubbed	Hand rubbed with stainless steel wool lubricated with neutral soap solution. Final rubbing with No. 00 steel wool.
	M35	Brushed	Brushed with rotary stainless steel wire brush, wire diameter 0.24 mm (0.0095 in.); peripheral wheel speed 30 m/s (6,000 feet per min.); or various proprietary satin finishing wheels or satin finishing compounds with buffs.
	M3X	Other	To be specified.
Nondirectional textured	M40	Unspecified	
	M41	Extra fine matte	Air blasted with finer than 200 mesh washed silica or aluminum oxide. Air pressure 310 kPa (45 psi); gun distance 203 to 305 mm (8 to 12 inches) from work at 90° angle.
	M42	Fine matte	Air blasted with 100 to 200 mesh silica sand if darkening is not a problem; otherwise aluminum oxide type abrasive. Air pressure 207 to 621 kPa (30 to 90 psi) (depending upon thickness of material); gun distance 305 mm (12 inches) from work at angle of 60° to 90°.
	M43	Medium matte	Air blasted with 40 to 50 mesh silica sand if darkening is not a problem; otherwise aluminum oxide type abrasive. Air pressure 207 to 621 kPa (30 to 90 psi) (depending upon thickness of material); gun distance 305 mm (12 inches) from work at angle of 60° to 90°.
	M44	Coarse matte	Air blasted with 16 to 20 mesh silica sand if darkening is not a problem; otherwise aluminum oxide type abrasive. Air pressure 207 to 621 kPa (30 to 90 psi) (depending upon thickness of material); gun distance 305 mm (12 inches) from work at angle of 60° to 90°.
	M45	Fine shot blast	Shot blasted with cast steel shot of ASTM size 70 to 170 applied by air blast or centrifugal force. To some degree, selection of shot size is dependent on thickness of material since warping can occur.
	M46	Medium shot blast	Shot blasted with cast steel shot of ASTM size 230 to 550 applied by air blast or centrifugal force. To some degree, selection of shot size is dependent on thickness of material since warping can occur.
	M47	Coarse shot blast	Shot blasted with cast steel shot of ASTM size 660 to 1320 applied by air blast or centrifugal force. To some degree, selection of shot size is dependent on thickness of material since warping can occur.
	M4X	Other	To be specified.

[a]The complete designation must be preceded by AA—signifying Aluminum Association.
[b]Examples of methods of finishing are intended for illustrative purposes only.
Courtesy of the Aluminum Association

may be rolled with one side having very smooth rollers and the back side uncontrolled. Sheets may be embossed by using rollers that have a design on the surface. Aluminum siding that looks like it has a wood grain is an example of an embossed finish.

Mechanical Finishes

Mechanical finishes are used to alter the appearance of the surface of the aluminum product. They are usually performed before the surface is cleaned to receive other finishing processes. The surface may be *buffed* with an abrasive or *polished* to a high luster with very fine abrasive. It may be ground with a dry *grinding* wheel. This is often necessary to clean up ridges on castings. It produces a rough, scratched finish. Rotating wire brushes produce a *scratched surface*. The scratches can vary from very fine to coarse depending on the size of the wire in the brushes. A *matte surface* is produced by blasting the

Table 16.12 Chemical Finishes on Aluminum

Type of Finish	Designation[a]	Description	Examples of Methods of Finishing[b]
Nonetched cleaned	C10	Unspecified	
	C11	Degreased	Organic solvent treated.
	C12	Inhibited chemical cleaned	Inhibited chemical type cleaner used.
	C1X	Other	To be specified.
Etched	C20	Unspecified	
	C21	Fine matte	Trisodium phosphate, 22-45 g/l (3-6 oz per gal) used at 60-71°C (140-160°F) for 3 to 5 min.
	C22	Medium matte	Sodium hydroxide, 30-45 g/l (4-6 oz per gal) used at 49-66°C (120-150°F) for 5 to 10 min.
	C23	Coarse matte	Sodium fluoride, 11 g/l (1.5 oz) plus sodium hydroxide 30-45 g/l (4-6 oz per gal) used at 54-66°C (130-150°F) for 5 to 10 min.
	C2X	Other	To be specified.
Brightened	C30	Unspecified	
	C31	Highly specular	Chemical bright dip solution of the proprietary phosphoric-nitric acid type used, or proprietary electrobrightening or electropolishing treatment.
	C32	Diffuse bright	Etched finish C22 followed by brightened finish C31.
	C3X	Other	To be specified.
Chemical coatings[c]	C40	Unspecified	
	C41	Acid chromate-fluoride	Proprietary chemical treatments used producing clear to typically yellow colored surfaces.
	C42	Acid chromate-fluoride-phosphate	Proprietary chemical treatments used producing clear to typically green colored surfaces.
	C43	Alkaline chromate	Proprietary chemical treatments used producing clear to typically gray colored surfaces.
	C44	Non-chromate	Proprietary chemical coating treatment employing no chromates.
	C45	Non-rinsed chromate	Proprietary chemical coating treatment in which coating liquid is dried on the work with no subsequent water rinsing.
	C4X	Other	To be specified.

[a]The complete designation must be preceded by AA—signifying Aluminum Association.
[b]Examples of methods of finishing are intended for illustrative purposes only.
[c]Includes chemical conversion coatings.
Courtesy the Aluminum Association

surface with sand or steel shot. Round pieces of abrasive cloth or steel wool can be rotated against the surface to produce a series of concentric circles. Code designations for mechanical finishes are in Table 16.11.

Chemical Finishes

Chemical finishes are produced by the reaction of the aluminum surface to various chemicals. *Conversion coatings* are a major chemical finish. They prepare the surface for the bonding of paints, organic coatings, and laminates. The natural oxide film does not always provide an adequate bonding surface. The surface can be *etched* with a chemical that produces a frosty surface. Designs can be etched on the surface by protecting all areas except those to be etched. Chemical *oxide films* can be used to produce a surface having greater resistance to corrosion than the natural oxide films. Aluminum can be plated with zinc in a process called *zincating*. This produces a thin coating of zinc. The zinc coating protects the aluminum from galvanic action and also prepares the surface for electroplating. Aluminum can be used to produce surfaces that are highly reflective of heat and light. These mirrorlike surfaces are produced

by chemical *brightening*. Code designations for chemical finishes are in Table 16.12.

Anodic Finishes

A very important, widely used finish on aluminum is an *anodized electrolytic oxide layer. Anodized films* can be used as the finish on surfaces to be painted. *Anodized coatings* are most often used as the finished protecting layer. A film is less than 0.1 mil thick. A coating is a 0.1 mil or thicker.

Following is a typical procedure for the anodizing process.

1. The surface is cleaned using alkaline and/or acid cleaners to remove grease and surface dirt.

2. Next the surface is given a pretreatment that may be either etching or brightening. Etching involves producing a matte surface with hot solutions of sodium hydroxide. This removes minor surface imperfections. A thin layer of the aluminum is removed by this process. Brightening involves producing a mirrorlike surface with a concentrated solution of phosphoric and nitric acids. These chemically smooth the surface.

Table 16.13 Anodic Coatings on Aluminum

Type of Finish	Designation[a]	Description	Examples of Methods of Finishing[b]
General	A10	Unspecified	
	A11	Preparation for other applied coatings	3 μm (0.1 mil) anodic coating produced in 15% H_2SO_4 at 21°±1°C (70°F±2°F) at 129 A/m² (12 A/ft².) for 7 min, or equivalent.
	A12	Chromic acid anodic coatings	To be specified.
	A13	Hard, wear and abrasion resistant coatings	To be specified.
	A1X	Other	To be specified.
Protective & Decorative Coatings less than 10 μm (0.4 mil) thick	A21	Clear coating	Coating thickness to be specified. 15% H_2SO_4 used at 21°±1°C (70°F±2°F) at 129 A/m² (12 A/ft².).
	A211	Clear coating	Coating thickness—3 μm (0.1 mil) minimum. Coating weight—6.2 g/m² (4 mg/in².) minimum.
	A212	Clear coating	Coating thickness—5 μm (0.2 mil) minimum. Coating weight—12.4 g/m² (8 mg/in².) minimum.
	A213	Clear coating	Coating thickness—8 μm (0.3 mil) minimum. Coating weight—18.6 g/m² (12 mg/in².) minimum.
	A22	Coating with integral color	Coating thickness to be specified. Color dependent on alloy and process methods.
	A221	Coating with integral color	Coating thickness—3 μm (0.1 mil) minimum. Coating weight—6.2 g/m² (4 mg/in².) minimum.
	A222	Coating with integral color	Coating thickness—5 μm (0.2 mil) minimum. Coating weight—12.4 g/m² (8 mg/in².) minimum.
	A223	Coating with integral color	Coating thickness—8 μm (0.3 mil) minimum. Coating weight—18.6 g/m² (12 mg/in².) minimum.
	A23	Coating with impregnated color	Coating thickness to be specified. 15% H_2SO_4 used at 27°C±1°C (80°F±2°F) at 129 A/m² (12 A/ft².) followed by dyeing with organic or inorganic colors.
	A231	Coating with impregnated color	Coating thickness—3 μm (0.1 mil) minimum. Coating weight—6.2 g/m² (4 mg/in².) minimum.
	A232	Coating with impregnated color	Coating thickness—5 μm (0.2 mil) minimum. Coating weight—12.4 g/m² (8 mg/in².) minimum.
	A233	Coating with impregnated color	Coating thickness—8 μm (0.3 mil) minimum. Coating weight—18.6 g/m² (12 mg/in².) minimum.
	A24	Coating with electrolytically deposited color	Coating thickness to be specified. Application of the anodic coating, followed by electrolytic deposition of inorganic pigment in the coating.
	A2X	Other	To be specified.
Architectural Class II[c] 10 to 18 μm (0.4 to 0.7 mil) coating	A31	Clear coating	15% H_2SO_4 used at 21°C±1°C (70°F±2°F) at 129 A/m² (12 A/ft².) for 30 min, or equivalent.
	A32	Coating with integral color	Color dependent on alloy and anodic process.
	A33	Coating with impregnated color	15% H_2SO_4 used at 21°C±1°C (70°F±2°F) at 129 A/m² (12 A/ft².) for 30 min, followed by dyeing with organic or inorganic colors.
	A34	Coating with electrolytically deposited color	Application of the anodic coating followed by electrolytic deposition of inorganic pigment in the coating.
	A3X	Other	To be specified.
Architectural Class I[c] 18 μm (0.7 mil) and thicker coatings	A41	Clear coating	15% H_2SO_4 used at 21°C±1°C (70°F±2°F) at 129 A/m² (12 A/ft².) for 60 min, or equivalent.
	A42	Coating with integral color	Color dependent on alloy and anodic process.
	A43	Coating with impregnated color	15% H_2SO_4 used at 21°C±1°C (70°F±2°F) at 129 A/m² (12 A/ft².) for 60 min, followed by dyeing with organic or inorganic colors, or equivalent.
	A44	Coating with electrolytically deposited color	Application of the anodic coating followed by electrolytic deposition of inorganic pigment in the coating.
	A4X	Other	To be specified.

[a]The complete designation must be preceded by AA—signifying Aluminum Association.
[b]Examples of methods of finishing are intended for illustrative purposes only.
[c]Aluminum Association Standards for Anodized Architectural Aluminum.
Courtesy the Aluminum Association

3. The third step is the actual anodizing process in which the anodic film is built and combined with the aluminum by passing an electric current through an acid electrolyte bath in which the aluminum is immersed. The coating thickness and finished surface characteristics can be carefully controlled.

4. Coloring the anodized surface can occur in several ways. One method is to combine the coloring with the actual anodizing process (step 3), which simultaneously forms and colors the oxide cell wall in bronze and black shades. This produces a more abrasion resistant coating than other methods but is more expensive because it requires more electricity.

A second coloring procedure involves a two-step electrolytic coloring process. After the aluminum is anodized (step 3) it is immersed in a bath containing an inorganic metal salt and an electric current is passed that deposits the metal salt at the base of the pores. Color depends on the metal salt used (tin, copper, nickel, cobalt). This method provides the greatest variation of colors.

A third method involves organic dyeing. This produces vibrant colors that are highly weather resistant.

One other coloring process is described as interference coloring. It involves modification of the pore structure produced in sulfuric acid. It produces light-fast colors ranging from blue, green, and yellow to red.

Anodic coatings are classified into four groups—General, Protective and Decorative, Architectural I, and Architectural II.

General coatings are those less than 0.1 mil thick. *Protective and Decorative* coatings are less than 0.4 mil thick. These two classes are used for general industrial applications. The architectural classes are used on materials to be exposed to weather and wear. Details can be found in Table 16.13.

Architectural I coatings are recommended for exterior use where they will receive no regular maintenance. They are also used for interior purposes where extra protection is needed. They must be over 0.7 mil in thickness and weigh more than 27 mg per in.^{2}.

Architectural II coatings are recommended for interior applications not expected to receive heavy wear and for exterior uses where the product will be regularly maintained. They must have a thickness from 0.4 to 0.7 mils and weigh 17 to 27 mg per in.^{2}. Code designations for architectural classes can be found in Table 16.13.

Coatings

Many aluminum products are *painted* to provide additional protection or for surface decoration. These products, such as aluminum gutters, are painted in a factory and delivered to the job site finished, ready to install. The factory-applied finish is electrostatically sprayed or roller applied with an organic paint and is then passed through an oven where it is baked to a hard, uniform finish. The paint is flexible and will not crack if the product is bent or formed on the job.

Porcelain enamel coatings are produced by using a vitreous inorganic material that is bonded to the metal by fusing it at high temperatures. The coating is very resistant to corrosion, is durable, and is available in a wide range of colors. It is widely used on curtain wall panels as seen on high-rise buildings where it is difficult to provide regular maintenance.

Aluminum can be *electroplated* with chromium, copper, and other materials. If chromium is electroplated, the aluminum must first be zincated or plated with copper, brass, or nickel. Chromium plating gives a mirrorlike surface and provides resistance to abrasion. Copper plating requires the aluminum to be zincated. It is mainly used where things, such as electrical connections, must be soldered to the aluminum product. Tin and brass plating also provide good soldering surfaces. Zinc and cadmium plating improve corrosion resistance.

Aluminum surfaces can also have *laminated* finishes. This involves bonding another material, such as vinyl or polyvinyl chloride films, to the surface with an adhesive. Usually the aluminum surface must be chemically cleaned to provide the strongest bond.

Aluminum surfaces that are to serve as high-quality mirrors are treated by electropolishing. This is an anodic smoothing of the surface and requires high-purity aluminum. The finished surface usually has a final anodic protective coating applied.

Protecting the Finished Aluminum Product

Aluminum products on the construction site need to be protected from damage during delivery, storage, and installation and until the job is finished. Following are several coatings used for this purpose.

The best and most expensive protective coatings are some form of *paper or plastic sheet material* bonded to the product with an adhesive that permits them to be removed easily, leaving little or no sticky residue behind. If on-the-job protection is needed, masking tape, as used by painters, does a good job.

Sometimes a coat of clear *lacquer* is sprayed on the surface of the aluminum. Methacrylate lacquers will chalk off the surface after several years. Protection of aluminum is afforded by *waxing* the surface. This is a good way to maintain the surface after the job is completed. Any type of automobile wax or polish is good. This will not provide protection from damage by mortar and other abrasive materials during construction.

Routine Maintenance of Aluminum Surfaces

When cleaning aluminum surfaces, the type of protective film must be considered. A natural aluminum will develop its own oxide film. If this is scrubbed or rubbed with abrasive cleaners or steel wool the film will be damaged. It will eventually form another oxide layer. However, unless the surface is stained, abrasive cleaners should be avoided. Likewise, anodized surfaces, plated surfaces, and other coatings also can be damaged and cannot be repaired easily.

The simplest cleaner is clear *water*. The surfaces can be washed with water and liquid soaps and rinsed. If some discoloration needs to be removed a *mild polish*, such as liquid auto polish, is satisfactory. Liquid non-scratch cleaners used on plastic bath fixtures are satisfactory. Cleaners containing abrasive materials, such as kitchen cleansers, will scratch the surface. Pads of very fine *steel wool* will remove stains but may damage the finish. *Liquid or paste wax* is a good final protective coating.

Joining Aluminum Members

Aluminum members can be joined by any of the standard fastening techniques. *Mechanical fasteners* include screws, bolts, rivets, and a variety of specially designed products. Sheet stock can be joined by *stitching*, which involves sewing the sheets together with aluminum wire. Aluminum can be joined by *welding, brazing,* and *soldering.* This includes gas, arc, resistance, and inert gas shielded arc welding. An ever-increasing joining technique is *adhesive bonding.* A variety of adhesives

are used, depending on the materials and design considerations. Bonding produces a strong joint and increases the design possibilities for joining aluminum and forming laminates.

Aluminum Products

Aluminum is a major construction material. In its many alloys it finds use in all types of construction. Following are examples of some of the most common uses.

One major use for aluminum is the production of maintenance-free *windows* (Fig. 16.4). The members used for this purpose are extruded from an aluminum alloy such as 6063. Typical window types available are shown in Fig. 16.5.

Sliding glass doors are made with two or more framed glass doors encased in an extruded aluminum frame. One or more of the panels moves (Fig. 16.6). In addition, a wide range of interior and exterior aluminum doors are made. These can be panelized, flush, louvered, and have glass lights (Fig. 16.7). Doors and windows often have two or three layers of glass and weather stripping to prevent air infiltration. In addition, the exterior and interior metal parts are separated by plastic strips, forming a thermal break. This reduces the heat loss or gain through the aluminum parts.

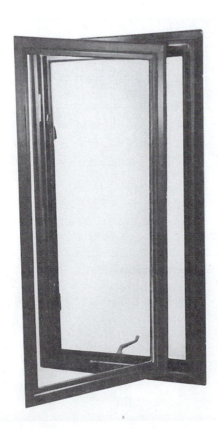

Figure 16.4 Aluminum windows are widely used in residential and commercial construction. *(Courtesy Acorn Building Components)*

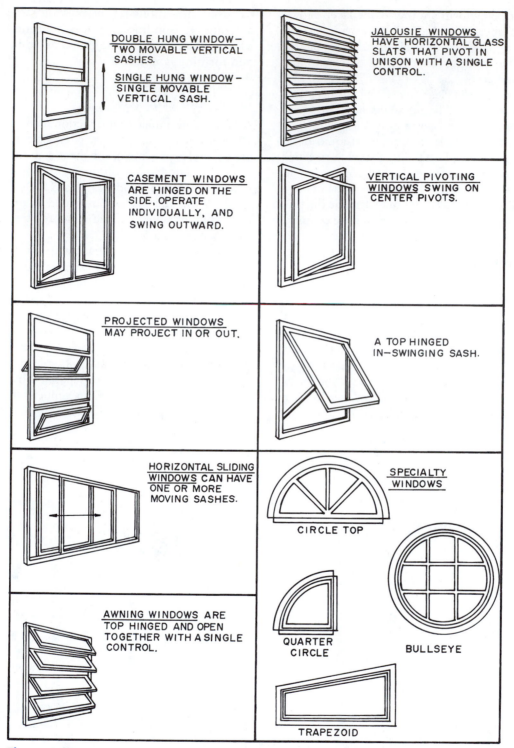

Figure 16.5 These are the most commonly available types of aluminum windows.

Figure 16.6 Aluminum sliding doors with anodized coatings and energy-efficient glass are widely used. *(Courtesy Acorn Building Components)*

Figure 16.7 Aluminum doors resist weathering and are energy efficient. *(Courtesy Endure-A-Lifetime Products, Inc.)*

Specifications developed by the Architectural Aluminum Manufacturers Association (AAMA) have been adopted by the American National Standards Institute (ANSI). These specify minimum frame strength, thickness, corrosion resistance, air infiltration, water resistance, wind-load capacity, and condensation resistance. Units meeting these specifications have the AAMA seal attached.

Aluminum is used for residential and commercial siding and roofing. The specifications are developed by the AAMA. Residential siding is cold-rolled to shape and is available in several widths. Horizontal siding is most popular, although vertical siding panels are available. The panels may or may not have a backer board. Backer boards are either fiberboard or foamed plastic, which provide some insulation value (Fig. 16.8).

The same material is used for soffits, fascia, flashing, gutters, and frieze boards. Panels are secured to the wood framing with aluminum nails (Fig. 16.9).

Aluminum panels for siding and roofing commercial buildings are made in a variety of designs. Some of the designs available are in Fig. 16.10. They are one-third the weight of steel. Aluminum alloy 3004 is often used. Common finishes include mill finish, unpainted, painted, and stucco embossed.

Figure 16.8 Aluminum siding is weather resistant and adds insulation value with backer boards. *(Courtesy the Aluminum Association and Berle Cherney, photographer)*

Figure 16.9 Aluminum soffits, fascias, and gutters require no maintenance. *(Courtesy the Aluminum Association and Berle Cherney, photographer)*

Figure 16.10 Aluminum siding and roofing materials are widely used on commercial buildings. *(Courtesy the Aluminum Association and Berle Cherney, photographer)*

Aluminum curtain wall systems include preformed insulated wall panels that are integrated with aluminum windows to form a weathertight, maintenance-free exterior surface. They are available with anodized or factory-applied baked enamel finishes. You can find more information on these in Chapter 31.

A wide array of *roof accessories,* such as skylights, roof hatches, and smoke and fire vents and louvers, are made with extruded and sheet aluminum. For example, a smoke hatch dome is made with 6063-T-5 aluminum extrusions and an acrylic dome.

Aluminum *structural shapes* are rolled in much the same configurations as discussed for structural steel members. These include shapes such as S and W beams, channels including a number of special shapes, tees, zees, bulb angles, round and square tubes, pipe, plate, and rods of various shapes.

Aluminum is used for large-diameter electrical conductors, various sheet metal applications, and for applications where corrosion exists, as in food-processing and chemical plants.

COPPER

Copper (chemical symbol Cu) is a nonmagnetic reddish brown metal that has excellent electrical and thermal conductivity. It has the highest conductivity properties of all commonly used metals except silver. It is ductile and malleable and easily worked. When alloyed it offers a wide range of properties making it a very valuable and widely used material in construction.

Copper and copper alloys are used to produce a wide range of products used in construction. Although copper products are initially more expensive than aluminum, they have properties that make them less costly in the long run because of their resistance to corrosion and other damaging conditions.

HISTORY

Copper is one of the first metals humans used to make tools and utensils and for various art and decorative purposes. In some parts of the world it is found naturally in almost pure form, and early humans could beat it into the shapes desired in a cold condition. Copper is available in many parts of the world, and early in the history of humankind it was used in widely spread areas.

Archaeological discoveries indicate that copper likely was used as early as 4500 B.C. or possibly earlier. Hammered copper specimens have been found that are believed to be from this time. It is clear that by 3000 B.C. the Egyptians were making utensils, simple tools, and ornaments from copper. Archaeological finds indicate it was also used by other Near Eastern peoples at about the same time. About 2500 B.C. the Egyptians had discovered how to make bronze, and the use of copper and bronze spread over most of the Mediterranean area.

Copper was named after the island of Cyprus, where large quantities were produced by 3000 B.C. This was a major supply for the Romans, and the island was frequently conquered by other peoples because of the copper. Copper was first called cuprium, then cuprum, and in English became copper.

Copper artifacts have been found in China and India dating from about 2500 B.C., and bronze items were made around 1770 B.C. Bronze, an alloy of copper and tin, became so widely used that the period was referred to as the *Bronze* Age. Copper and bronze were being used in Central America and parts of South America around 100 A.D. Native Americans used copper for tools and decorative beads long before Europeans discovered the continent.

Through the years, with improvements in mining, refining, and alloying, copper has become one of the major metals used in construction products and tools.

ITEM   OF   INTEREST

DEAD SEA SCROLLS

In 1948 on the northwest shore of the Dead Sea in Jordan the first of eight ancient manuscripts now known as the Dead Sea Scrolls were found in a cave where they had remained from the period about 200 B.C. to 70 A.D. These were copies of books of the Torah, which forms the Christian Old Testament, and non-Biblical texts from this time. These scrolls were written by scribes and most were on leather. However, one found in 1952 was written on thin copper sheet material and rolled. The copper scroll was too brittle to unroll and eventually was read by cutting it into paper-thin strips through which the letters could be read. It can be seen that copper was produced and used several thousand years ago in the Mediterranean area.

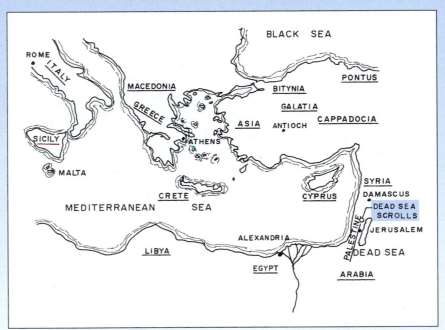

The Mediterranean world at the time of the Dead Sea Scrolls. Copper was known and used throughout this area.

Properties of Copper

The most important properties of copper and copper alloys as used in the construction industry are electrical and thermal conductivity, resistance to corrosion, wear resistance, ductility, and high temperature performance. Copper has a relatively low tensile strength, about 32,000 psi, that can be improved by heat treating, cold-working, and alloying. High electrical conductivity makes it a good material for electrical wiring and parts in devices that conduct electricity. Its good thermal conductivity makes it useful in heat transfer situations, and its corrosion resistance properties make it useful for parts exposed to the atmosphere or to such corrosive elements as chemicals, for plumbing pipe, and for gas lines. Its ductility properties make it a material easily bent, stretched, stamped, machined, and otherwise formed into useful products. Copper has a melting point of 1981°F (1083°C) and a coefficient of thermal expansion of 0.0000168/°F (0.0000093/°C).

A unique characteristic of copper is that if exposed unprotected to the atmosphere it will develop a green coating naturally over a period of years. This coating, called *patina*, provides a natural protection from additional corrosion. It takes many years of exposure to the atmosphere to build up a fully developed patina coating.

Production of Copper

Much of the copper mined in the United States is low-grade sulfate ore. There are various processes for producing the metal copper from the ore. You can see one example detailed in Fig. 16.11. This shows that the mined ore is crushed and pulverized in a *ball mill*. The crushed ore is ground to a powder by a *grinder* and flows to a concentrator where the *flotation method* is used. Here the pulverized ore is mixed with water, oil, and a foaming agent and is agitated by air. The copper sulfate particles collect in the foam on the surface, and other materials settle to the bottom and are removed as waste. The froth flows to a *reverberatory furnace* where it is smelted (roasted) to eliminate sulfur and certain metal

impurities by oxidation. This produces a layer of copper on top that is tapped off and moved to a *converter*.

The converting processes is much like the Bessemer process used to produce steel. In the converter, the iron in the remaining material is oxidized, forming a layer of slag that is removed. The sulfur is oxidized forming sulfur dioxide gas and drawn off the converter. The remaining material, called *blister copper*, is about 99 percent pure. The blister copper is processed in a *refining furnace*, reducing its oxygen content and forming fire-refined copper. This fire-refined copper can be used to produce copper products by rolling, extruding, drawing, and casting. It does not produce as pure a copper as needed for some applications.

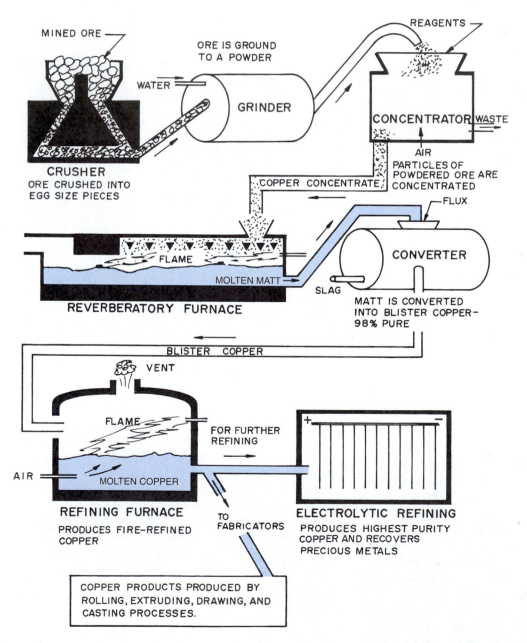

Figure 16.11 These are the steps needed to produce copper from copper ore.

The purest copper is produced by the *electrolytic process.* Fire-refined copper can be cast in molds that become the anodes in the electrolytic process. The electrolytic process utilizes an acid-proof tank containing an electrolyte, warm diluted sulfuric acid, and copper sulfate. The fire-refined copper anodes are suspended in the tank. Thin sheets of pure copper are suspended between them and serve as the cathodes. An electric current is passed through the system, causing the copper in the anodes to dissolve and bond to the cathodes. This produces copper that is about 99.9 percent pure. The cathodes are melted and used to produce the base metal for a wide variety of copper alloys.

After the copper has been fire refined or purified by the electrolytic process it is formed into sheets, bars, tubes, and wires used to produce various products. Examples of these products are shown in Fig. 16.12. Some views from within the mill producing these products can be seen in the following photos. Finished copper tubing is shown moving on a conveyor in Fig. 16.13. It is being prepared for shipment. Copper plumbing tube is wound on large-diameter reels, as shown in Fig. 16.14. The flat copper sheet in Fig. 16.15 is being cut to specified widths for forming into gutters, downpipes, and for use as flashing. The machine shown with the large coils of copper sheet material in Fig. 16.16 alternately rolls and anneals the sheet to the desired thickness and temper.

Classifying Copper and Copper Alloys

Copper and copper alloys are specified by the Unified Numbering System for Metals and Alloys (UNS). They are identified by a five-digit number code preceded by the letter C. Wrought materials are assigned UNS numbers from C10000 to C79999. Cast alloys are numbered from C80000 to C99999. A summary of these can be found in Table 16.14.

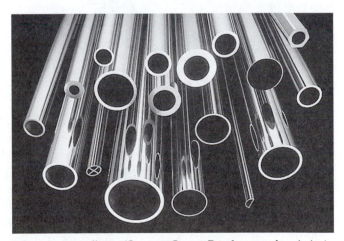

Figure 16.12 These are wrought copper products produced at the mill from copper alloys. *(Courtesy Copper Development Association)*

Figure 16.13 Finished copper tubing moving on a conveyor and being prepared for shipping or storage. *(Courtesy Copper Development Association)*

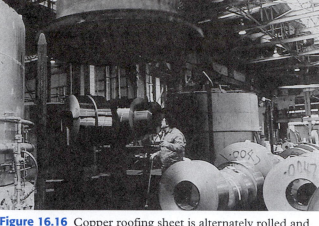

Figure 16.16 Copper roofing sheet is alternately rolled and annealed to achieve the desired thickness and temper. *(Courtesy Copper Development Association)*

Figure 16.14 Copper plumbing tube wound on large-diameter reels for storage and shipment. *(Courtesy Copper Development Association)*

Coppers are numbered from C10100 to C15999 and are pure or nearly pure, having a minimum copper content of 99.3 percent or higher. The coppers in this series are very similar in chemical composition. The numbering system identifies the production or refining processes used to produce the copper. For example oxygen-free copper is C10200, electrolytic tough pitch copper is C11000, and phosphorus-deoxidized high residual phosphorus copper is series C12200. Coppers containing less than a total of 0.7 percent of the specified alloying constituents include tellurium-bearing copper (C14500) and zirconium copper (C15000).

The numerical designations indicate the type of copper or copper alloy and the alloying elements and impurities. Detailed handbooks from the Copper Development Association give specific details.

You must identify the copper or copper alloy by its UNS number. Over the years some frequently used types of copper were given trade names, as you can see in the tables that follow. The trade name does not indicate the composition of the material, and sometimes is used to describe similar materials that have slightly different amounts of various elements.

Deoxidized copper (UNS C122200) contains 99.90 percent copper and 0.025 percent phosphorus. It has better forming and bending qualities than electrolytic copper and resists embrittlement at high temperatures. It is used for water and refrigeration piping, oil burner service, and in sheets and plates where welding is the major joining method. *Electrolytic tough pitch* copper (UNS C11000) is 99.90 percent copper and has high electrical and thermal conductivity. It can be easily formed into useful shapes. Major uses include electrical conductors of all types as well as flashing, gutters, roofing, and forgings (Fig. 16.17).

Figure 16.15 Copper sheet being cut to the desired width for forming flashing, gutters, and downspouts. *(Courtesy Copper Development Association)*

Table 16.14 UNS Designations for Coppers, Brasses, and Bronzes

Wrought Coppers and Copper Alloys	
C10000–C15760	Copper
C16200–C19900	High copper alloys

Cast Coppers and Copper Alloys	
C80100–C81200	Copper
C81400–C82800	High copper alloys

Wrought Brasses and Brass Alloys	
C21000–C28000	Brasses
C31200–C38500	Copper-zinc-lead alloys
C40400–C48600	Copper-zinc-tin alloys

Cast Brass and Brass Alloys	
C83300–C83810	Brasses
C84200–C84800	Copper-tin-zinc and copper-tin-zinc-lead alloys
C85200–C85800	Copper-zinc and copper-zinc-lead alloys
C86100–C86800	Manganese bronze and leaded manganese bronze alloys
C87300–C87800	Copper-silicon alloys

Wrought Bronzes and Bronze Alloys	
C50100–C54400	Copper-tin-phosphous alloys
C55180–C55284	Copper-phosphorus and copper-silver-phosphorus alloys
C60800–C64210	Copper-aluminum alloys
C64700–C66100	Copper-silicon alloys
C66400–C69710	Other copper-zinc alloys
C70100–C72950	Copper-nickel alloys
C73500–C79800	Copper-nickel-zinc alloys (nickel silvers)

Cast Bronzes and Bronze Alloys	
C90200–C91700	Copper-tin alloys
C92200–C92900	Copper-tin-lead alloys (leaded tin bronzes)
C93100–C94500	Copper-tin-lead alloys (high leaded-tin alloys)
C94700–C94900	Copper-tin-nickel alloys
C95200–C95900	Copper-aluminum-iron and copper-aluminum-iron-nickel alloys
C96200–C96900	Copper-nickel-iron alloys
C97300–C97800	Copper-nickel-zinc alloys (nickel silvers)
C98200–C98840	Copper-lead alloys
C99300–C99750	Special alloys

Courtesy Society of Automotive Engineers

Figure 16.17 Copper flashing is used on high-quality construction.

Copper Alloys

The major copper alloying elements are tin, aluminum, zinc, nickel, silicon, manganese, lead phosphorus, and beryllium. Following is a brief discussion of each. Additional information about brasses and bronzes is given later in this section. The UNS designations shown are for wrought products. Designations for cast products are in Table 16.14.

High copper alloys (UNS C16200–C19199) are wrought alloys having specified copper contents from 96 to 99.3 percent copper. Cast high copper alloys have a minimum of 94 percent copper. Wrought high copper alloys have very high electrical and thermal conductivity and are almost as high as pure copper. However, they are much stronger than pure copper, which increases the number of possible uses. They also have good corrosion resistance.

Brasses (UNS C20000–C49999) are copper alloys with zinc as the major alloying element. Other elements, such as lead, phosphorus, nickel, silicon, iron, and aluminum, may be added in small specified amounts. They are extremely useful and find application in many products.

Bronzes (UNS C50000–C66399) are copper alloys in which neither nickel nor zinc is used as the major alloying element. The various types are used for electrical contacts and corrosion-resistant applications.

Miscellaneous copper-zinc alloys (UNS C66400–C69999) are often referred to as manganese or nickel bronzes. Typically, the major alloying element is zinc, so they are much like some of the brasses.

Copper-nickel alloys (UNS C70000–C72999) contain from 3 to 33 percent nickel. Other elements may be added to improve corrosion resistance and strength. They are used in marine applications because of their outstanding capacity to resist corrosion. Typical applications include use in heat exchangers, condensers, piping, valve and pump parts, and relay and switch springs. Alloys with more than 50 percent nickel are called Monel® metals and form a separate class. *Monels* have high strength at elevated temperature.

Copper-nickel-zinc alloys (UNS C73000–C79999) are referred to as silver nickels because of their silver color. Zinc is the principle alloying element and nickel is secondary. Other elements may be added to alter the properties. They contain no silver. They have good electrical and mechanical properties and good corrosion resistance. They are used for fasteners and various electrical components.

Copper Alloy Finishes

Copper alloys are available with a variety of finishes. Some are supplied by the mill that manufactures the copper alloy, and others are supplied by the company fabricating the copper into a product or into stock shapes. These finishes use the same three classifications as used on aluminum products. These include mechanical, chemical, and coatings. You can see a summary in Table 16.15.

Table 16.15 Copper Alloy Finishes

Finish	Copper Development Assn. Finish Designation
Mechanical	
As fabricated	M10 series
Buffed	M20 series
Directional textured	M30 series
Non-directional textured	M40 series
Patterned	M4X (specify)
Chemical	
Cleaned only	C10 series
Matte dipped	
Bright dipped	
Conversion coatings	C50 series
Coatings	
Organic:	
Air dry	060 series
Thermo-set	070 series
Chemical cure	080 series
Vitreous	
Laminated	L90 series
Metallic	

Courtesy Copper Development Association

Notice the Copper Development Association finish designation. An M before the identifying digits denotes mechanical finishes, a C designates chemical finishes, and coatings use three-digit numbers.

Care of Copper and Copper Alloys

New copper products, especially sheet stock, have a bright, shiny, light brownish color. If this is to be maintained it must be covered with a protective coating such as a clear lacquer. When left exposed to the atmosphere it will first turn to a darker brown and eventually take on a permanent light green patina. This is considered highly desirable from an appearance standpoint and can be produced artificially if the natural aging is too slow.

Copper can be washed with liquid soap and water. If scrubbed with an abrasive cleaner the brown or green color will be damaged and will take a period of time to return.

Uses of Copper

Copper and copper alloys are excellent for outdoor uses. Some uses in construction include siding, roofing, flashing, guttering, and screen wire (Fig. 16.18). The alloys are used extensively for plumbing pipe in residential and commercial structures as well as in the manufacture of plumbing fittings, such as valves, drains, and faucets (Fig. 16.19). Sewage treatment plants and industrial plants, such as chemical processing installations, utilize copper for many purposes, including to line vessels that will contain corrosive materials. Various types of hardware and fasteners, such as nails, screws, and bolts, are made from copper alloys. A major use is in electrical wire, electrical conductors, and parts in electrical appliances that conduct electricity.

BRASS

Brasses (UNS C20000–C49999) are copper alloys having zinc as the principal alloying element, but variations are produced by adding small quantities of other elements. *Zinc* improves strength and ductility and produces changes in color. *Lead* is added to improve machinability, and *tin* improves strength, hardness, workability, and ductility.

Brasses are hardened by cold-working. However, hardness is also influenced by the alloy composition. The compositions of several wrought brasses and brass alloys used in construction are shown in Table 16.16. Similar information is available for cast brasses and brass alloys. Notice that zinc is the major alloying element with lead and iron in much smaller amounts. The chemical symbols used to identify the various elements in all metals are shown in Appendix E.

Brasses fall into three general classes. *White brasses* contain less than 55 percent copper and are hard and

Figure 16.18 Copper is used for a wide range of construction applications. Here it is used over a bay window and for a dormer roof.

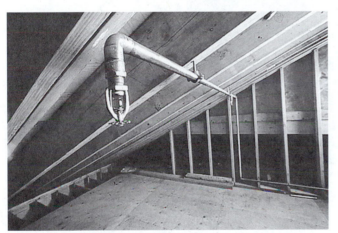

Figure 16.19 Copper is used for a wide range of piping systems, such as this fire protection sprinkler. *(Courtesy Copper Development Association)*

In Table 16.14 you can find the Unified Numbering System designation for wrought and cast brasses. Brasses are very important metals and are used in construction almost as much as copper.

Plain Brasses (Copper-Zinc Alloys)

Copper-zinc alloys are sometimes referred to as plain brasses. Several that find use in products related to construction are red brass, commercial bronze, cartridge brass, and muntz metal. The alloying elements for each can be found in Table 16.16.

Red brass contains 85 percent copper, 15 percent zinc, and small amounts of lead and iron. It has excellent resistance to corrosion and has higher ductility and strength than copper. It is used for plumbing pipe, handrails, balusters, stair posts, tubing, and hardware (Fig. 16.21).

Commercial bronze contains 90 percent copper and 10 percent zinc. It has good ductility and good cold-working properties. It is used for screws, forgings, and some types of hardware.

Cartridge brass contains 70 percent copper and 30 percent zinc. It has the best strength and ductility of all the brasses and is easily worked cold. It is widely used where copper products require extensive fabrication, such as stamping or deep drawing. It finds use in the manufacture of electric sockets, reflectors, rivets, and heating units.

brittle. They are used for cast products and cannot be hammered or worked without breaking.

Alpha brasses contain 63 to 95 percent copper and are the easiest type to work. They are used to make radiator parts, springs, and grilles. See the product in Fig. 16.20.

Alpha-beta brasses contain 55 to 63 percent copper. They are stronger than alpha brasses and can be worked hot. They are used where strength is important, such as for rivets and screws.

Table 16.16 Composition of Selected Wrought Brasses and Brass Alloys[a]

UNS Designation	Descriptive Name	Major Alloying Elements in Percent[b]					Other Named Elements
		Cu	Zn	Pb	Fe	Sn	
Copper-Zinc Alloys (Brasses)							
C22000	Commercial bronze	89.0–91.0	REM[c]	0.05 max.	0.05 max.		
C23000	Red brass	84.0–86.0	REM	0.05 max.	0.05 max.		
C26000	Cartridge brass	68.5–71.5	REM	0.07 max.	0.05 max.		
C28000	Muntz metal	59.0–63.0	REM	0.03 max.	0.07 max.		
Copper-Zinc-Lead Alloys (Leaded Brasses)							
C31400	Leaded commercial bronze	87.5–90.5	REM	1.3–2.5	0.10 max.		0.7 Ni
C35000	Medium-leaded brass	60.0–63.0	REM	0.8–2.0	0.15 max.		
C37700	Forging brass	58.0–61.0	REM	1.5–2.5	0.30 max.		
C38500	Architectural bronze	55.0–59.0	REM	2.5–3.5	0.35 max.		
Copper-Zinc-Tin Alloys (Tin Brasses)							
C44300	Admiralty, arsenical	70.0–73.0	REM	0.07 max.	0.06 max.	0.8–1.2	0.02–0.06 As
C46400	Naval brass, uninhibited	59.0–62.0	REM	0.20 max.	0.10 max.	0.5–1.0	
C48500	Naval brass, high lead	59.0–62.0	REM	1.3–2.2	0.10 max.	0.5–1.0	

[a]These are only a few of the many types of brass and brass alloys available.
[b]Cu copper, Zn zinc, Pb lead, Fe iron, Sn tin, Ni nickel, As arsenic
[c]Remainder for the difference between elements specified and 100 percent.
Reproduced with permission from *Standards Handbook,* parts 5 and 6, Copper Development Association.

Figure 16.20 This product was cast in yellow brass, which falls in the alpha brass class. *(Courtesy OMC Industries, Inc.)*

Figure 16.21 Valves are commonly cast from brass.

Muntz metal contains 60 percent copper and 40 percent zinc, which you will notice is the greatest percent of zinc alloyed with the copper. Muntz metal has low ductility but high strength and is used for things such as sheet stock and exposed architectural features.

Leaded Brasses (Copper-Zinc-Lead Alloys)

Lead is added to brass to make it easier to machine. Those types used in some products used in construction include architectural bronze, forging brass, and medium-leaded brass. The alloying elements for each can be found in Table 16.16.

Architectural bronze has the least copper and most lead of the three mentioned above. It contains 55 to 59 percent copper, 41 to 45 percent zinc, and 2.5 to 3.5 percent lead. It is widely used for forgings and products produced by machining. On a building it can be found in decorative grilles, handrails, architectural trim, door parts, and other such uses.

Forging brass contains about 60 percent copper, 38 percent zinc, and 2 percent lead. It has great plasticity when hot and is therefore widely used for forgings. Because it has good corrosion resistance, it is used for plumbing and hardware items.

Medium-leaded brass contains about 62 percent copper, 34 percent zinc, and 2 percent lead. It is used when good machining properties are required, such as for keys, parts of locks, plaques, and various scientific instruments.

Tin Brasses (Copper-Zinc-Tin Alloys)

When tin is alloyed with copper, zinc, and other elements the alloy has additional properties not present in plain brasses. The two types of tin brasses are admiralty and naval brasses.

Admiralty brass contains about 71 percent copper, 28 percent zinc, 1 percent tin, and traces of lead, iron, and arsenic. These alloying elements improve strength and ductility and, most important, increase resistance to corrosion. Admiralty brass is widely used in the manufacture of condenser and heat-exchanger plates, various tubes, and in equipment in chemical and electrical power plants as well as in products that must have resistance to seawater.

Naval brasses fall into several categories used for wrought products. These include uninhibited, arsenical, medium-leaded, and high-leaded. The copper content for all ranges from 59 to 62 percent. There is a wide range in the amounts of the lead and tin alloying elements. Naval brasses are used in chemical, steam power plant, and marine equipment. They can withstand the corrosive effects of the materials to which they are typically exposed.

HISTORY

Brass is a mixture of copper and zinc. It most likely was produced by accident since zinc and copper occur together naturally in some ores. The smelting of this ore produced what is now called brass. It appears that the Romans may have been the first to deliberately add zinc to copper to produce brass. They used it to produce coins that may be seen today. In the Middle Ages the production of brass in Europe was a major industry. It was widely used to cast religious objects, candlesticks, locks, and plates to remember the dead. These were engraved and contained words and various figures. Companies started producing brass utensils, lamps, and other household items. The cannons of the nineteenth century were massive brass castings that were eventually replaced by steel. Brass is a widely used material today and is used for the very best quality items.

BRONZE

Technically, bronze has been used to identify a product 90 percent copper and 10 percent tin. However, other alloying elements are now added to produce a wider range of materials called bronze that have variations in the properties. Bronze now refers to alloys of copper having alloying elements of silicon, aluminum, manganese, and other elements. They may or may not have zinc. Some of the products in Table 16.16 are referred to as bronze but are actually brass alloys. These include commercial bronze, architectural bronze, and Muntz metal.

The composition of a few selected wrought bronzes and bronze alloys are shown in Table 16.17. The phosphor bronzes contain about 89 percent copper. C51800 has 4 to 6 percent tin and 0.10 to 0.35 percent phosphor. C53400 has more lead and is referred to as leaded phosphor bronze. Aluminum bronzes have 6 to 7.5 percent aluminum, while the low silicon bronze has no tin or lead but 0.8 to 2.0 percent silicon. Cast bronzes have similar elements. In Table 16.14 you can find the Unified Numbering System designations for bronzes.

Bronze is used for various cast products and hardware. Screws, washers, nuts and bolts, and weather stripping are other uses. See the art piece in Fig. 16.22, which was cast from a statuary bronze that has strong and excellent casting qualities. The medical symbol in Fig. 16.23 and the items in Figs. 16.24 and 16.25 were all cast in bronze. They are attractive and very durable.

Table 16.17 Composition of Selected Bronzes and Bronze Wrought-Type Alloys[a]

UNS Designation	Pevious Trade Name	Major Alloying Elements in Percent[b]										Other Named Elements
		Cu	Pb	Fe	Sn	Zn	P	Al	Mn	Si	Ni	
Copper-Tin-Phosphorus Alloys (Phosphor Bronzes)												
C51800	Phosphor bronze	REM[c]	0.02 max.	—	4.0–6.0	—	0.10–0.35	0.01				
Copper-Tin-Lead-Phosphorus Alloys (Leaded Phosphor Bronzes)												
C53400	Phosphor bronze B-1	REM	0.8–1.2	0.10 max.	5.0–5.8	0.30 max.	0.03–0.35					
Copper-Aluminum Alloys (Aluminum Bronzes)												
C61300	Aluminum bronze	REM	0.01 max.	2.0–3.0	0.20–0.50	0.10 max.	—	6.0–7.5	2.0 max.	0.10 max.	0.15 max.	0.15 P max.
C64200	———	REM	0.05 max.	0.30 max.	0.20 max.	0.50 max.	—	6.3–7.6	0.10 max.	1.0–2.2	0.25 max.	0.15 As max.
Copper-Silicon Alloys (Silicon Bronzes)												
C65100	Low silicon bronze B	REM	0.05 max.	0.8 max.	—	1.5 max.	—	—	0.7 max.	0.8–2.0		

[a]These are only a few of the many alloys available.
[b]Cu copper, Pb lead, Fe iron, Sn tin, Zn zinc, P phosphorus, Al aluminum, Mn manganese, Si silicon, Ni nickel, As arsenic
[c]Remainder for differences between elements specified and 100 percent
Reproduced with permission from *Standards Handbook,* parts 5 and 6, Copper Development Association.

Figure 16.22 This piece of art was cast using statuary bronze. It is a strong alloy and has excellent working qualities.

Figure 16.23 This bronze casting shows great detail, indicating that the metal has excellent casting qualities. *(Courtesy Metal Arts)*

Figure 16.24 These letters were cast in bronze with the finish protected by a baked-on polyurethane coating. *(Courtesy OMC Industries, Inc.)*

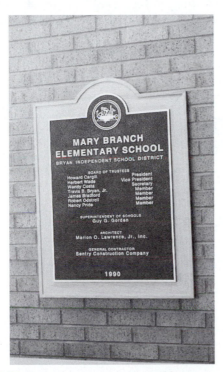

Figure 16.25 A cast bronze plaque provides an attractive, permanent means for giving recognition. *(Courtesy OMC Industries, Inc.)*

HISTORY

Bronze was used by early humankind after copper found widespread use. Bronze is an alloy of copper and tin. Since tin deposits were often mixed with copper ore deposits, it is believed that the early discovery or use of bronze was accidental. People probably noted that some ores (those with some tin ore) were harder and more useful for making tools and weapons. Sometime through the years it was discovered that mixing tin with copper produced this improved product.

LEAD

Lead (chemical symbol Pb) is a soft, heavy metal that is easily worked, with good corrosion resistance. A special feature is its ability to resist penetration from radiation.

Properties of Lead

Other advantageous properties are its high density and weight, softness and malleability, low melting point, 620°F (327°C), and good electrical conductivity. Lead is low in strength and elasticity.

HISTORY

Archaeological findings indicate that lead may have been used as early as 5000 to 6000 B.C. Through the centuries it found use for glazing pottery and as a solder for joining copper. It was used in China for money and as an alloying element in early bronzes. The Romans made wide use of lead pipes, which we now know can cause lead poisoning in those drinking the water. By 800 B.C. lead was used in the manufacture of glass, and by 1100 A.D. it was used as roofing and as cames used to assemble the glass pieces in stained-glass windows. Early American settlers made lead bullets.

Production of Lead

The major source of lead is the mineral galena or lead sulfide. Other sources are cerussite (lead carbonate) and anglesite (lead sulfate). Lead ores frequently contain zinc, and some have gold, silver, and other metals.

The lead-bearing ore is first crushed and ground into fine particles. The metal-bearing material is separated from the rock particles by flotation. Flotation involves mixing water, oils, and chemicals with the ground ore. As air is blown into the mix from the bottom, the lead-bearing particles are wetted by the oil and float to the top in a froth of air bubbles. The lead particles are drawn off and the waste material (gangue) settles to the bottom and is removed.

The concentrated ore is roasted in the air, which changes the lead sulfide to lead oxide. Sulfur escapes as the gas sulfur dioxide and is recovered and made into sulfuric acid. The lead oxide is then smelted in a blast furnace in which the lead settles to the bottom. Any gold or silver settles with it. Waste materials float to the top forming a slag and are removed. The lead is then processed to remove the gold and silver.

Grades of Lead

Of the several grades of lead, chemical lead, desilverized lead, and corroding lead are used in some way in construction. *Chemical lead* and *desilverized lead* are used for pipes, sheets, and alloys. *Corroding lead* is used for white lead, red lead, and litharge (used in the manufacture of batteries, pottery, lead glass, and ink).

Lead Alloys

Lead is alloyed with antimony to improve hardness and strength. However, many other elements are also added. Among these are arsenic, nickel, zinc, copper, iron, and manganese.

Unified Numbering System Designations

Lead and lead alloys are designated by the UNS numbers L50001 through L59999. For example, L50121 is described as a solder alloy containing 98.0 percent lead, and L50770 is a battery grid alloy, lead-calcium, containing 99.6 percent lead.

Uses of Lead

Lead pipes and tank liners are used in installations processing highly corrosive materials, but they are never used for piping to carry drinking water. Since lead is a good self-lubricant, it is used when high-pressure lubricating is necessary. Lead solder is used for electrical connections because it is a good conductor, but it is not used on water pipe connections. Hard solders have antimony added. High-temperature solders are alloyed with silver and are generally referred to as silver solder.

The use of lead in solder for joining copper water pipes has been banned because of the possibility of increasing the lead present in the water, which will cause lead poisoning. Lead pipes are not used to move drinking water for the same reason. This danger is especially present when soft or distilled water is used. However, lead pipe and lead-lined tanks have high corrosion resistance and can find use in industrial production applications, such as in the chemical manufacturing industry. Sheet lead is used for roofing, flashing, and spandrels in areas where there is severe industrial air contamination or in areas along the seacoast.

Another interesting use for lead is in the form of lead azide. Lead azide is easily exploded by an electrically heated wire, so it is used in the manufacture of blasting caps, which are used to set off other explosives.

Lead is used in products such as adhesives, caulking, pigments, glazing, pipes and tanks, compounds, and protective coatings over steel and copper. It is an element added to produce products such as brass, bronze, certain asphalt products, glazes, various fusible alloys, glass, porcelain enamel, iron, steel, certain plastics, sol-

der, rust-resistant prime coatings before painting, tin, and as an additive to certain wood preservative preparations. As mentioned in Chapter 32, the use of lead in paint has been completely stopped because of the danger of children eating pieces of peeling paint.

Architects specify lead for waterproofing, soundproofing, reduction of vibration, and radiation shielding. Following are more detailed examples of two lead products, which will give you a closer look at its elements and properties.

Lead strip and sheet products are made from lead that is almost 100 percent pure and from an alloy with about 7 percent antimony, which improves the strength and stiffness. Seams can be folded, soldered, or folded and soldered, as shown in Fig. 16.26. You can see several other applications in Figs. 16.27 and 16.28. Soldering is done with a 50 percent lead–50 percent tin solder.

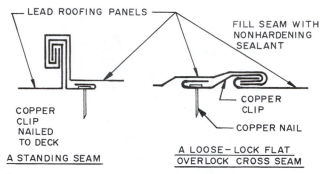

Figure 16.26 Typical seams used when installing lead and terneplate roofing.

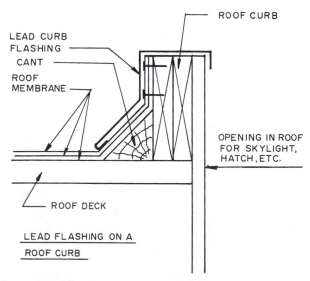

Figure 16.27 Lead is an excellent flashing material, especially in areas where severe air contamination exists or along the seacoast.

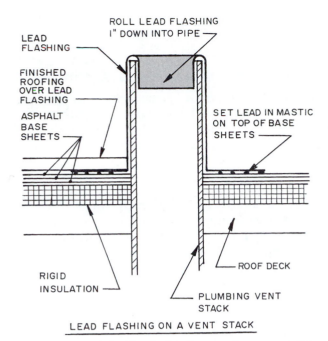

LEAD FLASHING

ROLL LEAD FLASHING 1" DOWN INTO PIPE

FINISHED ROOFING OVER LEAD FLASHING

ASPHALT BASE SHEETS

SET LEAD IN MASTIC ON TOP OF BASE SHEETS

RIGID INSULATION

ROOF DECK

PLUMBING VENT STACK

LEAD FLASHING ON A VENT STACK

Figure 16.28 Lead flashing is ductile enough to permit it to be rolled over the top of vent pipes, sealing the flashing at the top.

Red lead is a lead oxide that is a widely used primer applied generally to steel before it receives the final coats of finish paint. It is available as a paste, dry, or as a liquid paint. Red lead is available dry in grades A, B, and C. Grade C has the highest lead content. The red lead paste in an oil is available in grades B and C. Again grade C has a higher percentage of lead.

SOLDERS

Solders are nonferrous metals used to join metals in a waterproof joint and to make secure electrical connections. Many people do soldering without thinking about the composition of the metal. However, the effectiveness of the completed job depends on using the proper metal composition.

Properties

Solders have low melting temperatures, typically around 375°F (192°C) to 595°F (315°C). However, some high-temperature solders range up to about 740°F (396°C). The low melting point enables the solder to join metals without raising their temperatures to their melting points; therefore, the solder has no alloying action with the metals being joined.

Solders have little shear, tensile, or impact strength, so they are used where there will be no load on the joint or on a joint that has other fasteners, rivets, bolts, or interlocking seams to carry a load or stress.

Types of Solder

Solders are mainly an alloy of lead and tin with small amounts of other elements included. The four major classifications of solder are tin-lead, tin-lead-antimony, silver-lead, and a variety of special alloys.

The *tin-lead solders* are general-purpose alloys used for joining metals. The tin-lead composition varies from an alloy with 70 percent lead and 30 percent tin through a series of about ten combinations with the amount of lead increasing and the amount of tin decreasing. The maximum lead solder has 90 percent lead and 10 percent tin. Tin-lead solders have less than one percent antimony. Typical 50-50 tin-lead solder is most commonly used.

Tin-lead-antimony solders have from 1 to 2 percent antimony, and of the five commonly available types all have more lead than tin. Tin content varies from 20 to 40 percent.

Silver-lead solders contain more than 97 percent lead, from 0 to 1.25 percent tin, traces of antimony, and silver content of 1.5 to 2.5 percent. They have melting points of about 580°F (307°C) and produce fairly strong corrosion-resistant joints. They are used for soldering copper and brass using a torch heating method.

Special-purpose solders are designed for a specific application such as soldering copper roofing or when a high-temperature solder is required. They typically contain various amounts of lead, zinc, silver, and cadmium.

Fluxes

If solder is to bond to a metal surface, the surface must be clean and free of oxides. In addition to mechanically cleaning the surface (washing, buffing, brushing, etc.), any oxides on the surface must be removed. *Fluxes* are materials used to remove these oxides. Three general types of fluxes are available—neutral, corrosive, and noncorrosive.

Neutral fluxes are mild and are used on metals that can be soldered easily, such as copper, lead, brass, and tin plate. They typically are a form of a mild acid that is wiped on the surface. Often they do not need to be wiped off before you begin to solder.

Corrosive fluxes include salt-type and acid-type. They are more effective in cleaning the surface than the neutral fluxes, but it is important that they, along with the oxide residue, be removed. Otherwise their corrosive action will continue to attack the base metal being soldered. Corrosive fluxes (often called acid fluxes) cannot be used on electrical connections.

Noncorrosive fluxes tend to be used only on metals that are easily soldered and on electrical connections. Typically they contain rosin as the main flux. Noncorrosive fluxes are good to use when soldering electrical connections because they will not cause the wires to corrode.

TIN

Tin (chemical symbol Sn) is produced from the ore containing the mineral *cassiterite,* which is a tin oxide. Since there is little cassiterite in North America, tin has to be imported from Malaysia, Brazil, Russia, Indonesia, Thailand, China, and Bolivia. The ore usually contains little tin, so an extensive refining process is required.

HISTORY

The use of tin can be traced back to around 2000 B.C. when it was used as an element in producing bronze in Egypt. The Chinese and the peoples of the Malay Peninsula used tin several hundred years B.C. Some tin was mined in France, Spain, and England around 500 B.C. Around 100 A.D. tin was used as an alloying element to produce a lead-tin mixture used as a solder to join metals. The Romans coated copper with tin, and by about 1500 A.D. tinplate was being used in Europe. Tinplate is iron or steel sheets coated with tin.

Properties of Tin

Tin is a soft metal that is malleable and ductile and blue-white in color. It is corrosion resistant when exposed to air and moisture. It has a low melting point of 450°F (232°C) and can be cast. It will take a high polish and has properties enabling it to coat other metals.

Working Characteristics

Because tin is soft and malleable, it can be worked by rolling, spinning, extrusion, and casting.

Production of Tin

After the ore is mined the impurities must be removed to get a concentrated ore. This involves a series of mechanical and chemical processes for crushing and cleaning the ore. Magnetic separators and screening operations are used from which the ore and the tailings are separated.

The concentrated cassiterite is refined, producing a high-purity metal of at least 99.8 percent tin. Several different processes are used but they all involve smelting (melting) the ore in a furnace in which a crude tin is separated from the slag. The crude tin is then heated to a predetermined temperature at which impurities with higher melting temperatures than tin remain. This process is called liquidation. This partly pure tin is drawn off and heated above its boiling point. As it is stirred, additional impurities rise to the surface and are drawn off. This continues until impurities are removed, leaving an almost pure tin.

Unified Numbering System Designations

Tin alloys are designated in the Unified Numbering System by the numbers L13001 through L13999. For example, UNS L13630 is a lead-tin solder containing 37 percent lead and 63 percent tin.

Uses of Tin

Since tin and tin alloys have high corrosion resistance and excellent coating ability, they are used extensively to provide protective coatings on other metals, especially steel. One example is the coating on cans used to store food for retail sale. Tin is also used as an alloying element in other metals. Its low melting point makes it useful in some solders. It finds applications in mirrors, hardware, and fusible alloys. Tin compounds are used in the production of glazes, glass, and porcelain enamel.

TERNEPLATE

Terneplate is a mixture of lead and tin applied to copper-bearing steel sheet or stainless steel sheet to produce a corrosion-resistant coating. Tin is added because lead alone will not alloy with the iron. Terneplate sheets are available in two types, short terne and long terne.

HISTORY

The development of terneplate can be traced to Wales, a division of the United Kingdom in southwest Great Britain, around 1650 to 1700. It was made from sheets or strips of iron coated with the lead-tin alloy. Eventually steel plate and strip materials were coated and the product progressed to its present configuration.

In the early days of the United States terneplate was used on the roofs of homes of the wealthy, such as The Hermitage, home of Andrew Jackson (1835).

Short terne, used for roofing, is available with stainless steel or copper-bearing steel bases. The stainless steel type uses UNS S30400 stainless steel in 26 and 28 gauge thicknesses. The copper-bearing steel terneplate is made in 26, 28, and 30 gauge thicknesses. Both types use a coating on both sides consisting of 75 to 80 percent lead and 20 to 25 percent tin.

Long terne is used for various industrial purposes and is available in 14 to 30 gauge low-carbon steel with a coating on both sides consisting of 75 to 87½ percent lead and 12½ to 25 percent tin.

Uses of Terneplate

Short terne, often called roofing terne, is used for finished roofing, gutters, downspouts, and flashing. Terneplate roofing is installed with batten and standing seams. Flat locked and horizontal seams can be used when it is installed over wood sheathing. Figure 16.26 has some examples.

Long terne is used for fireproof doors and frames, other fireproofing items, and roofing. The various types are available in sheets and rolls. Terne on steel should be painted after installation. Both sides require painting to seal any pinholes. If pinholes are allowed to exist in terne-coated steel, corrosion will result. Most of the time the mill applies a red iron-oxide primer, but a high-quality finish coat of paint is required over this. Terne-coated stainless steel does not require a primer or painting.

Production of Terne Sheets

The sheets are chemically cleaned, usually by passing them through a dilute solution of sulfuric or hydrochloric acid. Then they are fluxed by passing them through a heated solution of zinc chloride. Finally, the sheets are passed through a molten solution of lead and tin. The terne-coated sheet is then passed between smooth metal rollers to produce the finished surface. After it has cooled it is cleaned to remove any traces of oil and is ready to be primed.

TITANIUM

Titanium (chemical symbol Ti) is found in large quantities in the earth's surface. It is a very light, strong, ductile, silvery metal and is one of the most common elements.

HISTORY

The element titanium was discovered in England about 1790. It was in the form of black sand, and upon analysis it yielded a white metallic oxide. This work was with titanium dioxide, a compound found in nature that produces a white powder when refined. This powder is used today in the paint industry. It was not until about 1910 that titanium in metallic form was produced. At this time it was in the form of a laboratory experiment. It was not until 1946 that a process was patented that enabled titanium to be produced in commercial quantities. By 1950 an industrial type titanium was available. Since then it has been used for an ever-increasing number of applications.

Properties

Titanium has low electrical conductivity and a low coefficient of thermal expansion. It is also paramagnetic, meaning that when it is placed in a magnetic field it possesses magnetization in direct proportion to the field strength. It has a melting point of 3300°F (1820°C) and a coefficient of thermal expansion of 0.0000085/°C.

Two important characteristics of titanium are its high strength-to-weight ratio and its ability to resist corrosion by salt water and the atmosphere. These have an important impact on its use in various products.

Working Characteristics

Pure titanium is easier to use than titanium alloys; however, in general all types can be fabricated using standard manufacturing processes. These include machining, welding, riveting, drilling, punching, hot- and cold-rolling, extruding, forging, and drawing.

Production of Titanium

Titanium is produced by the Kroll process, named after the person who developed it. The process is chemically driven, using magnesium in an inert atmosphere of helium or argon to produce the reduction of the titanium tetrachloride. These elements react releasing magnesium chloride, which is distilled off. Left is a sponge metal that is crushed and melted into ingots. The ingots (similar to pig iron ingots) are used to produce titanium and titanium alloys that are processed into various products.

Titanium Alloys

The strength of titanium varies depending on the purity of the metal. The higher the purity the weaker the metal; therefore, various elements are added to produce titanium alloys with greatly increased strength. Typically vanadium, molybdenum, aluminum, iron, chromium, and manganese are used as alloying elements.

Unified Numbering System Designations

Titanium alloys are designated in the Unified Numbering System by the numbers UNS R50001 through R59999. For example, R56210 is a titanium alloy containing 90.2 percent titanium.

Uses of Titanium

The major use for titanium is in aircraft and aerospace industries and other military applications. Its strength and light weight make it a desirable material for these applications. It can also be formed in sheet, strip, pipe, and tube products, and it can be forged and cast. Future uses in construction could be as doors, curtain walls, and other exterior applications such as flashing and guttering. Since titanium has a high melting point it will also find use as a structural material.

NICKEL

Nickel (chemical symbol Ni) is a silver-colored metal mainly used as an alloying element. It provides increased resistance to atmospheric and chemical corrosion and increases the strength of the alloy.

HISTORY

Nickel in various alloyed forms was used by prehistoric man (even though its presence was not known) because it was in the meteoric iron used to produce tools. Other metallic items from China and Asia Minor (a peninsula between the Black and Mediterranean Seas) have been found to contain nickel as well as copper and other alloying elements. In the mid-1700s a crude nickel was first isolated from an ore in England and by the early 1800s refined nickel was being produced. Canada and New Caledonia (an island in the South Pacific) are the largest producers of nickel-bearing ore.

Properties

Nickel is resistant to strong alkalis and many acids. It has good resistance to corrosion and oxidation and is strong and tough. It has a melting point of 2651°F (1455°C) and a coefficient of thermal expansion of 0.000013/°C. It is magnetic up to 680°F (360°C).

Working Characteristics

Nickel can be fabricated using most of the commonly used processes, such as hot- and cold-rolling, extruding, bending, forging, and spinning. It can be joined by some welding processes, soldered, brazed, or joined with mechanical fasteners.

Production of Nickel

Several processes are used to produce nickel from the ore. The most recent is the Hybinette process developed in Canada. The ore is processed using the Bessemer process from which the molten metal goes to a cooling chamber. From here it is crushed and ground and mag-netically separated. The result is a nickel-copper platinum alloy that is treated by the electrolysis process to separate the nickel, copper, and platinum.

Nickel Alloys

A major use for nickel is as an alloying element. The alloying of nickel to other metals provides increased ductility, corrosion resistance, strength, hardness, and toughness. Nickel alloyed to nonferrous metals improves electrical resistance and magnetism and helps control expansion. An examination of steel alloys in Chapter 15 will show you that nickel is a commonly used alloying element. It is also widely used in Monel® metals and aluminum alloys. Monel alloy is about 66 percent nickel and 34 percent copper. Inconel 600® is a special nickel alloy containing about 75 percent nickel, 15 percent chromium, and 7 percent iron. Since it has excellent oxidation resistance, it is used in food-processing and chemical industries. Other nickel alloys include those used for electrical resistance coils, magnetic and nonmagnetic alloys containing iron, alloys designed to have a high coefficient of thermal expansion that are used in the production of glass, and copper-nickel alloys used for products exposed to marine conditions.

Unified Numbering System Designations

Nickel and nickel alloys are designated in the Unified Numbering System by the numbers N02001 through N99999. For example, UNS N02250 is a commercially pure nickel alloy having 99.0 percent minimum nickel.

Uses of Nickel

In addition to its use as an alloying element in ferrous and nonferrous metals, nickel is also an excellent material to use for electroplating and electroless plating. Electroplating is the process of depositing a coating of metal on another metal by electrolysis. Electrolysis is a method of plating a material by chemical means in which the piece to be plated is immersed in a reducing agent that, when catalyzed by certain materials, changes metal ions to metal, forming a deposit on the surface of the piece. Nickel is used in electric heating elements, lamp filaments, and plumbing fittings.

ZINC

Zinc (chemical symbol Zn) is a bluish white metal that is brittle and has low strength. It is often referred to as a white metal and is widely used as a protective coating over steel.

An accurate knowledge of zinc as a basic metal has existed for about 200 years. This is because zinc as found in ores in nature is difficult to convert to the metal.

It found use as an element in brass by the Romans and other Middle and Far Eastern civilizations about 2000 years ago.

The process for isolating zinc as a metal occurred in 1721 by a German, Johann Henckel. It is believed that some form of zinc was produced in China about 500 years ago. These discoveries led to the smelting of zinc in England about 1740. The first successful smelting operation was begun in Belgium about 1807. Zinc was first produced in the United States in 1835 at the Washington, D.C., arsenal. Commercial production was underway by about 1860 in LaSalle, Illinois, and Bethelehem, Pennsylvania.

Properties

Zinc has low strength and is brittle. Although it can be damaged by alkalis and acids, it resists corrosion by water and forms a protective oxide when exposed to air. Zinc has a melting point of 787°F (419°C). It is also subject to creep. The tensile strength can be greatly increased by cold-working and alloying.

Working Characteristics

Since zinc is a soft material, it can be hot- and cold-rolled, drawn, extruded, cast, and machined. It can be joined by welding, soldering, and various mechanical fasteners.

Production of Zinc

Zinc is extracted from zinc blende (sphalerite) ore. The mined ore is crushed and ground, and the ore particles are separated from the rock. Elements such as copper, lead, and iron sulfides tend to be in the ore and are separated by a flotation process. This involves separating the elements in the finely grained ore by floating them on a liquid. The floating capacity of the various elements varies. The lead and copper sulfates will float off the top of the liquid. Chemicals are added to cause the zinc sulfide to float and be taken off. This concentrated zinc material is dried and ready to be refined into the metal zinc.

The *electrolytic process* involves roasting the zinc concentrate and removing soluable parts with a weak sulfuric acid solution. The solution is filtered to remove some of the other metals. Finally, this solution is moved to electrolytic tanks where cathodes of pure aluminum and anodes of lead or lead-silver are lowered into the tank and an electric current is passed between them through the solution. Pure zinc is attracted to and plates the cathodes. The layer is removed, melted, and poured in slabs for processing into various applications.

The *vertical furnace* method involves mixing coking coal briquettes and the dried zinc concentrate into the top of a vertical furnace where they are heated until the zinc concentrate vaporizes. The vapor is removed and condensed at the temperature at which the vapor turns into a solid. The solid zinc slabs are ready for use in producing various products.

Zinc Alloys

Zinc alloys used for die casting consist of about 95 percent zinc and 4 percent aluminum and magnesium. Some copper may be present.

Unified Numbering System Designations

Zinc and zinc alloys are designated in the Unified Numbering System by the numbers Z00001 through Z99999. For example, Z13001 is identified by the name zinc metal and contains 99.90 percent zinc minimum.

Uses of Zinc

The major use for zinc is to form a protective coating over steel. This is referred to as galvanizing. Galvanizing involves placing the steel to be coated into a bath of molten zinc, which bonds to the surface. It is important that the coating be free of imperfections such as pinholes that would permit moisture to reach the steel and cause it to rust. Both galvanized sheet and strip material are available.

Since zinc and zinc alloys have low melting temperatures, they are easy to cast and are used for some types of hardware and plumbing items. They are usually die cast and finished by polishing or plating with chromium, brass, or other materials.

Zinc also finds use as an alloying element in brasses. Various zinc compounds find use in the production of paper, plastics, ceramics, rubber, abrasives, paint, and other products. Zinc is also used for specialized products in which corrosion resistance is important, such as anchors, flashing, screws, nails, expansion joints, and corner beads. Solid zinc strip material, as shown in Fig. 16.29, is used to produce a wide range of products, such as low-voltage buss bars, cavity wall ties, electric cable binders, electric motor covers, grading screens, and roofing and fascia material.

Zinc is high on the galvanic table of electrolytic potentials. This means it can be used to coat a material

Figure 16.29 Solid zinc strip material is used to produce a wide range of products for construction and other industries. *(Courtesy Alltrista Zinc Products Co.)*

lower on the table to protect the material if galvanic action does occur. The zinc will be sacrificed, thus protecting the coated metal.

Zinc Galvanizing Processes

Zinc sheet, strip, coils, and wire are galvanized in a continuous process. The processes in use include electrogalvanizing, hot-dip galvanizing, metallic spraying, and **sherardizing.**

Electrogalvanizing involves placing cleaned steel or iron in an electrolyte solution of zinc sulfate. The electrolytic action deposits a layer of zinc on the material. The thickness of the coating can be controlled, but the thickness is limited. Typical thicknesses range from 0.0001 to 0.0005 in. (0.0025 to 0.0127 mm).

Hot-dip galvanizing involves immersing clean steel in a bath of molten zinc. It is a semiautomatic process.

Metallic spraying involves coating the sheet iron or steel by applying a fine spray of molten zinc. It can be applied after an installation is complete, thus coating the bolts, rivets, and welds.

Sherardizing is a process in which the cleaned iron or steel is placed in a container filled with zinc dust. The temperature in the container is raised and the objects to be coated are tumbled in the dust. The heated zinc bonds to the metal, forming a thin coating. This is usually limited to coating small parts and provides minimum protection.

REVIEW QUESTIONS

1. How does the specific gravity of aluminum compare with that of steel?

2. How does aluminum compare with copper as a conductor of electricity?
3. How is alumina produced from the processed aluminum hydrate?
4. What is the purity of newly produced aluminum?
5. What do the aluminum temper designations F, O, and H mean?
6. What does solution heat-treating mean?
7. When galvanic corrosion occurs, what forms the electrolyte?
8. How can galvanic corrosion be reduced or eliminated?
9. What types of finishes are used on aluminum products?
10. What is a natural, controlled finish?
11. What is an anodized finish?
12. What are the types of anodic coatings used on industrial and architectural applications?
13. How are aluminum surfaces painted in a factory?
14. What is a porcelain enamel coating?
15. What is laminated finish?
16. How can aluminum members be joined?
17. What are the major properties of wrought copper alloys?
18. What are the two most widely used copper alloys?
19. What is copper patina?
20. What type of ore contains lead sulfide?
21. What are the major properties of lead?
22. What are the major properties of tin?
23. What are the major properties of zinc?

KEY TERMS

alumina A hydrated form of aluminum oxide from which aluminum is made.

anodize To apply an electrolytic oxide coating to an aluminum alloy.

bauxite Ores containing high percentages of aluminum oxide.

cast products Products formed by pouring molten aluminum into a mold.

galvanic corrosion Corrosion that develops by galvanic action when two dissimilar metals are in contact in the atmosphere.

galvanize To coat steel or iron with zinc.

heat-treatable alloys Aluminum alloys whose strength characteristics can be improved by heat treating.

heat treating Heating and cooling a solid metal to produce changes in its physical and mechanical properties.

non-heat-treatable alloys Alloys that do not gain strength if heat treated but do gain strength by the addition of alloying elements.

nonferrous metals Metals containing little or no iron.

oxide layer A very thin protective layer formed naturally on aluminum due to its reaction with oxygen.

reduction The electrolytic process used to separate molten aluminum from alumina.

sherardize To coat steel with a thin cladding of zinc by heating it in a mixture of sand and powdered zinc.

strain hardening Increasing the strength of a metal by cold-rolling.

temper designation A specification of the temper or metallurgical condition of an aluminum alloy.

wrought products Products formed by any of the standard manufacturing processes—drawing, rolling, forging, or extruding.

SUGGESTED ACTIVITIES

1. Collect samples of each of the commonly used nonferrous metals. Identify and list the properties of each.

2. How many uses of nonferrous metals can you find around your campus? Make a list of the location and type of material.

3. Test samples of a variety of nonferrous metals for tensile strength and hardness. Compare your findings with those of others who made the same tests.

4. Cite the location and name of the product made from nonferrous metals that were extruded, drawn, rolled sheet, and cast.

ADDITIONAL INFORMATION

Merritt, F.S., and Rickett, J.T., *Building Design and Construction Handbook*, McGraw-Hill, New York, 1994.

Zahner, L.W., *Architectural Metals: A Guide to Selection, Specification, and Performance*, John Wiley and Sons, New York, 1994.

Other resources are

Numerous related publications from the organizations listed in the "Professional and Technical Organizations" section in Appendix B.

Steel Frame Construction

This chapter will help you to:

1. Become acquainted with the framing drawings prepared for structural steel buildings.

2. Describe the procedure for erecting the structural steel frame of a building.

3. Be familiar with the fastening techniques used to join structural steel members.

4. Describe the fire protection procedures for structural steel members required by building codes.

5. Discuss steel framing systems using manufactured components.

Steel framed buildings are a type of skeleton-frame construction in which the walls, floors, and roof are supported by a structural framework of steel beams, columns, girders, and related structural elements (Fig. 17.1). Some designs rely in part on the skeleton frame for support and utilize other methods, such as wall-bearing construction, where the walls (masonry, panelized, etc.) form part of the structural system. The interior walls in a skeleton-framed building are non–load bearing, which permits great freedom for using the interior space. The members of the frame are manufactured to design sizes, transported to the construction site, and erected.

The design of the structural frame requires careful engineering analysis. Design loads include live and dead loads as well as loads due to forces such as earthquake, wind, rain, and snow. Soil and hydrostatic pressures act horizontally below grade and must be considered. The stresses to which steel is subjected in use must be considered as the design process continues. Provisions also must be made for temporary stresses that occur during construction, such as supporting a crane. The allowable working stresses are regulated by building codes. Detailed information is available from the American Institute of Steel Construction (AISC). Design information for light-gauge steel structural members can be obtained from the Steel Joist Institute and the American Iron and Steel Institute.

Another design consideration is the economic use of structural steel members. The spacing of columns influences the span of beams and girders. Spans that are too short or too long produce uneconomical use of framing members. Another engineering consideration is the design of connections and the method of securing these. Typically rivets, bolts, or welding is used.

The procedure for the design and construction of structural steel framed buildings includes the engineering design of the structure, the preparation of shop drawings, the manufacture of the structural members, and erection of the structural members on the site.

Figure 17.1 This steel framed building has a cellular steel floor through which electrical wiring can be run. Notice the sprinkler system and suspended ceilings. (*Reproduced with permission from* The Building Systems Integration Handbook. *Richard Rush, ed., Butterworth-Heinman Publishers, Newton, Mass., 1986*)

STRUCTURAL STEEL DRAWINGS

Several types of drawings are developed, including engineering design drawings, shop drawings, and erection plans. The **design drawings** are prepared by the structural engineer. They indicate the type of construction and give data on shears, loads, moments, and axial forces that must be resisted by each member and all of the connections. A partial drawing is shown in Fig. 17.2. This drawing shows the elevation of the beam as (75′-0″) above an established on-site datum. The sizes of the beams are given, and the forces are indicated in k (kips). Notes are used to provide additional information. From the information given, the structural detailer can prepare shop drawings for the various members.

Shop drawings have all the information needed for the fabrication of the member. It includes the size of the member and the exact location of holes for connections. Provision must be made to allow for clearances so erection can proceed without interference between joining members. You can see a typical example in Fig. 17.3. Notice the use of notes and the mark (B1) identifying the beam.

Erection plans are assembly drawings that are very much like the design drawings. They are assembly-type drawings used on the site to place each member in its desired location. One example is in Fig. 17.4. They show each piece or any subassembly of pieces and their assigned shipping or assembly mark. The erection plan also includes details showing the anchor bolt locations in the foundation. Anchor bolts are used to secure columns to the foundation and are placed in the concrete as the foundation is poured, so their location is critical. The engineering design drawings are basically

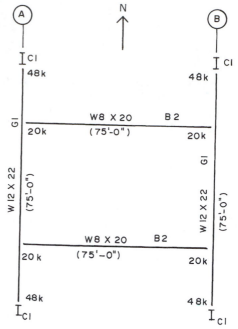

A PARTIAL DESIGN DRAWING

GENERAL NOTES
SPECIFICATIONS: LATEST AISC EDITION
MATERIAL: ASTM A36
FASTENERS: 3/4 Ø A325 IN BEARING TYPE CONNECTIONS
THREADS IN SHEAR PLANES
CONNECTIONS: PER AISC MANUAL AND DEVELOP
INDICATED END REACTIONS

Figure 17.2 A partial engineering design drawing contains the data needed by the structural detailer.

the same as the erection drawings but have more design detail and in some cases can be used as erection drawings. Typically they are small and crowded with details, so erection drawings are usually provided at a larger scale and with only the information needed to identify and locate each member.

Each steel member is identified by an *erection mark*, usually a letter and number giving the part a unique identification. For example, B2 means beam number 2, C2 means column number 2, and G2 means girder number two. All members that are identical will use the same mark. The mark can also include a number indicating the floor where it is to be used. For example, B2(3) means beam B2 is to be erected on the third floor. A column designation as C2(2-4) means that column 2 is in the second tier of the building including the second to fourth floors.

The structural steel members are delivered to the job site with a prime coat and numbered as shown on the erection plan. Connectors are also primed and numbered as required. If a steel member is to be encased in concrete it usually is not primed. This helps the concrete

to bond to the member. After the steel is erected, some field painting is required, such as in areas where welding occurred. The welds are chipped free of any slag formed on the surface before they are painted.

THE ERECTION PROCESS

The finished members are delivered to the job site in the order in which they are needed as erection occurs. The erecting company is responsible for setting in place each member of the frame and securing it as required by the design engineer. As the steel arrives on the site, the members must be placed as near their point of erection as possible and in the order they will be needed.

The erection crew lifts the first level of columns with a ground-level crane and places them over anchor bolts cast in the foundation. Columns in multistory buildings are generally two stories high (Fig. 17.5). The columns have steel baseplates that distribute the load on the column over a larger area of the foundation. The baseplate is usually welded to the column when it is manufactured. There are several methods for setting columns. Smaller columns have a steel plate leveled over a bed of grout before the column is erected (Fig. 17.6). Larger columns have leveling nuts on the baseplate. These are adjusted until the column is plumb. Then grout is worked below the baseplate (Fig. 17.7). A column with stiffener plates is shown in Fig. 17.8. Very large baseplates required for large, heavy columns may have the baseplate leveled and grouted before the column is welded to it. Large-diameter holes may be specified near the center of the plate so grout can be forced below the center of the plate (Fig. 17.9). For additional examples consult the publications of the American Institute of Steel Construction.

Once the columns are in place the crane lifts the girders and beams and they are secured to the columns with the required connections (Fig. 17.10). Now the next level of columns is erected. The second level of two-story columns is lifted with a crane and connected to the columns below with some type of splice plate (Fig. 17.11). Welded splices are also used. A crane sets the second-level columns, beams, and girders (Fig. 17.12). As the floors are framed, the specified decking is installed (Fig. 17.13). In all cases safety regulations are observed, including safety nets, railings, and protection of openings.

As the height of the building gets beyond the reach of the ground-based crane a tower crane can be used to reach the upper levels. This might be installed outside the building frame as shown in Fig. 17.14 or be located inside the structure in an elevator shaft opening or a special temporary opening planned for this purpose. This type of crane will have some means of increasing its height as shown in Fig. 17.15.

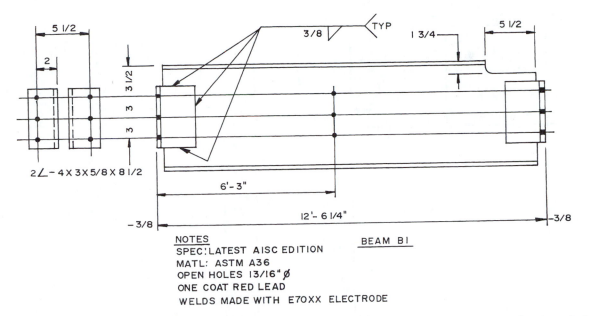

BEAM BI

NOTES
SPEC: LATEST AISC EDITION
MATL: ASTM A36
OPEN HOLES 13/16" ∅
ONE COAT RED LEAD
WELDS MADE WITH E70XX ELECTRODE

Figure 17.3 A typical shop drawing of a beam that has shop welded connectors that are prepared for bolting on the site. *(Courtesy Bethlehem Steel Corporation)*

Notice the identification marks on this column.

Another lifting device for multistory construction is a guy derrick. It consists of a mast, a boom that pivots at the base of the mast, hoisting tackle, and supporting guy lines. The guy derrick can lift itself to floors constructed above it. The hoisting operation is controlled by a power winch located on grade. As the derrick is raised to higher levels the winch remains on the ground. A typical construction procedure is shown in Fig. 17.16. Since the derrick is mounted on the structural frame, the loading must be considered as the design is developed. When the frame is completed, the derrick is disassembled and lowered to the ground with smaller winches. Additional information can be found in books on rigging.

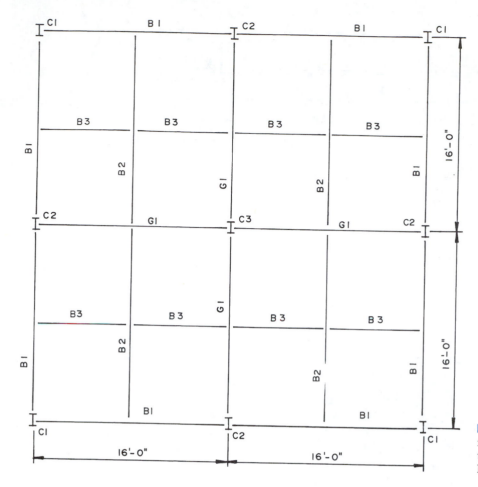

Figure 17.4 An erection drawing of the roof framing that identifies each structural member by a mark and shows their locations.

Figure 17.5 Columns on multistory buildings are usually two stories high. (Courtesy Vulcraft)

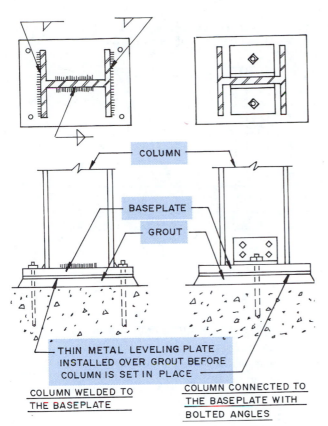

COLUMN

BASEPLATE

GROUT

THIN METAL LEVELING PLATE
INSTALLED OVER GROUT BEFORE
COLUMN IS SET IN PLACE

COLUMN WELDED TO
THE BASEPLATE

COLUMN CONNECTED TO
THE BASEPLATE WITH
BOLTED ANGLES

Figure 17.6 Typical base connections for small columns.

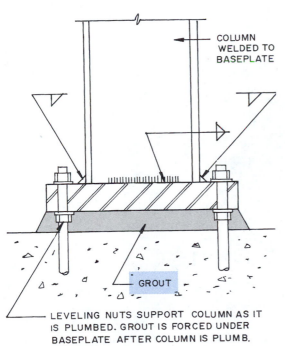

COLUMN
WELDED TO
BASEPLATE

GROUT

LEVELING NUTS SUPPORT COLUMN AS IT
IS PLUMBED. GROUT IS FORCED UNDER
BASEPLATE AFTER COLUMN IS PLUMB.

Figure 17.7 Some columns are plumbed with leveling nuts before grouting.

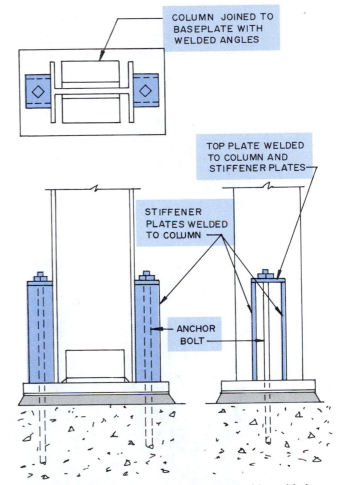

COLUMN JOINED TO
BASEPLATE WITH
WELDED ANGLES

TOP PLATE WELDED
TO COLUMN AND
STIFFENER PLATES

STIFFENER
PLATES WELDED
TO COLUMN

ANCHOR
BOLT

Figure 17.8 This is one type of stiffener used in welded construction.

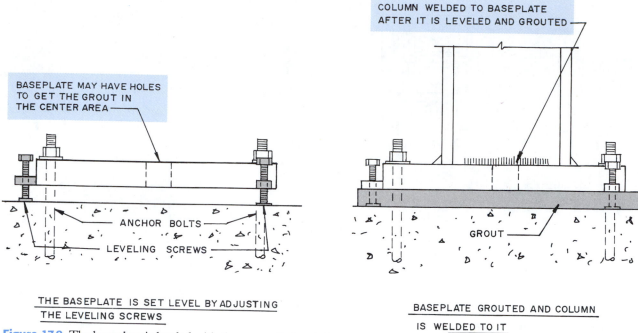

BASEPLATE MAY HAVE HOLES TO GET THE GROUT IN THE CENTER AREA

COLUMN WELDED TO BASEPLATE AFTER IT IS LEVELED AND GROUTED

ANCHOR BOLTS

LEVELING SCREWS

GROUT

THE BASEPLATE IS SET LEVEL BY ADJUSTING THE LEVELING SCREWS

BASEPLATE GROUTED AND COLUMN IS WELDED TO IT

Figure 17.9 The baseplate is leveled with the screws before it is grouted or has the column welded to it.

Figure 17.10 A steel beam is being welded to a steel column. (*Courtesy Bethlehem Steel Corporation*)

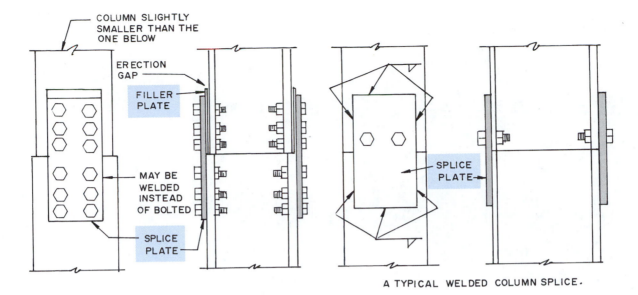

COLUMN SLIGHTLY SMALLER THAN THE ONE BELOW

ERECTION GAP

FILLER PLATE

MAY BE WELDED INSTEAD OF BOLTED

SPLICE PLATE

SPLICE PLATE

A TYPICAL WELDED COLUMN SPLICE.

SPLICE PLATE

MAY BE WELDED INSTEAD OF BOLTED

SAME SIZE COLUMNS MAY BE SPLICED BY BOLTING OR WELDING THE SPLICE PLATE.

SPLICE PLATE

FILLER PLATES

BUTT PLATE

A BUTT PLATE IS USED WHEN SPLICING COLUMNS OF DIFFERENT WIDTHS WHEN THE FLANGES WOULD HAVE NO SUPPORT BY THE COLUMN BELOW.

Figure 17.11 Typical W-column splices.

Figure 17.12 These ground-based cranes set the first columns and the connecting beams and purlins. *(Courtesy Bethlehem Steel Corporation)*

Figure 17.13 Steel decking is welded to the beams and purlins, forming the base for the finished floor or roof. *(Courtesy Vulcraft)*

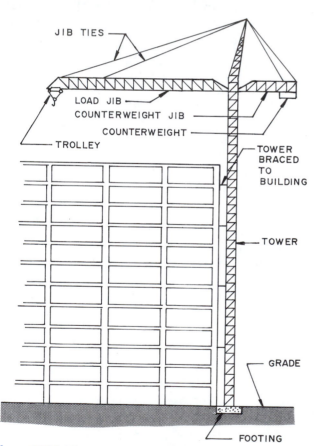

Figure 17.14 Tower cranes are used to reach the upper levels of multistory buildings.

FASTENING TECHNIQUES

The steel members in a building frame can be joined with rivets or bolts or by welding. In some cases a combination of these is used.

Rivets used to connect structural steel members must have properties that enable them to handle the shear, bending, and other stresses developed by the frame. Rivets are made from high-strength carbon and alloy steels according to ASTM dimensional and chemical specifications.

Rivets are installed in holes drilled or punched in the framing members. They are heated to a white heat, inserted into the holes of the members to be joined, and a head is formed by a pneumatic hammer that produces a head on the other side (Fig. 17.17). When the rivet cools it shrinks, shortens in length, and pulls the members together. The compression and friction developed between the members being joined assist in resisting shear and tensile stresses in the joint.

Bolts are used more commonly than rivets (Fig. 17.18). Those used to join steel frame members are either carbon steel, as specified by ASTM A449, or ASTM A325 and A490 high-strength bolts. ASTM A449 bolts are used in bearing connections (shear) in which their lower strength is adequate. ASTM A325 and A490 high-strength bolts are used for friction connections. They have a higher shear strength than carbon A449 bolts and high tensile strength. They are identified by markings on the head and nut (Fig. 17.19). Bolts used in friction connections must be tightened to a minimum of 70 percent of their ultimate tensile strength.

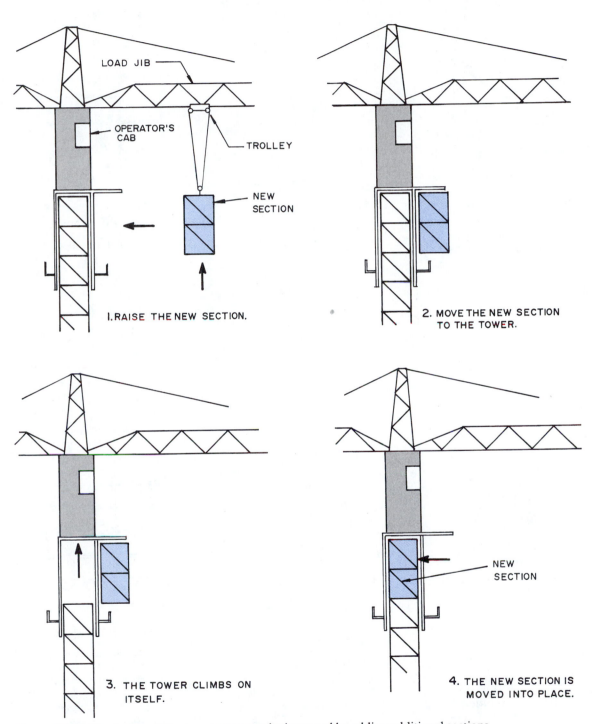

LOAD JIB

OPERATOR'S
CAB

TROLLEY

NEW
SECTION

1. RAISE THE NEW SECTION.

2. MOVE THE NEW SECTION
TO THE TOWER.

3. THE TOWER CLIMBS ON
ITSELF.

4. THE NEW SECTION IS
MOVED INTO PLACE.

NEW
SECTION

Figure 17.15 The height of the tower crane can be increased by adding additional sections.

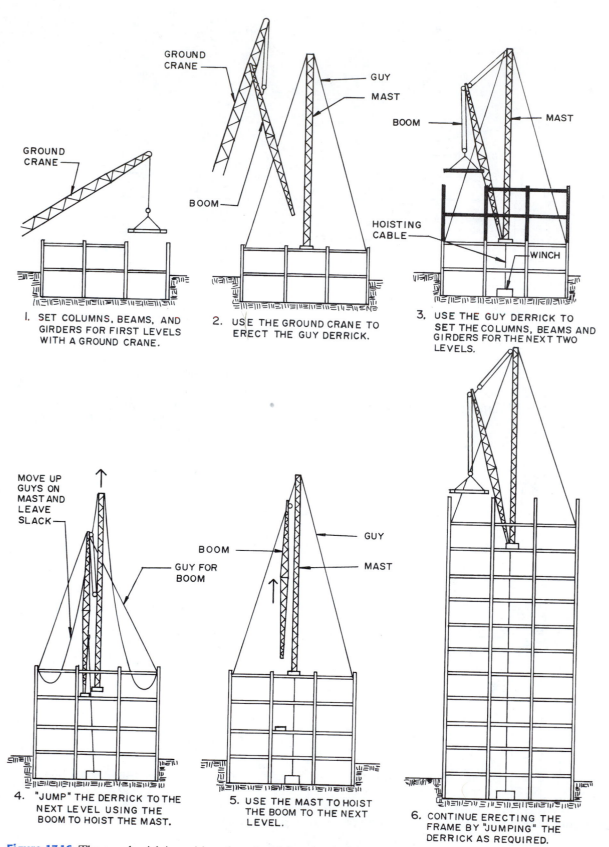

Figure 17.16 The guy derrick is positioned on the highest level of the building and can be raised to higher levels as the height of the building increases.

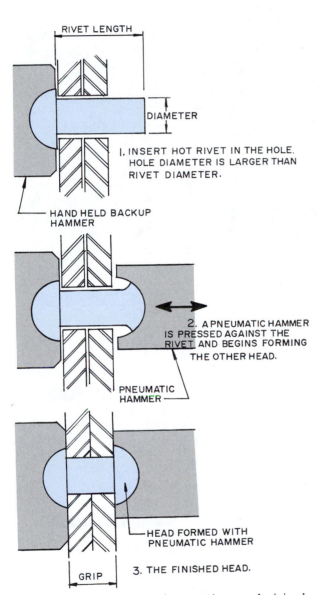

RIVET LENGTH

DIAMETER

1. INSERT HOT RIVET IN THE HOLE. HOLE DIAMETER IS LARGER THAN RIVET DIAMETER.

HAND HELD BACKUP HAMMER

2. A PNEUMATIC HAMMER IS PRESSED AGAINST THE RIVET AND BEGINS FORMING THE OTHER HEAD.

PNEUMATIC HAMMER

HEAD FORMED WITH PNEUMATIC HAMMER

GRIP

3. THE FINISHED HEAD.

Figure 17.17 Structural steel connections can be joined using rivets.

Figure 17.18 High-strength steel bolts are tightened with a pneumatic impact wrench. *(Courtesy Bethlehem Steel Corporation)*

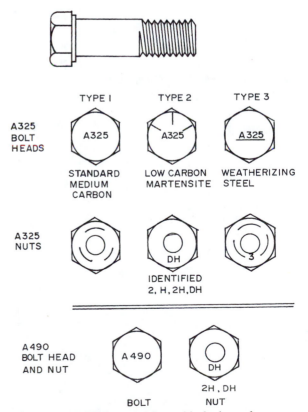

	TYPE I	TYPE 2	TYPE 3
A325 BOLT HEADS	A325	A325	A325
	STANDARD MEDIUM CARBON	LOW CARBON MARTENSITE	WEATHERIZING STEEL
A325 NUTS		DH	3
		IDENTIFIED 2, H, 2H, DH	

A490 BOLT HEAD AND NUT

A490 — BOLT

DH — 2H, DH NUT

Figure 17.19 High-strength steel bolts have the ASTM identification markings on the heads and nuts.

I T E M   O F   I N T E R E S T

FASTENERS FOR METAL BUILDING CONSTRUCTION

Two commonly used fasteners in the construction of metal buildings are designed to join sheet metal panels metal-to-metal and metal-to-wood. Metal-to-metal fasteners include drilling fasteners and self-tapping fasteners. The *drilling fasteners*, shown in Fig. A, have a tip that drills the correct size hole through the sheet metal followed by threads that tie the pieces together. The fastener must have a high surface hardness and a head that will withstand the use of a power impact driver. Drilling fasteners are available in carbon steel and several classifications of stainless steel and in many sizes and lengths. The manufacturer should be consulted for this information.

The *self-tapping fasteners*, shown in Fig. B, are installed in holes that are drilled or punched in the sheet metal. The size of the hole must be that specified by the manufacturer for each of the screws they manufacture. Self-tapping fasteners are available in carbon steel and stainless steel. Consult a manufacturer for the sizes available.

Typical *sheet-metal-to-wood fasteners* are shown in Fig. C. They are available in carbon steel and are power driven through predrilled holes in the metal into the wood. They are available with protective coatings such as a polymer film and zinc-and-chromate finish. The manufacturer will paint the heads the desired color. These fasteners are available in lengths up to 2 in. (50.8 mm).

An examination of the types of head shows the hexagon head is used for those fasteners to be driven with socket type power impact tools. Smaller types have slotted heads and also are power driven. A variety of washer faces are available. Some have a metal washer that gives metal-to-metal contact. Others have a watersealing gasket that is typically polyvinyl chloride (PVC), which is a plastic, and ethylene-propylene-diene (EPDM), which is a synthetic rubber. Other materials are used also.

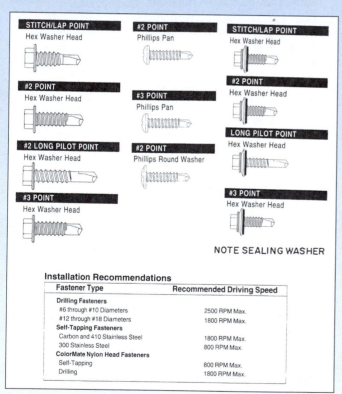

Figure A

Installation Recommendations

Fastener Type	Recommended Driving Speed
Drilling Fasteners	
#6 through #10 Diameters	2500 RPM Max.
#12 through #18 Diameters	1800 RPM Max.
Self-Tapping Fasteners	
Carbon and 410 Stainless Steel	1800 RPM Max.
300 Stainless Steel	800 RPM Max.
ColorMate Nylon Head Fasteners	
Self-Tapping	800 RPM Max.
Drilling	1800 RPM Max.

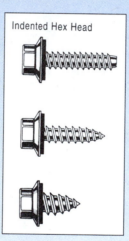

Figure B

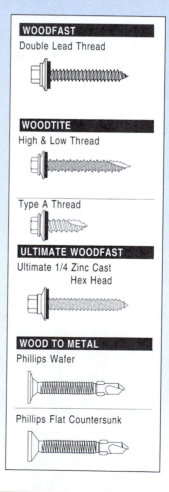

Figure C

I T E M   O F   I N T E R E S T

FASTENERS FOR METAL BUILDING CONSTRUCTION

Continued

Corrosion-Resistant Heads

Corrosion-resistant systems are essential to the life expectancy of a metal screw fastener. *Mechanical plating* is a process used to apply a thick film of zinc plating. Zinc powder is held in solution and peened onto the surface by mechanical force. Color sidings on metal buildings brought a demand for *paint finishes.* However, these do not hold up when the screws are installed with power impact tools. A *nylon-headed fastener* has the head enclosed in nylon by an injection molding process (Fig. D). These fasteners offer color and long life. *Metal-capped screws* have a thin stainless steel or zinc-alloy cap crimped on the head. Although these protect the head, there is the possibility of the cap coming off as the screw is installed or of moisture seeping between the cap and the head.

Another technique being used is to *post-plate* the heads with a long-lasting film applied over conventionally plated films. Corrosion resistance is enhanced if the heads are then painted. A development by the Atlas Bolt and Screw Company uses a *zinc-alloy cast head.* The head is a die-cast zinc-alloy unit that is permanently fastened to the head of the fastener. It has a polyvinyl chloride (PVC) washer to form a waterproof seal with the sheet metal (Fig. E). Typical head styles and drives available are shown in Fig. F.

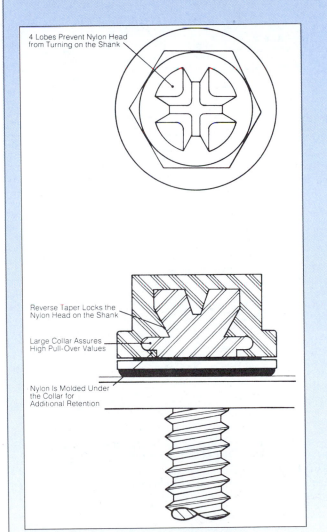

Reverse Taper Locks the Nylon Head on the Shank

Large Collar Assures High Pull-Over Values

Nylon Is Molded Under the Collar for Additional Retention

4 Lobes Prevent Nylon Head from Turning on the Shank

Figure D

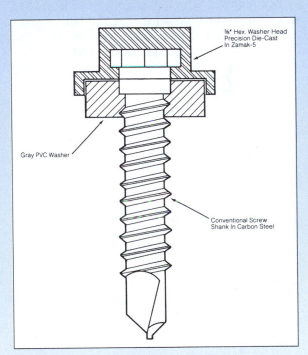

⅜" Hex. Washer Head Precision Die-Cast In Zamak-5

Gray PVC Washer

Conventional Screw Shank In Carbon Steel

Figure E

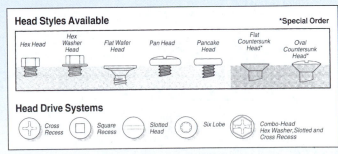

Head Styles Available   *Special Order

Hex Head	Hex Washer Head	Flat Wafer Head	Pan Head	Pancake Head	Flat Countersunk Head*	Oval Countersunk Head*

Head Drive Systems

Cross Recess   Square Recess   Slotted Head   Six Lobe   Combo-Head Hex Washer, Slotted and Cross Recess

Figure F

The bolts can be tightened with *calibrated torque wrenches,* which develop the specified tension values in the connector. They have an automatic cutoff and a gauge that registers the tensile stress developed. Another way to determine the tension is the turn-of-the-nut method. To do this, the nut is turned until it is tight and then turned an additional specified amount, such as one-third to one full turn. Another method is the *direct tension indicator.* Load indicator washers—metal washers with rounded protrusions on one face—are placed under the bolt head. As the nut is tightened they are flattened. The amount of flattening that occurs is measured with a feeler gauge and correlated with the amount of tension developed (Fig. 17.20). Still another method is to use *tension-control bolts,* as shown in Fig. 17.21. The bolt has a splined end that is inserted in a power wrench that grips the nut and the splined end. When the specified torque is reached the splined end twists off at the torque control groove.

Plain smooth hardened washers may be used to distribute the load of the bolt or nut over a larger area, but they are not required. The hardened washer is used to prevent **galling,** which could render the tension calibration ineffective (Fig. 17.20). Galling is the wearing or abrading of one material against another under extreme pressure. The bolt holes are usually ¹⁄₁₆ in. (1.6 mm)

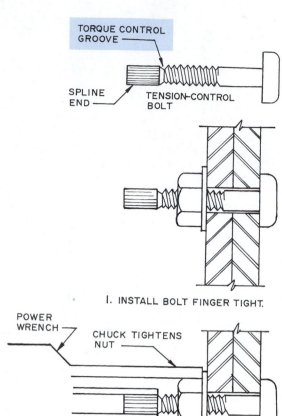

I. INSTALL BOLT FINGER TIGHT.

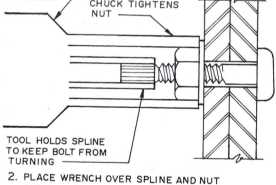

2. PLACE WRENCH OVER SPLINE AND NUT AND TIGHTEN.

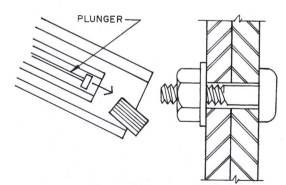

3. TIGHTEN UNTIL SPLINE BREAKS OFF FROM BOLT. A PLUNGER EJECTS SPLINE FROM TOOL.

Figure 17.21 Connections can be tensioned as specified by using tension-control bolts, which have a spline that shears when the proper tension is reached.

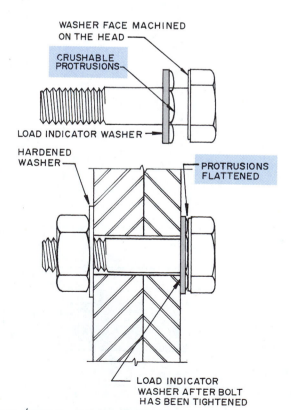

Figure 17.20 Load indicator washers can be used to verify that the bolt is under the proper tension.

larger in diameter than the bolt, but some field conditions may require a larger allowance. The size of oversize holes must be approved by the structural engineer.

The amount of threads that fall within the inside of the hole in the connection influences the strength of the connection. High-strength bolts have shorter threads than other types, so there is a minimum encroachment of threads into the grip. However, the bolt length must also provide sufficient protrusion beyond the connection to hold a washer and the nut. To achieve higher strengths in bearing connections, the amount of threads that fall within the grip must be less than the thickness of the outside metal piece. Some examples are shown in Fig. 17.22.

Welding is the third method used to fasten structural steel members. Occasionally welding and bolting are used together on some connections. Welded beam connections are of the same general types used for bolted and riveted connections. The connections may be riveted or bolted to the member in the fabrication shop and welded to the connecting member on the site or welded in the shop and bolted in the field or welded in the shop and the field (Fig. 17.23). When field connections are welded, the members are held with several bolts that hold them together as the field welding occurs. Welds in the fabrication shop are made by clamping the connector to the member and welding it in place. A beam detail showing a shop welded, field bolted connection is in Fig. 17.24. Notice the symbols indicating the size and type of weld. Weld symbols are made up of several parts, as shown in Fig. 17.25. Some frequently used penetration welds are shown in Fig. 17.26. To keep the molten welding rod material from running out the bottom of the connection, a backing bar or backing bead is placed below it before the weld is started. Sometimes bars are welded on the open ends of an area to be welded so the weld thickness is the same its entire length.

Although there are several types of welding processes, *electric arc welding* is generally used for structural steel connections. This involves passing an electric current through a metal electrode (the welding rod) to the piece to be welded. The electrode is kept slightly above the piece, which causes an arc to jump the gap. This produces considerable heat, which melts the electrode and a small area of the members being welded. As the electrode is moved along the line of the weld it leaves a continuous bead of metal that fuses the members together as it cools (Fig. 17.27). Large welds require several beads to be laid on top of each other.

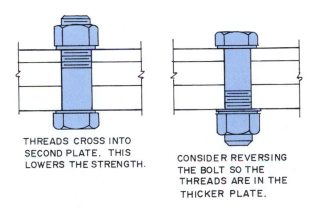

THREADS CROSS INTO SECOND PLATE. THIS LOWERS THE STRENGTH.

CONSIDER REVERSING THE BOLT SO THE THREADS ARE IN THE THICKER PLATE.

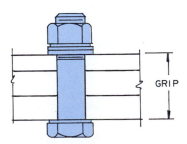

GRIP

OR USE A LONGER BOLT WITH ADDITIONAL WASHERS TO PERMIT THE FULL TIGHTENING OF THE NUT.

Figure 17.22 The amount of threads falling within the connection can influence the strength.

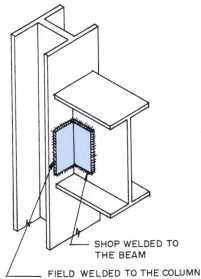

SHOP WELDED TO THE BEAM

FIELD WELDED TO THE COLUMN

Figure 17.23 This angle connector has been welded to the beam during fabrication and welded to the column during erection.

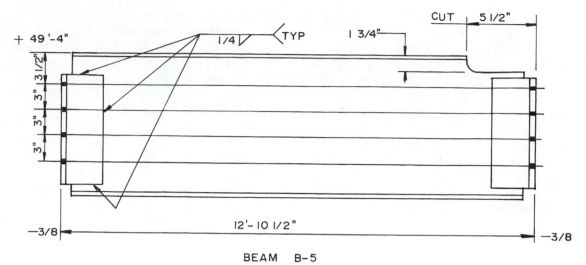

BEAM   B–5

Figure 17.24 A detailed drawing of a beam with connections shop welded to the beam and prepared to be field bolted to a column or girder.

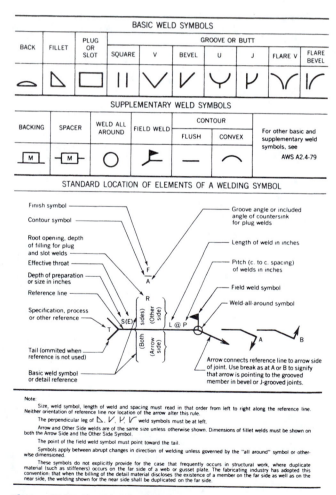

Figure 17.25 Standard symbols for welded joints. *(Courtesy American Institute of Steel Construction)*

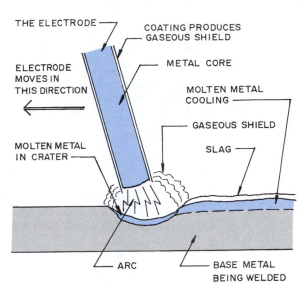

Figure 17.27 The welding arc merges the molten electrode core with the molten base material.

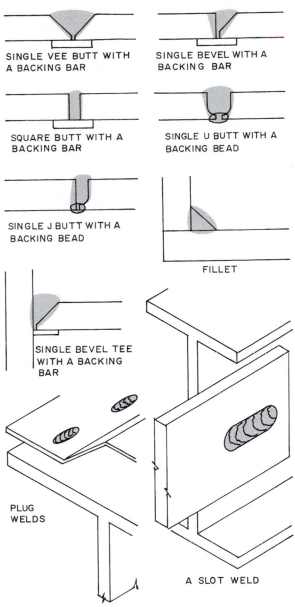

Figure 17.26 Penetration welds commonly used on structural steel construction. Penetration welds extend all the way through the material.

The structural engineer determines the size, length, and type of weld to be used. Some welds that are in critical locations (such as in a nuclear plant) require that each weld be inspected. There are a number of techniques available for checking welds for flaws.

THE STEEL FRAME

The members making up the steel frame are joined with metal connections and riveted, bolted, or welded. The connections typically are angles, tees, or plates. The connections transmit vertical (shear) forces and bending moments (rotational forces) from one member to the

other. **Shear stress** is produced when vertical loads are applied to a member, creating a deformation in which parallel planes slide relative to each other so as to remain parallel (Fig. 17.28). A **moment** is the property by which a force tends to cause a body to which it is applied to rotate about a point or line as shown in Fig. 17.29. **Bending moment** is the moment that produces bending at a section of a structural member. The various connections illustrated in the next section of this chapter are identified as shear connections or moment connections or connections controlling both.

Frame Stability

A building designed using structural steel framing (or concrete or wood) must be designed to be stable against the forces acting upon it, such as wind, live loads, and earthquakes. This can be accomplished by adding shear walls on the exterior or interior. They may be partitions or form a core, as used for an elevator shaft, or a series of stairwells. The frame can also be braced using *diagonal bracing* that triangulates some of the bays on the exterior and interior, installing *surface cladding* that is designed to provide structural bracing, and by using *connections* designed to control shear stresses and bending moments. This produces a rigid frame designed to prevent any change in the angle between the members (Fig. 17.30).

Types of Beam Connections

Beam connections may be framed or seated. **Framed connections** join the beam to an adjoining member with connectors, such as angles, that are secured to the beam web. **Seated connections** have the flange of the beam

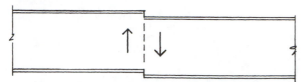

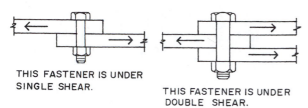

SHEAR STRESS APPLIES A FORCE TO THIS BEAM
PRODUCING A TENDENCY IN ONE SECTION TO "SLIDE"
PAST THE ADJACENT SECTION.

THIS FASTENER IS UNDER
SINGLE SHEAR.

THIS FASTENER IS UNDER
DOUBLE SHEAR.

Figure 17.28 Structural members and connections are subject to shear forces.

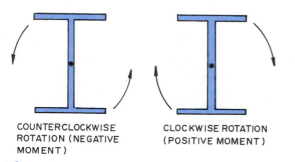

COUNTERCLOCKWISE
ROTATION (NEGATIVE
MOMENT)

CLOCKWISE ROTATION
(POSITIVE MOMENT)

Figure 17.29 A bending moment can be positive or negative.

resting on a seat, such as a metal angle (Fig. 17.31). When the connections are designed, the type, strength, and size of fasteners, strength of the base material, and the end reaction of the beam must be considered. Usually one side of the framed connector is connected to the beam in the fabrication shop and the other side is field-connected to the column or butting beam.

Simple, Rigid, and Semirigid Connections

Moment connections are used to transfer the forces in the beam flanges to the column. This transfer is in addition to the need for shear connections used to support the beam. Framed, seated, and end plate connections are shear connections. Moment connections have flange stresses developed independently of the shear connections.

Simple connections provide for shear only and require additional bracing to stabilize the structural frame, such as diagonal bracing or shear panels (Fig. 17.30).

Rigid connections provide for shear and have sufficient rigidity to hold the original angles between the connected members (moment). Rigid-frame construction does not require additional bracing.

Semirigid connections provide for shear, and although they are not as rigid as rigid connections, they do provide a known moment-resisting capacity.

Welded Connections

Examples of welded beam-to-column and combinations of welded and bolted connections are shown in Fig. 17.32. The load capacity depends on the connector, type of weld, strength of the electrode material, and base material. AISC tables indicating the sizes and capacities of angle connections are based on the use of E70 electrodes. A typical welded beam-to-girder or beam-to-beam connection is shown in Fig. 17.33. Welded column splicing is illustrated in Fig. 17.11.

Bolted and Riveted Connections

Two types of connections are possible using bolts or rivets. A *bearing-type connection* is one in which the fasteners bear against the sides of the holes in the connections. The rivet or bolt bears the stress. All riveted and some bolted connections are bearing type. *Friction-type connections* are those in which the fastener can be tightened enough to clamp the connected parts together under high pressure. The shearing force on the connection is resisted by the friction between the connected parts rather than on the bolts. Only high-strength bolts are used in this type of connection.

A partial design drawing for a bolted beam-to-column connection is in Fig. 17.34. It shows the type of construction and all information on loads, shears, moments, and axial forces that must be resisted by the steel members and the connections.

The size and strength of the structural steel members illustrated in the following figures are not indicated. This will vary with the design conditions. Standards for design, fabrication, and erection are available from the American Institute of Steel Construction. The type of steel is indicated by its ASTM number. See Chapter 15 for additional details.

Examples of bolted beam-to-column details are shown in Figs. 17.35 and 17.36. Riveted connections use similar details. Bolted column splicing details are shown in Fig. 17.11. A partial framing drawing indicating beam-to-beam connections and design information is in Fig. 17.37. A typical beam-to-beam connection is in Fig. 17.38. Notice the top of the beam butting the girder has a clearance cut so they are flush on top. The notch (called a cope, block, or cut) allows clearance all around the flange.

Channels are often used between beams to support decking, as on a roof. A typical connection between channels is shown in Fig. 17.39. Many special connections require careful analysis by the structural engineer. Not all designs fit the rather standard connections illustrated to this point. A few examples are shown in Fig. 17.40.

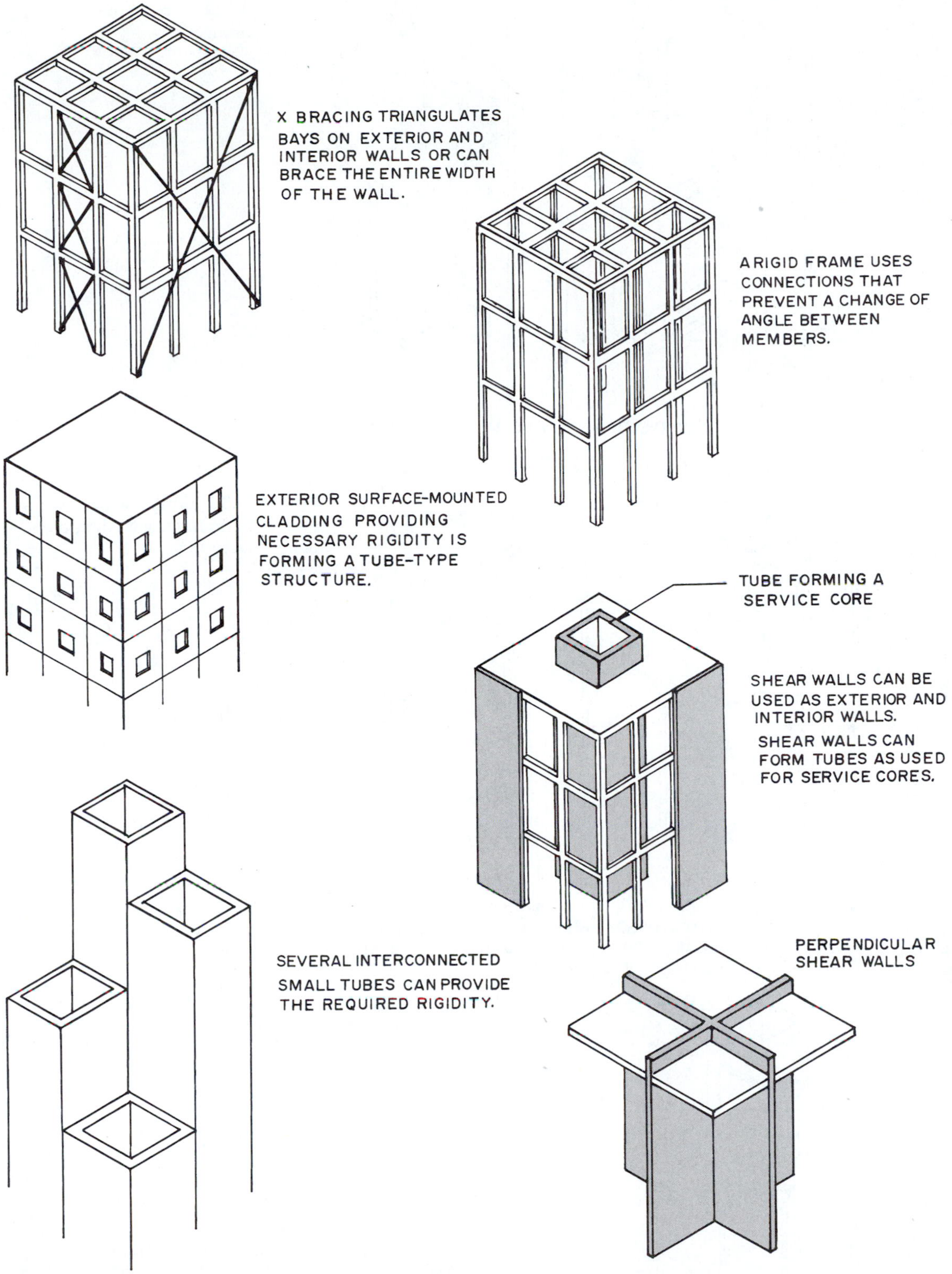

X BRACING TRIANGULATES BAYS ON EXTERIOR AND INTERIOR WALLS OR CAN BRACE THE ENTIRE WIDTH OF THE WALL.

A RIGID FRAME USES CONNECTIONS THAT PREVENT A CHANGE OF ANGLE BETWEEN MEMBERS.

EXTERIOR SURFACE-MOUNTED CLADDING PROVIDING NECESSARY RIGIDITY IS FORMING A TUBE-TYPE STRUCTURE.

TUBE FORMING A SERVICE CORE

SHEAR WALLS CAN BE USED AS EXTERIOR AND INTERIOR WALLS.

SHEAR WALLS CAN FORM TUBES AS USED FOR SERVICE CORES.

SEVERAL INTERCONNECTED SMALL TUBES CAN PROVIDE THE REQUIRED RIGIDITY.

PERPENDICULAR SHEAR WALLS

Figure 17.30 Typical methods for stabilizing the structural frame.

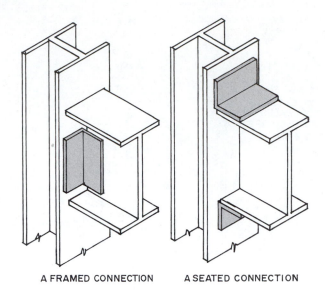

Figure 17.31 Typical framed and seated connections.

A FRAMED CONNECTION   A SEATED CONNECTION

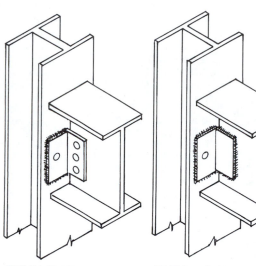

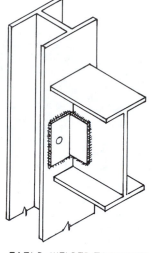

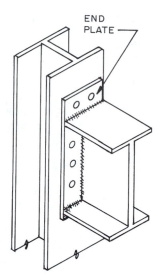

END PLATE

Figure 17.32 Typical welded beam-to column framed, seated, and end-plate connections.

FIELD WELDED TO COLUMN.
SHOP BOLTED TO BEAM.
A FRAMED CONNECTION.

FIELD WELDED TO COLUMN.
SHOP WELDED TO BEAM.
A FRAMED CONNECTION.

END PLATE SHOP WELDED TO
BEAM AND FIELD BOLTED TO
COLUMN.

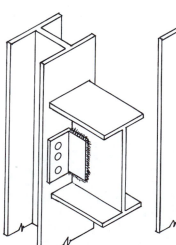

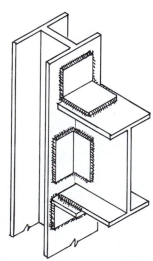

FIELD BOLTED TO COLUMN.
SHOP WELDED TO BEAM.
A FRAMED CONNECTION.

SHOP AND FIELD WELDED
SEATED CONNECTION.

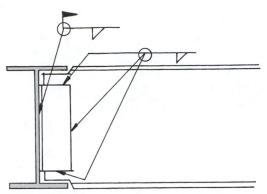

Figure 17.33 A typical welded beam-to-beam connection.

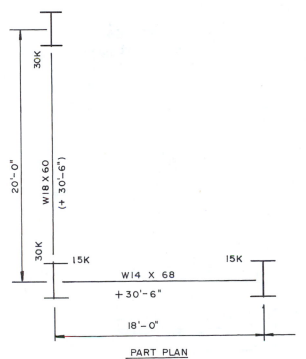

PART PLAN

ELEVATION TOP OF STEEL + 30'- 6"

GENERAL NOTES

SPECIFICATIONS: AISC LATEST EDITION.
MATERIAL: ASTM A36
FIELD FASTENERS: 3/4" Ø BOLTS IN BEARING
   TYPE CONNECTIONS, THREADS IN SHEAR
   PLANES.
SHOP FASTENERS: 3/4" Ø A502-1 RIVETS.
CONNECTIONS MUST CONFORM TO AISC
   MANUAL AND DEVELOP END REACTIONS
   SHOWN ON DESIGN DRAWINGS.
ONE COAT SHOP PAINT.

Figure 17.34 A partial design drawing for a bolted beam-to-column connection.

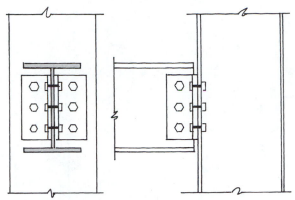

SHOP BOLTED TO THE BEAM. FIELD BOLTED TO
THE COLUMN. THIS IS A FRAMED SHEAR
CONNECTION.

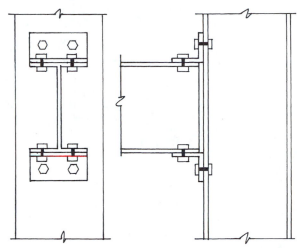

SHOP BOLTED TO THE BEAM. FIELD BOLTED TO
THE COLUMN. THIS IS A SEATED SHEAR
CONNECTION.

Figure 17.35 A typical bolted beam-to-column framed shear connection.

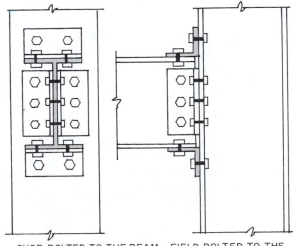

SHOP BOLTED TO THE BEAM. FIELD BOLTED TO THE
COLUMN. THIS IS A MOMENT-RESISTING
CONNECTION.

Figure 17.36 One type of bolted moment-resisting connection.

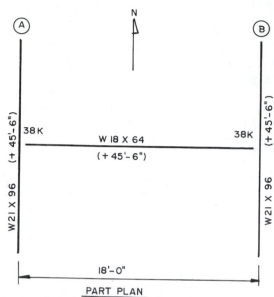

PART PLAN
ELEVATION TOP OF STEEL + 45'-6"

GENERAL NOTES

SPECIFICATIONS: AISC LATEST EDITION.
MATERIAL: ASTM A36
FIELD FASTENERS: 3/4"Ø BOLTS IN BEARING
   TYPE CONNECTIONS, THREADS IN SHEAR
   PLANES.
SHOP FASTENERS: 3/4" Ø A502-1 RIVETS.
CONNECTIONS MUST CONFORM TO AISC
   MANUAL AND DEVELOP END REACTIONS
   SHOWN ON DESIGN DRAWINGS.
   ONE COAT SHOP PAINT.

Figure 17.37 Design information for a beam-to-beam connection.

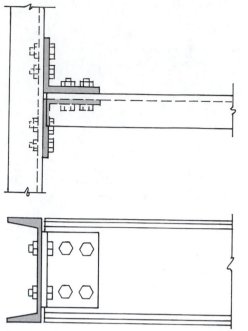

Figure 17.39 A typical bolted channel connection.

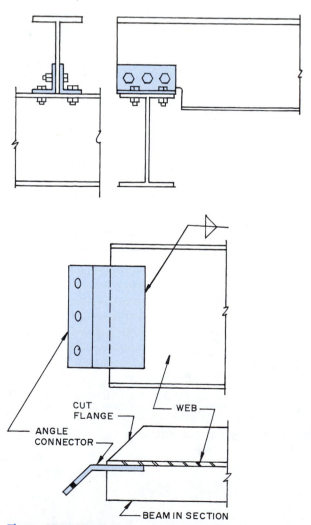

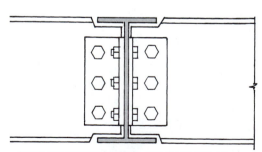

SHOP BOLTED TO BEAMS, FIELD BOLTED TO
ADJOINING BEAM.

Figure 17.38 A typical bolted beam-to-beam connection.

CUT
FLANGE

WEB

ANGLE
CONNECTOR

BEAM IN SECTION

Figure 17.40 Examples of special connections designed by structural engineers.

FIRE PROTECTION OF THE STEEL FRAME

Structural steel is an incombustible material and will not melt even during a building fire, but sustained extreme heat applied to it affects its properties, including its strength. This can lead to failure of steel columns and beams during a prolonged building fire. To protect the occupants and the structural integrity of the building, codes require certain steel frames to be protected with a fire-resistant material. The requirements depend on the type of construction, building height, floor area, occupancy, the fire protection system to be installed, and the location of the building. Structural steel may be protected by many materials, including concrete, tile, brick, stone, gypsum board, gypsum blocks, fire-resistant plasters, sprayed-on mineral fibers, intumescent fire-retarding coatings, liquids, and flame shields.

Unprotected mild steel loses about half its room temperature strength at temperatures exceeding 1000°F (542°C). Fire resistance ratings are given as the number of hours a material can withstand fire exposure, as specified by standard test procedures. Most standard fire tests on structural steel members are conducted by the National Institute of Standards and Technology, Gaithersburg, Maryland, and the Underwriters Laboratories, Northbrook, Illinois.

Insulating concrete can be used to protect steel members. The amount of protection depends on the thickness of the cover, the concrete mix, and the method of support. Lightweight concrete (aggregates such as perlite, vermiculite, expanded shale and slag, pumice, and sintered flyash) have greater fire resistance than normal concrete because it has greater resistance to heat transfer and has a higher moisture content (Fig. 17.41). Other insulating materials provide greater protection and are much lighter than concrete.

Masonry units, such as concrete block, brick, gypsum block, and hollow clay tile, are used to enclose structural steel members. In addition to the protection offered by the material, those with hollow cores can be filled with insulating materials (such as vermiculite) or mortar to increase their fire protection qualities (Fig. 17.42). Masonry protection is heavy, and this limits where it is used. Information on fire resistance of masonry units is available from the National Concrete Masonry Association, Herndon, Virginia.

Lightweight plaster on metal lath is a lightweight fire-resistant protection material. It is also applied over gypsum block (Fig. 17.43). The plaster is troweled over the lath or block. Lightweight plaster using perlite or vermiculite aggregates has good insulating qualities.

Fire-resistant sheet products are secured mechanically to the structural steel members. Those with a hard surface, such as gypsum board, can also be painted and serve as the finished surface when used inside a building (Fig. 17.44).

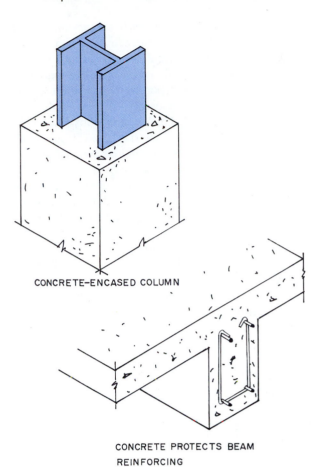

CONCRETE-ENCASED COLUMN

CONCRETE PROTECTS BEAM REINFORCING

Figure 17.41 Structural steel can be protected from fire with concrete encasements.

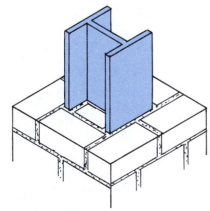

Figure 17.42 Masonry can be used to provide fire protection for steel members.

Sprayed cementitious materials, such as gypsum plaster with perlite, vermiculite, or other insulating materials, are applied directly to the structural steel (Fig. 17.45). The steel must be clean and primed to receive the coating. Since these coatings can be easily damaged by abrasion they are used where they will not be exposed and are covered with a durable protective material. Thick applications (over 2 in. or 50 mm) require that a metal mesh be used (Fig. 17.46).

Mineral fiber slabs are used to enclose structural steel members. Mineral fibers have excellent heat flow retardation properties and can withstand temperatures above 1000°F (542°C). The coating is easily damaged and requires a protective covering if exposed to possible impact or the weather (Fig. 17.47).

Intumescent coatings are a sprayed-on mastic fire-retarding coating. They dry to a hard durable finish much like a paint. When exposed to heat, the coating expands increasing its thickness and forming an insulation blanket. These coatings are available in colors and can serve as the finished surface.

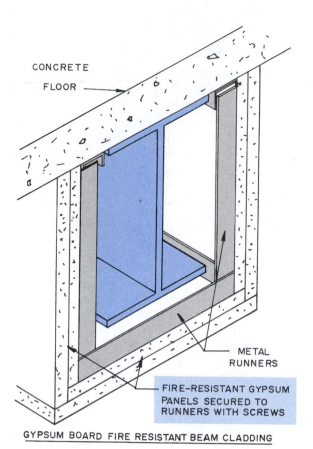

GYPSUM BOARD FIRE RESISTANT BEAM CLADDING

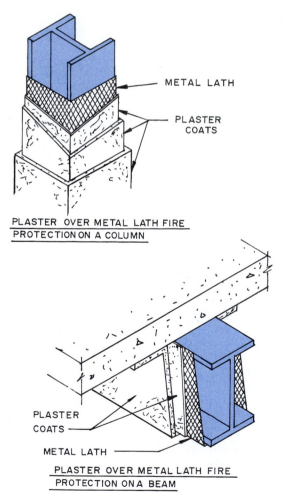

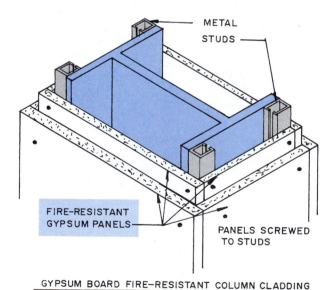

GYPSUM BOARD FIRE-RESISTANT COLUMN CLADDING

Figure 17.44 Gypsum board products are used to provide fire protection for steel members.

Figure 17.43 Lightweight plaster over metal lath provides fire protection for structural steel.

Courtesy Gomaco Corporation

As the project nears completion the road construction begins. After the grade and base are set, the concrete curbing can be laid.

Courtesy Peri Formwork Systems, Inc.

Extensive framework is required for the construction of this foundation. Notice the huge size of the required excavation. The heavy steel form structure is painted red.

Installing temporary utilities is part of the preliminary work required on the site.

Courtesy Velux

Skylights provide considerable natural lighting and a feeling of spaciousness and enhance the colors of the materials used on the inside of the building.

Courtesy Tubelite, Inc.

This multistory building was constructed using a custom-designed and -fabricated ribbon window system. Notice the use of light blue spandrel panels to enhance the horizontal lines of the building.

A series of metal-framed doors creates an attractive entrance to this building, which is clad with a glazed curtain-wall system. Notice the use of colored glazing to provide a dramatic contrast with the white facing panels.

Courtesy EFCO Corporation

This entrance to a commercial building used heavy-duty metal framed doors with wide glazed side panels providing considerable natural light into the lobby. The anodized aluminum frames blend harmoniously with the stone side panels.

Figure 17.45 A cementitious material is being sprayed on structural steel members to provide fire protection. *(Courtesy W.R. Grace and Co.)*

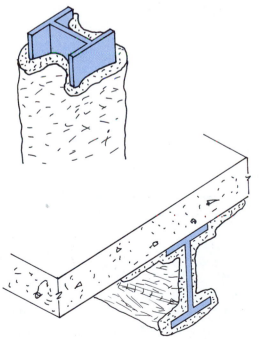

Figure 17.46 Sprayed cementitious materials form a fire-protective covering.

Liquid-filled columns are used to reduce the heat of the steel. The tube or box-type columns are filled with water, which is supplied from a central source, such as a water main. This replaces water lost due to heat from the fire. Vent valves release the steam produced. Pumps are sometimes used to maintain circulation within the system (Fig. 17.48). Antifreeze is added if the columns are exposed to freezing temperatures.

Flame shields are metal barriers that deflect the flames and reflect the heat from a fire away from exterior structural steel members. In Fig. 17.49 the shield deflects the heat from the flanges of a beam. Notice the insulation on the inside face of the beam.

The installation of *ceilings* that provide the needed fire protection for floors and the roof is another procedure used. The design depends on the situation, but plastered ceilings, fire-rated dropped ceilings, acoustic tiles of various types, and drop-in ceiling panels all provide various degrees of fire protection (Fig. 17.50).

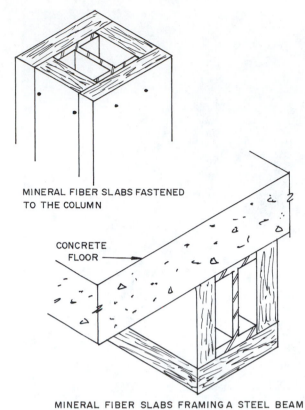

MINERAL FIBER SLABS FASTENED
TO THE COLUMN

CONCRETE
FLOOR

MINERAL FIBER SLABS FRAMING A STEEL BEAM

Figure 17.47 Mineral fiber slabs have excellent resistance to heat flow.

Figure 17.48 Water in tubular columns can carry away heat by convection.

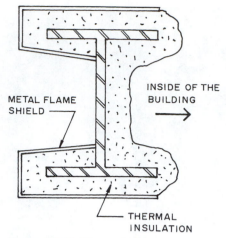

METAL FLAME
SHIELD

INSIDE OF THE
BUILDING

THERMAL
INSULATION

Figure 17.49 Flame shields can protect exposed sections of structural steel members.

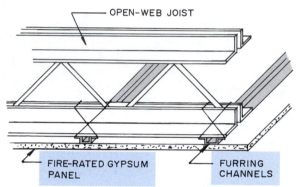

OPEN-WEB JOIST

FIRE-RATED GYPSUM
PANEL

FURRING
CHANNELS

FIRE-RATED CEILING TIED TO AN OPEN-WEB JOIST

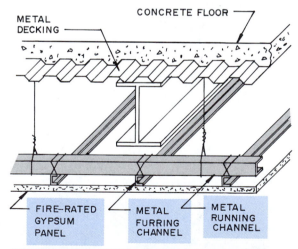

METAL
DECKING

CONCRETE FLOOR

FIRE-RATED
GYPSUM
PANEL

METAL
FURRING
CHANNEL

METAL
RUNNING
CHANNEL

FIRE RATED SUSPENDED CEILING PROTECTING LONG
SPAN STEEL BEAMS

Figure 17.50 Typical fire-rated ceilings used to protect structural steel framing above them.

DECKING

Several types of floor and roof decking are used with structural steel framing, metal decking and precast concrete units.

Metal Decking

Sheet and cellular steel decking are illustrated in Chapter 15. *Sheet decking* is available in various thicknesses of metal and depths of corrugation (Fig. 17.51). The designer considers the span between supporting members and the loads to be carried as the product is selected. Metal roof decking is often covered with rigid insulation and the finish roofing material. Under these conditions it must carry all of the imposed loads.

Composite metal decking provides a base on which a concrete slab is poured (Fig. 17.52). The reinforced slab provides most of the structural strength. The metal decking provides tensile reinforcing to the concrete slab because it is bonded to the slab by perforations in the metal or by welded wire fabric that is tack welded to the decking. Sometimes the designer specifies the use of metal **shear studs** that are welded to the beams and protrude up into the concrete slab (Fig. 17.53). These tie the steel beam and the concrete slab together, which increases the load-carrying capacity of the steel beam. Properly installed metal roof decking can serve as shear diaphragms under lateral loads such as wind and seismic forces.

Cellular steel floor raceway systems consist of metal decking that has wiring raceways and a structural concrete slab on top (Fig. 17.54). The cellular raceways provide space to run electrical wiring, telephone lines, and data cables. These connect to a large main header duct that has a removable cover for lay-in wiring. Inserts provide access through the concrete slab to the wiring in the raceway (Fig. 17.55). The cellular system provides a fire-resistant barrier between floors, serves as a form for pouring the concrete deck, and provides tensile reinforcement for the concrete floor slab.

Metal decking is generally plug welded to the steel beams, girders, and joists. The edges are usually joined by welding or self-drilling screws.

Concrete Decking

A number of precast concrete decking units are available for use on structural steel framing. These are shown in Chapter 9. They include hollow-core units and precast slabs and channels of various designs. Long-span precast concrete roof and floor units are generally prestressed. Short-span members have reinforcement, as do cast-in-place members.

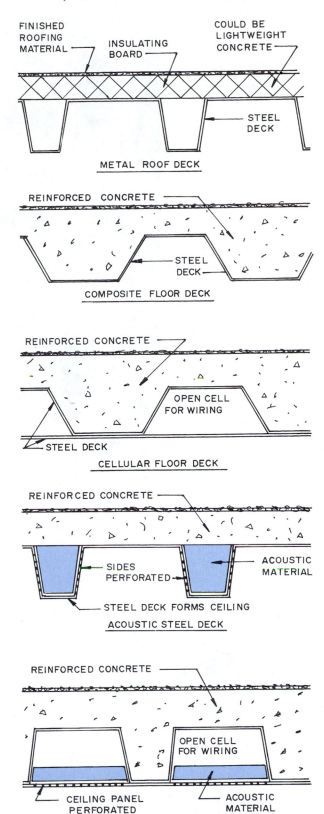

Figure 17.51 Typical types of metal decking.

Figure 17.52 Composite metal decking serves as a base for a concrete slab floor or roof. *(Courtesy Vulcraft)*

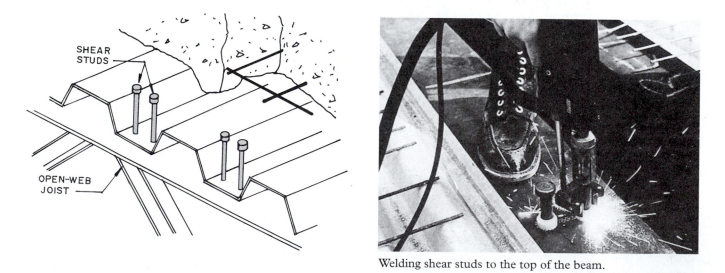

Welding shear studs to the top of the beam.

Figure 17.53 Shear studs tie the steel structural member to the concrete slab. *(Drawing Courtesy Vulcraft) (Photo Courtesy Bethlehem Steel Corporation)*

Precast slabs and channels are placed on the structural steel framing. They form the deck and, in some cases, the ceiling below (Fig. 17.56). They are secured to beams and joists by welding metal plates that are cast in them to the steel structure or by using clips, as shown in Fig. 17.57.

Precast hollow-core decking units are used for long spans and are placed on the steel framing with a crane

(Fig. 17.58). Generally the top surface is left rough so the topping placed over it will bond (Fig. 17.59). The topping helps tie the precast units together and adds to the structural integrity of the roof or floor. Reinforcing can be cast in the topping.

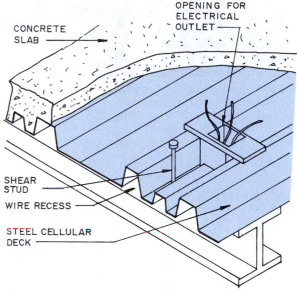

Figure 17.54 A steel cellular floor system designed for welding to a structural steel frame as manufactured by H.H. Robertson.

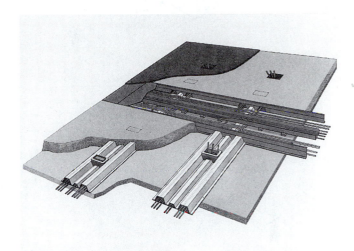

Figure 17.55 A steel cellular floor wire distribution system for installation on a slab-on-grade. *(Courtesy H.H. Robertson)*

Precast channel roof decking is secured to the structural frame.

Precast channel roof decking can provide an attractive finished ceiling.

Figure 17.56 These lightweight precast concrete roof deck channels provide a roof deck. They can be made with an acoustical treatment on the underside that forms an attractive ceiling. *(Courtesy Federal Cement Products)*

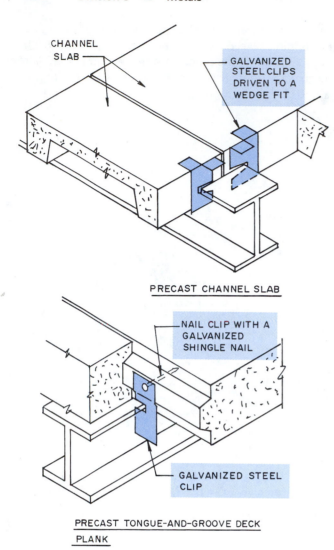

CHANNEL SLAB

GALVANIZED STEEL CLIPS DRIVEN TO A WEDGE FIT

PRECAST CHANNEL SLAB

NAIL CLIP WITH A GALVANIZED SHINGLE NAIL

GALVANIZED STEEL CLIP

PRECAST TONGUE-AND-GROOVE DECK PLANK

Figure 17.57 Precast deck planks can be secured to the steel framing with metal clips.

Figure 17.58 Precast concrete hollow-core floor and roof decking can be used with steel or concrete structural frames. *(Courtesy Flexicore Manufacturing Association)*

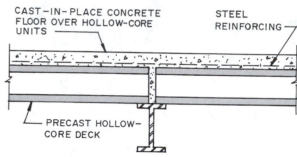

CAST-IN-PLACE CONCRETE FLOOR OVER HOLLOW-CORE UNITS

STEEL REINFORCING

PRECAST HOLLOW-CORE DECK

Figure 17.59 Precast hollow-core decking units can be topped with a finished concrete floor and reinforced to help increase the structural integrity of the structural steel frame.

STEEL TRUSSES

A number of steel trusses are used for construction purposes. Included are roof trusses, open-web joists, and space frames. A *truss* is a coplanar (forces operating in the same plane) assembly of structural members that are joined at their ends. The diagonals form triangles, producing a rigid framework. Triangulation is the concept behind the design of a truss. A triangle is the only multisided geometric figure that is rigid and will not move or rock unless a structural member bends or a connection fails. This principle is illustrated in the design of the bridge in Fig. 17.60.

Roof trusses act like beams and support the roof and all imposed loads. The common types of roof trusses are in Fig. 17.61. A typical steel roof truss consists of top and bottom chords and a system of webs (Fig. 17.62). The slope of the top chord can vary as desired, and it can be parallel to the bottom chord for flat roofs. Parallel chord trusses are used for floors and flat roofs, and as major structural beams and girders. A typical trussed roof framing structure includes the truss, purlins to support a deck, the deck, and sometimes a system of rafters, as shown in Fig. 17.63.

Figure 17.60 The structure of this bridge uses the principle of triangulation. *(Courtesy West Virginia Highway Department)*

In Fig. 17.64 large floor-to-floor steel trusses carry the loads of the floors, roofs, and ceilings. This is often referred to as a staggered truss system because trusses are one story high and erected at every other floor.

Open-web joists are another type of truss. They are lightweight structural members made from steel angles and bars (Fig. 17.65). They, like other trusses, are end bearing and can span long distances and are used for floor and roof construction. One installation is shown in Fig. 17.66.

Space frames are another truss type member. They provide a structural frame that can span in several directions and over large areas with a minimum number of vertical supports. Various designs are used. The space frame is generally made from tubular members (Fig. 17.67). However, wide flange beams and tee members are also used. The attractiveness of the assembly is lost if it must be covered to meet fire codes. Some codes do not require fire protection if the space frame is more than 20 ft. (6.1 m) above the floor.

TENSILE STRUCTURES

Tensile structures, also referred to as tensioned fabric structures, are framed using high-strength steel cables suspended between supporting members (Fig. 17.68). The structures shown in Fig. 17.69 and 17.70 are covered with a tensioned membrane made of a woven electrical-grade fiberglass substrate with a Teflon PTFE coating. These structures can cover large areas. The engineer must design the structure to resist wind uplift and flapping of the membrane.

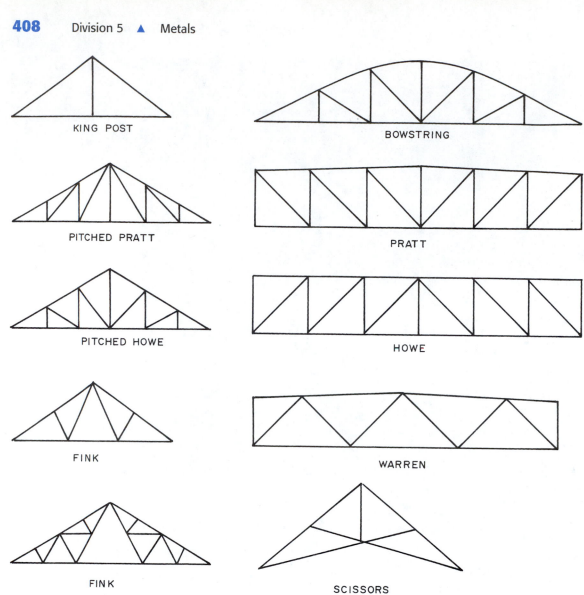

Figure 17.61 Commonly used roof trusses.

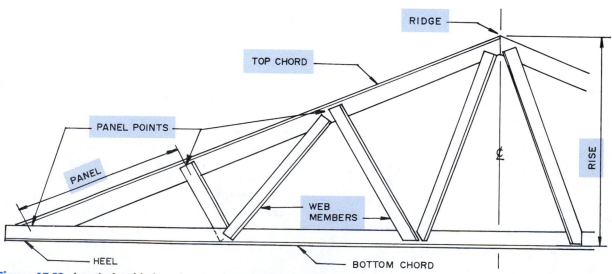

Figure 17.62 A typical welded steel truss used for light construction.

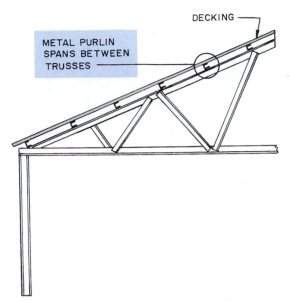

Figure 17.63 The roof deck is supported by purlins spanning the distance between the metal trusses.

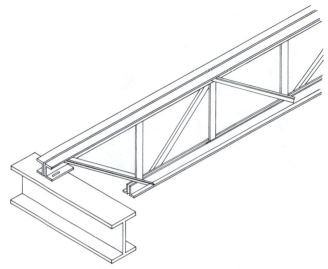

Figure 17.65 Open-web joists are lightweight truss-type structural members that can span long distances.

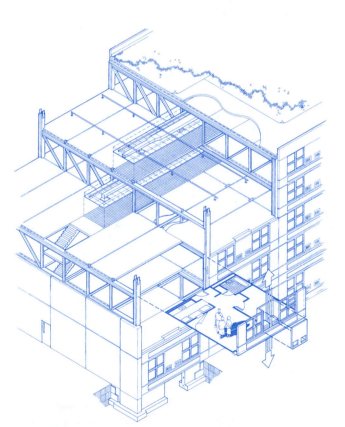

Figure 17.64 Staggered floor-to-floor steel trusses serve as load-bearing members for floors, roofs, and ceilings. *(Reproduced with permission from* The Building Systems Integration Handbook, *Richard Rush, ed., Butterworth-Heinmann Publishers, Newton, Mass., 1986)*

Figure 17.66 Open-web joist girders connect the columns and support the long span open web joists. Notice the metal decking being installed. *(Courtesy Vulcraft)*

Figure 17.67 Space frames span long distances and permit the integration of electrical and mechanical systems within the structure. (*Reproduced with permission from* The Building Systems Integration Handbook, *Richard Rush, ed., Butterworth-Heinmann Publishers, Newton, Mass., 1986*)

Figure 17.68 Tensile or tensioned fabric structures use a network of steel cables in sleeves in the fabric covering to form the structural support and exterior finish. (*Reproduced with permission from* The Building Systems Integration Handbook, *Richard Rush, ed., Butterworth-Heinmann Publishers, Newton, Mass., 1986*)

LIGHT-GAUGE STEEL FRAMING

Light-gauge structural steel shapes are formed from flat cold-rolled pieces of carbon steel. The gauge thickness ranges from No. 12 to No. 20. Some shapes are formed from a single steel sheet, but others have several forms shape welded together. They are available in nailable and nonnailable types and are either galvanized or primed with zinc chromate. Some of the commonly available shapes are in Fig. 17.71. Load-carrying capacities and spans should be obtained from the manufacturer. Light-gauge metal is widely used in commercial, industrial, and residential construction (Fig. 17.72).

Typical construction details for light-gauge steel structural members are shown in Fig. 17.73. The floors can be steel decking with a concrete topping or precast concrete planks. Roofs can be any decking material available. Stud walls can support lightweight steel or open-web joists and various roof trusses or metal rafters. The finished exterior can be any of the materials typically used. In Fig. 17.74 a steel framed building is enclosed with a brick veneer exterior. Additional light steel framing details can be found in Chapter 19.

PRE-ENGINEERED METAL BUILDING SYSTEMS

The building frames of pre-engineered metal building systems are constructed from a variety of structural steel and light-gauge steel framing members. The vertical members are supported on cast-in-place concrete footings. The structural components are designed to carry specified loads. The company that manufactures and sells the systems distributes them through a network of franchised dealers. The dealers work with the customer and architect to produce the structural system required. Figure 17.75 shows an assembly of a pre-engineered building using rigid frames as the main structure and enclosing materials supported by light-gauge steel members. The complete assembly, including the steel frame, girts, purlins, and bracing, provides the strength needed. The common types of pre-engineered building available include truss-type, rigid frame, post-and-beam, and sloped roof.

Truss-type buildings use steel columns and open-web joists and girders to form the roof (Fig. 17.76). Joist girders span the long distances between steel columns.

Figure 17.69 Tensioned fabric structures provide dramatic structural enclosures. *(Courtesy Birdair, Inc.)*

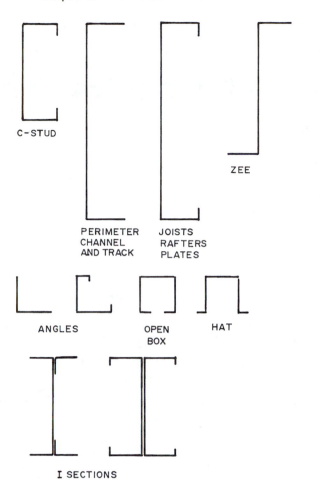

Figure 17.71 Typical cold-formed lightweight structural steel shapes.

Figure 17.70 This tensioned fabric structure covers a large area over a shopping mall. *(Courtesy Birdair, Inc.)*

Figure 17.72 Lightweight steel studs and joists are widely used in commercial and residential construction.

Construction details

Load bearing exterior wall

Load bearing interior wall

Non-bearing exterior wall

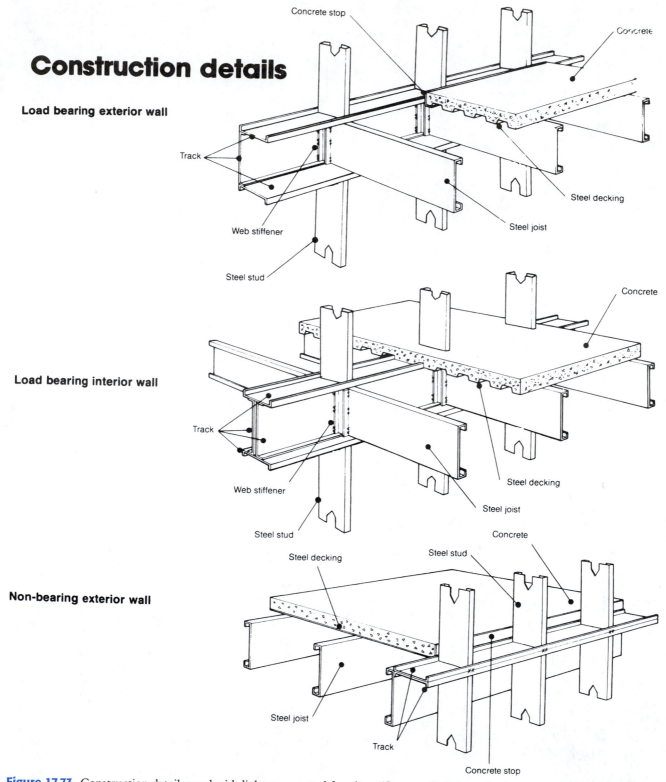

Figure 17.73 Construction details used with light-gauge steel framing. *(Courtesy Dale/Incor, Inc.)*

Construction details

Typical load bearing wall panel

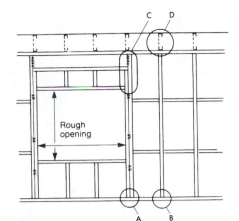

Rough opening

Detail A
Multiple stud and stud to track attachment (single stud to track identical)

Load bearing studs must be seated tight to track web

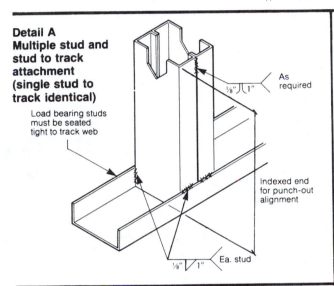

⅛" ⌐ 1"

As required

Indexed end for punch-out alignment

⅛" ⌐ 1"   Ea. stud

Detail B
Stud to track (alternate to welding)

NOTE: In curtain wall application, only (1) screw required at both sides.

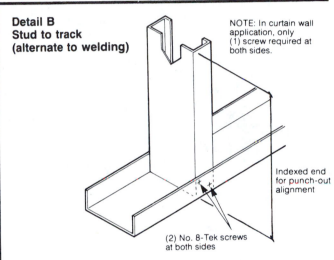

Indexed end for punch-out alignment

(2) No. 8-Tek screws at both sides

Detail C
Header detail

Track

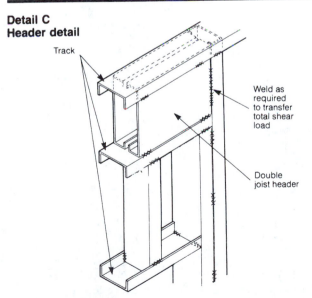

Weld as required to transfer total shear load

Double joist header

Detail D
Joist to track attachment

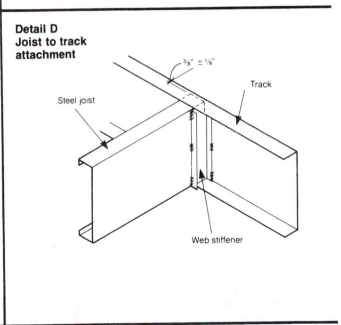

⅜" ± ⅛"

Track

Steel joist

Web stiffener

Figure 17.73 *Continued*

Curtainwall Construction

Spandrel conditions

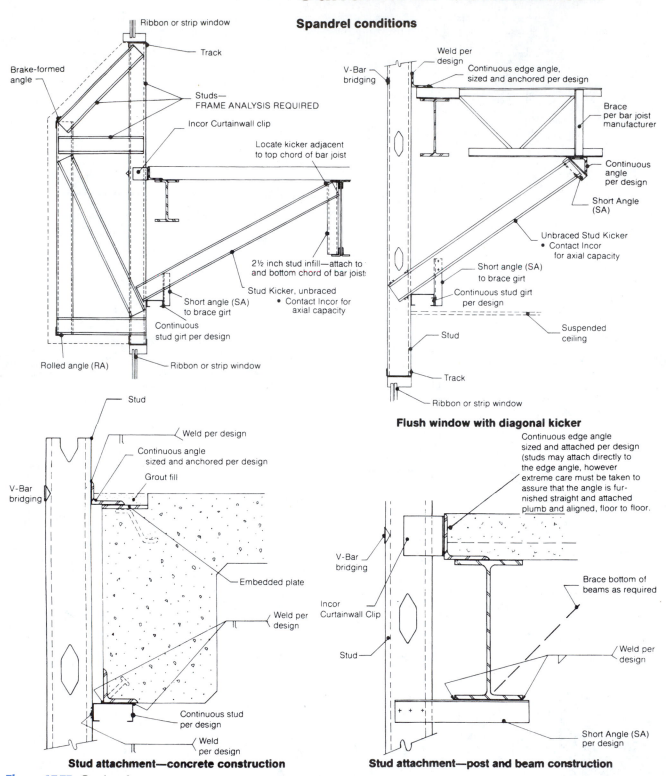

Ribbon or strip window

Track

Brake-formed angle

Studs—FRAME ANALYSIS REQUIRED

Incor Curtainwall clip

Locate kicker adjacent to top chord of bar joist

2½ inch stud infill—attach to and bottom chord of bar joist

Stud Kicker, unbraced
• Contact Incor for axial capacity

Short angle (SA) to brace girt

Continuous stud girt per design

Rolled angle (RA)

Ribbon or strip window

Weld per design

V-Bar bridging

Continuous edge angle, sized and anchored per design

Brace per bar joist manufacturer

Continuous angle per design

Short Angle (SA)

Unbraced Stud Kicker
• Contact Incor for axial capacity

Short angle (SA) to brace girt

Continuous stud girt per design

Suspended ceiling

Stud

Track

Ribbon or strip window

Flush window with diagonal kicker

Stud

Weld per design

Continuous angle sized and anchored per design

Grout fill

V-Bar bridging

Embedded plate

Weld per design

Continuous stud per design

Weld per design

Stud attachment—concrete construction

Continuous edge angle sized and attached per design (studs may attach directly to the edge angle, however extreme care must be taken to assure that the angle is furnished straight and attached plumb and aligned, floor to floor.

V-Bar bridging

Incor Curtainwall Clip

Stud

Brace bottom of beams as required

Weld per design

Short Angle (SA) per design

Stud attachment—post and beam construction

Figure 17.73 *Continued*

Incor steel stud bearing applications

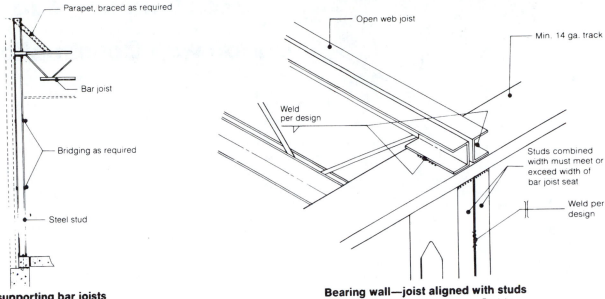

Studs supporting bar joists

Bearing wall—joist aligned with studs

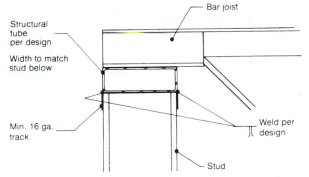

Alternative A—Bearing wall—joist not aligned with studs

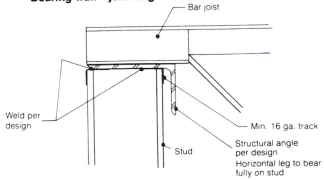

Alternative B—bearing wall—joist not aligned with studs

Curtainwall Construction

Stub wall (base connection to resist overturning due to lateral forces)

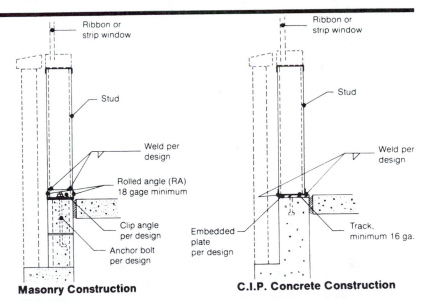

Masonry Construction

C.I.P. Concrete Construction

Figure 17.73 *Continued*

Brick Veneer—Stud Curtainwall Construction

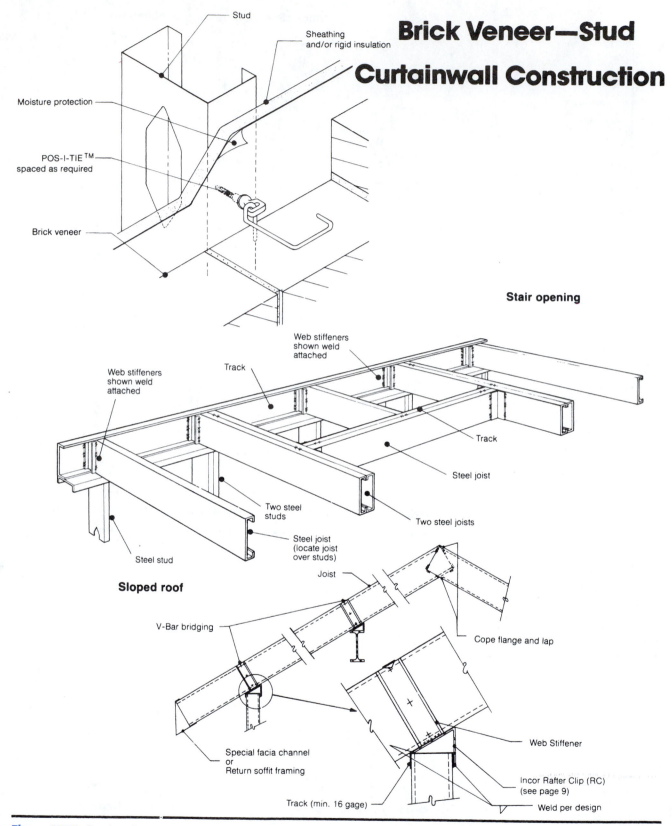

Stud

Sheathing and/or rigid insulation

Moisture protection

POS-I-TIE™ spaced as required

Brick veneer

Stair opening

Web stiffeners shown weld attached

Track

Web stiffeners shown weld attached

Track

Steel joist

Two steel studs

Two steel joists

Steel joist (locate joist over studs)

Steel stud

Sloped roof

Joist

V-Bar bridging

Cope flange and lap

Special facia channel or Return soffit framing

Web Stiffener

Incor Rafter Clip (RC) (see page 9)

Track (min. 16 gage)

Weld per design

Figure 17.73 *Continued*

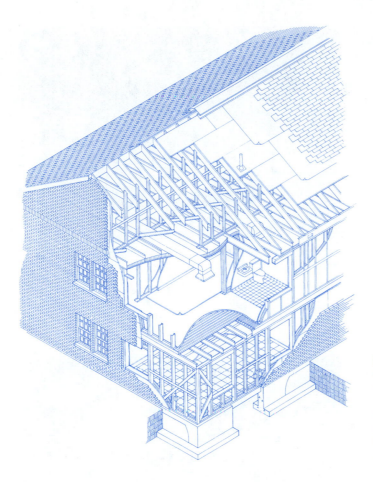

Figure 17.74 This multistory building is framed using lightweight steel studs, joists, rafters, and bracing. The exterior is sheathed, and a brick veneer provides the finished exterior. *(Reproduced with permission from* The Building Systems Integration Handbook, *Richard Rush, ed., Butterworth-Heinmann Publishers, Newton, Mass., 1986)*

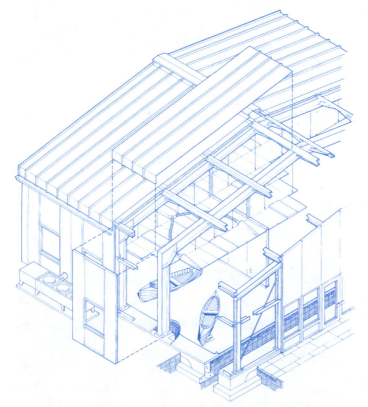

Figure 17.75 This pre-engineered metal building uses factory-assembled rigid frames for the wall and roof structural members. Z-purlins running perpendicular to the rigid frames support the metal roof. *(Reproduced with permission from* The Building Systems Integration Handbook, *Richard Rush, ed., Butterworth-Heinmann Publishers, Newton, Mass., 1986)*

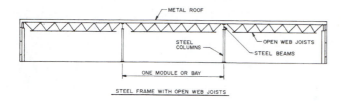

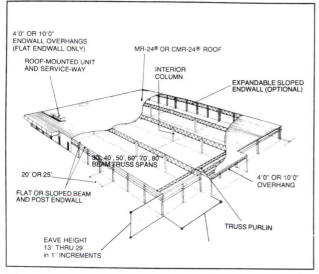

Figure 17.76 A pre-engineered truss-type building uses steel columns, joist girders, and open-web joists. *(Artwork provided by and reflecting the products, components, and systems of Butler Manufacturing Company, P.O. Box 917, Kansas City, Mo. 64141)*

Figure 17.77 Open-web roof joists are supported by a joist girder carried on steel columns. *(Courtesy Vulcraft)*

Figure 17.78 These rigid frames support metal purlins upon which the metal roof decking will be installed.

Figure 17.79 This roof is insulated with fiberglass insulation encased in a vinyl fabric covering.

Smaller open-web joists serve as truss purlins between the joist girders (Fig. 17.77).

Rigid frame buildings, as shown in Fig. 17.75, are available with low sloped and high sloped roofs. The roof decking is supported by light-gauge steel Z-shaped purlins spanning the rigid frames (Fig. 17.78). The ceiling is insulated with long insulating blankets laid over the purlins. The insulation is typically a fiberglass blanket encased in a vinyl fabric covering (Fig. 17.79).

Post-and-beam framed buildings are built using posts on the interior to support the roof. These divide the building into standard-size bays. A large building would be made up of several bays with posts inside the enclosed area (Fig. 17.80). This roof has a small amount of slope. If greater slope is desired the roof is framed with trusses or a rigid frame is used. The roof decking is supported by light-gauge metal purlins that span the roof beams, trusses, or rigid frames.

A roof framing system that uses joists similar to space frame construction is shown in Fig. 17.81. These *triangular joists* span long distances, providing a large covered area free of columns. Electrical and mechanical systems can pass through the joists. They provide an attractive ceiling if left exposed.

Tubular steel trusses are used to carry loads over long distances and provide an attractive, massive appearance.

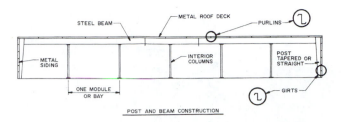

POST AND BEAM CONSTRUCTION

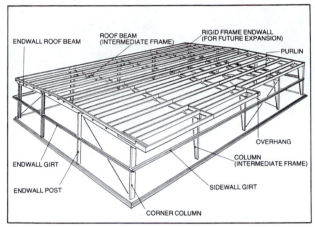

Figure 17.80 This is a four-bay post-and-beam pre-engineered structural frame. *(Artwork provided by and reflecting the products, components, and systems of Butler Manufacturing Company, P.O. Box 917, Kansas City, Mo. 64141)*

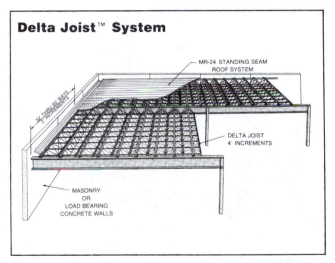

Figure 17.81 These triangular joists span long distances and may be supported by load-bearing walls or interior beams. Those shown are identified as Delta Joists™ by the manufacturer. *(Artwork provided by and reflecting the products, components, and systems of Butler Manufacturing Company, P.O. Box 917, Kansas City, Mo. 64141)*

The one shown in Fig. 17.82 provides a roof over a large covered area without posts. This particular application uses a tensioned fabric roof membrane.

Pre-engineered buildings can also be several stories. An example of a three-story building is in Fig. 17.83. A wide range of metal roofing products is available. These are covered in Chapters 15 and 27.

Figure 17.82 This large tubular truss provides a large open area free of columns and supports a tensioned fabric roof. *(Courtesy Birdair, Inc.)*

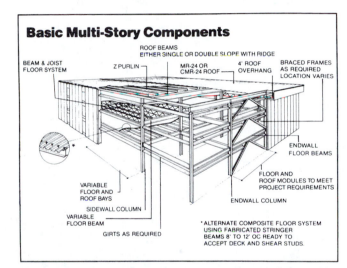

Figure 17.83 Pre-engineered components are used to provide rapid erection of multistory buildings. *(Artwork provided by and reflecting the products, components, and systems of Butler Manufacturing Company, P.O. Box 917, Kansas City, Mo. 64141)*

EXTERIOR WALL SYSTEMS

A number of metal exterior wall systems are available. Some are single-thickness metal panels with ribs formed in them for strength and decorative purposes. Other types have insulation in a hollow-core panel. The insulation may be a foamed core, rigid insulation board, or some form of fibrous insulation blanket. Methods for joining panels vary depending on the manufacturer (Fig. 17.84).

The exposed exterior surface can be a colored coating, such as fluoropolymer over a smooth or embossed metal face (Fig. 17.85); a stucco-type finish, such as applying a fiber-reinforced polymer to the metal; panels with a stone aggregate bonded to the metal (Fig. 17.86); or a kiln-fired clay brick surface.

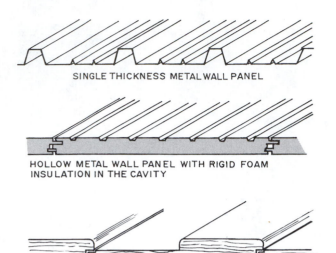

SINGLE THICKNESS METAL WALL PANEL

HOLLOW METAL WALL PANEL WITH RIGID FOAM INSULATION IN THE CAVITY

HOLLOW METAL PANEL WITH FIBERGLASS INSULATION

Figure 17.84 Several types of metal wall panels used on pre-engineered buildings.

Figure 17.86 A metal wall panel with stone aggregate bonded to the metal. *(Courtesy Butler Manufacturing Company)*

Figure 17.85 A metal wall panel with a colored fluoropolymer coating. *(Courtesy Butler Manufacturing Company)*

REVIEW QUESTIONS

1. What are the purposes of the various types of drawings used for the design and erection of steel framed buildings?
2. How are structural steel members lifted for installation as the erection process proceeds?
3. What are the commonly used fastening methods for joining structural steel members?
4. How can the specified tension values be ensured when bolting structural steel members?
5. What techniques can be used to provide for the stability of the structural steel frame?
6. What are the two types of beam connections?
7. What is the difference between bearing-type and friction-type riveted connections?
8. What techniques are used to provide fire protection of structural steel framing?
9. What types of decking are used with structural steel framing?
10. What is a truss?
11. What type of structural system is used to support tensile structures?

KEY TERMS

bending moment The moment that produces bending at a section of a structural member.

design drawings Structural drawings prepared by the structural engineer that give all the information needed for the detailing and fabrication of each steel member.

erection plan An assembly drawing where each structural steel member is located on a building frame.

framed connections Connections joining structural steel members with a metal connector, such as an angle, that is secured to the web of the beam.

galling The wearing or abrading of one material against another under extreme pressure.

moment The property by which a force tends to cause a body to which it is applied to rotate about a point or line.

seated connections Connections that join structural steel members with metal connectors, such as an angle, upon which one member, such as a beam, rests on top of the connector.

shear stress A force on a member that creates a deformation in which parallel planes slide relative to each other so as to remain parallel.

shear studs Metal studs welded to the steel frame that protrude up into the cast-in-place concrete deck.

shop drawings Working drawings giving the information needed to fabricate structural steel members.

SUGGESTED ACTIVITIES

1. Contact an architect or general contractor and try to secure drawings of steel framed buildings. Examine the framing and connection drawings. Notice the system used to identify members and connections.

2. Select one method for fireproofing steel framing listed in your local building code and prepare a sketch showing the details for beam and column fireproofing.

3. Visit a site where some type of steel framed building is under construction. This can be a multistory building or a factory manufactured metal building. Prepare a report with necessary sketches to show the types of connections used. If possible, examine the erection drawings and note the system used to identify each member and its location.

4. Examine the local building code and describe each type of construction in which structural steel framing may be used.

ADDITIONAL INFORMATION

Detailing for Steel Construction, American Institute of Steel Construction, Chicago, Ill., 1983.

Manual of Steel Construction, American Institute of Steel Construction, Chicago, Ill., 1990.

Merritt, F.S., and Ricketts, J.T., *Building Design and Construction Handbook*, McGraw-Hill, New York, 1994.

Newman, M., *Structural Details for Steel Construction*, McGraw-Hill, New York, 1988.

Standard Specifications, Load Tables, and Weight Tables for Steel Joists and Girders, Steel Joist Institute, Myrtle Beach, S.C., 1994.

Courtesy Southern Forest Products Association

Wood

This chapter will help you to:

1. Understand the structural composition of trees and identify the species used in construction.

2. Make decisions pertaining to the influence of defects in lumber on various applications.

3. Know the standard inch and metric lumber sizes.

4. Use information about lumber grades to secure products that will serve the intended purpose.

5. Use the data from the In-Grade Testing Program.

6. Use information about the properties of wood when selecting species to be used.

7. Use the structural properties of wood when designing structural members.

8. Take action to protect the wood portions of a structure from damage by insects and moisture.

Wood is a natural organic material. It is unique among construction materials in that it is a renewable resource. Carefully managed tree farms and natural wild growth provide a continuing source of wood. Wood in the form of lumber and timbers is one of the most familiar construction materials. In addition, wood is used to produce a variety of reconstituted products such as plywood, particleboard, and hardboard. Wood is also used in the manufacture of paper and cardboard. Wood fibers also provide a source of nitrocellulose for the manufacture of explosives.

Because there are many different species of trees, there is a wide variation in the properties of wood. As a result, wood is useful for many applications, both structural and decorative. It is important to understand the properties of a species of wood before selecting it for a particular application.

TREE SPECIES

Several hundred species of wood grow over the world. Some are abundant, while others are rather rare. Some species, because of their properties, location, and abundance, are used commonly in construction. Woods are divided into two classes, hardwoods and softwoods. This division is based on botanical differences not on their actual hardness or softness. The major forest areas providing construction lumber in the United States and Canada are shown in Fig. 18.1.

Softwoods

Much of the production of wood for commercial use is in the class **softwoods.** These are used for framing, sheathing, roofing, subflooring, siding, trim, and millwork. Softwoods are referred to as *coniferous* because they bear cones and, with a few exceptions, have needle-like leaves that stay green all year.

Hardwoods

Hardwood trees are broad-leaved and shed their leaves in the winter. They are called *deciduous.* They are more expensive than softwoods and find use in cabinets, furniture, paneling, interior trim, and flooring.

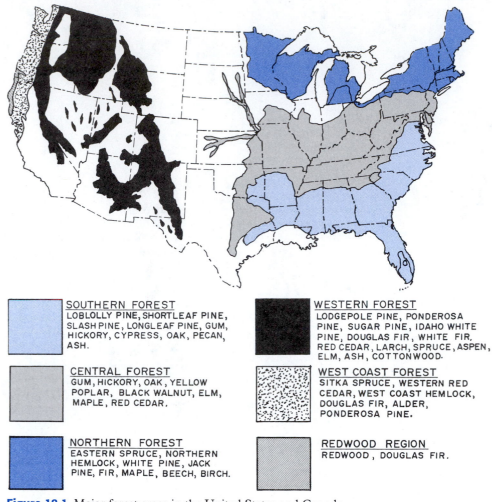

SOUTHERN FOREST
LOBLOLLY PINE, SHORTLEAF PINE,
SLASH PINE, LONGLEAF PINE, GUM,
HICKORY, CYPRESS, OAK, PECAN,
ASH.

WESTERN FOREST
LODGEPOLE PINE, PONDEROSA
PINE, SUGAR PINE, IDAHO WHITE
PINE, DOUGLAS FIR, WHITE FIR,
RED CEDAR, LARCH, SPRUCE, ASPEN,
ELM, ASH, COTTONWOOD.

CENTRAL FOREST
GUM, HICKORY, OAK, YELLOW
POPLAR, BLACK WALNUT, ELM,
MAPLE, RED CEDAR.

WEST COAST FOREST
SITKA SPRUCE, WESTERN RED
CEDAR, WEST COAST HEMLOCK,
DOUGLAS FIR, ALDER,
PONDEROSA PINE.

NORTHERN FOREST
EASTERN SPRUCE, NORTHERN
HEMLOCK, WHITE PINE, JACK
PINE, FIR, MAPLE, BEECH, BIRCH.

REDWOOD REGION
REDWOOD, DOUGLAS FIR.

Figure 18.1 Major forest areas in the United States and Canada.

THE STRUCTURE OF WOOD

The root system anchors the tree to the ground. The *tap root* draws minerals and water from the ground. *Side roots* grow out around the tree, and *feeder roots* grow off the side roots. They catch soil water and minerals in the ground and feed the tree. The water and minerals flow up into the sapwood (xylem). The leaves use sunlight and carbon dioxide to change water and minerals into food. This food flows down into the tree through the inner bark (phloem).

Each part of a tree serves a specific purpose in the tree's growth and development (Fig. 18.2). At the center is a small core called the *pith*. During the early years of growth the pith helps support the stem and feed the tree. As the tree matures it ceases to function. Next to the pith is the **heartwood,** which is hard, mature wood that forms the largest part of the trunk. It strengthens the tree and has the color associated with the particular species. The next layer, **sapwood,** is a living layer that carries water and food throughout the tree. It is soft, usually light in color, and contains more moisture than heartwood. The heartwood and sapwood form growth rings that can be seen clearly when a tree is cut down (Fig. 18.3). The light, soft ring develops in the spring when the tree grows rapidly and is referred to as *springwood*. The dark, hard ring is formed in the summer when growth is slow and is called *summerwood*. When the tree is cut down its approximate age can be determined by counting the hard summerwood annual rings.

Next to the sapwood is the *cambium layer*. It forms new cells that are either xylem or phloem. Sapwood and heartwood are formed with xylem cells. On the outside of the cambium layer are the phloem cells, which form the inner bark layer. Phloem cells in the inner bark carry food to the roots. As more phloem cells are formed, the outer layer of the inner bark changes into outer bark. The *outer bark* is the layer that protects the tree.

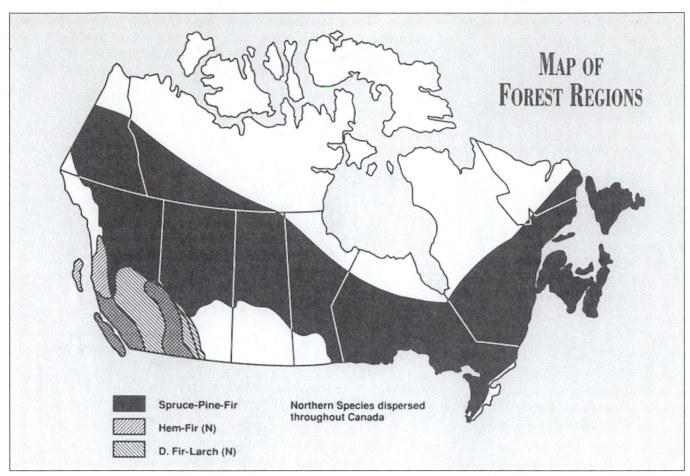

Spruce-Pine-Fir

Hem-Fir (N)

D. Fir-Larch (N)

Northern Species dispersed throughout Canada

Figure 18.1 *Continued*

Within the tree are *medullary rays* that run perpendicular to the growth rings. They carry food and water from the cambium layer to the interior of the tree.

WOOD DEFECTS

A piece of wood may have *natural defects* that occurred while the tree was growing or have *seasoning defects* that are produced as the wood is dried for use.

The natural defects include knots, shake, wane, insect holes, and pitch pockets (Fig. 18.4). *Knots* occur when a branch that is imbedded in the trunk as the tree grows is cut. Knots weaken the wood and are one factor a lumber grader considers as wood is graded. *Pitch* is the accumulation of sap or resin in pockets in the tree. It occurs in softwoods and usually does not cause major problems. *Shake* occurs when a tree is racked or bent, as in a wind storm. Shake appears as small cracks running with or across the annual rings. *Wane* refers to the absence of wood or the presence of bark along the edge of a board. *Insect holes* are caused by boring insects that eat their way into the wood.

Seasoning defects include warp, checks, stain, honeycombing, and case-hardening. **Warp** refers to any variation of shape of a board other than a flat, true surface. Warp develops as the wood loses moisture and can be controlled by proper seasoning. Some of the common types of warp are cup, bow, crook, and twist (Fig. 18.5).

Checks are small cracks on the surface or ends of the boards that are perpendicular to the growth rings. Wood sometimes develops a *stain* on its surface after it is cut into lumber. Stains may be brown, green, or blue. They do not influence the wood's strength. *Honeycombing* occurs when cracks occur on the interior of a board. You will notice these when you cut into a board. They occur if the lumber is not seasoned properly. *Case-hardening* occurs when the outer surface is drier than the interior of the board and has greater stress than the inner section. Again, this is caused by improper seasoning.

SEASONING LUMBER

Seasoning is the process of drying wood from the **green lumber** state to the recommended moisture content de-

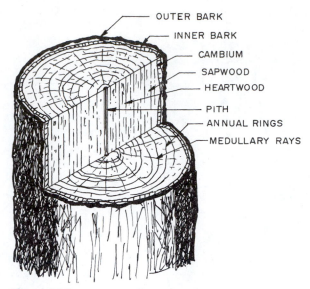

Figure 18.2 The parts of a tree.

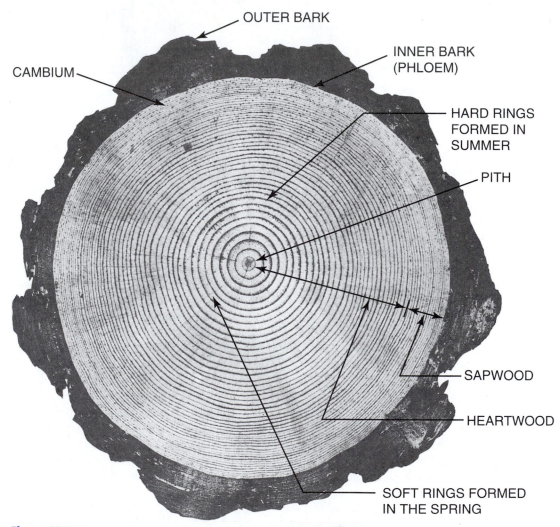

Figure 18.3 The sawed log shows the springwood and summerwood rings. *(Courtesy Forest Products Laboratory, USDA Forest Service, Madison, Wis.)*

Figure 18.4 Typical knots in softwood lumber. These often are loose and drop out. *(Courtesy Forest Products Laboratory, USDA Forest Service, Madison, Wis.)*

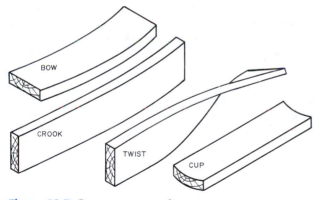

Figure 18.5 Common types of warp.

sired for finished lumber. The methods of seasoning lumber include air drying, kiln drying, dehumidification, and solar drying. Air drying and kiln drying are used most commonly. Seasoning requirements for softwoods as specified by the Southern Pines Inspection Bureau are in Table 18.1.

Air-Dried Lumber

Air drying involves stacking the boards with stickers (wood strips) in between the layers. This permits air to circulate between the layers, aiding the drying process. Air drying is a slow process and does not provide control over how fast the drying occurs. On hot, dry, windy days the wood will season too fast, and checking and warping will occur. On cool days with high humidity the wood will dry very little. It is difficult to air dry lumber to the percentage of moisture desired (Fig. 18.6). Some dimension lumber and lower grades of softwood lumber are often air dried. Structural timbers are so large they would take a very long time to air dry, so they often are shipped green.

The moisture content of air-dried lumber for construction purposes should be in the 15 to 19 percent range in the United States. In some parts of the country, such as the Southwest, it could be somewhat lower than 15 percent, while in the moist Pacific Northwest it may be closer to 19 to 20 percent. Lumber grading rules for the various softwoods specify the maximum moisture content for each grade regardless of how it is dried.

Hardwoods are usually air dried for a while then dried in a kiln. It is difficult to get the moisture content of hardwoods low enough for use in furniture and cabinets by air drying only. Normally a moisture content of 6 to 8 percent is required for these purposes, so hardwoods have to be kiln dried.

Kiln-Dried Lumber

Kiln drying involves stacking and sticking the lumber as described for air drying. It is stacked in a kiln, which is an enclosed building. In the kiln the temperature, humidity, and air circulation are controlled to carefully reduce the moisture content of the wood. Air temperatures in the kiln reach 180° F (82° C) with an equally high relative humidity (Fig. 18.7). Since the temperature and humidity are controlled, the lumber can be quickly dried to any desired moisture content. Most woods can be dried in less than two weeks in a kiln. Kiln drying reduces the defects in the wood and produces a product that will not expand or contract as much as air-dried wood.

Dehumidification and Solar Kilns

These are two relatively new types of kilns. The *dehumidification method* uses electricity to dry the lumber. *Solar kilns* use the energy of the sun to produce the heat needed, and they are the most economical. Those in current use can handle only small amounts of wood.

Unseasoned Lumber

Most lumber more than 2 in. thick is air dried. The thicker the stock the longer the drying time required to achieve the desired moisture content of 19 percent for this type of material. Most stock in these thicknesses is used in a green condition (more than 19 percent moisture content). It continues to dry and shrink after it has been used, which sometimes causes problems. The moisture content can be checked on the job site with a battery-operated moisture meter.

LUMBER SIZES

After the rough sawed wood has been dried, it is taken to a planing mill to be surfaced and shaped into products such as boards, dimension stock, timbers, molding, and trim. Softwood lumber is produced in standard sizes. Hardwoods are often **dressed** to thickness but not to any specific width or length. After the lumber is planed or shaped, it must be stored in weatherproof sheds so the moisture content is not increased. Some manufacturers wrap it in plastic sheets.

Table 18.1 Seasoning Requirements for Softwoods[a]

Items (Nominal)	Moisture Content Limit	
	Maximum (Dry)	Kiln-Dried (KD or MC15)
D&btr grades		
1″ & 1¼″	15%	12% on 90% of pieces 15% on remainder
1½″, 1¾″ & 2″	18%	15%
Over 2″ not over 4″	19%	15%
Over 4″	20%	18%
Paneling		
1″		12%
Boards 2″ and less and dimension 2″ to 4″	19%	15%
Decking		
2″ thick	19%	15%
3″ and 4″ thick		15% on 90% of pieces 18% on remainder
Heavy dimension		
Over 2″ not over 4″	19%	15%
Timbers		
5″ and thicker	23%	20%

[a]As specified by the Southern Pine Inspection Bureau

Figure 18.6 Lumber stacked for air drying before being moved to the kiln. (*Courtesy Southern Forest Products Association*)

Figure 18.7 Lumber being loaded into a dry kiln. (*Courtesy Southern Forest Products Association*)

Softwood Lumber

Softwood lumber is sold by designating its nominal size. The nominal size is the size of the board after it has been rough cut at the sawmill. Inch and metric sizes are shown in Tables 18.2, 18.3, and 18.4. The dressed or finished size is the actual size of the board after it has dried and been surfaced. For example, a typical inch stud has a nominal size of 2 × 4 in. and a finished size of 1½ × 3½ in. A dry, dressed metric size stud is 38.10 × 88.90 mm. If stock is surfaced green it is made larger than the dressed size of dry stock so that when it dries and shrinks it will be the same size as the dried stock. The size of worked stock includes the overall size, as shown in Fig. 18.8. *Worked stock* refers to boards that have some machining operation, such as tongue-and-groove flooring.

Lumber shrinks in direct proportion to its loss of moisture as it seasons from a green to a dry condition. This makes it possible to establish separate sizes for green and dry surfaced lumber so that when the green lumber used in a building eventually reaches the required moisture content it will have essentially the same size, strength, and stiffness as kiln-dried lumber over the same spans and loading conditions. The dressed sizes of green and dry softwood lumber are specified in American Softwood Lumber Standard PS 20–70 and the Canadian National Lumber Grades Authority standard CSA 0141. American and Canadian standards are coordinated and accepted in both countries. They are based on dry lumber having a moisture content of 19 percent or less and green lumber more than 19 percent. The standard sizes for finish lumber, boards, siding, and other wood products are in Tables 18.5 and 18.6.

Softwood lumber is sold in standard lengths of two-foot multiples ranging from 6 ft. to 24 ft. Some special types, such as precut studs, are cut to the exact desired length. Metric lumber lengths are specified in meters (m) and decimal parts of a meter. Metric lengths are shown in Table 18.4.

Table 18.3 Standard Sizes of Surfaced Timbers

Thickness and Width of Stock					
Customary Units (in.)			Metric Units (mm)		
Nominal	Actual Dry	Actual Green	Nomen-clature[a]	Actual Dry	Actual Green
5	-	4½	114	-	114.3
6	-	5½	140	-	139.7
7	-	6½	165	-	165.1
8	-	7½	191	-	190.5
9	-	8½	216	-	215.9
10	-	9½	241	-	241.3
12	-	11½	292	-	292.1
14	-	13½	343	-	342.9
16	-	15½	394	-	393.7
18	-	17½	445	-	444.5
20	-	19½	495	-	495.3

Courtesy Canadian Wood Council and the Southern Forest Products Association
[a]Nomenclature means the size used to describe the member. Actual size may be slightly larger or smaller.

Table 18.2 Standard Sizes of Surfaced Dimension Lumber

Thickness of Stock					
Customary Units (in.)			Metric Units (mm)		
Nominal	Actual Dry	Actual Green	Nomen-clature[a]	Actual Dry	Actual Green
2	1½	1⁹/₁₆	38	38.10	39.69
2½	2	2¹/₁₆	51	50.80	52.39
3	2½	2⁹/₁₆	64	63.50	65.09
3½	3	3¹/₁₆	76	76.20	77.79
4	3½	3⁹/₁₆	89	88.90	90.49
4½	4	4¹/₁₆	102	101.60	103.19
Width of Stock					
2	1½	1⁹/₁₆	38	38.10	39.69
3	2½	2⁹/₁₆	64	63.50	65.09
4	3½	3⁹/₁₆	89	88.90	90.49
5	4½	4⅝	114	114.30	114.47
6	5½	5⅝	140	139.70	142.87
7	6½	6⅝	165	165.10	168.28
8	7¼	7½	184	184.15	190.50
10	9¼	9½	235	234.95	241.30
12	11¼	11½	286	285.75	292.10
14	13¼	13½	337	336.55	342.90
16	15¼	15½	387	387.35	393.70

Courtesy Canadian Wood Council and the Southern Forest Products Association
[a]Nomenclature means the size used to describe the members. Actual size may be slightly larger or smaller.

Table 18.4 Metric Lengths for Softwood Lumber

Nominal Length (ft.)	Metric Length (m)
3	0.91
4	1.22
5	1.52
6	1.83
7	2.13
8	2.44
9	2.74
10	3.05
11	3.35
12	3.66
13	3.96
14	4.27
15	4.57
16	4.88
17	5.18
18	5.49
19	5.79
20	6.10
21	6.40
22	6.71
23	7.01
24	7.32

Courtesy Canadian Wood Council

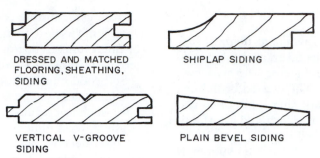

DRESSED AND MATCHED
FLOORING, SHEATHING,
SIDING

SHIPLAP SIDING

VERTICAL V-GROOVE
SIDING

PLAIN BEVEL SIDING

Figure 18.8 Examples of worked stock.

Table 18.5 Standard Sizes for Softwood Products

	Thickness (in.)		Width (in.)		
	Nominal	Worked	Nominal	Face	Overall
Bevel siding	$\frac{1}{2}$	$\frac{3}{16} \times \frac{7}{16}$	4	$3\frac{1}{2}$	$3\frac{1}{2}$
	$\frac{5}{8}$	$\frac{7}{16} \times \frac{9}{16}$	5	$4\frac{1}{2}$	$4\frac{1}{2}$
	$\frac{3}{4}$	$\frac{3}{16} \times \frac{11}{16}$	6	$5\frac{1}{2}$	$5\frac{1}{2}$
	1	$\frac{3}{16} \times \frac{3}{4}$	8	$7\frac{1}{4}$	$7\frac{1}{4}$
Drop siding	$\frac{5}{8}$	$\frac{9}{16}$	4	$3\frac{1}{8}$	$3\frac{3}{8}$
Rustic and drop	1	$\frac{23}{32}$	5	$4\frac{1}{8}$	$4\frac{3}{8}$
siding (dressed)			6	$5\frac{1}{8}$	$5\frac{3}{8}$
and matched)			8	$6\frac{7}{8}$	$7\frac{1}{8}$
			10	$8\frac{7}{8}$	$9\frac{1}{8}$
Rustic and drop	$\frac{5}{8}$	$\frac{9}{16}$	4	3	$3\frac{3}{8}$
siding	1	$\frac{23}{32}$	5	4	$4\frac{3}{8}$
(shiplapped)			6	5	$5\frac{3}{8}$
			8	$6\frac{5}{8}$	$7\frac{1}{8}$
			10	$8\frac{5}{8}$	$9\frac{1}{8}$
			12	$10\frac{5}{8}$	$11\frac{1}{8}$
Flooring	$\frac{3}{8}$	$\frac{5}{16}$	2	$1\frac{1}{8}$	$1\frac{3}{8}$
	$\frac{1}{2}$	$\frac{7}{16}$	3	$2\frac{1}{8}$	$2\frac{3}{8}$
	$\frac{5}{8}$	$\frac{9}{16}$	4	$3\frac{1}{8}$	$3\frac{3}{8}$
	1	$\frac{3}{4}$	5	$4\frac{1}{8}$	$4\frac{3}{8}$
	$1\frac{1}{4}$	1	6	$5\frac{1}{8}$	$5\frac{3}{8}$
	$1\frac{1}{2}$	$1\frac{1}{4}$			
Ceiling	$\frac{3}{8}$	$\frac{5}{16}$	3	$2\frac{1}{8}$	$2\frac{3}{8}$
	$\frac{1}{2}$	$\frac{7}{16}$	4	$3\frac{1}{8}$	$3\frac{3}{8}$
	$\frac{5}{8}$	$\frac{9}{16}$	5	$4\frac{1}{8}$	$4\frac{3}{8}$
	$\frac{3}{4}$	$\frac{11}{16}$	6	$5\frac{1}{8}$	$5\frac{3}{8}$
Partition	1	$\frac{23}{32}$	3	$2\frac{1}{8}$	$2\frac{3}{8}$
			4	$3\frac{1}{8}$	$3\frac{3}{8}$
			5	$4\frac{1}{8}$	$4\frac{3}{8}$
			6	$5\frac{1}{8}$	$5\frac{3}{8}$
Paneling	1	$\frac{23}{32}$	3	$2\frac{1}{8}$	$2\frac{3}{8}$
			4	$3\frac{1}{8}$	$3\frac{3}{8}$
			5	$4\frac{1}{8}$	$4\frac{3}{8}$
			6	$5\frac{1}{8}$	$5\frac{3}{8}$
			8	$6\frac{7}{8}$	$7\frac{1}{8}$
			10	$8\frac{7}{8}$	$9\frac{1}{8}$
			12	$10\frac{7}{8}$	$11\frac{1}{8}$
Shiplap	1	$\frac{3}{4}$	4	$3\frac{1}{8}$	$3\frac{1}{2}$
			6	$5\frac{1}{8}$	$5\frac{1}{2}$
			8	$6\frac{7}{8}$	$7\frac{1}{4}$
			10	$8\frac{7}{8}$	$9\frac{1}{4}$
			12	$10\frac{7}{8}$	$11\frac{1}{4}$
Dressed and	1	$\frac{3}{4}$	4	$3\frac{1}{8}$	$3\frac{3}{8}$
matched	$1\frac{1}{4}$	1	5	$4\frac{1}{8}$	$4\frac{3}{8}$
	$1\frac{1}{2}$	$1\frac{1}{4}$	6	$5\frac{1}{8}$	$5\frac{3}{8}$
			8	$6\frac{7}{8}$	$7\frac{1}{8}$
			10	$8\frac{7}{8}$	$9\frac{1}{8}$
			12	$10\frac{7}{8}$	$11\frac{1}{8}$

Courtesy Southern Forest Products Association

Table 18.6 Metric Sizes for Surfaced Boards, Shiplap Siding, and Centre Match Siding

Actual Thickness (mm) Dry, Dressed	Actual Width (mm) Dry, Dressed
17	38
19	64
25	89
32	114
	140
	165
	184
	210
	235

Courtesy Canadian Wood Council

Hardwood Lumber

Hardwood lumber is sold in random widths and lengths and often is not surfaced to standard thicknesses. Since hardwoods are used in the manufacture of cabinets and furniture, the boards are cut into many sizes and thicknesses. Cutting to standard widths and lengths would cause great waste. Standard rough thickness in customary units range from ³⁄₈ in. to 1¹⁄₂ in. in ¹⁄₄ in. increments. Hardwood lumber is also available in 2, 3, and 4 in. thicknesses.

Hardwood lumber in Canada is sized so the metric units and customary units are within 1 percent of each other. The standard thicknesses of rough lumber are 15 mm to 60 mm in 5 mm increments and 70 mm to 100 mm in 10 mm increments. Surfaced lumber thicknesses are 5 mm less than the rough size for green or air-dried lumber and 5 to 8 mm for kiln-dried lumber. Standard lengths are 1.2 m through 4.8 m in increments of 30 cm. Hardwood lumber is sold by volume in cubic meters (m³).

Buying Lumber

Most lumber in the United States is sold by the **board foot.** A board foot is equal to a piece of lumber with an actual size of 1 in. thick, 12 in. wide, and 1 ft. long. For example a board 1 in. thick, 12 in. wide, and 8 ft. long contains 8 board feet. These sizes are the nominal sizes. The actual size is ³⁄₄ in. × 11¹⁄₄ in. × 8 in. Usually stock under 1 in. thick is figured as 1 in. To calculate board feet multiply the thickness in inches by the width in inches by the length in feet and divide by 12 (which converts the width to feet.) (Fig. 18.9).

Metric lumber is sold by the cubic meter (m³). The volume is based on the actual size. This gives the actual volume of the piece in cubic meters. It is computed by the formula

Volume = thickness (mm) × width (mm) × length (m)

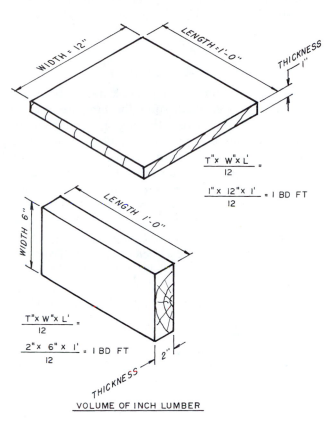

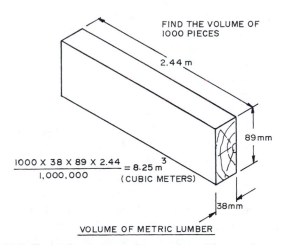

Figure 18.9 Examples representing one board foot and metric volume of lumber.

Thickness and width are given in millimeters (mm) and the length is in meters (m). An example is shown in Fig. 18.9. The Canadian Wood Council provides tables for converting metric volume to board feet.

Other materials, such as molding and trim, are sold by the linear foot. Posts and pilings are sold by the piece. Wood shingles are sold by the square (enough to cover 100 ft.²).

LUMBER MANUFACTURING

The production of lumber of all kinds begins in the forest where trees are felled. After the branches are cut away the log is moved to a loading site. Logs may be skidded to the site by a tractor or hauled down a steep hillside on a long cable called a choker line. Once they reach the loading site they are generally loaded on trucks (Fig. 18.10). The trucks operate on dirt roads built into the forest especially for the logging operation. At the mill the logs are stacked into large piles and often sprayed with water to keep them from drying out and splitting. Most mills use a mechanical peeler to grind off the bark (Fig. 18.11). After the log has been peeled it is ready to be cut into lumber of various types.

A log is hauled onto the carriage of a large machine called a *heading*. This machine has a saw blade that is most often a band type. Small mills use circular saw blades 3 to 4 ft. in diameter. The carriage moves the log into the blade, which cuts off slabs of wood. The saw operator, called the sawyer, judges how best to get the most marketable wood from each log and adjusts the machinery to rotate and advance the log into the saw. Some mills use computer controls to assist with this process (Fig. 18.12). The slabs cut from the log fall on a conveyor and move to an *edger*, where the slabs are cut to desired widths and edges are cut square. Then the slab goes to a *trim saw*, which cuts the rough, square-edged slabs to standard lengths. Hardwoods are not edged or cut to length.

Most lumber used for construction purposes is *plain-sawed*. This method produces the maximum yield from the log and the widest stock (Fig. 18.13). Plain-sawed lumber produces boards that have a broad grain running the length of the board. Another common sawing method, **quartersawing,** produces boards with the edges of the annual rings showing on the face. This is often done with flooring because the annual rings are hard and withstand wear better than the wide areas of springwood exposed on plain-sawed boards. Examples of commonly used softwoods showing plain-sawed, quartersawed, and end grain are in Fig. 18.14.

Next the lumber moves on a conveyor past a **lumber grader.** The grader usually grades each board and marks it as it passes from the trim saw. The mark indicates the **grade.** This is a very technical job because of the extensive specifications for each of the many grades

Figure 18.10 Southern pine logs being moved to a loading area where they will be put on trucks and moved to a mill. *(Courtesy Southern Forest Products Association)*

Figure 18.11 This log is having its bark removed by a mechanical debarking machine. *(Courtesy Southern Forest Products Association)*

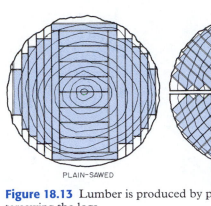

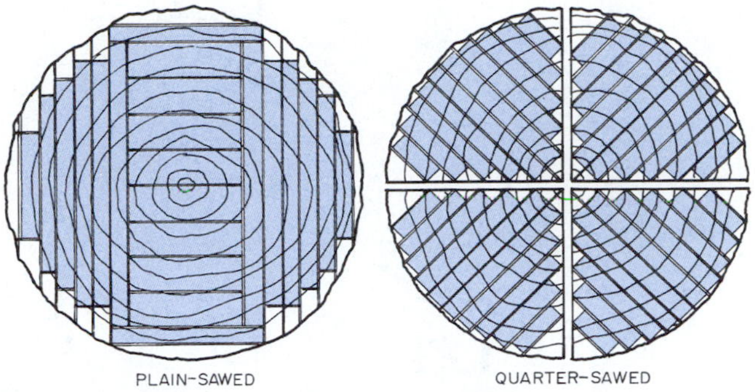

Figure 18.13 Lumber is produced by plain-sawing or quartersawing the logs.

Figure 18.12 This log is being cut into lumber. *(Courtesy Southern Forest Products Association)*

and types of lumber. Finally, the green lumber goes to an air-drying area where it is stacked with sticks between each layer. After a number of weeks the lumber is usually moved, still stacked, to a kiln for the final reduction of moisture.

After the lumber has been dried it is dressed to its finished size in a planing mill. All dimensional lumber (up to but not including 5 in. or 114 mm) must be marked S-Dry (surfaced dry) if it was surfaced dry. If it was surfaced green (moisture content more than 19 percent), it must be grade stamped S-GRN. If the lumber is surfaced at 15 percent moisture content it is often stamped MC15. Hardwood lumber is generally shipped dried but unsurfaced to manufacturers that use it for furniture, cabinets, and other products.

LUMBER GRADING AND TESTING

Lumber Grades

The grades of softwood lumber are based on American Softwood Lumber Standard PS 20–70 published by the U.S. Department of Commerce. Under the provisions of this standard a National Grading Rule Committee (NGRC) was established to develop uniform nationwide grade requirements for dimensional lumber. The functions of the committee were to "establish, maintain, and make fully and fairly available grade strength ratios, nomenclature, and descriptions of grades of dimension lumber conforming to American Lumber Standards." The specifications developed are known as the National Grading Rule for Dimensional Lumber (NGRDL). They form a part of the grading rules of all wood association grading rules, such as those of the Western Wood Products Association, the Southern Pine Inspection Bureau, West Coast Lumber Inspection Bureau, and the Redwood Inspection Service. The grading rules of these

individual associations pertain to those species produced in their geographic areas and contain the National Grading Rules for Dimensional Lumber specifications plus additional specifications unique to that association (Table 18.7).

Grades of Canadian lumber are identical with those used in the United States, or the same as the requirements of American Softwood Lumber Standard PS 20–70. The grades are published by the Canadian National Lumber Grades Authority (NLGA) and appear in the publication Standard Grading Rules for Canadian Lumber (Table 18.7). The mills that manufacture lumber according to these standards place a grade stamp on each piece (Fig. 18.15).

The softwood lumber grading standards cover primarily the nomenclature and bending strength ratios for structural light framing, light framing, structural joists and planks, appearance framing, and stud grades. The bending stress ratios indicate the relationships between basic stresses of clear wood free of strength-reducing defects (knots, splits) and the stresses developed in a particular grade of lumber of the same species. Some examples using customary units are in Table 18.8. Data for Canadian species are given in kilograms per square meter (kg/m^2) and are available from the Canadian Wood Council. A description of the most commonly used lumber grades for softwoods as used by the Southern Pine Inspection Bureau is shown in Table 18.9. Other U.S. regional and Canadian associations have established similar grades for the species harvested in their regions. The higher the grade the higher the allowable stresses. However, the lower grades of lumber cost less.

Lumber Classifications as to Manufacture

Rough lumber has not been dressed, and it shows saw marks on all four surfaces.

Blanked lumber is dressed to a size larger than standard dressed sizes but smaller than the nominal size. It may be surfaced on one surface (S1S), one edge (S1E), on to surfacing the four sides (S4S).

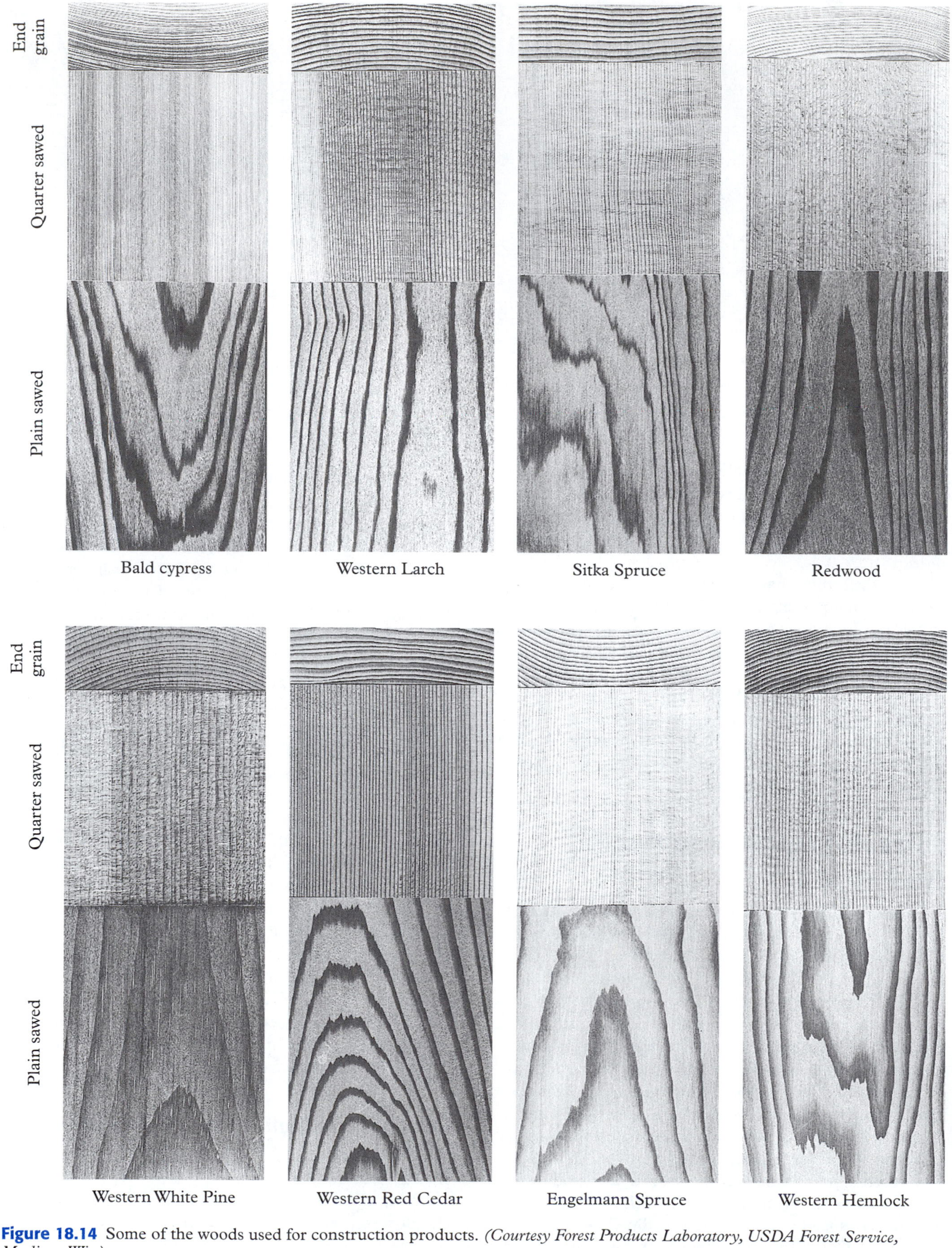

Figure 18.14 Some of the woods used for construction products. *(Courtesy Forest Products Laboratory, USDA Forest Service, Madison, Wis.)*

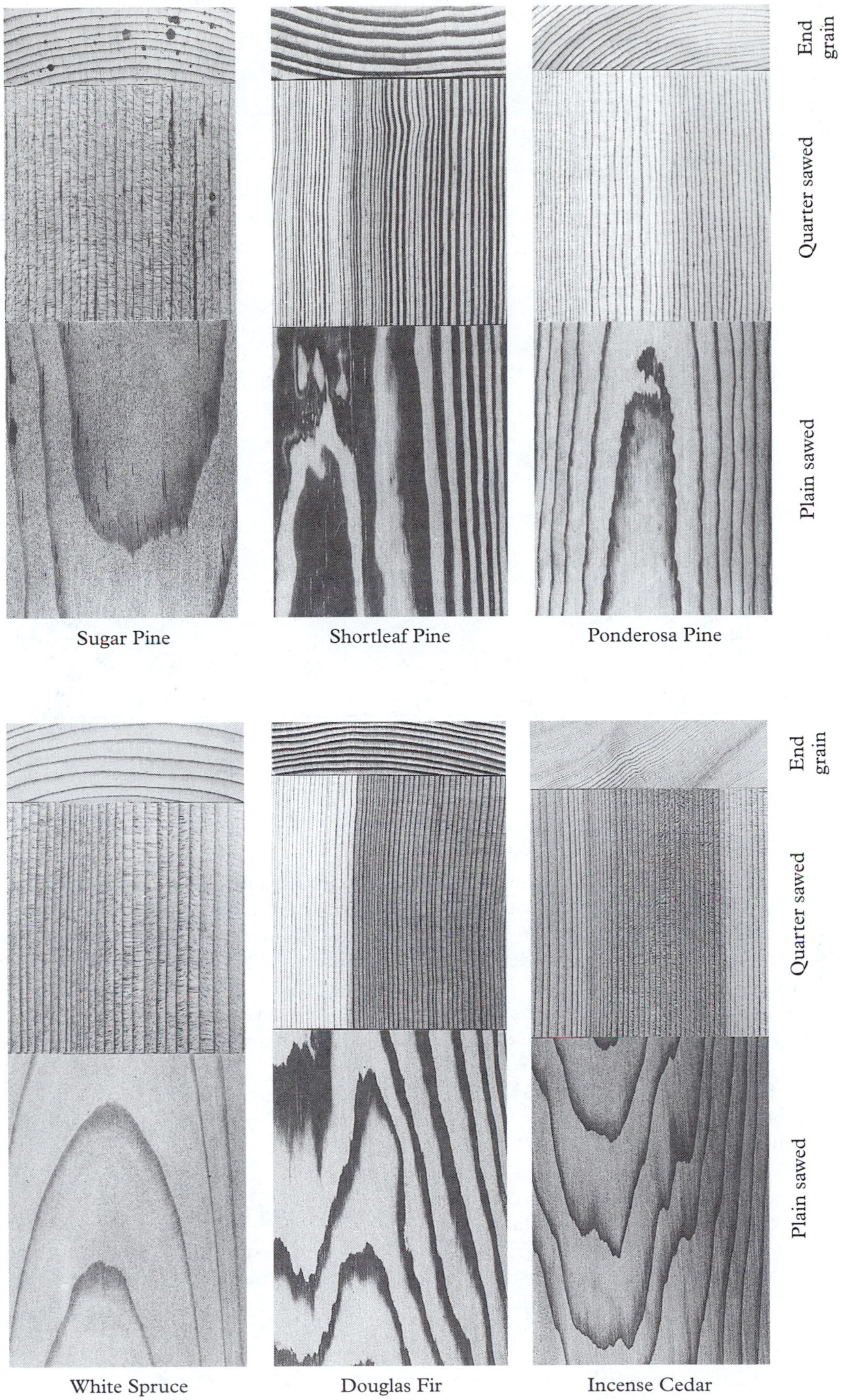

End grain
Quarter sawed
Plain sawed

Sugar Pine   Shortleaf Pine   Ponderosa Pine

End grain
Quarter sawed
Plain sawed

White Spruce   Douglas Fir   Incense Cedar

Figure 18.14 *Continued*

Table 18.7 U.S. and Canadian Softwood Grading Authorities by Region

United States			
Softwood Region	**Species**	**Lumber Association**	**Grading Authority**
Western wood region	Western red cedar Ponderosa pine Douglas fir White fir Western hemlock Englemann spruce Western larch Sitka spruce Lodgepole pine Idaho white pine Sugar pine	Western Wood Products Association	Western Wood Products Association West Coast Lumber Inspection Bureau
Southern pine region	Shortleaf pine Longleaf pine	Southern Forest Products Association	Southern Pine Inspection Bureau
Redwood region	Redwood Douglas fir	California Redwood Association	Redwood Inspection Service
Canada			
	Spruce Pine, several types Fir, several types Larch Douglas fir Hemlock Western red cedar Aspen Poplar	Canadian Wood Council and 12 lumber grading authorities	Canadian National Lumber Grades Authority

Dressed lumber has one or more surfaces smoothed in any combination of surfaces and edges, such as S1E, S2E, S1S1E, S1S2E, S4S, etc. For example S1E means surface one edge, S4S means surface four sides, and S1S1E means surface one side and one edge.

Worked lumber has been dressed and has been matched, shiplapped, or patterned. *Matched lumber* has a tongue on one edge and a groove on the other providing a tongue-and-groove joint. *Shiplapped lumber* has been worked or rabbeted on both edges of each piece to provide a lapped joint when the pieces fit together. *Patterned lumber* has been worked to a shaped or molded form in addition to being dressed, matched, or shiplapped.

Resawn lumber is produced by resawing any thickness of lumber into thinner lumber. *Ripped lumber* is produced by sawing any width lumber in narrower pieces.

Size Classifications

The following size classifications are used by the Southern Pine Inspection Bureau. A summary of these grades and classifications is in Table 18.10.

Boards are less than 2 in. in nominal thickness and 1 in. or more in width. If they are less than 6 in. wide they are classified as strips.

Dimension lumber is from 2 in. up to but not including 5 in. thick and 2 in. or more in width. It is subdivided into five classes: structural light framing, light framing, structural joists and planks, appearance framing, and stud.

Timbers are 5 in. or more in their least dimension and are subdivided into classes such as beams, posts, girders, etc.

Stress-Rated Lumber

Each piece is **stress rated** at the mill. The lumber is evaluated by mechanical stress-rating equipment that subjects each piece to bending stress. The modulus of elasticity (E), which is a measure of stiffness, is measured by instruments on the machine. The stress grade is electronically calculated and includes consideration of the effect of the slope of the grain, knots, growth rate,

Typical Grade Marks Used in the United States

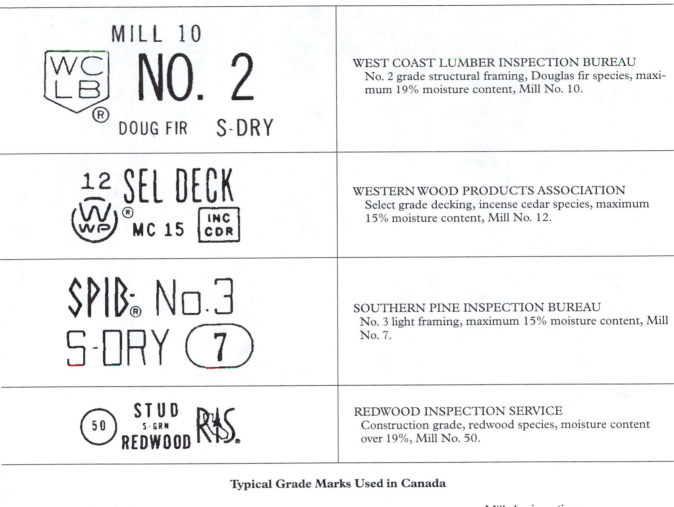

WEST COAST LUMBER INSPECTION BUREAU
No. 2 grade structural framing, Douglas fir species, maximum 19% moisture content, Mill No. 10.

WESTERN WOOD PRODUCTS ASSOCIATION
Select grade decking, incense cedar species, maximum 15% moisture content, Mill No. 12.

SOUTHERN PINE INSPECTION BUREAU
No. 3 light framing, maximum 15% moisture content, Mill No. 7.

REDWOOD INSPECTION SERVICE
Construction grade, redwood species, moisture content over 19%, Mill No. 50.

Typical Grade Marks Used in Canada

PARTS OF THE GRADE STAMP

CARIBOO LUMBER MANUFACTURERS ASSOCIATION
205–197 North 2nd Ave
Williams Lake, B.C. V2G 1Z5

Figure 18.15 Typical grade stamps for dimension lumber used by various grading authorities in the United States and Canada.

Table 18.8 Working Stresses for Selected Softwood Species

Species and Commercial Grade	Size Classification	Allowable Unit Stresses[a] in lb/in.²						
		Extreme Fiber in Bending "F_b"[b]		Tension Parallel to Grain "F_t"	Horizontal Shear "F_v"	Compression Perpendicular to Grain "F_c" ⊥	Compression Parallel to Grain "$F_{c\,//}$"	Modulus of Elasticity "E"
		Single-Member	Repetitive Member					
Douglas Fir-Larch (Surfaced dry or surfaced green. Used at 19% max. m.c.)								
Select Structural		2100	2400	1200	95	385	625	1,800,000
No. 1		1750	2050	1050	95	385	625	1,800,000
No. 2	2" to 4" thick	1450	1650	850	95	385	625	1,700,000
No. 3	2" to 4" wide	800	925	475	95	385	625	1,500,000
Appearance		1750	2050	1050	95	385	625	1,800,000
Hem-Fir (Surfaced dry or surfaced green. Used at 19% max. m.c.)								
Select Structural		1650	1900	975	75	405	1300	1,500,000
No. 1		1400	1600	825	75	405	1050	1,500,000
No. 2	2" to 4" thick	1150	1350	675	75	405	825	1,400,000
No. 3	2" to 4" wide	650	725	375	75	405	500	1,200,000
Appearance		1400	1600	825	75	405	1250	1,500,000
Southern Pine (Surfaced at 15% maximum moisture content, K.D. 15. Used at 15% max. m.c.)								
Select Structural		2150	2500	1250	105	565	1800	1,800,000
No. 1	2" to 4" thick	1850	2100	1050	105	565	1450	1,800,000
No. 2	2" to 4" wide	1550	1750	900	95	565	1150	1,600,000
No. 3		850	975	500	95	565	675	1,500,000

[a]Values shown are for normal loading conditions at maximum 19% m.c., as in most covered structures.
[b]Values in bending for "repetitive member uses" are intended for design of members spaced not over 24" o.c., in groups of not less than three members and when joined by floor or roof sheathing elements.

moisture content, and density. The machine then puts the grade stamp on the piece.

The grade stamp on machine stress-rated lumber indicates that the stress-rating system used to make the test meets certification requirements. The grade stamp shows the agency trademark, the mill name or number, the phrase "Machine Rated," the species and the "E" rating in millions of pounds per square inch (Fig. 18.16).

The Southern Pine Inspection Bureau has established fifteen categories of stress-graded lumber 2 in. or less in thickness. These categories can be used for most structural purposes, such as for trussed rafters. There are five classes with lower allowable bending stresses in relation to the modulus of elasticity. The classes can be used when lower bending stress can be used, such as in floor joists.

The Southern Pine Inspection Bureau also has two grades of stress-rated timbers 5 in. × 5 in. and larger, Select Structural (SR) and Dense Select Structural (SR). They are divided into No. 1 SR, No. 1 Dense SR, and No. 2 SR, No. 2 Dense SR. There is also a series of grades for stress-rated industrial lumber. Other regional associations in the United States and Canada have similar grading procedures.

In-Grade Testing Program

The In-Grade Testing Program was a twelve-year research program conducted by the U.S. Department of Agriculture Forest Products Laboratory in cooperation with U.S. and Canadian lumber industry associations. The program was initiated to verify the softwood lumber design values for visually graded lumber and to provide a scientific basis for wood engineering similar to that used for steel and concrete. Thousands of lumber specimens of many species, grades, and sizes were tested and design values were applied.

The In-Grade Testing Program developed new *species groupings*. Six species groups were developed for Western lumber species and Eastern lumber species. (Table 18.11). The Canadian species groupings are in Table 18.12. Southern pine is a separate grouping for woods in the south and southeastern United States. The design values are available in the Supplement to the National Design Specifications for Wood Construction. The various lumber associations also have publications pertaining to the species in their areas.

Table 18.9 Southern Pine Softwood Lumber Grades

Product	Grade	Character of Grade and Typical Uses
Finish	B&B	Highest recognized grade of finish. Generally clear, although a limited number of pin knots permitted. Finest quality for natural or stain finish.
	C	Excellent for painted or natural finish where requirements are less exacting. Reasonably clear but permits limited number of surface checks and small tight knots.
	C&Btr	Combination of B&B and C grades; satisfies requirements for high-quality finish.
	D	Economical, serviceable grade for natural or painted finish.
Boards S4S	No. 1	High quality with good appearance characteristics. Generally sound and tight-knotted. Largest hole permitted is $1/16''$. A superior product suitable for wide range of uses, including shelving, form, and crating lumber.
	No. 2	High-quality sheathing material, characterized by tight knots. Generally free of holes.
	No. 3	Good, serviceable sheathing, usable for many applications without waste.
	No. 4	Admit pieces below No. 3 which can be used without waste or contain usable portions at least 24″ in length.
Dimension Structural light framing 2″ to 4″ thick 2″ to 4″ wide	Select Structural Dense Select Structural	High quality, relatively free of characteristics that impair strength or stiffness. Recommended for uses where high strength, stiffness, and good appearance are required.
	No. 1 No. 1 Dense	Provide high strength; recommended for general utility and construction purposes. Good appearance, especially suitable where exposed because of the knot limitations.
	No. 2 No. 2 Dense	Although less restricted than No. 1, suitable for all types of construction. Tight knots.
	No. 3	Assigned design values meet wide range of design requirements. Recommended for general construction purposes where appearance is not a controlling factor. Many pieces included in this grade would qualify as No. 2 except for single limiting characteristic. Provides high-quality, low-cost construction.
Studs 2″ to 4″ thick 2″ to 6″ wide 10′ and shorter	Stud	Stringent requirements as to straightness, strength, and stiffness adapt this grade to all stud uses, including load-bearing walls. Crook restricted in 2″ × 4″-8′ to 1/4″, with wane restricted to $1/3$ of thickness.
Structural joists and planks 2″ to 4″ thick 5″ and wider	Select Structural Dense Select	High quality, relatively free of characteristics that impair strength or stiffness.
	Structural	Recommended for uses where high strength, stiffness, and good appearance are required.
	No. 1 No. 1 Dense	Provide high strength; recommended for general utility and construction purposes. Good appearance; especially suitable where exposed because of the knot limitations.
	No. 2 No. 2 Dense	Although less restricted than No. 1, suitable for all types of construction. Tight knots.
	No. 3 No. 3 Dense	Assigned stress values meet wide range of design requirements. Recommended for general construction purposes where appearance is not a controlling factor. Many pieces included in this grade would qualify as No. 2 except for single limiting characteristic. Provides high-quality, low-cost construction.
Light framing 2″ to 4″ thick 2″ to 4″ wide	Construction	Recommended for general framing purposes. Good appearance, strong, and serviceable.
	Standard	Recommended for same uses as Construction grade, but allows larger defects.
	Utility	Recommended where combination of strength and economy is desired. Excellent for blocking, plates, and bracing.
	Economy	Usable lengths suitable for bracing, blocking, bulkheading, and other utility purposes where strength and appearance not controlling factors.
Appearance framing 2″ to 4″ thick 2″ and wider	Appearance	Designed for uses such as exposed-beam roof systems. Combines strength characteristics of No. 1 with appearance of C&Btr.
Timbers 5″ × 5″ and larger	No. 1 SR No. 1 Dense SR No. 2 SR No. 2 Dense SR	No. 1 and No. 2 are similar in appearance to corresponding grades of 2″ dimension. Recommended for general construction uses. SR in grade name indicates Stress Rated.
Structural lumber	Dense Str. 86 Dense Str. 72 Dense Str. 65	Premier structural grades from 2″ through and including timber sizes. Provides some of the highest design values in any softwood species with good appearance.

Courtesy Southern Forest Products Association

Table 18.10 Classifications, Sizes, and Grades of Softwood Lumber

	Standard Units		
Classification	**Thickness Nominal (in.)**	**Width Nominal (in.)**	**Grades**
Finish	$^3/_8$-4	2-16	B&B, C, C&Btr, D
Boards	1½	2-12+	No. 1, No. 2, No. 3, No. 4
Structural light framing	2-4	2-4	Select Structural, No. 1, No. 2, No. 3
Light framing	2-4	2-4	Construction, Standard, Utility
Studs	2-4	2-6	Stud
Structural joists and planks	2-4	5 and wider	Select Structural, No. 1, No. 2, No. 3
Appearance framing	2-4	2 and wider	Appearance
Timbers, nonstress	5 and larger	5 and larger	Square-edge and sound, No. 1, No. 2, No. 3
Timbers, stress-rated	5 and larger	5 and larger	Sel, Str, SR, DNS, Sel Str SR, No. 1 SR, No. 1 DNS SR, No. 2 SR, No. 2 DNS SR

	Metric Units		
Classification	**Thickness Nomenclature[a] (mm)**	**Width Nomenclature[a] (mm)**	**Grades**
Finish	8–64	38–286	B&B, C, C&Btr,D
Boards	17–32	38–387	No. 1, No. 2, No. 3, No. 4
Structural light framing	38–102	38–387	Sel Str, No. 1, No. 2, No. 3
Light framing	38–102	38–387	Construction, Standard, Utility
Studs	38–89	38–140	Stud
Structural joists and planks	38–89	114–387	Sel Str, No. 1, No. 2, No. 3
Appearance framing	38–89	38 and wider	Appearance
Timbers, nonstress	114 and larger	114 and larger	Square edge and sound, No. 1, No. 2, No. 3
Timbers, stress-rated	114 and larger	114 and larger	Sel Str SR, DNS Sel Str SR, No. 1 SR, No. 2 DNS SR, No. 3 DNS SR

[a]Nomenclature means the size used to describe the members. Actual size may be slightly larger or smaller.
Courtesy Southern Forest Products Association (standard) and Canadian Wood Council (metric)

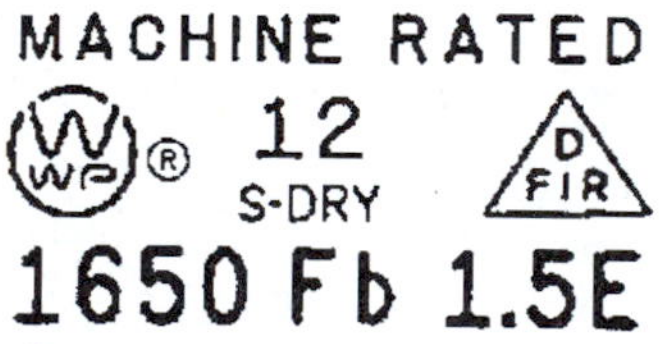

```
MACHINE RATED
(W)(WP)®  12   /D\
      S-DRY  /FIR\
1650 Fb 1.5E
```

WESTERN WOOD PRODUCTS ASSOCIATION
Douglas fir species, machine rated to 1650 psi extreme fiber stress in bending, modulus of elasticity 1,500,000 psi, maximum moisture content 19%, Mill No. 12.

Figure 18.16 Grade stamp for machine stress-rated lumber. (*Courtesy Western Wood Products Association*)

Table 18.11 Western Lumber Species Groups

Douglas Fir-Larch
  Douglas fir
  Western larch
Douglas Fir-South
  Douglas fir grown
  in AZ, CO, NV, NM, & UT
Hem-Fir
  Western hemlock
  Noble fir
  California red fir
  Grand fir
  Pacific silver fir
  White fir

Spruce-Pine-Fir (South)
  Engelmann spruce
  Sitka spruce
  Lodgepole pine
Western Woods
  Ponderosa pine
  Sugar pine
  Idaho white pine
  Mountain hemlock
Western Cedars
  Incense cedar
  Western red cedar
  Port orford cedar
  Alaska cedar

Table 18.12 Canadian Lumber Species Groups

Douglas Fir-Larch
  Douglas fir
  Western larch
Hem-Fir
  Western hemlock
  Amabilis fir
Spruce-Pine-Fir
  White spruce
  Red spruce
  Black spruce
  Engelmann spruce
  Lodgepole pine
  Jack pine
  Alpine fir
  Balsam fir
Northern Species
  All species graded in accordance with the NLGA standard grading rules for Canadian lumber

Base Values

The design data are presented in the form of *Base Values* for the various species groupings. The Base Values can be adjusted for a particular application in which the structural lumber is to be used. The Southern Forest Products Association refers to these as empirical values. Base Values are assigned to six *Basic Properties* of wood. The six Basic Properties are (1) extreme fiber stress in bending (F_b) (bending strength), (2) tension parallel to the grain (F_t) (3) horizontal shear (F_v), (4) compression parallel to the grain ($F_{c//}$), (5) compression perpendicular to the grain ($F_{c\perp}$) (side-grain crushing), and the modulus of elasticity (E or MOE) (stiffness) (Table 18.13).

Adjustment Factors

The Base Values are adjusted for various *Conditions of Use*. The Conditions of Use and their application to the Base Values are shown in Table 18.14. The seven Conditions of Use are

Size Factors (C_f)—Applied to dimension base values

Repetitive Member Factors (C_r)—Applied to size-adjusted F_b (bending stress)

Duration of Load Adjustment (C_d)—Applied to size-adjusted values

Horizontal Shear Adjustments (C_h)—Applied to F_v (horizontal shear) values

Flat Use Factors (C_{fu})—Applied to size-adjusted F_b (bending stress)

Adjustments for Compression Perpendicular to Grain ($C_{c\perp}$) (compression perpendicular) values

Wet Use Factors (C_m)—Applied to size-adjusted values

How to Apply Adjustment Factors

Suppose that you want to find the adjusted bending stress, F_b, for a Select Structural 2 × 6 in. member from the Douglas Fir-Larch group. The Base Value for DF-L in SS grade is 1450 psi. Multiply it by the Size Value, which is 1.3: $1450 \times 1.3 = 1885$ psi (size adjusted F_b). This can now be adjusted for other conditions of use,

such as for a *Repetitive Member Factor,* which is 1.15: $1885 \times 1.15 = 2167$ psi (adjusted F_b).

The various lumber associations have published new span tables for members such as joists and rafters. These tables include some applications of conditions of use. Before using the span tables it is necessary to note which conditions of use have been applied.

PHYSICAL AND CHEMICAL COMPOSITION OF WOOD

Since wood is a naturally occurring material, it has considerable variation in its properties. For example, it varies in color, density, weight, and strength. The physical and chemical composition of wood determine its properties and, therefore, its uses.

Porosity of Wood

Wood is a cellular material, as shown in Figs. 18.17 and 18.18. Softwood cellular structure contains large longitudinal cells called *tracheids* and smaller radial cells called *rays*. These cells store and transfer nutrients. In addition, the annual rings are cellular. The structure of hardwoods is more complex. They have two different types of longitudinal cells, small-diameter *fibers* and larger diameter *vessels* or *pores,* which transport the sap of the tree. They also have a higher percentage of rays than found in softwoods.

The surface area of these cells is very large and gives wood several important properties. First, it makes it possible for wood, when dry, to absorb toxic chemicals needed to prevent decay and insect attack. Second, it can absorb moisture repellents to minimize moisture exchange and therefore control shrinkage. Third, the cells are air pockets that provide insulating qualities. Fourth, it enables wood to shrink and swell as moisture content varies. Fifth, it contributes to the ease of adherence of paint, adhesives, and other synthetic resins used on wood surfaces.

Table 18.13 Base Values for Western Dimension Lumber[a]

Species or Group	Grade	Extreme Fiber Stress in Bending "F_b" Single	Tension Parallel to Grain "F_t"	Horizontal Shear "F_v"	Compression Perpendicular "$F_{c\perp}$"	Compression Parallel to Grain "$F_{c//}$"	Modulus of Elasticity "E"
Douglas Fir/Larch	Select Structural	1450	1000	95	625	1700	1,900,000
	No. 1 & Btr.	1150	775	95	625	1500	1,800,000
	No. 1	1000	675	95	625	1450	1,700,000
	No. 2	875	575	95	625	1300	1,600,000
	No. 3	500	325	95	625	750	1,400,000
	Construction	1000	650	95	625	1600	1,500,000
	Standard	550	375	95	625	1350	1,400,000
	Utility	275	175	95	625	875	1,300,000
	Stud	675	450	95	625	825	1,400,000

[a]Sizes: 2″ to 4″ thick by 2″ and wider
Courtesy Western Wood Products Association

Table 18.14 Adjustment Factors for Base Values for Western Dimension Lumber

Size Factors (C_F) (Apply to Dimension Lumber Base Values)

Grades	Nominal Width (depth)	F_b 2" & 3" thick nominal	F_b 4" thick nominal	F_t	$F_{c//}$	Other Prop-erties
Select Structural, No. 1 & Btr., No. 1, No. 2 & No. 3	2", 3" & 4"	1.5	1.5	1.5	1.0	1.0
		1.15	1.0			
	5"	1.4	1.4	1.4	1.1	1.0
	6"	1.3	1.3	1.3	1.1	1.0
	8"	1.2	1.3	1.2	1.05	1.0
	10"	1.1	1.2	1.1	1.0	1.0
	12"	1.0	1.1	1.0	1.0	1.0
	14" & wider	0.9	1.0	0.9	0.9	1.0
Construction & Standard	2", 3", & 4"	1.0	1.0	1.0	1.0	1.0
Utility	2", & 3"	0.4	—	0.4	0.6	1.0
	4"	1.0	1.0	1.0	1.0	1.0
Stud	2", 3" & 4"	1.1	1.1	1.1	1.05	1.0
	5" & 6"	1.0	1.0	1.0	1.0	1.0

Repetitive Member Factor (C_r) (Apply to Size-Adjusted F_b)

Where 2" to 4" thick lumber is used repetitively, such as for joists, studs, rafters, and decking, the pieces side by side share the load and the strength of the entire assembly is enhanced. Therefore, where three or more members are adjacent or are not more than 24" apart and are joined by floor, roof, or other load distributing elements, the F_b value can be increased 1.15 for repetitive member use.

Repetitive Member Use

$$F_b \times 1.15$$

Duration of Load Adjustment (C_d) (Apply to Size-Adjusted Values)

Wood has the property of carrying substantially greater maximum loads for short durations than for long durations of loading. Tabulated design values apply to normal load duration. (Factors do not apply to MOE or $F_{c\perp}$).

Load Duration

Load Duration	Factor
Permanent	0.9
Ten years (normal load)	1.0
Two months (snow load)	1.15
Seven day	1.25
One day	1.33
Ten minutes (wind and earthquake loads)	1.6
Impact	2.0

Confirm load requirements with local codes. Refer to Model Building Codes or the National Design Specification for high-temperature or fire-retardant treated adjustment factors.

Flat Use Factors (C_fu) (Apply to Size-Adjusted F_b)

Nominal Width	Nominal Thickness 2" & 3"	4"
2" & 3"	1.00	—
4"	1.10	1.00
5"	1.10	1.05
6"	1.15	1.05
8"	1.15	1.05
10" & wider	1.20	1.10

Horizontal Shear Adjustment (C_H) (Apply to F_v Values)

Horizontal shear values are based upon the maximum degree of shake, check or split that might develop in a piece. When the actual size of these characteristics is known, the following adjustments may be taken.

2" Thick Lumber

For convenience, the table below may be used to determine horizontal shear values for any grade of 2" thick lumber in any species when the length of split or check is known and any increase in them is not anticipated.

3" and Thicker Lumber

Horizontal shear values for 3" and thicker lumber also are established as if a piece were split full length. When specific lengths of splits are known and any increase in them is not anticipated, the following adjustments may be applied.

When Length of Split on Wide Face is:	Multiply Tabulated F_v Value by:	When Length of Split on Wide Face is:	Multiply Tabulated F_v Value by:
No split	2.00	No split	2.00
½ of wide face	1.67	½ of narrow face	1.67
¾ of wide face	1.50	1 of narrow face	1.33
1 of wide face	1.33	1½ of narrow or more	1.00
1½ of wide face or more	1.00		

Adjustments for Compression Perpendicular to Grain (C_c⊥)
(For Deformation Basis of 0.02" Apply to F_c⊥ Values)

Design values for compression perpendicular to grain ($F_{c\perp}$) are established in accordance with the procedures set forth in ASTM Standards D 2555 and D 245. ASTM procedures consider deformation under bearing loads as a serviceability limit state comparable to bending deflection because bearing loads rarely cause structural failures. Therefore, ASTM procedures for determining compression perpendicular to grain values are based on a deformation of 0.04" and are considered adequate for most classes of structures. Where more stringent measures need to be taken in design, the following formula permits the designer to adjust design values to a more conservative deformation basis of 0.02":

$$Y_{02} = 0.73 Y_{04} + 5.60$$

Example: Douglas Fir-Larch: $Y_{04} = 625$ psi
$Y_{02} = 0.73 (625) + 5.60 = 462$ psi

Wet Use Factors (C_M) (Apply to Size-Adjusted Values)

The design values shown in the accompanying tables are for routine construction applications where the moisture content of the wood does not exceed 19%. When use conditions are such that the moisture content of dimension lumber will exceed 19%, the Wet Use Adjustment Factors below are recommended:

	Property	Adjustment Factor
F_b	Extreme fiber stress in bending	0.85*
F_t	Tension parallel to grain	1.0
F_c	Compression parallel to grain	0.8**
F_v	Horizontal shear	0.97
$F_{c\perp}$	Compression perpendicular to grain	0.67
E	Modulus of elasticity	0.9

*Fiber Stress in Bending Wet Use Factor 1.0 for size-adjusted F_b not exceeding 1150 psi.
**Compression Parallel to Grain in Wet Use Factor 1.0 for size-adjusted F_c not exceeding 750 psi.
Courtesy Western Wood Products Association

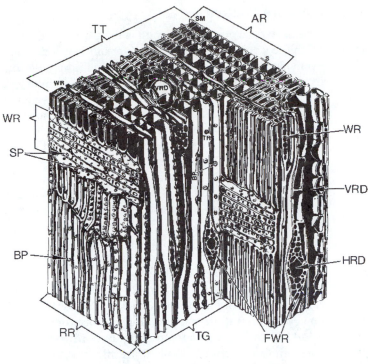

Cell structure of a softwood

Figure 18.17 A greatly enlarged example of wood cellular structure. Tracheids (TR) are vertical cells that make up the major part of the structure. Rays (WR) are cells that run radially from the center to the outside of the tree. Annual rings (AR) are made up of small hard summerwood cells (SM) and larger, softer springwood cells (S). Resin is in horizontal resin ducts (HRD) and vertical resin ducts (VRD) centered in fusiform wood rays (FWR). Simple pits (SP) allow sap to pass back and forth between the ray cells and the longitudinal cells. Border pits (BP) transfer sap between longitudinal cells. Face RR indicates a radial cut through the wood and TG indicates a tangential cut. *(Courtesy Forest Products Laboratory, USDA Forest Service, Madison, Wis.)*

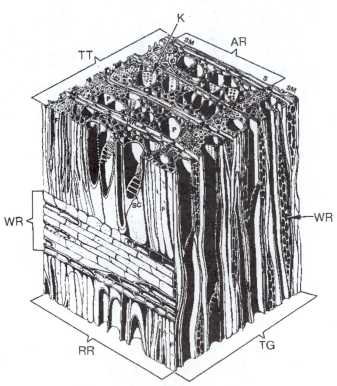

Cell structure of a hardwood

Figure 18.18 Hardwood trees have a more complex structure. The rays (WR) make up the major mass and produce the grain features. Large vertical cells with pores (P) move the sap. Wood fibers (F) give the tree structural strength. Pits (K) transfer sap from one cell to another, and rays (WR) run radially from the center of the tree. A radial cut is indicated by RR and a tangential cut by TG. An annual ring is shown by AR. *(Courtesy Forest Products Laboratory, USDA Forest Service, Madison, Wis.)*

Composition of Wood

The cells are made of fibers that are mainly *cellulose* and *hemicellulose* that are bonded together with an organic substance called **lignin.** Cellulose and hemicellulose are complex glucose compounds. Glucose is a sugar that is useful to fungi and insects as a food. Although the exact amounts of cellulose, hemicellulose, and lignin vary with different species of wood, the general composition in kiln-dried woods is shown in Table 18.15. These elements are what give wood its strength, susceptibility to decay, and hygroscopic properties. **Hygroscopic** refers to the property of wood that permits it to absorb and retain moisture. Cellulose provides strength in tension, toughness, and elasticity. Lignin, because it bonds the fibers together in fiber bundles, gives wood its compressive strength. The remaining substances in wood do not contribute to its structure but give each species its unique color, odor, taste, density, resistance to decay, and flammability.

Paper products are made using the cellulose and hemicellulose in wood. The wood is cut into chips and cooked to separate the fibers from the lignin. The fibers form the pulp used to make paper and cardboard products. Hemicellulose plays an important part in fiber-to-fiber bonding in the papermaking process. See Chapter 23 for additional information on paper products.

Cellulose is also used in some textiles, plastics, and other products requiring cellulose derivatives. *Extractive* materials derived from wood include tannins, coloring matters, essential oils, fats, resins, waxes, gums, starch, and simple metabolic intermediates.

HYGROSCOPIC PROPERTIES OF WOOD

Wood expands when it absorbs moisture and shrinks when it loses moisture. Therefore, wood is hygroscopic. During its life a tree contains considerable water. In this condition, wood is called green. To be useful in construction, furniture, and other products it must be in a dry condition. The moisture content of wood must be reduced to a level acceptable for its intended purpose.

Table 18.15 Composition of Kiln-Dried Wood

Material	Softwoods Percent	Hardwoods Percent
Cellulose	40–50	40–50
Hemicellulose	20	15–35
Lignin	23–33	16–25
Extraneous materials	5–10	5–10

Moisture Content

The **moisture content** of wood is the weight of the water in the wood divided by the weight of the wood, when oven dry, expressed as a percent. One method to do this in a laboratory is to weigh the green sample, oven dry it, and weigh it again. For example, if the sample weighed 4 oz. green and 3 oz. dry it would have had a moisture content of 1 oz. of water or a moisture content of 33 percent. On the job the moisture content is checked with a moisture meter. This battery-operated unit shows the moisture content on a meter when a probe is placed in contact with the wood (Fig. 18.19). The points on the probe are driven into the wood. The moisture content is read on the dial.

The moisture content of lumber from newly cut trees varies by species but ranges from 30 percent to as much as 200 percent. The moisture content of sapwood is higher than that in the heartwood. For example, in the various species of pine the heartwood moisture content ranges from 30 to 98 percent, and the sapwood ranges from 100 to 220 percent. In general, hardwoods have lower moisture contents when freshly cut than softwoods.

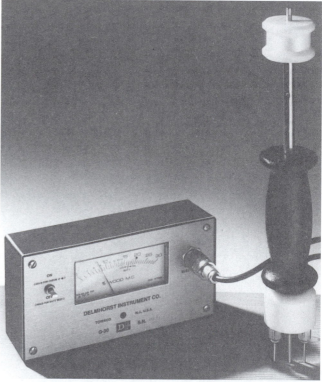

Figure 18.19 A moisture meter used to determine the amount of moisture in wood. *(Courtesy Delmhorst Instrument Co.)*

Fiber Saturation Point

Moisture in the living tree or in freshly cut wood (green) is present in the *cell fibers* as absorbed water and in the *cell cavities* as free water. When the wood begins to dry the cavities in the cell cavities begins to disappear. Once all free water in the cell cavities is gone the cell fibers making up the cell walls contain any remaining water. This is the **fiber saturation point.** Most softwoods have a moisture content of about 30 percent at this point. As moisture is removed from the cell walls they begin to shrink. Therefore, the fiber saturation point is the point at which shrinkage begins (Fig. 18.20).

The total shrinkage in lumber occurs between the fiber saturation point, 30 percent, and the desired moisture content, such as the 15 percent required for most construction lumber. Shrinkage is approximately proportional to the amount of moisture loss. Wood shrinks about $1/30$ of the total shrinkage with each 1 percent reduction in moisture below the fiber saturation point. The reduction mentioned above, 30 percent to 15 percent, is about half the possible shrinkage.

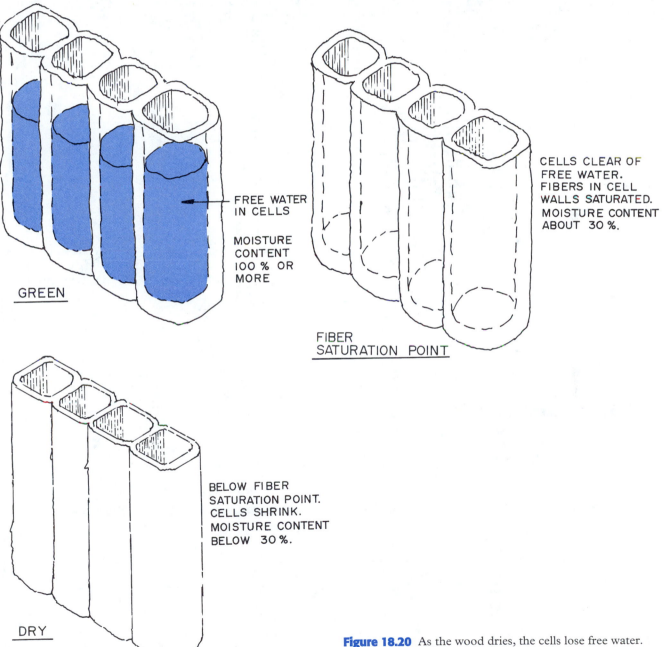

FREE WATER
IN CELLS

MOISTURE
CONTENT
100 % OR
MORE

GREEN

CELLS CLEAR OF
FREE WATER.
FIBERS IN CELL
WALLS SATURATED.
MOISTURE CONTENT
ABOUT 30 %.

FIBER
SATURATION POINT

BELOW FIBER
SATURATION POINT.
CELLS SHRINK.
MOISTURE CONTENT
BELOW 30 %.

DRY

Figure 18.20 As the wood dries, the cells lose free water. Then the cell wall fibers dry and the cells shrink.

I T E M   O F   I N T E R E S T

HANDLING AND STORING WOOD PRODUCTS

Wood is a hygroscopic material—it absorbs and retains moisture from the air. Likewise, under certain conditions it can lose moisture. These properties are a concern for both solid wood and engineered wood products, such as plywood and laminated veneer lumber (LVL). When wood is stored it must be protected from moisture so the moisture content is not increased beyond that of the finished product. Excess moisture can be released in the building after wood members are installed and can cause long-term problems, such as drywall cracking. Also, as wood is stored it must be protected from crushing and twisting. The figure shows some suggested ways to store lumber and panel products before they are used in construction. Proper storage will reduce possible waste due to damaged materials, thus helping to control the on-site costs of a project.

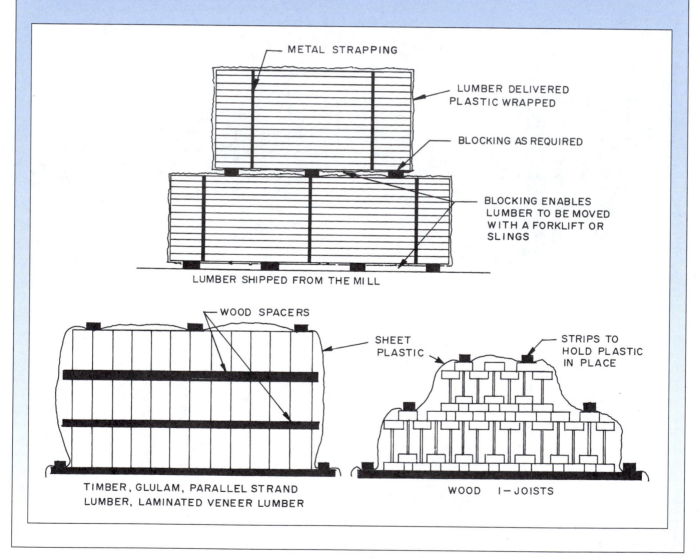

Equilibrium Moisture Content

Green lumber gives off moisture to the air, and if it is left long enough it will lose moisture until it has the same moisture content as the surrounding air. It will continue to take on and give off moisture as the moisture content of the air varies. When this point is reached the wood has reached its **equilibrium moisture content.**

Since wood reacts to changes in humidity and temperature of the air and these vary constantly, wood is constantly seeking to reach equilibrium with the surrounding air. Variations tend to be seasonal, and there is less variation inside heated buildings than on the exterior.

As the wood seeks equilibrium it gradually shrinks or expands. Excessive expansion or contraction can cause problems in a building. Doors and windows stick if expansion occurs. Cracks occur in the interior wall finish if studs and plates have excessive shrinkage. To minimize the amount of expansion and contraction, wood should be installed at a moisture content as close as possible to the equilibrium moisture content it will have after it is installed. The recommended moisture content for interior wood, such as flooring, trim, cabinets, windows, and doors, is between 6 and 12 percent. This varies with the region from the dry Southwest to the more moisture-laden areas of the Northwest and Southeast (Fig. 18.21). Since wood shipped to a job can have its moisture content increased or decreased during shipping or storage, it is recommended that wood items be stored in the area where they will be used for several days so they can approach equilibrium before use.

The moisture content for exterior wood products is usually 12 to 15 percent for most areas of the country. These products, such as wood siding, are exposed to greater variations in humidity and temperature and therefore reach equilibrium at a higher moisture content. For example, surfaced framing lumber having a moisture content at maximum 15 percent is stamped MC15. If it is surfaced and has a moisture content at a maximum of 19 percent it is stamped S-DRY (surfaced dry). If it has more than 19 percent moisture it is stamped S-GRN (surfaced green).

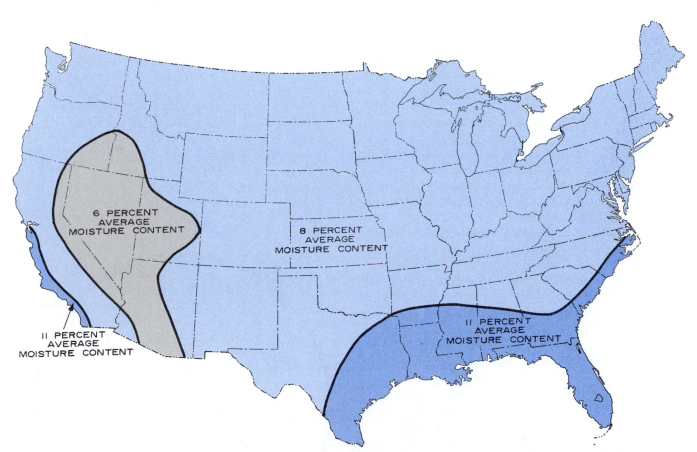

Fig. 18.21 The equilibrium moisture content varies in different sections of the country. *(Courtesy Forest Products Laboratory, USDA Forest Service, Madison, Wis.)*

To be sure that the lumber does not exceed the moisture content stamped on it, mills sometimes dry it several percentages below that required. In dry regions framing lumber should have a moisture content of not more than 15 percent when interior finishes are installed. In most of the country, a maximum moisture content of 19 percent is satisfactory.

How Moisture Affects Wood Properties

The moisture content of wood affects its size, dimensional stability, strength, stiffness, decay resistance, glue bonding, and paintability.

Size and *dimensional stability* are directly affected by the moisture content of wood. Shrinkage and swelling occur only after the moisture content falls below the fiber saturation point. Most woods shrink and swell very little *parallel with the grain* (longitudinal). This shrinkage has very little influence on construction uses. Wood shrinks and swells a great deal in thickness and width *across the grain*. Shrinkage is greatest in the direction parallel to the annual growth rings (tangential) and is about twice as much as the shrinkage across the rings (radial) (Fig. 18.22).

Figure 18.23 shows the combined effects of tangential and radial shrinkage as wood dries from its green condition. Notice that the shrinkage is affected by the direction of the growth rings. Tangential shrinkage is about twice as great as radial shrinkage. Tangential shrinkage is that which is tangent to the circumference of the tree. Radial shrinkage is that which occurs parallel to a line passed through the center of the tree. This produces distortions in the lumber. The nature of the distortion depends on the location the piece of lumber occupied in the tree. These distortions due to shrinkage are greater in plain-sawed lumber.

Most hardwoods shrink more than softwoods. Heavier species shrink more than lighter species. Stock with a large cross-sectional area, such as 6 in. × 6 in. timbers, do not shrink as much proportionately because the inside does not dry at the same rate as the outside. The outer layers dry faster, become set, and keep the inner area from shrinking normally as it dries.

Softwood lumber shrinks about $\frac{1}{32}$ in. per inch of face width while drying from a green condition to about 19 percent moisture content. This means a rough cut 2 in. × 10 in. board shrinks about $\frac{1}{4}$ in. in width as it dries to 19 percent moisture content and about $\frac{5}{16}$ to $\frac{3}{8}$ in. when it reaches 15 percent moisture content.

As the moisture content of wood decreases, its *strength* increases. This is caused by the stiffening and strengthening of the fibers in the cell walls and the fact that as the wood shrinks it becomes a denser material. Crushing strength and bending strength increase a great deal in dry woods. Stiffness increases moderately. Shock resistance

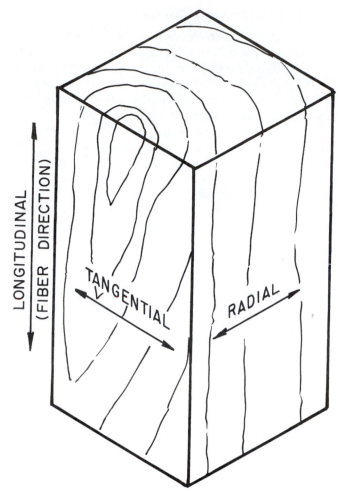

Figure 18.22 Wood shrinks in three directions in relationship to the direction of the grain and growth rings.

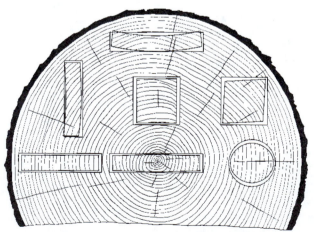

Fig. 18.23 Typical shrinkage and distortion of various wood shapes in relation to the direction of growth rings. *(Courtesy Forest Products Laboratory, USDA Forest Service, Madison, Wis.)*

depends on the pliability of the material and is less for dry wood than for the original green stock. Green wood also bends farther before it breaks than dry wood.

Wood that maintains a moisture content below 20 percent will be free from *decay;* it is important to control the moisture content. This can be accomplished in any of several ways, including painting and treating it with wood-preserving chemicals. Green wood should not be *painted.* The recommended moisture content for wood to be painted is 16 to 20 percent. Also, if the wood has surface moisture, as from a rain, it should not be painted until the surface is thoroughly dry.

Reducing the moisture content of wood also improves the *glue bond.* Wood used for interior purposes is glued at 5 to 6 percent moisture content. Exterior wood can be glued satisfactorily at 10 to 12 percent moisture content. Veneers, such as those used to make plywood, usually have 2 to 6 percent moisture content. Woods glue satisfactorily up to about 15 percent moisture content, but they should be in equilibrium with the air in which they will be used. Interior wood at 12 to 15 percent moisture can be glued satisfactorily, but the glued members will check and warp as the wood loses moisture and seeks equilibrium with the air.

Specific Gravity

Specific gravity is a ratio of the weight of a certain volume of a material to the weight of an equal volume of water at 39.2°F (4°C). The specific gravity of wood is found by weighing a piece of wood of known volume, such as 6 in.^{3}, and dividing this by the weight of an equal volume of water. Water has a specific gravity of 1. If a material has a specific gravity less than this it will float. If it is greater than 1 it will sink. You can see some examples for selected woods in Table 18.16.

The specific gravity of wood fibers forming the cells is about 1.5, so the fibers would sink. However, most woods float because much of the volume is made up of air-filled cells. The size of the cells and the thickness of the cell walls influence the specific gravity (Figs. 18.17 and 18.18). Specific gravity is an approximate indicator of the density of wood and an indication of strength. In general, woods that are dense weigh more, have a higher specific gravity, and can handle higher stresses.

STRUCTURAL PROPERTIES OF WOOD

Wood is an *orthotropic* material. Orthotropic pertains to a mode of growth that is more or less vertical. Wood has unique and independent mechanical properties in the directions of three mutually perpendicular axes—longitudinal, radial, and tangential. The longitudinal axis is parallel to the fibers (grain). The radial axis is normal to the

Table 18.16 Specific Gravity and Weight of Woods Commonly Used In Construction[a]

Softwoods	Specific Gravity	Density[b]
Cedar, western	0.32	20.0
Cypress	0.46	28.7
Douglas fir, coast	0.48	30.0
Larch, western	0.52	32.4
Pine, western white	0.38	23.7
Pine, lodgepole	0.41	25.5
Pine, ponderosa	0.40	25.0
Pine, shortleaf	0.51	31.6
Redwood	0.40	25.0
Spruce, Englemann	0.35	21.6
Spruce, Sitka	0.40	25.0
Hardwoods		
Birch, yellow	0.62	38.7
Cherry, black	0.50	31.2
Maple, sugar	0.63	39.3
Oak, red	0.61	37.9
Oak, white	0.64	39.9

[a]Oven-dry samples
[b]Density in points per cubic foot based on specific gravity as shown and 0% moisture.
Courtesy Forest Products Laboratory, USDA Forest Service, Madison, Wis.

growth rings (perpendicular to the grain in the radial direction). The tangential axis is perpendicular to the grain but tangent to the growth rings. Refer to Fig. 18.22.

Wood is a *fibrous* material. The fibers are bonded together with lignin forming the walls of the cells making up the material. The fibers in hardwoods are about $\frac{1}{25}$ in. long and from $\frac{1}{8}$ to $\frac{1}{3}$ in. in softwoods. The strength of wood does not depend on the length of the fibers but on the thickness of the cell walls and the direction of the fibers in relation to applied loads. Most fibers are oriented with their lengths parallel with the vertical dimension of the tree. Strength of wood parallel with the fibers (parallel with the grain) is greater than the strength perpendicular to the fibers (perpendicular with the grain).

Structural members under an *external load,* such as wind, furniture, or people, produce internal forces called stresses in a member to resist these external forces. *Tensile stresses* result when an external force tends to stretch a member. *Compressive stresses* occur when a member is under a squeezing force. When a member, such as a **beam,** is loaded so that the applied force is acting approximately perpendicular to the member, the load produces a *bending stress.* When a member is under bending stresses it develops compressive stresses in the upper part and tensile stresses in the lower part.

The strength of wood under various stresses is found by testing samples in a laboratory. Tests include bending, shear, stiffness, tension, and compression. Samples free of any defects are tested to find the *basic stresses* for each species used in construction. Samples of the various grades in each species are tested to find realistic stresses for each. These are lower than the basic stresses and are called *working stresses*. The working stress takes into account the things that lower the load a member can carry, such as knots, pitch pockets, or checks.

The working stresses for several species of wood used in construction are in Table 18.17. These are for normal loading conditions, which include dead loads (roof or floor materials) and live loads (occupants or furniture). They are based on the normal duration of loading, which assumes a fully stressed member under the full maximum design load for 10 years and the application of 90 percent of this maximum normal load continuously through the remainder of the life of the structure.

The working stresses are given in pounds per square inch and include working stresses for bending (F_b), **horizontal shear** (F_v), **vertical shear** (V), modulus of elasticity (E), compression parallel to the grain (F_c), fiber stress in tension (F_t), and compression perpendicular to the grain ($F_{c\perp}$).

Bending

Wood beams under load deflect, and bending stresses are produced in the fibers. Wood has high fiber strengths in bending. However, if a beam is loaded enough to produce stresses greater than the fiber strength of the wood, the beam will break. The stresses developed in the fibers are greater the farther they are from the central axis of the beam. Fibers located twice as far from the central axis of the beam as other fibers have twice the stress. Therefore, when a beam bends, the maximum stresses are developed in the outer fibers of the top and bottom

of the beam. This is called the *extreme fiber in bending*, F_b (Fig. 18.24).

Shear

A beam is subject to vertical and horizontal shear (Figs. 18.25 and 18.26). Vertical shear refers to the tendency for one part of a beam to move vertically in relation to an adjacent part, allowing the beam to slip down between supports. Horizontal shear refers to the tendency of the sapwood fibers to move horizontally in relation to the bottom fibers. Horizontal shear is more of a factor in beam failure than vertical shear. The fiber stress in horizontal shear must be kept below the working stress values indicated by the symbol F_v.

Modulus of Elasticity

The modulus of elasticity (E) is a measure of a beam's resistance to deflection, or its stiffness. Stiffness is important when selecting columns, beams, rafters, joists, and other structural members (Fig. 18.27).

Tension

Wood under tension has good tensile strength. The allowable unit stress in *tension parallel to the grain* (F_t) is applied to the entire cross section of a wood member. Knots, checks, holes, and splits reduce the allowable unit stress in a member. Wood is weak in *tension perpendicular to the grain* (Fig. 18.28).

Table 18.17 Working Stresses for Selected Species of Structural Light Framing Lumber[a]

Species	Grade	Extreme Fiber in Bending "F_b"	Tension Parallel to the Grain "F_t"	Compression Parallel to the Grain "F_c"	Horizontal Shear "F_v"	Compression Perpendicular to the Grain "$F_{c\perp}$"	Modulus of Elasticity "E"
Southern Pine	Dense Select Structural	2500	1500	2100	105	475	1,900,000
	Select Structural	2150	1250	1800	105	405	1,800,000
	No. 1 Dense	2150	1250	1700	105	475	1,900,000
	No. 1	1850	1050	1450	105	405	1,800,000
	No. 2 Dense	1800	1050	1350	95	475	1,700,000
	No. 2	1550	900	1150	95	405	1,600,000
	No. 3 Dense	1000	575	800	95	475	1,500,000
	No. 3	850	500	675	95	405	1,500,000
	Stud	850	500	675	95	405	1,500,000
Hem-Fir	Select Structural	1650	975	1300	75	405	1,500,000
	No. 1/Appearance	1400	825	1050/1250	75	405	1,500,000
	No. 2	1150	675	825	75	405	1,400,000
	No. 3	650	375	500	75	405	1,200,000
Douglas Fir-Larch	Select Structural	2100	1200	1600	95	625	1,800,000
	No. 1/Appearance	1750	1050	1250/1500	95	625	1,800,000
	No. 2	1450	850	1000	95	625	1,700,000
	No. 3	800	475	600	95	625	1,500,000

[a]Structural light framing 2″ to 4″ thick 2″ to 4″ wide MC 15%

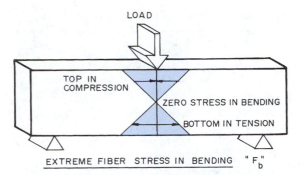

EXTREME FIBER STRESS IN BENDING "F_b"

LOADS APPLIED TO A BEAM CAUSE IT TO BEND, PRODUCING TENSION IN FIBERS ALONG THE FACE FARTHEST FROM THE LOAD AND COMPRESSION IN FIBERS ALONG THE FACE NEAREST THE LOAD. THESE INDUCED STRESSES ARE DESIGNATED AS "EXTREME FIBER STRESSES IN BENDING, F_b".

Figure 18.24 Forces designated as extreme fiber stresses in bending occur along the faces of the beam.

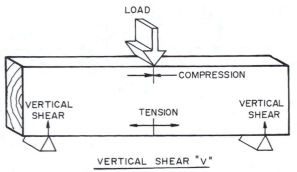

VERTICAL SHEAR "V"

VERTICAL SHEAR STRESSES TEND TO CAUSE ONE PART OF A MEMBER TO MOVE VERTICALLY IN RELATION TO THE ADJACENT PART.

Figure 18.25 Horizontal shear stress tends to cause fibers to slice horizontally much the same as when you bend a deck of cards.

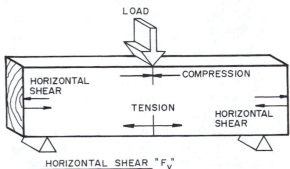

HORIZONTAL SHEAR "F_v"

HORIZONTAL SHEAR STRESSES TEND TO SLIDE FIBERS OVER EACH OTHER HORIZONTALLY. INCREASING THE BEAM CROSS SECTION DECREASES SHEAR STRESSES.

Figure 18.26 Vertical shear tends to cause fibers in one part of a beam to move vertically in relation to those adjacent to them.

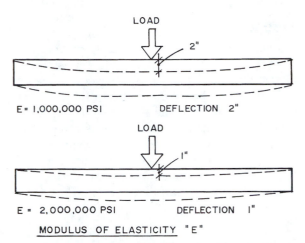

E = 1,000,000 PSI   DEFLECTION 2"

E = 2,000,000 PSI   DEFLECTION 1"

MODULUS OF ELASTICITY "E"

THE MODULUS OF ELASTICITY (E) IS A RATIO OF THE AMOUNT A MATERIAL WILL DEFLECT IN PROPORTION TO THE APPLIED LOAD.

Figure 18.27 The modulus of elasticity is a measure of the stiffness of a beam.

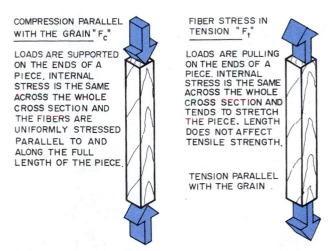

COMPRESSION PARALLEL WITH THE GRAIN "F_c"

LOADS ARE SUPPORTED ON THE ENDS OF A PIECE. INTERNAL STRESS IS THE SAME ACROSS THE WHOLE CROSS SECTION AND THE FIBERS ARE UNIFORMLY STRESSED PARALLEL TO AND ALONG THE FULL LENGTH OF THE PIECE.

FIBER STRESS IN TENSION "F_t"

LOADS ARE PULLING ON THE ENDS OF A PIECE. INTERNAL STRESS IS THE SAME ACROSS THE WHOLE CROSS SECTION AND TENDS TO STRETCH THE PIECE. LENGTH DOES NOT AFFECT TENSILE STRENGTH.

TENSION PARALLEL WITH THE GRAIN.

Figure 18.28 These are typical tension and compression stress applied parallel to the grain of a member.

Compression

The unit of stress in compression parallel to the grain is several times greater than that perpendicular to the grain (Fig. 18.29). Selected values for compressive strength perpendicular to grain ($F_{c\perp}$) and parallel to grain (F_c) are in Table 18.17.

Gluing Properties

As mentioned earlier, the glue bond is improved as moisture content decreases, with 10 to 12 percent moisture effective for exterior products. In addition, some species of wood bond better than others (Table 18.18). A satisfactory glue joint is one that is as strong as the wood itself. Properly made joints are often stronger than the wood, and the wood will break before the joint will. The density and structure of the wood influence the strength of the bond. In general, heartwood does not bond as well as sapwood, and hardwoods do not bond as well as softwoods.

OTHER PROPERTIES OF WOOD

Wood has other properties, some unique to it as a material, that need to be considered. These include thermal properties, decorative features, decay resistance, and insect damage.

Thermal Properties

Softwoods have a *thermal conductivity* of approximately 1 Btu/in. of thickness. Therefore, wood is a fairly good insulator but not as good as other materials used specifically for insulation. As the moisture content of the wood increases it becomes less of an insulator because the

Table 18.18 Gluing Properties of Selected Woods

Bond easily	Alder
	Aspen
	Fir: white, grand, noble, pacific
	Pine: eastern, white, western white
	Red cedar, western
	Redwood
	Spruce, Sitka
Bond well	Elm, American rock
	Maple, soft
	Sycamore
	Walnut, black
	Yellow poplar
	Douglas fir
	Larch, western
	Pine: sugar, ponderosa
	Red Cedar, eastern
	Mahogany: African, American
Bond satisfactorily	Ash, white
	Birch: sweet, yellow
	Cherry
	Hickory: pecan, true
	Oak: red, white
	Pine, southern
Bond with difficulty	Persimmon
	Teak
	Rosewood

Courtesy Forest Products Laboratory, USDA Forest Service, Madison, Wis.

moisture increases thermal conductivity. The lighter (less dense) woods are better insulators because of their larger cell structure.

Wood also experiences *thermal expansion*. It expands when heated and contracts when cooled. For most construction purposes this is not a factor because the amount of change is so small. Expansion due to moisture increases is much greater and is the factor to consider.

Decorative Features

Wood has unique decorative features that make it a prized material. Various species have unique *colors*, including the mature heartwood and the new growth sapwood. Species vary in density, producing wood with *smooth*, closed grain to *rough*, very open grain producing beautiful decorative appearances. Wood can also be stained or bleached to alter the color, or a paste filler can be worked into the grain.

Wood also takes on decorative colors and texture when it is permitted to *weather* with no protective coating. Some woods, such as cedar and cypress, are decay resistant and weather to a silver gray. Some woods develop a rough, checked surface when weathered while others do not. All woods tend to warp as they weather, some more than others. Decay-resistant woods, such as redwood, cedar, and cypress, warp less than most. Narrow boards, such as 1 × 6 and 1 × 8 siding, warp less than wider boards.

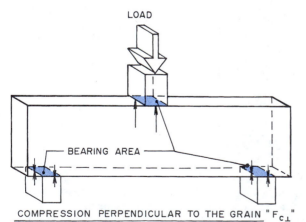

LOAD

BEARING AREA

COMPRESSION PERPENDICULAR TO THE GRAIN "$F_{c\perp}$"

WHERE A MEMBER BEARS ON SUPPORTS, THE LOADS TEND TO COMPRESS THE FIBERS. THE BEARING AREA MUST BE SUFFICIENT TO PREVENT SIDE GRAIN CRUSHING.

Figure 18.29 The working compression stresses perpendicular to the grain are considerably less than those parallel to the grain.

Decay Resistance

Some species of wood have *natural decay-resistant substances* in their cell structures that enable the heartwood to resist attack by decay or by insects, such as termites. Decay-resistant woods include cedars, redwood, cypress, black walnut, and black locust. Insect-resistant woods include cypress, some cedars, and redwood. These species may be used for all above-ground uses that are exposed to the weather. To be most effective they should be 100 percent heartwood. Wood used in the ground as foundations for permanent structures should be pressure treated with a wood preservative.

Wood can be attacked by *fungi* (microscopic plants) that cause decay, molds, and stains. Fungi develop in wood when the moisture content is above 20 percent, temperatures are mild (40° to 100°F), there is sufficient oxygen, and the wood provides an adequate food supply. Wood in the ground is exposed to moisture, oxygen, and a rather uniform mild temperature. Therefore, it is ideally situated for fungi attack (decay). Wood submerged in water lacks an adequate supply of oxygen.

Often parts of a building under renovation are dry but found to be heavily decayed. Although this condition is often called "dry rot," it occurs because the wood over the years has been intermittently wet (as from a leaking roof) and dry. The moisture permits the growth of fungi, and even though the wood may be dry at the time it is removed it had to have moisture frequently for decay to occur.

A harmless type of fungi called *white pocket* is found in living softwood trees. When the tree is cut into lumber the fungi ceases to develop. It will not spread in the cut lumber or transfer to other lumber stored on it. It is an acceptable defect and accepted in most grades of lumber except the very best.

Molds and stains are also caused by fungi but do not damage the wood. The wood may be discolored, but its strength is not impaired. If the wood is used where appearance is important the staining may be undesirable. Many times molds can be removed by brushing the surface. Stains appear as blue, blue-black, or brown specks, as streaks or as spots.

Insect Damage

Insects bore holes in living trees and cut lumber. Some holes are very small and are called *pinholes*. Others are larger and are *grub holes*. Lumber graders watch for this damage and take it into account because it can influence the strength of the stock. When structural integrity is important, do not use insect-damaged lumber.

Completed buildings are also subject to attack by insects. The *termite* is the major offender. Most parts of the country have termites, but they occur in larger numbers in the milder climates. (Fig. 18.30). Most termites are the *subterranean type*. They live in nests in the ground and build tunnels through the earth to get to material containing cellulose, which is their food. This type of termite must have a moist atmosphere to exist. When they attack the wood in a house they must build mud tunnels up the foundation to reach the floor joists. The tunnels protect them from the atmosphere and keep them moist. Termites will also enter a building through wood that has direct contact with the earth or through very small cracks in the foundation or concrete floor slab. They eat the inside of the wood members, leaving a thin outer layer to protect them from drying. You can check for termites by sticking an ice pick into the joists. It will easily penetrate an infested wood member.

Following are things a builder can do to reduce termite attack.

1. Install a metal termite shield on top of the foundation. Termites have been known to build tunnels around these shields.
2. Remove all wood scraps, paper, and cardboard from the construction site. Any scraps buried, as when backfilling, will provide a food supply and attract a colony of termites.
3. Chemically treat the soil to form a barrier to repel termites. Many cities require this as part of the construction process.
4. Use chemically treated, pressure impregnated wood for sills, posts, and other parts near the soil.
5. Use poured concrete foundations rather than concrete block. They are less likely to crack and they have no open cells through which termites can build tunnels sight unseen inside the foundation.
6. If a concrete block foundation is used it should be capped with a 4 in. solid concrete cap.

A few areas have a *nonsubterranean type* of termite. They can live in damp or dry wood and do not require contact with the ground. They tend to leave fine wood particles through openings where they entered the wood. If you see "sawdust" you will know you have this type of termite.

WOOD PRESERVATIVES

Wood **preservatives** are used to protect wood from decay and insect attack. Preservative chemicals are forced into the wood to provide needed protection and are a must for wood to be in contact with the ground. Nonpressure treatments, such as dipping, brushing, or soaking, are used only on exterior wood that will not be in contact with the ground.

The major types of wood preservatives are pentachlorophenol, inorganic arsenic, creosote, and ACQ Preserve®. Pentachlorophenol and creosote are oil-borne,

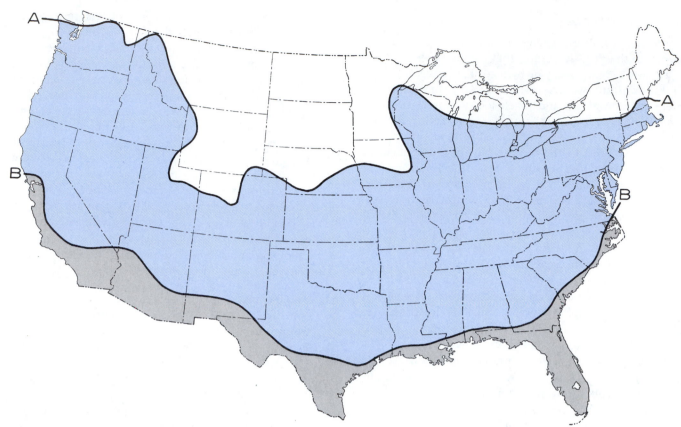

Figure 18.30 Line A indicates the northern limit of recorded damage by subterranean termites. Line B indicates the northern limit of damage by dry-wood (nonsubterranean) termites. *(Courtesy Forest Products Laboratory, USDA Forest Service, Madison, Wis.)*

and inorganic arsenic is water-borne. Other types finding some applications include copper naphthenate, zinc naphthenate, and chromated zinc chloride.

Inorganic Arsenic

Inorganic arsenic is a water-borne pesticide registered with the Environmental Protection Agency for use as a wood preservative. It protects against insects and decay. The types included are ammoniacal copper arsenate (ACA), chromated copper arsenate (CCA), and ammoniacal copper zinc arsenate (ACZA). Wood treated with inorganic arsenic should be used only where protection from insect attack and decay are important. It is widely used on decks, gazebos, utility poles, and piles on land and in fresh water. It is used on manufactured products such as plywood and millwork.

Water-borne preservatives leave the surface relatively clean, paintable, and free from objectionable odor. The chemicals are forced into the wood and the inorganic arsenic penetrates deeply. The wood product should be dried after treatment to the moisture content required for untreated wood.

There are certain hazards connected with using wood treated with inorganic arsenic. It must never be used where it will come in contact with food or be burned. Those cutting or handling treated wood products should wash thoroughly, wear a mask while cutting it, and wash work clothes separately from the other clothes.

ACQ Preserve®

ACQ Preserve is wood treated with ACQ. ACQ is a formulation of alkaline copper (AC) and a quaternary (Q) compound. The alkaline copper is the primary biocide, eradicating insects that attack wood. The quaternary or "quat" eliminates copper tolerant fungi. Both components are effective wood preservers, yet they have very low mammalian toxicity. ACQ contains no arsenic or chromium, and it carries lower environmental burdens over its lifetime than similar preservatives, such as CCA and ACZA (Fig. 18.31).

Figure 18.31 Logs being moved into a pressure chamber for treatment by ACQ Preserve®. *(Courtesy Chemical Specialties, Inc.)*

Pentachlorophenol

Pentachlorophenol is an oil-borne pesticide registered with the Environmental Protection Agency for use in protecting wood from insects and decay. It penetrates deeply into the wood and remains in the pressure-treated wood for a long time. It is used for a wide range of products, such as utility poles and glulams to be used in areas with moisture. A major use is in treating pilings for use on land and in fresh water. They are not generally used in salt water. Workers should observe the same precautions mentioned for inorganic arsenic.

Creosote

Creosote is an oil-borne pesticide registered with the Environmental Protection Agency to be used to pressure treat wood to protect it from insects and decay. The most effective is coal-tar creosote but water-tar, oil-tar, and water-gas-tar creosotes are used. Materials treated with creosote have a black to dark brown appearance and are not paintable.

Applying Preservatives

Wood products are treated by pressure and nonpressure processes.

Pressure Treatment

Pressure treatment involves impregnating the wood with the desired preservative by placing the wood in closed vessels and raising the pressure considerably above atmospheric pressure. The pressure forces the preservative into the wood until the desired amount has been absorbed. Considerable preservative is absorbed with relatively deep penetration. There are several different pressure processes in use, but the basic principles are the same. Pressure processes provide a closer control over preservation retentions and penetrations and generally better protection than nonpressure methods.

Nonpressure Processes

A wide range of nonpressure processes are used. These differ widely in penetration and retention; therefore, the degree of protection afforded varies. One method is *brushing.* This is the easiest and least effective. *Dipping* for a few minutes in a preservative gives greater assurance that all faces and checks have been coated. This is often used for assembled window sashes, frames, and other exterior millwork. A third method is *cold soaking.* Well-seasoned wood soaked for several hours to several days in low-viscosity oil preservatives. Some woods, such as pine, are successfully treated this way and have a long life when in contact with the ground. The *diffusion process* is used with green and dry wood. It uses a water-borne preservative that diffuses out of the treating solution or paste into the wood. The *double diffusion process* is more effective. It involves steeping the wood first in one chemical and then in another. Other diffusion processes include injecting the preservative at the ground line of a pole, applying a paste to the surface, and pouring the chemical into holes bored in the pole at ground line.

Another nonpressure process is a *thermal process.* In the hot bath, the air in the wood expands and some is forced out. In the cooling bath that follows, the air contracts, which creates a partial vacuum that forces liquid into the wood. It is used for fence posts and lumber. One last nonpressure process is the *vacuum process.* It utilizes a quick, low initial vacuum in a chamber followed by a brief immersion in the preservative followed by a final high vacuum. It is used to treat millwork with water-repellent preservatives and construction lumber with water-borne and water-repellent preservatives.

Effect of Treatment on Strength

Coal-tar, creosote, creosote-coal-tar mixtures, creosote-petroleum mixtures, and pentachlorophenol dissolved in petroleum oils are practically inert to wood and have no chemical influence affecting its strength. Chemicals in water-borne salt preservatives, such as chromium, copper, arsenic, and ammonia, are reactive with wood. They are potentially damaging to mechanical properties and can cause corrosion in mechanical fasteners. At retention levels required for ground contact, mechanical properties are essentially unchanged, but maximum load in bending, impact bending, and toughness are reduced somewhat. Heavy salt loadings for marine use may reduce bending strength by 10 percent and work properties by 50 percent.

Quality Certification

The American Wood Preservers Association (AWPA) is a professional society responsible for establishing consensus standards for the wood preserving industry. The American Lumber Standards Committee certifies wood preservation inspection agencies to provide quality control services to individual wood preserving plants. The accrediting agencies permit the use of their quality stamp, which indicates to the consumer that a product meets established standards (Fig. 18.32).

The National Wood Window and Door Association (NWWDA) has a testing, plant inspection, and certification program for non-pressure-treated products. They provide standards for water-repellent preservatives, wood window units, wood flush doors, wood sliding patio doors, wood skylight and roof windows, and wood swinging doors. Their quality certification seals are shown in Fig. 18.33.

FIRE RETARDANT TREATMENTS

In many applications wood must be treated with fire retardant chemicals. The two general methods for applying fire retardant chemicals are *pressure impregnating* the wood with water-borne or organic solvent–borne chemicals or applying fire retardant *chemical coatings* to the wood surface. Pressure treating chemicals include inorganic salts and more complex chemicals. Salts are most commonly used. They react to temperatures below the ignition point of wood, causing the combustible vapors generated in the wood to break down into nonflammable water and carbon dioxide. After treatment the wood should be dried to its original required moisture content.

Coatings have low surface flammability, and when they are exposed to fire they form an expanded low-density film. The film insulates the surface from high temperatures.

Most fire retardant chemicals do not resist exposure to weather so it is necessary to use leach-resistant types for exterior use, such as on wood shakes. Fire retardant treatment results in some slight reduction in the strength properties of wood, so design values for allowable stresses are reduced. Most chemicals cause fasteners to corrode. Designers must select a combination of chemicals and metal fasteners that can coexist without corrosion. Crystal salts in wood have an abrasive effect on cutting tools. Carbide-tipped cutting tools should be used when working with fire retardant wood. Gluing is also a problem. Special resorcinol-resin adhesives have proven acceptable. Fire-retardant wood can be painted if the moisture content has been reduced sufficiently.

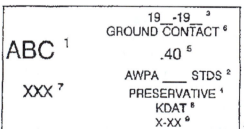

Interpreting a Quality Mark

1 - The identifying symbol, logo or name of the accredited agency.
2 - The applicable American Wood Preservers' Association (AWPA) commodity standard.
3 - The year of treatment if required by AWPA standard.
4 - The preservative used, which may be abbreviated.
5 - The preservative retention.
6 - The exposure category (e.g. Above Ground, Ground Contact, etc.).
7 - The plant name and location; or plant name and number; or plant number.
8 - If applicable, moisture content after treatment.
9 - If applicable, length, and/or class.

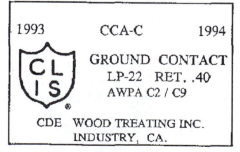

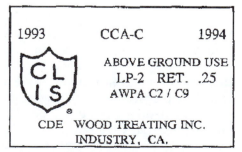

Figure 18.32 The quality mark indicates that the wood has been pressure treated following industry standards. (*Courtesy American Lumber Standard Committee and the California Lumber Inspection Service*)

WATER-REPELLENT
PRESERVATIVE   CONFORMS TO NWWDA I.S.-4

WOOD WINDOW UNIT   CONFORMS TO NWWDA I.S.-2

WOOD FLUSH DOOR   CONFORMS TO NWWDA I.S.-1

WOOD SLIDING PATIO DOOR   CONFORMS TO NWWDA I.S.-3

SKYLIGHT/ROOF WINDOW   CONFORMS TO NWWDA I.S.-7

WOOD SWINGING PATIO DOOR CONFORMS TO NWWDA I.S.-8

Figure 18.33 The seal of the National Wood Window and Door Association certifies that millwork has been treated with a nonpressure preservative following NWWDA standards. *(Courtesy National Wood Window and Door Association)*

REVIEW QUESTIONS

1. How can you tell a coniferous tree from a deciduous tree?
2. What natural defects are found in wood?
3. What are the two ways wood is seasoned?
4. What is the recommended moisture content for lumber used for framing?
5. Why is softwood lumber that is surfaced green produced larger than softwood lumber that is surfaced when it is dry?
6. What standard is followed by those producing softwood lumber?
7. How many board feet are in a $2'' \times 10'' \times 12'-0''$ piece of stock?
8. What is the volume of 250 pieces 38 mm $\times$ 140 mm $\times$ 3.66 m?
9. How does a contractor know the lumber received on the job has been dried to a 15 percent moisture content?
10. How can you differentiate between dimension lumber and timbers?
11. What is stress-rated lumber?
12. What is meant by the equilibrium moisture content?
13. In what ways does moisture effect wood?
14. What will be the size of a 5 in. $\times$ 8 in. rough cut timber after it dries?
15. What is the recommended moisture content for wood to be painted?
16. Explain why wood floats even though the wood fibers have a specific gravity of 1.5.
17. What is a good moisture content for producing a satisfactory glue bond?
18. Which species of wood have natural resistance to decay?
19. What types of preservatives are used on wood?
20. What are the two processes used to apply fire retardant chemicals to wood?

KEY TERMS

air drying Allowing lumber to dry by exposure to the air.

beam A horizontal structural member supporting a load.

board Lumber less than 2 inches thick and 1 inch or more wide.

board foot The measure of lumber having a volume of 144 cubic inches.

dimension lumber Lumber from 2 inches to, but not including, 5 inches thick and 2 inches or more wide.

dressed lumber Lumber having one or more sides planed smooth.

equilibrium moisture content The point at which the moisture content of the wood is the same as that of the surrounding air at a given relative humidity and temperature.

fiber saturation point The point in the drying or wetting of lumber at which the cell walls are saturated with water, but the cell cavities are free of water.

grade The designation of the quality of a board.

green lumber Lumber having a moisture content over 19 percent.

hardwood A botanical group of trees that are broad-leaved and deciduous.

heartwood The mature, nonliving central core of a tree.

horizontal shear The tendency of the top wood fibers to move horizontally in relationship to the bottom fibers.

hygroscopic Able to absorb moisture and expand or lose moisture and shrink.

kiln drying Drying lumber in a kiln where temperature and moisture are controlled.

lignin An organic substance that bonds together the fibers in wood.

lumber grader A person who inspects each piece of lumber and assigns a grade to it.

moisture content The weight of water in a piece of wood compared with the weight of the wood oven dry expressed as a percent.

preservatives Chemicals forced into the wood to provide protection from decay, insects, and other damaging conditions.

quartersawing Sawing lumber so the hard annual rings are nearly perpendicular to the surface.

rough lumber Lumber that has not been dressed.

sapwood The newly formed wood in a living tree.

seasoning The process of drying wood from its original green condition to a moisture content making it suitable for use.

softwood A botanical group of trees that, with a few exceptions, have needles and are evergreen.

stress-rated lumber Lumber that has its modulus of elasticity determined by actual tests on mechanical stress rating equipment.

timbers Lumber that is 5 inches or more in its least dimension.

vertical shear The tendency of one part of a member to move vertically in relationship to the adjacent part.

warp A variation in a board from a flat, plane condition.

SUGGESTED ACTIVITIES

1. Collect a variety of samples of wood species used in construction and conduct various tests, such as
    a. Tension and compression tests.
    b. Moisture content tests. Place samples outdoors for several weeks and test again. What happened to the moisture content?
    c. Drive common nails into the samples and devise a way to measure the pounds of pull required to pull them out. Which woods had the best holding characteristics?

2. Prepare a visual aid showing actual samples of species used in construction. Label each and list its properties and characteristics. Try to find samples having commonly found defects and identify these.

3. Collect samples of freshly cut (green) woods used in construction. Machine to a carefully selected thickness, width, and length. Measure the moisture content. In an oven or with some other method (air drying) let the samples dry until they reach a moisture content of 15 percent, then measure the size of the samples. Record your findings in a written report.

4. Visit the local lumber yard and examine and make a list of the sizes and species available. Pay special attention to the type of treatment used on pressure-treated rot-resistant woods.

5. Invite a local exterminator to address the class to cite the chemicals and applications used to treat a building for termites and other insects. Ask questions about the hazards and safety precautions taken.

ADDITIONAL INFORMATION

Chen, W.F., *The Civil Engineering Handbook*, CRC Press, Boca Raton, Fla., 1996.

Merritt, F.S. and Ricketts, J.T., *Building Design and Construction Handbook*, McGraw-Hill, New York, 1994.

Timber Construction Manual, American Institute of Timber Construction, John Wiley and Sons, New York, 1985.

Wood Handbook: Wood as an Engineering Material, Forest Products Laboratory, Forest Service, U.S. Department of Agriculture, 1987. Available from the Superintendent of Documents, U.S. Government Printing Office, Washington, DC 20402.

Wood Reference Handbook, Canadian Wood Council, Ottawa, 1991.

19

Products Manufactured from Wood

This chapter will help you to:

1. Learn the technical information about industrial plywood needed to make construction-related product decisions.

2. Select the proper reconstituted wood panel products for various applications.

3. Identify and select hardwood plywood for construction applications.

4. Identify and select manufactured wood structural components used for building construction.

5. Be aware of the types, sizes, and properties of many wood products manufactured from wood.

Wood is a major construction material. For example, high-quality wood is used to manufacture products such as doors and windows. In addition, many products are manufactured using wood by-products, such as wood chips and logs too small to produce usable solid lumber. This chapter presents information on many of these wood products.

PLYWOOD AND OTHER PANEL PRODUCTS

A variety of veneered and reconstituted panels are made from wood. These include plywood, composite panels, waferboard, oriented strand board, particleboard, and hardboard.

Plywood Panel Construction

Plywood panels are made by bonding together thin layers or plies of wood. The grain in each ply is perpendicular to the ply below it. Panels always have an odd number of plys, such as three, five, or seven. Each ply may be a single thickness veneer or two veneers glued together to form a thicker ply.

The plywood constructions commonly used in building construction are the veneer core, lumber core, particleboard core, and medium-density fiberboard core (Fig. 19.1). The *veneer core panel* has three to nine plies. The *lumber core panel* has strips of solid wood glued together as the core and two veneers glued to each side. *Particleboard core* and *medium-density fiberboard core* panels have a single sheet of one of these materials on the core and a single ply glued on each side. Notice that each type has the same number of plies on each side of the core. This provides *balanced construction*.

Specifications for Plywood Panels

Some grades of veneered plywood panels are manufactured under specifications or performance testing standards of U.S. Product Standard PS 1-83, Construction and Industrial Plywood. This is a manufacturing specification developed cooperatively by members of the plywood industry and the Office of Product Standards Policy of the National Bureau of Standards. Other

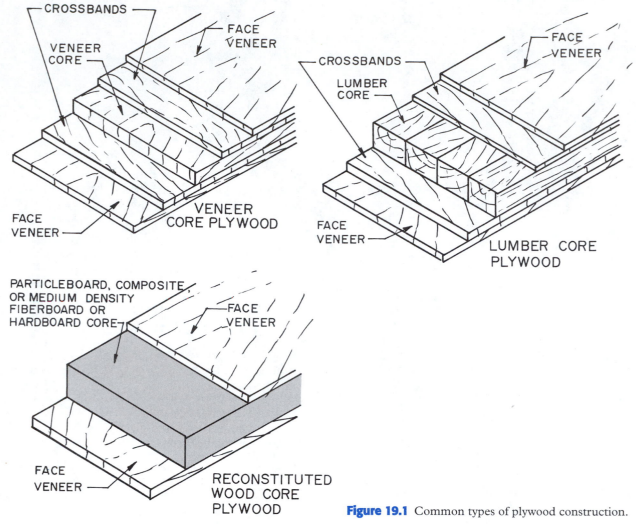

Figure 19.1 Common types of plywood construction.

veneered panels, including a number of performance rated composite and nonveneered panels, are manufactured under provisions of APA - The Engineered Wood Association performance standards. APA-rated panels that meet PS 1-83 requirements have the designation "PS 1-83" in the APA trademark.

In Canada there are three standards for softwood plywood, all of which are in the metric terms: CSA 0121-M Douglas Fir Plywood, CSA 0151-M Canadian Softwood Plywood, and CSA 0153-M Poplar Plywood. Allowable stresses and section properties for Douglas fir plywood are in CAN3-086-M.

Construction and Industrial Plywood

Construction and industrial plywood are manufactured according to the specifications in U.S. Product Standard PS 1-83. Canadian plywood is made following the standards detailed earlier in this chapter.

Grades

The outer veneers of construction and industrial plywood are classified in five appearance groups, N, A, B, C, and D. N is the best and D is the poorest grade (Fig. 19.2).

Species

Construction and industrial plywood is made using about seventy species of wood. These may be mixed within a panel and may be hardwoods and softwoods. The species used are divided into five groups, depending on their strength and stiffness. Group 1 includes the strongest and stiffest species. Group 5 includes those with the lowest properties (Table 19.1). The inner plies in Groups 1, 2, 3, and 4 may be any of the species in Groups 1, 2, 3, and 4. Inner plies in Group 5 may be any of the species listed. The front and back plies are of the same species and must be from the group indicated by the group number.

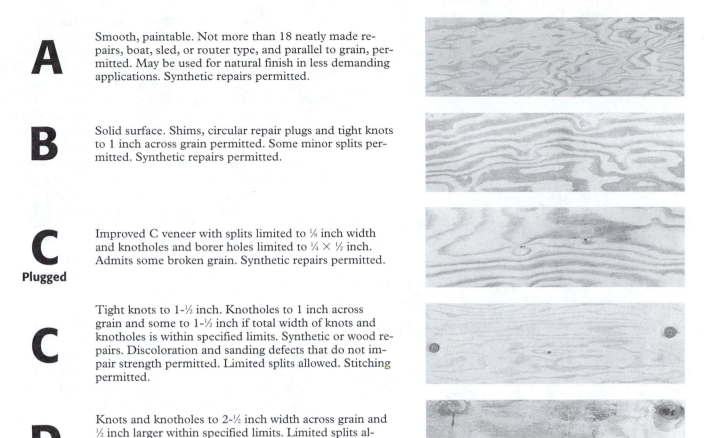

A Smooth, paintable. Not more than 18 neatly made repairs, boat, sled, or router type, and parallel to grain, permitted. May be used for natural finish in less demanding applications. Synthetic repairs permitted.

B Solid surface. Shims, circular repair plugs and tight knots to 1 inch across grain permitted. Some minor splits permitted. Synthetic repairs permitted.

C
Plugged Improved C veneer with splits limited to ⅛ inch width and knotholes and borer holes limited to ¼ × ½ inch. Admits some broken grain. Synthetic repairs permitted.

C Tight knots to 1-½ inch. Knotholes to 1 inch across grain and some to 1-½ inch if total width of knots and knotholes is within specified limits. Synthetic or wood repairs. Discoloration and sanding defects that do not impair strength permitted. Limited splits allowed. Stitching permitted.

D Knots and knotholes to 2-½ inch width across grain and ½ inch larger within specified limits. Limited splits allowed. Stitching permitted. Limited to Interior, Exposure 1 and Exposure 2 panels

Figure 19.2 Plywood veneer grades. *(Courtesy APA - The Engineered Wood Association)*

Table 19.1 Plywood Outer Veneer Group Numbers

Group 1	Group 2		Group 3	Group 4	Group 5
Apitong	Cedar, Port Orford	Maple, black	Alder, red	Aspen	Basswood
Beech, American	Cypress	Mengkulang	Birch, paper	Bigtooth	Poplar, balsam
Birch	Douglas fir 2	Meranti, red	Cedar, Alaska	Quaking	
Sweet	Fir	Mersawa	Fir, subalpine	Cativo	
Yellow	California red	Pine	Hemlock, eastern	Cedar	
Douglas fir 1	Grand	Pond	Maple, bigleaf	Incense	
Kapur	Noble	Red	Pine	Western red	
Keruing	Pacific silver	Virginia	Jack	Cottonwood	
Larch, western	White	Western white	Lodgepole	Eastern	
Maple, sugar	Hemlock, western	Spruce	Ponderosa	Black (western	
Pine	Lauan	Red	Spruce	poplar)	
Caribbean	Almon	Sitka	Redwood	Pine	
Ocote	Bagtikan	Sweetgum	Spruce	Eastern white	
Pine, southern	Mayapis	Tamarack	Engelmann	Sugar	
Loblolly	Red	Yellow poplar	White		
Longleaf	Tangile				
Shortleaf	White				
Slash					
Tanoak					

Courtesy APA-The Engineered Wood Association

Table 19.2 Plywood Panel Thicknesses and Dimensions

Nominal Thickness	
In.	**mm^a**
1/4	6.4
5/16	7.9
11/32	8.7
3/8	9.5
7/16	11.1
15/32	11.9
1/2	12.7
19/32	15.1
5/8	15.9
23/32	18.3
3/4	19.1
7/8	22.2
1	25.4
1 3/32	27.8
1 1/8	28.6

Nominal Dimensions (width × length)		
ft.	**mm**	**m^a**
4 × 8	1219 × 2438	1.22 × 2.44
4 × 9	1219 × 2743	1.22 × 2.74
4 × 10	1219 × 3048	1.22 × 3.05

^{a}Soft converted metric sizes
Courtesy APA-The Engineered Wood Association

Sizes of Panels

In customary units, standard nominal thicknesses of sanded construction and industrial plywood panels are 1/8 through 1 1/4 in., in 1/8 in. increments. Unsanded panels are 5/16 in. to 1 1/4 in., in increments for panels over 3/8 in. thick. Panel widths are 36, 48, and 60 in., and lengths are 60 to 144 in., in 12 in. increments (Table 19.2).

Soft converted metric thicknesses are also shown in Table 19.2. Panels sized 1200 × 2400 mm are designed for use in hard metric designed buildings while panels 1219 × 2438 mm are soft conversions and can be used in buildings designed using customary units.

APA Performance-Rated Panels

APA Performance-Rated Panels are manufactured to performance standards of APA - The Engineered Wood Association, which establishes performance criteria for designated construction applications. The four panels specified by APA are plywood, **oriented strand board, composite** (Com-Ply®), and APA Rated Siding (Fig. 19.3).

Plywood panels are made using all veneer plies, producing the strongest type of plywood. *Oriented strand board* (OSB) is manufactured from strands or wafers oriented in general in one direction. These layers of ori-

ented strands are bonded together at right angles to each other. Usually three to five layers are bonded to form a panel. Most OSB panels are textured on one side to produce a nonslick surface.

Composite or *Com-Ply*® panels have a core of reconstituted wood bonded between solid wood veneers. This produces a panel that allows for efficient use of wood materials yet has a wood grain surface on the front and back.

APA Rated Siding is made by bonding wood veneers in the same manner as plywood. It is available in a variety of surface textures and designs. Panels are available in 4 × 8, 4 × 9, and 4 × 10 ft. dimensions. A guide to APA Performance-Rated Panels is in Fig. 19.4.

APA Rated Sheathing is used for subfloor, wall, and roof sheathing. It is used where strength and stiffness are required. Nominal panel thicknesses range from 5/16 to 3/4 inch. *Structural I* is a type of rated sheathing used where increased resistance to wracking and cross-panel strength are needed. It is used on structural diaphragms and panelized roofs. *APA Rated Sturd-I-Floor*® is a single-layer flooring for use under carpets. It may eliminate the need for installing an underlayment. Panels are available with square and tongue-and-groove edges. It is available in thicknesses from 19/32 to 1 1/8 inch (Fig. 19.4).

Exposure Durability Classifications

APA Performance-Rated Panels are manufactured in four exposure durability classifications—Exterior, Exposure 1, Exposure 2, and Interior. These classifications pertain to how well the bonding agent used to join the veneers can resist exposure to moisture.

Exterior panels use a waterproof adhesive and are designed for use on applications subject to permanent exposure to moisture and weather.

Exposure 1 panels use the same waterproof adhesive as Exterior panels and are designed to be used where they will be exposed to moisture or the weather for long periods before they are finally protected. They are also used where the panels may be subjected occasionally to moisture after they are in service. They differ from Exterior panels in some compositional aspects and are not recommended for permanent exposure to the weather.

Exposure 2 panels are used in applications in which they will be briefly exposed to the weather during construction before being permanently protected. They are actually an interior-type panel with an intermediate adhesive.

Interior panels are manufactured with an interior glue and are intended for interior use only. They are identified by abbreviation INT-APA and the omission of glue-line information on their trademark.

Veneer grades used on APA Performance-Rated Panels are A, B, C, C plugged and D (Fig. 19.2).

Plywood
All-veneer panels consisting of an odd number of cross-laminated layers, each layer consisting of one or more piles. Many such panels meet all of the prescriptive or performance provisions of U.S. Product Standard PS 1-83/ANSI A199.1 for Construction and Industrial Plywood.

APA Rated Siding
All-veneer panels constructed in the same manner as plywood panels. Available in a variety of panel sizes and thicknesses. Various surface textures and designs available.

Composite (Comply)®
Panels of reconstituted wood cores bonded between veneer face and back plies.

Oriented Strand Board
Panels of compressed strand-like particles arranged in layers (usually three to five) oriented at right angles to one another.

Figure 19.3 APA Performance-Rated Panels. *(Courtesy APA - The Engineered Wood Association)*

Product Identification

APA trademarks on the types of plywood used in construction are in Fig. 19.5. The *sanded* panels with B-grade or better veneer panels are used for various construction applications. The *specialty grades* include panels designed for a specific use, such as Plyform® for concrete forms and underlayment. Most of the information on the trademark is self-explanatory except for span rating. A span rating of $^{32}/_{16}$ means the panel can be used as roof sheathing with rafters spaced up to 32 in. on-center and as subflooring with floor joists spaced up to 16 in. on-center.

Specialty Plywoods

Many plywood manufacturers make various plywood products that are not included in the national standards for plywood manufacture. These are classified as specialty plywoods. The user must refer to information supplied by the manufacturer when selecting these products because they meet no standard. Several of the most frequently used are overlaid plywood, siding panels, and interior paneling. The APA does have standards and trademarks for several specialty panels.

Overlaid plywood is a high-grade exterior-type panel that has a resin-impregnated fiber ply bonded to one or both sides. It is made in two types, high density and medium density. *High-density panels* have a hard, smooth chemically resistant surface. They are made in several colors and require no additional finish. *Medium-density panels* have a smooth, opaque, nonglossy surface that hides the grain of the veneers below it. They are used when a high-quality paint finish is needed.

Siding panels are used as finished exterior siding and are available in a variety of surface finishes, such as rough sawed, V-grooved, and reverse batten grooves.

Paneling is used on interior walls as the finish material. It is usually prefinished and only requires installation. It is available in a wide variety of wood species and surface fixtures.

Hardwood Plywood

Hardwood plywood is plywood made using various species of hardwoods for the outer veneer. It is manufactured following the standards in ANSI/HPMA HP 1983, American National Standards for Hardwood and Decorative Plywood. These standards are published by the Hardwood Plywood Manufacturers Association, P.O. Box 2789, Reston, VA 22090. Canadian hardwood plywoods are made according to metric standard CSA 0115-M, Hardwood and Decorative Plywood.

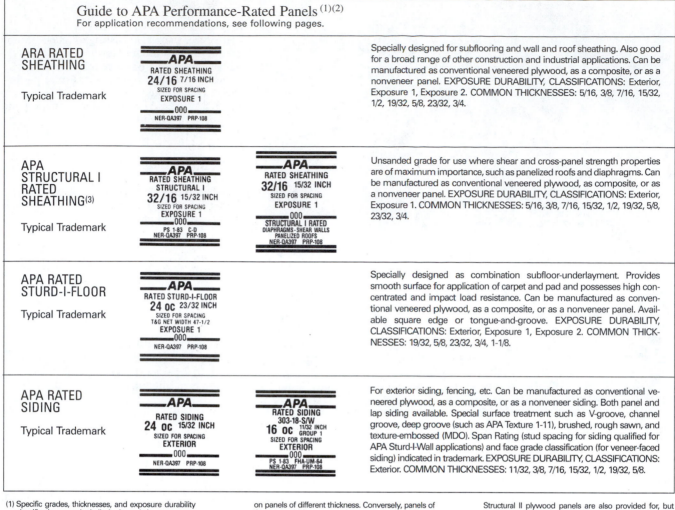

Figure 19.4 APA - The Engineered Wood Association trademarks and uses of APA Performance-Rated panels. *(Courtesy APA - The Engineered Wood Association)*

Species

Species used in hardwood plywood are divided into four categories (Table 19.3). These categories reflect the modulus of elasticity (stiffness) of each species and its specific gravity. Category A is the stiffest and D has the lowest rating.

Grades of Veneers

Hardwood plywood offers six grades of hardwood veneers. These grades are summarized in Table 19.4. There are also specific requirements for softwoods used in hardwood plywood panels.

A grade (A) is the best. The face is made of hardwood veneers carefully matched as to color and grain.

B grade (B) is suitable for a natural finish, but the face veneers are not as carefully matched as on the A grade.

Sound grade (2) provides a face that is smooth. All defects have been repaired. It is used as a smooth base for a paint finish.

Industrial-grade (3) face veneers can have surface defects. This grade permits knotholes up to 1 in. (25 mm) in diameter, small open joints, and small areas of rough grain.

Backing grade (4) uses unselected veneers having knotholes up to 3 in. in diameter and certain types of splits. No defect is permitted that affects the strength of the panel.

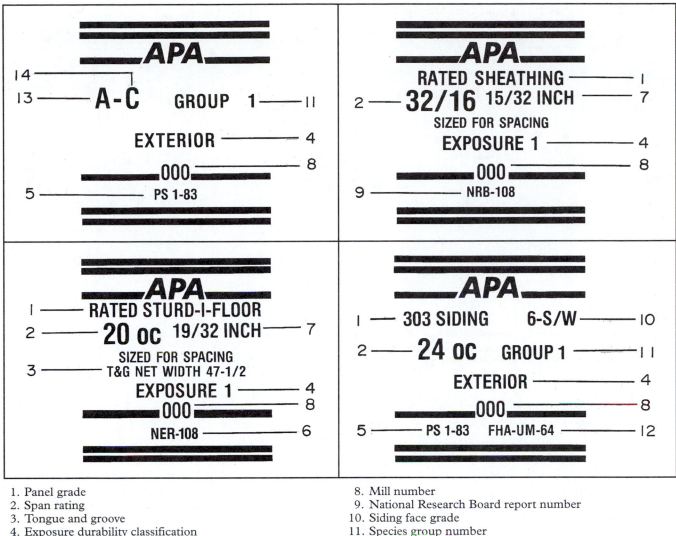

1. Panel grade
2. Span rating
3. Tongue and groove
4. Exposure durability classification
5. Product standard
6. Code recognition of APA as a quality assurance agency
7. Thickness
8. Mill number
9. National Research Board report number
10. Siding face grade
11. Species group number
12. FHA recognition
13. Grade of face veneer
14. Grade of back veneer

Figure 19.5 A few of the trademarks of the APA - The Engineered Wood Association. *(Courtesy APA - The Engineered Wood Association)*

Specialty grade (SP) includes veneers having characteristics unlike any of those in the other grades. The characteristics are agreed upon between the manufacturer and the purchaser. For example, species such as wormy chestnut or bird's-eye maple are considered as specialty grade.

Types of Hardwood Plywood

There are four types of hardwood plywood rated according to their water-resistance capacity. The Technical type is the most water resistant, and Type III is the least (Table 19.5).

Construction of Hardwood Plywood

The construction of hardwood plywood is described by identifying the core. The commonly available constructions are

Hardwood veneer core: Has an odd number of plies, such as 3-ply, 5-ply, and so on.

Softwood veneer core: Has an odd number of plies, as 3-ply, 5-ply, and so on.

Hardwood lumber core: Used in 3-ply, 5-ply, and 7-ply constructions.

Softwood lumber core: Used in 3-ply, 5-ply, and 7-ply constructions.

Table 19.3 Categories of Commonly Used Decorative Species in Hardwood Plywood[a]

Category A	Category B	Category C	Category D
Ash, white	Ash, black	Alder, red	Aspen
Apitong	Avodire	Basswood, American	Cedar, eastern red
Beech, American	Birch, paper	Butternut	Cedar, western red
Birch, yellow, sweet	Cherry, black	Cativo	Fuma
Bubinga	Cypress	Chestnut, American	Willow, black
Hickory	Elm, rock	Cottonwood, black	
Kapur	Fir, Douglas	Cottonwood, eastern	
Keruing	Fir, white	Elm, American (gray, red, or white)	
Oak (Oregon, red, or	Gum, sweet	Gum, black	
white)	Hemlock, western	Hackberry	
Paldao	Magnolia, cucumber	Hemlock, eastern	
Pecan	Sweetbay	Lauan	
Rosewood	Maple, sugar (hard)	Maple, red (soft)	
Sapele	Mahogany, African	Maple, silver (soft)	
	Mahogany, Honduras	Meranti, red	
	Maple, black (hard)	Pine, ponderosa	
	Pine, western white	Pine, sugar	
	Poplar, yellow	Pine, eastern white	
	Spruce, red, Sitka	Prima-vera	
	Sycamore	Redwood	
	Tanoak	Sassafras	
	Teak	Spruce (black, Engelmann, white)	
	Walnut, American	Tupelo, water	

[a]Based on an evaluation of published modulus of elasticity (MOE) and specific gravity values.
Courtesy Hardwood Plywood Manufacturers Association

Table 19.4 Hardwood Veneer Grades

Grade	Symbol
A grade	A
B grade	B
Sound grade	2
Industrial grade	3
Backing grade	4
Specialty grade	SP

Table 19.5 Types of Hardwood Plywood[a]

Technical	(Exterior use)
Type I	(Exterior use)
Type II	(Interior use)
Type III	(Interior use)

[a]Based on water-resistance capacity

Particleboard core: Used in 3-ply and 5-ply constructions.

Medium-density fiberboard core: Used in 3-ply construction.

Hardboard core: Used in 3-ply construction.

Special cores: Used in 3-ply construction. Special cores are those made of any other material than those listed above.

Sizes and Thicknesses of Panels

Hardwood plywood is available in panels 48 in. wide and 96 and 120 in. long. Standard thicknesses are $\frac{1}{4}$, $\frac{3}{8}$, $\frac{1}{2}$, and $\frac{3}{4}$ in. Lumber core panels and particleboard core panels are available in only the $\frac{3}{4}$ in. thickness.

Canadian hardwood plywood is available in incremental thicknesses from 4.8 to 32 mm.

Product Identification

The trademark stamp is placed on the back of the panel (Figs. 19.6 and 19.7). Most of the information shown is self-explanatory. The flame spread rating shows the rating given the panel following tests. Flame spread ratings are set by ASTM E162. For example, codes require cabinets to have a flame spread rating of 200 or less. The bonding agents include formaldehyde, which tends to filter into the air. Some codes require the plywood to meet formaldehyde emission requirements.

Reconstituted Wood Products

In addition to the APA Performance-Rated Panels discussed earlier, a number of reconstituted wood panels are used, primarily in cabinet and furniture construction. These include hardboard, particleboard, fiberboard, and waferboard.

Hardboard is made from wood chips converted into fibers and bonded into panels under heat and pressure. It is manufactured following standards developed by the American Hardboard Association. It is available in thicknesses from $\frac{1}{12}$ up to $1\frac{1}{8}$ in. The most used panel size is 4 × 8 ft., but other sizes are available on special order.

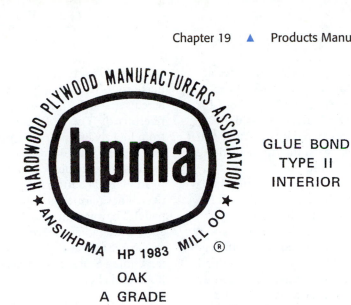

Figure 19.6 An HPMA trademark stamp for hardwood plywood. *(Courtesy Hardwood Plywood and Veneer Association)*

HARDWOOD PLYWOOD MANUFACTURERS ASSOCIATION		
FORMALDEHYDE EMISSION 0.2 PPM CONFORMS TO HUD REQUIREMENTS ——— LAY UP 16 3.6MM THICK HP-SG-86	SIMULATED DECORATIVE FINISH ON PLYWOOD ————— **hpma** ® MILL SPECIALTY GRADE	FLAME SPREAD 200 OR LESS ASTM E84 GLUE BOND TYPE II ANSI/HPMA HP 1983

PREFINISHED PLYWOOD BACKSTAMP

HARDWOOD PLYWOOD MANUFACTURERS ASSOCIATION		
FORMALDEHYDE EMISSION 0.3 PPM CONFORMS TO HUD REQUIREMENTS	**hpma** ® MILL	FLAME SPREAD 200 OR LESS ASTM E84 ——— SIMULATED DECORATIVE FINISH ON PARTICLEBOARD

PREFINISHED PARTICLEBOARD BACKSTAMP

Figure 19.7 HPMA trademark backstamps for prefinished plywood and prefinished particleboard. *(Courtesy Hardwood Plywood and Veneer Association)*

Metric hardboard thicknesses for standard hardboard are 2.1, 2.5, 3.2, 4.8, 6.4, 7.9, and 9.5 mm. Panels are 1200 × 2400 mm.

Hardboard is available in five classes (Table 19.6), which are defined as follows:

Class 1: Tempered is impregnated with siccative material and is stabilized by heat and special additives to impart substantially improved properties of stiffness, strength, hardness, and resistance to water and abrasion, as compared with the Standard Class.

Class 2: Standard is the form of the material as it comes from the press. It has high strength and water resistance.

Class 3: Service-tempered is impregnated with siccative material and is stabilized by heat and additives. Its properties are substantially better than Service Grade.

Class 4: Service is basically the same form as it comes from the press but has less strength than Standard Class.

Class 5: Industrialite is a medium-density hardboard that has moderate strength and lower unit weight than the other classes.

Table 19.6 Classifications of Hardboard Panels

Class 1	Tempered
Class 2	Standard
Class 3	Service-tempered
Class 4	Service
Class 5	Industrialite

Hardboard is manufactured in the United States according to the following standards: ANSI/AHA A135.4-1988, Basic Hardboard; ANSI/AHA A135.5-1988, PF Hardboard Paneling; and ANSI/AHA A135.6-1990, Hardboard Siding. These are promoted by the American Hardboard Association. Panels manufactured to these standards have an AHA grade stamp.

Canadian hardboard products are manufactured following CGSB 11-GP-3M, Hardboard, and CGSB 11-GP-5M, Hardboard for Exterior Cladding.

Particleboard is made from wood chips, water, and a synthetic resin binder. The particles are bonded with heat and pressure. It is manufactured to standards developed by the National Particleboard Association. It is mainly used for furniture and cabinet construction (Fig. 19.8).

The general uses and grades are in Table 19.7. The grades are identified by a letter designation followed by a hyphen and a digit or letter.

The second digit or letter designation indicates the grade identification within a particular density or product description. For example, M-2 indicates a medium-density particleboard, Grade 2.

If there is a third designation it indicates a special characteristic, such as panel M-3 Exterior Glue. This is a medium-density panel, Grade 3, with exterior glue.

Grades 1, 2, and 3 relate to the relative magnitude of the mechanical properties as shown in Fig. 19.9. For example, Grade 3 has the highest modulus of elasticity, modulus of rupture, and hardness. Grade 1 has the lowest of these properties. Canadian particleboard is made following ANSI 208.1. Particleboard standards.

Waferboard is made by bonding large wood flakes 1½ in. or longer into panels having the same thicknesses and panel sizes as particleboard.

In the United States *fiberboard* consists of two products, softboard and hardboard. In Canada these are listed separately, and softboard panels are identified as fiberboard. The softboard panels are manufactured from loosely bound paper pulp and other types of fibers into panels that have insulating properties. They are used as rigid insulation on walls and roofs.

In Canada softboard panels have no standard thickness. These are established by the manufacturer of the panel. They are cut into 1200 × 2400 mm and 4 × 8 ft. panels. Canadian fiberboard is manufactured according to CSA Standard A247-M, Insulating Fiberboard. In the United States thicknesses available are ¾, 1, 1½, and 2 in., and sheet sizes are 24 × 48 in., 48 × 48 in., and 4 × 8 ft.

STRUCTURAL BUILDING COMPONENTS

Trusses are used for floor and roof construction. They are carefully engineered to carry known loads over specified distances. Most trusses are made from 2 × 4 in. and 2 × 6 in. lumber joined with wood or metal gusset plates (Fig. 19.10).

Glued laminated wood members are made by bonding ¾ in. thick wood strips for curved members and 1½ in. thick wood strips for straight members to form a beam or column. Laminations in Canadian products are 19 and 38 mm. These are commonly called **glulam** members. Members of any size can be made, but manufacturers produce a wide range of standard sizes. Inch size widths range from 3⅛ in. to 10¾ in. and depths range from 6 in. to 81 in. Standard Canadian widths range from 80 to 365 mm and depths from 114 to 2128 mm.

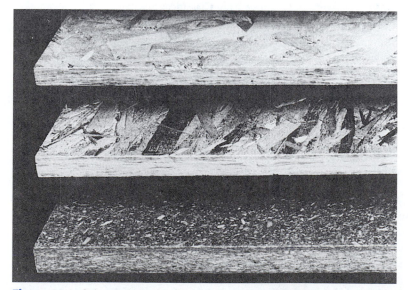

Waferboard
Panels of compressed wafer-like particles or flakes randomly or directionally oriented.

Oriented Strand Board
Panels of compressed strand-like particles arranged in layers (usually three to five) oriented at right angles to one another.

Structural Particleboard
Panels comprised of small particles usually arranged in layers by particle size, but not usually oriented.

Figure 19.8 Other panels composed of reconstituted wood that are designed to carry known loads over known distances.

Table 19.7 Grades and Uses of Particleboard

Type	Grade	Use
High density	H-1, H-2, H-3	High-density industrial
	H-1, H-2, H-3 Exterior Glue	High-density exterior industrial
Medium density	M-1	Commercial
	M-2, M-3	Industrial
	M-1, M-2, M-3 Exterior Glue	Exterior construction
		Exterior industrial
Medium density—specialty grade	M-S	Commercial
Low density	LD-1, LD-2	Door core
Underlayment	PBU	Underlayment
Manufactured home decking	D-2, D-3	Flooring in manufactured homes

Courtesy National Particleboard Association

Table A
Requirements for Grades of Particleboard [1,2]

ANSI A208.1-1993

Grade[3]	Length & Width mm (inch)	Thickness Tolerance[4] Panel Average from Nominal mm (inch)	Thickness Tolerance[4] Variance from Panel Average mm (inch)	Modulus of Rupture N/mm² (psi)	Modulus of Elasticity N/mm² (psi)	Internal Bond N/mm² (psi)	Hardness N (pounds)	Linear Expansion max. avg. (percent)	Screw-holding Face N (pounds)	Screw-holding Edge N (pounds)	Formaldehyde Maximum Emissions (ppm)
H-1	±2.0 (0.080)	±0.200 (0.008)	±0.100 (0.004)	16.5 (2393)	2400 (348100)	0.90 (130)	2225 (500)	NS[5]	1800 (405)	1325 (298)	0.30
H-2	±2.0 (0.080)	±0.200 (0.008)	±0.100 (0.004)	20.5 (2973)	2400 (348100)	0.90 (130)	4450 (1000)	NS	1900 (427)	1550 (348)	0.30
H-3	±2.0 (0.080)	±0.200 (0.008)	±0.100 (0.004)	23.5 (3408)	2750 (398900)	1.00 (145)	6675 (1500)	NS	2000 (450)	1550 (348)	0.30
M-1	±2.0 (0.080)	±0.250 (0.010)	±0.125 (0.005)	11.0 (1595)	1725 (250200)	0.40 (58)	2225 (500)	0.35	NS	NS	0.30
M-S[6]	±2.0 (0.080)	±0.250 (0.010)	±0.125 (0.005)	12.5 (1813)	1900 (275600)	0.40 (58)	2225 (500)	0.35	900 (202)	800 (180)	0.30
M-2	±2.0 (0.080)	±0.200 (0.008)	±0.100 (0.004)	14.5 (2103)	2250 (326300)	0.45 (65)	2225 (500)	0.35	1000 (225)	900 (202)	0.30
M-3	±2.0 (0.080)	±0.200 (0.008)	±0.100 (0.004)	16.5 (2393)	2750 (398900)	0.55 (80)	2225 (500)	0.35	1100 (247)	1000 (225)	0.30
LD-1	±2.0 (0.080)	+0.125 (0.005) -0.375 (0.015)	±0.125 (0.005)	3.0 (435)	550 (79800)	0.10 (15)		NS	0.35	400 (90)	NS	0.30
LD-2	±2.0 (0.080)	+0.125 (0.005) -0.375 (0.015)	±0.125 (0.005)	5.0 (725)	1025 (148700)	0.15 (22)		NS	0.35	550 (124)	NS	0.30

1) Particleboard made with phenol formaldehyde based resins do not emit significant quantities of formaldehyde. Therefore, such products and other particleboard products made with resin not containing formaldehyde are not subject to formaldehyde emission conformance testing.

2) Grades listed in this table shall also comply with the appropriate requirements listed in Section 3. Panels designated as "exterior glue" must maintain 50% MOR after ASTM D 1037 accelerated aging (paragraph 3.3.3).

3) Refer to Annex C for general use and grade information.

4) Thickness tolerance values are only for sanded panels as defined by the manufacturer. Unsanded panels shall be in accordance with the thickness tolerances specified by agreement between the manufacturer and the purchaser.

5) NS - Not Specified

6) Grade "M-S" refers to medium density, "special" grade. This grade was added to the Standard after grades M-1, M-2, and M-3 had been established. Grade "M-S" falls between M-1 and M-2 in physical properties.

Table B
Requirements for Grades of Particleboard Flooring Products [1,2]

Grade[3]	Length & Width Tolerance mm (inch)	Thickness Tolerance[4] Panel Average from Nominal mm (inch)	Thickness Tolerance[4] Variance from Panel Average mm (inch)	Modulus of Rupture N/mm² (psi)	Modulus of Elasticity N/mm² (psi)	Internal Bond N/mm² (psi)	Hardness N (pounds)	Linear Expansion max. avg. (percent)	Formaldehyde Maximum Emissions (ppm)
PBU	+0 (0) -4.0 (0.160)	±0.375 (0.015)	±0.250 (0.010)	11.0 (1595)	1725 (250200)	0.40 (58)	2225 (500)	0.35	0.20
D-2	±2.0 (0.080)	±0.375 (0.015)	±0.250 (0.010)	16.5 (2393)	2750 (398900)	0.55 (80)	2225 (500)	0.30	0.20
D-3	±2.0 (0.080)	±0.375 (0.015)	±0.250 (0.010)	19.5 (2828)	3100 (449600)	0.55 (80)	2225 (500)	0.30	0.20

1) Particleboard made with phenol formaldehyde based resins do not emit significant quantities of formaldehyde. Therefore, such products and other particleboard products made with resin not containing formaldehyde are not subject to formaldehyde emission conformance testing.

2) Grades listed in this table shall also comply with the appropriate requirements listed in Section 3. Panels designated as "exterior glue" must maintain 50% MOR after ASTM D 1037 accelerated aging (paragraph 3.3.3).

3) Refer to Annex C for general use and grade information.

4) Thickness tolerance values are only for sanded panels as defined by the manufacturer. Unsanded panels shall be in accordance with the thickness tolerances specified by agreement between the manufacturer and the purchaser.

Figure 19.9 Requirements for grades of particleboard and particleboard flooring products. *(Courtesy National Particleboard Association)*

Wood can be laminated into various shapes, including curved, and be stronger than solid wood the same size. Glued laminated wood members can be made to a high level of quality because wood defects can be cut away. They are manufactured in the United States according to ANSI/AITC A190.1-1983, Structural Glued Laminated Timber. In Canada the standards include CSA 0177-M, Qualification Code for Manufacturers of Structural Glued-Laminated Timber; CSA 0122, Structural Glued-Laminated Timber; and CSA3-086-M, Code for Engineering Design in Wood.

A typical glulam member is shown in Fig. 19.11. The laminations can be butted in the lightly stressed center laminations. The outer laminations, subject to tension and compression forces, must be joined with a *scarf or finger joint*.

Laminated veneer lumber is a manufactured structural member made by bonding wood veneers with an exterior adhesive. It has almost no shrinking, checking, twisting, or splitting. It has more load-bearing capacity per pound than sawed lumber. It has a modulus of elasticity of about 2.0 and an allowable bending stress of 3100 psi. It is available in depths from 9¼ to 18 in. and thicknesses of 1½ and 1¾ in. (Fig. 19.12).

A comparison of the physical properties of glued laminated lumber, laminated veneer lumber, and solid lumber is shown in Table 19.8. Although actual data vary for species, grades, and manufacturing requirements, this comparison shows that glued manufactured wood structural members exhibit properties that are greater than those of most grades and species of solid woods.

Plywood-lumber beams (box beams) are made according to carefully engineered specifications as to the size of members and nailing requirements. They are glued and built in factories under controlled conditions. One example is shown in Fig. 19.13.

Stressed skin panels are prefabricated panels using solid lumber stringers and headers and plywood skins. They are used for floors, walls, and roofs. They enable a constructor to rapidly erect a building because they cover a large area and span fairly long distances. For example, on a floor the joists would be replaced with beams, possibly box beams, spaced 4 ft. on-center. See Fig. 19.14.

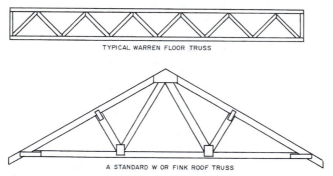

Figure 19.10 Two types of wood trusses.

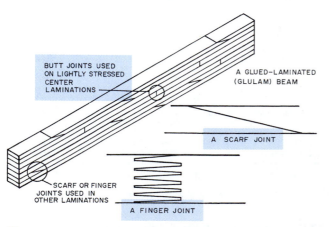

Figure 19.11 Glued laminated beams can span long distances and carry heavy loads. *(Courtesy Southern Forest Products Association)*

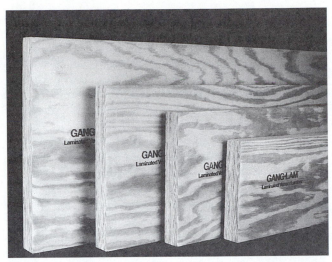

Figure 19.12 Laminated veneer lumber is made by bonding together layers of wood veneer with an exterior adhesive. *(Courtesy Louisiana-Pacific)*

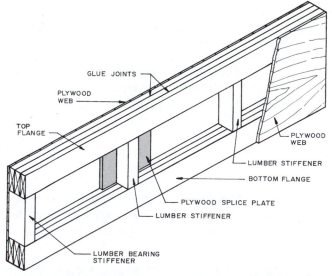

Figure 19.13 A typical plywood-lumber beam (box beam) with plywood webs and solid wood flanges and stiffeners.

Table 19.8 Properties of Laminated Veneer, Glued Laminated Members, and Solid Lumber

Property	LVL[a]	Glued Laminated[b]	Solid Lumber[c]
Extreme fiber in bending (F_b) psi	3000	1600–2400	760–2710
Horizontal shear (H_v) psi	290	140–650	95–120
Compression perpendicular to grain ($F_{c\perp}$) psi	3180	375–650	565–660
Compression parallel with grain (F_c) psi	—	1350–2300	625–1100
Tension parallel with grain (F_t) psi	2300	1050–1450	450–1200
Modulus of elasticity (E) million psi	2.0	1.2–1.8	1.3–1.9

[a]Limited to one size member
[b]Data from one manufacturer
[c]Data for one species of lumber

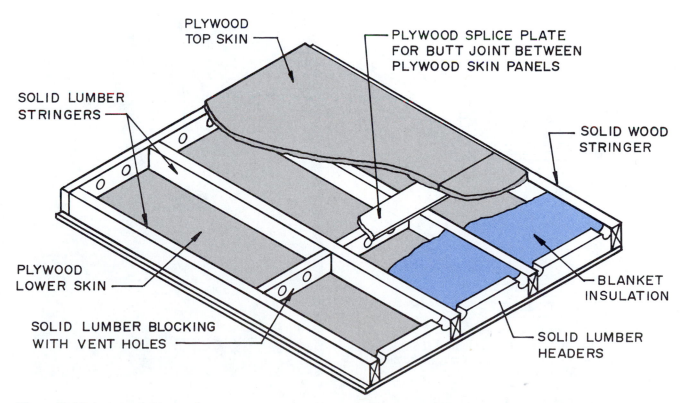

Figure 19.14 A stressed skin panel.

A proprietary type of joist is an **I joist** made of softwood veneers bonded together to make the top and bottom flanges with a composite panel core (Fig. 19.15). Another type of manufactured structural member is made from **parallel strand lumber** (PSL). This type is manufactured under the registered name Parallam®. Parallam structural products include beams, headers, posts, and columns. They are made from Douglas fir or southern pine. The logs are peeled to produce a veneer that is dried and screened to remove strength-reducing defects. Then the sheets of veneer are clipped into strands up to 8 ft. in length and $\frac{1}{10}$ and $\frac{3}{8}$ in. thick. Small defects are removed and the strands are coated with a waterproof adhesive. The long oriented strands are fed into a rotary belt press and cured under pressure using microwave energy. This produces a PSL billet that can be cut to standard sizes. It is available in lengths up to 66 ft. (Fig. 19.16).

Parallam products have been tested for fire resistance, fastener-holding ability, moisture response, long-term loading, flexural strength, stiffness, and internal bond. They have proved to exceed the performance of competing wood products.

Parallam beam and column sizes are in Table 19.9. They are cut, drilled, and installed in the same manner as solid wood and laminated wood structural members.

There are a wide variety of *prefabricated wall panels* available from manufacturers. Several of these are shown in Figs. 19.17 and 19.18. The framed panel is much like the conventional framed wall, floor, or roof. The stressed

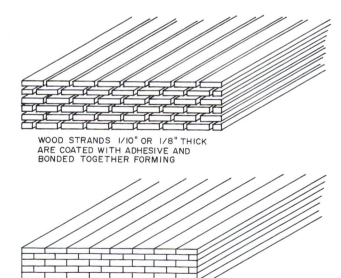

Figure 19.15 This I joist is made with flanges made of solid lumber and a composite wood core. Some are made with laminated veneer flanges. *(Courtesy Louisiana-Pacific)*

Table 19.9 Stock Sizes of Parallam® Parallel Strand Lumber

Parallam Beams and Headers	
Thickness (in.)	**Depth (in.)**
$1\frac{3}{4}$, $3\frac{1}{2}$, $5\frac{1}{4}$, 7	$7\frac{1}{4}$, $9\frac{1}{4}$, $9\frac{1}{2}$, $11\frac{1}{4}$, $11\frac{1}{2}$, $11\frac{7}{8}$, 12, $12\frac{1}{2}$, 14, 16, 18
Parallam Columns	
Dimensions (in.)	
$3\frac{1}{2} \times 3\frac{1}{2}$, $3\frac{1}{2} \times 5\frac{1}{4}$, $3\frac{1}{2} \times 7$, $5\frac{1}{4} \times 5\frac{1}{4}$, $5\frac{1}{4} \times 7$, 7×7	

Courtesy MacMillan Bloedel Limited

WOOD STRANDS 1/10" OR 1/8" THICK
ARE COATED WITH ADHESIVE AND
BONDED TOGETHER FORMING

A STRONG WOOD STRUCTURAL MEMBER

Figure 19.16 Parallam® parallel strand lumber (PSL) is made by bonding thin wood strands into structurally sound members. *(Courtesy MacMillan Bloedel Limited)*

Figure 19.17 This wall panel is made by bonding wood composition panels to a rigid foam core. *(Courtesy AFM Corporation)*

skin panel has the skin adhesive bonded to interior wood members, forming a stronger panel. The sandwich panel has facing material, such as plywood, bonded to an insulating foam core. This is highly energy efficient.

American Wood Systems

American Wood Systems (AWS) was created by APA - The Engineered Wood Association to serve the needs of the engineered wood systems industry. The AWS created a new trademark, APA-EWS, (Engineered Wood Systems). This trademark appears on glued laminated beams and other engineered products such as wood I beams, structural composite lumber, and other engineered wood products. These products receive the same technical and promotional services APA provides to the manufacturers of structural wood panels. A typical APA-EWS trademark is shown in Fig. 19.19.

WOOD SHINGLES AND SHAKES

Wood shingles and shakes are popular as a finished roof covering and are also used as exterior siding on wood framed buildings (Fig. 19.20). Red cedar *shingles* are

smooth sawed on both faces and are uniform in thickness at the butt end. They are tapered from the thick butt end to the thin top edge. They are available in grades No. 1 (blue label), No. 2 (red label), No. 3 (black label), and in an undercoursing grade (green label). They come in lengths of 16, 18, and 24 in. Widths range from a 3 in. minimum to a 14 in. maximum.

Red cedar *shakes* are available in two types, Certi-Sawn® and Certi-Split®. Certi-Sawn shakes are sawed on both faces, as are shingles, but they are not as precisely manufactured. They are thicker on the butt end and vary more in thickness than shingles, which produces a rougher, more textured appearance. They are available in 18 and 24 in. lengths and widths from 4 to 14 in. Certi-Split shakes have a rough, split face and a smooth sawed back. They are tapered from a thick butt end. The thickness varies with the grade. Certi-Sawn cedar shakes are available in grades 1, 2, and 3, as described for shingles, but these grades have different specifications from those for the same grade number used for shingles. Certi-Split shakes are available in grade No. 1 and premium grade, which is the best. They are available in 18 and 24 in. lengths and widths from 4 to 14 in.

Cedar shakes and shingles are available impregnated with fire-retardant polymers that penetrate the innermost cells of the wood. They meet the requirements for class C and B shake and shingle roof systems and class A shake roof systems. They have been tested with the intermittent-flame test, spread-of-flame test, burning brand test, flying brand test, rain test, and the weathering test.

Southern pine taper-sawed shakes are available in grade No. 1 (the best) and No. 2. Special hip and ridge units are made from No. 1 shakes. The shakes are $\frac{3}{16}$ in. thick and 18 or 24 in. long.

WOOD WINDOWS, DOORS, AND CABINETS

Windows

Windows made of wood are available in a wide range of sizes, styles, and quality. Some are covered with vinyl or aluminum to eliminate the need for painting. There have been great improvements in window design and construction, and windows are more energy efficient and easy to operate. Figure 19.21 shows some of the more frequently used types. The National Wood Window and Door Association standards IS-2 and IS-3 established specifications for wood windows and sliding glass doors. Their seal on a product certifies that it meets these standards (Fig. 19.22). NWWDA also has standard IS-4, which pertains to water-repellent preservative treatment.

Single-strength glass is limited to sizes less than 76 united inches (width plus height). Double-strength glass is limited to a maximum of 100 united inches. Tinted

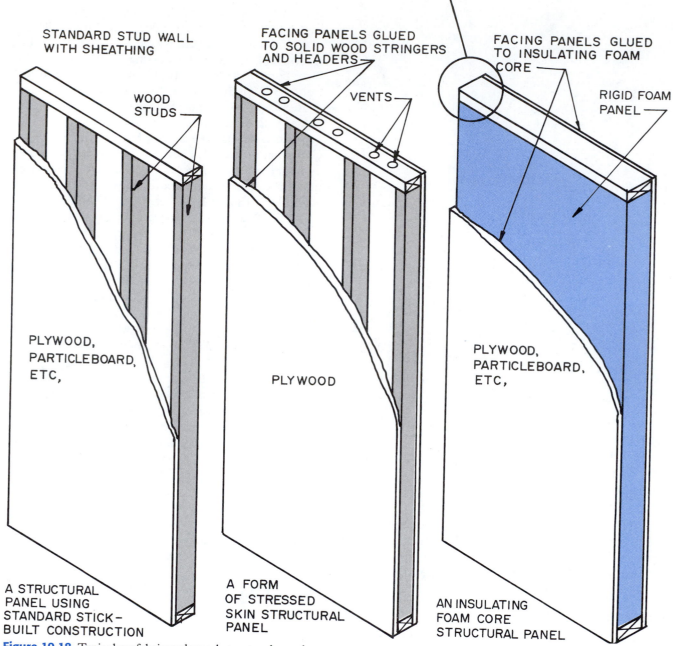

STANDARD STUD WALL
WITH SHEATHING

WOOD
STUDS

PLYWOOD,
PARTICLEBOARD,
ETC,

A STRUCTURAL
PANEL USING
STANDARD STICK-
BUILT CONSTRUCTION

FACING PANELS GLUED
TO SOLID WOOD STRINGERS
AND HEADERS

VENTS

PLYWOOD

A FORM
OF STRESSED
SKIN STRUCTURAL
PANEL

FACING PANELS GLUED
TO INSULATING FOAM
CORE

RIGID FOAM
PANEL

PLYWOOD,
PARTICLEBOARD,
ETC,

AN INSULATING
FOAM CORE
STRUCTURAL PANEL

Figure 19.18 Typical prefabricated wood structural panels.

glass to reduce heat and glare is available. Insulating glass, consisting of two layers of glass with a dehydrated air space between, is available with metal or glass edges. Reflective coated glass has thin films of metal or metal oxide on the surface. The film turns away solar energy by reflecting it away from the building. Detailed information on glass is in a later chapter.

Doors

Doors are made from wood, metal, fiberglass, and hardboard fiber material. Regardless of the material, the types are the same. The common types of wood doors are shown in Fig. 19.23.

Flush doors have wood veneers or hardboard face panels bonded to a core of solid wood strips (called a solid-core door) or a core made of wood or kraft paper forming a hollow, honeycomb interior (called a hollow-core door). They can have openings cut for windows and louvers (Fig. 19.24). Stile-and-rail doors have solid wood vertical and horizontal members enclosing panels made of wood, glass, or louvered sections. Some are made from pressed wood fibers, forming a paneled hardboard door (Fig. 19.25). Fire-rated doors use a

Figure 19.19 A typical APA-EWS trademark. 117-93 is the number of the manufacturing standards and specification for glulam beams published by the American Institute of Timber Construction. 24F means the allowable bending stress in fiber is 24,000 psi, and V4 is the lamination layup for Douglas fir. *(Courtesy APA - The Engineered Wood Association)*

(1) Indicates structural use:
   B - Simple span bending member.
   C - Compression member.
   T - Tension member.
   CB - Continuous or cantilevered span bending member.
(2) Mill number.
(3) Identification of ANSI Standard A190.1, Structural Glued Laminated Timber.
(4) Applicable laminating specification.
(5) Applicable combination number.
(6) Species of lumber used.
(7) Designates appearance grade, INDUSTRIAL, ARCHITECTURAL, PREMIUM.

Figure 19.20 Cedar shingles and hand-split shakes are widely used for finished roof coverings. *(Courtesy Cedar Shake and Shingle Bureau)*

SHAKE

SHINGLE

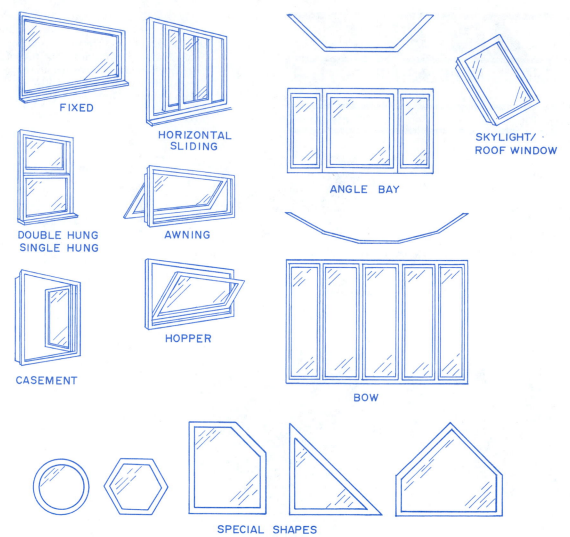

Figure 19.21 Wood windows in common use.

Figure 19.22 Seals of the National Wood Window and Door Association related to windows and preservation treatments. *(Courtesy National Wood Window and Door Association)*

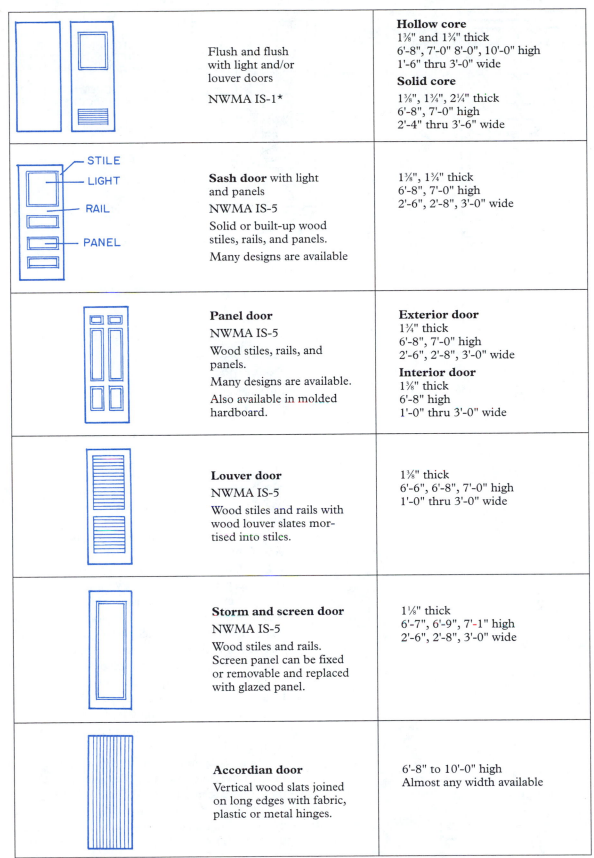

	Flush and flush with light and/or louver doors NWMA IS-1*	**Hollow core** 1⅜" and 1¾" thick 6'-8", 7'-0" 8'-0", 10'-0" high 1'-6" thru 3'-0" wide **Solid core** 1⅜", 1¾", 2¼" thick 6'-8", 7'-0" high 2'-4" thru 3'-6" wide
STILE — LIGHT — RAIL — PANEL	**Sash door** with light and panels NWMA IS-5 Solid or built-up wood stiles, rails, and panels. Many designs are available	1⅜", 1¾" thick 6'-8", 7'-0" high 2'-6", 2'-8", 3'-0" wide
	Panel door NWMA IS-5 Wood stiles, rails, and panels. Many designs are available. Also available in molded hardboard.	**Exterior door** 1¾" thick 6'-8", 7'-0" high 2'-6", 2'-8", 3'-0" wide **Interior door** 1⅜" thick 6'-8" high 1'-0" thru 3'-0" wide
	Louver door NWMA IS-5 Wood stiles and rails with wood louver slates mortised into stiles.	1⅜" thick 6'-6", 6'-8", 7'-0" high 1'-0" thru 3'-0" wide
	Storm and screen door NWMA IS-5 Wood stiles and rails. Screen panel can be fixed or removable and replaced with glazed panel.	1⅛" thick 6'-7", 6'-9", 7'-1" high 2'-6", 2'-8", 3'-0" wide
	Accordian door Vertical wood slats joined on long edges with fabric, plastic or metal hinges.	6'-8" to 10'-0" high Almost any width available

*National Woodwork Manufacturers Association

Figure 19.23 Common types of wood doors.

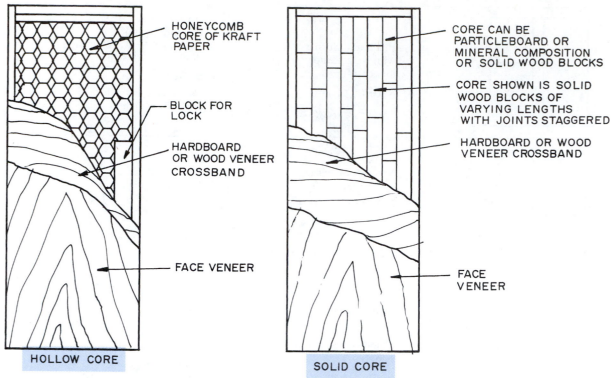

HONEYCOMB
CORE OF KRAFT
PAPER

BLOCK FOR
LOCK

HARDBOARD
OR WOOD VENEER
CROSSBAND

FACE VENEER

HOLLOW CORE

CORE CAN BE
PARTICLEBOARD OR
MINERAL COMPOSITION
OR SOLID WOOD BLOCKS

CORE SHOWN IS SOLID
WOOD BLOCKS OF
VARYING LENGTHS
WITH JOINTS STAGGERED

HARDBOARD OR WOOD
VENEER CROSSBAND

FACE
VENEER

SOLID CORE

Figure 19.24 Construction of hollow-core and solid-core flush doors.

Figure 19.25 A panel door made of molded hardboard.

mineral composition core. Acoustical doors use sound-dampening cores or sheets of lead over the core. Solid wood core doors reduce sound transmission and retard fire better than hollow-core doors.

Doors can be mounted with various hardware to provide a variety of operating methods. These are shown in Fig. 19.26. *Swing doors* have hinges secured to the side jamb. Very large and heavy swinging doors use a pivot at the floor and head jamb rather than hinges. *Sliding doors* are suspended on a metal track secured to the head jamb using nylon rollers attached to the top of the door. A plastic guide is secured to the floor to keep the door in line. *Folding doors* are on a pivot at the floor and head jamb. A pin at the top of the end door runs in an overhead track to keep the door operating in a straight line. *Accordian folding doors* operate in the manner described for folding doors. Very long and heavy folding doors may require a floor track.

Many companies manufacture doors to the specifications of the National Wood Window and Door Association. These standards establish minimum requirements for material, design, construction, and pressure treatment. Doors meeting these standards are recognized by the NWWDA seal (Fig. 19.27).

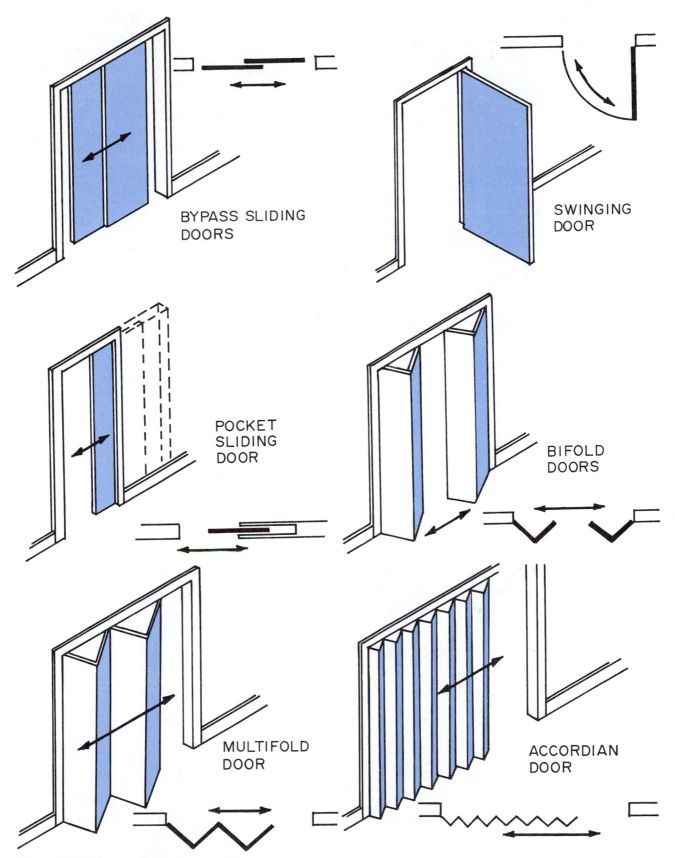

BYPASS SLIDING DOORS

SWINGING DOOR

POCKET SLIDING DOOR

BIFOLD DOORS

MULTIFOLD DOOR

ACCORDIAN DOOR

Figure 19.26 Commonly used types of door operation.

Cabinets

Cabinets are generally made from wood and wood products, although some steel cabinets are available. They are made in a wide variety of styles, such as colonial, contemporary, and provincial, and use a wide range of woods, such as birch, cherry, and walnut. Some are covered with a plastic laminate over a particleboard substrate.

In the kitchen are base cabinets and wall cabinets. Bath cabinets, such as lavatory units, are built the same as kitchen cabinets.

Cabinet manufacturers build their units to meet the standards of the Kitchen Cabinet Manufacturers Association set forth in ANSI A161.1-1980. The standard sizes are shown in Figs. 19.28 and 19.29. Cabinets manufactured to these standards have the KCMA seal (Fig. 19.30). Cabinets are constructed in several ways. The basic difference is that some use a faceframe and some do not. The parts of a basic cabinet are identified in Fig. 19.31. Construction using a faceframe is shown in Fig. 19.32, and that without a faceframe is shown in Fig. 19.33.

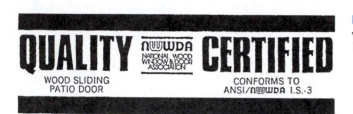

Figure 19.27 Quality certification seal for wood doors. *(Courtesy National Wood Window and Door Association)*

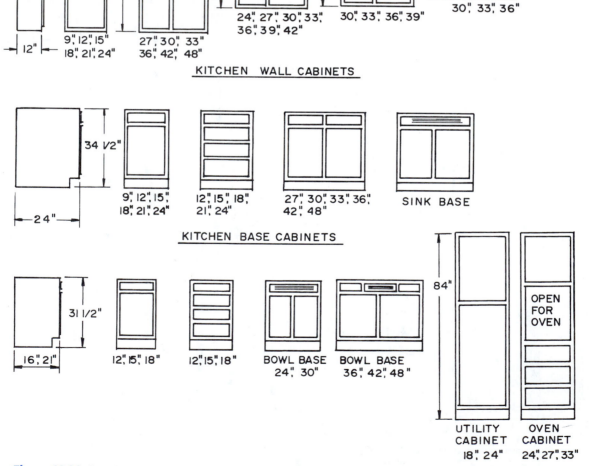

Figure 19.28 Standard modular sizes for kitchen cabinets and bath lavatories. *(Courtesy Architectural Woodwork Institute)*

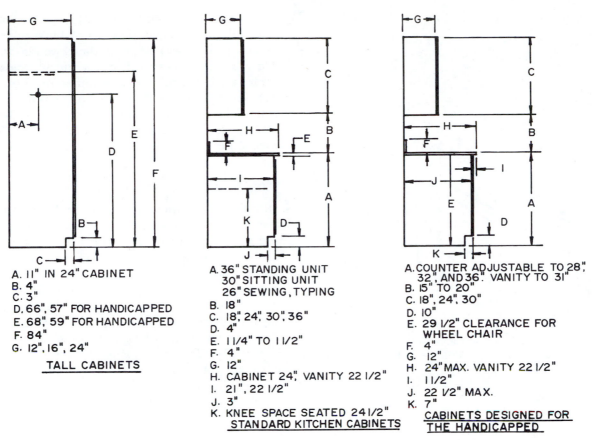

A. 11" IN 24" CABINET
B. 4"
C. 3"
D. 66", 57" FOR HANDICAPPED
E. 68", 59" FOR HANDICAPPED
F. 84"
G. 12", 16", 24"

TALL CABINETS

A. 36" STANDING UNIT
   30" SITTING UNIT
   26" SEWING, TYPING
B. 18"
C. 18", 24", 30", 36"
D. 4"
E. 1 1/4" TO 1 1/2"
F. 4"
G. 12"
H. CABINET 24", VANITY 22 1/2"
I. 21", 22 1/2"
J. 3"
K. KNEE SPACE SEATED 24 1/2"
STANDARD KITCHEN CABINETS

A. COUNTER ADJUSTABLE TO 28", 32", AND 36". VANITY TO 31"
B. 15" TO 20"
C. 18", 24", 30"
D. 10"
E. 29 1/2" CLEARANCE FOR WHEEL CHAIR
F. 4"
G. 12"
H. 24" MAX. VANITY 22 1/2"
I. 1 1/2"
J. 22 1/2" MAX.
K. 7"
CABINETS DESIGNED FOR THE HANDICAPPED

Figure 19.29 Standard cabinet heights and widths. *(Courtesy Architectural Woodwork Institute)*

Figure 19.30 This seal certifies the cabinet has been manufactured to ANSI A161.1-1980 requirements and certified by the Kitchen Cabinet Manufacturers Association. *(Courtesy Kitchen Cabinet Manufacturers Association)*

WOOD FLOORING

Both hardwoods and softwoods are used for finish flooring. They are popular because of the beauty of the grain and the color. The most commonly used hardwoods are oak, beech, birch, pecan, and maple. Softwoods include southern pine, western larch, bald cypress, eastern Englemann, eastern spruce, red pine, ponderosa pine, eastern hemlock and Douglas fir.

Grading rules for wood flooring vary with the species and sections of the country. These associations are identified in Table 19.10. They specify requirements for kiln drying, grading, and control of moisture and establish standard sizes. Many manufacturers produce additional sizes for special applications, such as very thick flooring for industrial use.

Wood flooring is available quartersawed and plainsawed. It is produced in four basic types: strips, planks, parquet (thin wood blocks), and solid end grain blocks (Fig. 19.34). The construction of these products varies widely and manufacturers should be consulted.

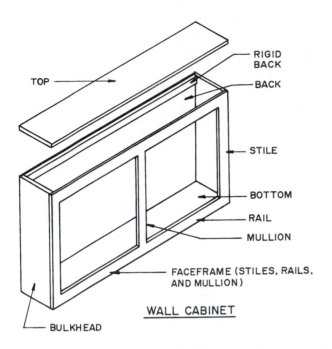

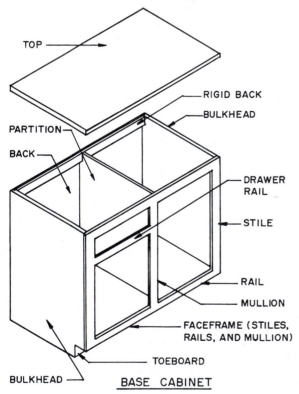

Figure 19.31 Names of parts used in cabinet construction.

Oak flooring is made in both plain-sawed and quartersawed types. Maple, beech, and birch flooring are usually plain-sawed because these woods are hard, dense, and close grained. Southern pine flooring is graded under the rules of the Southern Pine Inspection Bureau. It is available in both flat-grain and edge-grain. The most commonly used type is flat-grain. Douglas fir flooring is available in both flat-grain and edge-grain.

Strip Flooring

Strip flooring is available in soft and hard woods. It is manufactured with square edges, side matched (tongue and groove), and side and end matched (tongue and groove on sides and ends). It usually has a small hollow relief area cut on the bottom. It is available in solid wood and laminated wood strips. Strip flooring is supplied either unfinished or prefinished. Unfinished flooring usually has square, tight-fitting edge joints, which produces a smooth floor. Prefinished strip flooring usually has a V-joint on the edges and ends of each strip. A wide range of woods and finishes are available. Sizes of strip flooring are in Table 19.11.

Planks

Planks may be solid wood or laminated. Solid wood planks are usually ¾ in. (18 mm) thick and from 3½ to 8 in. (87 to 204 mm) wide. Laminated planks are usually thinner ranging from ⁷⁄₁₆ to ⅜ in. thick. They are used to produce a floor representative of the wide flooring used in the early days of this country. Some are secured with screws covered with wood plugs.

Parquet Flooring

Parquet flooring is a thin block flooring made from solid wood or laminated wood blocks. These blocks are joined on the floor to form a parquet floor. They are available unfinished and prefinished. Generally the blocks are square, but other units are available from various manufacturers (Fig. 19.35). The American Parquet Association and NBS-PS 27-70, Mosaic Parquet Hardwood Flat Flooring, grading rules are used for the manufacture of the flooring. Laminated hardwood block flooring is manufactured following ANSI/HPMA LHF 1982. Some of the typical patterns used for parquet floors are in Fig. 19.36.

Solid block flooring for use in areas where heavy traffic occurs, as in a factory, and where easy maintenance is required, is cut so the face exposes the end grain. Southern pine is frequently used.

Canadian hardwood strip flooring is manufactured in the customary sizes compatible with those used in the United States and in metric sizes. Strip and parquet

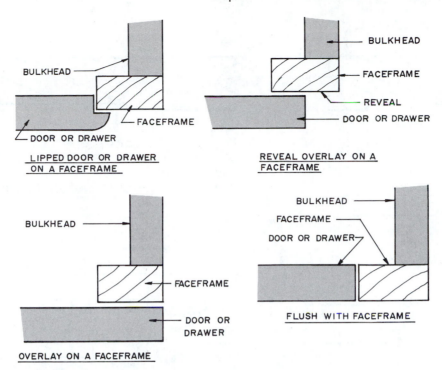

Figure 19.32 Cabinet construction with a faceframe.

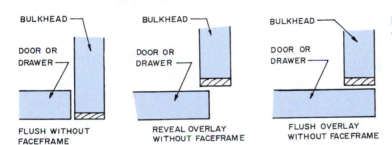

Figure 19.33 Cabinet construction without a faceframe.

Table 19.10 Grades of Unfinished Hardwood and Softwood Flooring

Unfinished Oak[a]	Prefinished Oak[a]
Clear Plain or Clear Quartered (best appearance)	Prime (excellent)
Select Plain or Select Quartered (excellent appearance)	Standard and Better (mix of Standard and Prime)
Select and Better (mix of Clear and Select)	Standard (variegated)
No. 1 Common (variegated)	Tavern and Better (mix of Prime, Standard, and Tavern)
No. 2 Common (rustic)	Tavern (rustic)
Beech, Birch, and Hard Maple[a]	**Pecan[a]**
First Grade White Hard Maple (face all bright sapwood)	First Grade Red (face all heartwood)
First Grade Red Beech (face all red heartwood)	First Grade White (face all bright sapwood)
First Grade (best)	First Grade (excellent)
Second and Better (excellent)	Second Grade Red (face all heartwood)
Second (variegated)	Second Grade (variegated)
Third and Better (mix of First, Second, and Third)	Third Grade (rustic)
Maple[b]	**Softwood[c]**
First (highest quality)	B and Btr (best quality)
Second and Better (mix of First and Second)	C (good quality, some defects)
Second (good quality, some imperfections)	C and Btr (mix of B & Btr and C)
Third and Better (mix of First, Second, and Third)	D (good economy flooring)
Third (good economy flooring)	No. 2 (defects but serviceable)

[a]Hardwood flooring grades of the National Oak Flooring Manufacturers Association
[b]Maple flooring grades of the Maple Flooring Manufacturers Association
[c]Softwood flooring grades of the Southern Pine Inspection Bureau

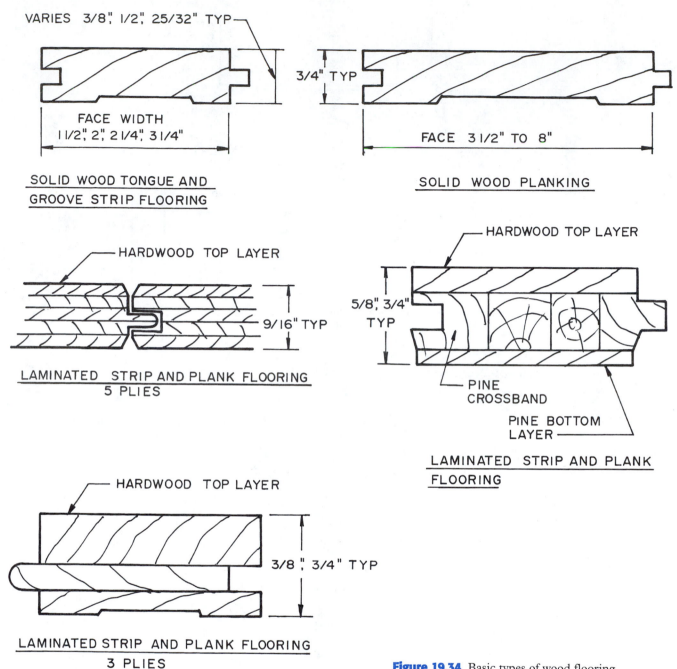

VARIES 3/8", 1/2", 25/32" TYP

3/4" TYP

FACE WIDTH
1 1/2", 2", 2 1/4", 3 1/4"

FACE 3 1/2" TO 8"

SOLID WOOD TONGUE AND
GROOVE STRIP FLOORING

SOLID WOOD PLANKING

HARDWOOD TOP LAYER

9/16" TYP

LAMINATED STRIP AND PLANK FLOORING
5 PLIES

HARDWOOD TOP LAYER

5/8", 3/4"
TYP

PINE
CROSSBAND

PINE BOTTOM
LAYER

LAMINATED STRIP AND PLANK
FLOORING

HARDWOOD TOP LAYER

3/8", 3/4" TYP

LAMINATED STRIP AND PLANK FLOORING
3 PLIES

Figure 19.34 Basic types of wood flooring.

flooring are measured by surface area in square meters (m^2). The two Canadian industry standards for strip and hardwood flooring are published by the Canadian Lumbermen's Association.

Cushioned Flooring

Hardwood floors used in gymnasiums, auditoriums, dance studios, and other locations needing cushioning use special methods for installation. Hardwood flooring 1 1/4 to 2 1/2 in. thick is used. Some is square edged and some uses tongue-and-groove edges. It can be installed over any strong, firm, level surface. Typical installation details are shown in Fig. 19.37.

Table 19.11 Stock Sizes for Wood Strip Flooring

Southern Pine Flooring[a]				
Species	Actual Thickness in.	mm	Actual Width[b] in.	mm
Longleaf pine, slash pine, shortleaf pine, loblolly pine	$5/16$	8	$1^3/8$	35
	$7/16$	11	$2^3/8$	60
	$9/16$	14	$3^3/8$	86
	$3/4$	76	$4^3/8$	101
	1	101	$5^3/8$	136
	$1^1/4$	107		

[a]Courtesy Southern Pine Inspection Bureau
[b]Widths available in all thicknesses

Hardwood Strip Tongue-and-Groove Flooring[c]				
Species	Actual Thickness in.	mm	Actual Width in.	mm
Oak, beech, birch, hard maple, hickory	$3/4$	19	$1^1/2$, 2, $2^1/4$, $3^1/4$	38, 51, 57, 82.5
	$11/32$	8.7	$1^1/2$, 2	38, 51
	$15/32$	12	$1^1/2$, 2	38, 51
	$33/32$	26	2, $2^1/4$, $3^1/4$	51, 57, 82.5

[c]Courtesy National Oak Flooring Manufacturers Association

Maple Strip Tongue-and-Groove Flooring			
Actual Thickness in.	mm	Actual Width in.	mm
$25/32$, $33/32$	20, 26	$1^1/2$, $2^1/4$, $3^1/4$	38, 57, 82
$41/32$	32.5	$2^1/4$, $3^1/4$	57, 82

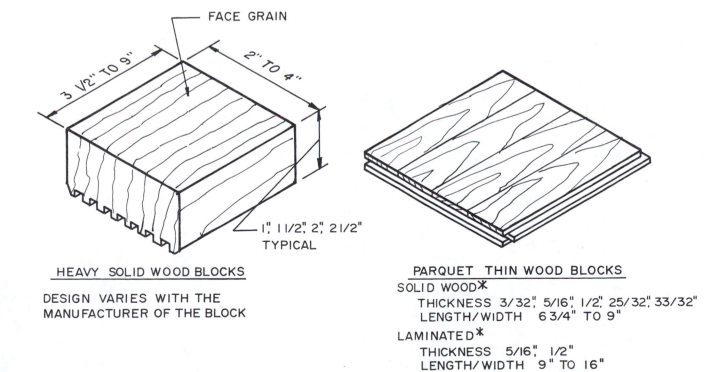

FACE GRAIN

$3^1/2$" TO 9"

2" TO 4"

1", $1^1/2$", 2", $2^1/2$" TYPICAL

HEAVY SOLID WOOD BLOCKS

DESIGN VARIES WITH THE MANUFACTURER OF THE BLOCK

PARQUET THIN WOOD BLOCKS
SOLID WOOD*
   THICKNESS 3/32", 5/16", 1/2", 25/32", 33/32"
   LENGTH/WIDTH 6 3/4" TO 9"
LAMINATED*
   THICKNESS 5/16", 1/2"
   LENGTH/WIDTH 9" TO 16"

* REFER TO MANUFACTURER'S DATA

Figure 19.35 Typical parquet flooring blocks and heavy wood block flooring.

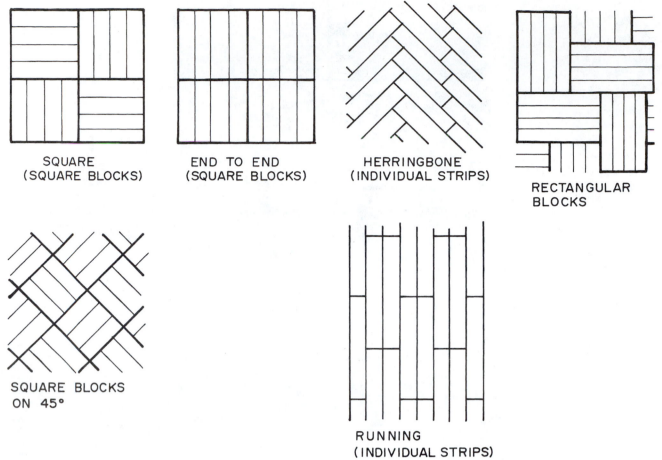

SQUARE
(SQUARE BLOCKS)

END TO END
(SQUARE BLOCKS)

HERRINGBONE
(INDIVIDUAL STRIPS)

RECTANGULAR
BLOCKS

SQUARE BLOCKS
ON 45°

RUNNING
(INDIVIDUAL STRIPS)

Figure 19.36 Parquet flooring patterns.

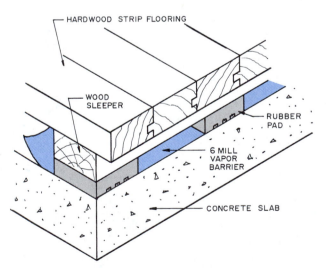

HARDWOOD STRIP FLOORING

WOOD
SLEEPER

RUBBER
PAD

6 MILL
VAPOR
BARRIER

CONCRETE SLAB

Figure 19.37 A typical cushioned floor installation. Other designs are available.

REVIEW QUESTIONS

1. What are the differences between plywood, oriented strand board, and composite panels?
2. What are the plywood exposure durability classifications? Give an example of where each could be used.
3. How does hardwood plywood differ from industrial plywood?
4. What are the differences between laminated veneer lumber and glued laminated members?
5. How do wood shingles and shakes differ?
6. What types of wood flooring are available?
7. What are the four plywood constructions commonly used in building construction?
8. What are the appearance groups established for the outer veneers of construction plywood?
9. How is the strength of the woods used in plywood indicated?
10. What is a performance-rated wood panel?
11. What are the grades of outer veneers for hardwood plywood?
12. What are the classes of hardboard?

KEY TERMS

APA performance-rated panels Panels whose structural properties meet the standards of APA - The Engineered Wood Association.

composite panels Panels having a reconstituted wood core bonded between layers of solid veneer.

glulam A structural wood member made by bonding wood strips to form a solid member; short for "glued laminated."

hardboard Panels made by applying heat and high pressure to bond wood fiber.

hardwood plywood Plywood made with various species of hardwoods as the outer veneers.

I joist A joist made of an assembly of wood top and bottom flanges and a composite wood panel core.

laminated veneer lumber Lumber made by bonding wood veneers to form a solid member.

oriented strand board Panels manufactured by bonding wood fiber strands all oriented in the same direction.

overlaid plywood Plywood panels whose exterior surfaces are covered with a resin-impregnated fiber ply.

parallel strand lumber Lumber produced by bonding lengths of wood veneer.

particleboard Panels made from wood chips, water, and a resin binder.

plywood Panels made by bonding layers of wood veneer with each layer at right angles to the one next to it.

trusses Structural members made by assembling components to form a rigid framework.

waferboard Panels made by bonding large waferlike wood chips in random positions.

SUGGESTED ACTIVITIES

1. Collect samples of as many kinds of plywood products as you can find. Prepare a display. Identify each one with as much technical information about each as you can gather.

2. Immerse samples of interior and exterior plywood in water. Continue until you begin to notice delamination. Keep a record of the length of time it takes each to fail. Did some types never fail?

3. Cut 2-inch-wide strips of ½-inch-thick fir plywood 4 feet long. Support a strip with the flat side down with each end resting on chairs. Slowly add weights and note the amount of deflection until the strip breaks or slides off the chair. Then stand the strip on its edge and repeat the test. In which position did it carry the greatest load? Why?

4. Build scale models of the commonly used trusses.

5. Collect samples of the various types of wood flooring. Your building supply dealer may have manufacturers samples for you to use. Label each.

ADDITIONAL INFORMATION

The Wood Truss Handbook, Gang-Nail Systems, Inc., 1981, P.O. Box 59-2037 AMF, Miami, Fla. 33159.

Wood Building Technology, Canadian Wood Council, Ottawa, 1993.

Wood Handbook: Wood as an Engineered Material, Forest Products Laboratory, Forest Service, U.S. Department of Agriculture, available Superintendent of Documents, U.S. Government Printing Office, Washington, D.C. 20402.

Wood Reference Handbook, Canadian Wood Council, Ottawa, 1991.

Other resources are

Numerous publications from the APA-The Engineered Wood Association, Tacoma, Wash.

Numerous related publications from the organizations listed in the "Professional and Technical Organizations" in Appendix B.

Wood and Metal Light Frame Construction

This chapter will help you to:

1. Develop an understanding of the methods used to construct light frame buildings.

2. Understand the differences in framing when using the various materials and products available.

3. Be aware of the influence codes and ordinances have on the design of light frame buildings.

Wood light frame construction is the most widely used system for the construction of residences and small apartments in the United States and Canada. Wood, a renewable resource, is in abundance in these countries, and it is easily harvested, dried, and processed into structural members. The exterior finish can be wood, stucco, masonry, or other available siding products.

Wood light frame construction is a very flexible system that permits an almost unlimited range of design possibilities. The architect can produce designs that are contemporary (Fig. 20.1) or classic (Fig. 20.2), simple (Fig. 20.3) or complex, low cost or very expensive, and that can accommodate almost any electrical, heating, air conditioning, plumbing, and security system desired. With the vast array of products on the market, the building can be insulated, sealed, and waterproofed to produce a building with long life and low maintenance. It can be built in almost any climate and on any site that will accept an adequate foundation.

The system has evolved from basically a solid wood structure to the use of a variety of reconstituted wood products, which are discussed in Chapter 19, and some details follow in this chapter. Many of the problems that developed with early wood light frame construction—wood decay, swelling and shrinking of members, sticking doors and windows, and creaking floors—have been remedied by improved construction techniques and materials. In addition, considerable effort has been made to utilize factory assembled panels and modules, thus reducing the on-site labor costs associated with what is often referred to as "stick built" construction. Light wood frame construction will continue to be used because of its many advantages and the improvements in materials and construction techniques.

ORDINANCES AND CODES

Zoning and Light Frame Construction

Among other things, zoning ordinances set minimum sizes for buildings and restrict where they can be placed on a lot. Ordinances usually establish minimum distances from each side of the lot, called setbacks. The building can be located anywhere within the area clear of the setbacks. Look at Fig. 20.4 to see how these apply. As a general contractor you should be very careful to be certain the corners of the house used for construction of the foundation are accurately located. If they are at all near the setback lines you will probably want to employ a land surveyor to stake the corners. Some localities require that a surveyor do the staking.

Building Restrictions and Covenants

In many localities you will find that there are limitations on the building, such as exterior appearance of the building or the materials used. The architect in the locality will be aware of these and the zoning requirements as the building is designed. For example, flat or butterfly roofs may be prohibited, the height of the building to the ridge or even the top of the chimney can be limited, and the choice of siding, shingles, windows, and other features limited. These and other similar requirements will influence the design and structure of wood light frame buildings. Since this system is so flexible, it can be used for the construction of almost any design that falls within the building codes.

Building Codes

Wood light frame building construction is carefully regulated by the various building codes. As you study methods and materials of construction it would be helpful for you to have copies of the local building code available for consultation. Codes specify such features as accessibility, interior environment, means of egress, structural loads, and many other features of the building.

One major concern building codes have with wood light frame construction is the danger of fire. These buildings are typically made almost completely of combustible materials and this calls for restrictions on their occupancy, size, and height. Most codes have a classification for all wood framed buildings with minimal fire ratings of assemblies. In the BOCA National Building Code this is Type 5, Combustible, and in the Standard Building Code, Southern Building Code Congress International, it is Type VI. These codes specify the parts that can be wood and include all aspects of wood light framed construction.

More restrictive types of construction permit wood framing but with greater fire resistance requirements. The fire resistance is specified by the number of hours the particular assembly, such as an exterior wall, must resist penetration by fire. If the architect plans to use wood light frame construction in the more restrictive types of construction, provision will be made to increase the fire resistance rating of the walls, floor, ceiling, and other related assemblies. Typically this is done by using additional gypsum board on interior walls and in some

Figure 20.1 This contemporary residence utilizes a heavy timber wood frame. (*Courtesy American Forest and Paper Association*)

Figure 20.2 The Isaiah Thomas Printing Office built in 1784. Wood was plentiful and supplied the framing, sheathing, siding, and finished roofing. (*Courtesy Old Sturbridge Village, Sturbridge, Mass.*)

Figure 20.3 This is a typical single-family wood light frame house. (*Courtesy American Forest and Paper Association*)

cases by adding sheathing. Sprinkler systems can also be used to help meet fire requirements. One example can be noted in Fig. 20.5.

PREPARING THE SITE

After the building is designed, bids are accepted, and building permits are issued, it is time to prepare the site for construction. This begins by removing unwanted brush and trees that are in the way of the building, sidewalks, and driveway. In some areas tree removal is carefully regulated and each tree must be marked and its removal approved before the site is cleared.

Next the surveyor stakes the corners of the foundation, being certain they are located as shown on the site plan. The builder then sets up batter boards on each corner as shown in Fig. 20.6. Batter boards are usually 1×6 lumber held by 2×4 posts driven into the ground.

A chalk line is pulled across the boards on each side of the proposed building. Where they cross is the center of the stake in the ground. Once the lines are located they may be tied to a nail driven into the batter board or run in a saw cut in the board. Figure 20.7 is an enlargement of a typical corner layout. The corner can be checked for squareness with the 3-4-5 method. When the base of a right triangle measures 3 ft. and the vertical side measures 4 ft. the hypotenuse will measure 5 ft. If it does not, the corner is not square. Some prefer to use longer measurements in multiples of three such as 9, 12, and 15 ft.

Excavation

Excavation for a basement can be accomplished with a bulldozer or backhoe. Basements are dug to the depth specified. A plumb line is dropped from the batter board chalk lines, and the corner of the foundation is established in the excavation, as you can see in Fig. 20.8. Now the excavation for the footing is dug, usually by hand, to the required width and depth. If it is dug too deep, soil cannot be put back. The footing is poured thicker so it rests on undisturbed soil. If the soil crumbles on the sides of the footing excavation, wood forms must be installed. Now the footings for columns, piers, and the fireplace are located and dug. The specified reinforcing is placed in the footing excavation and the concrete is poured. Excavations for footings for buildings with crawl spaces or slab floors follow a similar procedure, but the amount of excavation is much less. Figure 20.9 shows one example.

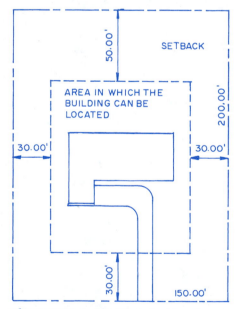

Figure 20.4 A simplified site plan showing the setbacks and the location of the building.

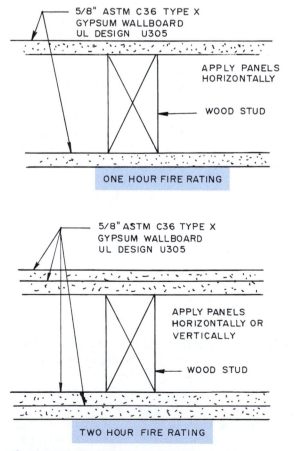

Figure 20.5 Typical types of interior partition construction and their fire ratings.

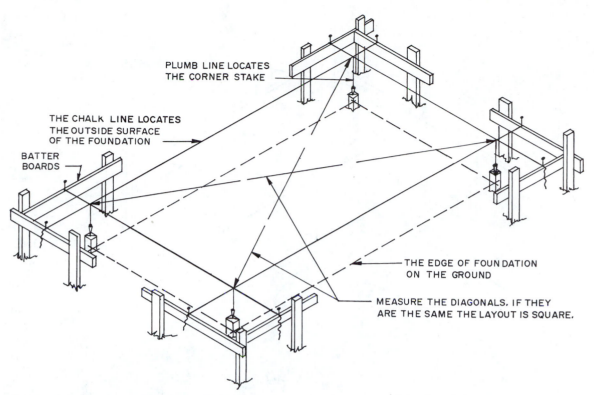

PLUMB LINE LOCATES
THE CORNER STAKE

THE CHALK LINE LOCATES
THE OUTSIDE SURFACE
OF THE FOUNDATION

BATTER
BOARDS

THE EDGE OF FOUNDATION
ON THE GROUND

MEASURE THE DIAGONALS. IF THEY
ARE THE SAME THE LAYOUT IS SQUARE.

Figure 20.6 Batter boards and chalk lines are used to locate the corners of the foundation.

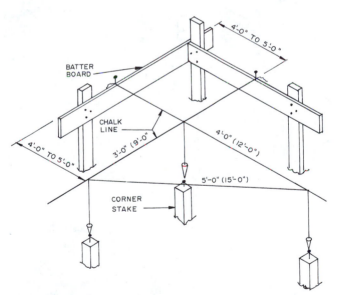

BATTER
BOARD

CHALK
LINE

4'-0" TO 5'-0"

4'-0" TO 5'-0"

3'-0" (9'-0")

4'-0" (12'-0")

5'-0" (15'-0")

CORNER
STAKE

Figure 20.7 The squareness of the corner marked by the chalk line can be checked by 3-4-5 method. Notice the batter boards are set back about 4 ft. from the corner stake.

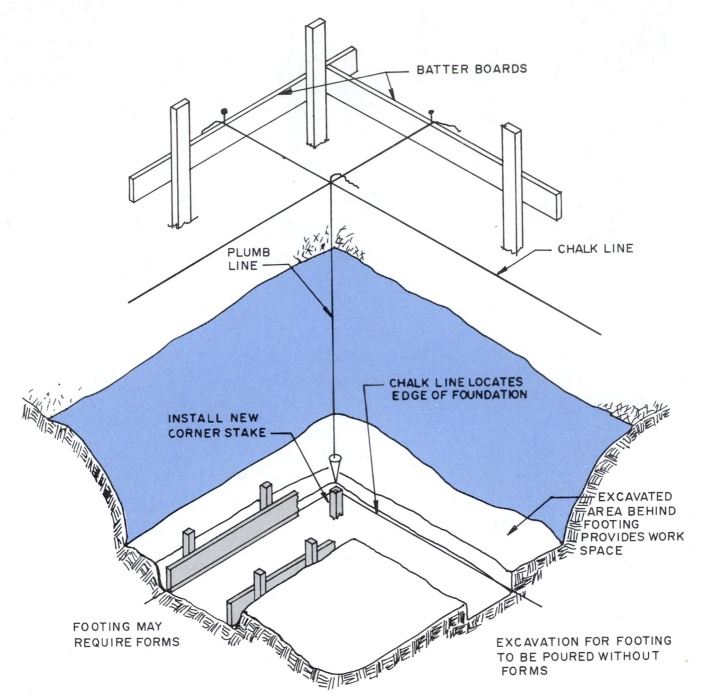

Figure 20.8 The basement excavation is dug to the specified depth and footings are dug. Footing forms are used if the sides of the footing excavation crumble.

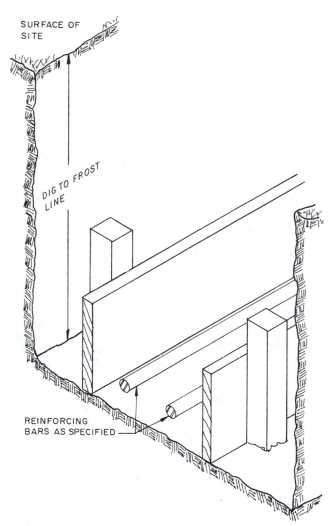

Figure 20.9 The excavation for a footing for a building with a crawl space extends below the frost line.

FOUNDATIONS

The footing is cast-in-place reinforced concrete. The foundation wall may be concrete block, brick, cast-in-place concrete, or pressure-treated wood. The building may have a basement, crawl space, or a concrete slab floor. Additional information on foundations is in Chapter 6. Here we will review those commonly used when building wood light framed buildings. These include buildings with basements, crawl spaces, and concrete slab-on-grade floors.

Basements are used in some parts of the country, typically in the colder climates. They provide considerable living space for a small cost per square foot, especially when you consider that a crawl space requires a footing and some foundation wall and provides no living space.

The building needs a foundation and footing to the frost line so in many areas it does not take much more to get a full 8 ft. (2.4 m) deep basement.

A typical basement detail can be found in Fig. 20.10. The walls can be cast-in-place reinforced concrete or concrete block. The footings are poured first. When they have reached adequate strength, the concrete block foundation can be laid or the forms for the poured concrete foundation installed. After the reinforcing bars are installed, the foundation can be poured.

After the basement foundation has been built, the soil should not be backfilled against it until the floor platform is built because the floor is needed to provide protection against horizontal stress created by the weight of the soil. If this is not done the foundation may crack or collapse.

The foundation in Fig. 20.11 is typical for a building with a crawl space and wood exterior siding or a brick veneer. It requires that the foundation have a brick ledge that is usually near the grade. Building codes limit the minimum height between the ground and the bottom of the floor joists. Since plumbing, heating, air conditioning, and electrical service are often run below the joists, you will want to try to make this height well above the minimum. Even a 2 ft. (610 mm) high crawl space is claustrophobic and very difficult to work in and makes servicing utilities a problem.

Since it is important to keep the ground in the crawl space dry, the exterior of the foundation below grade should be waterproofed, and gravel and a French drain should be placed at the footing and drain to a drywell or to daylight. The ground in the crawl space should be covered with plastic sheet material. This will retard the passage of moisture from the soil into the air in the crawl space.

Several foundations for houses with *concrete slab-on-grade floors* are illustrated in Fig. 20.12. The monolithic poured foundation and floor is used in areas where freezing does not occur. The others are used where the footing has to go below grade. The anchor bolts for the bottom plate are set in the concrete before it hardens.

Wood Foundations

Wood foundations can be used for buildings with basements or crawl spaces. They use pressure-treated plywood and pressure-treated solid wood members (Fig. 20.13). The materials are stress graded so it is known that they will withstand lateral soil and subsurface water pressures. The wood materials must have the stamp of the American Wood Preservers Association (AWPA).

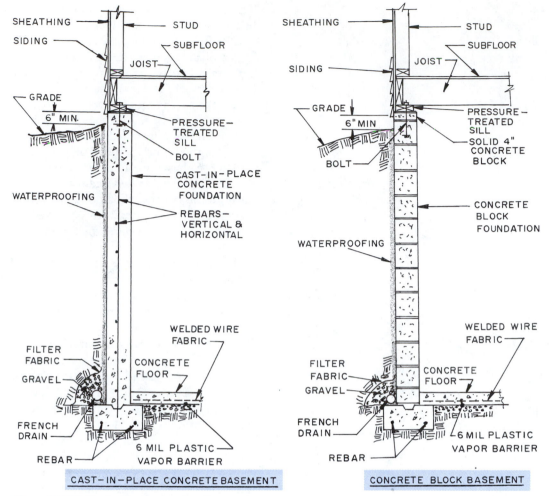

Figure 20.10 Typical basement details.

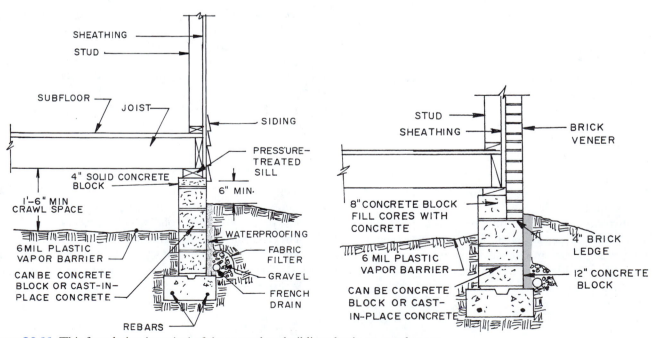

Figure 20.11 This foundation is typical of those used on buildings having a crawl space.

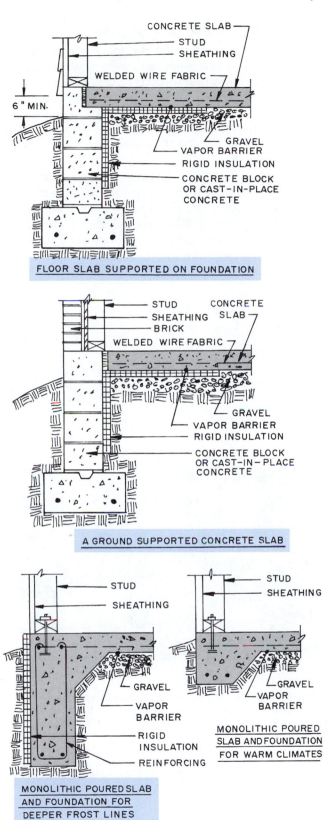

CONCRETE SLAB
STUD
SHEATHING
WELDED WIRE FABRIC
6" MIN.
GRAVEL
VAPOR BARRIER
RIGID INSULATION
CONCRETE BLOCK
OR CAST-IN-PLACE
CONCRETE

FLOOR SLAB SUPPORTED ON FOUNDATION

STUD   CONCRETE
SHEATHING   SLAB
BRICK
WELDED WIRE FABRIC
GRAVEL
VAPOR BARRIER
RIGID INSULATION
CONCRETE BLOCK
OR CAST-IN-PLACE
CONCRETE

A GROUND SUPPORTED CONCRETE SLAB

STUD
SHEATHING
GRAVEL
VAPOR
BARRIER
RIGID
INSULATION
REINFORCING

STUD
SHEATHING
GRAVEL
VAPOR
BARRIER

**MONOLITHIC POURED
SLAB AND FOUNDATION
FOR WARM CLIMATES**

**MONOLITHIC POURED SLAB
AND FOUNDATION FOR
DEEPER FROST LINES**

Figure 20.12 Types of concrete slab floor construction used with wood light frame construction.

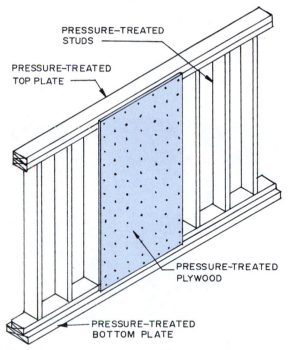

PRESSURE-TREATED
STUDS
PRESSURE-TREATED
TOP PLATE
PRESSURE-TREATED
PLYWOOD
PRESSURE-TREATED
BOTTOM PLATE

Figure 20.13 The assembly of a section of a permanent wood foundation.

The panels are often built in a shop and shipped to the site. This makes their assembly easier and more accurate because jigs can be used. The studs forming the panel face the inside of the building, and they can be insulated just as you would insulate an exterior above-grade wall (Fig. 20.14).

A typical basement foundation panel is shown in Fig. 20.15. The panels must be carefully assembled using bronze, copper, silicon, or stainless steel nails. They must follow the designs engineered by the American Forest and Paper Association. Notice that the bottom plate is set on a bed of gravel rather than on a concrete footing. The floor joists rest on the double top plate. A detail for a foundation for a building with a crawl space is in Fig. 20.16. Notice in Fig. 20.17 the wood foundation can be used to support a brick veneer exterior wall.

Piers and Columns

In addition to constructing the foundation walls, the contractor must pour footings for piers or columns to support the beams that will carry the floor joists. A typical pier used on wood light frame buildings with crawl space is often built with concrete blocks, as you can see in Fig. 20.18. The size depends on the loads to be carried. The one in Fig. 20.18 is typical for small residential buildings.

If the building has a basement, steel columns are used. One example is shown in Fig. 20.19. Again, the size of the footing and the type of column depend on the load to be carried. When the foundation is ready for the carpenters to install the first floor platform it will look like the one shown in Fig. 20.20.

Figure 20.14 Permanent wood foundations are installed using pressure-treated lumber over a gravel bed for the footing. *(Courtesy APA - The Engineered Wood Association)*

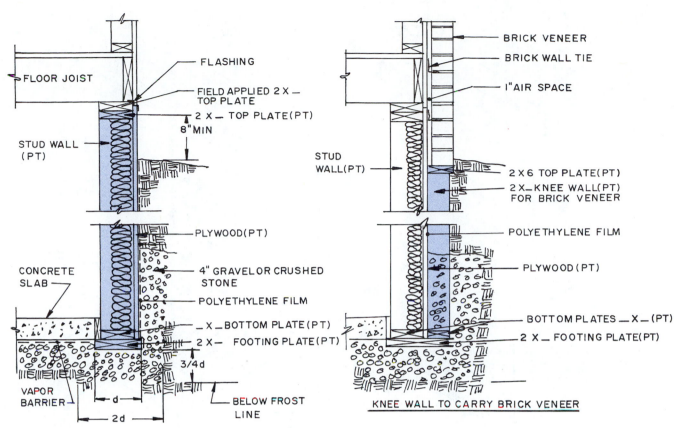

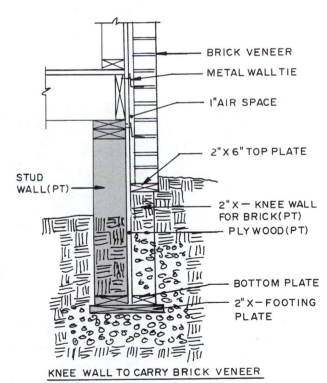

Figure 20.15 Typical details for a permanent wood foundation basement wall. *(Courtesy American Forest and Paper Associations (formerly NFPA), Washington, D.C.)*

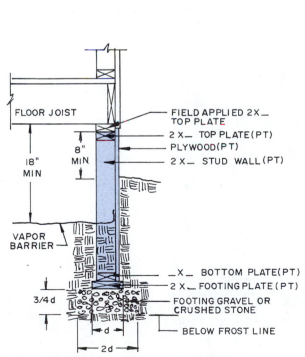

Figure 20.16 The permanent wood foundation can be used on buildings having a crawl space. *(Courtesy American Forest and Paper Associations (formerly NFPA), Washington, D.C.)*

Figure 20.17 This detail shows the use of a brick ledge to carry the brick veneer over a wood frame wall. It is used on crawl space and basement foundations.

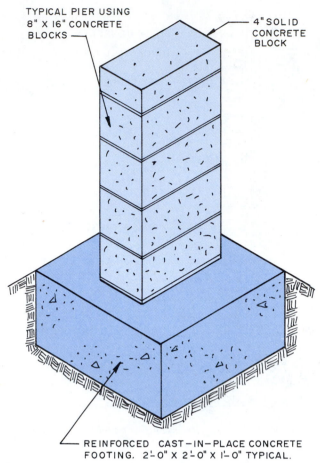

TYPICAL PIER USING 8" X 16" CONCRETE BLOCKS

4" SOLID CONCRETE BLOCK

REINFORCED CAST—IN—PLACE CONCRETE FOOTING. 2'-0" X 2'-0" X 1'-0" TYPICAL.

Figure 20.18 A typical concrete block pier used in residential construction.

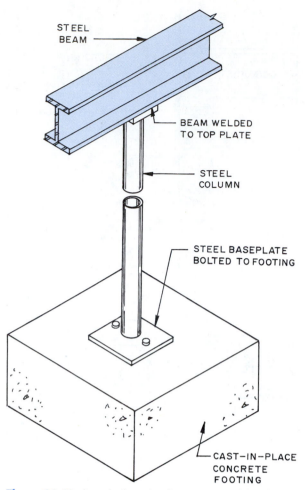

STEEL BEAM

BEAM WELDED TO TOP PLATE

STEEL COLUMN

STEEL BASEPLATE BOLTED TO FOOTING

CAST—IN—PLACE CONCRETE FOOTING

Figure 20.19 A typical steel column supporting a beam that will carry the wood floor joists.

Figure 20.20 A completed foundation for a wood light frame building.

EVOLUTION OF WOOD LIGHT FRAME CONSTRUCTION

Early wood frame construction in the U.S. colonies involved building a structural frame of heavy timbers that were fitted together with mortise and tenon joints. The walls below this frame were filled in with smaller wood members or masonry. This construction required a large group of workers to hand cut and fit the joints and lift the timbers into place. The walls below the horizontal timbers were non–load bearing.

Today this type of construction is not in common use and has been replaced by wood light frame construction. However, there are companies that do produce the beams and columns, cut the joints, and assemble the frames. Examples can be seen in Chapter 21. You can see in Figs. 20.21 and 20.22 the structure itself is a work of art produced by skilled craftsmen. It is left exposed inside the building and becomes part of the ambience of the interior.

A system that was the forerunner of the current widely used platform framing system of wood light frame construction was developed in the early 1800s. It involved using wall studs to carry the roof and second floor and eliminated the heavy timber frame. The studs not only carried the loads but also served to enclose the building as the non-load-bearing walls did on the heavy timber construction.

This design used wall studs that ran the full two stories and eventually was called **balloon framing.** You can see from the detail in Fig. 20.23 that the studs rested on the sill on top of the foundation and ran continuously to the double top plate. The floor joists rested on a ribbon let into the stud and were nailed to the stud. You can find additional details in Figs. 20.24 and 20.25. Balloon framing is seldom used now and has been replaced by platform framing.

PLATFORM FRAMING

Platform framing involves building the first-floor deck on top of the foundation (Fig. 20.26). Upon this, the walls for the first floor are assembled, erected, and braced (Fig. 20.27). The joists for the second floor are laid on top of the double plate of the exterior first-floor walls and supported by load-bearing interior walls. The second-floor joists are covered with sub-flooring, forming the second floor platform, and the walls for the second floor are assembled, erected, and braced (Fig. 20.28). The ceiling joists are laid on the

Figure 20.21 Heavy timber wood framing for a recently built structure is similar to early framing techniques. *(Courtesy Canadian Wood Council)*

Figure 20.22 The exposed heavy timber roof framing provides a very attractive ceiling. *(Courtesy Canadian Wood Council)*

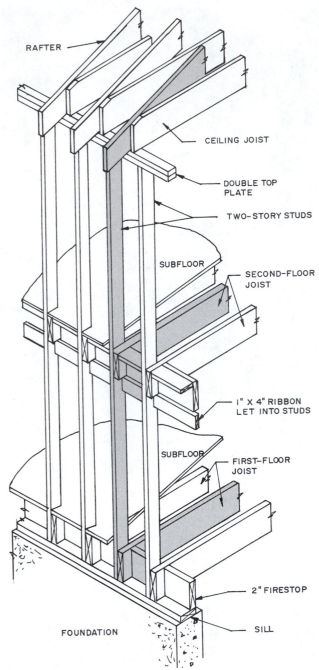

Figure 20.23 A typical framing detail for balloon framing.

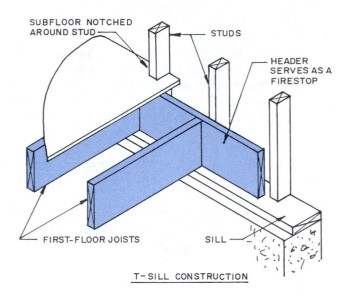

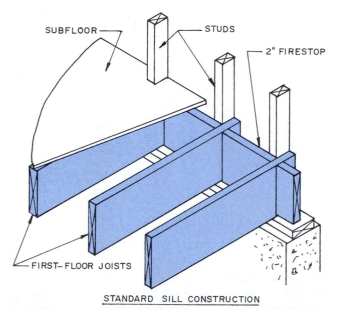

Figure 20.24 Construction details for the floor joist at the sill commonly used with balloon framing. Notice that the studs rest on the wood sill.

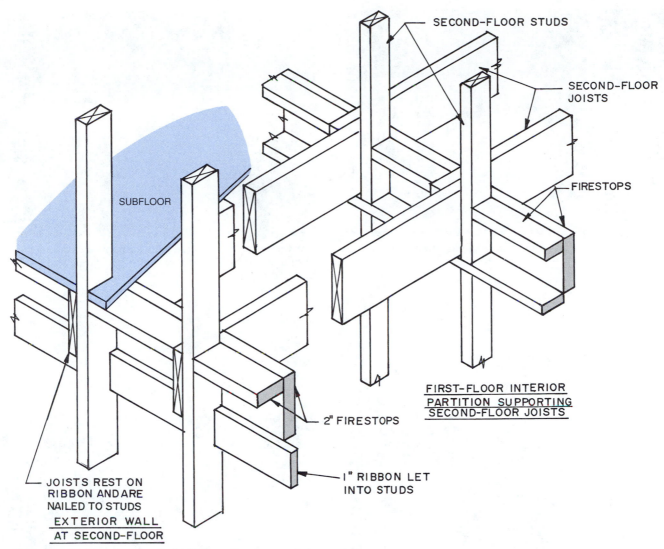

Figure 20.25 Framing details for balloon framing on the second floor.

double plate of the second-floor exterior walls and supported by interior load bearing walls. Finally, the carpenters install the rafters, which rest on the double plate of the second-floor walls (Fig. 20.29). After the walls are sheathed, the roof sheathing is installed. This greatly strengthens the roof structure and the rigidity of the entire structure (Fig. 20.30).

If roof trusses are used there is no need for ceiling joists because the bottom chord of the truss serves as a ceiling joist. The trusses are erected bearing on the double top plate of the second-floor walls. Trusses are discussed later in this chapter.

In Fig. 20.31 the first floor has been built on the foundation using **box sill** construction. Box sill construction has the floor joists resting on a sill that is bolted to the foundation. The joists butt a header, which in effect forms the floor framing into a box (Fig. 20.32). The header is end nailed to the joists. The joists are toenailed to the sill. Sometimes metal connectors are used.

The floor joists usually cannot span from one side of the foundation to the other, so some form of a beam is run through the foundation. The beam may be steel, wood, or a manufactured wood product. It is supported at intervals by piers or columns (Fig. 20.33).

The floor joists are assembled on the foundation.

Figure 20.26 Building the first-floor platform.

The floor joists are covered with subflooring, forming a platform upon which the interior and exterior walls are assembled and raised.

The walls are assembled on the platform.

The assembled wall is raised and nailed in place. It must be braced with wood members to the floor.

Figure 20.27 Assembling and erecting the walls. *(Courtesy APA - The Engineered Wood Association)*

Figure 20.28 This structure has the second-floor platform and walls in place. Notice the diagonal bracing let into the studs. *(Courtesy Southern Forest Products Association)*

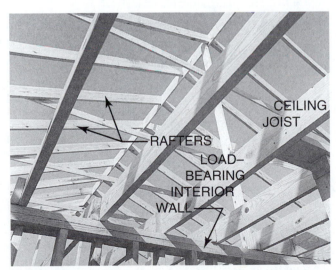

Figure 20.29 The carpenters have installed the ceiling joists, which are shown resting on an interior load-bearing wall. The stick-built roof rafters are also in place. *(Courtesy Southern Forest Products Association)*

Figure 20.30 The joints between the plywood sheathing panels are staggered as the panels are laid. *(Courtesy APA - The Engineered Wood Association)*

I T E M   O F   I N T E R E S T

FRAMING TO CONTROL WIND AND SEISMIC LOADS

Winds and earthquakes subject the structural frame to lateral loading, forcing the building to move sideways as shown in Fig. A. To prevent this movement wood frame buildings can be constructed using the floors or roof to act as horizontal beams or diaphragms. The diaphragm transfers the load to the end walls, which are constructed as shear walls. The shear walls transfer the load to the foundation.

Diaphragm and shear wall construction contain intermediate framing members, chords, struts, and sheathing. The sheathing can be oriented strand board or plywood. The size and spacing of the nails must be indicated on the structural drawings along with information about the other materials used. Thicker sheathing and closer spacing of the nails increase the capacity of the diaphragm or shear wall.

Diaphragms work as horizontal beams, with the sheathing acting as the web of the beam (Fig. B). Blocking is used to carry the unsupported edges of the sheathing. The top wall plates on shear walls and the header joist on floor or roof diaphragms act as a chord and serve as the flange on a beam. If the chord is not continuous, it must be spliced.

The shear wall is constructed much like the diaphragm (Fig. C). Chord members are secured to the foundation with special metal tiedowns (Fig. D).

If there are openings in a diaphragm or shear wall, additional framing and connections around the openings are required to transfer the loads.

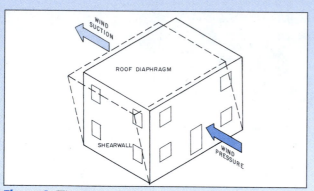

Figure A The structural frame is subject to lateral forces.

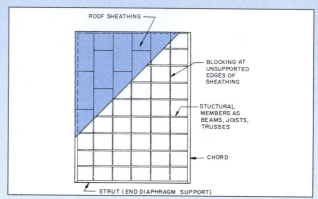

Figure B A typical diaphragm detail.

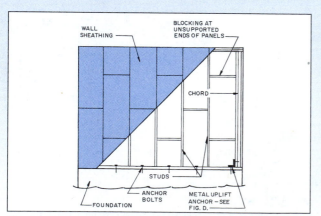

Figure C An elevation of a typical shear wall.

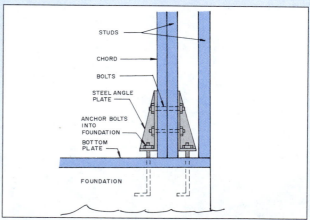

Figure D A typical method for anchoring a shear wall to the foundation.

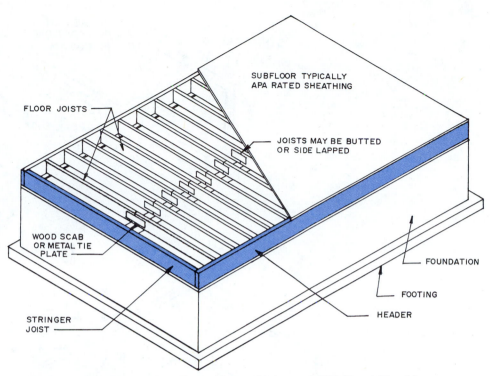

Figure 20.31 The framing for a floor with wood joists and APA Rated Sheathing.

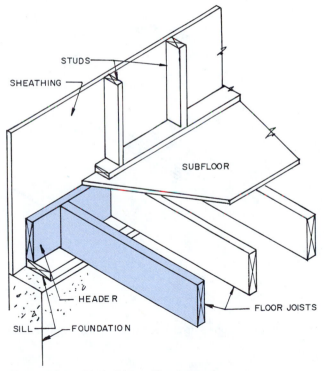

Figure 20.32 Typical box sill construction.

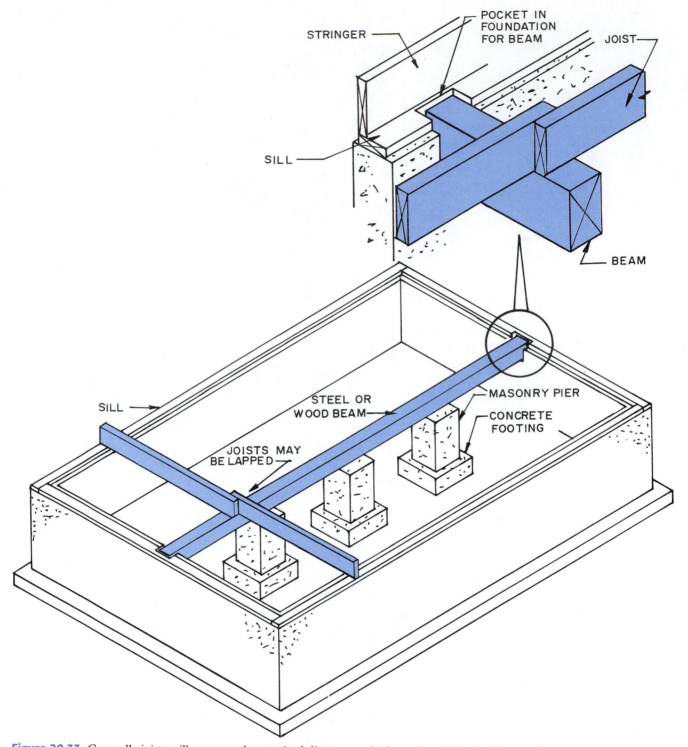

Figure 20.33 Generally joists will not span the required distance and a beam is required to shorten the span.

After the first-floor joists are in place and the subfloor glued and nailed to them, the first-floor walls are assembled on this platform and raised into position, as shown in Fig. 20.34. Next, the floor joists for the second floor are installed and covered with subflooring, and the walls for the second floor are built and raised on this platform, as you can see in Fig. 20.35. The structure is now ready to have the ceiling joists installed. Finally, the rafters can be set in place on the top plate of the exterior walls, and the roof can be sheathed. The completely assembled two-story wood light framed building is shown in Fig. 20.36. This is a typical example of platform framing.

Stud Wall Construction

In Fig. 20.37 you can see the framing for a typical exterior wall with an opening. The studs are typically 2 × 4 stock, but 2 × 6 studs provide space for additional insulation. The studs are generally spaced 16 in. (406 mm) on-center. However, other spacings, such as 12 and 24 in. (305 and 610 mm) are used. The load-carrying capacity of the wall and the size of the studs provides for these spacings. They are always on the even foot because panels of sheathing are 4 × 8 ft. (1.2 × 2.4 m), and the panel edges will rest on a stud.

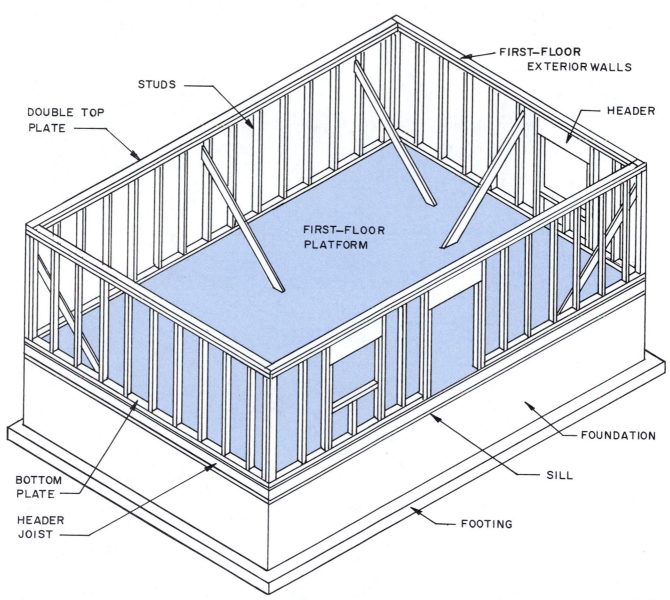

Figure 20.34 The first-floor exterior walls are assembled on the platform and raised in place. The interior partitions are built and raised next.

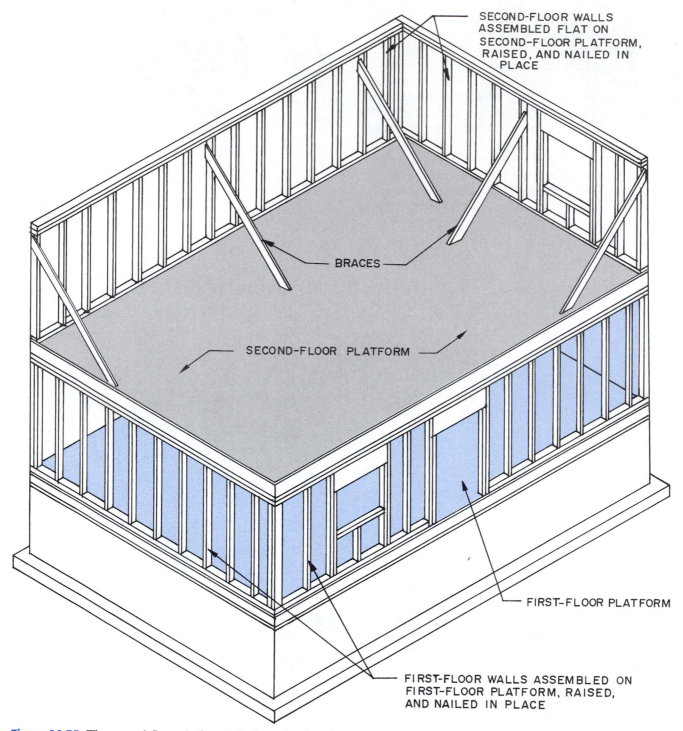

Figure 20.35 The second-floor platform is built on the first-floor walls and the second-floor walls are assembled and raised.

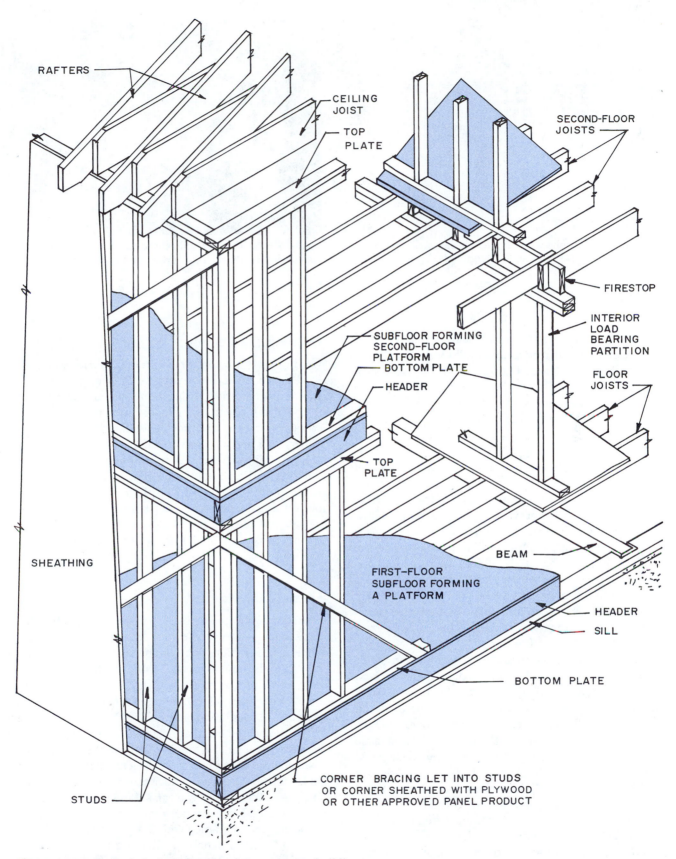

Figure 20.36 Typical platform framing for a two-story building.

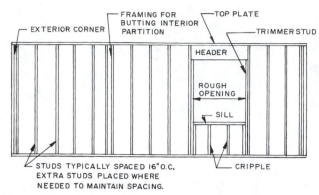

Figure 20.37 A framing detail for an exterior wall. Notice that the large header spanning the opening is supported by trimmer studs.

Figure 20.38 This opening is framed with a header and small cripples between it and the top plate.

Framing an opening in a load-bearing wall requires that a header span the opening and be large enough to carry the imposed loads. Figure 20.38 shows how an opening is framed. The header can be built in many ways, and you can see some of them in Fig. 20.39. Openings in interior non-load-bearing partitions are framed as shown in Fig. 20.40. Framing for a door opening in a load-bearing wall is shown in Fig. 20.41. Notice the rough opening door size includes the size of the door plus space for the door frame and a little room to allow the carpenter to plumb the door frame.

The corners of exterior walls have to be built to allow for the side wall to butt the front or rear wall and to leave a flange to which the gypsum board is nailed. You can see several examples in Fig. 20.42. Where interior partitions meet exterior walls, the top plate overlaps the lower plate of the double top plate, as shown in Fig. 20.43. Where interior partitions butt an outside wall or another partition, they can be joined by installing blocking or extra studs, as shown in Fig. 20.44.

When the walls are in place the sheathing can be installed. Sheathing may be plywood, oriented strand board, Com-Ply, or asphalt-impregnated fiber or rigid foam plastic sheets. When plywood, oriented strand board, or Com-Ply are used, they also provide the required bracing of the wall. When rigid foam plastic sheets are used, the wall is braced with let-in diagonal braces (Fig. 20.45), or sheets of plywood, oriented strand board or Com-Ply are nailed at each corner and at intervals along the wall as specified by the architect. In Fig. 20.46 you can see the nailing pattern used for rigid panel sheathing. Figure 20.47 shows the rigid panels on the corner and the nonstructural asphalt-impregnated sheets on the remainder of the wall.

Framing the Ceiling

Ceiling joists are supported on the exterior wall and on load-bearing interior partitions. They carry the gypsum wallboard or other ceiling materials and insulation. If they are subject to other loads, such as a second-floor storage area, the architect will size them accordingly. When they become joists for the second floor of a two-story building, sizing is even more important. A typical detail is in Fig. 20.48. The first ceiling joist is set in from the outer edge of the plate on the gable end to permit construction of the framing for the gable end, as you can see in Fig. 20.49. The ceiling joists can be trimmed as shown in Fig. 20.50 to follow the rafter shape.

Framing the Roof

The commonly used types of roofs are shown in Fig. 20.51. Roof framing is one of the most difficult parts of wood light frame construction. The following discussion will be limited to the basic stick built gable and hip roofs. To learn how to lay out the rafters you should consult a carpentry book. The types of rafters used are identified on the drawings in Figs. 20.52 and 20.53. The rafters rest on top of the double plate and are nailed to it. In areas with high winds a metal strap is used to connect them to the plate. Examine the layout in Fig. 20.54. Overhangs on the gable end can be built using a ladder construction for overhangs (12 in. or 305 mm), as you can see in Fig. 20.55, and lookouts are used for overhangs beyond this (Fig. 20.56). The gable end can be framed with 2 × 4 in. studs as shown in Fig. 20.57. Information on finishing the cornice can be found in Chapter 22.

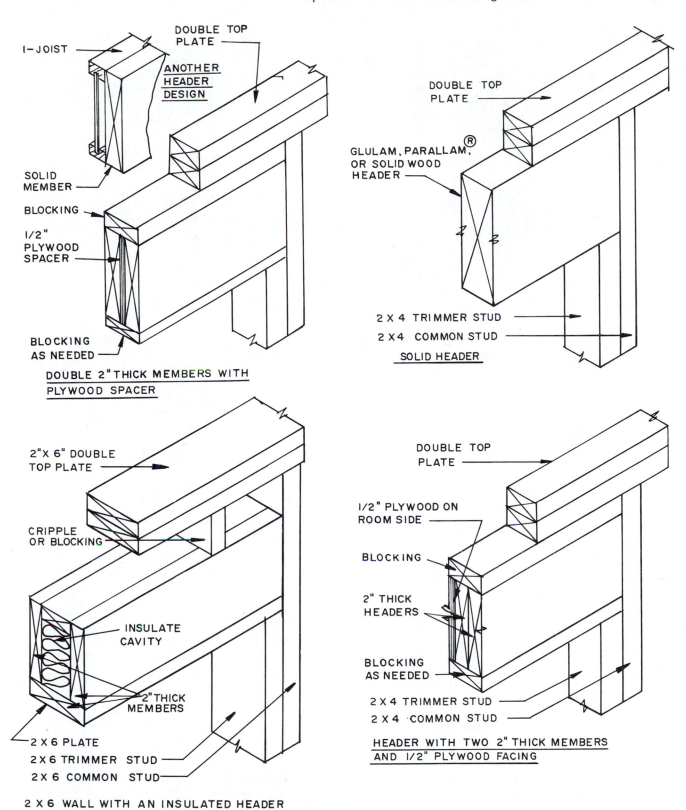

Figure 20.39 Some ways headers in wood light frame construction can be built.

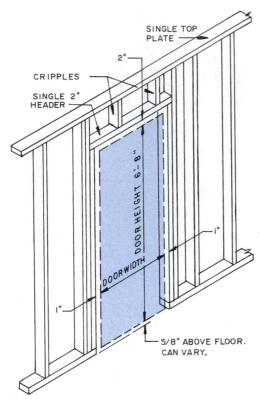

Figure 20.40 How to frame a door opening in an interior non-load-bearing wall.

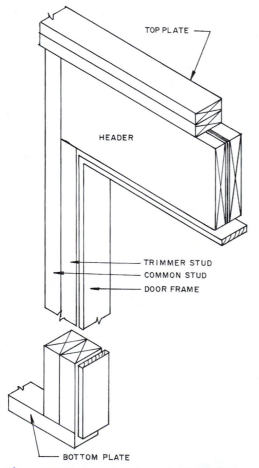

Figure 20.41 A section through the framing for a door opening in an exterior wall.

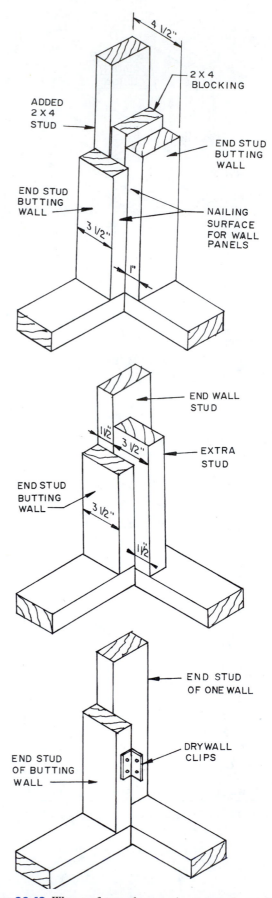

Figure 20.42 Ways to frame the exterior corner formed by butting exterior walls. It is important to provide nailing flanges for the interior wall finish materials.

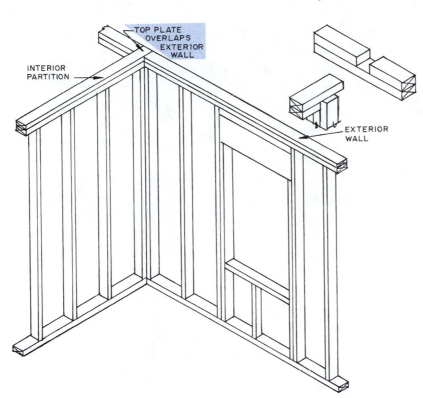

Figure 20.43 When interior partitions butt exterior walls the top plate overlaps, providing a horizontal tie.

TOP PLATE OVERLAPS EXTERIOR WALL

INTERIOR PARTITION

EXTERIOR WALL

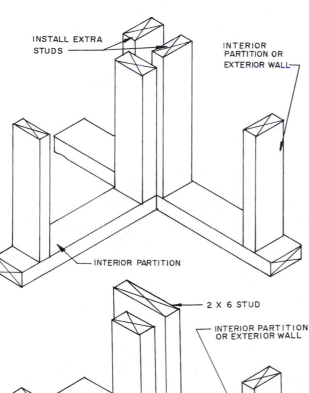

INSTALL EXTRA STUDS

INTERIOR PARTITION OR EXTERIOR WALL

INTERIOR PARTITION

2 X 6 STUD

INTERIOR PARTITION OR EXTERIOR WALL

INTERIOR PARTITION

Figure 20.44 Two ways that butting partitions can be joined. These techniques can also be used when butting exterior walls.

WOOD DIAGONALS LET INTO STUDS

Figure 20.45 One way to brace exterior walls is to let in wood diagonals.

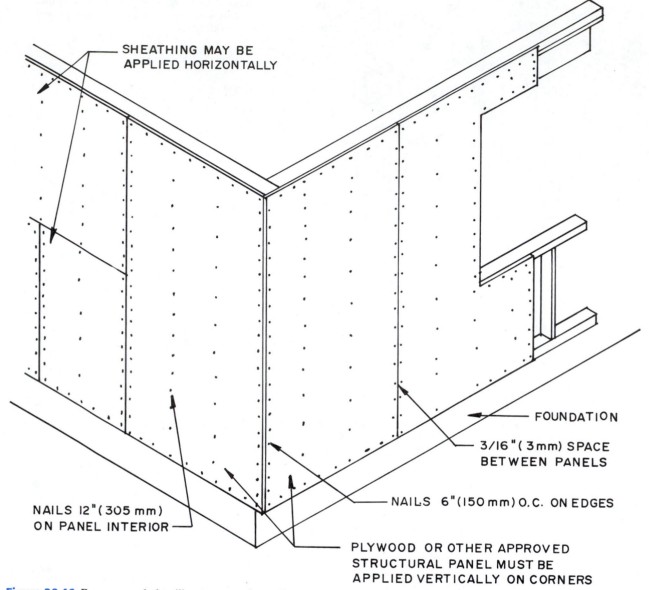

SHEATHING MAY BE
APPLIED HORIZONTALLY

FOUNDATION

3/16" (3 mm) SPACE
BETWEEN PANELS

NAILS 6" (150 mm) O.C. ON EDGES

NAILS 12" (305 mm)
ON PANEL INTERIOR

PLYWOOD OR OTHER APPROVED
STRUCTURAL PANEL MUST BE
APPLIED VERTICALLY ON CORNERS

Figure 20.46 Recommended nailing patterns for wall sheathing when using rigid panels, such as plywood. Diagonal bracing is not needed when rigid sheets are used as sheathing.

Figure 20.47 This building has the exterior wall braced with sheets of rigid sheathing, such as plywood, on each corner. Long walls require additional sheets at intervals along the wall.

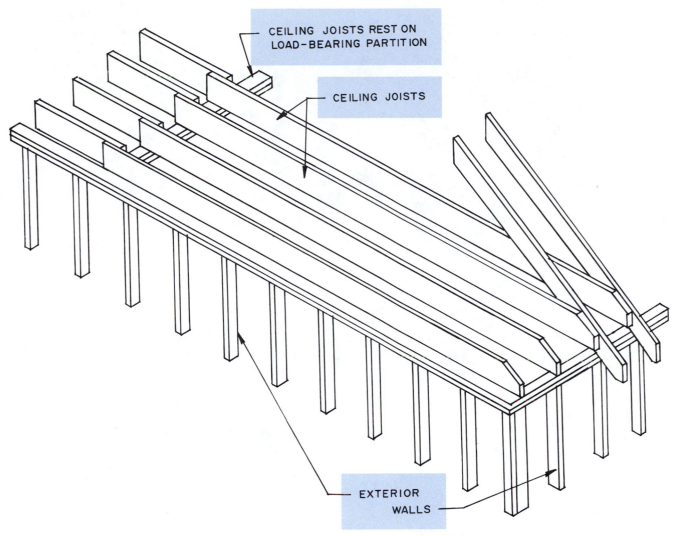

Figure 20.48 Ceiling joists span from an exterior wall to a load-bearing interior partition.

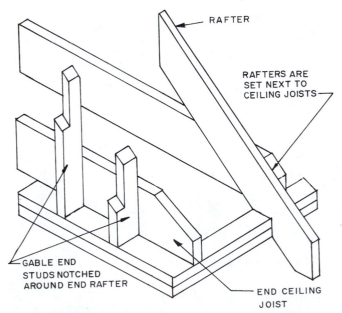

Figure 20.49 How to place the ceiling joists. When the rafters are installed they are placed next to the ceiling joists and are nailed to them.

WOOD TRUSSES AND JOISTS

Wood Roof Trusses

Wood roof trusses are widely used for framing wood light frame buildings. A roof truss is a structural unit, usually in a triangular form, made by assembling structural wood members into a rigid frame. A typical example is in Fig. 20.58.

The *advantages* of wood trusses are that they speed up on-site erection time and can span the width of most buildings without interior load-bearing walls, and since they are carefully engineered and factory built, they have a consistent, reliable quality. One *disadvantage* is that they make it difficult to use the attic for storage. However, trusses can be designed to allow the center of the building to be open for storage or second-floor rooms. A few of the many types of wood trusses available are shown in Fig. 20.59. Many other designs are available.

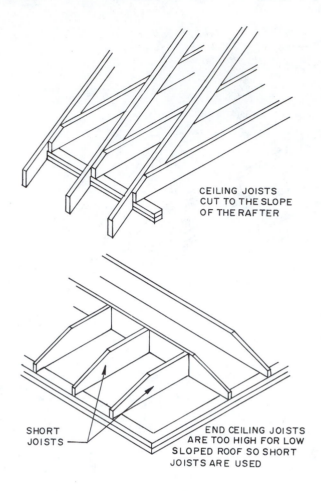

CEILING JOISTS
CUT TO THE SLOPE
OF THE RAFTER

SHORT
JOISTS

END CEILING JOISTS
ARE TOO HIGH FOR LOW
SLOPED ROOF SO SHORT
JOISTS ARE USED

Figure 20.50 Ceiling joists have to be cut to the slope of the roof.

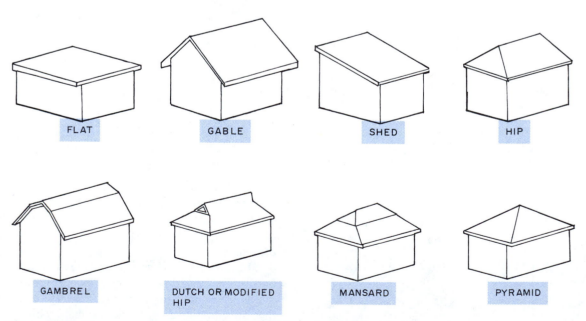

FLAT   GABLE   SHED   HIP

GAMBREL   DUTCH OR MODIFIED HIP   MANSARD   PYRAMID

Figure 20.51 The most commonly used types of roofs used with light frame construction.

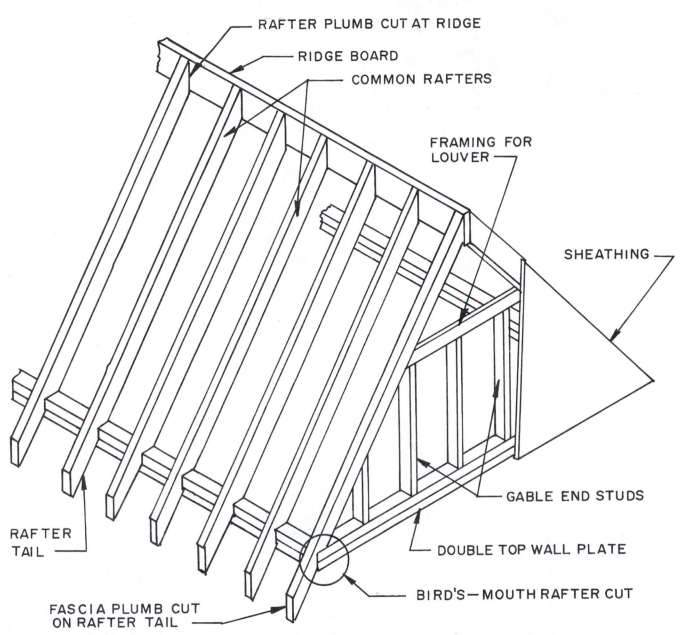

RAFTER PLUMB CUT AT RIDGE

RIDGE BOARD

COMMON RAFTERS

FRAMING FOR LOUVER

SHEATHING

RAFTER TAIL

FASCIA PLUMB CUT ON RAFTER TAIL

GABLE END STUDS

DOUBLE TOP WALL PLATE

BIRD'S—MOUTH RAFTER CUT

Figure 20.52 The framing members of a typical gable roof and gable end. Notice the names of the cuts on the common rafter.

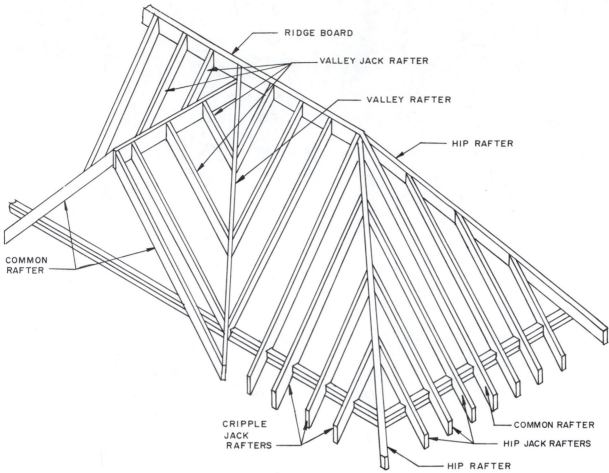

Figure 20.53 Typical hip roof construction. Notice that the intersection of a gable roof with the section of the roof built with common rafters forms a valley.

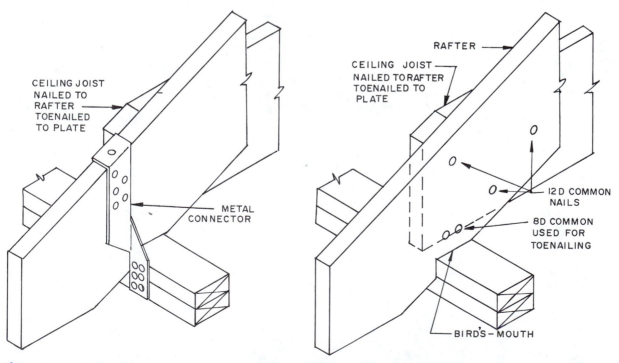

Figure 20.54 Rafters, trusses, and ceiling joists can be toenailed to the top plate. Metal connectors increase the strength of the connection.

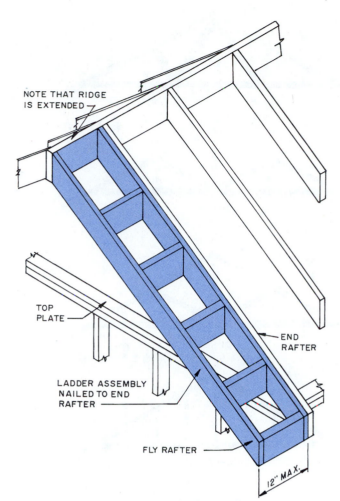

Figure 20.55 Small roof overhangs on the gable end can be built with a ladder-type construction. The roof sheathing strengthens the extension.

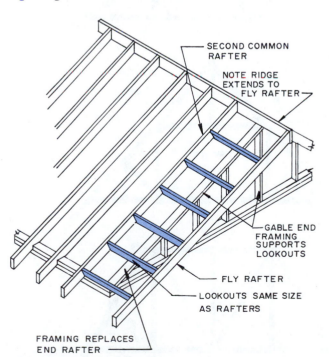

Figure 20.56 Gable end roof overhangs over 12 in. (305 mm) are built using lookouts.

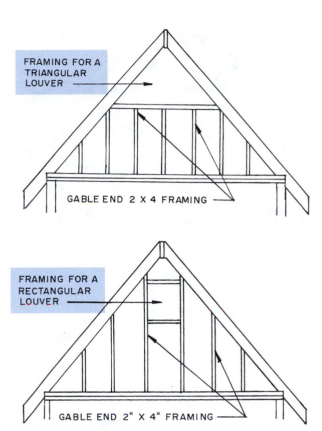

Figure 20.57 Gable ends frequently are framed to hold some form of louver.

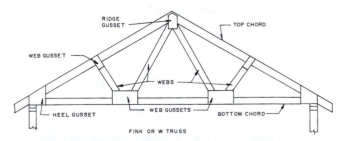

Figure 20.58 The parts of a typical wood frame truss.

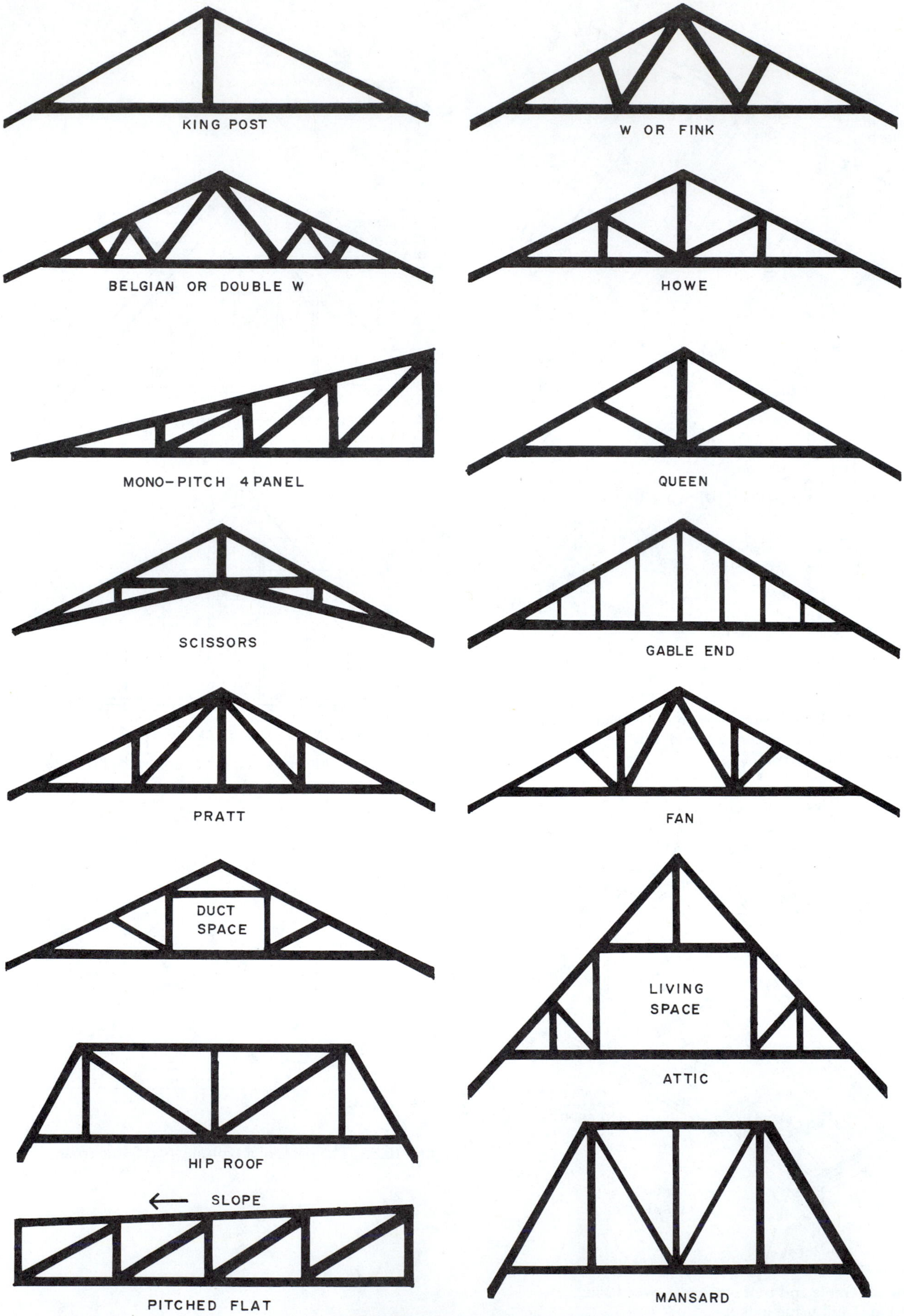

KING POST

W OR FINK

BELGIAN OR DOUBLE W

HOWE

MONO–PITCH 4 PANEL

QUEEN

SCISSORS

GABLE END

PRATT

FAN

DUCT SPACE

LIVING SPACE

ATTIC

HIP ROOF

← SLOPE

PITCHED FLAT

MANSARD

Figure 20.59 Some of the commonly used wood roof trusses.

Small trusses can be hung between the exterior walls with the ridge point down. These can be raised by workers standing on the platform using wood poles to spin the point into an upright position. Workers on the top plate set it in position, secure it to the top plate, and brace it (Fig. 20.60). Generally, trusses are set by a crane that raises them over the building and lowers them to the top plate, and workers on the plate secure and brace them, as shown in Fig. 20.61.

As the trusses are erected they are secured by braces to the ground and platform, and braces are nailed between them until the sheathing is installed. You see a typical example in Fig. 20.62. There are other techniques that carpenters use to raise, brace, and install frames. These can be found in general carpentry books.

Wood Floor Trusses

Wood floor trusses are frequently used instead of solid wood joists. They provide long spans and often eliminate the need for one or more beams to support the floor in long spans. Since they have open webs it is easy to run electrical, mechanical, and plumbing systems through them. An all-wood floor truss is shown in Fig. 20.63, and one with metal webs is in Fig. 20.64. Wood floor trusses must be installed following the manufacturer's instructions. You can see some typical details in Fig. 20.65. A framing detail for a two-story house using wood floor and roof trusses and platform framing can be seen in Fig. 20.66.

Figure 20.60 Small trusses can be raised by workers standing on the platform and lifting it with a long pole.

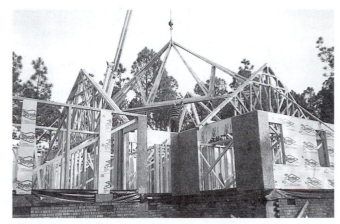

Figure 20.61 Large roof trusses are lifted into place with a crane.

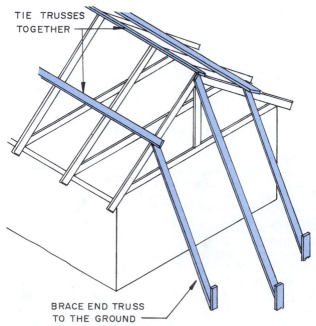

TIE TRUSSES TOGETHER

BRACE END TRUSS TO THE GROUND

Figure 20.62 As trusses are erected they must be temporarily braced until permanent bracing and roof sheathing are in place.

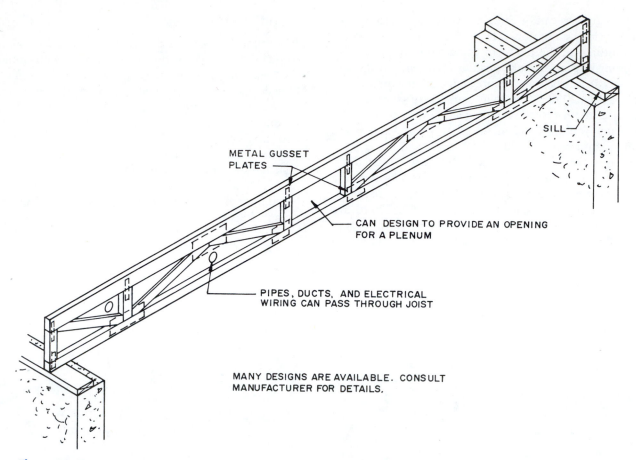

METAL GUSSET
PLATES

CAN DESIGN TO PROVIDE AN OPENING
FOR A PLENUM

PIPES, DUCTS, AND ELECTRICAL
WIRING CAN PASS THROUGH JOIST

SILL

MANY DESIGNS ARE AVAILABLE. CONSULT
MANUFACTURER FOR DETAILS.

Figure 20.63 A typical wood frame truss joist is used for floor and ceiling construction.

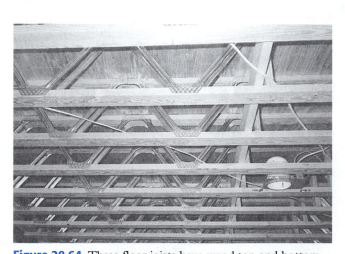

Figure 20.64 These floor joists have wood top and bottom chords and metal webs nailed to them.

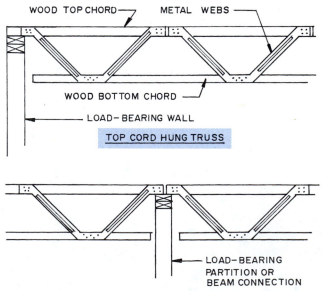

WOOD TOP CHORD — METAL WEBS

WOOD BOTTOM CHORD

LOAD-BEARING WALL

TOP CORD HUNG TRUSS

LOAD-BEARING
PARTITION OR
BEAM CONNECTION

Figure 20.65 A typical wood floor truss with metal webs designed to be suspended from the top chord.

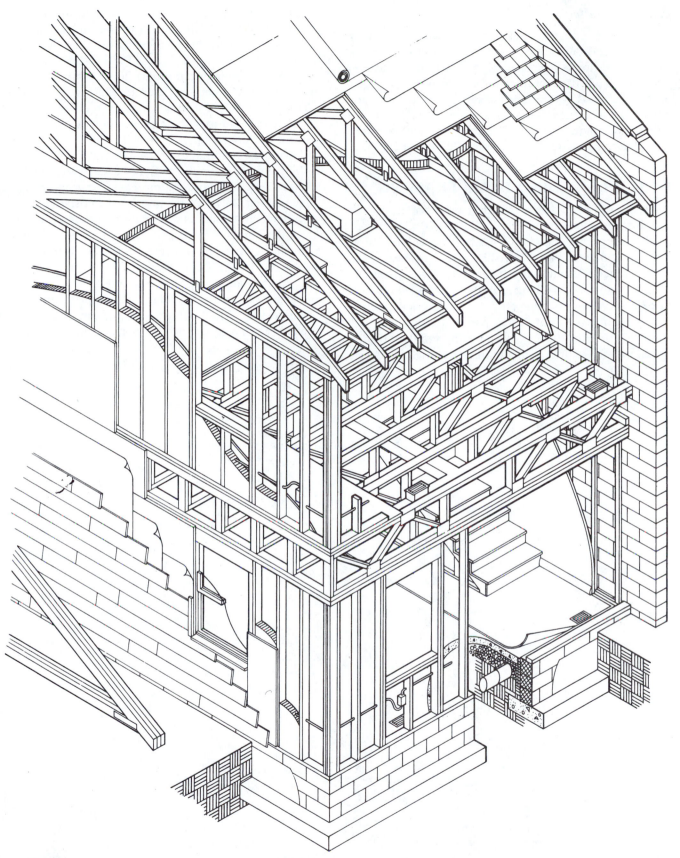

Figure 20.66 A wood light frame building using wood roof and floor trusses and platform construction. *(Reproduced with permission from* The Building Systems Integration Handbook, *Richard D. Rush, ed., Butterworth–Heinmann Publishers, Newton, Mass., 1986)*

Wood I-Joists

Wood **I-joists** are structural members made in a factory using plywood or oriented strand board webs and laminated veneer lumber or flanges, as shown in Fig. 20.67. They are widely used for floor and ceiling joists and sometimes as rafters. They must be installed following the manufacturer's directions. They must not be installed using only the techniques common for solid wood joists. Some typical installation details are in Figs. 20.68 and 20.69.

POST, PLANK, AND BEAM FRAMING

Post, plank, and beam framing uses wood posts to support the second floor or roof (Fig. 20.70). Beams are used to support roof and floor loads. A typical structural frame is shown in Fig. 20.71. The posts and beams are widely spaced, usually in the range of 4 ft. on-center. The floor and roof decking are either thick wood planks or plywood designed to span these distances and carry the design loads. Composition panels are used for roof sheathing. In Fig. 20.72 the roof beams and floor beams run across the width of the building and the planks run perpendicular to them. Another framing method is to run the roof beams the length of the building and the decking perpendicular to them in the direction of the width of the building (Fig. 20.73).

Roof and floor decking may be solid wood or laminated wood. You can find examples in Fig. 20.74. Each plank is face nailed to the beam and toenailed through the tongue. A composite stressed-skin panel can be used as roof decking (Fig. 20.75). These must be installed following the manufacturer's directions.

The posts are secured to floor beams with metal connectors, and post and column caps are used on top as shown in Fig. 20.76. At the bottom, the posts are secured to concrete foundations with metal column bases such as shown in Fig. 20.77. Interior partitions that run parallel with the floor beams must have a beam placed below or on top of the floor decking. Two examples are in Fig. 20.78. Rafters are secured to posts with some form of metal strap, and at the ridge metal tie plates and straps are used (Fig. 20.79). Butting beams can be hung with metal beam hangers. One type is shown in Fig. 20.80.

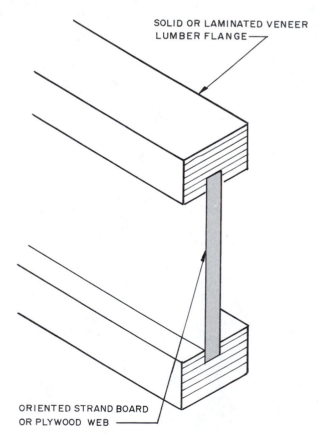

SOLID OR LAMINATED VENEER
LUMBER FLANGE

ORIENTED STRAND BOARD
OR PLYWOOD WEB

Figure 20.67 A wood I-joist.

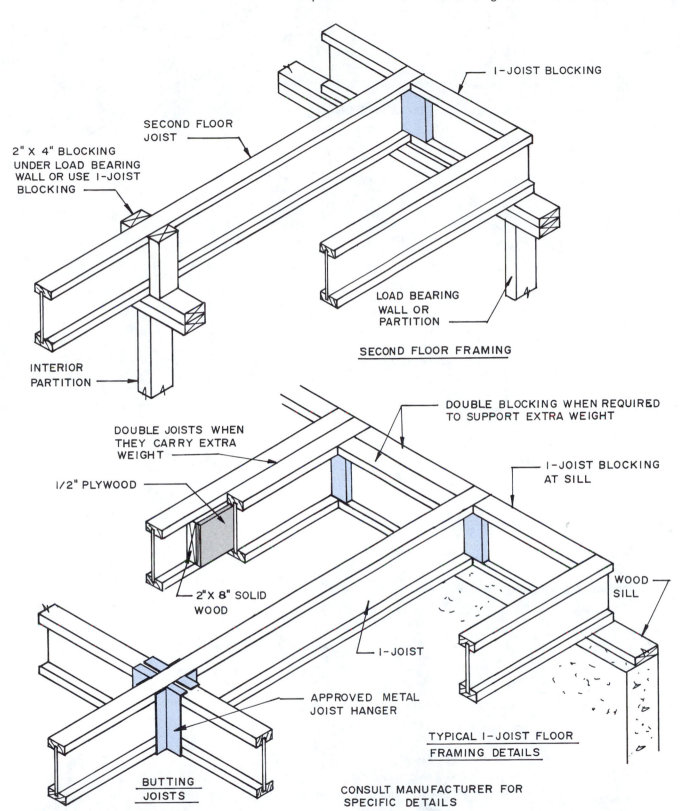

I-JOIST BLOCKING

SECOND FLOOR JOIST

2" X 4" BLOCKING UNDER LOAD BEARING WALL OR USE I-JOIST BLOCKING

LOAD BEARING WALL OR PARTITION

SECOND FLOOR FRAMING

INTERIOR PARTITION

DOUBLE BLOCKING WHEN REQUIRED TO SUPPORT EXTRA WEIGHT

DOUBLE JOISTS WHEN THEY CARRY EXTRA WEIGHT

I-JOIST BLOCKING AT SILL

1/2" PLYWOOD

2" X 8" SOLID WOOD

WOOD SILL

I-JOIST

APPROVED METAL JOIST HANGER

TYPICAL I-JOIST FLOOR FRAMING DETAILS

BUTTING JOISTS

CONSULT MANUFACTURER FOR SPECIFIC DETAILS

Figure 20.68 Typical framing details for use with manufactured wood I-joists.

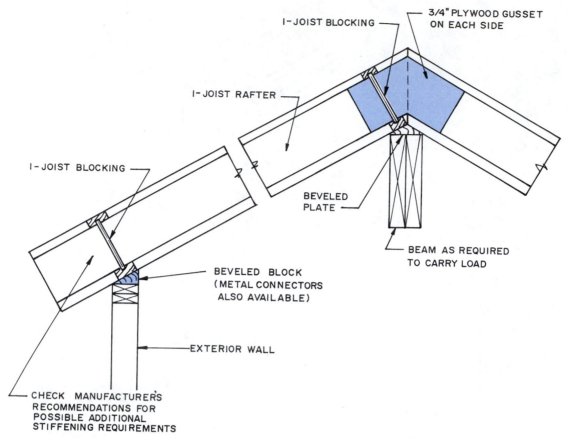

I-JOIST BLOCKING

3/4" PLYWOOD GUSSET
ON EACH SIDE

I-JOIST RAFTER

I-JOIST BLOCKING

BEVELED
PLATE

BEAM AS REQUIRED
TO CARRY LOAD

BEVELED BLOCK
(METAL CONNECTORS
ALSO AVAILABLE)

EXTERIOR WALL

CHECK MANUFACTURER'S
RECOMMENDATIONS FOR
POSSIBLE ADDITIONAL
STIFFENING REQUIREMENTS

Figure 20.69 One of several ways used to frame a roof using wood I-joists as rafters.

Figure 20.70 This one-story commercial building is framed with glued laminated columns and roof beams. The columns are anchored to a concrete foundation with metal column bases that are set in the concrete, and the column is bolted to them. *(Courtesy Timber Structures, Inc.)*

Figure 20.71 Today a crane is used to raise heavy timber members. In early times human labor was used thus presenting great difficulty. The structural frame is a thing of beauty. *(Courtesy The Beamery)*

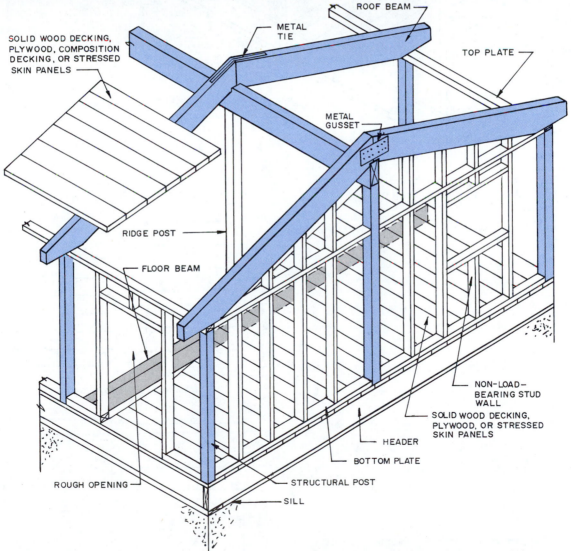

ROOF BEAM

METAL TIE

TOP PLATE

SOLID WOOD DECKING, PLYWOOD, COMPOSITION DECKING, OR STRESSED SKIN PANELS

METAL GUSSET

RIDGE POST

FLOOR BEAM

NON-LOAD-BEARING STUD WALL

SOLID WOOD DECKING, PLYWOOD, OR STRESSED SKIN PANELS

HEADER

BOTTOM PLATE

ROUGH OPENING

STRUCTURAL POST

SILL

Figure 20.72 The roof framing on this post, plank, and beam structure runs across the width of the building.

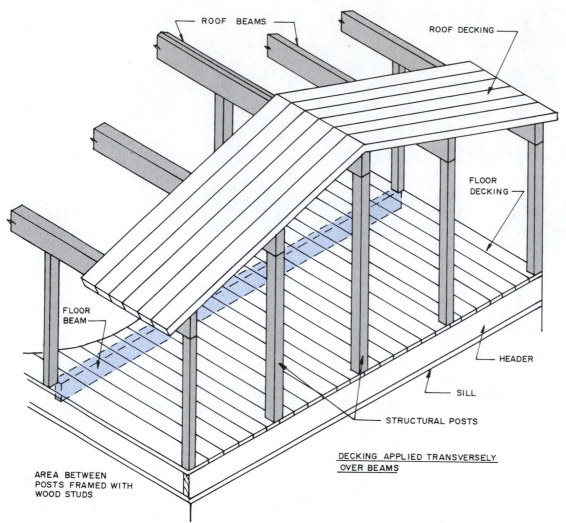

Figure 20.73 The roof on this post, plank, and beam structure is supported by beams running the length of the building.

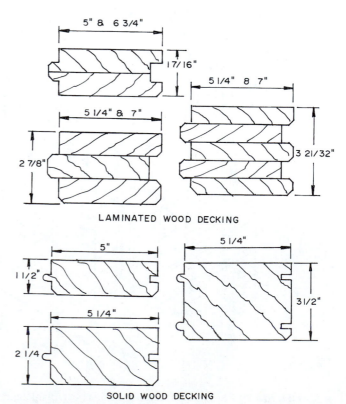

Figure 20.74 Samples of some of the heavy wood decking used on the floors and roofs of post, plank, and beam structures.

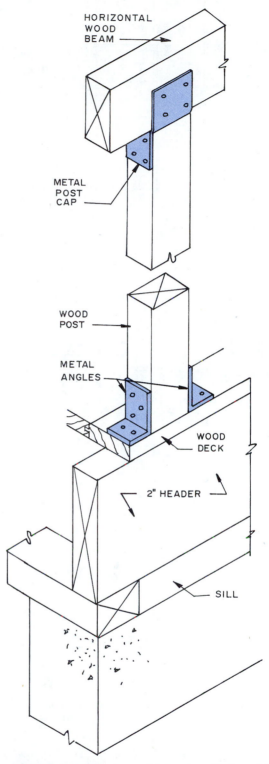

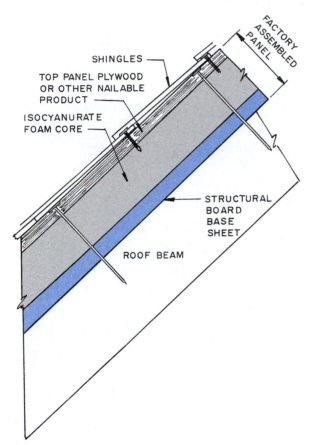

Figure 20.75 Roofs on post, plank, and beam structures can be decked with factory assembled stressed skin panels like this example.

Figure 20.76 A typical post and beam installation with the column supported by the header on the floor. Notice the use of metal connectors.

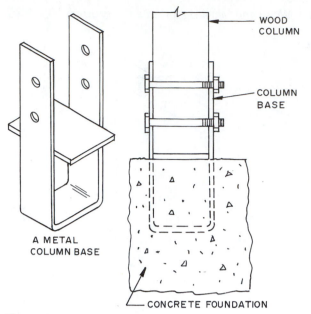

WOOD COLUMN

COLUMN BASE

A METAL COLUMN BASE

CONCRETE FOUNDATION

Figure 20.77 Metal column bases are set in the concrete foundation and the wood column bolted to them.

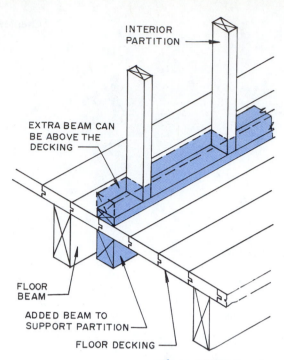

INTERIOR PARTITION

EXTRA BEAM CAN BE ABOVE THE DECKING

FLOOR BEAM

ADDED BEAM TO SUPPORT PARTITION

FLOOR DECKING

Figure 20.78 Interior partitions that run parallel with the floor beams need an extra beam to provide support.

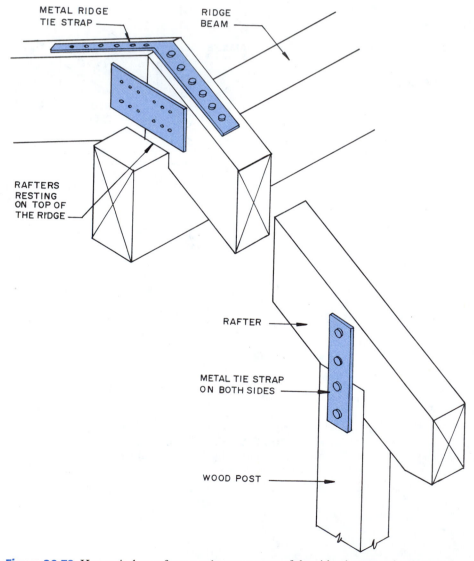

METAL RIDGE TIE STRAP

RIDGE BEAM

RAFTERS RESTING ON TOP OF THE RIDGE

RAFTER

METAL TIE STRAP ON BOTH SIDES

WOOD POST

Figure 20.79 Heavy timber rafters can be set on top of the ridge beam and connected with metal plates and straps. Metal tie straps connect the rafters to the columns.

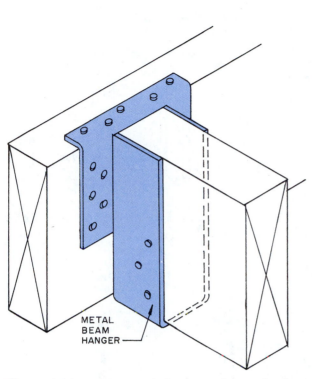

Figure 20.80 When heavy timber beams butt each other heavy-gauge metal beam hangers can be used.

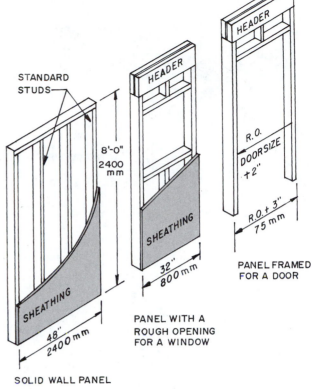

Figure 20.81 Examples of prefabricated wall panels.

PANELIZED FRAME CONSTRUCTION

Panelized construction uses a series of standard factory assembled wall, floor, and roof units that are assembled on the job. In the United States the units are made in 16 in. units based on a 4 in. module. Metric units are based on a 100 mm module and probably in 400 mm units. Some typical wall panels are shown in Fig. 20.81. They include provisions for doors and windows. The actual length of the units available depends on the manufacturer, but they are typically 16, 32, 64, 80, 96, and 144 in. based on a 4 in. module. Comparable metric units are 400, 800, 1600, 2000, 2400, and 3600 mm based on a 100 mm module.

When the units are erected on the subfloor, they are tied together with a top plate as shown in Fig. 20.82. Manufacturers install the sheathing and sometimes windows in the wall in the factory. The design of the panels must include ways to join exterior and interior corners. This is accomplished by adding extra studs and providing an overlapping flange of sheathing as shown in Fig. 20.83. Interior partitions also butt the exterior walls, and the framing detail in Fig. 20.84 is typical. It provides an extra stud in the panel. Another type of structural wall panel is made using a rigid foamed plastic core and oriented strand board skins (Fig. 20.85.)

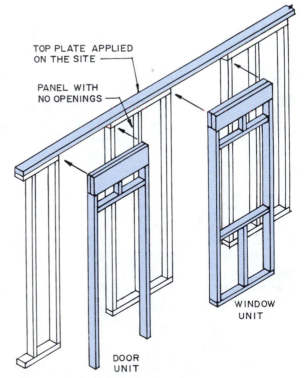

Figure 20.82 Prefabricated wall units are secured to the subfloor and tied together with a field-applied top plate.

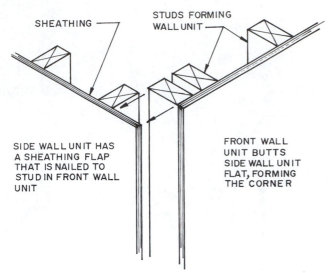

Figure 20.83 This is one way prefabricated wall units can form exterior corners.

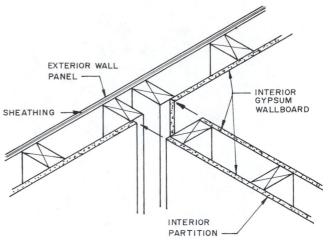

Figure 20.84 Exterior wall panels are prepared to receive an interior partition.

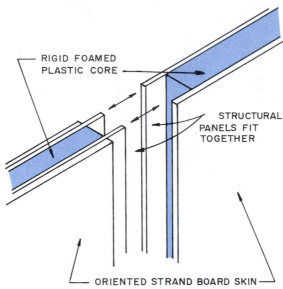

Figure 20.85 A manufactured wall panel with a rigid foam plastic core.

The roof is framed with factory assembled gable ends and trusses. They are set in place with a crane and braced until the sheathing is applied. The gable end is sheathed and has the siding applied (Fig. 20.86). The floor can be built with conventional joists and plywood subflooring. However, large factory assembled floor panels can be used, and widely spaced beams can replace the floor joists (Fig. 20.87).

STEEL FRAMING

Several manufacturers offer a complete **lightweight steel framing** system for walls, floors, and roofs. Specific details should be obtained from the manufacturer.

The following discussion is typical of those available. It should be noted that because quality solid wood framing is getting more expensive, steel framing will find increasing use.

The steel members are perforated to lighten them and to permit the passage of utilities such as plumbing and electrical wiring. They are noncombustible; various types of exterior sheathing and siding can be applied; they are easy to insulate; and gypsum and other interior finish materials can be secured to them. Steel members can be preassembled into wall panels to speed erection. Steel framing is widely used for interior and exterior wall construction in commercial and, more recently, in residential construction.

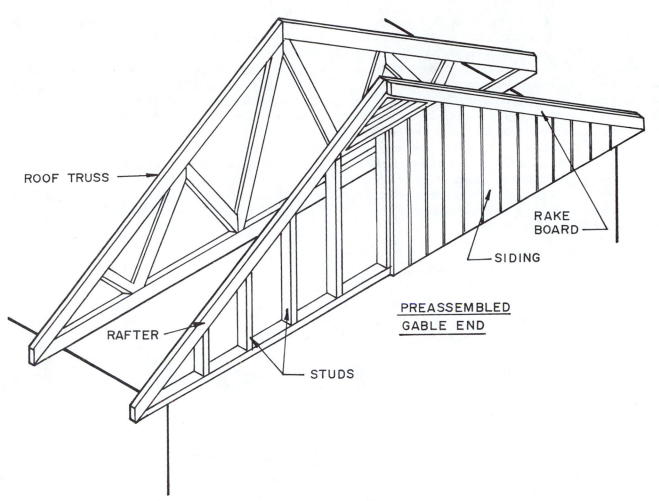

ROOF TRUSS

RAKE BOARD

SIDING

RAFTER

PREASSEMBLED GABLE END

STUDS

Figure 20.86 The gable ends are preassembled and set in place on the end wall.

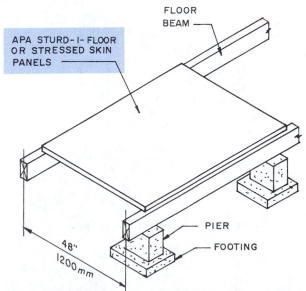

FLOOR BEAM

APA STURD-1-FLOOR OR STRESSED SKIN PANELS

PIER

FOOTING

48"
1200 mm

Figure 20.87 Floor panels can be used instead of thick wood decking to span the distance between floor beams.

The lightweight steel framing system includes studs, tracks, joists, bracing, hangers, and other accessories needed to assemble the unit. Bearing walls can be single and double studs as shown in Fig. 20.88. The studs can be doubled to increase the load-bearing capacity of the wall. The studs sit in a metal track that is screwed to the subfloor. The studs are usually joined to the track with special power-driven screws. Some connections such as bracing or joining beams to the wall's top track must be welded. Manufacturer's specifications must be followed. Posts are made by assembling studs inside tracks as shown in Fig. 20.89. Headers over door and window openings are built up using joists secured inside track material. Refer back to Fig. 20.88.

Assembled walls require metal bridging to be welded and spaced as specified. Two types are channel and strap (Fig. 20.90).

Floors and ceilings are assembled using lightweight steel joists. Bridging may be straps or sections of joists forming solid bridging (Fig. 20.91).

In a multistory building the second-floor joists rest directly above a first-floor stud that may be single or double. Short sections of joist are welded in on the end of each floor joist to stiffen it and help carry the load of the second-floor wall (Fig. 20.92). Roofs are built using trusses assembled from metal channels. Refer to Chapter 17 for additional details.

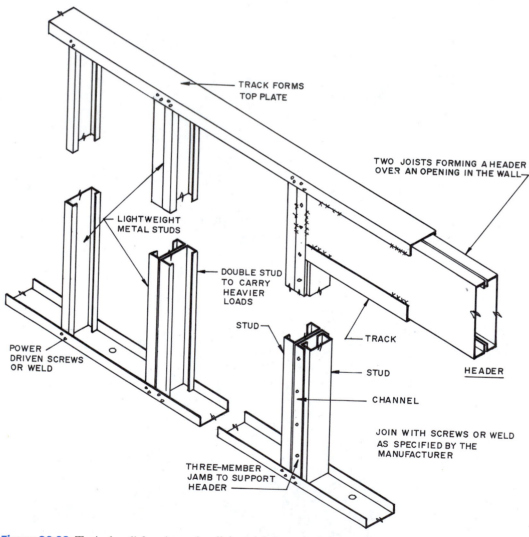

Figure 20.88 Typical wall framing using lightweight metal studs.

Figure 20.89 A post built up from metal studs and track.

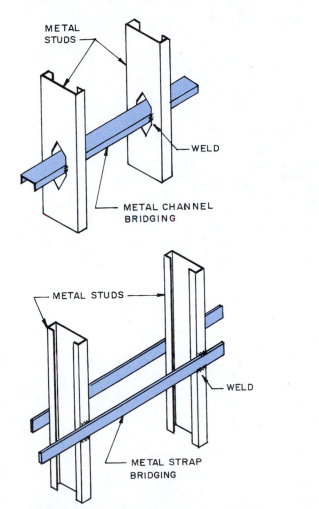

Figure 20.90 Metal studs are braced with metal bracing.

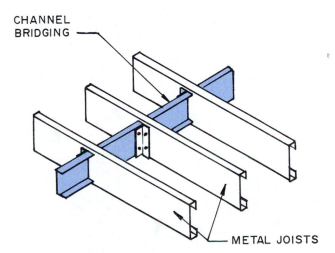

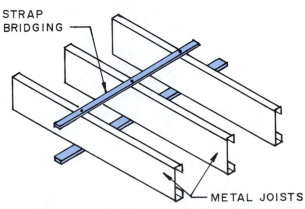

Figure 20.91 Metal bridging is used with metal floor and ceiling joists.

Steel studs designed to be nailed to wood top and bottom plates provide rapid assembly of wall and partitions (Fig. 20.93). The wall is assembled by laying out the studs on the subfloor and nailing them to the wood plates in the same manner as used with wood studs. The wall is then lifted into place (Fig. 20.94).

Additional lightweight steel framing details for commercial construction can be found in Chapter 17.

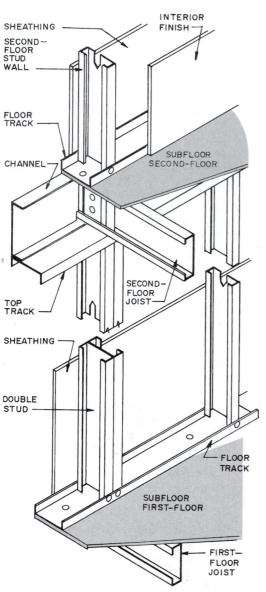

Figure 20.92 A typical frame for a two-story building using lightweight steel framing.

Figure 20.93 Some types of metal studs are nailed to wood top and bottom plates. *(Courtesy H.L. Stud Corporation)*

Figure 20.94 The assembled wall is raised into position. *(Courtesy H.L. Stud Corporation)*

REVIEW QUESTIONS

1. What are the commonly used framing methods used in light frame construction?
2. What is box sill construction?
3. When is a beam required under the floor joists?
4. How do the studs in balloon framing differ from those in platform framing?
5. How are second-floor joists supported when using balloon frame construction?
6. What are the two ways to frame a post-and-beam structure?
7. What components make up an assembled truss unit used for truss framed construction?
8. What is meant by panelized construction?
9. What members are used to make a wood I joist?
10. What are the advantages of using wood floor and roof joists?

11. What are the members used to construct a wall with lightweight steel members?
12. How are posts built using lightweight steel members?

KEY TERMS

balloon framing A wood framing system with the exterior studs extending the full height of the frame and the floor joists secured to the studs.

box sill A type of sill used in frame construction in which the floor joists butt and are nailed to a header joist and all rest on the sill.

firestop Material inserted in a wall, floor, or ceiling to block the natural draft within the assembly that may occur during a fire.

I-joist A wood joist made of an assembly of laminated veneer wood top and bottom flanges and a web of plywood or oriented strand board.

lightweight steel framing Structural framing members made from cold-formed lightweight steel sheet material.

panelized construction Preassembled panels for wall, floor, and roof construction.

platform framing A wood framing system in which the wall studs are one story high and the floor joists of the story above rest on the top plate of the walls below.

post, plank, and beam framing A wood framing system using beams for horizontal structural members, which rest upon posts that form the vertical members.

T-sill A type of sill construction used in balloon framing in which the header joist is placed inside the studs and is butted by the floor joists.

SUGGESTED ACTIVITIES

1. Invite the local building official to speak to the class about how plans are checked to see if they meet the building code.

2. Lay out the corners of the foundation for a small building and set the batter boards.

3. Visit as many construction sites as possible and observe the procedures used to lay out and construct light frame buildings.

4. Construct a two-story scale model of a small building using platform framing.

ADDITIONAL INFORMATION

Lorre, E.N., *Residential Steel Framing Construction Guide,* Technical Publications, Las Vegas, 1994.

Newman, M., *Structural Details for Wood Construction,* McGraw-Hill, New York, 1988.

Ramsey, C.G., Sleeper, H.R., and Ambrose, J., eds., *Residential and Light Construction,* John Wiley and Sons, New York, 1991.

Spence, W.P., *Residential Framing,* Sterling Publishing Co., New York, 1993.

Spence, W.P., *Finish Carpentry,* Sterling Publishing Co., New York, 1995.

Thallon, R., *Graphic Guide to Frame Construction,* Taunton Press, Newtown, Conn., 1991.

Wood Building Technology, Canadian Wood Council, Ottawa, 1993.

Wood Reference Handbook, Canadian Wood Council, Ottawa, 1991.

CHAPTER 21

Heavy Timber Construction

This chapter will help you to:

1. Understand the requirements of the building code as it pertains to heavy timber construction.

2. Become familiar with the methods of framing heavy timber buildings and the types of connections used.

Heavy timber construction is used for both residential and commercial construction. In Chapter 20, post, plank, and beam construction commonly used for residential construction is illustrated. The illustrations in this chapter relate to larger buildings used for commercial purposes.

The wood products commonly used include timber and dimension lumber, glulams, parallel strand lumber, laminated veneer lumber, and prefabricated wood I joists. The selection of the members depends on the load-bearing capacity, appearance of the member, and availability of the product. Various types of metal connectors are available. Sheathing and decking can also be thick wood members capable of spanning the distances between the wood structural members (Fig. 21.1).

When properly designed, heavy timber construction will meet established engineering standards and building code, fire safety, and energy efficiency requirements.

BUILDING CODES

The designer must observe the requirements of the building code when designing any building, including those with heavy timber construction. Heavy timber construction requirements are specified in the various building codes. They typically specify the size and type of approved structural members, such as columns, floor framing, roof framing, floors, and roof decks. The approved types of connections are detailed, and the fire resistance rating of the structural elements is specified. Codes also require seismic analysis in areas where earthquakes are a known factor. This includes such things as diaphragms, shear panels, and selection and spacing of fasteners. The building codes also limit the height of heavy timber buildings, depending on the design and occupancy.

Heavy timber construction in the Building Officials and Code Administrators (BOCA) International Code falls under Type 4 and Type 5. Under Type 4 the exterior walls are constructed of approved noncombustible materials and internal structural members may be of solid or laminated wood without concealed spaces or approved noncombustible materials. Under Type 5 the exterior walls, load-bearing walls, partitions, floors, and roofs may be any approved material that includes heavy

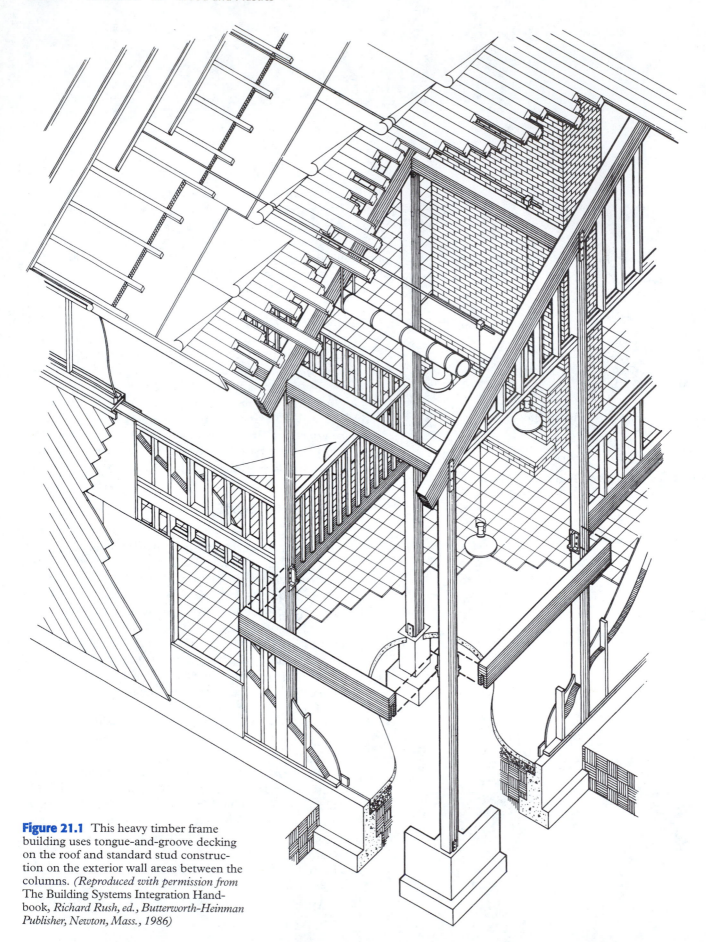

Figure 21.1 This heavy timber frame building uses tongue-and-groove decking on the roof and standard stud construction on the exterior wall areas between the columns. *(Reproduced with permission from* The Building Systems Integration Handbook, *Richard Rush, ed., Butterworth-Heinman Publisher, Newton, Mass., 1986)*

timber members. This permits the use of heavy timber construction in small buildings, such as residences, commercial, and religious buildings. There are no restrictions on the minimum sizes of material as long as the code requirements for Type 5 construction are met.

FIRE RESISTANCE

Heavy timber frame structural systems are able to absorb heat and char when exposed to a fire. They retain much of their structural strength until the cross-sectional area becomes so small that it will not carry the load. The char actually tends to protect the wood in the member from sustained fire damage. Unprotected steel structures will weaken when exposed to heat and collapse.

GLUED LAMINATED CONSTRUCTION

Columns, beams, joists, rigid frames, arches, domes, and decking are made from glued laminated wood. Information about these products is available from The American Institute of Timber Construction (AITC) and in Chapter 19.

Glued laminated members (called glulams) are engineered, stress-rated members made by laminating

wood strips, usually 1⅜ in. (35 mm) and 1½ in. (38 mm) thick for straight members and ¾ in. (19 mm) for curved members, such as arches. Commonly available glulam timber members are shown in Fig. 21.2. The actual sizes of commonly available rectangular glued laminated members are shown in Table 21.1. In addition, rectangular, tapered, and spaced columns are manufactured (Fig. 21.3).

Arches and Domes

Glued laminated arches and domes may be two-hinged, with hinges at each base, or three-hinged, with an additional hinge at the crown (Fig. 21.4). A **hinge joint** is any joint that permits action similar to a hinge but in which there is no appreciable separation of adjacent members. These members produce considerable horizontal thrust at their base, which is controlled by the foundation and tie rods (Fig. 21.5).

Laminated arches span long distances, some spanning 70 ft. (21.4 m) or more. The span depends on the type of arch, loads, roof pitch, and species of wood. Arches are usually left exposed inside the building, forming an attractive ceiling (Fig. 21.6).

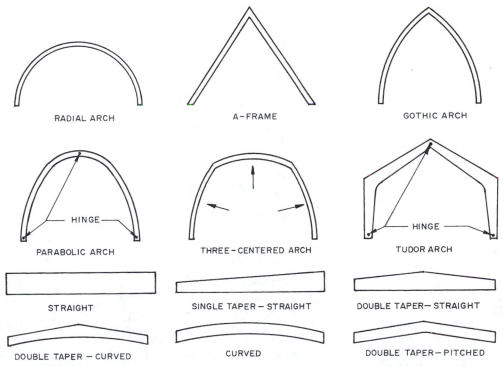

Figure 21.2 Some of the glued laminated structural timber members available from various manufacturers.

Table 21.1 Actual Sizes of Commonly Available Rectangular Glued Laminated Members

Laminations 1½ in. (38 mm) Thick				Laminations 1⅜ in. (35 mm) Thick			
width (in.)	depth (in.)	width (mm)	depth (mm)	width (in.)	depth (in.)	width (mm)	depth (mm)
3⅛	7½	79.3	190.5	3	6⅞	76.2	174.6
5⅛	6	130.1	152.4	5	6⅞	127	174.6
5⅛	9	130.1	228.6	5	8¼	127	209.5
5⅛	10½	130.1	266.7	5	11	127	279.4
6¾	9	171.4	228.6	6¾	8¼	171.4	209.5

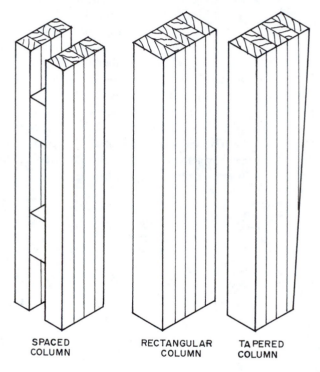

SPACED COLUMN   RECTANGULAR COLUMN   TAPERED COLUMN

Figure 21.3 Several types of glued laminated columns.

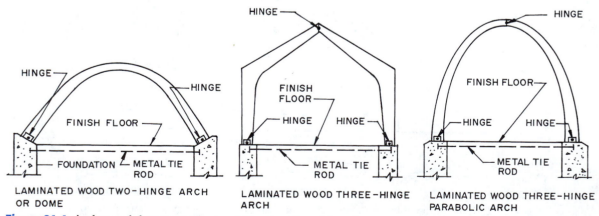

LAMINATED WOOD TWO-HINGE ARCH OR DOME

LAMINATED WOOD THREE-HINGE ARCH

LAMINATED WOOD THREE-HINGE PARABOLIC ARCH

Figure 21.4 Arches and domes may be two-hinged or three-hinged.

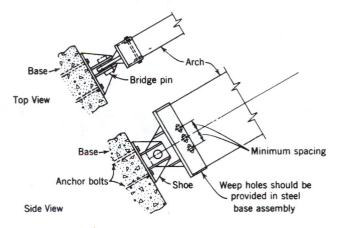

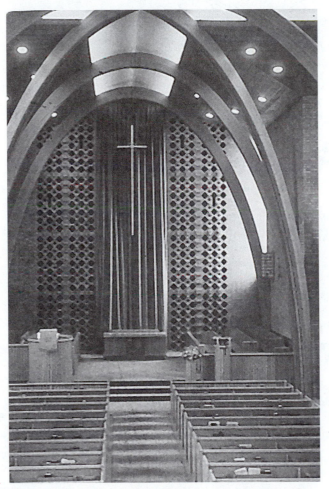

Figure 21.5 The horizontal thrust at the base of the glued laminated arch may be controlled with metal tie rods extending from one base to the other.

Figure 21.6 These parabolic glued laminated wood arches provide a lofty atmosphere and a beautiful ceiling. (*Courtesy Canadian Wood Council*)

Typical construction details for arches are shown in Figs. 21.7 through 21.10. In Fig. 21.7 are details for connecting parabolic arches and domes to a buttress foundation. The shoe plate is anchored to the foundation and bolted to the metal connection on the end of the arch. Connections for wood three-hinged arches are shown in Fig. 21.8. This shows several ways to anchor the arch to steel and concrete bases. Several types of crown connections are shown in Fig. 21.9. Long-span arches may require a section be added on the site because the assembled arch would be too large to transport. One connection used is shown in Fig. 21.10.

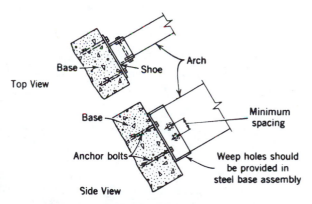

TRUE HINGE ANCHORAGE FOR ARCHES.

ARCH ANCHORAGE WHERE TRUE HINGE IS NOT REQUIRED.

Figure 21.7 Dome-type arches are anchored to concrete foundations with steel anchor plates secured to them. (*Courtesy American Institute of Timber Construction, 7012 S. Revere Parkway, Suite 140, Englewood, Colo. 80112*)

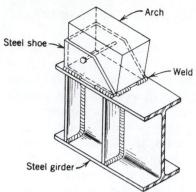

ARCH ANCHORAGE TO STEEL GIRDER. Vertical uplift load and thrust are taken through the weld.

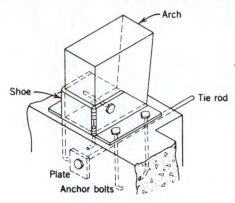

TIE ROD IN CONCRETE. Thrust is taken by anchor bolts in shear into the concrete foundation and tie rod.

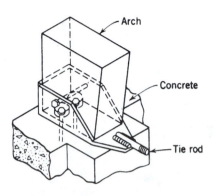

TIE ROD TO ARCH SHOE. Thrust due to vertical load is taken directly by the tie rod welded to the arch shoe.

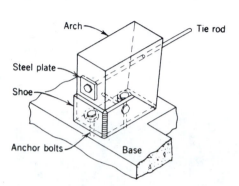

TIE ROD TO ARCH. Thrust due to vertical load is taken directly by the tie rod.

Figure 21.8 Base connections for glued laminated arches use some type of steel shoe. Notice the tie rods shown. *(Courtesy American Institute of Timber Construction, 7012 S. Revere Parkway, Suite 140, Englewood, Colo. 80112)*

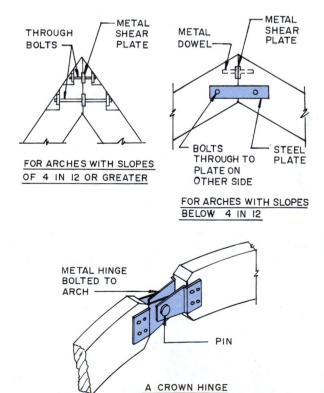

Figure 21.9 Typical connections for the crown of arches.

Glued laminated wood domes are designed as a radial arch or a triangulated system (Fig. 21.11). The triangulated arch can span greater distances than the radial arch. The one shown in Fig. 21.12 spans an entire football field. Another type of dome construction, shown in Fig. 21.13, produces a dramatic curved ceiling. Arched laminated wood structural members are also used for a variety of other projects, such as the pedestrian bridge over a freeway shown in Fig. 21.14.

Columns and Beams

Several ways to secure glued laminated beams to foundations are shown in Fig. 21.15. The heavy steel clips and bolts resist vertical and horizontal forces. It is recommended that the beam rest on a metal plate. Column-to-foundation connections are shown in Fig. 21.16. The column should rest on a metal plate and be at least 3 in. (76 mm) above grade or the finished floor. Figure 21.17 shows steel hold-down anchors used to meet structural tie-down and tilt-over requirements specified by the building codes. Steel connectors are used to secure beams and girders to the columns (Fig. 21.18). The design engineer must indicate the type of fastener to be used and the size and number of bolts. A connection for a large laminated girder and column is shown in Fig. 21.19. Several beam-to-beam and beam-to-girder connections are shown in Fig. 21.20.

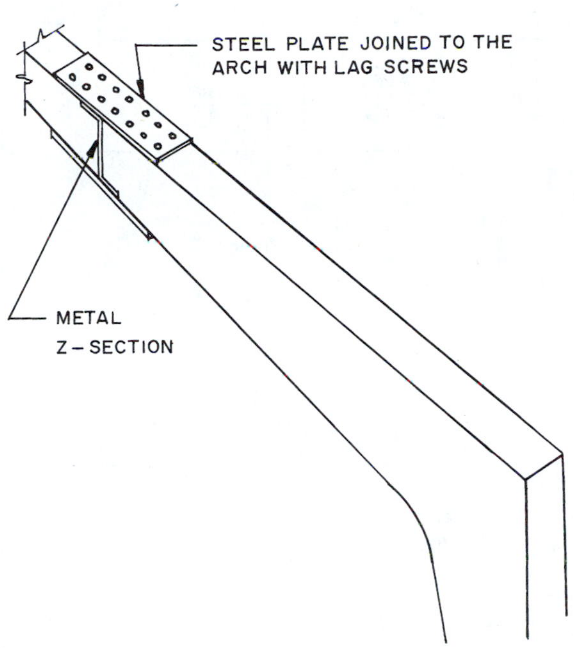

STEEL PLATE JOINED TO THE
ARCH WITH LAG SCREWS

METAL
Z – SECTION

Figure 21.10 Sections may be spliced on long-span glued laminated arches.

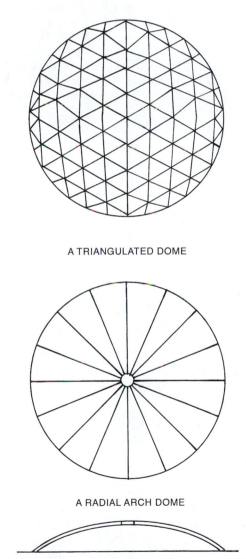

A TRIANGULATED DOME

A RADIAL ARCH DOME

Figure 21.11 Glued laminated wood domes use either radial arch or triangulated construction.

I T E M   O F   I N T E R E S T

TERMINOLOGY FOR METAL FASTENERS USED TO JOIN WOOD MEMBERS

There are a number of ways and types of metal fasteners used to join wood members. These range from the nails and staples used for light frame construction to bolts, side plates, and other types of hardware. The effectiveness of metal fasteners depends on their being large enough to carry the loads and transfer them over a large area so the wood fiber that is in contact with the fastener is not deformed. The spacing between metal fasteners and between the ends and edges of the wood member and the fasteners is critical to a successful union (Fig. A).

Metal fasteners are subject to various types of loads, as shown in Fig. B. The architect and engineer design connections so the calculated loads are sustained by the connectors and their placement in the wood members.

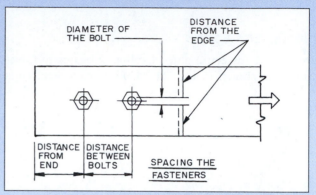

Figure A

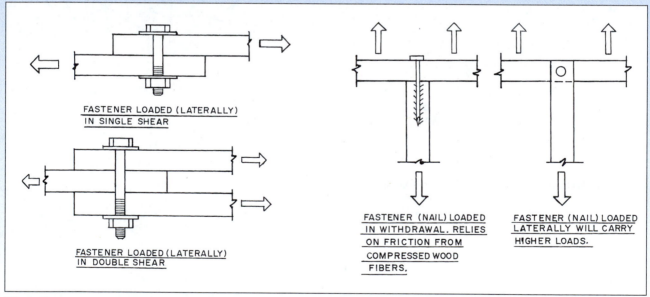

Figure B

Figure 21.12 This triangulated glued laminated wood dome spans a large area. *(Courtesy American Institute of Timber Construction, 7012 S. Revere Parkway, Suite 140, Englewood, Colo. 80112)*

Figure 21.13 These glued laminated wood domes produce a dynamic architectural feature. *(Courtesy Canadian Wood Council)*

Figure 21.14 Glued laminated arches were used to construct this bridge. *(Courtesy American Institute of Timber Construction, 7012 S. Revere Parkway, Suite 140, Englewood, Colo. 80112)*

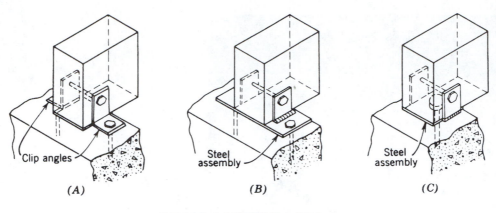

BEAM ANCHORAGES

Figure 21.15 Typical glued laminated connections used to secure a beam to the foundation. *(Courtesy American Institute of Timber Construction, 7012 S. Revere Parkway, Suite 140, Englewood, Colo. 80112)*

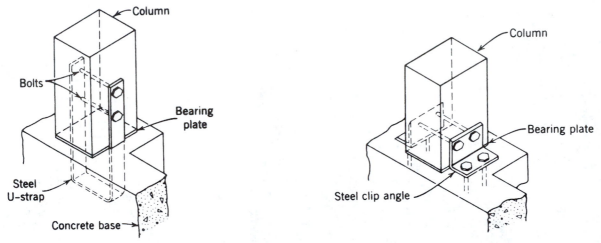

Figure 21.16 Examples of glued laminated column-to-foundation connections. *(Courtesy American Institute of Timber Construction, 7012 S. Revere Parkway, Suite 140, Englewood, Colo. 80112)*

Figure 21.17 Steel hold-downs are used to meet code tilt-over and tie-down requirements. *(Courtesy Canadian Wood Council)*

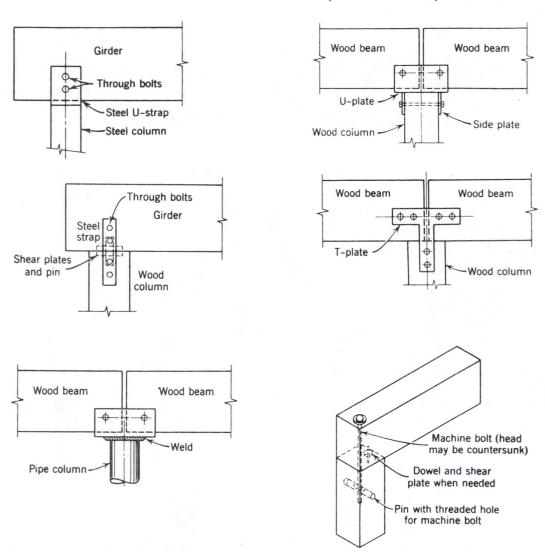

Figure 21.18 Typical beam-to-column and girder-to-column connections. *(Courtesy American Institute of Timber Construction, 7012 S. Revere Parkway, Suite 140, Englewood, Colo. 80112)*

Figure 21.19 A heavy steel column-to-girder connection joins two girders butting over the column. *(Courtesy Canadian Wood Council)*

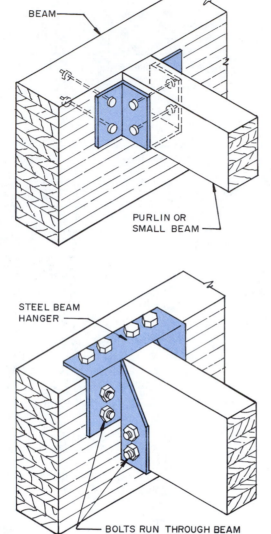

BEAM

PURLIN OR
SMALL BEAM

STEEL BEAM
HANGER

BOLTS RUN THROUGH BEAM

Figure 21.20 Typical beam-to-girder connections.

Decking

Decking systems include laminated and solid wood decking, stressed skin panels, and $1\frac{1}{8}$ in. (29 mm) plywood panels, as shown in Fig. 21.21. Typical glulam assemblies are shown in Fig. 21.22. The sizes of laminated decking vary depending on the species of wood used. Typical actual sizes include $2\frac{7}{8} \times 5\frac{3}{8}$ in. and $3 \times 7\frac{1}{8}$ in. Solid wood sizes include $1\frac{1}{2} \times 5\frac{1}{4}$ in., $2\frac{1}{2} \times 5\frac{1}{4}$ in., and $3\frac{1}{2} \times 5\frac{1}{4}$ in. (Fig. 21.23).

Several ways for installing laminated and solid wood decking are shown in Fig. 21.24. Notice that in the single span and continuous two span the end joints meet on a beam or purlin. Manufacturer's directions should be followed when installing decking. The decking forming

the roof and ceiling of the church shown in Fig. 21.25 is the major design feature of the building. Design details for decking on arched, flat, and sloped roofs are shown in Figs. 21.26 and 21.27.

SOLID HEAVY TIMBER CONSTRUCTION

A wide range of techniques can be used when constructing heavy timber buildings. In Fig. 21.28 is a building using specially designed roof truss construction supported by columns. Some of the connections are made by mortising them to receive a tenon. This type of construction makes it possible to erect a building with minimum intrusion on the surroundings, as shown in Fig. 21.29.

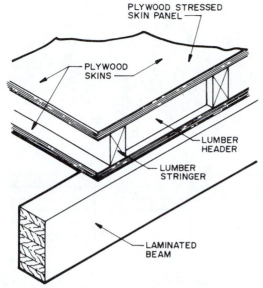

STRESSED SKIN PANELS ON LAMINATED BEAM SYSTEM.

Stressed skin panels, which have a practical span range of 32 ft, are fastened directly to the main laminated timber beams by lag screws or gutter spikes.

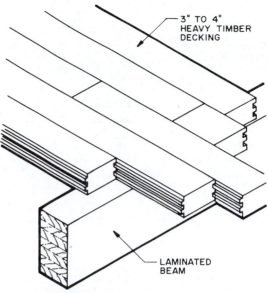

HEAVY TIMBER DECKING ON A LAMINATED BEAM SYSTEM.

Heavy timber decking either laminated or solid 3 or 4 in. nominal thickness is nailed directly to the main laminated beams. The economical span range for the heavy timber decking is 8 to 20 ft depending upon the thickness and loading conditions.

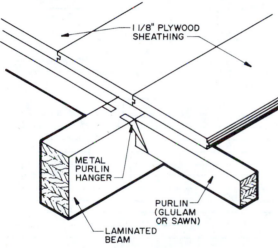

ONE AND ONE-EIGHTH-INCH PLYWOOD ON A LAMINATED BEAM AND PURLIN SYSTEM.

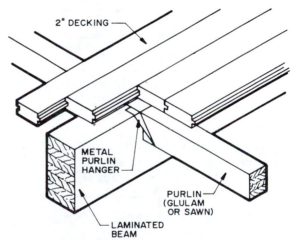

TWO-INCH DECKING ON A LAMINATED BEAM AND PURLIN SYSTEM.

Two-inch nominal thickness decking with an economical span range of 6 to 12 ft is nailed directly to glulam or sawed wood roof purlins, typically on 8 ft centers. Purlins are connected to the main laminated timber beams by metal purlin hangers.

Figure 21.21 Types of decking used with glued laminated framing. *(Courtesy American Institute of Timber Construction, 7012 S. Revere Parkway, Englewood, Colo. 80112)*

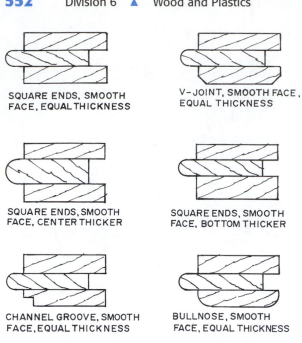

SQUARE ENDS, SMOOTH
FACE, EQUAL THICKNESS

V-JOINT, SMOOTH FACE,
EQUAL THICKNESS

SQUARE ENDS, SMOOTH
FACE, CENTER THICKER

SQUARE ENDS, SMOOTH
FACE, BOTTOM THICKER

CHANNEL GROOVE, SMOOTH
FACE, EQUAL THICKNESS

BULLNOSE, SMOOTH
FACE, EQUAL THICKNESS

GROOVED FACE, EQUAL
THICKNESS

SQUARE ENDS, SMOOTH
FACE, DOUBLE TONGUE
AND GROOVE, EQUAL
THICKNESS

Figure 21.22 Examples of the types of glued laminated decking available.

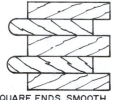

ACTUAL SIZES OF
DOUGLAS FIR, LARCH,
AND SOUTHERN PINE
GLUED LAMINATED
DECKING

2 3/16" (55.5 mm)
2 7/8" (73 mm)
3 21/32" (93 mm)
5 3/8" (136.5 mm)
EXPOSED FACE

1 1/2" (38 mm)
5" (127 mm)
EXPOSED FACE

2 1/2" (64 mm)
3 1/2" (89 mm)
5 1/4" (133 mm)
EXPOSED FACE

ACTUAL SIZES OF DOUGLAS FIR AND LARCH SOLID
DECKING.

NOTE: EXPOSED FACE MAY BE SMOOTH, BRUSHED,
GROOVED, OR STRIATED.

Figure 21.23 Sizes of glue laminated wood decking made from Douglas fir, larch, and southern pine.

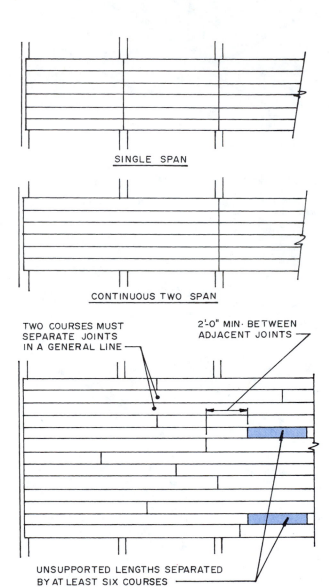

SINGLE SPAN

CONTINUOUS TWO SPAN

TWO COURSES MUST
SEPARATE JOINTS
IN A GENERAL LINE

2'-0" MIN. BETWEEN
ADJACENT JOINTS

UNSUPPORTED LENGTHS SEPARATED
BY AT LEAST SIX COURSES

CONTROLLED RANDOM PATTERN

Figure 21.24 Typical ways glued laminated wood decking may be installed.

Figure 21.25 Wood decking that is left exposed can form an attractive ceiling. *(Courtesy American Institute of Timber Construction, 7012 S. Revere Parkway, Englewood, Colo. 80112)*

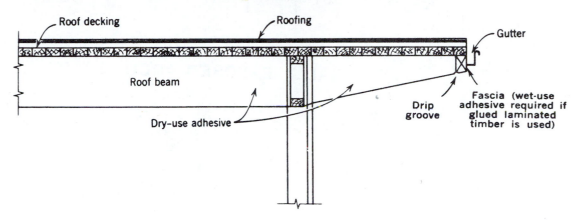

Roof decking

Roofing

Gutter

Roof beam

Dry-use adhesive

Drip groove

Fascia (wet-use adhesive required if glued laminated timber is used)

BUILDING WITH COVERED OVERHANG.

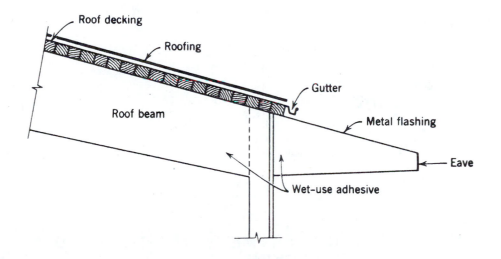

Roof decking

Roofing

Gutter

Roof beam

Metal flashing

Eave

Wet-use adhesive

BUILDING WITH UNCOVERED OVERHANG.

Figure 21.26 Details for using glued laminated beams and decking on sloped and flat roofs. *(Courtesy American Institute of Timber Construction, 7012 S. Revere Parkway, Englewood, Colo. 80112)*

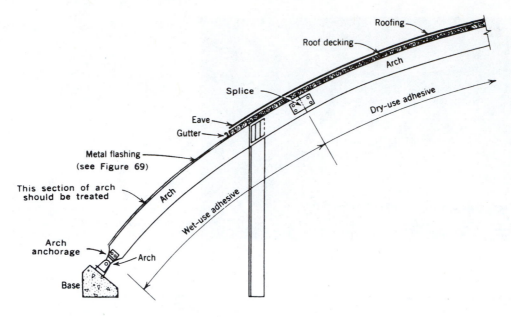

PARTIALLY EXPOSED ARCH.

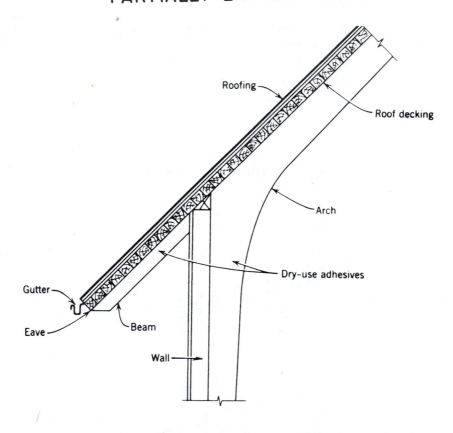

RIGID FRAME WITH PROTECTED OUTLOOKER.

Figure 21.27 Details for installing glued laminated wood decking on arches. *(Courtesy American Institute of Timber Construction, 7012 S. Revere Parkway, Englewood, Colo. 80112)*

Figure 21.28 This heavy timber framing uses mortised connections, producing an exposed interior framing free from metal connectors. *(Courtesy The Beamery)*

Figure 21.29 Heavy timber framing can be erected with a minimum of disturbance of the surrounding area. *(Courtesy The Beamery)*

Figure 21.30 This building is framed with solid wood columns, beams, and rafters. *(Courtesy Canadian Wood Council)*

Solid heavy timber construction is used for commercial and religious buildings. The details are much the same as used for glued laminated construction. The structural framing in Fig. 21.30 shows typical beam and column construction and a heavy timber frame roof. Figure 21.31 shows details for setting beams on a masonry wall. The beam is set into a steel pocket anchored into the masonry bearing wall. The columns are set in the same manner as laminated columns. Beam-to-column connections may be made with metal connectors or wood bearing blocks bolted to the column with split rings to control vertical forces (Fig. 21.32). Beam-to-girder or beam- or purlin-to-beam connections generally use some type of metal connector. However, wood ledgers with metal lateral ties are used in some cases, as shown in Fig. 21.33. Notice the wood decking shown in this illustration. In Fig. 21.34 a beam has been installed between two wood columns forming a spaced column. A carefully engineered connection using metal connectors and wood angled supports bolted to the connector is shown in Fig. 21.35. The work of the structural engineer is critical when designing structural systems to carry the imposed loads.

Heavy timber trusses are used to span the width of a building and support purlins that carry the roof decking. The bolted, multimember truss in Fig. 21.36 is supported on corbelled masonry pilasters that are an integral part of the masonry wall. The bolted connections have split rings.

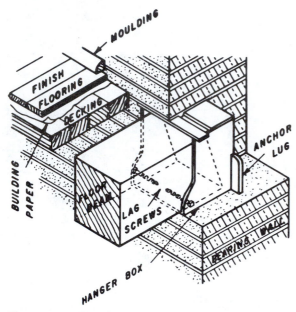

Figure 21.31 Wood beams are set in metal anchors that are tied to the foundation or a masonry wall. *(Courtesy American Forest and Paper Association (formerly NFPA), Washington, D.C.)*

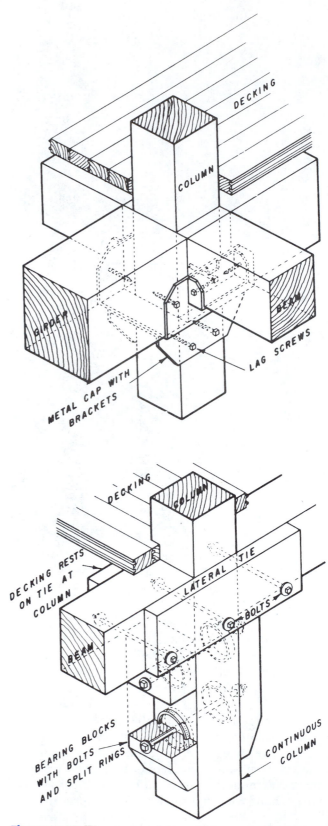

Figure 21.32 Ways to tie solid wood beams to columns. *(Courtesy American Forest and Paper Association (formerly NFPA), Washington, D.C.)*

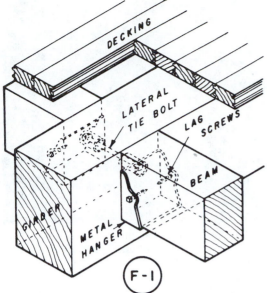

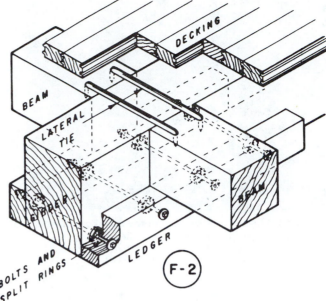

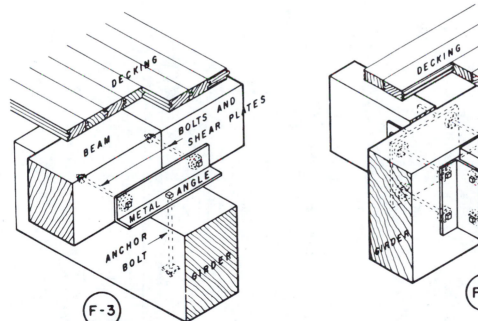

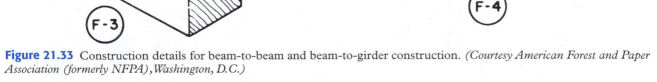

Figure 21.33 Construction details for beam-to-beam and beam-to-girder construction. *(Courtesy American Forest and Paper Association (formerly NFPA), Washington, D.C.)*

Figure 21.34 This beam is secured to the members making up a spaced column. *(Courtesy Canadian Wood Council)*

Figure 21.35 This connection has angled wood members tied to overhead beams, providing additional support for the lower beam. *(Courtesy Canadian Wood Council)*

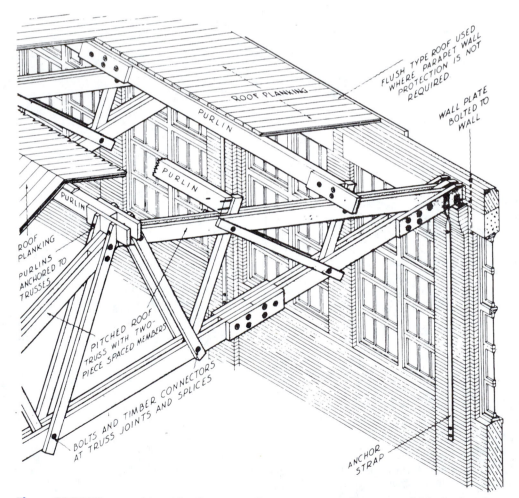

Figure 21.36 These multimember heavy wood trusses carry purlins upon which the wood decking is secured. *(Courtesy American Forest and Paper Association (formerly NFPA), Washington, D.C.)*

Connectors

Two connectors used to provide additional strength to a bolted joint are **split ring connectors** and **shear plate connectors.**

Split ring connectors are ring-shaped metal inserts placed in grooves cut into mating wood pieces that are to be secured with bolts (Fig. 21.37). They are used on wood-to-wood connections to distribute lateral forces over a larger bearing area than the bolts alone provide. They greatly increase the load-bearing capacity of the joint (Fig. 21.38). They are available in $2\frac{1}{2}$ in. (60 mm) and 4 in. (102 mm) diameters. The $2\frac{1}{2}$ in. split ring is used with $\frac{1}{2}$ in. (12 mm) bolts and the 4 in. split ring requires a $\frac{3}{4}$ in. (19 mm) bolt. They are also used when lag screws are used instead of bolts.

Another type of connector used in heavy timber construction is a shear plate (Fig. 21.39). Shear plates are used for metal-to-wood connections. The example in Fig. 21.39 shows a shear plate and bolts securing a steel plate forming a connection between two wood members. Shear plates are available in $2\frac{5}{8}$ in. (52 mm) and 4 in. (102 mm) diameters. A $\frac{3}{4}$ in. (19 mm) bolt is used with the $2\frac{5}{8}$ in. shear plate and a $\frac{3}{4}$ in. or $\frac{7}{8}$ in. (22 mm) bolt is used with the 4 in. shear plate. They also are used when lag screws are used instead of bolts.

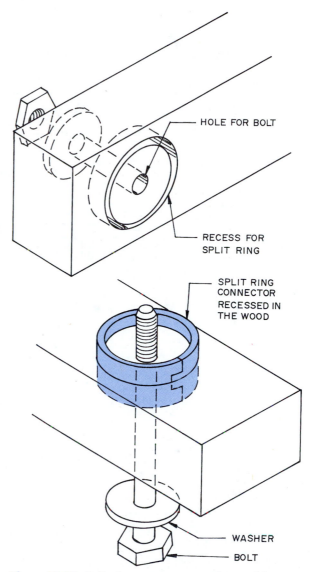

Figure 21.38 Split ring connectors are inserted in recesses cut in the joining members.

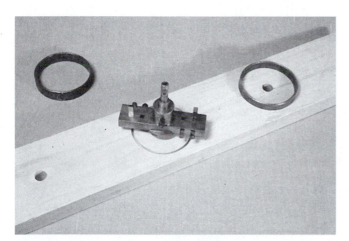

Figure 21.37 A hole is bored for a bolt and a circular recess is cut by the tool shown in the center. The installed split ring is shown on the right. *(Courtesy Canadian Wood Council)*

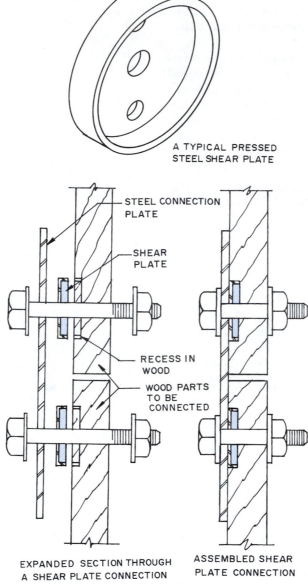

A TYPICAL PRESSED
STEEL SHEAR PLATE

STEEL CONNECTION
PLATE

SHEAR
PLATE

RECESS IN
WOOD

WOOD PARTS
TO BE
CONNECTED

EXPANDED SECTION THROUGH
A SHEAR PLATE CONNECTION

ASSEMBLED SHEAR
PLATE CONNECTION

Figure 21.39 Shear plates distribute lateral forces over a
larger bearing area in metal-to-wood connections.

Timber joinery is another method of connecting
heavy timber structural members. It does not use metal
connectors but uses machined interlocking wood joints
cut into the wood members. These are much like the
joints used in furniture construction. Some of the connections used are shown in Fig. 21.40. Although these
connections take longer to make and assemble than using metal connections, the finished job shows the natural beauty of the wood with no metal plates or bolts
intruding.

POLE CONSTRUCTION

Pole construction uses round wood poles for the vertical load-carrying members and standard framing for
floors, walls, and roofs. The poles must be pressure
treated with a preservative to inhibit rot. A typical structural frame is shown in Fig. 21.41. The poles carry the
floor and roof loads. Generally the poles are set in holes
in the ground that are filled with concrete. The simplest
connection is to use bolts through the dimension lumber and the post. The post could be notched to provide
a flat surface and a seat for the joist (Fig. 21.42).

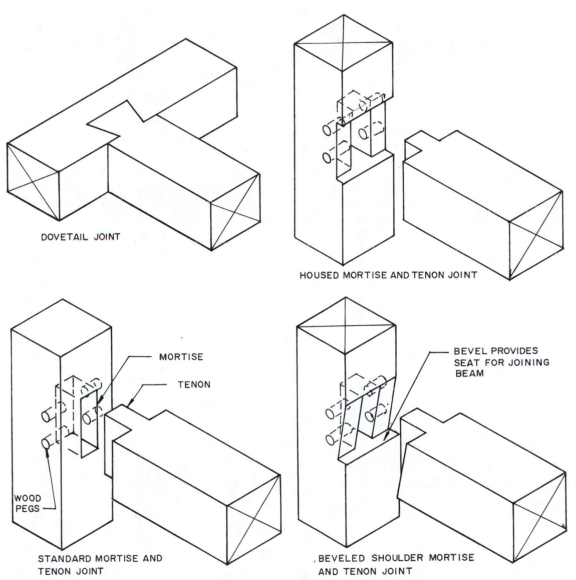

DOVETAIL JOINT

HOUSED MORTISE AND TENON JOINT

MORTISE

TENON

WOOD PEGS

STANDARD MORTISE AND TENON JOINT

BEVEL PROVIDES SEAT FOR JOINING BEAM

BEVELED SHOULDER MORTISE AND TENON JOINT

Figure 21.40 Typical joints used in joining solid wood framing members. These are typical of construction used in years before metal connectors were developed.

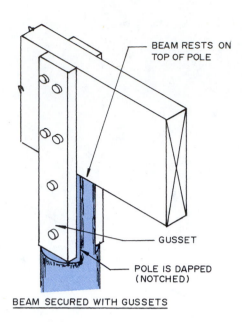

BEAM RESTS ON TOP OF POLE

GUSSET

POLE IS DAPPED (NOTCHED)

BEAM SECURED WITH GUSSETS

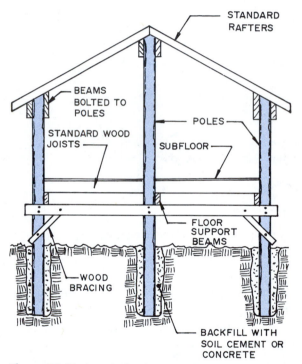

STANDARD RAFTERS

BEAMS BOLTED TO POLES

POLES

STANDARD WOOD JOISTS

SUBFLOOR

FLOOR SUPPORT BEAMS

WOOD BRACING

BACKFILL WITH SOIL CEMENT OR CONCRETE

Figure 21.41 A typical pole construction building.

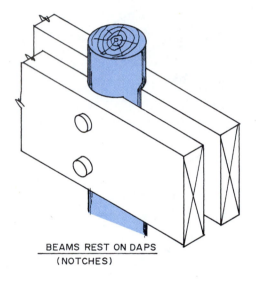

BEAMS REST ON DAPS (NOTCHES)

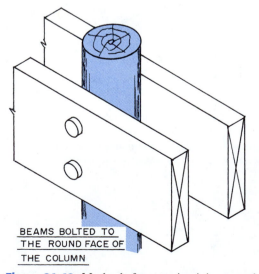

BEAMS BOLTED TO THE ROUND FACE OF THE COLUMN

Figure 21.42 Methods for securing joists to poles.

REVIEW QUESTIONS

1. What are the types of wood products used to frame heavy timber buildings?
2. What is the difference in reaction to fire between timber structural systems and steel systems?
3. What is meant by a hinged arch?
4. What two types of design are used for glued laminated dome construction?
5. Why do codes require steel hold downs?
6. What are the types of span details used with glued laminated wood decking?
7. Why are split ring connectors used?
8. When are shear plates used?

KEY TERMS

glued laminated members Wood members made by bonding together laminations of dimension lumber.

heavy timber construction A type of construction using wood columns, beams, girders, purlins, and decking.

hinge joint A joint that permits some action similar to a hinge but in which there is no appreciable separation of the joining members.

pole construction Construction using large-diameter log poles in a vertical position to carry the loads of the floors and roof.

shear plate connector A circular metal connector recessed into the wood member that is to be bolted to a steel member.

split ring connector A ring-shaped metal insert placed in circular recesses cut in joining wood members that are held together with a bolt or lag screw.

timber joinery The joining of structural wood members using wood joints such as the dovetail and mortise and tenon.

SUGGESTED ACTIVITIES

1. Build a scale model of the structural frame of a heavy timber building.

2. Make sketches of the various types of metal connectors.

3. Copy from the local building code the requirements that apply specifically to heavy timber construction.

ADDITIONAL INFORMATION

Merritt, F.S., and Ricketts, J.T., *Building Design and Construction Handbook*, McGraw-Hill, New York, 1994.

Newman, M., *Structural Detailing for Wood Construction*, McGraw-Hill, New York, 1988.

Spence, W.P., *Residential Framing*, Sterling Publishing Co., New York, 1993.

Timber Construction Manual, American Institute of Timber Construction, John Wiley and Sons, New York, 1985.

Wood Building Technology, Canadian Wood Council, Ottawa, 1993.

Wood Reference Handbook, Canadian Wood Council, Ottawa, 1991.

CHAPTER 22

Finishing the Exterior and Interior of Light Wood Frame Buildings

This chapter will help you to:

1. Have a complete understanding of the many materials and design possibilities for finishing the exterior of light wood frame buildings.

2. Be familiar with the many things required to finish the interior of light wood frame buildings and some of the installation procedures.

After the building is framed and sheathed, the exterior finish work can begin. The contractor tries to close up the building from the weather as rapidly as possible so that interior work may begin. After the exterior is complete, the drive and landscape work can proceed, although some contractors delay this until the interior work is complete because it could sustain considerable damage from the trades doing the interior work.

FINISHING THE EXTERIOR

The exterior finish work includes installing the windows and exterior doors, finishing the eave, shingling the roof, installing the siding and gutters, and doing the exterior painting. With the installation of the windows and finishing the roof, the building is weathertight.

Framing the Eaves and Rake

The **eaves** and **rake** are framed after the roof is sheathed. There are many ways these can be constructed. The carpenter will find the required design on the working drawings. The architect designs the eave and rake to suit the architectural style of the house. Typical designs for sloped roofs are in Figs. 22.1 through 22.3 and a flat roof design is in Fig. 22.4. The rough fascia is $1\frac{1}{2}$ in. (38 mm) thick. It provides spacing support for the ends of the rafters and a means of securing gutters and the finished fascia. The finished fascia can be painted wood or wood covered with aluminum or vinyl. The **soffit** can be plywood, hardboard, or another exterior reconstituted wood product, vinyl, or aluminum. Vinyl and aluminum soffits are perforated, providing a flow of air into the attic (Fig. 22.5). Solid soffits require that metal vent strips be installed (Fig. 22.6).

The rakes are finished similar to the eave. The rake fascia may be set flush over the siding or extend from the gable end as shown in Figs. 22.7 and 22.8.

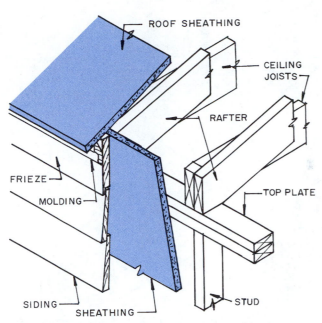

Figure 22.1 This is a close cornice that provides no overhang and uses a frieze board at the intersection of the roof and wall sheathing.

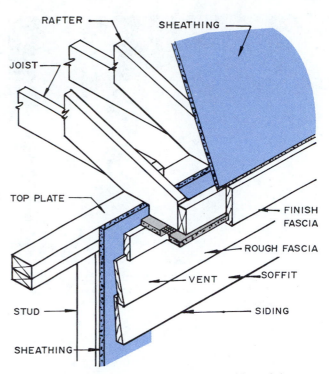

Figure 22.3 This narrow box cornice provides minimum overhang and a vent strip.

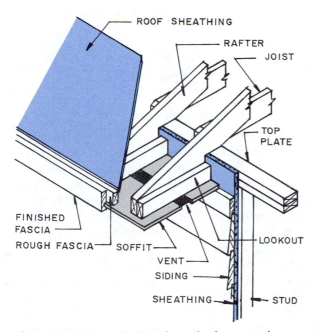

Figure 22.2 This wide boxed cornice has a continuous attic vent strip.

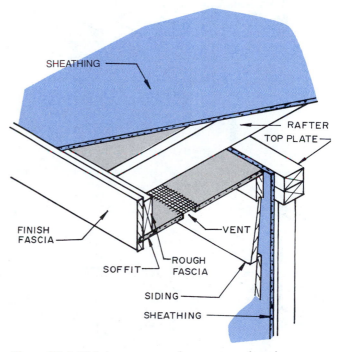

Figure 22.4 This is one way to frame an overhanging cornice on a flat roof.

Figure 22.5 This vinyl soffit is perforated to provide attic ventilation. The fascia is also vinyl clad. *(Courtesy of Alcan Building Products)*

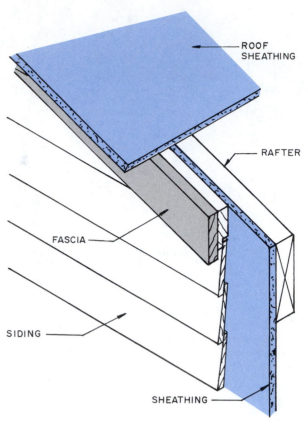

ROOF SHEATHING

RAFTER

FASCIA

SIDING

SHEATHING

Figure 22.7 This framing is for a close fascia on the rake.

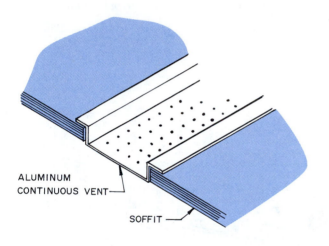

ALUMINUM CONTINUOUS VENT

SOFFIT

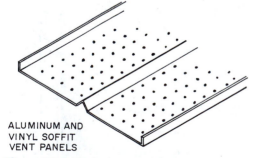

ALUMINUM AND VINYL SOFFIT VENT PANELS

Figure 22.6 Typical types of soffit vents.

Ventilating the Attic

Attic ventilation is required to remove high-temperature air in the summer and the accumulation of moisture in summer and winter. Moisture in the attic in the winter will freeze on the rafters and sheathing. As the air temperature gets above freezing, the moisture will melt and drip on the ceiling insulation and damage it. Attic ventilation is also required to keep the air temperature in the winter about the same as the outside air. If the ceiling is poorly insulated or ventilation is inadequate, as the air temperature rises and causes snow on the roof to melt, water will run to the uninsulated eave and freeze again. This causes an ice dam on the eave, and moisture will back up under the shingles and possibly leak through into the exterior wall and on the ceiling insulation. In cold climates extra layers of builder's felt are laid from the eave up the roof to provide additional waterproofing (Fig. 22.9). To get adequate ventilation, the soffit must have some form of vent allowing air to flow up into the attic. (Refer to Figs. 22.2, 22.3, and 22.4.)

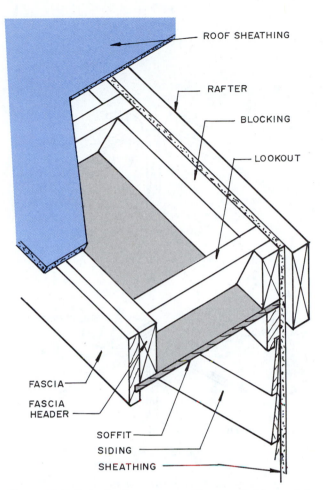

Figure 22.8 Rake fascias are generally extended beyond the end wall and have a soffit returning to the wall.

It is necessary to install some type of baffle at the exterior wall so the ceiling insulation does not block the flow of air from the soffit into the attic (Fig. 22.10). Vents near the ridge are used to allow the flow of air to exit the attic. Power fans, ridge vents, and gable end vents are commonly used (Figs. 22.11 and 22.12). Typical venting details for flat roofs are in Fig. 22.13. They use soffit vents and require that a space be left between the sheathing and insulation.

The Finished Roof Material

After the roof sheathing is covered with builder's felt, the finished roofing may be installed. These materials are covered in Chapters 27 and 28. A commonly used roofing material is composed of a fiberglass base that is saturated with asphalt and has mineral granules embedded in the surface to form a protective coating (Fig. 22.14). Wood shingles and shakes (Fig. 22.15), clay, perlite, and concrete tile (Fig. 22.16), and metal roofing materials are also used. Flat roofs have a built-up or single-ply sheet membrane (Fig. 22.17).

Shingles are usually installed by a roofing contractor. The crews are trained to properly install flashing and the shingles. It is important they be installed according to the manufacturer's recommendations. The fire resistance of roofing materials is an important consideration, and they are classified by their ability to resist exposure to fire.

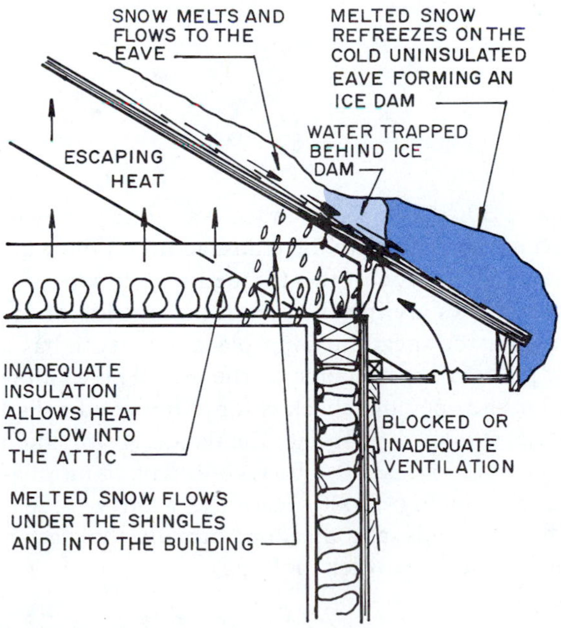

Figure 22.9 Ice dams are formed when melted snow flows over the cold eave and refreezes.

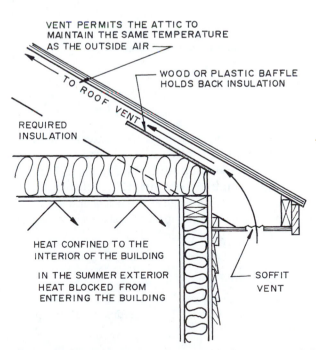

Figure 22.10 Baffles are used to provide for passage of air at the eave into the attic.

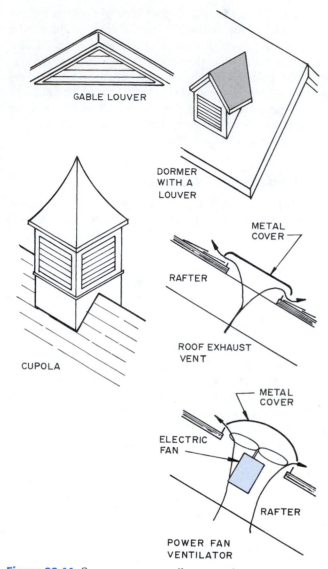

GABLE LOUVER

DORMER WITH A LOUVER

CUPOLA

METAL COVER

RAFTER

ROOF EXHAUST VENT

METAL COVER

ELECTRIC FAN

RAFTER

POWER FAN VENTILATOR

Figure 22.11 Some ways to ventilate an attic.

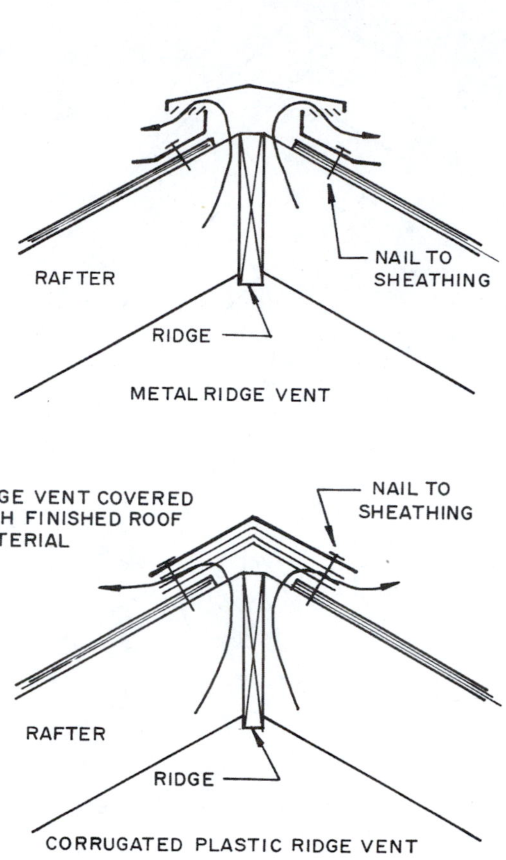

RAFTER

NAIL TO SHEATHING

RIDGE

METAL RIDGE VENT

RIDGE VENT COVERED WITH FINISHED ROOF MATERIAL

NAIL TO SHEATHING

RAFTER

RIDGE

CORRUGATED PLASTIC RIDGE VENT

Figure 22.12 Ridge vents are widely used for attic ventilation.

After the shingles are in place the gutters may be installed, although this is often delayed until the exterior is finished to keep them from being damaged. The gutters are either aluminum or vinyl. The roof shingles extend about 1 in. over the edge of the fascia and direct the water into the gutter. The gutter also serves as a decorative finish on the fascia (Fig. 22.18).

Installing Windows

Many types and designs of windows are used in wood light frame construction. These are described in Chapters 19 and 30. The recommended way to install windows varies somewhat with the manufacturer. Following is a general description of two frequently used types.

Most windows are installed from the outside of the building after the sheathing is in place. One type has a metal or plastic flange secured to the unit. The window is placed in the opening, checked for plumb, and nailed to the studs and header through the flange. A simplified illustration is in Fig. 22.19. The second type of installation is shown in Fig. 22.20. Here the window unit is nailed to the studs and header through the casing (sometimes called the brick molding).

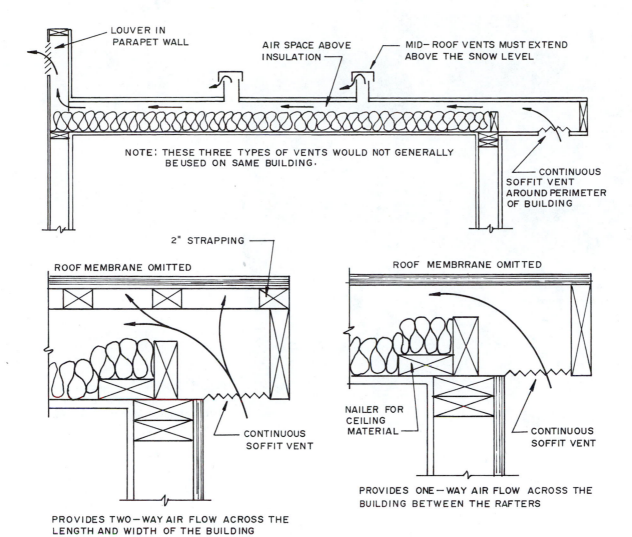

Figure 22.13 Techniques for ventilating flat roofs.

Figure 22.14 Fiberglass asphalt shingles are widely used as a finish roofing material.

Figure 22.15 Wood shingles and shakes produce a rustic roof with attractive shadow lines. *(Courtesy Cedar Shake and Shingle Bureau)*

Figure 22.16 These fire-resistant roofing tiles are made from concrete. *(Courtesy Monier, Orange, Calif.)*

Figure 22.17 Flat roofs use some form of built-up membrane using builder's felt, tar, and gravel. *(Courtesy National Roofing Contractors Association)*

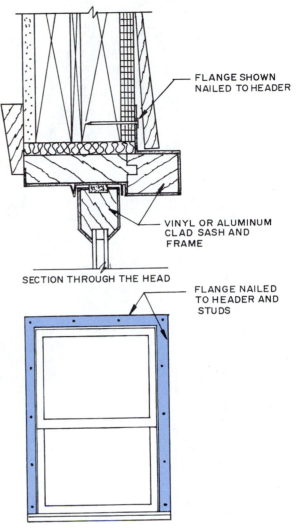

FLANGE SHOWN
NAILED TO HEADER

VINYL OR ALUMINUM
CLAD SASH AND
FRAME

SECTION THROUGH THE HEAD

FLANGE NAILED
TO HEADER AND
STUDS

Figure 22.18 Gutters and downspouts are usually installed after all exterior finish work is complete. *(Courtesy Genova Products, Inc.)*

Figure 22.19 Windows made with flanges on the frame are installed by nailing through the flanges into the sheathing and studs.

The installation procedure begins by checking the opening to see that the sill is level. Usually the sides of the rough opening are covered with builder's felt or plastic sheeting. Techniques used to install the unit are in Fig. 22.21. The wood, aluminum, or vinyl siding or the brick veneer is installed after the windows are securely in place.

Exterior Door Installation

A variety of exterior doors and frames are available. These are described in Chapters 19 and 30. The doors may be wood, metal, or fiberglass.

The door and frame come in an assembled unit. It is important that the frame be kept square. Typical exterior door frame construction at the sill is shown in Fig. 22.22. One type has the sill resting flat on the subfloor. Another has it resting on the header, and the subfloor fits into a rabbet in the sill. Still another type has a sloped sill that requires that the floor header and any butting joists be cut.

To install an exterior door, cover the edges of the rough opening with builder's felt or sheet plastic. Set the door frame in the opening and check to be certain the sill is level. Shim it level if necessary. Check the sides for plumb. Place wedge-shaped blocking between the studs and the door jamb. Drive one casing nail through the casing into the studs on both sides. Adjust the blocking, keeping the jambs plumb, and nail through the blocking into the studs (Fig. 22.23).

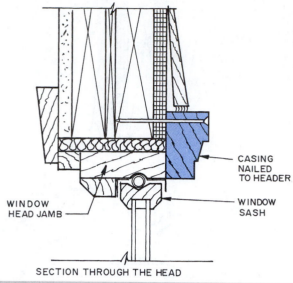

SECTION THROUGH THE HEAD

CASING
NAILED
TO HEADER

WINDOW
HEAD JAMB

WINDOW
SASH

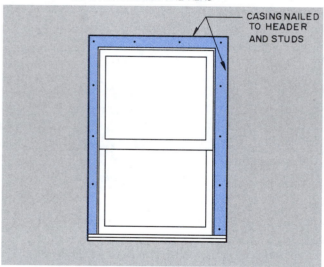

CASING NAILED
TO HEADER
AND STUDS

Figure 22.20 Windows made with wood brick casing attached are installed by nailing through the casing into the sheathing and framing.

Installing the Siding

Exterior siding may be wood, plywood, wood shingles, hardboard, plastic, vinyl, stucco, brick, or stone. The methods for installing wood siding are shown in Figs. 22.24 and 22.25. Notice that the nails in the lower end of the horizontal siding do not penetrate the piece of siding directly below. This allows each piece to expand and contract independently of the other. Hardboard siding is available in lap siding and panel siding. Lap siding installation is shown in Fig. 22.26. Hardboard panel siding is available in 4 × 8 ft. (1220 × 2440 mm)

and 4 × 9 ft. (1220 × 2745 mm) sheets. It typically has batten strips nailed over it.

Plywood siding is available in lap siding and panel siding. It is made with a variety of surface textures. Lap siding installation is shown in Fig. 22.27. Plywood panel siding is available in 4 ft. (1220 mm) widths and 8, 9, and 10 ft. (2440, 2745, 3050 mm) lengths and in a variety of panel patterns (Fig. 22.28). Sheathing and bracing are not required because the panel provides the required structural strength (Fig. 22.29). Wood products are covered in Chapters 18 and 19.

ITEM OF INTEREST

ENERGY-EFFICIENT LAMPS

In 1995 the Energy Policy Act of 1992 went into effect, banning an array of lamps considered too inefficient to permit continued use. The ban includes many of the standard 4 ft. (12 192 mm) fluorescent tubes, and a number of incandescent reflector and PAR (parabolic aluminized reflector) flood lamps. The legislation requires the packaging of lamps to give the light output in lumens, energy usage in watts, and the expected hours of life of the lamp (see figure). The new labels are not required to give the color temperature or the color rendering index (CRI). Color temperature is an expression of the visual warmth or coolness of a light source. It is expressed in degrees Kelvin (°K). The lower the temperature the "warmer" the light will appear visually. The color rendering index pertains to the rendition of color as it occurs in areas where the color of the light is very important, such as lighting works in an art museum. Here lights having high CRI continuous-spectrum sources are used.

Lamp manufacturers have various codes used to describe the lamps. Typically this includes the size and shape of the lamp, the wattage, and other features, such as special materials and color temperature. Following are several examples.

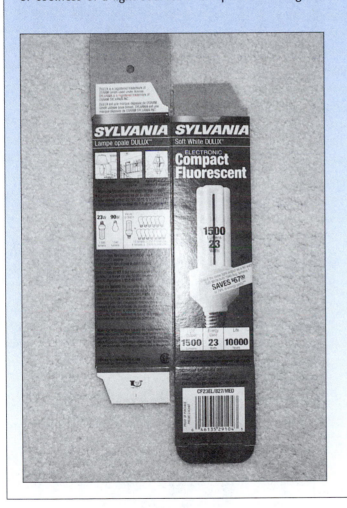

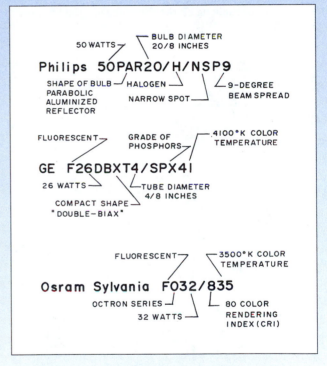

1. Lay a bead of caulking on the sheathing around the window opening.

2. Set the window unit in the opening.

3. Check for levelness and plumb. Then nail through one corner. Recheck for levelness and plumb and nail the other corner.

4. Check the diagonals to be certain the unit is square before nailing the rest of the flange. Also check to see that the windows open easily. Then nail the flange on all sides.

Figure 22.21 Procedures for installing windows getting the unit level and plumb before final nailing. *(Courtesy Pella Corporation)*

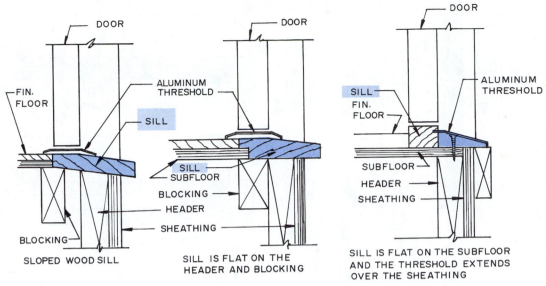

Figure 22.22 Typical exterior door frame construction.

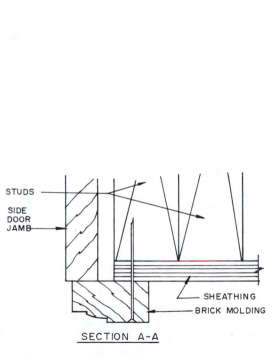

SECTION A-A

Figure 22.23 The exterior door frame is plumbed with wedges and nailed to the studs through the jamb and wedges. The brick casing is nailed through the sheathing into the frame.

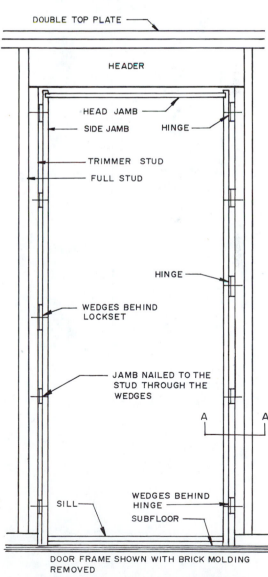

DOOR FRAME SHOWN WITH BRICK MOLDING REMOVED

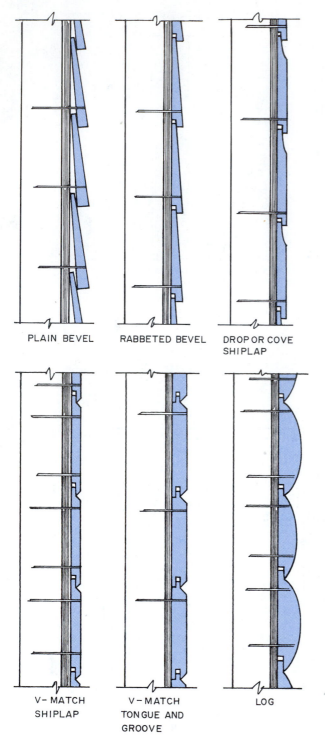

PLAIN BEVEL   RABBETED BEVEL   DROP OR COVE SHIPLAP

V-MATCH SHIPLAP   V-MATCH TONGUE AND GROOVE   LOG

Figure 22.24 Types of horizontal wood siding with nailing patterns.

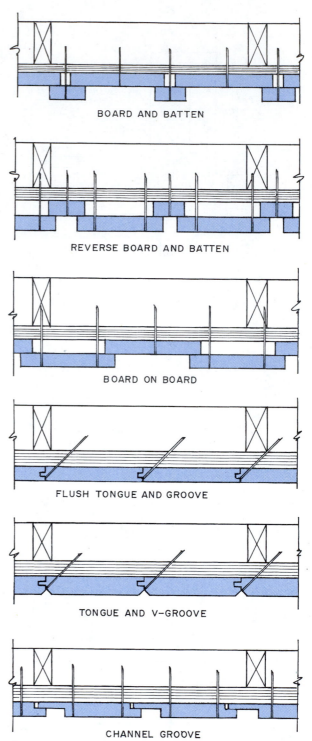

BOARD AND BATTEN

REVERSE BOARD AND BATTEN

BOARD ON BOARD

FLUSH TONGUE AND GROOVE

TONGUE AND V-GROOVE

CHANNEL GROOVE

Figure 22.25 Types of vertical wood siding with nailing patterns.

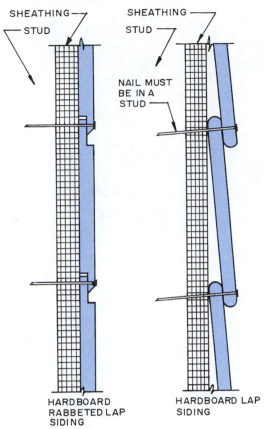

Figure 22.26 Techniques for installing hardboard siding.

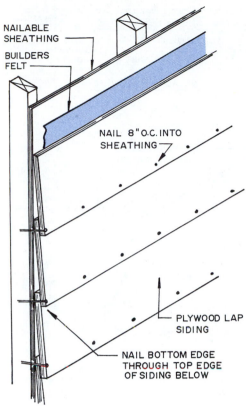

Figure 22.27 Plywood lap siding can be installed over structural or nonstructural sheathing.

Aluminum and vinyl siding are available in a variety of colors and surface textures (Fig. 22.30). They must be nailed to nailable sheathing or directly into the studs because they have no structural strength. Specially designed channels are placed at internal and external corners, and where the siding butts a surface, such as the wood frame of a door or window. A typical installation detail is shown in Fig. 22.31. Wood shingles used as siding are applied over nailable sheathing, such as plywood, and may be installed in a single or double course (Figs. 22.32 and 22.33).

Stucco exterior surfaces may be constructed using a finish coat of a portland cement, lime, and sand mixture (Fig. 22.34). The stucco is troweled over a wire mesh that has been nailed to the sheathing (Fig. 22.35). Another exterior finish similar to stucco is finding increasing use (Fig. 22.36). It is composed of a glass-fiber-reinforced portland cement mixture containing special bonding agents. It is applied over a glass fiber mesh bonded to the sheathing. This system has no means of draining water that may penetrate into the wall assembly from around poorly flashed and caulked windows and other openings. This moisture causes the sheathing and wood wall framing to deteriorate. Figure 22.37 shows a wall assembly that has the same exterior appearance but is flashed and permits water to weep from the wall. This is the USG Corporation Water-Management Finish System.

Brick and stone veneers are also used as exterior siding on light frame construction. They rest on a brick ledge constructed as part of the foundation. Metal ties are nailed to the studs and inserted in the mortar joints between the masonry units. Wall ties are generally spaced every 16 in. (406 mm) vertically (Fig. 22.38). Masonry and stone materials are covered in Chapters 11 through 14.

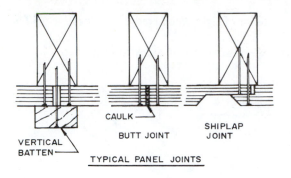

VERTICAL BATTEN

CAULK

BUTT JOINT

SHIPLAP JOINT

TYPICAL PANEL JOINTS

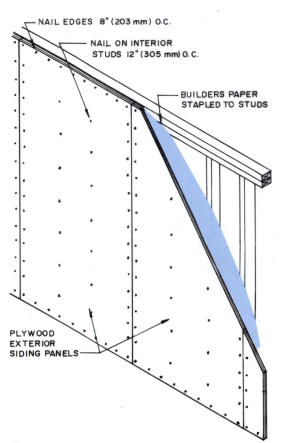

NAIL EDGES 8" (203 mm) O.C.

NAIL ON INTERIOR STUDS 12" (305 mm) O.C.

BUILDERS PAPER STAPLED TO STUDS

PLYWOOD EXTERIOR SIDING PANELS

Figure 22.28 Plywood sheet siding is available in a variety of patterns and can be nailed to the studs without sheathing.

Figure 22.29 Plywood siding panels are power nailed directly to the studs. *(Courtesy APA - The Engineered Wood Association)*

Figure 22.30 Vinyl siding, soffits, and fascia complete this low-maintenance exterior. *(Courtesy Wolverine Technologies)*

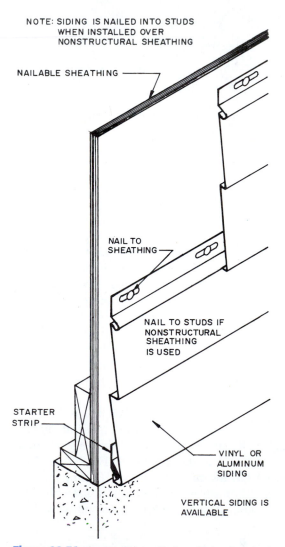

NOTE: SIDING IS NAILED INTO STUDS WHEN INSTALLED OVER NONSTRUCTURAL SHEATHING

NAILABLE SHEATHING

NAIL TO SHEATHING

NAIL TO STUDS IF NONSTRUCTURAL SHEATHING IS USED

STARTER STRIP

VINYL OR ALUMINUM SIDING

VERTICAL SIDING IS AVAILABLE

Figure 22.31 A typical installation design for vinyl and aluminum siding. The connection system will vary depending on the manufacturer of the siding.

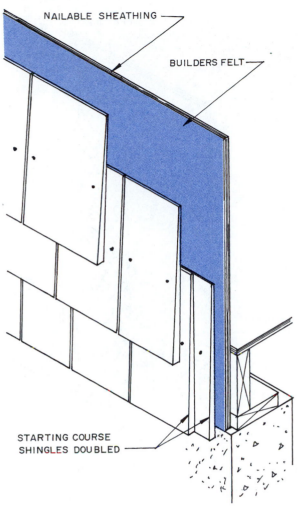

Figure 22.32 These wood shingles are applied in a single course.

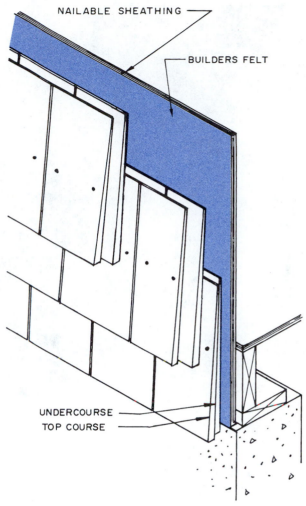

Figure 22.33 Wood shingles and shakes can be used as siding. These are applied in a double course.

Before the siding is installed, the sheathing is usually covered with asphalt-impregnated builder's paper or a watertight, vapor-permeable plastic sheet made from synthetic fibers. This provides a waterproof layer that permits water vapor in the wall cavity to pass through to the outside, thus preventing it from being trapped inside the wall cavity. Some types of sheathing have a cover of perforated aluminum foil that serves the same purpose (Fig. 22.39).

FINISHING THE INTERIOR

After the building is weathertight, the electricians, plumbers, and air conditioning/heating contractor may begin their work. The electricians run the wires and mount the required boxes (Fig. 22.40). The plumbers run the water and waste disposal lines (Fig. 22.41). The heating contractor places the ducts or runs the hot water pipes to the locations of registers. After they have completed this, the local building inspector checks the work for compliance. The electrician installs the light fixtures, switches, and outlets after the interior walls and ceilings are completed. The plumber sets the fixtures, except the tub or shower, which is installed as the interior partitions are built. The heating contractor covers the duct openings until the carpet is installed (Fig. 22.42) or the hardwood floors are finished (Fig. 22.43). Then the ducts are uncovered and vacuumed carefully and the registers are set in place.

Figure 22.34 Stucco coats are troweled over the base. *(Courtesy Georgia-Pacific)*

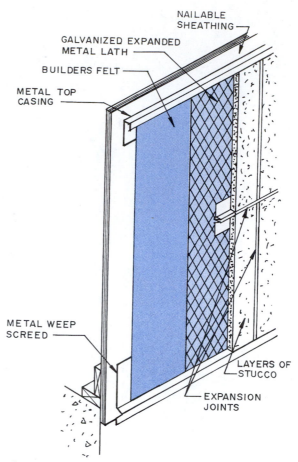

NAILABLE SHEATHING

GALVANIZED EXPANDED METAL LATH

BUILDERS FELT

METAL TOP CASING

METAL WEEP SCREED

LAYERS OF STUCCO

EXPANSION JOINTS

Figure 22.35 Typical construction for a portland cement stucco installation.

Figure 22.36 This building has an exterior finish using a synthetic stucco. *(Courtesy Dryvit Systems, Inc.)*

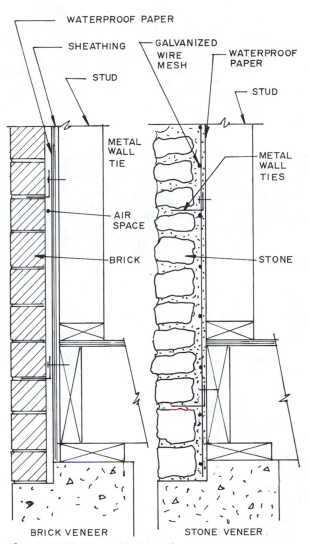

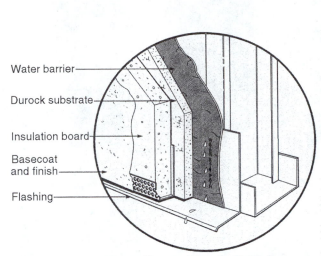

Insulscreen 2100 Water-Managed Exterior Finish System

Figure 22.37 USG's Water-Management Finishing System uses a cement board (Durock) as the substrate for the insulation over which the synthetic stucco is applied. Note the use of flashing and the provision for the wall to weep at the bottom. *(Courtesy USG Corporation)*

Figure 22.38 Typical brick and stone veneer over frame construction.

Figure 22.39 The roof is covered with builder's felt before the shingles are applied, and the wall is covered with a vapor-permeable plastic sheet made from synthetic fibers.

Figure 22.40 These main electric panels are installed after the building is weathertight.

Figure 22.41 The plumbing is installed after the framing carpenters enclose the building.

Figure 22.42 The carpet is installed after the walls are finished, the base is in place, and the painting is complete. *(Courtesy No-Muv Corporation, Inc.)*

Installing Insulation

Thermal insulation and vapor barriers are discussed in Chapter 25. Once the interior is protected from the weather and the electrical and plumbing systems are installed, the insulation can be installed in the walls, floors, and ceilings. Blankets and batts may have a water-resistant paper covering that forms a vapor barrier. If so, it is placed facing the interior of the building. The paper provides tabs used to staple the insulation to the studs (Fig. 22.44). When unfaced blankets are used, the entire wall is covered with plastic sheets, forming a vapor barrier (Fig. 22.45). The blankets should fit snugly against the top and bottom plates. They are cut to fit tightly around plumbing and electrical boxes. Another form of insulation is a rigid batt. It is pressed between the studs and is held in place by friction. A plastic sheet vapor barrier is stapled over these batts.

Insulation is placed wherever the interior is exposed to exterior temperatures. A typical example is in Fig. 22.46. Notice the placement at the floor headers and, in cold climates, on the interior of the foundation of a crawl space or basement. The ceiling can be insulated after the drywall has been installed by laying the blankets on top of the gypsum. Floor insulation is often friction fit. Blankets can be supported with wires placed between the joists (Fig. 22.47). Ceilings are also insulated by spraying in loose fiber insulation (Fig. 22.48).

The insulation properties of a wall can be improved by using a rigid plastic foam insulation-type sheathing. This can be run down over the foundation into the ground (Fig. 22.49). It must be protected by a hard material where it is exposed to the weather. Some sheets are made with a protective coating already applied.

Figure 22.43 These hardwood floors were installed after the interior walls and ceiling were finished and the electrical and plumbing fixtures were set. *(Courtesy Harris-Tarkett, Inc.)*

Figure 22.44 These insulation blankets have a water-resistant paper vapor barrier that must face the side of the wall. *(Courtesy Gold Bond Building Products)*

Figure 22.45 Unfaced insulation blankets are placed between the studs and covered with a plastic sheet. *(Courtesy Gold Bond Building Products)*

Insulation may also be used to reduce the passage of sound through walls, floors, and ceilings. Various types of sound-control batts are available. More information on acoustical treatments is in Chapter 35.

Interior Wall and Ceiling Finishes

Interior walls and ceilings are generally covered with gypsum board after the insulation has been installed. Other finishes include plaster, plywood, and hardboard paneling and solid wood paneling. Gypsum and plaster products are discussed in Chapter 34 and wood products in Chapter 19. Gypsum board and plaster finishes provide a hard, durable surface that can be covered with a wide range of decorative materials. In addition, they provide excellent fire ratings of the areas covered. The installation of these materials is illustrated in Chapter 36. The gypsum panels usually are installed by subcontractors specializing in this work (Fig. 22.50). Plaster finishes require the services of highly skilled plasterers, and these craftsmen are not generally available in some areas (Fig. 22.51). Wood, plywood, and hardboard paneling are usually installed by finish carpenters. Usually wood paneling is applied over a substrate of gypsum wallboard. Masonry partitions are used in many commercial projects but seldom in light frame construction because they require a footing. They can be conveniently used on buildings with concrete slab floors.

Ceilings are typically finished with gypsum board, plaster, some form of composition tile, or panels set in a suspended metal frame (Fig. 22.52). See Chapter 36 for additional details.

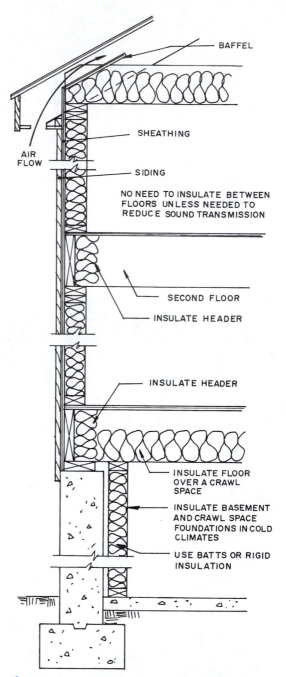

Figure 22.46 Insulation is placed wherever the interior walls, ceilings, or floors are exposed to exterior temperatures.

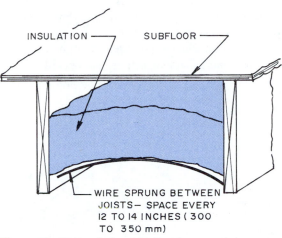

Figure 22.47 Insulation blankets between joists can be supported with wires strung between the joists.

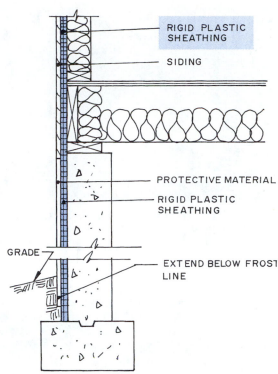

Figure 22.49 Rigid foam plastic sheathing can be used to insulate the foundation.

Figure 22.48 Loose-fill insulation is blown into attic. *(Courtesy Louisiana-Pacific)*

Figure 22.50 Gypsum wallboard is a widely used material for finishing interior walls. *(Courtesy United States Gypsum Corporation)*

Figure 22.51 Plaster walls have the base coat keyed into a wire mesh, producing a finish that has high resistance to cracking. *(Courtesy United States Gypsum Corporation)*

Figure 22.52 Suspended ceiling panels are held by a metal grid. *(Courtesy Celotex Corporation)*

INTERIOR FINISH CARPENTRY

Interior finish carpentry includes projects such as installing interior molding, wainscoting, cabinets, built-in shelves and other units, and stairs. It requires the efforts of the most skilled craftsmen and cabinetmakers.

Installing Interior Doors

One task performed by the finish carpenter is to install interior doors and jambs. Many varieties and types of interior doors are available. Manufacturers often supply installation instructions with each unit. Additional information about doors is in Chapters 19 and 30.

Most interior swinging doors arrive prehung on a one-piece door jamb. The carpenter places the jamb into the rough opening, gets the sides plumb by driving wood wedges between the frame and the wall stud, and nails through the side jamb and wedges into the stud. A long level is used to check for plumb (Fig. 22.53). The door is closed and the spacing between it and the jambs is checked to be certain it is uniform and that the door does not rub on the jamb. Several types of split door jambs are available. They are made with the casing attached to each half as shown in Fig. 22.54. The frames are slid into the rough opening from opposite sides and nailed to the studs.

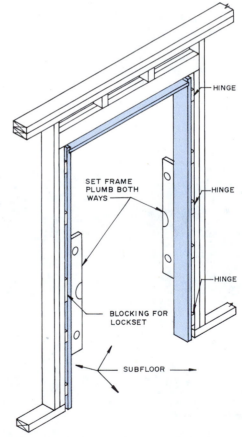

HINGE

SET FRAME PLUMB BOTH WAYS

HINGE

HINGE

BLOCKING FOR LOCKSET

SUBFLOOR

Figure 22.53 An interior door frame is set plumb using wood wedges to position it in the rough opening.

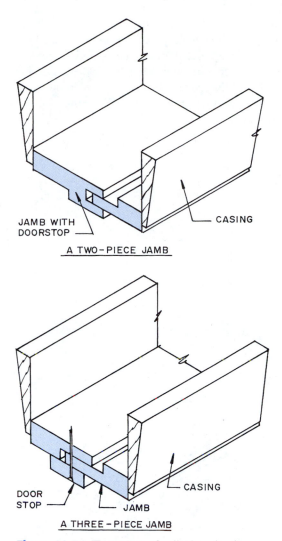

JAMB WITH
DOORSTOP

CASING

A TWO-PIECE JAMB

DOOR
STOP

CASING

JAMB

A THREE-PIECE JAMB

Figure 22.54 Two types of split door jambs.

Installing Casings and Base Molding

After the door jambs are in place and the gypsum wall-board is installed, the finish carpenter can install the casings and baseboards. The casings are available in a wide range of sizes and patterns. The architect specifies the type to use. A typical cased interior door opening is shown in Fig. 22.55. This particular example shows mitered corners, but other types of corners may be used.

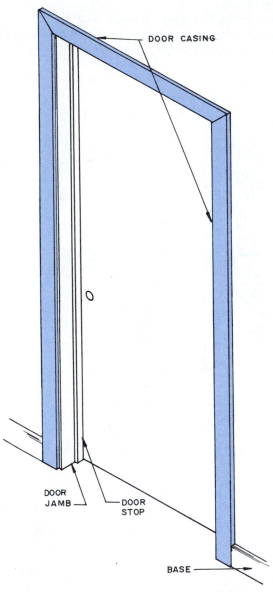

DOOR CASING

DOOR
JAMB

DOOR
STOP

BASE

Figure 22.55 Interior door openings are trimmed on both sides with a casing after the finish wall material is installed.

Window casings are usually like those shown in Figs. 22.56 and 22.57. One type shows the casing on all sides of the opening. The other has a stool with an apron below.

Base molding is used where the wall meets the floor. It is usually mitered at outside corners and coped at inside corners. Coping refers to cutting the ends of the base molding to the shape of the piece it will butt (Fig. 22.58). If the floor is to be carpeted, the base is usually placed a little above the subfloor. If hardwood flooring is to be used, the base is installed after the flooring is in place, and a small molding covers the crack between the flooring and the base molding (Fig. 22.59). Flooring is discussed in Chapters 19 and 37.

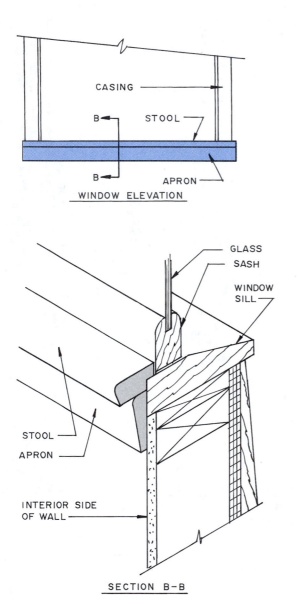

Figure 22.56 This window is cased on three sides and has a stool and apron on the bottom.

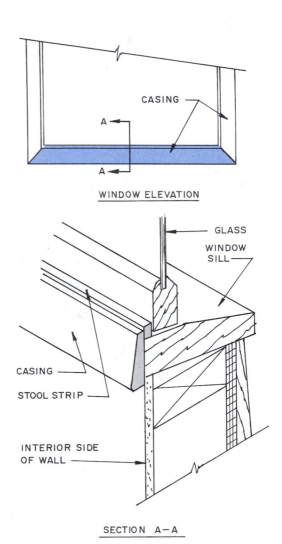

Figure 22.57 The casing on this window is on all four sides and the stool is omitted.

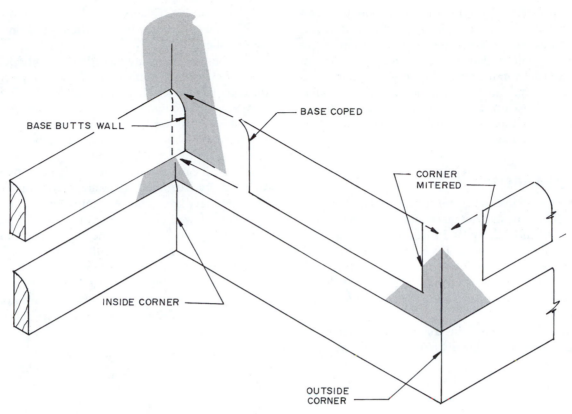

Figure 22.58 The base is coped at the inside corners and mitered at the outside corners.

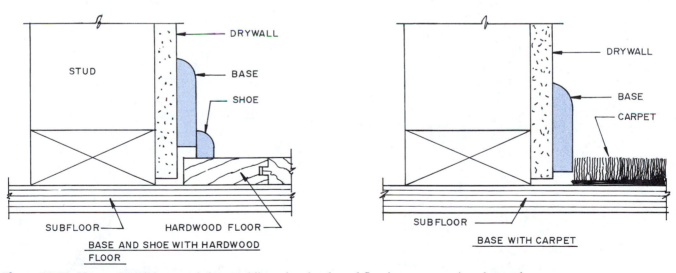

Figure 22.59 How to install base and shoe molding when hardwood flooring or carpet is to be used.

Other types of molding are often shown on the architectural drawings. For example, some form of crown molding is often used where the wall and ceiling meet (Fig. 22.60). Chair rails are used to protect the wall from damage, especially in formal dining rooms, and to add a decorative feature. Often the wall below the chair rail is paneled or at least finished in a different way from the wall above it. Moldings are often mounted on the wall to form decorative panels (Fig. 22.61).

Installing Cabinets

Cabinets are available in several styles and types of construction. Standard cabinet sizes and construction details are shown in Chapter 19. Most cabinets used today are made in a factory and shipped to the site in strong cardboard boxes. The finish carpenter installs them following the architectural drawings (Fig. 22.62). Many times the local company supplying the cabinets will also do the installation.

Some installers prefer to install the wall cabinets first because they can stand directly below the cabinet while they are working. Others set the base cabinets first and use them as a support to hold the wall cabinets as they are installed.

The base cabinets are placed on the subfloor and leveled by shimming them at the floor. They are screwed through the back rail to the wall studs (Fig. 22.63). Some types have metal legs that can be adjusted to level the cabinet. The wall must be checked for straightness. If it has a bow, the base cabinets must be shimmed out so they are straight (Fig. 22.64).

The wall cabinets are installed the specified distance above the base cabinets. They are often supported on the base cabinets and adjusted until they are level. Then they are screwed through the back rails into the studs (Fig. 22.65). They are shimmed to get them plumb.

The countertop is usually plywood or oriented strand board with a plastic laminate top. Ceramic tile is also used over the wood base. The carpenter cuts the openings needed for the sinks and lavatories. The plumber installs these later on as the building nears completion.

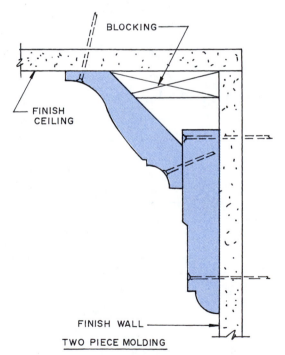

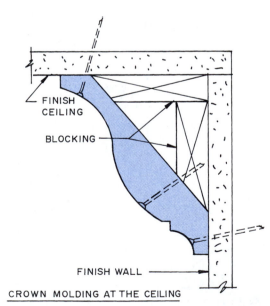

Figure 22.60 Moldings are often used at the intersection of the wall and ceiling.

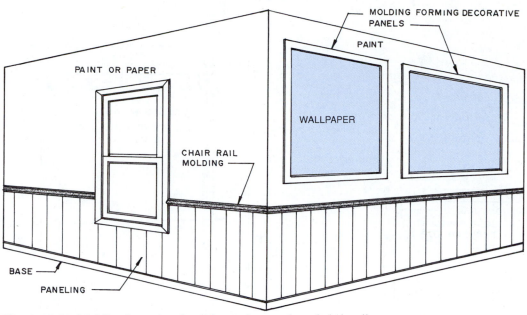

Figure 22.61 Molding is used to form decorative panels and chair rail.

Figure 22.62 A wide variety of manufactured cabinets are available for use in residential and commercial buildings. *(Courtesy Amera Cabinetry)*

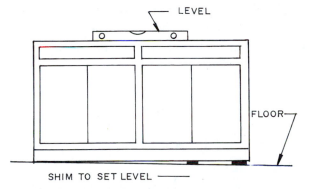

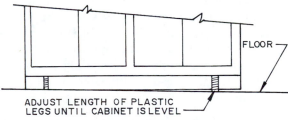

Figure 22.63 The base cabinet can be leveled with wood shims or by adjusting metal legs, if they were used.

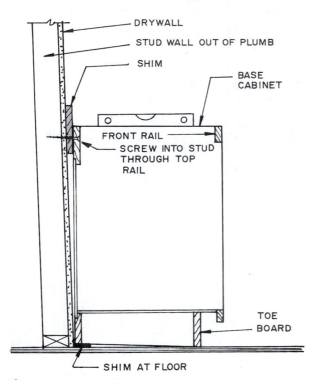

Figure 22.64 Shim the base cabinet against the wall when the wall is not plumb.

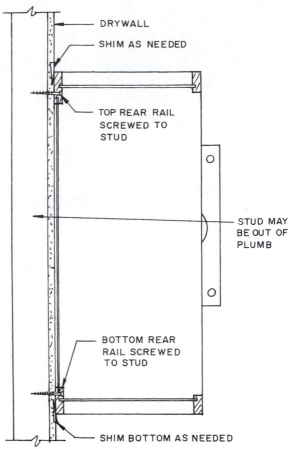

Figure 22.65 Wall cabinets are plumbed and leveled and screwed to the studs.

Installing Stairs

The stair design is shown on the architectural drawings. The type and quality of the treads, handrail, newel posts, balusters, and other parts are specified. Some stairs are built by the carpenter on the site. Usually they build the stringers and buy manufactured parts. Others buy completely manufactured stairs and assemble them on the site. As the stair was designed, the architect had to observe the requirements in the local building codes that regulate riser height, tread width, stair width, and handrail requirements.

The design of a typical wood frame stair is illustrated in Fig. 22.66. The load is carried by the carriage. Generally a stair has three or more carriages. The riser is a board covering the vertical distance between treads. The rough framing for a stair with a landing is in Fig. 22.67.

Some stairs have the treads and risers set in housed carriages. They are held in place with glue and wedges (Fig. 22.68). There are many factory manufactured stairs that provide an interesting and attractive appearance (Fig. 22.69).

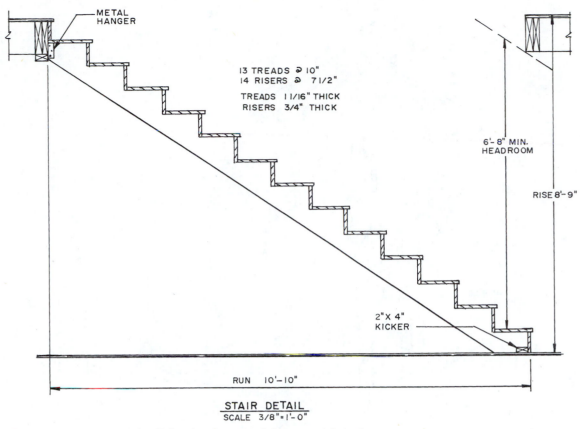

Figure 22.66 A typical detail drawing for a wood frame straight stair.

Other Finish Items

The finish carpenter has to do many things as the building nears completion. These include installing locks, door stops, bath accessories, and fireplace mantels. Some items, such as mirrors and closet shelving, are installed by the subcontractor supplying the item (Fig. 22.70). Finally, the building is thoroughly cleaned, the subfloor is scraped clean and sanded if necessary, and the carpet, composition floor covering, or tile, marble, or slate floor covering is installed. If prefinished hardwood floors are installed, they need to be cleaned and waxed. Unfinished floors are sanded and finished as specified.

Other trades have to phase in their work at the proper time. This includes painting and installing wallpaper and other wall coverings. Window shades and curtains are installed after the carpet is down and all interior finish is complete.

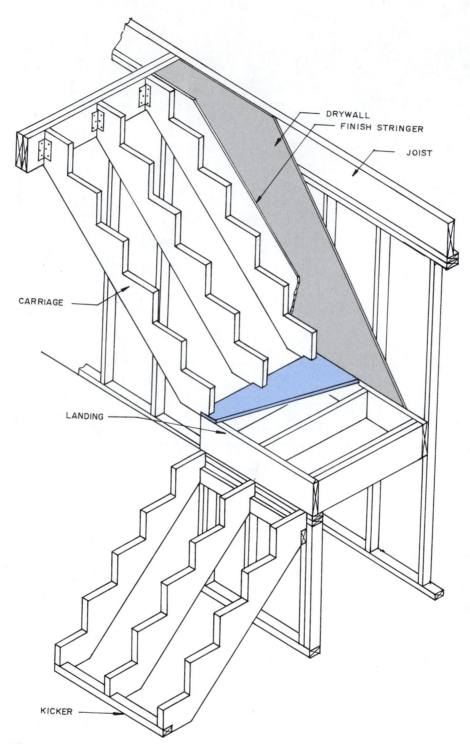

DRYWALL
FINISH STRINGER
JOIST
CARRIAGE
LANDING
KICKER

Figure 22.67 Framing for a stair with a landing.

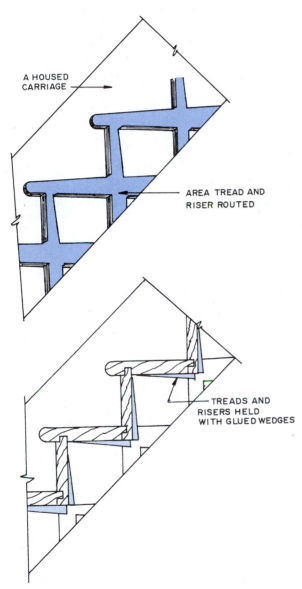

Figure 22.69 This stair was built using commercially available parts. *(Courtesy L.J. Smith Stair Systems)*

Figure 22.68 Some stairs use housed carriages to hold the treads and risers.

Figure 22.70 Among the last things the finish carpenter has to do is install mantels, locks, and other finish hardware. *(Courtesy Driwood Molding Co.)*

REVIEW QUESTIONS

1. What materials can be used for the finished soffit and fascia?
2. Why is attic ventilation important?
3. What are some of the ways a roof can be vented?
4. What types of exterior siding are in common use?

KEY TERMS

eave   The lower edge of a sloping roof that projects beyond the exterior wall.

rake   The board along the sloping edge of a gable.

soffit   The exposed undersurface of any overhead component of a building.

SUGGESTED ACTIVITIES

1. Visit a construction site regularly and follow the steps from finishing the exterior through the work of the trades involved in interior finishing. Keep a log indicating when each part was started, thus producing a construction schedule for the job.

2. Sketch several ways used for framing the eave and rake.

3. Carefully examine the construction of the stair in a building and see if it meets recommended construction standards. If it does not, prepare a report that includes sketches citing what you would do to improve the construction.

ADDITIONAL INFORMATION

Gypsum Construction Handbook, United States Gypsum Company, Chicago, 1992.

Spence, W.P., *Finish Carpentry,* Sterling Publishing Co., New York, 1995.

Spence, W.P., *Residential Framing,* Sterling Publishing Co., New York, 1993.

Wood Building Technology, Canadian Wood Council, Ottawa, 1993.

23

Paper and Paper Pulp Products

This chapter will help you to:

1. Understand the technical processes used to produce paper and paper pulp.

2. Be able to use the knowledge of the properties of paper and paper pulp products when making material selections.

3. Specify the required chemical treatments of paper and paper pulp products to meet the needs of a construction application.

4. Be aware of the many paper and paper pulp products used in construction.

A wide variety of paper and paper pulp products are used in the construction industry. New products appear regularly as paper and paper pulp products are combined with other materials such as synthetic resins, plastic materials, and metal foils.

MANUFACTURE OF PAPER

Paper is manufactured from paper pulp, which is produced using fibers from materials containing large amounts of **cellulose.** The main source of pulp is wood from coniferous (softwood) and deciduous (hardwood) trees. Coniferous trees supply the major source of cellulose. Other pulp materials include straw, cornstalks, jute, wastepaper, rags, and bagasse fibers. **Jute** is a strong, coarse fiber taken from two types of East Indian tiliaceous plants. **Bagasse** fibers are the refuse left over from the crushing of sugarcane or beet refuse from sugar making. Wood pulp fibers are used to produce paper and fiber panels, such as wallboard, acoustical tile, fiberboard, paneling, and roof decking.

In addition to cellulose, paper pulp may include **carbohydrates** and **lignin** from the wood and *coloring materials,* such as dyes, vegetable colors, and mineral pigments. A wide range of *chemicals* are used to produce the type of pulp needed for the finished products. *Rosin, glues,* and *synthetic resins* also are added to provide sizing, and a number of *filler* and *coating materials,* such as gypsum, clay, talc, and chalk, may be added to the pulp.

The manufacturer of paper pulp begins by removing the bark from the logs. The logs then go to a chipper where they are ground to a pulp or into small pieces. If ground into pulp, the wood goes directly into the paper-making process. If ground into chips, it goes into a chemical process where it is digested into a soft wood-pulp mass (Fig. 23.1).

After the pulp is digested it goes to a bleacher where, using chemicals, the pulp is purified and the chlorine is neutralized. The pulp is rinsed between stages. Some products, such as brown wrapping paper and pulp for paperboard, are not bleached. Bleaching increases the brightness of the paper.

Once the pulp is out of the bleacher it goes to the beater, where the fiber and added materials are beaten into a mass of the proper consistency. The pulp leaves

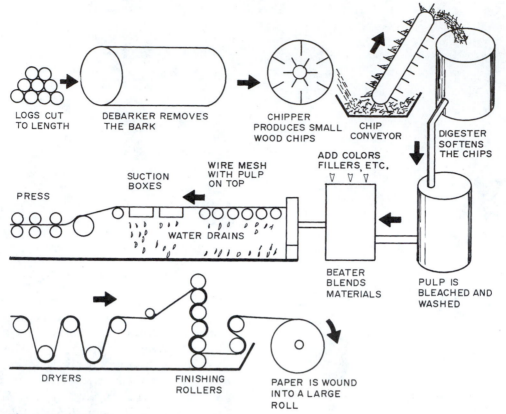

Figure 23.1 The chemical process used to produce paper from wood chips.

the beater and flows to the papermaking machine. Here it passes over a wire mesh screen where some of the water drains away and a sheet of fibers is formed. The sheet still contains considerable water, so it is passed over a series of suction boxes and through a series of smooth rollers that remove additional moisture. The sheet then goes to a series of rollers in a dryer, where the moisture is reduced to 5 to 10 percent. The dry sheet passes through a set of finishing rollers, which gives it the final surface finish. The finished paper is wound in large rolls and is ready to be processed into useful products.

PROPERTIES OF PAPER PRODUCTS

The properties of paper and paper pulp products vary with the composition of the pulp, the cellulose fibers used, the manufacturing process, and the finishing operations involved.

Water absorption is a factor in determining where a paper pulp product can be used. Although some forms of pulp product are waterproof, most require an environment containing little moisture.

Fire resistance properties influence where they can be used in a building. Fire codes spell out these situations.

Tensile strength is important when paper products are under a load, such as in sheathing paper or backing on insulation batts.

Porosity influences acoustical and insulation properties of paper pulp products as well as their water absorption characteristics.

Chemical resistance properties have a direct relationship to specific uses including possible reactions with materials touching the paper product.

CHEMICAL TREATMENTS

Paper products can be treated with chemicals that enable them to resist fire or flame, water and water vapors, bacteria, mold, fungi, and insects.

Papers that are **flame-retardant** are not totally **fireproof,** but they will not burst into flames, which would release deadly combustible gases. **Fire-resistant** paper is difficult to ignite, will not support combustion, and will self-extinguish after the source of the heat is removed. In Table 23.1 are chemicals commonly used for providing flame-retardant and fire-resistant properties in paper and paper pulp products. Some fire-resistant and flame-retardant chemicals are destroyed by wetting, so they should not be used on products to be in areas exposed to moisture.

Paper and paper pulp products are *waterproofed* by mixing synthetic resins with the pulp. Some types, such as wallpaper, are moisture resistant because they have a plastic film on the exposed surface. Waterproof membranes used on walls and floors below the water level in the soil are coated with tar, asphalt, or a synthetic resin. Waterproof paper with a laminated metal foil is used for concealed flashing. Papers used for vapor barriers are also water resistant.

Table 23.1 Flame Retarding and Fire-Resisting Chemicals Commonly Used on Paper Products

Material	Effectiveness	Moisture Resistance
Antimony trioxide	Excellent	Are moisture resistant
Zinc borate	Good	
Antimony oxide	Excellent	
Ammonium sulfate	Excellent	Are not moisture resistant
Ammonium chloride	Good	
Sodium phosphate	Good	
Boric acid–sodium borate	Good	

PAPER PRODUCTS USED IN CONSTRUCTION

A summary of selected paper products used in construction is in Table 23.2. Various types of building papers and felts are used for sheathing, reduction of dust and air filtration, vapor barriers, and for sound-deadening purposes. Some have a pulp made from cotton, old rags, and paper. Others use a kraft paper base or jute pulp.

There are papers for the preparation of architectural drawings, reproduction of drawings, correspondence, billing, and many other of the business operations involved in construction.

Other paper products are used to make bags, such as protective wrappings of products, and to encase materials such as fiber-type insulation batts. It is important when selecting one of these products to examine the recommendations of the manufacturer pertaining to its best use, relationship to building codes, and installation techniques.

Cardboard

Cardboard is a pulp product that varies widely in its thickness and properties. It is a relatively stiff, rigid product. In construction it is used for boxes and cartons in which products are shipped. Corrugated board is probably the most widely used. Corrugated board consists of outer plies of a coarse, stiff brown paper bonded to the face and back of a corrugated core (Fig. 23.2). Corrugated paper

Table 23.2 Selected Paper Products Used in Construction

Paper Product	Uses	Coating
Asphalt felt	Sheathing paper, built-up roofing	Cotton and rag pulp paper saturated with asphalt
Asphalt sheathing paper	Between sheathing and siding	Asphalt saturated
Floor lining paper	Between subfloor and finish floor	Asphalt or tar saturated and coated
Carpet felt	Carpet underlayment	No special coating
Copper foil laminated	Concealed flashing, vapor barrier, waterproofing	Copper foil applied to one side of kraft paper
Deadening felt	Sound deadening in walls and floors	Made from rags and paper with no special surface coating
Plastic-coated paper	Vapor barriers	Laminated kraft paper coated with plastic on one side
Building paper reinforced with glass fibers	Vapor barriers	Laminated paper reinforced with glass fibers
Rolled roofing	Final (exposed) roof layer	Felts saturated with asphalt and coated with mineral aggregate
Sheathing paper	Protection against dust and air infiltration in walls and floors	No special coating
Wallpaper	Cover finished interior walls	Clay or casein coating
Resin-coated wallpaper	Washable finished interior wall coating	Synthetic waterproof coating, such as plastic film
Kraft paper	Bags, protective wrapping, concrete forms, plastic laminate base material, concealed flashing, cover gypsum sheets	Wood pulp sized with resin
Vellum	Drafting paper	100% rag impregnated with synthetic resin
Mineral-fiber papers	Where high temperatures exist, for electrical insulation	Glass, silica, and ceramic fibers in the pulp

honeycomb cores are also used in the construction of some wall panels and hollow-core doors.

Paper Pulp Sheets

Paper pulp sheets are available with a variety of surface treatments, structural characteristics, chemical properties, and colors. The simplest form is a single homogeneous ply of wood pulp with no special additives or surface coatings. Another form has the ply treated with resin. Some sheets are made by laminating sheets of paper and impregnating them with resin. Other sheets have an aluminum foil or plastic film bonded to the surface. The pulp forming the sheet may be reinforced with glass fibers or have a thin core of plastic foam or corrugated paper. Sheets may be impregnated with asphalt or synthetic resins and can be waterproof. Some are rigid, durable, and strong. They are available with plain, perforated, and corrugated surfaces. The basic sheet is usually brown, but various surface finishes and colored finished surfaces are available. Paper pulp sheets are available in 4 × 8 ft. (1219 × 2438 mm) sheets in thicknesses of $^1/_8$, $^3/_{16}$, and $^1/_4$ in. (3, 5, and 6 mm).

Laminated paper sheet material, made by laminating sheets of paper, is used for finished surfaces on interior walls. It is available in 4 ft. (1220 mm) width and lengths to 16 ft. (4880 mm).

Fiberboard Wood Pulp Panel Products

Fiberboard wood pulp products are made from pulpwood. A wide variety of products are manufactured. They differ from pulp sheets in that they are thicker.

They are available as laminated layers of paper pulpboard or panels and as a single homogeneous layer of wood pulp fibers.

Single layer *fiberboard* (often called insulation board) is a high-density asphalt-impregnated panel used for wall sheathing. One-inch fiberboard has an R-value of about 2.6 and is available in 4 × 8 ft. (1219 × 2438 mm) sheets and thicknesses of $^1/_2$, 1, $1^1/_2$, and 2 in. (12, 25, 38, and 51 mm). Longer sizes are also available. Standard fiberboard will not hold nails. A nail base fiberboard is available. Fiberboard is also used as backer boards behind shingles and some types of siding. Fiberboard sheets are also used for roof insulation. Some panels have plastic foam or reflective foil bonded to them.

Fiberboard finds some use as interior finished wall panels with one surface prefinished. Acoustical ceiling tiles are made with a variety of surface treatments, such as plain, patterned, textured, and perforated. Tiles are usually 12 × 12 in. (305 × 305 mm) and 16 × 16 in. (406 × 406 mm) square and $^1/_2$ and $^3/_4$ in. (12 and 19 mm) thick. Larger size panels are available from some manufacturers.

Mineral fiberboard is made of mats of rock wool or fiberglass with stiff paper faces on both sides giving it rigidity. These are used for insulation.

Fiberboard panels are installed by nailing or gluing them to studs or other surfaces. Joints are not laid tight, so there is allowance for expansion. Acoustical ceiling tiles are installed as recommended by the manufacturer. This may be with suspended tracks or by stapling or gluing them to another surface or to wood furring. The joints between the panels on decorative interior walls are covered with some type of strip. If high moisture

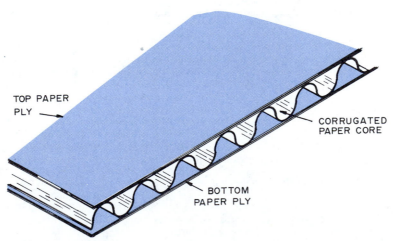

Figure 23.2 Corrugated board has coarse outer paper plies bonded to a corrugated paper core.

conditions exist, fiberboard panels are not the best choice. If they are used under these conditions, use plastic-coated panels and seal them on all edges with a waterproof caulking. Be certain to compare their fire-resistant and flame spread ratings with those required by the building code.

Wood Pulp and Fiber Decking

A variety of roof decking products are available made from layers of pulpboard or fiberboard bonded with a waterproof adhesive. These products have a variety of surface treatments, such as a coating of fibrous cement, plastic film, plywood, metal foil, asphalt, roofing felt, paper, or plastic foam sheets. These coatings increase water resistance, insulation value, and strength. The bottom surface is often exposed to view and becomes the finished ceiling. It can be treated to provide acoustical value. This surface may be painted, covered with a textured fibrous sheet, or have a prefinished plywood veneer. Most contain a vapor barrier in the assembled deck panel (Fig. 23.3).

When choosing a product, it is important to know the distance it will span with known loads (weight of roofing, snow, wind, etc.). The span varies with the known load and thickness. Typical panel thicknesses range from 2 to 4 in. (51 to 102 mm), widths from 20 to 48 in. (508 to 1219 mm), and lengths from 48 to 144 in. (1219 to 3657 mm). Spans in general range from 16 to 48 in. (406 to 12,192 mm). Manufacturers' catalogs must be consulted for specific data.

Decking is manufactured with various edge treatments. These commonly include all four edges square, all four edges tongue and groove, and tongue and groove on two edges and square on the ends. Those used with tees have rabbeted edges (Fig. 23.4).

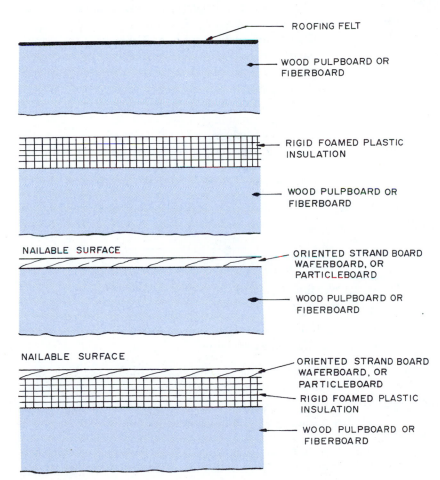

ROOFING FELT

WOOD PULPBOARD OR FIBERBOARD

RIGID FOAMED PLASTIC INSULATION

WOOD PULPBOARD OR FIBERBOARD

NAILABLE SURFACE

ORIENTED STRAND BOARD WAFERBOARD, OR PARTICLEBOARD

WOOD PULPBOARD OR FIBERBOARD

NAILABLE SURFACE

ORIENTED STRAND BOARD WAFERBOARD, OR PARTICLEBOARD

RIGID FOAMED PLASTIC INSULATION

WOOD PULPBOARD OR FIBERBOARD

Figure 23.3 Examples of wood pulp and fiber roof decking.

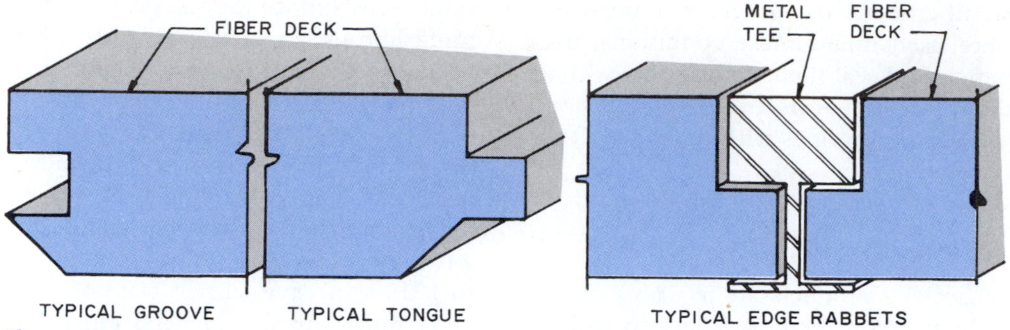

Figure 23.4 Typical edge treatments for wood pulp and fiber roof decking.

The atmospheric conditions in which wood fiber decking is used should be considered. Again, observe the manufacturer's limitations. For example, some require that the temperature be kept 40°F (4.4°C) or higher and that the relative humidity be kept below 50 percent. The type of finished roofing material to be applied must also be considered. If it is to have a built-up roof on which hot asphalt or tar is applied, the joints between panels must be sealed to keep the asphalt or tar from leaking through and damaging the finished ceiling below. If the finished roofing material, such as shingles, requires nailing or stapling, the pulp roof decking must have a plywood, oriented strand board, particleboard, or waferboard veneer to hold the nails. It must also be thick enough so the nails do not break through the finished ceiling below.

Following is a typical example of the design information provided by wood pulp and fiber decking manufacturers:

Diaphragm resistance—can be used in seismic regions and areas of severe wind loads
Meets U.L. wind uplift resistance requirements
Flame spread 20
Smoke developed 5
Roof and ceiling assembly 1½ hr. fire rating
Linear expansion 0.2%
Noise reduction coefficient—2" NRC = .65
   3" NRC = .80
Light reflectance 60%
Sound transmission class 2½ in. thick is STC 41
R value 2"-4, 3"-6, 4"-8
Compressive strength 300 psi
Fastener withdrawal loads with oriented strand board
   #14 all-purpose screw—380 lb.
   #12 steel decks—330 lb.
   #10 sheet metal screw—180 lb.
   #11 gauge roofing nail—51 lb.

REVIEW QUESTIONS

1. What is the main source of wood pulp for papermaking?
2. In addition to wood pulp, what other materials are used in papermaking?
3. What types of construction products are made from wood pulp fibers?
4. What sizing materials are added during the papermaking process?
5. What filler materials are added during the papermaking process?
6. What is the main difference between the two methods of papermaking?
7. How does bleaching affect the paper?
8. What properties of paper products directly influence where they are used?
9. How are paper products waterproofed?

KEY TERMS

bagasse Crushed sugar cane or beet refuse from sugar making.

carbohydrates Organic compounds that form the supporting tissues of plants.

cellulose An inert carbohydrate that is the chief constituent of the cell walls of plants, wood, and paper.

fireproof A material or construction that will not burn or is almost unburnable.

fire-resistant The capacity of a material or construction to withstand fire or give protection from it.

flame-retardant Having resistance to the propagation of flame and therefore not readily ignited. Indicates a lower resistance than fire-resistant.

jute A coarse fiber obtained from two East Indian tiliaceous plants.

lignin An organic substance forming a major part of woody tissue.

SUGGESTED ACTIVITIES

1. Collect samples of the various paper and paper pulp products used in construction and identify their properties.

2. Visit a building supply dealer and make a list of all the paper and paper pulp products you can observe.

ADDITIONAL INFORMATION

Hornbostel, C., *Construction Materials, Types, Uses, and Applications,* John Wiley and Sons., New York, 1991. Other resources are
Publications of the American Forest and Paper Association, Washington, D.C. 20036.

Plastics

This chapter will help you to:

1. Learn the properties and characteristics of plastic materials used in construction.

2. Select the proper plastic material for various construction applications.

Although the use of plastics is relatively new in the construction field, it has rapidly become a major material and is being used for structural and non-structural purposes (Fig. 24.1). Plastic is replacing conventional materials, such as wood and metal, in many parts of a building. This is occurring because extensive research in plastic development has resulted in a variety of plastic materials that have properties not available in conventional materials. In general, plastics resist corrosion and moisture, are tough, are light in weight, and are easily formed into useful products. Since plastics are chemically derived, a wide range of plastics with special properties can be developed. This makes them particularly useful as a construction material because of the extensive range of applications that exist.

Vinyl siding is available in a variety of colors, surface textures, and sizes, and it provides a tough, maintenance-free exterior. Wood windows are clad in vinyl, eliminating the need for painting, and some windows are framed entirely with solid plastic extruded parts. Some types of plastics are replacing glass glazing because the plastic is lighter and more resistant to shattering. Plastic lavatories, showers, and bathtubs have largely replaced those of ceramic-coated metal with cast iron fixtures. A high percentage of items used in electrical systems, such as boxes and wiring insulation, are plastic. Plastic film is used for vapor barriers and to reduce air infiltration, and plastic foams are used for insulation.

Since there are so many different kinds of plastics and these have varying properties, the designer and constructor must be careful to use a plastic for the purpose for which it was designed. For example, using plastic pipe designed to carry cold water for hot water lines will eventually cause problems.

There are many families of plastic materials, and within each the properties can be varied widely. Plastics from several families may be able to serve the same purpose, but others exhibit unique, special characteristics.

The term **plastics** is used today to describe man-made **polymers** that contain carbon atoms covalently bonded with other elements. Plastic is obtained by breaking down materials found in nature, such as petroleum,

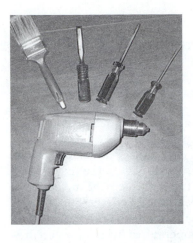

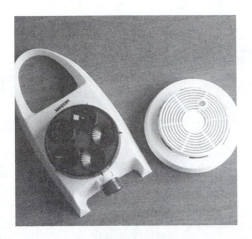

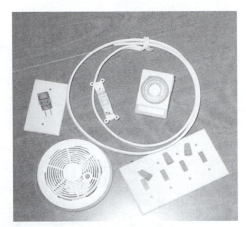

Figure 24.1 Plastic products are widely used in the construction industry.

coal, and natural gas (Fig. 24.2). Plastics are **synthetic materials,** resulting from chemical manipulations of natural materials. In other words, they are man-made and not found in nature. In their formulation they are soft, can be formed into the shapes desired, and exhibit **plastic behavior.** Plastics are **organic materials.** They are made up of molecules built around a *carbon atom.* An exception to this is a group of materials considered a plastic that is built around the *silicon atom.*

MOLECULAR STRUCTURE OF PLASTIC

You can understand how the properties of plastics can be made to vary by examining the molecular structure of the material. Most materials accepted in the plastics family are based on the carbon atom. The carbon atom

bonds with other atoms through **covalent bonding.** Covalent bonding is the process in which small numbers of atoms are bound into *molecules.* A single molecule is known as a **monomer.** When several monomers link together to form a chain they form a *polymer.* This chain formation is called *polymerization.* Plastic polymers are called *macromolecules* because they are composed of many smaller molecules ("macro" means large). Since the small molecules are joined together in a chainlike condition, they are often called chains. Plastics are made up of these polymers (chains). Note that many plastics are described by the prefix "poly," such as polyvinyl and polystyrene. Plastic is a polymer. Polymers are moldable materials sold in the form of granules, powder, flakes, liquids, or pellets for processing into useful products.

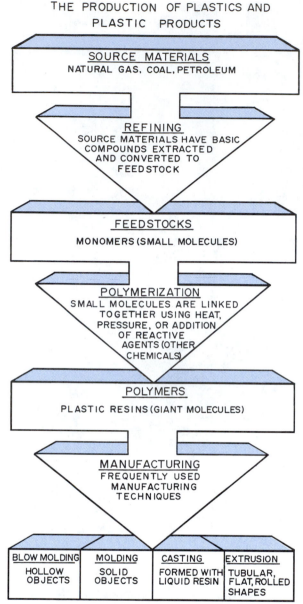

THE PRODUCTION OF PLASTICS AND PLASTIC PRODUCTS

SOURCE MATERIALS
NATURAL GAS, COAL, PETROLEUM

REFINING
SOURCE MATERIALS HAVE BASIC COMPOUNDS EXTRACTED AND CONVERTED TO FEEDSTOCK

FEEDSTOCKS
MONOMERS (SMALL MOLECULES)

POLYMERIZATION
SMALL MOLECULES ARE LINKED TOGETHER USING HEAT, PRESSURE, OR ADDITION OF REACTIVE AGENTS (OTHER CHEMICALS)

POLYMERS
PLASTIC RESINS (GIANT MOLECULES)

MANUFACTURING
FREQUENTLY USED MANUFACTURING TECHNIQUES

BLOW MOLDING	MOLDING	CASTING	EXTRUSION
HOLLOW OBJECTS	SOLID OBJECTS	FORMED WITH LIQUID RESIN	TUBULAR, FLAT, ROLLED SHAPES

Figure 24.2 The production of plastics and plastic products. *(Courtesy the Society of the Plastics Industry, Inc.)*

The molecular bonding process can be shown by the following illustration. The carbon atom has a **valence** of four, which means there are four points at which other atoms can attach themselves to the carbon atom to form covalent bonds. Other elements have different valences. For example, oxygen atoms have a valance of two. Two oxygen atoms can bond to the four carbon valence points creating carbon dioxide (CO_2) (one carbon atom and two oxygen atoms).

CLASSIFICATIONS

Plastics can be divided into two basic classifications, thermoplastics and thermosetting materials.

Thermoplastics

Thermoplastics are plastic materials that can be softened or remelted by application of heat and reformed. Some thermoplastic materials will experience contamination and chemical degradation if they are reheated frequently.

Thermoplastic materials are composed of long chainlike molecules that are unattached to each other. These molecules can slide past each other and change shape. At normal temperatures, 70°F (21°C), the material retains its shape because the movement of molecules is slight. However, when heat is applied, the bonding between molecules weakens and the material expands and becomes soft. At high temperatures (which vary according to the composition) the plastic softens and has sufficient flow to permit it to be molded. As the molded part cools, the plastic holds the new shape.

Thermosetting Plastics

Thermosetting plastics (also called **thermosets**) are plastics that cannot be reheated and reformed once they have been softened, formed, and cured. They are formed with a chemical process that produces a strong bond between the molecules, and they cannot slide by each other. The final form of the material is irreversible.

ADDITIVES

Most plastic resins in pure form do not have the properties needed for particular applications. Therefore, **additives** are mixed with the resin to modify the properties. Frequently more than one additive is needed to obtain the properties desired. The amount added also has great influence on the final result. The commonly used additives include plasticizers, fillers, stabilizers, and colorants.

Plasticizers

Plasticizers are added to the plastic resin to reduce brittleness and increase flexibility, resiliency, and moldability and, in some cases, to improve impact resistance. A plasticizer lessens the bond between the chainlike molecules that make up a plastic. Since they can slide past each other, the plastic can bend and flex. In doing this the strength, heat resistance, and dimensional stability are lessened (Fig. 24.3).

Fillers

Some fillers are used to provide bulk, thus reducing the cost of the plastic. Others are used to improve a specific property. **Fillers** to increase *bulk* or *ease of molding* include finely ground hardwood and nutshells. **Hardness** is improved by adding mineral oxides and mineral powders. *Heat resistance* is improved by adding inorganic fillers, such

Chapter 24 ▲ Plastics 607

as clay, silica, ground limestone, or asbestos. *Electrical resistance* is increased by adding quartz or mica. *Toughness* is improved by adding fibers, such as hemp, cotton, sisal, rayon, polyester, or nylon. *Mechanical properties* are strengthened by adding glass fibers or metal fibers.

Stabilizers

Stabilizers, or lead compounds, are used to stabilize plastic by helping it resist heat, loss of strength, and the effect of radiation on the bonds between the chains.

Colorants

Colorants added to the resin include *organic dyes* and *inorganic pigments* (metal-based). Some of these also have a *stabilizing* influence. Inorganic pigments disperse throughout the resin rather than dissolve as do the dyes, so they reduce the transparency of the material. They also withstand high temperatures without fading.

PROPERTIES

The important properties of plastics are illustrated in Tables 24.1 through 24.5. Notice the variances within the two classifications, thermosets and thermoplastics.

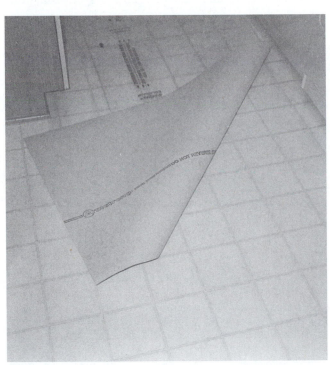

Figure 24.3 Floor products have plasticizers added to increase flexibility.

The limitations on these materials for certain uses in construction can be seen by comparing them with other materials on tensile strength and other properties. Following is a general discussion of the major properties of plastics. The composition of plastics can be varied considerably, so the specifics of their properties do also. To make proper use of plastic materials, the properties of the specific formulation of resin and additives must be known.

Mechanical Properties

The mechanical properties that are generally most important to consider for plastics used in construction are tensile strength, stiffness, toughness (or impact strength), hardness, and creep.

As shown in Table 24.1, the *tensile strength* of most plastics is lower than steel and more closely related to wood. Plastic laminates and fiberglass reinforced plastics do have tensile strengths approaching steel.

Stiffness is measured by the modulus of elasticity (E). The greater the E value the stiffer the material. Overall, plastics are not as stiff as steel, aluminum alloys, and concrete (Table 24.2), so they deflect more under load. One advantage plastic structural members

Table 24.1 A Comparison of the Tensile Strengths of Plastics with Other Commonly Used Construction Materials

Material	Tensile Strength, psi
Thermoset plastics	175–35,000
Thermoplastics	1000–20,000
Reinforced plastics	800–50,000
Plastic laminates	800–50,000
Glass (annealed)	3,000–6,000
Cast iron	20,000–100,000
Hot-rolled carbon steel	43,000–130,000
Southern pine	500–1,250
Wrought aluminum	5,000–82,000

Table 24.2 A Comparison of the Modulus of Elasticity of Plastics with Other Commonly Used Construction Materials

Material	Modulus of Elasticity (E), 10^6 psi
Thermoset plastics	0–2.4
Thermoplastics	0–1.8
Reinforced plastics	0–5.00
Cast iron	15–25
Wood	0.5–2.6
Aluminum alloys	9–11
Glass	9–11
Concrete	1–5
Steel, carbon, high strength, and stainless	29.0

do have is that they are lighter than conventional materials.

Impact strength (toughness) is the ability of a material to resist impact from an object striking it. In general, thermoplastics exhibit better impact strength than thermosets. Polyurethane is good and is used on car bumpers. Polyethylene and polyvinyl chloride (PVC) have wide ranges of impact strength and find uses ranging from bottles to pipe. Compared to other construction materials, plastics are quite high in impact strength. This is why acrylic and polycarbonate sheet material is used for glazing in many situations.

Hardness is a measure of abrasion resistance. In general, plastics are not as good as glass or steel in resisting abrasion. This can be noted by the scratches that can be seen on plastic glazing if it is improperly cleaned.

Creep is the term used to describe the tendency of a material to flow while cold (in normal temperatures) and to have a permanent change in size and shape when under continued stress. Although all materials creep, it is insignificant in materials such as steel. Creep is a significant factor for many plastics, even at room temperature. It is most apparent in the thermoplastics because they contain unattached molecules that can begin to slip past each other when under stress. When the stress is for a short time, the plastic may return to its original condition. If the stress is long-term it could produce permanent deformation.

Laminates, fiberglass reinforced plastics, and some thermosets do resist creep better than thermoplastics, but they do not resist stress as effectively as steel and concrete.

Electrical Properties

Plastics have excellent electrical insulating properties and have enabled the development of greatly improved electrical equipment. In addition, they have good heat resistance, which is necessary for the use of a material in many electrical devices. Silicones and fluorocarbons have these properties and are widely used. Epoxies are used to encapsulate electric components subjected to temperatures as high as 300°F (150°C), and phenolics are used as housings for electrical parts, switches, and receptacles. In other words, plastics have excellent **dielectric strength** (maximum voltage a dielectric can withstand without rupture). A *dielectric* is a nonconductor of electric current. Since electrical components are subject to damage by arcing, the arc resistance is important. **Arc resistance** is measured by the total elapsed time in seconds an electric current must arc to cause a part to fail. Plastic parts carbonize and become conductors, burn, develop thin lines between electrodes, or become glowing hot. Plastics used for applications in which arcing is a possibility have an arc resistance time from 100 to 300 seconds.

Thermal Properties

Two major factors to consider when examining thermal properties are the influence of temperature on strength and expansion and contraction of the material. Many plastics lose strength at comparatively low temperatures. The **service temperature** (maximum temperature at which a plastic can be used without affecting its properties) of plastics is low when compared with other construction materials (Table 24.3). Some plastics soften below 200°F (94°C), so they soften if placed in boiling water. In general, thermosets have higher service temperatures than thermoplastic materials. Polyamide thermosetting resins resist intermittent heat up to 930°F (500°C) under low stress and up to 480°F (250°C) under continuous stress.

All building materials *expand and contract*, and this must be taken into consideration as a designer selects materials and prepares the design drawings. Most plastics have a higher *coefficient of thermal expansion* than other construction materials (Table 24.4). In general, thermosets have a lower rate of expansion than thermoplastics.

Plastics are poor *conductors* of heat. Unmodified plastics compare favorably with brick, concrete, and glass (Table 24.5). Notice that foamed plastics have the lowest thermal conductivity and are therefore *excellent insulators*. They are widely used for all types of insulation.

Table 24.3 A Comparison of the Service Temperatures of Plastics with Other Commonly Used Construction Materials

Material	Service Temperature, °F
Plastics	80–500
Plastic foam (polystyrene)	140–150
Carbon steel	900–1350
Wood	225–400
Glass	350–1600
Concrete	450–1350
Aluminum alloys	300–575

Table 24.4 A Comparison of the Coefficient of Thermal Expansion of Plastics with Other Commonly Used Construction Materials

Material	Coefficient of Thermal Expansion 10^{-6} in./in./°F
Thermoset plastics	7–167
Thermoplastics	19–195
Reinforced plastics	7–38
Cast iron	5.9–6.6
Carbon steel	6.5
Wood	7–38
Glass	1
Aluminum alloys	9
Concrete	7

Table 24.5 A Comparison of the Thermal Conductivity of Plastics with Other Commonly Used Construction Materials

Material	Thermal Conductivity BTU/hr./ft.²/in./°F
Thermoset plastics	1–2
Thermoplastics	0.5–5.0
Plastic foams	0.15–0.26
Wood	0.5–1.5
Glass	5–6
Brick	3–6
Concrete	5.5–9.5
Steel alloys	300–500
Aluminum alloys	500–1500
Copper	440–2550

When comparing plastics in general with other materials it can be seen that plastics have the lowest coefficient of thermal conductivity.

Flammability is the ability of a material to resist burning. Since plastics are organic materials, they will burn. The range of flammability, however, is great. Cellulosics are highly flammable and are often banned by building codes. Polyurethane and polyvinyls give off toxic fumes when burning, and building codes specify protective measures for many uses. Vinyl siding, for example, begins to burn at 698°F (370°C), but wood ignites at about 400°F (206°C). In general, thermoplastics are more flammable than thermosets, though these properties can be changed with additives. Some plastics can develop a char layer that serves as a shield to deter burning. The National Aeronautics and Space Administration uses nylon and phenolic resins on the exterior of spacecraft. These are exposed to the searing heat of reentry.

Although plastics are no more combustible than wood, they can produce toxic fumes, which are more likely to cause death in a fire than the flames. Most building codes have a separate chapter devoted to the use of plastics in building construction.

Chemical Properties

Plastics do not corrode like metals, but they can deteriorate and be damaged by chemical attack. When a metal corrodes, its surface is damaged to the extent that it loses weight. Deteriorated plastics gain weight because the attacking chemicals combine with the plastic resin. This causes swelling, crazing, and discoloration. In addition, there is loss of impact, flexural, and tensile strengths. Usually the *chemical resistance* of a plastic decreases as the temperatures increase.

The consideration of the **weatherability** of a plastic involves consideration of moisture, ultraviolet light, heat, and chemicals found in the air as ozone and hydrochloric acid. High-density polyethylene (HDPE) has great resistance to acids, water absorption, and weathering. It is used for electrical wire insulation. Acrylics also have great resistance to weathering (and excellent optical qualities), which makes them widely used for glazing.

Density

The density of plastic materials is in general lower than other commonly used construction materials. Glass reinforced plastics (GRP) are lighter than steel and aluminum.

Specific Gravity

The specific gravity of plastics varies considerably, ranging from 0.06 for foams to 2.0 for fluorocarbons. *Specific gravity* is the ratio of the mass of a volume of material to the mass of an equal volume of water at a standard temperature. The specific gravity of water is 1, so some plastics will float (lighter than water) and others will sink. The specific gravity for softwoods is 0.5, about 2.7 for aluminum, and about 8.0 for steel.

Optical Properties

Some plastics have optical properties equal to that of glass. Acrylics are as transparent as fine optical glass and have a light transmission of 92 percent. Polystyrene, polypropylene, and polycarbonates are also 90 percent or better.

PLASTIC CONSTRUCTION MATERIALS

The following discussion gives a brief description of the plastics most commonly used in construction and some of their major uses. They are divided into thermoplastics and thermosets.

Since plastic materials are frequently identified in technical literature and professional magazines by approved abbreviations, the designer and constructor must be familiar with them. A list of these abbreviations as compiled by the American Society for Testing and Materials is in Table 24.6.

Thermoplastics

More products used in construction are made from thermoplastic materials than thermoset. Following are commonly used thermoplastic materials and products using them.

Acrylonitrile-Butadiene-Styrene

Acrylonitrile-butadiene-styrene (ABS) plastics are a combination of high-impact polystyrene, which is very tough, and acrylonitrile, which improves its rigidity, tensile strength, and chemical resistance. ABS plastics are widely used for pipe and pipe fittings for water lines with water up to 180°F (83°C) (nonpressure), gas supply lines, waste, drain, and sewage vent systems. It also is used for hardware, such as handles and knobs (Fig. 24.4).

Acrylics

The most widely used acrylic is polymethyl methacrylate (PMMA). It has excellent optical clarity and is used for glazing. Typical uses include door and window lights and roof domes and skylights. It is also used to make lighting fixtures, but it will soften at 200°F (94°C). It finds use in outdoor signs, corrugated roofing, and molded hardware.

Acrylics are also dispersed as fine particles in a liquid producing a *latex* that is used to make latex paints.

Cellulosics

Two common plastics based on the cellulose molecule are cellulose acetate (CA), and cellulose acetate-bu-tyrate (CAB). Cellulose acetate is not used for construction purposes, but CAB can be made resistant to weathering and is used for pipe for gas and chemicals. Cellulosics are also used in coating compounds and adhesives.

Fluorocarbons

The most widely used fluorocarbon is polytetrafluoroethylene (PTFE). It has high resistance to chemical degradation and has a service temperature from −450°F (−234°C) to +500°F (262°C). It has a low coefficient of friction and is marketed under the trade name Teflon. It is used for pipe that must handle corrosive liquids at high temperatures and for parts that require easy sliding surfaces.

Table 24.6 ASTM[a] Abbreviations for Plastics

Thermoplastics	
Type	**Abbreviation**
Acrylonitrile-butadiene-styrene	ABS
Acrylic:	
Polymethyl methacrylate	PMMA
Cellulosics:	
Cellulose acetate	CA
Cellulose acetate-butyrate	CAB
Cellulose acetate-propionate	CAP
Cellulose nitrate	CN
Ethyl cellulose	EC
Fluorocarbons:	
Polytetrafluoroethylene	PTFE
Nylons:	
Polyamide	PA
Polyethylene	PE
Polypropylene	PP
Polycarbonates	PC
Styrene:	
Polystyrene	PS
Styrene-acrylonitrile	SAN
Styrene-butadiene plastics	SBP
Vinyl:	
Polyvinyl acetate	PVAc
Polyvinyl butyral	PVB
Polyvinyl chloride	PVC

Thermoset Plastics	
Type	**Abbreviation**
Epoxy, epoxide	EP
Fiberglass reinforced plastics	FRP
Melamine-formaldehyde	MF
Phenolic:	
Phenol-formaldehyde	PF
Polyester	—
Polyurethane:	
Urethane plastics	UP
Silicone plastics	SI
Urea-formaldehyde	UF

[a]American Society for Testing and Materials

Figure 24.4 ABS plastics are used extensively for plumbing installations.

Nylons

Polyamides (PA), also known as nylon, are tough, high strength, and have good chemical resistance. Since they resist abrasion well they are used for molded parts such as locks, rollers, gears, and cams. They do not weather well.

Polycarbonates

Polycarbonates (PC) have high impact strength and good heat resistance. They are dimensionally stable and are transparent. They are used for light fixtures, molded parts, and signs. They are used in place of glass in areas where damage is likely, such as in skylights.

Polycarbonate plastic sheets such as Lucite and Lexan have high impact strength, reaching up to 250 times the strength of glass and 30 times that of acrylic. They are suitable for glazing openings in areas where high security is needed. They are laminated to produce bullet-resistant panels. The light transmission varies with the thickness but runs 75 percent for 0.50 in. (12 mm) panels to 86 percent for 0.125 in. (3 mm) panels. Some companies make these sheets in transparent solar tints, translucent white, and a variety of colors.

Double-skinned units with internal ribs and dead air spaces are available for use on vertical and sloped glazing.

Polyethylene

Polyethylene (PE) is light, strong, and flexible even at low temperatures. It has good water resistance and low vapor transmission. Its major use in construction is as a vapor barrier on walls, floors, and ceilings. It is also used on basement walls as part of the waterproofing application (Fig. 24.5).

Polypropylene

Polypropylene (PP) is much like polyethylene but is more heat resistant and stiffer. It is used for pipes to carry hot water and for waste disposal systems. It is used to make strong fibers for carpeting.

Polystyrene

Polystyrene (PS) is a water-resistant, dimensionally stable, transparent plastic that maintains its properties at low temperatures but begins to soften around 200°F (94°C). It is brittle and has poor weathering qualities. When produced in a foamed condition, it is widely used as an insulation. The foam is also used as the core for insulated doors and sandwich panels and for roof insulation (Fig. 24.6).

Vinyls

The term vinyl describes a large group of plastics developed from the ethylene molecule. Those used in construction are polyvinyl chloride (PVC) and polyvinyl butyral (PVB).

Figure 24.5 Polyethylene sheets can be used to protect construction materials.

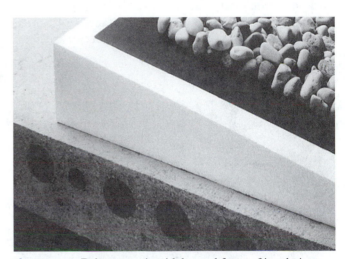

Figure 24.6 Polystyrene is widely used for roof insulation panels. *(Courtesy Associated Foam Manufacturers)*

Polyvinyl chloride is the most widely used vinyl in the manufacture of products used in construction. It has high impact resistance, high abrasion resistance, and good aging qualities, and it is dimensionally stable. A long list of products are made from PVC, including siding, gutters, floor tile, pipe, and window frames (Fig. 24.7). It is bonded to other materials, such as plywood panels, to give a protective skin. PVC is available as a rigid or flexible foam and used as a core material in panel construction. It is copolymerized with other plastics to produce a binder for terrazzo floors and a number of adhesives.

Polyvinyl butyral (PVB) is used as an inner layer in safety glass and as a protective coating on fabrics. Another type of vinyl, *polyvinyl acetate* (PVAc) is used in mortars, paints, and adhesives.

Thermosets

Thermoset materials are used for products requiring greater heat resistance and stiffness than that afforded by thermoplastics. However, they find limited use because they are brittle and harder to form.

Epoxies

Epoxies have good chemical and moisture resistance but are mainly used because of their excellent adhesive qualities. They are widely used in coating compounds and as adhesives. They bond to almost any material and are used in the assembly of panels, the bonding of veneers and overlays, and as protective coatings. They are used as mortars for bonding concrete block and in patching material for damaged concrete. Epoxies are also used to produce some types of fiberglass reinforced plastic.

Formaldehyde

Formaldehyde plastics are incapable of plastic deformation and are hard, strong, heat resistant, and brittle. There are three types that find some use in construction.

Phenol-formaldehyde (PF) generally referred to as *phenolics,* has fillers such as glass fibers added to improve impact resistance and strength. It has good electrical and thermal properties. Phenolic plastics are the most widely used of the thermosets. They are used to mold electric parts, such as switches, boxes, and circuit breakers, and hardware.

An important use is the coating of kraft paper that forms the base of high-pressure plastic laminates used for countertops and other surfaces. The cardboard interior structure of hollow-core doors and panels is impregnated with phenolic resin. It is also used to make some types of adhesives, protective coating materials, and foamed insulation (Fig. 24.8).

Melamine-formaldehyde (MF), generally referred to as melamines, are hard, scratch-resistant plastics that withstand chemical attack. They are used in the production of high-pressure plastic laminates such as those used for countertops. They are used as an adhesive in the production of plywood, and they are used to mold hardware and electrical fixtures.

Urea-formaldehyde (UF) has the same uses as melamine-formaldehyde. Urea-formaldehyde is not as hard and does not have as good heat-resisting properties as melamine-formaldehyde.

Polyesters

A large number of plastics fall under the generic name *polyester.* Polyesters include the Mylars from which drafting film is made, alkyds used for paints and enamels, and fiberglass reinforced plastics. Saturated and unsaturated

Figure 24.7 Polyvinyl chloride is used to produce exterior products such as siding, gutters, soffits, and fascias. *(Courtesy Wolverine Technologies)*

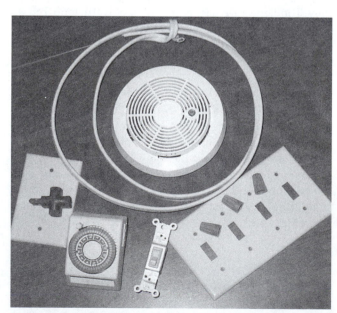

Figure 24.8 Phenolics are used to mold electrical parts.

polyesters are produced. The most important group is the *unsaturated polyesters.* They can be linked to a monomer, such as styrene, forming a strong, hard thermosetting plastic. Their major use is in the production of reinforced plastics to which glass fibers are added. This produces a stable, tough, impact-resistant, high-strength material. Major uses include molded bathtubs, showers, sheets for roofing and partitions, curtain wall exterior laminates, and window frames and sash. Reinforced polyesters have ultimate strengths up to 50,000 psi (344 500 kPa) and E values as high as 3×10^6 psi. The strength of the plastic is proportional to the amount of glass reinforcement used because the glass fiber has a tensile strength approaching 150,000 psi (1 033 500 kPa).

Polyurethanes (UP)

Polyurethanes are used to produce low-density foams that can be varied from soft, open-cell types that are flexible to a tough, closed-cell, rigid material. These foams have very low thermal conductivity and make excellent insulation. They have good chemical and heat resistance and good tensile strength. Polyurethane foams are used to fill wall, ceiling, and floor cavities by spraying them as a creamy foam on the surface or in the cavity, after which they solidify. They are also used to insulate pipes, ducts, and wall panels. Rigid insulation sheets are widely used for many purposes, such as insulating flat roofs over which a hot tar and gravel roof is applied.

Polyurethanes find use in the manufacture of **elastomers** (synthetic rubbers), caulking, adhesives, and glazing materials.

Silicones (SI)

Silicone plastics are not based on the carbon atom, as all those previously discussed, but on the silicon atom. They have good corrosion resistance and are good electrical insulators. Heat resistance properties enable them to have a service temperature ranging from $-80°F$ ($-27°C$) to $500°F$ ($260°C$). An important property is that they withstand exposure to the elements.

Since silicons are very water repellent, they (in liquid form) are applied to exterior masonry materials to provide a water-resistant coating. This stops water from penetrating a wall, such as a brick exterior wall, yet has the permeability to allow internally developed moisture to pass through as a vapor (Fig. 24.9).

Silicone rubber is soft, heat resistant, and does not harden at low temperatures. It is used in sealants and gaskets where watertightness is desired.

MANUFACTURING PROCESSES

Plastic resins in the form of liquid, powder, or beads are produced by chemical companies and sold to companies that manufacture plastic products. The manufacturer com-

Figure 24.9 Silicone is commonly used in sealants and waterproofing materials.

bines these polymers with additives to modify the properties of the polymer to meet the needs of the product to be produced. The modified polymer is then processed by equipment of various types to produce the products.

Common manufacturing processes include blow molding, calendaring, compression molding, casting, extrusion, expandable bead molding, form molding, injection molding, laminating, rotational casting, transfer molding, and thermoforming.

Blow molding involves placing a heated, preformed plastic tube called a paison in a forming die. Air pressure is raised inside the plastic tube, forcing it to conform to the shape of the die. When it is cool the die is opened and the part is removed (Fig. 24.10). Most bottles are formed this way.

Calendaring involves moving a plastic material in a liquid state through a series of rollers, which creates a thin plastic film as the material solidifies (Fig. 24.11). Vapor barrier and floor covering materials are typical products.

Compression molding involves placing plastic resin in powder form in a heated mold to which heat and pressure are applied. The resin melts and fills the mold (Fig. 24.12).

Casting involves pouring plastic in a liquid state into a cavity in a mold. The liquid plastic fills the cavity and hardens. *Extrusion* is a process in which a semiliquid plastic is forced under pressure through an opening in a die. The shape of the die opening determines the shape of the extruded member. This is much the same as squeezing toothpaste out of a tube (Fig. 24.13).

Expandable bead molding is a process in which small granules of resin, such as polystyrene, are mixed with an expanding agent and placed in a steam-heated rolling drum. When the granules have expanded, they are cooled and are transferred to a mold in which they are heated until they fuse together (Fig. 24.14).

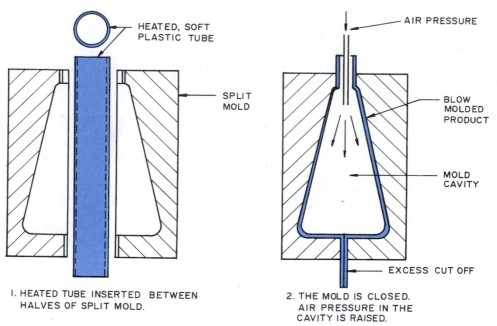

Figure 24.10 Products formed by blow molding are shaped by air pressure forcing the plastic to the contours of the mold.

Form molding requires that an expanding agent be mixed with plastic granules or powder and injected into a mold, where it is heated, melting the resin and forming a gas that expands the resin to fill the mold cavity.

Injection molding uses granules or powder resin. This is fed into the heated cylinder of the injection molding machine and forced by a ram into a cold mold where it solidifies (Fig. 24.15).

Laminating is a process in which several layers of material are bonded together to form a single sheet. An example is laminating a colored plastic veneer to a plywood backing for use as wall paneling. The materials to be laminated are impregnated with the plastic resin, placed together, and bonded by the application of heat and pressure.

Rotational molding forms hollow one-piece items from polyethylene powders. The resin is placed inside the mold,

which is heated as it rotates about two axes. The resin is distributed to the surfaces of the mold by centrifugal force and is fused by the heat (Fig. 24.16).

Transfer molding is a combination of compression and injection molding. The resin is made liquid in the transfer chamber outside the mold and is then injected into the mold, where it fills the cavity and solidifies (Fig. 24.17).

Thermoforming involves two commonly used procedures, vacuum forming and pressure forming. *Vacuum forming* involves placing a heated sheet of plastic over a mold cavity and pulling a vacuum below it. Atmospheric pressure forces the sheet to the shape of the mold (Fig. 24.18). *Pressure forming* involves placing a heated sheet of plastic over a mold cavity and increasing the pressure behind the sheet, forcing it into the cavity of the mold.

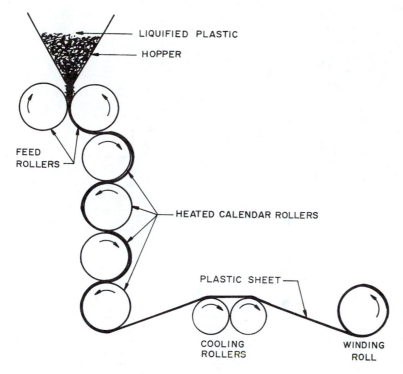

LIQUIFIED PLASTIC

HOPPER

FEED ROLLERS

HEATED CALENDAR ROLLERS

PLASTIC SHEET

COOLING ROLLERS

WINDING ROLL

Figure 24.11 Plastic sheet materials such as vapor barriers and floor covering are produced by the calendaring process.

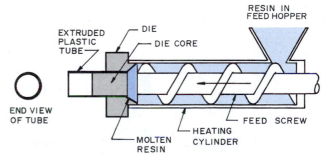

RESIN IN FEED HOPPER

EXTRUDED PLASTIC TUBE

DIE

DIE CORE

END VIEW OF TUBE

MOLTEN RESIN

FEED SCREW

HEATING CYLINDER

Figure 24.13 The extrusion process involves forcing molten plastic through an opening in a die.

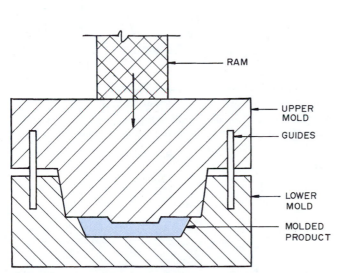

RAM

UPPER MOLD

GUIDES

LOWER MOLD

MOLDED PRODUCT

Figure 24.12 Compression molding applies pressure and heat on a powdered resin, forcing it into a mold cavity.

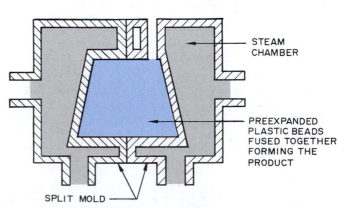

STEAM CHAMBER

PREEXPANDED PLASTIC BEADS FUSED TOGETHER FORMING THE PRODUCT

SPLIT MOLD

Figure 24.14 Expanded plastic beads are fused together to form a solid using the expandable bead molding process.

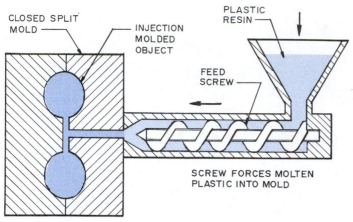

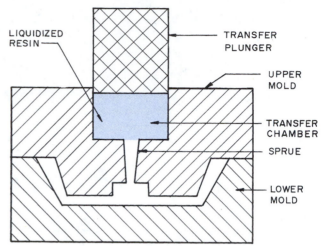

Figure 24.15 Injection molded parts are formed by forcing a molten plastic resin into a die cavity.

I. RESIN IS LIQUIDIZED IN THE TRANSFER CHAMBER.

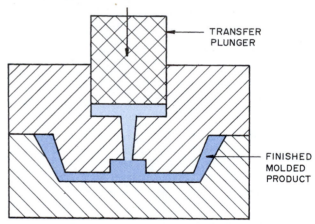

2. RESIN IS INJECTED INTO THE MOLD CAVITY.

Figure 24.17 Transfer molding liquifies the plastic resin in a transfer chamber and injects it into a mold cavity.

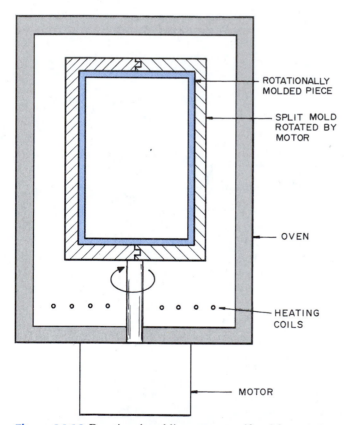

Figure 24.16 Rotational molding uses centrifugal force to spread the plastic resin to the surface of the mold, where it is fused by heat.

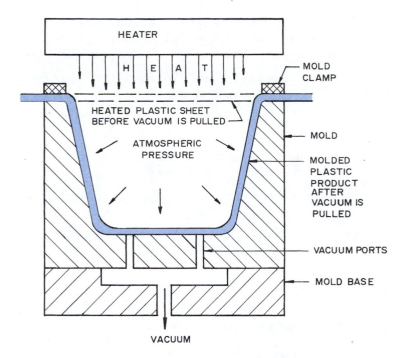

HEATER

H E A T

MOLD CLAMP

HEATED PLASTIC SHEET
BEFORE VACUUM IS PULLED

ATMOSPHERIC
PRESSURE

MOLD

MOLDED
PLASTIC
PRODUCT
AFTER
VACUUM IS
PULLED

VACUUM PORTS

MOLD BASE

VACUUM

Figure 24.18 Pressure forming involves pulling a vacuum in the mold cavity, permitting atmospheric pressure to force the heated plastic sheet to the surface of the cavity.

REVIEW QUESTIONS

1. What two atoms form the nucleus for the derivation of the material plastics?
2. What is the process called that enables carbon atoms to bond with other atoms?
3. What is a single molecule called?
4. What is formed when several monomers link together?
5. What is the valence of the carbon atom?
6. What are the two major classifications of plastics?
7. Why is it possible to heat and reform thermoplastic materials?
8. Why is it that thermoset plastics cannot be reheated and reformed into new shapes?
9. What are the major groups of additives used in formulating plastics?
10. How do plasticizers change the properties of plastics?
11. What is a common way to reduce the cost of plastics?
12. What is added to plastic to increase its strength?
13. What colorants are used to produce desired colors?
14. What are the most important mechanical properties of plastics used in construction?
15. What are the two major considerations when considering the thermal properties of plastics?
16. Which plastics have the lowest thermal conductivity?
17. What is the major danger produced by burning plastics?
18. When plastics deteriorate due to chemical action, what type of damage occurs?
19. Which plastics have good optical properties?
20. Identify the following plastics by their standard abbreviation: ABS, PMMA, PA, PE, PVC.
21. What is the major use for ABS plastics?
22. What type of plastic would you choose for door and window glazing?
23. What type of plastic is used in the manufacture of latex paints?
24. The vapor barriers used in walls and floors most likely are made from what type of plastic?
25. What type of plastic weathers well and is used for exterior purposes, such as siding, gutters, and window frames?
26. What thermoset plastic is known for its use as a strong adhesive?
27. What plastics are used in the manufacture of high-pressure plastic laminates?
28. What type of plastic is used in the manufacture of showers and bathtubs?
29. What thermoset plastic is widely used to produce foamed insulation as commonly sprayed on roofs and in wall cavities?
30. What uses are made of silicones in building construction?

KEY TERMS

additive Material mixed with the basic resin to alter its properties.

arc resistance The total elapsed time in seconds an electric current must arc to cause a part to fail.

covalent bonding A process in which small numbers of atoms are bonded into molecules.

creep A permanent dimensional change due to stress applied over a long period of time without the use of heat.

dielectric strength The maximum voltage a dielectric (nonconductor) can withstand without fracture.

elastomers Synthetic rubbers that can at room temperature be stretched twice their original length and upon release return instantly to their original length.

fillers An inert material added to a resin to alter the strength, working properties, and to lower the cost.

flammability The ability of a material to resist burning.

hardness A measure of abrasion resistance.

monomer A molecule of low molecular weight that can react with other molecules of low molecular weight to form a polymer.

organic material A class of compounds comprised only of those existing in plants and animals.

plastic An organic material that is solid in its finished state but is capable of being molded or of receiving form.

plastic behavior The ability of a material to become soft and formed into desired shapes.

plasticizers Materials added to the plastic resin to increase the flexibility and workability.

polymer A chemical compound formed by the union of simple molecules to form more complex molecules.

service temperature The maximum temperature at which a plastic can be used without altering its properties.

stabilizers Additives used to stabilize the plastic by helping it resist heat, loss of strength and resist the effect of radiation on the bonds between the molecular chains.

synthetic materials Materials formed by the artificial building up of simple compounds.

thermoplastic A plastic that can be repeatedly softened by heating and solidified by curing.

thermosets A plastic that hardens upon curing and cannot be softened again by heating.

valence Points on an atom at which other elements can bond.

weatherability The ability of a plastic to resist deterioration due to moisture, ultraviolet light, heat, and chemicals found in the air.

SUGGESTED ACTIVITIES

1. Collect samples of various plastic products used in construction. Try to identify the type of plastic.

2. Develop a series of tests you can perform with the facilities available and test plastics samples. For example, see which will ignite the quickest when exposed to a continuous flame and which will burn after the flame is removed. Run tensile tests, expose samples to acids, hot water, etc., and record what happens to each.

ADDITIONAL INFORMATION

Hornbostel, C., *Construction Materials: Types, Uses, and Applications,* John Wiley and Sons, Inc., New York, 1991.

Ramsey, C.G., Sleeper, H.R., and Hoke, J.R., eds., *Architectural Graphic Standards,* John Wiley and Sons, New York, 1988.

Structural Plastics Selection Manual, American Society of Civil Engineers, New York, 1990.

PART V

THERMAL AND MOISTURE PROTECTION, DOORS, WINDOWS, AND FINISHES

E nergy conservation is a major consideration when designing a building and selecting materials. In this period of increasing energy costs and potential shortages, energy conservation is of increasing importance.

Structures are exposed to the weather and below-ground water. New and improved products to control moisture inside and outside of buildings appear on the market frequently. Energy conservation and moisture control are necessary to protect the building and the comfort of the occupants.

CSI Division 7 chapters cover information on thermal and moisture protection. Chapter 25 describes various products used to provide thermal insulation and to prevent vapor penetration through assembled materials, as a wall. Chapter 26 cites the types of bonding agents available and their appropriate applications. It also covers the very important topic of joint sealants for use on exterior applications. Protection from moisture penetration is critical to a successful job. Finally, the chapter presents information on above- and below-grade waterproofing coatings and membranes.

Chapter 27 gives technical information on bituminous materials and the many products that use them as a major ingredient, including those used for waterproofing and finished roofs.

Chapter 28 gives detailed information about roofing systems used on residential, commercial, and industrial buildings. These include bituminous materials, as well as metal, clay, concrete, wood, and plastic products.

CSI Division 8 chapters include information on glass, doors, windows, entrances, and cladding systems. Chapter 29 tells you about the types of glass used in construction and their appropriate applications. Doors and windows serve a number of functions, such as access, light, ventilation, and emergency egress, and are a major exterior design feature. Chapter 30 illustrates the wide choices available and cites the materials used in their construction.

Chapter 31 presents examples of the very important subject of cladding the exterior. It will help you understand what materials are used, how building codes influence choices, and how various systems are installed.

The remainder of Part V is devoted to finishes, the topic of CSI Division 9. Chapters 32 through 38 cover an extensive array of materials and installation techniques—gypsum, plaster, various paints and other coatings, acoustical materials, flooring, and carpeting. ▲

D I V I S I O N 7

THERMAL AND MOISTURE PROTECTION

CSI MASTERFORMAT™

Photo courtesy the Asphalt Institute

Courtesy Gerard Roofing Technologies

Thermal Insulation and Vapor Barriers

This chapter will help you to:

1. Select appropriate types of insulation for various applications in the design of a building.

2. Understand the way heat is transferred and how insulation can be used to control it.

3. Be aware of the problems caused by moisture penetrating the insulation and learn how you can design an assembly of materials to reduce moisture transmission.

A wide variety of materials are in use as thermal insulation. These include wood, plastics, and metal products. Considerable attention is now given to the design and construction of energy-efficient buildings. Manufacturers have responded with an array of products for use in almost any environment or situation. A key to energy-efficient design is understanding how heat is transferred.

THERMAL INSULATION

Thermal insulation is manufactured from a variety of materials. *Metallic* insulation is in the form of aluminum or copper and other metallic foil or as an organic insulation material with a metallic laminate. *Organic fibrous* insulation materials include cane, cotton, wood, cellulose, and synthetic fibers. *Organic cellular* materials include polyurethane, polystyrene, cork, and foamed rubber. *Mineral cellular* insulation materials include perlite, vermiculite, and foamed glass. *Mineral fibrous* materials include rock, glass, slag, and asbestos melted and spun into fibers.

Thermal insulation is available as granular or fibrous *loose fill*, *flexible* woollike blankets and batts, *rigid* sheet material, *liquid spray* using a mineral fiber or insulating concrete, *cast-in-place* using insulating concrete, and *foamed-in-place*, such as polyurethane foams.

METHODS OF HEAT TRANSFER

Heat is transferred through materials from the warm side to the cool side. To avoid a loss of heat in a building in the winter, insulation is used to break this transfer. Likewise, in the summer high exterior temperatures cause heat to pass into a building unless the transfer is retarded by insulation.

The amount of heat that can be transferred depends on the characteristics of the material and its thickness. Porous and fibrous materials that have many air pockets, such as wood, transfer less heat than solid materials, such as brick. A thick material transfers less heat than a thin section of the same material.

Heat is transferred by radiation, convection, or conduction. **Radiation** involves the transmission of heat by electromagnetic waves. The heat energy passes through the air between the source and the body to be heated without heating the intervening air. Some materials accept radiant energy and others reject it. Shiny materials reject radiant energy and dark materials absorb it. A white shingle roof reflects more radiant energy than a brown or black roof, thus increasing the energy efficiency of a building.

Convection heat transfer involves the transfer of heat by the circulation or movement of heated liquids or gases. The heat is moved by natural or forced (fan) means by currents of air that absorb heat brought to the space, such as by a hot water heating system convector (Fig. 25.1). As warm air passes over a cool surface it transfers some of its heat to the cool surface. This convection current is a means for heat transfer and occurs in every building.

Conduction heat transfer involves the transmission of heat by its passing through a solid or a liquid. For example, the sun can heat a brick wall and the heat is transferred to the other side through the brick by conduction.

DESIGNATING THERMAL PROPERTIES

Heat is the result of the movement of molecules of a substance. Heat quantity is measured in **British thermal units (Btu).** One Btu is the amount of heat needed to raise the temperature of 1 lb. of water 1°F. In the metric system heat quantity is measured in calories. One *calorie* is the heat required to raise the temperature of 1 gram of water 1°C. Calories can be converted to joules (J) by multiplying calories by 4.18. A **joule** is a derived metric unit used to describe work, energy, and quantity of heat. The terms used to describe the thermal properties of materials include thermal conductivity (k), thermal resistance (R), thermal conductance (C), and thermal transmittance (U).

Thermal conductivity, k, is a measure used to indicate the amount of heat that will be conducted through a square foot of area of a material per a specified unit of thickness. The lower the k value the better the insulating qualities.

In the U.S. customary system of measurement thermal conductivity is indicated as Btu · in./ft.^{2} · hr. · °F or Btu per inch of thickness per square foot per hour per degree Fahrenheit. In the metric system it is W/m · K or watts per meter-kelvin.

Thermal resistance, R, is a measure used to indicate the ability of a material to resist the flow of heat through it. The larger the R-value the greater the resistance and, therefore, the better the insulating value. Thermal resistance, R, is the reciprocal of conductance, C, or R = ¹⁄c. In the U.S. customary system of measurement thermal resistance R = (hr · ft.^{2} · °F)/Btu (hours per square foot per degree Fahrenheit per Btu). In the metric system thermal resistance is specified as

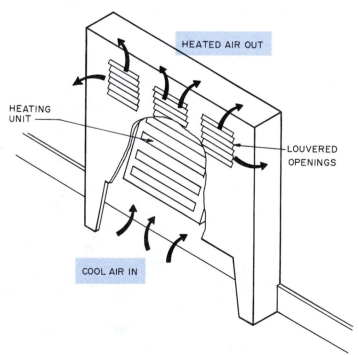

Figure 25.1 This convector has a hot-water heating unit that heats the air that passes over it.

R SI. In this system R SI = $(K \cdot m^2)/W$ or kelvin-meter squared per watt.

Total thermal resistance, R_t, is the resistance to heat flow through an assembly of materials.

Thermal conductance, C, is a measure used to indicate the amount of heat that will pass through a specified thickness of material. It is the reciprocal of thermal resistance: $C = \frac{1}{R}$. In the U.S. customary system of measurement thermal conductance C = Btu (hr. $\cdot$ ft.$^2 \cdot$ °F) or Btu per hour per square foot per degree Fahrenheit. In the metric system thermal conductance is W/(m$^2 \cdot$ °C) or watts per square meter-degree celsius.

Thermal transmittance, U, is a measure of the amount of heat that would pass through an assembly of various materials, such as an exterior wall. It is the reciprocal of the total resistance, R, of the assembly: $U = \frac{1}{R}$. The smaller the U-value the greater the resistance to the transmission of heat.

INSULATION MATERIALS

A wide array of insulating materials are available (Table 25.1). The decision of which to use depends on their location in the building, the environment to which they will be exposed, their effectiveness in resisting heat transfer, and their cost relative to the savings expected from improving the energy efficiency of the building. The major classifications of insulation include loose fill, batts and blankets, rigid, reflective, foamed-in-place, and sprayed. Insulation materials are rated by their ability to resist heat flow, which is indicated by their R-value. Typical R-values are in Table 25.2.

Loose-Fill Insulation

Loose-fill insulation is available as a granular or loose fibrous material. Granular insulation is poured and fibrous is machine-blown into the areas to be insulated. *Granular* insulation includes perlite (expanded volcanic rock), vermiculite (expanded mica), cork, and expanded polystyrene. These are available in different densities, which influences the R-value (Table 25.3). Granular insulation materials are also used to produce a lightweight concrete roofing base material that is pumped and leveled on the roof deck. Expanded polystyrene roof insulation boards are placed on this and are bonded to the deck. The roof base material is flowed over the insulation board, forming a base for the application of the finished roofing material (Fig. 25.2).

Fibrous loose insulation is made by blowing a jet of air through molten glass, slag, or rock to form thin fibers that form a woollike substance when gathered. One type is made from cellulosic fibers obtained from wood chips, newsprint, and other organic fibers. Cellulosic fibers must be fireproofed. This treatment is noted by the Underwriters Laboratory rating on the label. Granular insulation is used when filling vertical cavities, such as cores in concrete block and wall cavities (Fig. 25.3). Fibrous loose insulation is difficult to work into such places. It is widely used in large areas, such as insulating ceilings, where it can be sprayed. It rapidly covers the area with a uniform layer.

Batts and Blankets

Batts and blankets are flexible insulation mats made from fiberglass, mineral wool, cotton fibers, or wood fibers. Batts are 48 in. (1219 mm) long. Blankets are in rolls up to 8 ft. long (2438 mm). Some thinner blankets are available in longer rolls. Common thicknesses are 3½, 6½, 9½, and 12 in. (89, 158, 241, 305 mm) and widths 15 and 23 in. (380 and 584 mm). Other sizes are available for special applications, such as noise barrier insulation, 2¾ and 4 in. (70 and 100 mm), and furred out masonry walls, ¾ in. (18 mm). Batts and blankets are available for a variety of applications, such as low-density insulation used in residential construction, extra-low-flame-spread types used in commercial construction in which the insulation will remain exposed, special batts for use over ceiling panels in suspended ceilings, and one type that forms a noise barrier, increasing the sound transmission class performance of the wall, floor, or ceiling.

Batts and blankets are available unfaced, faced on one side with moisture-resistant kraft paper that forms a vapor barrier, and faced with aluminum foil that forms a fire-resistant facing. Some types have a facing on both sides and are used on vertical applications, such as walls, and horizontal applications, such as floors and ceilings in commercial buildings. They can be used to wrap around items that need to be insulated, such as water heaters.

Some types are designed to be placed between studs or joists and held by friction. Other types are stapled to the face or side of the stud. The vapor barrier side faces the inside of the building (Fig. 25.4).

Table 25.1 Forms of Building Insulation

Materials
Metallic, organic fibrous, organic cellular, mineral cellular, mineral fibrous
Forms
Loose fill, flexible blankets, rigid sheets, liquid spray, cast-in-place, foamed-in-place

Table 25.2 R-Values of Insulation Materials[a]

Fiberglass Board Panels			
Thickness (in.)	Thickness (mm)	R-Value (customary units)[b]	R = Value (SI units)[c]
1	25	4.30	30
1.5	38	6.50	45
2	50	8.70	60
2.5	63	10.90	75
3	75	13.00	90
4	100	16.50	114

Extruded Polystyrene Rigid Panels			
Thickness (in.)	Thickness (mm)	R-Value (customary units)[b]	R = Value (SI units)
0.75	18	3.80	26
1	25	5.00	35
1.5	38	7.50	52
2	50	10.00	69
2.6	66	16.67	115
3.1	78	20.00	138

Polyisocyanurate Rigid Panels			
Thickness (in.)	Thickness (mm)	R-value (customary units)[b]	R = Value (SI units)
1.2	30	7.14	49
1.6	40	10.00	69
2.0	50	12.50	86
2.6	66	16.67	115
3.1	79	20.00	138

Polyurethane Rigid Panels			
Thickness (in.)	Thickness (mm)	R-Value (customary units)[b]	R = Value (SI units)
1.0	25	6.25	43
1.5	38	10.00	69
2.0	50	14.30	98
2.7	68	20.00	138

Wood Fiber Rigid Panels			
Thickness (in.)	Thickness (mm)	R-Value (customary units)[b]	R = Value (SI units)
0.5	12	1.40	10
0.75	18	2.10	14
1	25	2.80	19
1.5	38	4.20	29
2	50	5.60	38

Granular Insulation			
	Thickness (mm)	R-Value (customary units)[b]	R = Value (SI units)
Vermiculite per inch	per 25 mm	2.1 to 2.3	14 to 16
Perlite per inch	per 25 mm	2.6 to 3.5	18 to 24

Aluminum Reflective Insulation			
	Thickness (mm)	R-Value (customary units)[b]	R = Value (SI units)
Multilayer foil batts with three quarter inch air spaces	18 mm air space	summer 17 winter 13	117 89
Multilayer foil batts with one inch air spaces	25 mm air space	summer 30 winter 18	207 124

[a]Data obtained from various manufacturer catalogs. Consult manufacturer's specification sheets for specific data.
[b]R-value (customary units) hr. · ft.^{2} · °F/Btu
[c]R SI (metric units) m^2 · K/W

Table 25.3 Properties of Loose-Fill Insulation

Material	Density (lb./ft.^{3})	R-Value (hr. · ft.^{2} · °F/Btu)	Water Vapor Permeability
Cellulose	2.2–3.0	3.1–3.7	High
Expanded polystyrene	0.9–1.8	3.8–5.0	High
Fiberglass	0.6–1.0	2.9–3.7	High
Mineral wool	1.5–2.5	2.9–3.7	High
Perlite	2–11	2.7–2.9	High
Vermiculite	4–10	2.1–2.3	High

Figure 25.2 Casting an insulating roof base coat using Zono-lite®, which uses granular vermiculite aggregate in a Type I or Type III portland cement. *(Courtesy W.R. Grace and Co.)*

Figure 25.3 Granular insulation is a good choice for filling cavities. *(Courtesy W.R. Grace and Co.)*

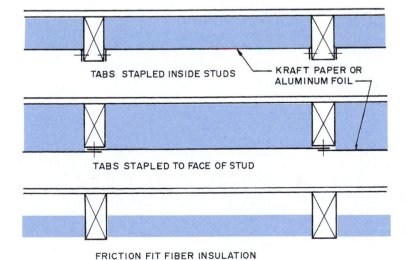

TABS STAPLED INSIDE STUDS — KRAFT PAPER OR ALUMINUM FOIL

TABS STAPLED TO FACE OF STUD

FRICTION FIT FIBER INSULATION

Figure 25.4 Insulation batts and blankets are installed between studs and joists in several ways.

I T E M   O F   I N T E R E S T

INSULATING AROUND RECESSED LIGHTS

Codes require that insulation be kept away from recessed light fixtures, usually 3 to 4 in. (76.2 to 101.6 mm). Incandescent lamps get very hot and can be the source of fire. Metal or other fire-resistant material can be used to form a baffle around the light. It should extend at least 4 in. (101.6 mm) above the insulation and be fastened to a ceiling joist. The only disadvantage to having baffles larger than the 4 in. (101.6 mm) clearance is that it increases the uninsulated ceiling area and therefore the heat loss. The baffle can be altered to allow it to be fastened to the joist as shown in the accompanying illustration.

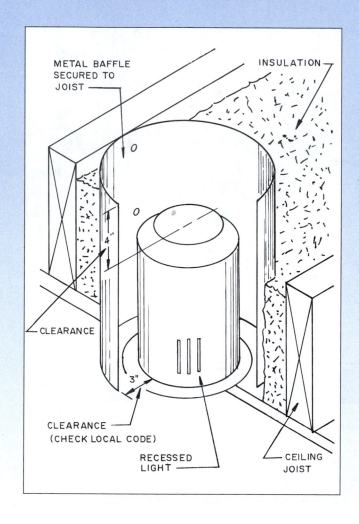

METAL BAFFLE SECURED TO JOIST

INSULATION

CLEARANCE

CLEARANCE (CHECK LOCAL CODE)

RECESSED LIGHT

CEILING JOIST

Rigid Insulation

Rigid insulation board is made using organic fibers, such as wood or cane; mineral wool fibers; glass fibers; corkboard; several forms of expanded plastics, such as expanded and extruded polystyrene and polyisocyanurate foam; and some forms of cellular hard rubber. These products are used in all parts of a building, including roof, wall, and floor insulation (Fig. 25.5).

Wood and cane fiberboard are commonly used for exterior sheathing and shingle backer boards and are asphalt impregnated. They are also used for roof insulation. Rigid insulation sheets made from *granulated cork* are used for roof, wall, and floor insulation and are available in thicknesses from 2 to 12 in. (50 to 305 mm) and sheet sizes 24 × 36 in. (610 × 915 mm). *Mineral wool wall panels* have the insulation mat bonded to a rigid back sheet and are commonly used for roof insulation.

Expanded and extruded plastic rigid sheet insulation is used on walls, floors, roofs, and foundations (Fig. 25.6). Some are made tapered so a flat roof will have some slope. Usually a built-up roof membrane is placed over the rigid insulation panel. One type is bonded to a particleboard sheet that provides a nailing surface for shingles (Fig. 25.7). Rigid polystyrene insulation panels are also used on roof decks to be covered with insulating concrete. The variation in panel thicknesses allows the roof to slope to drainage sources (Fig. 25.8). The insulating concrete has a cellular structure that reduces weight and increases the insulation properties.

Flat panels containing a fiber insulation core may be unfaced or faced with an aluminum foil vapor barrier. Common thicknesses are 1, 1½, 2, 2½, and 3 in. (25, 38, 50, 63, and 75 mm). Standard fiber core panel sizes are 24 × 48 in. (610 × 1220 mm). Extruded and expanded plastic panels typically are 16 and 24 in. (406 and 610 mm) in width and 48 to 96 in. (1220 to 2438 mm) in length. Common thicknesses are 1, 1½, 2, 2½, 3, and 4 in. (25, 38, 50, 63, 75, and 100 mm). Other sizes are available. Manufacturers should be consulted for sizes and R-values.

Most rigid insulation panels are manufactured from flammable materials. Building codes require they be covered with a fire-resistant material, such as gypsum board. Polystyrene products deteriorate when exposed to ultraviolet rays from the sun and must be covered to avoid exposure. Molded polystyrene and expanded perlite board must be protected from water. Many other forms of insulating board products are available, primarily for use on commercial buildings. A few of these are shown in Fig. 25.9.

Figure 25.6 Rigid insulation board can be used on foundations below grade. *(Courtesy U.C. Industries, Inc.)*

Figure 25.7 This roof decking is made by laminating rigid foamed insulation sheets to particleboard, which provides a nailing surface. *(Courtesy Tectum, Inc.)*

Figure 25.5 Rigid insulation board is used to insulate concrete roof decks. *(Courtesy U.C. Industries, Inc.)*

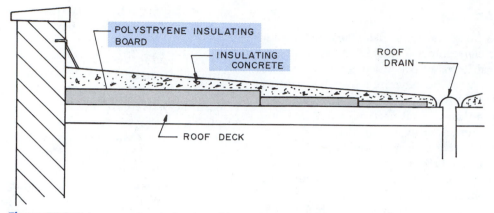

Figure 25.8 Polystyrene insulating board is used to insulate concrete roofs and to provide a sloping base on which insulating concrete can be poured.

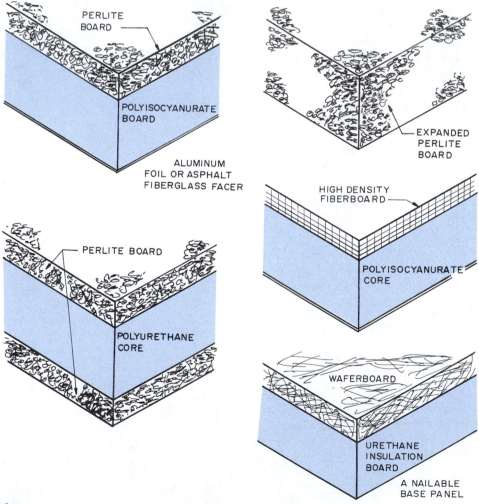

Figure 25.9 Common forms of insulating sheet products.

Reflective Insulation

Reflective insulation is usually aluminum or copper foil in sheets or rolls (Fig. 25.10). Rolls are usually 24 and 48 in. (610 and 1220 mm) wide and 500 ft. (152.5 m) long. It is available in single thickness layers or in a multilayer batt that has dead air spaces between the layers. The reflective foil utilizes the reflective properties to reject the passage of heat and increase the effectiveness of the dead air spaces (Fig. 25.11). The foil may be bonded onto a heavy kraft paper, insulation board, or gypsum lath. Some forms of fiber insulation batt have reflective foil bonded to one side. The foil also serves as a vapor barrier. Reflective insulation is used in residential and commercial construction in walls, floors, ceilings, and roofs (Fig. 25.12). It can reduce heat flow by as much as 25 percent, reducing the amount of conventional insulation needed and in some cases reducing the size of the air-conditioning units required. It is easily cut with a knife and installed by stapling.

Foamed-in-Place and Sprayed Insulation

Foamed-in-place insulations are generally polyurethane or phenol-based compounds that provide excellent insulation. When mixed they are pumped through hoses into cavities, such as wall cavities, and sprayed in layers on flat and sloping surfaces, such as roof decks. The ingredients are carefully measured and mixed by special equipment. The equipment meters the isocyanate and polyol at a one-to-one ratio. These are pumped as separate materials into the proportioning unit, which heats each and pumps them into separate hoses that are heated. The two components are mixed in a spray gun and sprayed on the substrate. The mixture also can be applied by power or hand rollers. Since the material expands as it hardens, the amount injected into a cavity must be carefully measured. If a residential wall cavity is overfilled the pressure could break the gypsum. Polyurethane will burn and must be covered with a fire-resistant finish.

Figure 25.10 Reflective insulation is available in rolls. *(Courtesy Advanced Foil Systems)*

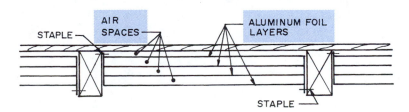

Figure 25.11 Some forms of reflective insulation have multilayers of foil with dead air spaces between them.

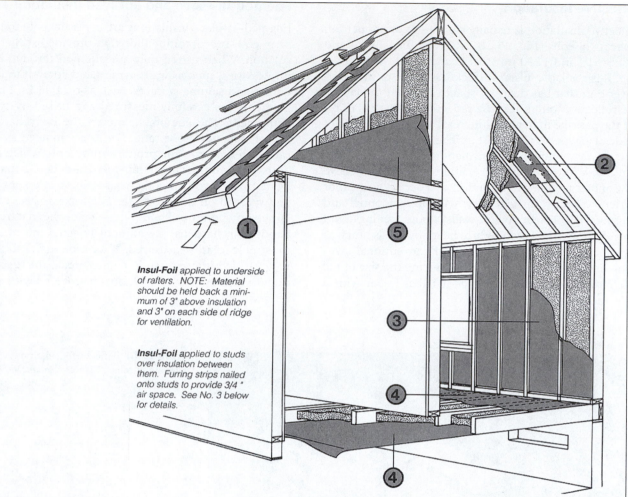

Insul-Foil applied to underside of rafters. NOTE: Material should be held back a minimum of 3" above insulation and 3" on each side of ridge for ventilation.

Insul-Foil applied to studs over insulation between them. Furring strips nailed onto studs to provide 3/4 " air space. See No. 3 below for details.

Insul-Foil Applications:

1. Attics: New construction – **Insul-Foil** draped between rafters or trusses just prior to sheathing. Or, stapled between, or across the face of top chord of truss or bottom of rafters.

2. Cathedral Ceilings: New Construction – In 2" x 10"s or greater with R-19, **Insul-Foil** should be stapled between trusses or rafters approximately 1-1/2" below deck. Note: A vapor barrier should be used below fiberglass.

3. Sidewalls: Apply **Insul-Foil** on studs after batts have been installed between studs. Nail 3/4" x 1-1/2" furring strips over radiant barrier into studs. This provides a 3/4" space between radiant barrier and drywall. This also makes an excellent vapor retarder. Because this is an enclosed air space it would provide an additional R-value of R-2.77.

4. Floors: Under floors over crawl spaces, in addition to being draped over joists prior to sheathing, **Insul-Foil** can be stapled between joists 1" to 1-1/2" below flooring. Excellent for reducing winter heat losses.

5. Kneewalls: Apply **Insul-Foil** to studs on attic side prior to installing fiberglass. **Insul-Foil** is an excellent backer for wet spray-on cellulose.

Figure 25.12 Recommended locations for placing reflective radiant barriers. Insul-Foil® is a registered trademark. *(Courtesy Advanced Foil Systems)*

After the sprayed polyurethane foam insulation is in place it must be protected from exposure to moisture and ultraviolet radiation. Commonly used protective coatings include acrylics, butyls, chlorinated synthetic rubber, modified asphalts, silicones, and urethanes.

Sprayed-on insulations are used on ceilings, walls, tanks, and other items. Although polyurethane is widely used, vermiculite or perlite aggregate mixed with a gypsum or portland cement binder and sometimes fibers are mixed with an inorganic binder and used as insulation. Severe restrictions are placed on insulation with asbestos fibers. Sprayed-on insulation is also used to increase fire resistance and moisture resistance and to improve acoustical properties. It has the advantage of being able to bond to irregular shaped and sloped surfaces. Consult manufacturers for R-value, fire resistance, flame spread, smoke developed and compressive and impact strength.

VAPOR BARRIERS

Another part of the insulation of a building is the installation of vapor barriers. A vapor barrier is used to keep water vapor that is generated inside a building, such as by cooking, from penetrating the wall and condensing as moisture on the insulation. When circumstances warrant, it can be used to reduce the penetration of moisture from outside sources into the building (Fig. 25.13).

Various types of insulation are made with a vapor barrier as part of the sheet or blanket. *Kraft paper* coated with wax or asphalt is commonly used on one side of fibrous insulation batts and blankets. Aluminum foil is a more expensive coating that has heat reflective values as well. *Polyethylene film* is sometimes stapled to the studs on the side facing inside the building and placed under the subfloor and on the ceiling. It also is used to cover the ground in a crawl space to retard moisture from leaving the

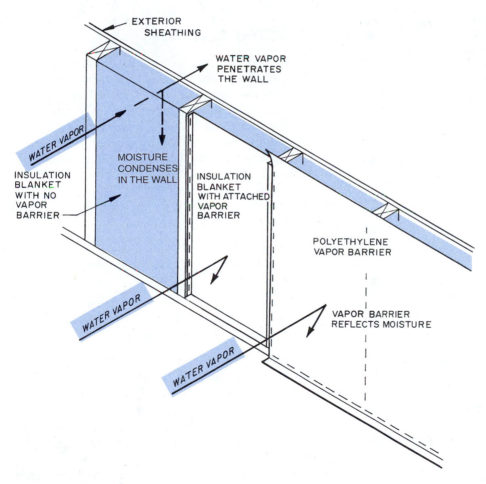

Figure 25.13 Polyethylene vapor barriers reflect moisture generated inside a building, and keeps the moisture from penetrating the insulation.

ground into the air in the crawl space. It is laid below welded wire fabric before concrete slabs are poured. It is an excellent vapor barrier and also assists in reducing air infiltration. Polyethylene film is available in thicknesses of 2, 3, 4, and 6 mils and rolls 3 to 20 ft. (0.9 to 6 m) wide.

Air infiltration from the exterior of a building is also reduced by applying a plastic barrier on the exterior over the sheathing (Fig. 25.14). This material is a spunbound olefin formed into a sheet of very fine high-density polyethylene fibers. It resists tearing and puncture and will not rot. It has a perm of 94, so moisture vapors that get into the wall from inside the building can pass through it to the exterior. In all cases vapor barriers and air infiltration wrap must be carefully installed, overlapped at joints, and sealed to reduce penetration.

Various liquid materials also serve as vapor barriers. Enamels, primers, latex paints, and oil-based paints are used.

Figure 25.14 This building is being wrapped with DuPont Tyvek® Housewrap to reduce air infiltration.

REVIEW QUESTIONS

1. What types of organic fibers are used for insulation?
2. What kinds of organic cellular materials are used for insulation?
3. What kinds of mineral cellular materials are used for insulation?
4. What kinds of mineral fibrous materials are used for insulation?
5. In what forms is thermal insulation available?
6. In what three ways can heat be transferred?
7. What are the four terms used to describe thermal properties of a material?

8. How are insulation materials rated on their ability to resist heat flow?
9. What materials are used for vapor barriers?

KEY TERMS

British thermal unit (Btu) A measure of the amount of heat needed to raise the temperature of one pound of water one degree Fahrenheit.

conduction The transfer of heat by passing it through a solid or liquid.

convection Transmission of heat by currents of air resulting from differences in density due to temperature differences in the heated space.

joule A meter-kilogram-second unit of work or energy.

radiation Transfer of heat through space by means of electromagnetic waves.

SUGGESTED ACTIVITIES

1. Try to set up an experiment to get a rough evaluation of the relative effectiveness of various types of insulation. For example, you might build a box from $\frac{3}{4}$ in. plywood, leaving the top open. Place a heat source in the box, such as a number of high-wattage incandescent lamps. Cover the open top with an insulation product and measure the temperature of the exterior surface after a set number of hours. Repeat with various types and thicknesses of insulation. Be alert for the possibility of fire and do not leave the experiment unattended.

2. Collect samples of various insulation products and label them, identifying their composition and properties.

ADDITIONAL INFORMATION

Brand, R., *Architectural Details for Insulated Buildings*, Van Nostrand Reinhold, New York, 1990.

Insulation Manual, National Association of Home Builders, Washington, D.C., 1990.

Ramsey, C. G., Sleeper, H. R., and Hoke, J. R., eds., *Architectural Graphic Standards*, John Wiley and Sons, New York, 1995.

Wing, C., *The Visual Handbook of Building and Remodeling*, Rodale Press, Emmaus, Pa., 1990.

Wood Building Technology, Canadian Wood Council, Ottawa, Ontario, 1993.

Bonding Agents, Sealers, and Sealants

This chapter will help you to:

1. Select the proper bonding agents for various construction applications.

2. Understand how the various bonding agents cause adherence.

3. Select the correct sealers for use on exterior applications.

4. Decide the best way to protect joints in exterior construction.

5. Select waterproofing materials best suited for various applications.

Bonding agents, sealers, and sealants are made from natural and man-made substances, many of which are described in other chapters. For example, bitumens and various synthetic resins are used to produce many of these products. These materials are discussed in Chapters 24 and 27.

BONDING AGENTS

A **bonding agent** is a compound that holds materials together by bonding to the surfaces to be joined. In many parts of a building and in many products bonding agents are used to permanently fasten things together, such as wood doors and plywood veneers. Bonding agents used in the fabrication of construction products and in on-site applications typically join materials by **mechanical action** or by **specific adhesion.** Those that join by mechanical action are best represented by the bonding of wood. The bonding agent enters the pores of the wood, hardens, and forms a mechanical link. Bonding agents that join by specific adhesion are used to bond dense materials without pores, such as glass and metal. The bonding is caused by the attraction of unlike electrical charges. The positive $(+)$ and negative $(-)$ charges in the bonding agent are attracted by the electrical charges on the surface of the material to be bonded. This molecular attraction provides a strong holding force.

Bonding agents can have their properties varied for specific conditions or for use on a particular material. Some are a combination of two or more types, such as a phenol and resorcinal resin combination or a urea resin blended with a melamine resin. They are available as powders, solids, liquids, and pastes. Some require the addition of a catalyst.

A summary of the commonly used bonding agents, their uses, and the materials they will join are in Table 26.1.

Curing of Bonding Agents

The curing of bonding agents is accomplished by loss of solvents, anaerobic environments, catalysts in two-part mixtures, and cooling of hot melts. *Loss-of-solvent bonding agents* cure by the loss of volatile liquids, water,

Table 26.1 Commonly Used Adhesives and Their Applications

Adhesive	Bonded Material	Typical Uses
Acrylic	Plastics to metal, plastics to plastics, rubber to metal	Curtain walls
Casein	Wood to paper, wood to wood	All interior wood-joining needs
Cyanoacrylate (anaerobic)	Acrylics, phenolic, rubber, glass, polycarbonates, ceramics, steel, copper, aluminum	Any use (known as "super glue"); electronic and electrical devices
Epoxy	Almost any material except a few plastics and silicones	Interior and exterior uses, panels, glass to metal, curtain walls
Melamine formaldehyde	Paper, textiles, hardwood, interior plywood	Interior uses, plywood manufacturing
Natural rubber	Leather, paper, cork, foam rubber	Pressure-sensitive tape
Neoprene rubber (contact cement)	Many plastics, ceramics, aluminum	Plastic laminates, other interior uses
Nitrile rubber	Many plastics, ceramics, glass, aluminum	General uses
Phenol	Wood, cardboard, cork	Exterior plywood, any exterior use
Polyvinyl acetate	Porous materials (paper and textiles)	Various interior applications
Resorcinol	Rubber, paper, cork, asbestos, wood	Furniture, wood beams, columns
Silicone	Glass, ceramics, aluminum, polyester, acrylics, phenolic, rubber, steel, textiles	Sealant, gasket material
Polyurethane	Many plastics, glass, copper, aluminum, ceramics	Bonding dissimilar materials (e.g., on steel and glass sun roofs)
Urea	Many plastics, glass, copper, aluminum, ceramics	Particleboard, furniture, cabinets
Vinyl butyral	Glass	Laminating glass

or organic solvents that were used to dissolve the base material. The solvents evaporate or soak into porous materials to be bonded. Sometimes a solvent may damage the materials on which it is applied, so the manufacturer's instructions should be consulted. **Anaerobic bonding agents** maintain a fluid condition when exposed to oxygen but set hard when oxygen is omitted, as occurs when two parts are clamped together. *Two-part mixtures* have the resin and catalyst in separate containers. When they are mixed the catalyst causes cross-linking of the resin. **Hot-melt adhesives** are in solid form, become liquid when heated, and set rapidly when heat is removed.

Types of Bonding Agents

Bonding agents may be divided into three major classes—adhesives, glues, and cements. Following are some of the more commonly used types.

Adhesives

Adhesive bonding agents are made from synthetic materials. They fall into two types, *thermoplastic* and *thermosetting*.

Thermoplastic Adhesives Thermoplastic adhesives are moisture resistant but are not used where they will be exposed to moist conditions.

Aliphatic resins are a type of polyvinyl resin and are stronger and more heat resistant than other polyvinyls. They are yellow in color, but other polyvinyls are white. They are used for furniture and general carpentry work (Fig. 26.1).

Alpha-cyanoacrylate, often called "superglue," is used to bond metals, plastics, and other dense material. It is not recommended for use with porous materials, such as wood.

Hot-melts are a mixture of polymers sold in solid form, such as rods, pellets, ribbons, or films. They are placed in an electric hot-melt applicator that melts the adhesive as it is applied to the surface. There are various types for bonding plastics, particleboard, softwoods, and hardwoods. They set up fast but are not very strong. They should be used for interior purposes where not much stress is expected (Fig. 26.2).

Polyvinyl acetate adhesives, generally referred to as "white glue," have moderate moisture resistance but high dry strength and are recommended for interior use only. They are used with paper, wood, and vinyl plastics and sometimes with metal. They are one of the most widely used adhesives. Typical uses include furniture assembly, bonding plastic laminates, and flush doors (Fig. 26.1).

Thermoset Adhesives Thermosetting adhesives are available in a variety of forms. *Epoxy adhesives* are produced in a two-part liquid form. An epoxy resin and an epoxy hardener are mixed immediately before being used (Fig. 26.3). Epoxy resins will bond to almost any

material and produce a strong joint. Epoxies are used in products such as curtain walls and for bonding steel and concrete in applications such as bridge construction. Epoxies are being used for more and more structural applications. Epoxies are available in a variety of types for special uses.

Melamine adhesives are produced as a powder with a catalyst to be added when used. They have good bond to paper and wood. They are cured by applying heat of about 300°F (150°C). Melamine adhesives are used as fortifiers for urea resins that are used for hardwood plywood, gluing lumber, and scarf joining softwood plywood.

Phenol resin adhesives have good bonding qualities to paper and wood, have good shear strength, and are resistant to moisture and temperature. They are cured in a hot press at about 300°F (150°C). They are widely used in the manufacture of plywood and particleboard.

Resorcinol resin adhesives are used when a waterproof bond in wood products is required. They are two-part adhesives consisting of a resin and a catalyst. The mixture must be used within eight hours. One important use is the manufacture of glued laminated wood structural

Figure 26.1 Polyvinyl acetate (white glue) and aliphatic resin types of polyvinyl resin (yellow or carpenter's glue) are used for furniture and general carpentry.

Figure 26.2 Hot melts are applied with this type of electric hot melt gun. *(Courtesy Arrow Fastener Co., Inc.)*

Figure 26.3 Epoxies are two-part bonding agents. The resin and catalyst are in separate tubes.

members. Some adhesives are a mixture of resorcinol and phenol resins.

Urea resin adhesives are available in powder form and are mixed to proper consistency with water. They are moisture resistant but not waterproof. They bond well to paper and wood and have good resistance to heat and cold. Normal clamp time is about sixteen hours, but this can be reduced to minutes or seconds using a radio-frequency glue-drying machine. Typical uses include interior hardwood plywood, interior particleboard, flush doors, and furniture.

Glues

Glues are bonding agents made from animal and vegetable products, such as bones, hides, fish, and milk. Those finding some use are animal, blood albumin, vegetable, and fish glues.

Animal glue is produced as flakes or dry powder. It is mixed with water and heated. The heated mixture is applied to the wood surfaces. It is primarily used for furniture manufacture. It has excellent bonding and shear strength but has been largely replaced by newer products.

Another form, *liquid hide glue,* is more widely used. It is a ready-mixed form of animal glue. It is not moisture resistant and is used only on interior products, such as furniture, paper, and textiles.

Blood albumin glue is a form of animal glue used to bond paper products. It finds limited use on some forms of interior plywood. It has moderate bonding power with wood, has poor resistance to moisture, and has only fair resistance to heat and cold. *Fish glues* are used for sealing cardboard boxes and on packing tape.

Casein glue is made from dried milk curds. It is in powdered form and is mixed with water. It is water re-sistant and used for interior wood and on exterior products that will be sheltered and not be directly subjected to the weather, such as laminated timbers. It is used on wood and paper and provides a strong joint. It resists heat well but has only moderate resistance to cold.

Vegetable glues are made from soybeans, starch, and dextrin. They are used for bonding interior plywood, paper, and wallpaper.

Cements

Cements are made from synthetic rubber, such as neoprene, nitrile, and polysulfide, suspended in a liquid.

Cellulose cements, such as cellulose acetate, cellulose nitrate, and ethyl cellulose, are used for interior purposes and mainly for bonding plastics and glass and for porous materials, such as wood and paper. Cellulose nitrate is more commonly known as a general household cement and is sold in tubes to make application easy. Cellulose cements have moderate resistance to temperature change and good moisture resistance.

Contact cements are neoprene cements that stick immediately upon contact and require no clamping time. They are used to bond plastic laminates on countertops, plastics to plastics, plastics to wood, and wood to wood. Although they have good resistance to moisture, they are generally used for interior purposes. They have a very high bonding strength.

Mastic is a elastomeric construction cement. It is sold in tubes and is applied with a caulking gun. Typical uses include bonding plywood subfloors to joists; bonding wall paneling to studs; laminating gypsum wallboard, styrene, and similar materials; and assembling wall, floor, and roof panels. Mastic is water resistant. It develops full strength after several weeks (Fig. 26.4).

Figure 26.4 Mastic is an elastomeric construction cement applied with a caulking gun.

Buna N resins are a form of acrylonitrile butadiene rubber. They are liquid cements that have good moisture resistance and strength. They are good general-purpose interior cements.

A summary of some of the properties of these bonding agents is in Table 26.2.

SEALERS FOR EXTERIOR MATERIALS

Sealing materials are applied to the surface of a material to seal it against penetration by water or to keep water from passing through the material. Brick walls are often sealed to keep moisture from penetrating the mortar joints and into the inside of the wall. **Sealers** have **adhesive** properties and are closely related to bonding agents. In addition to bonding to the surface, they must form an unbroken film over it and fill any minute pores, cracks, or other minute openings. Large cracks or defects must first be filled with some form of sealant before a sealer is applied because the sealer will not span most cracks.

Acrylic sealer is a high-solid-content clear acrylic sealer. It maintains the original appearance of masonry or concrete while protecting it from moisture, airborne

dirt, and other pollutants. It is used on products such as exposed aggregate panels, brick, and stone. Acrylic sealer is virtually unaffected by prolonged exposure to moisture, common acids, ultraviolet rays, oils, and aliphatic solvents. It reduces damage from the freeze-thaw cycle, efflorescence, and stains. It is available in a number of variations for different applications, such as methyl methacrylate acrylic polymer.

Asphalt driveway sealer is a thin, quick-drying sealer that gives a black protective coating. It is applied by brush or roller.

Coal tar coatings are used to seal surfaces and provide protection against corrosive conditions encountered in industrial plants, water and sewage systems, chemical plants, refineries, and other such industries. They resist moisture and most acids, alkalies, corrosive vapors, and atmospheric corrosion. They can be applied to metal, concrete, and masonry surfaces.

Silicone sealers are in liquid form and are used on brick, concrete block, stucco, cement plaster, and concrete. They produce a water-repellent coating that protects the surface but does not prevent water vapor from escaping. They seal the surface, minimize efflorescence, and protect from absorbing staining materials, such as

Table 26.2 Characteristics of Bonding Agents

	Form	Moisture Resistance	Application Temperature	Clamp Time
Thermoplastic Adhesives				
Aliphatic	Liquid	Low	45°F (7°C)	1 hr.
Alpha-cyanoacrylate	Liquid	Low	75°F (21°C)	Minutes
Hot melts	Solid	Low	Electronic welder 300°F (150°C)	None
Polyvinyl acetate	Liquid	Low	60°F (16°C)	1 hr.
Thermosetting Adhesives				
Epoxy	Liquid	Water resistant	60°F (16°C)	None
Melamine resin	Liquid	Water resistant	Hot press 250°F (122°C)	Minutes
Phenol resin	Liquid and powder	Waterproof	300°F (150°C)	Minutes
Resorcinol resin	Liquid	Waterproof	70°F (21°C)	16 hr.
Urea resin	Powder and liquid	Water resistant	70°F (21°C)	1 to 3 hr., or seconds with high-frequency glue curing machine
Glues				
Animal	Flakes and powder	Low	70°F (21°C)	2 to 3 hr.
Liquid hide	Liquid	Low	70°F (21°C)	2 to 3 hr.
Blood albumin	Liquid	Moderate	70°F (21°C)	2 hr.
Vegetable or fish	Liquid	Low	70°F (21°C)	2 hr.
Casein	Powder	Water resistant	32°F (0°C)	2 hr.
Cements				
Contact cement	Liquid to thick	Water resistant	70°F (21°C)	None
Mastic	Paste in tubes	Water resistant	70°F (21°C)	15 min.
Cellulose	Liquid	Low	70°F (21°C)	Minutes
Buna N	Liquid	Water resistant	70°F (21°C)	Minutes

dirt and soot. Since protection against moisture penetration is provided, damage due to freeze-thaw cycles is reduced (Fig. 26.5).

Epoxies are used to provide a waterproof coating. They are often used on floors in areas where moisture and chemicals exist, such as food, meat processing, and other industrial plants. They are also used on floors around swimming pools, decks, showers, restrooms, loading docks, and on stair treads. When mixed with an aggregate, such as emery, epoxies provide a slip-resistant surface. These are available in a variety of formulations.

Polyurethane sealers are used to minimize concrete problems such as scaling, spalling, chloride penetration, and damage due to the freeze-thaw cycle. Some types have an ultraviolet stabilizer added. They do not inhibit the transmission of vapor out of the concrete.

Wood Sealers

Sealers on wood products are used to keep the various layers of finish, such as stains and fillers, from bleeding through the final topcoat. The sealer also provides the required base for the finished topcoat. The sealer used depends on the recommendations of the manufacturer of the topcoat. If the wood is to be stained and varnished, *shellac* is often used as a sealer over the

Figure 26.5 Silicone sealers are used on wood, brick, stone, concrete, and plaster.

stain. After this, the wood filler can be applied and another sealer, often shellac, is used. Then the varnish topcoat is applied. Synthetic resin sealers are also used.

Lacquer topcoats require that a *lacquer sanding sealer* be applied. After careful sanding the lacquer topcoats can be applied. Other finishes, such as polyesters and polyurethanes, may or may not require a sealer. Always follow the instructions of the topcoat manufacturer.

For exterior and interior paints the manufacturer will have specific recommendations for sealers. Following is an example for one manufacturer:

Drywall sealer—latex primer
Plaster sealer—wall and wood primer
Wood sealer—alkyd enamel undercoat
Concrete block sealer—block filler
Masonry sealer—latex primer
Ferrous metal sealer—water-based acrylic paint

Walls to receive wallpaper or fabric coverings also must have sealers. Usually some form of casein glue is used.

SEALANTS

A **sealant** is a material used to seal joints between members in construction materials and protect materials against the penetration of moisture, air, corrosive substances, and foreign objects. Examples include expansion joints in large masonry walls, spaces between glass and frames, and openings between exterior siding and door and window units. The sealing of the joints between these and other parts of the building is essential to ensure the integrity of the entire structure.

Joints in the exterior walls are needed to allow materials to expand and contract. These are called **working joints.** The joint must allow for this movement. If a foreign object, such as a rock, gets into the joint, contraction may be blocked and the rock may cause the materials on each side to spall or crack (Fig. 26.6).

The two basic methods for protecting joints are with *sealants* and with *prefabricated covers*. A sealant is a flexible adhesive material that is worked into the joint, bonds to the sides, and sets into a firm but rubbery stretchable material. Prefabricated covers are usually metal and are made so they can allow for the movement. The sealant filling a joint is a substitute for the material that was removed. As such, it must meet the performance requirements of the basic material. It must maintain the integrity of the assembly of materials but allow for the movement between them.

Moldable sealants may be deformable or elastic/elastomeric. The *deformed sealant* is installed in its natural stretched shape. It stretches as the joint widens and de-

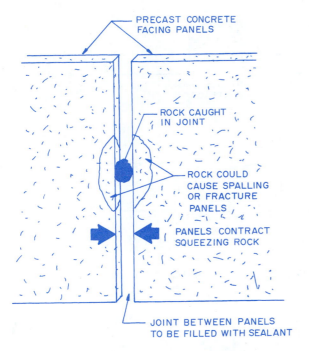

PRECAST CONCRETE FACING PANELS

ROCK CAUGHT IN JOINT

ROCK COULD CAUSE SPALLING OR FRACTURE PANELS

PANELS CONTRACT SQUEEZING ROCK

JOINT BETWEEN PANELS TO BE FILLED WITH SEALANT

Figure 26.6 Joints between masonry materials to be sealed must be free of foreign objects.

forms as the joint width is reduced. The elastic/elastomeric sealant stretches as the joint widens and shrinks back to its normal size as the joint width is reduced (Fig. 26.7).

Sealant Performance Considerations

A key factor is the *percent of elongation* the sealant can safely stretch and still give expected protection. Some high-performance sealants have a 50 percent elongation, but intermediate types are usually rated up to 25 percent. The sealant must have excellent *adhesiveness* to the material to which it is expected to bond. It must be *flexible* and have minimum internal shrinkage through the years of use, must resist *staining* material around it, and must have a tough *nontacky* elastic surface skin so dirt and solid objects do not stick to it or penetrate it.

Types of Sealants

Sealants are of two basic types: (1) a flexible, moldable adhesive compound and (2) a solid, flexible preformed shape.

Preformed sealants are available in a variety of materials and shapes. Examples of several are in Fig. 26.8.

Moldable sealants are manufactured in three performance levels. *Low-performance sealants* are less costly, have a short life (four to seven years), and are used where limited joint movement is expected. *Intermediate-performance sealants* are more expensive but last seven to

fourteen years. They are used where joint movement is greater than where low-performance sealants are used. *High-performance sealants* are the most expensive, but they have a life expectancy of twenty to thirty years. They are used in joints where the movement is the greatest. Manufacturers' recommendations on joint size and the allowable percent of elongation must be observed.

Moldable bulk sealants are available in pourable form, in knife grade (such as glazing compounds), gunable form for manual or pneumatic caulking guns, and as preformed tapes that may be cured or not cured.

Following are descriptions of some of the commonly used sealants. The actual makeup of these can vary considerably, so many special-purpose sealants are available.

Polyurethane sealants may be one part or two part. They are general-purpose sealants used for areas such as precast concrete and masonry joints, for glazing, and for sealing around door and window openings, in swimming pools, and in waste treatment plants. Some are designed for exterior concrete joints in roads and sidewalks. This type is resistant to damage from fuels and oils. Some are formulated to be applied by a gun, and others are liquid and are poured.

An *epoxy* crack filler is available for use on very small cracks that occur in concrete, concrete block, and brick walls.

Silicone rubber sealants are available in a variety of compounds. One type is for interior use on nonporous surfaces where high humidity and temperature extremes exist, such as around bathtubs. Another type is to seal exterior building joints, and it will bond to nonporous materials such as glass, most metals, ceramics, and most plastics. It can be gunned in below freezing temperatures, is available in a variety of colors, and has a life expectancy of up to thirty years.

Latex sealant adheres to wood, metal, concrete, masonry, marble, porcelain, ceramic tile, glass, and many types of plastics. Some types permit joint movement up to 25 percent.

Polysulfide polymer sealants are elastomeric and bond to all masonry, concrete, wood, glass, and metal surfaces. They are available in a variety of formulations, including those with one and two components. They are used in applications such as sealing joints between curtain wall panels, precast concrete, and various window assemblies. They cure in about twenty-four hours to a rubberlike material with excellent stretch capabilities.

A *urethane bitumen sealant* is a two-part catalyzed 100 percent solid polyurethane-coal tar elastomeric compound. It bonds to concrete, stone, brick, glass, wood, metal, cement asbestos, concrete block, and most plastics with the exception of polyethylene. It offers excellent resistance to acids and commonly used solvents.

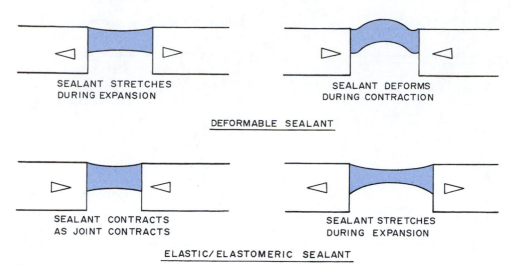

Figure 26.7 Elastic/elastomeric and deformable sealants allow for expansion and contraction at each joint.

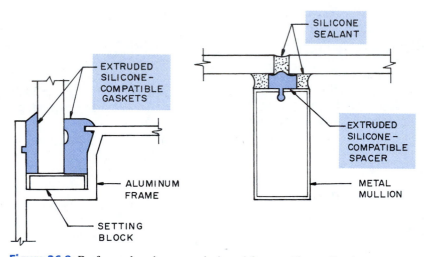

Figure 26.8 Preformed sealants are designed for specific applications.

Sealants and Joint Design

The width of the joint between two members, such as precast concrete facing panels, must be carefully determined. For example, one type of sealant available tolerates a joint movement of +100 to −50 percent. The manufacturer requires the joint width be two times the expected joint movement and at least ¼ in. (6 mm) wide. A backup rod is used to control the depth (Fig. 26.9). Other types of sealer joints are shown in Fig. 26.10.

Backup Materials

Proper installation of sealants depends on the use of proper backup materials installed to control the depth of the sealant, as shown in Fig. 26.9. The backup material not only controls the depth of the sealant but also can serve as a bond breaker to keep the sealant from bonding to the back of the joint. Some of these are installed in the factory as the unit, such as a window, is made. Others are applied to joints in the field, such as in control joints in a concrete floor slab.

Backup materials vary depending on the location and use. Most frequently used are rods or tubing made from polyethylene, butyl, urethane, and neoprene that control the depth of the sealant. Other rod types form a primary water seal to keep out moisture during construction until the sealant can be installed. Typical uses for rods and tubes are in joints between precast curtain wall panels and expansion joints in long masonry walls.

Tapes are another form of backup material. They keep the sealant from adhering to joint surfaces where bonding is not wanted. Polyethylene film is also used as a bond breaker.

Other materials, such as resin-impregnated fiberboard, corkboard, and dense plastic foamed strips, are used in places such as isolation and control joints in concrete and masonry construction. Rods and tubes can also be used for these purposes. Some examples are in Fig. 26.11.

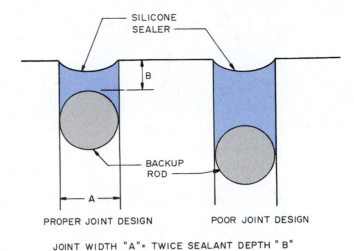

Figure 26.9 A backup rod is used to control the depth of the sealant.

Caulking and Glazing

Caulking is a procedure for sealing joints, cracks, or other small openings with **caulking compound** (sealer). Caulking compound is a resilient mastic material.

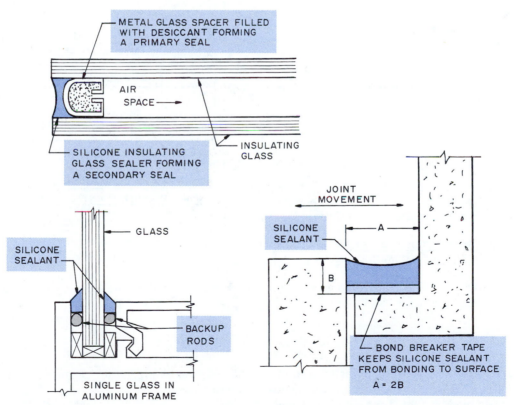

Figure 26.10 Examples of other types of joints requiring backup rods or the use of bond breaker tape.

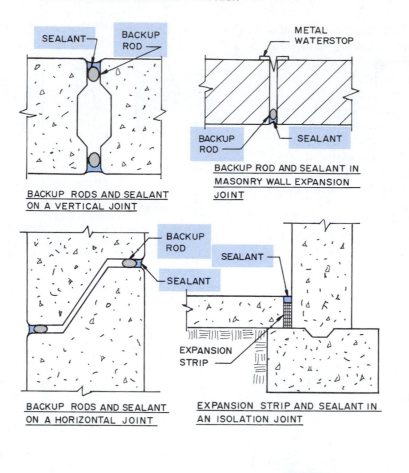

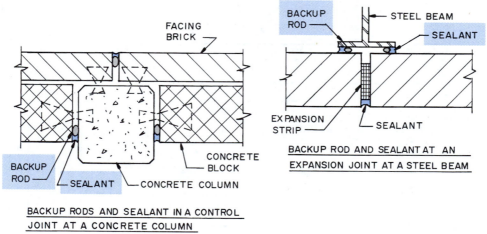

Figure 26.11 Typical applications of the use of backup rods and expansion strips.

Glazing compounds are a form of sealer used to set glass in place in frames. They serve to seal out water and air and form a cushion allowing for expansion and contraction of the glass and frames.

Caulking and glazing compounds are commonly silicone, acrylic, butyl, polysulfide, or polyurethane. These five types are gunable, meaning they are of a consistency to be applied by a gunlike device. Other types are re-ferred to as "knife grade" because they are applied with a putty knife.

WATERPROOFING MEMBRANES AND COATINGS

Waterproofing involves applying a material on the surface of an assembly of materials, such as a foundation,

ITEM OF INTEREST

WATERPROOFING TIPS

There are a number of ways to waterproof a foundation wall. The manufacturer of a system usually requires the contractor to employ a certified applicator if the manufacturer's guarantee is to be valid. Common systems include liquid membranes, sheet membranes, cementitious coating, built-up systems, and bentonite.

Liquid membranes are applied with a roller, trowel, or spray. The liquid solidifies into a rubbery coating. Different materials are available, such as polymer-modified asphalt and various polyurethane liquid membranes.

Sheet membranes tend to be self-adhering rubberized asphalt sheets, typically an assembly of multiple layers of bitumen and reinforcing materials. Some companies manufacture PVC and rubber butyl sheet membranes.

Cementitious products are available from building supply outlets. They are mixed on the site and are applied with a brush. Some have an acrylic additive available that improves bonding and makes the cementitious coating more durable. One disadvantage is that these coatings will not stretch if the foundation cracks, thus opening the possibility for leakage through cracks.

Built-up systems may be like the widely used hot tar and felt membrane. Alternate layers of hot tar and felt are bonded to the foundation. Usually at least three layers of felt are specified.

Bentonite is a clay material that expands when wet. It is available in sheets that are adhered to the foundation. As groundwater penetrates the clay, it swells many times its original volume providing a permanent seal against water penetration.

Surface Preparation

Regardless of the type of system used it is important to prepare the surface before application. This includes (1) drying the wall and footings, (2) removing the concrete form ties, making certain they break out inside the foundation so they do not penetrate the waterproof membrane, (3) cleaning the wall so it is free of all dirt or other loose material, and (4) sweeping the wall free of dust and mud film residue. A residue left when wet mud is wiped off and left to dry on the foundation can inhibit bonding. Finally, any openings around pipes or other items that penetrate the wall must be grouted.

Safety

Waterproofing presents some hazards that must be controlled. First is a possible cave-in of the soil, burying the workers. Normal shoring procedures should be observed. Many of the materials used are flammable and solvent-based, presenting a fire hazard. Workers should not smoke or use any tools that might cause ignition. Solvent fumes can be very harmful and workers must wear respirators. Fumes are usually heavier than air and settle around the foundation in the excavated area. The solvents, asphalt, and other materials used may cause skin problems, so protective clothing, including gloves, is required with many products. As always, wear eye protection. When in doubt consult the manufacturer of the product.

to make it impervious to water. Many parts of a building need waterproofing. Most common among these are foundation walls, roofs, exterior wood or masonry walls, and exposed structural components, including steel.

Waterproofing must be adequate to resist the forces that tend to force water through the assembly of materials. These forces include

▲ *Gravity,* which forces water through horizontal areas such as a roof or deck
▲ *Hydrostatic pressure* on one side of a horizontal or vertical assembly (most commonly due to subsoil water)
▲ A difference in *air pressure* on one side of a horizontal or vertical assembly

For surfaces not subjected to hydrostatic, gravity, or air pressure differences, waterproofing can be a light-duty coating, such as silicone or coal tar pitch. This provides a moisture-resistant membrane that *dampproofs* the assembly. Surfaces under pressure require a heavy-duty membrane, such as tar and felt or a synthetic membrane. Waterproofing is most effective if applied on the surface directly facing the source of moisture.

Waterproofing can be accomplished by:

1. Applying a built-up bituminous membrane of felt and hot or cold tar pitch.
2. Applying a heavy coating, such as portland cement plaster or a trowelable asphalt.

3. Bonding an elastomeric membrane to the wall.
4. Applying a thin film or coating to the exterior of the wall, such as liquid silicone or coal tar pitch.
5. Adding waterproofing admixtures to the concrete as it is mixed.
6. Applying a dry coating that will emulsify in place, such as bentonite clay.

A summary of the most frequently used types of waterproofing is in Table 26.3.

Waterproofing coatings and membranes are not self-supporting but must be bonded to the surface to be treated. In addition, waterproofing must be able to adjust to the stresses caused by movements of the assembly and any cracks or deterioration without losing the waterproofing capabilities. It should be noted that waterproofing coatings and membranes can be damaged during installation and construction. One example is damage to roof membranes by someone walking on them. Protective pads are available for installation on top of the roof membrane to protect the areas where workers may walk. Another frequently occurring damage occurs during backfilling a foundation. A protective material can be placed over the waterproofing to keep rocks from piercing it.

Bituminous Coatings

Hot-applied and emulsified coal tar pitch, hot-applied and cold-applied asphalt, and emulsified asphalt can be applied to the foundation by brush, roller, or spray. These are effective only for situations in which hydrostatic pressure is not a factor.

A waterproofing system that will resist hydrostatic pressure consists of alternate layers of hot mopped coal tar pitch, asphalt, or cold mopped emulsified asphalt over layers of mineral or glass fiber felts in much the same way as laying a built-up roof. The number of layers of felt and asphalt depends on the hydrostatic conditions. Manufacturers of these systems have established specifications for various conditions.

Another form of light-duty waterproofing is a bituminous binder with asbestos fibers forming a trowelable mix. It is hand troweled on the concrete foundation and is often used to bond foamed plastic insulation boards to the foundation. It is only good for dampproofing a wall not subject to hydrostatic pressure. A thinner version can be sprayed using a mastic pump (Fig. 26.12). Refer to Chapter 27 for additional information.

There are a variety of solvent-based asphalt dampproofing compounds. They are available as a thick mastic, a semisolid mastic, and a spray mastic. They give dampproofing properties to interior and exterior above-grade and below-grade surfaces. They are used on metal to prevent corrosion.

Liquid Coatings

An acrylic copolymer waterproof coating is available in a variety of colors. It may be obtained with fillers and texturing aggregates that are fused onto the concrete or masonry surface. This offers waterproofing protection and an attractive finish coating. Important uses are on above-grade exterior concrete walls, columns, and spandrels. The coatings are also used on portland cement plaster and stucco walls, giving a textured, sandlike finish in color that minimizes surface defects. They are applied by brush, roller, or spray.

Clear silicone dampproof liquids are widely used on exterior concrete, masonry, and wood surfaces. They are not a surface coating but penetrate into the surface, carrying solids into the pores. They do not color the wall but do reduce efflorescence. They can be applied with a brush, roller, or spray.

A liquid waterproofing material that is self-curing is available in a polyurethane rubber with a coal tar additive. It is applied to decks, roof slabs, and floors with a brush, roller, or squeegee. It will cover small cracks and is flexible enough to cover surfaces of irregular shapes. It can be covered with a concrete slab, hardboard, or mineral surface roll roofing (Fig. 26.13).

Table 26.3 Types of Waterproofing

| Sheet membranes | Composite membranes | Built-up Membranes | | Liquid membrane | Applied coating |
		Hot applied	Cold applied		
Butyl	Elastomeric, backed	Asphalt, type I, II, III	Bitumen emulsion	Butyl	Acrylic, silicone
Ethyene propylene	Polyethylene and	Coal tar pitch, type B	Bitumen, fiberated	Urethane	Asphalt emulsions,
Neoprene	rubberized bitumen	Felts, saturated and	cement	Polychlorene	cut backs
Polyethylene	Polyvinyl chloride backed	coated	Felts, coated	(neoprene)	Cementitious with
Polyvinyl chloride	Saturated felts and		Bentonite clay	Polyurethane,	admixtures
	bitumen coated		Fabric, saturated	coal tar	Epoxy, bitumen
			Glass fiber mesh,		Urethane, bitumen
			saturated		Bitumen, rubberized
			Cementitious		
			membrane		

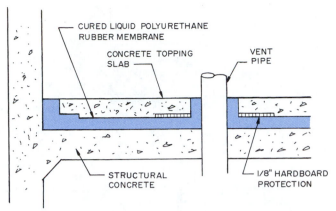

Figure 26.13 Self-curing polyurethane rubber forms a seamless waterproof membrane.

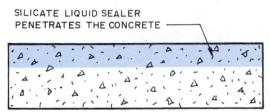

Figure 26.14 Silicate liquid gel forms a waterproof coating by reacting with the soluble calciums in the concrete.

Figure 26.12 This trowelable dampproofing material has a nonvolatile bituminous binder dispersed in water by means of selected mineral colloids. *(Courtesy Karnak Corporation)*

Another form of liquid waterproofing uses a silicate liquid gel that reacts with the soluble calciums in concrete to form a glass gel in the microscopic pores (Fig. 26.14). The gel will penetrate 1.5 to 2.0 in. (38 to 50 mm), providing a strong moisture-proof barrier that also reduces efflorescence.

Synthetic Sheet Membranes

Synthetic membranes are available made from neoprene, polyvinyl chloride, polyethylene, butyl, and ethylene propylene. These sheet materials are bonded to the foundation wall using adhesives recommended by the system manufacturer. Since they are flexible membranes, they tend to adjust to settling and compaction of the soil and are not likely to rupture. They are available

as single materials or composite membranes. Following are examples of a few of these products.

Composite Membranes

Composite membranes are sheet products made by laminating two or more waterproofing materials. One frequently used composite membrane is made of a rubberized asphalt layer with a polyethylene film bonded to the outer surface. It remains stable below ground and below water. When the sheets are lapped the rubberized asphalt back bonds to the polyethylene face of the sheet beside it. Under most conditions it will bridge gaps up to ¼ in. (6 mm). It is available in rolls 48 in. (121 mm) wide and 60 ft. (18.3 m) long. Since the back surface is very tacky, it is covered with a peelable paper covering (Fig. 26.15).

Another composite membrane is made by laminating a chlorinated polyethylene (CPE) film to a nonwoven polyester fabric. This composite is bonded with water-based acrylic adhesives, avoiding problems that occur from fumes of solvent-based adhesives. It can be installed on surfaces that are damp at the time of installation.

Sheet Membranes

Sheet membranes are waterproofing products composed basically of one major waterproofing material.

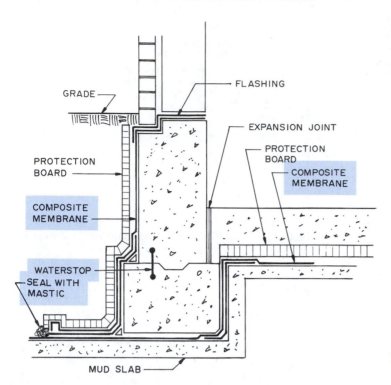

GRADE

FLASHING

EXPANSION JOINT

PROTECTION BOARD

COMPOSITE MEMBRANE

PROTECTION BOARD

COMPOSITE MEMBRANE

WATERSTOP SEAL WITH MASTIC

MUD SLAB

Figure 26.15 Composite membranes provide high-quality waterproofing.

One such product is made from polyvinyl chloride alloyed with high-density polymer resins. It is not affected by aging, mildew, or corrosion. It remains flexible at low temperatures and has good abrasion and tear resistance. It is bonded, and laps are sealed with a special adhesive. Some forms of PVC membrane can be formulated to be resistant to gasoline and oil. The seams can be welded with hot air or bonded with an adhesive specified for that purpose.

Another type of sheet membrane is a chlorinated polyethylene (CPE) product. It is available in a range of thicknesses with and without integral reinforcement. It is suitable for use above and below grade and on horizontal and vertical surfaces. Laps can be joined by chemical or thermal fusion. It will bridge cracks up to $\frac{1}{4}$ in. (6 mm).

Cement-Based Waterproofing

A number of cement-based heavy-duty waterproof coatings are available. These have carefully graded aggregates that produce a high-density and high-strength waterproof coating. Some types are applied with a trowel, and others require special procedures. They are used to waterproof concrete masonry and stone in interior and exterior locations. They are used on water reservoirs, swimming pools, basements, parapet walls, and other heavily exposed surfaces.

Lead Waterproofing

Lead waterproofing sheets are frequently used for projects in which waterproofing must be of uncompromised security. Examples include areas of grass, fountains, and reflecting pools that may be built over underground facilities, such as parking or stores, or on the roof of a building. The gauge of the lead sheet varies for different purposes, but it must be at least 6 lb. lead ($\frac{3}{32}$ in. or 2.3 mm) thick to allow for the burning (welding) necessary to join the sheets. If lead is in contact with cementitious materials it must be protected with a bituminous coating. Lead wool is also used to waterproof joints in above-ground installations.

Bentonite Clay Waterproofing

Bentonite is a clay formed from decomposed volcanic ash, with a high content of the mineral montmorillonite. It can absorb large amounts of water, which causes it to swell many times its original volume, forming a waterproof barrier. The dried, finely ground particles are usually applied as a waterproofing membrane in three ways.

Bentonite Panels

Panels, usually 4 × 4 ft. (1200 × 1200 mm), consist of a biodegradable paper covering over bentonite clay particles. One type is $\frac{3}{16}$ in. (2.5 mm) thick with a corrugated kraft board core. It is used on vertical walls and under structural slabs. Another type is $\frac{5}{8}$ in. (16 mm) thick composed of the layers of corrugated kraft board with the center layer holding the bentonite clay. The hollow outer layers allow space for the expansion of the bentonite clay and reduce upward pressure against a thin nonstructural slab (Fig. 26.16).

The panels are ready to apply when received on the job and can be applied at all temperatures and over moist substrates. They can be nailed to green concrete walls. The panels are overlapped $1\frac{1}{2}$ in. (38 mm). A hydrated sodium bentonite gel is used to fill gaps around pipes and fittings. If the backfill contains rocks that may pierce the panels, they should be covered with a protective material (Fig. 26.17).

Sprayed Bentonite

A bentonite clay mixed with a modified asphalt that serves as an adhesive to bond the clay to the surface is applied by spraying. A $\frac{3}{8}$ in. (10 mm) membrane is built up for normal applications, but thicker layers can be used for situations in which hydrostatic pressure is severe.

Bentonite-Sand Mixture

Bentonite clay mixed with sand is used to produce a waterproof barrier under concrete slabs. The mix is carefully measured and spread over the area to be covered with concrete. It is then covered with a polyethylene sheet to protect it from the moisture in the concrete. The reinforcing is placed on top and the slab is poured. If too much bentonite is used the forces of expansion could crack the slab.

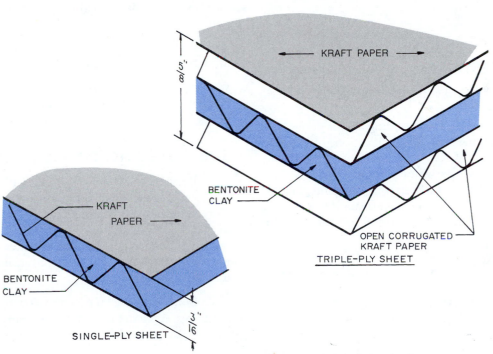

Figure 26.16 Bentonite clay is held in a corrugated paper core bonded to kraft paper outer layers.

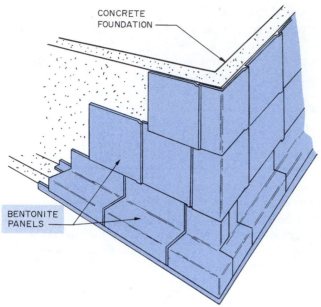

CONCRETE
FOUNDATION

BENTONITE
PANELS

Figure 26.17 Bentonite panels can be nailed to concrete foundations and are overlapped at the joints between panels.

REVIEW QUESTIONS

1. What are the two ways bonding agents hold materials together?
2. In what ways do various bonding agents cure?
3. What are the three major classifications of bonding agents?
4. What are the two classes of adhesives?
5. What is the source of materials for making glues?
6. From what materials are cements made?
7. What is the difference between a sealer and a sealant?
8. What is a working joint?
9. What are the two basic methods of protecting exterior joints?
10. What are the two types of sealants?
11. What are the performance levels of sealants?
12. What purpose does a sealant backup rod serve?
13. Why are tapes sometimes used with sealants in joints?
14. What forces tend to force water through an assembly of materials?
15. What is bentonite?

KEY TERMS

adhesive A substance that is sticky and causes something to adhere.

adhesive bonding agents Bonding agents made from synthetic materials.

anaerobic bonding agents Bonding agents that set hard when not exposed to oxygen.

bonding agent A compound that holds materials together by bonding the surfaces to be joined.

caulking compound A resilient material used to seal cracks and prevent leakage of air or water.

cements Bonding agents made from synthetic rubber suspended in a liquid.

glues Bonding agents made from animal and vegetable products.

hot-melt adhesives Adhesives that bond when they are heated to a liquid form.

mechanical action Action that bonds materials by adhesives that enter the pores and harden, forming a mechanical link.

sealants Materials used to close joints between abutting members.

sealers Materials applied to a material to seal it against penetration by water.

specific adhesion Bonding dense materials using the attraction of unlike electrical charges.

working joints Joints in exterior walls that allow for expansion and contraction of materials in the wall.

SUGGESTED ACTIVITIES

1. Bond identical wood samples with various types of adhesives recommended for use on porous materials. Cure following the manufacturers' recommendations. Test each sample for bond strength with a tensile testing machine. Soak other samples in water for an identical number of days and test these for tensile strength. Write up your findings.

2. Bond identical metal samples with various types of adhesives. After the recommended curing time, test for strength of bond with a tensile testing machine.

3. Visit several commercial buildings and note how various expansion joints and other areas needing to be sealed are protected. Observe any evidence of failure of the sealant. What would you recommend be done to repair any failures?

ADDITIONAL INFORMATION

Joint Sealants, Structural Sealant Glazing Systems, and Fenstration Sealants Guide Manual, American Architectural Manufacturers Association, Palatine, Ill., 1982.

The NRCA Roofing and Waterproofing Manual, National Roofing Contractors Association, Rosemont, Ill., 1992.

Panek, J. A., and Cook, J. P., *Construction Sealants and Adhesives,* John Wiley and Sons, New York, 1992.

Weismantel, G. E., *Paint Handbook,* McGraw-Hill, New York, 1981.

Another resource is *A Professional's Handbook on Grouting, Concrete Repair, and Waterproofing,* available from Five Star Products, Inc., 425 Stillson Rd., Fairfield, CT 06430.

CHAPTER 27

Bituminous Materials

This chapter will help you to:

1. Understand the properties of bitumens and how these properties influence your decision as to how bituminous materials might be used.

2. Interpret the results of laboratory tests and use these for decision making.

3. Be aware of the array of bituminous products available and the best applications for each.

Bitumen is a mixture of complex hydrocarbons that occur naturally or are heat produced from materials such as coal and wood. Bitumens may be in a gaseous, liquid, semisolid, or solid state.

Asphalt, tar, and coal tar pitch are those most commonly used in construction. Asphalt is found in natural deposits, but most is produced from petroleum. Tar is produced by the distillation of wood and coal. Fractional distillation of tar produces coal tar pitch.

PROPERTIES OF BITUMENS

Possibly the most important property of bitumens is *water resistance*. The water resistance of bitumens is excellent. This varies with the different products produced, but overall bitumen is a good waterproof material.

Bitumens *adhere well* to dry solid surfaces. Bitumens bond well because they are in a semifluid state needed by adhesives to bond. They will not bond to wet surfaces.

Bitumens are *flammable* and will ignite when heated to their *flash point*. This influences the temperatures used to heat bituminous materials for various applications.

Another property to consider is the *softening point*. Although this varies with the composition of the product, it is a very important property to consider in relation to the temperatures to be experienced by various applications. For example, the softening point for roofing asphalts ranges from 200 to about 220°F (95 to 104°C), but coating-grade asphalts, such as those used for waterproofing, have a softening point of 50 to 55°F (11 to 13°C).

Bitumens exhibit *cold flow* properties. This means they tend to flow or spread or lose their shape. This is more pronounced when temperatures rise. It does offer the advantage of causing a check or crack on a roof membrane to seal or heal when the sun heats up the asphalt.

The **viscosity** of asphalt is an important property when considering applications. Viscosity is the ability to stay in place when subjected to heat. Asphalts have good viscosity properties up to 135°F (58°C), and some can withstand temperatures up to 275°F (136°C).

Asphalts rate high in their **ductility** properties. Their molecules hold fast even when they are extended by heat and pressure. Asphalts can expand and still remain bonded to the materials upon which they have been placed.

ASPHALT

Asphalt is a dark brown to black cementitious material in semisolid or solid form made up of bitumens found in deposits of natural asphalt. A similar product is a residue from the distillation of petroleum. Petroleum provides most of the asphalts used today (Fig. 27.1).

Asphalt is an ingredient in many products used in construction. Most apparent are its uses in roofing and siding and as a paving material. It is used in some paints, adhesives, acid- and alkali-resistant coatings, and damp-proofing and waterproofing solutions. It is used to coat organic fiber and fibrous plastic panels, to coat papers of various types, in adhesives and cementitious materials, and in the manufacture of drain and sewer pipe. Some types, such as waterproof coatings, are applied on the site. Others, such as a coating for fibrous panels, are applied in the factory (Table 27.1).

The grades and types of asphalt used for coatings, cements, paints, and adhesives are in Tables 27.2 and 27.3. The grade refers to the liquefaction of the asphalt.

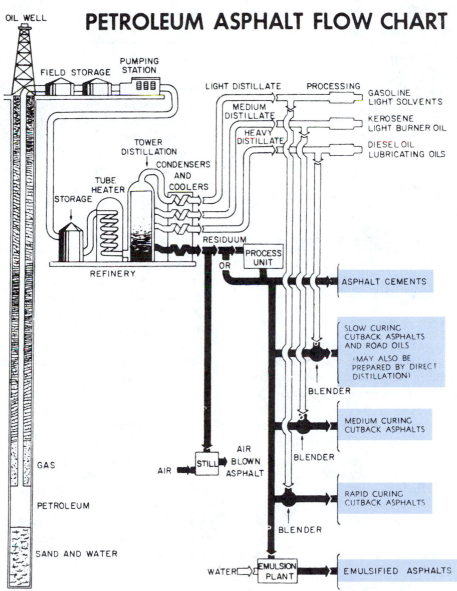

Figure 27.1 The process for producing various asphalt products from petroleum. *(Courtesy The Asphalt Institute)*

Table 27.1 Major Classes and Uses of Bituminous Mixtures

Class	Use
Liquid	Alleviate dust (spraying) Waterproofing Impregnation or saturation of materials
Medium consistency	Sealing compound Road surface binder Adhesive compound for roofing Expansion joint caulking
Solid	Electrical insulation compound Molded products

Table 27.2 Grades of Asphalt Used for Paints, Coatings, Adhesives, and Cements

Grades of Liquefaction
0 (Thinnest)
1
3
4
5 (Thickest)

Table 27.3 Types of Asphalt Used for Paints, Coatings, Adhesives, and Cements

Type	
Steam Refined with Petroleum Solvent	**Emulsified in Chemically Treated Water**
Slow curing (SC)	Slow setting (SS)
Medium curing (MC)	Medium setting (MS)
Rapid curing (RC)	Rapid setting (RS)

Table 27.4 Liquid Asphalt Products and Their Solvents

Classification	Solvent
Rapid curing (RC)	Gasoline or naphtha
Medium curing (MC)	Kerosene
Slow curing (SC)	Slowly volatile or nonvolatile oils
Asphalt emulsions	Water and emulsifiers

Table 27.5 Physical Characteristics of Roofing Asphalts

Type	Softening Point	Flash Point
Type I, Dead level	135–151°F	475°F
	58–67°C	248°C
Type II, Flat	158–176°F	475°F
	70–80°C	248°C
Type III, Steep	185–205°F	475°F
	85–97°C	248°C
Type IV, Special steep	210–225°F	475°F
	100–108°C	248°C

Grade 0 is the thinnest, and grade 5, which is much like a thick paste, has the thickest consistency. The types are slow curing (SC), medium curing (MC), and rapid curing (RC). These refer to steam refined or oxidized asphalt having a petroleum solvent (Table 27.4). Asphalt emulsified in chemically treated water has grades. These are slow setting (SS), medium setting (MS), and rapid setting (RS).

Roofing asphalt is available in four types:

Type I – Dead level
Type II – Flat
Type III – Steep
Type IV – Special steep

Physical characteristics for these four types of roofing asphalts are in Table 27.5.

Asphalt cements are binders used to produce high-quality asphalt pavements. They are highly viscous and are made in several grades based on consistency. Asphalt cements are semisolid at normal ambient temperatures. They are tested for viscosity at 140°F (60°C).

Asphalt cements are graded from the softest, AR 1000, to the hardest, AR 16000. AR 4000 is a general-purpose grade, and AR 8000 is used in hot climates. The grade chosen depends on the quality of the aggregate and on the climate.

Asphalt cements are blended with aggregates graded into a range of sizes. Typical aggregates include crushed stone, gravel, and sand. Aggregates compose about 90 percent of the weight of the paving mix. The aggregates give the mix its strength, and the asphalt is the binder. In some cases tars are added because they increase the resistance to damage from gasoline spilled on the paved surface. Asphalt cement paving is used for paving roads, drives, and parking lots (Fig. 27.2).

Cutback asphalt is a broad classification of residual asphalt materials left after the petroleum has been processed to produce gasoline, kerosene, diesel oil, and lubricating oils. The *residual asphalt* is then blended with various solvents to produce cutback asphalt. There are three classifications—rapid curing (RC), medium curing (MC), and slow curing (SC). Cutback asphalt often is mixed with granular soil to stabilize the roadbed before paving or to serve as the binder for a finished road surface for light traffic, such as on secondary roads.

The *emulsion group* of asphalts consists of emulsified asphalt cement mixed with water. This type is used in road construction. The emulsion is made by adding heated, fluid asphaltic cement into water to which an emulsifying agent such as soap or bentonite clay has been added. A stabilization agent, protein, is added to prevent the particles from blending together within the mix. The suspended particles of asphalt blend with the aggregate or soil particles as the water drains away and evaporates.

Figure 27.2 Asphalt paving cement is used to pave roads, driveways, and parking lots. *(Courtesy The Asphalt Institute)*

Table 27.6 Grades of Emulsified Asphalts

Anionic
RS asphalt—rapid setting
MS asphalt—medium setting
SS asphalt—slow setting
Cationic
CRS asphalt—rapid setting
CMS asphalt—medium setting
CSS asphalt—slow setting

Emulsified asphalts are either anionic or cationic, depending on the emulsifying agent used. Anionic emulsions are those in which the asphalt globules in the mix are negatively charged. Cationic emulsions have asphalt globules that are positively charged. Each comes in three grades (Table 27.6). They are used for the same purposes as cutback asphalts.

Emulsified asphalts have the advantage of being applied cold, which eliminates the need to heat the asphalt to bring it to a fluid condition. They may be applied to damp aggregates, as in the base for a road, because water is already a part of the mix. Water is not volatile, as are the solvents in cutback asphalt, so danger from fire and the production of hazardous fumes is not present. Emulsified asphalts work best when weather conditions are such that water evaporation will not occur, such as in a rainy season.

Plastic asphalt cements are made using asbestos fibers and asphalts that have good plasticity and elastic properties. This black cement is used to bond flashing and to make roof repairs. It does not flow at high temperatures or become brittle at cold temperatures. It allows for expansion and contraction of materials without breaking.

Quick-setting asphalt cement is much like plastic asphalt cement, but it has greater adhesive properties and sets up rapidly. It is used to cement down the free tabs on shingles and the laps between layers of roll roofing.

Asphalt roofing tape is a porous fabric strip saturated with asphalt. It is available in rolls 4 to 36 in. (100 to 914 mm) wide and 50 yd. (45.5 m) long. It is used with plastic asphalt cements to patch holes in roofs, to patch seams, and to seal flashing.

Asphalt Paving

Asphalt is widely used for paving because it is a cementitious substance and will bind other ingredients, such as aggregates, together. It experiences a minimum of cracking because it is flexible, plastic, and ductile. It withstands exposure to the weather and has resistance to salt, acids, and alkalis.

A bituminous binder is used to hold together and protect the mineral aggregate used on paving surfaces. Asphalt cements are used as binders for quality paving. Cutback asphalt is used to stabilize the roadbed before paving and as a binder on the surface of light duty roads. Emulsion asphalts may be applied cold to damp aggregates.

Asphalt paving mix designs are tested by a variety of methods. The purpose of these tests is to get a blend of the asphalt and aggregates of various sizes to give a product that will withstand the traffic and still be easy to place.

Laboratory Tests of Asphalt

There are many standard types and grades of asphalt. Since they are used where temperatures may vary considerably, standard tests are used to ascertain their flow properties. Various testing procedures have been developed by the Asphalt Institute, the American Association of State Highway and Transportation Officials, and the American Society for Testing and Materials.

Viscosity Test

Viscosity tests are used to find the flow properties of asphalts at the temperature at which they will be applied and their temperature in service. One testing method uses a capillary tube viscometer (Fig. 27.3). The viscometer is mounted in a constant temperature bath and preheated asphalt is poured into the large side of the viscometer until it reaches the fill line. The viscometer is left in the bath for a prescribed length of time. A partial vacuum is pulled on the small tube, and the time it takes the asphalt to move between two timing marks is recorded. This is an indication of flow or viscosity.

Penetration Test

The penetration test is, in general, being replaced by the viscosity test. It involves measuring the depth that a weighted needle penetrates into an asphalt sample over a specified time.

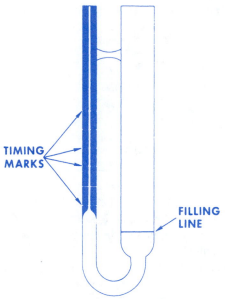

Figure 27.3 This device is used to make viscosity tests to find the flow properties of asphalts. *(Courtesy The Asphalt Institute)*

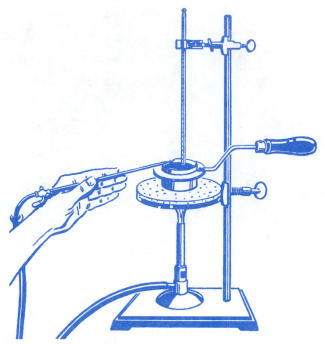

Figure 27.4 This apparatus is used to find the flash point of asphalt samples. *(Courtesy The Asphalt Institute)*

Flash Point Test

The flash point test determines the temperature to which an asphalt may be heated without the danger of an instantaneous flash occurring when the asphalt is exposed to an open flame. The flash point temperature is well below the temperature at which the asphalt will burn. The burn temperature is called the "fire point."

The test is made by placing the sample in a brass cup and heating it at a prescribed rate. A small flame is played over the surface periodically. The temperature at which the vapor produces a flash is the flash point temperature (Fig. 27.4). There are other tests available for determining the flash point.

Thin Film Oven Test

This test subjects a sample of asphalt to hardening conditions similar to those that occur in a hot-mix plant operation. A 50 mil asphalt sample is placed in a cylindrical flat-bottom pan 5.5 in. (140 mm) in diameter and $\frac{3}{8}$ in. (9.5 mm) deep. An asphalt layer about $\frac{1}{8}$ in. (3 mm) deep is placed in the pan. This is heated at 325°F (163°C) in an oven with a revolving shelf for five hours. The sample is then subjected to the penetration test.

Ductility Test

The ductility test is used to find out how much an asphalt sample will stretch at various temperatures below its softening point. A standard briquette is molded. It is brought to the specified test temperature and pulled at a specified rate until the sample completely separates. The elongation, in centimeters, at which the final thread of asphalt breaks is used to indicate ductility (Fig. 27.5).

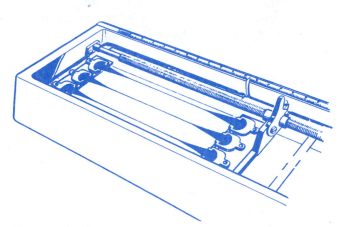

Figure 27.5 The ductility test is made to find out how much an asphalt sample will stretch at various temperatures. *(Courtesy The Asphalt Institute)*

Solubility Test

This test is used to ascertain the purity of the asphalt cement. A predetermined amount of asphalt cement is dissolved in trichloroethylene. The soluble portion represents the active cementing constituents. Inert matter is filtered out and measured. The result is given as a percent of the soluble content.

Specific Gravity Test

The specific gravity of a material is the ratio of the weight of a given volume of the material to the weight of an equal volume of water. Water has a specific gravity of 1. The specific gravity of a material will vary with the temperature because the volume of a material changes with temperature changes. An asphalt with a specific gravity of 1.1 is 1.1 times as heavy as water.

Softening Point Test

The various grades of asphalt soften at different temperatures. The softening point is found by the ring and ball test. The heated asphalt is poured into a brass ring. The sample is suspended in a water bath, and a steel ball of specified weight and diameter is placed in the center of the sample. The bath is heated at a controlled rate until the ball reaches the bottom of the glass bath container. The temperature of the water at this point is the softening point (Fig. 27.6).

Distillation Test

The distillation test is used to find the proportions of asphalt and diluent present in a sample. It is also used to find how much diluent distills off the sample at various

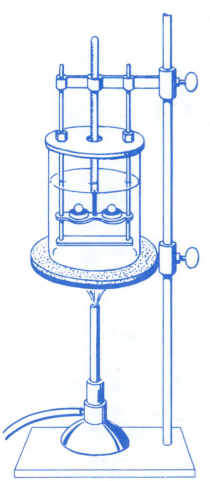

Figure 27.6 The softening point of asphalt is found by this ring and ball test. (*Courtesy The Asphalt Institute*)

temperatures. This information is useful to ascertain the asphalt's evaporation characteristics, which influence the rate the asphalt, as used in road construction, will cure after it is applied.

OTHER PRODUCTS MADE WITH BITUMINOUS MATERIALS

In addition to asphalt, a wide range of products are manufactured using bituminous materials as the major components. Most common among these are coal tar pitch, felts, stabilizers, and surfacing materials.

Coal Tar Pitch

Coal tar pitch is a dark brown to black hydrocarbon obtained by the distillation of coke-oven tar. It is available in several grades and is used as the basis for a number of paints, roofing products, and waterproofing materials. It has a softening point near 150°F (65°C).

Coal tar enamel is made from coal tar pitch by adding mineral fillers to the pitch. It is used to protect pipe in pipeline work. Cold-applied coal tar products have a solvent added to liquify them. Hot applied coal tar coatings give better protection than cold coal tar coatings.

Felts

Felt is a sheet material made from cellulose fibers from organic materials such as wood, paper and rags, or glass fibers or asbestos.

Saturated felts are made with an organic mat saturated with coal tar pitch or asphalt and coated with a surface coat of thin asphalt. This is sometimes called tar paper. It is used as an underlayment for shingles, as sheathing paper, and as laminations in built-up roof construction. Tar paper is also used to produce roll roofing and shingles.

Saturated felt is available in three weights, Type 15, 20, and 30. The type number indicates the weight per square of felt. For example, Type 15 weighs 15 pounds per square (100 sq. ft.).

Saturated felt is available in rolls 36 in. (914 mm) wide. Type 15 felt rolls are 144 ft. (44 m) long, Type 20 rolls are 108 ft. (11 m) long, and Type 30 rolls are 72 ft. (22 m) long.

Fiberglass Sheet Material

Fiberglass mats are said to be impregnated with asphalt. However, they are not "saturated" because the glass fibers will not absorb the asphalt. The asphalt forms a coating on the surface and fills the spaces between the fibers. Fiberglass mats are available in rolls 36 in. (914 mm) wide and 108 ft. (11 m) long.

Fireproofing Paper

Fireproofing paper is made using asbestos fibers either in a pressed mat like felt or in woven sheets. The various sheet products are used as underlayment for finished roofing materials, as vapor barriers in walls and floors, and for other such applications. They are not left exposed to the weather because coal tar pitch oxidizes rapidly when exposed to ultraviolet rays from the sun.

Waterproof Coatings

Asphalt waterproofing is used on masonry walls above and below grade. Below grade it must resist the pressure of subsurface water and keep it from passing through the foundation. Since it is not subjected to high temperatures below grade, it is possible to use asphalts with lower softening points. Above grade it resists passage through a wall or roof deck. Here it might be exposed to sunlight and should have a higher softening point.

A bituminous waterproofing coating consists of one or more coats mopped on either hot or cold. Cold applied can be reinforced by the addition of glass, plastic, or asbestos fibers. Cutbacks and emulsions are used extensively for this purpose. They may be covered with a plastic membrane or a felt membrane.

A dampproofing board product is made with an asphalt core covered on both sides by layers of asphalt-impregnated paper or by felt and covered with a weather resisting coating. Additional information on waterproof coatings can be found in Chapter 26.

ROOF COVERINGS

Building codes classify roof covering requirements. They vary with the type of building, size, occupancy, and fire district. The following pages discuss the various types of bituminous roof covering materials in common use. It should be noted that coal tar pitch and asphalt are not compatible and are not used where they will be in physical contact (Fig. 27.7).

Fire Ratings

In choosing asphalt roof coverings, one factor to consider is the fire rating classification of the products available. The following fire ratings of roofing materials are based on specifications of the Underwriters Laboratory Inc. (UL). UL labels indicating the fire rating appear on packaged asphalt roofing products (Fig. 27.8).

Class A: Highest rating. The covering is effective against severe fire exposure. Roof covering is not readily flammable and offers a high degree of fire protection for the roof deck.

Class B: Moderate protection against fire exposure. Roof covering is not readily flammable and offers a moderate degree of fire protection for the roof deck.

Class C: Minimal protection. Covering provides protection against light fire exposure. Covering is not readily flammable and offers a measurable degree of fire protection for the roof deck.

Figure 27.7 A variety of roof covering materials use asphalt. *(Used by permission of Georgia-Pacific Corporation. All rights reserved.)*

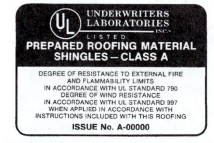
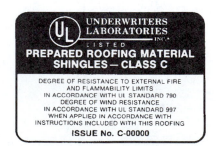

Figure 27.8 Manufacturers of shingles whose products meet the safety standards of Underwriters Laboratories Inc. (UL) may be authorized by UL to use these labels which include the UL Mark on those products. *(Reproduced with permission of Underwriters Laboratories Inc.)*

Roll Roofing

Roll roofing uses either organic felt or fiberglass mats as the base. A viscous bituminous coating is applied on this base, forming the exposed surface. Roll roofing is made in four types—smooth-surfaced, mineral-surfaced, mineral-surfaced selvage-edged, and pattern-edged (Fig. 27.9). *Smooth-surfaced roll roofing* has both sides covered with a fine talc or mica to keep the surfaces from sticking as it is made into rolls.

Mineral-surfaced roll roofing has mineral granules in a wide range of colors rolled into the surface, giving a surface that is attractive and protecting the bitumen from damage from the ultraviolet rays from the sun. The minerals also increase the fire resistance of the product. Mineral-surfaced roll roofing is available in rolls 36 in. (915 mm) wide and from 36 to 72 ft. (8 to 22 m) long. Smooth-surfaced rolls come in weights of 50 to 65 lb. per square, mineral-surfaced rolls 90 lb. per square, and fiberglass-reinforced mineral fiber rolls 75 lb. per square.

Mineral-surfaced selvage-edge roofing is of the same construction as mineral-surfaced roll roofing except that only 17 in. of the 36 in. wide surface is covered with granules. The 19 in. selvage edge is used for lapping with the next layer, forming a two-ply covering.

Pattern-edged roll roofing is a mineral-surfaced product that has a 4 in. uncoated band in the center. The roll is semicut along this strip to form two 18 in. wide patterned roofing strips that are thin lapped 2 in. over the layer below.

Roll roofing can be installed in a single or double thickness. The double thickness provides much better protection over a period of time (Fig. 27.10).

Conventional Hot Built-Up Roof Membranes

A built-up roof consists of alternate plies of organic or fiberglass roofing felt with a hot bitumen coating mopped over each layer (Fig. 27.11). The design of the roof varies with the situation, but generally it consists of three or more layers of felt with a bitumen layer over each and a bitumen topcoat with some form of aggregate rolled on top. A conventional hot built-up roof system is shown in Fig. 27.12.

The felts give the needed reinforcements to keep the bitumens in each layer from alligatoring. Alligatoring refers to surface cracking due to oxidation and shrinkage stresses, which gives a repetitive mounding of the asphalt surface similar to that on an alligator hide.

The aggregate surface is usually gravel, marble chips, or slag. Gravel and marble chips are usually applied 400 pounds per 100 sq. ft. and slag 300 pounds per 100 sq. ft. Conventional built-up roofs can be designed for slopes up to 3 in. (76 mm) per 12 in. (305 mm). Slopes above ½ in. (12 mm) per 12 in. (305 mm) usually require the use of steep asphalt.

Modified Asphalt Roofing Systems

Modified asphalt roll roofing is composed of polymer modified bitumen (modified asphalt) reinforced with one or more plies of fabric, such as polyester glass fiber (Fig. 27.13). These membranes are of uniform thickness and have consistent physical properties throughout the membrane.

A variety of modifiers and types of reinforcing plies are designed for use on almost every type of construction, such as new roofing, reroofing, domes, and spires.

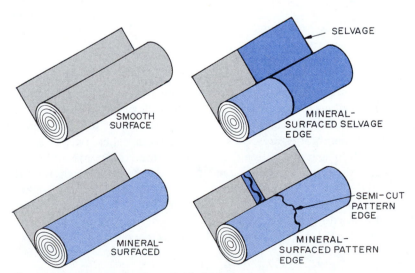

Figure 27.9 Roll roofing is available in four types.

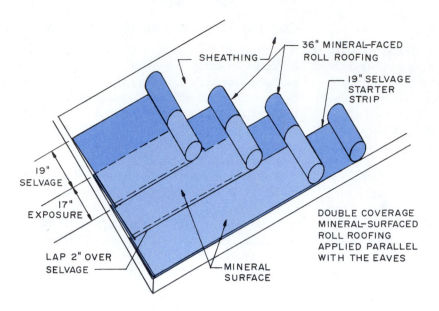

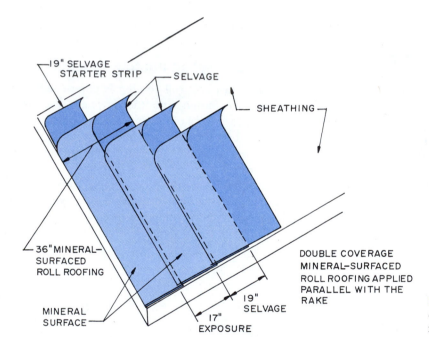

Figure 27.10 Two ways to apply double coverage mineral surfaced roll roofing.

Modified membranes may be used below grade for waterproofing canals, water reservoirs, and landfills.

Most modified bitumen membranes are made using either styrene-butadiene-styrene (SBS) or atactic polypropylene (APP). APP modified membranes are generally applied using a propane torch to heat and soften the underside of the membrane. This surface becomes a molten adhesive that is placed on the substrate and rolled. When it cools it bonds to the substrate.

SBS modifies the bitumen by forming a polymer lattice within the bitumen. When this is cooled the membrane acts somewhat like rubber. Since the membrane is more flexible than the APP membrane, SBS is used where flexibility is needed, such as where the substrate may be subject to movement or deflection. SBS modified membranes are generally mopped in hot asphalt, self-adhered, or adhered with cold process adhesives. Some types can have their joints heat welded. Some have factory-applied mineral aggregate surfaces to protect them from ultraviolet damage. Those without this covering need to be protected with some form of ultraviolet protective coating.

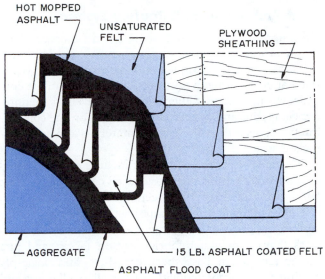

Figure 27.12 A conventional hot asphalt built-up roof system on a nailable deck.

Figure 27.11 Tar can be applied by hot mopping. *(Used by permission of Georgia-Pacific Corporation. All rights reserved.)*

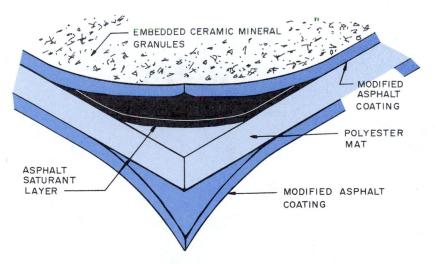

Figure 27.13 A form of a modified asphalt roll roofing installation. *(Courtesy Asphalt Roofing Manufacturers Association)*

Modified bitumen membranes are supplied in rolls generally 36 in. wide and in varying lengths. Examples of the construction of typical modified bitumen roofing systems are in Fig. 27.14.

Cold-Applied Asphalt Roofing Systems

Cold-applied systems use some form of coated base sheet, fabric, or similar reinforcement over which the principal waterproofing agent, which is a liquid, is applied at ambient temperatures. The selection of the cold process application depends on the levels of maintenance, repair, and service expected and on the compatibility between the cold-applied materials and any existing substrate.

The *reinforcing felts and fabrics* include organic coated base sheets, organic mineral-surfaced cap sheets, organic felt sheets, fiberglass ply sheets, fiberglass base sheets, fiberglass mineral-surfaced cap sheets, single-ply smooth-surfaced sheets, single-ply mineral-surfaced sheets, fiberglass fabrics, polyester fabrics, cotton fabric, and jute burlap fabric. The base sheet, cap sheet, and single-ply roofing are incorporated with cold-applied adhesives and cold-applied surface coatings to form a roof membrane.

The *coatings and adhesives* used are designed to be brushed or sprayed at normal room temperatures. These include filled and nonfilled asphalt cutbacks, filled and nonfilled asphalt emulsions, filled and nonfilled coal tar coatings, and filled and nonfilled aluminum pigmented asphalt. Toppings include gravel and aluminum chips placed in the topcoat while it is still wet. They block ultraviolet light, can reflect heat, and are decorative.

Asphalt and Fiberglass Shingles

Asphalt shingles are made using an organic felt base saturated with asphalt and coated with a mineral stabilized coating of asphalt on both sides. The top side to be exposed to the weather is coated with mineral granules. They are available in a wide range of colors. The bottom is relatively smooth and is coated with talc to prevent it from sticking to the shingle below it as they are bundled for shipping (Fig. 27.15).

Fiberglass shingles are made using a fiberglass mat as the base. The mat does not have to be saturated with asphalt before it receives the asphalt top coating. Mineral aggregates are rolled into the top coating. These shingles are lighter weight than asphalt shingles (Fig. 27.16).

Shingles are made in a variety of styles. Most commonly they are in strips 12 × 36 in. (305 × 915 mm) with the exposed surface cut to resemble three smaller shingles. They are available in weights of 100 to 300 lb. per square (45 to 135 kg per square). Commonly available forms of asphalt and fiberglass shingles are shown in Fig. 27.17. Fiberglass shingles are more commonly used today than asphalt shingles.

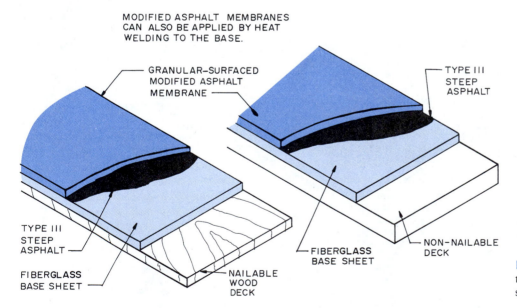

MODIFIED ASPHALT MEMBRANES CAN ALSO BE APPLIED BY HEAT WELDING TO THE BASE.

GRANULAR–SURFACED MODIFIED ASPHALT MEMBRANE

TYPE III STEEP ASPHALT

TYPE III STEEP ASPHALT

FIBERGLASS BASE SHEET

NAILABLE WOOD DECK

FIBERGLASS BASE SHEET

NON–NAILABLE DECK

Figure 27.14 Typical applications of modified asphalt roofing systems.

Figure 27.15 Two styles of the many asphalt/fiberglass shingles available. *(Used by permission of Georgia-Pacific Corporation. All rights reserved.)*

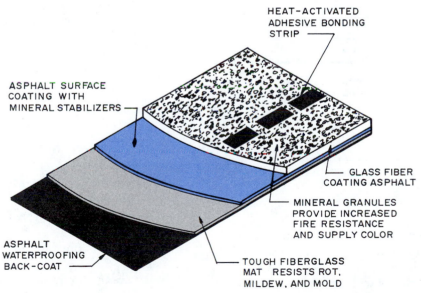

HEAT–ACTIVATED
ADHESIVE BONDING
STRIP

ASPHALT SURFACE
COATING WITH
MINERAL STABILIZERS

GLASS FIBER
COATING ASPHALT

MINERAL GRANULES
PROVIDE INCREASED
FIRE RESISTANCE
AND SUPPLY COLOR

ASPHALT
WATERPROOFING
BACK-COAT

TOUGH FIBERGLASS
MAT RESISTS ROT,
MILDEW, AND MOLD

Figure 27.16 The layers forming the construction of a fiberglass shingle. *(Courtesy GAF Building Materials Corporation)*

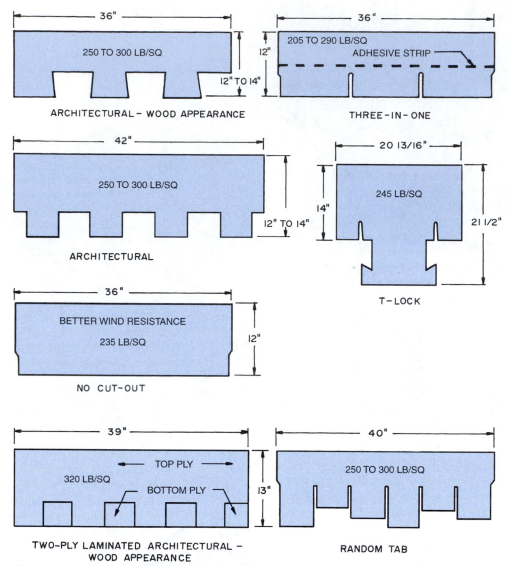

Figure 27.17 Commonly available forms of asphalt and fiberglass shingles.

REVIEW QUESTIONS

1. What will prevent bitumens from adhering to a concrete wall?
2. What is the point at which bitumens will burn?
3. What property of bitumens allows them to spread after they are in place?
4. What is the binder used with aggregate to produce quality asphalt pavements?
5. From what source are most asphalts obtained?
6. How do cutback asphalts differ from asphalt emulsions?
7. What are the two major uses for cutback asphalts?
8. What can happen that will cause asphalt emulsions on a roadbed not to harden?
9. What asphalt product is used to make roof repairs?
10. What are the big differences between saturated felts and fiberglass felts?
11. Why are fiberglass felt mats not saturated with asphalt?
12. What types of fiber are used in making fireproofing paper?
13. What are the fire rating classes used to rate asphalt roofing materials?
14. Which fire rating class indicates the highest fire protection?
15. How does a built-up roof differ from one covered with roll roofing?
16. What are the various materials used to make fiberglass shingles?

KEY TERMS

asphalt A dark brown to black cementitious material derived from natural asphalt deposits and by refining petroleum.

bitumen A mixture of complex hydrocarbons derived from petroleum or natural sources.

coal tar pitch A dark brown to black hydrocarbon produced by the distillation of coke-oven tar.

ductility A measure of the capability of a material to be stretched or deformed without breaking.

felt A sheet material made using a fiber mat, saturated and topped with asphalt.

viscosity A measure of the resistance exhibited by a fluid in resisting a force that tends to cause it to flow.

SUGGESTED ACTIVITIES

1. Collect samples of as many bituminous products as possible.

2. Try to prepare a setup to make some tests on a variety of types and grades of asphalt. You may not have the required laboratory equipment, but you might be able to improvise. The results will not be usable, but you can compare what you found about the general characteristics of the material. Since the material may flash when exposed to heat or flame, be especially careful and wear protective face and body guards.

3. Visit construction sites when the finished roofing is being applied and observe the installation techniques and safety protection used.

4. Visit a road paving project and observe the procedures for preparing, delivering, and laying the asphalt final coat. Try to find out what was done to prepare the base. If possible, find the local or state requirements for street and parking area paving.

ADDITIONAL INFORMATION

Chen, W. F., *The Civil Engineering Handbook,* CRC Press, Boca Raton, Fla., 1995.

The NRCA Roofing and Waterproofing Manual, National Roofing Contractors Association, Rosemont, Ill., 1992.

Watson, J. A., *Commercial Roofing Systems,* Prentice Hall, Englewood Cliffs, N.J., 1984.

Weismantel, G. E., *Paint Handbook,* McGraw-Hill, New York, 1981.

Other resources are numerous related publications from the associations listed in the Professional and Technical Organizations in Appendix B.

CHAPTER

28

Roofing Systems

This chapter will help you to:

1. Be familiar with the vast array of roofing systems available.

2. Select the proper roofing system for various applications.

3. Be able to discuss with clients the merits of each roofing system.

4. Understand the way each roofing system should be installed.

A roof system consists of a number of interacting materials that when properly combined provide a weather-resistant top surface to a building. In some systems roofs also insulate the building.

Many types of decks, insulation, and roof covering materials are available. Manufacturers of these materials provide technical information and installation instructions. A roofing system can be designed to serve a variety of purposes, such as resistance to fire, water penetration, and wind. In addition, it can provide acoustical and thermal insulation and contribute to energy conservation.

Roof systems are generally divided into two categories, low-slope and steep-slope. **Low-slope roofs** are flat or nearly flat and provide little runoff of rain. **Steep-slope roofs** permit the rapid runoff of rain, which reduces the likelihood of penetration through the water-resistant surface.

Low-slope roofs are generally cheaper to build and can be extended over large buildings, but steep-sloped roofs are limited by the spans possible by the roof structural materials. Water tends to pond on flat roofs even though drains are provided. As the temperatures change or if there is movement in the deck, the finished membrane may crack or pull loose from parapets and drains. Expansion joints are used to try to diminish this problem. If water vapor gets under the finished membrane, solar heat will cause it to get hot and may form blisters in some roofing materials.

Steep-slope roofs can be finished with materials such as metal or fiberglass shingles, clay tiles, or standing seam metal roofing. They may have an attic space below that, when properly ventilated, reduces possible damage due to vapor or high temperatures under the roofing material. The finished roofing material is highly visible on a steep-slope roof and forms an important part of the overall design of the building.

BUILDING CODES

Building codes govern the materials, design, construction, and quality of the roof and roof covering. This includes the ability to resist rain and wind, durability specifications, compatibility of materials with each other, physical characteristics, and fire protection. Fire protection classifications are specified by ASTM E108.

There are four classifications of roof covering. *Class A roof coverings* are effective against severe fire test exposure. They include clay tile, mineral fiber, concrete, slate, some fiberglass asphalt shingles, and other materials that have been so certified by an approved testing agency. They may be used on buildings of all types.

Class B roof coverings are effective against moderate fire test exposure. They include metal sheets and shingles, some composition shingles, and other materials certified by an approved testing agency.

Class C roof coverings are effective against light fire test exposure. They include materials certified as Type C by an approved testing agency.

Nonclassified roof coverings are not permitted on any buildings covered by most codes. In some cases these may be approved on various types of storage buildings.

ROOFING SYSTEM MATERIALS

Roofing systems are composed of decks, vapor retarders, insulation, and finished roofing materials. Materials for most of these components have been discussed in earlier chapters, which are referenced below.

Roof Decks

Materials commonly used for roof decks are detailed in other chapters related to materials and types of construction. A properly functioning roof depends on a structurally sound deck that is compatible with the roofing system. Following are the frequently used decking materials.

Cement-wood-fiber panels are made by bonding treated wood fibers with portland cement or some other binder and compressing them into structural panels.

Concrete decks are made by using cast-in-place techniques and precast concrete structural members. (Review Chapters 8 and 9.)

Gypsum concrete decks are produced by mixing gypsum, wood fibers or mineral aggregates, and water and casting it on formboards. Precast gypsum planks are also used.

Lightweight insulating concrete is made by mixing insulating aggregates, such as perlite, portland cement, and water and casting it on top of metal or bulb-tee and formboard decking systems.

Reconstituted wood panels are produced by bonding wood veneers or wood chips. Plywood and oriented strand board are the two commonly used materials.

Steel decks are produced by cold-rolling sheet steel into various structural shapes.

Wood planks are solid wood decking or glued laminated members made by bonding solid dimensional lumber into a structural decking member.

Vapor Retarders

It is important to stop the flow of water vapor from inside the building up into the roofing system. This flow will cause some decking to deteriorate and will damage some types of finished roofing surfaces. The moisture will also damage insulation in the exposed roofing assembly. Changes in temperature and humidity cause a difference in water vapor pressure between the inside and outside of the building, causing water vapor to migrate into the insulation (Fig. 28.1). The vapor retarder must resist the passage of water vapor. Any tears or holes in the material must be sealed. The protected roof membrane also serves as a vapor retarder. Vapor retarders should usually be applied to the warm side of the insulation in roof decks in cold climates. In warm, humid climates, winter condensation in the roof assembly is not a problem, and vapor retarders are often not used. Air-conditioning in warm climates may actually cause a reverse vapor migration. Vapor retarders must be used on top of concrete and poured gypsum decks.

The performance of a vapor retarder is indicated by its **water-vapor-transmission rate (WVTR)**, which is specified in *perms*. The perm is a measure of the porosity of a material to the passage of water vapor. Materials must have a perm rating of 0.00 to 0.50 perms. A perm rating indicates the number of grains of water vapor that will pass through one square foot of a material per hour when the vapor-pressure differential between the two sides equals 1 in. of mercury (0.49 psi). The metric equivalent is expressed in terms of grams/m²/hr./mm of mercury pressure difference between the two faces.

Some frequently used vapor retarders are

Bituminous materials, such as layers of asphalt roofing felt covered with hot asphalt.

Kraft paper layers bonded with asphaltic adhesive and a glass-fiber reinforcement. They are bonded to the decking with cold-applied asphalt adhesive.

Polyethylene sheet material laid loose on the deck or attached with mechanical fasteners. The sheets are overlapped and taped or joined with an adhesive.

Aluminum foil on the face of insulation batts and rigid insulation. It provides a vapor barrier and serves as a reflective insulation. The joints between insulation panels should be covered with aluminum tape.

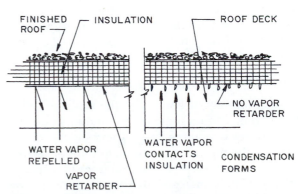

Figure 28.1 The vapor retarder is placed on the warm side of the roof assembly below the insulation.

Other vapor retarders contain a combination of materials. Examples of products available include combinations such as a polypropylene scrim, kraft paper sheet, vinyl scrim, aluminum foil, or polypropylene scrim and aluminum foil. The scrim used is a fiberglass fabric. (Scrim is an open-weave fabric.)

Insulation

The location of roof insulation varies with the design of the roof. In a typical steep-slope roof with an attic no roof insulation is required. Instead the ceiling is insulated. A low-slope roof must have some form of insulation and ventilation. These are illustrated in Chapter 22. All of the types of insulation can find some use in the design of the roof system. These are detailed in Chapters 22 and 25.

Finished Roofing

The finished roofing is exposed to the weather. A wide range of products are available, providing the designer with considerable choice. The specific properties of the products provided by the various manufacturers must be known and their installation recommendations followed. Table 28.1 describes many of the finished roofing materials available.

LOW-SLOPE ROOF ASSEMBLIES

The following discussion and illustrations detail some of the construction used for roofing on low-slope roofs. A low-slope roof system is one with a slight slope and upon which water may collect and then slowly flow to drainage outlets. The actual slope permitted varies with the type of roofing system used. The manufacturer's recommendations must be observed. Typically these will fall in a range of ¼ in. rise per 12 in. run (6.4 mm per 305 mm) to 2 in. rise per 12 in. of run (50.8 mm per 305 mm). Special provisions can be made for roofs with a 3 in. per 12 in. (76.2 mm per 305 mm) slope.

Low-slope roofing systems include built-up, thermoset single-ply coverings, thermoplastic single-ply coverings, modified bitumen coverings, spray-applied polyurethane foam coverings, liquid-applied coatings, and metal sheet coverings. An important feature in the design of low-slope roofs is adequate insulation.

Insulation for Low-Slope Roofs

Roof insulation may be applied to the top side of the decking. The insulation must meet building code and insurance bureau requirements. Roofs insulated this way increase the possibility of condensation occurring within the roof system, so a vapor retarder is required. Since the insulation increases the roof temperature on hot days, the roof materials age faster and the expansion and contraction stresses on the membrane are greater.

Rigid roof insulation includes cellular glass, composite boards, glass fiber, perlitic boards, polyisocyanurate foam boards, polystyrene boards, polyurethane foam boards, and wood fiber boards.

Roof insulation may be placed as a single or double layer. The double layer eliminates any leakage of heating or cooling energy that may occur between the joints in the first layer. The first layer may be mechanically joined to the deck, such as when steel decking is used. The second layer is bonded with hot bitumen or an adhesive (Fig. 28.2). The first layer on concrete decks is bonded to the deck with hot bitumen (Fig. 28.3).

Built-Up Roofing

The traditional **built-up roofing** system used on low-slope roofs consists of bitumen (asphalt or coal tar) usually applied hot over felts, which may be glass fiber, organic, or polyester, and a finished top surface, such as an aggregate (gravel or slag) or a cap sheet. Various manufacturers offer a variety of compositions of felts and roofing asphalts. Their recommended methods and application systems also vary. The following examples are typical of those in use.

Typical built-up roof construction on uninsulated nailable roof decks is shown in Fig. 28.4. The deck is covered with one ply of sheathing paper (a vapor retarder) nailed to the deck. Nailable decks include wood, plywood, structural wood fiber panels, lightweight insulating concrete, and precast and poured gypsum. Next, three to five layers of an asphalt-coated base felt are applied, bonded with coatings of a hot mopped bitumen. The top coating is covered with roofing asphalt and gravel or slag. Usually 400 lb. of gravel or 300 lb. of slag are applied per 100 sq. ft. of surface area. Manufacturers recommend that dead flat roofs be given a slight slope such as ½ in. per linear foot.

Nonnailable decks, such as steel, precast concrete, and poured concrete, have the insulation bonded with hot bitumen or an approved adhesive. This is followed by layers of asphalt-saturated roofing felt and hot roofing asphalt. The layers of felt are laid in a full bed of hot asphalt and broomed in place (Fig. 28.5). The roofing asphalt is brought to the site in a tank truck and heated in an asphalt kettle (Fig. 28.6). The heated asphalt is pumped to a tank on the roof and moved to the area where it is needed (Fig. 28.7).

Roof penetrations have to be flashed (Fig. 28.8), as do roof drains (Fig. 28.9) and parapets and other places where the roof butts against a wall (Fig. 28.10). Finally, the gravel forming the top protective coating is lifted to the roof (Fig. 28.11) and spread in a bed of hot roofing asphalt (Fig. 28.12). A typical construction detail is in Fig. 28.13.

Another type of built-up roofing is an assembly using an asphalt glass fiber roof membrane covered with a mineral-surfaced inorganic cap sheet. A *mineral-surfaced roof material* consists of a base felt that is coated on one or both sides with asphalt and surfaced with mineral granules. A typical assembly is shown in Fig. 28.14. An

Table 28.1 Materials Used for Finished Roofing

Material	Type of Roof	Descriptive Factors	Weight per Square (lb./100 ft.²)	Weight per Square Meter (kg/m²)
Aluminum (sheet, shingles)	Steep-slope	Fire resistant, long life, range of colors	5–90	2.44–4.39
Asphalt (built-up)	Steep-slope, low-slope	Granular topping applied influences fire class, life 20–30 years	100–600	4.88–29.3
Asphalt shingles (fiberglass, asphalt)	Steep-slope	Fire resistance varies with product, range of colors, life 20–30 years	235–325	11.47–15.86
Cement-fiber tile	Steep-slope	Fire resistance, long life, heavy, use in warm climates	950	46.4
Clay tile	Steep-slope	Fire resistant, long life, heavy	800–1600	39–78
Copper (sheet)	Steep-slope, low-slope	Fire resistant, long life, can be soldered	0.019″ thick 160 0.040″ thick 320	7.8 15.6
Lead, copper coated	Steep-slope, low-slope	Fire resistant, long life	$\frac{1}{32}$″ thick 200 $\frac{1}{16}$″ thick 400	9.76 19.52
Monel (Ni-Cu)	Steep-slope	Fire resistant, long life	22 gauge 1424 26 gauge 827	69.5 40.4
Perlite-portland cement	Steep-slope	High fire rating, lightweight, long life	900–1000	43.9–48.8
Plastic (single-ply membrane)	Low-slope	Long life, requires careful installation, limited fire classification, several types available	Loose laid Ballasted, 1000–1200 Fully adhered, 30–55	48.8–58.6 1.5–2.7
Plastic (liquid applied)	Low-slope	Limited fire classification follow manufacturer's directions	20–50	0.98–2.4
Slate	Steep-slope	Fire resistant, heavy, long life	$\frac{3}{16}$″ thick 800 $\frac{1}{4}$″ thick 900 $\frac{3}{8}$″ thick 1100 $\frac{1}{2}$″ thick 1700 $\frac{3}{4}$″ thick 2600	39.0 43.9 53.7 83.0 126.9
Stainless steel, terne coated	Steep-slope	Fire resistant, long life	90	3.89
Steel (sheet, shingles)	Steep-slope, low-slope	Fire resistant, long life, durable colors	Copper coated 130 Galvanized 130	6.3 6.3
Wood (shingles, shakes)	Steep-slope	No fire resistance unless treated, limited life	200–450	9.8–22.0
Zinc	Steep-slope	Fire resistant, long life, can be painted	9 gauge 670 12 gauge 1050	32.7 51.2
Terneplate copper-bearing sheet	Steep-slope	Fire resistant, long life	30 gauge 540 26 gauge 780	26.4 38.1
Mineral-surfaced cap sheet	Low-slope	Limited fire resistance, limited life	55–60	2.68–2.9
Modified bitumen	Low-slope	Fire resistance, 10 years or more	100	4.9

asphalt glass fiber membrane is nailed to the nailable deck and additional layers are bonded with hot asphalt. The mineral-surfaced inorganic cap sheet is bonded to the asphalt glass fiber base with hot asphalt. It is recommended that this system be used on roofs with a slope of ¼:12 in. (6.25:305 mm) or greater.

Modified Bitumen Membranes

Modified bitumen membranes combine polymer-modified asphalt and a polyester or fiberglass mat, resulting in a product of exceptional strength. The two membranes available are SBS (styrene-butadiene-styrene) and APP (atactic polypropylene). SBS sheets have a reinforcement mat coated with an elastomeric blend of asphalt and SBS rubber. APP membranes have a reinforcement mat coated with a blend of asphalt and APP plastic. They are available from various manufacturers with several reinforcements. However, fiberglass reinforcement is most frequently used.

The major difference between SBS and APP products is the blended asphalt used. The blends create a product that has greater elongation, strength, and flexibility than traditional roofing asphalts. Elongation, recovering after elongation, flexibility, and cold weather performance of SBS membranes are superior to those of APP membranes.

Figure 28.2 The first layer of insulation is fastened to the deck with mechanical fasteners. The second layer is bonded with hot asphalt. *(Courtesy Schuller International, Inc.)*

Figure 28.3 Rigid insulation is placed over hot asphalt on a concrete deck. *(Courtesy National Roofing Contractors Association)*

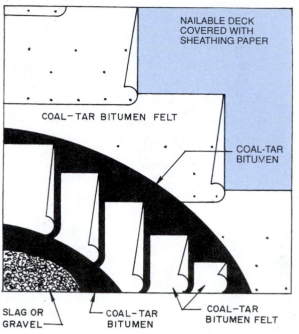

NAILABLE DECK COVERED WITH SHEATHING PAPER

COAL–TAR BITUMEN FELT

COAL-TAR BITUVEN

SLAG OR GRAVEL

COAL–TAR BITUMEN

COAL–TAR BITUMEN FELT

Figure 28.4 One of several ways built-up roof constructions are applied over a nailable roof deck.

Figure 28.5 Overlapping sheets of asphalt-saturated felt are bonded to the rigid insulation with hot roofing asphalt. *(Courtesy National Roofing Contractors Association)*

Figure 28.6 Roofing asphalt is heated in an asphalt kettle and pumped to the roof. *(Courtesy National Roofing Contractors Association)*

Figure 28.9 A roof drain flange is sealed with cold-applied asphalt mastic. *(Courtesy National Roofing Contractors Association)*

Figure 28.7 The hot roofing asphalt is pumped up to a tank on the roof. *(Courtesy National Roofing Contractors Association)*

Figure 28.10 Layers of asphalt-saturated organic felt are mopped with hot roofing asphalt to build up a multilayer membrane along the edge of the roof. *(Courtesy National Roofing Contractors Association)*

Figure 28.8 Roof penetrations are flashed with hot roofing asphalt and asphalt-saturated organic felt. *(Courtesy National Roofing Contractors Association)*

Figure 28.11 Gravel is raised to the roof by a hoist and dumped into a buggy. *(Courtesy National Roofing Contractors Association)*

Figure 28.13 A construction detail for an insulated low-slope roof deck with built-up roofing.

Figure 28.12 Stone ballast is spread in a bed of hot roofing asphalt to help hold the membrane in place and to protect it. *(Courtesy National Roofing Contractors Association)*

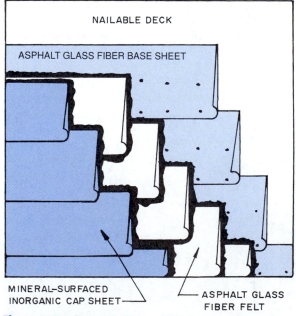

Figure 28.14 One installation detail for using mineral-surfaced cap sheets.

SBS products are generally installed using hot asphalt as the bonding material. They are applied as cap sheets over a base of hot asphalt and roofing felts as shown in Fig. 28.15. The cap sheet (SBS membrane) may have a ceramic granule surfacing to protect it from ultraviolet light, or it may be unsurfaced. The unsurfaced type must be coated with asphalt and gravel to give it ultraviolet protection (Fig. 28.16).

APA products are applied by a method called "torching." This is made possible by the unique properties of the modified bitumen. The back coating of modified asphalt is heated with a propane torch to the point at which it becomes able to bond the sheet to the substrate. APP products cannot be installed with hot mopped asphalt.

Single-Ply Roofing Systems

Single-ply roofing systems can be applied over almost any commonly used roof decking and over existing asphalt or built-up roofs when reroofing a building. Roofing insulation is required over the roof deck except over very smooth concrete, plywood, and splinter-free solid wood decks. The roof must drain freely and have sufficient drain outlets to carry away the water.

Single-ply roofing membranes are available in a number of materials, and the manufacturer's specifications and installation instructions should be carefully observed. Single-ply membranes are either thermoset or thermoplastic materials. Thermoset materials cure during manufacture and can only be bonded to themselves with an adhesive. Thermoplastic materials do not completely cure during manufacturing and can be welded together, usually with high-temperature air.

Single-ply membranes may be secured to the deck with ballast. The membrane is laid loose on the deck and covered with ballast to keep it from being lifted by the wind. The ballast is usually large, smooth aggregate or concrete pavers (Fig. 28.17). A variation of the ballasted roof membrane is shown in Fig. 28.18. The membrane is covered with insulation and a protective mat. The ballast is placed on the mat. In both systems the abutting edges of the membrane are spliced and all edges, penetrating pipes, and other penetrations are flashed as specified by the manufacturer.

Single-ply membranes can also be mechanically fastened to the deck. This can be accomplished using metal batten bars placed at intervals on top of the membrane and then screwed to the deck. The bars are then covered with a batten cover strip (Fig. 28.19). Another method uses screws and metal discs placed over the edge of a layer of membrane and screwed into the deck (Fig. 28.20). The adjoining membrane is lapped over this attachment. A third method is to place large-diameter discs on top of the membrane spaced as required and screw them to the deck. A waterproof cover is placed over the discs (Fig. 28.21).

Following are some of the commonly used single-ply roofing membranes.

Figure 28.15 These modified SBS bitumen membranes can be laid in hot roofing asphalt. *(Courtesy Schuller International, Inc.)*

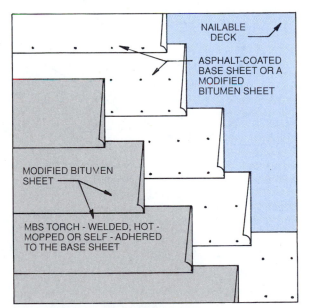

Figure 28.16 A typical assembly for a partially adhered single-ply modified bitumen roof system.

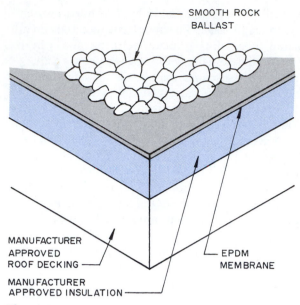

Figure 28.17 Single-ply membranes are sometimes covered with aggregate, which serves as ballast to hold it to the deck.

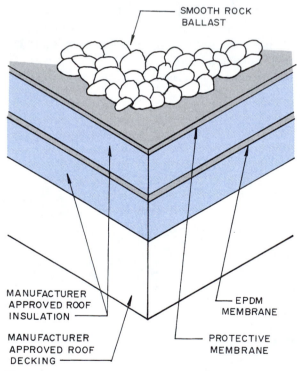

Figure 28.18 A single-ply protected membrane is covered with a layer of rigid insulation and ballast.

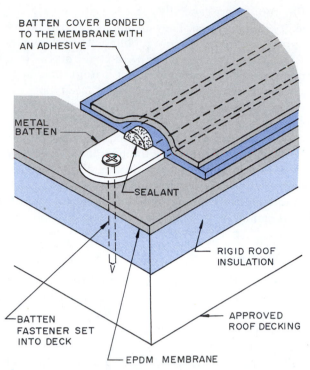

Figure 28.19 This single-ply membrane is mechanically fastened to the deck. The metal batten strip is covered with a plastic batten cover.

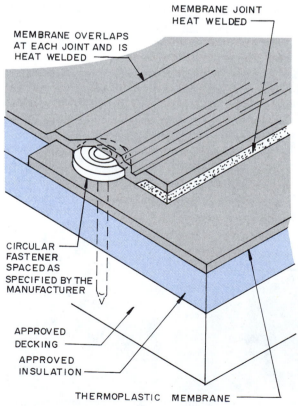

Figure 28.20 Some membranes are fastened to the deck with metal discs and sealed with a heat-welded lap joint.

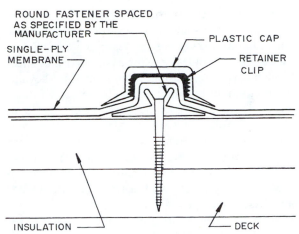

Figure 28.21 This circular mechanical fastener joins the membrane to the deck without putting a hole through it.

Figure 28.22 The EPDM single-ply membrane is unrolled on the deck. *(Courtesy Schuller International, Inc.)*

Chlorosulfated polyethylene (CSPE) is a thermoset membrane that completes its cure after it is installed. It has excellent weathering qualities and is resistant to ozone, sunlight, and most chemicals.

Ethylene propylene diene monomer (EPDM) is a thermoset elastomeric compound produced from propylene, ethylene, and diene monomer. It has good resistance to weathering, ultraviolet rays, abrasion, and ozone (Figs. 28.22 and 28.23).

Polyvinyl chloride (PVC) is a thermoplastic membrane produced by the polymerization of vinyl chloride monomer, stabilizers, and plasticizers. PVC membranes are resistant to weather and chemical atmospheres, have good fire resistance, and are easy to bond. They do not work well with bituminous materials.

Styrene-butadiene-styrene (SBS) is a thermoplastic membrane made by blending SBS with a high-quality asphalt over a fiberglass mat. It has good fire resistance, and it can be applied with hot or cold asphalt or be torched. The method used depends on the specific composition of the membrane.

Spray-Applied Roof Coatings

Spray-applied roof coatings consist of a polyurethane foam insulation layer applied to the deck and then topped with a protective coating. The protective coating is usually acrylic, polyurethane, or silicone. Sometimes mineral granules or aggregate applied to the wet top coating are used to provide additional protection. An application using a top coating consisting of several layers of elastomeric polyurethane is shown in Fig. 28.24.

Figure 28.23 The EPDM single-ply membrane is unfolded and laid over the surface of the deck. *(Courtesy Schuller International, Inc.)*

Liquid-Applied Roof Systems

Liquid-applied coverings are available from various manufacturers as one- or two-component elastomeric materials. They are applied by spraying, brushing or rolling over the roof decking, which is typically plywood or concrete. Any joints require special attention as recommended by the manufacturer.

Metal Roof Assemblies

Various manufacturers have metal roofing systems designed for low-slope roof assemblies (Fig. 28.25). They have developed panel profiles and seaming methods to be used to produce a watertight roof. The panels are available in a wide range of colors and may be galvanized or Galvalume steel, aluminum, copper, or terne-coated stainless steel. Galvalume is a trade name for a patented steel sheet coated with a corrosion-resistant aluminum-zinc alloy applied by a continuous hot-dip process. Terne-coated stainless steel panels are nickel-chrome stainless steel coated on both sides with an alloy of 80 percent lead and 20 percent tin. They are recommended for use in severe chemical and marine environments. Metal roofing panels are available embossed to create a stuccolike finish. Various types of other coatings are available, such as siliconized polyester and epoxy-based coatings.

Low-slope standing seam metal roofing systems require some slope. Typically a slope of $\frac{1}{4}$:12 is minimum. This type of roofing is generally held to the structure with some form of metal clip. The clip is joined to the metal frame as a bar joist or Z purlin with metal fasteners. One such design is shown in Fig. 28.26. The hold-down clip hooks onto the leading edge of the panel. The clips and edges of the butting panels are interlocked during the seaming process. The standing seam is formed on the roof with a portable seaming machine that runs along the joint and folds the members into a watertight joint (Fig. 28.27). The clip fits into a slot in the clip base that permits the clip to move with the roof during expansion and contraction. Metal roofing panels also are used extensively for reroofing flat conventional roofs and steep-slope roofs. This requires the installation of a metal subframe that is secured through the old deck to the structural frame and supports the new metal roof deck (Fig. 28.28).

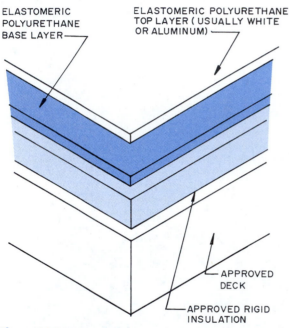

ELASTOMERIC POLYURETHANE BASE LAYER

ELASTOMERIC POLYURETHANE TOP LAYER (USUALLY WHITE OR ALUMINUM)

APPROVED DECK

APPROVED RIGID INSULATION

Figure 28.24 A spray-applied double-layer elastomeric polyurethane membrane on a low-slope roof.

Figure 28.25 This metal roofing system is designed for use on low-slope roofs.

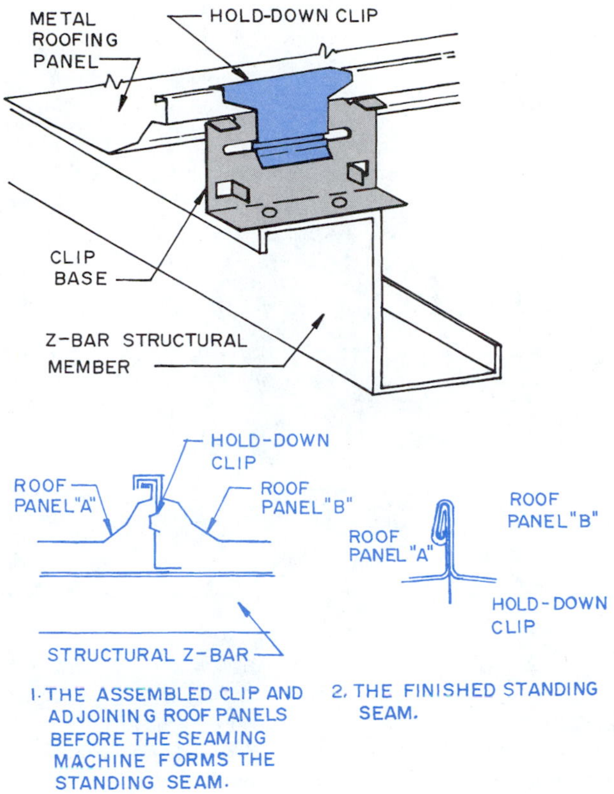

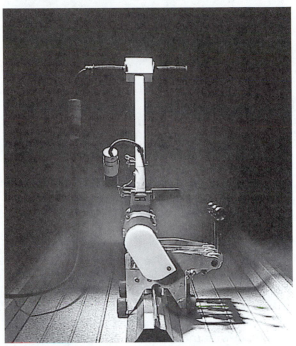

Figure 28.26 This metal roofing panel is held to the structural frame with a metal clip that is rolled into a standing seam. *(Courtesy CECO Building Systems)*

Figure 28.27 The standing seam is formed with a seaming machine that moves along the butting panel edges. *(Courtesy CECO Building Systems)*

ROOFTOP WALKWAYS

Rooftop walkways are placed over standing seam metal roofs and other flat roof membranes to provide maintenance workers access to mechanical units on the roof and to other parts of the roof where access is desired. One type consists of lightweight perforated steel planks designed to fit over the standing seams. They are anchored with clips and supports attached directly to the standing seam (Figs. 28.29 and 28.30).

Walkways used on aggregate-ballasted roof systems use concrete pavers. The pavers may be set on plastic or concrete blocks, or a bed of gravel is laid over the membrane and the concrete pavers or slabs are laid on it with cracks left open between the pieces. In both types drainage below and between the pavers is provided. In no situation should the roof membrane be pierced to support a walkway.

FLASHING

Flashing is made of a thin material impervious to penetration by water. It is placed where needed to prevent water penetration and provide for water drainage. Typical places in roof construction include joints between the roof and a wall, at expansion joints, and around objects, such as pipes, that penetrate the roof membrane.

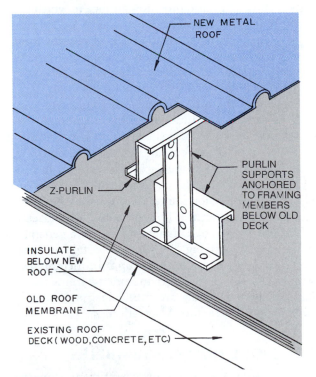

Figure 28.28 Typical construction for reroofing a building with metal standing seam roofing panels. *(Courtesy CECO Building Systems)*

Figure 28.29 These lightweight perforated steel planks are anchored to the standing seam. *(Courtesy Unistrut Corporation)*

Figure 28.30 Steel plank walkways provide access to various parts of the roof, protecting it from damage. *(Courtesy Unistrut Corporation)*

Materials used for flashing include copper, galvanized steel, lead, aluminum, stainless steel, bituminous sheet material, and plastics. In some cases a combination of these is used, such as galvanized steel covered with bitumen, which prevents corrosion of the steel when in contact with mortar.

The materials used in flashing have different thermal expansion and contraction characteristics from those of the roofing membrane. These differences can cause cracking and tears in the roof membrane unless allowance is made for this movement.

Wood nailers are used at roof edges and wherever roof insulation ends. They also provide a surface to which the roof membrane and flashing can be nailed. Wood nailers must be securely fastened to the deck to prevent wind uplift of the flashing. The wood nailers are treated with a wood preservative. Care must be taken to be certain the preservative will not damage the roofing materials.

As the size of the roof increases, the danger of tears from expansion and contraction of the membrane also increases. The designer locates expansion joints to provide allowances for movement. Different designs are available, and one for a built-up roof is shown in Fig. 28.31. Also included in the design are area dividers located between expansion joints. They must be designed so they do not restrict the flow of water. The spacing varies depending on the climate, but they are generally every 150 to 200 ft. (Fig. 28.32). The details in Figs. 28.33 through 28.39 are for built-up roof membranes. Those for EPDM, PVC, and modified bitumens are similar. Additional details are available from the National Roofing Contractors Association.

STEEP-SLOPE ROOF ASSEMBLIES

Steep-slope roof assemblies have sufficient slope to permit water to drain and flow to the ground. Usually gutters and downspouts are used to control the flow. A variety of roofing materials are available for use on steep slopes. The choice depends on the architectural appearance and the slope. Minimum slopes are recommended for the various products available.

For example, if a sloped roof is below the minimum acceptable slope for various types of shingles it must be roofed with a built-up membrane or metal standing seam roof as described for low-slope roofs. Generally accepted slopes for steep-slope roofs are listed in Table 28.2.

Asphalt Shingles

Asphalt shingles may have an organic or fiberglass base. Organic asphalt shingles have a wood fiber base that is saturated with asphalt and coated with colored mineral granules. Fiberglass asphalt shingles have a fiberglass mat that has top and bottom layers of asphalt. It is also coated with mineral granules. Asphalt shingles are die cut from the mat, are typically three tab, and have square edges. Other shapes are available and are illustrated in Chapter 27.

A typical installation starts with the underlayment. The material is doubled at the eave as shown in Fig. 28.40. In cold climates this double layer should extend up the roof until it is 24 in. (610 mm) inside the exterior wall.

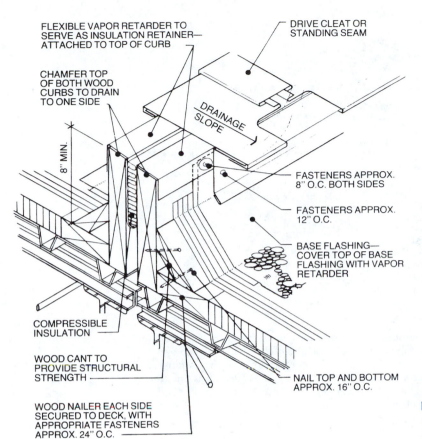

FLEXIBLE VAPOR RETARDER TO
SERVE AS INSULATION RETAINER—
ATTACHED TO TOP OF CURB

DRIVE CLEAT OR
STANDING SEAM

CHAMFER TOP
OF BOTH WOOD
CURBS TO DRAIN
TO ONE SIDE

DRAINAGE
SLOPE

8" MIN.

FASTENERS APPROX.
8" O.C. BOTH SIDES

FASTENERS APPROX.
12" O.C.

BASE FLASHING—
COVER TOP OF BASE
FLASHING WITH VAPOR
RETARDER

COMPRESSIBLE
INSULATION

WOOD CANT TO
PROVIDE STRUCTURAL
STRENGTH

NAIL TOP AND BOTTOM
APPROX. 16" O.C.

WOOD NAILER EACH SIDE
SECURED TO DECK, WITH
APPROPRIATE FASTENERS
APPROX. 24" O.C.

Figure 28.31 The expansion joints allow for expansion and contraction in the roof. *(Courtesy National Roofing Contractors Association)*

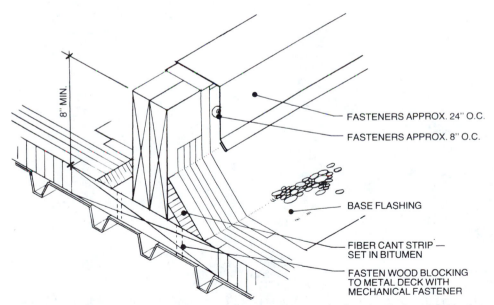

8" MIN.

FASTENERS APPROX. 24" O.C.

FASTENERS APPROX. 8" O.C.

BASE FLASHING

FIBER CANT STRIP—
SET IN BITUMEN

FASTEN WOOD BLOCKING
TO METAL DECK WITH
MECHANICAL FASTENER

Figure 28.32 An area divider is a raised wood member attached to the deck and flashed. *(Courtesy National Roofing Contractors Association)*

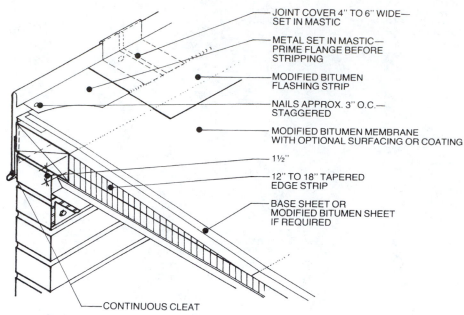

JOINT COVER 4" TO 6" WIDE—
SET IN MASTIC

METAL SET IN MASTIC—
PRIME FLANGE BEFORE
STRIPPING

MODIFIED BITUMEN
FLASHING STRIP

NAILS APPROX. 3" O.C.—
STAGGERED

MODIFIED BITUMEN MEMBRANE
WITH OPTIONAL SURFACING OR COATING

1½"

12" TO 18" TAPERED
EDGE STRIP

BASE SHEET OR
MODIFIED BITUMEN SHEET
IF REQUIRED

CONTINUOUS CLEAT

Figure 28.33 This roof edge detail shows the wood edge boards protecting the insulation and providing a nailing surface for the membrane. The plys of the membrane lap over the wood and are nailed to it. Two plys of felt are placed over these, forming the flashing. A metal cleat serves a finished fascia and covers the felt flashing. This edge turns the water toward a scupper or roof drain, keeping it from running over the edge of the roof. *(Courtesy National Roofing Contractors Association)*

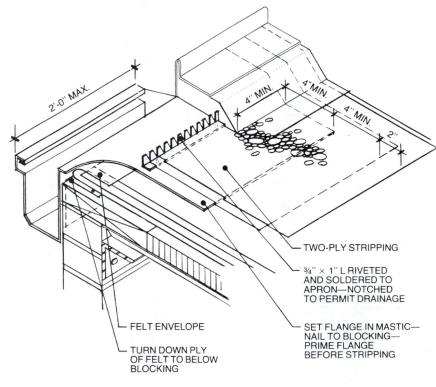

2'-0" MAX.

4" MIN.

4" MIN.

4" MIN.

2"

TWO-PLY STRIPPING

¾" × 1" L RIVETED
AND SOLDERED TO
APRON—NOTCHED
TO PERMIT DRAINAGE

FELT ENVELOPE

TURN DOWN PLY
OF FELT TO BELOW
BLOCKING

SET FLANGE IN MASTIC—
NAIL TO BLOCKING—
PRIME FLANGE
BEFORE STRIPPING

NOTES:
THIS DETAIL SHOULD BE USED ONLY WHERE THE DECK IS SUPPORTED BY THE OUTSIDE WALL.

Figure 28.34 This is a scupper, which is an opening in the edge of the roof that permits the water to drain to the ground, usually through a downpipe. Notice the layers of felt laid over the sheet metal from the scupper that is secured to the deck. *(Courtesy National Roofing Contractors Association)*

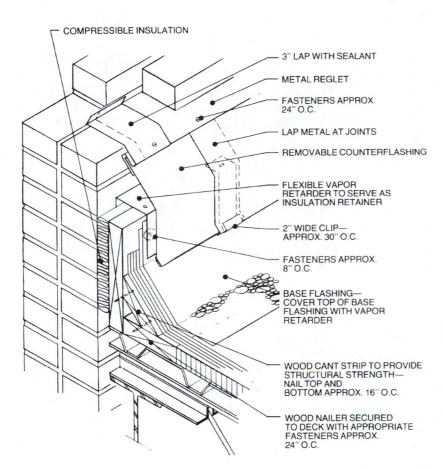

COMPRESSIBLE INSULATION

3" LAP WITH SEALANT

METAL REGLET

FASTENERS APPROX. 24" O.C.

LAP METAL AT JOINTS

REMOVABLE COUNTERFLASHING

FLEXIBLE VAPOR RETARDER TO SERVE AS INSULATION RETAINER

2" WIDE CLIP— APPROX. 30" O.C.

FASTENERS APPROX. 8" O.C.

BASE FLASHING— COVER TOP OF BASE FLASHING WITH VAPOR RETARDER

WOOD CANT STRIP TO PROVIDE STRUCTURAL STRENGTH— NAIL TOP AND BOTTOM APPROX. 16" O.C.

WOOD NAILER SECURED TO DECK WITH APPROPRIATE FASTENERS APPROX. 24" O.C.

Figure 28.35 This detail shows a roof butting a masonry wall. A wood nailer protects the edge of the insulation. A wood header butts the vertical wall and has a layer of insulation between it and the wall. The corner is sloped with a wood cant strip. Metal flashing is set into a mortar joint and overlaps the felt laid up over the cant strip. This assembly allows movement between the roof and the masonry wall. *(Courtesy National Roofing Contractors Association)*

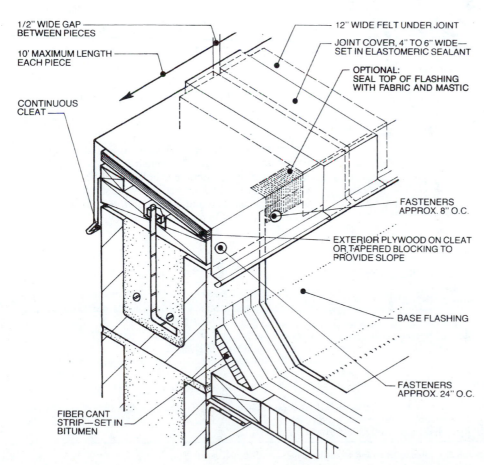

1/2" WIDE GAP BETWEEN PIECES

10' MAXIMUM LENGTH EACH PIECE

CONTINUOUS CLEAT

12" WIDE FELT UNDER JOINT

JOINT COVER, 4" TO 6" WIDE— SET IN ELASTOMERIC SEALANT

OPTIONAL: SEAL TOP OF FLASHING WITH FABRIC AND MASTIC

FASTENERS APPROX. 8" O.C.

EXTERIOR PLYWOOD ON CLEAT OR TAPERED BLOCKING TO PROVIDE SLOPE

BASE FLASHING

FASTENERS APPROX. 24" O.C.

FIBER CANT STRIP—SET IN BITUMEN

Figure 28.36 This light-metal parapet cap uses a wood nailer and cant strip. The base flashing is overlapped with a metal parapet cap. This construction is used only when the deck is supported by the wall. *(Courtesy National Roofing Contractors Association)*

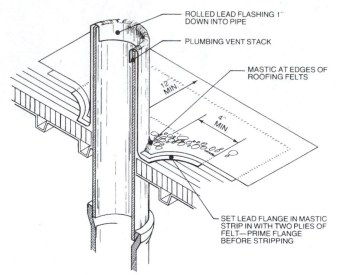

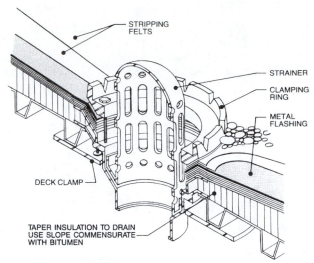

Figure 28.37 Lead flashing used to seal a pipe penetrating the roof membrane. Several layers of roofing felt are laid over the lead flashing and sealed to the pipe with mastic. The built-up roof membrane is laid over this assembly. *(Courtesy National Roofing Contractors Association)*

Figure 28.38 Roof drains are located in various parts of the roof and carry away water through pipes running down inside the building. A variety of products are manufactured. Installation details are provided by the manufacturer. *(Courtesy National Roofing Contractors Association)*

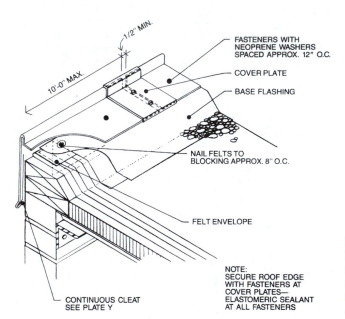

Figure 28.39 This detail shows the use of a heavy-metal roof edge. The metal roof edge may be steel, aluminum, or stainless steel. The cover plate seals the joint between pieces and allows for expansion and contraction. *(Courtesy National Roofing Contractors Association)*

Table 28.2 Typical Minimum Slopes for Roofing on Steep-Slope Roofs[a]

Material	Allowable Slope
Asphalt shingles	4:12 if one layer of asphalt-saturated felt underlayment is used.
	2:12 if two layers of asphalt-saturated felt underlayment are used.
Clay and concrete tile	4:12 if interlocking tile are used with one layer of 4 lb. cap sheet underlayment.
	Less than 3:12 when noninterlocking tile are used with two layers of No. 40 asphalt-saturated felt underlayment.
Slate	4:12 with one layer of 30 lb. asphalt-saturated felt underlayment.
	2:12 with double underlayment.
Wood shingles and shakes	3:12 for shingles and 4:12 for shakes with 30 lb. felt underlayment and interlayment with shakes or shingles.
Metal shingles and shakes	3:12 and 4:12 depending on the material and style.
	Requires 30 lb. asphalt-saturated felt underlayment.

[a]Consult local building codes and observe the manufacturer's recommendations on acceptable slope and installation details.

A metal drip edge is placed along the edge of the roof at the eave and rake. The underlayment goes below the drip edge on the rake and on top at the eave (Fig. 28.40). The placement of shingles can be done in several ways. One way is to keep the cutouts centered over the tabs as shown in Fig. 28.41.

The nails or staples are placed just above the notch in each shingle (Fig. 28.42). Nails should be 11 or 12 gauge hot-dipped galvanized roofing nails having shanks ⅞ to 1 in. (25 mm) long. Codes specify staple sizes. Typical recommendations are in Table 28.3.

The valleys can be flashed as shown in Figs. 28.43 through 28.45. When a roof meets a wall, step flashing is used. Each piece is nailed to the wall and rests on top of each shingle. The wood siding is laid over the flashing (Fig. 28.46). When the roof meets a brick wall, the flashing is set into the mortar joint. Finally, hips and ridges are capped with single tabs overlapping each other, as shown in Fig. 28.47.

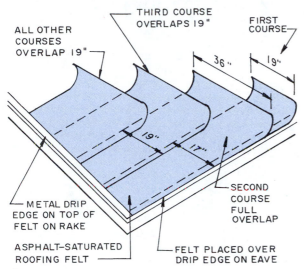

Figure 28.40 This illustrates a double-layer asphalt-saturated roofing felt underlayment with drip edge flashing.

Figure 28.42 Asphalt shingles can be fastened with a roofing coil nailer. *(Courtesy ITW Paslode)*

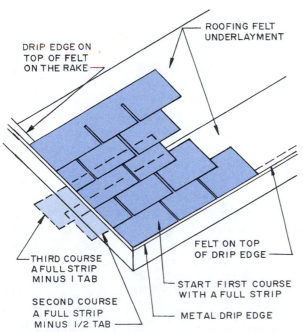

Figure 28.41 Asphalt shingles are spaced so the slots between the tabs on overlapping shingles do not line up.

Table 28.3 Typical Staple Sizes for Installing Asphalt Shingles

Wood Deck Thickness	Minimum Staple Leg Length	Staple Width
⅜″	⅞″	⅞″
½″	1″	1″
⅝″ and thicker	1¼″–1½″	1¼″–1½″

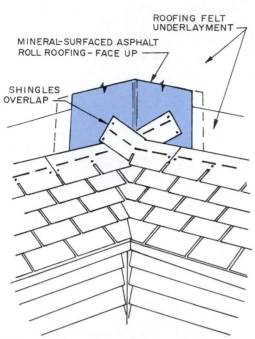

Figure 28.43 The shingles are woven over roll roofing valley flashing, forming the valley.

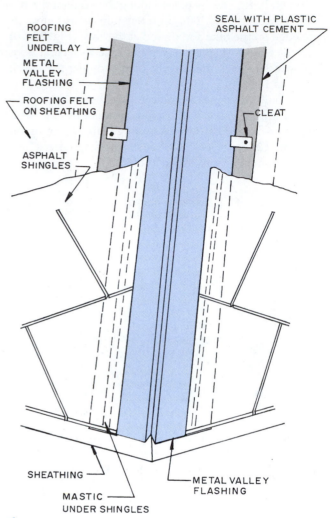

Figure 28.44 Metal valley flashing is placed over asphalt felt that is bonded to the felt on the roof with plastic asphalt cement.

Wood Shingles and Shakes

Wood shingles and shakes are made from cedar, redwood, southern pine, and other woods. Shakes are hand-hewn and shingles are sawed. Building codes in many areas require they be treated with fire-retardant chemicals.

Usually wood shingles are applied over spaced 1 × 4 in. or 1 × 6 in. wood sheathing boards. The size used depends on the amount of shingle to be exposed to the weather. Wood shakes are placed over 1 × 6 in. spaced wood sheathing wood. Unspaced sheathing is used at the eave and runs up until it is 24 in. (610 mm) inside the exterior wall. It is covered with roofing felt (Fig. 28.48).

Shingles are doubled or tripled at the eave; shakes are usually doubled. A layer of felt is laid over the top edge of each row of shakes (Fig. 28.49). The hips and ridges are capped with shingles cut as shown in Fig. 28.50. These may be purchased ready-made. Valleys use metal flashing placed over roofing (Fig. 28.51).

Shakes can be applied on roofs with slopes less than 4:12 if special construction is used. A typical example is in Fig. 28.52. Each wood shingle or shake should be applied with two corrosion-resistant nails, such as aluminum, stainless steel, or hot-dipped zinc-coated steel. Staples should be stainless steel or aluminum (Fig. 28.53).

Slate Shingles

Slate shingles are available in a variety of colors depending on the rock from which it was quarried. They are split to thickness and trimmed to size from larger slabs of slate, then holes for fasteners are drilled or punched. Additional details are in Chapter 13. Slate is a heavy material, and the roof structure must be designed to carry the weight. It is fire-resistant and has a long life but is more costly than most roofing materials. It produces a roof that enhances the architectural beauty of the building (Fig. 28.54).

Slate roofing is divided into two classifications, textural (or random) and standard commercial. The textural slate is delivered to the job in a range of thicknesses and sizes and must be sorted on the job by the slaters. Standard commercial slate is graded at the quarry by thickness, length, and width. Thickness is usually about ¼ in. (6 mm).

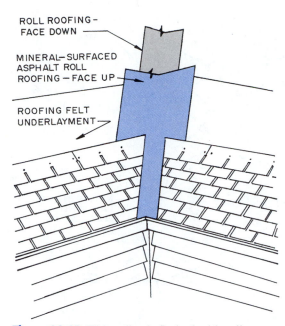

Figure 28.45 This valley is flashed with roll roofing material.

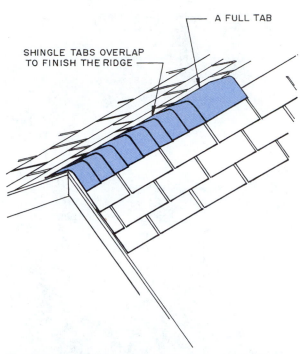

Figure 28.47 The ridges and hips are capped with single-tab asphalt shingles.

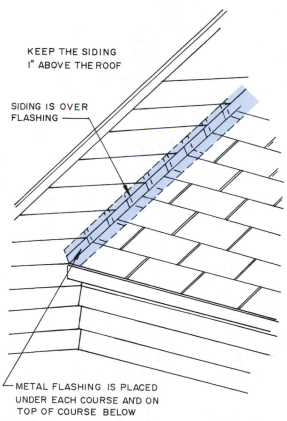

Figure 28.46 The siding is placed over the flashing and kept 1 in. above the surface of the shingles.

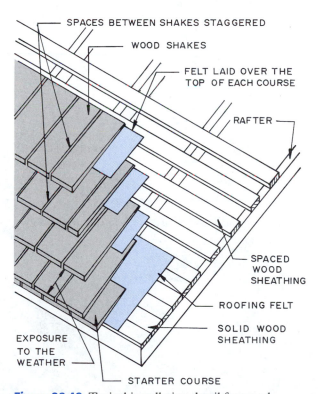

Figure 28.48 Typical installation detail for wood shakes.

Figure 28.49 Wood shakes are double coursed at the eave and a layer of roofing felt is laid between the courses. *(Courtesy Cedar Shake and Shingle Bureau)*

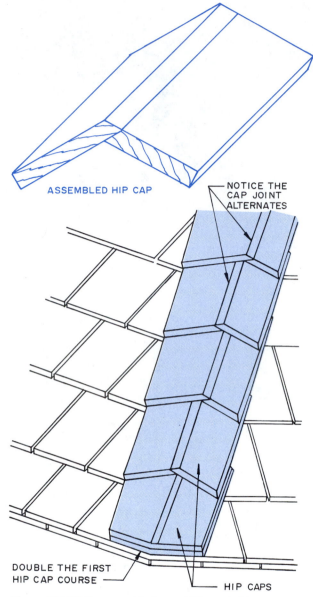

Figure 28.50 Preassembled wood hip and ridge caps are used to seal the shingles over hips and ridges.

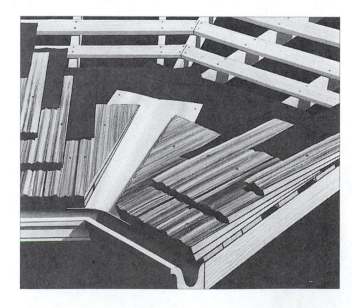

Figure 28.51 Valleys are flashed with metal flashing over a layer of roofing felt. *(Courtesy Cedar Shake and Shingle Bureau)*

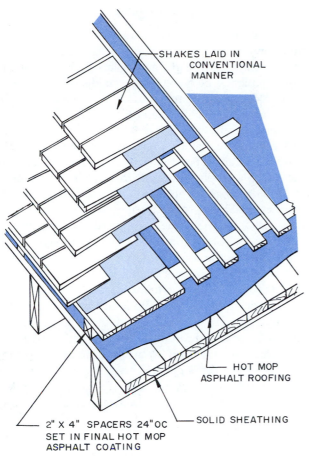

SHAKES LAID IN
CONVENTIONAL
MANNER

HOT MOP
ASPHALT ROOFING

2" X 4" SPACERS 24"OC
SET IN FINAL HOT MOP
ASPHALT COATING

SOLID SHEATHING

Figure 28.52 Recommended construction for applying wood shakes on roofs with less than 4:12 slope.

Figure 28.53 Wood shakes can be fastened with corrosion-resistant nails or staples. *(Courtesy Cedar Shake and Shingle Bureau)*

Figure 28.54 Slate shingles produce an attractive, durable, fire-resistant roof covering. *(Courtesy Evergreen Slate Co., Inc.)*

ASTM C406 Standard Specification for Slate Roofing addresses material characteristics, physical requirements, and sampling procedures for the selection of roofing slate. Roofing slate is in three grades. Grade S_1 has an expected life of 75 years; S_2, 40 to 75 years; and S_3, 20 to 40 years.

Slate installation details are in Fig. 28.55. It is applied over solid sheathing covered with asphalt-saturated roofing felt. Codes require double underlayment for roofs with less than certain specified slopes. Each course of slate covers the joints in the course below. Plastic cement is applied over the joint to be covered. Large-head copper slater's nails are used. Each slate must have at least two fasteners.

The slates are laid in the same manner as other shingles. The exposed ends may be in even rows or in a random edge. Hips and ridges cannot be nailed and are held in place with a waterproof elastic slater's cement that is colored to match the slate.

Cement-Fiber Shingles

Cement-fiber shingles are made by combining portland cement and various fibrous materials. One such product is a composition of portland cement, organic and inorganic fibers, perlite, and iron oxide pigments for color.

Clay and Concrete Roofing Tile

Tile is made from concrete or clay. It is durable, fire resistant, heavy, and expensive. Some forms of concrete tile are made using lightweight aggregates. It is important that the aggregates be carefully screened so the sizes are controlled. They are manufactured in the same basic types as clay tile. The colors can be varied by adding iron oxides (Fig. 28.56).

Clay roofing tile may be unglazed or glazed. Unglazed tile ranges from an orange-yellow to a dark red, depending on the clay. Glazed tile is available in a wide range of solid colors. Interlocking tiles are used on roofs with slopes of 3:12 or more. Shingle tiles, often called Norman tiles, require a 5:12 slope. Clay tile is discussed in Chapter 11.

The specific installation details for clay and concrete tile vary by the tile manufacturer. The following are typical examples. Products should be installed following the directions from the manufacturer.

Flat interlocking clay tile installation is shown in Fig. 28.57. The tiles are on solid sheathing that is covered with asphalt-saturated roofing felt. A flat under-eave tile or an under-eave tile with an apron is used along the eave. The gable rake is covered with special right and left rake tiles. The ridge and hips are covered with V-shaped tiles. These tiles are set in a cement mortar and nailed.

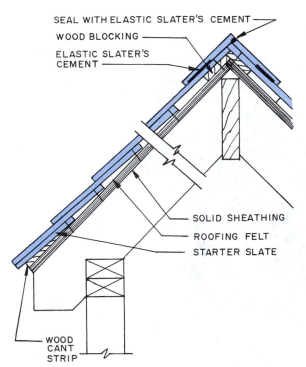

SEAL WITH ELASTIC SLATER'S CEMENT
WOOD BLOCKING
ELASTIC SLATER'S CEMENT
SOLID SHEATHING
ROOFING FELT
STARTER SLATE
WOOD CANT STRIP

Figure 28.55 Slate shingle installation is started with a wood cant strip at the eave. Wood blocking is used at the ridge.

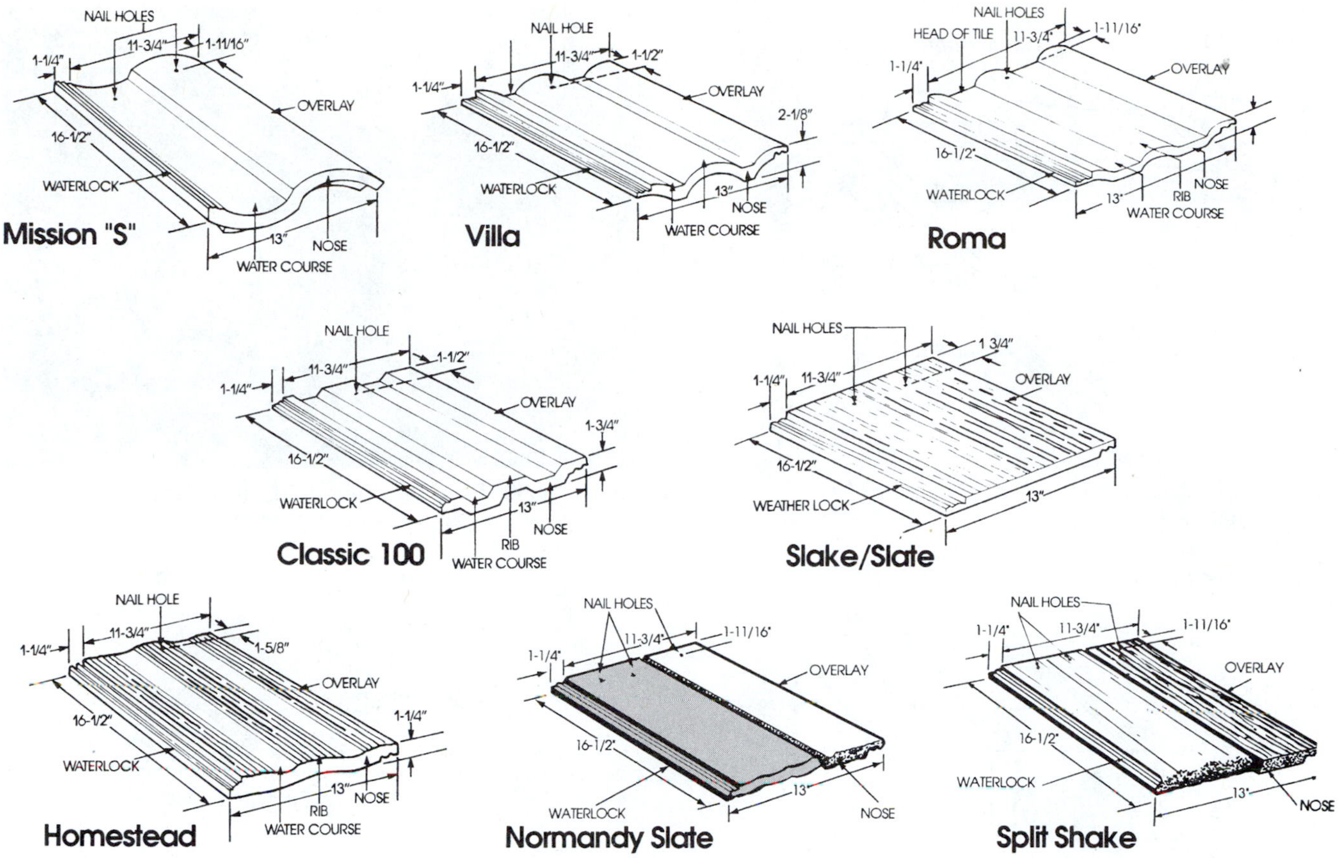

Figure 28.56 The shapes of concrete roofing tile available from one manufacturer. *(Courtesy Monier)*

An installation detail for flat cement tiles is shown in Fig. 28.58. They are installed much like slate shingles. Spaced sheathing can be used for roofs over 4:12 slope. Roofing felt is laid between the layers. A finished installation is shown in Fig. 28.59.

A number of circular profile tiles are available. Two common ones are shown in Fig. 28.60. These tiles have a gable rake tile and a curved ridge and hip tiles. In addition to being nailed to the sheathing, they are set in a cement mortar. Valleys are flashed as described for other roofing products. A typical installation is shown in Fig. 28.61. Some scenes showing the installation of concrete roof tiles are in Fig. 28.62.

Metal Roof Systems

Some commonly used seams on metal roofing on steep-slope roofs are the standing, capped, and batten. Some metal roofing is installed over decking, and other types

have sufficient strength to span between properly spaced purlins. Examples of purlin-supported metal roofing are in the low-slope section of this chapter (see Fig. 28.26).

A typical batten seam is shown in Fig. 28.63. The roofing panels have an L-flange that butts a cleat. The cleat is fastened to the deck and folded over the flanges. A metal batten cap is snapped over this assembly.

A capped seam is shown in Fig. 28.64. A T-shaped cleat is fastened to the deck and the roofing flanges butt it. The cleat is folded over the flanges and a metal cap is snapped over the assembly.

One type of standing seam is shown in Fig. 28.65. A cleat is fastened to the deck. The cleat is folded around one roof panel flange. Then the butting panel is placed next to the cleat and folded over the assembly and the entire assembly is folded over again. In all cases it is important that the cleats and fasteners provided by the manufacturer of the roofing panels be used. A finished standing seam metal roof is in Fig. 28.66.

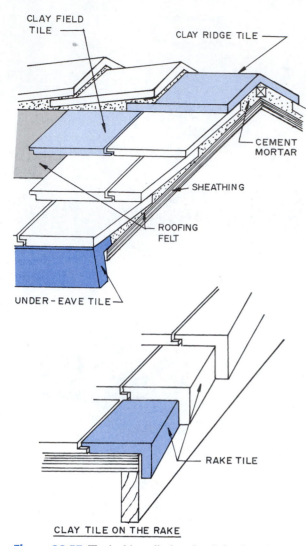

CLAY FIELD TILE

CLAY RIDGE TILE

CEMENT MORTAR

SHEATHING

ROOFING FELT

UNDER-EAVE TILE

RAKE TILE

CLAY TILE ON THE RAKE

Figure 28.57 Typical installation detail for flat clay tile using special eave, rake, and ridge tile.

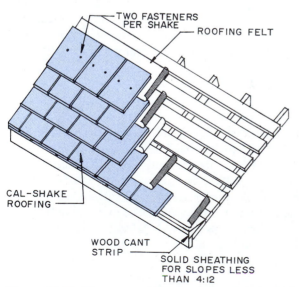

TWO FASTENERS PER SHAKE

ROOFING FELT

CAL-SHAKE ROOFING

WOOD CANT STRIP

SOLID SHEATHING FOR SLOPES LESS THAN 4:12

Figure 28.58 These extruded cement flat roof tiles are installed on spaced sheathing with layers of roofing felt between courses. (*Courtesy Cal-Shake, Inc.*)

Figure 28.59 This concrete roof tile installation produces an attractive, durable, fire-resistant finished roof. (*Courtesy Cal-Shake, Inc.*)

Figure 28.60 Two of the commonly manufactured circular tile profiles.

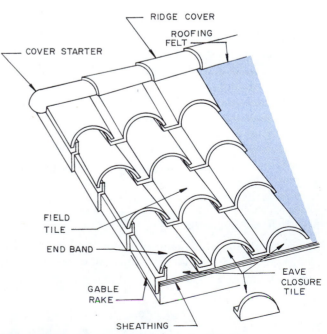

RIDGE COVER

ROOFING FELT

COVER STARTER

FIELD TILE

END BAND

GABLE RAKE

SHEATHING

EAVE CLOSURE TILE

Figure 28.61 A typical installation detail for curved clay roof tiles.

Figure 28.62 Procedures for installing concrete roofing tiles. *(Courtesy Monier)*

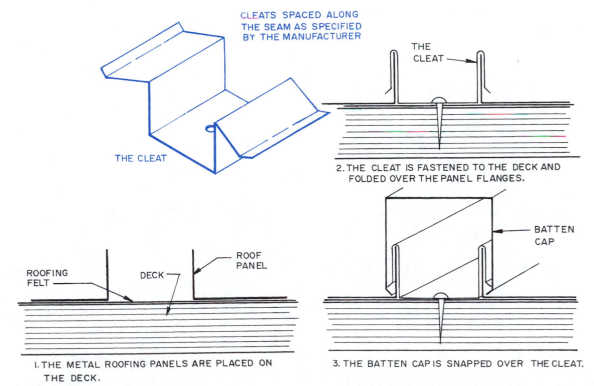

CLEATS SPACED ALONG THE SEAM AS SPECIFIED BY THE MANUFACTURER

THE CLEAT

THE CLEAT

2. THE CLEAT IS FASTENED TO THE DECK AND FOLDED OVER THE PANEL FLANGES.

ROOFING FELT

DECK

ROOF PANEL

BATTEN CAP

1. THE METAL ROOFING PANELS ARE PLACED ON THE DECK.

3. THE BATTEN CAP IS SNAPPED OVER THE CLEAT.

Figure 28.63 A batten seam is constructed by snapping a metal batten cap over the cleats that secure the edge of the metal roofing to the deck.

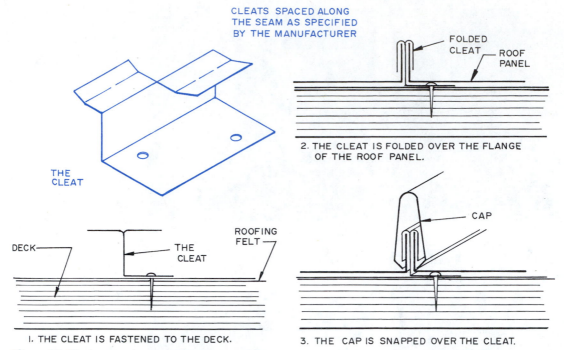

Figure 28.64 A capped seam provides a watertight seal over the union of metal roofing panels.

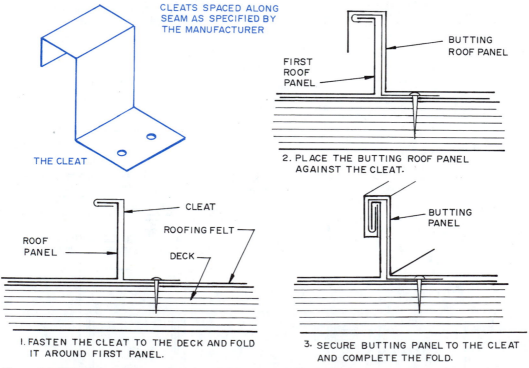

Figure 28.65 This standing seam is formed by rolling over the edges of each metal roof panel and the cleat that secures them to the deck.

Metal shingles and shakes are available that have the surface embossed with a range of patterns and textures. A typical construction detail is in Fig. 28.67. This shows the shingles locked together and secured to the deck with metal clips. The fastening system used varies with the manufacturer of the shingles. A close-up view of a finished metal shake roof is in Fig. 28.68. Notice the surface texture stamped into the metal. A smooth, slightly concave metal shingle was used on the building in Fig. 28.69.

Figure 28.66 This durable aluminum standing seam roof provides a major architectural feature for this building. *(Courtesy Atlas Aluminum Corporation)*

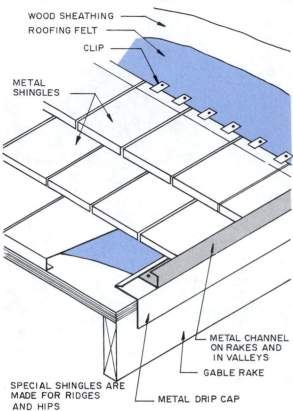

WOOD SHEATHING
ROOFING FELT
CLIP
METAL SHINGLES
METAL CHANNEL ON RAKES AND IN VALLEYS
GABLE RAKE
SPECIAL SHINGLES ARE MADE FOR RIDGES AND HIPS
METAL DRIP CAP

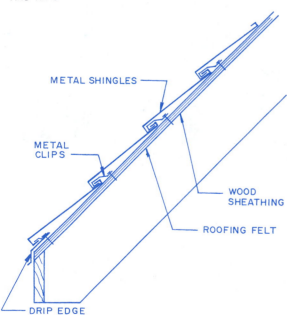

METAL SHINGLES
METAL CLIPS
WOOD SHEATHING
ROOFING FELT
DRIP EDGE

Figure 28.67 A typical construction detail showing the installation of metal shingles and shakes.

Figure 28.68 A finished metal shake roof. (*Courtesy Berridge Manufacturing Co.*)

Figure 28.69 This metal shingle closely resembles a slightly concave clay tile. (*Courtesy Atlas Aluminum Corporation*)

REVIEW QUESTIONS

1. List the four fire classifications for roof coverings.
2. What types of roof decking are used for low-slope roofs?
3. Why are vapor retarders an important part of a roof assembly?
4. What materials are used for vapor retarders?
5. Describe the process for installing a built-up roof covering.
6. What is a modified bitumen?
7. How are single-ply roofing membranes secured to the roof deck?
8. What types of single-ply roofing membranes are available?
9. Describe the makeup of a spray-applied roof coating.
10. How are the standing seam metal roof panels joined to form a watertight joint?
11. How are standing seam metal roof panels secured to the deck?
12. Why are rooftop walkways needed?
13. List several places flashing is required.
14. What roof covering materials are used on steep-slope roofs?

KEY TERMS

built-up roofing A continuous roof covering made up of plies of saturated roofing felt alternated with layers of asphalt or coal tar pitch with a topcoat of aggregate.

flashing A thin waterproof material installed as needed to prevent water penetration and provide drainage.

low-slope roofs Roofs that are flat or nearly flat.

modified bitumens A roofing membrane composed of a polyester or fiberglass mat saturated with a polymer-modified asphalt.

single-ply roofing A roofing membrane composed of a sheet of waterproof material secured to the roof deck.

steep-slope roofs Roofs with sufficient slope to permit rapid runoff of rain.

water-vapor transmission rate (WVTR) A measure of the porosity of a material to the passage of water vapor.

SUGGESTED ACTIVITIES

1. Collect samples of as many roofing products as you can find. Label each, giving the properties, fire rating, and advantages and disadvantages.

2. Set up a partial roof deck outside the classroom and install as many types of roofing systems as you can make available.

3. Ask a roofing contractor to visit the classroom and discuss bidding procedures, relative costs of various systems, and safety requirements that must be met during installation.

4. Visit various construction sites and observe the procedures being used for the installation of the roofing system. Back in the classroom, discuss what you saw. What improper practices did you observe?

ADDITIONAL INFORMATION

Griffin, C. W., *Manual of Built-Up Roofing Systems*, Prentice Hall, Englewood Cliffs, N.J., 1982.

Hornbostel, C., *Construction Materials: Types, Uses, and Applications*, John Wiley and Sons, New York, 1991.

McElroy, W. C., *Roof Builder's Handbook*, Prentice Hall, Englewood, Cliffs, N.J., 1993.

Merritt, F.S., and Ricketts, J.T., *Building Design and Construction Handbook*, McGraw-Hill, New York, 1994.

The NRCA Roofing and Waterproofing Manual, National Roofing Contractors Association, Rosemont, Ill., 1992.

Scharff, Robert, *Roofing Handbook*, McGraw-Hill, New York, 1988.

Spence, W. P., *Finish Carpentry*, Sterling Publishing Co., New York, 1995.

Watson, J. A., *Commercial Roofing Systems*, Prentice-Hall, Englewood Cliffs, N.J., 1984.

Other resources are numerous related publications from the organizations listed in Appendix B.

DIVISION 8

DOORS AND WINDOWS
CSI MASTERFORMAT™

Courtesy H.H. Robertson

29

Glass

This chapter will help you to:

1. Be familiar with the various types of glass products used in building construction.

2. Select appropriate glass products for various applications.

Glass is an inorganic mixture that has been fused at a high temperature and cooled without crystallization. It has an unusual internal structure because mechanically it is rigid and has the characteristics of a solid, yet the atoms in glass are arranged in a random order similar to those in a liquid. Glass is technically a supercooled liquid.

TYPES OF GLASS

There are six basic types of glass. They are classified by their ability to resist heat (Table 29.1). *Soda-lime-silica* is the type commonly used for door and window glazing and for bottles. About 90 percent of all glass produced is this type. Its general composition includes 74 percent silica, 15 percent soda, 10 percent lime, and 1 percent alumina. It is easy to form and cut. It has fair chemical resistance and does not resist high temperatures or rapid thermal changes. It is easily shattered into small, sharp pieces.

Lead-alkali-silica glass is more expensive than soda-lime and has the same properties.

Fused silica glass is composed of about 99 percent silicon dioxide and is the most expensive. It has the highest resistance to heat, handling temperatures ranging from 1650 to 2190°F (900 to 1200°C). It also has the highest corrosion resistance and affords excellent transmission of ultraviolet rays.

Ninety-six percent silica is used when high thermal hardness and the ability to withstand thermal shock (going from hot to cold rapidly) are required.

Borosilicate glass is a thermally hard glass that has been used for many years. It is best recognized by its Corning trade name, Pyrex®.

Aluminosilicate glass is more costly than borosilicate but can withstand high service temperatures and is similar in its ability to withstand thermal shock.

The materials in this chapter are devoted to the uses of soda-lime-silica glass because it is the most widely used glass for construction applications.

THE MANUFACTURE OF GLASS

Although several glassmaking processes have been used for many years, most glass today is produced by the

Table 29.1 Thermal Properties of Commonly Used Glass

Categories of Glass	Thermal Expansion	Heat Resistance
Soda-lime-silica	High	Low
Lead-alkali-silica	High	Low
Fused silica	Low	High
96% silica	Low	High
Borosilicate (Pyrex)	Medium	Medium
Aluminosilicate	Medium	Medium

float process. The first production of glass by this method was in 1959 by the English firm, Pilkington Brothers, Ltd. This process is now used worldwide.

Float Glass

The float process involves producing the molten glass in a furnace from which it is conveyed to a float bath. Here the molten glass is floated across a bath of molten tin (Fig. 29.1). The molten tin gives a very flat surface that supports the glass as it is polished by the application of heat from above. The heat melts out any irregularities in the glass. The ribbon of glass moves on to a cooling zone where heat is reduced, permitting the glass to solidify enough to be conveyed on to the **annealing** lehr. After the glass has been annealed, it is moved to a section where it is cut into lengths, inspected, and packed. The sheets of glass produced by this method have parallel surfaces, a smooth, clear finish, and high optical clarity.

Float glass is a flat glass that is available as *regular float glass* or *heavy float glass*. Thicknesses range from $3/32$ to $1/2$ in. (2.5 to 12 mm). Regular float glass is made in three qualities: *silvering*, which is used for selected high-quality pieces for optical uses and mirrors; *mirror glazing*, which is for general-purpose mirrors; and *glazing*, which is for door and window glazing. Heavy float glass is available in glazed type. Float glass is used when clarity and visual transparency with a minimum of distortion are desired. Float glass is used for many products, such as reflective glass, mirrors, tinted glass, laminated glass, and insulating glass. Selected data are in Table 29.2.

Sheet Glass

Sheet glass is a type of flat glass that is less expensive than float glass. It is made by older methods that involve drawing a ribbon of molten glass along a series of rollers where its thickness is established and it is annealed, cooled, and cut to size. Sheet glass has more distortion than float glass and is not as widely used as in the past. It is available in single strength, $3/32$ in. (2.3 mm) thick; double strength, $1/8$ in. (3.1 mm) thick; and heavy sheet, $3/16$ in. (4.7 mm) and $7/32$ in. (5.6 mm) thick. Picture glass is a thinner version, $3/64$, $1/16$, and $5/64$ in.

(1.2, 1.6, and 2.0 mm) thick, that is used for covering pictures and charts, and for other purposes for which strength is not a factor. Sheet glass is available in three grades, AA (best), A (good), and B (general glazed) and as clear, tinted, reflective, tempered, or heat-treated products.

PROPERTIES OF SODA-LIME-SILICA GLASS

The following discussion presents general properties of soda-lime-silica glass. These properties can be varied by altering the composition of the ingredients.

Mechanical Properties

A major consideration is the ability of glass to withstand breakage. Glass is brittle yet remains elastic up to its ultimate tensile strength. This means that glass can be bent up to the breaking point and if released will return to its original position. Glass breaks by bending or stretching; therefore, *tensile strength* is a major mechanical property to be considered.

Glass does not have a clearly defined tensile strength because the actual strength depends on the condition of the surface of the glass. A small scratch or nick is sufficient to cause tensile stresses to concentrate at that weaker point. Therefore, the actual tensile strength is less than the theoretical tensile strength. The mechanical strength can be increased by chemical and heat treatments.

Thermal Properties

Heat can be transferred by conduction, convection, and radiation. The thermal conductivity of glass is high, but it is lower than most metals. The U-value for a single-strength sheet of glass is about 1.04 to 1.10. Double glazing with a $1/2$ in. (12 mm) air space reduces the U-value to about 0.49 to 0.56. Typical thermal properties for selected flat glass products are in Table 29.3.

For single glazed glass most of the thermal resistance is borne at the outdoor and indoor surfaces. Indoors about two-thirds of the heat flows by radiation to room surfaces and one-third flows by convection. The heat transfer of the inner surface of glazing can be reduced a great deal by adding low-emissivity metallic films (referred to as **Low-E**) to the glass. This also reduces the U-value for glazing having air spaces. A typical double glazed opening with suspended Low-E film has a U-value of 0.31 to 0.32. Other glass products, such as tinted glass, reflective glass, and insulating glass, reduce heat gain and loss.

The thermal expansion of glass must be considered as units containing glass are designed. The coefficient of expansion of soda-lime glass is about 4.5×10^{-6}, and the coefficient for aluminum is about 13×10^{-6}. An

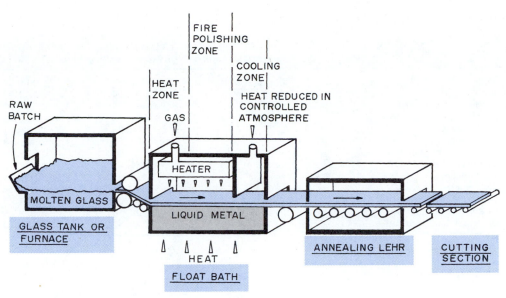

Figure 29.1 The process for making float glass.

Table 29.2 Thicknesses of Commonly Used Float Glass Sheets[a]

Thickness	
in.	mm
$3/32$	2.5
$1/8$	3.0
$5/32$	4.0
$3/16$	5.0
$1/4$	6.0
$3/8$	10.0
$1/2$	12.0

[a]Sheet size depends on the type of glass and design pressures.

aluminum window frame contracting under low temperatures imposes stress on the edge of the glass unless sufficient clearance is allowed for the difference in thermal movement.

Chemical Properties

Glass used in typical applications in building construction is a very durable material and more resistant to corrosion than most other materials. Since it is not porous, it will not absorb moisture or chemical elements in the ground or atmosphere. There are few exceptions that may occur in isolated industrial applications. Hot concentrated alkali solutions and superheated water can cause soda-lime-silica glass to dissolve. Hydrofluoric acids will cause corrosion.

Electrical Properties

Glass is a good electrical insulator. It is widely used for applications where this property plus its other properties make it useful. Light fixtures often use glass, and the envelope on a light bulb is glass because it is strong, heat resistant, and an electrical insulator.

Optical Properties

When light falls on glass some is absorbed and some is reflected. Clear sheet glass permits the passage of about 86 to 89 percent of visible light. Double glazed lights (window panes) permit about 80 to 82 percent light passage. Typical daylight transmittance figures are in Table 29.4.

When glass is treated to reflect or absorb light the percent of light transmitted is greatly affected. Tinted glass frequently transmits less than 50 percent of the daylight striking it. Glass with reflective coatings transmits 6 to 50 percent of the visible light.

The optical quality is also influenced by the lack of distortion. If the front and back surfaces of a sheet of glass are not parallel the image will form some angles, appearing wavy or distorted. With the increased use of float glass, distortion has been reduced and optical properties improved.

HEAT TREATING GLASS

The properties of glass can be improved by various heat treating methods. These include annealing, tempering, and heat strengthening.

Table 29.3 Thermal Properties of Flat Glass

Glass	Thickness (in.)	Winter Nighttime U-value/R-value	Summer Daytime U-value/R-value
Clear single	¼	1.3/0.78	1.04/0.96
Clear double	⅛	0.49/2.04	0.52/1.92
Clear double	¼	0.49/2.04	0.56/1.79
Clear double with Low-E film	¼	0.31/3.32	0.32/3.03
Light brown single	¼	1.13/0.88	1.10/0.91
Dark brown single	¼	0.89/0.88	0.89/1.88
Double glass light brown/clear	¼	0.31/2.04	0.33/1.75
Double glass dark brown/clear	¼	0.41/2.04	0.47/1.72
Light green single	¼	1.13/0.88	1.10/0.91
Dark green single	¼	0.95/1.14	0.98/1.12
Double light green/clear	¼	0.50/2.04	0.59/1.75
Double dark green/clear	¼	0.42/2.50	0.50/2.13
Clear insulating glass with suspended Low-E film	¼	0.23/4.30	0.37/2.70

Table 29.4 Daylight Transmittance for Selected Glass Units

Glass	Percent Daylight Transmittance
Clear single	86–89
Clear double	80–82
Clear double with Low-E film	72
Clear triple	72–74
Clear insulating with Low-E film	37–69
Light brown single	52
Light green single	75
Light gray single	41

Annealing

Annealing occurs in the normal production of glass. After the glass ribbon has been formed it is passed through an annealing lehr where the temperatures are carefully controlled. The temperature of the glass is raised high enough to relieve strains developed during the forming in the float bath. Then the temperature is slowly lowered, allowing all parts of the glass to cool uniformly.

Tempering

Tempering is used to increase the strength of glass. Tempered glass is used when strength beyond that of standard annealed glass is required.

Tempering involves raising the temperature of the glass to near the softening point and then blowing jets of cold air on both sides, suddenly chilling it. The surfaces harden and shrink and the interior is still fluid. As the interior cools and shrinks the exterior remains unchanged in size. This causes the surfaces and edges to be in compression and the interior in tension. The opposing compressive and tension forces balance each other, resulting in a stronger glass.

Tempered glass is three to five times as resistant to damage as annealed glass. When the thin tempered skin on the glass is broken, the entire sheet disintegrates into small pebblelike particles instead of sharp slivers. Tempered glass is made in accordance with Federal Specification DD-G-1403B, which requires a minimum compression of 10,000 lb. per sq. in. (surface) or 9,700 lb. per sq. in. (edge). Fully tempered glass can meet safety standards required by ANSI Z97.1-1975 and Federal Standard CPSC 16 CFR 1201, Safety Standard for Architectural Glazing Materials.

Heat Strengthening

Heat strengthened glass is heated and cooled much like tempered glass. Heat strengthened glass has a compression range of 3,000 to 10,000 lb. per sq. in. (surface) or 5,500 to 9,700 lb. per sq. in. (edge). It is about twice as strong as annealed glass.

Chemical Strengthening

Glass can also be **chemically strengthened** by immersing the glass in a molten salt bath, causing larger potassium ions in the salt to replace the smaller sodium ions already in the glass. This crowds the surface, causing compressive stresses.

FINISHES

The finish on glass depends on its end use. Although most glass produced is clear and transparent, finishes make it useful for other applications.

Fire Finish

As glass leaves the annealing lehr it has a smooth, transparent finish. This is referred to as a fire finish.

Etching

Etching produces a surface that can range from almost opaque to rather smooth but translucent. The degree of opacity depends on the length of time the glass is exposed to hydrofluoric acid or some other etching compound. The glass may be dipped or sprayed with the etching fluid. The finished surface is opaque but rather smooth to the touch.

Sandblasting

Glass is sandblasted by bombarding the surface with coarse-grained sand particles blown by compressed air. This produces a translucent finish that is generally a bit rougher than etched glass. Etching and sandblasting abrade the surface, which greatly lowers the strength of the sheet.

Patterned Finish

Patterned glass has texture or pattern rolled into the surface as the glass is drawn through a furnace. It may be colorless or tinted and is available in a wide variety of patterns. It may be relatively transparent from one side and translucent from the other.

Silvering

Mirrors are produced by spraying silver nitrate and tin chloride on the surface of the glass as it moves on a con-veyor. For additional protection this layer can be covered with shellac, varnish, or paint, or a layer of copper can be electroplated over it.

Float glass mirrors are the highest quality and are available in silvering quality, mirror glazing quality, and glazing quality. Silvering quality is the very best. Sheet glass used for window glazing can be used for low-quality mirrors. These are in two grades, A and B.

Ceramic Frit

Glass surfaces are sometimes coated with a ceramic frit that is then fired at a high temperature. Ceramic frit is composed of small powdered particles produced by quenching a molten glossy material. This produces a colored layer on the surface that is highly weather resistant. Both translucent and opaque colored layers are available. A major use is a finished exterior surface on wall spandrels (Fig. 29.2).

Tinting

Another way to infuse color in glass is by tinting. Tinting is done by adding color-producing ingredients to the glass at the beginning of the glass production process. Therefore, the color is not a surface finish but is infused in the glass.

This type of glass is used to reduce glare. The blue-green glass gives a moderate reduction in glare and brightness. The bronze tint is used primarily for heat

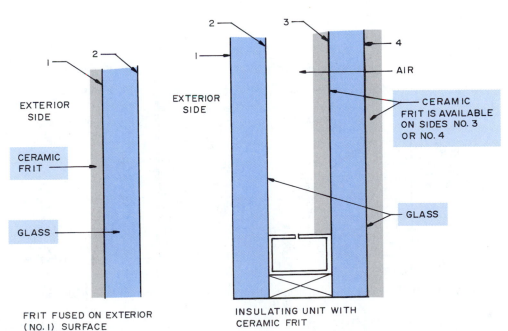

FRIT FUSED ON EXTERIOR (NO. 1) SURFACE

INSULATING UNIT WITH CERAMIC FRIT

Figure 29.2 Glass surfaces, such as used on building exteriors, may be coated with a colored ceramic frit.

reduction. The gray tint gives a wide range of light transmittance. Translucent glass diffuses most of the direct sunlight yet permits transmission of considerable diffused light.

GLASS PRODUCTS

Following are examples of commonly used glass products. Specifications vary with the manufacturer.

Low-E Glass

Low-E (low-emissivity) glass reduces energy costs by creating a heat barrier that helps keep heat outside in the summer and inside in the winter. It is used on residential and commercial buildings, on greenhouses, and on buildings using passive solar design. It is suitable for use in all climates and is available in double and triple insulating glass units. Tinting the outer pane increases the energy efficiency.

Low-E glass insulating units have a thin metallic coating on the inside glass surface, so the coating is fully protected. This coating selectively reflects the ultraviolet and infrared wavelengths of the energy spectrum. Low-E coatings permit the use of natural daylighting techniques because they are virtually invisible and have a high visible light transmission (Fig. 29.3).

Laminated Glass

Laminated glass is used in areas where the glazing is subject to possible breakage or where security of glazed openings is required. Typical applications from a safety standpoint include residential and commercial door glazing and windows in a gymnasium. Security applications occur in prisons and banks.

Laminated glass is made by bonding layers of float glass with interlayers of plasticized polyvinyl butyral (PVB) resin or polycarbonate (PC) resin. The glass is chemically strengthened by immersing it in a molten salt bath. Laminated glass is made to meet ANSI Z97.1-1975 and Consumer Products Safety Commission Regulation 16 CFR1201 Categories 1 and 11.

Many variations and thicknesses are available from various manufacturers. Laminated glass is available with ultraviolet filtering laminates and wire glass laminates. Laminated mirrors are also available. The thicknesses of the various products vary with the design, but thickness from 1/4 to 3 in. (6 to 75 mm) are common. Typical designs are shown in Fig. 29.4.

A laminated glass for use in a high-traffic well-guarded area is composed of three sheets of annealed glass and two interlayers. It will withstand attack by baseball bats and other small lightweight objects. When more security is needed, a five-glass panel with four plastic interlayers is used to provide protection against attack by bricks, bottles, and chairs. A laminate for high-risk areas has a polycarbonate core of 1/4 to 3/8 in. (6 to 9 mm) thick with special interlayer materials plus chemically strengthened glass. It is made with multiple cores for greater strength. It will withstand pounding by sledge hammers and is used where maximum security is needed. Bullet-resistant laminated glass has the same construction as those already shown. It has additional layers of glass and polyvinyl butyral interlayers. It is available in thicknesses from 1³/₁₆ to 2¹/₂ in. (30 to 63 mm). Greater thicknesses are available on special order.

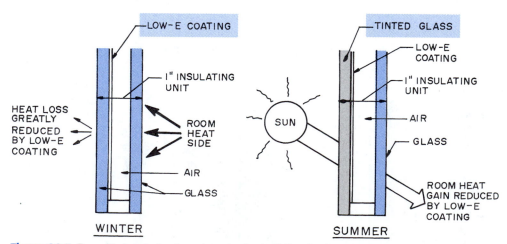

Figure 29.3 Low-E glass helps keep heat in the building in the winter and out of the building in the summer.

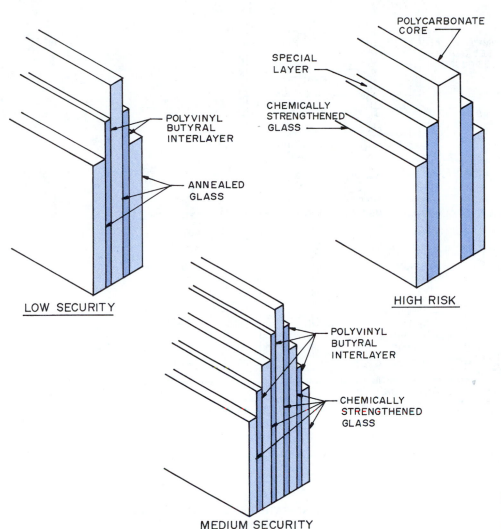

Figure 29.4 Several types of laminated glass used to increase the security of glazed openings.

Another type of laminated glass is **acoustical glass.** It is a construction of sound-absorbing plastic between two or more pieces of glass. The sound reduction is achieved because this soft interlayer allows the glass unit to bend in response to pressure from the sound waves. It is also available in insulated constructions, thus reducing heat loss and gain as well as reducing sound transmission. It is available in thicknesses of $\frac{1}{4}$, $\frac{1}{2}$, and $\frac{3}{4}$ in. (6, 12, and 18 mm). Insulated units are thicker. Sound class ratings are 36, 40, and 41. Common uses are in office partitions and in radio, TV, and recording studios.

Insulating Glass

Insulating glass is a manufactured glazing unit composed of two layers of glass with an airtight, dehydrated air space between them. They are made with hermeti-cally sealed glazing with edges fused together or set in gaskets. A wide range of systems are designed for use on commercial and residential buildings. Examples of these are in Fig. 29.5. Additional designs are available from various manufacturers. Insulating glass lowers heating and cooling costs by reducing air-to-air heat transfer.

When the glass is installed where it will be subjected to unusual thermal stresses or high wind loads, heat strengthened or tempered glass is used. Insulating glass units also use tinted, reflective, clear laminated, and Low-E glass. The sizes of typical units are in Table 29.5.

Reflective Glass

To understand how **reflective glass** reduces the amount of solar energy transmitted through a glass opening, consider the following explanation.

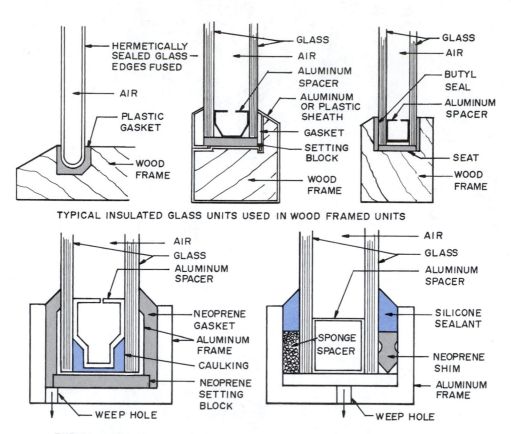

TYPICAL INSULATED GLASS UNITS USED IN WOOD FRAMED UNITS

TYPICAL INSULATED GLASS UNITS USED FOR LARGE COMMERCIAL WINDOWS

Figure 29.5 Some of the types of insulating glass units available.

Table 29.5 Sizes of Insulating (Float) Glass

| Maximum size | 90 × 144 in. (2 286 × 36 576 mm) |
| Minimum size | 15 × 20 in. (380 × 508 mm) |

| Glass Thicknesses | | Air Space | | Unit Thickness | |
in.	mm	in.	mm	in.	mm
3/32	2.5	1/4	6.0	7/16	11.0
1/8	3.0	1/4	6.0	1/2	12.7
5/32	4.0	1/4	6.0	9/16	14.2
3/16	5.0	1/2	12.0	7/8	22.2
1/4	6.0	1/2	12.0	1	25.4

When solar energy strikes a glass surface it can be (1) transmitted through the glass, (2) reflected, or (3) absorbed in the glass. Of that portion that is absorbed in the glass, part of the energy will be reradiated and convected (transmitted) inward and part of it will be reradiated and convected outward. The solar heat gain consists of that portion that is transmitted through the glass plus the portion of the absorbed energy that is reradiated and convected to the inside. The solar heat rejected is that portion of the original solar incidence that is returned to the exterior. This includes the energy reflected from the glass surface plus any absorbed energy that is reradiated and convected outward.

Another source of energy transfer through a glass window is due to convection. Convection is the transfer of heat by currents of air resulting from differences in air temperature and density in the heated space. The total heat gain of a window is the sum of solar heat gain and convective heat gain (Fig. 29.6). It is these factors designers consider as they choose the type of glazing opening.

Reflective glass has one surface covered with thin, transparent layers of metallic film. Several metals and mineral oxides are used, producing a variety of colors and heat reflection ratings. On single sheet glazing the metallic film is on the surface facing the inside of the building. On insulating units it is on the outside face of the glass facing the inside of the building (Fig. 29.7). Reflective glass is available in colors such as silver, green, blue, bronze, copper, and gold. Variations of these colors are available from various manufacturers (Fig. 29.8).

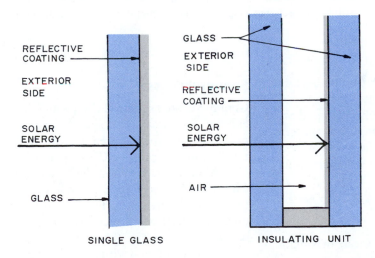

Figure 29.6 Reflective glass reduces the amount of solar energy transmitted through a glazed opening.

Figure 29.7 Insulating glass units use a transparent layer of metallic film to reduce energy transfer through the glass by convection.

Figure 29.8 The facade is covered with high-performance reflective glass that lets daylight enter the building but reflects the rays of the sun. *(Courtesy PPG Industries, Inc.)*

Visible light transmittance varies with the coating but ranges in general from 8 to 45 percent. Coatings are available on clear or tinted heat-absorbing glass. Most reflective glasses are ¼ in. (6 mm) thick, but thicker sheets are available. A summary of the solar-optical properties of clear, tinted, and reflective float glass is in Table 29.6.

Wired Glass

Wired glass is safety glass that has a mesh of small-diameter wires rolled into a sheet of molten glass. When wired glass breaks the wires hold the glass shards together. It also maintains its integrity as a fire barrier and is used in fire doors and window glazing. Tempered and laminated glass have replaced wire glass in many applications.

Spandrel Glass

Ceramic-coated spandrel glass is an effective way to clad exterior wall areas. The spandrel panel eliminates the problem of corrosion, greatly reducing deterioration of the building (Fig. 29.9). Glass spandrel panels use ¼ in. (6 mm) heat strengthened glass with an opaque ceramic frit fired on the surface. Some companies fire the frit on the exposed exterior surface, and others fire it on the protected interior surface. The spandrel is insulated, usually with 1 to 2 in. (25 to 50 mm) fiberglass or other acceptable insulation with a ½ in. (12 mm) air space and an aluminum foil vapor barrier facing the interior side of the panel. Panel sizes vary with the design, but typical sizes available include minimum panels of 18 × 33 in. (457 × 838 mm) and maximum panels of 84 × 144 in. (2130 × 3660 mm).

Table 29.6 Solar Optical Properties of Glass

	Light (%)		Solar Energy (%)		U-value Btu/hr./ft.²/°F	
	Transmitted	Reflected	Transmitted	Reflected	Winter	Summer
Single Glazed						
Sheet clear	91	8	86–89	7	1.13	1.02–1.03
Float clear	79–90	8	58–86	7	1.00–1.13	.98–1.03
Float blue-green	74–83	7	48–64	6	1.10–1.13	1.08–1.09
Float bronze	26–68	5–6	24–65	6	1.08–1.13	1.08–1.09
Float gray	19–62	4–6	22–63	5	1.06–1.13	1.08–1.09
Float reflective coating	8–34	6–44	6–37	14–35	0.90–1.11	0.89–1.12
Double Glazed						
Clear	78–82	14–15	60–71	14	0.49–0.57	0.55–0.62
Tinted	37–70	6–12	34–55	6–11	0.49–0.57	0.57–0.64
Reflective coating	7–30	6–44	5–29	14–35	0.41–0.49	0.47–0.58
Triple Glazed						
Clear	70–73	20	46–60	19–20	0.31–0.38	0.40–0.46
Tinted	11–56	8–17	25–46	8–16	0.31–0.38	0.40–0.46
Reflective coating	16–22	8–44	5–22	8–44	0.28–0.31	0.34–0.41

Figure 29.9 This building is sheathed with silver spandrel glass. *(Courtesy PPG Industries, Inc.)*

Bent Glass

Bent glass provides the designer the opportunity to produce a building with a unique and dramatic look (Fig. 29.10). It is available in clear float, tinted float, obscure, wire, Pyrex, patterned, and other types of glass. Some types of reflective glass can be bent. Laminated safety glass and double glazed thermal insulating units can be fabricated. Thickness and size specifications must be developed with the help of the manufacturer.

Architectural Beveled Glass

Architectural beveled glass provides a unique transition from outside to inside and is an outstanding architectural detail. The glass is made in a wide range of standard sizes and designs as well as custom designs. The pieces of beveled glass are assembled into panels with lead caming (Fig. 29.11). Energy-efficient triple glazed panels are available.

Glass Blocks

Glass blocks are made as solid and cavity units and are laid in a mortar bed similar to masonry units. Various designs permit light to be diffused, reduced, or reflected. The translucence or transparency is varied by design (Fig. 29.12).

Figure 29.10 Bent laminated glass provides a strong transparent exterior partition. *(Courtesy Laminated Glass Corporation)*

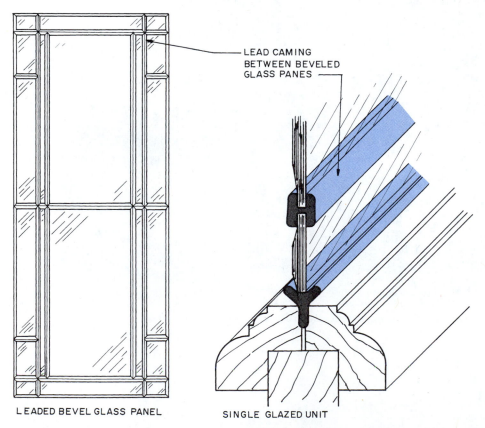

LEAD CAMING BETWEEN BEVELED GLASS PANES

LEADED BEVEL GLASS PANEL

SINGLE GLAZED UNIT

Figure 29.11 Beveled glass provides a unique decorative appearance.

Glass blocks are made by fusing two halves of pressed glass together, which creates a space with a partial vacuum. This gives glass block an insulating value equal to a 12 in. (305 mm) thick concrete wall (U-value 0151, R-value 1.96). Glass blocks are very strong and provide excellent security.

Glass blocks serve well as solar glazing in walled areas facing in a southerly direction. Blocks are available with a variety of patterns on the surface. These help to provide privacy and brightness and to control light transmission. Blocks also can have a highly reflective thermally bonded oxide surface coating that reduces heat gain and transmitted light. Selected properties of glass block are in Table 29.7.

PLASTIC GLAZING MATERIALS

Chapter 24 covers the types of plastic materials used in construction. Reference can be made to that chapter for technical information under the headings polycarbonates and acrylics. Chapter 31 contains additional information about glazing applications.

Figure 29.12 Glass blocks are available in a wide range of sizes and styles. *(Courtesy Pittsburgh Corning Corp.)*

Table 29.7 Properties of Single Cavity Glass Block

	Nominal Sizes 3" thick	4" thick	Compressive Strength (psi)	Light Transmission (%)	U-value
Clear	3 × 6 6 × 6 4 × 8 8 × 8 4 × 12 6 × 8 12 × 12	6 × 6 8 × 8 12 × 12	400–600	75	.52
Clear with reflective coating		8 × 8 12 × 12	400–600	5–21	.56
Light-diffusing pattern		8 × 8 12 × 12	400–600	39	.60
Decorative pattern	3 × 6 6 × 6 4 × 8 4 × 12 8 × 8 6 × 8 12 × 12		400–600	20–75	.65

REVIEW QUESTIONS

1. What are the basic types of glass?
2. What type of glass is used in most construction applications?
3. What are the two types of float glass?
4. What are the grades of sheet glass?
5. How does a small scratch or nick in glass reduce the tensile strength of the sheet?
6. How does the thermal conductivity of glass compare with that of metal?
7. Why is glass resistant to corrosion?
8. How does glass rate on optical clarity?
9. Which types of soda-lime-silica glass are annealed?
10. List the types of heat treated glass from the weakest to the strongest.
11. In addition to heat treating glass, how can it be strengthened?
12. List the types of surface finishes commonly used on glass.
13. What type of glass helps reduce heat transfer into a building in the summer and heat loss to the outside in the winter?
14. How does tinting affect the performance of glass?
15. What are the main reasons for using laminated glass?
16. How does insulating glass differ in construction from laminated glass?
17. What three things occur to produce solar heat gain when solar energy strikes a glass surface?
18. How much visible light is transmitted through reflective glass?
19. What are the two important features of wired glass?
20. What type of finish is used on spandrel glass?

KEY TERMS

acoustical glass A glazing unit used to reduce the transmission of sound through the glazed opening by bonding a soft interlayer between layers of glass.

annealing Freeing from internal stress by heating and gradually cooling.

chemical strengthening A process for strengthening glass. It involves immersing the glass in a molten salt bath.

float process A glass manufacturing process in which the molten glass ribbon flows through a furnace supported on a bed of molten metal.

insulating glass A glazing unit used to reduce the transfer of heat through a glazed opening by leaving an air space between layers of glass.

laminated glass A glazing unit made by bonding two or more sheets of glass with interlayers of plastic resin.

Low-E glass Low-emissivity glass has a thin metallic coating that selectively reflects ultraviolet and infrared wavelengths of the energy spectrum.

reflective glass A glazing unit used to reduce the amount of solar energy transmitted through a glazed opening by having one surface of the glass covered with a thin, transparent metallic film.

tempering A process used to strengthen glass. It involves raising the glass to near the softening point and then blowing jets of cold air on both sides, suddenly chilling it.

SUGGESTED ACTIVITY

Collect samples of glass products used for glazing, identify each, and describe their properties. Glass manufacturers often have small sample pieces available.

ADDITIONAL INFORMATION

Hornbostel, C., *Construction Materials: Types, Uses, and Applications*, John Wiley and Sons, New York, 1991.

Ramsey, C.G., Sleeper, H.R., and Hoke, J.R., eds., *Architectural Graphic Standards*, John Wiley and Sons, New York, 1995.

Other resources are numerous related publications from the organizations listed in Appendix B.

Doors, Windows, Entrances, and Storefronts

This chapter will help you to:

1. Be familiar with the types of doors, windows, entrances, and storefronts available.

2. Apply the information to the decision-making processes as these units are selected for a building.

3. Understand the properties of various units as they relate to fire, security, privacy, and operation.

DOORS

Doors are used to provide security, privacy, and fire protection to access openings in interior and exterior walls. The choice of a door depends on the traffic passing through the opening and the door's appearance. In residential work interior doors provide privacy and some degree of security, and exterior doors provide protection from the weather, security, and, often, an important part of the exterior design. In commercial buildings doors must permit ingress and egress as required by building codes. They are subject to considerable use, so they must be strong and have hardware to withstand constant use. Other doors provide access for large equipment, such as trucks and aircraft. They are usually mechanically opened and closed because they are large and heavy. Doors are also used to divide large interior spaces into smaller rooms.

Doors are available manufactured from wood, hardboard, plastic, metal, and glass. The typical types are shown in Figs. 30.1 and 30.2. These types are used in residential and commercial construction.

Hollow-Core Metal Doors

Hollow-core metal doors are available in steel and aluminum. They are used on interior and exterior openings and are available in a wide range of designs (Fig. 30.3). The exterior surface may be flush or paneled, have windows or louvers, and be finished with a baked enamel finish on a smooth or textured surface or be covered with plastic veneers that may be wood grain or a variety of colors and patterns.

The exact construction of hollow metal doors varies with the manufacturer, but two types of construction are common. One uses a tubular framework over which the face panels are welded. Insulation and sound-deadening material are placed in the hollow core. Another type uses some type of core material, such as kraft paper honeycomb, polyurethane, polystyrene, a steel grid, mineral fiberboard, or vertical steel stiffeners, over which metal-faced panels are bonded (Fig. 30.4). The door may be seamless, full-flush, flush-panel, or industrial-tube construction

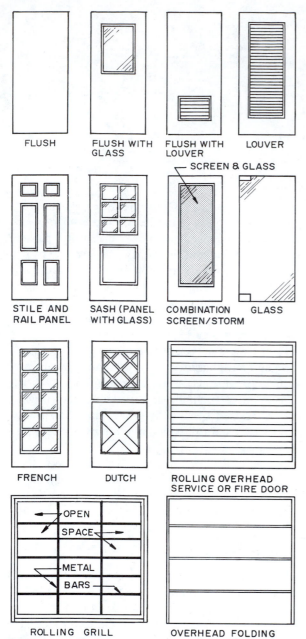

Figure 30.1 Some of the doors available for residential and commercial applications.

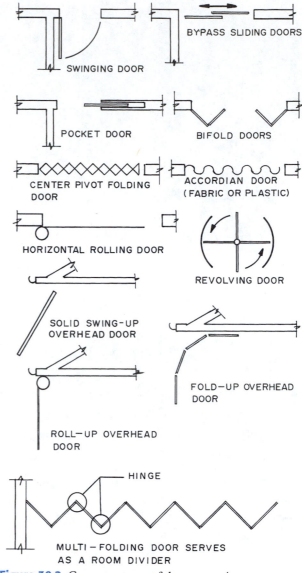

Figure 30.2 Common types of door operation.

(Fig. 30.5). Seamless doors have no seams showing on the face or edges. The edge seams are hidden by welding them and grinding the weld smooth. Full-flush doors have visible seams on the edges but no seams on the face. Flush-panel doors use either a stile and panel or a stile and rail. The face panel may be slightly recessed. Industrial-tube doors are made using tubular steel stiles and rails and have recessed panels. Metal doors are classified by grades and models as shown in Table 30.1. The higher the grade the thicker the metal specified. The nomenclature for hollow steel door designs is shown in Fig. 30.6. The letter symbols are used as the basis for the description of steel doors. Some typical construction details for the top and bottom edges of the doors are shown in Fig. 30.7. Some

of the commonly used meeting stiles for hollow metal doors are shown in Fig. 30.8.

Hollow Metal Frames

Hollow metal frames are available in steel or aluminum. They are supplied knocked down or preassembled. A wide range of designs are available providing for many types of interior and exterior wall construction (Fig. 30.9). Hollow metal frames are anchored to masonry walls with some type of strap or wire loop. Usually three anchors are required on each jamb. Several strap designs are used to secure the hollow metal door frames to wood or metal studs. Brackets are used to secure the base to the floor (Fig. 30.10). The installation of some of these frames is detailed in Fig. 30.11.

Figure 30.3 A typical hollow-core metal door.

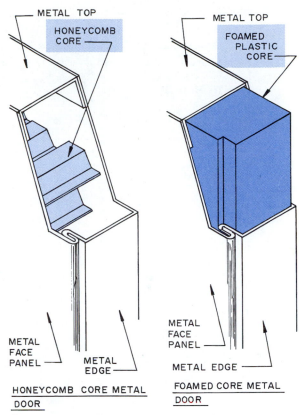

Figure 30.4 Typical metal door construction.

The frame and the door are prepared to receive a lockset and strike plate. The holes, recesses, and screw holes are cut. Assembled door frames frequently have the door installed before the unit is shipped to the job (Fig. 30.12).

The required steel gauge for door frames is related in the standards to the door grade and model. The higher the door grade the thicker the steel used for the door frame.

Metal Fire Doors and Frames

Fire doors are an assembly of a door and frame manufactured and installed to give protection against the passage of fire through an opening in a wall. The design, location, and installation are controlled by national standards and local building codes. Approved fire door assemblies are required to meet the test requirements specified in ASTM E152, Methods of Fire Tests of Door Assemblies. Standards for fire doors and windows are in the publication ANSI/NFPA 80. This publication was prepared by a committee of the National Fire Protection Association and has been approved by the American National Standards Institute.

All fire door assemblies must bear the label of an approved agency and include the name of the manufacturer,

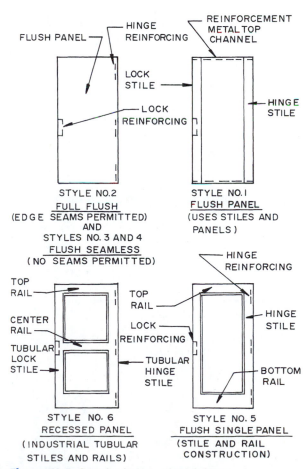

Figure 30.5 Standard types of steel doors.

Table 30.1 Grades and Models of Standard Steel Doors

Grade	Model
Grade I—Standard duty, 1⅜″ and 1¾″ thick	Model 1—Full flush design, hollow metal and composite Model 2—Seamless design, hollow metal and composite
Grade II—Heavy duty, 1¾″ thick	Model 1—Full flush design, hollow metal and composite Model 2—Seamless design, hollow metal and composite
Grade III—Extra heavy duty, 1¾″ thick	Model 1 and 1A[a]—Full flush design, hollow metal and composite Model 2 and 2A[a]—Full flush design, hollow metal and composite Model 3—Stile and rail, flush panel

[a]1A and 2A are made with a heavier gauge metal than models 1 and 2.
Courtesy The Steel Door Institute

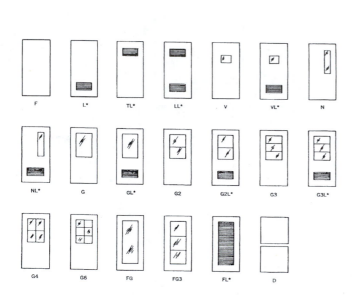

NOMENCLATURE LETTER SYMBOLS

F   -- Flush

L* -- Louvered (bottom)

TL* -- Louvered (top)

LL* -- Louvered (top
        and bottom)

V   -- Vision Lite

VL* -- Vision Lite
        and Louvered

N   -- Narrow Lite

NL* -- Narrow Lite
        and Louvered

G   -- Half Glass (options
        G2, G3, G4, and G6)

GL* -- Half Glass and
        Louvered (options
        G2L* and G3L*)

FG -- Full Glass
        (option FG3)

FL* -- Full Louver

D   -- Dutch Door

* Louvered door designs : specify design, louver size and/or free area requirements
ADD SUFFIX I TO INDICATE INSERTED LOUVER
ADD SUFFIX P TO INDICATE PUNCHED LOUVER
ADD SUFFIX A TO INDICATE AIR CONDITIONING GRILLE

Figure 30.6 Nomenclature for steel doors and frames.
(Courtesy The Steel Door Institute)

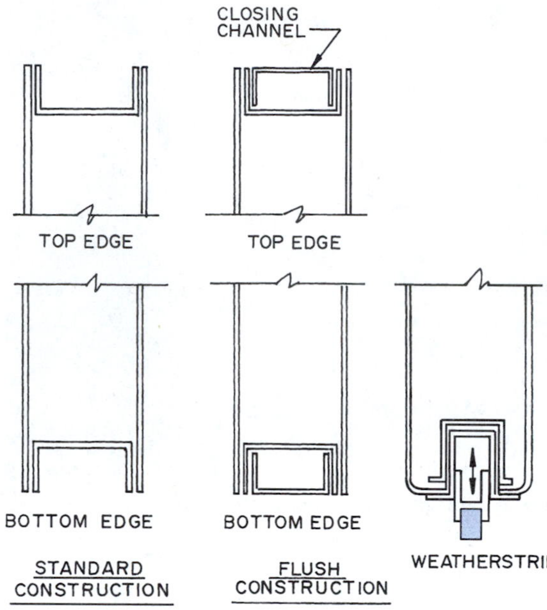

Figure 30.7 Some commonly used top and bottom construction details for hollow metal doors.

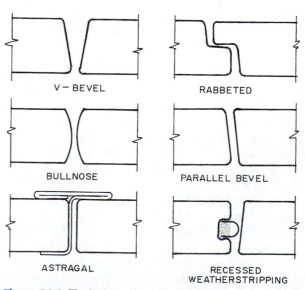

Figure 30.8 Typical meeting stile profiles used on metal doors.

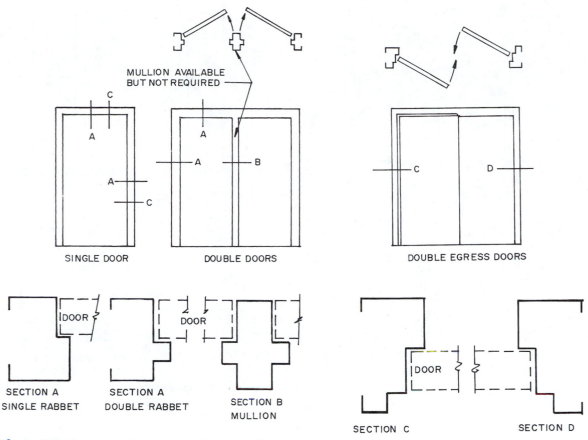

MULLION AVAILABLE
BUT NOT REQUIRED

SINGLE DOOR   DOUBLE DOORS   DOUBLE EGRESS DOORS

SECTION A
SINGLE RABBET

SECTION A
DOUBLE RABBET

SECTION B
MULLION

SECTION C   SECTION D

Figure 30.9 Some typical metal door frame details.

the fire protection rating, and the maximum transmitted temperature end point. Some assemblies are automatic and keep the door open (as on a long hall) and will close automatically if a fire raises the temperature or if they are exposed to smoke.

Fire doors may be classified by the following designation systems:

1. Hourly rating designation.
2. Alphabetical letter designation.
3. A combination of hourly and a letter designation.
4. Horizontal access doors use a special listing indicating the fire rated floor, floor-ceiling, or roof-ceiling assembly for which the door will be used.

The *hourly designation* indicates the duration of the fire test exposure in hours and is called the fire protection rating. The alphabetical *letter designation* in use follows:

Class A—Openings in fire walls and in walls dividing a single building into fire areas

Class B—Openings in enclosures of vertical communications through buildings and in two-hour rated partitions providing horizontal fire separations

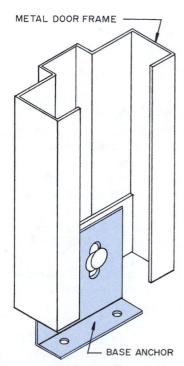

METAL DOOR FRAME

BASE ANCHOR

Figure 30.10 Metal door frames are anchored to the floor with some form of metal angle.

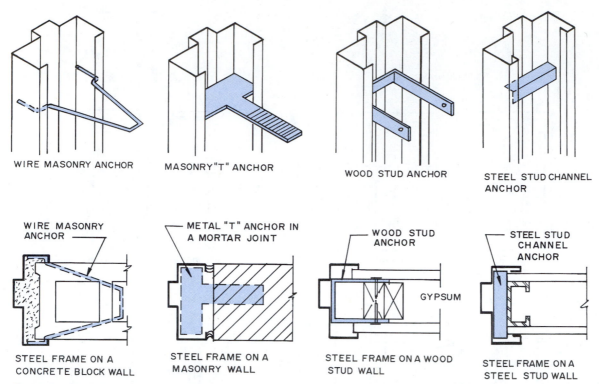

WIRE MASONRY ANCHOR   MASONRY "T" ANCHOR   WOOD STUD ANCHOR   STEEL STUD CHANNEL ANCHOR

WIRE MASONRY ANCHOR   METAL "T" ANCHOR IN A MORTAR JOINT   WOOD STUD ANCHOR   STEEL STUD CHANNEL ANCHOR

GYPSUM

STEEL FRAME ON A CONCRETE BLOCK WALL   STEEL FRAME ON A MASONRY WALL   STEEL FRAME ON A WOOD STUD WALL   STEEL FRAME ON A STEEL STUD WALL

Figure 30.11 Some of the types of anchors used to secure steel door frames to the wall.

Class C—Openings in walls or partitions between rooms and corridors having a fire resistance rating of one hour or less

Class D—Openings in exterior walls subject to severe fire exposure from outside of the building

Class E—Openings in exterior walls subject to moderate or light fire exposure from outside of the building

Special listings are tested in accordance with NFPA 251, Standard Methods of Fire Tests of Building Construction and Materials. It indicates the fire rated assembly and its hourly rating. Fire door rating details are shown in Table 30.2. Labels on the doors specify the class of the door, which indicates the time interval the door will meet. For example, a Class A door has a 3 hr. time interval. Metal door frames have fire ratings of ¾, 1½, and 3 hr. The hardware used must be fire rated to maintain the fire rating of the door and frame.

The typical types of construction for fire doors includes

Wood core fire doors have wood, hardboard, or plastic laminate faces bonded to solid wood core or a wood particleboard core.

Hollow metal fire doors may be of flush or panel design with the steel face being 20 gauge or thicker.

Metal clad fire doors have wood cores or stiles and rails with insulated panels. The face is 24 gauge steel or lighter.

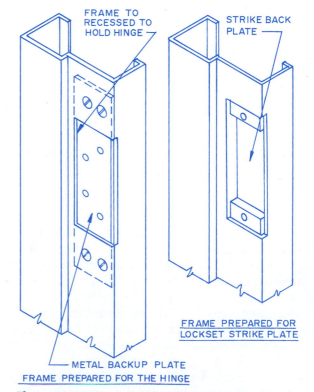

FRAME TO RECESSED TO HOLD HINGE

STRIKE BACK PLATE

METAL BACKUP PLATE
FRAME PREPARED FOR THE HINGE

FRAME PREPARED FOR LOCKSET STRIKE PLATE

Figure 30.12 The metal door frame has recessed hinge and lockset back plates to receive the hinges and strike plate.

Table 30.2 Fire Door Ratings

Class	Hour	Glazing	Location
A	3	No glazing	In fire wall and walls that divide a building into fire areas
B	1½	100 sq.in. of glazing	In enclosures of vertical communication through buildings
C	¾	1296 sq.in. of glazing per light	Openings in walls requiring a fire resistance rating of 1 hr. or less
D	1½	No glazing	Openings in exterior walls subject to fire exposure from the outside of the building
E	¾	720 sq.in. glazing	Openings in exterior walls subject to moderate or light fire exposure from outside the building

Sheet metal fire doors are made using 22 gauge or lighter steel.

Tinclad fire doors have a solid wood core covered with a 24 to 30 gauge terneplate or galvanized steel facing.

Composite fire doors consist of some combination of wood, steel, or plastic laminate bonded to a solid core material.

Rolling steel fire doors are steel doors that move on an overhead barrel that is enclosed in a hood. It may have an automatic closing mechanism (Fig. 30.13).

Curtain-type fire doors have interlocking steel blades forming a steel curtain in a steel frame.

Special-purpose fire doors include a fire door and frame assembly. These include acoustical fire doors, security fire doors, armored attack–resistant fire doors, radiation-shielding fire doors, and pressure-resistant fire doors. Also, a variety of automatic fire vents are manufactured (Fig. 30.14).

Wood and Plastic Doors

The commonly available types of wood doors are illustrated in Chapter 19. They are available in solid wood and with wood veneer, hardboard, and plastic laminate faces. Another door has wood stiles and rails covered with a fiberglass skin. These are all manufactured in the same basic styles (Fig. 30.15). Wood and plastic doors can have glass or louvered openings.

Glass Doors

Glass doors may be frameless and have metal channels to hold a pivot hinging apparatus or have a very narrow metal frame on all sides or a wider frame that provides a stronger door for use in areas where heavy traffic is expected (Fig. 30.16). The glass must meet building code safety requirements.

Special Doors

Many doors are manufactured to meet special needs other than providing human access within a building.

Figure 30.13 A rolling steel fire door. *(Courtesy Cornell Iron Works, Inc.)*

These include units that provide access to an area such as a roof (Fig. 30.17), various types of sliding doors (Fig. 30.18), doors to resist explosions, and doors to provide an airtight or a watertight closure. Special doors and glazing are used for security purposes. Various types of folding and accordion doors are widely used in residential and commercial construction. There are many variations of rolling doors, such as those used on aircraft hangers and openings in manufacturing plants. Many are power activated (Figs. 30.19 and 30.20). Various types of rolling grille doors are used to provide security to contents of stores opening onto a public mall (Fig. 30.21).

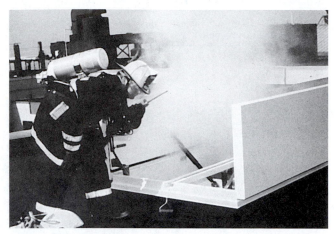

Figure 30.14 This automatic fire door opens during a fire to let the rising smoke and gases clear from the building. *(Courtesy The Bilco Company)*

Figure 30.15 A wood frame door with fiberglass outer skin.

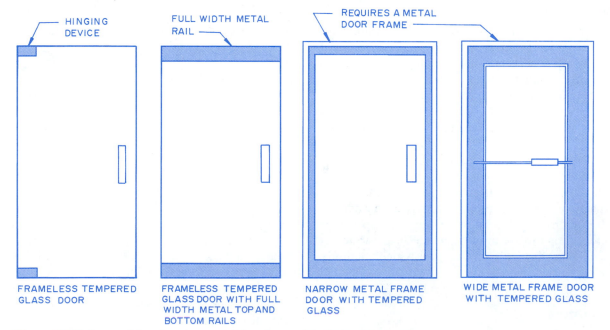

FRAMELESS TEMPERED GLASS DOOR

HINGING DEVICE

FRAMELESS TEMPERED GLASS DOOR WITH FULL WIDTH METAL TOP AND BOTTOM RAILS

FULL WIDTH METAL RAIL

NARROW METAL FRAME DOOR WITH TEMPERED GLASS

REQUIRES A METAL DOOR FRAME

WIDE METAL FRAME DOOR WITH TEMPERED GLASS

Figure 30.16 Some of the glass doors used in commercial construction.

Figure 30.17 A roof scuttle provides safe and easy access to the roof. *(Courtesy The Bilco Company)*

Figure 30.19 These steel rolling service doors provide a fire-resistant closure between areas separated by this interior wall in a manufacturing plant. *(Courtesy Cornell Iron Works, Inc.)*

Figure 30.20 These insulated rolling service doors have either steel or aluminum interiors and exterior skins separated by a closed cell urethane foam insulation core. *(Courtesy Cornell Iron Works, Inc.)*

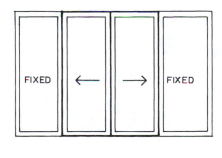

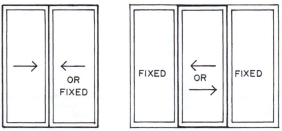

Figure 30.18 Some of the available configurations of sliding doors.

Figure 30.21 During the hours this business is open it is totally exposed to the public, and rolling grille doors provide security during closed hours yet permit the interior to be visible. *(Courtesy The Overhead Door Company)*

WINDOWS

The basic types of windows used in residential construction are illustrated in Chapter 19. In addition, various windows are manufactured primarily for use in commercial buildings. Windows are available made from solid wood, wood clad with plastic or aluminum, solid plastic, steel, stainless steel, aluminum, bronze, and **composite** materials.

A major consideration in the selection of windows is energy savings. Various types of energy-efficient glazing are available, such as energy-efficient glass, double glazing, and glass with louver blinds set between the panes of glass. Some double glazed units have a gas inserted in the space between the panes. See Chapters 29 and 31 for additional information on glass and glazing. Windows must have airtight unions between both fixed and moving parts to reduce air infiltration. Metal units transmit heat and cold rapidly through the material, so the design must provide insulation between touching interior and exterior parts. These are called thermal breaks.

Windows are also a major source of light to the inside space. From a design standpoint they are a major feature of the appearance of the exterior. They used to be a major source of ventilation, but this need has been reduced to some degree by year-round interior temperature control and air-filtering systems.

Quality windows are tested and certified against standards for air leakage, water infiltration, uniform wind load structural requirements, and uniform load deflection. These tests are made following ASTM standards.

The energy performance of **fenestration** products is rated by the National Fenestration Rating Council (NFRC). Fenestration refers to the design of windows and other exterior openings of a building. The NFRC combines U-value, solar heat gain factors, optical properties, air infiltration, condensation resistance, and other characteristics into a uniform rating system to reflect annual energy performance. Products are certified and labeled, providing a means for comparing the products manufactured by various companies. The NFRC certification label is shown in Fig. 30.22.

A wide range of locking and operating devices are available and should be considered as window units are selected. These include manual and electric operating devices, hinges and other hardware, and safety devices, such as a keyed lock or safety bar that limits the amount a window can be opened.

Local fire codes also influence the choice of windows. These codes typically control the sizes of windows, height of sill above the floor, and the ability to open the window from the inside without the need of a special tool. Windows used in fire walls must be fire rated and labeled to restrict the spread of fire and smoke. Security windows are tested to be certain they meet the standards required to provide resistance to forced entry. Finally, the material and finish must fit in with the surrounding area and be able to withstand any corrosive elements to which it will be exposed.

Wood Windows

Wood frame windows are available with fixed and operable sashes. They may be clad with plastic or aluminum or left natural and painted. Wood windows provide better insulation than metal and plastic windows. However,

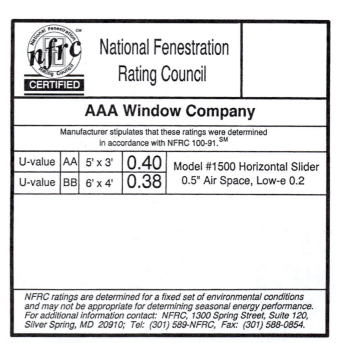

Figure 30.22 A generic example of the certification label of fenestration products rated by the National Fenestration Rating Council. The blank space is used to indicate solar heat gain, air infiltration, long-term energy performance, and other energy performance attributes. *(Courtesy The National Fenestration Rating Council)*

they do swell and shrink as the moisture content changes. The cladding of wood window exteriors has reduced this considerably.

Wood windows are installed in wood frame walls by nailing through a metal or plastic flange through the sheathing into the wood studs and headers in the rough opening (Fig. 30.23). Some are designed with a wood molding, and this is nailed to the sides of the wood frame rough opening (Fig. 30.24). An example of another installation method is the use of metal clips to secure the window in a wall framed with metal studs, as shown in Fig. 30.25. Windows come with various types of glazing and insect screens.

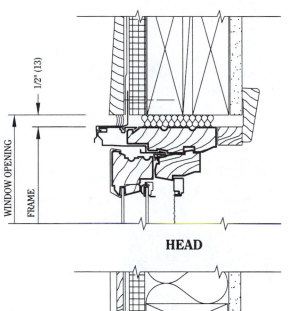

HEAD

HEAD DRIP FIN #7290

SUGGESTED USE OF INSULATION
TO FILL ALL VOIDS AT WINDOW
PERIMETER BY OTHERS

PERIMETER SEALANT
BY OTHERS

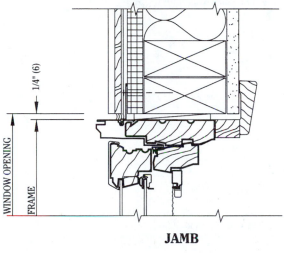

JAMB

INSTALLATION FIN #4205

SHIM AND PLUMB UNITS
AS REQUIRED

WOOD TRIM BY OTHERS

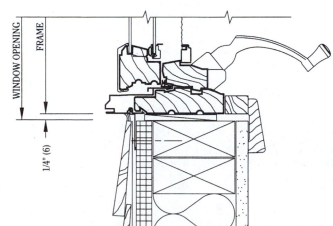

SILL

INSTALLATION FIN #4205

WHEN STYROFOAM OR OTHER UNSTABLE
SHEATHING MATERIAL IS USED, PROVIDE
SOLID BLOCKING FOR FIN ATTACHMENT

LEVEL UNITS AS REQUIRED

Figure 30.23 An installation detail for a clad casement window secured in a wood frame wall by nailing through the plastic flange. *(Courtesy Pella Windows and Doors)*

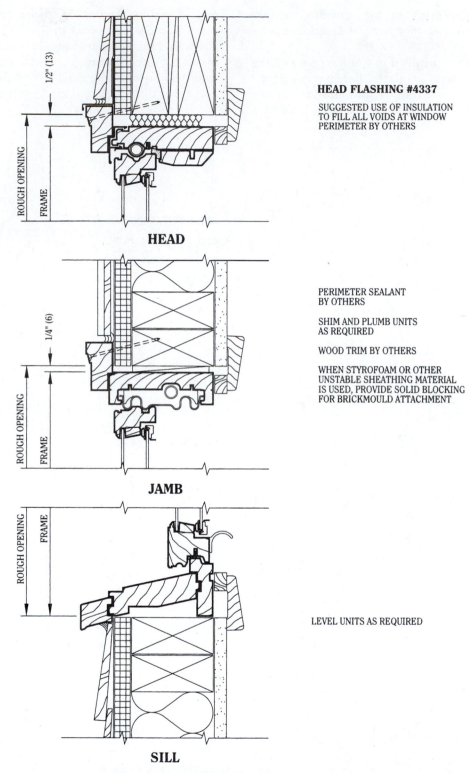

HEAD FLASHING #4337

SUGGESTED USE OF INSULATION
TO FILL ALL VOIDS AT WINDOW
PERIMETER BY OTHERS

HEAD

PERIMETER SEALANT
BY OTHERS

SHIM AND PLUMB UNITS
AS REQUIRED

WOOD TRIM BY OTHERS

WHEN STYROFOAM OR OTHER
UNSTABLE SHEATHING MATERIAL
IS USED, PROVIDE SOLID BLOCKING
FOR BRICKMOULD ATTACHMENT

JAMB

LEVEL UNITS AS REQUIRED

SILL

NOTE: THESE DETAILS ARE FOR TYPICAL SINGLE PUNCH OPENINGS.

Figure 30.24 An installation detail for a wood frame double-hung window in a wood frame wall installed by nailing through the exterior molding. *(Courtesy Pella Windows and Doors)*

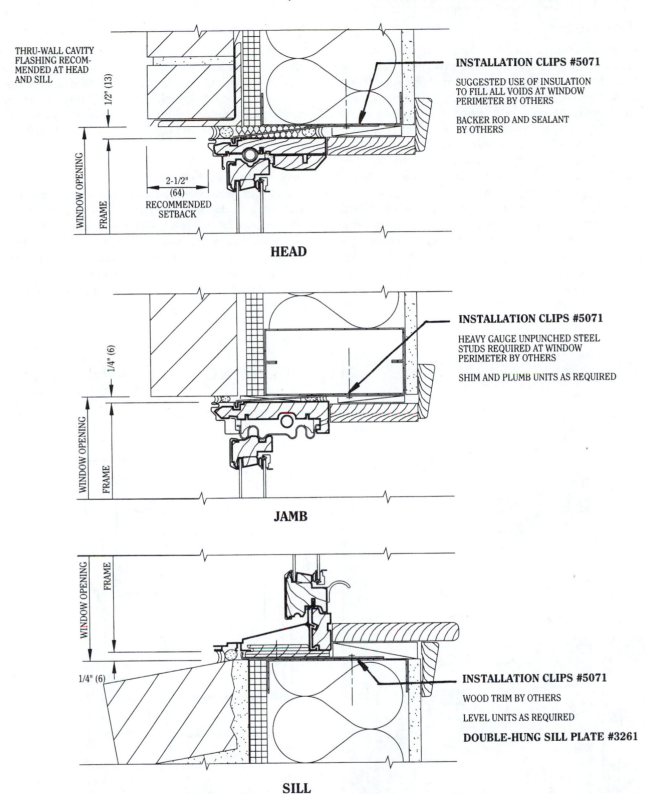

Figure 30.25 An installation detail for a clad double-hung window installed in a steel stud frame wall with clips that are secured to the metal studs. *(Courtesy Pella Windows and Doors)*

Plastic Windows

Plastic windows are available in the same basic types as described for wood windows. The structural frame and casement frame are made from polyvinyl chloride (PVC) and glass fibers and a polyester resin. The material will not rust, swell, pit, peel, or corrode, and it never needs painting. It is a fairly good insulator, and the frame is made with dead air pockets that increase the insulation value. PVC also has excellent sound insulation value, and members can be produced with the accuracy needed to provide airtight fits (Fig. 30.26).

PVC windows are available in a range of colors. Some color the exterior members and leave the interior exposed surfaces an off-white. A section through a plastic window unit is shown in Fig. 30.27. Windows are fastened to the wall using steel wall anchors with zinc coat screws applied through slots in the frame as specified by the manufacturer. Units are available with various types of glazing and insect screens.

Metal Windows

Metal windows used in residential construction are most often aluminum. Although some types of steel windows are designed for use on residential buildings, they mainly are used for commercial, industrial, and monumental buildings. Monumental buildings include structures such as schools, public buildings, hospitals, and churches.

Steel Windows

Steel windows are made from structural grade steel. *Residential grade* windows are made from lighter gauge steel than windows designed for use in monumental, industrial, and commercial buildings, which are classified as *intermediate grade*. *Heavy-duty intermediate grade* windows are used where the size of the window requires additional strength or it must meet other conditions, such as resisting high winds. Manufacturers' catalogs show the gauges and recommended uses for their products.

Figure 30.26 This vinyl window has low air infiltration, is maintenance free, and is assembled by fusion welding. *(Courtesy Georgia-Pacific Corporation)*

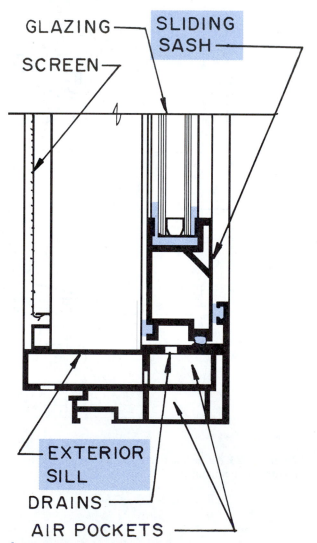

Figure 30.27 A section through the sill of a typical plastic double-hung window.

Steel window units are rigid and strong and require narrower members than other materials, producing a smaller sight line. They are available primed or galvanized and ready for painting. Some types have a polyvinyl chloride or urethane plastic coating that gives permanent protection. The units are weatherstripped and available with single and double glazing. Manufacturers produce a range of stock sizes but will produce custom-designed units.

Steel windows used in residential construction are typically casement type. In commercial, industrial, and monumental type buildings other designs are available, as shown in Fig. 30.28. A section through a typical steel window is shown in Fig. 30.29.

Stainless Steel Windows

Stainless steel windows are usually more expensive than steel or aluminum; however, they have a long life and may save money over a period of years. They are made from various types of stainless steel formed into structural members that are welded to form the unit. The sizes and types of window vary by the manufacturer and many are custom built to the architect's design. Typical types available include folding, awning, casement, and various hinged types. Some are like those shown for commercial and industrial steel windows.

Aluminum Windows

Aluminum windows are made from extruded aluminum structural members. Since aluminum is easy to work, complex designs on these members are readily produced. Aluminum has good structural properties, and it will resist corrosion and damage from weather. If used in an atmosphere where damage may occur it can be given various organic protective coatings. It can also be colored using anodized finishes. Additional information on aluminum is in Chapter 16.

Aluminum windows are assembled using both mechanical connections and welding. Any connectors, fasteners, hardware, or anchors must be aluminum or a material such as stainless steel that is compatible with aluminum so corrosion does not occur because of galvanic action.

Several aluminum alloys are used in the manufacture of aluminum windows. These affect the properties of the product and vary with the purpose for which the window will be used. Standards for the quality of aluminum windows are established by the Architectural Aluminum Manufacturers Association. These vary with the type of window and its intended use. The windows are of three types—residential, commercial, and monumental. They are available in the same types as described for plastic and steel windows. Some manufacturers have other special designs and will produce custom-designed units for special applications. A section through a typical aluminum window is in Fig. 30.30. Additional window details are in Chapter 31.

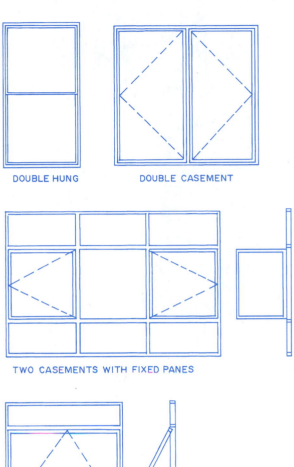

DOUBLE HUNG   DOUBLE CASEMENT

TWO CASEMENTS WITH FIXED PANES

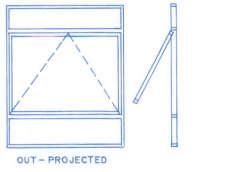

OUT – PROJECTED

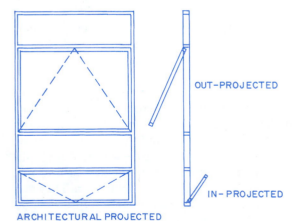

OUT–PROJECTED

IN – PROJECTED

ARCHITECTURAL PROJECTED

Figure 30.28 Typical types of steel windows available. Other designs are available.

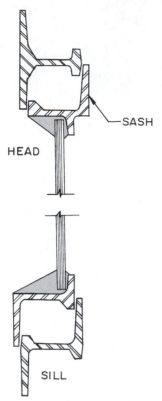

Figure 30.29 A section through a typical steel window.

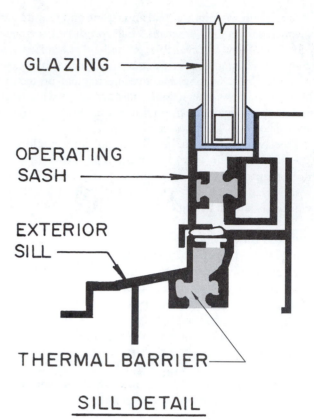

Figure 30.30 A section through the sill of an aluminum frame window showing the thermal barriers.

Composite Materials

Increasing use is being made of composite materials in construction. Composites are materials made by combining several different materials. One such product is shown in Fig. 30.31. This window is made entirely of composites (a cellular PVC material) and wood by-products, producing a product that looks like wood but will not split, warp, or rot. It has twice the insulation value of wood and is available in several colors.

Special Windows

Special windows are those designed for a special application, such as skylights, security windows, pass windows, and storm windows. Skylights and roof windows perform the same basic function as windows, but they are in the roof. Skylights have fixed glazing, but the sash on roof windows can be opened to provide ventilation. They may be installed as single units, as is common in residential construction (Fig. 30.32), or in strings or clusters, admitting considerable light (Fig. 30.33). They are glazed with glass or plastic lights. Some operable units are opened manually with a crank, and other types are electrically opened. They may be designed to blow out the glazing if

an explosion occurs or to open automatically if the temperature increases, serving as a fire vent.

Most skylights used in commercial, industrial, and monumental work come with aluminum frames (Fig. 30.34). The small units used in residential work have wood frames. Small domed and barrel skylights may have single or double glazing (Fig. 30.35).

Skylights are available for enclosing large spaces, such as a mall, reception area, swimming pool, or restaurant. They are also used to enclose walkways, protecting the interior from the weather and providing abundant light. They are glazed with glass and plastic materials. Some examples are in Figs. 30.36 and 30.37.

Security windows (also called detention windows) are tested to ascertain their resistance to forced entry. This includes testing the locking device, the impact resistance of the sash and frame, the resistance of security bars, and the resistance of the glazing. The degree of security is specified as one of four classes:

Class 1: Minimal
Class 2: Moderate
Class 3: Medium
Class 4: Relatively high

Figure 30.31 These windows are made from a composite material composed of wood by-products and a cellular PVC material. *(Courtesy Jeld-Wen, Inc.)*

Figure 30.32 Single skylights are used to provide natural light in residential construction. *(Courtesy Velux Roof Windows and Skylights)*

Figure 30.33 These skylights are installed as a string to enhance the lighting of the interior of this room. *(Courtesy Eagle Window and Door, Inc.)*

Figure 30.34 Commercial buildings make extensive use of skylights. *(Courtesy Lin-El, Inc.)*

Figure 30.36 Large skylights form the roof structure and are available in a variety of shapes. *(Courtesy Lin-El, Inc.)*

Figure 30.35 This barrel skylight provides considerable light into the interior. *(Courtesy O'Keeffe's Inc.)*

Figure 30.37 Walkway covers protect those moving between buildings. *(Courtesy Lin-El, Inc.)*

In some cases window security is specified in building codes that require a high-grade window in certain areas of the wall. If windows are protected by devices such as bars, screens, or shutters, provisions for meeting exit requirements in the fire code must be met. Manufacturers can indicate if the windows meet ANSI/ASTM security standards. In addition to a secure window, the window frame, lock, hinges, glazing, and design to prohibit insertion of tools through the unit are considered. Laminated glass used on security windows is discussed in Chapter 29.

Security windows are used in places where maximum security is required, such as detention centers, correctional, and psychiatric facilities. They are designed so horizontal members do not deflect and are tied with steel members in the frame. Some have fixed glazing, and others permit a limited amount of ventilation (Fig. 30.38). They are typically made of maximum security tool-resisting steel. The frame is usually welded. Laminated security glass is used and some units are double glazed. Bullet-proof glass is also available. Finally, screens made of tool-resistant steel encased in a steel frame complete the unit (Fig. 30.39). Some companies also make aluminum-framed security windows. The hardware also must be of highest quality.

Top pivoted awning security window

Medium Security Guard window

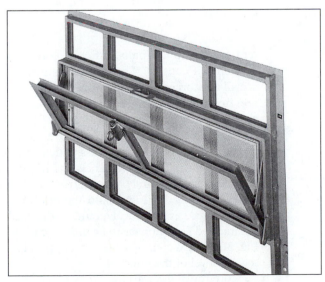

Protected out awning security window

Figure 30.38 Detention windows are available with fixed or operating glazing. (*Courtesy The William Bayley Company*)

Figure 30.39 Security windows may have a screen over the glazing to increase the security of the installation. This is a psychiatric window with a guard screen. *(Courtesy The William Bayley Company)*

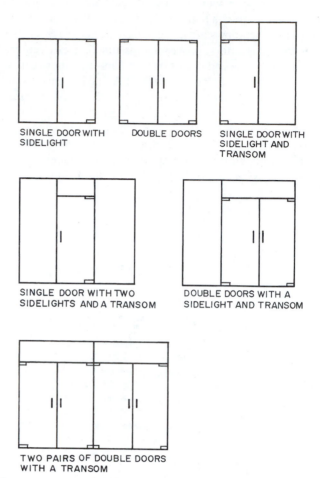

Figure 30.40 A few of the tempered glass frameless entrances with one or more doors, sidelights, and transoms. The same configurations are available in metal frame entrances.

ENTRANCES AND STOREFRONTS

Entrances and storefronts are made using flat architectural glass products such as tempered and heat-strengthened glass, laminated glass, and insulating glass. Detailed information on glass products is in Chapter 29. Typically, tempered glass doors are available with door closers, top and bottom pivots, locks, and push-pulls. The unit may include one or more doors, sliding glass panels, and transoms (Fig. 30.40). The doors may be manually or automatically opened. The metal frame, typically aluminum, is usually anodized, and a variety of colors are offered. Stainless steel frames are also available. The glazing is available in clear, obscure, and several colors. Some door units are frameless (Fig. 30.41).

Entrances may be custom built for a particular opening using stock door sizes and custom-built frames. The frames are mechanically joined and then welded (Fig. 30.42). Figure 30.43 shows a high-traffic entrance system for shopping malls, schools, concert centers, and convention centers. The doors, frames, and hardware are all reinforced heavy-duty assemblies. The door walls are heavier, and the hinges and pivots are directly mounted into the reinforced frame (Fig. 30.44).

Figure 30.45 shows the entrance to a high-rise building that provides reliable access and an impressive architectural feature. The glass panels and hidden mullions reflect the architectural solution to the cladding of the entire building. In Fig. 30.46 the entrance shelters the walkway and is the focal point of the building. It was painted red for emphasis. The unique circular entrance in Fig. 30.47 also contains a stair that provides access to the second level of the building.

Storefronts use various types of entrance units plus glazing and solid panels of various types (Fig. 30.48). These large glass units are backed by a metal rolling grille door to provide security. The grille can be raised during business hours to permit a clear view of the store interior.

Figure 30.41 This entrance to a multistory building uses frameless glass doors. *(Courtesy EFCO Corporation)*

Figure 30.42 These aluminum frame entrance doors were custom designed and manufactured. They are engineered to withstand the heavy use and weather conditions to which they are exposed. Notice the use of large transom panels above the doors. *(Courtesy Tubelite, Inc., Reed City, Michigan)*

Figure 30.43 These heavy-duty entrance doors and frames are designed to provide years of service under severe conditions. *(Courtesy Kawneer Company, Inc.)*

Figure 30.44 This impressive entrance uses metal frame glass doors with a marble surround. *(Courtesy EFCO Corporation)*

Figure 30.45 This entrance forms the focal point to the architectural design of this building. It uses glass panels with concealed mullions and exposed rails. *(Courtesy Bruce Wall Systems Corporation)*

Figure 30.46 This striking entrance provides a focal point for the building, provides shelter from the weather, allows natural light to enter the walkway, and is reflected by the designer in the construction of the curtain wall mullions. *(Courtesy Kawneer Company, Inc.)*

Figure 30.47 This custom-designed entrance contains stairs to provide access to a second level and includes a short covered walkway to the right. It shelters the area from the weather and allows natural light to enter the area. *(Courtesy Lin-El, Inc.)*

Figure 30.48 This storefront provides excellent visibility of the store interior yet can be secure when desired by lowering the rolling grill security door. *(Courtesy Cornell Iron Works, Inc.)*

When designing a glazed storefront it is essential to include wind load calculations. Many window manufacturers have this information for their products. In Fig. 30.49 is a storefront typical of those used on many small retail establishments. It provides access to the store and an area in which the retailer can display merchandise. These stores open onto the street.

In malls the stores face a wide pedestrian walkway. These stores use a number of entrances that not only provide access but also a clear view of the products inside the store (Fig. 30.50).

Building Codes

Entrances and storefronts are subject to a variety of requirements in local building codes. Among these are consideration of fire resistance, wind and snow loads, hazards such as human impact loads, the means for supporting the doors and related glazing, hardware, locking arrangements, requirements for power-operated doors, and the size and number of doors required.

Figure 30.49 A typical storefront used on many small retail shops to provide access and a merchandise display area.

Figure 30.50 This glass wall and entrance provides those passing by on the mall a view of the interior of the store or office and those inside a degree of security. *(Courtesy EFCO Corporation)*

REVIEW QUESTIONS

1. Describe the construction of various types of hollow-core metal doors.
2. Explain how hollow metal door frames are secured to the wall and floor.
3. What designations are used to identify metal fire doors?
4. Identify each of the letter designations used on metal fire doors.
5. List some of the special doors available.
6. What are some of the types of energy-efficient glazing available?
7. What association has established energy performance rating for fenestration products?
8. What two methods are commonly used to secure wood windows to the wall framing?
9. What are the grades of steel windows?
10. How do windows made of composite materials compare with solid wood windows?
11. What are the classes of security windows?
12. What are typical building code requirements placed upon entrances and storefronts?

KEY TERMS

composite materials Materials made by combining several different materials.

fenestration Pertains to the design and disposition of windows and other exterior openings of a building.

SUGGESTED ACTIVITY

Walk through a commercial development and list the various doors, windows, entrances, and storefronts used. Record the materials and glazing used. Check your local codes to see if those you observed meet the requirements specified.

ADDITIONAL INFORMATION

Hornbostel, C., *Construction Materials: Types, Uses, and Applications,* John Wiley and Sons, New York, 1991.

Ramsey, C. G., Sleeper, H. R., and Hoke, J. R., eds., *Architectural Graphic Standards,* John Wiley and Sons, New York, 1995.

Spence, W. P., *Finish Carpentry,* Sterling Publishing Co., New York, 1995.

Spence, W. P., *Residential Framing,* Sterling Publishing Co., New York, 1993.

Other resources are numerous related publications from the organizations listed in Appendix B.

31

Cladding Systems

This chapter will help you to:

1. Understand the forces that must be considered when designing cladding systems.

2. Be alert to building code requirements pertaining to cladding design and installation.

3. Supervise the erection of various cladding systems.

Cladding is a non-load-bearing exterior wall enclosing a building. It may be brick, aluminum, steel, bronze, plastic, glass, stone, or other acceptable material. Cladding is exposed to the weather and must resist the forces generated by the wind, rain, earthquakes, and temperature changes. It must control penetration by rain, condensation, high winds, and transfer of heat into and out of the building, and it must meet building code requirements, including weather resistance, structural requirements, and fire resistance. Cladding therefore enables the indoor environment to be controlled year-round so it maintains conditions required by the occupancy of the building. A building may be clad with precast concrete panels, masonry panels, or some type of lightweight curtain wall (Fig. 31.1).

CLADDING DESIGN CONSIDERATIONS

The design and construction details of cladding systems require careful engineering analysis and the consideration of a wide range of factors. The various elements and the final assembled materials must be tested to ensure they will meet the design requirements and building codes. Following are some of the factors to be considered when selecting and designing cladding systems.

Structural Performance

Panels can run continuously one or more stories in height and may be made up of spandrels and vision glass or one solid continuous panel. They must carry their own weight, have connectors to secure the panels to the structure, resist wind forces, provide for movement after they are installed, and meet fire codes.

Stiffness and Weight

The structural design of the panel must enable it to be transported to the site, be lifted into place, and support its own weight after installation. Since cladding is non–load bearing, it is not designed to carry additional loads. Stiffness requirements are also a major structural factor enabling the panel to resist wind loads.

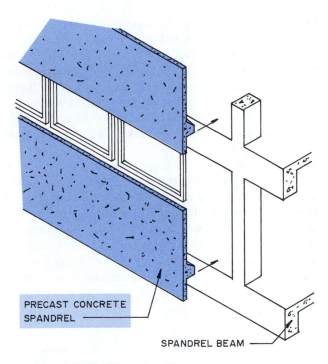

PRECAST CONCRETE
SPANDREL

SPANDREL BEAM

PRECAST CONCRETE CURTAIN WALL

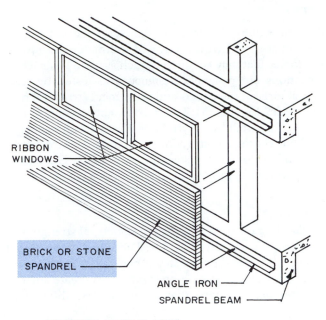

RIBBON
WINDOWS

BRICK OR STONE
SPANDREL

ANGLE IRON

SPANDREL BEAM

MASONRY CURTAIN WALL

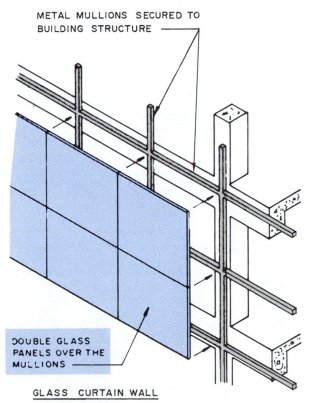

METAL MULLIONS SECURED TO
BUILDING STRUCTURE

DOUBLE GLASS
PANELS OVER THE
MULLIONS

GLASS CURTAIN WALL

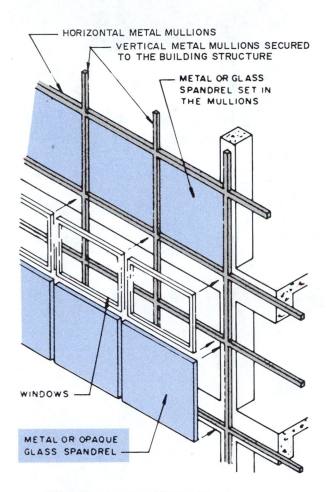

HORIZONTAL METAL MULLIONS

VERTICAL METAL MULLIONS SECURED
TO THE BUILDING STRUCTURE

METAL OR GLASS
SPANDREL SET IN
THE MULLIONS

WINDOWS

METAL OR OPAQUE
GLASS SPANDREL

METAL AND GLASS CURTAIN WALL

Figure 31.1 Some of the commonly used cladding systems.

Connections

The connections securing the panels to the structural frame must support the weight of the panel and resist other forces that may act upon it, such as wind loads. The connections are generally designed and tested by the manufacturer of the panels and must meet structural requirements. Figures 31.2 and 31.3 show connectors designed for installing a crystallized glass curtain wall system. One shows a connection to a concrete structural frame and the other to a steel frame building. Examples of other connections are shown later in this chapter.

Wind Forces

The cladding must have sufficient structural strength to resist forces produced by winds. These forces may create positive pressure against the panel that may cause deflection. In some situations wind forces create a negative pressure (suction) that tends to blow the panel from the structure. Wind blowing by a building is likely to create a vacuum (negative pressure). On high-rise buildings these negative pressures are usually greatest near the corners of the building. The indoor air pressure, due to the heating and cooling system, may be higher than the outdoor air pressure, thus putting a stress on the curtain wall from the inside of the building. If the panel lacks adequate stiffness and/or connectors it could blow off (Fig. 31.4).

Wind forces are more severe on the upper floors of high-rise buildings because they are subject to winds of greater velocity than lower floors or low-rise buildings. In an area with surrounding multistory buildings, the wind pressures are influenced by factors such as the direction of the prevailing winds, the shape of the building and neighboring buildings, the positioning of the building on the site, and the topography of the surrounding area. Detailed information is available from the American Architectural Manufacturers Association.

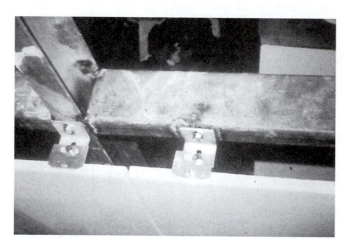

Figure 31.2 These connections were designed to secure crystallized glass curtain wall panels to a structural steel framework. *(Courtesy N.E.G. America, Inc.)*

Figure 31.3 These connections were designed to secure crystallized glass curtain wall panels to a structural concrete framework. *(Courtesy N.E.G. America, Inc.)*

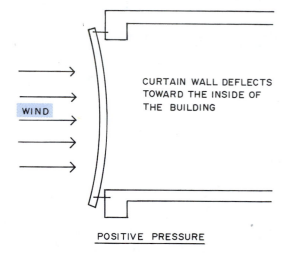

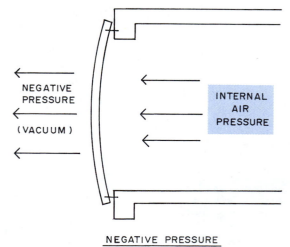

Figure 31.4 Wind forces subject the curtain wall to positive and negative forces.

Movement

Since a building is constantly subject to various forces, the cladding system must be designed to allow for movement within the structural system. For example, wind forces and earthquakes may cause the structural frame to deflect or twist, putting the cladding panels, windows, and connections under stress (Fig. 31.5). In addition, the cladding may be subject to forces created by gravity, thermal expansion and contraction, and moisture penetration and condensation. The weight of construction materials and differential (uneven) settling or heaving (lifting) of a foundation can cause beams to deflect. Creep that may occur over a long period of time can cause beams, columns, and girders to produce stress on the cladding. Moisture on the curtain wall can cause movement because of swelling and drying. The designer must produce a product that will serve to shelter the interior of the building and remain structurally intact (Fig. 31.6).

Cladding panels must have expansion joints that permit some type of sliding overlap or other method of accommodating movement. Large glass panes must be glazed into frames that provide a watertight seal yet permit movement between the glass and the frame. Should the design prove inadequate, windows could break, cladding attachments could pull loose, panels could rupture, and parts could even fall off the building.

Air Infiltration

The control of air infiltration is closely related to the methods of controlling water penetration. Air leaks permit unconditioned air to enter and require energy to heat or cool the unwanted air. Water vapor can enter the cladding panel through air leaks and condense inside the panel. Air leaks also permit pollution to enter and, if extensive, permit the entrance of outside noise.

Water Penetration

A typical curtain wall is an assembly of different materials and has a number of joints between similar and dissimilar materials. The watertightness of the exterior cladding is critical to the success of the installation. The sealing material must bond to the surfaces of the joint, withstand temperature changes, rain, wind, and the stress due to expansion, contraction, and wind loads. All of these directly influence the watertightness of the cladding.

Since the cladding is directly exposed to the weather, it is subjected to wind-driven rain, snow, and hail. The upper levels of high-rise buildings are subject to greater possible penetration because of higher wind velocities at these levels. In some areas water penetration seals must withstand chemical attack by polluted air. Variations in temperatures also attack seals.

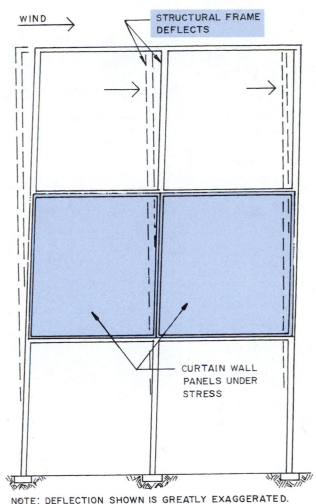

NOTE: DEFLECTION SHOWN IS GREATLY EXAGGERATED.

Figure 31.5 Earthquakes cause wracking and twisting of the structural frame. Wind causes movement in the frame. This puts stress on the curtain wall panels, windows, and connections, possibly leading to a failure.

In multistory buildings large amounts of wind-driven rain cascade down the face of the building. Provisions must be made to carry away this water and maintain the integrity of the water penetration seals. The major factor in controlling water penetration is the design of the joints and the proper installation of the seals. In some types of panels, such as aluminum-faced panels, provision is made for internal drainage of water that may have penetrated a joint. This consists of flashing, drainage channels, and weep holes.

Water penetration of cladding may be caused by wind-driven rain and other forces as shown in Fig. 31.7. *Gravity* works constantly to force water through improperly designed joints. Proper joint design can reduce this type of penetration. Wind-driven rain has sufficient **kinetic energy** to carry the water through openings in joints and wall surfaces. This can be reduced by covering the joints with some type of batten or internal baffle.

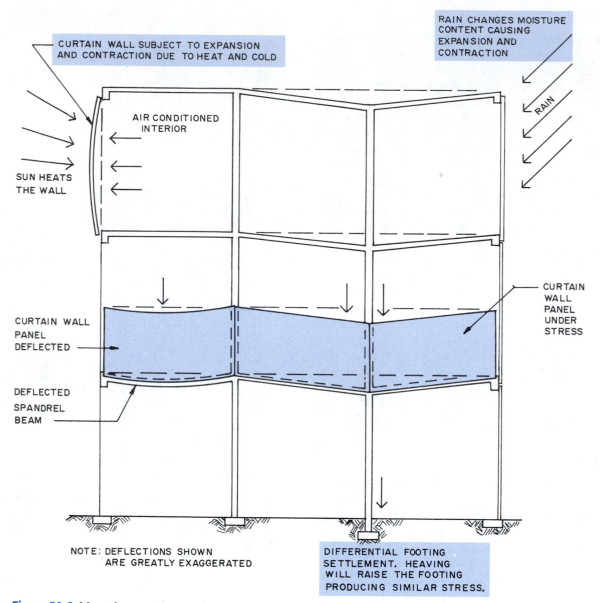

Figure 31.6 Many forces acting on the structural frame can put excessive stress on the curtain wall, possibly leading to a failure.

Water also can penetrate poorly designed joints due to the *surface tension* of the water, which allows it to cling to and flow along soffit areas. This can be prevented by adding a drip on the outer edge. Another force to consider when designing joints is **capillary action.** This can occur when the butting surfaces of a joint provide a very small opening. One way to control this is to have an air gap in the joint, which breaks the capillary path. These four forces can be easily controlled by proper joint design (Fig. 31.8).

Wind forces may also cause a difference in pressure on the outside face and inside the cladding panel. When the air pressure is greater on the outside than the inside, water may be forced through the joint into the panel

(Fig. 31.9). These penetrations can be controlled by the **rainscreen principle.**

The rainscreen principle is a pressure equalization method. The rainscreen is the exposed outer surface of the wall. It is backed by an air space. The joints are designed to prevent water from penetrating and are not sealed. This permits the air pressure on the face of the panel to equalize with the air pressure inside the panel.

The air space behind the rainscreen must be limited in size by subdividing it into relatively small subdivisions, and each area should have one opening to the exterior. A structural air barrier must be provided on the interior side of the panel. This is necessary to prevent leakage to the interior of the building. The use of the

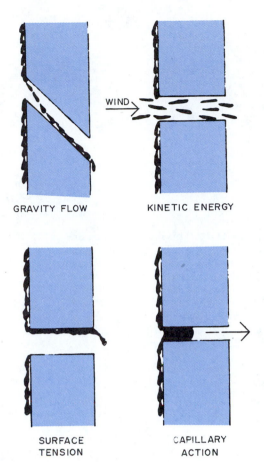

Figure 31.7 These forces can cause water penetration in joints and other openings in exterior walls.

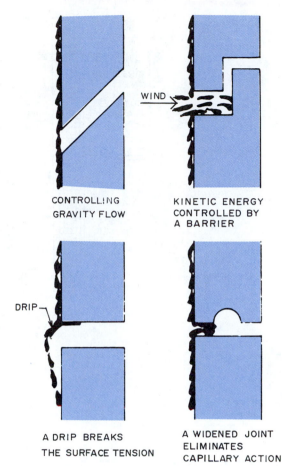

Figure 31.8 Techniques for controlling water penetration at joints. *(Courtesy American Architectural Manufacturers Association, from Curtain Wall Manual CW 1–9 The Rainscreen Principle)*

rainscreen principle with aluminum curtain wall panels is shown in Fig. 31.10 and with precast concrete curtain wall panels in Fig. 31.11.

Pressure equalization can also be used with glazing details. The units have a water-resistant seal, a pressure equalization chamber along the edge of the glass, and an inner vapor seal (Fig. 31.12). The rainscreen principle can also be used to resist water penetration by air pressure differences in traditional wood and masonry walls.

ASTM tests for water penetration and wind infiltration are made on curtain wall units with glazing sealants or gaskets installed. The actual conditions can vary during a storm and may, under certain circumstances, exceed those normally specified.

PERFORMANCE FACTORS

Cladding must be designed to last. It must control condensation, radiation and conduction, sound and pollution, and the amount of natural light entering the building. Cladding also must meet building code requirements for fire resistance.

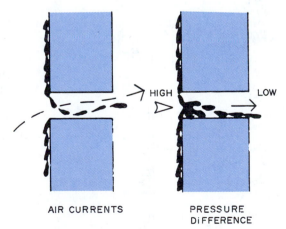

Figure 31.9 Wind currents can drive water through openings in the wall and cause a difference in the air pressure between the outside face and the air space inside the panel.

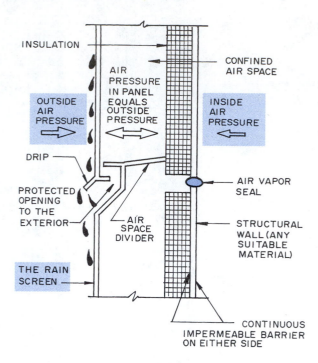

Figure 31.10 This illustrates the rain-screen principle. This aluminum curtain wall panel has divided air spaces that are vented with protected openings to the exterior. This equalizes the air pressure in the panel with the outside air. (*Courtesy American Architectural Manufacturers Association, from Curtain Wall Manual CW 1–9* The Rainscreen Principle)

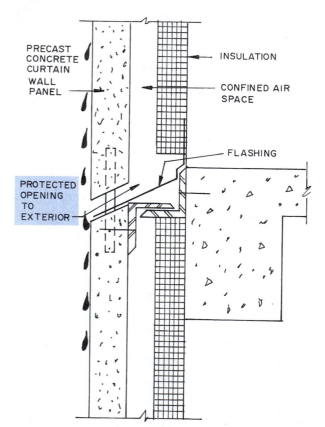

Figure 31.11 A generalized example of how the rain-screen principle can be used with precast concrete curtain wall construction. (*Courtesy American Architectural Manufacturers Association, from Curtain Wall Manual CW 1–9* The Rainscreen Principle)

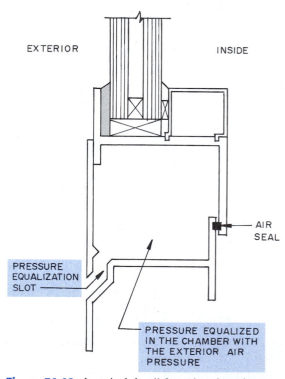

Figure 31.12 A typical detail for using the rain-screen principle with glazed units. (*Courtesy American Architectural Manufacturers Association, from Curtain Wall Manual CW 1–9* The Rainscreen Principle)

Durability and Maintenance

Cladding should last over the expected life of the building. Materials used should have a proven record of durability, including the connections, seals, and joints. If a reduced durability is accepted, then maintenance over the life of the building will be greater. The decision may involve choosing a more expensive but durable material that has higher initial cost but a lower maintenance cost or a less costly but less durable material, providing a building at lower initial cost but one that will have increasing maintenance costs over the years.

Generally, masonry coverings require little maintenance and tend to show weather streaking less than impervious skins. The exterior of a building can be routinely cleaned using vertical ladders or a scaffold or cradle hanging from the gantry arms on a trolley (Fig. 31.13). Some windows are designed so the outside can be cleaned by someone from inside the building. However, codes may prohibit operating windows. In many buildings the windows are all fixed and must be cleaned from the exterior.

Condensation Control

As is true with any type of wall construction, the cladding must have a vapor barrier on the inside wall facing to prevent water vapor from passing into the wall assembly. Water vapor will condense and run down the surface of masonry and concrete cladding. If it penetrates into a hollow assembly, such as an aluminum panel assembly, it will condense, reduce the value of the insulation, and cause damage due to freezing and thawing. Some form of drainage is provided to remove any condensation that may occur. Provision for the escape of moisture vapors to the exterior of the building can be provided. Interior surfaces of the panel should be insulated so they are warmer than the dew point of the air within the panel, thus reducing condensation.

Radiation and Conduction Control

The cladding must control the influx of radiant heat from the sun into the building that could cause physical discomfort and large air-conditioning costs. Radiant heat is actually radiant energy and not heat. Heat is induced when the radiant energy strikes something. The energy moves from a hot body to a cold body. For example, a person sitting near a cold wall radiates heat to the wall, making the person feel cold. Radiant energy from the sun passes through windows and heats up people and materials it strikes inside the building.

Heat is also transferred by thermal conduction. The cladding must prevent the conduction of heat out of and into the building through the wall. Thermal conduction is the process of heat transfer through a material. The cladding must use insulation and thermal breaks in ma-

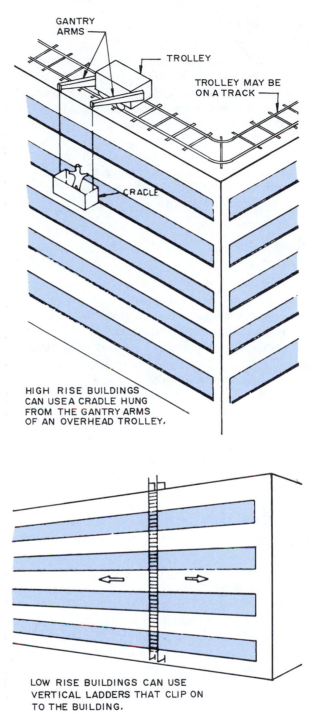

Figure 31.13 Two methods for performing maintenance on the exterior of multistory buildings.

terials that are good thermal conductors. A **thermal break** is an insulating material placed between thermal conducting materials that touch, retarding the passage of heat or cold. The use of multiple glazing and energy-efficient glass also offers radiant controls as discussed in Chapter 29.

Sound and Pollution Control

The importance of controlling the influx of exterior sounds and air pollution into the building varies with the type of occupancy. For example, a hospital would require greater control than many other occupancies. Likewise, in some cases it is necessary to keep the noise or pollution inside the building, such as with some types of manufacturing or chemical processing operations. The cladding system would have to be airtight and well insulated and, in many cases, have sound-deadening material as part of the assembly. Generally, glazed wall areas admit more sound than spandrel panels.

Natural Light

Natural light is used to provide some illumination and relieve the closed-in feeling. In actual practice, artificial lighting is on all the time regardless of the amount of natural light available. Sunlight must be controlled to prevent distracting glare and unwanted radiant energy (heat). Although some control can be had using various types of solar screening and energy-efficient glass, the amount of glazing used is a major factor.

Windows are a part of the exterior cladding systems. They are exposed to the same factors as the actual panels, such as wind load, thermal movement, air infiltration, and water penetration. Their design must enable them to meet the structural requirements as well as have adequate fastening systems and durability.

Fire Resistance

Cladding is subject to the requirements of the building code. Of great importance is its ability to meet the fire resistance requirements. The fire resistance ratings of cladding are determined by tests of the assembly and framing following ASTM test procedures. These involve the combustibility of materials in the cladding, the fire resistance ratings of the assembled panels and spandrels, and the nearness of other buildings. The required fire resistance of non-load-bearing cladding varies with the occupancy of the building and the fire separation distance. The *fire separation distance* is the distance between buildings. It is used to reduce the risk of fire spreading from one building to another. The distances specified in the code typically might be measured in feet between the building and the closest lot line, from the centerline of the street, or from a halfway distance between two buildings on the same site.

The *connectors* should hold the cladding in place during a fire for at least the time specified for the fire resistance rating of the assembly. Exterior trim and other wall finish materials are also subject to established fire resistance ratings. Codes also contain requirements for the fire protection of openings in the wall.

Another requirement is to install fire-stopping. *Fire-stopping* uses approved materials to prevent the movement of flame and gases through small vertical and horizontal concealed openings in building components (Fig. 31.14). Materials commonly approved for use as fire-stops include masonry set in mortar, concrete, mortar or plaster on metal lath, plasterboard, sheet metal of approved gauge, asbestos-cement board, mineral, slag, or rock wool, if solidly compacted into a confined space. These materials must be permanently fastened in place. They are commonly referred to as **safing.**

CURTAIN WALL CLADDING

A curtain wall is defined as a non-load-bearing exterior building wall supported by the building framework. A metal curtain wall is an exterior non-load-bearing building wall consisting of metal, glass, or other surfacing materials supported by a metal framework. A window wall

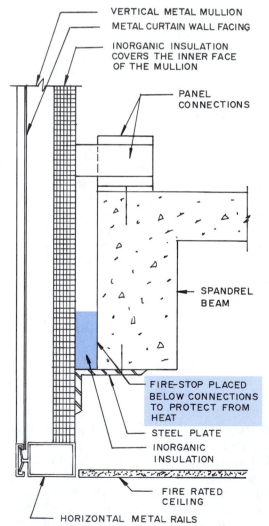

Figure 31.14 One example of fire-stopping a curtain wall at the spandrel beam.

is a type of metal curtain wall composed of metal framing members containing operable sash, fixed lights, ventilators, or opaque glass panels. All of these systems are the result of years of engineering development and testing. Various manufacturers supply completely designed systems engineered for a wide range of applications (Fig. 31.15).

A curtain wall system typically includes panels that form the exposed exterior walls, glazing, mullions, connections, gaskets and sealants. Curtain walls are lightweight and non–load–bearing, are anchored to columns, spandrel beams, and floors, and are generally supported at their bottom edge. Therefore, the materials making up the panel are not under tension. The facing panels are made from metal, glass, modified stucco, molded glass fiber–reinforced polyester or other code approved materials (Fig. 31.1). Since the curtain wall panels are non–load bearing, they can be rather thin and lightweight, thus reducing the loads on the foundation. This can be a significant saving in weight and construction costs on high-rise buildings. Curtain walls range from a simple single-thickness material, such as metal siding, to a multilayer sandwich panel containing insulation. The choice depends on the desired appearance and the requirements for the interior environment. Following are general examples of details for commonly used curtain walls.

Masonry Curtain Walls

Masonry curtain walls that cover the structural frame are usually laid on a steel shelf angle secured to the frame. The shelf may be anchored to the concrete span-

drel beam or bolted or welded to a structural steel spandrel beam (Figs. 31.16 and 31.17). The wall may be laid up a brick at a time as is done with low-rise masonry construction. Provisions are made for horizontal and vertical expansion joints to permit the panel to expand and contract as temperatures vary. Masonry curtain wall panels can be prefabricated on the ground and the assembly lifted into place with a crane. High-bond mortars are used because they have excellent bonding characteristics and high compressive and tensile strength.

The masonry curtain wall has a backup wall, typically concrete masonry units or steel studs (Fig. 31.18). It is non–load-bearing and supports sheathing, insulation, air and vapor barriers, and interior finish materials. The curtain wall is tied to the backup wall with ties. The ties are stiff, but they can flex enough to accommodate differential movements between the veneer and the backup wall.

Figure 31.15 This building is clad with several types of curtain walls. On the left is an all-glass wall with exposed mullions and rails. On the right it has ribbon windows with a solid material forming the spandrel below. *(Courtesy H.H. Robertson)*

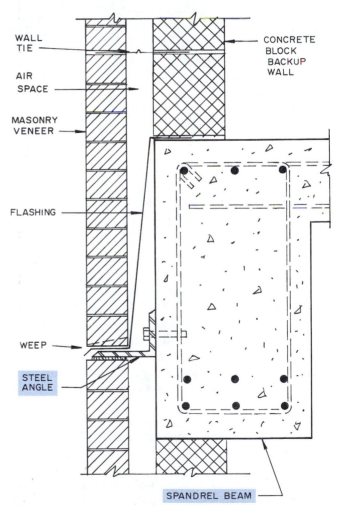

Figure 31.16 A typical detail for a brick masonry curtain wall placed over a concrete structural system. The brick is mounted on a steel angle and tied to a masonry backup wall.

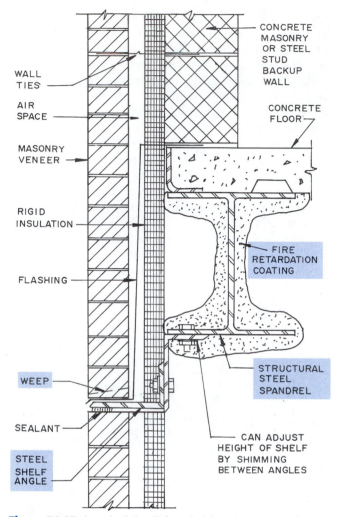

WALL TIES
AIR SPACE
MASONRY VENEER
RIGID INSULATION
FLASHING
WEEP
SEALANT
STEEL SHELF ANGLE

CONCRETE MASONRY OR STEEL STUD BACKUP WALL
CONCRETE FLOOR
FIRE RETARDATION COATING
STRUCTURAL STEEL SPANDREL
CAN ADJUST HEIGHT OF SHELF BY SHIMMING BETWEEN ANGLES

Figure 31.17 A typical detail for a brick masonry curtain wall placed over a steel frame building. The brick is mounted on a steel shelf and is tied to a backup wall.

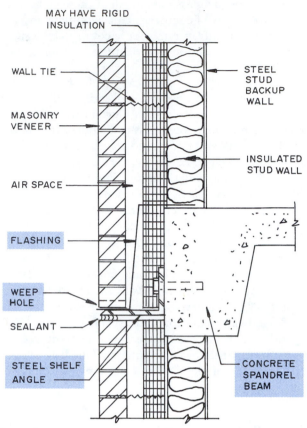

MAY HAVE RIGID INSULATION
WALL TIE
MASONRY VENEER
AIR SPACE
FLASHING
WEEP HOLE
SEALANT
STEEL SHELF ANGLE

STEEL STUD BACKUP WALL
INSULATED STUD WALL
CONCRETE SPANDREL BEAM

Figure 31.18 A typical detail for a brick masonry curtain wall placed over a structural concrete frame building. It is tied to a steel stud backup wall. The weep hole permits drainage of moisture and equalizes the air pressure (the rainscreen principle) on both sides of the masonry.

Most single-width masonry walls will experience some penetration by driving rain. This moisture is handled by the rainscreen principle, which provides a cavity behind the masonry and weep holes to drain the moisture. The cavity acts as a pressure equalization chamber and provides some degree of pressure equalization between the exterior and interior surfaces of the masonry.

Some architectural designs permit the structural system to be visible on the finished building. The masonry curtain wall is laid on top of the spandrel beam as shown in Fig. 31.19.

A typical detail showing a ribbon window installation on top of a masonry curtain wall is in Fig. 31.20. The aluminum frame has a sill extending into the room and a slot to hold the top of the interior wall finish material.

Masonry curtain walls can also be made using prefabricated panels that are hoisted into place. Two methods are commonly used. One uses a plastic gasket form liner inside a form into which the bricks are placed (Fig. 31.21, photo 1). This spaces the bricks and holds them in place. Reinforcement is placed over the bricks (Fig. 31.21, photo 2). A concrete panel is poured over the assembly (Fig. 31.21, photo 3). When cured, the panel is lifted and the liner is removed.

Another method uses steel studs faced with sheathing such as plywood or exterior gypsum wallboard (Fig. 31.22). The thin bricks are factory applied to Styrofoam panels that are secured to the studs with metal clips. The assembled prefabricated panels are lifted into place and secured to the structural frame (Fig. 31.23). Clay masonry products are discussed in Chapter 11 and mortar in Chapter 10.

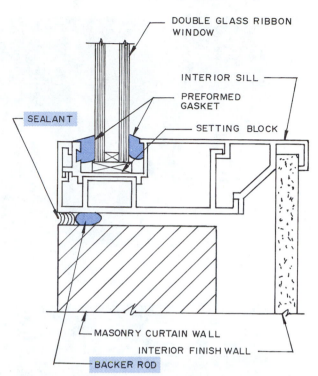

Figure 31.20 A typical installation of a metal window on a masonry curtain wall panel.

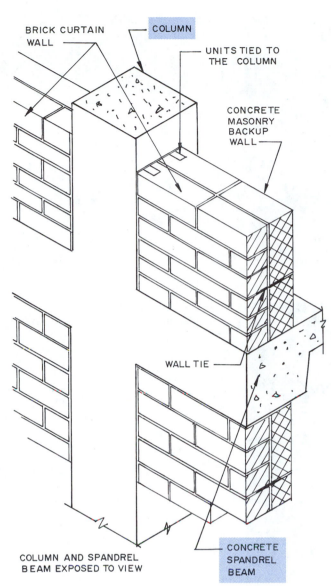

Figure 31.19 Masonry curtain walls can be supported on the spandrel beam, leaving the beam and column exposed to form part of the architectural design of the exterior.

Stone Curtain Walls

Stone is a widely used material for facing low-rise, mid-rise, and high-rise buildings. This facing material is secured to the structural frame of the building. Commonly used stones include granite, slate, marble, and limestone. The material is cut into panels of various thicknesses ranging generally from 1¼ in. to 4 in. (32 to 102 mm), depending on the material and the size of the panel.

The architect must consider the specifications related to the physical characteristics of the material. For example, the characteristics for granite relating to absorption by weight, density of material, compressive strength, and

the modulus of rupture are established by ASTM standards. The stone panels are available in a variety of finishes ranging from smooth, highly polished to sandblasted rough surfaces. Some manufacturers produce panels with the rough sawed surface exposed. Joints between panels are typically ¼ in. (6 mm), but this can vary with the product. Stone products are discussed in Chapter 13.

Installation details are recommended by the manufacturer of the panels. Manufacturers often supply or recommend engineering approved connections. A typical way to support stone panels is with some type of metal subframe (Fig. 31.24). They can be mounted to the structural concrete frame or to steel or masonry backup walls. A number of connection systems are shown in Figs. 31.25 through 31.27. One method of cladding a soffit, such as may occur at a window opening, is shown in Fig. 31.28 and typical parapet construction is shown in Fig. 31.29.

Preassembled stone panels are made by mounting stone panels on a steel frame. The assembly is lifted into place and secured to the structural frame of the building. Other types join stone panels to a reinforced concrete backing with steel anchors. Preassembled panels reduce the time required for plumbing and leveling the facing because several panels are installed in one operation. A building with stone veneer cladding is shown in Fig. 31.30.

1. Place the thin brick units in the form.

2. Place the required reinforcing.

3. Pour a concrete backing over the brick units.

Figure 31.21 The steps to cast a prefabricated brick curtain wall panel. *(Courtesy The Scott System, Inc.)*

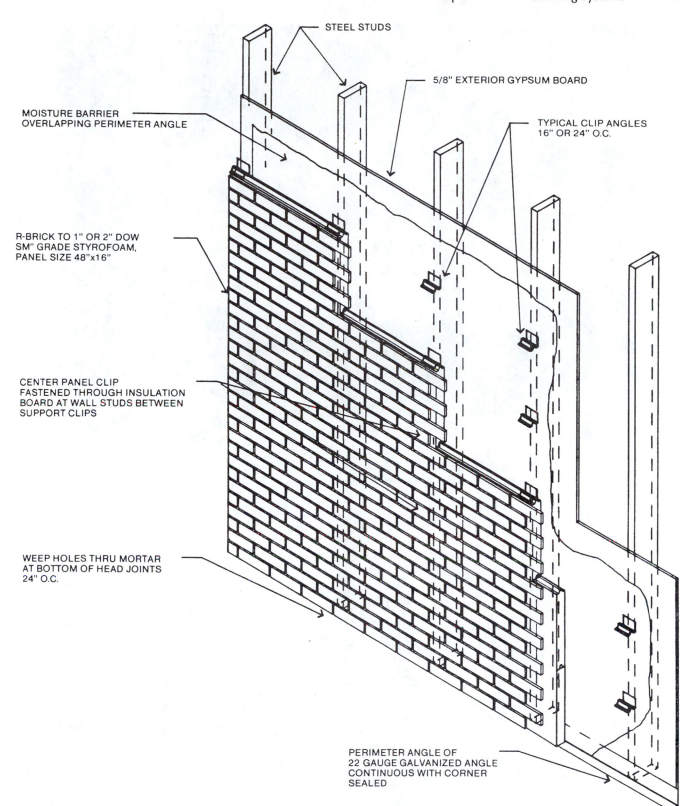

STEEL STUDS

5/8" EXTERIOR GYPSUM BOARD

MOISTURE BARRIER
OVERLAPPING PERIMETER ANGLE

TYPICAL CLIP ANGLES
16" OR 24" O.C.

R-BRICK TO 1" OR 2" DOW
SM" GRADE STYROFOAM,
PANEL SIZE 48"x16"

CENTER PANEL CLIP
FASTENED THROUGH INSULATION
BOARD AT WALL STUDS BETWEEN
SUPPORT CLIPS

WEEP HOLES THRU MORTAR
AT BOTTOM OF HEAD JOINTS
24" O.C.

PERIMETER ANGLE OF
22 GAUGE GALVANIZED ANGLE
CONTINUOUS WITH CORNER
SEALED

Figure 31.22 This prefabricated brick masonry curtain wall panel is assembled over metal studs and an approved sheathing. *(Courtesy American Brick Company)*

Figure 31.23 A preassembled curtain wall panel is hoisted into place. *(Courtesy American Brick Company)*

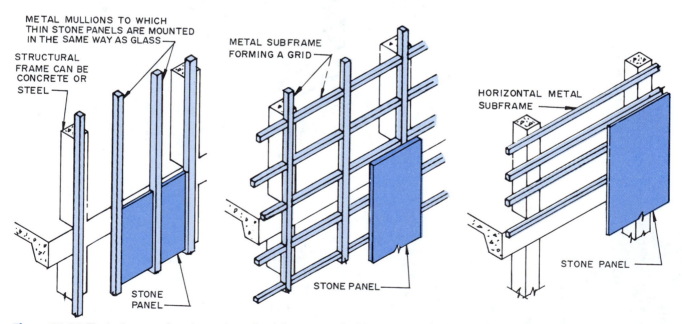

Figure 31.24 Typical types of steel curtain wall subframes used with stone panels. The subframes may be secured to steel or concrete structural systems.

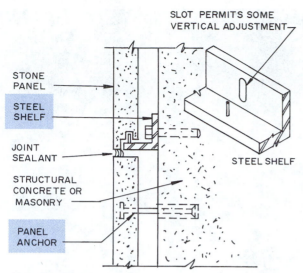

Figure 31.25 The stone curtain wall panel is mounted on a metal angle seat and tied to the masonry backup wall.

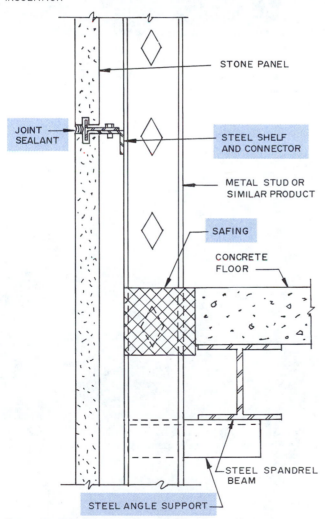

Figure 31.26 A stone curtain wall panel mounted over a steel structural system and steel studs.

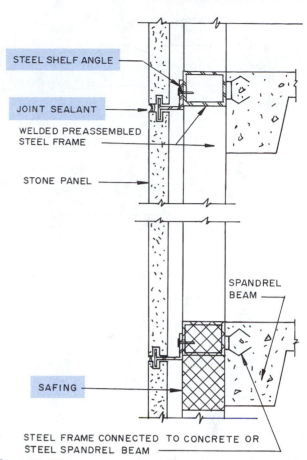

Figure 31.27 A stone curtain wall panel mounted on a tubular steel frame that is connected to the structural system of the building.

Precast Concrete Curtain Walls

Precast concrete curtain wall panels are available with a variety of surface finishes and thicknesses (Fig. 31.31). The surface finish may be produced by the surface texture of the casting form or after casting and prior to hardening by methods such as brooming, stippling, floating, troweling, or exposing aggregate. The surface may be recessed or panelized. The design of the reinforced precast concrete curtain wall panels can vary depending on the exterior appearance specified by the architect and the structural design of the engineer. Several typical panels are in Fig. 31.32. They may be made using conventional reinforcing or as prestressed units. The use of glass fiber reinforcing along with steel reinforcement enables lighter and thinner panels to be produced than those using only steel reinforcement. The glass fibers reduce the need for the secondary steel reinforcing used to resist cracking and to help control thermal expansion and contraction.

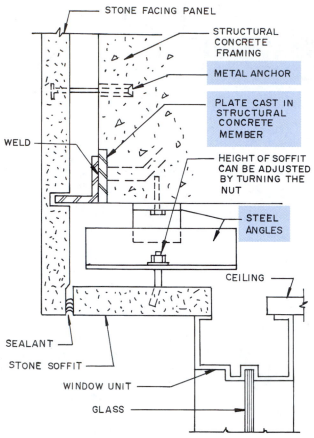

STONE FACING PANEL

STRUCTURAL CONCRETE FRAMING

METAL ANCHOR

PLATE CAST IN STRUCTURAL CONCRETE MEMBER

HEIGHT OF SOFFIT CAN BE ADJUSTED BY TURNING THE NUT

WELD

STEEL ANGLES

CEILING

SEALANT

STONE SOFFIT

WINDOW UNIT

GLASS

Figure 31.28 A typical stone soffit detail as used at window returns.

Figure 31.30 This high-rise building is clad with stone spandrels. The roto-vent window strip has triple glazing and venetian blinds between two layers of glass. It allows for ventilation and is excellent for high security. *(Courtesy EFCO Corporation)*

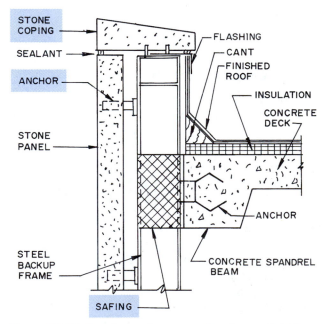

STONE COPING

SEALANT

ANCHOR

STONE PANEL

STEEL BACKUP FRAME

SAFING

FLASHING

CANT

FINISHED ROOF

INSULATION

CONCRETE DECK

ANCHOR

CONCRETE SPANDREL BEAM

Figure 31.29 One design for a stone-faced parapet with a stone coping.

Figure 31.31 This building has a standard ribbon curtain wall with precast concrete spandrel panels. *(Courtesy Tube-lite, Inc.)*

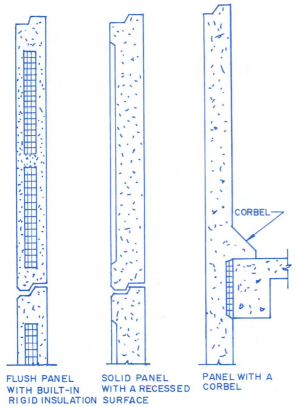

FLUSH PANEL WITH BUILT-IN RIGID INSULATION | SOLID PANEL WITH A RECESSED SURFACE | PANEL WITH A CORBEL

Figure 31.32 Some of the types of precast concrete curtain wall panels available.

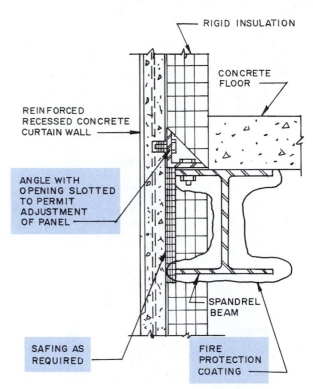

RIGID INSULATION

CONCRETE FLOOR

REINFORCED RECESSED CONCRETE CURTAIN WALL

ANGLE WITH OPENING SLOTTED TO PERMIT ADJUSTMENT OF PANEL

SPANDREL BEAM

SAFING AS REQUIRED

FIRE PROTECTION COATING

Figure 31.33 A typical precast concrete curtain wall panel over a steel structure frame.

Precast panels are installed basically the same as stone panels. Typically, anchors are cast into the panel and angles are bolted to the panel and the structural frame. The panels can be insulated by bonding rigid insulation to the inside surface of the panel or by sandwiching rigid insulation in the center of the panel. Allowances must be made between panels for expansion and contraction. An adequate sealant is required to provide a waterproof wall. Typical details are shown in Figs. 31.33 and 31.34. They are installed over steel and structural concrete frames. Typical parapet and window opening details are shown in Fig. 31.35.

Modified Stucco Curtain Walls

Modified stucco curtain wall panels are an assembly of materials providing insulation and a weather-tight rainscreen. The system is referred to as an exterior insulation and finish system (EIFS). An installation designed for use on residential buildings is shown in Fig. 31.36. Plywood or OSB sheathing is nailed to the wood studs. An airtight, water-resistant material is troweled over the substrate. Expanded polystyrene (EPS) insulation board

is secured to the substrate with special mechanical fasteners. A base coat is troweled over the insulation board; then a reinforcing mesh is embedded into it. The synthetic plaster finish coat is troweled over the base coat (Fig. 31.37).

When EIFS is used over metal studs in commercial building construction, the assembly appears as in Fig. 31.38. The substrate in this design is Dens-Glas® Gold. It has inorganic glass mat facings, a water-resistant, silicone-treated gypsum core, and an alkali-resistant surface coating. It is secured to the metal studs with approved self-tapping metal screws. All joints are reinforced with a mesh material. The rest of the assembly is the same as that described for the residential system. The manufacturer of EIFS should be consulted for specifications.

Curtain wall panels one or more stories high are an assembly of steel studs covered with the EIFS materials. The assembled panels are lifted into place and bolted or welded to connectors on the structual frame of the building as described for other types of curtain wall construction (Fig. 31.39). Additional details for commerical EIFS curtain wall construction are shown in Figs. 31.40 and 31.41. A finished commercial building is shown in Fig. 31.42.

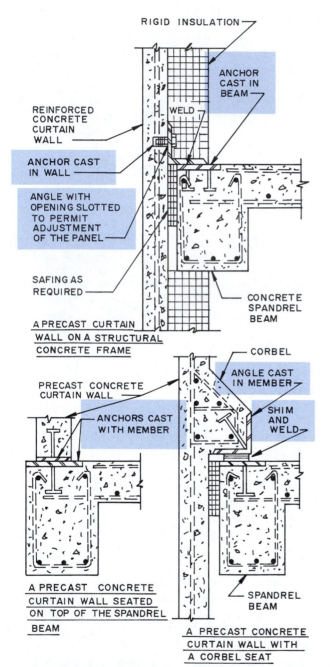

RIGID INSULATION

ANCHOR CAST IN BEAM

WELD

REINFORCED CONCRETE CURTAIN WALL

ANCHOR CAST IN WALL

ANGLE WITH OPENING SLOTTED TO PERMIT ADJUSTMENT OF THE PANEL

SAFING AS REQUIRED

CONCRETE SPANDREL BEAM

A PRECAST CURTAIN WALL ON A STRUCTURAL CONCRETE FRAME

CORBEL

ANGLE CAST IN MEMBER

PRECAST CONCRETE CURTAIN WALL

SHIM AND WELD

ANCHORS CAST WITH MEMBER

A PRECAST CONCRETE CURTAIN WALL SEATED ON TOP OF THE SPANDREL BEAM

SPANDREL BEAM

A PRECAST CONCRETE CURTAIN WALL WITH A CORBEL SEAT

Figure 31.34 Several ways precast concrete curtain walls are secured to concrete spandrel beams.

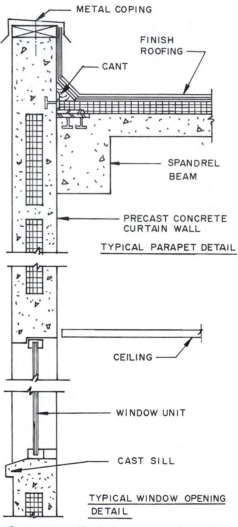

METAL COPING

FINISH ROOFING

CANT

SPANDREL BEAM

PRECAST CONCRETE CURTAIN WALL

TYPICAL PARAPET DETAIL

CEILING

WINDOW UNIT

CAST SILL

TYPICAL WINDOW OPENING DETAIL

Figure 31.35 Typical parapet and window opening details for precast concrete curtain walls.

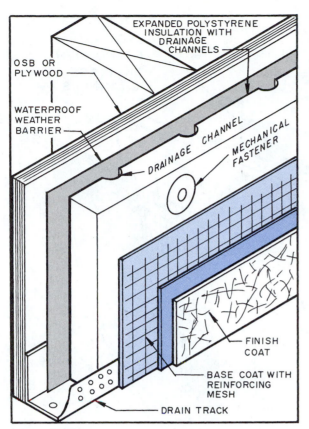

Figure 31.36 This EIFS is designed for use on residential and other wood framed buildings. It illustrates the Dryvit® Residential MD (moisture drainage) System. *(Copyrighted and used with the permission of Dryvit Systems, Inc.)*

Figure 31.37 The finish coat of the specified color is applied over the base coat. *(Courtesy Dryvit® Systems, Inc.)*

A number of similar panel systems are manufactured for lightweight exterior wall cladding. For example, one has a composite panel similar to the one just described, but it has a hard polymer exterior finish. Another provides a stucco, exposed aggregate, or simulated brick or stone exterior surface.

Metal and Glass Curtain Wall Systems

Glass and metal curtain walls are widely used because they provide a lightweight exterior wall and are rapidly erected (Fig. 31.43). Various types of glass panels are used. The metal framing is usually aluminum. The glass panels are typically insulating glass, laminated glass, heat-reflective glass, and spandrel glass. Spandrel glass is a heat-strengthened glass with a ceramic frit fired on it. Frit is a molten glossy material. It is recommended that rigid insulation or some other rigid backing material be bonded to the back of the glass spandrel panel. This helps

hold the glass in place should a panel fracture due to high internal temperatures or storm damage.

Metal and glass curtain wall systems are available in two basic types, custom and standard. Custom walls are those designed especially for a particular building and built by the manufacturer for that application. Standard walls are those that are standardized by the manufacturer and can be assembled from stock parts.

Methods of Installation

Custom and standard curtain wall systems can be classified by the method of installation. The following classifications are detailed in *Aluminum Curtain Wall Design Guide Manual* published by the American Architectural Manufacturers Association.

In the *stick system* mullions are installed followed by horizontal rails into which glazing and spandrel panels are placed (Fig. 31.44). Figure 31.45 shows a typical aluminum frame stick system with vision glazing and ceramic tile–faced spandrel panels.

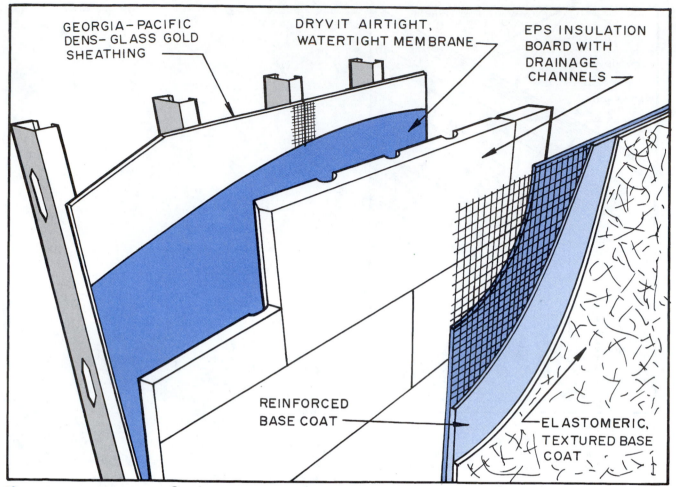

GEORGIA-PACIFIC DENS-GLASS GOLD SHEATHING

DRYVIT AIRTIGHT, WATERTIGHT MEMBRANE

EPS INSULATION BOARD WITH DRAINAGE CHANNELS

REINFORCED BASE COAT

ELASTOMERIC, TEXTURED BASE COAT

Figure 31.38 This is the Dryvit® Infinity EIFS. It provides a water-tight membrane and an insulation board with drainage channels to capture, control, and discharge any incident moisture that may enter the system. This system is typically used on commercial buildings. *(Copyrighted and used with the permission of Dryvit Systems, Inc.)*

Figure 31.39 Preassembled modified stucco curtain wall panels can be several stories high. *(Courtesy Dryvit® Systems, Inc.)*

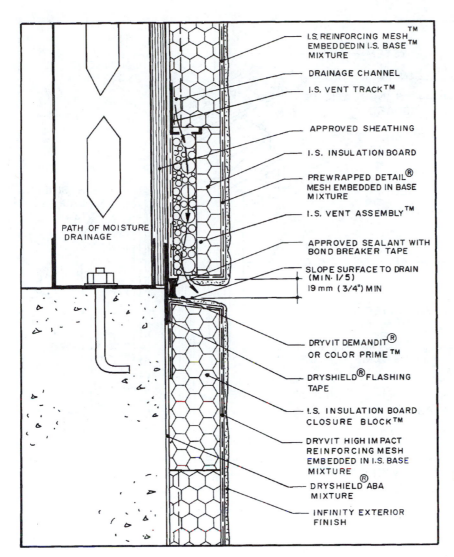

Figure 31.40 This detail shows the recommended construction of horizontal joints for the Dryvit® Infinity EIFS applied to curtain wall construction. *(Copyrighted and used with the permission of Dryvit Systems, Inc.)*

The *unit system* uses large preassembled wall panels. The vertical edges of the units are joined, serving as a mullion, and the bottom of one panel joins the top of the one below, forming a horizontal rail (Fig. 31.46).

The *unit-and-mullion* system uses mullions secured to the building structure, and preassembled wall panels are installed between the mullions (Fig. 31.47).

The *panel system* uses homogenous wall panels that can be precast concrete, some form of molded plastic, or stamped from sheet metal. They may or may not have openings for windows. They are connected to the structural frame as shown in Fig. 31.48. A typical building is shown in Fig. 31.49.

The *column-cover-and-spandrel system* uses long spandrel panels that span the column covers. Glazing may be installed above the spandrel and the columns are enclosed with a hollow cover (Fig. 31.50). A building with a column-cover-and-spandrel system can be seen in Fig. 31.51.

Many other system designs are being developed, including some with little or no exposed framing.

Design Factors

There are many factors to consider when designing glass curtain walls and glass spandrels. For example, when glass is exposed to sunlight, the glass temperatures rise. If the glass panels are glazed directly to a material that easily absorbs heat, such as concrete, the edge of the glass is cooler than the center of the panel. The hotter center expands more than the edges, causing increased stresses at the edges. This may cause the panel to fracture, and proper glazing is required to reduce this

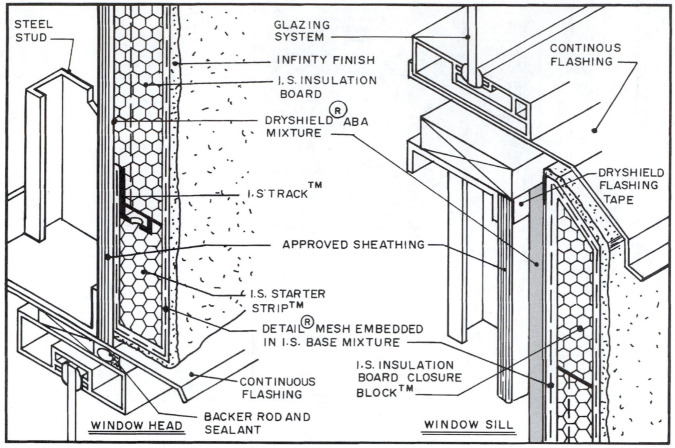

Figure 31.41 Installation details for the Dryvit® Infinity EIFS cladding at the head and sill of glazing units. *(Copyrighted and used with permission of Dryvit Systems, Inc.)*

danger. Interior heat traps, as mentioned for spandrel panels, can cause temperature differences in the panel, leading to fracture.

The design must allow for air circulation. This is also a consideration when interior shades are used. They must be hung several inches from the glass curtain wall to permit air to circulate between the shade and the glass (Fig. 31.52). Exterior sun shading of glass curtain walls also can produce uneven temperatures in the panel because often only a part of the glass is shaded, producing differences in temperature. If most of the panel is shaded, the stresses are minimal (Fig. 31.53).

The large glass panels used in curtain wall construction are heavy. They must have the strength to span a distance equal to the width of the panel and to resist wind loads. They must be isolated from the frame so expansion, contraction, and possible bending from wind loads will not fracture the glass. The curtain wall system is designed so the frame provides the amount of grip around the edge of the glass to hold it in place when under stress. The frame and mullions are designed to withstand the combined stresses of weight and wind loads.

Glass manufacturers frequently specify the minimum compressive pressure of the gasket on the glass surface. Typically, a pressure of at least 4 lb. per linear in. (700 N/m) is needed to provide support for the glass edge and a watertight seal. Excessive pressure, usually more than 10 lb. per linear in. (1750 N/m) can increase mechanical stress and may contribute to glass breakage.

Figure 31.42 A building clad with modified stucco curtain wall panels. *(Courtesy Dryvit® Systems, Inc.)*

Figure 31.43 A glass-to-the-front, stick fabrication framing system has the advantage of fast field erection and significant savings. Used effectively in ribbon-type applications, this multi-purpose framing system may be inside or outside glazed and has an option for structural silicone glazing. It accepts infills of ¼ in. or 1 in. with a minimum front mullion projection. *(Courtesy Kawneer Company, Inc.)*

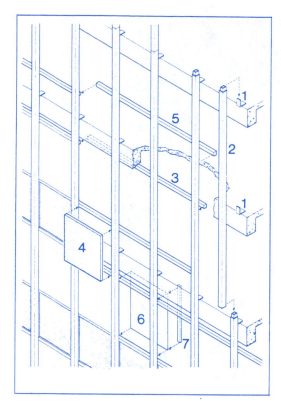

Figure 31.44 The stick curtain wall system:
1. anchors, 2. mullion, 3. horizontal rail, 4. spandrel panel, 5. horizontal rail, 6. vision glass, 7. interior mullion trim. *(Courtesy American Architectural Manufacturers Association, from* Aluminum Curtain Wall Design Guide Manual)

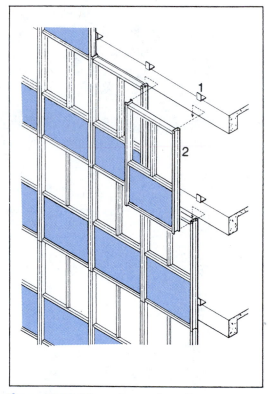

Figure 31.46 The unit curtain wall system:
1. anchor, 2. preassembled framed curtain wall unit. *(Courtesy American Architectural Manufacturers Association, from* Aluminum Curtain Wall Design Guide Manual)

Figure 31.45 A typical stick system curtain wall with vision glazing and spandrels faced with ceramic tile. *(Courtesy United States Ceramic Tile Company)*

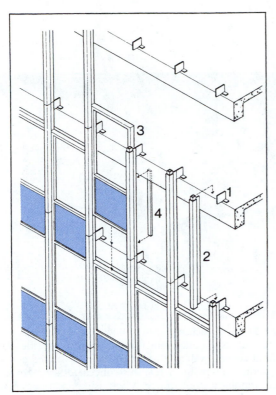

Figure 31.47 The unit-and-mullion curtain wall system: 1. anchors, 2. mullion, 3. preassembled curtain wall unit lowered into place behind the mullion from the floor above, 4. interior mullion trim. *(Courtesy American Architectural Manufacturers Association, from* Aluminum Curtain Wall Design Guide Manual)

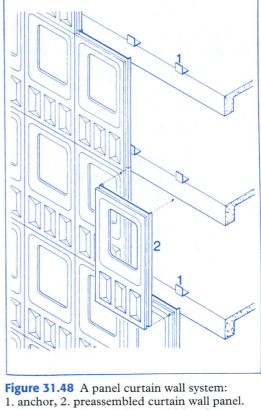

Figure 31.48 A panel curtain wall system: 1. anchor, 2. preassembled curtain wall panel. *(Courtesy American Architectural Manufacturers Association, from* Aluminum Curtain Wall Design Guide Manual)

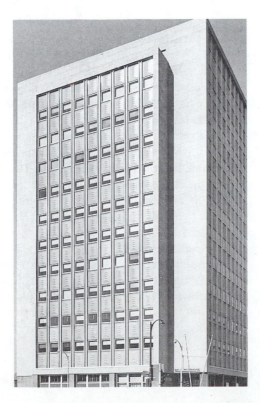

Figure 31.49 This curtain wall system used precast concrete wall panels with windows placed between the mullions. These are custom-made panels that provide character to the appearance of the building.

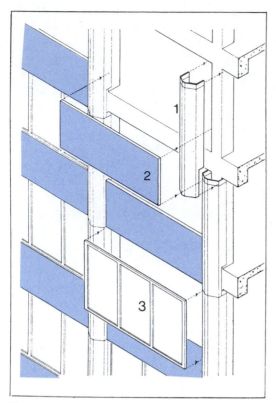

Figure 31.50 The column-cover-and-spandrel curtain wall system: 1. column cover, 2. spandrel panel, 3. glazing unit. *(Courtesy American Architectural Manufacturers Association, from* Aluminum Curtain Wall Design Guide Manual*)*

Figure 31.51 Pressure glazed hollow mullion curtain wall for low- to mid-rise applications is an integrated, outside-glazed captured curtain wall and structural silicone glazed curtain wall product. The joinery features concealed fasteners for a clean, monolithic appearance. When using open-back horizontal mullions, the fillers snap at the edge, producing an uninterrupted seam line. *(Courtesy Kawneer Company, Inc.)*

Design Examples

Construction details for a glass curtain wall using structural silicone sealant to attach the panels to metal mullions that are hidden behind the glass are shown in Fig. 31.54. This design clads the entire wall with glass curtain wall panels that admit some natural light and permit those inside to see out through the panel. They are referred to as vision-type glass ribbon windows. The glass may be any of several types, such as reflecting or heat resisting. A building with a total glass curtain wall is in Fig. 31.55. In Fig. 31.56 a similar construction is used, but opaque glass spandrel panels are used to cover interior structural members and other installations, such as ceiling panels, heating, plumbing, and electrical runs.

Neopariés are another type of glass curtain wall. Neopariés are unique glass ceramic products used for interior and exterior wall cladding. They are harder and lighter than natural stone, will not absorb water, and are available in flat and curved panels. They are secured to

the structural frame by connectors shown in Figs. 31.2 and 31.3. Flat panel installation is shown in Fig. 31.57 and curved panels in Fig. 31.58.

Possibly the most commonly used metal curtain wall has exposed mullions and rails. The panels held by these metal frames are shown in Fig. 31.59. This is called the stick system. The mullions are often larger than the rails, but this varies considerably. This system is shipped disassembled, and mullions are usually installed first followed by the horizontal rails. Then the glazing and spandrel panels are installed. This system requires more on-site time to install than other factory prepared panel systems (Fig. 31.60).

Spandrel panels may be materials other than glass. Manufactured spandrel panels with aluminum, stainless steel, stone aggregate set in a plastic matrix, ceramic tile panels, and fiberglass reinforced plastic sheet materials are available. Aluminum spandrel panels are most commonly used. They may be left natural or finished with an

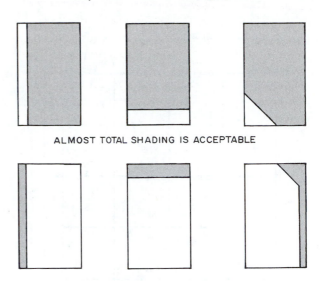

ALMOST TOTAL SHADING IS ACCEPTABLE

VERY SMALL SHADED AREAS ARE HARMFUL

Figure 31.53 The amount of shading of exterior glass influences the degree of stress in the panel.

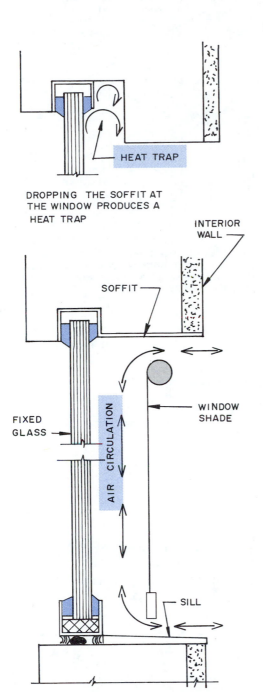

DROPPING THE SOFFIT AT THE WINDOW PRODUCES A HEAT TRAP

HEAT TRAP

INTERIOR WALL

SOFFIT

FIXED GLASS

AIR CIRCULATION

WINDOW SHADE

SILL

Figure 31.52 Provision must be made to allow for air circulation to prevent a heat trap from developing between the shade and the glass.

organic coating (enamel, lacquer), vinyl or other plastic coatings, clear and color anodic finishes, and porcelain enamel. These spandrel panels are installed in aluminum frames much the same as described for glass spandrel panels. Aluminum panels are also used to cover the entire wall when windows are not wanted. The construction of the panels varies with the manufacturer. A typical metal curtain wall panel is illustrated in Fig. 31.61.

A number of manufacturers provide wall systems for sloped glazing. The sloped glazing serves as the roof and is integrated with the vertical curtain wall. It produces a naturally lighted area and a dramatic setting (Fig. 31.62).

The building in Fig. 31.63 illustrates the freedom of architectural design a curtain wall system can provide. Some of the structural system is left exposed, and other parts are covered by the glass curtain wall. The exposed columns and spandrel beams create an interesting architectural feature. A close-up showing the materials covering the exposed structure is in Fig. 31.64.

Curtain Wall Joint Sealants

The choice of joint sealant depends on the materials to be sealed, the design of the joints, and possible changes that may occur in the joints after the sealant has been installed. Joints may be *working joints*, which are designed to allow movement, or *nonworking joints* that are joined by a fastener so they do not move (Fig. 31.65). Sealants may be flowable compounds or solid materials.

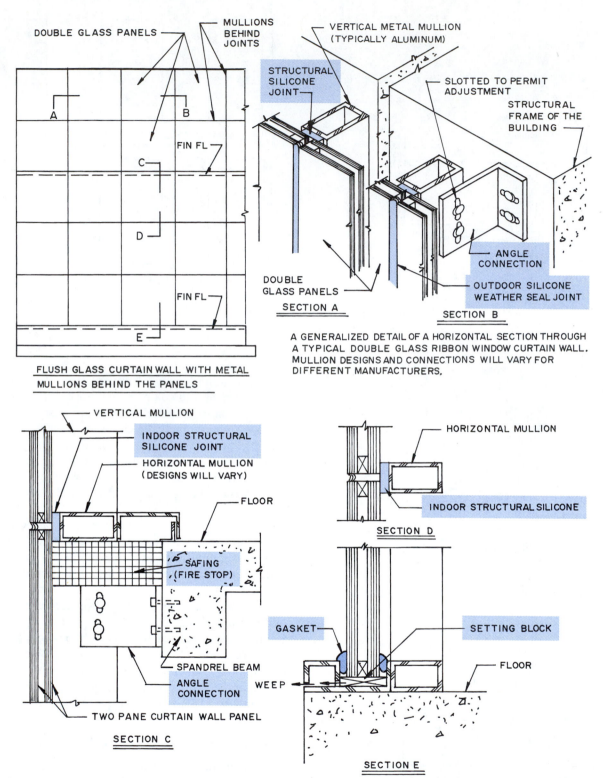

DOUBLE GLASS PANELS

MULLIONS BEHIND JOINTS

A

B

FIN FL

C

D

FIN FL

E

FLUSH GLASS CURTAIN WALL WITH METAL MULLIONS BEHIND THE PANELS

VERTICAL METAL MULLION (TYPICALLY ALUMINUM)

STRUCTURAL SILICONE JOINT

SLOTTED TO PERMIT ADJUSTMENT

STRUCTURAL FRAME OF THE BUILDING

ANGLE CONNECTION

OUTDOOR SILICONE WEATHER SEAL JOINT

DOUBLE GLASS PANELS

SECTION A

SECTION B

A GENERALIZED DETAIL OF A HORIZONTAL SECTION THROUGH A TYPICAL DOUBLE GLASS RIBBON WINDOW CURTAIN WALL. MULLION DESIGNS AND CONNECTIONS WILL VARY FOR DIFFERENT MANUFACTURERS.

VERTICAL MULLION

INDOOR STRUCTURAL SILICONE JOINT

HORIZONTAL MULLION (DESIGNS WILL VARY)

FLOOR

SAFING (FIRE STOP)

SPANDREL BEAM

ANGLE CONNECTION

TWO PANE CURTAIN WALL PANEL

SECTION C

HORIZONTAL MULLION

INDOOR STRUCTURAL SILICONE

SECTION D

GASKET

SETTING BLOCK

FLOOR

WEEP

SECTION E

Figure 31.54 These are generalized details for installing a double glazed ribbon window curtain wall. Specific details such as mullion design and connections vary depending on the manufacturer of the system.

Figure 31.55 This glass curtain wall system has concealed mullions and capped horizontal rails. *(Courtesy Bruce Wall Systems Corporation)*

Flowable Sealant Materials

Flowable sealant materials have adhesive qualities and are applied either by a **sealant gun** or a knife (Fig. 31.66). Some types are in an extruded preformed strip. They adhere to the surfaces on which they are applied and cure into a rubbery material, sealing the joint. They allow for expansion and contraction in the joint because they can stretch and contract without fracturing. Flowable sealant materials are grouped into three classes, determined by the amount of change in joint size they can accommodate.

Low performance sealants have minimum movement capacity and are used in stable joints. Typical materials include oil-and-resin-based compounds, bituminous-based caulks and mastics, and polybutene compounds.

Medium performance sealants can accommodate elongations in the range of ±5 to ±12.5 percent. Typical materials in this group include acrylics, butyls, and neoprene.

High performance sealants are used when the cyclic movement is expected to be between ±12.5 to ±25 percent. Typical materials in this group include polymercaptans, polysulfides, polyurethanes, silicones, and some solvents that release acrylics.

Detailed information on sealants and their applications is in Chapter 26.

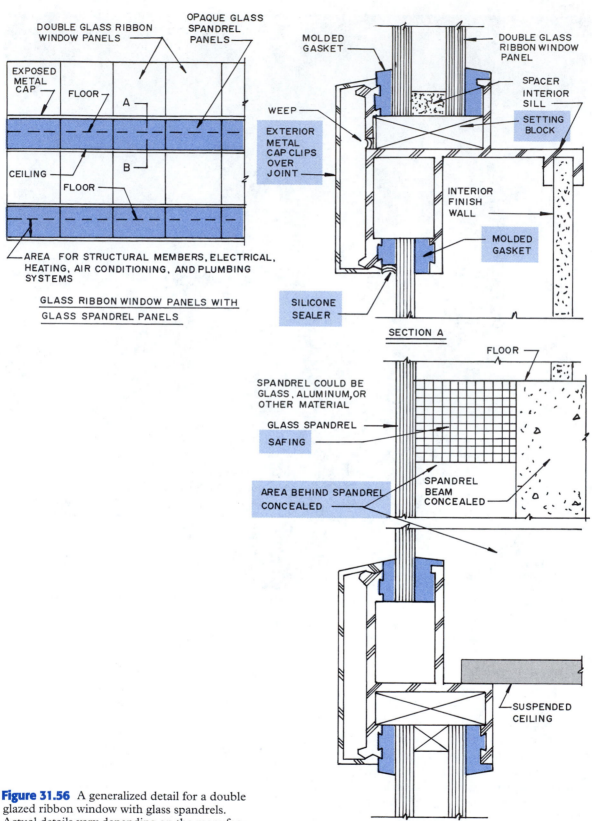

DOUBLE GLASS RIBBON
WINDOW PANELS

OPAQUE GLASS
SPANDREL
PANELS

EXPOSED
METAL
CAP

FLOOR

A

CEILING

B

FLOOR

AREA FOR STRUCTURAL MEMBERS, ELECTRICAL,
HEATING, AIR CONDITIONING, AND PLUMBING
SYSTEMS

GLASS RIBBON WINDOW PANELS WITH
GLASS SPANDREL PANELS

MOLDED
GASKET

DOUBLE GLASS
RIBBON WINDOW
PANEL

WEEP

SPACER
INTERIOR
SILL

SETTING
BLOCK

EXTERIOR
METAL
CAP CLIPS
OVER
JOINT

INTERIOR
FINISH
WALL

MOLDED
GASKET

SILICONE
SEALER

SECTION A

FLOOR

SPANDREL COULD BE
GLASS, ALUMINUM, OR
OTHER MATERIAL

GLASS SPANDREL

SAFING

AREA BEHIND SPANDREL
CONCEALED

SPANDREL
BEAM
CONCEALED

SUSPENDED
CEILING

SECTION B

Figure 31.56 A generalized detail for a double
glazed ribbon window with glass spandrels.
Actual details vary depending on the manufacturer of the system.

Figure 31.57 These crystallized glass curtain wall panels are secured to the structural frame with metal connectors. Soffits are set with an adhesive and a metal channel. *(Courtesy N.E.G. America, Inc.)*

Figure 31.58 Crystallized glass curtain wall panels are available in curved units. Here they are cladding a round concrete tower. *(Courtesy N.E.G. America, Inc.)*

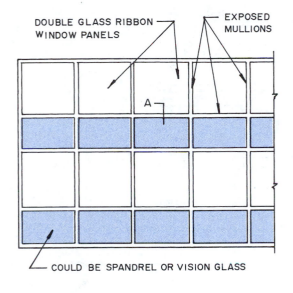

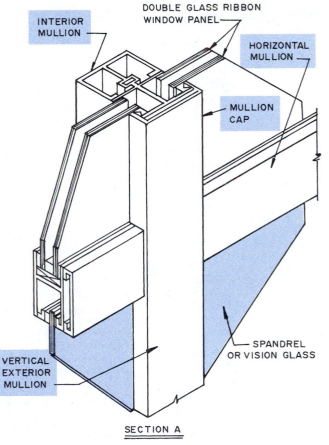

Figure 31.59 Glass curtain walls may have exposed mullions and rails.

Figure 31.60 This mid-rise building has exposed mullions and rails and glass curtain wall panels. *(Courtesy Tubelite, Inc.)*

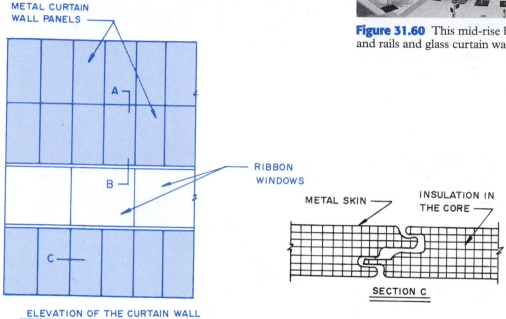

METAL CURTAIN WALL PANELS

RIBBON WINDOWS

A

B

C

ELEVATION OF THE CURTAIN WALL

METAL SKIN

INSULATION IN THE CORE

SECTION C

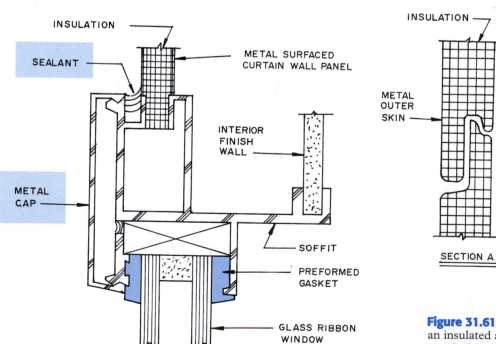

INSULATION

SEALANT

METAL SURFACED CURTAIN WALL PANEL

INTERIOR FINISH WALL

METAL CAP

SOFFIT

PREFORMED GASKET

GLASS RIBBON WINDOW

SECTION B

INSULATION

METAL OUTER SKIN

SECTION A

Figure 31.61 A generalized detail for an insulated aluminum curtain wall panel with ribbon windows. Details vary depending on the manufacturer of the system.

Figure 31.62 Standardized aluminum wall system for sloped glazing is installed as a stick system. Screw spline fastening at purlins to rafter joints substantially reduces erection labor and eliminates unsightly bolts or exposed fasteners. The standardized system can be integrated with a vertical storefront or curtain wall, terminated on a parapet wall or curb, applied to a steel grid subframe, and used on inside or outside corner applications. *(Courtesy Kawneer Company, Inc.)*

Figure 31.63 The high-rise building on the right has a fixed glass curtain wall with stone facings on the structural members exposed to view. The mid-rise building on the left has ribbon windows and precast concrete spandrels. *(Courtesy Bruce Wall Systems Corporation)*

Figure 31.64 A close-up view of the glass curtain wall and faced exposed building structural system of the high-rise building in Fig. 31.63. *(Courtesy Bruce Wall Systems Corporation)*

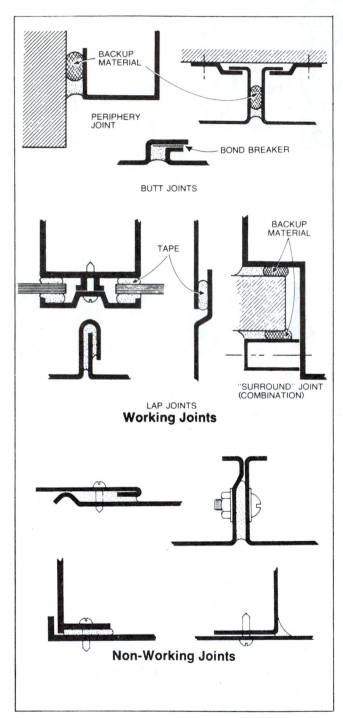

Figure 31.65 Some of the frequently found shapes used as edge joints of joining members. *(Courtesy American Architectural Manufacturers Association, from* AAMA *Joint Sealants Manual)*

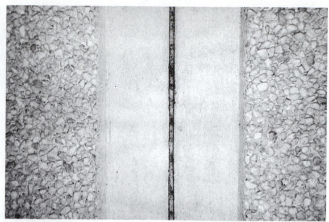

Figure 31.66 These precast concrete curtain wall panels have their joints filled with a flowable sealant. Notice the exposed surface of the panel is exposed aggregate.

Solid Sealant Materials

Solid sealant materials include tapes and gaskets. *Preformed solid tapes* are made from polybutene or polyisobutylene and have adhesive on one or both sides. They are available in rectangular, square, circular, and wedge shapes. They are completely cured when manufactured and seal by being compressed. They are mainly used on lap joints. Refer to Fig. 31.50.

Preformed cellular tapes are made from polyurethane, neoprene, vinyl nitrile, and polyvinyl chloride. They contain a chemical solvent and are delivered in a tightly compressed condition. When installed they expand, filling the joint and bonding to the contacting surfaces as they cure. Refer to Fig. 31.50.

Gaskets are preformed solid elastomeric materials designed to fit the configuration of materials and panels to be sealed. They are compressed and forced into the joint, forming a watertight seal (Fig. 31.67).

BUILDING CODE REQUIREMENTS

Building code requirements consider factors such as fire resistance, structural characteristics, and connections. Fire requirements specify the ability of the cladding to resist fire and to prevent the spread of fire to or from neighboring buildings. Critical is the inclusion of fire-stops between the cladding and each floor. Fire-stops are also required on other vertical openings such as covers on columns. Fire-stops are usually mineral wool safing, gypsum on metal lath, or steel plates with grout. All fire-stops (safing) must be securely fastened to the edge of the floor or other structural member (Fig. 31.30).

Structural requirements include factors such as the strength of the cladding to support its own weight over the distance it spans between connectors, accommodating forces of cyclical expansion and contraction, resisting wind loads, and meeting seismic requirements. The

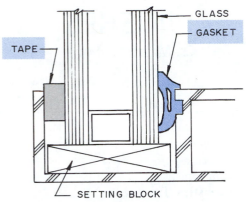

Figure 31.67 Typical gasket and tape sealant applications on a glass unit set in an aluminum frame.

method of connecting the cladding to the structural frame is also subject to analysis. One consideration is whether the designer should rely on local and national standards pertaining to wind loads or conduct wind tunnel tests of the building. Some cities require a wind analysis of buildings above a specified height.

REVIEW QUESTIONS

1. What types of cladding are in general use on multistory buildings?
2. What are the major considerations when designing a cladding system?
3. What problems do wind forces cause on cladding?
4. What forces can cause movement within the cladding system?
5. What is the rainscreen principle?
6. What forces may cause water to penetrate the joints in cladding?
7. What materials are used as fire-stopping?
8. Define a curtain wall.
9. Describe each of the systems used for metal and glass curtain wall design.
10. What is the difference between working joints and nonworking joints?
11. Describe the classes of flowable sealants.
12. Describe the types of solid sealant materials used in curtain wall construction.

KEY TERMS

capillary action The movement of a liquid through small openings in porous or fibrous material as a result of surface tension.

cladding A non-load-bearing exterior wall enclosing a building.

kinetic energy The energy of a body with respect to the motion of the body.

rainscreen principle The theory used to improve the watertightness of cladding by dividing the air space inside the cladding into air chambers vented to the exterior to eliminate air pressure differentials between the exterior air and the air inside the cladding panel so water is not moved into the panel through the joints.

safing Fire-resistant material used to seal open spaces in wall construction to block the passage of fire, heat, and gases.

sealant gun A tool used to extrude flowable sealant into a joint.

thermal break Material with a low thermal conductivity inserted between materials, such as metal with a high thermal conductivity used to retard the passage of cold or heat through the highly conductive material.

SUGGESTED ACTIVITIES

1. Try to find buildings with various cladding materials and examine the finished installation. Prepare a report detailing how the architect handled the various factors, such as meeting fire resistance requirements, controlling condensation, and resisting water penetration.

2. Examine manufacturers' brochures (as in Sweet's Catalogs). Sketch the designs proposed and indicate the materials to be used, such as glazing materials, sealants, and thermal breaks. Show and label construction details, such as how the units are installed, insulated, and vented, and how condensation drainage is provided.

3. Invite a local architect to address the class on how decisions are made related to various cladding systems and to relate personal experiences with the actual performance of systems.

ADDITIONAL INFORMATION

Aluminum Curtain Wall Design Guide Manual, Architectural Manufacturers Association, Palatine, Ill., 1979.

Exterior Wall Construction in High-Rise Buildings, Canada Mortgage and Housing Corporation, Ottawa, Ontario, 1990.

Hornbostel, C., *Construction Materials: Types, Uses, and Applications,* John Wiley and Sons, New York, 1991.

Howard, P. J., ed., *Precast Concrete Cladding,* Halsted Press, New York, 1992.

The Metal Curtain Wall Manual, American Architectural Manufacturers Association, Palatine, Ill., 1989.

Rush, R. D., ed., *The Building Systems Integration Handbook,* Butterworth-Heinemann, Stoneham, Mass., 1986.

D I V I S I O N 9

FINISHES
Csi MasterFormat™

Photo courtesy Harris Tarkett, Inc.

Interior Finishes

This chapter will help you to:

1. Understand the relationship between the various interior finishing processes and other aspects of the construction of the building.

2. Be able to apply the various considerations pertaining to interior finishes to the selection of interior finish materials.

Interior finishes refers to the wall, ceiling, and floor finishes; to other material applied to them, such as paneling, wainscoting; and to materials used for acoustical treatment, insulation, decoration, and other features. Interior finish installation is carefully regulated by building codes, which vary depending on the proposed occupancy of the building. Interior finishes include a wide range of materials, as shown by the MasterFormat outline for Division 9, which is on the page introducing this section of the book. These systems and materials are described in other chapters.

In addition to meeting code requirements, the selection of interior finish materials also depends on factors such as cost, appearance, durability, coatings, acoustical considerations, and fire requirements. The architect and the owner work to select materials that meet the requirements of the building and the building code.

Interior finishing begins after the building has been enclosed, protecting the interior from the weather. The roof must be finished, cladding installed, and doors and windows set. The electrical and mechanical trades can begin to install the electrical, communications, telephone, and computer systems as well as waste and potable water lines, the automatic sprinkler system, and the heating, cooling, and ventilation systems, including the air handlers, boilers, pumps, chillers, fans, cooling towers, and related ductwork. Installation of transportation systems, such as elevators, escalators, and moving walks and ramps, can also be started.

ELECTRICAL AND MECHANICAL SYSTEMS AND THE INTERIOR FINISH

The installation of electrical and mechanical systems is discussed in Chapters 44 through 47. The building must be designed to facilitate the installation of these systems. Horizontal installations, such as electrical, plumbing, and ducts, are often placed below the floor and concealed with a suspended ceiling. Some buildings have cellular floor construction using cellular metal decking or precast hollow-core concrete decking through which wires and sometimes plumbing can be run (Fig. 32.1). Some designs use a raised floor system that provides access below the finished floor. Restrooms usually have a

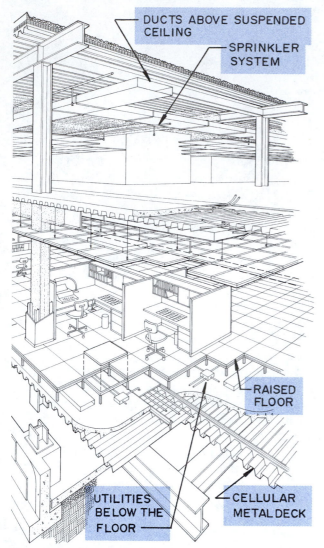

Figure 32.1 Electrical and mechanical systems may be run horizontally below the floors or through cellular metal decking or hollow-core concrete decking. Raised flooring systems also provide space for horizontal runs. *(Reproduced with permission from* The Building Systems Integration Handbook, *Richard D. Rush, ed., Butterworth-Heinmann Publishers, Newton, Mass., 1986.)*

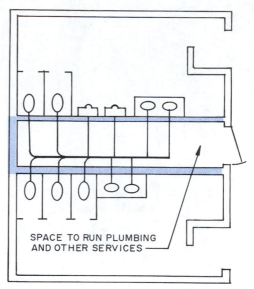

Figure 32.2 Back-to-back restrooms can have a plumbing wall built between them to accommodate plumbing and provide access to it.

plumbing wall installed so the plumbing can be placed out of sight yet accessible for servicing when needed (Fig. 32.2). The restrooms may be part of the **central service core,** which permits the plumbing walls on each floor to be directly above the one below. This facilitates the vertical run of the plumbing.

In multistory buildings, areas to house the major electrical service entrance, sewer connections to the central sewage system, potable water, elevator-operating equipment and the heating, cooling, and ventilation units are generally placed in a basement or subbasement (Fig. 32.3). These are run up through the building to each floor through a vertical central service core

(Fig. 32.4) from which they are distributed as required to each floor. The service core may house stairs, elevators, escalators, restrooms, electrical and mechanical chases, fire protection equipment storage areas, and ventilation air shafts. Each floor will have an area allotted to the services provided by the central service core. A typical example is in Fig. 32.5. Areas such as electrical and mechanical rooms are enclosed on each floor with a floor and ceiling. Areas such as the air shaft and elevator shaft are often open the entire height of the building. This influences the interior finish. The enclosed rooms must meet fire codes and must be designed to contain sounds generated within them. For example, the mechanical room may contain air-handling fans and pumps that could cause sound vibrations in the floor to be transmitted to the occupied space outside the core (Fig. 32.6). The interior finish must handle these and other disturbances. Likewise, the air shaft handles ventilation ducts carrying fresh air into the building and exhausting stale air. The interior finish on the walls of air and elevator shafts is carefully controlled by codes. These shafts can serve as a passage for fire, smoke, and gases from a fire on one floor to all the floors above. The interior finish is critical to the design of the service core.

It should be noted that in very high multistory buildings major electrical and mechanical equipment could be placed on one or more floors. For example, major equipment on the tenth floor could service the core areas on floors ten through nineteen, and another major equipment area on floor twenty could service core areas on floors twenty through twenty-nine. This could mean

Figure 32.3 Major units of the mechanical system are usually located in a basement or subbasement. This is a gas-fired steam double-effect absorption chiller. *(Courtesy Carrier Corporation)*

that some areas, such as air shafts, might not extend all the way to the roof. Also, banks of elevators might serve only the first ten floors and a second bank serve floors eleven to twenty.

COST

The cost of interior finish materials involves consideration of the price of the material plus the labor to install it plus the contractor's overhead and profit. Other cost factors include the expected life of the installation before replacement is required, regular maintenance, minor repairs, and the possibility of increased replacement costs years later because of inflation and increased labor costs. Using life-cycle costing, calculations can reveal if it would be less costly over the long run to use higher quality, more expensive materials. This is frequently the case if the owner plans to retain title to the building over a long time period.

APPEARANCE

The interior finish generally hides the mechanical and electrical systems and provides an attractive interior surface. In some cases, portions of the structural system are left exposed and become part of the interior design (Fig. 32.7). The materials used are directly influenced by the desired appearance of the interior

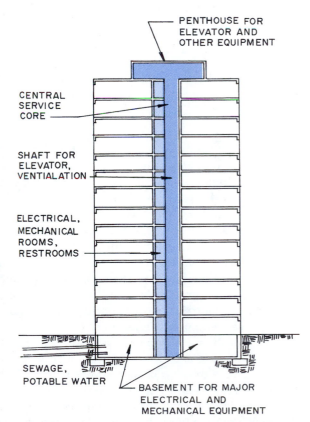

PENTHOUSE FOR ELEVATOR AND OTHER EQUIPMENT

CENTRAL SERVICE CORE

SHAFT FOR ELEVATOR, VENTILATION

ELECTRICAL, MECHANICAL ROOMS, RESTROOMS

SEWAGE, POTABLE WATER

BASEMENT FOR MAJOR ELECTRICAL AND MECHANICAL EQUIPMENT

Figure 32.4 The services required for multistory buildings are usually distributed vertically using a central service core.

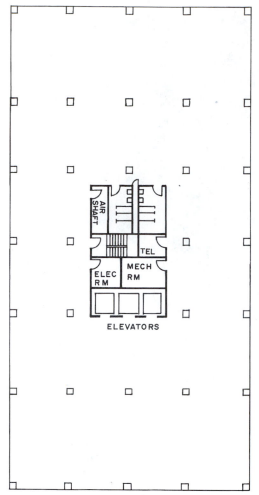

Figure 32.5 The central service core provides space for the utilities and the vertical transportation system. Each floor will have an area like that shown for distributing services to that floor.

spaces and their proposed use. For example, a hotel lobby may have darker textured walls, carpeted floors and stairs, and elaborate wall hangings. The lighting system can provide soft general illumination yet spotlight features such as paneled walls and paintings. The interior of a health and fitness center will be in sharp contrast to this by having light, durable wall surfaces, acoustical ceilings, considerable natural light, and floor coverings to suit the activity in the area. The halls and lobby may have a durable clay tile floor, the aerobics room a soft resilient floor covered with carpet, and a basketball area a composition floor.

The color of the interior finish establishes the mood and reactions to the area. Warm colors, such as those ranging from reds through orange and yellow, create a feeling of friendliness and warmth. Cool colors, such as those ranging from greens through blues and purples, create a feeling of coldness. Lighter colors increase the level of illumination, and darker colors absorb some light, reducing the amount of reflective light. Generally, interior finish uses a range of both warm and soft colors.

Figure 32.6 The vibrations from these pumps in a mechanical room could cause sound vibrations to penetrate adjacent rooms.

The texture of the finished wall and ceiling surfaces also influences the appearance and other qualities, such as acoustical properties and light reflection.

Coatings

Protective and decorative coatings are applied over many types of interior walls, ceilings, and floors. These are discussed in detail in Chapter 33. Coatings are layers of material in liquid form applied to a surface to decorate, preserve, protect, seal, or smooth it. When the coat solidifies, it may leave a flat, a semigloss, or a gloss finish. The coatings may be transparent, semitransparent, or opaque. They are made from a wide range of materials, which determine the color, hardness, and opacity of the coating. The type of coating used is regulated by building codes. Fire resistance and the support of combustion are important. Codes specify the flame spread, smoke, and fuel contributed ratings of the coatings. See Chapter 33 for information on coatings.

Other surface coverings may be paper, plastic, fabric, or thin wood veneers. They are subject to the same code restrictions as liquid coatings.

Figure 32.7 These exposed glued laminated wood arches and roof purlins provide a major architectural detail to the interior of the room. The ceiling is exposed finished wood decking. *(Courtesy American Institute of Timber Construction)*

DURABILITY

The durability of interior finishes includes the ability of the exposed material and any protective or decorative coating to resist damage due to bumps, abrasion, or soil. Hard materials or hard-surfaced materials provide considerable protection and are used in areas where wear and tear are to be expected. In some areas, such as commercial kitchens and restrooms, the ability to withstand water and high humidity is important. Floor coverings are especially vulnerable to wear and damage. A lobby of a busy building will require very durable floor covering, but other areas may be able to use something less expensive. The ability to easily clean the materials with minimum damage to them is also a factor. Finally, the expected life of the product and replacement costs must be a factor when considering the degree of durability.

ACOUSTICAL CONSIDERATIONS

The acoustical properties of the finish materials on walls, ceilings, and floors are another major consideration. These are discussed in Chapter 35. The type of material, texture of the surface, and any coatings influence the acoustical properties. Under consideration is the ability of a finish to control the sound created within an area as well as restrict its flow through walls, ceiling, and floor to adjoining areas. Sound transmitted through materials may be due to vibrations in the material caused by actions such as someone walking on the floor above or by machinery operated on the floor. Assemblies can be tested to ascertain their **impact noise rating.** Impact noise can be reduced by insulating the floors and covering them with soft materials such as a soft underlayment, carpet padding, and carpet. This must be considered as the finish is planned.

Sound transmission of airborne noise through a material or assembly of materials is indicated by its **sound transmission class** (STC). This is a single number that indicates the effectiveness of the material to reduce the transmission of airborne sound through the material into the next area. The higher the STC rating the more efficient is the material in blocking the transmission of sound. Building codes have regulations for the transmission of sound between areas for some types of construction, such as noise from a hall passing into apartments. The manufacturer of various finishing materials has them tested and can provide the needed information.

The amount of airborne sound energy absorbed by a material is indicated by its **noise reduction coefficient** (NRC). It is a single number rating used to calculate the quantity of sound-absorbing material required. Various materials can be compared using this rating.

Sound can also be transmitted between areas by such things as poorly fitting doors, openings around electrical outlets and piping, partitions that are not sealed at the floor or ceiling, air ducts, and ceilings that have inadequate soundproofing.

FIRE CONSIDERATIONS

Interior finish and trim are subject to a wide range of building code requirements pertaining to fire. The combustibility of an interior finish material is rated by testing the flame spread of the surface of the material. *Flame spread* is the rate at which combustion will spread across a material. The **flame spread rating** is a single number that designates the ability of a material to resist flaming combustion over its surface. The rate of flame travel is measured under ASTM testing procedures. Noncombustible cement-asbestos board has a rating of 0. An untreated specified species of wood has a designated rating of 100. The lower the rating the slower the flame will spread across its surface. Flame spread ratings are Class I flame spread, 0–25; Class II flame spread, 26–75; Class III flame spread, 76–200. The acceptable flame spread ratings for various types of buildings and occupancies are specified by the building code (Table 32.1).

Smoke development ratings classify materials by the amount of smoke they will give off as they burn. Most codes prohibit materials having a rating of 450 or more to be used inside a building. Materials are tested using ASTM test procedures. Class I, Class II, and Class III ratings permit smoke developed to range from 0 to 450.

The **fuel contributed rating** is a measure of the amount of combustible substances in a given material. The **fire resistance rating** indicates the capacity of a material to withstand fire for a specified time and under conditions of standard intensity so it will not fail structurally and will not permit the side away from the fire to become hotter than a specified temperature. The ratings are given in hours.

Table 32.1 Maximum Flame-Spread Class[a]

Occupancy Group[b]	Enclosed Vertical Exitways	Other Exitways[c]	Rooms or Areas
A	I	II	II[d]
E	I	II	III
I	I	I[e]	II[f]
H	I	II	III[g]
B, F, M, and S	I	II	III
R–1	I	II	III
R–3	III	III	III[h]
U		No restrictions	

[a]Foam plastics shall comply with the requirements specified in Section 1001.5. Carpeting on ceilings and textile wall coverings shall comply with the requirements specified in Sections 804.2 and 805, respectively.
[b]A—Assembly building
B—Factories, office buildings, retail stores
E—Educational buildings
H—Hazardous materials
I—Nursing homes
M—Private garages, agricultural buildings
R–1—Hotels, apartment houses
R–3—Residences, lodging houses
[c]Finish classification is not applicable to interior walls and ceilings of exterior exit balconies.
[d]In Group A, Divisions 3 and 4 Occupancies, Class III may be used.
[e]In Group I, Divisions 2 and 3 Occupancies, Class II may be used or Class III when the Division 2 or 3 is sprinklered.
[f]In rooms in which personal liberties of inmates are forcibly restrained, Class I material only shall be used.
[g]Over two stories shall be of Class II.
[h]Flame-spread provisions are not applicable to kitchens and bathrooms of Group R, Division 3 Occupancies.

Reproduced from the 1994 edition of the *Uniform Building Code*™, copyright© 1994, with the permission of the publisher, the International Conference of Building Officials.

Codes also specify regulations pertaining to **fire-retardance.** Curtains, draperies, and other decorative materials are required by code to be fire-retardant.

Floor finish materials, such as wood, composition tile, and carpet, are also regulated by building codes. Doors and windows are also controlled by code. Typical fire doors are discussed in Chapter 30.

REVIEW QUESTIONS

1. What materials are included in Division 9, Finishes, of the MasterFormat?
2. What factors are considered as the interior finish is planned?
3. How are electrical and mechanical services distributed in multistory buildings?
4. How do flame spread, smoke contributed, and fuel contributed considerations influence interior finish?
5. What acoustical measurements are considered when selecting interior finish materials?
6. What purpose do coatings serve on interior finish?

KEY TERMS

central service core A fire-resistant vertical shaft through a multistory building used to route electrical, mechanical, and transportation systems.

fire resistance rating The time in hours a material or assembly of materials can withstand exposure to fire developed from data of standard tests.

flame retardance The ability of a material to resist ignition when exposed to flames.

flame spread rating A single numerical designation used to present a comparative measure of the ability of a material to resist flaming combustion over its surface.

fuel contributed rating A measure of the amount of combustible substance in a material.

impact noise rating A single numerical designation used to express the effectiveness of a floor construction to provide insulation against the transmission of sound due to impact.

noise reduction coefficient The average of the sound absorption coefficients of acoustical materials tested at several predetermined frequencies.

smoke developed rating A numerical classification of a building material as determined by an ASTM test that indicates its surface burning characteristics.

sound transmission class A single numerical designation of the sound insulation of a material or assembly of materials.

Protective and Decorative Coatings

This chapter will help you to:

1. Be aware of the range of protective and decorative coatings available and the properties of each.

2. Select appropriate protective and decorative coatings for interior and exterior work.

Coatings are layers of materials in liquid form applied to surfaces to decorate, preserve, protect, seal, or smooth them. When the liquid coat solidifies it leaves a thin layer over the surface (**substrate**) to which it was applied.

Coatings are used to protect materials from heat, soiling, **abrasion,** solar radiation, moisture, chemicals in the air or in a solution, and corrosion. They also are used to provide decorative surfaces.

A coating consists of a primer coat, an intermediate coat (sometimes called an undercoat), and a finish coat, also referred to as a topcoat (Fig. 33.1).

The **primer coat** is applied directly to the substrate (such as steel or wood). It must have good **adhesion** to the surface, have appropriate flexibility, permit the intermediate coat to bond to it, retard corrosion of the substrate, and resist weathering long enough to permit the application of the intermediate and finish coats.

The **intermediate coat** is applied over the primer coat. It must provide an adequate film thickness and structural strength, bond to the primer coat and permit the finish coat to bond to it, and be an excellent barrier to chemicals and other environmental contaminates.

The **topcoat** (finish coat) is applied over the intermediate coat. It provides the initial barrier to action of the environment, gives a pleasing appearance, and may be required to serve other functions, such as providing a nonskid surface or resisting fouling agents as found in marine situations.

The total coating system must be one that is compatible with the substrate, and each coating must be compatible with the next one to be applied and/or the previous one applied. If this does not occur, the coating will deteriorate and have to be completely removed before the substrate can be recoated.

Coatings may be clear, semitransparent, or opaque. **Clear coatings** let most of the natural color and texture of the substrate show through. They are used when it is desired to show the natural appearance, such as the color and grain of wood or the color and texture of colored concrete or the color of the exposed concrete aggregate.

Semitransparent coatings allow some of the color and texture to show but obscure much of it. They are

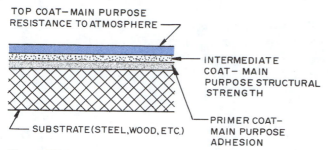

Figure 33.1 A coating typically is composed of three layers of finishing material.

used when the appearance of the substrate is to be changed, such as adding a bit of color yet permitting the material, such as the grain in wood, to show.

Opaque coatings completely obscure the color and much of the texture of the substrate. They are used when a solid, uniformly colored surface is desired. The material composing the substrate is often not identifiable.

FEDERAL REGULATIONS

Federal regulations that affect the use of paint and other coatings are the 1992 Residential Lead-Based Paint Hazard Reduction Act and the 1970 Clean Air Act.

Lead in Coatings

Lead is now prohibited from being used in paints. An ingestion of loose paint dust or chips can cause lead poisoning. Typical hazards include harm to workers on remodeling projects where paint containing lead was used in previous years and harm to children living in older houses and apartments who may eat loose paint chips pulled from the trim and cabinets. See the Items of Interest section of this chapter for additional details.

Air Quality Regulations

The 1970 Clean Air Act mandates National Air Quality Standards that set health-based limits on ozone and carbon monoxide. This includes regulating emissions from automobile exhausts, *construction, manufacturing,* and other areas, including all products that emit **volatile organic compounds (VOCs),** which includes paints and other coatings. Emissions from solvents in coatings are one of the largest sources of air pollution.

VOCs include any solvent, propellant, or other substance, except water, that evaporates from a coating as it dries. VOC is calculated as the combined weight of *all solvents* in a given volume of coating, excluding certain exempted materials. VOC is specified in *grams per liter* (g/l), which can be converted to pounds per gallon (lb./gal.) by dividing g/l by 119.8.

Architectural coatings are limited to not more than 250 g/l (2.09 lb./gal.) and *industrial maintenance primers and topcoats* to no more than 420 g/l (3.5 lb./gal.). Architectural coatings are used on surfaces exposed to mild environments. Industrial maintenance coatings are used to protect surfaces against chemical corrosion, high temperatures, abrasion, immersion in water, and solvents.

Some waterborne coatings that currently are available meet VOC requirements. However, just because a product is water based does not mean it meets the requirements. Some water-based coatings contain sufficient cosolvent to raise the VOC above acceptable limits. Many waterborne coatings cannot meet the severe conditions presented by the industrial maintenance environment. Currently some coating manufacturers have solvent and water-based products that meet VOC requirements and provide the required protection. For example, the Rust-Oleum Corporation has a water-based two-compound amine epoxy coating having a VOC level of 250 g/l that meets emission requirements. Contractors must observe the VOC ratings of the coatings they use or see if an exempt product is available. Manufacturers include VOC ratings of their products in their sales literature.

The implications of the Clean Air Act as it is enforced by various states means that many of the coatings mentioned in this chapter may cease to be used as newly formulated coatings that meet the limitations are developed. New coatings will have fewer solvents, be water-based with little or no cosolvent, or be composed of 100 percent solids. This will also influence how these new materials will be applied. Brush application may become less viable and new application equipment, including newly designed spray systems, will be developed. It is possible that more coatings will be factory applied and that less on-the-job application will occur. The constructor must be constantly aware of these changes as they take place.

COMPOSITION OF COATINGS

Many different kinds of coatings are available, and their composition varies widely. However, they all have the same basic composition. This includes a binder, a solvent, and pigments. The **binder** is a nonvolatile natural or synthetic **resin** that forms the base of the hardened coating. The **solvent** is the volatile part of the coating in which the binder is dispersed. The mixture of the binder in the solvent forms the **vehicle** of the coating. **Pigments** are insoluble particles that are suspended in the vehicle and give it color and opacity. The binder plus the pigments form the *solids* of the coating and produce the layer remaining after the solvent has evaporated.

Clear coatings are made of binder and solvent but contain no pigments. Semitransparent coatings have a small amount of pigment, so they provide some opacity. Opaque coatings contain considerable pigment and totally obscure the face of the substrate. The binder bonds itself to the substrate. It also must have the properties needed to meet the requirements of the coating. For example, it must have plasticizers to make it flexible, stabilizers to enable it to resist solar radiation or sources of heat, and driers to control the rate of curing.

Coatings are either solvent-based or water-based. *Solvent-based coatings* are volatile, and **water-based coatings** have the binders and pigments dissolved in water. Solvent-based coatings require the use of a solvent, such as paint thinner, to thin the mix and to clean brushes and spray guns. The solvent used depends on the composition of the coating. Water-based coatings are thinned and tools are cleaned with water.

Coatings are specified for specific uses. Manufacturer's directions clearly tell whether a coating can be used for exterior or interior purposes or, in some cases, both. They specify the substrate (wood, metal, masonry, etc.) to which the coating can be applied and whether it will resist attack by chemicals, heat, and ultraviolet rays. In addition, the final appearance is indicated as clear, semitransparent, or opaque and high gloss, satin gloss, or flat (no gloss). Coating specifications may include wet film thickness and/or dry film thickness.

Wet Film Thickness

To ensure expected performance the proper thickness of paint must be applied. This is specified by giving the *minimum wet film (MWF)* thickness. This is stated as "applied at the rate of xx mils per coat minimum wet film thickness (MWF)." A mil is a unit of measure equal to 0.001 in. (0.025 mm).

The wet film thickness varies depending on the **spreading rate** per gallon. The wet film thickness of a coating can be determined by measuring the wet film thickness with special gauges designed for that purpose or by using an industry-developed formula that relies on the spreading rate. Table 33.1 shows some typical wet film thicknesses found using the formula.

The minimum wet film thickness for various coatings can also be found by noting the recommended spreading rates indicated by the manufacturer of the coating material.

Dry Film Thickness

Specifications state the *minimum dry film* (MDF) thickness per coat for the coating material. Dry thickness determines the amount of protection the coating will provide. Sufficient wet mils must be applied to get a specified dry film thickness.

The dry film thickness is found by multiplying the wet film thickness in mils by the percent of solids in the paint. This gives a theoretical dry film thickness. The percent of solids in a coating is determined by the manufacturer of the material.

Table 33.1 Wet Film Thickness Based on the Spreading Rate

Spreading Rate (sq. ft./gal.)	Wet Film Thickness (mils)
1600	1.0
1000	1.6
700	2.3
400	4.0
200	8.0
160	10.0

FACTORS AFFECTING COATINGS

After a coating has been applied to a substrate and has hardened it is subject to many external factors. These must be considered as a choice of coatings is considered.

Water is one of the most common factors. Rain on external surfaces may cause a thermal shock if the coating is very hot. It may penetrate the coating through checks and cracks and freeze or become very hot, causing blisters. *Water vapor* may penetrate the coating from the outside, resulting in damage to the substrate that causes the coating to peel and crack. Water vapor may penetrate the substrate from behind the coating, again causing blisters and peeling. This is one reason why vapor barriers are used on the interior side of exterior walls. The substrate, such as wood, may contain moisture before the coating is applied. The moisture content of the substrate must be within specified limits for successful coating. Excess moisture in the substrate will cause poor adhesion of the coating.

Solar radiation bombards the coating with ultraviolet radiation, which can cause fading of the pigment and some chemical reactions in the solvents and binders, leading to coating deterioration. This permits ultraviolet radiation to reach the substrate, causing possible damage to it. Solar radiation causes an increase in the *temperature* of the coating and the substrate, which causes both to expand. The coating must be flexible enough to expand and contract without cracking. This includes handling expansion of the substrate, which may exceed the normal expansion of the coating at the same temperature. Likewise, freezing temperatures cause damage, especially if moisture has penetrated the coating.

Coatings are subjected to *dust and dirt*. Their location will indicate the extent of possible damage. A ceiling, for example, usually has minimum exposure, but a baseboard or door has greater exposure. Location also dictates exposure to *abrasion*. Impact due to natural causes (such as hail), normal wear (as on doors), and vandalism must be considered.

In some situations coatings are affected by *chemical fumes, solutions, and reactions*. Chemical fumes are generated by many sources, including power plants and automobile emissions. Sea water, oils, solvents, and other chemical solutions impact heavily on coatings. Soluble alkaline salts in mortar and concrete can dissolve and crystallize on the surface, damaging the coating. This is called **efflorescence.** Sealants may react unfavorably, and rust may streak the coating surface. Wood knots heavy with resin may bleed through the coating. These and other possible reactions to coatings used must be carefully considered.

The *absorption* of the surface of the substrate will affect the coating. Some surfaces are hard and smooth and absorb none of the coating. Such surfaces may require roughing by sanding, sandblasting, or etching. Various porous surfaces have different rates of absorption, providing different levels of adhesion. This may cause some cracking of the coating.

New coatings applied over an *old coating* must be compatible. The surface of the old coating must be stable and have good adhesion to the substrate or the new coating will fail. Old coatings *chalk* as they age. If this is the only deterioration, a new coating can be applied over it. If there are cracks and loose peeling sections, the old coating must be removed before recoating.

Following is a discussion of commonly used field-applied coatings. There are many products finished in the factory, and they use mass production application and drying techniques. One example is polyvinyl chloride film used for factory-finished metal wall panels.

CLEAR COATINGS

In addition to the clear coatings that have been used for years, such as shellac, varnishes from natural resins, and lacquer, a number of products using synthetic resins are available. Clear coatings used in exterior locations in general do not have the durability of opaque coatings because they lack the pigments that protect against ultraviolet damage from the sun. They let the natural color and texture of the substrate show as well as protect it from moisture, abrasion, and other forms of damage.

Natural Resin Varnishes

Varnishes are one of the oldest finishes used to coat wood surfaces. Several types currently are available, and their properties and composition vary considerably. Varnishes fall into two broad classifications, natural-resin varnishes and synthetic varnishes.

There are three basic types of natural resin varnish—linseed oil varnishes, tung oil varnishes, and spirit varnishes. Natural varnishes are made from fossil resins or resins obtained from a variety of trees in tropical countries. The vehicle is some form of drying oil into which the resin is dissolved. The oil-to-resin ratio determines the classification of the varnish and is expressed as the number of gallons of oil that are mixed per 100 lb. (45 kg) of resin. Varnishes made with natural resins and oil are called oleoresinous varnishes.

Turpentine is usually the solvent for varnish, although mineral spirits, naphtha, and benzene can be used. The solvent evaporates, causing the varnish to harden. The drying oils cure by oxidation and polymerization following the loss of the solvent. This is why varnish is a slow-drying coating.

Linseed oil varnishes are available in three types—long-oil, medium-oil, and short-oil. Long-oil varnish is sold under the name spar varnish. It is used on exterior surfaces where moisture may be present but that are not constantly wetted. Some of the synthetic resin varnishes are better for very moist conditions. Medium-oil varnishes dry faster than long-oil types and have a harder film, but they are not as water resistant. Synthetic binders, alkyd and phenolic, are used to produce a modified spar varnish. Short-oil varnishes contain the least oil, dry rapidly, are rather brittle, and do not resist abrasion very well.

Tung oil varnishes are used in areas where heavy use is expected, such as on school furniture. They are usually factory applied.

Spirit varnish, also known as *shellac,* uses a resin obtained from the exudation of the lac insect, which is found in Southeast Asia and India. The resin is dissolved in denatured alcohol and is orange in color (referred to as orange shellac). It can be bleached, producing white shellac.

Shellac is available in various grades depending on the amount of resin dissolved in a gallon of denatured alcohol. The grades are referred to as cuts, such as a 4 lb. (1.8 kg) cut, which is 4 lb. of resin dissolved in one gallon of denatured alcohol.

Shellac dries rapidly but does not resist moisture. Its main use in construction is to seal knots and other resinous places in wood over which water-resistant coatings are applied. It will not withstand exposure to sunlight.

Synthetic Resin Varnishes

Synthetic resins are plastic materials suspended in a solvent. The most commonly used resins are **alkyds,** polyurethane, silicone, epoxy, acrylics, and phenolics. The vehicle may be the same drying oils used for natural resins, although synthetic materials have been developed. A listing of some clear synthetic resin varnishes and their solvents are in Table 33.2.

Acrylic resin varnishes produce a thermoplastic film that is resistant to yellowing with age, ultraviolet rays, and oxidation. They have good gloss retention and are almost colorless. They are used on metal, wood, plastics, textiles, and paper. Acrylic resin coatings are made with a solvent base or a water base. The solvent-based formulation has a high gloss and is impermeable to water vapor. The water-based formulation has a semigloss finish and is permeable to water.

Table 33.2 Clear Finishes

Binder	Base	Uses
Acrylic	Solvent or water	Waterproofing, sealing surface against dirt Used on concrete, masonry, stucco
Alkyd (spar varnish)	Solvent	Used on interior and exterior protected surfaces
Phenolic (spar varnish)	Solvent	Exterior wood exposed to moisture, marine applications
Silcone	Solvent	Waterproofing, sealing surface against dirt Used on concrete, masonry, stucco, wood
Polyurethane (one part)	Solvent	Resists chemical attack, abrasion, heavy foot traffic

Table 33.3 Polyurethane Resin Coatings

ASTM Type Designation	Curing Agent	Chemical Resistance
One Component		
Type 1	Oxygen	Good
Type 2	Humidity in air	Very good
Type 3	Heat	Excellent
Two Components		
Type 4	Amine	Excellent
Type 5	Polyester	Excellent

Alkyd resin varnishes are made by a chemical reaction between an alcohol and an organic acid. They are used for waterproofing and for reducing dirt retention on concrete, masonry, and stucco walls. Some types are used on exterior metal surfaces.

Phenolic resin varnishes include phenol formaldehyde and modified phenolic types. Phenol formaldehyde varnishes are excellent for materials exposed to the weather and can be used on wood marine products. They are resistant to caustic substances and acids. They tend to yellow and darken with age and lose gloss with exposure to the sun. Modified phenolic varnishes do not resist weathering as well, but they have an abrasion-resistant quality that makes them useful for interior applications such as furniture and floors.

Epoxy varnishes have excellent resistance to caustic materials, excellent adhesion qualities, and excellent resistance to abrasion. Epoxy coatings come in a variety of forms, each with special characteristics, such as being highly flexible. They can be used on a variety of materials, such as concrete, metal, and wood. Epoxies are often used as primers. Some are formulated to resist solvents, heat, chemicals, and salt water.

Polyurethane resin varnishes provide better abrasion resistance than natural varnish and offer resistance to solvents, chemicals, and oxidation. They are fast curing and are used on concrete floors and walls and places where a hard gloss surface is needed. Three types are one-component coatings, and two types have two components (Table 33.3). Type 1 cures by exposure to oxygen in the air. Type 2 requires 30 percent humidity in the air to cure. Types 1 and 2 can be applied on the job site. Type 3 cures with heat. Types 3, 4, and 5 are factory applied (Fig. 33.2).

Silicone varnishes have excellent heat resistance, water resistance, and resistance to corrosive atmospheric conditions. A silicone-acrylic coating has a high gloss and high resistance to blistering and crazing. Silicones are

Figure 33.2 Polyurethane clear coatings are used where a hard, durable finish is required.

also used to modify alkyds, improving their durability. Silicone-polyester coatings will resist damage by heat up to 550°F (288°C) if the mix contains 75 percent silicone.

Silicones bond well to wood, but steel must be cleaned and bonderized to get proper adhesion. Silicone solutions are formulated to be applied to concrete and masonry materials to provide a water-repellent coating by penetrating into the pores, leaving no film on the surface.

Lacquer

Lacquer is made from synthetic materials and is quick drying due to the rapid evaporation of the solvent. Lacquer has a nitrocellulose base used in combination with various resins and plasticizers. Drying oils are added to improve adhesion and elasticity. The *natural and synthetic resins* improve adhesion and hardness and give the desired gloss. There are many different resins added. Those used depend on the end use of the lacquer. Commonly used are alkyd resins, epoxy resins, acrylic resins, and cellulose acetate.

Solvents commonly used in lacquer include diethylene glycol, acetone, and amyl, ethyl, butyl, and isopropyl acetate. Formulating lacquer requires several different solvents to dissolve the synthetic and natural materials used. The blending of solvents used affects the gloss, ease of flow, setting time, and amount of bubbling as it is applied.

Thinners are sometimes added to lacquer before it is applied. This is especially important when using sprayed lacquers. Thinners adjust the consistency and rate of drying. In general, lacquers dry to the touch in five to ten minutes and form a firm film in thirty minutes to three or four hours, depending on the formulation. Some can have a second coat applied just fifteen to twenty minutes after the first coat was sprayed. Commonly used lacquer thinners are toluol, xylol, benzol, and ethyl, amyl, butyl, and isopropyl alcohol. Lacquer thinner spilled on a hardened lacquer finish will dissolve the finish.

Most lacquers are applied by spraying. A *brushing lacquer* is available and can be applied with a pure bristle brush. It is slower drying than spray lacquers. Wood surfaces to be lacquered need a washcoat of shellac or lacquer sealer before applying the finish coats. Several finish coats are required because the lacquer film is very thin.

Lacquer is used mainly on interior products and is widely used on cabinets and furniture. Pigments are added to give color and opacity.

OPAQUE COATINGS

Opaque coatings obscure the natural color of the surface, give it protection, and add a decorative feature by providing a wide range of colors. The type of coating material used depends on the material in the surface to be coated and its intended use. Exposure to the weather, the ground, chemical fumes, salt water, and other conditions must be considered when selecting the type of coating. Some of the frequently used coatings for various materials are in Table 33.4.

Before a surface is coated it must first be cleaned and prepared to receive a primer, if required. After priming, the required number of finish coats are applied. Manufacturer's specifications must be followed.

Primers for Opaque Coatings

A primer is a coating applied to the surface to seal it and prepare it to receive the finish topcoat. It helps hide discoloration and stains and seals areas that may cause a bleed through the finish topcoat. Resin in wood is a common bleeding problem. Primers are also used to smooth porous surfaces, such as those on concrete blocks. The use of the proper primer is vital to the successful use of the finish topcoat, so the manufacturer's recommendations must always be followed. In Table 33.4 are examples of some commonly recommended primers and associated topcoats.

Cleaning Metal Surfaces for Priming

The preparation of *steel* surfaces depends on the primer and finish topcoat. The directions of the manufacturer should be followed. In some cases cleaning with a detergent or solvent is adequate. Wire brushes, sandpaper, and sandblasting also clean the surface and roughen it. Galvanized metal can be cleaned with a dilute acetic acid, and steel can be cleaned with phosphoric acid. After the surface is washed and dried, the proper primer can be applied. Some prefer to let it weather for six to nine months before trying to prime it.

On-site *aluminum* to be painted should be allowed to weather for at least a month. Then it should be wiped with mineral spirits or any solution recommended by the coating manufacturer. Most aluminum products arrive on the site with a factory-applied finish. See Chapter 16 for more information.

Copper, bronze, and their alloys can be cleaned by wiping with mineral spirits or a dilute solution of hydrochloric or acetic acid, which must be completely washed away. Corrosion and material stuck on the surface can be removed by brushing or light sanding.

An etching primer is available for cleaning ferrous and nonferrous metals and some alloys. It cleans and chemically etches the surface and leaves a very thin protective film. Generally, a standard prime coat is applied over this material.

Preparing Wood Surfaces for Priming

Exterior wood to be painted should have a moisture content of 12 to 15 percent. Interior wood should have a moisture content of about 6 percent. Knots that might bleed through should have a coat of shellac or other **sealer** applied before the wood is primed. Mold, fungus, and other stains should be removed.

Preparing Concrete Surfaces for Priming

It is advisable to let concrete completely cure before priming. Concrete aged less than thirty days is generally considered not suitable for painting. Paint manufacturers' recommendations on curing time should be observed. The surface should be free of dirt, oil, and other substances. Loose material can be removed by brushing or sanding.

Table 33.4 Commonly Used Primers and Topcoats

Material	Primer	Topcoat	Remarks
Aluminum	Vinyl red lead	Vinyl	Exposure to weather
	Zinc chromate	Chlorinated rubber	Exposure to rain, salt water spray
	Zinc chromate	Alkyd or acrylic	Used on trim, flashing
	Self-priming	Epoxy ester	Exposure to fumes
Ferrous metal	Self-priming	Phenolic	Exposure to weather, high humidity
	Zinc silicate	Silicate, alkyd	Exposure to weather
	Zinc silicate	Silicone, aluminum pigmented	Exposure to weather
	Self-priming	Vinyl	Exposure to rain, salt-water spray
	Red lead	Acrylic	Will not resist abrasion
	Self-priming	Urethane	Exposure to corrosion, chemicals, abrasion
	Self-priming	Epoxy	Exposure to acids, alkalis, chemicals
	Self-priming	Coal tar	Apply hot to metal to be below ground
	Chlorinated rubber with red lead or zinc chromate	Chlorinated rubber Oil-based paints	Exposure to rain, salt-water spray, chemicals Normal exterior conditions
	Red lead		Not abrasion resistant, exposure to
	Red lead and alkyd-based red lead	Alkyd	severe weather
	Zinc-polystyrene	Polystyrene	Chemical fumes, exposure to fresh and salt water
Ferrous metal (galvanized)	Zinc dust or zinc chromate-zinc dust	Alkyd	Does not require topcoat
	Zinc dust or zinc oxide	Chlorinated rubber	Exposure to rain, salt-water spray
Ferrous metal in ground	Self-priming	Coal-tar-epoxy	Used on pipelines, buried structural steel
Gypsum wallboard	Vinyl	Alkyd	Light duty
	Self-priming	Acrylic	Heavy duty
Gypsum plaster	Self-priming	Acrylic	Plaster must be dry
Concrete and concrete masonry (dry), brick masonry	Self-priming	Acrylic	Interior locations, scrubbable
	Self-priming	Vinyl	Dry locations
	Self-priming	Epoxy esters	Exterior use, resists fumes, scrubbing
	Self-priming	Polychloroprene	Resists water, solvents, impact, exterior uses
	Self-priming	Urethane	Washable, interior locations
Concrete floors, no moisture exposure	Self-priming	Urethane	Light to moderate traffic
	Self-priming	Epoxy	Moderate to high traffic
Concrete, heavy moisture	Self-priming	Alkali-resistant chlorinated rubber	Water reservoirs, swimming pools
Portland cement plaster	Self-priming	Vinyl	Dry locations
	Styrene-butadiene	Alkyd	Dry locations
Wood, interior	Self-priming	Vinyl	Walls and floors
	Self-priming	Alkyd	Doors, paneling, trim light-duty floors
	Self-priming	Urethane	Surfaces subject to impact, scrubbing, heavy-duty floors
	Self-priming	Acrylic	Surfaces subject to impact, scrubbing
Wood, exterior	Self-priming	Urethane	Porch decking, exterior stairs
	Self-priming	Alkyd	Siding, plywood, cedar shakes, trim
	Self-priming	Acrylic	Siding, plywood
	Self-priming	Phenolic	Siding, plywood, trim
	Oil-based primer	Oil-based vehicle	Wood siding, exterior trim, plywood siding
	Self-priming	Epoxy	Any exterior wood

Priming Coats

A primer is a first coat applied to the substrate. It seals and fills the pores of the surface, inhibits rust in ferrous metals, and improves adhesion of subsequent coats of paint. Following are some of the frequently used primers.

Gypsum plaster can be primed with latex (acrylic), alkyd, or oil-based primer after a thirty-day waiting period. Latex flat interior finish may be used as a primer if the topcoat is a gloss or semigloss coating. Damp new plaster requires an **alkali**-resistant primer.

Gypsum wallboard and other paper-covered products are primed with polyvinyl acetate if an alkyd topcoating is used. Latex (acrylic) topcoatings are self-priming. *Portland cement plaster* often is painted with a vinyl topcoating, which is self-priming.

A wide range of primers and topcoatings are used on *concrete, concrete masonry,* and *brick masonry.* Most of the topcoats used are self-priming. The surfaces of concrete and masonry are usually rough and porous. They can be coated with a latex-portland cement grout or latex blockfiller to produce a smoother surface. This results in a better, watertight coverage because the topcoat will not always bridge the pores in the surface. A thinned coat of catalyzed-epoxy coating material can also be used. Concrete walls can be primed with a latex primer-sealer after which a topcoat of latex or alkyd paint can be applied.

Aluminum is often primed with zinc chromate. Bare *ferrous metals* use a variety of primers depending on the topcoat. Since ferrous metals rust rapidly, they must be coated with a rust-inhibiting primer. Red lead is the oldest of the primers and is almost exclusively used as a corrosion-inhibiting metal primer, usually on bridges and structural steel. Zinc silicate and zinc chromate are used when there will be considerable exposure to the weather. Coatings designed for special purposes often have a special primer designed to be used with that product. Etching primers that clean and seal the surface also are effective.

Galvanized metal can be chemically etched before painting to provide needed adhesion. This can be omitted if a latex metal primer designed for galvanized metal is used. Another primer is a varnish-based material pigmented with zinc dust, zinc oxide, or portland cement.

Types of Opaque Coatings

Opaque coatings are used to hide the color of the previous coating or substrate and obscure much of the grain or texture of the substrate. Commonly used opaque coatings are listed in Table 33.4. The following discussion gives basic details about these.

Alkyd Coatings

Alkyd coatings are currently a major type of organic coating. However, because of requirements relating to the amount of volatile organic compounds (VOCs) that alkyd coatings release to the environment, their use is becoming limited. Water-based products are replacing alkyd coatings.

Alkyd emulsions are formulated using a synthetic alkyd resin. Many types of alkyds are used in coating formulation. Alkyd coatings have only mild alkali resistance, but they have excellent water resistance and weather well. This makes them useful for exterior applications, such as enamels for porches. Alkyd coatings retain their color well and permit the formulation of a range of light colors. With some reformulation alkyd emulsions are used in making baking enamels, such as those used on kitchen appliances.

Alkyd resins are added to other coatings to improve adhesion and durability. For example, when modified with phenolic resins, water resistance and alkali resistance are improved and the coating will penetrate rusted surfaces. When alkyd resins are formulated with rust-inhibiting pigments, such as red lead, iron oxide, or zinc chromate, they are used as rust-inhibiting primers. A gloss enamel and resistance to chalking are produced by adding vinyl chloride acetate to alkyd resins. Silicone aids in color and gloss retention.

Chlorinated Rubber

Chlorinated rubber–based coatings are solvent thinned and have excellent resistance to alkalis and acids and some resistance to salt-water and salt-air exposure. They have excellent water and water vapor resistance and are used on swimming pools and basement walls. They have good abrasion resistance and can be used on masonry, plastic, concrete, and metal surfaces. Chlorinated rubber coatings are not recommended for use on wood because they are not permeable and blistering could occur, but they do resist attack by microorganisms. They do bond to metal and are used to stop corrosion due to galvanic action by separating dissimilar metals.

Enamel Coatings

Enamels are a form of pigmented coating using varnish as the vehicle. They form a gloss or semigloss film that is hard and durable. Oil-based and resin-based paints form similar coatings and can be classified as enamels.

Baked enamels are formulated to be applied in a factory by spraying. They are generally thermosetting materials that cure to a hard coating at 200 to 300°F (93 to 149°C). The coating becomes insoluble in the solvent used in its formulation. Enamels are available in a wide range of colors, are hard and washable, and resist alkalis and acids (Fig. 33.3).

Epoxy

Epoxy coatings are available in a variety of formulations. *Epoxy-ester* is an epoxy resin reacted with a drying oil. It has properties similar to phenolic varnishes and alkyd resins, but it offers better resistance to chemical fumes and exposure to water. *Epoxy-polyamide* offers excellent

Figure 33.3 Latex semigloss enamels are excellent for interior walls and trim.

resistance to chemical fumes, oils, atmospheric acids, and alkalies. It resists abrasion, adheres to concrete, metal, or wood, and will cure even if wet. *Epoxy-bitumens* may be formulated with coal tar or asphalt. They are used on items buried in the soil, such as tanks and piping. *Epoxy-polyester* coatings are heavy-bodied two-part systems used for protection of masonry and concrete. They have a high-solids vinyl filler that is applied to the concrete surface followed by a high-solids epoxy-polyester pigmented topcoat.

Epoxy resins are chemically setting and solventless. They do not harden by the evaporation of solvent but by the application of heat. When exposed to the weather they may fade and chalk, but the film remains undamaged.

Latex Coatings

Latex is a term applied to emulsions containing synthetic resins that are thinned with water. Since they contain no flammable solvent, they present no fire hazard when in storage or when being applied, and they release a minimum of volatile organic compounds into the atmosphere. Latex emulsion coatings dry rapidly. The common types available are acrylic, polyvinyl acetate,

and styrene-butadiene. They are used for coating interior and exterior vertical surfaces.

Acrylic Coatings

Acrylic coatings are thermoplastic resins that have a range of properties varying from a rather hard coating to a softer finish. They are water based and available in clear and pigmented coatings. They offer excellent protection to concrete against weathering. They are also used as factory-applied coatings on aluminum and steel wall panels because they have good **durability** and resistance to salt spray and chemicals. Acrylic coatings remain flexible and do not lose their color as they age. Some acrylic coatings are solvent based and are therefore not latex emulsions. They use solvents such as xylol or toluol.

One-part acrylic emulsions (latex) are used on interior and exterior vertical surfaces, such as wood, masonry, gypsum wallboard, plaster, and metal. They are permeable to water vapor (Fig. 33.4).

Two-part epoxy modified acrylic coatings are water based and used for interior and exterior vertical surfaces. They are tough coatings that resist stains and will withstand scrubbing.

Styrene-Butadiene Coatings

These water-based coatings are formulated using styrene and butadiene, producing a rubberlike film. They have good resistance to alkali and are permeable to water vapor. They have some exterior use as fillers over porous concrete surfaces. Some formulations are used on interior masonry, plaster, and gypsum wallboard, producing a film that resists abrasion and is washable. They generally are not used on wood.

Vinyl Coatings

Vinyl and polyvinyl acetate emulsions are available as water-based coatings. One formulation is used for interior application, and another is used for exterior use. *Polyvinyl chloride* copolymerized with polyvinyl acetate is an opaque solvent-based coating that has poor adhering properties and requires a special primer. It has good durability and resistance to oils, alkalies, acids, and salt water.

Oil-Based Coatings

Oil-based coatings are formulated by combining the body, pigment if needed, and a vehicle (drying oil, a thinner, and a drier). The paint *body* is a solid, fine material that provides the **hiding power.** White lead, lithopone, titanium white, and zinc oxide are used for bodies of white oil-based paint. *Pigments* are added to give the coating color. Both natural and synthetic pigments are used. Natural pigments are obtained from minerals, animal products, and vegetable products. For example, red lead is a typical red pigment. Synthetic pigments are generally derived from coal tar.

ITEM OF INTEREST

LEAD PAINT

Lead paint found on older buildings can poison construction workers doing remodeling or building additions on an old building. Common symptoms of lead poisoning include stomach cramps, nausea, loss of appetite, headache, and joint and muscle aches. The use of lead pigments in paint was banned in 1978, but some was used up until 1980. Most houses built before 1980 used lead paint.

The Residential Lead-Based Paint Hazard Reduction Act of 1992 sets forth detailed instructions on what is required of a contractor before beginning work in which lead paint exists and how the workers and occupants of the building must be protected. For example, workers must have respiratory protection and protective clothes, and work site facilities for their cleanup and clothes changing must be provided. Those who regularly do this work must have periodic blood tests. OSHA has detailed instructions they follow when inspecting work where lead pigmented paints are known to exist.

Before taking a job where lead paints may be expected the contractor should have tests made on the painted surfaces. Three tests are used—x-ray fluorescence, laboratory analysis, and chemical spot tests.

A portable x-ray fluorescence analyzer is used to measure the amount of lead on a painted surface expressed in milligrams per square centimeter. The analyzer can read through multiple layers of paint but cannot indicate which of the layers contain lead pigment.

Laboratory analysis is the most accurate test method for lead. It is known as atomic absorption spectrophotometry and gives results as a percentage of lead by weight. The paint samples are removed from the surface. It is important to get all the layers of paint down to the substrate in each chip to be tested. None of the substrate should be adhered to the backs of the samples. The chips are sent to laboratories equipped to run this test.

Spot tests for lead use a chemical test kit containing a swab or dropper used to apply a chemical reagent that reacts to lead by changing color if the paint contains 0.5 percent or more lead by weight. These kits are sold by many retail paint stores. See Fig. A.

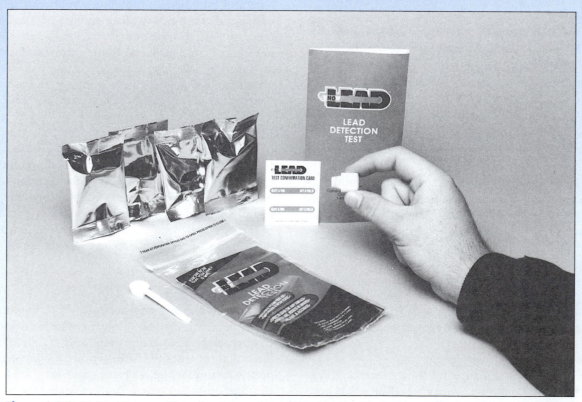

Figure A UF-33.01. (*Courtesy Know Lead, manufactured by Carolina Environment, Charlotte, N.C.*)

ITEM OF INTEREST

LEAD PAINT

Continued

Since the test only checks the top layer of paint, it is necessary to scrape down and test each layer below the top. Generally this test is used for preliminary analysis because it is inexpensive and fast. If it appears to indicate the presence of lead, paint chip samples should then be sent to a laboratory for a more accurate determination. The steps to make a test are in Fig. B.

INSTRUCTIONS FOR KNOW LEAD™ HOME TEST KIT

Test kit includes: 4 test daubers, 1 measuring spoon, 1 test confirmation card and instructions.

Please read instructions completely prior to actual use of Know Lead™.

Know Lead™ test daubers contain non-toxic chemicals required for the test. To prepare for testing, remove one of the inner pouches and remove test dauber from pouch. Fill measuring spoon with water.

DINNERWARE, CERAMICS, GLASS-WARE, TOYS, SOLDERED FOOD CANS, SOLDERED WATER PIPES:

STEP 1 DIP
Dip white non-abrasive end of test dauber in the spoon of water. White end of test dauber should turn pale yellow.

STEP 2 RUB
Firmly rub white non-abrasive end of test dauber on surface being tested for 10 seconds.

PAINTED WOOD OR PAINTED METAL SURFACES (Walls, Window or Door Trim, Baseboards, Radiators, Porch Railings, etc.)

STEP 1 CLEAN
Remove all dust and dirt from area to be tested.

STEP 2 SCRATCH
With gray abrasive end of the test dauber, vigorously scratch the painted surface. Note: Lead may be present underneath the first layer of paint.

STEP 3 DIP
Dip white non-abrasive end of test dauber in the spoon of water. White end of test dauber should turn pale yellow.

STEP 4 RUB
Firmly rub white non-abrasive end of test dauber on surface being tested for 10 seconds.

RESULTS: Allow 2 minutes for color change to appear. If non-abrasive end of test unit or test surface turns a pink color, test surface contains lead. If non-abrasive end of test unit or test surface does not change color, confirm negative test by rubbing underneath flap on the test confirmation card. If dauber or test spot on confirmation card turn pink, the test was performed correctly, and indicates a negative test. If the dauber or test spot on confirmation card do not turn pink, repeat the test with new unused test dauber. If you are uncertain about test results, or if you have concerns regarding a positive outcome of your test, contact your local Health Department for further information regarding lead. KNOW LEAD detects presence of lead in paint at .06% or greater. A negative test does not guarantee 0% presence of lead.

POSITIVE

NEGATIVE

WASH HANDS AFTER USE • DISPOSE OF PROPERLY • Not intended for use by children

Figure B UF-33.02

Figure 33.4 Latex flat wall paint is widely used on interior walls.

The *vehicle* is a nonvolatile fluid that suspends the body particles in solution. It is composed primarily of drying oil with small amounts of thinner and dryer. *Thinners* are volatile and they evaporate. They are used to regulate the flow of the paint. Turpentine, a product formed by distilling gum from pine trees, is a high-quality thinner. Naphtha and benzene are used in some formulations. *Driers* accelerate the oxidation and hardening process. Organic salts of iron, zinc, cobalt, and manganese are commonly used driers.

Oil-based paints are permeable to water vapor and thus minimize blistering over porous surfaces, such as wood. They should not be used in corrosive or alkaline conditions. Since they have excellent wetting properties, they are widely used as primers.

Phenolic Coatings

Phenolic coatings are made by polymerization of phenol and a formaldehyde reactant. They are solvent based and dry by the evaporation of the solvent, leaving a strong flexible coating. They are used on exterior concrete, plaster, wood, metal, and gypsum wallboard. Phenolic coatings are used where resistance to acids, alkalies, and some solvents is required as well as immersion in hot distilled water.

Two-part phenolic coatings have a catalyst added at the site and harden due to a chemical reaction. This coating is useful to resist very harsh conditions, such as chemical fumes.

Urethane or Polyurethane Coatings

Urethane resin coatings are available as one-part and two-part formulations. One-part coatings are moisture-cured (reaction to atmospheric moisture) and are generally clear (no pigment). They have better abrasion resistance than alkyd enamels. An oil-modified one-part urethane coating is available. Although it has better gloss retention, it has lower resistance to chemical attack.

Two-part formulations range from a hard to a rubberlike surface film. Adhesion to steel and concrete is poor, and the surfaces must be carefully prepared. It has good resistance to abrasion, water, and solvents.

Both one- and two-part coatings are used for heavy-duty wall coatings and surfaces subjected to heavy traffic as gymnasium floors. They resist scrubbing, abrasion, and impact. They have better abrasion resistance than regular varnishes.

SPECIAL-PURPOSE COATINGS

Special-purpose coatings are formulated to meet a specific need and are therefore not suitable for most coating situations. Several widely used types include bituminous, asphalt, reflective, and fire-retardant coatings.

Bituminous Coatings

Bituminous coatings are formulated by dissolving natural bitumens, such as coal tar or asphaltic products, in an organic solvent. They are most effective when used in below-ground applications because they do not react well when exposed to sunlight. One special weather-resistant type is a solvent-based asphalt roof repair coating. It has fillers to give it body and prevent it from running. Another type has *coal tar emulsified* in water, which enables it to be used on damp roofs. It has good resistance to sunlight. *Coal-tar paints* are made with a coal-tar-produced solvent, such as coal-tar naphtha or xylol. They are often mixed with other synthetic resins to produce high-quality water-resistant coatings for metal.

Asphalt Coatings

Asphalt coatings are produced from petroleum and are found in a variety of emulsions, cold-applied paints, and enamels. They have good moisture resistance.

Reflective Coatings

Reflective coatings absorb the ultraviolet band of solar radiation and reflect it as visible light. The life expectancy of reflective coatings exposed to sunlight is about one year.

Pigmented Fire-Retardant Coatings

A variety of pigmented fire-retardant coatings are available from coating manufacturers. Their products are rated for surface burning characteristics on combustible and noncombustible substrates. Class 1 flame spread (0–25) is required by codes for many applications. Pigmented fire-retardant paints meet this requirement. Although these coatings retard the spread of flame on a surface, they do not protect the substrate from fire or heat. If substrate protection is required, an **intumescent coating** must be used.

Intumescent Fire-Retardant Coatings

Intumescent coatings develop a thick, rigid foam protective layer that insulates the substrate and prevents the spread of fire. They are applied to wood, hardboard, cellulose board, and other wood-based products. They are noncombustible and do not produce toxic fumes. One type uses a urea-formaldehyde resin with an intumescent agent. It is a water-based material that dries rapidly by evaporation of the water. When exposed to fire or heat of about 350°F (178°C) it expands and develops a thick, insulating mat hundreds of times thicker than the original paint film. Usually one or two coats are sufficient. The important factor is not the number of coats but the amount of coating applied per square foot. The insulating layer delays contact between the flames and the combustible material below it. This impedes flame spread on the surface, holds down smoke, and, since it retards heat transfer, delays the ignition of the substrate, giving occupants additional time to evacuate the building (Table 33.5). These coatings should not be applied over other coating materials, and they generally are not suited for exterior use because they are water sensitive.

STAINS

Stains used on exterior wood color the wood and protect it from the weather. Those designed for use on interior wood provide color, and protection is given by a transparent topcoat, such as varnish or lacquer.

Exterior Stains

Exterior stains are blends of oil, driers, resins, a coloring pigment, a wood preservative (such as creosote or pentachlorophenol), a **mildewcide,** and a water repellent. They are low in film formation. Low-pigmented stains are penetrating types made to soak into the wood. Heavily pigmented stains have the same formulation but contain more pigment. Heavily pigmented stains are frequently used on cedar shakes and shingles. They usually do not have the durability of paints. Both types are used on any type of exterior wood, such as siding, plywood, fencing, decks, and trim.

Exterior wood stains are available in solid and semitransparent types. Solid stains hide the color of the wood but allow the texture of the wood to be seen. Semitransparent stains allow the natural wood color to be seen (Fig. 33.5).

Stains that are oil- and water-based are available for application on wood. Oil-based alkyd stains are solvent thinned and are available in opaque and semitransparent types. Acrylic latex stains are water soluble.

Interior Stains

Interior stains are used on doors, trim, cabinets, and other wood products that require a furniture quality finish. Many types of stain are available for quality interior wood finishing. A summary of these is in Table 33.6. Notice the various vehicles used and that some are penetrating stains while others are pigmented stains. Penetrating stains are made with dyes and do not obscure the grain. Pigments in pigmented stains stay on the surface of the wood and obscure some of the grain.

Fillers

Woods fall into two general groups—close-grained, such as maple, and open-grained, such as oak and walnut. When finishing these woods, a filler is applied to get a smoother surface. Close-grained woods are rather smooth and use a liquid filler to seal the pores on the surface. Some form of varnish is often used as a liquid filler. Open-grained woods have visible open pores that have to be filled with a paste wood filler to get a smooth surface. Paste wood fillers contain a vehicle such as linseed oil, a solvent such as mineral spirits, silex (ground quartz), and a drier. The silex mixed with the linseed oil makes a paste. It is forced into the pores and the excess is wiped off before it hardens, leaving the pores full. The silex is a neutral-colored material, and color pigment can be added so the filled pores match the color of the wood.

FLAME SPREAD, SMOKE, AND FUEL CONTRIBUTED RATINGS

The speed at which flames spread over a surface is indicated by the **flame spread** rating of the material. Coatings manufacturers have their products tested, and they

Table 33.5 Fire Retardant Coatings[a]
Surface Burning Characteristics (Based on 100 for Untreated Red Oak)

Coating Type[b] Surface	DS - Clear Douglas Fir	PR - Clear Douglas Fir	PR - White Douglas Fir	PR - White Cellulose Board	"DS 11" - Clear Douglas Fir	"DS 11" - Clear Cellulose Board
Flame spread	5	5	5	5	10	10
Smoke developed	0	0	0	0	30	20
Number of preliminary coats	None	None	None	None	None	None
Number of fire-retardant coats	2	2	2	2	2	2
Rate per coat (sq. ft./gal.)	200	200	200	200	200	200
Number of overcoats	None	None	None	None	None	None

[a]Tests conducted in accordance with ASTM E/84 (UL 723 and ULC-S-102)
[b]Coatings tested are Exolit Fire Retardant Coatings manufactured by American Vamag Company, Inc.
Courtesy American Vamag Company, Inc.

Figure 33.5 Exterior stains color and preserve wood shingles and siding.

Table 33.6 Wood Stains Commonly Used for Interior Application

Type of Stain	Vehicle Solvent	Staining Action	Remarks
Alcohol	Alcohol	Penetrating	Dries quickly but fades easily Sold in powder form Will raise grain
Gelled wood stain	Mineral spirits or turpentine	Pigmenting	Slow drying and does not fade Sold in gelled form
Latex stain	Water	Pigmenting	Slow drying and does not fade Sold in liquid form
Non-grain-raising stain	Alcohol Glycol	Penetrating	Dries quickly and does not fade or raise the grain Sold in liquid form
Oil stain (penetrating)	Mineral spirits or turpentine	Penetrating	Dries quickly and fades easily but does not raise the grain Sold in liquid form
Oil stain (pigmenting)	Mineral spirits or turpentine	Pigmenting	Slow drying and does not fade Sold in liquid form
Penetrating resin stain	Mineral spirits	Penetrating	Sold in liquid form Contains stain and protective coating in one coating
Water	Water	Penetrating	Dries quickly but fades easily Sold in powder form Will raise the grain

provide these data to the consumer. Flame spread is generally specified for different situations by building codes.

Flame spread is measured by following ASTM tests that give noncombustible cement-asbestos board a value of 0 and untreated red oak a value of 100.

Flame spread ratings are specified as Class I (0–25), Class II (26–75), and Class III (76–200).

Other considerations when selecting a coating are the **fuel contributed rating,** which is an indication of the amount of combustible material in the coating, and the **smoke developed rating,** which classifies the coating by the amount of smoke produced as it burns. This rating is on a scale of 0 to 100. The nearer the rating is to 0 the less smoke is developed. Coatings manufacturers can supply these data. Fuel contributed and smoke developed ratings use the same scale as flame spread.

REVIEW QUESTIONS

1. What is the purpose of regulations related to volatile organic compounds in coatings?
2. What are the VOC limits for architectural and industrial maintenance coatings?
3. What are the major parts of a coating?
4. What are the two types of coating bases?
5. What do MWF and MDF mean when used in coating specifications?
6. What are the major things that affect the life of coatings?
7. What are the major natural resin varnishes?
8. What are the major synthetic resin varnishes?
9. What is a primer?
10. What methods are used to clean metal surfaces for painting?
11. What is the recommended moisture content for wood to be painted?
12. What are the major types of opaque coatings?
13. What are some special-purpose coatings?
14. What is meant by a flame spread rating?
15. What are the flame spread classifications?
16. What does a smoke developed rating indicate?
17. What does a fuel contributed rating indicate?
18. How do semitransparent exterior stains affect wood siding?

KEY TERMS

abrasion resistance Resistance to being worn away by rubbing or friction.

adhesion The ability of a coating to stick to a surface.

alkali A substance such as lye, soda, or lime that can be destructive to paint films.

alkyd Synthetic resin modified with oil for good adhesion, gloss, color retention, and flexibility.

binder Film-forming ingredient in paint that binds the pigment particles together.

bituminous coatings Coatings formulated by dissolving natural bitumens in an organic solvent.

clear coating A transparent protective and/or decorative film.

coating A paint, varnish, lacquer, or other finish used to create a protective and/or decorative layer.

durability The ability of a coating to hold up against destructive agents such as weather and sunlight.

efflorescence White soluble salts that are deposited on masonry and mortar surfaces caused by free alkalies leached from the mortar or concrete as moisture moves through it.

enamel A classification of paint that dries to a hard, flat semigloss or gloss finish.

epoxy varnish A clear finish having excellent adhesion qualities, abrasion and chemical resistance, and water resistance.

flame spread rate The rate at which flames will spread across the surface of a material.

fuel contributed rating A rating of the amount of combustible material in a coating.

hiding power The ability of a paint to hide the previous color or substrate.

intermediate coat A layer of coating between the primer coat and the topcoat whose main purpose is to provide structural strength.

intumescent coatings Coatings that expand to form a thick fire-retardant coating when exposed to flame.

lacquer A fast-drying clear or pigmented coating that dries by solvent evaporation.

latex A water-based coating such as styrene, butadiene, acrylic, and polyvinyl acetate.

mildewcide An agent that helps prevent the growth of mold and mildew on painted surfaces.

oil-based paints Paints composed of resins requiring a solvent for reduction purposes.

opaque coatings Coatings that completely obscure the color and much of the texture of the substrate.

pigments Paint ingredients mainly used to provide color and hiding power.

primer coat A first coat of paint that seals and fills the surface, inhibits rust, and improves adhesion of subsequent coats of paint.

resin A natural or synthetic material that is the main ingredient of paint and binds the ingredients together.

sealer A thin liquid coat applied to a surface to prevent previous coatings from bleeding through or to prevent unfinished wood from absorbing excessive amounts of the topcoat.

semitransparent coatings Coatings that allow some of the texture and color of the substrate to show through.

smoke developed rating A rating classifying the material by the amount of smoke developed as it burns.

solvent A thin liquid used to dissolve the resin and keep the ingredients in solution.

spreading rate The area to which a paint can be spread, expressed in square feet per gallon.

stain A solution of coloring matter in a vehicle.

substrate The material (wood, metal, masonry, etc.) to which the coating is applied.

topcoat The final coat of paint.

varnish A transparent coating that dries on exposure to air, providing a protective coating.

vehicle The liquid portion of a paint composed mainly of solvents, resins, or oils.

VOCs Volatile organic compounds released to the atmosphere as a coating dries.

water-based coatings Coatings formulated with water as the solvent.

SUGGESTED ACTIVITIES

1. Visit a paint store and collect color charts and samples. Prepare a display citing the properties and some suggested uses for each type of material collected.

2. Secure several lead-sampling kits and test finishes in several old local residences and campus buildings for lead. Write up your findings and submit a report to the class.

ADDITIONAL INFORMATION

Hornbostel, C., *Construction Materials: Types, Uses, and Applications*, John Wiley and Sons, New York, 1991.
Weismantel, G. E., *Paint Handbook*, McGraw-Hill, New York, 1981.

34

Gypsum, Lime, and Plaster

This chapter will help you to:

1. Learn the properties and applications of various gypsum, lime, and plaster products.

2. Make appropriate choices when selecting interior finish materials and designing to meet codes.

Gypsum is a mineral found in rock formations and is a hydrated calcium sulfate. It is identified by the chemical formula $CaSO_4 \cdot \frac{1}{2}H_2O$, which means it is a compound of lime, sulfur, and water. The mined rock is crushed, ground, and calcined (heated). Calcining drives off most of the chemically combined water. The **calcined gypsum,** called **plaster of Paris,** is then ground into a fine powder that is used to produce products such as wallboard and gypsum plaster. Various materials are added to produce the properties needed for the different gypsum products.

When calcined gypsum (plaster of Paris) is mixed with water, the plaster absorbs the water and hardens into a solid state. This bonds the gypsum and any additives into a solid, hardened mass.

GYPSUM PLASTER PRODUCTS

A plaster finish consists of a supporting base, such as gypsum or metal lath, and one or more coats of plaster. The main ingredient in plaster is usually gypsum, but some types use portland cement. Gypsum plasters meet the requirements of ASTM C28, Standard Specification for Gypsum Plasters, or ASTM C587, Standard Specification for Gypsum Veneer Plaster.

Gypsum plasters include gauging, wood-fiber, neat, and ready-mixed plasters. Following are plaster products related to interior finishing.

Plaster of Paris is a calcined gypsum mixed with water to form a thick pastelike mixture. It sets in fifteen to twenty minutes and is used in construction for ornamental plastering and repairing plaster walls. If mixed with lime putty, it can be used for a **finish coat** on plaster walls. It hardens rapidly and has little shrinkage.

Unfibered gypsum is a neat gypsum that has no aggregate or filler added. It is used to form the plaster **scratch coat** (first), **brown coat** (second coat), and any required leveling coats. There are three kinds. *Regular* is used with sand aggregate and is for hand application. *LW* is used with lightweight aggregates for hand application. *Machine application* gypsum is used with either sand or lightweight aggregates.

Gypsum neat plaster is calcined gypsum plaster mixed at the mill with other ingredients to control working quality and setting time. It may be fibered or unfibered. Aggregates such as sand, perlite, and vermiculite are used.

Fibered gypsum is a neat gypsum with cattle hair or organic fibers, such as sisal, added. The fibers hold the plaster together. It is not generally used for machine application. It is used for the base coats.

Wood fibered gypsum is a factory-made product requiring the addition of water on the job. There are two types: *regular* for use over gypsum lath and *masonry* for use over masonry surfaces.

Bond coat gypsum is prepared for use as a base coat on monolithic concrete surfaces that are smooth and dense and do not have sufficient suction for standard plaster base coats. Finish plaster is applied over this coat.

Gauging plaster is a coarsely ground gypsum plaster composed of screened particles that regulate the set time at definite time periods. It is available in quick set (30 to 40 min.) and slow set (50 to 70 min.) mixtures. It is mixed with slaked lime to make a finishing coat.

Keene's cement is a double calcined gypsum (almost all water removed). The two types are regular (slow setting) and quick-setting. It has a high strength and is highly resistant to moisture. It is used with lime and sand for a float or sprayed finish.

Casting plaster is made from plaster of Paris that has been ground to a finer powder than regular plaster of Paris. It is used for molding ornamental plaster such as cornices. Lime putty is added to increase plasticity. It sets up slower than regular plaster of Paris, which allows more time for forming the ornamental work.

Finish plaster is used to cover the brown (second) coat with a hard, smooth finish or putty coat that can be painted or papered. It is mixed with hydrated lime putty and water in proportions of one part plaster to three parts lime putty. *Prepared finish plaster* has no lime and only requires the addition of water. It does not dry white as finish plaster, but it does dry faster.

Acoustical plaster is a calcined gypsum mixed with a lightweight mineral aggregate. Several compositions of acoustical plaster are in use. These use either gypsum, lime, or Keene's cement with aggregates such as rock wool, pumice, perlite, or vermiculite. Some mixtures include a foaming agent that forms small voids, producing a porous plaster.

Texture plaster is a type of finish plaster used to produce a rough, textured finished surface. It is applied in two coats over the plaster base coat or gypsum lath. The final coat is applied by sponge, brush, or trowel, depending on the desired texture.

Texture spray is a gypsum-based material that is mixed with water and applied with a spray machine as the final coat. The texture of the finished surface can be varied by adjusting the air pressure or spray orifice size or by varying the amount of water, thus changing the consistency of the mix by the addition of aggregate (Fig. 34.1). It is used to put a textured finish on gypsum wallboard plaster and monolithic concrete walls and ceilings.

Veneer plasters are specially formulated high-strength gypsum plasters that may be a thin monolithic base coat plaster over which a finish is applied or may be formulated for application as a finish plaster. Setting time and compressive strength are controlled by the composition prepared by the manufacturer. They may be applied over gypsum plaster base, masonry, or concrete surfaces.

Fireproofing plasters contain asbestos (where permitted), inorganic binders, or lightweight aggregates. They are sprayed directly onto bare steel shapes, providing fire protection as specified by codes. The finished coating is easily damaged and is not intended to be the exposed finish coat.

Joint compound is used to cover the heads of nails and the joints between sheets of gypsum wallboard. It is sold ready mixed for immediate use (Fig. 34.2). It is available with various setting times ranging from thirty minutes to six hours. The rapid chemical hardening and low shrinkage permit same-day finishing and next-day decorating.

GYPSUM BOARD PRODUCTS

Gypsum board is the generic name for a series of panel products having a noncombustible core of calcined gypsum with paper surfacing on the face, back, and long edges. It is often called drywall or wallboard. These products are manufactured to ASTM specifications.

Advantages of Gypsum Board

Gypsum board is highly *fire resistant* and is the major material used on walls, ceilings, floors, and other parts of buildings. Type X is formulated to meet fire code requirements. As it is exposed to fire, the chemically combined water is slowly released, retarding the transfer of heat through the panel. In addition, it has a *low flame spread index* and *smoke density index*. Type X panels meet the requirements of ASTM C36 and tests made by the Underwriters Laboratories.

Gypsum board is also an effective *barrier to the transfer of sound*, as through a partition or floor-ceiling construction. It is a *durable* material that produces high-quality walls and ceilings and maintains excellent dimensional stability. Gypsum products are inexpensive, easy to install, and accept a wide range of finishes.

Manufacturing Gypsum Board

Gypsum board products are made by mixing calcined gypsum with water and additives, forming a slurry. This is fed between continuous layers of paper on a board machine. As the board moves down a conveyer, the calcium sulfate recrystallizes or rehydrates, forming a solid core.

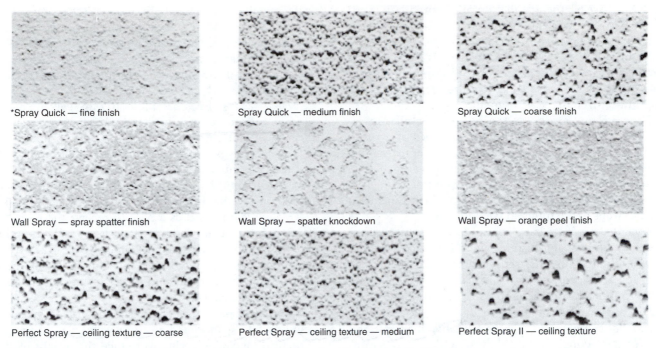

*Spray Quick — fine finish

Spray Quick — medium finish

Spray Quick — coarse finish

Wall Spray — spray spatter finish

Wall Spray — spatter knockdown

Wall Spray — orange peel finish

Perfect Spray — ceiling texture — coarse

Perfect Spray — ceiling texture — medium

Perfect Spray II — ceiling texture

Figure 34.1 Textures typically used to finish interior drywall ceilings and walls. The gypsum product can also be used on plaster and monolithic concrete walls. *(Courtesy Gold Bond Building Products, National Gypsum Company)*

The gypsum bonds to the paper. Finally, the board is cut to size and passed through dryers to remove excess moisture.

Types of Gypsum Board Products

Regular gypsum wallboard is used as the finished surface of walls and ceilings. It is highly fire resistant because it has a gypsum core. It has an off-white paper on the finish side and a gray paper on the back side of the gypsum core. It is available with an aluminum foil covering on the back that acts as a vapor barrier. The standard sheet is 4 × 8 ft. Metric sizes are approximately 1200 × 2400 mm. Sheets 10, 12, and 14 ft. long are available. It is available in thicknesses of ¼, ⁵⁄₁₆, ⅜, ½, and ⅝ in. (6.4, 8, 9.5, 12.7, and 16 mm). The edges manufactured are tapered, square, beveled, rounded, or tongue and groove (Fig. 34.3).

Fire-resistant gypsum board (Type X) has improved fire-resistance qualities due to the addition of fire-resistive materials in the core. It is used in assemblies that must meet fire code ratings. It is available in ½ in. (12 mm) and ⅝ in. (16 mm) thicknesses. Some types have an aluminum foil back layer that serves as a vapor barrier.

Predecorated gypsum board has the finish surface covered with a printed, textured, painted, coated, or vinyl film. It needs no further treatment after installation.

Water-resistant gypsum board has a water-resistant gypsum core and a green water-repellent paper. It is

Figure 34.2 Drywall joint compound is available ready mixed for immediate use.

used as a base for the installation of tile and plastic panels in bath, shower, kitchen, and laundry areas. It is available with a regular or fire-resistant core and in ½ in. (12 mm) and ⅝ in. (16 mm) thickness. It is not recommended for use on ceilings with joist spacing greater than 12 in. because the weight of the tile ceiling could exceed the strength of the panel.

Backer board is used as the base ply or plies in assemblies that require more than one layer of gypsum board. It is also used as a backing behind acoustical tile, plywood paneling, and other decorative wall paneling. It is available in regular and fire-resistant panels.

Gypsum coreboard is a 1 in. (25 mm) thick panel used in shaft walls and laminated gypsum panels.

Gypsum liner board has a special fire-resistant core enclosed in a moisture-resistant paper. It is used as a liner panel on shaft walls, stairwells, chaseways, corridor ceilings, and area separation walls. It is in ¾ in. (18 mm) and 1 in. (25 mm) thicknesses and 4 in. (610 mm) and 48 in. (1220 mm) widths.

Exterior gypsum soffit board is available in regular and fire-resistant core in ½ in. (12 mm) and ⅝ in. (16 mm) thicknesses. It is used in exterior areas such as soffits, canopies, and carport ceilings where it has indirect exposure to the weather.

Gypsum sheathing panels have a fire-resistant and water-resistant gypsum core faced with specially treated water-repellent paper on both faces and the long edges. It is used on wood and steel framing in residential and commercial construction. Panels are available with beveled, square, and V-shaped tongue-and-groove edges. It is available in ½ in. (12 mm) and ⅝ in. (16 mm) thicknesses.

Sound-deadening board is a ¼ in. (6 mm) gypsum panel used in connection with fire-resistant gypsum panels to meet the requirements for sound and fire resistance.

Lay-in ceiling panels are made from ½ in. (12 mm) fire-resistant gypsum wallboard. They are cut into 2 × 2 ft. (610 × 610 mm) and 2 × 4 ft. (610 × 1220 mm) panels for easy installation in suspended ceiling grids (Fig. 34.4).

Gypsum lath is used as a base to receive the layers of hand-laid or machine-laid plaster. It is available in 16 in. (406 mm) and 24 in. (610 mm) panels 48 in. (1220 mm) long. It is made in ⅜ in. (9 mm) and ½ in. (12 mm) thicknesses. It may have an aluminum foil backing that provides heat reflective qualities.

Gypsum fiberboard is a composite material of gypsum, cellulose fibers, and fibers from recycled newspapers. It sometimes includes perlite to reduce the weight of the panel. The cellulose fiber reinforces the gypsum that surrounds it, making a product that works more like particleboard than the typical gypsum wallboard. It is a solid material and does not have the paper facing used on conventional wallboard. This increases its fire resistance.

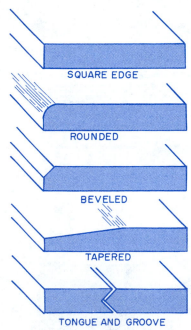

Figure 34.3 Regular gypsum wallboard is available with a variety of edge shapes.

Figure 34.4 Gypsum panels are used on suspended ceilings. *(Courtesy Chicago Metallic Corporation)*

It can be installed with a pneumatic stapler, nail gun, or drywall screws in wood stud applications. It is strong enough to permit mirrors and towel racks to be anchored to it rather than to studs. It is more stable than conventional gypsum wallboard, is moisture, fire, impact, mildew, and sound resistant. The seams are not taped as are conventional gypsum wallboard but are filled and smoothed with two layers of a specially formulated joint compound. The panels are available in light weight and a heavier, high-density panel. They are available 4 ft. wide and in stock lengths of 8, 10, and 12 ft. (2440, 3050, and 3660 mm). Widths to 8 ft. (2440 mm) and lengths to 20 ft. (6100 mm) are available by special order. Panels are made in thicknesses of ⅜, ⁷⁄₁₆, ½, and ⅝ in. (10, 11, 13, and 16 mm). Panels are available with four square edges or two or four tapered edges. It is available for use as interior wall finish, floor underlayment, tile backing, and as gypsum sheathing.

Poured gypsum roof decks are formed by pouring gypsum concrete over permanently installed *gypsum formboards*. The formboards are supported by structural steel roof framing. The gypsum is in a thick, creamy condition and is pumped through a hose to the deck. A construction detail for a poured gypsum roof deck is in Fig. 34.5.

APPLICATIONS

Various plaster products are used for finish wall and ceiling construction in residential and commercial construction and for cast ornamental features. Plaster is used for sound attenuation and fire protection of structural members, such as steel beams and columns (Fig. 34.6). It is applied over gypsum plaster base panels and metal and wire lath that has been secured to the structural framing.

Gypsum wallboard products are used for finishing interior walls and ceilings (Figs. 34.7 and 34.8), providing sound attenuation, fire resistance on walls and ceiling, and fire protection of structural members, such as steel beams and columns. Some types have water-resistant coverings and are used on walls to be tiled. Other types are used for exterior sheathing. Gypsum wallboard is also used over subfloors to increase the fire resistance of the assembly. Installation details and applications on interior walls and ceilings are illustrated in Chapter 36.

LIME

Lime is produced by burning limestone, marble, coral, or shells in a kiln at 2000°F (1100°C) to remove the carbonic acid gas. The purest lime (97 percent) is used in plaster and is a *finishing lime*. The lime produced by the kiln is called quicklime. It has the capacity to **slake** or **hydrate** when allowed to soak up to two or three times its weight in water. Slaking or hydration is a chemical reaction that causes the temperature of the mix to rise rapidly. This mixture is allowed to cool and sit three weeks before it can be used. Contractors buy lime already slaked so they do not have this waiting period.

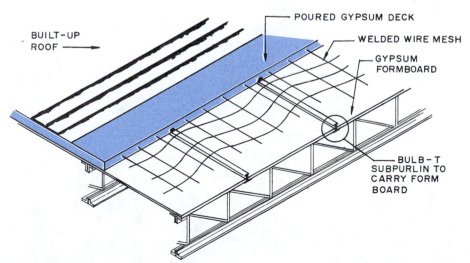

Figure 34.5 Gypsum formboard is supported on the structural members and provides a floor for pouring the gypsum roof deck. The formboard remains in place after the roof deck has hardened.

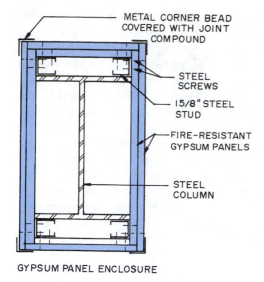

METAL CORNER BEAD
COVERED WITH JOINT
COMPOUND

STEEL
SCREWS

15/8" STEEL
STUD

FIRE-RESISTANT
GYPSUM PANELS

STEEL
COLUMN

GYPSUM PANEL ENCLOSURE

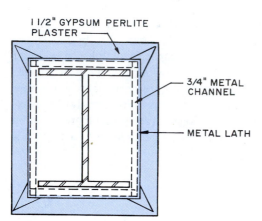

1 1/2" GYPSUM PERLITE
PLASTER

3/4" METAL
CHANNEL

METAL LATH

GYPSUM PERLITE PLASTER
ENCLOSURE

Figure 34.6 Gypsum panels are used to provide fire protection to structural members.

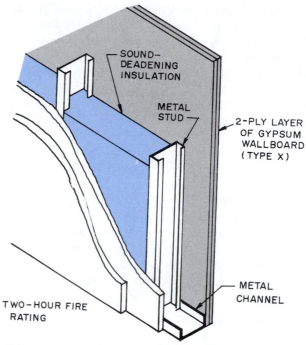

SOUND-
DEADENING
INSULATION

METAL
STUD

2-PLY LAYER
OF GYPSUM
WALLBOARD
(TYPE X)

METAL
CHANNEL

TWO-HOUR FIRE
RATING

Figure 34.7 A typical interior wall with a two-ply gypsum wallboard finish.

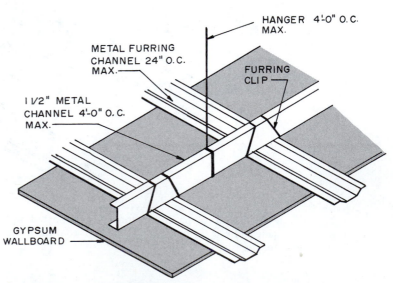

HANGER 4'-0" O.C.
MAX.

METAL FURRING
CHANNEL 24" O.C.
MAX.

FURRING
CLIP

1 1/2" METAL
CHANNEL 4'-0" O.C.
MAX.

GYPSUM
WALLBOARD

Figure 34.8 Gypsum wallboard can be secured to suspended steel channels, forming a fire-resistant ceiling.

The slaked quicklime is transformed by the chemical reaction into **hydroxide of lime,** a fine dry powder. This is the lime product used in plaster. There are two types, Type N (normal) and Type S (special). Type N must be soaked twelve to sixteen hours before it can be added to the plaster. Type S can be added to the plaster as soon as it is mixed with water.

CEMENT BOARD

Cement board is a product made with an aggregated portland cement core reinforced with polymer-coated glass-fiber mesh embedded in both surfaces (Fig. 34.9). Although it is not a gypsum product, it is used for some of the same applications. Cement board is a durable, fire-resistant, and water-resistant panel used on load-bearing and non-load-bearing walls on interior and exterior surfaces. It can be attached to wood or steel framing with special screws for each application. Holes must be drilled to accept fasteners. Galvanized roofing nails can be used to secure it to wood framing. Joints can be covered with an open glass-fiber mesh tape. The panels can be finished with a portland cement mortar containing dry latex polymers. These materials are supplied by the manufacturer, and the manufacturer's recommendation should be carefully followed.

Cement board may be used in areas with high humidity, such as baths, kitchens, and pools. It is also used for soffits, fences, and chimney enclosures. It is used as a substrate for the application of ceramic tile, slate, and quarry tile on interior surfaces. It can be used as a heat shield behind stoves and heaters.

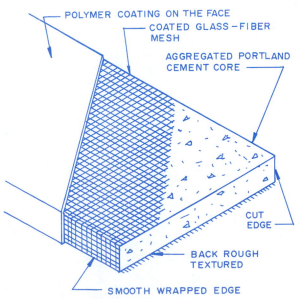

Figure 34.9 Cement board is used when the wall or ceiling will be exposed to water or high humidity.

REVIEW QUESTIONS

1. What use is made of casting plaster?
2. Where is perlite gypsum plaster used?
3. What gypsum plaster product is used as a base coat on dense, smooth concrete surfaces?
4. What gypsum product would you use in areas subject to high moisture?
5. What material is used to cover the heads of nails or screws that hold gypsum wallboard to the studs?
6. How does gypsum wallboard retard the spread of fire?
7. What are the advantages of gypsum wallboard?
8. What sizes of regular gypsum wallboard are available?
9. Where is water-resistant gypsum board used?
10. What is a gypsum backer board?
11. Where is gypsum liner board used?
12. Where can exterior gypsum soffit board be used?
13. What two types of gypsum sheathing are available?
14. What is the major use for gypsum board substrate?
15. What are the differences between the uses for gypsum base and gypsum lath?
16. How are poured gypsum roof decks constructed?

KEY TERMS

acoustical plaster Calcined gypsum mixed with lightweight aggregates.

brown coat The second coat of gypsum plaster. It is applied over the scratch coat.

calcined gypsum Ground gypsum that has been heated to drive off the water content.

cement board A panel product made with an aggregated portland cement core reinforced with polymer-coated glass fiber mesh on each surface.

fibered gypsum A neat gypsum with cattle hair or organic fillers added.

finish coat The third or final coat of gypsum plaster. It is the finished, exposed surface.

finish plaster The final top coat of plaster on a wall or ceiling.

fire-resistant gypsum A gypsum product that has increased fire resistant properties due to the addition of fire resistive materials in the gypsum core.

gypsum A mineral, hydrated calcium sulfate.

gypsum board A panel product having a gypsum core and covered on both sides with special papers.

gypsum neat plaster A gypsum plaster with no aggregates or fillers added. Sometimes called unfibered gypsum.

hydrate Capacity of lime to soak up water several times its weight.

hydroxide of lime The product produced by the chemical reaction during the slaking or hydrating of lime.

lime A white to gray powder produced by burning limestone, marble, or coral or shells.

plaster of Paris A calcined gypsum mixed with water forming a thick paste-like mixture.

scratch coat The first coat of gypsum plaster. It is applied to the lath.

slake Capacity of lime to soak up water several times its weight.

texture plaster A finish plaster used to produce rough, textured finished surfaces.

unfibered gypsum A neat gypsum. It has no additives.

SUGGESTED ACTIVITIES

1. Lay up some gypsum panels on a mock stud wall and practice taping the seams. Use various joint compounds and note the drying time and sanding properties.

2. Build a stud wall and cover it with gypsum lath and metal lath. Apply a three-coat plaster finish on each.

3. Try to visit a building under construction and observe the drywall crew installing and taping panels. If possible, try to be there when they are applying a textured finish.

ADDITIONAL INFORMATION

Gorman, J. R., Jaffe, S., Pruter, W. F., and Rose, J. J., *Plaster and Drywall Systems Manual,* BNI Books Division of Building News, Inc., Los Angeles, and Mc-Graw-Hill, New York, 1988.

Gypsum Construction Handbook, United States Gypsum Company, 125 South Franklin St., P.O. Box 806278, Chicago, Ill. 60680-4124.

Van Den Braden, F., and Hartsell, T. L., *Plastering Skills,* American Technical Publisher, Homewood, Ill., 1984.

Other resources are numerous related publications from the organizations listed in Appendix B.

35

Acoustical Materials

This chapter will help you to:

1. Understand the factors involved when considering the specifications and recommendations for the acoustical treatment of a building.

2. Select appropriate materials to provide acoustical control of various interior spaces.

Acoustics is the science of sound, including the generation, transmission, and effects of sound waves. Acoustical materials are those used to control sound. They are used to (1) reduce the *levels* of sound within an area, such as a room, by absorption and (2) to control the *transmission* of sound from one area to an adjacent area caused by the transmission of vibrations through the building structure.

SOUND

Sound is the sensation produced by the stimulation of the organs of hearing by vibrations transmitted through the air. This includes vibrations traveling in the air at a speed of about 1130 ft./sec. (345 m/sec.) at sea level and mechanical vibrations transmitted through an elastic medium, such as steel. Sound is the movement of air molecules moving in a wavelike motion. A sound wave produces changes in the atmospheric pressure above and below the existing static pressure. This deviation in atmospheric pressure is called *sound pressure*.

Sound Waves

Sound waves move in a spherical direction. That means that they move from the source in all three dimensions (Fig. 35.1). There is a delay between the time a sound is created and the time it is heard. Since it takes audio sound about a second to travel 1130 ft. (345 m) in the air, a person 2260 ft. (690 m) away will hear the sound two seconds after it is generated. When you see lightning in the distance you do not hear the thunder until several seconds later due to the differing speeds of sound and light. Sound travels at different speeds in various materials. In wood it travels about 11,000 ft./sec. (3355 m/sec.) and in steel about 16,000 ft./sec. (4880 m/sec.).

Frequency of Sound Waves

The **frequency** of sound is the number of *cycles* of like waveforms per second. Sound travels in sine waves as shown in Fig. 35.2. Wavelength equals the *velocity* of the sound (1130 ft./sec. in air), divided by the frequency (number of cycles). For example, a sound having a frequency of 15 cps would have a wavelength of 1130/15 or

801

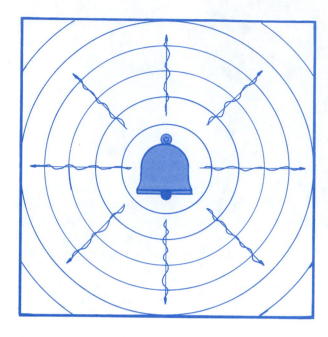

Figure 35.1 Sound travels in all directions from a generating source.

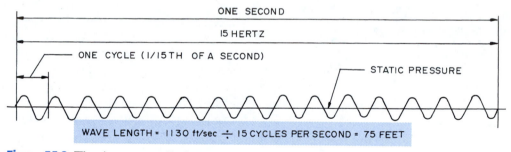

Figure 35.2 The sine wave audio frequency.

75 ft. The frequency, cycles per second, is expressed in units of hertz (Hz). A hertz is one cycle per second (cps), so in the above example the frequency is expressed as 15 hertz (15 cps = 15 Hz).

The frequency of sound determines the pitch. Low, deep sounds have low frequencies, and high-pitched sounds have high frequencies. The human ear can receive sounds ranging from a low of about 20 Hz to a high of about 20,000 Hz (or 20 KHz).

The frequency of sound to be considered varies, often rapidly and constantly. For example, the sound produced by music from a radio often ranges from low to high frequencies rapidly and often. Frequency from some other sources, such as a dishwasher, is more limited in range.

Sound Intensity

Sound intensity depends on the strength of the force that sets off the sound vibrations. Sound intensity is measured in **decibels** (dB). The decibel scale for normal applica-

tions ranges from 0 dB, which is just below the lowest audible sound (about 20 Hz), to 120 dB, which can produce a feeling of vibration in the ear. Intensities above this can cause actual damage to human hearing and structural damage to buildings. Representative sound levels in decibels for selected situations are in Table 35.1.

The intensity of sound varies in much the same way as described for frequency. Music could produce very high decibel levels, but a dishwasher produces lower, steady decibel intensity.

SOUND CONTROL

Sound generated within a space such as a room is controlled by acoustical materials that have sound-absorbing properties and by materials that reduce sound transmission through assemblies of materials, such as a wall or floor.

Sound-absorbing materials absorb some of the sound waves that strike them and reflect the rest into the area. The acoustical engineer selects materials to control the sound in the manner desired. Some materials, such as plaster walls, reflect most of the sound striking them into the room. As the sound hits other walls it is reflected again (Fig. 35.3). Sound-absorbing materials are porous and have openings in and out of which the vibrating air particles move. This movement causes friction, which generates heat. Some of the sound energy is lost as heat, which may be reflected into the room or transmitted through the material.

Sound Transmission Class

Sound transmission class (STC) is a single number rating that indicates the effectiveness of a material or an assembly of materials to reduce the transmission of airborne sound through the unit. The larger the STC number the more effective the material is as a sound

transmission barrier. Materials are tested following the specifications in ASTM E90-70.

Typical building code STC requirements for residences, apartments, and hotels are

1. All separating walls and floor-to-ceiling assemblies must provide an STC of 50.
2. All penetrations in assemblies, such as openings for piping and electrical wiring, must be sealed to maintain the STC 50 rating.
3. Entrance doors and their seals must have an STC rating of 26 or more.

Manufacturers of materials used to reduce sound transmission indicate the STC rating for each material and the various thicknesses. Figure 35.4 shows STC ratings for several assemblies of materials.

Table 35.1 Sound Pressure Levels of Selected Sounds

Sound Levels (decibels)	Source of Sound	Sensation
140	Near a jet aircraft	Deafening
130	Artillery fire	Threshold of pain
120	Elevated train, rock band, siren	Threshold of feeling
110	Riveting, air-hammering	Just below threshold of feeling
80–100	Power mower, thunder close by, symphony orchestra, noisy industrial plant	Very loud
60–80	Noisy office, average radio or TV, loud conversation	Loud
40–60	Conversation, quiet radio or TV, average office	Moderately loud
20–40	Average residence, private office, quiet conversation	Quiet
0–20	Whisper, normal breathing, threshold of audibility	Very faint

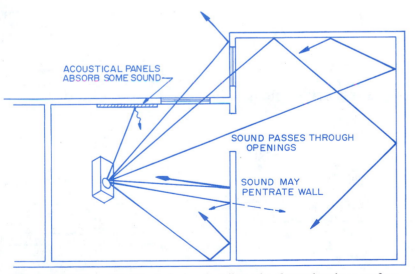

Figure 35.3 Sound is reflected when it strikes a hard nonabsorbent surface.

Impact Isolation Class

Impact isolation class (IIC) is a single number giving an approximate measure of the effectiveness of floor construction to provide isolation against the sound transmission from impacts. Impacts include walking, skidding, or dropping items on the floor, all of which set up vibrations that radiate to the area below. Floors with low IIC ratings can have improved performance by applying carpet or other sound-absorbing materials on them. Several design suggestions are in Figs. 35.4 and 35.5. IIC ratings of 45 to 65 are common for floor-ceiling assemblies in multifamily dwellings. Building codes typically require that floor-ceiling assemblies have an IIC rating of 50.

Noise Reduction Coefficient

The **noise reduction coefficient (NRC)** is an indication of the amount of airborne sound energy absorbed by a material. The single number rating is the average of the sound absorption coefficients of an acoustical material at frequencies of 250, 500, 1000, and 2000 Hz. The larger the NRC number the greater the efficiency of the materials to absorb sound. Some typical examples are in Table 35.2.

ACOUSTICAL MATERIALS

The control of sound transmission and the absorption of sound can be accomplished using a wide variety of materials. Some standard construction materials, such as brick or concrete block, are used to control sound transmission. Other materials are especially designed for acoustical purposes. Many of these materials are discussed in detail in other chapters.

Floor Coverings

The installation of *carpet* and *vinyl composition floor covering* reduce impact noise and dampen airborne noise. Carpeting is much more effective than vinyl floor covering and is even better when installed over a cushion. See Chapter 38 for additional information on carpets.

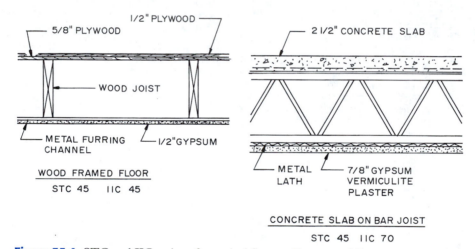

Figure 35.4 STC and IIC ratings for typical floor-ceiling assemblies.

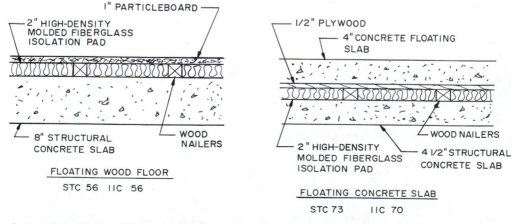

Figure 35.5 STC and IIC ratings for typical floating floor construction.

Table 35.2 Noise Reduction Coefficients for Selected Materials

Material	NRC
Unpainted brick wall	0.02–0.05
Painted brick wall	0.01–0.02
Glazed clay tile	0.01–0.02
Concrete wall	0.01–0.02
Lightweight concrete block	0.45
Heavyweight concrete block	0.27
Standard plaster wall	0.01–0.04
Gypsum wall	0.01–0.04
Acoustical plaster wall	0.21–0.75
Glass	0.02–0.03
Fiberglass	0.50–0.95
Wood panel	0.10–0.25
Mineral wool	0.45–0.85
Acoustical tile	0.55–0.85
Carpeting	0.45–0.75
Vinyl floor covering	0.01–0.05

Acoustical Plaster

Acoustical plaster is composed of a plaster made with perlite or vermiculite aggregate. It may be applied with a brush, but it usually is sprayed on the surfaces. It has the advantage of uniformly covering curved and irregular shapes. The plaster is applied in several layers, with the finished job about ½ in. (12 mm) thick. It has an NRC of about 0.21 to 0.75. See Chapter 34 for more information about plaster.

Another form of spray-applied acoustical wall and ceiling covering is not a plaster but is composed of cellulosic fibers in a bonding agent. It will bond to almost any surface and provides acoustical control and thermal insulation.

Acoustical Ceiling Tile and Wall Panels

A variety of ceiling tiles and wall panels are available. They are made from various materials such as wood fibers, sugarcane fibers, mineral wool, gypsum, fiberglass, aluminum, and steel. Most ceiling tiles have some form of perforation in their surface. Drilled or punched holes in a variety of patterns are common. Some tiles have slots, fissures, or striations (Fig. 35.6). Other types use a sculptured, irregular molded surface. Following are brief descriptions of some of those available.

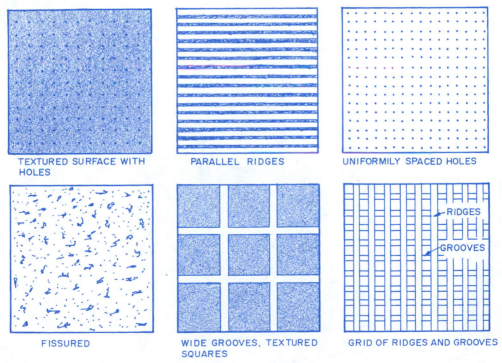

TEXTURED SURFACE WITH HOLES

PARALLEL RIDGES

UNIFORMILY SPACED HOLES

FISSURED

WIDE GROOVES, TEXTURED SQUARES

GRID OF RIDGES AND GROOVES

RIDGES

GROOVES

Figure 35.6 A few of the acoustical surface treatments used on ceiling tile and panels.

Wood fiber and *sugar cane fiber ceiling tiles* are available in thicknesses ranging from ⅝ in. (16 mm) to 1½ in. (38 mm). Tile sizes commonly available include 12 × 12 in. (305 × 305 mm), 12 × 24 in. (305 × 610 mm), 24 × 24 in. (610 × 610 mm), and 24 × 48 in. (610 × 1220 mm).

Ceiling tiles and wall panels are also made from *molded mineral fibers* and are cast having a wide choice of surface textures. Some manufacturers use *mineral fiber* or *fiberglass* sound-deadening panels covered with a fabric. This provides an attractive finished ceiling or wall panel. This type has STC ratings of 35 to 40, NRC ratings of 0.60 to 0.80, and flame spread ratings of 0 to 25.

Some ceiling tile and wall panels can resist damage from moisture and bumps. One type is a *ceramic ceiling tile* made from mineral fibers in a ceramic bond. It is fire re-sistant and is used in areas of high humidity. It has STC ratings of 40 to 44, NRC ratings of 0.50 to 0.60, and flame spread rating 0. Another type uses *perforated aluminum or steel tiles.* One type applies a vinyl coating over a perforated aluminum tile that is backed by a mineral fiber substrate. The sound passes through the holes and is absorbed by the mineral fiber. Metal tiles have STC ratings of 35 to 45, NRC ratings of 0.60 to 0.70, and flame spread rating of 0. Another type that resists damage is made by bonding wood fibers into a porous panel resembling particleboard. It resists moisture and heavy blows.

Lightweight panels are made that have a vinyl covering over a laminate of high-density molded glass fiber bonded to a core of 1 or 2 in. (25 to 50 mm) sound-absorbing fiberglass (Fig. 35.7). Wall panels

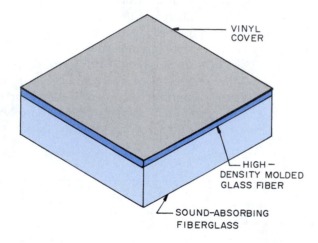

Figure 35.7 A fabric-covered acoustical panel with a tough layer that will resist damage from impact.

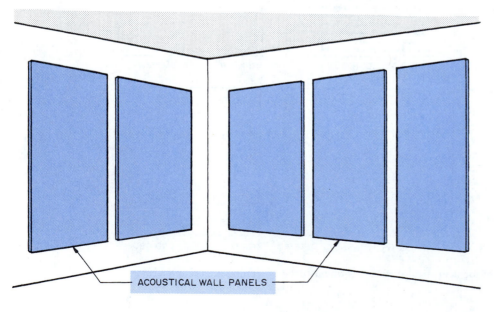

Figure 35.8 Acoustical wall panels of molded glass fiber are used to control interior sound reverbera-tions.

made from these materials are applied to walls in halls, restaurants, gymnasiums, offices, and other places where sound control is necessary (Fig. 35.8). One type uses a perforated zinc-coated steel or aluminum panel bonded to a 2 in. (50 mm) thick fiberglass panel. These panels are mounted several inches off the wall (Fig. 35.9).

Another type of ceiling panel used is baffles. *Baffles* are acoustical panels hung from the ceiling to reduce airborne sound in a large space such as a factory, restaurant, or auditorium (Fig. 35.10). One type uses a lightweight panel of vinyl-covered fiberglass. Panels are 1 in. (25 mm) thick, 10 in. (254 mm) high, and from 2 to 4 ft. (610 to 1220 mm) long (Fig. 35.11). Another type uses a flame-resistant polyethylene cover over a 1½ in. (38 mm) fiberglass core. It is flexible, like a blanket,

and is hung from the ceiling. Another type of baffle uses rigid wood fiber sound-deadening panels. These baffles can be joined to form an attractive grid on the ceiling (Fig. 35.12).

Sculptured acoustical wall units made from a high-density molded fiberglass layer bonded to a sound-absorbing glass fiber blanket are used to absorb sound and provide a decorative feature. They are covered with a wide range of fabrics. Typically they are in round, octagonal, and triangular shapes (Fig. 35.13).

Some ceiling tiles are designed to be glued to a ceiling substrate and some can be nailed or stapled to the ceiling. Tile to be installed with an adhesive has nut-size daubs of adhesive placed on the back. The tile is pressed against the ceiling substrate and slid into place (Fig. 35.14).

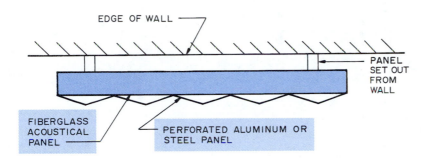

Figure 35.9 A perforated metal acoustical panel.

Figure 35.10 Acoustical materials are used on ceiling baffles to reduce sound reverberations. Notice also the use of acoustical wall panels. *(Courtesy Metal Building Interior Products Co.)*

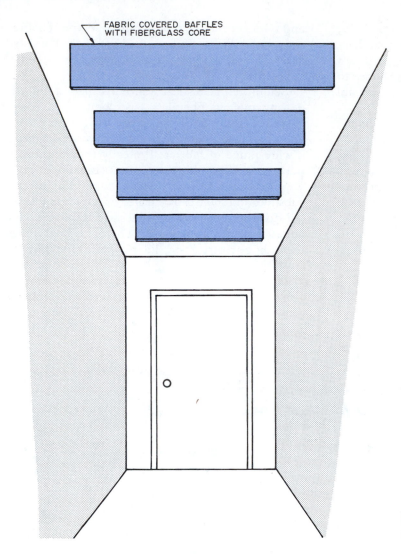

FABRIC COVERED BAFFLES
WITH FIBERGLASS CORE

Figure 35.11 Baffles at the ceiling used to control sound in a hall.

If ceiling tiles are to be installed to the bottom of floor joists without a substrate, wood 1 × 3 in. (25 × 75 mm) strips are nailed perpendicular to the joists at 12 in. (305 mm) on center spacing. The ceiling tiles are nailed or stapled through their tongues to the wood strips (Fig. 35.15).

Ceiling tiles and panels are made with a variety of edge shapes, which produces different appearances. Some of these are shown in Fig. 35.16.

The suspended ceiling system is widely used in commercial construction and finds some use in residential work. It consists of a grid of metal runners hung from overhead on wires into which acoustical ceiling panels fit. Light fixtures are also designed to fit into this grid. A typical system is shown in Fig. 35.17. The space be-

tween the metal grid and the overhead structure can be used to run mechanical and electrical systems.

A number of grid systems are available, but basically they fall into three types—exposed grid, semiexposed grid, and concealed grid. The manufacturer produces acoustical tiles with edges designed to fit their grid, producing the type of exposure desired. *Exposed systems* have the main runner and cross runner exposed. This produces a square or rectangular grid appearance on the ceiling. The acoustical units rest on the flange of the T-shaped runner. *Semiexposed systems* have the main runner exposed and the cross runner concealed. This gives a series of long parallel lines in the ceiling. The *concealed system* has no metal runners showing (Fig. 35.18).

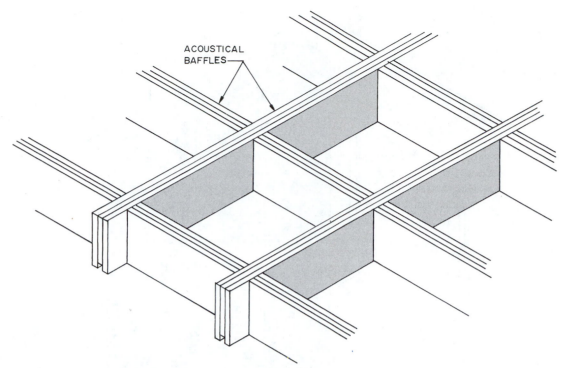

ACOUSTICAL
BAFFLES

Figure 35.12 Decorative baffles in a grid mounted on the ceiling.

Sound Barriers

A wide range of material can be used to block the transmission of sound through an assembly, such as a wall. Many of these, such as brick, concrete, concrete block, and gypsum, are discussed in earlier chapters. These units provide mass. Mass blocks the transmission of sound. For example, a 4 in. (100 mm) brick or concrete masonry unit has an STC of about 40. A 4 in. (100 mm) concrete floor has an STC of 44.

Lead sheet material is used in commercial buildings to block sound transmission. For example, a lead sheet can be laid on top of a suspended ceiling or hung from the bottom of the floor above, blocking the transmission of sound that has penetrated the ceiling. Lead is available in sheets and foils as thin as 0.0005 in. (0.013 mm). It is used in walls, doors, and other areas where a thin but effective sound barrier is needed.

Acoustical sealant is available as a pumpable material and is applied to all openings through which sound may penetrate. For example, space around electrical boxes and pipes that pierce a wall or floor must be sealed. Openings, even very small ones, reduce the effectiveness of an otherwise efficient sound barrier.

CONSTRUCTION TECHNIQUES

There are ways to construct walls, floors, and ceilings to reduce sound transmission. Those that follow show some of the basic ways in common use.

Several wall assemblies are shown in Fig. 35.19. It shows three frequently used methods for increasing the STC rating of wood and metal frame walls. All of these could have sound-deadening insulating batts installed to increase the STC value.

Several floor-ceiling assemblies are shown in Figs. 35.4 and 35.5. The floating floor technique shown is used in commercial construction. The same type of construction can be used with wood frame floors as used in residential work.

CONTROLLING SOUND FROM VIBRATIONS

Mechanical equipment is mounted on the structural frame of buildings. This produces vibration sounds that travel for great distances to other parts of the building.

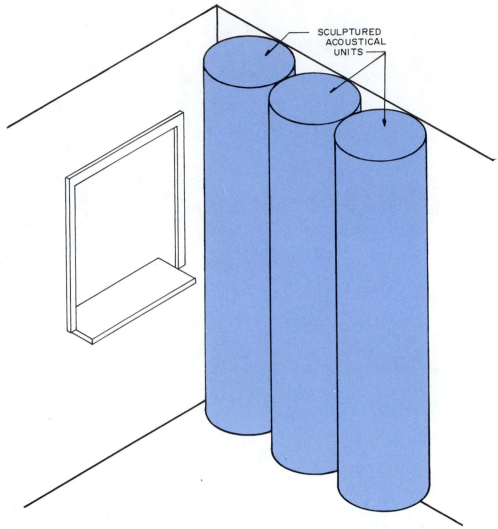

Figure 35.13 Sculptured acoustical shapes used along a wall to control sound and serve as a decorative feature.

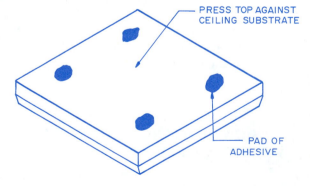

PRESS TOP AGAINST
CEILING SUBSTRATE

PAD OF
ADHESIVE

Figure 35.14 Ceiling tile can be adhered to a solid ceiling substrate with pads of adhesive.

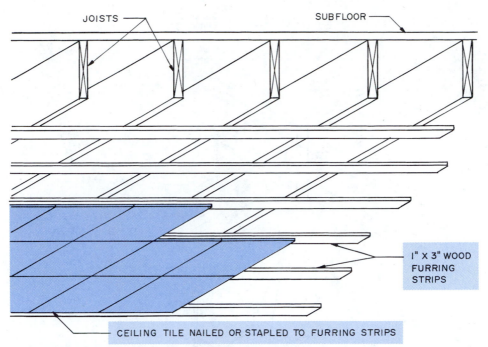

JOISTS

SUBFLOOR

I" X 3" WOOD FURRING STRIPS

CEILING TILE NAILED OR STAPLED TO FURRING STRIPS

Figure 35.15 Ceiling tile can be installed by nailing or stapling to wood furring strips.

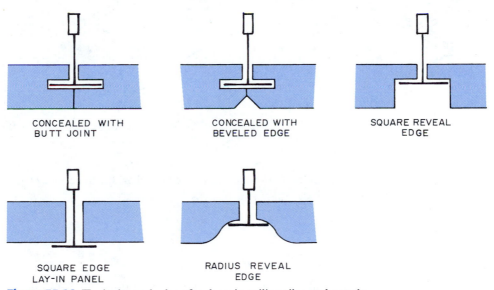

CONCEALED WITH BUTT JOINT

CONCEALED WITH BEVELED EDGE

SQUARE REVEAL EDGE

SQUARE EDGE LAY-IN PANEL

RADIUS REVEAL EDGE

Figure 35.16 Typical panel edges for drop-in ceiling tiles and panels.

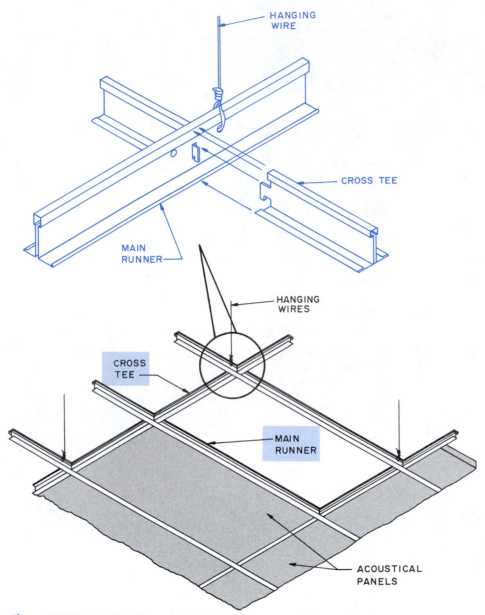

Figure 35.17 A view looking up from below at a suspended metal grid with acoustical panels resting on the flanges.

Vibrating water and sewer pipes also create sounds that need to be controlled. Much of the sound can be dampened by mounting the mechanical equipment on various types of *isolator pads*. Rubber, neoprene, cork, and fiberglass pads are used for small pieces of equipment or larger pieces located in a basement area (Fig. 35.20).

For heavier duty units mounted within the building or on the roof, a steel spring mounting is used. The base of the spring pad must itself be isolated from the structure with an isolator pad to prevent the transmission of audible high-frequency vibration through the spring to the structure.

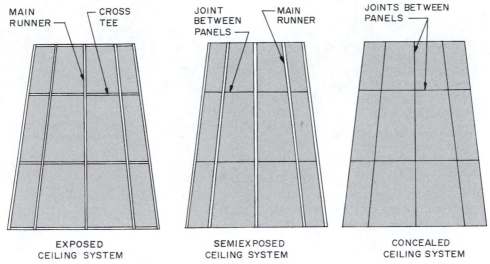

MAIN RUNNER — CROSS TEE

JOINT BETWEEN PANELS — MAIN RUNNER

JOINTS BETWEEN PANELS

EXPOSED CEILING SYSTEM

SEMIEXPOSED CEILING SYSTEM

CONCEALED CEILING SYSTEM

Figure 35.18 Types of suspended ceiling systems.

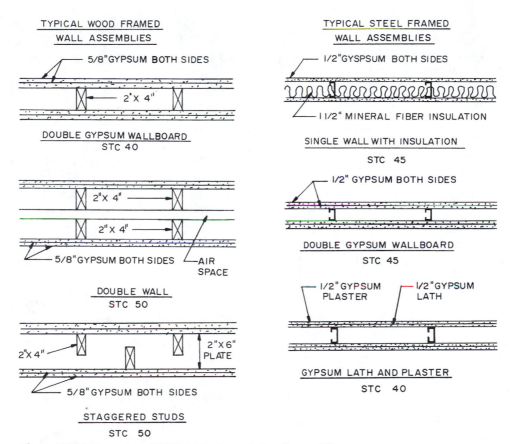

TYPICAL WOOD FRAMED WALL ASSEMBLIES

5/8"GYPSUM BOTH SIDES

2'X 4"

DOUBLE GYPSUM WALLBOARD
STC 40

2"X 4"

2"X 4"

5/8"GYPSUM BOTH SIDES — AIR SPACE

DOUBLE WALL
STC 50

2"X 4"

2"X 6" PLATE

5/8"GYPSUM BOTH SIDES

STAGGERED STUDS
STC 50

TYPICAL STEEL FRAMED WALL ASSEMBLIES

1/2"GYSPSUM BOTH SIDES

1 1/2" MINERAL FIBER INSULATION

SINGLE WALL WITH INSULATION
STC 45

1/2" GYPSUM BOTH SIDES

DOUBLE GYPSUM WALLBOARD
STC 45

1/2"GYPSUM PLASTER   1/2"GYPSUM LATH

GYPSUM LATH AND PLASTER
STC 40

Figure 35.19 Examples of STC ratings for typical wall assemblies.

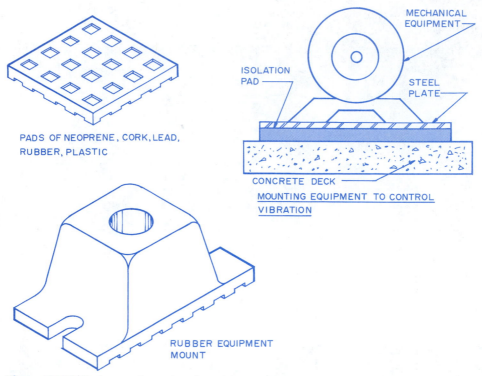

PADS OF NEOPRENE, CORK, LEAD, RUBBER, PLASTIC

MECHANICAL EQUIPMENT

ISOLATION PAD

STEEL PLATE

CONCRETE DECK

MOUNTING EQUIPMENT TO CONTROL VIBRATION

RUBBER EQUIPMENT MOUNT

Figure 35.20 Various pads used to damp the sound produced by vibrations of mechanical equipment.

REVIEW QUESTIONS

1. In what two ways do acoustical materials control sound?
2. What is sound pressure?
3. In what directions do sound waves move?
4. How long will it take a sound to travel one mile from its source?
5. What is the wavelength of a sound having a frequency of 25 cps?
6. What is the frequency in hertz of a sound having a frequency of 35 cps?
7. How does sound frequency influence the pitch of the sound?
8. What is the normal range of sound that can be heard by the human ear?
9. How many decibels of sound will damage a person's hearing?
10. What is the difference between sound-absorbing materials and sound-reflective materials?
11. What are typical STC and IIC requirements for apartment construction?
12. How does acoustical plaster differ from standard gypsum plaster?
13. What acoustic product is used to reduce airborne sound at the ceiling in large areas, such as an auditorium?
14. What are the three types of grids used on suspended ceilings?
15. How is sound transmission of vibrations from mechanical machinery controlled?

KEY TERMS

acoustics The science of the generation, transmission, and effects of sound waves.

decibel A unit used to express the intensity of a sound wave.

frequency The number of times per second sound pressure cycles above and below atmospheric pressure.

IIC (impact isolation class) A single number indicating the effectiveness of floor construction to provide isolation against sound transmission from impact.

NRC (noise reduction coefficient) A single number indicating the amount of airborne sound energy absorbed by a material.

sound A sensation produced by the stimulation of the organs of hearing by vibrations transmitted through the air. It is the movement of air molecules moving in a wavelike motion.

STC (sound transmission class) A single number indicating the effectiveness of a material or an assembly of materials to reduce the transmisfion of airborne sound.

SUGGESTED ACTIVITIES

1. Visit rooms used for various types of activities, such as a classroom, music room, auditorium, and mechanical room in your school. List the materials on the ceiling, walls, and floor plus any special acoustical treatments. Write a report explaining what might be done to improve the acoustics of the space.

2. Ask the director of the physical plant of your school to show you the devices used on mechanical equipment to reduce the transmission of vibrations into the structure. Write a report and make sketches of the various devices.

ADDITIONAL INFORMATION

Egan, M. D., *Architectural Acoustics,* McGraw-Hill, New York, 1989.

Hornbostel, C., *Construction Materials: Types, Uses, and Applications,* John Wiley and Sons, New York, 1991.

Ramsey, C. G., Sleeper, H. R., and Hoke, J. R., Jr., eds., *Architectural Graphic Standards,* John Wiley and Sons, New York, 1995.

Rettinger, M., *Handbook of Architectural Acoustics and Noise Control,* McGraw-Hill, New York, 1990.

Wood Building Technology, Canadian Wood Council, Ottawa, 1993.

Other resources are numerous related publications from the organizations listed in Appendix B.

CHAPTER

36

Interior Walls, Partitions, and Ceilings

This chapter will help you to:

1. Be aware of the code requirements for various interior walls, partitions, and ceilings.

2. Select appropriate designs and materials for interior walls, partitions, and ceilings.

3. Supervise the installation of the various types of finish walls and ceilings.

Interior walls and partitions may be load bearing or non–load bearing and constructed with wood or steel studs or masonry. Various types of finish material can be applied over the structure. Walls and partitions divide a building into areas having various types of occupancy and hide electrical and mechanical systems that may be run inside the wall cavity. They also provide privacy, security, sound control, insulation, and fire and smoke protection.

Ceilings provide an attractive finished surface as well as hide electrical and mechanical systems suspended above them. They also provide sound control and fire protection.

Interior finish on walls, partitions, and ceilings may be a gypsum board product, plaster, solid wood, plywood or hardboard panels, some type of fiberboard, or ceramic tile. Gypsum and plaster products are described in Chapter 34, wood products in Chapter 19, and tile products in Chapter 11. Procedures for finishing wood frame interiors are discussed in Chapter 22. Various interior finish materials and coatings are presented in Chapters 32 and 33.

INTERIOR WALLS AND PARTITIONS

In most cases in commercial construction interior walls and partitions are non–load bearing and serve to divide the open space into usable areas and to enclose openings such as stairs, shafts, and exits. The building structural system carries the loads of roof and floors, interior walls and partitions, and furniture and other items to be placed in the building. The requirements an interior wall or partition must meet are specified by the building code. In Table 36.1 are minimum fire resistance requirements of walls, partitions, and opening protective devices as specified in the Standard Building Code.

Non–load-bearing non–fire-resistant partitions are those not used as fire separation walls. They must meet the code requirements specifying the combustibility of materials for the type of construction in which they are to be used. In Table 36.2 are minimum interior wall finish classifications based on the occupancy of the building. They are based on the acceptable flame spread and smoke development classifications.

Table 36.1 Minimum Fire Resistance of Walls, Partitions, and Opening Protectives[1] (hrs)

Protectives	Walls and Component	Opening Partitions
SHAFT ENCLOSURES (including stairways, exits & elevators)		
4 or more stories	2	$1\frac{1}{2}$B
less than 4 stories	1^2	1B[2]
all refuse chutes	2	$1\frac{1}{2}$B
WALLS AND PARTITIONS		
fire walls[3]	4	3A
within tenant space	See 704.2.3	
tenant space (see also 704.3)	1	$\frac{3}{4}$C
horizontal exit	2	$1\frac{1}{2}$B
exit access corridors[4,5]	1	20 min.
smoke barriers	See 409.1.2	
refuse and laundry chute access rooms	1	$\frac{3}{4}$C
incinerator rooms	2	$1\frac{1}{2}$B
refuse and laundry chute termination rooms	1	$\frac{3}{4}$C
hazardous occupancy control areas	1	$\frac{3}{4}$C
high rise buildings	See 412	
covered mall buildings	See 413	
assembly buildings	See Note 2	
bathrooms & restrooms	See Note 6	
OCCUPANCY SEPARATIONS[7]	Required Fire Resistance	
	4	3A
	3	3A
	2	$1\frac{1}{2}$B
	1	$\frac{3}{4}$C
EXTERIOR WALLS[8]	All	$\frac{3}{4}$E

Notes:
See Standard Building Code for references.
1. Table 600 may require greater fire resistance of walls to insure structural stability.
2. All exits and stairways in Group A and H occupancies shall be 2 hours with $1\frac{1}{2}$–hour B door assemblies.
3. See also 503.1.2.
4. See 704.2.3.
5. See 409 for sprinkled Group 1- buildings.
Standard Building Code© 1994. Used with permission from the Standard Building Code, Southern Building Code Congress International, Inc.

Table 36.2 Minimum Interior Finish Classification

Occupancy	Unsprinklered			Sprinklered		
	Exits[1]	Exits Access	Other Spaces	Exits[1]	Exit Access	Other Spaces
A	A	A	B	B	C	C
B	B	B	C	C	C	C
E	A	B	C	B	C	C
F	C	C	C	C	C	C
H		Sprinklers required		B	C	C
I Restrained	A	A	C	A	A	C
I Unrestrained		Sprinklers required		B	B	B[3]
M	B	B	C	C	C	C
R[2]	B	B	C	C	C	C
S	C	C	C	C	C	C

Notes:
1. In vertical exitways of buildings three stories or less in height of other than Group I Restrained, the interior finish may be Class B for unsprinklered buildings and Class C for sprinklered buildings.
2. Class C interior finish materials may be used within a dwelling unit.
3. Rooms with 4 or less persons require Class C interior finish.
   1. Class A Interior Finish, Flamespread 0–25, Smoke Developed 0–450. Any element thereof when so tested shall not continue to propagate fire.
   2. Class B Interior Finish. Flamespread 26–75, Smoke Developed 0–450.
   3. Class C Interior Finish. Flamespread 76–200, Smoke Developed 0–450.
   Standard Building Code© 1994.
   Used with permission from the Standard Building Code, Southern Building Code Congress International, Inc.

Fire separation walls or *fire partitions* are installed to provide enclosure of shafts, floor openings, and exits and to subdivide an area. The construction and selection of materials are specified by codes for the type of construction involved. They are used to control the spread of fire between areas on a floor. The fire separation wall or fire partition should have a fire-resistance rating equal to code requirements specified for the occupancy of the fire areas that are separated (Table 36.3). They must extend from the top of the fire-resistant floor to the bottom of the fire-resistant floor or roof slab above and must be securely fastened to them. The sizes of openings in fire separation walls are restricted by code, and these openings must be protected by fire doors, windows, or shutters.

Fire walls are **fire-rated walls** that restrict the spread of fire and extend continuously from the foundation to or through the roof. If the fire wall stops at the roof, the roof assembly must be of noncombustible material. If it extends above the roof, the height above the roof is specified by the code. The fire wall should be able to remain intact even when the construction on either side collapses due to fire damage. Fire-resistance ratings of fire walls are usually four hours. The wall must be smoke-tight where it meets the exterior wall. The sizes of openings in fire walls are restricted by code, and the openings require protection by fire doors or wire glass.

Party walls are fire walls used jointly by two parties under an easement agreement. They are erected on an interior lot line dividing two parcels of land each of which is a separate real estate entity. A party wall is actually a common wall between two buildings that meet on a lot line.

Smoke barriers are fire-resistant continuous membranes used to resist the movement of smoke. They have a fire-resistance rating specified by the building code.

Table 36.3 Occupancy Separation Requirements[1]

Large or Small Assembly	2 hour
Business	1 hour
Educational	2 hour
Factory-Industrial	2 hour
Hazardous	See 704.1.4
Institutional	2 hour
Mercantile	1 hour
Residential	1 hour
Storage, Moderate Hazard S1	3 hour
Storage, Low Hazard S2	2 hour
Automobile Parking Garages[2]	1 hour
Automobile Repair Garages	2 hour

Note:
See Standard Building Code for references.
1. The minimum fire resistance of construction separating any two occupancies in a building of mixed occupancy shall be the higher rating required for the occupancies being separated.
2. See 411.2.6 for exceptions.
Standard Building Code© 1994. Used with permission from the Standard Building Code, Southern Building Code Congress International, Inc.

They form a continuous membrane from one outside wall to the other and from the floor slab to the roof or floor slab above. Doors in smoke barriers must meet special code requirements and have door closers that cause them to close when smoke detectors sense smoke. They typically have a one-hour minimum fire-resistance requirement.

Interior wall, partitions, and ceilings are also subject to specifications relating to the transfer of sound into adjacent areas and the control of reverberating sound within the rooms. Acoustical principles and construction details are in Chapter 35.

INTERIOR GYPSUM WALL SYSTEMS

Metal Stud Nonbearing Interior Walls

Metal studs are formed from light-gauge steel and are available in a range of sizes. Some are manufactured from heavier gauge steel and are used for load-bearing walls. Other types are designed for use in curtain wall construction. Metal studs for non-load-bearing interior walls and partitions are typically made from 18 gauge steel sheet and may have solid or perforated webs (Fig. 36.1). The punch-outs in the web facilitate installation of bridging, pipe, and electrical conduit (Fig. 36.2).

Another type of metal stud uses round rod bent on a diagonal truss design between vertical cords (Fig. 36.3). These studs are used primarily for walls covered with gypsum lath or metal lath. The lath is joined to the studs with clips. The studs are secured to floor and ceiling runner tracks supplied by the manufacturer of the stud. The solid and perforated web studs are secured to the runners with self-drilling self-tapping screws. They are made of hardened steel and are driven with an electric screwdriver. The tracks are secured to the concrete floor and overhead material with power-driven fasteners. The gun uses a gunpowder cartridge to drive the steel fastener through the track into the concrete floor (Fig. 36.4). Figure 36.5 shows the large area that can be covered with a single sheet of gypsum wallboard. The sheets are carefully positioned and screwed to the metal studs (Fig. 36.6).

Typical Metal Stud Wall Assemblies

The actual assembly of gypsum wallboard walls depends on the desired sound transmission level, thermal insulation, and building code fire resistance specifications. In Fig. 36.7 is a typical assembly for a *cavity-type wall* that, depending on the circumstances, could serve as a fire, separation, or party wall. This is a continuous vertical non-load-bearing wall assembly with metal studs and furring, sound attenuation, fire blankets, gypsum liner panels, and water-resistant fire-resistant gypsum facing

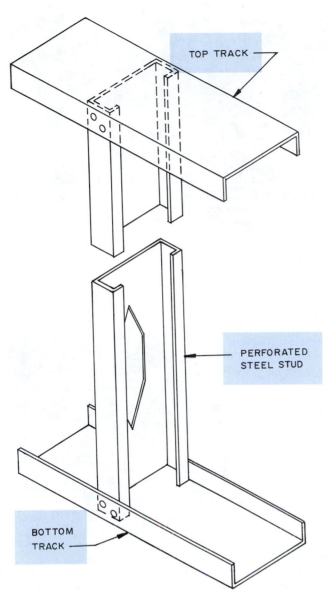

Figure 36.1 Light-gauge steel studs are used for interior walls and partitions.

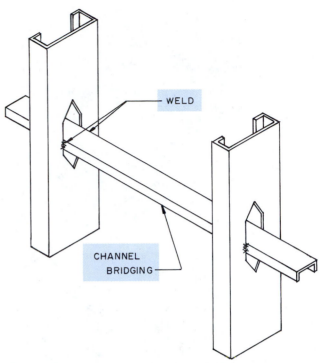

Figure 36.2 One type of bridging used with perforated steel studs.

panels. The studs used are C-H steel studs with steel E-studs at the end of the wall. A wall with this construction can have a two-hour fire rating.

A typical assembly for a fire-rated partition designed to offer effective sound control and a high fire rating is shown in Fig. 36.8. It uses steel studs, gypsum lath on steel furring, and a gypsum finish coat. A sound attenuation blanket is in the wall cavity.

Fire-resistant drywall partitions are used to enclose shafts in multistory buildings (Fig. 36.9). Shafts include those for elevators, mechanical equipment, stairwells, and air returns. A typical assembly for a non-load-bearing fire-resistant gypsum board partition for enclosing shafts, air ducts, and stairwells is shown in Fig. 36.10. It uses a C-H stud system, a gypsum panel liner, and a sin-

gle layer of fire-resistant gypsum board on each side. This assembly will give a two-hour fire rating.

Gypsum fire-resistance wall construction is considerably lighter than masonry fire-resistant walls. Additional layers of fire-resistant gypsum board will increase the fire rating. Figure 36.11 shows several steel frame partition systems with gypsum board, lath and plaster, and veneer plaster cladding. These contain no fire-resistant insulation or gypsum liner panels. The manufacturers of gypsum products have a wide range of wall designs for various fire ratings. Additional information can be found in Chapter 34.

Installing Gypsum Wallboard

Gypsum wallboard is fastened to wood studs and ceiling joists with special nails or power-driven screws (Fig. 36.12) and to metal studs and metal furring with self-drilling self-tapping screws (Fig. 36.13). Panels can be installed with the long edge perpendicular to or parallel to the studs. Perpendicular application is usually used because it reduces the amount of joint to be taped (Fig. 36.14). Perpendicular application also places the strongest dimension of the panel across the studs.

Generally the ceiling panels are installed first and are run perpendicular or parallel to the joists, depending on which method produces the fewest joints. Panels may be single or double nailed. Nails on single-nailed ceilings are usually spaced 7 in. (178 mm) O.C. and 8 in. (203 mm)

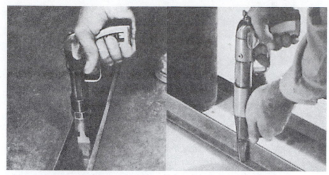

Figure 36.4 The metal runner is fastened to the concrete floor with steel fasteners driven by a gun that uses a small charge of gunpowder. *(Courtesy USG Corporation)*

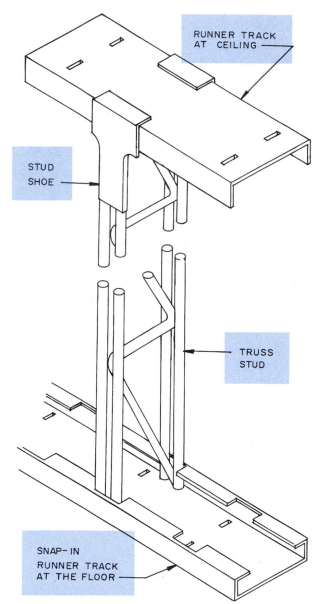

Figure 36.3 A truss stud is set in snap-in runners and secured to the top runner with a stud shoe.

Figure 36.5 Gypsum panels cover a large wall area and are screwed to the metal studs. *(Courtesy Georgia-Pacific Corporation)*

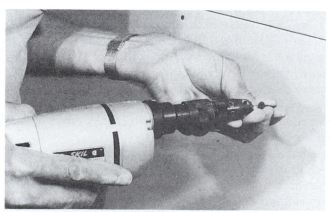

Figure 36.6 Gypsum wallboard is secured to wood and metal studs with screws driven with a power screwdriver. *(Courtesy USG Corporation)*

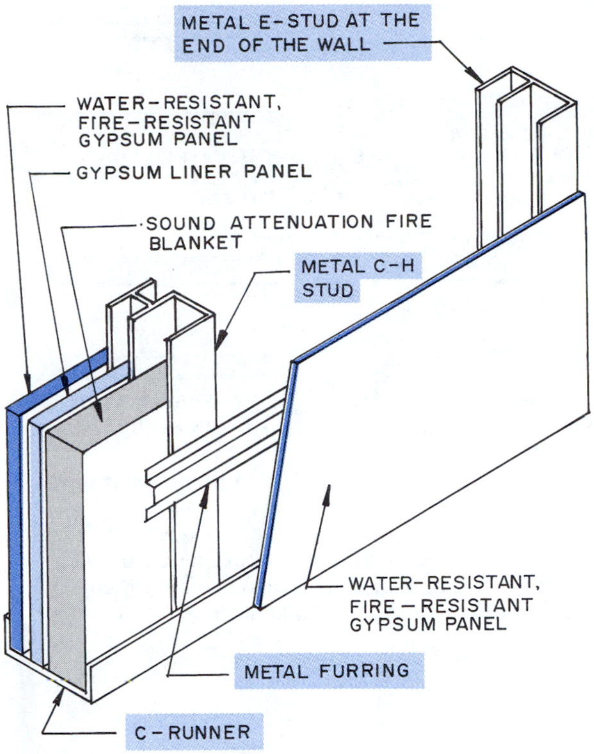

Figure 36.7 A cavity-type separation wall utilizing a gypsum liner panel and a sound attenuation fire blanket.

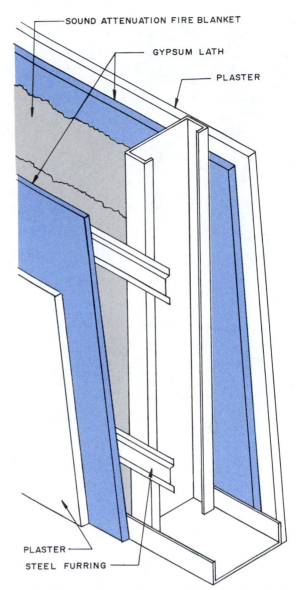

Figure 36.8 A sound attenuation steel frame wall with fire-resistant properties.

O.C. on side walls (Fig. 36.15). Screws are spaced 12 in. (305 mm) on the ceiling and 16 in. (406 mm) on the side walls. To reduce the possibility of nail popping after the wall is finished, the panel can be double nailed, as shown in Fig. 36.16. The fasteners are driven until the head is set in a shallow dimple but the paper covering is not broken (Fig. 36.17). Nails are driven with a drywall hammer that has the nailing face the required shape and diameter (Fig. 36.18).

Gypsum panels are often installed in a double layer. The first layer is nailed or screwed to the framing, and the second layer is bonded to it with an adhesive plus a few nails or screws. The adhesive is applied to the back of the panel as shown in Fig. 36.19. This produces a wall with greater sound attenuation and fire resistance and reduces the possibility of nail popping.

Gypsum panels are cut by scoring the surface along the edge of a metal T-square with a utility knife, as shown in Fig. 36.20. The panel is bent to break the core, and the paper is scored on the back to separate the parts of the panel.

The joints between the panels are covered with layers of **joint compound** and tape. The long edges are made with a taper, allowing the compound and tape to fill it flush with the surface. The short ends of the panels are not tapered and produce a slight bulge if taped.

Therefore, end joints are to be avoided when possible (Fig. 36.21). The joint is covered with a thin layer of compound and the tape is pressed into it with a finishing knife (Fig. 36.22). A thin skim coat is applied over the tape and the compound is allowed to dry. A layer of joint compound is placed over all nail and screw heads. Additional layers of compound are applied and sanded. Each layer is wider than the one before and is feathered out. The tape and joint compound can be mechanically applied to the wall and ceiling joints as shown in Figs. 36.23 and 36.24. The compound may be hand sanded, but a pole sanding unit with a vacuum greatly reduces the dust in the air (Fig. 36.25).

Elevator shafts

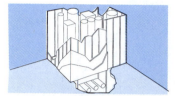

Mechanical shafts (HVAC, plumbing, electrical, etc.)

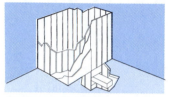

Air return shafts (unlined)

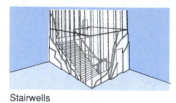

Stairwells

Figure 36.9 Examples of commonly found shaft walls that must meet fire code requirements. *(Courtesy USG Corporation)*

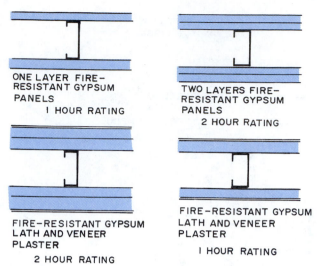

ONE LAYER FIRE-RESISTANT GYPSUM PANELS
I HOUR RATING

TWO LAYERS FIRE-RESISTANT GYPSUM PANELS
2 HOUR RATING

FIRE-RESISTANT GYPSUM LATH AND VENEER PLASTER
2 HOUR RATING

FIRE-RESISTANT GYPSUM LATH AND VENEER PLASTER
I HOUR RATING

Figure 36.11 Typical fire ratings for various types of metal frame gypsum wallboard and plaster partitions. Consult the manufacturer for specific data for recommended stud sizes and panel thicknesses.

RING SHANK NAIL

CUPPED HEAD NAIL

COLOR HEAD PIN FOR PREDECORATED PANELS

Figure 36.12 Nails used to secure gypsum wallboard to wood studs.

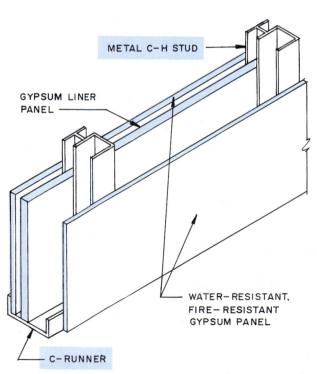

METAL C-H STUD

GYPSUM LINER PANEL

WATER-RESISTANT, FIRE-RESISTANT GYPSUM PANEL

C-RUNNER

Figure 36.10 Typical construction of a single-layer fire-resistant non-load-bearing cavity shaft wall as used on elevator and mechanical shafts, air ducts, and stairwells.

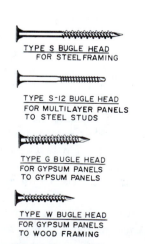

TYPE S BUGLE HEAD FOR STEEL FRAMING

TYPE S-12 BUGLE HEAD FOR MULTILAYER PANELS TO STEEL STUDS

TYPE G BUGLE HEAD FOR GYPSUM PANELS TO GYPSUM PANELS

TYPE W BUGLE HEAD FOR GYPSUM PANELS TO WOOD FRAMING

Figure 36.13 Screws used to secure gypsum wallboard to wood and metal studs and for joining panels.

Figure 36.14 Gypsum wallboard panels are often installed with the long edge perpendicular to the studs. *(Courtesy USG Corporation)*

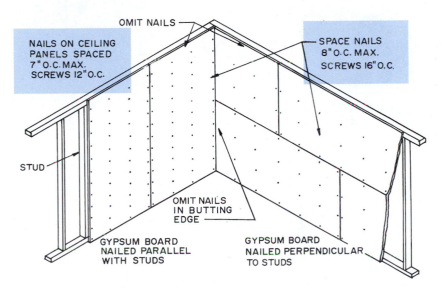

OMIT NAILS

NAILS ON CEILING
PANELS SPACED
7" O.C. MAX.
SCREWS 12" O.C.

SPACE NAILS
8" O.C. MAX.
SCREWS 16" O.C.

STUD

OMIT NAILS
IN BUTTING
EDGE

GYPSUM BOARD
NAILED PARALLEL
WITH STUDS

GYPSUM BOARD
NAILED PERPENDICULAR
TO STUDS

Figure 36.15 Suggested single nailing pattern for gypsum wallboard panels.

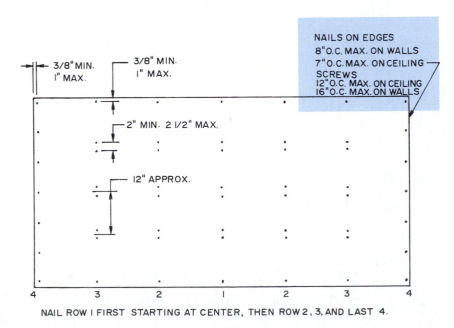

NAILS ON EDGES
8" O.C. MAX. ON WALLS
7" O.C. MAX. ON CEILING
SCREWS
12" O.C. MAX. ON CEILING
16" O.C. MAX. ON WALLS

3/8" MIN.
1" MAX.

3/8" MIN.
1" MAX.

2" MIN. 2 1/2" MAX.

12" APPROX.

4   3   2   1   2   3   4

NAIL ROW I FIRST STARTING AT CENTER, THEN ROW 2, 3, AND LAST 4.

Figure 36.16 Suggested double nailing pattern for gypsum wallboard panels.

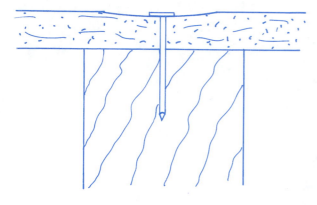

Figure 36.17 Gypsum board fasteners are set with a shallow dimple without breaking the paper cover.

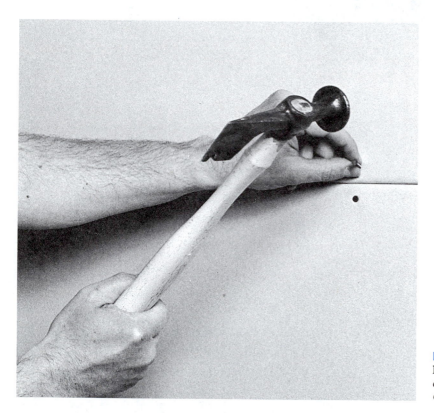

Figure 36.18 The face on this drywall hammer is crowned to produce a shallow dimple around the nail head. *(Courtesy Georgia-Pacific Corporation)*

Figure 36.19 Adhesive is applied in ridges on the back of a panel that is to be bonded to a panel that is nailed or screwed to the studs. *(Courtesy USG Corporation)*

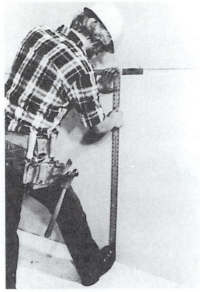

1. Score the panel along the straight-edge with a utility knife.

2. Bend the panel forward, breaking the core.

3. Cut the paper on the back of the panel.

4. Pull the parts of the panel apart.

Figure 36.20 Gypsum wallboard is cut by scoring the panel along a straightedge, bending it to break the core, and cutting the paper on the back side. *(Courtesy USG Corporation)*

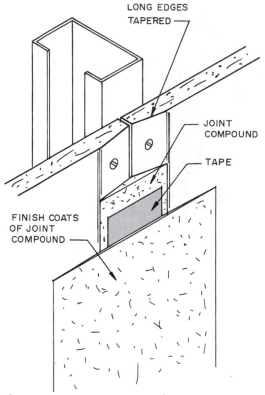

Figure 36.21 The joints between panels are covered with layers of joint compound and tape.

Figure 36.22 The tape is pressed into the first layer of joint compound. *(Courtesy Georgia-Pacific Corporation)*

Internal corners are finished with paper or fiberglass tape bonded in joint compound and finished as described for joints (Fig. 36.26). External corners are finished with metal or plastic corners nailed or screwed to the studs and covered with joint compound (Fig. 36.27). Other finishing accessories include edge trim, control joints, and trim for round corners such as found on arches (Figs. 36.28 and 36.29).

Door frames on walls with metal studs can be installed as shown in Fig. 36.30. These details show the installation of metal and wood door jambs.

PLASTER WALL FINISHES

The types of plaster and gypsum base materials used to finish wall surfaces are detailed in Chapter 34. **Plaster** is applied over expanded metal lath, gypsum lath, or veneer **plaster base. Metal lath** is made by slitting sheets of thin metal alloy and stretching it to form a diamond-shaped mesh or by punching and forming the openings in the metal sheet. Typical types are riblath, diamond-mesh lath, sheet lath, and wire lath (Fig. 36.31). They are usually wired to steel studs or secured with self-tapping screws (Fig. 36.32).

Figure 36.23 This taping tool applies the tape to the joint after the first coat of joint compound is in place. *(Courtesy USG Corporation)*

Figure 36.24 This tool is being used to apply joint compound over the heads of nails. *(Courtesy USG Corporation)*

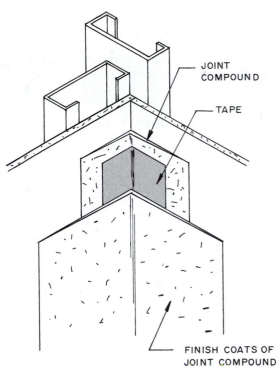

Figure 36.26 Internal corners are covered with paper or fiberglass tape and joint compound.

Figure 36.25 A dustless drywall sanding unit uses a vacuum to collect the joint compound dust. *(Courtesy Hyde and Meeks Industries, Inc., 56 Dudley St., Arlington, Va 02174)*

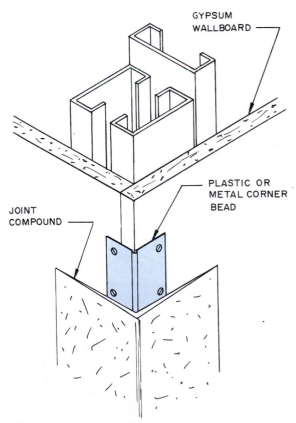

Figure 36.27 External corners are covered with a metal or plastic corner bead and joint compound.

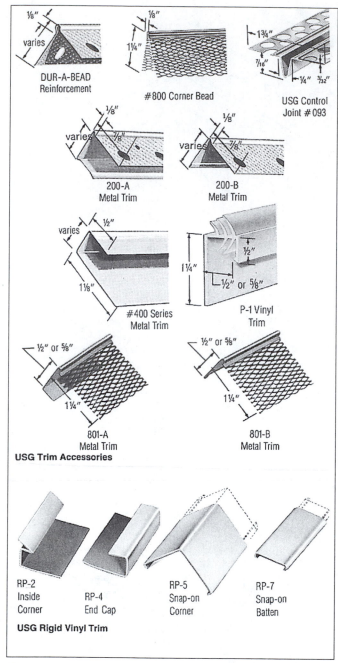

Figure 36.28 Drywall finishing accessories are available in metal and plastic. (*Courtesy USG Corporation*)

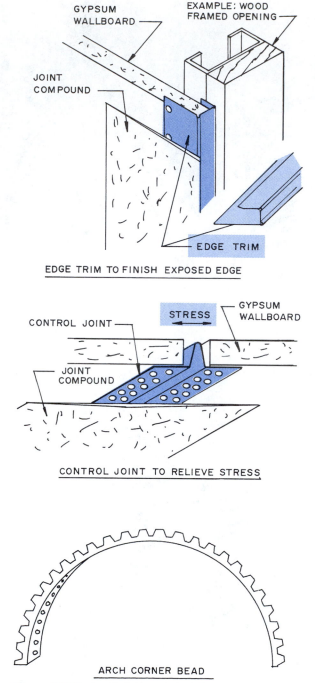

Figure 36.29 Applications of drywall edge trim, control joint, and arch finishing accessories.

Gypsum lath is a rigid fire-resistant base to which gypsum plasters are applied. The gypsum lath is available in several thicknesses and as two different products. One is a standard gypsum lath that is used for nailing or stapling to wood studs or screwed to metal studs and furring. A second type has the addition of a fire-resistant core. **Veneer gypsum base** has a gypsum core and is faced with paper. It is used as the base for **veneer plaster.**

Plaster over expanded metal lath is a three-coat system (Fig. 36.33). The first coat (the scratch coat) is applied to the metal lath with enough pressure to force it into the mesh and form a good key. It is cross-raked, leaving a flat but textured surface. After the scratch coat has hardened, the second coat (the brown coat) is applied over the scratch coat. It is leveled and allowed to harden. The third coat (the finish coat) is applied over

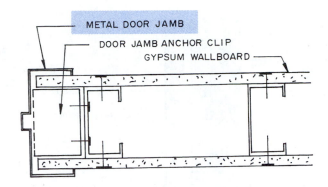

METAL DOOR JAMB INSTALLATION

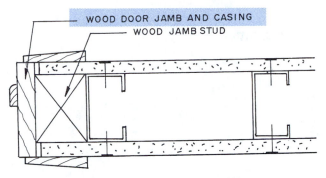

WOOD DOOR JAMB INSTALLATION

Figure 36.30 Typical installation details for metal and wood door jambs.

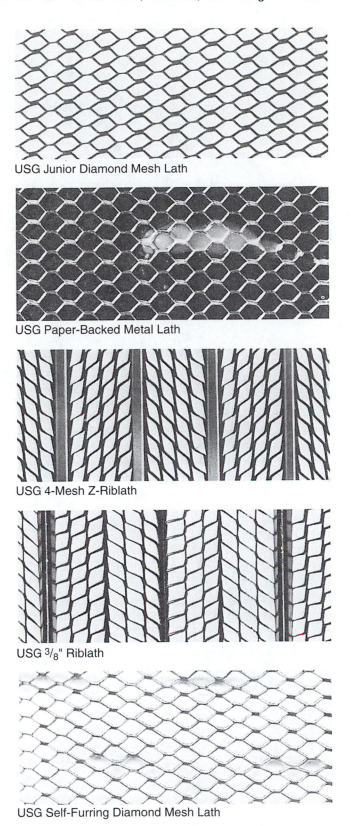

USG Junior Diamond Mesh Lath

USG Paper-Backed Metal Lath

USG 4-Mesh Z-Riblath

USG $\frac{3}{8}$" Riblath

USG Self-Furring Diamond Mesh Lath

Figure 36.31 Types of metal lath. *(Courtesy USG Corporation)*

the brown coat and the surface is finished to the desired texture.

Plaster over gypsum lath usually is a two-coat system, but three coats may be used (Fig. 36.34). The first coat is applied with pressure so it bonds to the lath and provides a level coat with a rough surface. The second coat is applied over the first coat and is finished to the desired texture.

Veneer gypsum base is secured to the studs and is covered with one or two coats of veneer plaster (Fig. 36.35). The first coat is a thin layer that is covered by a second thin coat before the first coat has hardened. The second coat is troweled to the desired texture. Veneer plaster usually is completely hardened within twenty-four hours.

Plaster is installed at door openings as shown in Fig. 36.36. Other methods of assembly are used depending on the weight of the door and the method of securing the door frame to the steel wall stud. The door frame is grouted at the anchors to increase rigidity of the frame and improve resistance to frame rotation. The grouting may be located only at each frame anchor or the frame may be fully grouted.

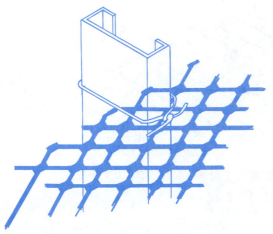

Figure 36.32 Metal lath is tied to metal studs with 18 gauge wire.

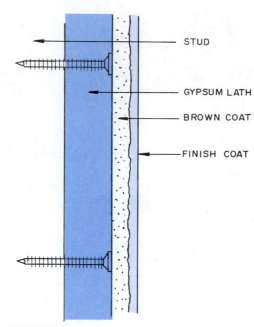

Figure 36.34 This is a two-coat plaster wall finish over gypsum lath.

STUD

GYPSUM LATH

BROWN COAT

FINISH COAT

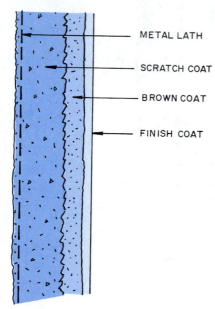

METAL LATH

SCRATCH COAT

BROWN COAT

FINISH COAT

Figure 36.33 This is a three-coat plaster wall finish over metal lath.

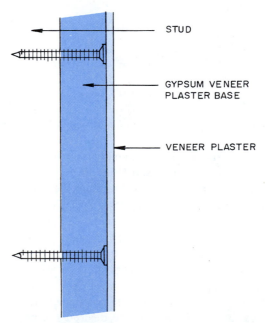

STUD

GYPSUM VENEER PLASTER BASE

VENEER PLASTER

Figure 36.35 Veneer plaster is a single-coat plaster applied over a gypsum veneer plaster base.

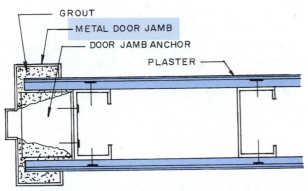

Figure 36.36 One way to install metal door frames when the finish wall is plaster.

Plaster over Masonry and Concrete Walls

Plaster may be applied directly to masonry, concrete, and clay tile walls. Concrete block walls are porous and provide a satisfactory base for plaster. Clay tile and brick walls may be used as a base for plaster walls, but they must be sufficiently porous to provide suction for the plaster and be scored to increase the mechanical bonding. Smooth-surface glazed or semiglazed tile cannot be covered. It requires that the wall be furred and a metal or gypsum plaster base be installed. Monolithic concrete walls are porous, but they usually require an application of a bonding agent to produce the adhesive bond necessary for direct application of gypsum plasters. Generally, exterior masonry walls are not plastered directly to the masonry because of the likelihood of water seepage and condensation that will wet the plaster.

Masonry concrete and clay tile walls can have gypsum lath installed by securing metal furring to the surface. On exterior walls this provides an air space, keeping the plaster base and plaster away from the masonry wall. Furring can be shimmed to help bring a wall with minor irregularities into a flat surface. The furring may be placed horizontally or vertically (Fig. 36.37).

Plastering

Plaster base may be metal lath or gypsum lath. The lath may be nailed to wood studs as shown in Fig. 36.38. Notice that the ends of the lath rest on a stud and are nailed to it. Gypsum lath may also be secured with self-drilling self-tapping screws that are driven with a power screwdriver (Fig. 36.39). In Fig. 36.40 long lengths of gypsum lath are being secured to metal runners with screws.

Plaster may be applied to the base with a hand trowel, as shown in Fig. 36.41. In one hand the plasterer holds a square flat tool called a hawk. The hawk holds a supply of plaster, which is removed by a trowel and applied to the wall. In Fig. 36.41 the gypsum lath base has metal

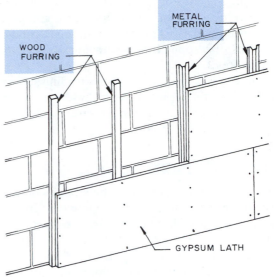

FUR THE WALL WITH WOOD OR METAL FURRING

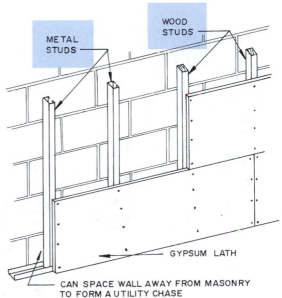

CAN SPACE WALL AWAY FROM MASONRY TO FORM A UTILITY CHASE
BUILD A WALL OVER THE MASONRY WALL

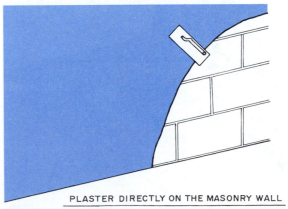

PLASTER DIRECTLY ON THE MASONRY WALL

Figure 36.37 Masonry walls can be covered with a plaster finish.

clips supporting the ends of the panels because they do not rest on a vertical support. Small wire clips are used to tie the gypsum panels to the truss-type metal studs. Plaster may also be applied by a spraying process, as shown in Fig. 36.42.

Various accessories are used with plaster finishes, as described for drywall. Typical among these are casing beads, external corner beads, flexible corner beads, and expansion joints (Fig. 36.43).

SOLID GYPSUM PARTITIONS

Solid gypsum partition systems may be studless or use channel studs. The solid studless partition has a vertical core of metal mesh lath. The lath specifications vary with the height of the wall. The lath is connected to the ceiling with a runner that is attached to the overhead and a metal runner that is fastened to the floor (Fig. 36.44).

The lath is covered with plaster scratch, brown, and finish coats. Another assembly for a studless partition is shown in Fig. 36.45. Long length gypsum lath is secured to runners at the overhead and floor. It is plastered in the conventional manner.

Solid gypsum partitions are also assembled using C-studs and metal lath. The studs are secured to the overhead and floor with metal runners. Metal lath is tied to the studs, and plaster scratch, brown, and finish coats are applied (Fig. 36.46).

Similar construction is used to assemble a double-channel stud hollow partition. Vertical studs are secured to the overhead and floor with metal runners. Wire mesh is tied to each row of studs, and conventional plastering procedures are used (Fig. 36.47). The studs are stiffened and braced with horizontal channels. This type of partition is useful for running electrical and mechanical system components.

Figure 36.38 This gypsum plaster base is being nailed to wood studs. *(Courtesy USG Corporation)*

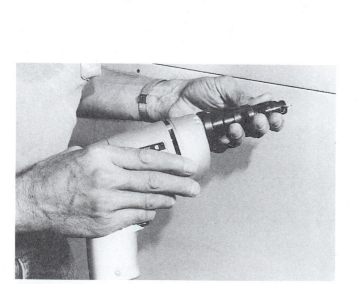

Figure 36.39 Gypsum plaster lath is secured to studs with screws driven with a power screwdriver. *(Courtesy USG Corporation)*

Figure 36.40 Long lengths of gypsum lath are being screwed to steel runners around a column, forming a fire-resistant enclosure. *(Courtesy USG Corporation)*

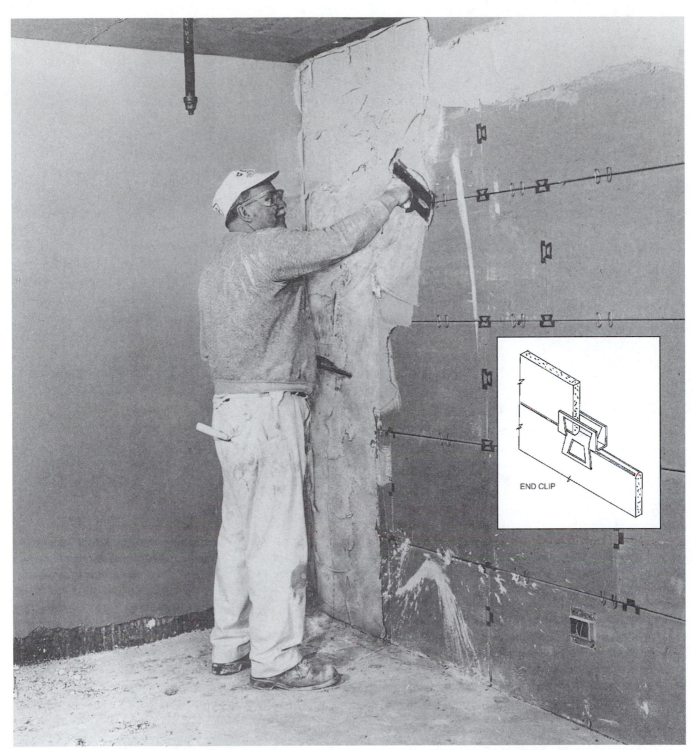

Figure 36.41 The plasterer is hand-troweling the scratch coat to the gypsum lath base that is secured to metal studs with wire clips. The large metal clips support the ends of the panels that do not rest on a stud. *(Courtesy USG Corporation)*

Figure 36.42 The plasterer on the right is applying a scratch coat by spraying it on the base. The plasterer in the center is hand-troweling the brown coat while the person on the left is leveling the coat by pulling a darby across it. After it has been leveled it may be troweled again to smooth the surface. *(Courtesy USG Corporation)*

STRUCTURAL CLAY TILE PARTITIONS

Structural clay tile partitions are made with hollow clay masonry units that are glazed on one or both sides. The glazes provide a smooth surface that is impervious to penetration by water and is available in a wide range of colors (Fig. 36.48). The tiles resist abrasion, wear, and mildew and can be routinely scrubbed and sanitized. Information about tile sizes and wall construction are in Chapter 11.

CERAMIC TILE WALL FINISHES

Ceramic tile used to finish walls and partitions are available in a wide range of sizes and colors. They provide a hard, water-resistant facing and are durable and easily cleaned (Fig. 36.49). Information about tile sizes and installation is in Chapter 11.

Ceramic tile is generally bonded to walls faced with gypsum or cement panels designed to support the tile and resist damage if moisture penetrates the joints between the tiles. Ceramic tile may be bonded to concrete or masonry that has a surface prepared by sandblasting or scarifying to provide a true uncontaminated surface. Ceramic tile may also be applied to wall surfaces consisting of metal lath covered with a cement mortar bed (Fig. 36.50). Portland cement mortar is used for setting tile in a thick bed. Other mortars and adhesives are used for thin beds.

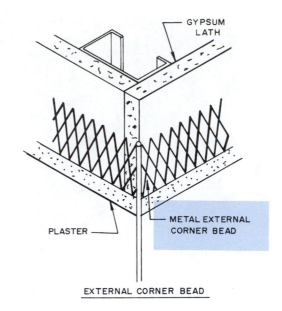

EXTERNAL CORNER BEAD

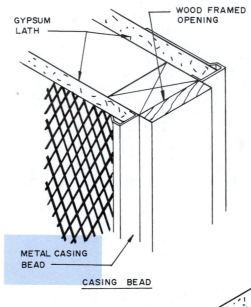

CASING BEAD

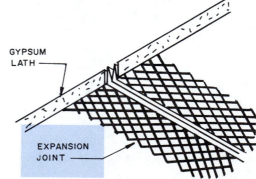

Figure 36.43 Accessories used with plaster finishes.

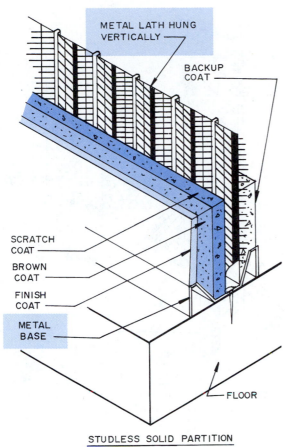

Figure 36.44 Typical construction of a studless solid gypsum partition built around vertical metal lath.

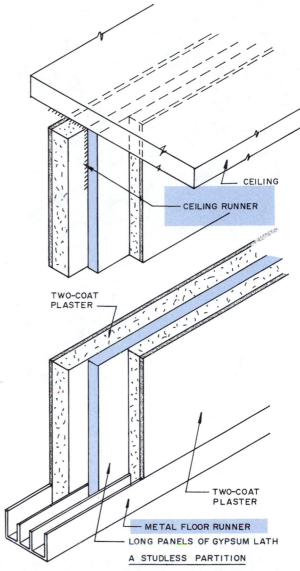

Figure 36.45 A studless solid partition built around a gypsum lath core set in metal tracks.

WOOD PRODUCT WALL FINISHES

Interior wall finish may be some form of solid wood paneling or a reconstituted wood panel such as plywood or hardboard. They are available finished including some types with vinyl wall coverings. Details about these products is in Chapter 19. The use of wood interior wall finishes may be severely limited or prohibited by code regulations for certain building types and occupancies.

CEILING CONSTRUCTION

The ceiling serves a variety of functions. It provides a finished appearance to the overhead area, contributes to the acoustical treatment of the room, and provides a light-reflecting surface to enhance illumination. Some types provide a space below the floor above in which electrical and mechanical systems are run out of view.

The ceiling may support lights, provide a measure of fire protection to the floor or roof above, and permit the use of sprinkler heads in a fire control system. Ceilings may be flat or curved or constructed on any of a wide variety of sloping surfaces that provide for the architectural enhancement of the area (Fig. 36.51).

Ceiling finish materials may be plastic, metal, plaster, gypsum panels, wood, or one of the many fibrous panels available. The floor or roof decking may be exposed and used as the finished ceiling. This could be concrete, steel, or wood. The types of ceilings include suspended, furred, and contact.

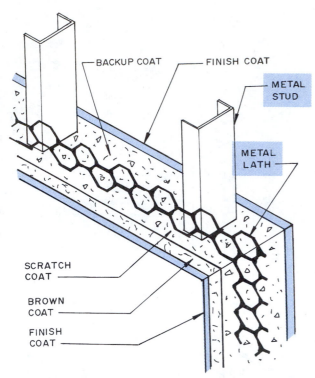

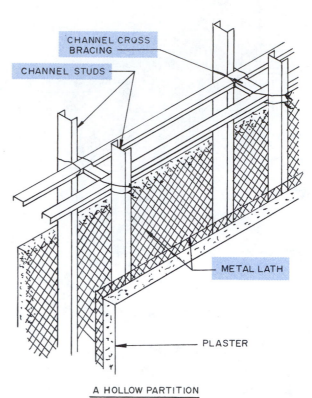

Figure 36.46 Typical construction details for a solid gypsum partition built with metal tracks and a metal lath core.

Figure 36.47 A double-channel hollow partition can be used to provide space to run utilities.

Figure 36.48 Partitions and walls are constructed with glazed structural ceramic tile units, providing a fire-resistant wall that is easily cleaned and resists damage. (*Courtesy Stark Ceramics, Inc.*)

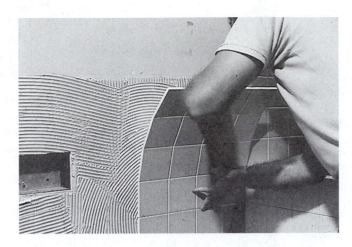

Figure 36.49 Ceramic tile wall facing provides a fire-resistant, durable surface that is easily cleaned. *(Courtesy American Olean Tile Company)*

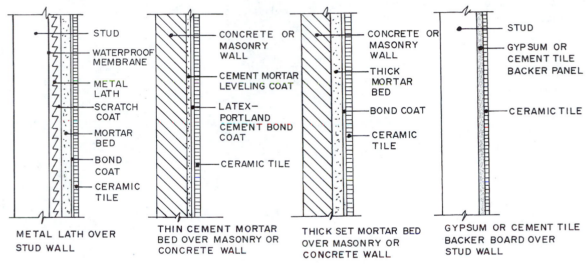

STUD
WATERPROOF MEMBRANE
METAL LATH
SCRATCH COAT
MORTAR BED
BOND COAT
CERAMIC TILE

METAL LATH OVER STUD WALL

CONCRETE OR MASONRY WALL
CEMENT MORTAR LEVELING COAT
LATEX–PORTLAND CEMENT BOND COAT
CERAMIC TILE

THIN CEMENT MORTAR BED OVER MASONRY OR CONCRETE WALL

CONCRETE OR MASONRY WALL
THICK MORTAR BED
BOND COAT
CERAMIC TILE

THICK SET MORTAR BED OVER MASONRY OR CONCRETE WALL

STUD
GYPSUM OR CEMENT TILE BACKER PANEL
CERAMIC TILE

GYPSUM OR CEMENT TILE BACKER BOARD OVER STUD WALL

Figure 36.50 Some of the constructions used when building a wall to have a ceramic tile face.

Figure 36.51 This flat aluminum ceiling will withstand the moisture in this atmosphere. It accommodates lights and heating system vents.

Suspended Ceilings

Suspended ceilings are possibly the most widely used in commercial construction. They offer considerable advantages because they allow electrical and mechanical systems to be hidden and are available with a variety of finished surface materials (Figs. 36.52 and 36.53).

Suspended ceilings are hung from the floor or roof overhead by a series of wires that support a metal grid that carries the finished ceiling. The length of the wires can vary, so a flat horizontal ceiling can be installed below a structural floor or roof that has members of various sizes (Fig. 36.54). Typically, the lighting system is set in the grid and is flush with the ceiling (Fig. 36.55). Some types of panels have acoustical properties and others provide a degree of fire protection below the floor or roof above (Fig. 36.56).

Suspended plaster ceilings are constructed by hanging channel runners to which furring channels are tied with wire. The spacing is determined by the conditions. The wire or metal lath is wired to the furring. This construction produces a hard, durable, highly fire-resistant ceiling. Typical construction details are in Fig. 36.57.

Various types of suspended acoustical systems are available that use a suspended metal grid into which lightweight panels and lights are placed. The grids are assembled with some form of locking connection (Fig. 36.58). The drop-in panels may have an exposed grid or a concealed grid (Fig. 36.59). These systems also can accommodate heat ducts, sprinkler heads, and other required penetrations. Some grid systems have recessed panels, and a large number of surface textures are available. Ceiling panels are available in materials such as aluminum, fiberglass-reinforced polymer gypsum cement, gypsum panels faced with vinyl and fabrics, fiberglass, and wood fibers.

Figure 36.52 The exposed metal grid of this suspended ceiling is finished a dark color to emphasize the grid pattern. *(Courtesy The Celotex Corporation)*

Figure 36.53 The exposed metal grid on this suspended ceiling is a neutral color, permitting it to blend with the acoustic ceiling panels. *(Courtesy The Celotex Corporation)*

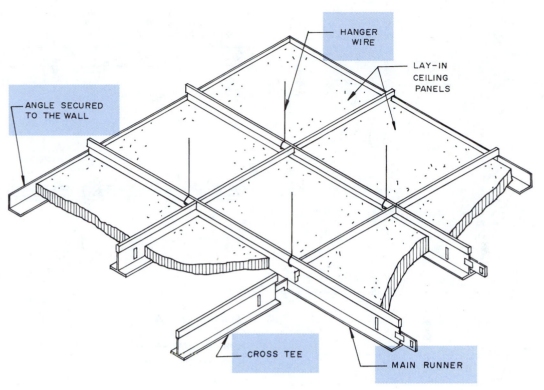

Figure 36.54 A suspended ceiling is hung from overhead by wires. The wire carries the main runners, which are joined by cross tees to form a metal grid. In this design, ceiling panels are laid into the grid, forming the ceiling.

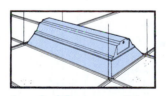

A vaulted fluorescent light fixture.

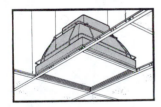

A high-intensity discharge lamp.

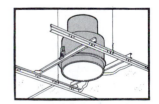

An incandescent fixture.

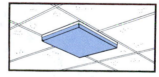

A surface-mounted fixture.

Figure 36.55 Some of the types of lights that are used with suspended ceilings. *(Courtesy Chicago Metallic Corporation)*

Figure 36.56 This fire-resistant acoustical ceiling has recessed incandescent and fluorescent light fixtures.

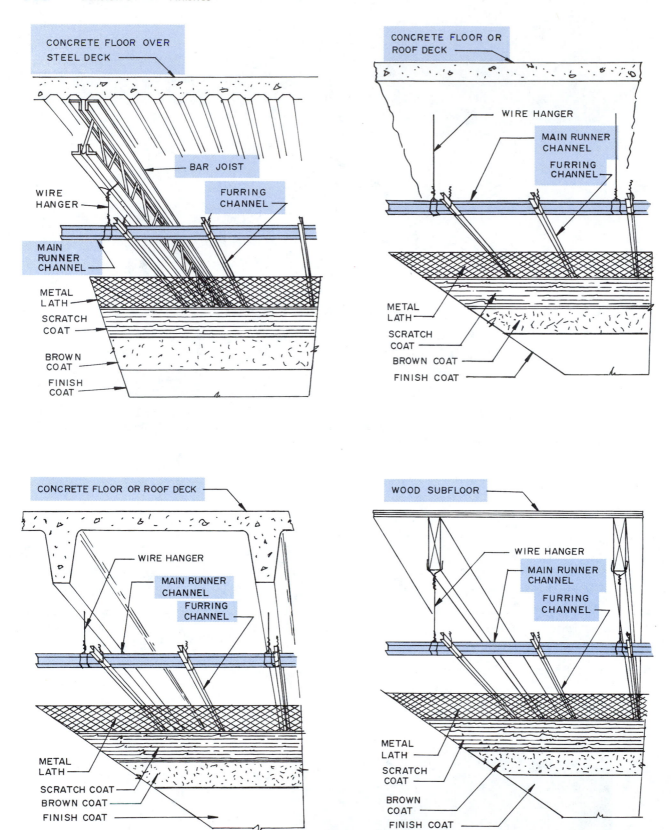

Figure 36.57 A suspended plaster ceiling has the main channel hung from overhead by wires, and metal furring channels are wired to the main channel. The plaster coats are as described for finished walls.

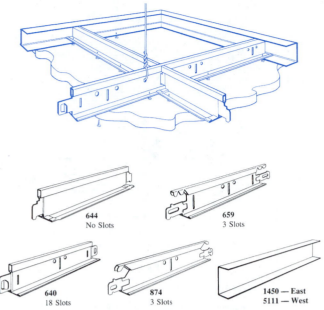

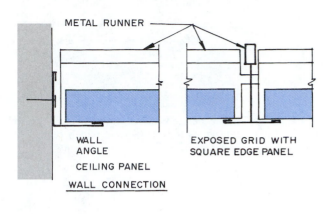

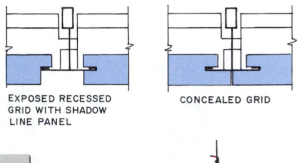

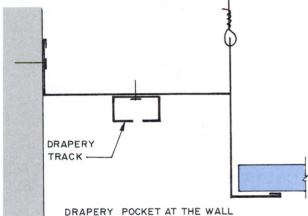

Figure 36.58 This suspended ceiling system has a variety of grid members. It is designed to support a gypsum ceiling. *(Courtesy Chicago Metallic Corporation)*

Figure 36.59 Drop-in ceiling panels may have an exposed, recessed, or concealed grid.

Another suspended ceiling system uses a metal grid with gypsum panels secured to it with screws. The system in Fig. 36.60 may be fire-rated or nonrated construction. In this system the main runners are spaced 48 in. (1219 mm) O.C. and the furring cross channels either 16 in. (406 mm) or 24 in. (610 mm) O.C. The gypsum drywall panels are attached to the grid with conventional drywall screws. The ceiling grid is secured to the wall with a wall track (36.61), and partitions are secured to the ceiling by screwing them to a drywall furring cross-channel or main runner (36.62). The system has provision for the introduction of ducts through the ceiling.

Furred Ceilings

Furred ceilings are constructed using metal furring strips secured to the metal or wood structural system (Fig. 36.63). Concrete overhead floors and roofs often have a C-channel hanger hung below them, and the furring strips are wired to the hanger. The metal lath is wired to the furring strip (Fig. 36.64). The spacing and size of the hangers depend on the design of the system.

Contact Ceilings

Ceilings can be attached directly to the floor or ceiling joists, bar joists, or other structural members. In wood frame construction, gypsum wallboard panels are screwed directly to the joists as shown in Chapter 20. Plaster ceilings can have the metal lath secured directly to the steel, wood, or concrete structural system as shown in Fig. 36.65. The ceilings are then finished with several layers of plaster as described for plaster wall construction.

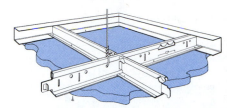

Figure 36.60 This suspended ceiling system uses C-channels at the wall and is designed to have gypsum panels secured to it with screws. *(Courtesy Chicago Metallic Corporation)*

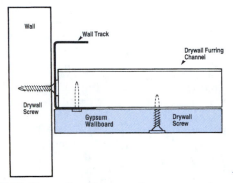

Figure 36.61 Drywall ceiling panels are secured to the metal grid with drywall screws. *(Courtesy Chicago Metallic Corporation)*

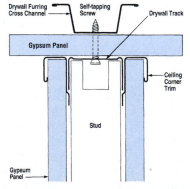

Figure 36.62 Partitions are secured to the suspended ceiling grid with metal drywall track and ceiling corner trim. *(Courtesy Chicago Metallic Corporation)*

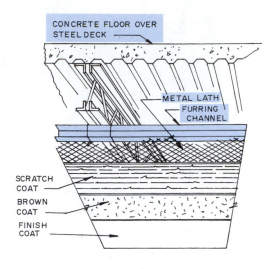

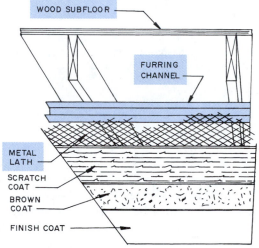

Figure 36.63 Furred ceilings have the furring channels secured to the floor or roof structure.

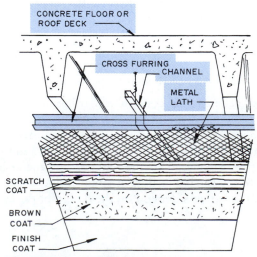

Figure 36.64 This furred ceiling has a C-channel supporting the cross furring.

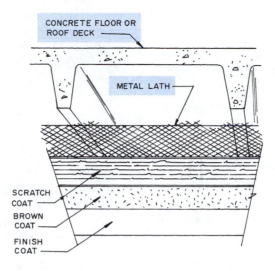

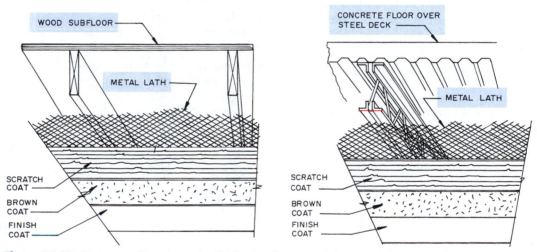

Figure 36.65 Contact ceilings have the finished ceiling material secured directly to the overhead floor or roof structural system.

REVIEW QUESTIONS

1. What purposes do interior partitions serve?
2. What is the difference between a fire partition and a fire wall?
3. How can electrical wiring be run inside partitions having metal studs?
4. What assembly of materials is often used in a fire-rated partition that also requires effective sound control?
5. How is gypsum wallboard secured to metal studs?
6. What can be done to reduce nail popping in gypsum wallboard installation?

7. How are external corners on gypsum wallboard finished?
8. What base materials are used when constructing a plaster interior finish?
9. What are the plaster coats applied over metal lath?
10. Why can concrete block walls have plaster applied directly to them?
11. How can you plaster over a glazed tile wall?
12. Describe the construction of typical solid gypsum partitions.
13. What are the merits of using glazed hollow clay tiles for partition construction?
14. What is a suspended ceiling?

KEY TERMS

fire-rated partition A partition assembly that has been tested and given a rating indicating the length of time it will resist a fire in hours.

fire wall A construction of noncombustible materials that subdivides a building or separates adjoining buildings to retard the spread of fire.

gypsum lath A gypsum board product designed to serve as the base for a gypsum plaster wall finish.

joint compound A plastic gypsum mixture used to cover the joints and fasteners in gypsum wallboard installations.

metal lath Perforated sheets of thin metal secured to studs and serving as the base for a finished plaster wall.

plaster A thick plaster mixture of gypsum, lime, or portland cement with water and, if specified, sand, fibers, or hair.

plaster base Any material suitable for the application of plaster.

smoke barriers Continuous membranes used to resist the movement of smoke.

veneer gypsum base A gypsum board product designed to serve as the base for the application of gypsum veneer plaster.

veneer plaster A calcined gypsum plaster formulated to be applied in thin coats over veneer gypsum base.

SUGGESTED ACTIVITIES

1. Examine catalogs of companies that manufacture interior walls, partitions, and ceiling systems. For each, list the materials used, fire rating, sound transmission class, and sketch a section of the assembly.

2. Tour various buildings on your campus and record the location of different types of interior wall assemblies, partitions, and ceilings. Do those in older buildings meet the current codes? What should be done to upgrade each system found lacking?

ADDITIONAL INFORMATION

Gorman, J. R., Jaffe, S., Pruter, W. F., and Rose, J. J., *Plaster and Drywall Systems Manual*, BNI Books Division of Building News, Inc., Los Angeles, and McGraw-Hill, New York, 1990.

Gypsum Construction Guide, Gold Bond Building Products, National Gypsum Company, 2001 Rexford Road, P.O. Box 25884, Charlotte, N.C. 28211.

Gypsum Construction Handbook, United States Gypsum Company, 125 South Franklin St., P.O. Box 806278, Chicago, Ill. 60680–4124.

Hornbostel, C., *Construction Materials: Types, Uses, and Applications*, John Wiley and Sons, New York, 1991.

Remodeler's Guide to Suspended Ceilings, Chicago Metallic Corporation, 4849 S. Austin Ave., Chicago, Ill. 60638.

Spence, W. P., *Finish Carpentry*, Sterling Publishing Co., New York, 1995.

Spence, W. P., *Residential Framing*, Sterling Publishing Co., New York, 1993.

Van Den Braden, F., and Hartsell, T. L., *Plastering Skills*, American Technical Publisher, Homewood, Ill., 1984.

Other resources are numerous related publications from the organizations listed in Appendix B.

37

Flooring

This chapter will help you to:

1. Be aware of the importance of consulting building codes before selecting flooring.

2. Select flooring materials appropriate for various applications.

3. Understand the types of flooring available and their sizes and properties.

Finish flooring materials are installed over substrates of various kinds and in a variety of situations. The designer must consider factors such as concrete slabs on grade or above grade, wood substrate, cellular steel floors with a concrete slab, the use of radiant heating, the traffic loads on the floor, required maintenance, fire resistance, building code requirements, acoustical requirements, color, and texture. The finish flooring must meet the prescribed conditions at an acceptable cost. Some areas require that special conditions be observed, such as a need for sanitation or unique situations as found in hospitals.

BUILDING CODES

Building codes have sections devoted to floor finish. Codes are related to rooms or enclosed spaces, to vertical exits and passageways, and to corridors that provide access to exits. The requirements can vary depending on the classification and occupancy of the building. The most commonly used finish flooring materials, such as wood, vinyl, terrazzo, and clay tile, do not present unusual hazards and are generally not subject to building code interior finish requirements. Carpeting does present a hazard and must meet the requirements specified by U.S. Department of Commerce standard DOC FF1 or the National Fire Protection Association Standard NFPA 253. Carpeting is discussed in Chapter 38.

WOOD FLOORING

The various types of wood flooring are described in detail in Chapter 19. Both hardwoods and softwoods are used to form strip flooring and various types of parquet and heavy wood-block flooring (Fig. 37.1). Some types have a factory-applied finish, and others are sanded and finished after installation. Some types are nailed to a wood subfloor, and others are bonded with a mastic adhesive. Many are available as a very thin veneer and cannot be sanded.

Standard wood strip flooring is manufactured in several thicknesses. Several methods for installing it are shown in Fig. 37.2. The recommendations of the manufacturer for method, nails, and adhesives should be observed. Parquet

Figure 37.1 Hardwood strip flooring protected with a durable coating provides a beautiful finished floor. *(Courtesy Harris-Tarkett, Inc.)*

flooring is generally installed with adhesives. It is available in a wide variety of species of wood and patterns (Fig. 37.3).

Other wood flooring products include a durable hardwood plank or parquet flooring that is acrylic impregnated. This prefinished laminated product is ⅜ in. (915 mm) thick (Fig. 37.4). It meets slip-resistance standards set forth by the Americans with Disabilities Act of 1990 and has a Class B flame spread (ASTM E-84). Still another product is composed of a hardwood veneer covered with a 20 mil vinyl top layer. The wood veneer is bonded to a base made up of several layers of vinyl and fiberglass (Fig. 37.5).

RESILIENT FLOORING

Resilient flooring in a polyvinyl chloride material is available in sheets and individual tiles. Sheet vinyl has a top vinyl layer and a composition backing. Vinyl composition tiles are composed of vinyl resins, plasticizers, stabilizers, fibers, and pigments (Fig. 37.6) and are available in 9, 12, 18, and 36 in. (22.8, 30, 45.7, and 91.4 mm) squares and some rectangular shapes. Available thicknesses are ⅛ and 3⁄32 in. (2.4 and 3.2 mm).

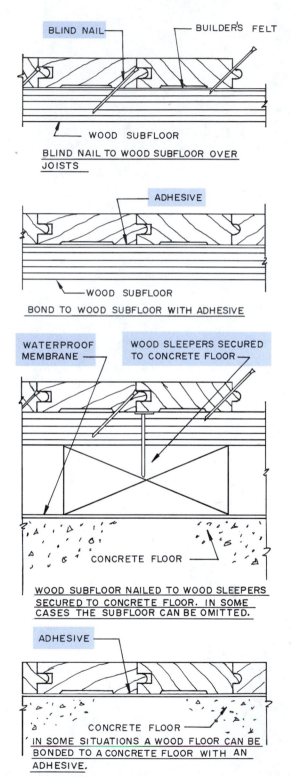

Figure 37.2 Several ways wood flooring can be installed.

Figure 37.3 Parquet flooring has been widely used for many years and provides a wide variety of patterns. *(Courtesy Harris-Tarkett, Inc.)*

Figure 37.4 This wood flooring is acrylic-impregnated hardwood suitable for commercial and other high-traffic uses. *(Courtesy PermaGrain Products, Inc.)*

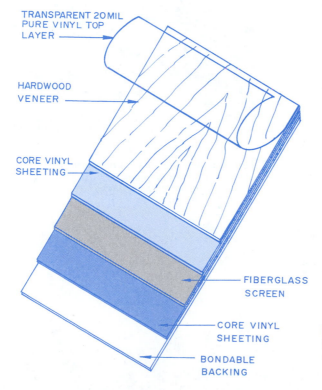

TRANSPARENT 20 MIL
PURE VINYL TOP
LAYER

HARDWOOD
VENEER

CORE VINYL
SHEETING

FIBERGLASS
SCREEN

CORE VINYL
SHEETING

BONDABLE
BACKING

Figure 37.5 This GenuWood 11 flooring is a laminated product with a hardwood veneer and a pure vinyl top layer. *(Courtesy PermaGrain Products, Inc.)*

Sheet vinyl flooring is available in rolls 6, 9, and 12 ft. (1.8, 2.7, and 3.7 m) wide and up to 50 ft. (15.2 m) long and in a wide range of thicknesses from about 0.069 to 0.224 in. (1.75 to 5.69 mm). Also available are solid vinyl commercial vinyl (Fig. 37.7) and PVC resilient flooring. Solid vinyl floor covering offers the maximum wear potential. PVC flooring is used when a heat-welded surface is required. Plastic materials are discussed in detail in Chapter 24. Vinyl composition floor coverings are tested for fire spread and smoke produced. Data is available from the manufacturer. They can be installed over wood or concrete subfloors with a mastic recommended by the manufacturer.

Asphalt resilient flooring is available in tiles usually 9 or 12 in. (2745 or 3660 mm) square. They are a composition of an asphalt binder for standard tile or a resinous binder for greaseproof tile, inert fillers, fibers, and pigments for color. They are available in grades A, B, C, and D, based on color and are usually ⅛ or ³⁄₁₆ in. (3 or 4.8 mm) thick. They are durable and fire resistant. They are bonded to wood or concrete subfloors with mastic recommended by the manufacturer.

Rubber flooring is a form of resilient flooring available in tiles and sheets (Fig. 37.8). Tile sizes vary by the manufacturer, but 12 and 36 in. (305 and 915 mm) square tiles are common. Sheets range from 36 to 50 in. (915 to 1270 mm) wide. Flooring is typically ⅛ and ³⁄₁₆ in. (3 and 4.7 mm) thick. Rubber flooring is comfortable for walking, wears well, and resists damage from oils, solvents, alkalis, acids, and other chemicals. Most types have some form of gridded surface that aids traction and a roughened or recessed back grid. Some typical designs are shown in Fig. 37.9. They are available in a wide range of colors.

Installation techniques depend on the situation, and manufacturer's recommendations should be observed. Typically, rubber flooring is bonded to concrete or wood subfloors with an approved adhesive. They usually are not adhesive-attached to below-grade slabs. On-grade slabs require the use of a moisture-resistant adhesive. Manufacturers also have rubber flooring products such as stair tread, base, entrance mats that are set in a recess in the floor, and long runners that are not bonded to the subfloor.

Figure 37.6 Resilient vinyl composition tile is available in a wide range of patterns and colors.

Figure 37.7 This inlaid commercial vinyl flooring has a long life and provides an attractive, durable, easy-to-clean surface. (*Courtesy Azrock Industries, Inc.*)

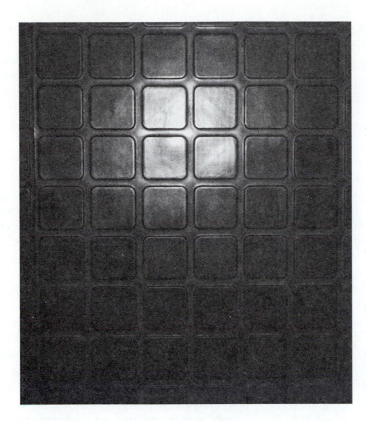

Figure 37.8 Rubber finish flooring is available in tiles and sheets.

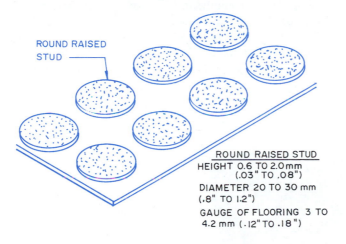

ROUND RAISED STUD

ROUND RAISED STUD
HEIGHT 0.6 TO 2.0 mm
(.03" TO .08")
DIAMETER 20 TO 30 mm
(.8" TO 1.2")
GAUGE OF FLOORING 3 TO
4.2 mm (.12" TO .18")

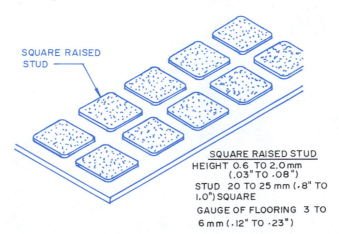

SQUARE RAISED STUD

SQUARE RAISED STUD
HEIGHT 0.6 TO 2.0 mm
(.03" TO .08")
STUD 20 TO 25 mm (.8" TO
1.0") SQUARE
GAUGE OF FLOORING 3 TO
6 mm (.12" TO .23")

Figure 37.9 Typical stud sizes and shapes used on rubber flooring tiles.

CLAY TILE

Clay floor tile includes quarry tile, paver tile, and ceramic tile. Information on these is in Chapter 11. *Quarry tile* is unglazed and made from shales and fire clays. It is very durable, cleans easily, resists damage due to freezing, thawing, and abrasion and is used on floors subject to heavy and extra heavy use (Fig. 37.10). Proper installation is essential. Some types are made with an abrasive grain surface to provide slip resistance. Although various manufacturers have a variety of sizes and shapes, square, rectangular, and hexagonal are most common. Various trim units such as stair nosing and cove base units are available.

Paver tiles may be glazed or unglazed and are much like ceramic tiles but larger. Typical sizes are 4 × 4 in. (102 × 102 mm), 4 × 8 in. (102 × 203 mm), and 8 in. (203 mm) hexagon shape. They are weather resistant and resist abrasion. Typical uses include residential and moderate-duty commercial floors such as shopping centers and restaurants and heavy-duty floors such as a commercial kitchen (Fig. 37.11). A variety of trim materials are available.

Ceramic mosaic tile is available glazed and unglazed and in squares, rectangles, and hexagon shapes. They are small tiles, typically 1 or 2 in. (25.4 or 50.8 mm) square or hexagon shapes or 1 × 2 in. (25.4 × 50.8 mm) rectangles. They are used for interior floors, countertops, walls and swimming pools, and exterior floors and walls with special installation procedures. Some types have slip-resistant surfaces (Fig. 37.12). A variety of trim materials are available.

Glazed ceramic floor tile is available with smooth and textured surfaces. The textured surface provides greater slip resistance. The tiles are available in 4 and 8 in. (101 and 202 mm) squares, 8 in. (202 mm) hexagon, and 4 × 8 in. (101 × 202 mm) rectangles. They are used for interior residential and moderate-duty commercial floors such as in restaurants and shopping malls (Fig. 37.13).

Mortars and Adhesives for Tile Floors

Clay floor tile can be installed using a variety of bonding agents. Portland cement mortar is the only one recommended for a thick bed. Epoxies and furans do not contain portland cement and are more expensive.

Portland cement mortar is a mixture of portland cement, hydrated lime, sand, and water. It can be used to provide a leveling bed up to 1¼ in. (32 mm) thick. The tiles may be set into this bed while it is still plastic. After it cures, tiles may be bonded to the leveling bed with a thin-set coat such as dry-set or latex portland cement mortar.

Dry-set mortar is a mixture of portland cement, resinous materials, sand, and water. It is a thin-set mortar (about 3/32 in. or 2.4 mm thick). It may be applied over the hardened thick set portland cement mortar bed and used to bond the tiles in place.

Epoxy mortar contains no portland cement but is a mix of epoxy resin and a hardener. It has high bond strength, and it resists impact and chemical attack. It is a thin coat adhesive.

Latex portland cement mortar is a mixture of portland cement, a latex additive, and sand. It is a thin coat application, usually about ⅛ in. (3 mm) thick.

Figure 37.10 Quarry tile provides a clean, durable finish floor.

Figure 37.11 Paver tiles are weather resistant and resist abrasion. *(Courtesy American Olean Tile Company)*

Organic adhesives harden by evaporation. They are applied with a notched trowel, leaving a thickness of about 1/16 in. (1.6 mm). They are not used on exterior applications or on interior uses that are exposed to considerable water.

Furan mortar is a mixture of furan resin and a hardener. It has high resistance to chemicals.

Tile Grouts

Grouts typically are a mixture of portland cement, fine sand, and lime. They generally come premixed, requiring the addition of water. Latex, furan, epoxy, or silicone rubber can be added to influence the properties. The manufacturer should be consulted to get the most desirable grout for the situation at hand.

BRICK FLOORING AND BRICK AND CONCRETE PAVERS

Brick flooring and brick pavers are used to produce durable finish flooring. Brick pavers are thinner than standard brick, ranging from 1/2 to 2 1/4 in. (12 to 57 mm) thick and are usually preferred to standard brick. Pavers are also available in concrete and asphalt. Glazed brick is also used for a finished floor. Brick flooring is used both indoors and outside, as on a patio (Fig. 37.14). Some types are suitable for heavy use on industrial floors (Fig. 37.15). The various types and sizes of brick are discussed in Chapter 11.

Bricks can be laid with the large flat face exposed or on edge, exposing a narrow face. Brick pavers are laid with the flat face exposed. There are various ways to lay them.

Figure 37.12 Ceramic mosaic tile floors can be laid in a variety of colors and geometric patterns. *(Courtesy American Olean Tile Company)*

Figure 37.13 Textured glazed ceramic tile floors are used in areas subject to moderately heavy use.

Brick floor laid with minimum mortar joint.

Brick floor laid without mortar.

Brick floor laid with wide mortar joint.

Figure 37.14 Bricks provide an attractive, durable finish flooring inside and outside the building.

Figure 37.15 These hard-baked bricks provide a durable floor for heavy industrial applications.

The most common are shown in Fig. 37.16. Exterior brick patios are often laid on a bed of sand with no mortar between the units. This is not satisfactory when heavy loads are to be placed on the floor. Floors subject to heavy loads are laid over a concrete slab.

CONCRETE FLOORS

Finished concrete floors are widely used and provide a durable and economic solution to many building floor situations. The properties of concrete and placing techniques are discussed in Chapters 7 and 8.

The concrete floor surface may be finished in several ways. The slab can be poured, leveled, and screeded and the surface smoothed with a wood float. A smoother surface can be produced by smoothing the wood float surface with a steel trowel. These are done before the concrete has firmly set. Various *textured surfaces* can be had by techniques such as brushing the surface with a broom after it has been troweled but not set. An *exposed aggregate* finish is made by leveling the slab, embedding aggregates in the surface with a float to get a level surface, and, after the concrete has begun to set, flushing the concrete off the aggregate with water, leaving it exposed and forming the finished surface (Fig. 37.17). A high-quality smooth finished surface can be produced by covering the base concrete with a 1 in. (25 mm) thick specially formulated concrete layer. This is applied before the base concrete has hardened.

Colored surfaces are produced by adding color pigments, usually metallic oxides, to the concrete mix as the batch is prepared. Another technique is to pour and finish the floor and spread a dry shake coloring material prepared especially for this purpose. The dry shake is spread evenly over the surface before it has hardened and floated to smooth the surface and even out the color.

Other finishes include painting, applying an abrasive aggregate to the surface to produce a nonslip finish, adding metallic aggregate to the surface to produce a more wear-resistant finish, and the use of various chemical coatings to harden the surface (Fig. 37.18).

Heavy-duty concrete floors are used in many industrial applications. They are usually built as two-course construction. The reinforced floor base is poured and topped with a durable concrete layer made with special abrasion-resistant aggregate, such as emery or iron filings.

TERRAZZO

Terrazzo is a matrix consisting of marble or granite chips, portland cement, and water or a synthetic resin. It is placed as a top layer over a concrete underbed, steel decking, or a wood subfloor.

Portland cement terrazzo is divided into sections by 1¼ in. (32 mm) high metal or plastic dividing strips, which reduce the possibility of cracking. The three methods for casting portland cement terrazzo are monolithic, bonded, and sand cushion (Fig. 37.19). *Monolithic terrazzo* is a ⅝ in. (16 mm) thick layer placed as the topping on a green concrete slab. *Bonded terrazzo* is a topping 1¾ in. (44.5 mm) or thicker. The cured concrete slab is cleaned and coated with a neat portland cement, and a concrete underbed of about 1 in. (25 mm) is laid. Divider strips are placed on the underbed and pushed into it. The terrazzo topping is leveled over the underbed.

Sand cushion terrazzo has a ½ in. (12.7 mm) bed of dry sand laid over the concrete slab. This is covered with a waterproof membrane, and the concrete underbed is laid over it. Dividing strips are placed on the underbed and the terrazzo topping is poured and leveled. The sand cushion isolates the slab from the floor slab protecting it from damage due to possible movement of the floor slab.

Synthetic resin matrix terrazzo should be installed following the directions of the manufacturer of the resin. It is a thin coat material usually ranging from ⅛ to ¼ in. (3 to 6 mm) thick. The stone chips used in synthetic resin terrazzo are much smaller than those used in portland cement terrazzo. Dividing strips are bonded to the wood, metal, or concrete underbed with an adhesive (Fig. 37.20).

After the terrazzo topping has hardened, it is ground smooth and polished. The polishing brings out a gloss and the color of the chips. Sometimes a less glossy nonslip surface is ground. A protective sealer is applied over the finished floor (Fig. 37.21).

Another form of terrazzo is as precast units. Synthetic resin terrazzo is cast in square and rectangular shapes in thicknesses from ⅛ to ¼ in. (3 to 6 mm). Precast stair treads, risers, window stools, wall base, and

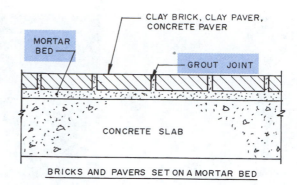

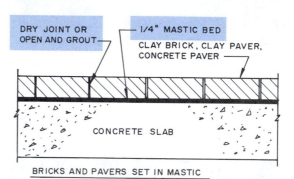

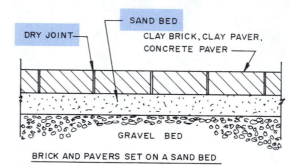

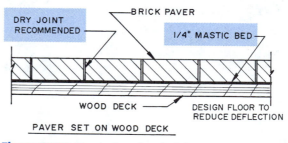

Figure 37.16 Examples of typical floor construction using clay brick, clay pavers, or concrete pavers.

other shapes are available precast in portland cement terrazzo (Fig. 37.22).

STONE FLOOR COVERING

Stone finish flooring is typically slate, marble, or granite (Fig. 37.23). When used on the interior of a building, stones are laid over a wood or a concrete slab in the same

Figure 37.17 Exposed aggregate concrete floors provide a durable and attractive finished surface.

Figure 37.18 This concrete floor in an industrial plant was treated with ArmorSeal®. (*Courtesy The Sherwin-Williams Company*)

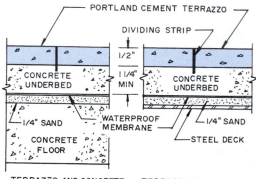

TERRAZZO AND CONCRETE FLOOR SEPARATED BY A SAND CUSHION

TERRAZZO AND METAL FLOOR SEPARATED BY A SAND CUSHION

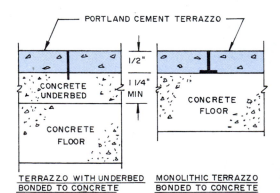

TERRAZZO WITH UNDERBED BONDED TO CONCRETE

MONOLITHIC TERRAZZO BONDED TO CONCRETE

TERRAZZO WITH UNDERBED OVER WOOD DECK

Figure 37.19 Typical portland cement terrazzo floor installations over various types of floor decks.

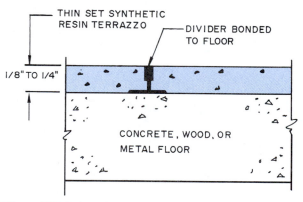

Figure 37.20 A typical synthetic resin terrazzo floor construction detail.

Figure 37.21 The glass block wall enhances the terrazzo floor. This is a very durable floor that is easily cleaned and presents a multicolored finish. *(Courtesy Pittsburgh Corning)*

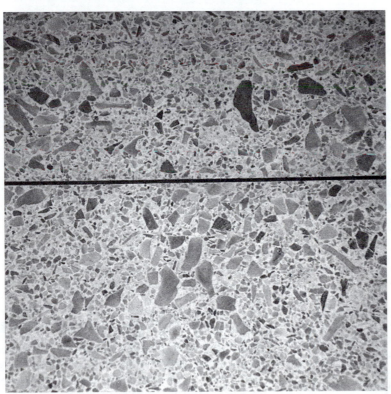

Close up of finished terrazzo floor. Note the metal divider strip.

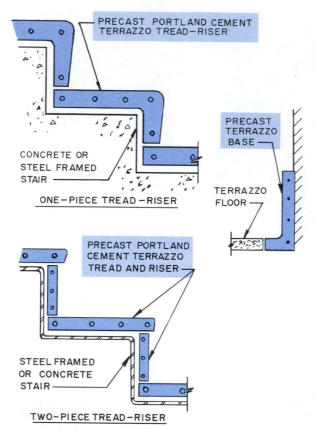

Figure 37.22 Precast terrazzo units are often used for stair treads and risers.

Marble

Slate

Figure 37.23 This marble floor is used on an entrance hall and is laid without mortar joints between the pieces. The slate floor is durable and attractive.

manner as brick pavers. Sand-laid stone is usually used for exterior work, and the joints are left open. Interior stone floors may have the joints filled with joint grout or be set tight with no joint grout (Fig. 37.24).

Another stone floor covering material is a cast stone tile flooring formed by impregnating and encapsulating the stone aggregate with high-technology resins that create a very dense surface. This flooring meets Underwriters Laboratory requirements for slip resistance (Fig. 37.25).

OTHER FLOORING APPLICATIONS

Industrial buildings are generally built with a concrete slab. The finish flooring used is mainly to protect the slab from corrosive materials and the movement of heavy traffic. The concrete can be treated with chemicals to harden the surface (Fig. 37.26) or covered with a flooring, such as wood blocks standing on end so the end grain forms the surface (Fig. 37.27). Asphalt mastic

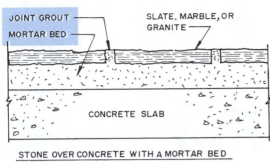

STONE OVER CONCRETE WITH A MORTAR BED

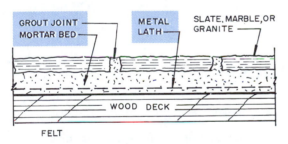

STONE OVER CONCRETE WITH MASTIC BED

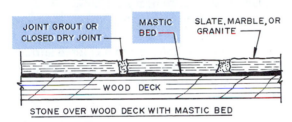

STONE OVER WOOD DECK WITH MASTIC BED

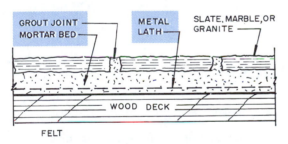

STONE OVER MORTAR BED ON WOOD DECK

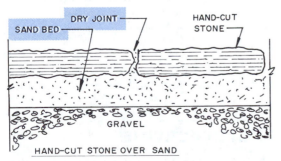

HAND-CUT STONE OVER SAND

Figure 37.24 Various ways stone flooring may be laid for interior and exterior applications.

Figure 37.25 This flooring is made using cast marble tiles formed with marble aggregate cast in a plastic resin. (*Courtesy PermaGrain Products, Inc.*)

Figure 37.26 This concrete garage floor has been given an epoxy primer and a single coat of Accoguard™, which is a chemical product (urethane) that cures to produce a durable, weather-resistant coating. (*Courtesy Environmental Technologies Company*)

floors also can be used. The entire floor, or at least certain work areas, may need treatment for slip resistance, which could be incorporated in the concrete top coating or be movable rubber mats.

Some floors must have a low electrical resistance to prevent sparks from static electricity. One place this *non-sparking conductive flooring* is used is in areas where volatile vapors are present. Other flooring materials must be grease resistant, slip resistant, fire resistant, acid and alkali resistant, static discharging, or x-ray protective.

Various types of seamless floor coverings are available. Some are formed from plastic, such as an acrylic, latex epoxy, resins, methyl methacrylate, and polyacrylate resins, and are spread over a concrete or wood subfloor.

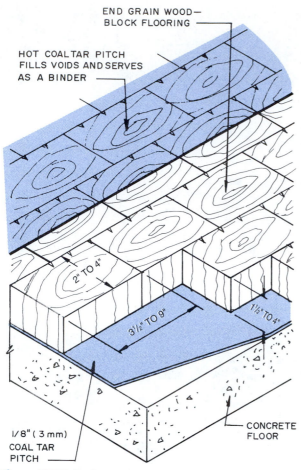

END GRAIN WOOD—
BLOCK FLOORING

HOT COAL TAR PITCH
FILLS VOIDS AND SERVES
AS A BINDER

2" TO 4"

3½" TO 9"

1½" TO 4"

1/8" (3 mm)
COAL TAR
PITCH

CONCRETE
FLOOR

Figure 37.27 End grain wood-block floors are used in industrial areas where heavy wear is expected.

Figure 37.28 Typical poured-in-place basketball floors are made from a two-component polyurethane elastomer.

A wide range of flooring for athletic use is available. This includes coatings for showers and saunas, basketball, tennis, and volleyball courts, tracks, and locker rooms. Materials used include elastomers, rubbers, polymers, and urethane products (Fig. 37.28).

REVIEW QUESTIONS

1. What types of finish flooring are in common use?
2. What type of finish flooring presents the greatest hazard in a fire?
3. What techniques are used to secure wood flooring to the substrate?
4. How can a designer find the fire spread and smoke produced data for various finish flooring products?
5. What type of floor tile is used for areas subject to heavy use?
6. How is the surface of rubber flooring designed to provide slip resistance?
7. What types of mortars and adhesives are used to install tile floors?
8. Why is grout used to finish the installation of ceramic tile floors?
9. What substrate should be used under brick floors that will be subject to heavy use?
10. How is an exposed aggregate concrete floor surface produced?
11. How are concrete floors given a colored surface?
12. What substrates can be used when constructing a terrazzo finished floor?
13. What are the commonly used types of terrazzo flooring?
14. Where are precast terrazzo products used?
15. What materials are used to form seamless floors?

KEY TERMS

clay tile A clay unit that is fired and sometimes glazed. Used as a finished surface on floors and walls.

grout A mortar used to fill spaces in masonry walls, floors, ceramic tile, and preformed roof deck units.

resilient floor covering A manufactured interior floor covering in either sheet or tile form that returns to its original form after being bent, compressed, or stretched.

terrazzo A finish flooring material made from aggregate and portland cement and water or a synthetic resin.

SUGGESTED ACTIVITIES

1. Visit floor covering dealers and obtain samples of as many different products as you can gather. Arrange the samples in a display and label each, citing the brand, material makeup, and places where it would be appropriate to use it.

2. Visit buildings under construction when the floor finishing installers are at work. Observe the tools and procedures used. Prepare a report on what you saw.

3. Have the managers of various flooring supply dealers visit the class and explain how they help the client select a covering material and how they prepare a bid on the job.

ADDITIONAL INFORMATION

Hornbostel, C., *Constructional Materials: Types, Uses, and Applications,* John Wiley and Sons, New York, 1991.
Spence, W. P., *Finish Carpentry,* Sterling Publishing Co., New York, 1995.

Other resources are numerous related publications from the organizations listed in Appendix B.

Carpeting

This chapter will help you to:

1. Understand the various types of carpet construction.

2. Choose from among the various carpet materials the best one for various applications.

3. Supervise the installation of carpeting.

Carpet is a widely used floor covering in residential and commercial construction. It competes well with other floor covering products in terms of cost, durability, and maintenance. In addition, it is available in a wide range of fibers, textures, and colors. It is made in widths of 6, 8, 12, and 25 ft. and in some cases greater widths, resulting in a material that produces few seams and rapidly covers large floor areas.

Although carpeting is often selected for its beauty and comfort, it also provides sound-deadening properties by reducing impact noise and sound reflection within a room (Fig. 38.1).

PILE FIBERS

Wool, a **natural fiber,** has been a major fiber in carpet construction for many years. The development of **synthetic fibers** has enabled the carpet manufacturer to produce products with a wider range of characteristics. Carpet piles often are a blend of fibers. For example, a wool carpet may have a blend of 10 to 30 percent nylon fibers to increase the toughness. Commonly used synthetic fibers are acrylic, modacrylic, nylon, polyester, and polypropylene (olefin).

Wool Fibers

Wool used in carpets is a natural fiber imported into the United States from wool-producing countries such as Australia, New Zealand, and Scotland. The product varies with the different breeds of sheep. The wool fibers range from 3½ to 7 in. (89 to 178 mm) long. The wool pile is made by spinning the fibers into yarn. Wool is relatively unaffected by weak alkalis and organic solvents. It has good resistance to abrasion and aging, and it resists damage to mildew and sunlight. Generally, wool is more expensive than synthetic fibers.

Acrylic Fibers

This synthetic fiber is similar to wool in texture and abrasion resistance. It is composed of more than 85 percent by weight of acrylonitrile. It has good resistance to mildew, aging, sunlight, moths, and chemicals. It is manufactured in a wide range of colors.

Figure 38.1 Carpeting adds beauty and comfort to the room in which it is installed. *(Courtesy Shaw Industries, Inc.)*

Modacrylic Fibers

This synthetic fiber is a member of the acrylic group. It contains at least 35 percent acrylonitrile but less than the 85 percent of acrylic fibers. It is soft, resilient, quick drying, and abrasion and flame resistant. It resists acids, alkalis, sunlight, and mildew.

Nylon Fibers

Nylon is a synthetic petrochemical product produced as a continuous filament, but it is often cut into short lengths varying from 1½ to 6 in. (38 to 152 mm). The short fibers are spun into yarn much the same as wool fibers. Nylon is a strong fiber that has antistain properties and can be dyed. It is resistant to mildew, aging, and abrasion, and it is resilient and low in moisture absorbency.

Polyester Fibers

Polyester fibers are synthetic fibers that have a high tensile strength, good resistance to mineral acids, and excellent mildew, aging, and abrasion resistance. Prolonged exposure to sunlight may cause some loss of strength. They are not as durable as nylon fibers.

Polypropylene Fibers (Olefin)

Polypropylene is a synthetic fiber that is a class of olefin that consists of 85 percent propylene by weight. It has the lowest moisture absorption rate of all fibers in use. It resists mildew, abrasion, aging, sunlight, and common solvents. It is not as resilient as nylon, but it is less expensive. Olefin is very lightweight, strong, and soil resistant.

CARPET CONSTRUCTION

Although the construction of carpets varies somewhat, the details in Fig. 38.2 will help show the terms used. The *pile yarns* form the exposed top surface that takes the wear and abrasion. Pile yarns are made from wool or one or more synthetic fibers. The *backing yarns* consist of *weft yarns* (running the width of the carpet) and *warp yarns* (running the length of the carpet). *Stuffer yarns* run lengthwise and give strength and stability to the carpet. The warp yarns pass over the weft yarns, pulling the pile yarns into place.

WOVEN CARPETS

Woven carpets include **Axminster, loomed, velvet, and Wilton construction.**

Axminster Construction

Axminster carpet is woven on a loom that inserts each tuft of pile yarn individually, providing the possibility of varying the color of each tuft. This gives great flexibility for design. Axminster is usually a cut pile that is even in height. The backing is heavily ridged and sized, making it so stiff that it must be rolled lengthwise. A typical construction detail is shown in Fig. 38.3.

Loomed Construction

Loomed carpet has the pile yarn bonded to a ³⁄₁₆ in. (5 mm) thick rubber cushion. It has a low-loop single-level pile. The pile back has a waterproof coating to which the rubber cushion is bonded. A typical construction detail is shown in Fig. 38.4.

Velvet Construction

The velvet weave is shown in Fig. 38.5. The pile, stuffer, and weft yarns are held together with the double warp yarns. The setting of the loom establishes the height of the piles, and at this point they are loops. The wire that sets the height has a knife edge that cuts the pile yarns to the desired height. A textured surface can be produced by setting the wires to form and cut loops of different heights. Patterned designs cannot be produced by this loom, but tweed effects can be created by varying the colors of the yarn.

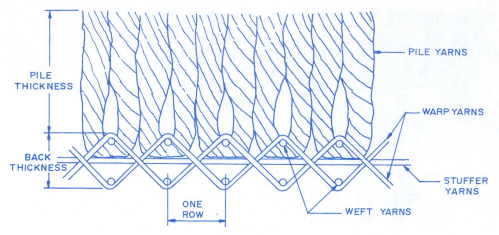

Figure 38.2 Terms used to identify the parts of carpeting.

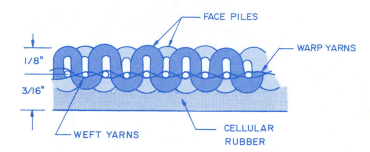

Figure 38.3 Construction details of Axminster carpet.

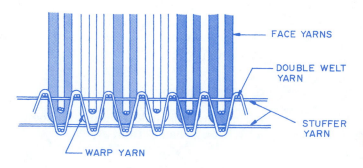

Figure 38.4 Loomed carpet construction bonds the pile yarn to a rubber cushion.

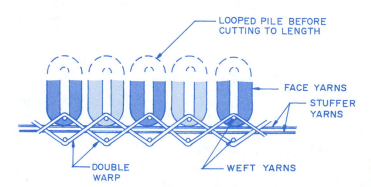

Figure 38.5 Velvet construction has double warp yarns.

Wilton Construction

Wilton construction is produced on a loom like the one used for velvet construction, but the Wilton loom has a mechanism to feed yarns of various colors. The various color yarns are selected by a series of punched cards. The cards regulate which color yarn is looped over the loop-forming wires. These become the visible pile tufts and the other colored yarns are buried in the body of the carpet. This produces a multicolored product with a stiff backing (Fig. 38.6).

Wilton looms can produce sculptured and embossed textures by varying the pile height and a combination of cut and uncut loops. Uncut loop Wilton weaves are also available.

TUFTED CONSTRUCTION

Tufted construction involves stitching the pile yarn through a backing material, which is much like sewing on a power sewing machine. The back is then coated with a latex bonding agent that holds the tufts to the base and bonds a second backing over it. The first backing is usually cotton, polypropylene olefin, or jute. The second backing is usually a coarse jute fabric (Fig. 38.7).

Tufted carpets are available in level loop, multilevel, cut, uncut, and a mixture of cut and uncut piles. Carved and textured surfaces are produced.

KNITTED CONSTRUCTION

Knitted carpet is made by looping together the pile yarn, stitching, and backing in a single operation. A coat of latex is spread on the back to bond these together and stiffen the carpet (Fig. 38.8). Knitted carpets are available in solid colors and tweeds and in single or multilevel uncut pile.

FLOCKED CONSTRUCTION

Flocked construction involves electrostatically spraying short strands of pile yarn onto an adhesive-coated backing sheet. The pile strands become vertically imbedded in the adhesive, a second backing is laminated to the first, and the unit is cured (Fig. 38.9).

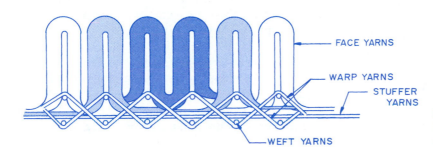

FACE YARNS
WARP YARNS
STUFFER YARNS
WEFT YARNS

Figure 38.6 Wilton construction is used to produce carpets containing multicolored yarns.

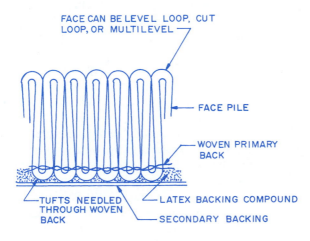

FACE CAN BE LEVEL LOOP, CUT LOOP, OR MULTILEVEL

FACE PILE

WOVEN PRIMARY BACK

TUFTS NEEDLED THROUGH WOVEN BACK

LATEX BACKING COMPOUND

SECONDARY BACKING

Figure 38.7 Tufted carpet is made by stitching pile yarns through a backing material and bonding two layers of backing over it.

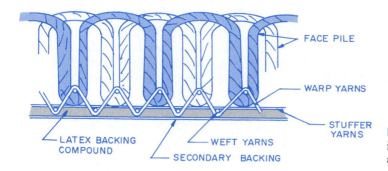

FACE PILE

WARP YARNS

STUFFER YARNS

SECONDARY BACKING

WEFT YARNS

LATEX BACKING COMPOUND

Figure 38.8 Knitted carpet construction involves looping the pile yarns, stitching, and backing in a single operation.

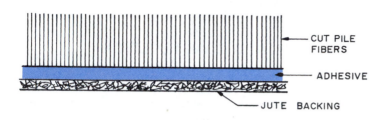

CUT PILE FIBERS

ADHESIVE

JUTE BACKING

Figure 38.9 Flocked carpet construction electrostatically sprays short strands of pile yarn onto an adhesive-coated backing.

FUSION BONDED CONSTRUCTION

Fusion bonded carpet is made by bonding the pile yarn between two parallel sheets of backing. These are coated with vinyl adhesive, making a sandwichlike product with the pile yarn in the center. After the adhesive hardens, the pile yarn is cut in the middle to form two pieces of cut pile carpet. A secondary backing is often bonded to the first backing sheet (Fig. 38.10).

QUALITY SPECIFICATIONS

Carpet is made in various qualities suitable for different locations. Following are things to consider when choosing carpet.

The Federal Housing Administration carpet quality specifications include pile yarn weight, pile density, and pile thickness. ASTM tests indicate how to measure these qualities. The General Services Administration has also issued carpet specifications.

Pile yarn weight (W) is the average weight of pile yarn in ounces per square yard, excluding the backing.

Pile thickness (t) is the height of pile tufts above the backing in inches.

Pile density (D) is the weight of pile yarn per unit of volume in the carpet, stated in ounces per cubic yard. It is 36 times the finished pile weight divided by the average pile thickness:

$$D = \frac{36W}{t} = \text{oz./cu.yd.}$$

The minimum acceptable pile density is 2800 oz./cu.yd.

Minimum requirements for *pile weight-density* (WD) and *pile yarn weight* (W) are in Table 38.1. Pile weight density equals pile weight (W) multiplied by the minimum pile density, 2800 oz./cu.yd.

Carpet can be specified by the number of tufts per square inch, rows, wires, and stitches per linear inch and by the pitch, gauge, and number of needles. These describe the spacing of the tufts. *Yarn size* and the number of piles could also be specified. *Piles* are the number of strands that are twisted together to form the pile yarn.

Tufts per square inch is an indication of the overall density of the carpet. The greater the number of tufts per square inch the denser the carpet.

Rows are the way of indicating the number of tufts per inch. Wiltons and velvets are described by the number of *wires* per inch (Fig. 38.11). This spacing of tufts indicates the pile density lengthwise. The more rows or wires per linear inch the denser the carpet. Since tufted carpets have no weft yarns, their density is indicated by the number of tufts per inch in the backing. These are referred to as *stitches* per inch.

Pitch, gauge, and needles are used to describe the number of tufts across the width of a carpet. **Pitch** is the number of tufts in a 27 in. (686 mm) width of carpet. The higher the pitch number the tighter the weave and the denser the carpet (Fig. 38.12). When no continuous warp exists, the term *gauge* is used to indicate the spacing of tufts across the width of the carpet in inches. For example, ⅛ gauge means 8 tufts per inch and ³⁄₁₆ gauge means 16 tufts per 2 inches. The term *needles* is used by some instead of gauge to indicate the number of tufts per inch.

Table 38.1 Minimum Physical Requirements for Carpeting in Moderate Traffic Areas[a]

Pile Fiber	Minimum Pile Yarn Weight (W) oz./yd.^{2}	Minimum Weight-Density Factor (WD)[b]
Acrylic Modacrylic Wool	25	70,000 (25 × 2800)
Nylon (staple and filament) Polypropylene (olefin)	20	56,000 (20 × 2800)

[a]Based on FHA Use of Materials Bulletin UM-44A.
[b]Pile density (D) = 2800

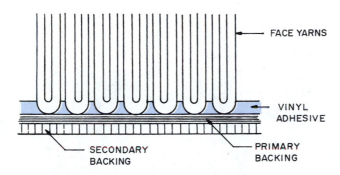

Figure 38.10 Fusion bonded carpet construction involves bonding pile yarn to backing sheets coated with a vinyl adhesive.

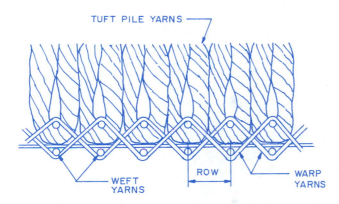

Figure 38.11 Rows are used to identify the number of tufts per inch.

Physical specifications for carpeting in moderate use areas are in Table 38.1. Physical specifications by the GSA for heavy traffic areas are in Table 38.2.

Building codes specify flammability requirements for carpets in various locations, such as in commercial buildings. The flame resistance of carpets is most commonly ascertained by the Flooring Radiant Panel Test. The carpet specimen is mounted on the floor of the test chamber and exposed to intense radiant heat from above. The rate of flame spread is measured.

CARPET CUSHIONS

Carpet is usually installed over a **cushion** because this increases the life of the carpet, absorbs considerable traffic noise, has insulation value, and provides a soft, resilient floor covering. There are three types of carpet installation. The carpet can be installed over a separate cushion, the carpet can be made with a foam cushion attached to the back, or the carpet may be laid without a cushion (Fig. 38.13).

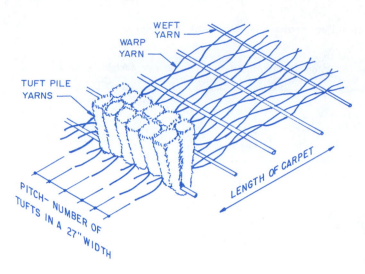

WEFT YARN

WARP YARN

TUFT PILE YARNS

LENGTH OF CARPET

PITCH— NUMBER OF TUFTS IN A 27" WIDTH

Figure 38.12 Pitch indicates the number of tufts in a 27 in. (686 mm) width of carpet.

Table 38.2 Minimum Physical Requirements for Carpeting in Heavy Traffic Areas[a]

Pile Description	Wilton Wool or Acrylic — Single level loop woven thru back	Velvet — Wool or Acrylic — Single level loop pile woven thru back	Velvet — Nylon[2] — Single level loop pile woven thru back	Velvet — Wool or Acrylic — Multilevel loop woven thru back	Tufted Wool or Acrylic — Single level loop	Knitted Wool or Acrylic — Single level loop
Tufts/sq.in.	52	60	32	34	60	26
Frames	3	—	—	—	—	—
Shots/wire	2	2	2	2	—	—
Weight (oz./sq.yd.) Pile	48	42	29	44	42	37
Total	72	60	50	64	—	58
Pile height	Min. 0.250 Max 0.320	Min. 0.210 Max. 0.310	Min. 0.200 Max. 0.290	Min. 0.190 Max. 0.370	Min. 0.250 Max. 0.320	Min. 0.230 Max. 0.290
Chain	Cotton and/or rayon	Cotton and/or rayon	Cotton and/or rayon	Cotton and/or rayon	—	Cotton, rayon or nylon
Filling	Cotton and/or rayon, or jute	Cotton and/or rayon, or jute	Cotton and/or rayon, or jute	Cotton and/or rayon, or jute	—	Jute or kraftcord
Stuffer	Cotton, jute or kraftcord	Cotton, jute or kraftcord	Cotton, jute or kraftcord	Cotton, jute or kraftcord	—	—
Back coating (oz./sq.yd.)	—	8	6	8	—	—
Tuft bind (oz.)	80	80	80	80	100	14
Ply twist turns (per inch)	Min. 1.5 Max. 3.5	Min. 1.5 Max. 3.5	Min. 1.0 Max. 3.0	Min. 1.5 Max. 3.5	Min. 2.5 Max. 4.5	Min. 1.5 Max. 3.5

[a]Based on General Services Administrative Federal Specifications DD-C-95.

Carpet cushions are available in three basic materials—urethane, fiber, and rubber (Table 38.3). They are manufactured in Class 1 (light and moderate traffic) and Class 2 (heavy-duty traffic). These recommendations meet U.S. government requirements for cushion used in FHA-financed housing. Rubber cushion is more expensive, but it lasts longer, resists mildew and mold, and is nonallergenic.

INSTALLING CUSHION AND CARPET

Carpet is installed over concrete, plywood, particleboard, or another firm smooth substrate. Carpet can be installed in several ways. One method uses tackless wood strips. These are thin wood strips with tack points sticking through them. Metal strips with hooked points are also used. They are nailed to the subfloor. The cushion is placed inside the strips and stapled to the floor. The

Table 38.3 Minimum Cushion Recommendations for Residential Applications

Type	Key Characteristics	Class 1	Class 2
Urethane			
A. Prime	Density lbs./cu. ft. min. Thickness, in. min.	2.2 0.375	Not Recommended for class 2
B. Grafted prime[a]	Density lbs./cu. ft. min. Thickness, in. min.	2.7 0.375	2.7 0.25
C. Densified prime	Density lbs./cu. ft. min. Thickness, in. min.	2.2 0.313	2.7 0.25
D. Bonded[b]	Density lbs./cu. ft. min. Thickness, in. min.	5.0 0.375	6.5 0.375
E. Mechanically frothed[a]	Density lbs./cu. ft. min. Thickness, in. min.	10.0 0.250	12.0 0.250
Fiber			
A. Rubberized hair jute	Weight, oz./sq. yd. min. Thickness, in. min. Density lbs./cu. ft. min.	40.0 0.27 12.3	50.0 0.375 11.1
B. Rubberized jute	Weight, oz./sq. yd. min. Thickness, in. min. Density lbs./cu. ft. min.	32.0 0.3125 8.5	40.0 0.375 8.9
C. Synthetic fibers[a]	Weight, oz./sq. yd. min. Thickness, in. min. Density lbs./cu. ft. min.	22.0 0.250 6.5	28.0 0.300 6.5
D. Resinated recycled textile fiber[a]	Weight, oz./sq. yd. min. Thickness, in. min. Density lbs./cu. ft. min.	24.0 .25 7.3	30.0 0.30 7.3
Rubber			
A. Flat rubber	Weight, oz./sq. yd. min. Thickness, in. min. Density lbs./cu. ft. min.	56.0 0.22 18.0	64.0 0.22 21.0
B. Rippled rubber	Weight, oz./sq. yd. min. Thickness, in. min. Density lbs./cu. ft. min.	48.0 .285 14.0	64.0 0.330 16.0

Maximum thickness for any product is 0.5 in.
Class 1: Light and moderate traffic (such as living rooms, dining rooms, bedrooms, recreational rooms, and corridors. Class 2 cushion may be used in Class 1 applications.)
Class 2: Heavy duty traffic (for heavy traffic use at all levels, but specifically for public areas such as lobbies and corridors in multi-family facilities.) The Carpet Cushion Council also recommends Class 2 for stairs and hallways.
[a]The Carpet Cushion Council has endorsed HUD UM72 as its recommended minimum standards. Products noted above are Carpet Cushion Council recommendations for inclusion in UM72a, HUD's proposed revision of UM72.
The UM72 standards include other technical characteristics for each type of carpet cushion. Please contact the Carpet Cushion Council for further details.
[b]Bonded 8 lb. density can qualify for Class 2 with minimum thickness of .25–5%.
This chart illustrates products that meet **minimum** guidelines. Better grades of carpet cushion than the minimum suggested are always recommended when possible to provide more support and cushion for carpet. In areas where heavy use is expected, the Carpet Cushion Council suggests using firmer grades of cushion. These areas include stairways, halls, and areas where heavy furniture is used (such as living rooms and dining rooms). Softer cushion may be used in bedrooms and lounge areas where use is lighter and a plusher "feel" is desired.

carpet is laid over the cushion, stretched with a special device, and pressed onto the tack points on the tackless strip (Fig. 38.14).

Another installation method involves bonding the cushion to the subfloor with an adhesive and bonding the carpet to the cushion (Fig. 38.15). The materials and techniques recommended by the manufacturer must be followed to achieve maximum results. In some cases carpet is bonded directly to the subfloor (Fig. 38.16). This often is done in a kitchen or on a porch. When this is done, the carpet will experience accelerated fiber failure. Following is an abbreviated description of how to install the cushion and carpet using adhesives.

After the subfloor has been made smooth and clean, the cushion is rolled out on it and cut to size. Then the adhesive is spread on the subfloor using a clean, sharp trowel with square notches. Since trowels wear down rapidly, they are replaced frequently (Fig. 38.17).

Next, the cushion is bonded to the floor (Fig. 38.18). The carpet is rolled out on top of the cushion, measured, and cut to the needed size. Then the carpet is folded back and the adhesive is laid down on the cushion (Fig. 38.19). The carpet is lifted and laid over the area with adhesive. The carpet must be moved, stretched, and rolled perfectly into place before the adhesive sets (Fig. 38.20).

A special adhesive is used for closing the seams between rows of carpet. It is applied to the seam areas and the cut edge of the carpet. Then the seams are pressed down with a carpet tractor (Fig. 38.21). Finally, the entire carpet is rolled thoroughly. Rolling helps work the carpet adhesive into the carpet backing and into the pores of the upper surface of the cushion. It also helps bond the pad to the subfloor (Fig. 38.22).

1. Carpet installed over separate cushion. This can be done by stretching the carpet in over tack strip, or by glueing the cushion to the floor and the carpet to the cushion.

2. Carpet with attached cushion.

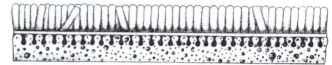

3. Direct glue down: carpet cemented directly to the floor without any cushion.

Figure 38.13 Types of carpet cushions. *(Courtesy Cushion Carpet Council)*

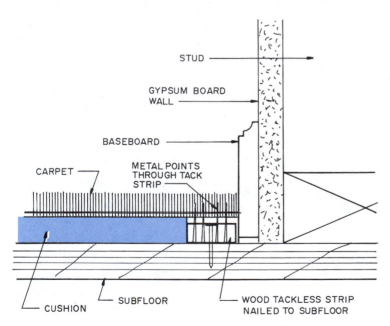

Figure 38.14 Some carpet is installed using wood tackless strips around the edge of the room to hold the carpet tightly in place.

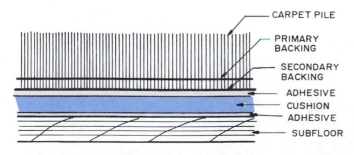

Figure 38.15 Carpet can be installed by bonding the cushion to the subfloor and bonding the carpet to the cushion.

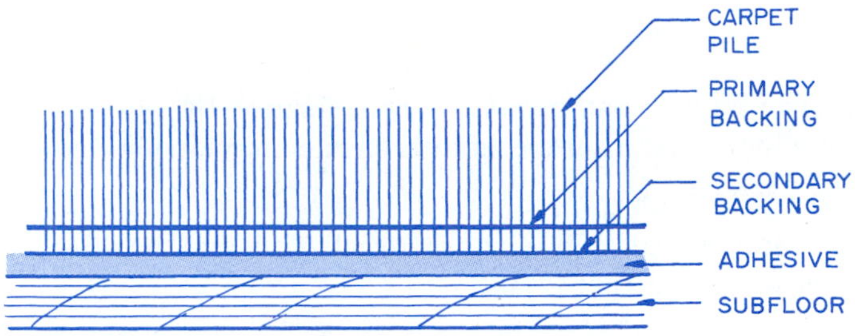

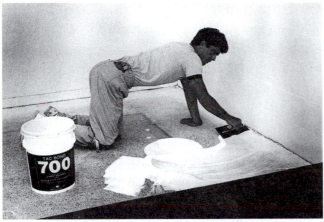

Figure 38.16 Carpet can be bonded directly to the subfloor.

CARPET PILE
PRIMARY BACKING
SECONDARY BACKING
ADHESIVE
SUBFLOOR

Figure 38.19 After cutting the carpet to size, adhesive is applied to the top of the cushion. *(Courtesy No-Muv Corporation, Inc.)*

Figure 38.17 Adhesive is applied to the subfloor with a notched trowel. *(Courtesy No-Muv Corporation, Inc.)*

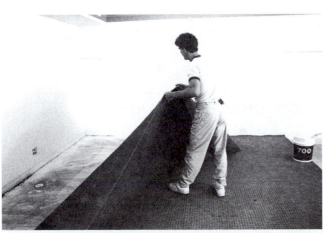

Figure 38.18 The cushion is laid over the adhesive-coated floor. *(Courtesy No-Muv Corporation, Inc.)*

Figure 38.20 The carpet is laid over the adhesive-coated cushion and stretched and rolled into place. *(Courtesy No-Muv Corporation, Inc.)*

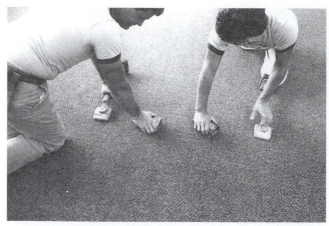

Figure 38.21 A special adhesive is applied at the seams, and the seams are pressed closed. *(Courtesy No-Muv Corporation, Inc.)*

Figure 38.22 The carpet is rolled to smooth it and to force the adhesive into the carpet backing. *(Courtesy No-Muv Corporation, Inc.)*

REVIEW QUESTIONS

1. In what widths is carpet generally available?
2. What is the major natural fiber used in carpet construction?
3. What are the commonly used synthetic fibers used in carpet construction?
4. What are the major types of carpet construction?
5. What is meant by pile yarn weight?
6. How is pile thickness measured?
7. How is pile weight density calculated?
8. What does a measure of tufts per square inch indicate?
9. What does the number of rows indicate?
10. From what materials are cushions made?

KEY TERMS

Axminster construction Carpet formed by weaving on a loom that inserts each tuft of pile individually into the backing.

cushion A layer of resilient material applied to the floor over which a carpet is laid.

flocked construction Carpet formed by electrostatically spraying short pile strands on an adhesive-coated backing material.

fusion bonded construction Carpet formed by bonding pile yarn between two sheets of backing material and cutting the pile yarn in the center, forming two pieces of carpet.

knitted construction Carpet formed by looping pile yarn, stitching, and backing together.

loomed construction Carpet formed by bonding the pile yarn to a rubber cushion.

natural fibers Fibers found in nature, such as wool.

pitch The number of tufts in a 27 in. width of carpet.

rows The number of tufts per inch.

synthetic fibers Fibers formed by chemical reactions.

tufted construction Carpet formed by stitching the pile yarn through the backing material.

velvet construction Carpet formed by joining the pile, stuffer, and weft yarns with double warp yarns.

Wilton construction Carpet formed on a loom capable of feeding yarns of various colors.

SUGGESTED ACTIVITIES

1. Collect carpet samples from manufacturers and local carpet retail outlets. Arrange a display and label the samples, detailing the fibers, methods of construction, and suitable applications.

2. Review the local building code pertaining to carpeting and prepare a brief summary.

ADDITIONAL INFORMATION

Ramsey, H. R., Sleeper, J. R., and Hoke, J. R., eds., *Architectural Graphic Standards,* John Wiley and Sons, New York, 1995.

Other resources are publications of the Carpet and Rug Institute, Dalton, Ga., the Carpet Cushion Council, Riverside, Ca. and the American Fiber Manufacturers Association, Washington, D.C.

PART VI

SPECIALTIES, EQUIPMENT, AND FURNISHINGS

Commercial and industrial buildings have widely varying requirements when it comes to items needed to prepare the building for use by the occupants. Chapter 39 presents examples and descriptions of many of the items specified by the Construction Specifications Institute as *specialty* items. Examples are compartments, cubicles, screens, cupolas, identifying signs, lockers, and fire protection equipment.

In addition to specialty items, the designer has to identify and specify items classified as *equipment*. Chapter 40 cites some of these items, which include things such as security, maintenance, solid waste handling, office, medical, and residential equipment.

Furnishings are specified as products that are decorative in nature, such as artwork and window treatments and items such as casework and multiple seating. These items are presented in Chapter 41.

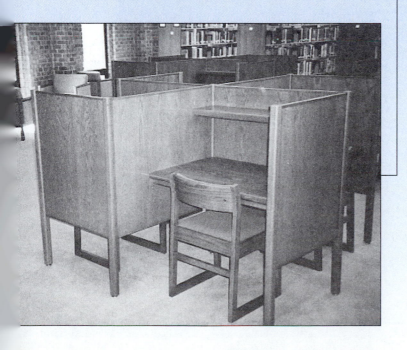

D I V I S I O N 10

SPECIALTIES
CSI MasterFormat™

CHAPTER 39

Specialties

This chapter will help you to:

1. Become aware of some of the specialty products as classified by the Construction Specifications Institute.

2. Make informed decisions when deciding on and specifying specialty items.

Products listed as specialties include a vast spectrum of items used to provide a limited special function. Following are brief descriptions of some of these. The entire CSI outline of specialties is given on the page introducing Division 10 of the MasterFormat.

VISUAL DISPLAY BOARDS

Various types of display board are shown in Fig. 39.1. **Chalkboards** are available with the exposed surface made from slate, porcelain enamel on steel, aluminum, or glass. The glass units use tempered float or plate glass coated with a colored, vitreous glaze that contains a fine abrasive. Chalkboards may be permanently mounted on a wall or may be portable. Some have panels that slide horizontally or vertically. They are available in sizes up to 6 by 12 ft. (1830 by 3660 mm). Natural slate chalkboards are typically from $\frac{1}{4}$ to $\frac{3}{8}$ in. (6 to 10 mm) thick, in widths to 4 ft. (1220 mm), and in random lengths to 6 ft. (1830 mm). Other chalkboards are available in widths to 4 ft. (1220 mm) and lengths to 10 ft. (3050 mm).

Marker boards may have a vinyl sheet surface or porcelain enamel on steel. The substrate may be particleboard, fiberboard, Sheetrock, hardboard, plywood, or a plastic honeycomb. Special ink markers are used to write on the surface. Some are a form of watercolor and wipe off the marker board with a wet cloth. Another type has semipermanent ink and requires a special liquid cleaner.

Pegboard panels are usually $\frac{1}{4}$ in. (6 mm) thick hardboard with holes spaced 1 in. (25 mm) on center. They typically have an aluminum frame. Special wire hanging clips are available.

Bulletin boards or **tackboards** typically have a cork surface laminated to a fiberboard substrate. Some have a woven fabric facing and can hold paper with pins or hook-and-loop fasteners that grip into the fabric.

Various types of manual and motorized *projection screens* are available. They may be integrated into a larger unit with chalkboards, tackboards, and marker boards.

COMPARTMENTS AND CUBICLES

These products include enclosures such as metal, plastic laminate, and stone toilet **compartments,** shower and

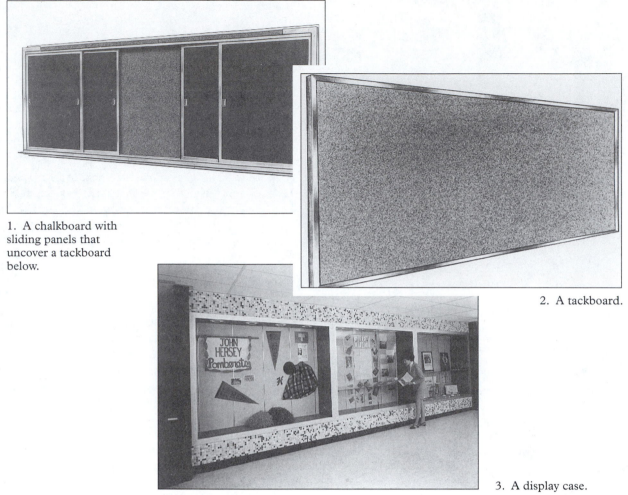

1. A chalkboard with sliding panels that uncover a tackboard below.

2. A tackboard.

3. A display case.

Figure 39.1 Some of the visual displays and chalkboards in common use. *(Courtesy Claridge Products and Equipment, Inc.)*

dressing compartments, and various **cubicles.** Also included are dividers, screens and curtains for showers, hospital cubicles, toilets, and urinals. The toilet enclosures may be supported by the floor, ceiling, and wall (Fig. 39.2). Toilet enclosures that are hung from the ceiling make cleaning the floor easier. Ceiling-hung units require a structural steel ceiling support.

Metal toilet compartments may be stainless steel or steel with a baked enamel finish. Plastic laminate compartments have a 1/16 in. (1.6 mm) plastic laminate sheet adhered to particleboard or high-pressure plastic laminate core. Stone compartments are typically marble.

Cubicle enclosures are hung from an overhead track as shown in Fig. 39.3. The track path can be varied to suit the conditions. It can be mounted directly to the ceiling or suspended (Fig. 39.4). Ceiling tracks are also used to support intravenous systems.

LOUVERS, VENTS, GRILLES, AND SCREENS

Some types of **grilles,** screens, louvers, and vents are specialty items. Louvers and other items for ventilation that are not an integral part of the mechanical system are specialty items. This includes items such as operable, stationary, and motorized metal wall louvers, louvered equipment enclosures, door louvers, and various types of wall and soffit vents. For example, a curtain wall louver is a specialty product. Interior and exterior grilles and screens made of any material are used for air distribution, sun screen, and other functions in addition to ventilation purposes (Fig. 39.5).

Grilles are open gratings or barriers used to cover, conceal, protect, or decorate an opening. They can be steel, aluminum, cast iron, or wood (Fig. 39.6). Various types of

FLOOR SUPPORTED

CEILING HUNG

OVERHEAD BRACED

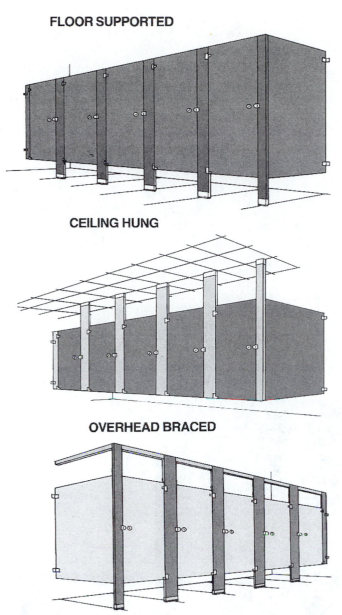

Figure 39.2 Toilet enclosures may be supported by the ceiling, wall, and floor. *(Courtesy Global Steel Products Corporation)*

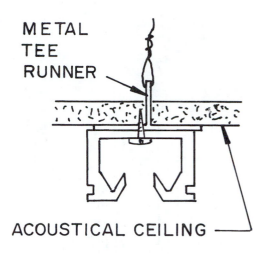

METAL TEE RUNNER

ACOUSTICAL CEILING

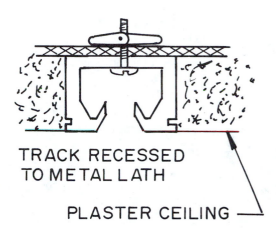

TRACK RECESSED TO METAL LATH

PLASTER CEILING

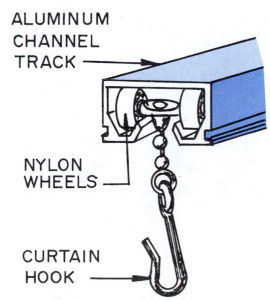

ALUMINUM CHANNEL TRACK

NYLON WHEELS

CURTAIN HOOK

Figure 39.3 Hospital cubicle curtains are hung from the ceiling on metal tracks.

grating are used to cover trench frames that control surface water flow. These are typically cast iron (Fig. 39.7). Street and parking lot construction must be designed to carry away surface water from rain and melting snow. Curb inlets are used as intakes for this purpose (Fig. 39.8). Tree grates are another form of grating used both inside and outside buildings. They provide protection yet permit the tree to receive water (Fig. 39.9).

SERVICE WALL SYSTEMS

Service wall systems incorporate services such as clocks, fire hoses, drinking fountains, fire extinguisher cabinets, telephones, and waste receptacles (Fig. 39.10).

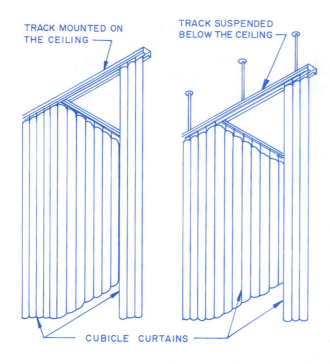

Figure 39.4 Cubicle curtains may be hung from ceiling-mounted or suspended tracks.

Figure 39.5 These metal louvers are part of the building ventilation system and are designed to be architecturally pleasing.

Figure 39.6 These cast metal grilles provide security and are a major architectural feature. *(Courtesy Lawler Foundry Corporation)*

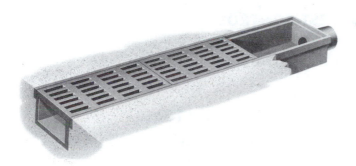

Figure 39.7 This cast iron trench frame is used to control surface water. *(Courtesy Neenah Foundry Company)*

Figure 39.8 This heavy-duty curb inlet frame and grate is manufactured in gray and ductile cast iron. *(Courtesy Neenah Foundry Company)*

Figure 39.9 A cast iron tree guard protects the tree roots from surface damage yet permits rain to reach them. *(Courtesy Neenah Foundry Company)*

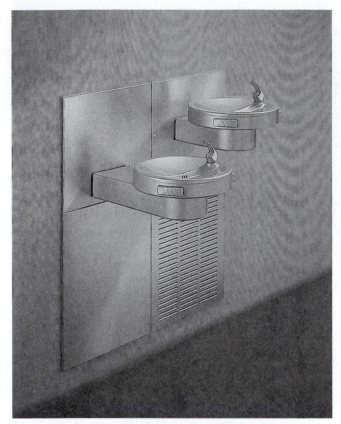

Figure 39.10 Drinking fountains are a part of the service wall system. *(Courtesy EBCO Manufacturing Company)*

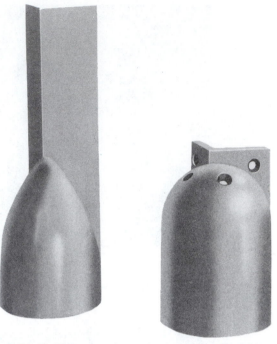

Figure 39.11 These cast iron wheel guards will withstand severe impact and abrasion as they protect the sides of garage door openings. *(Courtesy Neenah Foundry Company)*

WALL AND CORNER GUARDS

Wall and corner guards are protective devices made from metal, plastic, rubber, and other materials that will withstand impact and, for exterior use, exposure to the weather. Figure 39.11 shows cast iron wheel guards used to protect the sides of a large opening. Rubber bumpers are secured to walls, such as at a truck loading dock, to protect the wall and trucks from damage. Other types of corner guards are used to protect wall corners that may be damaged by passing traffic such as moving carts and forklifts. These typically are protected with molded rubber or plastic guards installed on the finished corners.

ACCESS FLOORING

Access flooring is freestanding flooring made in modular units and raised above the basic floor system to provide a space to run mechanical and electrical services. The systems commercially available have some type of metal adjustable pedestals on which the modular floor panels rest. The floor panels typically are covered with vinyl composition floor covering (Fig. 39.12), although carpet could be used.

PEST CONTROL

Various mechanical, chemical, and electrical pest-repellent systems and protective devices are available. These include control of rodents, birds, and insects.

FIREPLACES AND STOVES

Manufactured fireplaces and stoves, including dampers, metal chimneys, and fireplace screens and doors, are specialties. Manufactured fireplaces typically are metal-framed firebox and vent units designed to accommodate gas-fired logs. They may be vented vertically through a metal pipe or horizontally through the wall. Some types do not require venting. They are delivered completely assembled (Fig. 39.13). A wide range of gas and wood-burning stoves are available. Some are placed in the room and become not only a source of heat but a part of the decor (Fig. 39.14). Others are large units with ducts running to several rooms.

MANUFACTURED EXTERIOR SPECIALTIES

These items include stock and custom-designed steeples, spires, cupolas, and weather vanes. A **steeple** is an ornamental construction usually ending in a spire (Fig. 39.15). It is usually erected on a roof or tower and

Figure 39.12 This access floor permits computer and electrical runs to be located below the floor and easily accessed for changes as needed. *(Reproduced with permission from* The Building Systems Integration Handbook, *Richard D. Rush, Editor, Butterworth-Heinmann, 313 Washington St., Newton, Mass. 02158–1626)*

Figure 39.13 This manufactured direct vent gas fireplace is quickly installed and has high efficiency and heat output. *(Courtesy Heat-N-Glo Fireplace Products, Inc.)*

Figure 39.14 This freestanding top-vented gas-fired stove provides an attractive focal point in a room. *(Courtesy Heat-N-Glo Fireplace Products, Inc.)*

Figure 39.15 This beautiful steeple contains a clock and is crowned with a tall spire.

may hold a clock or bells. A **spire** is a tall pointed pyramidal roof built on a tower or steeple. A **cupola** is a light roofed structure mounted on the roof. Cupolas typically are circular or polygonal. In addition to the architectural beauty, a cupola often serves as a roof vent (Fig. 39.16).

FLAGPOLES

This specialty area includes complete flagpole assemblies, including the required accessories. Flagpoles typically are made from wood, metal, or fiberglass. They may be ground set or wall mounted (Fig. 39.17).

IDENTIFYING DEVICES

Identifying devices include directories, direction signs, bulletin boards, and various letters, signs, and plaques used to communicate or identify. These may be fixed devices, or they may be lighted, computerized or some type of electronic device.

Directories usually have replaceable letters and a locked glass door (Fig. 39.18). They are available as wall mounted or freestanding units. *Bulletin boards* may have a cork face or a cork face covered with nylon or vinyl fabric. The substrate is typically a fiberboard panel.

Various types of *accessibility signs* have become standardized for use across the country. Many building codes require accessible elements to be identified by the international symbol of accessibility (Fig. 39.19). The international symbol of accessibility is posted to indicate all elements and spaces of accessible facilities. This includes facilities such as parking spaces reserved for individuals with disabilities, accessible passenger-loading zones, accessible entrances, and accessible toilet and

Figure 39.16 This cupola has louvered sides so it can serve as a roof vent.

Figure 39.17 A flagpole flies the Stars and Stripes.

bathing facilities. In addition, signs are used to identify volume-control telephones, text telephones, and assistive listening systems (Fig. 39.20).

One example of code regulated signs is the identification of exits and signs pointing the direction to move to get to an exit. Exit signs are illuminated. They can point to stairs, elevators, or doors exiting directly from the building to the outside (Fig. 39.21). Signs also are used to identify refuge areas and the direction to move to reach them. A refuge area is an area that has protection against smoke and fire. The maximum occupant capacity for assembly rooms, restraints, and other areas where numbers of people gather is indicated by signs posted in the area. Control of parking and identification of facilities for the handicapped are other major parts of any sign system related to a building (Fig. 39.22). These and many other signs are required by the building codes.

Other *symbol signs* are widely used to present a message graphically and to help communicate with people using various languages. These are found extensively in places such as airports, hotels, and hospitals, and they are used as highway traffic control signs (Fig. 39.23).

Figure 39.18 This building directory gives room locations for the offices in the court house. (*Courtesy Claridge Products and Equipment, Inc.*)

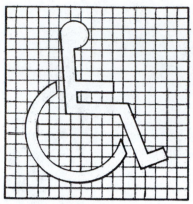

Proportions
International Symbol of Accessibility.

Display Conditions
International Symbol of Accessibility

Figure 39.19 The international symbol of accessibility.

International TDD Symbol

International Symbol of Access for Hearing Loss

Figure 39.20 Symbols used to identify telephones and assistive listening devices for those with hearing impairments.

Figure 39.21 Lighted exit signs are one of the more important identifying devices in a building.

Figure 39.22 Various signs are used to control parking and to identify facilities accessible by the handicapped.

Figure 39.23 Some of the symbols used to communicate graphically.

PEDESTRIAN CONTROL DEVICES

The products used for pedestrian control include portable posts and railings, rotary gates, turnstiles, electronic detection and counting systems, and items to limit access (Figs. 39.24 through 39.27). These are strong, durable products that can withstand exposure to the weather.

LOCKERS AND SHELVING

Lockers provide temporary storage in areas such as airports (Fig. 39.28) and athletic facilities (Fig. 39.29). Some are coin operated. All types of open manufactured metal and wood storage shelving used for general storage are specialty items. This includes open shelving, various manufactured shelving systems, and mobile storage systems.

FIRE PROTECTION SPECIALTIES

This specialty includes all portable firefighting devices and their storage facilities and excludes items directly connected to an installed fire protection system. Typical items include fire extinguishers, extinguisher cabinets, fire blankets and their storage cabinets, and any fire extinguishing units on wheels (Fig. 39.30).

PROTECTIVE COVERS

Various types of awnings, canopies, marquees, covered walkways, and sheltered bus and car stop structures are available as standard manufactured units or as custom designed and built structures. They are made from any of the commonly used materials that will withstand exposure to the weather (Fig. 39.31).

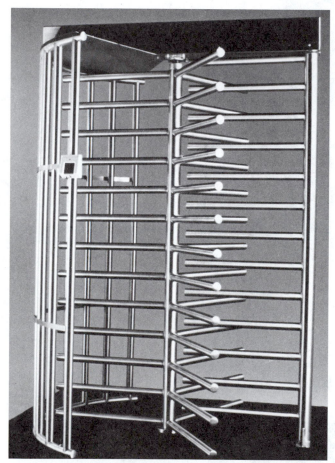

Figure 39.24 This is a 7 ft. high nonpenetrable security turnstile providing total passageway security. *(Courtesy Alvardo Manufacturing Company)*

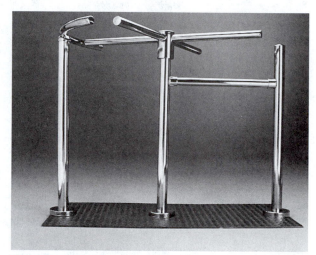

Figure 39.25 A four-arm turnstile that may be manually or electrically operated. It controls one-way ingress and egress. *(Courtesy Alvardo Manufacturing Company)*

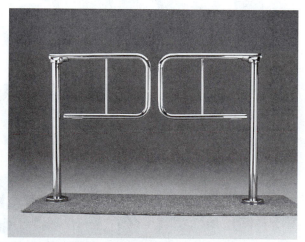

Figure 39.26 This self-closing double gate permits one-way passage. *(Courtesy Alvardo Manufacturing Company)*

Figure 39.27 A plastic-coated welded steel barrier used where control of equipment or people is required. *(Courtesy Alvardo Manufacturing Company)*

Figure 39.29 Single-tier lockers are widely used in schools and athletic facilities. *(Courtesy Republic Storage Systems Co., Inc., Canton, Ohio)*

Figure 39.28 Extensive banks of lockers are used in public places such as airports. These are coin-operated units. *(Courtesy American Locker Company, Inc.)*

Figure 39.30 A typical fire extinguisher stored in a protective cabinet.

Figure 39.31 This covered walkway provides safe access to buildings on opposite sides of a major thoroughfare. *(Courtesy Lin-El, Inc.)*

POSTAL SPECIALTIES

Postal specialty products include all types of postal service facilities and devices, such as view windows, letter slots, letter boxes, collection boxes, and chutes. In large commercial buildings these can become an extensive system used to speed the collection of mail (Fig. 39.32).

DEMOUNTABLE PARTITIONS

Demountable (movable) partitions include various dividers, screens, enclosures, and partitions. They may be freestanding post and panel units, hung from the wall or ceiling, or secured to the floor and ceiling. They may be solid, offering privacy and some sound control, or open mesh or screen panels, giving an area

security but permitting visual examination. Some types use folding gates or doors. They are made of a variety of materials including wood, plastic laminates, and metal (Figs. 39.33 through 39.36).

OPERABLE PARTITIONS

Operable partitions are those manually or power operated, allowing the moving partition to enclose or open up an area. They are hung from a track on the ceiling and are supported by a steel floor track (Fig. 39.37). Some are top hung and have a rubber seal at the floor. They include folding panel partitions, accordion folding partitions, and sliding and coiling partitions. Folding and accordion doors are not in this classification.

Figure 39.32 Post office boxes typically used in U.S. post offices.

Figure 39.34 Floor-to-ceiling partitions with glass panels block off noise from other areas, provide privacy and security of the contents of the room, yet do not give a closed-in feeling. *(Courtesy Glen O'Brien Partition Co., Inc.)*

Figure 39.33 Low demountable partitions provide privacy yet use the general illumination and the heating and air-conditioning systems. *(Courtesy Glen O'Brien Partition Co., Inc.)*

Figure 39.35 Counters can be built in as part of a demountable partition system. *(Courtesy Glen O'Brien Partition Co., Inc.)*

Figure 39.36 Floor-to-ceiling demountable partitions can be used to provide rooms with visual and sound protection. *(Courtesy Glen O'Brien Partition Co., Inc.)*

Figure 39.37 Operable partitions are used to divide large rooms into several smaller rooms.

EXTERIOR PROTECTION DEVICES FOR OPENINGS

An extensive range of protection devices are manufactured, including products such as sun control screens, shutters, louvers, screens to provide security, and panels to provide insulation and protection against storms. These may be fixed or may be manually or electrically movable (Fig. 39.38).

TELEPHONE SPECIALTIES

Telephone specialties include manufactured telephone enclosures that may be completely enclosed, supported on the floor (Fig. 39.39) or wall hung (Fig. 39.40). Telephone directory units and shelving are usually wall hung and are available in many sizes (Fig. 39.41). Telephone enclosures for outdoor use are weathertight and have lighting for use at night (Fig. 39.42). Some companies will manufacture the enclosures to the architect's design.

TOILET AND BATH ACCESSORIES

Toilet and bath accessories include all items manufactured for use in connection with toilets and baths. These include products used in most commercial restrooms (Fig. 39.43) plus special units designed for use in hospitals and areas where accessibility and security are important (Figs. 39.44 and 39.45). Items in this specialty area include mirrors, air fresheners, electric dryers (Fig. 39.46), grab bars (Fig. 39.47), waste receptacles, and dispensers for soap, toilet tissue, towels, razor blades, lotion, and paper cups (Fig. 39.48). Hospital products include special cabinets and bedpan storage units. Security items include mirrors that resist theft and breakage and specially designed soap and tissue dispensers.

Figure 39.39 A round floor-mounted telephone kiosk. *(Courtesy Sherron Division, RedyRef Pressed and Welded, Inc.)*

Figure 39.40 A bank of wall-hung telephones with privacy dividers. *(Courtesy American Specialties, Inc.)*

Figure 39.38 These windows have steel detention protection grids, providing maximum protection yet permitting light to enter the building. *(Courtesy The William Bayley Company)*

Figure 39.41 Telephone directory storage units with shelving provide security for the directories yet permit easy use. *(Courtesy American Specialties, Inc.)*

Figure 39.42 This stainless steel telephone booth provides the patron privacy and quiet. *(Courtesy Sherron Division, RedyRef Pressed and Welded, Inc.)*

Figure 39.45 The patient care module is an in-room combination lavatory and concealed water closet. *(Courtesy Bradley Corporation)*

Figure 39.43 This three-bowl lavatory provides multiple stations yet requires only one set of plumbing connections. *(Courtesy Bradley Corporation)*

Figure 39.46 This wall-mounted electric hand dryer is set into the wall to reduce the distance that it projects into the room. *(Courtesy World Dryer Corporation)*

Figure 39.44 This wash center provides barrier-free access. *(Courtesy Bradley Corporation)*

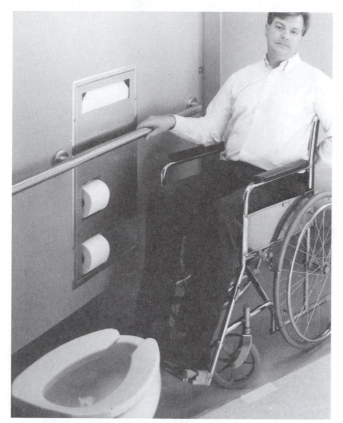

Figure 39.47 This wall unit provides the required grab bar for the handicapped plus the toilet paper and seat cover supply. *(Courtesy Bobrick Washroom Equipment, Inc.)*

SCALES

Any type of weighing device is considered a specialty item.

WARDROBE AND CLOSET SPECIALTIES

These specialties are units used to store clothing, such as hat and coat racks. Among these are wardrobes that provide hanging space as well as several drawers (Fig. 39.49). Other units typically used in hospitals and dormitories combine a wardrobe with a vanity area and lavatory (Fig. 39.50). Wardrobes often are constructed with a plastic laminate over a substrate such as particleboard. Racks are usually steel.

A wide variety of racks, wall mounted and floor supported, are used to store coats, hats and packages. Figure 39.51 shows an automatic coat and hat checking system used in restaurants, hotels, theaters, and other places where large numbers of people gather. It is activated by hand and foot remote controls or by an automatic selection device using the customer's check number to activate it and bring the customer's garments to the attendant.

Figure 39.48 This paper towel dispenser has a waste disposal container. *(Courtesy Bobrick Washroom Equipment, Inc.)*

Figure 39.49 This wardrobe, often used in hospitals and nursing homes, provides clothes hanging space plus shelf storage. *(Courtesy Hausmann Industries, Inc.)*

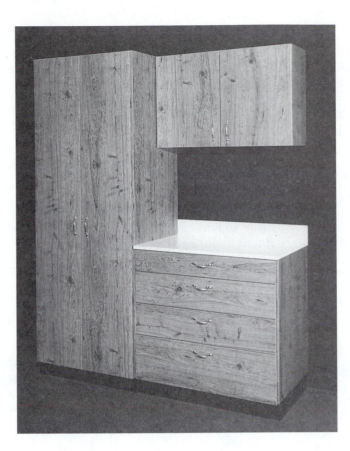

Figure 39.50 This wardrobe cabinet and drawer unit provides hanging and shelf space. A lavatory could be installed in the countertop. *(Courtesy Hausmann Industries, Inc.)*

Figure 39.51 This electric coat and hat check system speeds the storage and retrieval of the patrons' garments. *(Courtesy Railex Corporation)*

REVIEW QUESTIONS

1. What finished surfaces are available when selecting chalkboards?
2. What is a marker board?
3. What are the typical ways toilet enclosures are hung?
4. What materials are used for the panels on toilet enclosures?
5. What purposes do grilles serve?
6. What services are available with various service wall systems?
7. Why are access floors used?
8. How are gas-fired fireplaces vented?
9. What is the difference between a steeple, a spire, and a cupola?
10. What types of identifying devices are commonly used?
11. How do building codes reflect on identifying devices?
12. What are some frequently used pedestrian control devices?
13. Where are protective covers used?
14. What are the ways various partition systems may be installed?
15. List some exterior protection devices.

KEY TERMS

access floor A freestanding floor raised above the basic floor.

chalkboard A surface that can be written on with chalk and from which the chalk is easily removed.

compartment A small area within a larger area enclosed by partitions.

cubicle A very small enclosed space often large enough for just one person.

cupola A small roofed structure built on a roof, usually to vent the area below the roof.

grille An open grate used to cover, conceal, protect, or decorate an opening.

marker board A surface that can be written on with a water-based or semipermanent ink that can be removed.

spire A tall pyramidal roof built on a tower or steeple.

steeple A tall ornamental construction (like a tower), usually square or hexagonal, placed on the roof of a building and topped with a spire.

tackboard A surface used to display announcements and other material usually attached with tacks.

SUGGESTED ACTIVITY

Take a walk around your school or a major building and record the location and use of any specialty items you observe. Decide if the proper choice was made and suggest replacements for those you consider improperly used.

ADDITIONAL INFORMATION

Sweet's General Building and Renovation, Catalog File, Section 10, Specialties, McGraw-Hill, New York.

D I V I S I O N 11

EQUIPMENT
CSI MASTERFORMAT™

Residential and Commercial Equipment

This chapter will help you to:

Be aware of the extensive array of residential and commercial equipment the designer and contractor must be able to specify and install.

D ivision 11 of the CSI MasterFormat groups equipment related to residential and commercial construction. The sections are shown on the introductory page to this chapter. A few of these are discussed in the following pages.

MAINTENANCE EQUIPMENT

Maintenance equipment includes freestanding and built-in equipment used to maintain the building. This includes products such as window washing systems, vacuum systems, floor and wall cleaning equipment, and various types of housekeeping carts. Central vacuum systems have a centrally powered vacuum unit (Fig. 40.1) connected to pipes run through the wall to sealed outlets in the various rooms. The flexible plastic hoses with various end attachments are connected to the outlet when desired to vacuum an area (Fig. 40.2).

Window washing systems for high-rise buildings have hoists mounted on the roof, and the equipment that washes the windows on the exterior is controlled by equipment that usually rides on tracks on the roof. Some systems lower scaffolding from the roof from which workers wash the windows.

SECURITY AND ECCLESIASTICAL EQUIPMENT

Security and vault equipment is designed to store money or other valuables and includes vault doors (Fig. 40.3), day gates, security and emergency systems, safes, and safe deposit boxes.

Teller and service equipment is designed for use where money transactions occur. It includes service and teller windows, package transfer units, automatic banking systems (Fig. 40.4), and teller equipment systems (Fig. 40.5).

Ecclesiastical equipment includes items related to churches and includes baptisteries and various chancel items (Fig. 40.6). This varies according to the denomination and desires of the congregation. Some items are stock, and others are custom built. All types of materials are used, including stone, wood, and metal (Fig. 40.7).

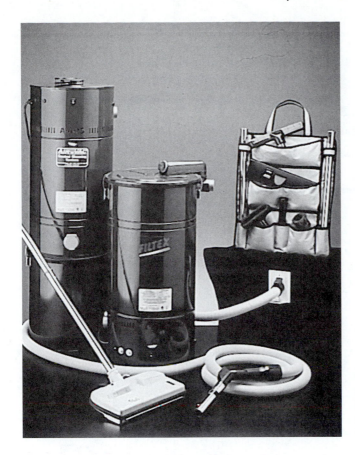

Figure 40.1 A central vacuum power unit that operates the system throughout the entire house. *(Courtesy M & S Systems, Inc.)*

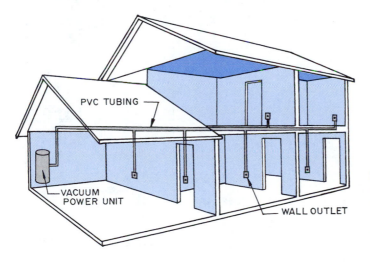

PVC TUBING

VACUUM
POWER UNIT

WALL OUTLET

Figure 40.2 A central power vacuum system operates off a power unit that is usually located in a garage, basement, or utility room. The tubing runs to each room and has an inlet connection for the vacuum hose.

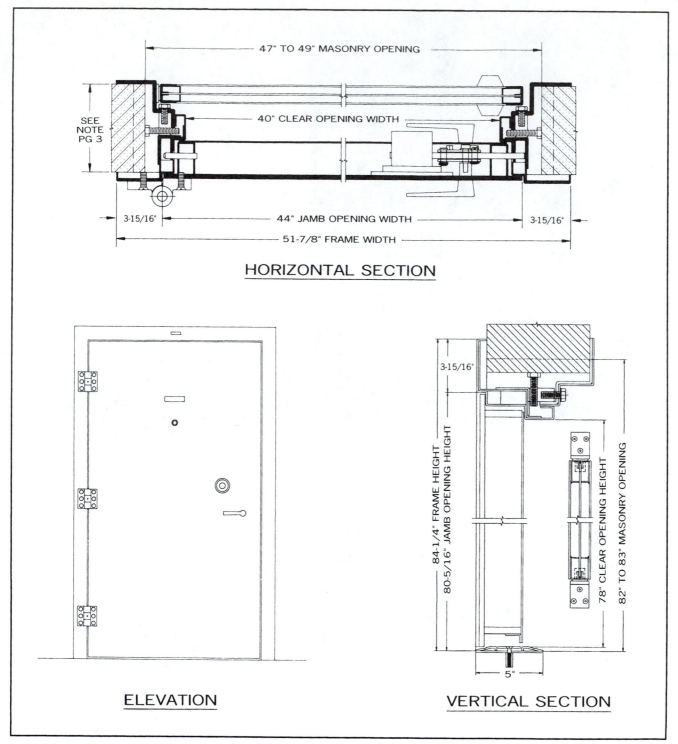

Figure 40.3 Security vault doors provide maximum security for valuables. Various types are made for different levels of security. The one in this illustration rates GSA Class 5. It is rated to protect against surreptitious entry (20 hours), against covert entry (30 minutes), against forced entry (10 minutes), against manipulation of the lock (20 minutes), and against radiological attack. *(Courtesy Overly Manufacturing Company)*

ITEM OF INTEREST

SECURITY OF VALUABLES

Security and vault equipment is a major item for banks, wholesale and retail stores, hotels, and many other business establishments. Available are both built-in and free-standing equipment especially designed for the storage of money and other valuables, such as paintings, jewelry, silver, and stock certificates. Large vaults are available for storing sizable items. Vaults require ventilation systems and specialized security systems.

Vault doors have fire ratings up to at least six hours (Fig. A). The controls include the timelock, dayguard lock,

alarm activator, lighting, emergency ventilator, and a connection for a telephone handset. Manufacturers offer centralized twenty-four-hour alarm systems for installations within the United States and maintain fleets of trucks manned by trained service technicians.

Safe deposit boxes come in stackable, interlocking sections to fit into various sizes of vaults (Fig. B). Various types of key systems are available, which greatly reduce the ability to improperly duplicate a key.

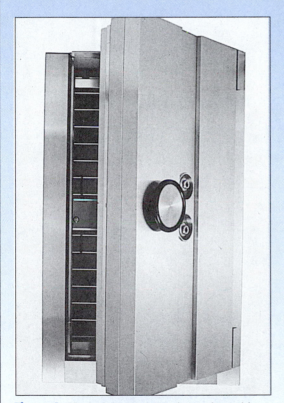

Figure A Vault doors are massive and provide maximum security. *(Courtesy Mosler, Inc.)*

Figure B Safe deposit boxes are installed in vaults and have a two-key security system. *(Courtesy Mosler, Inc.)*

Figure 40.4 Automatic teller machines are widely used to facilitate the banking needs of clients. *(Courtesy Branch Banking and Trust Company)*

Figure 40.5 Teller counters are used to provide individual service to customers in banks and related businesses.

Figure 40.6 The minister delivers the sermon from this custom-built pulpit.

Figure 40.7 The offering plates are kept on this stand beside the altar.

MERCANTILE EQUIPMENT

This includes equipment used in retail and service stores, such as display cases, cash registers and checking equipment, and food processing equipment. Mercantile equipment also includes items in specialty shops such as barber and beauty shops (Fig. 40.8).

SOLID WASTE HANDLING EQUIPMENT

Waste handling equipment includes units to collect, shred, compact, and incinerate solid waste. The systems manufactured include package incinerators, compactors, storage bins, pulping machines, pneumatic waste transfer systems, and various chutes and storage collectors. One such system is shown in Chapter 43.

OFFICE EQUIPMENT

Office equipment includes products such as copiers (Fig. 40.9), computers (Fig. 40.10), word processors, printers, fax machines, and modems. It does not include any office furniture, such as desks, files, and chairs. These are detailed in Division 12.

Office equipment is manufactured by a number of different companies. Since some of these products must communicate with each other, care should be exercised when different brands are mixed.

RESIDENTIAL EQUIPMENT AND UNIT KITCHENS

Residential equipment includes all appliances that may be built-in or freestanding, such as stoves, washers, dryers, freezers, microwaves, dishwashers, compactors, refrigerators, surface cooking units, ovens, range hoods, and indoor barbecue units. These are manufactured by a number of different companies, and if brands are mixed it is important to consider possible differences in color (Fig. 40.11).

Unit kitchens are manufactured units that combine a refrigerator, cooking unit, and sink plus cabinets into one unit. They are delivered totally assembled for rapid installation (Fig. 40.12).

This section does not include kitchen cabinets. They are detailed in Division 12.

ITEM OF INTEREST

ENTERTAINMENT CENTERS

Custom-built entertainment centers are finding increasing use in residential and commercial buildings. The system to be described is the RCA Custom Home Theatre, which is installed in a wall designed especially to house the components (Fig. A). A typical system centers around a large screen television. A high-quality sound system is used to provide the ultimate in clarity and performance. The side cabinets contain the speakers, VCR, CD player, and high-powered receivers. The system delivers brilliant pictures and CD-quality sound from a compact 18 in. satellite dish. The system is designed to be compatible with high-definition television.

The manufacturer provides detailed instructions to the carpenter on how to frame the interior wall. A partial example of the framing plan is shown in Fig. B.

Figure A The RCA Custom Home Theatre provides a complete array of high-quality visual and audio equipment. *(Courtesy RCA Custom Home Theatre)*

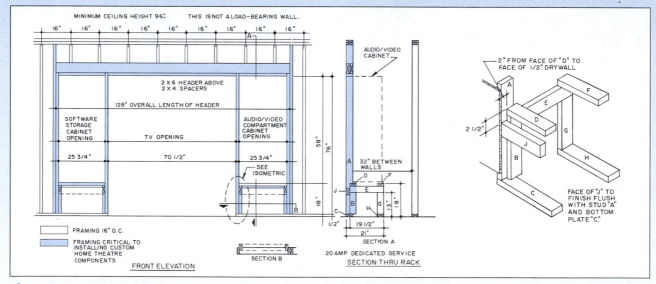

Figure B Part of the framing instructions provided by the manufacturers of the RCA Custom Home Theatre. *(Courtesy RCA Custom Home Theatre)*

Figure 40.8 Retail display fixtures are designed to exhibit a wide variety of merchandise of different sizes, weights, and values.

A small hand-feed office copier.

A large automatic-feed production copier that will collate copy.

Figure 40.9 Copiers are available from small hand-feed units to large automatic-feed, multiple-page units that collate pages and print in color.

A laser computer printer.

Figure 40.10 The personal computer has a large information storage and can operate an extensive range of programs, produce graphics, display information such as encyclopedia pages from a CD ROM, and play music using compact discs.

A personal computer.

Figure 40.11 Kitchen appliances include a wide range of electric and gas units in a range of sizes and colors. *(Courtesy General Electric Appliances)*

Figure 40.12 Unit kitchens are compact single units that fit along one wall. *(Courtesy Dwyer Products Corporation)*

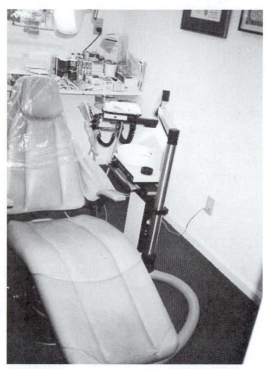

Figure 40.13 Typical equipment in a dental operatory.

FLUID WASTE TREATMENT AND DISPOSAL EQUIPMENT

This section includes a wide range of equipment used to treat and dispose of fluid waste. A few of the products include an oil/water separator, fluid pumping stations, sewage and sludge pumps, scum removal equipment, aeration equipment, and package sewer treatment plants. The selection of the units is a major part of the engineering design of a waste treatment system. The units must be durable and dependable.

MEDICAL EQUIPMENT

The number of items of *medical equipment* used for health care for humans and animals is extensive. It includes items such as sterilizing equipment, examination and treatment equipment, optical and dental equipment (Fig. 40.13), and items used in operating rooms and radiology facilities (Fig. 40.14). Technical changes and improvements occur rapidly in this area, and those planning these facilities must be constantly alert for new equipment. The items must be of the highest quality.

Laboratory equipment may be preassembled by the manufacturer or designed to be assembled on site. It includes the many items used in laboratories for testing, research, and developing new products. Typical items include fume hoods, incubators, sterilizers, refrigerators, and emergency safety appliances designed for use in laboratories. This includes a range of special service fittings and the accessories to use them.

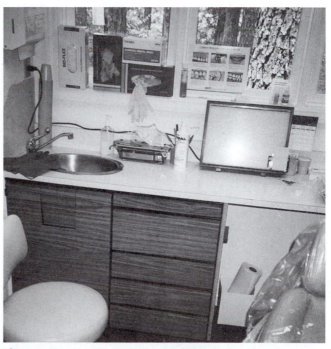

Figure 40.14 Medical facilities utilize a large number of small testing, treatment, and examination equipment.

REVIEW QUESTIONS

1. What types of equipment are used for building maintenance?
2. What types of ecclesiastical equipment can you name?
3. What operations are performed by solid waste handling equipment?
4. How does a unit kitchen differ from a typical residential kitchen?
5. What types of units are used for fluid waste treatment and disposal?

SUGGESTED ACTIVITY

Visit a hospital, a large restaurant, and a school and record the types of equipment that have been installed.

Discuss the merits of each with the person responsible for the operation and maintenance and record that person's comments on each item. Does your interviewee feel that any items are not adequate for the assignment?

ADDITIONAL INFORMATION

Sweet's General Building and Renovation Catalog File, Section 11, Equipment, McGraw-Hill, New York.

DIVISION 12 FURNISHINGS
CSI MASTERFORMAT™

Courtesy Stevens Industries, Inc.

CHAPTER 41

Furnishings

This chapter will help you to:

Have an insight into the range of furnishings used in residential and commercial buildings and be aware of the places that various art forms are used.

Furnishings include products that are decorative in nature such as artwork, rugs, and window treatments, and other items of a useful nature, such as casework, furniture for commercial buildings, and multiple seating. The interior decorator plays a major role in the selection of these items and sometimes becomes involved in the design of special products. The architect also is an important part of the overall design process. The CSI MasterFormat listing for furnishings is on the page introducing this section.

ARTWORK

Murals and Wall Hangings

A popular form of interior artwork is *photo murals*. They are available in a wide range of sizes and scenes. Murals painted on the wall are much more expensive but frequently become an important art item.

Mosaic murals are formed by bonding pieces of stone, such as marble, clay, or glass, together on heavy paper to form the desired design. The sheets are set against a mortar setting bed on the wall. When the setting bed has hardened, the paper is soaked off the face (Fig. 41.1).

All types of wall decorations are used, including paintings, prints, fabric wall hangings, and **tapestries** of various types (Fig. 41.2). Some high-quality woven rugs are used as wall hangings (Fig. 41.3).

Sculpture

Sculpture is another form of artwork used on interior and exterior locations. Some types are carved from stone (Fig. 41.4), and others are cast in bronze, lead, or aluminum (Fig. 41.5). Sculpture made by welding various shapes of metal is called reconstructed sculpture. It often has other materials, such as stone, integrated into the piece. *Relief* artwork is usually a panel-type piece with the images raised above the surface, producing a three-dimensional appearance. This gives the piece depth.

Figure 41.1 This mosaic mural was formed using small ceramic tiles and covers the entire wall of this restaurant. *(Courtesy American Olean Tile Company)*

Figure 41.2 Paintings are extensively used for interior decoration.

Figure 41.3 Woven rugs and tapestries are popular wall decorations.

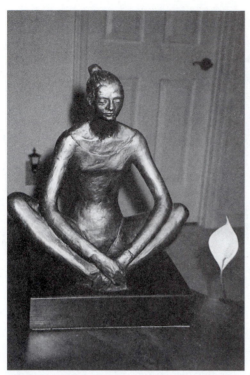

Figure 41.4 This ballerina was carved from stone.

Figure 41.5 Cast metal sculpture has been used for many years to produce striking art objects. This is Heloise, wed to the priest Pierre Abelard, 1079–1142.

Stained Glass Work

Stained-glass windows are mosaics of pieces of translucent glass cut to the desired shape. Sometimes the glass requires additional detailing, shading, or texturing. This is accomplished by painting it with special mineral pigments and firing it in a kiln. The two types of stained glass are leaded and faceted.

Leaded glass windows have the pieces of glass joined together with H-shaped lead strips called cames (Fig. 41.6). Large windows require round bracing bars be wired to the leaded glass. This provides bracing yet allows for thermal movement. Exterior leaded glass is pressed into a bed of glazing sealant or tape (Fig. 41.7).

Faceted glass uses colored glass slabs 1 in. (25 mm) thick. The glass is cut or faceted to reflect the rays of light in many directions. The glass is made in a wide range of translucency, opacity, and transparency. This influences the color and degree of reflected light (Fig. 41.8).

The faceted glass is set into a reinforced concrete or epoxy resin matrix much as bricks are set (Fig. 41.9). This forms a panel 1 in. (25 mm) thick. Light can pass through the faceted glass pieces bound together by the matrix. The panels are mounted in wood or metal frames or in a masonry wall.

Ecclesiastical Artwork

Ecclesiastical artwork includes pieces of religious significance to specific denominations. It can take the form of the types discussed in this chapter. Specific pieces could include altar pieces and religious symbols (Fig. 41.10).

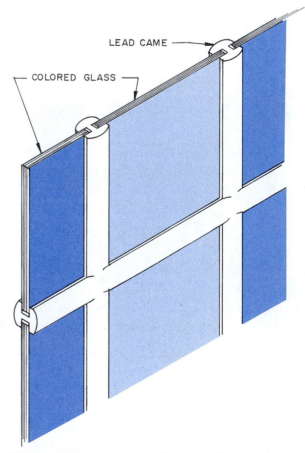

LEAD CAME

COLORED GLASS

Figure 41.6 Leaded glass is formed by joining glass pieces cut in various sizes and shapes with lead cames.

Gypsum wallboard can be formed to cover arches and curved walls, providing a strong, fire-resistant interior finish.

Plaster walls and ornamental plaster details impart a high quality to interior wall finish and decoration.

Poured polyurethane athletic floors provide a joint-free, durable, and resilient finished floor. Bands of color are used to mark off the various sections of the floor.

Clay brick pavers can be laid in a wide range of interesting patterns and colors.

Courtesy Chicago Metallic Corporation

The design installation and color of interior finish products are critical in setting the atmosphere of the interior spaces.

Courtesy Majestic

Manufactured fireplaces are available as wood-burning or gas-fired units. Gas-fired units may be ventless or may require venting.

Courtesy Phifer Wire Products Inc.

The windows on this building are protected by solar screens that reduce the heat load on the air-conditioning system. Notice how their color blends with that of the wall system.

Courtesy Alvarado Manufacturing Co., Inc.

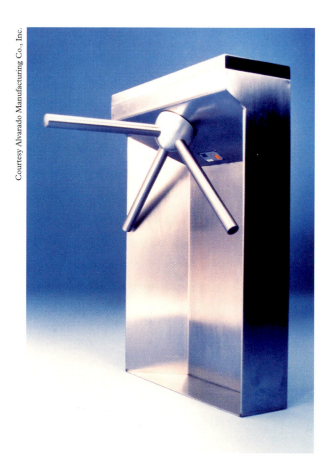

This pedestrian security turnstile allows traffic flow in only one direction. Notice the attractive finish of the stainless steel base.

A flagpole is one of many specialty items.

Courtesy Branch Banking and Trust Customers

Automatic teller machines have become
widely used for personal banking needs.
Brilliant colors are used to get the attention
of customers.

Courtesy Whirlpool Corporation

The brilliant white kitchen appliances blend in harmoniously
with the white cabinets, producing a unified whole.

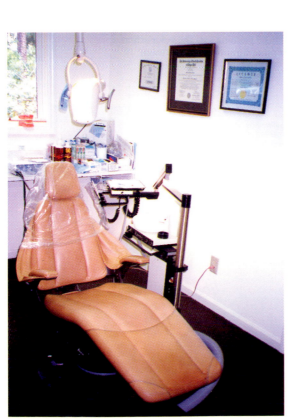

This dentist chair is one of an extensive array of medical
equipment manufactured.

Channel fittings are one type of built-in ecclesiastical
equipment. The colors are chosen to provide a soft,
restful atmosphere.

Courtesy Otis Elevator Company

Escalators are widely used to move people between floors rapidly and safely.

Courtesy Inclinator Company of America

Residential elevators provide easy access to two or more floors. The dark finish on the metal parts helps it blend into the room.

Courtesy Shuttlecraft, Inc.

Mobile hoists can carry large, heavy loads, as shown by this straddle crane. The bright color makes the unit highly visible and promotes safety in the work area.

Courtesy Universal-Rundle Corporation

An attractive, well-designed bath. Fixtures are available in a wide range of colors, and these colors are picked up in the tiles and other interior finishes.

Courtesy Durham-Bush, Inc.

An ACDX air-cooled package chiller with rotary screw compressors.

Courtesy Carrier Corporation

A Carrier 16DF direct-fired, double-effect absorption chiller.

Courtesy H. B. Smith Cast Iron Boilers

A large in-series gas-fired boiler heating system.

Mechanical

Electrical

A commercial multi-metering switchboard with plug-in meter sockets and fusible and circuit breaker disconnects.

Courtesy Square D Company

This electrical generation installation includes the Robinson Fossil Fuel Plant (on the right) and the Robinson Nuclear Plant (on the left) in Hartsville, S.C.

Courtesy Carolina Power and Light Company

This electrical substation is part of the electric power distribution system.

Figure 41.7 These dramatic leaded glass windows contribute to the atmosphere of the church interior. *(Courtesy Rambusch, Cunningham Werdnigg photographer)*

Figure 41.8 This faceted glass wall provides privacy, admits light, and is a major decorative feature.

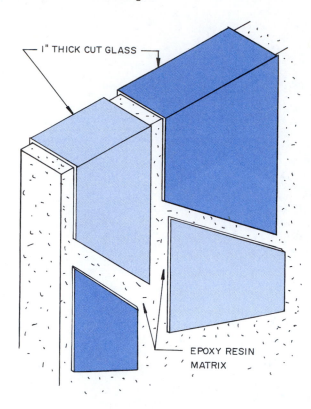

← 1" THICK CUT GLASS →

EPOXY RESIN
MATRIX

Figure 41.9 Faceted glass walls are formed by bonding 1 in. thick glass cut in various shapes and bonded with an opaque epoxy resin or reinforced concrete.

Figure 41.10 Ecclesiastical artwork reflects the beliefs of the denomination.

FABRICS

This section relates to fabrics, leathers, and furs for upholstery, curtains, draperies, and wall hangings and the various fabric treatments and fillers. Fire codes may require these products be fireproofed or their use will not be permitted (Fig. 41.11).

MANUFACTURED CASEWORK

Manufactured casework includes stock cabinets manufactured from wood, steel, and plastic laminates. It includes the countertops, sinks, and any accessories or fixtures mounted on the countertops. For residential applications it includes both kitchen and bath casework (Fig. 41.12).

Medical applications include casework designed for dental, hospital, and optical uses, for nurses stations, and for veterinary uses. All types of laboratory casework used in medical, research, and other laboratories are part of this section (Fig. 41.13).

Various manufacturers produce specialized casework that has a limited market. Included are casework for banks, dormitories, ecclesiastical uses, hotels, motels, schools, restaurants, and various display units (Fig. 41.14).

Figure 41.11 Draperies, curtains, and other wall hangings in commercial buildings must be noncombustible or be maintained flame resistant.

Figure 41.12 Residential casework includes kitchen and bath cabinets. *(Courtesy General Electric Appliances)*

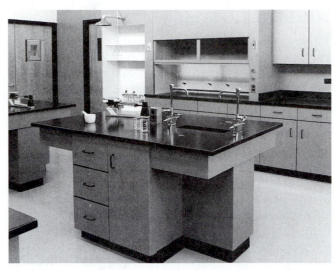

Figure 41.13 Laboratory casework frequently contains utility connections such as water, waste removal, electricity, gas, and computer connections. *(Courtesy Stevens Industries, Inc.)*

Figure 41.14 Commercial casework includes stock and custom-built units for commercial applications. This installation has storage, chalkboards, and bulletin boards. *(Courtesy Stevens Industries Inc.)*

WINDOW TREATMENT

Interior window treatment can be decorative, and it can be practical, blocking the intrusion of sunlight and providing privacy. *Metal blinds* with narrow horizontal slats (often called venetian blinds) can be adjusted to block the sun or to permit a view to the outside. They can be raised to the top of the window so they are completely out of view (Fig. 41.15).

Vertical blinds are made with fabric slats hanging from the top of the window. They can be adjusted to be open or closed (Fig. 41.16). *Shutters* made of wood or plastic are also used on the interior of the window. Some have adjustable louvers that will let in a little light. They permit very little view to the exterior.

A number of *shades* are manufactured, including insulating, lightproof, translucent, and woven wood or plastic types. They roll up to provide a view and are pulled down to block the view. An accordian type with a fiberglass fabric is also available (Fig. 41.17).

Curtains of various types can serve the same purpose as shades. Some have metal tracks across the top that permit the drapes to be open or closed over the windows. Other units remain on each side or over the top of the window and serve a decorative purpose. Again, fire codes must be observed when selecting shades and curtains.

FURNITURE AND ACCESSORIES

Furniture under CSI MasterFormat Division 12 includes all types of freestanding units for commercial and residential use. The listing of open office furniture includes office partitions, storage, work surfaces, shelving (Fig. 41.18), and light fixtures. General furniture includes all types from residential through commercial areas such as classroom, hotel, library (Fig. 41.19), medical, office (Fig. 41.20), and restaurant furniture.

Figure 41.15 Blinds with horizontal metal slats can be adjusted to block the sun or opened to permit the entrance of light.

Figure 41.16 Vertical blinds are widely used in residential and commercial construction.

Figure 41.17 Various types of shades are available for blocking the sun and providing privacy.

Figure 41.18 Libraries use large shelving units to store their inventory of publications.

Figure 41.20 Various types of upholstered furniture are specified under CSI Division 12 (Furnishings).

Figure 41.19 Library carrels provide quiet study areas.

Furniture accessories such as clocks, ashtrays, lamps, waste receptacles, and desk accessories are included (Fig. 41.21).

RUGS AND MATS

This section includes loose rugs and mats, gratings, and foot grilles. Carpet covering the floor area is included in Division 9. Typical products include chair pads, floor mats, entrance tiles, and floor runners. They are made from a variety of materials, including rubber, carpet, vinyl, aluminum, and stainless steel.

MULTIPLE SEATING

Multiple seating includes seating for areas in which audiences gather, such as schools, restaurants, theatres, churches, stadiums, and auditoriums (Fig. 41.22). It includes fixed, portable, and telescoping seating. Booths and various table and seat modules as found in restaurants also are included.

INTERIOR PLANTS AND PLANTERS

Extensive use of live and artificial plants is common in all types of commercial buildings. This section includes the plants, plant holders, and any landscaping accessories and maintenance materials required (Fig. 41.23).

Figure 41.21 Clocks are one type of furniture accessory.

Figure 41.22 Multiple seating includes products used in any area where large numbers of people gather. *(Courtesy Stevens Industries, Inc.)*

Figure 41.23 An extensive variety of live and artificial plants are used for interior decoration.

REVIEW QUESTIONS

1. What types of window blinds are commonly available?
2. How are mosaic murals formed?
3. What types of sculpture are commonly available?
4. How are leaded stained-glass pieces joined?
5. What are the differences between leaded and faceted glass?
6. What types of units are used for window treatment?
7. Where is multiple seating commonly found?

KEY TERMS

faceted glass window A window made by bonding 1 in. thick glass pieces with an epoxy resin matrix or reinforced concrete.

leaded glass window A window made of small glass pieces joined with lead cames.

mosaic A decoration made up of small pieces of inlaid stone, glass, or tile.

sculpture A three-dimensional work of art.

tapestry A fabric on which colored threads are woven by hand to produce a design.

SUGGESTED ACTIVITY

Visit commercial buildings of various types and shopping malls and record the types of furnishings used inside and outside.

ADDITIONAL INFORMATION

DeChiara, J., Panero, J., and Zelnik, M., *Time-Saver Standards for Interior Design and Space Planning*, McGraw-Hill, New York, 1991.

Sweet's General Building and Renovation Catalog File, Section 12, Furnishings, McGraw-Hill, New York.

P A R T VII

SPECIAL CONSTRUCTION AND CONVEYING SYSTEMS

The architect and the engineer frequently have to select, design, or specify products that are not typical or part of a conventional project. These are classified as *special construction*. Special construction can include a complete structure, such as an air-supported enclosure, or a building module, such as a prison cell. In Chapter 42 these and other projects are illustrated. The chapter runs from pre-engineered structures, such as a grandstand, to incinerators and dog shelters. This is a vast area of specification.

Part VII also includes considerable information on conveying systems. These are covered in Chapter 43 and include residential, freight, and passenger elevators, escalators, cart and tote box systems, lifts, moving walks and ramps, shuttle systems, conveyors, pneumatic systems, and various types of cranes and hoists.

DIVISION 13

SPECIAL CONSTRUCTION

CSI MASTERFORMAT™

CHAPTER 42

Special Construction

This chapter will help you to:

Be aware of the wide range of special construction facilities available and get a basic understanding of how they function.

Special construction encompasses an extensive array of structures, systems, and assemblies that serve a limited, specific purpose. Construction ranges from complex facilities, such as nuclear reactors, to simpler projects, such as site-constructed incinerators and specialized digestion tank covers and assemblies. Following are discussions of a few of these sections. A detailed listing is on the introductory page to Division 13.

AIR-SUPPORTED STRUCTURES

Air-supported structures have single- and multiple-wall enclosures made of a flexible material that is pneumatically supported. The structures are pre-engineered and prefabricated for delivery to the site. Typical materials include fiberglass fabric that is Teflon coated and polyester that is vinyl or neoprene coated.

Air-supported structures form an envelope that encloses a pressurized space. Some are unreinforced membranes in which the membrane is the primary structural element (Fig. 42.1). Another type uses a membrane that is reinforced by a network of cables or webbing. The webbing forms the primary structural system, and the membrane spans the area between the cables or webbing. Air locks are used as entrances, and emergency exits are counter-pressure balanced, are self-closing, and have panic hardware. The membrane is light admitting, and it must meet local fire codes.

INTEGRATED ASSEMBLIES

Integrated ceilings are a type of integrated assembly. They are pre-engineered and prefabricated for assembly on the site. A wide range of products are available, as discussed in Chapter 36. Some integrated systems include lighting, air supply registers, ducts, air return grilles, fire sprinkler systems, and communication linkage. These systems are coordinated with CSI MasterFormat Divisions 15 and 16 (Fig. 42.2).

SPECIAL-PURPOSE ROOMS

Many rooms are designed and constructed to meet specific performance requirements, such as various activities, fire protection, thermal control, or sound control. The specific

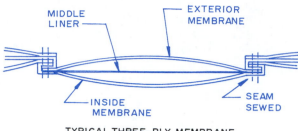

TYPICAL THREE-PLY MEMBRANE

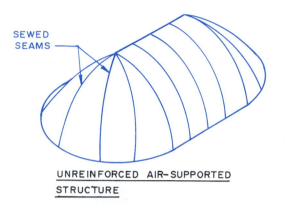

UNREINFORCED AIR-SUPPORTED STRUCTURE

Figure 42.1 This is an unreinforced air-supported structure constructed with a three-ply membrane.

Figure 42.2 This shows the installation of an integrated ceiling that, when completed, will cover the vaulted ceiling and include lighting, a sound system, and a sprinkler system. *(Courtesy Chicago Metallic Corporation)*

rooms in the CSI MasterFormat include athletic rooms, sound conditioned rooms, clean rooms, cold storage rooms, insulated rooms, shelters and booths, planetariums, prefabricated rooms, saunas, steam rooms, and vaults.

NUCLEAR REACTORS

The design and construction of nuclear reactors is a very demanding project and is regulated by a number of government agencies. In many areas nuclear fuels are used in a high-temperature high-concentration process to generate electric power. Figure 42.3 shows a fuel bundle being inspected and readied for placement in a storage pool. Figure 42.4 shows the massive construction required when building a nuclear reactor. See Chapter 47 for additional information.

PRE-ENGINEERED STRUCTURES

All types of pre-engineered prefabricated buildings and structures are classified as special construction, even if they are erected on temporary foundations. Some of the structures included are metal building systems, glazed structures (such as a greenhouse), portable buildings, grandstands and bleachers, cable supported, fabric, and log structures. Information on some of these is given in Chapters 1 and 17.

WASTE AND UTILITIES SECTIONS

A number of special constructions occur in the area of waste treatment and systems involved with utilities. *Filter underdrains and media* include the piping and filters used in water and fluid waste treatment. Filter media include anthracite, charcoal, diatomaceous earth, mixed, and sand media. *Digester covers and appurtenances* include special tank covers and assemblies used on digestion tanks.

Figure 42.3 Fuel bundles are being inspected and prepared for placement in the storage pool. *(Courtesy Tennessee Valley Authority)*

teorological information, such as solar and wind energy. *Transportation control instrumentation* describes systems used to monitor and control the various aspects in transportation systems, such as airport control, railway control, subway control, and transit vehicle control.

SOLAR ENERGY SYSTEMS

Solar energy systems may be active or passive. **Active solar systems** are assembled from manufactured components that convert solar energy to thermal energy and electrical power. The components include air and liquid flat plate collectors, concentrating collectors, and vacuum tube collectors (Fig. 42.5). Manufacturers have complete solar systems available. One example is a solar water heater (Fig. 42.6). A schematic of a warm water

Figure 42.5 Solar panels in a field arrangement provide services to a large industrial building. (*Courtesy Solar Development, Inc.*)

Figure 42.4 Massive concrete construction is used in construction of nuclear reactors, with some walls many feet thick. (*Courtesy Bureau of Reclamation, U.S. Department of the Interior*)

Oxygenation systems include site-assembled piping systems and related equipment for the dissolution and mixing of gaseous oxygen in liquid waste. This includes the oxygen generators, oxygen storage, and the dissolution system.

Utility control systems include the operating and monitoring systems for water supply, wastewater, and electrical power generation plants. They include metering devices, display panels, control panels, and sensing and communicating equipment.

MEASUREMENT AND CONTROL INSTRUMENTATION

Measurement and control instrumentation are covered in a section of Division 13. The types of controls vary with the industry and can include mechanical, electrical, fluid, pneumatic, and computer controlling devices. These systems may be totally automated or semi-automated. They can monitor situations such as temperatures or pressures, regulate the system so it operates at the required speed and quality, and control safety within a manufacturing, processing, or assembly plant.

Instruments listed as *recording instruments* are installed to measure and record such various occurrences as seismic information, stresses in structures, and me-

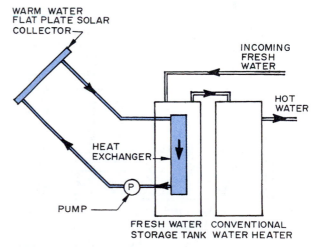

Figure 42.6 A simplified schematic of a solar hot water heater. The controls have been omitted. The solar collector heats the fluid that runs through the heat exchanger. It heats the water in the storage tank before it goes to the conventional water heater.

flat plate solar heating system is shown in Fig. 42.7. The flow is controlled by valves that are activated by thermostats. The liquid is moved by electrically operated pumps.

Passive solar systems are not a part of this section. They utilize natural means to collect, store, and distribute the heat through the building. Typical systems are shown in Figs. 42.8 and 42.9.

Solar photovoltaic collectors provide electricity when under the influence of light. Solar **photovoltaic cells** are thin flat semiconductors that convert light energy into direct-current electricity. They find many uses, especially in remote areas. An array of cells can be used to operate electrical equipment, such as a pump or lights, without a battery. They can be grouped in large numbers to add additional electrical power to an electric company grid (Fig. 42.10). Solar cells may be connected to a direct load for immediate use and used to charge a battery. The battery supplies the needed power when the sun is not shining. A typical solar photovoltaic system is shown in Fig. 42.11. The system also can be connected with the local utility as a backup.

WIND ENERGY CONVERSION SYSTEMS

Another source of electrical energy is derived from prefabricated systems using wind to drive turbines. These are tall steel towers with large propellers connected to an electrical generating device mounted on the top. They are placed in groups on sites known to have steady dependable winds. The power is sold to the local electric utility and fed into their power grid for delivery to the consumer. A typical system is shown in Fig. 42.12.

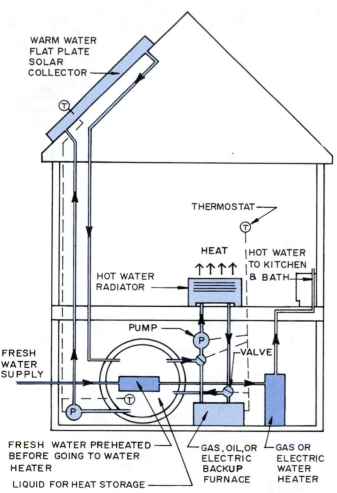

Figure 42.7 This schematic shows an active solar heating system using a warm water flat plate collector. Notice the heated liquid storage tank is also used to preheat the fresh water going to a conventional water heater.

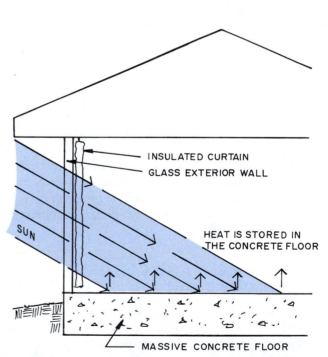

Figure 42.8 This passive solar heating design utilizes a massive concrete floor for heat storage.

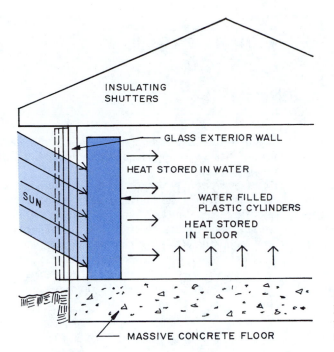

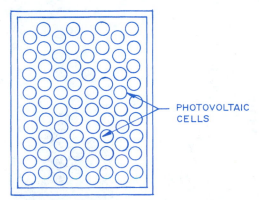

Figure 42.9 This type of passive solar heating uses plastic cylinders filled with water and a massive concrete floor for heat storage.

Figure 42.10 A typical solar photovoltaic panel.

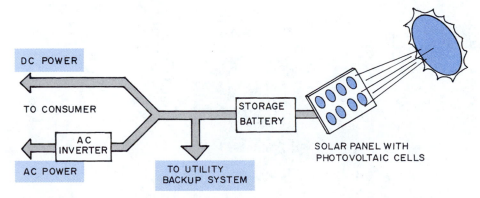

Figure 42.11 A schematic of a typical solar photovoltaic electric power producing system.

ITEM OF INTEREST

HOME SECURITY SYSTEMS

The use of home security systems is increasing rapidly. A basic system includes sensors on doors and windows, an alarm, and controls. The system is armed and disarmed by a keypad located near the door that is usually used for entry by the occupant (Fig. A). When the system is armed and an intruder opens a door or window the alarm is triggered. This may simply be an alarm inside and outside the building (Fig. B), or the system may be connected to a central monitoring station that verifies if it is a true entry and notifies the local police. Another system is a wireless system with a central control unit (Fig. C). Each door or window has a battery powered transmitter that sends a signal to the base station, which sounds the alarm. The system has a keypad that can be carried about the building (Fig. D).

Many types of protection sensors are available. Exterior sensors provide protection around the exterior of the building. These could include pressure detectors that detect a vehicle entering the driveway, and photoelectric beams and motion detectors that can sense a person in the yard.

Within the building a system can have alarms on window screens that sound when the screen is cut, glass breakage detectors, and sensors on doors and windows that activate when they are opened.

Figure A This wired security system is controlled by a keypad located near the entrance that is used most frequently. *(Courtesy Honeywell)*

Figure B This speaker sounds the alarm outside the building. An indoor alarm sounds at the same time.

I T E M   O F   I N T E R E S T

HOME SECURITY SYSTEMS

Continued

Other interior sensors include passive infrared sensors, which detect the intruder by the body heat, and motion sensors, which use sound waves to detect movement. Photoelectric sensors transmit an invisible beam of light from the transmitter to a receiver. When the beam is broken an alarm sounds. Some companies offer stress sensors that are attached to the floor joists in areas where there is a long span. These sense a deflection in the joists when someone walks over the floor.

In addition, security systems include fire and smoke alarms, which may sound an alarm within the building or be relayed to a monitoring service. Some manufacturers offer fixed temperature sensors, which sound when the temperature in an area gets above preset levels. These are often used in basements and attics. A cold temperature sensor is available that sounds an alarm when the temperature drops below a preset level, such as 40°F (4.5°C). This alerts the occupants to possible damage from freezing temperatures.

Carbon monoxide and natural gas sensors alert occupants to an unsafe buildup of these deadly gases.

Figure C The base station for a wireless security system contains the control buttons, receives the signal from the transmitters at each door or window, and sounds the alarm.

Figure D The wireless alarm system can be controlled by this portable keypad, which can be carried about the building.

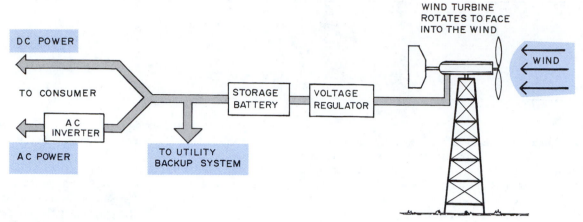

Figure 42.12 A schematic of a typical wind energy conversion system with an AC converter that converts the DC power generated into AC.

The turbine delivers DC electricity, which can be converted to AC by an inverter. Usually part of the power generated is used to charge batteries that are used when the wind is not blowing. This must be done with DC power. The system can also be connected to the local utility as a backup.

COGENERATION SYSTEMS

Cogeneration systems are systems using fossil fuel, geothermal, wind, or solar energy to produce electricity and heat as a system separate from the local utility. They are connected into the local electrical utility network so they can buy extra electricity when it is needed and sell power to the utility when it is developed.

BUILDING AUTOMATION SYSTEMS

Building automation systems monitor and control various types of equipment, special purpose devices, conveying systems, and mechanical and electrical systems. The automation system integrates all the systems within a building. Among those commonly found are systems to monitor and control energy and environmental systems, communications, security, clocks, alarms, detection devices, and door controls. Devices for moving people and materials, such as moving walks, moving ramps, elevators, and escalators, are also automated.

Figure 42.13 illustrates a possible security arrangement for a bank. It uses audio sensors, waterproof outlets, and an alarm control cabinet, which is usually located in the vault. Alarm actuators are at the teller windows, drive-in windows, and at the employees' desks, and surveillance cameras are placed in strategic locations (Fig. 42.14).

AUTOMATIC FIRE SUPPRESSION AND SUPERVISORY SYSTEMS

Fire detection and alarm systems include fire detectors, heat smoke detectors (Fig. 42.15), or a water flow switch, which indicates an automatic sprinkler system has been activated. Fire flame detectors are used where the materials that may burn, such as gasoline, do not generate smoke at first. They detect the presence of either infrared or ultraviolet radiation. Some systems also have manual pull stations that are activated by someone who sees a fire start (Fig. 42.16). Various types of fire alarms are in use. The most common are horns and bells (Fig. 42.17) and flashing lights (Fig. 42.18). These are often combined into one unit, providing sound and light signals.

Fire suppression systems include automatic sprinklers (Fig. 42.19), water fog generators that are used on highly flammable solids or liquids, and a liquid foaming agent introduced into the water in a sprinkler system. Several types of automatic gas suppression systems are available; carbon dioxide is commonly used. A high-expansion foam is sometimes used in small compartmentalized areas. The foam must completely cover the area to put out the fire. See Chapter 44 for additional information on fire suppression systems.

SPECIAL SECURITY CONSTRUCTION

Security construction includes the various systems and equipment to achieve a high level of security. This includes things such as crash barriers, blast-resistant products, gun ports, bullet-resistant security gates and doors, identification systems, and monitoring systems. A full alarm system uses closed-circuit television cameras that may be controlled by guards or automatically pan an area and zoom in when needed.

Specifications:

1. Audio sensors require junction box.
   Box shall be mounted flush with ceiling. Audio sensors shall not be mounted more than 25'-0" from any wall surface or more than 50'-0" apart. Minimum of one audio sensor per each enclosed area.
2. Waterproof floor outlet with bell nozzle.
3. Main junction box (12" x 12" x 4") complete with cover and located in accessible plan in equipment room or work area.
4. Outlet box (2"W x 3"H x 2-1/4"D) for transformer by Mosler. Mount as illustrated in main junction box.
5. Status control.
6. Alarm control cabinet must be in vault whenever possible.

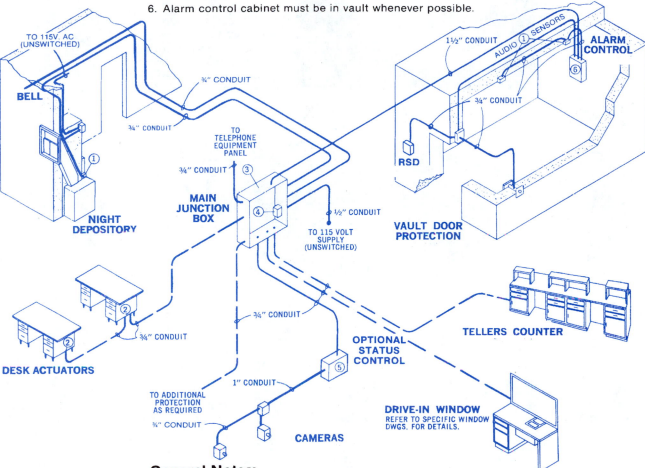

General Notes:

A. All conduit, outlet boxes and covers to be supplied and installed by others in accordance with national electric code and/or local codes without expense to Mosler.

B. No more than (2) 90° bends in a conduit run.

C. Conduit must have pull wires in place at time of installation.

——————— Must be separate conduit run.

——— ——— These conduits may be combined with each other wherever practical to eliminate and/or shorten conduit runs. Increase size as required.

Figure 42.13 A typical security system for a bank. *(Courtesy Mosler Inc.)*

Figure 42.14 Security cameras provide a wide coverage and record the activities within their area of surveillance. *(Courtesy Mosler Inc.)*

Figure 42.15 One type of fire detector that senses smoke and sounds an alarm.

Figure 42.16 Two types of manually activated fire alarm stations.

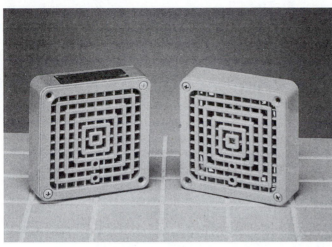

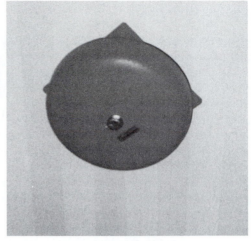

Horns   Bell

Figure 42.17 Horns and bells are used to provide a loud warning when the fire detection system senses a fire. *(Courtesy Federal Signal Corporation)*

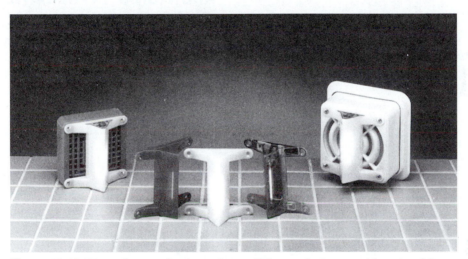

Figure 42.18 These fire warning horns have a light attached to provide a signal for the hearing impaired. *(Courtesy Federal Signal Corporation)*

Figure 42.19 One type of sprinkler head on an automatic sprinkler system.

REVIEW QUESTIONS

1. What types of covering materials are used for air-supported structures?
2. How are entrances to air-supported structures treated so the air pressure within the structure is maintained as people enter and leave?
3. What is meant by special-purpose rooms?
4. What types of structures are available as pre-engineered units?
5. What are some types of functions performed by process control systems?
6. What is the difference between active and passive solar energy systems?
7. What use is made of photovoltaic cells?
8. How is the DC current produced by a wind turbine converted to AC current?
9. What types of systems are controlled and monitored by building automation systems?

KEY TERMS

active solar systems Solar heating systems utilizing mechanical means to move and store solar energy.

air-supported structures Structures with the enclosing envelope made from a flexible material that is pneumatically supported.

cogeneration systems Systems using fossil fuel, geothermal, wind, or solar energy to produce electricity and heat.

passive solar systems Solar systems that utilize natural means to store solar energy.

photovoltaic cells Thin, flat semiconductors that convert light energy into direct current electricity.

SUGGESTED ACTIVITY

Locate any of the special construction areas mentioned and arrange a visit. Ask the person in charge to conduct a tour and explain how the system operates.

ADDITIONAL INFORMATION

Sweet's General Building and Renovation Catalog File, Section 13, Special Construction, McGraw-Hill, New York.

DIVISION 14 CONVEYING SYSTEMS

CSI MASTERFORMAT™

Courtesy Dover Elevator Systems

43

Conveying Systems

This chapter will help you to:

1. Be aware of the codes related to the design and installation of conveying systems.

2. Be able to inspect the installation of conveyor systems and pass judgment on the quality of the work.

3. Understand the various operating mechanisms used on elevators.

Many different types of conveying systems are available, each designed for a special purpose. Some move materials and equipment, and others move people. They range from a simple dumbwaiter used to carry mail between floors in a building to large, complex **elevator** systems carrying people or heavy materials and equipment.

ELEVATOR HOISTWAYS

The **hoistway** is a vertical fire-resistant enclosed shaft in which the elevator moves. It has a **pit** at the bottom and openings at each floor. The openings are protected by doors controlled by the operating system. Some types require a penthouse on the roof above the shaft. Codes typically require that the shaft have a 2 hr. fire rating and that the opening be protected with doors having a $1\frac{1}{2}$ hr. fire rating. A metal or concrete deck is required on the top of the hoistway (Fig. 43.1).

Hoistway Doors

The doors of the hoistway are controlled by the operating system and must have a $1\frac{1}{2}$ hr. fire rating. The codes specify the types of doors that are acceptable and the opening requirements. Passenger elevators generally use horizontal sliding doors, although swinging doors may be used. Vertical sliding doors are used on freight elevators.

The doors are closed automatically when the car leaves the landing zone. The **landing zone** is an area 18 in. (5490 mm) above or below the landing floor. The elevator car will not move unless all the doors on all levels are closed and locked. Doors have locks that prohibit them from being opened by someone on the landing side. Emergency access is available for maintenance and rescue personnel.

Machine Rooms

Machine rooms are part of the hoistway and provide a fire-resistant enclosure for the installation of required hoisting machinery, controls, pumps, and hydraulic oil storage. Since control systems are computerized, the area

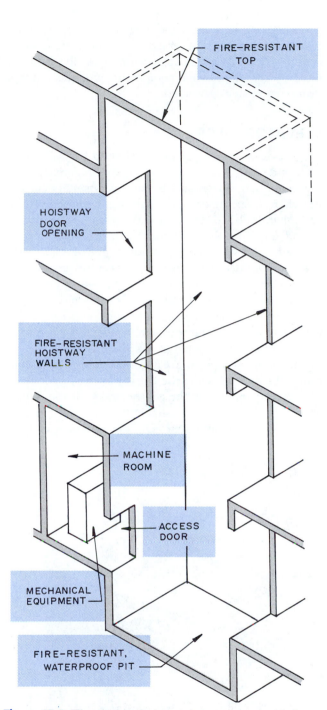

Figure 43.1 The elevator hoistway is constructed with fire-resistant walls, top, and pit.

should be air-conditioned to control the temperature. Nothing not involved with the operation of the elevator is permitted in the machine room. The size of the machine room will vary depending on the type of equipment.

Machine rooms for traction type elevators are generally in a penthouse on the roof over the hoistway (Fig. 43.2). The floor must be designed to carry the weight of the machinery plus loads required for lifting and lowering the car. In some cases traction elevator machine rooms can be located in the pit at the bottom of the shaft. Hydraulic elevator machine rooms are

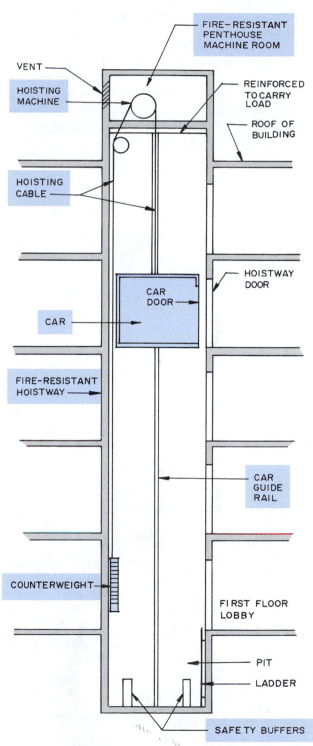

Figure 43.2 A traction electric elevator generally has the hoisting machine at the top of the hoistway.

usually located to the side of the hoistway (Fig. 43.3). They contain the hydraulic equipment and controls.

Venting Hoistways

In a fire, the elevator hoistway may become a vertical flue for gases and smoke and be used to carry away these noxious materials by venting the hoistway at the top. However, some building designs use other means of containing and venting smoke and gases from the site of the fire to prevent them from moving to other unaffected floors. Codes may require the hoistway to be vented, but other design considerations for controlling smoke must be considered as the hoistway is designed.

Hoistway Sizes

The clear inside dimensions for hoistways for standard size elevators are specified in the National Elevator Industry Standard, Elevator Engineering Standard Layouts. It is important that maximum and minimum clearances be-

tween the hoistway and cars, moving weights, and other moving equipment be observed. For specially designed elevators the hoistway must accommodate the larger or smaller car and moving equipment, and the required clearances must be observed.

ELEVATOR CARS

The **elevator car,** a platform enclosed by walls and a roof, is designed to carry passengers, materials, packages, and many special loads, such as hospital beds and stretchers. The car contains lighting, controls, venting units, handrails, telephones, and various types of wall, ceiling, and floor finish materials. *Passenger elevators* are finished to be attractive yet durable (Fig. 43.4). *Freight elevator* cars must have durable, abrasion-resistant interior finishes. Typically, they have steel floors and walls.

The car in a hydraulic elevator is mounted on top of the piston. Cars on electric elevators have a frame to which the wire hoisting ropes are connected. The design of elevator cars is detailed in ANSI/ASME A17.1.

The doors on the car are automatically operated and cannot be opened while the car is moving or is outside the landing zone. The car will not move if the door is open. Elevator cars are available that open to the front and to the front and rear. Double entrance cars are

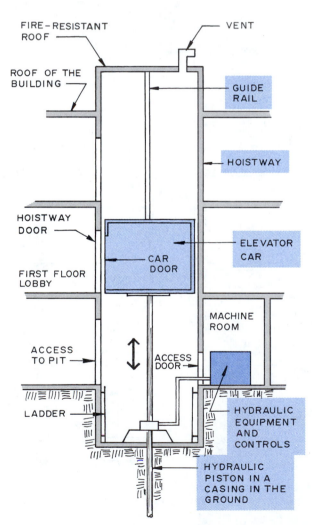

Figure 43.3 A hydraulic elevator typically has the machine room on the side of the hoistway.

Figure 43.4 A typical elevator car. *(Courtesy Dover Elevator Systems, Inc.)*

widely used on hospital and freight elevators (Fig. 43.5). Elevator car doors may be center opening, two-speed sliding doors, and single sliding doors (Fig. 43.6). The center-opening doors provide the quickest entry and exit. They are always used on high-speed systems. Two-speed sliding doors provide the widest possible opening but are slower to open and close than center-opening doors. Single sliding doors are the most economical and the slowest. They move right or left, depending on the car design, and the opening width is limited by the width of the car.

The car contains a panel with several controls. The control panel must be located so it can be reached by a person in a wheelchair. It has an alarm button that sounds an alarm outside the hoistway and an emergency stop switch. The telephone provides a means of communication with someone outside the hoistway. The panel has buttons used to indicate the floor desired and to hold the door open when the computer tries to close them (Fig. 43.7).

The capacity of a car is indicated by the maximum number of people it is designed to carry. The capacity is the total load in pounds divided by 150. For example, an elevator rated at 1500 lbs. would be marked to carry a maximum of 10 people.

Car sizes vary with the different manufacturers, but typical sizes for passenger elevators range from 6 ft. × 4 ft. (1.8 × 1.2 m) to 8 ft. × 6 ft. (2.4 × 1.8 m). Door widths are typically 3 ft. to 4 ft. (0.9 to 1.2 m) and 7 ft. to 9ft. (2.1 to 2.7 m) high. Hospital elevators are in the range of 9 ft. × 7 ft. (2.7 × 2.1 m). Freight elevators range from 5 ft. × 7 ft. (1.5 × 2.1 m) to 12 ft. × 16 ft. (3.7 to 4.9 m). They have clear door widths from 5 ft. to 12 ft. (1.5 to 3.7 m).

ELEVATOR CODE STANDARDS

Elevators, escalators, moving walks, and dumbwaiters are designed and installed following the code requirements of the American National Standard Safety Code for Elevators, Dumbwaiters, Escalators, and Moving Walks, ANSI/ASME A17.1, and local building codes. Standard elevator sizes and shapes have been developed by the National Elevator Industries, Inc. (NEII). It has standards such as Elevator Engineering Standard Layouts and Suggested Minimum Passenger Elevator Requirements for the Handicapped. These standards establish basic rules that require specific compliance and that should be coordinated with the design criteria of the complete elevator installation. Following are some examples.

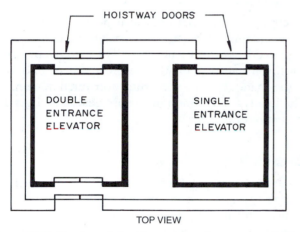

Figure 43.5 Single and double entrance doors are available.

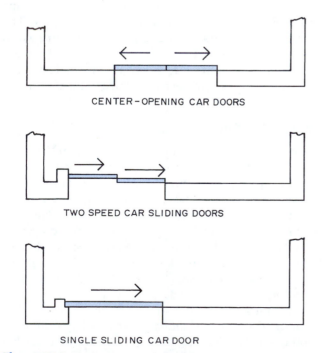

Figure 43.6 Typical types of car door action.

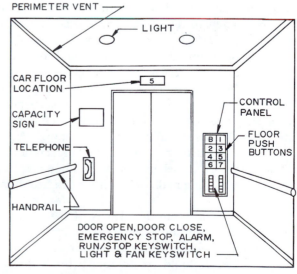

Figure 43.7 Typical elevator car controls.

The *hoistway design* criteria specify that the hoistway be enclosed its entire height with fire-resistant materials such as masonry, drywall, or concrete. Pits must also be of noncombustible material and waterproofed to prevent entry of groundwater. The top of the hoistway must be enclosed with a concrete or metal floor. The hoistway must be designed to prevent the accumulation of gases and smoke in case of a fire. No windows are permitted, and the hoistway and machinery space must be free of any pipes or ducts.

The *elevator machine room* must be enclosed with fire-resistant enclosures and doors. The only machinery allowed in the room is that required for the operation of the elevator. Since maintenance is required, permanent and easy access to the machine room is required.

The *electrical equipment and wiring* must conform to the National Electrical Code, ANSI/NFRA 70. All main electrical feeds are installed outside the hoistway. The only electrical equipment allowed in the hoistway are those things directly connected with the elevator. The machine room should have permanent lighting and natural or mechanical ventilation.

Buffers are required in the pit. Buffers are energy-absorbing units located in the pit at the bottom of the hoistway. They absorb any impact from a car that may descend below the normal lowest level.

Code requirements also include factors such as required tests, emergency condition operation, venting, opening protection, signals, and signs.

ELEVATOR TYPES

An elevator consists of a hoisting mechanism connected to a car or platform that slides vertically on guides on the sides of a fire-resistant hoistway. It is used to move passengers and materials between floors of a multistory building. *Passenger elevators* are designed to transport people between floors. *Freight elevators* carry materials between floors. Safety regulations permit only the operator or others needed to handle the materials being transported. *Hospital elevators* have special cars large enough to transport patients on stretchers or beds and their attendants. They can also serve as passenger elevators.

The Americans With Disabilities Act specifies the following for elevator requirements in new construction.

One passenger elevator should serve each level including mezzanines, in all multistory buildings. An exception is that elevators are not required in facilities that are less than three stories unless the building is a shopping center, shopping mall, or the office of a health care provider.

Elevators are not considered part of the system of egress for code purposes because they might not operate as required during an emergency, such as a fire or an earthquake. People could get stranded between floors. If the power fails, the elevator will automatically return to the lowest landing and the doors will open, allowing the passengers to exit. In the event of a fire the elevator fire service will be activated by the building's smoke alarm or by the fire service keyswitch located in the hall. When this happens, all car calls are canceled, the car returns to the main floor, and the doors open. Since elevators are vital for use by firefighters and other emergency personnel, they may be reactivated for use.

Elevators are either electric or hydraulic. The *electric elevator* uses an electric motor to supply the power. The mechanism includes the electric motor, a brake, a driving sheave or drum, gearing, belts, and wire rope. Electric elevators are used on low, medium, and high-rise buildings. They are faster than hydraulic elevators.

ELECTRIC PASSENGER ELEVATORS

Electric passenger elevators suspend the car from wire ropes and use weights to counterbalance the car. The car is guided by vertical guide rails. The electrically driven hoisting mechanism may be located at the top or bottom of the shaft. The wire rope runs over traction **sheaves.** The two types in common use are geared and gearless traction mechanisms.

Traction Driving Mechanisms

Electric elevators are powered by traction machines consisting of an electric motor connected to a driving sheave. This may be a direct connection or through a series of gears. A wire rope runs through grooves in the face of the sheave or the traction is provided by friction.

Gear-driven traction machines provide slower rising speeds and are used when slower speeds are desired. The gearing may be through a helical gearbox or with a worm gear. *Gearless direct drive* machines provide high speeds and are usually used on high-rise buildings. They have the traction sheave connected directly to the motor.

Geared Traction Elevators

A geared electric passenger elevator is shown in Fig. 43.8. This hoisting mechanism uses a worm gear to drive a large spur gear that is connected to the traction sheave. A sheave is a pulley that has a grooved rim for retaining a wire rope used to transmit force to the rope. The wire rope runs over the traction sheave and is moved as it rotates. This type of drive is used when low speeds and high lifting capacity are required. The speed can be changed by varying the size of the spur gear, which changes the gear ratio. Typical car rise speeds for geared traction *passenger elevators* range from 350 to 500 ft./min. (106 to 152 m/min.). The range of lifting capacity is typically from 2000 to 4500 lb. (900 to 2025 kg). The hoisting mechanism is usually housed in a penthouse on the roof. Geared traction *freight elevators* typically have a rise from 50 to 200 ft./min. (15 to 60 m/min.) and carry loads up to about 20,000 lb. (9072 kg).

Gearless Traction Elevators

A **gearless traction elevator** is shown in Fig. 43.9. The traction sheave and brake are mounted directly on the motor shaft. The speed of rotation of the traction sheave (which contains the wire rope) is the same as the speed of the motor. The speed can be varied by using a DC electric motor built to run at the speed required. Car rise speeds can range from 500 to 1200 ft./min. (152 to 366 m/min.). Gearless elevators also are run using a motor generator drive. The range of lifting capacity is typically from 2000 to 7000 lb. (900 to 3150 kg). The hoisting mechanism is usually housed in a penthouse on the roof.

Electric Elevator Control

The elevator control system regulates the starting, stopping, safety devices, speed of movement, direction of movement, acceleration, and deceleration of the car. Two types of controls are in general use—multivoltage and variable-voltage variable-frequency control (VVVF).

Multivoltage control is used with machines using DC motors. It controls the speed by varying the voltage supplied to the motor armature. It provides a smooth regulation of speed and is widely used on passenger elevators.

VVVF control is used to control AC motors and produces a smooth overall operation of the car. It is more efficient than multivoltage control and DC motor operation, and it is gaining popularity over DC controls and motors.

Car Safeties

Car safeties are used to stop the movement of the car and hold it in position. A governor monitors the speed of the car and activates a car safety if the car exceeds the safe speed. A car safety applies brake shoes against the guide rails, stopping the car. The governor also switches off the electrical power to the motor. Some types also activate a brake shoe to the motor drive shaft.

Roping

Traction type machine cars must be suspended from at least three hoisting ropes (ANSI 17.1). The wire rope consists of steel strands laid helically around a hemp core (Fig. 43.10). Each steel strand is made up of steel wires wrapped helically around a steel wire core (Fig. 43.11).

The roping of the traction type elevator greatly affects the speed of the car and the loads on the hoisting wires. Typical roping systems are shown in Fig. 43.12.

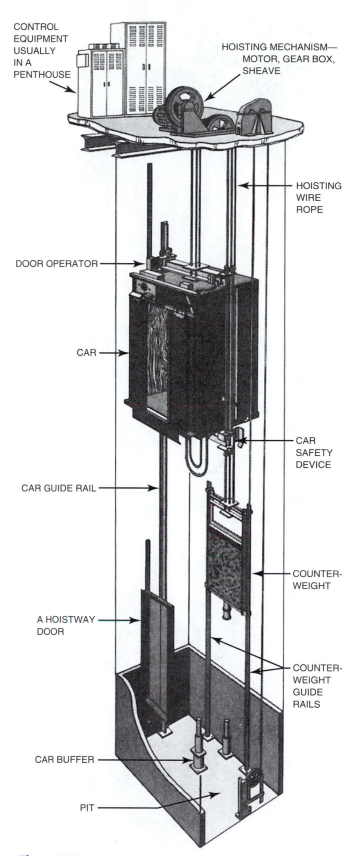

Figure 43.8 This is a geared electric passenger elevator that uses an electric motor to drive a gear box that regulates the speed of the traction sheave.

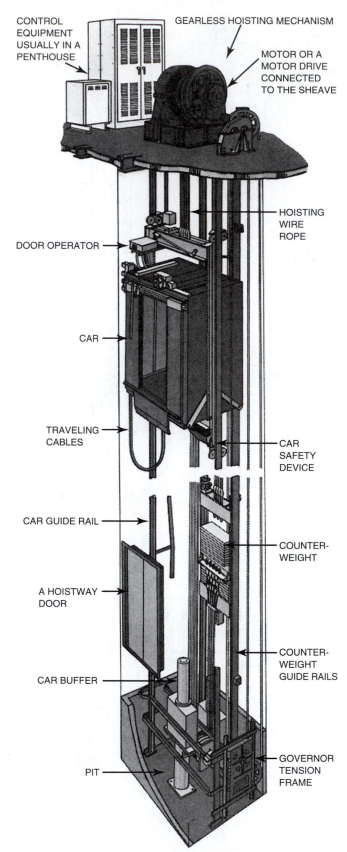

CONTROL EQUIPMENT USUALLY IN A PENTHOUSE

GEARLESS HOISTING MECHANISM

MOTOR OR A MOTOR DRIVE CONNECTED TO THE SHEAVE

HOISTING WIRE ROPE

DOOR OPERATOR

CAR

TRAVELING CABLES

CAR SAFETY DEVICE

CAR GUIDE RAIL

COUNTER-WEIGHT

A HOISTWAY DOOR

COUNTER-WEIGHT GUIDE RAILS

CAR BUFFER

PIT

GOVERNOR TENSION FRAME

Figure 43.9 This gearless electric passenger elevator has the traction sheave and brake connected directly to the motor shaft. *(Courtesy Otis Elevator Company)*

Single-wrap roping has the wire pass over the driving sheave one time and on to the counterweight. The deflection sheave moves the wire clear of the moving car. This provides a 1:1 ratio, which means it will move the car 1 ft. (305 mm) for every 1 ft. (305 mm) of travel along the circumference of the sheave.

Double-wrap roping has the wire wrapped twice around the driving sheave and a secondary sheave and on to the counterweight. Double-wrap roping can be used to provide a 1:1 or 2:1 ratio. The 1:1 double-wrap ratio causes less wear on the wire but increases the load on the sheave. The 2:1 double-wrap roping is used for heavily loaded elevators. This system moves the car 1 ft. (305 mm) for every 2 ft. (610 mm) of rope travel along the circumference of the sheave. Other roping systems are in use.

Winding Drum Machines

A winding drum hoisting mechanism has a gear-driven drum. The hoisting wire is attached to the drum and winds or unwinds around it as it rotates. This is used with dumbwaiters and some small residential elevators.

Counterweights

Counterweights are steel plates that slide on vertical steel guides. As the elevator rises they lower, and as the elevator lowers they rise. Counterweights generally equal the weight of the unloaded car and hoisting wires plus a percentage of the load capacity of the elevator. Since they rely on gravity, they help to raise the car thus reducing somewhat the power required.

Operating Systems

Elevator operating systems control the operation of the elevator. They range from simple systems to complex, automatic systems.

Typical *operator controlled systems* include the car-switch and signal systems. In the *car-switch system* the operator controls the direction of travel and initiates when the car is to move. In the *signal system* the operator presses buttons indicating the floor stops and then presses a start button. The car automatically stops at each floor for which a button was pressed.

Automatic operating systems do not require an operator. Signals are initiated by the passengers or by an automatic operating device. Typical automatic operating systems include the single, selective collective, and group systems.

Single automatic systems start when the passenger pushes a button to indicate the floor desired (Fig. 43.7). The car moves automatically to that floor, stops, and the doors open. The car is called to a floor by someone

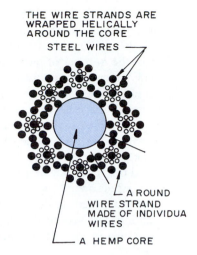

THE WIRE STRANDS ARE
WRAPPED HELICALLY
AROUND THE CORE

STEEL WIRES

A ROUND
WIRE STRAND
MADE OF INDIVIDUA
WIRES

A HEMP CORE

Figure 43.10 Typical wire rope construction.

ONE ROPE LAY

Left regular lay

Right regular lay

Figure 43.11 Typical types of wire rope. A lay is the length of a strand required to make a single wrap around the core. It is identified by the way the wires have been laid to form the strands and the way the strands are laid around the core. Left lay wire rope strands slant down to the left. Right lay strands slant to the right. Other types of wire rope constructions have special features. The type used depends on the application. *(Courtesy Bethlehem Steel Corporation)*

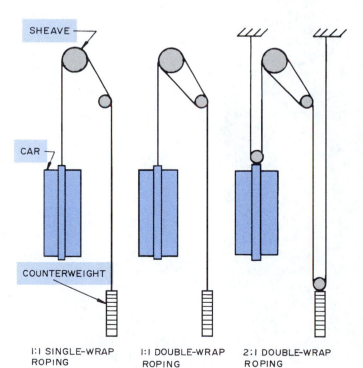

SHEAVE

CAR

COUNTERWEIGHT

1:1 SINGLE-WRAP
ROPING

1:1 DOUBLE-WRAP
ROPING

2:1 DOUBLE-WRAP
ROPING

Figure 43.12 Several types of roping systems used on electric traction elevators.

wanting to board by pressing a button on the wall by the doors to the elevator.

The *selective collective automatic system* has two call buttons, up and down (Fig. 43.13). When the up button in the car or on the hall wall has been pressed the car moves up but stops at every floor where the hall wall up button has been pressed. When it reaches the top up floor it descends, stopping at each floor on which a hall wall button was pressed or for which a floor button in the car was pressed (Fig. 43.14).

Group automatic systems control the operation of several cars that serve the same floors using a *supervisory control system.* The control system automatically decides which cars to send to various floors and coordinates the flow of traffic. The cars are dispatched from the first floor on predetermined intervals, and whichever car is nearest a floor where a passenger is waiting stops to make the pickup. This reduces waiting time and increases the number of passengers that may be carried.

The supervisory control system is able to make adjustments in car flow and direction to handle times when requirements vary, such as late afternoon in a high-rise office building when many workers are trying to leave at about the same time. The use of computers to make adjustments and control car movement has increased the efficiency of group car operations in heavily loaded situations.

Computerized systems vary, but many use a traffic information database to decide which pattern of car utilization will produce the shortest waiting time. For ex-

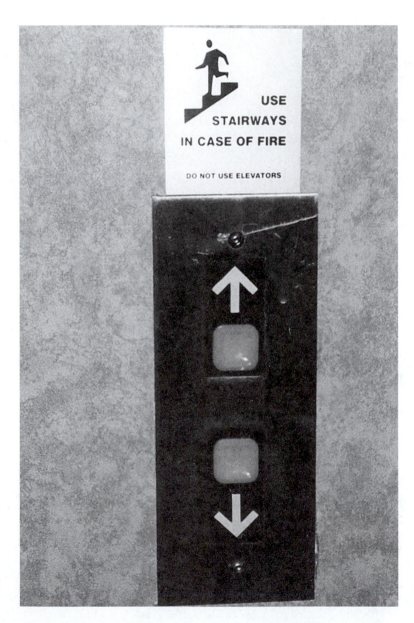

Figure 43.13 Call buttons signal the desired direction of travel of the person waiting for the elevator.

LIGHTED ARROW INDICATES
THE DIRECTION OF TRAVEL OF
THE ARRIVING CAR

CALL BUTTONS TO BRING
THE CAR MOVING IN THE
DESIRED DIRECTION TO
THIS FLOOR

Figure 43.14 Call buttons on the wall bring the elevator to the floor. The overhead direction arrows indicate the direction the arriving elevator is moving.

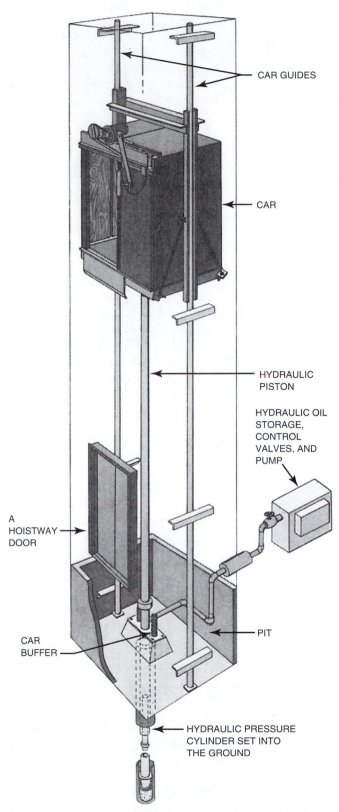

CAR GUIDES

CAR

HYDRAULIC PISTON

HYDRAULIC OIL STORAGE, CONTROL VALVES, AND PUMP

A HOISTWAY DOOR

PIT

CAR BUFFER

HYDRAULIC PRESSURE CYLINDER SET INTO THE GROUND

Figure 43.15 A hydraulic elevator moves the car with a long piston controlled by a hydraulic pump and valves. *(Courtesy Otis Elevator Company)*

ample, in peak up traffic times the cars may make fewer stops per round trip, returning to the lobby more often. A car may be assigned to serve only one group of floors and return quickly to the lobby. Another car can serve another group of floors. This reduces the stops for each car and gets them back to the lobby quickly for another trip. The reverse can occur when the down trips become heavily loaded. The control system display in the lobby indicates which floors each elevator will serve.

HYDRAULIC ELEVATORS

Hydraulic elevators are used for low-rise installations. The car is mounted on top of a piston that slides inside a hydraulic pressure cylinder. The hydraulic oil is pumped under pressure by an electric pump to the bottom of the piston, forcing it to rise and moving the car with it. When the car reaches the desired floor, the pump is automatically stopped and the oil under pressure holds the car at that level. The car is lowered by releasing oil from the pressure cylinder into the storage tank. The installation requires that the pressure cylinder be sunk into the ground a distance equal to the length of the cylinder (Fig. 43.15).

Hydraulic elevators are cheaper than electric and the mechanism is simpler. They do not require a penthouse on the roof to house the hoisting mechanism.

The speed of rise of *hydraulic passenger elevators* is typically in the range of 100 to 150 ft./min. (30 to 46 m/min.). The range of lifting capacity is typically 1500 to 4000 lb. (675 to 1800 kg). *Hydraulic freight elevators* typically rise at speeds of 50 to 100 ft./min. (15 to 30 m/min.) and carry loads from 3000 to 12,000 lb. (1350 to 5400 kg). Since gravity is at work, the down movement is faster than the rise.

FREIGHT ELEVATORS

Freight elevators are designed to carry loads of industrial trucks, general freight, motor vehicles, and heavy concentrated loads. Hydraulic elevators may be used in low-rise buildings and electric in mid- and high-rise buildings. Three classes of freight elevators are specified in the American National Standard Safety Code for Elevators, Dumbwaiters, Escalators, and Moving Walks, ANSI/ASME A17.1. *Class A* is limited to carrying a maximum of one-quarter of the rated load. It is for general freight that is hand loaded and unloaded with a lightweight hand truck. *Class B* handles only motor vehicles. *Class C* elevators are designed to handle heavy concentrated loads. A Class C1 elevator can carry materials plus industrial trucks (forklifts, pallet trucks, etc.). Class C2 elevators are used when materials are loaded by industrial trucks but the trucks are not carried with the materials. Class C3 elevators carry heavy concentrated loads other than trucks. These classifications are detailed in Fig. 43.16.

Some premanufactured freight elevators are available, but many have to be custom designed. The size of the car and the loading capacity vary widely with industrial requirements. The platform is designed to carry the anticipated load plus extra weight and movement when it is being loaded. For example, a truck loading materials can cause sideways movement of the car and shock if the load is not slowly lowered to the floor. A load placed off-center in the car produces eccentric loading that must be considered as the system is designed.

The doors on freight elevator cars are often an open steel mesh and lift vertically. The door to the hoistway is typically a vertical biparting door. It is a solid door with a locking system that will not allow the door to open until the car arrives at the floor (Fig. 43.17).

Generally freight elevators are automatically operated as described for passenger elevators. Some operator controlled systems are available.

OBSERVATION ELEVATORS

Observation elevators move up a hoistway secured to the exterior of a building. The cars are designed to have considerable glass in the walls so a view of the surrounding area is available as the elevator rises (Fig. 43.18). The hoisting machinery is placed out of view. The elevators may be geared, gearless, or hydraulic. The lower end of the hoistway (at the ground) must be enclosed with a safety barrier.

RESIDENTIAL ELEVATORS

Residential elevators are available in a range of sizes, including cars large enough to carry a wheelchair (Figs. 43.19 and 43.20). Designs vary, but residential elevators may use hydraulic or the electric traction system similar to that described for commercial passenger elevators. Typical examples are shown in Figs. 43.21 and 43.22. The car in the electric system is guided by channel guides. The doors have safety locking devices that prevent them from opening until the car is at the floor. The hydraulic unit uses a piston to raise the car.

The cars are steel reinforced and available with a variety of interior wall, floor, and ceiling finishes. Car sizes range from 36 in. (10,980 mm) to 42 in. (12,810 mm) square and 36 × 48 in. (10,980 to 14,640 mm) for wheelchairs. Most handle loads up to 450 lb. (202.5 kg).

AUTOMATED TRANSFER SYSTEMS

Automated transfer systems provide for vertical distribution of materials as well as horizontal distribution. These are widely used in hospitals, clinics, office buildings, hotels, and manufacturing plants. Among these are dumbwaiters, tote box and cart transfer systems, and horizontal tote box transfer systems.

Dumbwaiters

A dumbwaiter is a mechanism used to raise and lower a small car vertically within a building. Dumbwaiters are used in hospitals, restaurants, libraries, and office buildings to move mail, supplies, and materials (such as food, medicine, and books) from one floor to another. The sizes of the cars are controlled by local and national codes. Standard heights are 3, 3½, and 4 ft. (915, 1067, and 1220 mm) and the maximum platform size is 9 sq. ft. (0.837 m^2). The units range from light-duty lifts carrying 25 to 50 lb. (11.25 to 22.50 kg) to heavy-duty types carrying up to 500 lb. (225 kg). They may be manually or electrically powered. The *manually operated* units have an endless rope connected to a large pulley that is connected by gears to a pulley connected to the hoisting mechanism (Fig. 43.23). These have automatic braking mechanisms. They are usually limited to two-story applications.

Electrically powered dumbwaiters are used on buildings of any height. They are available as drum type, traction type, or hydraulic, as explained in the section on passenger elevators. The drum type has a maximum height of rise of 40 ft. (12.2 m), but the traction type height is unlimited.

DOVER. Capacity and Loading Requirements

All Dover freight elevators are designed and manufactured strictly in accordance with ASME A17.1 according to the following loading classifications:

CLASS A: GENERAL FREIGHT LOADING.

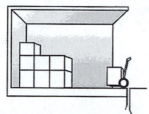

Where the load is distributed, the weight of any single piece of freight or any single hand truck and its load is not more than 1/4 the capacity of the elevator, and the load is handled on and off the car platform manually or by means of hand trucks. For this class of loading, the capacity shall be based on not less than 50 lb/ft2 (244.10 kg/m2) of inside net platform area.

CLASS B: MOTOR VEHICLE LOADING.

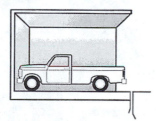

Where the elevator is used solely to carry automobile trucks or passenger automobiles up to the rated capacity of the elevator. For this class of loading, the capacity shall be based on not less than 30 lb/ft2 (146.46 kg/m2) of inside net platform area.

There are 3 Types of Class C Loading:

CLASS C1: INDUSTRIAL TRUCK LOADING.

Where truck is carried by the elevator.

CLASS C2: INDUSTRIAL TRUCK LOADING.

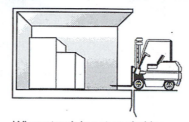

Where truck is not carried by the elevator but used only for loading and unloading.

CLASS C3: OTHER LOADING WITH HEAVY CONCENTRATIONS.

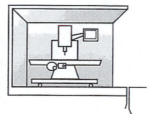

Where truck is not used. Determined on the actual loading conditions, but not less than that required for Class A loading. Consult your Dover representative to assist with the specific needs of this type of loading.

The following requirements shall apply to Class C1, C2, and C3:

The capacity of the elevator shall be not less than the load (including any truck) to be carried, and shall in no case be less than 50 lb/ft2(244.10 kg/m2) of inside net platform area. The elevator shall be provided with two-way automatic leveling.

For Class C1 and C2 the following additional requirements shall apply:

For elevators with a capacity of 20,000 lbs. (9,072 kg) or less, the car platform shall be designed for a loaded truck of weight equal to the capacity or for the actual weight of the truck to be used, whichever is greater. For elevators with a capacity exceeding 20,000 lbs. (9,072 kg), the car platform shall be designed for a loaded truck weighing 20,000 lbs. (9,072 kg) or for the actual weight of the loaded truck to be used, whichever is greater.

For C2 loading, the following requirements shall also apply:

The maximum load on the car platform during loading or unloading shall not exceed 150% of rated load.

Elevator size, capacity, and speed should be chosen to give you the most efficient and economical system possible. Your local Dover representative will be glad to work with you in designing an elevator system that will give you years of rugged use and dependable service.

Figure 43.16 Classifications, capacities, and loading requirements for freight elevators. (*Courtesy Dover Elevator Systems*)

Figure 43.17 Freight elevators are designed to carry the anticipated loads and resist actions from forklifts and other devices used to load them. *(Courtesy Dover Elevator Systems)*

Figure 43.19 This elevator is used in commercial and residential applications. It is large enough to handle a wheelchair. It is built into the wall structure. *(Courtesy Access Industries, Inc.)*

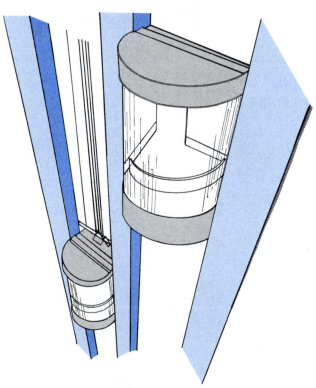

Figure 43.18 Observation elevators are glass enclosed and travel outside of a hoistway or in a hoistway open on one side.

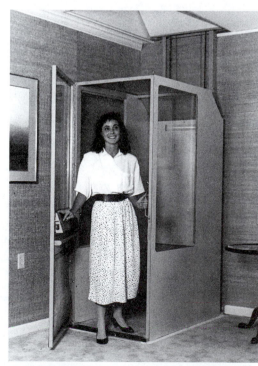

Figure 43.20 This single-passenger elevator is easily installed and especially useful in residential applications. *(Courtesy Access Industries, Inc.)*

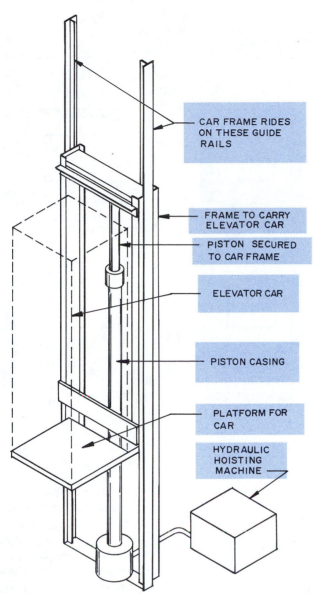

Figure 43.21 A simplified illustration of the structural frame and operating system for a hydraulic residential elevator.

CAR FRAME RIDES ON THESE GUIDE RAILS

FRAME TO CARRY ELEVATOR CAR

PISTON SECURED TO CAR FRAME

ELEVATOR CAR

PISTON CASING

PLATFORM FOR CAR

HYDRAULIC HOISTING MACHINE

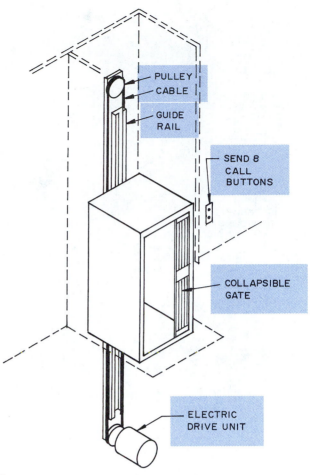

Figure 43.22 A simplified illustration of a cable-operated single-passenger residential elevator.

PULLEY

CABLE

GUIDE RAIL

SEND 8 CALL BUTTONS

COLLAPSIBLE GATE

ELECTRIC DRIVE UNIT

The electric traction and drum dumbwaiters ride on vertical rails secured to each floor with brackets forming the vertical structure (Fig. 43.27). The lifting mechanism is typically located on the top of the shaft.

Cart and Tote Box Transfer Systems

Cart and tote box transfer systems use high-performance dumbwaiter and elevator lift equipment that may be used in connection with a horizontal transfer system (Fig. 43.28). The systems available will lift up to 1000 lb. (450 kg). *Carts* are various sizes but may be up to 30 in. (9150 mm) wide, 55 in. (16,775 mm) deep, and up to 65 in. (19,825 mm) high (Fig. 43.25). *Tote boxes* are about 15 in. (4575 mm) wide, 20 in. (6100 mm) long, and 10 in. (3050 mm) deep. They can be carried on standard dumbwaiters. Small carts with weights not exceeding the capacity of a standard dumbwaiter can be carried by them. Tote box systems use counter-high doors.

Electric dumbwaiters have a speed of 50 to 150 ft./min. (15.25 to 45.75 m/min.). The higher speeds are typically used in buildings more than 50 ft. (15 m) high. Some types permit the standard speed to be reduced to as low as 25 ft./min. (7.6 m/min.) when carrying fragile items. The cars available can open on the front, front and rear, or front and side (Fig. 43.24). Standard doors are bi-parting, but slide-up, slide-down, and swinging doors are available (Fig. 43.25). Doors may be power or manually operated.

Dumbwaiters may be counter loading or floor loading. Floor loading types enable carts to be moved directly on the dumbwaiter (Fig. 43.26).

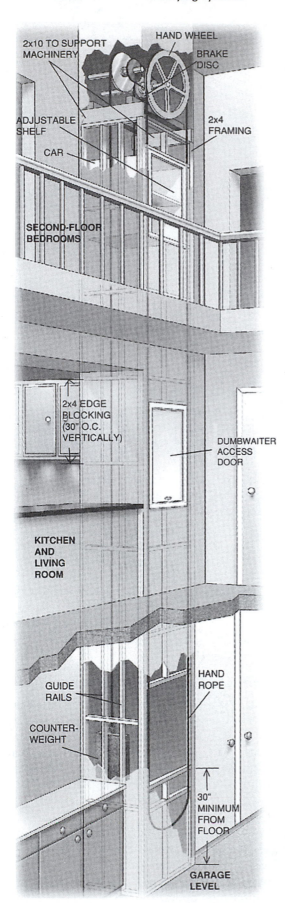

2x10 TO SUPPORT MACHINERY

HAND WHEEL

BRAKE DISC

ADJUSTABLE SHELF

2x4 FRAMING

CAR

SECOND-FLOOR BEDROOMS

2x4 EDGE BLOCKING (30" O.C. VERTICALLY)

DUMBWAITER ACCESS DOOR

KITCHEN AND LIVING ROOM

GUIDE RAILS

HAND ROPE

COUNTER-WEIGHT

30" MINIMUM FROM FLOOR

GARAGE LEVEL

Figure 43.23 Manually operated dumbwaiters use a hand rope over a pulley to operate the gears used to lift and lower the car. *(Courtesy Vincent Whitney Co.)*

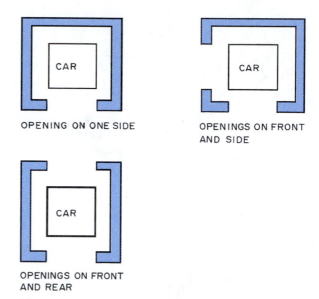

OPENING ON ONE SIDE

OPENINGS ON FRONT AND SIDE

CAR

CAR

CAR

OPENINGS ON FRONT AND REAR

Figure 43.24 Dumbwaiters and elevators are available with several door arrangements.

Horizontal transfer systems move tote boxes and other items into the vertical transfer system. Typically, items such as mail are placed in the tote box, which is placed on the conveyor. The conveyor system is installed in front of the vertical transfer system door. The boxes are placed on the conveyor and the operator presses the desired floor destination button. When a car arrives at the receiving floor the door opens automatically, and a car transfer device loads the tote box into the car of the tote box transfer system. The door closes and the car moves up or down to the desired floor, where the door opens and the box is automatically unloaded onto a conveyor (Fig. 43.29).

WHEELCHAIR LIFTS AND STAIR LIFTS

Wheelchair lifts are used on interior and exterior locations. They are made up of a steel platform with steel sides and a front gate that lowers to form a ramp to help load the wheelchair (Fig. 43.30). The ramp has a rubber skidproof surface. Wheelchair lifts are operated by an electric motor and have an automatic stop switch activated whenever the person in the wheelchair releases the control starting movement. The lift will not operate until all entry and exit doors are closed.

Typical maximum lifting capacity is in the range of 400 to 500 lb. (180 to 225 kg), and the speed of rise is gener-

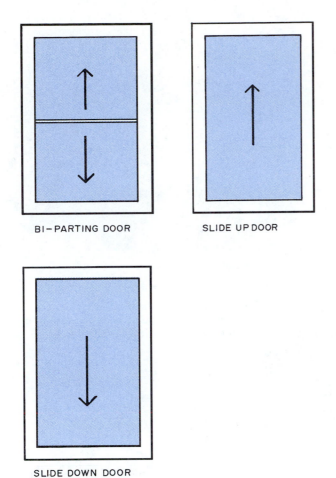

BI-PARTING DOOR

SLIDE UP DOOR

SLIDE DOWN DOOR

Figure 43.25 Dumbwaiter doors may slide up, slide down, or be bi-parting.

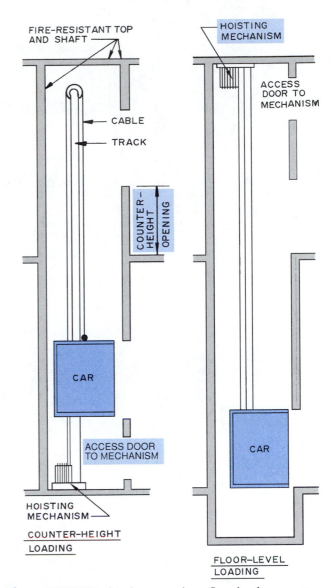

FIRE-RESISTANT TOP AND SHAFT

CABLE

TRACK

COUNTER-HEIGHT OPENING

CAR

ACCESS DOOR TO MECHANISM

HOISTING MECHANISM

COUNTER-HEIGHT LOADING

HOISTING MECHANISM

ACCESS DOOR TO MECHANISM

CAR

FLOOR-LEVEL LOADING

Figure 43.26 Dumbwaiters may have floor level or counter height loading.

ally about 8 ft./min. (2.8 m/min.). Various lifting heights are available, with 3 to 9 ft. (915 to 2745 mm) common. The platform is generally about 12 ft.^{2} (1.1 m^2).

Another type of wheelchair lift moves the passenger and wheelchair up the stair to another level. This system uses a platform with side enclosures to move the wheelchair up the stair on a rail system fastened to the wall or stair treads. It can travel up a multilevel straight stair. Two lifts are used when it is necessary to turn a 90° or 180° corner (Fig. 43.31).

Stair lifts are used to move people who do not use a wheelchair but have difficulty climbing stairs. The lift consists of a chair that runs along a track that may be mounted on the stair or along the wall of the stairwell. The chair can move around corners and across stair landings (Fig. 43.32). The rail can continue past the top of the stair, permitting the passenger to get off the chair a safe distance from the stairway (Fig. 43.33). The lifts are electrically powered, and a series of gears and shafts provide the motion.

ESCALATORS

Escalators are inclined, continuous, power-driven stairways used to move passengers up or down between floors. The escalator should be located where it will be most accessible to traffic. Adequate space must be allowed at each landing because people will be concentrated in these areas (Fig. 43.34). An alternate method for moving between floors, such as stairs parallel with the escalator, is necessary when codes require it. This provides access should the escalator fail or be under repair. Escalators may be used as a *required means of egress* if they meet all of the requirements for an emergency

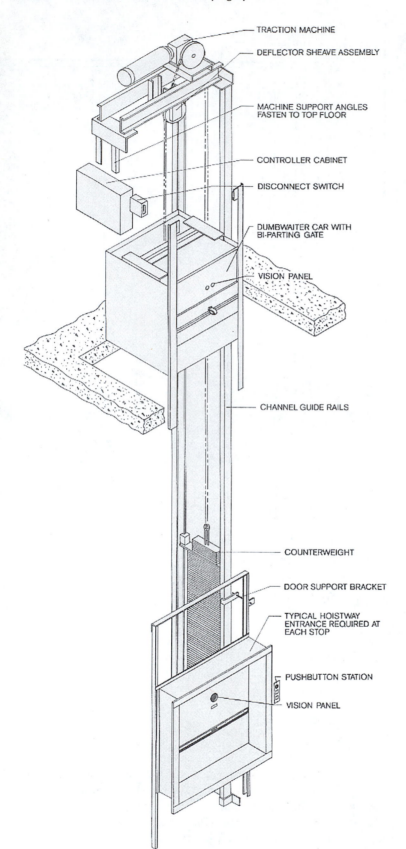

TRACTION MACHINE

DEFLECTOR SHEAVE ASSEMBLY

MACHINE SUPPORT ANGLES
FASTEN TO TOP FLOOR

CONTROLLER CABINET

DISCONNECT SWITCH

DUMBWAITER CAR WITH
BI-PARTING GATE

VISION PANEL

CHANNEL GUIDE RAILS

COUNTERWEIGHT

DOOR SUPPORT BRACKET

TYPICAL HOISTWAY
ENTRANCE REQUIRED AT
EACH STOP

PUSHBUTTON STATION

VISION PANEL

TRACTION (TR)

Figure 43.27 A traction electric powered dumbwaiter.
(Courtesy Matot, Inc.)

Figure 43.28 The cart transfer system uses
floor-level loading and carts designed to fit into
the dumbwaiter. *(Courtesy Matot, Inc.)*

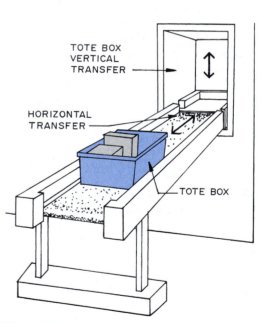

TOTE BOX
VERTICAL
TRANSFER

HORIZONTAL
TRANSFER

TOTE BOX

Figure 43.29 Horizontal transfer systems put
items to be moved in tote boxes that are automatically removed from the dumbwaiter when
they arrive at the specified floor then moved
onto a conveyor. *(Courtesy the Peelle Company)*

Figure 43.30 This porch lift is designed for exterior use on commercial and residential buildings. (*Courtesy Access Industries, Inc.*)

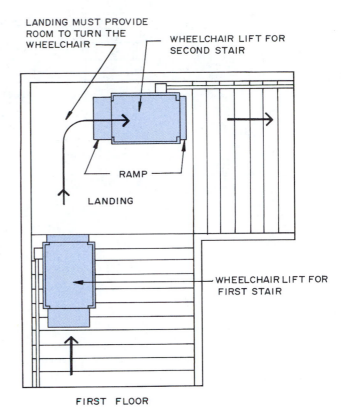

Figure 43.31 This stair-climbing wheelchair lift provides access to the upper floor by using two lifts.

egress stairway. This includes such things as having the floor openings enclosed to provide fire and smoke protection and having an approved sprinkler system. Escalators *not serving as a required exit* should have the floor openings enclosed or protected by one of the following systems: a *partial enclosure* that uses self-closing doors, an *automatic self-closing rolling shutter,* a system of high-velocity *water-spray nozzles,* or an *automatic water curtain* with an air-exhaust system.

Escalators can move large numbers of people much faster than elevators. Typical speeds are 90 to 100 ft./min. (27 to 30 m/min.) and can move 2000 to 4000 or more people per hour. However, they are seldom used to move passengers more than five or six stories.

Escalator Components

An escalator has a welded steel truss structural frame. The stair is a series of moving steps that are cast metal and grooved to provide a safe footing. The treads and risers are secured to a continuous chain that is moved by an electrically driven geared driving unit. Each side of the stair has a solid balustrade covering the ends of the stair and supporting a handrail. The handrail moves at the same speed as the stair. The escalator has electronic control devices for operation and emergency situations (Fig. 43.35).

Standards and Safety

Escalator standards are available in the publication, American National Standard Safety Code for Elevators, Dumbwaiters, and Escalators, ANSI/ASME A17.1, and the Life Safety Code of the National Fire Protection Association. In Canada the requirements are in CAN 3B-44. The safety features covered by the codes and the var-

Figure 43.32 A chair lift moves a person up a stair while they are comfortably seated. (*Courtesy Access Industries, Inc.*)

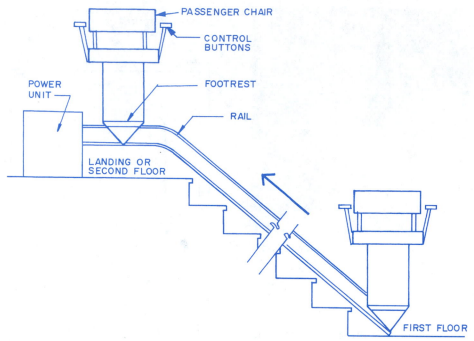

Figure 43.33 The stair lift moves the chair past the top step to provide a safe place to dismount.

Figure 43.34 This escalator has a large open area at each landing to provide needed space as people crowd on and off. (*Courtesy Otis Elevator Company*)

ious manufacturers include emergency stop buttons, broken step and drive chain switches, a brake that is electronically released when a power failure occurs, a switch to prevent the speed from exceeding the design speed, step lights, switches to stop the escalator if a handrail breaks, switches to control the handrail speed, a switch to shut it down if a step breaks, smooth balustrades that protect the sides of the step, and landing plates that protect against items getting caught as the steps flow under the floor.

Installation Examples

Escalators often are installed in pairs with one for the up direction and one heading down. They may be located in a parallel arrangement (Fig. 43.36) or in a parallel crisscross pattern (Fig. 43.37).

Escalator Sizes

The stair sizes established in ANSI/ASME A17.1 are shown in Fig. 43.38. The widths of escalator treads available are typically 24, 32, and 40 in. (0.6, 0.8, and 1.0 m). They are built on an angle of 30° and have various maximum rises, depending on the design. Rises of 20 to 30 ft. (6.1 to 9.2 m) are common for a single unit. The installation in Fig. 43.39 shows design sizes for one style of escalator.

MOVING WALKS AND RAMPS

Moving walks are horizontal conveyor belts designed to move people (Fig. 43.40). They may have a slight rise or fall, seldom exceeding 5°. **Moving ramps** move people up or down an incline with a maximum slope of 12°. They frequently connect with moving walks (Fig. 43.41). They are used where large numbers of people need to be moved over long distances, such as in an airport terminal.

Typical widths of the moving flexible rubber-covered endless belt are 24, 32, and 40 in. (0.6, 0.8, and 1.0 m).

Figure 43.35 Escalators are built using heavy structural components and drive mechanisms that can stand constant year-round operation. *(Courtesy Dover Elevator Systems)*

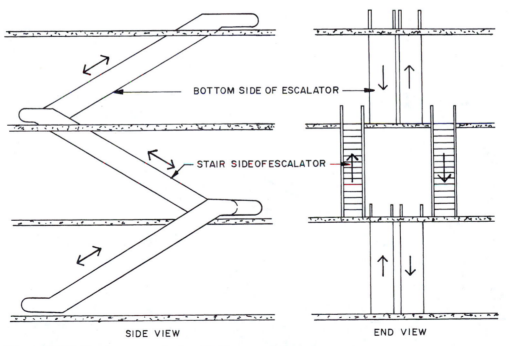

BOTTOM SIDE OF ESCALATOR

STAIR SIDE OF ESCALATOR

SIDE VIEW

END VIEW

Figure 43.36 Escalators may be installed in a parallel arrangement.

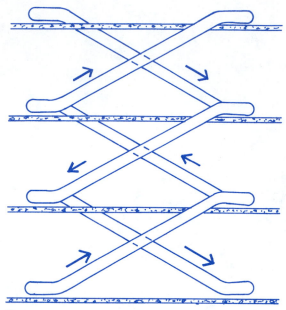

Figure 43.37 This is a typical crisscross escalator installation.

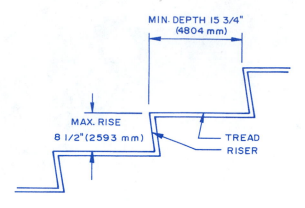

MIN. DEPTH 15 3/4"
(4804 mm)

MAX. RISE
8 1/2"(2593 mm)

TREAD
RISER

MIN. TREAD WIDTH 16" (4880 mm)
MAX. TREAD WIDTH 40" (12 200 mm)

Figure 43.38 Escalator stair size limitations.

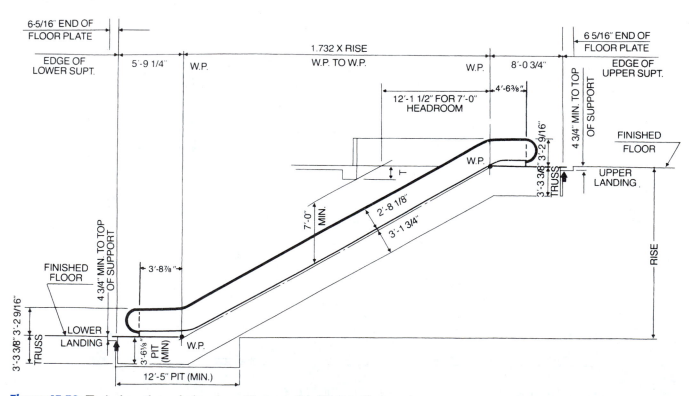

Figure 43.39 Typical escalator design sizes. *(Courtesy Otis Elevator Company)*

Figure 43.40 The pedestrians' view as they approach a moving walk. *(Courtesy Otis Elevator Company)*

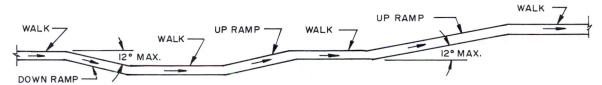

Figure 43.41 Moving ramps can move people and cargo up or down on an angle of 12° maximum.

The 24 in. belt accommodates one adult, the 32 in. belt provides room for an adult and a child or an adult with a shopping cart, and the 40 in. accommodates two adults or one adult with luggage.

The sides of the walks and ramps are enclosed with solid balustrades covered with an endless moving rubber handrail. The steel truss structural system is set in a concrete pit (Fig. 43.42). The electrically driven system has a mechanical and electrical safety system and controls similar to those used on escalators.

SHUTTLE TRANSIT

Shuttle transit systems provide horizontal transportation for distances beyond the practical limits of moving walks. They have many applications, such as linking office or retail areas to remote parking facilities or moving passengers between terminals in a large airport. Business and industrial parks can use them to provide rapid internal transportation (Fig. 43.43).

Shuttle transit systems operate much like elevators. The cars may be dispatched on a regular schedule or on call (Fig. 43.44). Each boarding station has a sign showing the scheduled arrival times. The cars operate on automatic controls. The system uses standard elevator gearless traction machine drives and cable equipment. The steel cable is attached along the side of the shuttle. The steel guide rails and power rails are located adjacent to the running surface. The vertical load is supported by a cushion of air developed between pads on the bottom of the car and the guideway running surface.

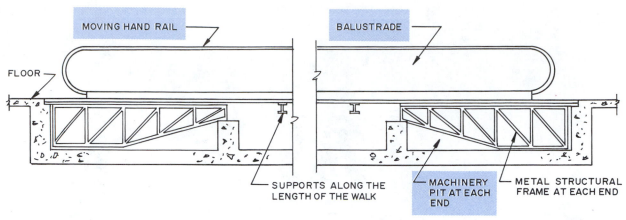

Figure 43.42 A generalized illustration showing a section through a moving walk.

Figure 43.43 This Otis Shuttle is one of a number of people movers used in areas where people must be moved over distances farther than they would like to walk. People are moved rapidly and comfortably. *(Courtesy Otis Elevator Company)*

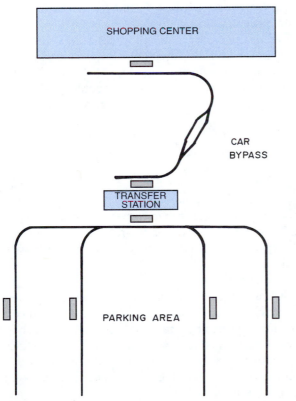

The cars may be operated singly or several can be joined to increase passenger capacity. The guideway running surface can be a single lane with dual lane bypasses located as needed (Fig. 43.45).

CONVEYORS AND PNEUMATIC TUBES

Various types of conveyor and pneumatic tube systems are used to move materials within a building. This might include a vertical transfer system as discussed earlier in this chapter under cart and tote box transfer systems (Fig. 43.28).

Another system, an electric track vehicle conveyor system, consists of self-propelled electric cars that travel over a network of tracks connecting stations within the building. The cars typically carry up to 50 lb. (22.5 kg) and move vertically and horizontally at speeds up to about 120 ft./min. (36.6 m/min.). They are used to transport paperwork, tools, test samples, and other small items (Fig. 43.46). The systems can be installed in a number of ways. The simplest is the single track system in which the car moves back and forth on one track. A dual track permits cars to go both ways at the same time, providing faster service. The loop system stores cars and sends them forward when they are called, after which they return to storage and wait to be called again (Fig. 43.47).

Pneumatic Tube Systems

Computerized pneumatic tube systems serve a variety of purposes. The system shown in Fig. 43.48 transmits small items placed in tubelike carriers, typically $3\frac{3}{4}$ to $5\frac{3}{5}$ in. (1144 to 1754 mm) in diameter and 15 to 16 in. (4575 to 4880 mm) long. They are moved through a system of transmission piping and controlled by a computerized control center. The system can be a simple single zone route or have multiple zones, which speed up the movement of the carriers.

Another pneumatic system is a large-diameter closed-pipe full vacuum system used to move trash and linens. One such installation is shown in Fig. 43.49. The mate-

Figure 43.44 The Otis Shuttle can connect widely spaced buildings or areas and follow a schedule. It operates unmanned and on automatic controls.

Figure 43.45 Cars running in different directions on the same track are scheduled to pass each other at a bypass.

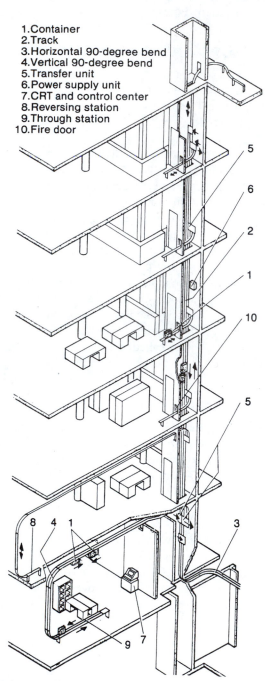

1. Container
2. Track
3. Horizontal 90-degree bend
4. Vertical 90-degree bend
5. Transfer unit
6. Power supply unit
7. CRT and control center
8. Reversing station
9. Through station
10. Fire door

Figure 43.46 This electric track vehicle conveyor system moves containers within the building following a single or dual track. *(Courtesy Translogic Corporation)*

rial to be transported is placed in bags that are inserted into the system through a door. The operator activates a key switch to open the door. The bag is placed in a holding chamber and the door is closed. The key is removed and the system is activated (Fig. 43.50). This system can be a single or multiple bag installation permitting 10 to 20 bags to be loaded in the collector before the cycling starts. The trash disposal system can have a shredding station. The trash collector system can handle loose trash consisting of paper, glass, plastic, and viscous liquids. The linen collector system, as used in hospitals, can handle plastic or cloth bagged linens or linens in loose form.

Material Conveyors

Material conveyors are used to move items, such as packages, luggage, parts, aggregate, and concrete, within a building or on the construction site. They include belt, roller, and segmented moving surfaces. The belt conveyor may be flat or troughed. Both types are powered by an electric motor. The *flat belt conveyor* is used to move items such as packages and manufactured parts within a building to stations where the items can be loaded or unloaded from the moving belt (Fig. 43.51). The system can turn corners using special power belt curves. It is able to move items horizontally or down small inclines. The *troughed conveyor belt* runs on rollers forming a U-shape and is used to move dry loose materials (Fig. 43.52).

Roller conveyors may have solid steel rollers across the unit or use a series of individual wheels, sometimes referred to as skate wheels (Fig. 43.53). *Solid roller units* may be gravity operated or power operated. They are used for medium- and heavy-duty work and may be permanently installed or movable. *Skate wheel conveyors* are used for light-duty use, such as unloading delivery trucks having light packages. They are gravity operated and usually are portable.

The *segmented conveyor* has a moving surface made up of flat sections joined with hingelike connectors (Fig. 43.54). Airport luggage conveyors are usually this type. They handle heavy loads and are power driven. Some types can be used to move solid and loose waste materials.

Portable belt conveyors are used to move materials on the construction site. The unit in Fig. 43.55 is to move concrete from a ready-mix truck. It uses a troughed conveyor and is powered by a gasoline engine. A loading hopper has been attached to receive the concrete (Fig. 43.56). This conveyor can move aggregate and other materials. Belt conveyors are available in a wide range of lengths and carrying capacities.

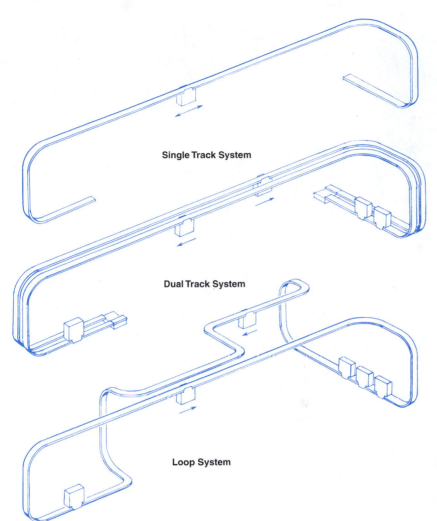

Single Track System

Dual Track System

Loop System

Figure 43.47 The electric track vehicle conveyor may have a single or double track on which containers move back and forth, or it may be a loop system. *(Courtesy Translogic Corporation)*

CRANES AND HOISTS

Cranes and hoists are used to move materials and heavy items within a building and on outdoor locations. The choice of the type to use depends on the applications required and is an important part of the design of the structure. Overhead, monorail, and underhung cranes run on rails generally supported by the structural frame of the building, so the load imposed by the crane must be carefully calculated. Some types use a structural system independent from the building structure.

Overhead cranes move along the top of fixed overhead rails. They often are used for specific jobs, such as moving steel members from storage to the fabrication area and moving the finished members to storage or a shipping area (Fig. 43.57). The lifting action is produced by a hoist that is part of the trolley. The trolley moves along a steel track. Large overhead cranes may have the operator seated in a cab connected to the crane bridge.

A *monorail* has the hoist slung below a single steel track as shown in Fig. 43.58. The *underslung crane* is much like the overhead crane except it travels along the bottom flange of the track (Fig. 43.59). A variation of this is the wall mounted *jib crane* shown in Fig. 43.60.

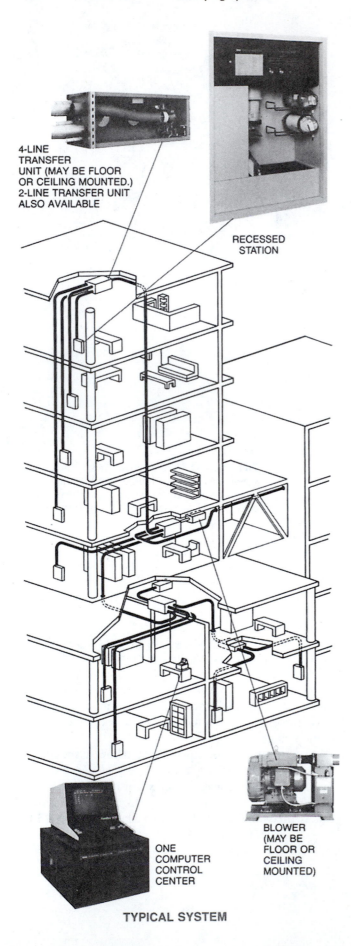

4-LINE
TRANSFER
UNIT (MAY BE FLOOR
OR CEILING MOUNTED.)
2-LINE TRANSFER UNIT
ALSO AVAILABLE

RECESSED
STATION

ONE
COMPUTER
CONTROL
CENTER

BLOWER
(MAY BE
FLOOR OR
CEILING
MOUNTED)

TYPICAL SYSTEM

Figure 43.48 A typical computerized pneumatic tube conveying system transmits items stored in tube-shaped containers through tube transmission piping. *(Courtesy Translogic Corporation)*

Typical Arrangement

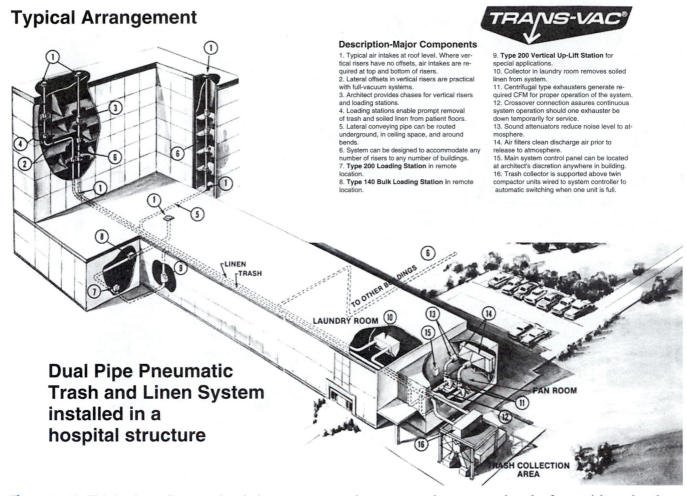

TRANS-VAC®

Description-Major Components

1. Typical air intakes at roof level. Where vertical risers have no offsets, air intakes are required at top and bottom of risers.
2. Lateral offsets in vertical risers are practical with full-vacuum systems.
3. Architect provides chases for vertical risers and loading stations.
4. Loading stations enable prompt removal of trash and soiled linen from patient floors.
5. Lateral conveying pipe can be routed underground, in ceiling space, and around bends.
6. System can be designed to accommodate any number of risers to any number of buildings.
7. **Type 200 Loading Station** in remote location.
8. **Type 140 Bulk Loading Station** in remote location.

9. **Type 200 Vertical Up-Lift Station** for special applications.
10. Collector in laundry room removes soiled linen from system.
11. Centrifugal type exhausters generate required CFM for proper operation of the system.
12. Crossover connection assures continuous system operation should one exhauster be down temporarily for service.
13. Sound attenuators reduce noise level to atmosphere.
14. Air filters clean discharge air prior to release to atmosphere.
15. Main system control panel can be located at architect's discretion anywhere in building.
16. Trash collector is supported above twin compactor units wired to system controller fo automatic switching when one unit is full.

Dual Pipe Pneumatic Trash and Linen System installed in a hospital structure

Figure 43.49 This is a large-diameter closed pipe vacuum conveying system used to move trash and soft materials, such as laundry, placed in bags. *(Courtesy Trans-Vac Systems)*

Figure 43.50 A loading station connected to the pipe transportation system. *(Courtesy Trans-Vac Systems)*

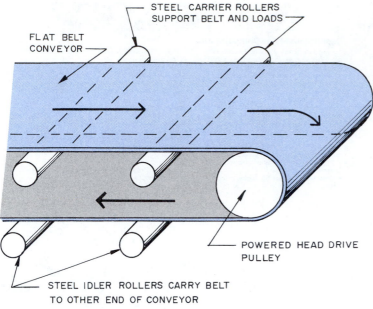

STEEL CARRIER ROLLERS
SUPPORT BELT AND LOADS

FLAT BELT
CONVEYOR

POWERED HEAD DRIVE
PULLEY

STEEL IDLER ROLLERS CARRY BELT
TO OTHER END OF CONVEYOR

Figure 43.51 Flat belt conveyors have a system of carrier and idle rollers to support the belt.

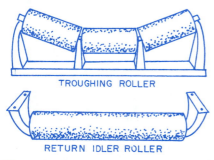

TROUGHING ROLLER

RETURN IDLER ROLLER

Figure 43.52 The troughing roller enables the conveyor belt to carry loose materials. The idle roller is below the troughing roller and supports the belt as it returns to the end of the conveyor.

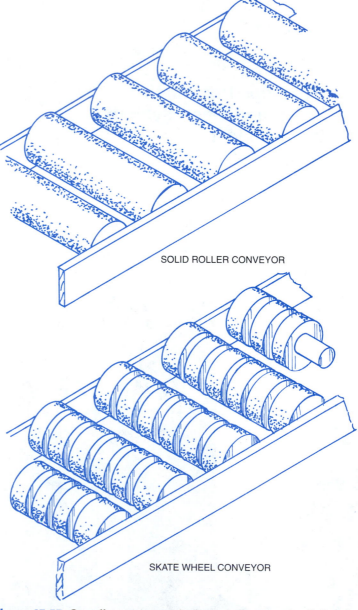

SOLID ROLLER CONVEYOR

SKATE WHEEL CONVEYOR

Figure 43.53 On roller conveyors the items being moved roll directly upon the solid roller or the skate wheels.

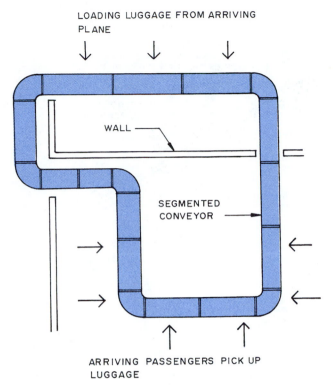

LOADING LUGGAGE FROM ARRIVING PLANE

WALL

SEGMENTED CONVEYOR

ARRIVING PASSENGERS PICK UP LUGGAGE

Figure 43.54 Segmented conveyor belts can move materials around corners.

Figure 43.56 A hopper is mounted on the end of the conveyor to receive the concrete. (*Courtesy Morgen Manufacturing Company*)

Figure 43.55 This high-capacity portable belt conveyor will move large amounts of concrete, aggregate, and other materials. (*Courtesy Morgen Manufacturing Company*)

These cranes are powered by electric motors and generally controlled by an operator on the floor. The control cable extends from the hoist to the floor. The operator can require the hoist to lift or lower a load and the trolley to move along the track.

A *gantry crane* has an overhead track and a trolley that runs on a horizontal track supported by legs that run on rails secured to the floor. Gantry cranes are used by industries that must move very heavy loads. They may operate inside or outside the building and do not require support from the structural system of the building (Fig. 43.61). On large cranes the operator is seated in a cab located below the bridge.

A diesel powered *mobile gantry* is shown in Fig. 43.62. It moves on wide-base radial tires, so it is more versatile than the track-bound gantry. It is available in a wide range of structural configurations. The cab is located to give the operator excellent visibility and is reinforced with steel guards. It also has a heater, defroster, windshield wipers, tinted glass, and a dome light. The control panel includes a load weight indicator so the weight being lifted is known. It is available with two-wheel and four-wheel drive.

A small *mobile hoist* is shown in Fig. 43.63. It is used for various material operations and is available in sizes with capacities from 15 to 400 tons. A *mobile straddle crane* is shown in Fig. 43.64. It is designed to lift long loads and place them in a new location, such as placing a freight container on a railroad car.

SPECIAL-PURPOSE SYSTEMS

Many material conveying systems are designed for special applications. One such unit is shown in Fig. 43.65. This unit is designed to move blowing-type insulation to areas where it must be placed. The unit is filled by insulation received on the site in bags. It will move the insulation over 100 ft. through a 3 in. (76 mm) hose.

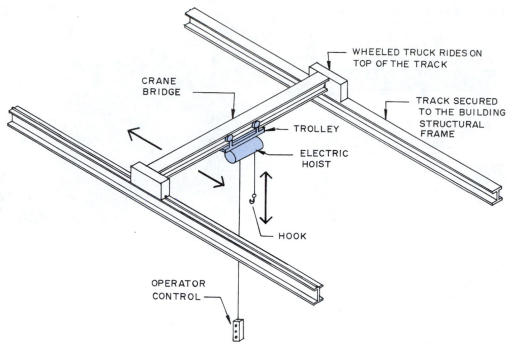

Figure 43.57 An overhead crane runs on the top of the track.

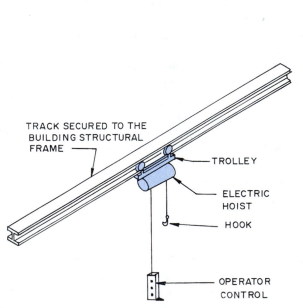

Figure 43.58 A monorail crane moves the hoist along a single track.

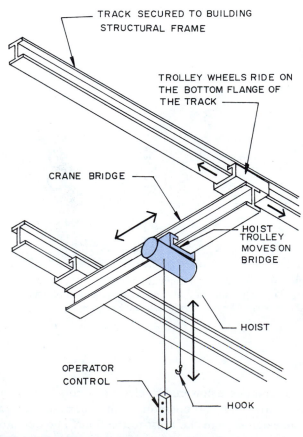

Figure 43.59 On an underslung crane the crane bridge trolleys ride on the lower track flange.

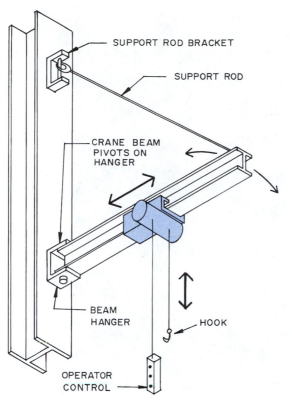

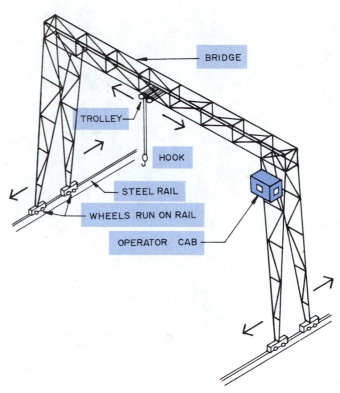

Figure 43.60 A wall-mounted jib crane rides on the horizontal track, which can rotate on the hanger pin, increasing the floor area it can serve.

Figure 43.61 This heavy-duty gantry crane rides on rails and can lift loads of several hundred tons.

Figure 43.62 A mobile gantry crane. *(Courtesy Shuttlecraft, Inc.)*

Figure 43.63 A small mobile hoist. It has the same features as the larger mobile gantry. *(Courtesy Shuttlecraft, Inc.)*

Figure 43.64 The mobile straddle crane can move long objects and raise them up for placement on surfaces above the ground. *(Courtesy Shuttlecraft, Inc.)*

Figure 43.65 One type of special-purpose material-moving equipment. This unit moves cellulose insulation at the rate of 1100 pounds per hour. *(Courtesy Intec)*

REVIEW QUESTIONS

1. What code is used to ascertain the safety standards of elevators and escalators?
2. Where can you get information on elevator size and shape standards?
3. What items are permitted in the elevator machine room?
4. What types of doors are used on various types of elevators?
5. What controls are normally located on the elevator car panel?
6. How is the capacity of an elevator car calculated?
7. What happens to elevator cars if there is a power failure?
8. What types of traction machines are used for hoisting elevators?
9. What are the major differences between gear-driven and gearless traction machines?
10. How is the speed of gearless traction elevator drive mechanisms varied?
11. What are the two types of elevator control systems?
12. What types of roping systems are used with traction elevators?
13. What types of elevator operating systems are available?
14. What mechanism is used for moving hydraulic elevators?
15. What are the classes of freight elevators?
16. What purposes do automated transfer systems serve?
17. What types of lifts are used to move the physically handicapped from floor to floor?
18. What escalator installation configurations are generally used?
19. What emergency features are part of the escalator system?
20. What is a moving walk? moving ramp?
21. Where is a shuttle transit system likely to be used?
22. What are the commonly used material conveyors?
23. What are the main differences between overhead cranes and monorails and mobile hoists and gantries?

KEY TERMS

buffer, elevator Energy-absorbing units placed in the elevator pit.

car, elevator The load-carrying unit of an elevator, consisting of a platform, walls, a ceiling, a door, and a structural frame.

car safeties, elevator Devices used to stop a car and hold it in position should it travel at excessive speed or go into a free fall.

elevator A hoisting and lowering mechanism equipped with an enclosed car that moves between floors in a building.

escalator A continuous moving stair used to move people up and down between floors.

gearless traction elevator An elevator with the traction sheave connected to a spur gear that is driven by a worm gear connected to the shaft of the electric motor.

hoistway, elevator A fire-resistant vertical shaft in which the elevator moves.

hydraulic elevator An elevator having the car mounted on top of a hydraulic piston that is moved by the action of hydraulic oil under pressure.

landing zone, elevator The area 18 in. (5490 mm) above or below the landing floor.

moving ramp A conveyor belt system used to move people or packages up an incline.

moving walk A conveyor belt system operating at floor level used to move people in a horizontal direction.

pit, elevator The part of the hoistway that extends below the floor of the lowest landing to the floor at the bottom of the hoistway.

sheave A pulley over which the elevator wire hoisting rope runs.

SUGGESTED ACTIVITY

Arrange a visit to buildings having a variety of conveying systems. Arrange to inspect the mechanical rooms and other aspects of the behind-the-scenes mechanism. Ask about the maintenance activities and schedule a visit for each system. Detail these in a written report.

ADDITIONAL INFORMATION

Ambrose, J. E., *Building Construction Service Systems,* Van Nostrand Reinhold, New York, 1990.

American National Standard Safety Code for Elevators, Dumbwaiters, Escalators, and Moving Walks, American National Standards Institute, New York.

Merritt, F. S., and Ricketts, J. T., *Building Design and Construction Handbook,* McGraw-Hill, New York, 1994.

Rush, R. D., ed., *The Building Systems Integration Handbook,* Butterworth-Heinemann, Stoneham, Mass., 1986.

Stein, B., and Reynolds, J. S., *Mechanical and Electrical Equipment for Buildings,* John Wiley and Sons, Inc., New York, 1992.

Other resources are

Publications of the National Association of Elevator Contractors, Conyers, Ga.

Publications of the National Elevator Industry, Inc., Fort Lee, N.J.

MECHANICAL AND ELECTRICAL

Fire protection is a major concern when designing commercial and industrial buildings, and the materials and structure are carefully regulated by codes. Chapter 44 identifies various fire codes and details commonly used fire suppression systems. This work requires the services of engineers who have special preparation in system design.

A basic need for a successful building is an adequate, well-functioning plumbing system. Moving potable water to where it is required is a considerable task in multistory buildings. Even more formidable is the removal of liquid waste. Chapter 45 cites the plumbing codes used across the country and describes the various piping and fittings used. Systems for potable water and sanitary piping are illustrated. A nonpotable water system and the Sovent System are described.

The comfort of the occupants of a building relies on the efficiency of the heating, ventilating, and air-conditioning (HVAC) system. The design of the system is a difficult task and, as with fire protection, requires the services of engineers prepared in system design. The basic concepts of HVAC and refrigeration are presented in Chapter 46 as are detailed illustrations of commonly used systems and the wide range of equipment available.

Part VIII concludes with the description of the multiple factors involved with planning and installing the electrical system. In most buildings this is probably the most detailed and extensive installation, from the introduction of electric power into the building to its distribution in multiple places with varying electrical requirements. The system is carefully regulated by code. Other systems within the building, such as HVAC and lighting, will not function without an adequate electrical system. The experienced electrical engineer is critical to the preparation of a design that meets codes and provides the services expected. Chapter 47 provides an introduction to electrical power, its generation, and its distribution. The materials required in a system are described and distribution systems are illustrated. The chapter concludes with a detailed description of the principles of lighting and lighting systems.

DIVISION 15 MECHANICAL

CSI MASTERFORMAT™

44

Fire Protection Systems

This chapter will help you to:

1. Have a record of the fire codes and the agencies providing them.

2. Develop a working knowledge of the various types of fire suppression systems.

A s a building is designed, consideration of the fire protection system becomes integrally involved with the design of the plumbing, mechanical, communications, and signaling systems. The fire protection system must provide for early detection of a fire and give adequate warning. The means of exiting the building is a part of the planning and is regulated by codes. This is especially critical in multistory buildings. The design should consider compartmentalization of the building, smoke control, and the type of fire control system to be used. A source of emergency electrical power is required.

FIRE PROTECTION STANDARDS

The standards of the *National Fire Protection Association (NFPA)* cover the wide range of fire protection requirements. The standards are typically adopted by local building inspection departments and become part of their code. The codes are published as individual books, such as *Fire Protection Systems* and the *National Fire Alarm Code*. Some publications are the result of a joint effort with the Society of Fire Protection Engineers. The entire code is available as a multivolume set titled *National Fire Codes*. NFPA also publishes the *National Electrical Code* and the *Life Safety Code*.

The Underwriters Laboratories, Inc. (UL) tests and approves fire protection equipment and issues reports in its Fire Protection Equipment List.

Other codes include

BOCA National Fire Prevention Code, Building Officials and Code Administrators
Standard Fire Prevention Code, Southern Building Code Congress International, Inc.
Uniform Fire Code, International Conference of Building Officials

FIRE SUPPRESSION SYSTEMS

Fire suppression systems include various types of sprinkler systems, foam and fog extinguishing systems, gas systems, and chemical systems. Manual water extinguishing systems are also used.

Automatic Water Sprinkler Systems

A typical automatic fire suppression sprinkler system uses a water supply from the city water system. Tall buildings require a backup supply, such as a storage tank on the roof (Fig. 44.1). The inflow is controlled by a valve. When water flow is detected, an alarm valve is activated (Fig. 44.2). The riser has a **siamese connection** (Fig. 44.3) that is located outside the building and is used by the fire department to pump additional water into the system from an outside source, such as a secondary water supply or a street hydrant. Each area on each floor that is protected by sprinklers has a similar system. In large buildings separate risers to various sections are used.

Sprinkler heads of various designs are available as shown in Fig. 44.4. They may be an upright type or a pendant type (Fig. 44.5). Automatic water sprinkler systems include wet pipe, dry pipe, deluge, preaction, and various types of water fog and liquid foams. These use some type of sprinkler head or nozzle.

Wet pipe systems keep water under pressure in the system of pipes at all times. When sprinkler heads are activated by heat from a fire, the water is immediately released. This is the most widely used water system (Fig. 44.6). Only the sprinkler heads over the fire area are activated, thus limiting damage from water. In areas where pipes may freeze, the system is filled with an antifreeze and water mixture.

Dry pipe system pipes are maintained with air or nitrogen under pressure. When a sprinkler head opens due to heat from a fire, the air pressure is released, causing the dry pipe to open. The pipes fill with water, which moves on through the open sprinkler heads (Fig. 44.7). Since the pipes are dry, this system is widely used in areas subject to freezing temperatures. Normally, the upright sprinkler head system is used because the pendant heads may hold water and freeze.

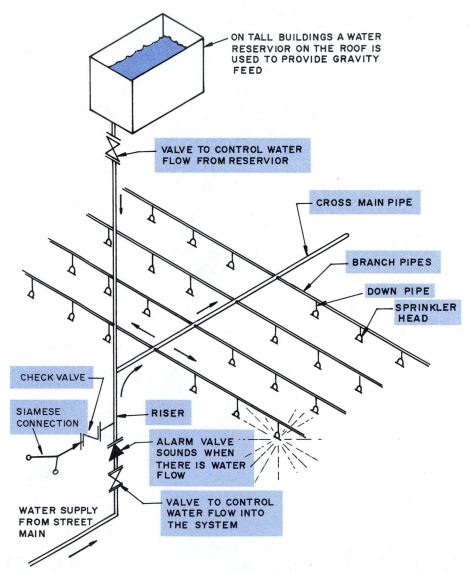

Figure 44.1 A simplified illustration of an automatic water fire suppression sprinkler system.

Figure 44.3 A siamese connection is used by the fire department to connect a supplemental source of water to the building fire suppression system.

Figure 44.2 An exterior alarm on an automatic water sprinkler system.

Figure 44.4 Typical sprinkler heads: (a) an upright head that is on the top of the exposed branch pipe and below the ceiling, (b) a recessed pendant that extends below the line of the ceiling, (c) an adjustable concealed head with a round plate that provides a decorative appearance, and (d) an adjustable pendant used in residential applications and standard commercial systems. *(Courtesy Central Sprinkler Company)*

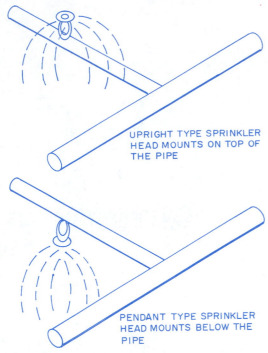

Figure 44.5 Sprinkler heads may be upright or pendant types.

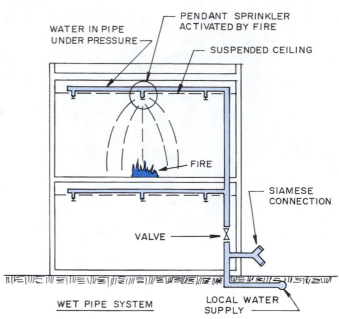

Figure 44.6 A wet pipe sprinkler system maintains water under pressure in the pipes.

The *deluge system* is designed to deliver as much water as possible as quickly as possible. It keeps the sprinkler heads or *spray nozzles* open at all times and the pipes are dry. It wets down the entire area because all the sprinklers are open. The system is activated by a sensitive fire detection system that opens the deluge valve rather than activating each of the sprinkler heads (Fig. 44.8).

Preaction systems have the sprinkler heads closed and the pipes dry. They use a fire detection system that is more sensitive than typical sprinkler heads. This detection system opens the preactive valve, allowing water to flow only to the sprinkler heads opened by heat from the fire (Fig. 44.9). The accidental opening of a sprinkler head will not cause the system to discharge water because the pipes are dry. This protects areas with sensitive equipment or expensive items from damage due to an accidental opening of a sprinkler head.

Water fog systems use a standard sprinkler piping system and have spray heads or nozzles instead of sprinkler heads. They are used in areas where highly flammable materials are stored. The fog tends to cool the material, keeping it below ignition temperature.

Foam Fire Suppressant Systems

Foam fire suppressant systems are used on fires such as gasoline in which water systems are ineffective. Foams are masses of air-filled or gas-filled bubbles formed by mechanical or chemical techniques. *Chemical foam* is

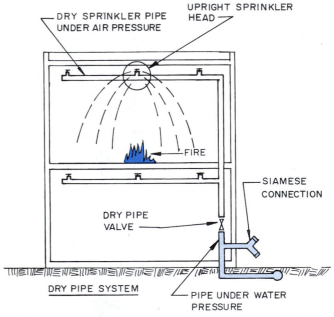

Figure 44.7 A dry pipe system maintains the sprinkler pipes dry and under air pressure.

formed by a reaction between water and several chemicals, producing a foam filled with bubbles produced by carbon dioxide. *Air foam* is formed with water and a chemical. It is moved through pipes and hoses and sprayed through discharge nozzles. *High-expansion foam* is formed by passing air through a screen that is constantly wetted with a chemical solution and a small

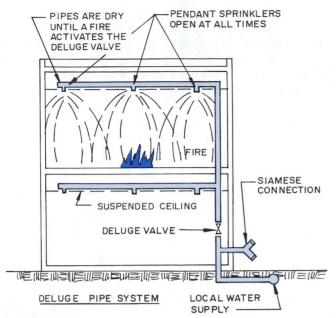

Figure 44.8 A deluge pipe system activates all the sprinkler heads in an area at the same time.

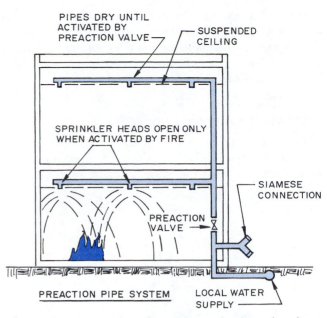

Figure 44.9 A preaction pipe system maintains a dry pipe until the preaction valve is opened by a sensitive fire detector. Water is sprayed only by sprinklers activated by the fire.

amount of water. It is moved to the fire area in large ducts and generally totally fills the compartment with the fire. Someone trapped in a compartment filled with foam will have a difficult time (Fig. 44.10).

Gas Fire Suppression Systems

Gas fire suppression systems have the advantage of being able to flood an area, suppressing the fire with little harm to the contents. This is especially important in areas with sensitive equipment, such as computer rooms, commercial aircraft, telephone exchanges, and libraries. Gas used for fire suppression is usually carbon dioxide or Halon 1301. Nonozone-depleting substitutes for Halon are being developed.

Carbon dioxide covers the fire with a blanket of heavy gas that reduces the oxygen content of the surrounding air so the combustion is extinguished. It is used in small compartmentalized areas, such as an electrical cabinet. It is not used in areas where people are present.

Carbon dioxide is stored in liquid form under great pressure and is released as a gas that cools and smothers the fire. A fire-sensing system activates a control valve that releases the gas through a series of pipes with nozzles directed at the point of a potential fire.

Halon 1301 extinguishes a fire by interfering with the combustion process, thus preventing combustion from occurring. Low concentrations of Halon 1301 extinguish flames rapidly, and higher concentrations completely diffuse into the atmosphere, preventing further burning or explosion. Halon 1301 is stored as a gas in cylinders un-

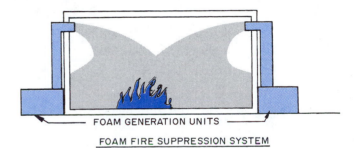

FOAM FIRE SUPPRESSION SYSTEM

Figure 44.10 Foam systems produce large amounts of air-filled or gas-filled bubbles and fill the compartment with foam from large-diameter ducts.

der pressure (Fig. 44.11). It vaporizes as it enters the fire area through discharge nozzles. The vapor diffuses into the surrounding atmosphere and leaves no residue. Most metals, rubber, and plastics are not affected by Halon 1301.

Dry Chemical Systems

Dry chemical systems do not penetrate the burning material but remain on the surface and smother the fire. Chemicals frequently used include sodium bicarbonate, potassium chloride, and monoammonium phosphate. They are effective on flammable liquid, electrical materials, and ordinary combustible materials. They have a high tolerance for extreme weather conditions and temperatures.

Automatic systems pipe chemicals directly to areas in which a fire might occur. These tend to be limited areas, such as a piece of equipment that is a constant fire hazard because of the materials used. Dry chemicals are also widely used in handheld fire extinguishers.

Manual Fire Suppression Systems

Manual fire suppression systems have fire hose stations on each floor of the building that are connected to a water piping system (Fig. 44.12). When water pressure is not enough to supply stations in the upper floors of a building or will not supply water in adequate quantity, standpipes are added to the system. A standpipe is a pipe or tank on the roof of a building that stores a supply of water. It provides an extra supply when normal water pressures fail. The standpipe system can include pumps to increase water pressure (Fig. 44.13). The standpipe also provides a reserve for the potable water needed for daily use in the building. Additional information on fire detection devices and alarm control systems is in Chapters 39 and 42.

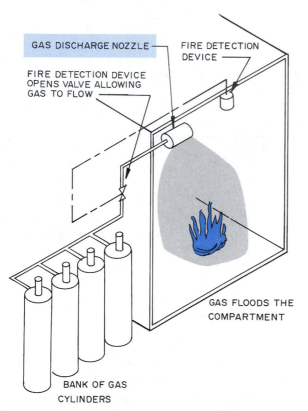

Figure 44.11 A simplified schematic of a Halon 1301 automatic fire detection and suppression system.

Figure 44.12 A typical manually operated fire hose station. *(Courtesy F.L. Industries)*

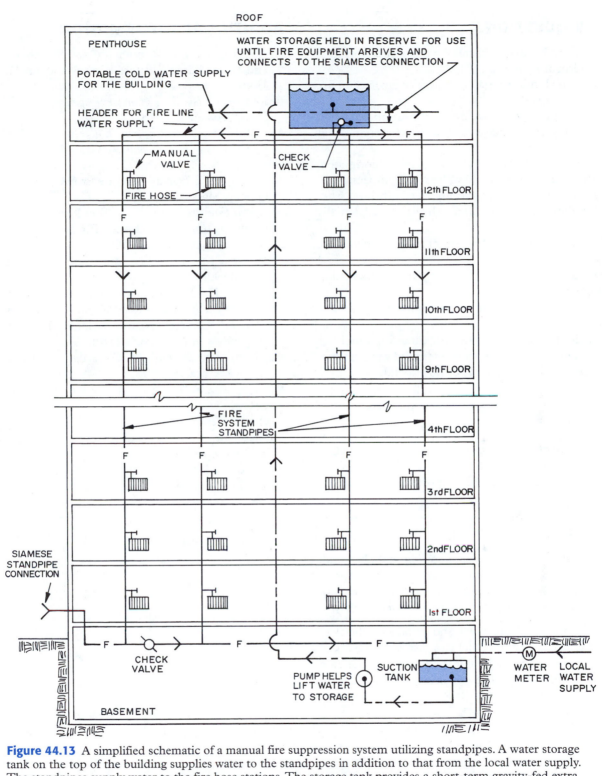

Figure 44.13 A simplified schematic of a manual fire suppression system utilizing standpipes. A water storage tank on the top of the building supplies water to the standpipes in addition to that from the local water supply. The standpipes supply water to the fire hose stations. The storage tank provides a short-term gravity-fed extra suppply to assist until the fire department can connect to the siamese connection.

REVIEW QUESTIONS

1. What organization publishes the major fire protection requirement manuals?
2. How is the Underwriters Laboratory involved in fire protection issues?
3. What purpose does a siamese connection serve?
4. How does the deluge fire protection system differ from the wet pipe system?
5. What is a good water-providing fire protection system to use if it is to be exposed to freezing temperatures?
6. Why does the preaction system protect the area from water damage if a sprinkler head is accidentally opened?
7. What types of foam fire suppression systems are available?
8. What is a major advantage of the gas fire suppression systems?
9. What gases are commonly used in gas fire suppression systems?
10. How do dry chemical systems extinguish a fire?
11. In a manual fire suppression system how is the water supplied to the individual fire hose stations?

KEY TERMS

siamese connection A connection outside the building to which firefighters connect an alternate water supply to boost the water used by the fire suppression system.

SUGGESTED ACTIVITIES

1. Tour your campus, local shopping mall, hospital, and other commercial buildings and record the type of fire suppression systems found in each. Take photos of the visible components and prepare a labeled display.

2. Invite the local fire chief and building official to address the class. The fire chief can discuss the work of the fire department in making inspections. The building official can address the fire requirements in the local code.

ADDITIONAL INFORMATION

BOCA National Fire Prevention Code, Building Officials and Code Administrators, Country Club Hills, Ill., 1996.

Bradshaw, V., *Building Control Systems,* John Wiley and Sons, New York, 1993.

Merritt, F. S., and Ricketts, J. T., *Building Design and Construction Handbook,* McGraw-Hill, New York, 1994.

National Fire Codes, National Fire Protection Association, Quincy, Mass.

Standard Fire Prevention Code, Southern Building Code Congress International, Inc., Birmingham, Ala.

Stein, B., and Reynolds, J. S., *Mechanical and Electrical Equipment for Buildings,* John Wiley and Sons, New York, 1992.

Traister, J. E., *Design and Application of Security Fire-Alarm Systems,* McGraw-Hill, New York, 1990.

Uniform Fire Code, International Conference of Building Officials, Whittier, Calif., 1996.

Wood Building Technology, Canadian Wood Council, Ottawa, Ontario, 1993.

Wood Reference Handbook, Canadian Wood Council, Ottawa, Ontario, 1991.

Other references are

Publications of the National Fire Sprinkler Association, Inc., Patterson, N.Y.

Numerous related publications from the organizations listed in Appendix B.

CHAPTER 45

Plumbing Systems

This chapter will help you to:

1. Relate the plumbing codes to plumbing system design.

2. Be familiar with the types of piping, tubing, and fittings used on plumbing systems.

3. Be knowledgeable about the sources of potable water and its distribution systems.

4. Be familiar with the parts of sanitary piping systems.

5. Be able to properly design facilities using various plumbing fixtures.

A typical building plumbing system includes hot and cold **potable** water distribution systems and a sanitary disposal system. Many other plumbing systems are designed to suit a special need, such as compressed air systems, fuel oil systems, natural gas systems, and oxygen gas systems. All of these are designed and installed as specified by plumbing codes.

PLUMBING CODES

The materials, plumbing design, and installation procedures are strictly regulated by local plumbing codes. Local governments usually adopt one of the model codes. These include the following:

▲ Uniform Plumbing Code, published by the International Association of Plumbing and Mechanical Officials
▲ National Standard Plumbing Code, published by the National Association of Plumbing, Heating, and Cooling Contractors
▲ Southern Standard Plumbing Code, published by the Southern Building Code Congress International, Inc.
▲ BOCA National Plumbing Code, published by the Building Officials and Code Administrators International

PIPING, TUBING, AND FITTINGS

Water is piped through buildings to a number of locations in which it serves various purposes. Potable (drinkable) water is delivered to sinks, lavatories, and water heaters. Typically it is also run to toilets (water closets) and exterior hose bibbs. Although these do not require potable water, a less pure and less expensive supply (known as gray water) is not generally available. Pipes for most situations are hidden in walls, floor, or ceiling. Pipes can be exposed in areas where their being seen is not important. However, they must be protected from freezing in all locations. The system must provide **valves** to control the flow of water into the building, within parts of large buildings, and to the individual fixtures.

The types of pipe used for potable water systems are detailed in Table 45.1. Decisions on which to use are based on cost, length of service expected, and the quality of the

Table 45.1 Pipe and Tubing Used for Water Supply Systems

Type	Connections	Diameters	Special Qualities
Galvanized steel	Threaded	⅛–4 in. (3–102 mm)	Strong, long life
Welded steel	Threaded	⅛–4 in. (3–102 mm)	Strong, long life
Copper tube	Soldered, brazed	¼–6 in. (6–152 mm)	Corrosion resistant
Red brass	Threaded, brazed	⅛–6 in. (3–152 mm)	Corrosion resistant
Plastic	Solvent joined, heat fusion, serrated inserts	¼–6 in. (6–152 mm)	Lightweight, corrosion resistant

water. Some water is loaded with minerals and is very corrosive. A pipe that carries hot water must withstand temperatures of 180°F (82°C) or higher. In special applications, the effect of the liquid or gas to be carried must be considered. Gasoline, liquified petroleum gas, and other such materials have a substantial effect on pipes.

Plastics, copper, and steel pipe are all made with the same diameters, but some types are made in larger diameters than others. Pipes also are made with a range of wall thicknesses. The thicker the pipe wall the greater the pressure the pipe can carry.

Galvanized steel pipe has been used for many years. It is strong, can be used for hot and cold water, and has a long life. The pipe and fittings are joined with threaded connections. It is available in three grades—standard weight, extra strong, and double extra strong.

Gas is piped to a meter outside the building by the utility company. The building gas piping is run inside the building to the fixtures, such as a gas-fired furnace, that requires the fuel. Codes regulate the materials and method of installation. Generally, the gas inlet to the building must be above grade. Gas piping within the building is *black steel pipe* meeting the requirements of ASTM A53 or ASTM A106. It uses steel or malleable iron fittings. Some codes permit the use of copper or brass pipe if the type of gas to be carried will not cause them to corrode. Some types of plastic pipe are permitted for underground outside installation. They must be given a protective coating to help resist corrosion from elements in the soil.

Welded steel pipe is made by rolling a flat hot steel strip into a circular shape and butt-welding the edges as they are pressed together. It is available in the same grades as galvanized steel pipe. The outside diameter is the same for all three grades and the difference in the thickness of the wall is taken up on the inside of the pipe. Welded steel pipe is joined by threaded fittings.

Red brass pipe is used for water lines, especially if the water contains corrosive elements. It is manufactured with threaded fittings and plain ends that are brazed to socket type fittings.

Copper water tubing is an excellent hot and cold water distribution material. It is resistant to corrosion and does not rust. The flexible type is easy to bend into various shapes, reducing the number of fittings needed. The joints fit together in a socket arrangement and are secured by soldering. The solder used must not contain lead because lead will contaminate the water causing those who use it to have health problems.

Copper water tubing has a high coefficient of expansion. When carrying heated liquids it could expand enough in length to cause damage to the pipe unless provision is made to allow for it. Typically a section of pipe is bent into a loop, allowing the pipe to expand and contract (Fig. 45.1).

The types of copper tubing are shown in Table 45.2. Copper water tubing is available in types K, L, and M. Type K has the heaviest wall and has a green stripe along its length. Because of its strength it can be used for underground runs. Type L has a medium wall and a blue stripe and type M has a light wall and a red stripe. All three are used for potable water, heating and air-

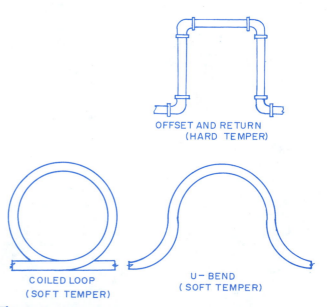

OFFSET AND RETURN
(HARD TEMPER)

COILED LOOP
(SOFT TEMPER)

U–BEND
(SOFT TEMPER)

Figure 45.1 Ways to allow for expansion and contraction of copper tubing. *(Courtesy Copper Development Association)*

Table 45.2 Copper Tubing, Types, Uses, and Sizes

Type	Color	Uses	Nominal Diameter
K	Green	Potable water, fire protection, solar, fuel/fuel oil, HVAC, snow melting	Straight lengths $\frac{1}{4}$–12 in. (6–305 mm) Coils $\frac{1}{4}$–2 in. (6–50.8 mm)
L	Blue	Potable water, fire protection, solar, fuel/fuel oil, HVAC, snow melting	Straight lengths $\frac{1}{4}$–12 in. (6–305 mm)
M	Red	Potable water, fire protection, solar, fuel/fuel oil, HVAC, snow melting	Straight lengths $\frac{1}{4}$–12 in. (6–305 mm)
DWV	Yellow	Drain, waste, vent, HVAC, solar	Straight lengths $1\frac{1}{4}$–8 in. (32–203 mm)
ACR	Blue Soft ACR not marked	Air-conditioning, refrigeration, natural gas, liquified petroleum gas	Straight lengths $\frac{3}{8}$–$4\frac{1}{8}$ in. (9.5–105 mm)
OXY, MED, OXY/MED, OXY/ACR, OXY/MED	K-Green L-Blue	Medical gas	Straight lengths $\frac{1}{4}$–8 in. (6–203 mm)
G	Yellow	Natural gas, liquified petroleum gas	Straight lengths $\frac{3}{8}$–$1\frac{1}{8}$ in. (9.5–28.5 mm) Coils $\frac{3}{8}$–$\frac{7}{8}$ in. (9.5–22 mm)

conditioning, and fuel and fuel oil systems. Type ACR is used for air-conditioning and refrigeration systems, DWV (yellow stripe) is used for drain, waste, and ventilation systems, and several other types identified as OXY and MED are used for medical gas applications.

Copper tubing is manufactured in hard and soft tempers. The soft-tempered pipe bends easily, reducing the need for many connections. The hard-tempered pipe requires soldered connectors to turn corners. Copper tube is available in diameters from $\frac{1}{4}$ to 8 in. (6 to 203 mm).

Plastic pipe is manufactured in several synthetic resins. The resin used greatly influences the strength and use of the pipe. Table 45.3 shows the commonly used resins for producing plastic pipe. Notice that most are not acceptable for hot water piping. Plastic pipe is lightweight, flexible, and available in long lengths. Various types are joined with solvent cement, elastomeric seals for bell-end piping and fittings, serrated insert fittings secured with stainless steel clamps, heat fusion used on plastics for which there is no solvent, and threaded fittings that can connect to threaded metal pipe.

Plastic piping and sanitary piping are manufactured in diameters from $\frac{1}{2}$ in. (12.7 mm) to 48 in. (1219 mm) and in lengths to 20 ft. (6 m). The wall thicknesses are identified by a schedule number, such as Schedule 40, 80, or 120. The larger the number the thicker the pipe wall.

Pipes made from glass, nickel, silver, and chrome are used for special applications. They are corrosion resistant. The metal pipes usually have threaded connections. Glass pipe is connected by neoprene gaskets secured with stainless steel straps over a stainless steel sleeve (Fig. 45.2).

Pipe Connections

The most common methods for connecting pipe are shown in Fig. 45.3. Water pipe uses butt-welded, socket, and threaded fittings. The bell and spigot connection is used on sewer lines and the flanged connection is used in applications such as petrochemical and power generation piping.

Water Pipe Fittings

The most commonly used water pipe fittings are shown in Fig. 45.4. The coupling connects two pipes end to end. The elbows change the direction of the pipe. A tee permits one pipe to intersect another. The end of a pipe can be closed by installing a coupling and screwing a plug into it. The cap screws over the end of the pipe.

VALVES

A wide variety of valves are available for controlling the flow of water, oil, gas, and chemicals. Some of the frequently used types are shown in Fig. 45.5. A *pressure regulator valve* limits the water pressure, preventing damage to piping and equipment. A *water hammer arrestor* has a hydraulic piston that absorbs shock waves produced by sudden changes in water flow; this reduces banging in the pipes.

Table 45.3 Major Types and Uses of Plastic Pipe

Type	Condition	Connections	Maximum Operating Temperature °F	°C	Typical Uses
Acrylonitrile butadiene styrene (ABS)	Rigid	Threaded, serrated fittings, solvent	100 pressure 180 nonpressure	38 82.9	Cold water, waste, vent, sewer, drain, conduit, gas
Chlorinated polyvinyl chloride (CPVC)	Rigid	Threaded, couplings, serrated fittings, solvent	180 at 100 psig (type 11)	82.9	Cold and hot water, chemical piping
Polyethylene (PE)	Flexible	Serrated fittings, fusion in socket, butt fusion	100 pressure 180 nonpressure	38 82.9	Cold water, gas, waste, chemicals
Polypropylene (PP)	Rigid	Mechanical couplings, butt fusion, socket fusion	100 pressure 180 nonpressure	38 82.9	Chemical piping, chemical drainage
Polyvinyl chloride (PVC)	Rigid	Solvent, threading, mechanical couplings, serrated fittings	100 pressure 180 nonpressure	38 82.9	Cold water, gas, waste, vents, drains, sewers, conduit
Styrene rubber plastic (SRP)	Rigid	Solvent, serrated fittings, elastomer seal	150 nonpressure (not used under pressure)	66	Sewage disposal field, storm drainage, soil drainage

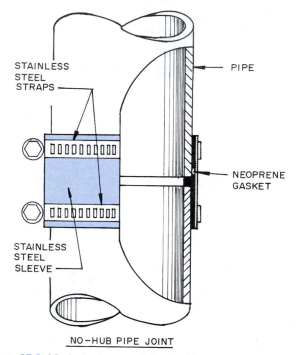

Figure 45.2 No-hub pipes are joined with gaskets and stainless steel straps.

The *backflow preventer* keeps water from backing up in the system, and the *expansion tank* absorbs excess pressure by absorbing the extra volume of water created when water is heated and expands. It has a rubber bladder that flexes against the water pressure. A *T and P valve* is a safety relief valve that senses a buildup of temperature or pressure and opens to release the excess pressure and allow cold water to enter to prevent an explosion.

Float valves are used to control water levels in tanks. As the water level rises the float rises and shuts off the water when a set capacity has been reached. *Strainers* typically have a 20 mesh screen that collects dirt and debris. They may be on the main line or on lines to a particular piece of equipment. *Saddle valves* are installed on piping that carries water under pressure. They clamp on the pipe and then pierce it.

Gate-style *stop and waste valves* were once commonly used to shut off water to the building on the owner's side of the meter. Now the more dependable *ball valve* is used for shutoffs. *Globe valves* are used to control the flow of hot and cold water, oil, and gas.

INSULATION

Condensation forms on the exterior of cold water pipes when warm humid air hits them. The condensation will drip down inside a wall cavity, ceiling, or floor, wetting the insulation and penetrating drywall or other wall finish.

To prevent damage from condensation, pipes are covered with preformed fiberglass or foam insulation that is usually ½ to 1 in. (12 to 25 mm) thick (Fig. 45.6). The insulation is fitted around the pipe and taped. Hot water pipes are insulated to reduce the loss of heat to the cooler atmosphere. Most plumbing units, such as water heaters, have insulation built inside their jackets. Extra batts can be taped on the outside.

Pipes carrying chilled water, brine, refrigerant, domestic hot water, commercial hot water and steam, steam condensate, and hot water heating systems require the use of minimum insulation shown in Table 45.4. If hot and cold piping run parallel they should be 6 to 8 in.(152 to 203 mm) apart to reduce the chance of heat and cooling exchange.

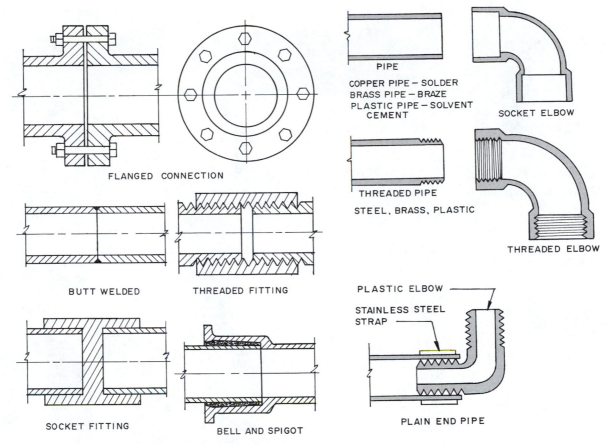

Figure 45.3 Types of pipe and fitting connections.

FLANGED CONNECTION

BUTT WELDED

THREADED FITTING

SOCKET FITTING

BELL AND SPIGOT

PIPE

COPPER PIPE – SOLDER
BRASS PIPE – BRAZE
PLASTIC PIPE – SOLVENT
CEMENT

SOCKET ELBOW

THREADED PIPE

STEEL, BRASS, PLASTIC

THREADED ELBOW

PLASTIC ELBOW

STAINLESS STEEL
STRAP

PLAIN END PIPE

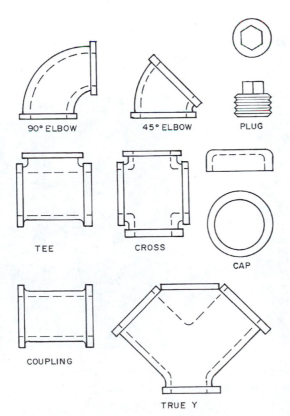

90° ELBOW

45° ELBOW

PLUG

TEE

CROSS

CAP

COUPLING

TRUE Y

Figure 45.4 Some of the commonly used pipe fittings.

Water Pressure Reducing Valve

Check Valve

Strainer

Ball Valve

Relief Valve

Gate Valve

Stop and Waste Valve

Water Hammer
Arrestor

Expansion Tank

Float Valve

Figure 45.5 Typical valves used in water distribution systems. *(Courtesy Webster Valve Company)*

POTABLE WATER SUPPLY

Water for public and privately owned central water systems may be obtained from rivers, lakes, ponds, surface runoff, and wells. Surface-collected water tends to contain a number of contaminates that must be removed to produce potable water. Quality problems include *hardness* caused by calcium and magnesium salts, *color* caused by manganese or iron, *corrosion* caused by acidity in the water, *pollution* by sewage and organic matter, *odor and taste* caused by organic matter, and *turbidity* caused by silt and

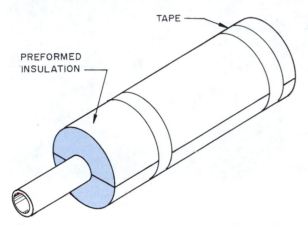

Figure 45.6 Hot and cold water pipes are insulated to maintain water temperature, prevent freezing, and stop condensation from dripping from the pipes.

other suspended matter. The water treatment plant processes the water to correct these problems (Fig. 45.7).

Water treatment includes processes such as *screening* the water at the intake, *sedimentation* (which allows particles to drop in a settling basin), *coagulation* (removing suspended matter with a chemical such as hydrated aluminum sulfate in a settling basin), *filtration* (removing suspended particles and some bacteria through a filter such as sand or diatomaceous earth or by chlorination filtration), *disinfection* (removing harmful organisms using bromine, iodine, ozone, or heat treatment), *softening* (removing calcium and magnesium), and *aeration* (exposing water to the air, such as by spraying, to improve taste and color).

In areas without a central water supply the water is obtained from wells. These are usually drilled, driven, or jetted. A *drilled well* is dug with a steel auger that drills into the earth into which a pipe is placed. A *driven well* is formed by putting a steel point on the end of sections of pipe and driving them into the earth. A *jetted well* has a well point from which a high-pressure stream of water is pumped, opening a hole for the well pipe to move behind it into the earth.

The water from wells tends to not require the extensive treatment needed for surface water. It must be tested for purity and treated as required to reduce hardness, iron, or magnesium.

Once the potable water is delivered to the building the quality must be maintained as the water is stored

Table 45.4 Minimum Insulation for Pipes (Inches)

Fluid Design Operating Temperature Range (°F)	Insulation Conductivity		Nominal Pipe Diameter (in.)					
	Conductivity Range [BTU-in./(h-ft³-°F)]	Mean Rating Temperature (°F)	Runouts[a] up to 2	1 and less	1–1¼ to 2	2½ to 4	5 and 6	8 and up
Heating Systems (Steam, Steam Condensate, and Hot Water)								
Above 350	0.32–0.34	250	1.5	2.5	2.5	3.0	3.5	3.5
251–350	0.29–0.31	200	1.5	2.0	2.5	2.5	3.5	3.5
201–250	0.27–0.30	150	1.0	1.5	1.5	2.0	2.0	3.5
141–200	0.25–0.29	125	0.5	1.5	1.5	1.5	1.5	1.5
105–140	0.24–0.28	100	0.5	1.0	1.0	1.0	1.5	1.5
Domestic and Service Hot Water Systems[b]								
105 and greater	0.24–0.28	100	0.5	1.0	1.0	1.5	1.5	1.5
Cooling Systems (Chilled Water, Brine, and Refrigerant)[c]								
40–55	0.23–0.27	75	0.5	0.5	0.75	1.0	1.0	1.0
Below 40	0.23–0.27	75	1.0	1.0	1.5	1.5	1.5	1.5

Source: Reprinted by permission from ASHRAE/IES Standard 90.1-1989, *Energy Efficient Design of New Buildings Except Low-Rise Residential Buildings,* © 1989 by the American Society of Heating, Refrigerating, and Air-Conditioning Engineers, Inc., Atlanta, Ga.
[a]Runouts to individual terminal units not exceeding 12 ft. in length.
[b]Applies to recirculating sections of service or domestic hot water systems and first 8 ft. from storage tank for nonrecirculating systems.
[c]The required minimum thicknesses do not consider water vapor transmission and condensation. Additional insulation, vapor retarders, or both, may be required to limit water vapor transmission and condensation.

Figure 45.7 Water treatment plant processes include settling, screening, sedimentation and coagulation, filtration, disinfecting, aeration and softening. *(Courtesy American Water Works Association)*

and flows through the building piping system. The potable water system must be kept isolated from all other plumbing systems.

WATER PIPE SIZING

The size of the required water supply pipe depends on the available pressure, the vertical and horizontal distances involved (friction in the pipe), the number of fittings (tees, elbows), and the demand of the use at the fixture. Experience has allowed values to be determined for flow of various fixtures under recommended minimum pressures. The values of all of these are considered as the pipe diameter is calculated. Several fixtures on one water line do not require the water flow to increase directly for each because not all of the fixtures will likely be used at the same time. Typical flow pressures and flow rates are shown in Table 45.5.

POTABLE WATER DISTRIBUTION SYSTEMS

Codes specify that all fixtures in a building must be supplied with potable water at adequate pressures. Potable water in residential buildings generally uses the water pressure from the central water system or the pump on a well to supply the fixtures (Fig. 45.8). This is an *up-feed system*, in which the water moves up from the source into the building and to the fixtures. The piping used must have the proper size to carry the amount of water needed by each fixture. The circulation pipes and risers usually feed more than one fixture, so the total expected flow must be figured.

The water enters through a meter and a shut-off valve. In cold climates the meter must be inside the building and the water service line must be below the frost line. The depth varies in different geographical areas. The water continues to a water softener if needed and then to a water heater and to horizontal circulation lines. The water moves to fixtures in the upper stories through risers and to individual fixtures through small-diameter pipe connectors. The hot water runs from the water heater through a similar piping system.

Low multistory buildings may also use an up-feed system if the demand on water is not excessive (Fig. 45.9). A series of pumps in the basement provides the needed pressure to raise the water to the desired height and to maintain it as fixtures are used.

Table 45.5 Minimum Flow and Pressure Required by Typical Plumbing Fixtures

Fixture	Flow Pressure (psi)	(kPa)	Flow Rate (gpm)	(L/s)
Ordinary basin faucet	8	55	2.0	0.13
Self-closing basin faucet	8	55	2.5	0.16
Sink faucet, ⅜ in. (9.5 mm)	8	55	4.5	0.28
Sink faucet, ½ in. (12.7 mm)	8	55	4.5	0.28
Bathtub faucet	8	55	6.0	0.38
Laundry tub faucet, ½ in. (12.7 mm)	8	55	5.0	0.32
Shower	8	55	5.0	0.32
Ball-cock for closet	8	55	3.0	0.19
Flush valve for closet	15	103	15–40	0.95–2.52
Flushometer valve for urinal	15	103	15.0	0.95
Garden hose (50 ft., ¾-in. sill cock) (15 m, 19 mm)	30	207	5.0	0.32
Garden hose (50 ft., ⅝-in. outlet) (15 m, 16 mm)	15	103	3.33	0.21
Drinking fountains	15	103	0.75	0.05
Fire hose 1½ in. (38 mm), ½-in. nozzle (12.7 mm)	30	207	40.0	2.52

Reproduced from *Manual of Individual Water Supply Systems,* U.S. Environmental Protection Agency

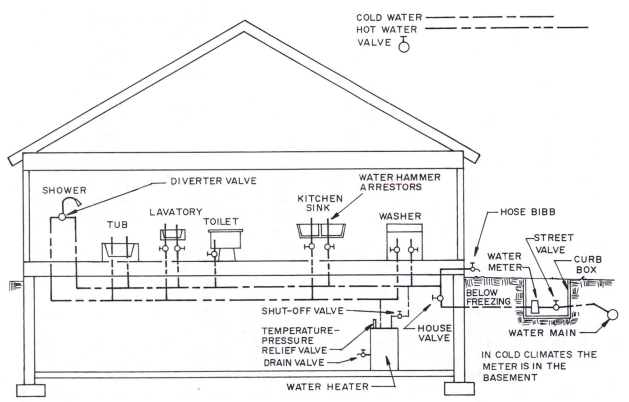

Figure 45.8 Components of a typical residential potable water distribution system.

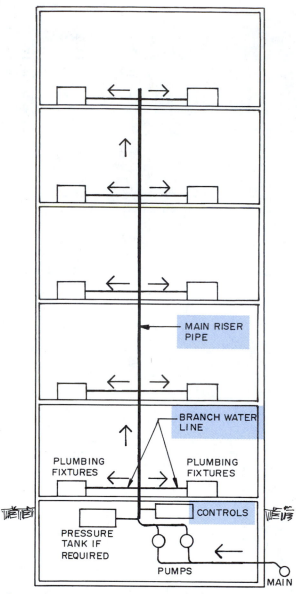

Figure 45.9 This is an up-feed potable water distribution system for a low-rise building.

Up-feed systems are useful on multistory buildings that are not high enough to warrant the expense of a rooftop water storage system. They do have height limitations. A system will have two or more pumps. As the demand for water increases, a second or third pump will come on line, increasing the flow and maintaining the pressure. It should be noted that the supply available from the central water system main should be large enough so that the demands of the building can be met without reducing water service to neighboring buildings. Notice that this system does not have a reserve water supply. A power failure can greatly reduce the water supply available.

Tall multistory buildings use a *down-feed* potable hot and cold water distribution system. The water from the central water system is pumped from a street main or a water storage suction tank to a roof storage tank with one or more pumps. This storage tank holds a reserve for the fire suppression system (see Chapter 44) and a supply of potable water for fixtures. The water flows from a header pipe to down-feed risers from which branch water lines run to fixtures on each floor (Fig. 45.10). Very tall buildings usually are divided into several zones each having its own pumps and storage tanks (Fig. 45.11).

SANITARY PIPING

Piping used for sanitary systems may be cast iron, copper, plastic, lead, glass, and clay. They are used for various **drainage, waste, and vent installations (DWV).**

Cast iron soil pipe and fittings are gray iron castings suitable for installation and service for storm drain, sanitary, waste, and vent piping. *Hub type cast iron* soil pipe and fitting specifications are detailed in ASTM A 74-87. They are available in Extra Heavy and Service classifications in diameters shown in Table 45.6. The pipes and fittings must be coated with a material to protect the surface. *Hubless cast iron* soil pipe and fittings are specified in ASTM A 888-90. Neither type is intended to be used on pressure systems. The selection of the proper design size allows the free air needed for gravity drainage. Both types are available in a wide range of fittings. A few are shown in Figs. 45.12 and 45.13. The hub type joints may be sealed with lead and **oakum** or with a neoprene compression gasket. Hubless connections use a gasket and a stainless steel casing and retaining clamps (Fig. 45.14).

Copper tubing used for sanitary waste systems is classified as DWV (Drainage, Waste, Vent). The available diameters are in Table 45.2. Copper tubing is used in residential, low-rise, and high-rise buildings for all parts of drainage plumbing, including soil and vent stacks and soil, waste, and vent branches. In high-rise buildings expansion must be considered as the system is designed. Changes of 1°F will cause the DWV pipe to change 0.001 in. (0.025 mm) for a 10 ft. section. One way to control thermal movement is to anchor the pipe to the floor as shown in Fig. 45.15. The number of anchors is a design decision but an example is to anchor it at every eight floors if the temperature rise is anticipated to be up to 50°F (10°C).

Joints in copper DWV tubing are made by properly cleaning and fluxing the joining parts and allowing the solder to flow between them by capillary action. The space between the parts is sized to allow the molten solder to flow into the joint (Fig. 45.16). Brazed connections are used if greater strength is needed.

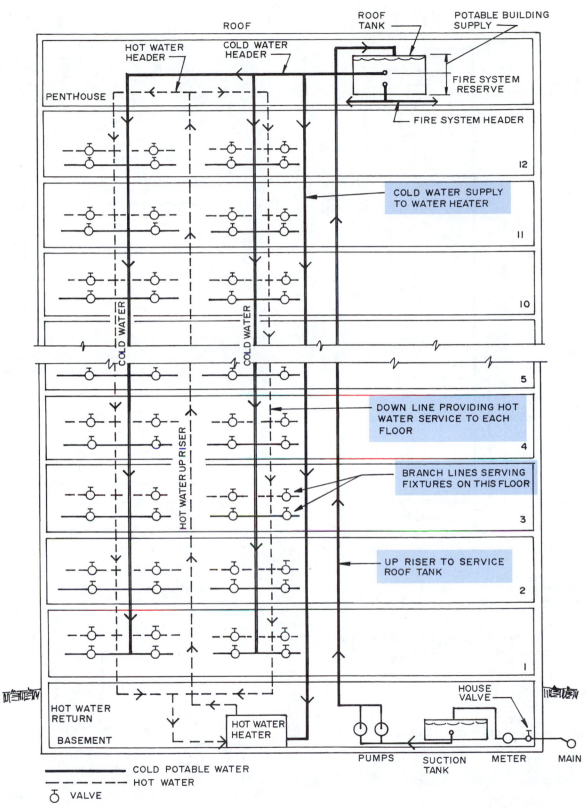

Figure 45.10 A simplified schematic of a down-feed potable hot and cold water distribution system typically used in high-rise buildings.

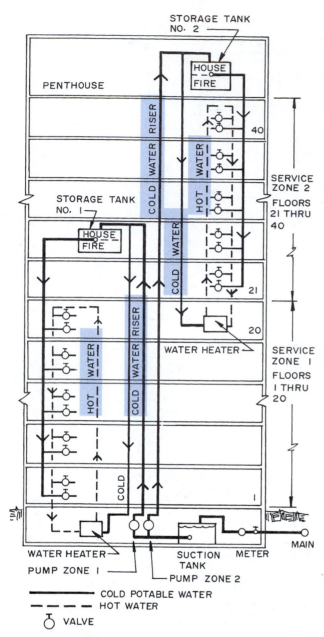

Figure 45.11 A simplified schematic of a two-zone down-feed hot and cold water distribution system.

Table 45.6 Cast Iron Soil Pipe Uses and Sizes

Type	Uses	Nominal Diameter
Hubless soil pipe, standard and extra heavy	Sanitary and storm drains, waste, vents	2–15 in. (50.8–381 mm)
Hub type soil pipe, service and extra heavy	Sanitary and storm drains, waste, vents	2–15 in. (50.8–381 mm)

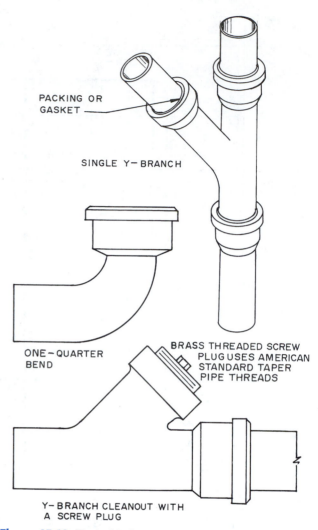

Figure 45.12 Typical hub type cast iron soil pipe fittings.

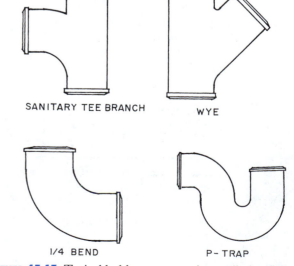

Figure 45.13 Typical hubless type cast iron soil pipe fittings.

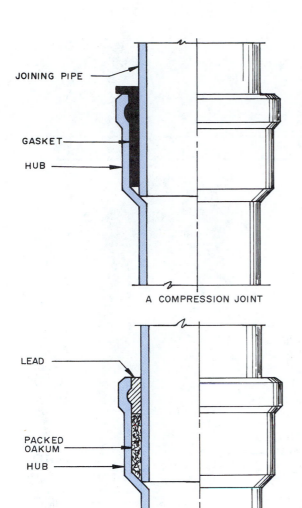

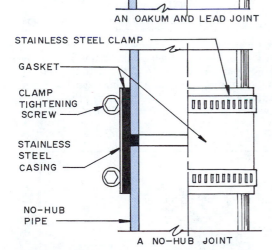

Figure 45.14 Joints used to join hub type and hubless cast iron soil pipe. *(Courtesy Cast Iron Pipe Institute)*

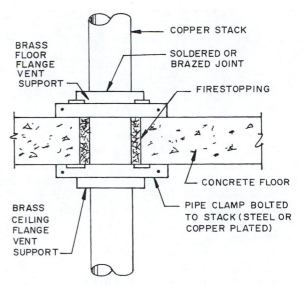

Figure 45.15 Thermal movement in long vertical runs of copper pipe can be controlled by anchoring the pipe to the floor as required by the design of the system. *(Courtesy Copper Development Association)*

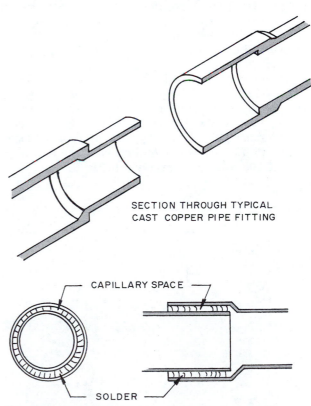

Figure 45.16 A typical soldered joint in copper tubing.

ITEM OF INTEREST

LOW PRESSURE SEWER SYSTEMS

Low pressure sewer systems offer an alternative to conventional gravity sewers and can replace failing septic systems in a community. Since they can move waste horizontally and up a grade under pressure they can be used on building sites on which gravity sewer systems cannot be economically installed or, in some cases, would be impossible to install. Examples include very flat terrain, very hilly or steep sites, and sites with high watertables, poor percolation characteristics, or rock. These systems enable unsuitable sites to be developed economically. The grinder pump can move waste horizontally thousands of feet and raise it 100 ft. (30.5 m) or more (Fig. A). One grinder pump may be able to serve more than one building.

Briefly speaking, the waste from the building flows through conventional size pipe to the grinder pump station, which may be installed outdoors or indoors. The waste is ground into a fine slurry by the grinder pump and discharged under low pressure through a small-diameter pressure piping. The pressure is normally under 60 psig. The slurry moves to a sewer main line, which is also under pressure, and then moves on to the wastewater treatment plant.

The heart of the system is the grinder pump illustrated in Fig. B. This is one of several sizes manufactured by Environment/One Corporation. It is a complete unit with all controls, valves, a pump, and a tank ready to be installed. The various components of the grinder pump are shown in Fig. C.

The inflow pipe comes from the waste system piping of the building and is the same diameter as used for gravity flow systems. However, the discharge pipe is much smaller, usually starting at 1¼ in. (32 mm) diameter. The main sewers running to the waste treatment plant will range from 2 to 6 in. (51 to 152 mm) compared to 8 to 24 in. (203 to 508 mm) for gravity systems (Fig. D).

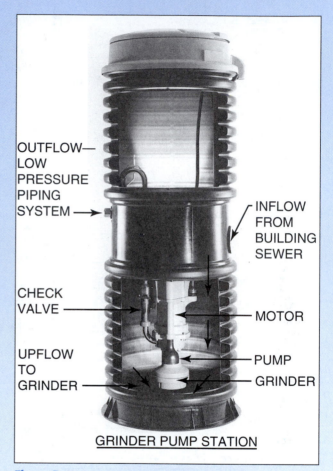

GRINDER PUMP STATION

Figure B This is the grinder pump station that receives the waste from the building sewer. The inflow waste settles to the bottom, is ground into fine particles, is pumped through a pipe, and is discharged under pressure at the outflow pipe. *(Courtesy Environment/One Corporation)*

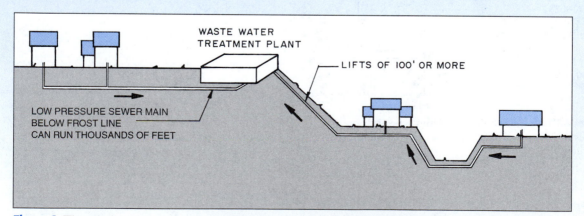

Figure A The grinder pump moves finely ground waste and liquids vertically and horizontally under pressure.

I T E M   O F   I N T E R E S T

LOW PRESSURE SEWER SYSTEMS

Continued

Since the system is under pressure, it can be run horizontally, uphill, or downhill because gravity is not required for flow. This means the small-diameter sewer mains can be installed in shallow ditches placed just below the frost line. This reduces the cost of installation when compared to the deep ditches required for large-diameter gravity flow systems.

The grinder pump can handle many of the items that should not but often do appear in domestic wastewater. For example, plastic, wood, rubber, and light metal objects can be routinely handled without jamming the grinder or clogging the pump or piping system. These can be ground into fine particles and become part of the flowing slurry.

The grinder pump is actuated when the depth of sewage in the tank reaches a predetermined "turn on" level and pumping is continued until the slurry reaches the "turn off" level. This means the pump running time is short, power consumption is low, and long pump life is ensured. The unit has protection against backflow from the discharge lines.

The grinder pump is designed with excess holding capacity in the tank to provide wastewater storage during electric power outages. How long the storage will last depends on the amount of waste flowing and the length of the power outage.

Companies that manufacture these units have extensive engineering design data and assist in designing the system.

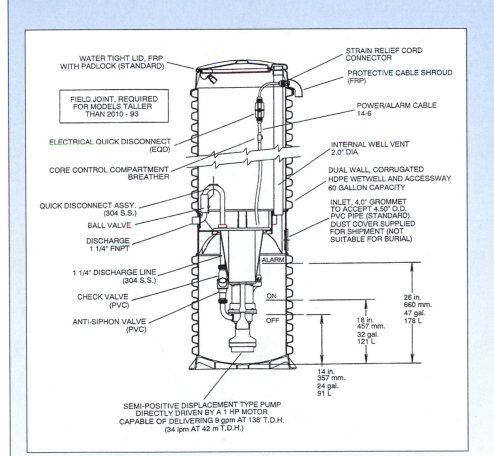

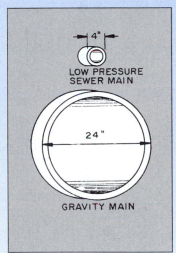

Figure D The low pressure waste disposal system uses sewer lines that are much smaller than those needed by gravity systems.

Figure C This drawing identifies the major components of the grinder pump station. *(Courtesy Environment/One Corporation)*

Plastic pipe suitable for DWV systems include acrylonitrile butadiene styrene (ABS), styrene rubber plastic (SRP), and polyvinyl chloride (PVC). All three can be used for sewer systems, and ABS and PVC can be used for drain, waste, and vent systems. Local plumbing codes should be checked to verify the approved uses.

Plastic pipe and fittings have identification symbols on each piece, as shown in Fig. 45.17. The pipe and fittings are joined with a solvent cement. Some plumbers prefer to clean the end of the pipe and the inside of the fitting by brushing on a coat of priming solvent. After about fifteen seconds a thick coat of solvent cement is applied to the end of the pipe and the inner surface of the fitting (Fig. 45.18). The pipe is inserted and twisted a quarter turn to spread the cement. You must wait at least three minutes before starting the next joint.

Lead and glass drain pipe is used for the disposal of special liquids such as in a chemical plant or research laboratory. Lead and glass resist attack by many chemicals. Lead pipe is joined by welding the joints. Glass pipe is connected with a gasket and stainless steel clamp as shown for hubless pipe in Fig. 45.14.

Clay pipe is used for waste disposal lines outside of a building. Its use is carefully controlled by building codes. The pipe is made from clay and burned in a kiln much like brick. It is available in diameters from 4 to 36 in. (102 to 914 mm) and has a variety of fittings. It uses

a hub joint that is filled with a packing compound. Clay pipe is impervious to acids and alkalines and does not deteriorate with age. It is not used where it might be subjected to shock or loads.

Ascertaining Waste Pipe Sizes

The diameter of the pipes used in a sanitary piping system depends on the amount of waste they are to carry. Plumbing codes typically establish the number of drainage fixture units running through each pipe size (Table 45.7). A **fixture unit** is a measure of the probable discharge into the drainage system by the various plumbing fixtures. It is expressed in units of cubic volume per minute. The value for a particular fixture depends on the volume rate of discharge, the time duration of a single discharge, and the average time between successive discharges. A fixture unit is generally equal to $7\frac{1}{2}$ gallons of flow.

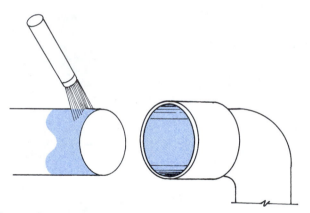

Figure 45.18 Adjoining surfaces of the plastic pipe are cleaned and primed before coating with a solvent cement.

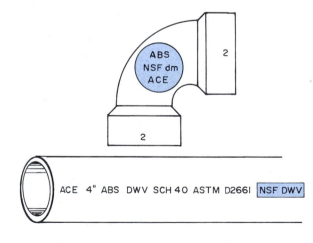

ACE	The name of the manufacturer.
4 in.	Diameter of the pipe.
ABS	Acrylonitrile-Butadiene-Styrene, the material.
DWV	Suitable for drainage waste and vent.
SCH 40	Schedule 40. This identifies the wall thickness of the pipe.
ASTM D2661	"Standards Number" assigned by the American Society for Testing and Materials.
NSF DWV	Tested by the National Sanitation Foundation Testing Laboratory. The pipe meets or exceeds the current standards for sanitary service.

Figure 45.17 Identification symbols used on plastic pipe. *(Courtesy Plastic Pipe Institute)*

Table 45.7 Drainage Fixture Units for Selected Plumbing Fixtures

Plumbing Fixture	Drainage Fixture Unit Value
Automatic clothes dryer	3
Bathtub with or without overhead shower	2
Clinic sink	6
Dental unit	1
Drinking fountain	$\frac{1}{2}$
Dishwasher, domestic	2
Floor drain, 2 in. waste pipe	3
Kitchen sink, $1\frac{1}{2}$ in. trap	2
Kitchen sink with dishwasher	3
Lavatory, $1\frac{1}{4}$ in. waste pipe	1
Shower stall, residential	2
Urinal, stall, washout	4
Water closet, tank operated	4
Water closet, valve operated	6

From *National Standard Plumbing Code*, National Association of Plumbing, Heating, and Cooling Contractors

Sanitary Piping Systems

The **sanitary piping** system removes the water and other waste materials discharged at the various fixtures. The system carries the material down through the building to the **building drain** from which it is discharged through the building sewer and on to the public **sanitary sewer** system or septic tank. The waste material flows from the building at a level below the lowest fixture and flows by gravity through the **building sewer.** If the public sewer is not below the sloped building sewer, waste will not drain and pumps will be used to lift the waste and move it to the public sewer.

Pipe diameters for sanitary piping systems are sized to carry the anticipated waste material flow rapidly without clogging the pipes or creating annoying water noises. Pipe diameters also are selected to produce minimum pressure variations where fixtures connect to **waste pipes** and waste pipes connect to **branch** soil pipes and where branch soil pipes connect to the **soil stack.**

The system relies on gravity flow, so it is not under pressure. The effluent flow increases in speed as it proceeds down the piping system. Friction within the pipe limits the speed and is another factor in sizing pipe.

Plumbing codes specify the required slope for horizontal piping. The requirement varies with pipe diameter but is typically from $\frac{1}{8}$ to $\frac{1}{2}$ in. (3 to 12.7 mm) per foot of run. Flow velocity increases with the amount of slope, and high velocities in horizontal pipes can increase siphonage. The capacity of a pipe should be increased by using a larger diameter pipe rather than by increasing the slope.

The sanitary drainage system receives waste matter that may be difficult for private or public waste treatment plants to process. Such waste will often clog the pipes in the sanitary drainage system.

Interceptors are used to catch large waste material near the point of origin, catching it before it moves very far into the system. Manufacturers have many different devices available to catch grease, hair, oil, glass and plastic grindings, plaster, and other undesirable materials. The interceptor is emptied regularly.

Indirect Wastes

Indirect waste such as wastes from food-handling equipment, dishwashers, sterilizers, and commercial laundries are discharged with an indirect waste pipe. The waste from the indirect waste pipe is not discharged directly into the building sanitary piping system but into a fixture connected to the building sanitary system (Fig. 45.19). This provides an **air gap** between the two systems. An air gap is used to prevent one system from accidentally backing up into another.

Toxic and corrosive waste discharges must be automatically diluted with water or chemically neutralized before being introduced into the building sanitary system.

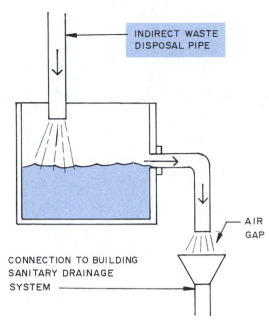

Figure 45.19 Indirect waste is transferred to the building sanitary system by transferring it from a vessel containing an air trap.

RESIDENTIAL SANITARY PIPING

A typical simple sanitary piping system for a residence is shown in Fig. 45.20. The waste from the fixtures drains through waste pipes or branch soil pipes into a soil stack. Each waste pipe is sized to carry the flow from a fixture. The branch soil pipe is sized to carry the flow from all the fixtures flowing into it. The soil stock extends below the building and connects to the building drain. The building sewer connects to the central sewer system or a septic tank. Each fixture has a **vent** pipe connected to a **vent stack** running through the roof. The vent pipe keeps the water in the **trap** of the fixture under atmospheric pressure, thus eliminating the chance of its being siphoned out when another fixture is used (Fig. 45.21). For example, if the waste system were a closed installation, a trap in a lavatory could have the water siphoned out when a connecting toilet is flushed. Some fixtures, such as toilets, have a trap built into the unit. The house drain will have one or more **cleanouts** (C.O.). A cleanout is a pipe fitting with a removable plug through which an auger may be run to dislodge an obstruction in the pipe.

Various types of traps are used (Fig. 45.22). Lavatories typically have a p-trap. A tub may have a drum trap. To keep plumbing costs down the designer attempts to place fixtures on each side of a wall. This reduces the piping needed and eliminates long horizontal runs to reach out-of-the-way fixtures (Fig. 45.23). Typically, the wall containing the plumbing uses at least 2 × 6 in. studs.

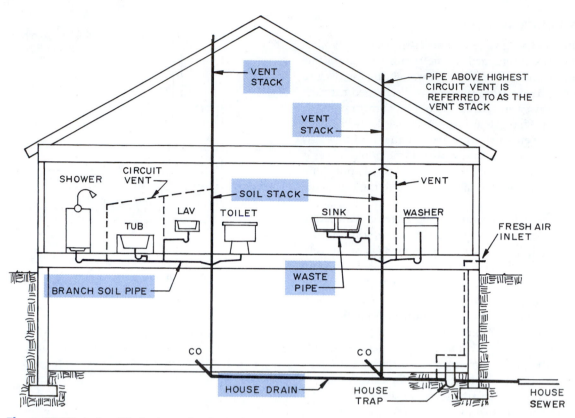

Figure 45.20 A simplified schematic of a typical residential sanitary piping system.

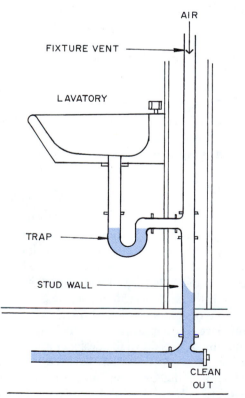

Figure 45.21 Fixtures are vented to keep water in the trap under atmospheric pressure so the water in it is not siphoned out when a nearby fixture discharges waste in the system.

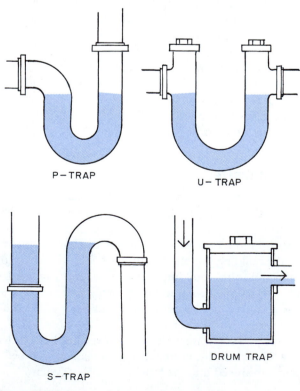

Figure 45.22 Some of the commonly used traps.

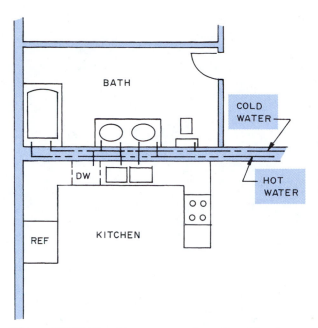

Figure 45.23 Plumbing costs can be reduced by clustering the plumbing fixtures.

MULTISTORY BUILDING SANITARY PIPING SYSTEMS

In multistory buildings it is common to design plumbing **chases** on each floor directly above the other to run water and waste disposal systems the height of the building (Fig. 45.24). Heating, air-conditioning, and fire suppressing systems can also use these chases. This consolidates the plumbing, reduces costs, and makes the planning of the uses of the space on each floor more flexible. Plumbing fixtures are typically located next to the sides of the vertical chase. However, horizontal runs to locations within the floor area can be made from the chase. In other cases, pipe **risers** are located inside a space around a structural column (Fig. 45.25). This enables fixtures to be placed in various locations within the building.

A schematic of a small multistory building sanitary piping system is shown in Fig. 45.26. The fixtures for the restrooms with the related piping are detailed on the second floor with notes indicating that they are the same on all other floors. The soil stack, vent stack, and building drain are shown for the entire building. Notice the venting of multiple fixtures to horizontal circuit vents that connect to the vent stack. In some cases the fixture waste pipes connect directly to the soil stack, but in other situations they feed into a horizontal branch soil pipe that connects to the vertical soil stack. The circuit vents, waste pipes, and branch soil pipes slope toward the soil stack, providing gravity flow. The pipe sizes are calculated depending on the flow, as explained earlier.

The system in Fig. 45.26 requires two sets of pipes, a vent system, and a waste disposal system. This two-pipe

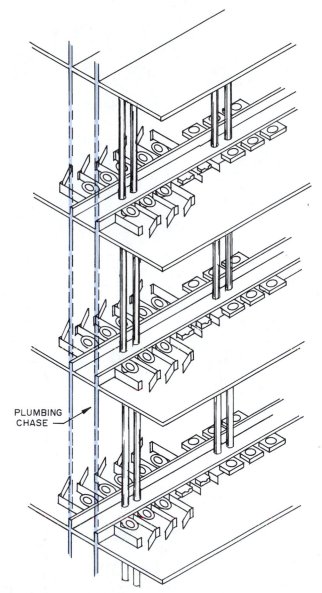

Figure 45.24 A plumbing chase is used to stack fixtures on floors above each other.

system is most commonly used in the United States. The venting controls the possibility of siphoning water from the traps, which would allow sewer gas to enter the building. Another system, the Sovent system, is used in Europe and Africa and has been approved for use in the United States by some of the national plumbing codes.

The Sovent System

The Sovent system is a vertical cast iron drainage and waste system conveying wastes from the upper levels of a building to the base of each Sovent stack. The stack begins just above the lowest deaerator fitting and continues up to just above the highest fixture connection (Fig. 45.27). This includes horizontal stack offsets located at intermediate levels. The stack uses traditional fittings and pipe made

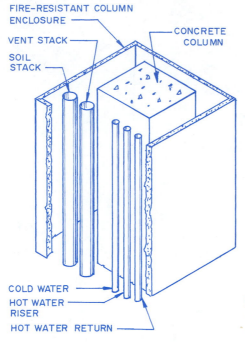

FIRE-RESISTANT COLUMN
ENCLOSURE

CONCRETE
COLUMN

VENT STACK

SOIL
STACK

COLD WATER

HOT WATER
RISER

HOT WATER RETURN

Figure 45.25 A column enclosure can be used to run pipe risers from floor to floor.

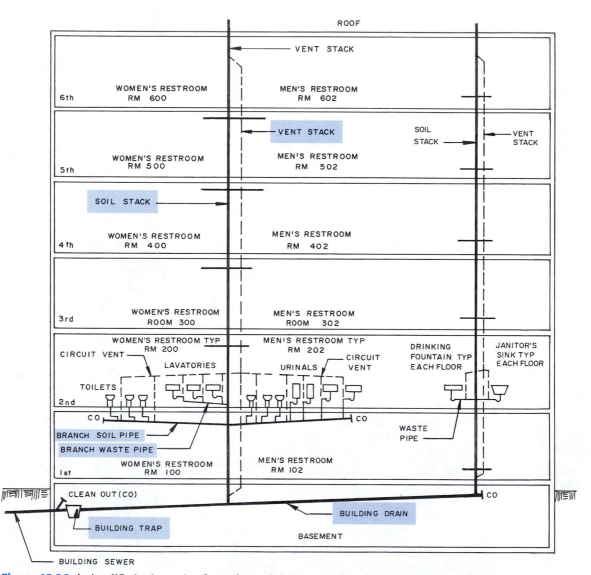

ROOF

VENT STACK

6th — WOMEN'S RESTROOM RM 600 — MEN'S RESTROOM RM 602

VENT STACK

SOIL STACK

VENT STACK

5th — WOMEN'S RESTROOM RM 500 — MEN'S RESTROOM RM 502

SOIL STACK

4th — WOMEN'S RESTROOM RM 400 — MEN'S RESTROOM RM 402

3rd — WOMEN'S RESTROOM ROOM 300 — MEN'S RESTROOM ROOM 302

WOMEN'S RESTROOM TYP RM 200

CIRCUIT VENT

LAVATORIES

MEN'S RESTROOM TYP RM 202

CIRCUIT VENT

URINALS

DRINKING FOUNTAIN TYP EACH FLOOR

JANITOR'S SINK TYP EACH FLOOR

TOILETS

2nd

CO

CO

BRANCH SOIL PIPE

BRANCH WASTE PIPE

WASTE PIPE

1st — WOMEN'S RESTROOM RM 100 — MEN'S RESTROOM RM 102

CLEAN OUT (CO)

CO

BUILDING DRAIN

BUILDING TRAP

BASEMENT

BUILDING SEWER

Figure 45.26 A simplified schematic of a sanitary piping system for a low-rise multistory building.

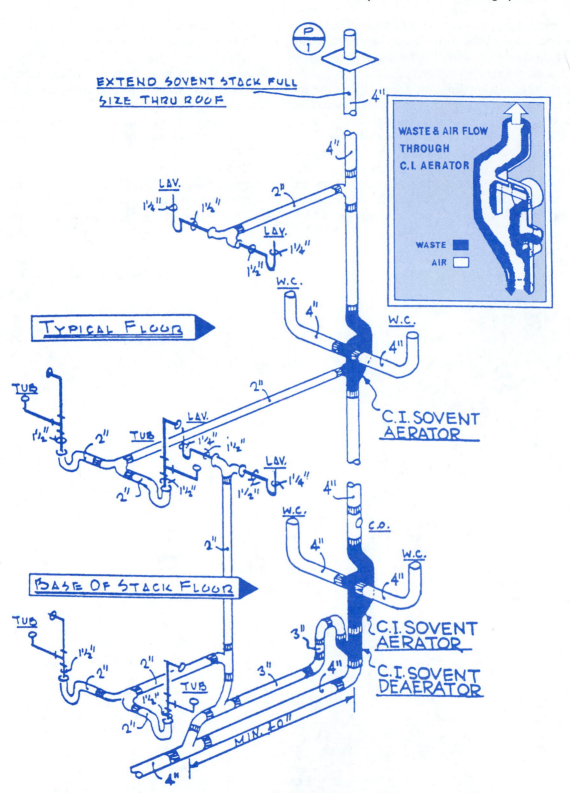

Figure 45.27 A vertical waste stack using cast iron Sovent aerators and deaerators and standard code approved drain, waste, vent (DWV) piping and fittings. *(Courtesy Conine Manufacturing Co., Inc.)*

from approved drain, waste, vent (DWV) materials. The choice is controlled by the manufacturer and codes. The Sovent stack penetrates the roof to the atmosphere much like the vent stack in traditional systems.

Waste flowing in a vertical pipe clings to the interior wall and moves down the pipe in a swirling motion. This leaves an open airway in the center needed for flow to continue. As the falling waste gathers speed it meets air resistance and flattens out and may form a complete blockage of the pipe. This blockage is eliminated by the Sovent aerator fitting (Fig. 45.28). Aerator fittings are available in many DWV materials and have no moving parts. The waste enters the offset chamber, which

breaks any attempt of the waste to form a plug and reduces its velocity. As the waste exits the offset chamber it clings to the interior pipe surfaces, leaving an open center air space. It then enters the mixing chamber. Here the incoming waste from horizontal branches hits the baffle, drops, and is mixed with the down-falling waste (Fig. 45.29). The waste flows down the stack to the lowest level that contains a deaerator fitting.

The deaerator fitting is designed to handle the pressure fluctuations that occur when falling water suddenly turns horizontal (Fig. 45.30). Here the velocity is slowed while additional waste continues to pile behind it, causing a wave (hydraulic jump) to form and possible develop-

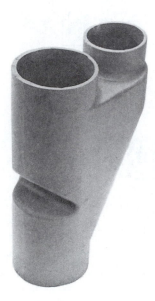

Figure 45.28 A cast iron Sovent aerator. *(Courtesy Conine Manufacturing Co., Inc.)*

STACK INLET

CROSS FLOW BAFFLE

VENT APERTURE

SEPARATION BAFFLE

OFFSET CHAMBER

MIXING CHAMBER

BRANCH INLETS

STACK OUTLET

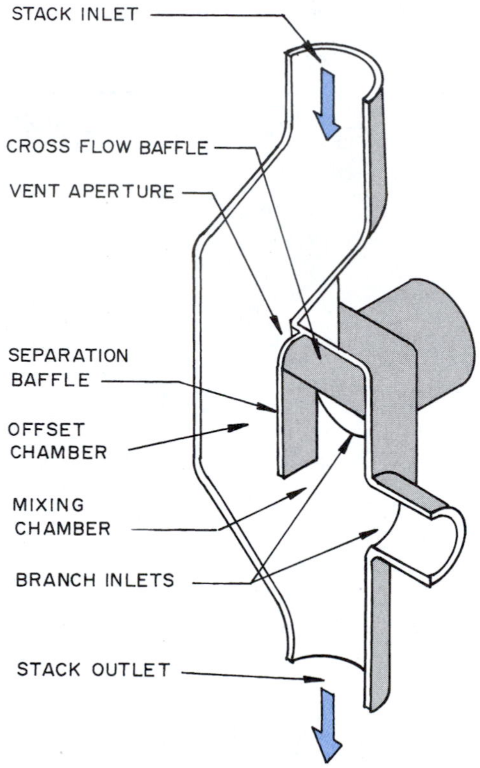

Figure 45.29 The cast iron Sovent aerator offsets the waste flow around the horizontal branch inlets and breaks up the hydraulic plug that occurs in unrestricted vertical flow and reduces the velocity of the falling waste. *(Courtesy Conine Manufacturing Co., Inc.)*

Figure 45.30 A cast iron Sovent deaerator. *(Courtesy Conine Manufacturing Co., Inc.)*

ment of positive and negative pressures. The water strikes the nosepiece, which reduces its velocity and allows the air and waste to separate and continue to flow down the horizontal pipe. The pressure relief line provides an outlet for pressures developed and eliminates any chance of the system's developing a vacuum, which would pull the water out of the traps at each fixture (Fig. 45.31).

NONPOTABLE WATER SYSTEMS

National awareness of the need for water conservation has produced a need for new technologies and alternate methods of handling water and wastewater. In a building in which large amounts of water are used, a considerable flow of wastewater is produced. Recyling this flow helps conserve the water supply and reduces the load on wastewater disposal facilities.

A schematic for an in-building wastewater treatment and recycling facility is shown in Fig. 45.32. Potable water is delivered to the drinking fountains, lavatories, and other fixtures where it is needed (1). This is a completely separate system from the nonpotable system. The toilet uses **nonpotable water.** The wastewater enters a pretreatment trash trap (2) and a sump (3) provides temporary storage in case of a mechanical failure of the system. Usually this will hold several days' supply of wastewater. A tank truck

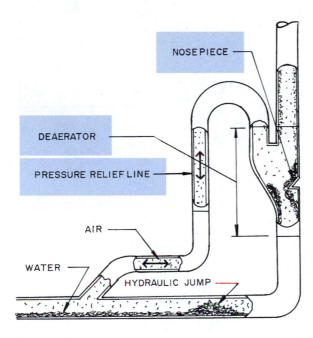

Figure 45.31 The cast iron Sovent deaerator fittings relieve excessive pressures that result from vertical flows changing to horizontal flows. (*Courtesy Conine Manufacturing Co., Inc.*)

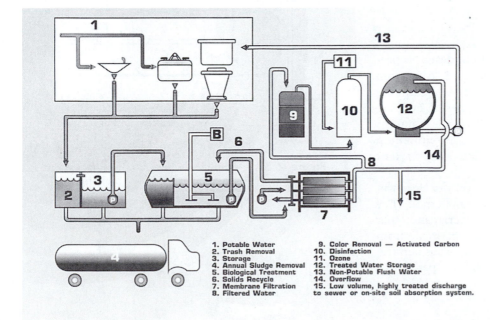

1. Potable Water
2. Trash Removal
3. Storage
4. Annual Sludge Removal
5. Biological Treatment
6. Solids Recycle
7. Membrane Filtration
8. Filtered Water
9. Color Removal — Activated Carbon
10. Disinfection
11. Ozone
12. Treated Water Storage
13. Non-Potable Flush Water
14. Overflow
15. Low volume, highly treated discharge to sewer or on-site soil absorption system.

Figure 45.32 A schematic for an in-building wastewater treatment and recycling facility. (*Courtesy SFA Enterprises, Inc., and John Irwin, Thetford Systems, Inc.*)

can be used to haul away excess wastewater if necessary. The wastewater proceeds automatically through the various processes, including biological treatment, filtering, color removal, disinfection, and ozone treatment. This produces treated water, which is kept in a storage reservoir (12). The water is moved under pressure to the piping serving the toilets (13).

It is vital that the potable and nonpotable systems are designed to be kept separate with no chance of accidentally becoming connected. All piping must be clearly marked "nonpotable water supply" or color coded and marked with colored tape. Valves, wall outlets, and other possible places of attachment are marked with warning tags.

The National Sanitation Foundation (NSF) has established a standard for recycled water quality in the publication Certification Standard No. 1.

ROOF DRAINAGE

Areas such as flat roofs or balconies on multistory buildings collect water during storms and require a drainage system. This system is separate from the sanitary sewer and must drain into the public storm sewer system (Fig. 45.33).

Flat roofs and other such surfaces usually are sloped toward interior roof drains. A typical example is shown in Fig. 45.34. The roof drains are raised cast iron, plastic, or aluminum strainers set in the center of a sloped area (Fig. 45.35). The water drains through downpipes, which may be hidden internally or mounted on the exterior of the building. The downpipes connect to a storm drain below the building, which connects to the storm sewer outside the building. Another technique is to slope the roof toward **scuppers** and drain the surface water through them into downpipes and on to the storm sewer (Fig. 45.36).

PLUMBING FIXTURES

One decision to be made as a plumbing design is prepared is how many fixtures are required. Minimum requirements are specified in the various plumbing codes. An example of some selected types of building occupancy requirements as specified by the Uniform Plumbing Code is in Table 45.8. These are minimum recommendations and known conditions may warrant the use of additional fixtures.

Plumbing fixtures require a steady supply of clean water to assist with the discharge of waste materials. They are under constant wear from water, bacteria, and other harmful elements. Therefore, they must be made from durable materials having a smooth, nonporous surface. Typical materials include stainless steel, copper, brass,

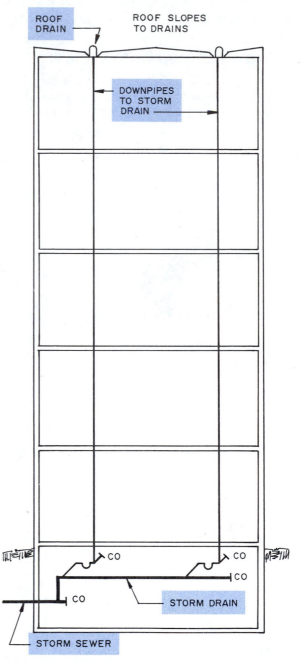

Figure 45.33 A simplified schematic showing a storm drain system for a roof on a multistory building.

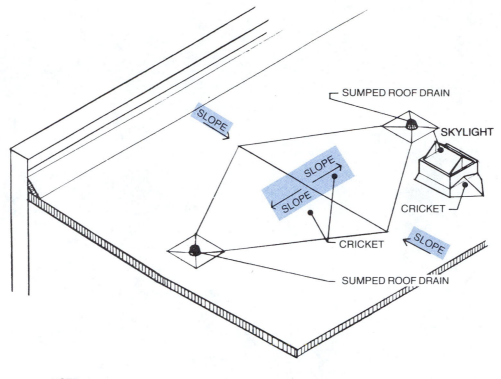

NOTE:
CRICKETS SHOULD BE LOCATED IN LOW VALLEYS BETWEEN ROOF DRAINS AND ON
THE HIGH SIDE OF ALL CURBS.

Figure 45.34 Flat roofs are often sloped toward interior storm drains. Crickets are used to direct the water toward the drains. *(Courtesy National Roofing Contractors Association)*

enameled cast iron, vitreous china, molded plastics, gel-coated fiberglass, and acrylic-faced fiberglass.

The selection of materials and fixture styles is usually made by the owner with the assistance of the architect and the plumbing supply dealer. They can give suggestions based on experience and technical expertise to assist the owner.

Special fixtures and design specifications are available in the publication, Americans with Disabilities Act Accessibility Guidelines. It records heights, spacings, grab bar requirements, and other features required for all aspects of a building's design to accommodate those with physical disabilities. Recommendations relating to fixture placement are shown in Figs. 45.37 through 45.42.

Water closets (toilets) are available as wall-hung or floor-mounted units. Floor-mounted tank type toilets are common in residential buildings (Fig. 45.43). Wall-hung water closets are widely used in industrial and public restrooms because it is easier to keep the floor clean around them. They do require special wall construction to carry the heavy weight involved (Fig. 45.44). Residential water closets typically use a flush tank while commercial establishments use a high-pressure flushing system (Fig. 45.45). Water closets have various methods for flushing. Some are quiet and use less water than others.

Urinals are wall-hung units used in men's restrooms in public facilities to reduce the number of water closets required. They are less expensive to install and take little space (Fig. 45.46). Some types have privacy shields between units. They may have a manual high-pressure flush valve or be connected to an automatic high-pressure flushing system. Sometimes one is set lower than the others for use by children.

Lavatories in residences are typically mounted in a base cabinet. The lavatory may be set into a top covered with plastic laminate or ceramic tile (Fig. 45.47). Others have a molded plastic top with the lavatory bowl and top as one integral piece (Fig. 45.48). This reduces the problems that occur around the stainless steel edge of those set into the top. Another popular lavatory is a pedestal type. The lavatory is actually wall hung and the pedestal covers up the plumbing below (Fig. 45.49). Wall-hung residential lavatories often have decorative metal legs instead of a pedestal.

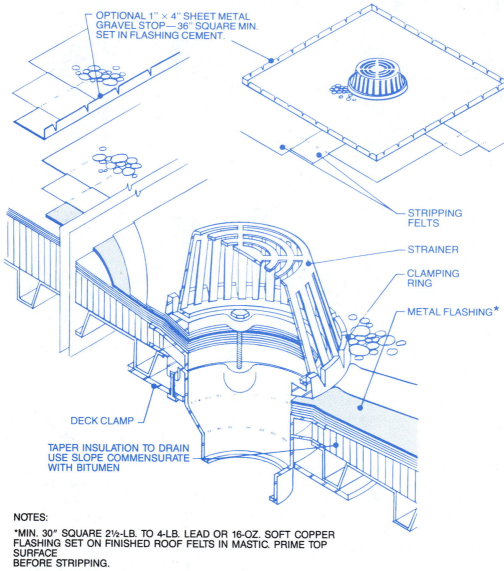

OPTIONAL 1" × 4" SHEET METAL
GRAVEL STOP—36" SQUARE MIN.
SET IN FLASHING CEMENT.

STRIPPING
FELTS

STRAINER

CLAMPING
RING

METAL FLASHING*

DECK CLAMP

TAPER INSULATION TO DRAIN
USE SLOPE COMMENSURATE
WITH BITUMEN

NOTES:

*MIN. 30" SQUARE 2½-LB. TO 4-LB. LEAD OR 16-OZ. SOFT COPPER
FLASHING SET ON FINISHED ROOF FELTS IN MASTIC. PRIME TOP
SURFACE
BEFORE STRIPPING.

MEMBRANE PLIES, METAL FLASHING, AND FLASH-IN PLIES EXTEND
UNDER
CLAMPING RING.

STRIPPING FELTS—EXTEND 4" AND 6" BEHOND EDGE OF FLASHING
SHEET, BUT NOT BEYOND EDGE OF SUMP.

THE USE OF METAL DECK SUMP PANS IS NOT RECOMMENDED.

Figure 45.35 A strainer protects the storm-water drain used on flat roofs. *(Courtesy National Roofing Contractors Association)*

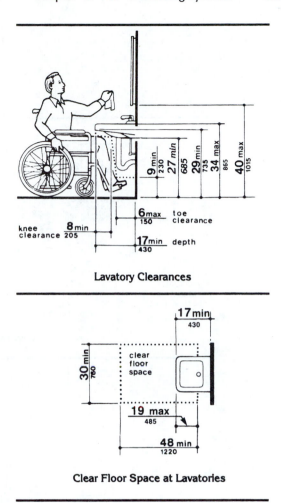

Lavatory Clearances

Clear Floor Space at Lavatories

Figure 45.37 Lavatories must be mounted with the top surface not more than 34 in. (865 mm) above the floor. They must be positioned so a person in a wheelchair will have the required knee room. Any exposed pipes must be located out of the way and any sharp edges must be covered. *(From Americans with Disabilities Act, U.S. Architectural and Transportation Barriers Compliance Board, Washington, D.C.)*

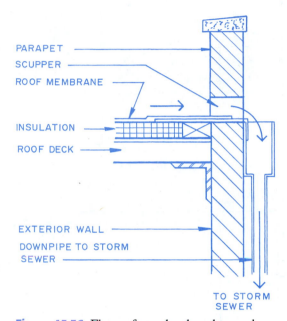

PARAPET
SCUPPER
ROOF MEMBRANE
INSULATION
ROOF DECK
EXTERIOR WALL
DOWNPIPE TO STORM SEWER
TO STORM SEWER

Figure 45.36 Flat roofs can be sloped toward scuppers in the exterior wall. The scupper permits roof water to flow into external downpipes.

Table 45.8 Minimum Plumbing Facilities

Type of Building or Occupancy	Water Closets	Urinals	Lavatories	Bathtubs or Showers	Drinking Fountains
Assembly places (theaters, etc.)	1 per 1–15 2 per 16–35 3 per 36–55 Over 55, 1 per additional 40	1 per 50 males	1 per 40	—	1 per 75

Hospitals
  Individual room
  Ward room
	
	1 per room
	1 per 8 patients
		
		—
		—
			
			1 per room
			1 per 10 patients
				
				1 per room
				1 per 20 patients
					
					—
					—

Restaurant	1 per 1–50 2 per 51–150 3 per 151–300	1 per 1–150 males	1 per 1–150 2 per 151–200 3 per 201–400	—	—
Worship place, assembly area	1 per 300 males 1 per 150 females	1 per 300 males	1 per toilet room	—	1 per 75
Worship place, educational and activities	1 per 250 males 1 per 125 females	1 per 250 males	1 per toilet room	—	1 per 75

Reprinted from the *Uniform Plumbing Code*™ with the permission of the International Association of Plumbing and Mechanical Officials © copyright 1994.

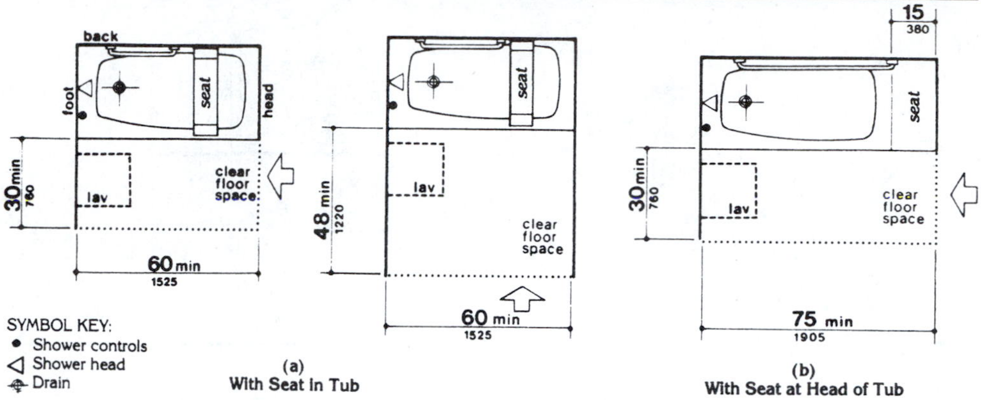

Clear Floor Space at Bathtubs

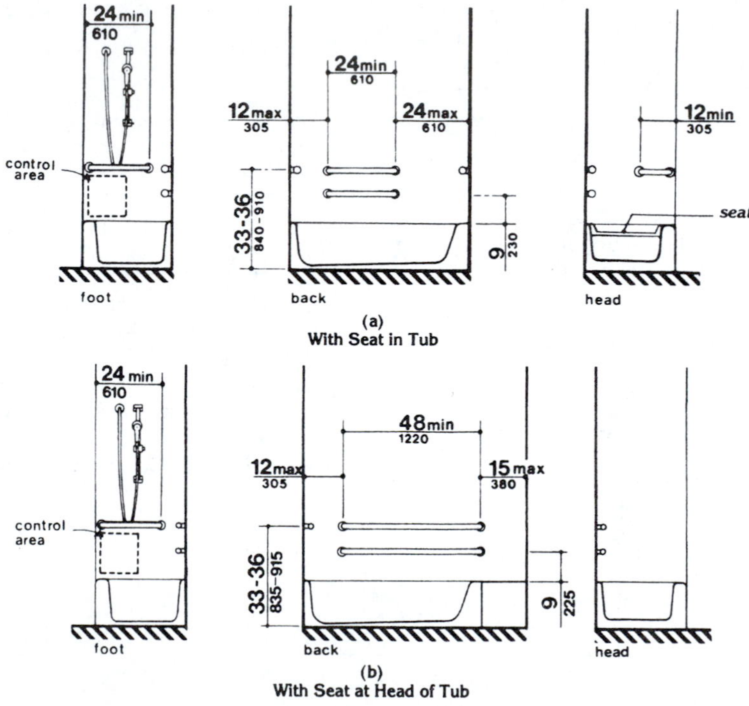

(a) With Seat in Tub

(b) With Seat at Head of Tub

SYMBOL KEY:
- Shower controls
- ◁ Shower head
- Drain

(a) With Seat in Tub

(b) With Seat at Head of Tub

Grab Bars at Bathtubs

Figure 45.38 Bathtubs should not be more than 9 in. (230 mm) high. They may have seats. Shower spray units must have a hose at least 60 in. (1525 mm) long and controls must be located so they are easily reached. Multiple grab bars are required. The tub may have an enclosure but it must not obstruct the controls or movement from a wheelchair into the tub. *(From Americans with Disabilities Act, U.S. Architecture and Transportation Barriers Compliance Board, Washington, D.C.)*

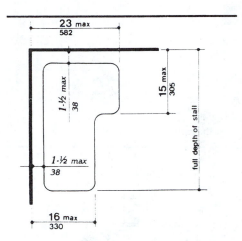

Shower Seat Design

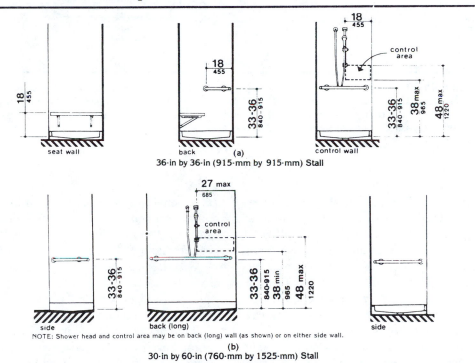

Grab Bars at Shower Stalls

Figure 45.39 Shower stalls 36 × 36 in. (915 × 915 mm) must have a seat and properly placed grab bars. Long stalls need not have a seat but must have grab bars. Controls must be placed so they can be reached from a seated position. *(From Americans with Disabilities Act, U.S. Architecture and Transportation Barriers Compliance Board, Washington, D.C.)*

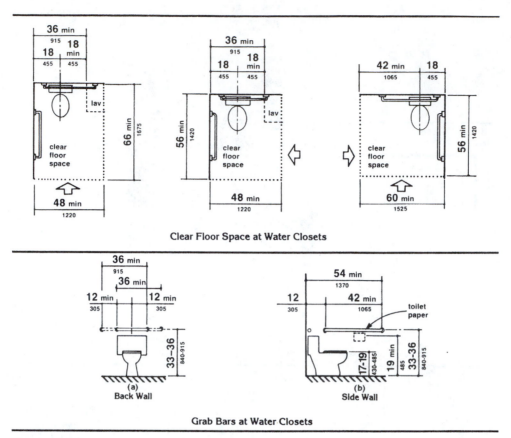

Clear Floor Space at Water Closets

Grab Bars at Water Closets

Figure 45.40 Water closets not in stalls require a minimum of 48 in. (1220 mm) clear of other items. The water closet must be 17 to 19 in. (430 and 485 mm) high. Grab bars to the side and rear are required. *(From Americans with Disabilities Act, U.S. Architectural and Transportation Barriers Compliance Board, Washington, D.C.)*

Lavatories in public facilities are typically wall-hung units, although some used lavatories set into a wall-hung countertop without a cabinet base below (Fig. 45.50). These usually have self-closing faucets that prevent someone from letting the water run after they leave. One type of faucet has an infrared control that turns on the water when hands are placed below the faucet and turns it off when the hands are removed. This saves on water use and reduces the cost of hot water.

Wash fountains are used in restrooms and dressing rooms in which a large number of people need to wash up at the same time, such as in an industrial plant. Wash fountains are typically half-round wall-hung units or round freestanding units. Large units can accommodate up to eight people at one time (Fig. 45.51).

Kitchen sinks are available in a range of types with the most typical being some form of two-bowl unit. How-ever, single- and triple-bowl units are available. One bowl is for general use and the second bowl contains the garbage disposal. Sinks are installed in plastic laminate or ceramic tile countertops (Fig. 45.52).

Service sinks are used by janitors to clean mops and for other cleaning activities. They are usually wall hung and deep. Mop service basins are floor-mounted units about 1 ft. (305 mm) high (Fig. 45.53).

Bathtubs are available in a range of sizes and designs. Some have two sides closed and fit in a corner. Others have one side closed and fit between end walls (Fig. 45.54). Square and rectangular units are available. Whirlpool tubs have pumps that circulate the water (Fig. 45.55). Some bathtubs made from gel-coated fiberglass have the tub and wall enclosure formed as a single unit. This leaves no cracks to form mold and the unit is easy to clean (Fig. 45.56).

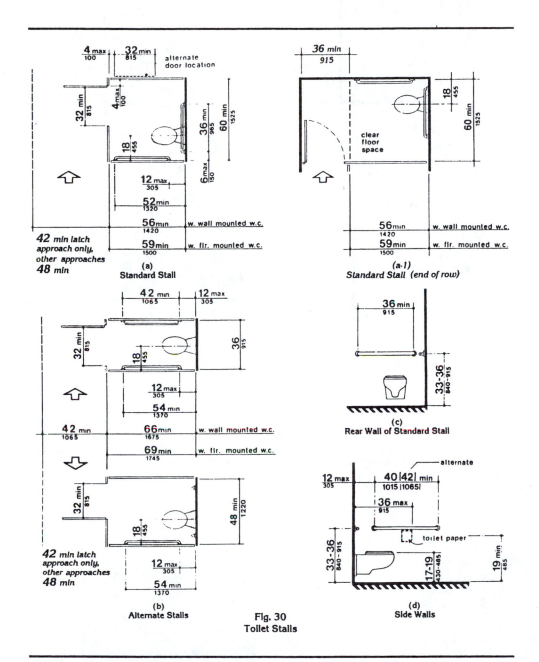

Figure 45.41 Water closets in stalls must provide more room than those without stalls. Space must be left for the wheelchair to enter and allow the door to close. Grab bars are required. *(From Americans with Disabilities Act, U.S. Architectural and Transportation Barriers Compliance Board, Washington, D.C.)*

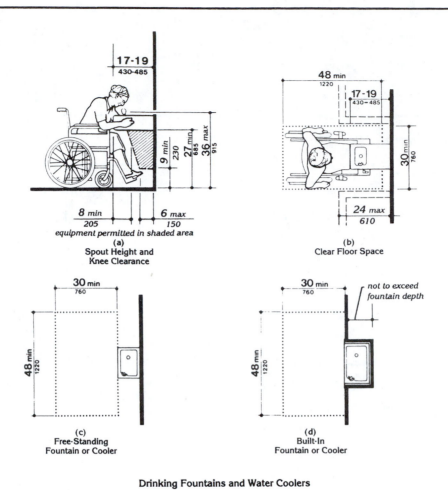

17-19
430-485

9 min
230

27 min
685

36 max
915

8 min
205

6 max
150

equipment permitted in shaded area

(a)
Spout Height and
Knee Clearance

48 min
1220

17-19
430-485

30 min
760

24 max
610

(b)
Clear Floor Space

30 min
760

48 min
1220

(c)
Free-Standing
Fountain or Cooler

30 min
760

not to exceed
fountain depth

48 min
1220

(d)
Built-In
Fountain or Cooler

Drinking Fountains and Water Coolers

Figure 45.42 Drinking fountains must have spouts no higher than 36 in. (915 mm) above the floor. The water flow must be projected so a person in a wheelchair can move below it. Controls must be front mounted or side mounted near the front edge.. *(From Americans with Disabilities Act, U.S. Architectural and Transportation Barriers Compliance Board, Washington, D.C.)*

Figure 45.43 A floor-mounted toilet typically used in residential construction.

Figure 45.44 This wall-hung toilet has an exposed flush valve. *(Courtesy Universal-Rundle Corporation)*

Figure 45.47 This bathroom lavatory is set into the countertop. *(Courtesy Universal-Rundle Corporation)*

Figure 45.45 A floor-mounted high-pressure toilet used in public restrooms. *(Courtesy Universal-Rundle Corporation)*

Figure 45.48 This molded plastic bathroom lavatory has the top and bowl cast as a single unit.

Figure 45.46 A wall-hung urinal with an exposed flush valve.

Figure 45.49 A pedestal lavatory is decorative and takes up very little space.

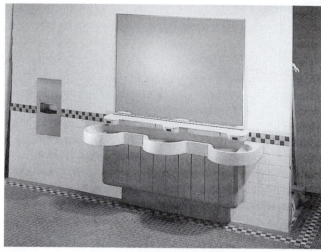

Figure 45.50 This three-basin integral cast lavatory has in-frared water control and allows access by the physically handicapped. Notice the relationship of the mirror and towel dispenser allowing use by the handicapped. *(Courtesy Bradley Corporation)*

Figure 45.52 A typical double-bowl stainless steel kitchen sink. The disposal is mounted below the bowl on the left.

Figure 45.51 A round freestanding stainless steel wash foun-tain. It measures 28 in. (711 mm) from the floor to the top of the rim. *(Courtesy Bradley Corporation)*

Figure 45.53 A deep service sink is used in residential utility rooms and janitor closets in commercial build-ings.

Figure 45.54 This bathtub fits between end walls and has one side exposed. *(Courtesy Universal-Rundle Corporation)*

Figure 45.55 This whirlpool bathtub is recessed in the floor. Provision must be made to install the pump below the floor. *(Courtesy Universal-Rundle Corporation)*

Figure 45.56 This fiberglass molded tub, shower, and wall-covering unit has smooth, easy to keep clean junctions between the various sections. *(Courtesy Universal-Rundle Corporation)*

Showers widely used are gel-coated fiberglass units that are in one piece. They are free from any joints, making them one solid, easy-to-clean unit (Fig. 45.57). Other types are available that have a cast terrazzo base. Walls are covered with ceramic tile or have a prefabricated enclosure made from galvanized-bonderized steel with an enamel finish or molded plastic panels.

Various types of *drinking fountains* are available. Some are wall hung and protrude into the room. Others are recessed or semirecessed into the wall. Freestanding and some recessed drinking fountains have a water cooling system in the base that produces temperature controlled water (Fig. 45.58). Units that provide access for the handicapped are wall mounted and have an electrically

Figure 45.57 Gel-coated fiberglass showers are lightweight, easy to install, and easy to keep clean.

Recessed drinking fountain

Semirecessed drinking fountain

Figure 45.58 Drinking fountains are available as semirecessed, recessed units, and freestanding. *(Courtesy EBCO Manufacturing Company)*

Free standing drinking fountain

activated push-button valve on the front that requires only a light touch (Fig. 45.59). Some pedestal types are designed for outdoor use and may have a foot-operated valve that is frost-proof. Stainless steel is the major material used on the surfaces exposed to water. Drinking fountains require a sanitary drain as well as a source of potable water.

Figure 45.59 This drinking fountain is designed to accommodate people in wheelchairs. *(Courtesy EBCO Manufacturing Company)*

REVIEW QUESTIONS

1. What types of pipe are used to carry natural gas?
2. How are copper pipe joints secured?
3. How can the expansion of copper pipe used to carry hot water be handled to prevent breakage?
4. How are the various types of copper water tubing identified?
5. What resins are used to produce plastic pipe?
6. What methods are used to join various types of plastic pipe?
7. Where would glass pipe be used?
8. Where would you use a relief valve?
9. Why are pipes insulated?
10. What processes are used at a water purification plant to produce potable water?
11. What is meant by an up-feed potable water distribution system within a building?
12. What type of potable water distribution system is used in large multistory buildings?
13. What materials are used for pipes in sanitary piping?
14. How does the plumbing contractor know what slope is required on sanitary piping installations?
15. What are indirect wastes?
16. How does the Sovent drainage and waste system operate without the venting piping used in traditional systems?
17. What uses are made of nonpotable water within a building?

KEY TERMS

air gap An unobstructed vertical distance between the lowest opening of pipe that supplies a plumbing fixture and the level at which the fixture will overflow.

branch A horizontal run of waste piping that carries waste material to a vertical riser.

building drain The part of the horizontal waste piping system that receives waste from soil, waste, and other drainage pipes within the building and moves it to the building sewer outside the building.

building sewer The pipe that receives the waste from the building drain and carries it to the public sewer or septic tank.

chase A recessed area in a wall for holding pipes and conduit passing vertically between floors.

cleanout Opening in the waste piping system that permits cleaning obstructions within the pipe.

DWV Drain, waste, vent pipe installation.

fixture unit A measure of the probable discharge into the waste disposal system from the various plumbing fixtures expressed in cubic volume per minute.

interceptor A trapping device designed to collect materials that cannot be handled by a sewage treatment plant, such as grease, glass chips, metal chips, and hair.

nonpotable water Recycled water that can be used by plumbing fixtures, such as toilets, when there is no possibility of it becoming intermingled with potable water used for human consumption.

oakum A caulking material made from hemp fibers treated with tar.

potable water Water that is safe to drink and meets standards of the local health authority.

riser A water supply pipe running vertically through the building to supply water to the various branches and fixtures on each floor.

sanitary piping Pipe that carries liquid and waterborne waste from plumbing fixtures and into which storm, surface, and groundwater are not admitted.

sanitary sewer A sewer to carry liquid or waterborne waste (sewage) from plumbing fixtures without the admittance of storm and surface water, street runoff, and groundwater.

scupper An opening in a parapet or wall allowing water to drain from the roof.

soil stack A vertical pipe into which waste flows through waste pipes from each fixture.

trap Device used to maintain a water seal against sewer gases that back up the waste pipe. Usually each fixture has a trap. Some are built into the fixtures, as with a toilet. Other fixtures, such as lavatories, have a trap installed in the waste line. Traps must be accessible so they can be cleaned.

valve A device used to regulate and stop the flow of water.

vent Pipe permitting the waste system to operate under atmospheric pressure. Vents allow air to enter and leave the system, preventing water in the traps from being siphoned off. If this occurs sewer gases can enter the building.

vent stack A vertical vent pipe used to provide circulation of air to and from any part of the building drainage system. It is the part of the soil stack above the highest vent branch.

waste pipe Horizontal pipe that connects a fixture to the soil pipe.

SUGGESTED ACTIVITIES

1. Arrange a tour of multistory buildings in which the plumbing system can be inspected. Attempt to view the hidden aspects such as in plumbing walls and vertical distribution systems.

2. Visit the local water treatment plant and sewage disposal plant. Request the process be explained in the order treatment occurs. Obtain information about regulations within which each facility must operate.

ADDITIONAL INFORMATION

Ambrose, J. E., *Building Construction: Service Systems,* Van Nostrand Reinhold, New York, 1990.

Americans With Disabilities Act Accessibility Guidelines, U.S. Architectural and Transportation Barriers Compliance Board, 1331 F Street, N.W. Suite 1000, Washington, D.C. 20004

BOCA Basic Plumbing Code, Building Officials and Code Administrators International, Country Club Hills, Ill.

Bradshaw, V., *Building Control Systems,* John Wiley and Sons, New York, 1993.

Chen, W. F., *The Civil Engineering Handbook,* CRC Press, Inc., Boca Raton Fla., 1995.

Merritt, F. S., and Rickets, J. T., *Building Design and Construction Handbook,* McGraw-Hill, New York, 1994.

National Standard Plumbing Code, National Association of Plumbing, Heating, and Cooling Contractors, Falls Church, Va.

Southern Standard Code, Southern Building Code Congress, Birmingham, Ala.

Spence, W. P., *Architecture: Design, Engineering, Drawing,* Glencoe/McGraw-Hill Educational Division, Mission Hills, Calif., 1991.

Spence, W. P., *Architectural Working Drawings: Residential and Commercial Buildings,* John Wiley and Sons, New York, 1993.

Stein, B., and Reynolds, J. S., *Mechanical and Electrical Equipment for Buildings,* John Wiley and Sons, New York, 1992.

Uniform Plumbing Code, International Association of Plumbing and Mechanical Officials, Whittier, Calif.

Other references are the numerous related publications from the organizations listed in Appendix B.

Heating, Air-Conditioning, Ventilation, and Refrigeration

This chapter will help you to:

1. Understand the many factors to be considered when designing heating and air-conditioning systems.

2. Make decisions concerning the need for and type of ventilation for acceptable indoor air quality.

3. Be familiar with the fuels used by heating, air-conditioning, and ventilation systems.

4. Select appropriate warm air heating and cooling systems.

5. Select appropriate designs for steam and hot water heating systems.

6. Discuss the systems and equipment available for use in cooling systems used in large buildings.

The design of a heating, air-conditioning and ventilation system involves many factors. A solution for one building may be insufficient for another structure. Solutions vary depending on the occupancy. For example, office building design gives major consideration to human comfort, but manufacturing plant design has a range of other considerations. The design of the building must be considered. Factors include type and amount of glazing, insulation, air infiltration and exfiltration, heat and cooling requirements due to machinery, industrial processes, solar load, materials used for walls, ceilings, floors, and roofs and their coefficient of thermal conductivity, and space within the building provided to house mechanical units and to access various parts of the structure with ducts, pipes, and other parts of a system. Some other factors, such as a need to control humidity, removal of chemicals, noxious gases, or dust, availability and cost of fuel, cost of the various systems, the expected maintenance expenses, possibility of down time, and climatic factors, illustrate a few additional things a HVAC engineer has to consider as a system is designed.

HEAT BALANCES

The HVAC engineer is concerned about getting the indoor temperature to a comfortable level and maintaining it there. In hot weather the system must remove excess heat from inside the building at the same rate it is gaining heat. In cold weather it must add heat to the interior at the same rate it is losing heat to the cold outside environment. This provides a balance so that comfortable conditions exist without large fluctuations in temperature.

HEAT TRANSFER

Heat flows from a hot material to a cold material. This flow occurs by *radiation, conduction,* or *convection.*

Thermal radiation transfers heat energy emitted by a source, such as waves through space, and it is absorbed by a material that it strikes. For example, the sun radiates heat energy through space to the earth where it warms anything it strikes. Radiant energy is used by solar collectors that receive it and through a transfer system that moves it to the interior of a building. The sun provides **radiant heat** through windows, which can help warm a room or even make it uncomfortably hot. Likewise, a cold exterior wall of a room absorbs heat from the bodies of the occupants, making them cold (Fig. 46.1). In the summer a hot exterior wall radiates heat, possibly making the occupant uncomfortable.

Thermal conduction involves the transfer of heat energy through a material by transmitting kinetic energy from one molecule to the next. The movement is caused by a temperature difference in the material. For example, if you hold a match on one end of a short piece of copper wire the other end will quickly get hot (Fig. 46.2). Heat gain or loss due to conduction occurs when your body has physical contact with a material. If you touch a metal object (a good thermal conductor) at room temperature, it will absorb body heat and feel cool to the touch. If you touch a wood object (a poor thermal conductor), it will not absorb as much heat and feel warm to the touch.

Thermal convection involves the transfer of heat by the circulation of heated air, gas, or liquid. Warmer air rises and cooler air settles, producing a circulatory air motion

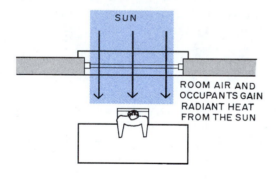

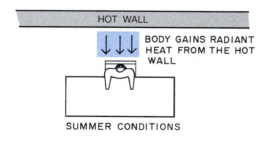

SUMMER CONDITIONS

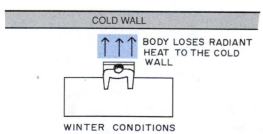

WINTER CONDITIONS

Figure 46.1 The HVAC engineer must consider various sources of radiant energy and how they influence human comfort.

in a room. The heat is transmitted to the ceiling, walls, and furnishings. Convection continues until they are the same temperature as the air. Hot air furnaces supply forced heated air through ducts and use convection to heat a room. Hot water and steam heating systems also use convection by heating the air with radiators.

If the air temperature is lower than your body temperature, your body loses heat to the air and you have a cooling sensation. If the air temperature is higher than your body temperature, your body absorbs heat from the air and becomes warmer.

SENSIBLE AND LATENT HEAT

Sensible heat is the type of heat that can be detected by the sense of touch when heat energy is taken away or added to a material. It is associated with a change of temperature.

Latent heat is involved with the action of changing the state of a substance, such as turning water to steam. This is an important factor to be considered by heating engineers. For example, if heat is added to a unit of water to raise the temperature to the boiling point, 212°F (100°C), the engineer can calculate the **Btu** of heat required to reach this point. If heat continues to be applied, the temperature of the water will not increase but it will begin to boil and change its state into water vapor (**steam**).

HUMAN COMFORT

The HVAC engineer designs the system to maintain recommended room temperatures. These vary by the season, with acceptable winter temperatures a few degrees below those acceptable for summer environments. For example, residences and similar occupancies such as offices and classrooms are typically most comfortable if kept in the range of 73 to 79°F (22.5 to 26°C) in the summer and 68 to 74°F (20 to 23.5°C) in the winter. The air temperature for most other occupancies, such as retail shops, medical facilities, and restaurants, is kept within this range. Some areas, such as showers, will have a higher design temperature. A factor that is difficult to handle is that various people feel comfortable at different temperatures. Also, different levels of activity, such as physical activity versus working at a desk, cause different temperature requirements. A space in which one area is cold and the other warm products discomfort. Rapid temperature changes, such as leaving an air-conditioned building and going out into a very hot day, may cause immediate discomfort. Some control can be had by trying to keep the inside temperature as near the outside temperature as possible yet provide an adequate inside temperature level.

Research by the American Society of Heating, Refrigerating and Air-Conditioning Engineers (ASHRAE) has established tolerances for high ambient temperature

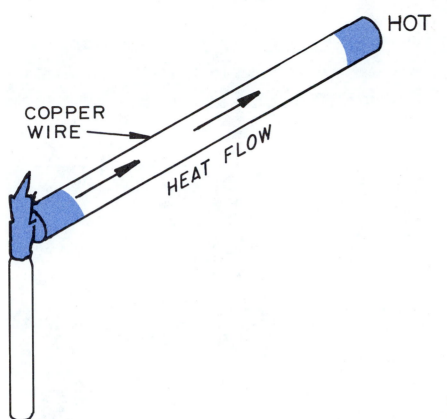

Figure 46.2 Heat can be transferred through some materials by thermal conduction.

(Table 46.1). (Ambient temperature is the temperature of the surrounding air.) When a person is exposed to these temperatures beyond these limits they are subject to heat stress and are in some danger.

Likewise, a person exposed to a very cold situation without adequate protection loses body heat faster than the body can replace it, which lowers the body temperature and produces discomfort. If not corrected, this creates a life-threatening situation.

ASHRAE Standard 55–1992, Thermal Environmental Conditions for Human Occupancy, establishes the indoor conditions and personal factors required for human comfort. The factors involved include clothing, air speed, temperature, thermal radiation, and the activity level of the occupants.

Table 46.1 High Ambient Temperature Tolerance Times

Air DB Temperature	Tolerance Time
180°F (82°C)	Almost 50 min.
200°F (93°C)	33 minutes
220°F (104°C)	26 minutes
239°F (115°C)	24 minutes

Reproduced with permission from the *ASHRAE Handbook of Fundamentals,* American Society of Heating, Refrigerating and Air-Conditioning

VENTILATION

Ventilation is used to control heat, condensation, humidity, odors, and contaminants in the air. In addition to exhausting air, it brings in a fresh air supply.

Air typically contains minute particles, such as dust raised by the wind and various pollens. The air in spaces in which large numbers of people assemble, such as a theater, becomes contaminated by carbon dioxide given off by the occupants. This carbon dioxide must be removed and replaced by fresh air. In some situations mist, fog, and various gases and vapors exist as well as a wide variety of bacteria and other microorganisms. These cause serious infections in humans and are a particular problem in medical facilities. Some business and industrial plants develop odors that must be removed. Unpleasant odors can cause physical discomfort such as headaches and nausea. Others are an indication that toxic substances are in the air that could cause physical harm to the occupants.

ASHRAE outdoor air ventilation specifications for various indoor spaces can be found in ASHRAE Standard 62–1989, Ventilation for Acceptable Indoor Air Quality. Selected examples from this publication for commercial facilities are listed in Table 46.2. In most examples the major contamination is in proportion to the number of persons to be in the space. For example, office space air requirements are based on the expectation that the maximum occupancy will be seven persons per

Table 46.2 Outdoor Air Requirements for Ventilation[a]—Commercial Facilities[b]

Application	Estimated Maximum[c] Occupancy (P/1,000 ft.² or 100 m²)	Outdoor Air Requirements				Comments
		cfm/ person[d]	L/s · person[e]	cfm/ft.²	L/s · m²	
Food Service						
Dining rooms	70	20	10			
Kitchens	20	15	8			Supplementary smoke removal equipment may be required
Retail Stores, Sales Floors, Show Room Floors						
Basement and street	30			0.30	1.50	
Upper floors	20			0.20	1.00	
Warehouses	5			0.05	0.25	
Specialty Shops						
Supermarket, hardware, drugs, fabric	8	15	8			
Beauty	25	25	13			
Workrooms						
Photo studio	10	15	8			
Pharmacy	20	15	8			
Offices						
Office space	7	20	10			Some office equipment may require local exhaust
Reception area	60	15	8			
		cfm/person[d]	L/s · person[e]	cfm/room[f]	Ls/ · room[g]	
Hotels, Motels, Resorts, Dormitories						
Bedroom				30	15	
Bath				35	18	
Conference room	50	15	8			

[a]This table prescribes supply rates of acceptable outdoor air required for acceptable indoor air quality. These values have been chosen to control CO_2 and other contaminants with an adequate margin of safety and to account for health variations among people, varied activity levels, and a moderate amount of smoking.
[b]This is only a partial listing.
[c]Column shows number of people per 1000 sq. ft. or 100 sq. meter.
[d]Cubic feet per minute per person.
[e]Liters per second per person.
[f]Cubic feet per minute per room.
[g]Liters per second per room.
Reprinted by permission from ASHRAE Standard 62–1989 Ventilation for acceptable Indoor Air Quality.

1,000 sq. ft. (100 m²). A hotel bedroom, on the other hand, is not based on the number of occupants but on the number of cubic feet per room.

Selected ASHRAE 62–1989 recommendations for institutional facilities are listed in Table 46.3. Most are based on the maximum number of persons to occupy 1000 sq. ft. (100 m²) of floor space.

ASHRAE 62–1989 recommendations for residential facilities are listed in Table 46.4. For each area of the residence the required outdoor air supply is specified in air changes per hour or cubic feet per minute.

Ventilated air may be brought in from the outside and filtered before entering the interior air distribution system. In some cases the interior air is recycled and has the contaminants removed and odors reduced. It is not always permissible to release air being removed from a building directly into the outside air without some form of treatment. Changing outside air with inside air can also increase heating and cooling costs, but heat exchanger units are available to reduce this cost. Heat exchangers use the heat in the outgoing air to warm the cooler incoming fresh air.

Table 46.3 Outdoor Air Requirements for Ventilation[a] of Institutional Facilities

Application	Occupation	Outdoor Air Requirements			
		cfm/person	L/s person	cfm/ft.^{2}	L/s • m^2
Education					
Classroom	50	15	8		
Laboratories	30	20	10		
Libraries	20	15	8		
Locker rooms				0.50	2.50
Corridors				0.10	0.50
Auditoriums	150	15	8		
Hospitals, Nurseries, Convalescent Homes					
Patient rooms	10	25	13		
Operating rooms	20	30	15		
Autopsy rooms				0.50	2.50

[a]This is only a partial listing for illustrative purposes.
Reprinted with permission from ASHRAE Standard 62–1989, Ventilations for Acceptable Indoor Air Quality.

Table 46.4 Outdoor Air Requirements for Ventilation[a] of Residential Facilities

Application	Outdoor Air Requirements
Living area	0.35 air changes per hour but not less than 15 cfm (7.5 L/s) per person
Kitchens	100 cfm (50 L/s) intermittent or 25 cpm (12 L/s) continuous or openable windows
Baths, toilets	50 cfm (25 L/s) intermittent or 20 cfm (10 L/s) continuous or openable windows

[a]This is only a partial listing for illustrative purposes.
Reprinted with permission from ASHRAE Standard 62–1989, Ventilation for Acceptable Indoor Air Quality.

Methods of Ventilation

Ventilation may be natural or mechanical. *Natural ventilation* may be through windows, louvered openings, roof ventilators, operable skylights, hoods, and jalousie-covered openings. *Mechanical ventilation* uses fans or blowers to move the air into the building through louvered exterior openings and directs the air to the areas where it is needed through ducts (Fig. 46.3). Mechanical ventilation is more reliable than natural ventilation because the fans, openings, and ducts can be sized to meet a specific need, and mechanical systems operate continuously to provide the specified ventilation. Natural ventilation has wide variations in performance due to such uncertain conditions as temperature and exterior weather conditions.

Some ventilation systems are used to remove smoke, gases, and heat developed during a fire. These are designed following the National Fire Protection Association Standard NFPA 90-A, Installation of Air-Conditioning and Ventilating Systems. Some ventilation systems, such as those in restaurant kitchens, remove fumes and heat. The air contains greasy materials that coat the ducts and fans and become a fire hazard. In addition to routine maintenance, these systems must have an automatic fire-control system. This typically includes fire dampers that close the duct when a fusible link is broken and activates some fire-smothering material such as a foam.

When designing a ventilating system, the engineer observes the many regulations in the building code. For example, in some situations it is not permissible to re-circulate the air. Areas such as restrooms, biology and chemistry laboratories, hospital areas such as operating rooms, and industrial facilities in which various flammable vapor or hazardous fumes are developed must exhaust the air and in some cases treat the exhausted air before it is released.

Types of Air Cleaners

The most commonly used air cleaners include fibrous panel filters, renewable filters, electronic filters, and combination air cleaners.

Viscous impingement fibrous panel filters are made of porous, coarse fibers that are coated with some type of adhesive substance that causes the particles in the air to stick to them. They are low cost and good for attracting lint type dust but do little with pollen or atmospheric dust. They are available in thickness from ½ to 4 in. (12.7 to 102 mm) and in a variety of standard sizes. A common application is in hot air residential furnaces and air-conditioning systems. Filter materials include coated animal hair, glass fibers, synthetic fibers, vegetable fibers, metallic wool, expanded metals and foils, crimped screens, random-matted wire, and synthetic open-cell

Figure 46.3 A fixed-bar louvered grille being used to move outside air into the building.

foams. Some types, such as the glass fiber residential filter, are disposable (Fig. 46.4). Metal filters can be cleaned and reused. They are typically washed with water or steam and may be recoated with the adhesive (Fig. 46.5).

Dry-type extended-surface air filters are made of random fiber mats or blankets using bonded glass fiber, cellulose fibers, synthetics, and wool felt that are supported by a wire frame forming pockets or pleats. Some types require the entire filter to be replaced, and others permit the mat to be replaced inside the wire frame. Generally, dry-type filters are more efficient and have higher dust-holding capacities than viscous impingement fiber panel filters.

Renewable filters are viscous impingement type filters in a roll form. They are either moved by hand or automatically unrolled across the filter area. As the clean filter material unrolls across the filter area the dirty portion is rolled on a roller at the bottom. When the roll is used up it is removed, thrown away, and a new roll is installed. *Random fiber dry filter* material is also used.

Electronic air cleaners use electrostatic precipitation to collect dust, pollen, and smoke. The filter has an ionization section and a collection plate section. The ionization section has small-diameter wires that have a positive direct current potential of 6 to 25kV DC. They are suspended between grounded plates as shown in Fig. 46.6. The wires create a positive ionizing field through which the particles in the air pass and are given a positive charge.

The collecting plate section has a series of equally spaced parallel plates. Alternate plates have a 4 to 10kV DC positive charge. The plates not charged (they are negative) are grounded. The charged particles in the air

Figure 46.4 A typical disposable glass fiber air filter panel.

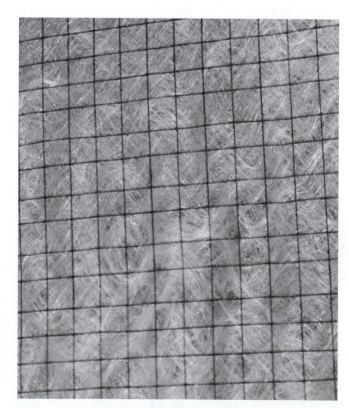

Figure 46.5 This metal mesh and plastic fiber mesh air filter can be washed and reused.

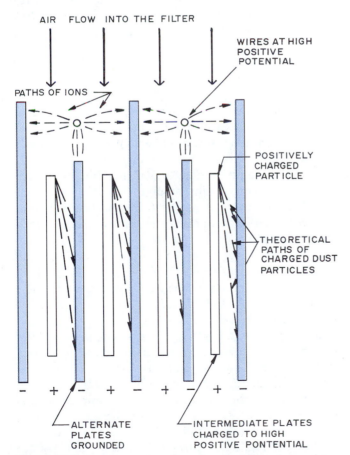

AIR FLOW INTO THE FILTER

WIRES AT HIGH
POSITIVE
POTENTIAL

PATHS OF IONS

POSITIVELY
CHARGED
PARTICLE

THEORETICAL
PATHS OF
CHARGED DUST
PARTICLES

− + − + − + − + −

ALTERNATE
PLATES
GROUNDED

INTERMEDIATE PLATES
CHARGED TO HIGH
POSITIVE PONTENTIAL

Figure 46.6 A cross section through an ionizing electronic air cleaner.

pass into the collecting section and are forced to the negative grounded plates by the electric field on the positively charged particles.

Electronic air cleaner cells are cleaned with a detergent and hot water. Large commercial installations may have an automatic washing system that cleans the cells in place. Small units, as used in residential systems, generally are removed and cleaned manually.

Combination air cleaners are an assembly of the systems just discussed. Typically a dry-type filter might be used in line after the electronic filter has processed that air. The dry filter can catch any particles that may come off the collector plates before they pass through the system.

FUEL AND ENERGY USED FOR HVAC

The major *fuels* include coal, fuel oil, natural gas, propane, butane, solid waste, and wood. All except natural gas require some form of storage. The *energy sources* include electricity, geothermal, ground heat, and solar. Solar requires a means for storing heat. The others depend on an outside source of supply.

Coal

Coal is a combustible material composed of elemental carbon, hydrocarbons, complex organic compounds, and various inorganic materials. The types commonly used for commercial and industrial heating and the production of electricity are anthracite, bituminous, lignite, semianthracite, and subbituminous. Coal is pulverized for burning in commercial furnaces. Fine coal is pressed into brickettes. It is converted into liquid and gaseous products that may be used as fuels and made into coke, which is used to produce steel. Systems that use coal as a fuel must have emissions controls to reduce air pollution and a means for disposing of the ash. Anthracite coal has a heat content of about 14,600 Btu/lb or 33,980 kJ/kg (kilojoules per kilogram).

Fuel Oil

Fuel oil is a fossil fuel derived from the refining of petroleum. It is a mixture of liquid hydrocarbons containing about 85 percent carbon, 12 percent hydrogen, and 3 percent other elements. The American Petroleum Institute (API) has a system for grading fuel oil. Kerosene is Grade 1, which means it is a very light and easily vaporized liquid. As the numbers increase the viscosity increases. Viscosity is a measure of the flowing quality of a liquid. Grades 1 and 2 fuel oils are frequently used in residential and small commercial furnaces, while Grade 3 is sometimes used. They flow easily and can be vaporized and mixed with air as required for burning. Grades 4, 5, and 6 are heavier and are used in large industrial and commercial burners. Grades 5 and 6 are very heavy and must be preheated before they can be pumped and burned. No. 2 fuel oil has a heat content of about 141,000 Btu/gal. or 39,300 kJ/L (kilojoules per liter).

Natural Gas

Natural gas is a fossil fuel. It is a combustible hydrocarbon gas consisting primarily of methane with small amounts of ethane and traces of butane, propane, pentane, and hexane. It is odorless in its natural form and has an odorant added as a safety measure so leaks can be detected by smell. Natural gas has a heat content of approximately 1050 Btu/ft.^{3} (39,100 kJ/m^3). It is obtained from wells drilled in strata in the earth containing pockets of gas. Natural gas is supplied to buildings through underground utility pipelines. The utility company is responsible for maintaining a continuous supply, so an on-site storage facility is not required.

Propane and Butane

Propane and butane are liquid petroleum products and are very volatile. They are colorless, flammable gases occurring in petroleum and natural gas. They are liquified under pressure and are transported in pressurized cylinders. They have a heat content of about 2,500 Btu/ft.^{3} (93,150 kJ/m^3).

Solid Waste

The collection of burnable solid waste is an ever-increasing activity and its use as a fuel is becoming more popular. It takes specialized incinerator equipment to produce a clean burn and get the desired heat energy. Typically, such a facility is built near a large city to ensure a steady supply of waste or near a commercial or industrial plant that has contracted to use the energy. It has the advantage of helping reduce the major problem of disposing of large amounts of waste materials while producing a usable end result. Typical systems are shown in Figs. 46.7 through Fig. 46.9.

The system shown in Fig. 46.7 was developed to serve a community of 30,000 people. It handles eighty-five tons of solid waste per day. Eighty percent of the waste processed is converted into steam for use by a local industry.

In the system shown in Fig. 46.8 the waste is shredded and electromagnets draw out the ferrous materials in the recycling process. The plant burns this *refuse-derived fuel (RDF)* with coal as a backup fuel. The fuel produced from the waste is the main fuel used to burn the incoming waste. About 50 percent of the waste entering the plant drives turbine generators that produce electricity. In this case, the electricity is used by the city to operate the sewer and water treatment plants and the street lights and to serve other public and private customers.

Another waste-to-energy system is illustrated in Fig. 46.9. The system accepts mixed waste and sorts it for recycling and composting before turning the remainder into refuse-derived fuel (RDF).

In all cases major effort is required to make waste-to-energy systems economical to operate. It requires cooperation from the local trash collectors and local and state governments.

Wood

Wood is a renewable resource, and the harvesting of trees is accompanied by a program of replanting. Wood has become a minor source of heat energy. Typically, it is used to heat small residences. Harvesting, cutting to size, storing, and handling wood are difficult. Large users require considerable storage space. Overall, wood has a heat content of about 7,000 Btu/lb. (16,290 kJ/kg).

A variety of wood-burning stoves are available, which many people choose instead of building a fireplace. Many newer designed stoves meet the Environmental Protection Agency's Phase II emission standards. These have the EPA approval label, and they consume less wood and burn 80 to 90 percent cleaner than the older wood-burning stoves.

The EPA Phase II stoves are available in two types: catalytic and noncatalytic. The *catalytic stove* has a cored combuster in the smoke path just past the main combustion chamber. The combuster has a catalytic coating that causes combustion gases to burn completely, thus reducing emissions. After the fire in the main combustion chamber reaches 500°F (262°C) a bypass damper is opened, directing the gases through the catalyst rather than straight up the chimney. The combustor does deteriorate with use and must be occasionally replaced.

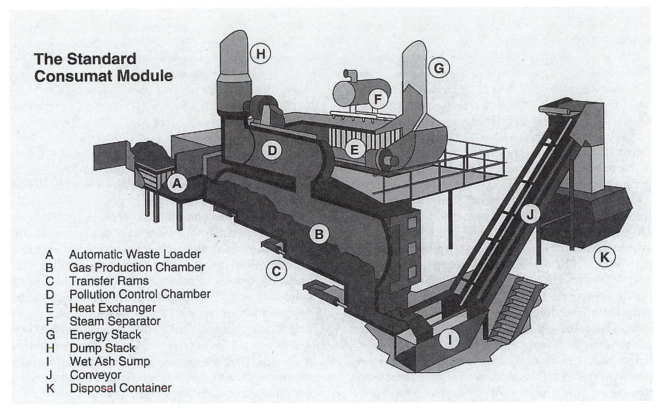

The Standard Consumat Module

A   Automatic Waste Loader
B   Gas Production Chamber
C   Transfer Rams
D   Pollution Control Chamber
E   Heat Exchanger
F   Steam Separator
G   Energy Stack
H   Dump Stack
I   Wet Ash Sump
J   Conveyor
K   Disposal Container

Figure 46.7 This waste-to-energy system is used to produce steam for a local industry *(Courtesy Great Lakes Regional Biomass Energy Program, Council of Great Lakes Governors)*

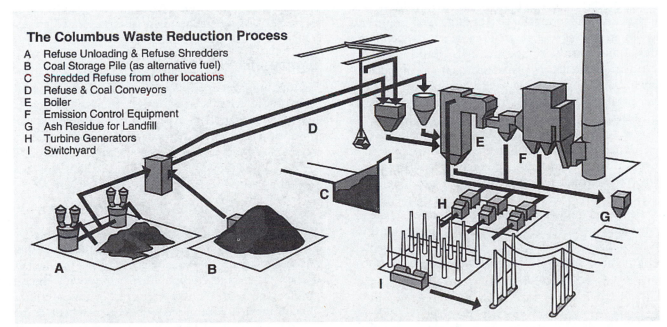

The Columbus Waste Reduction Process
A   Refuse Unloading & Refuse Shredders
B   Coal Storage Pile (as alternative fuel)
C   Shredded Refuse from other locations
D   Refuse & Coal Conveyors
E   Boiler
F   Emission Control Equipment
G   Ash Residue for Landfill
H   Turbine Generators
I   Switchyard

Figure 46.8 Refuse-derived fuel (RDF) is produced from waste in this plant and is burned in the plant to produce energy to drive turbine generators that produce electricity. *(Courtesy Great Lakes Regional Biomass Energy Program, Council of Great Lakes Governors)*

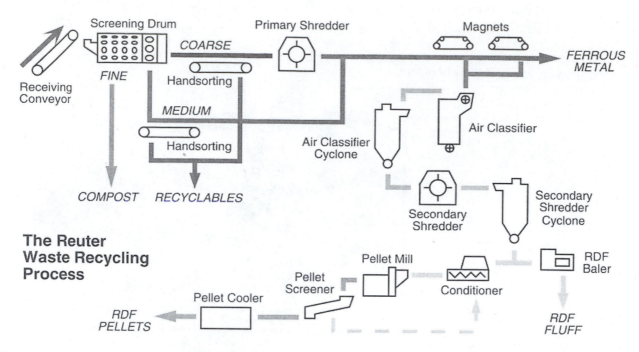

Figure 46.9 This waste-to-energy system accepts unsorted waste, sorts it, recycles it, composts it, and produces refuse-derived fuel (RDF). *(Courtesy Great Lakes Regional Biomass Energy Program, Council of Great Lakes Governors)*

The *noncatalytic stove* can also achieve low emissions, but it does this by keeping high temperatures in the main combustion chamber and adding turbulent secondary air at the top of the chamber where the hot gases leave the stove. This air provides additional oxygen so the combustion gases can burn more completely.

Pellet stoves are designed to burn a manufactured product made from wood and wood by-products. The pellets are small, about the size of large feed grain, and are sold in waterproof bags. They are a concentrated material and burn very hot and clean, producing very low emissions. Pellet fireplace inserts are also available. They use a motor-driven auger to feed the pellets from a fuel storage hopper. The typical hopper holds enough fuel to last several days.

Some types of fireplace inserts function much like stoves. The insert is a metal liner with pyroceramic glass doors that permit the heat generated to radiate into the room. They are airtight and burn hot, producing low emissions. This type of fireplace insert has emission qualities equal to the EPA Phase II woodstove. They can be installed in a wood-burning fireplace or have an approved brick or wood surrounding and are vented with a metal chimney. Some have an electric blower to increase the heat distribution into the room.

Electricity

Electricity is a major energy source for operating HVAC systems. It powers the blowers, pumps, and controls and supplies heat energy through various types of resistance heating units. It usually has the lowest installation cost and is easy to install anywhere in a building. It cannot always compete with other fuels on a cost basis. It requires no on-site storage, but large installations require space for transformers and switchgear. The power is usually supplied by a utility company that guarantees a steady supply, although some on-site power generation is used on large installations as discussed in Chapter 47.

Geothermal

Geothermal heat energy is derived from natural sources from the internal heat of the earth. Typical examples include geysers that send up jets of hot water and steam from within the earth. This source of heat is limited to a geyser basin (an area containing a group of geysers). The system can be tapped to supply heat to an entire village from this central source.

Ground Heat

The temperature of the ground fluctuates in the different seasons, but it remains fairly stable as the depth increases. The exact temperature varies with the geographic location. For example, in Minnesota's cold climate the temperature is relatively constant at 50°F (10°C) at a depth of 16 to 26 ft. (5 to 8 m). Earth-sheltered houses use this stable temperature to provide a steady source of heat through the exterior walls. If it maintains a 50°F (10°C) interior air temperature, very little supplemental heating is required. Another important use of ground heat is the use of ground source heat pumps. One type uses an earth coil that involves burying plastic pipe in either drilled holes or in a continuous ground loop. The water in the pipe absorbs the ground heat and is pumped to the water source heat pump, which extracts the heat and recirculates the water. Another source is to tap the temperature of water in a well. The well water heat pump removes water from a well, extracts the heat, and disposes of the water in a second well. This puts the water back into the aquifer. The groundwater temperatures vary by geographic area, but groundwater heat pumps in the range of 55 to 60°F (13 to 16°C) are commonly used.

Solar

Solar energy is derived from the rays of the sun that fall on the earth. It is a major source of heat energy that keeps all living things alive. It can be tapped by various types of collectors, stored, and used to heat and cool buildings and to produce electricity. It is a source of natural light. A variety of solar heating systems are available. Some require mechanical devices to gather, store, and distribute the heat energy as needed (referred to as active solar systems), but other methods do not require hardware of any type (referred to as passive solar systems).

WARM AIR HEATING EQUIPMENT

Warm air systems have some type of heat-generating device (furnace), controls, and a distribution system of ducts. The furnace may be fired by oil, gas, or electricity, and some solar systems use warm air distribution. An air-conditioning unit can be installed as a part of the furnace so the system can heat and cool the building.

Warm Air Furnaces

Warm air furnaces use blowers to move the heated air through ducts. These include upflow, downflow, and horizontal types. An *upflow warm air furnace* moves air out the top of the unit into the duct system. It is used in the basement when ducts are run under the floor or on the floor when the ducts are to be run in the attic (Fig. 46.10). A *downflow warm air furnace* moves the air out the bottom of the unit. This type typically is used when the ducts are in a concrete slab floor or below the floor in a crawl space (Fig. 46.11). *Horizontal warm air furnaces* are mounted in the attic or hung below the floor joists as shown in Fig. 46.12. Each of these may have an air-cooling unit that uses the furnace blower and ducts to provide cool air to the building spaces (Fig. 46.13).

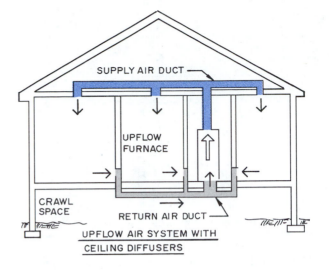

UPFLOW AIR SYSTEM WITH CEILING DIFFUSERS

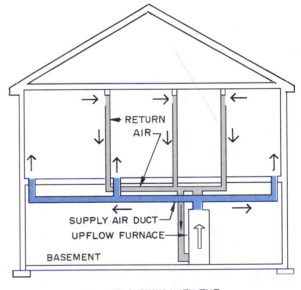

UPFLOW AIR SYSTEM WITH THE FURNACE IN THE BASEMENT

Figure 46.10 Typical upflow warm air heating systems.

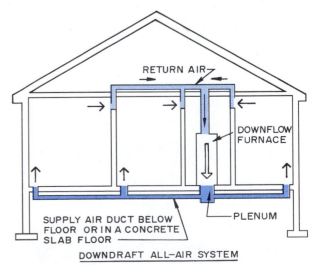

DOWNDRAFT ALL-AIR SYSTEM

Figure 46.11 Downflow warm air heating systems are used in a building with a crawl space or concrete floor.

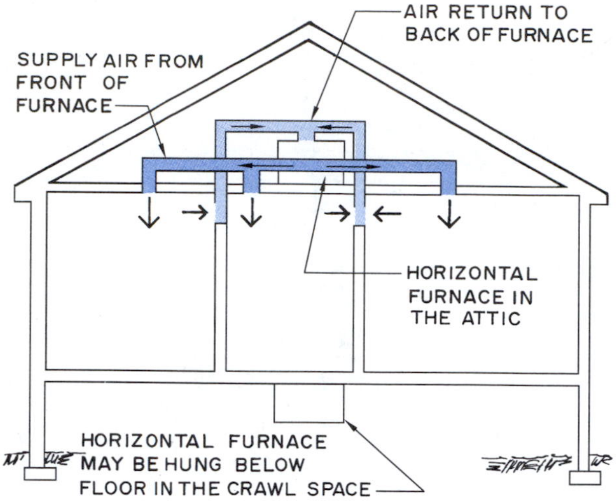

SUPPLY AIR FROM FRONT OF FURNACE

AIR RETURN TO BACK OF FURNACE

HORIZONTAL FURNACE IN THE ATTIC

HORIZONTAL FURNACE MAY BE HUNG BELOW FLOOR IN THE CRAWL SPACE

Figure 46.12 Horizontal warm air furnaces are hung below the floor or in the attic.

Figure 46.14 shows a horizontal gas-fired warm air furnace having both heating and cooling capabilities. It can be installed outdoors on a concrete slab for horizontal flow or on the roof for a downflow installation (Fig. 46.15). It can burn natural gas or be converted to propane.

Figure 46.16 shows the internal construction of a gas-fired warm air furnace. The blower operates on two speeds and also serves to circulate cool air if an air-cooling unit is mounted on top of the furnaces. These are available in a wide range of sizes and output (Btuh). The gas-fired furnace in Fig. 46.17 can be switched on the site from an upflow/horizontal configuration to a downflow mode.

Warm air furnaces may also use electric resistance coils to heat the air, which is then distributed through ducts by a blower. The furnace is similar to that shown for gas furnaces. A major problem is the high cost of electricity in many areas (Fig. 46.18).

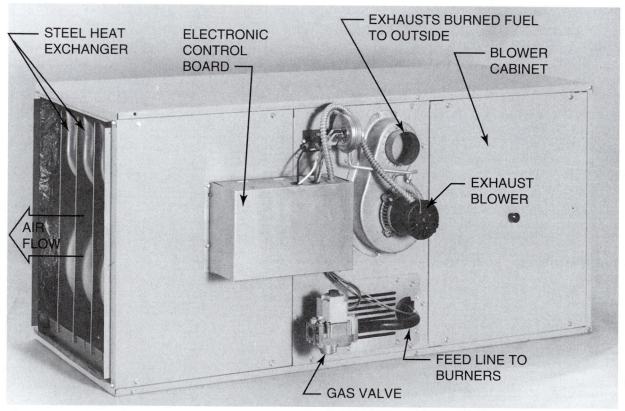

STEEL HEAT EXCHANGER

ELECTRONIC CONTROL BOARD

EXHAUSTS BURNED FUEL TO OUTSIDE

BLOWER CABINET

EXHAUST BLOWER

AIR FLOW

FEED LINE TO BURNERS

GAS VALVE

Figure 46.13 A horizontal gas-fired warm air furnace. *(Courtesy Bryant Air Conditioning)*

Figure 46.14 This single-package gas-fired warm air furnace has an electric cooling coil, thus providing summer cooling in addition to heating. It can be mounted on a concrete slab on the ground or on a roof. *(Courtesy The Trane Company)*

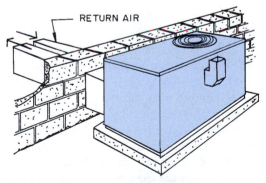

Figure 46.15 This horizontal gas-fired electric cooling unit is mounted on a concrete slab outside the building and feeds the conditioned air through ducts.

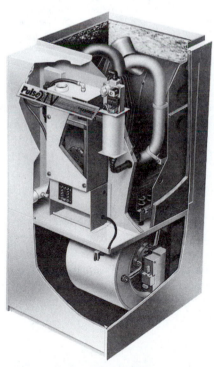

Figure 46.16 This is a high-efficiency gas-fired upflow furnace with a direct spark ignition that eliminates the pilot light. It is available in fixed and variable speed models. It uses pulse combustion. *(Courtesy Lennox International, Inc.)*

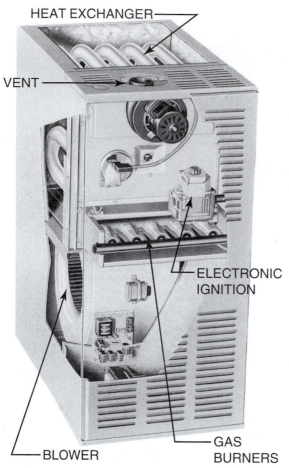

Figure 46.17 This gas-fired warm air furnace can be converted from the factory-shipped upflow/horizontal configuration to the downflow mode. *(Courtesy Lennox International, Inc.)*

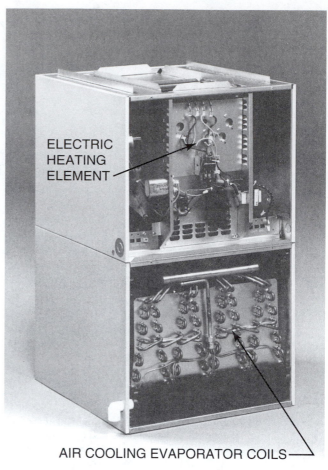

Figure 46.18 This electric central heating/cooling furnace has options of 5 to 25KW and up to 5 tons of cooling capacity. *(Courtesy Rheem Manufacturing Company)*

Oil-fired furnaces require a large storage tank outside the building from which the fuel oil is pumped to the burner in the furnace. A typical installation is shown in Fig. 46.19. The location and method of placement are regulated by codes. The oil tank construction must meet Underwriters Laboratory (UL) specifications. Tanks may be placed above or below ground. If below ground they should be below the frost line. Some codes permit oil storage tanks inside the building.

The oil-fired furnace (Fig. 46.20) is much like the gas-fired furnace but it has an oil burner unit with a nozzle that breaks the oil into a fine spray. The spray is ignited by a high-voltage ignition using spark electrodes to ignite the vapor.

Heat Pumps

A heat pump is a machine that can heat or cool a building using forced air through ducts. Basically, it is a machine that takes heat from one source, such as the outside air, and transfers it to another area, such as the air inside a building. The commonly available types include air-to-air, water-to-water, and air-to-water. Ground source heat pumps are finding increasing use.

The most frequently used heat pump is an *air-to-air unit*. The unit has a **compressor** similar to that used in a refrigerator. A **refrigerant** (a gas) is circulated in coils. When the unit is in heating mode, the coils in the outdoor unit enable the refrigerant to absorb heat from the outdoor air. The refrigerant is moved to the com-

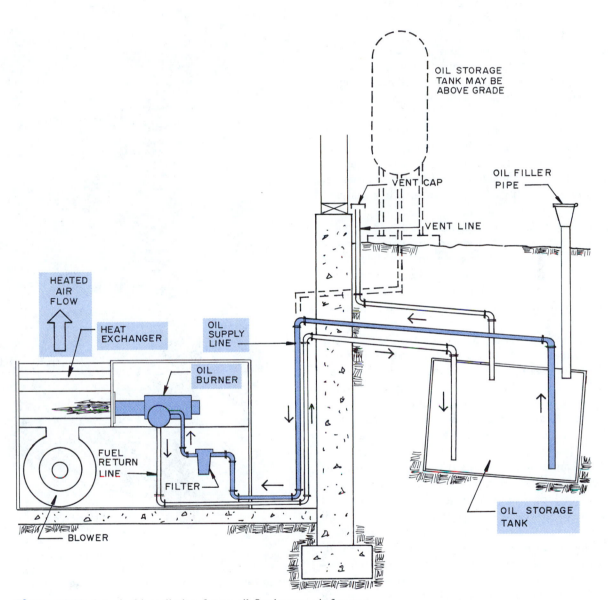

Figure 46.19 A typical installation for an oil-fired warm air furnace.

pressor where it is compressed, raising its temperature. It moves to indoor coils where indoor air is blown over them by a blower that removes the heat and moves it through ducts into the building (Fig. 46.21). In the cooling mode the reverse happens. Heat is absorbed by the inside coils and dispersed to the outside air by the outside coils.

An air-to-air heat pump compressor unit is shown in Fig. 46.22. The assembled outdoor unit rests on a concrete slab and the coils are fully exposed to the air. The fan on top pulls air through the coils (Fig. 46.23).

The *groundwater heat pump* (GWHP) is a water-to-air system. A water-to-air system pumps water from a well to the heat pump. The well water typically maintains a constant temperature, which varies with the geographic area but typically is from 55 to 65°F (13 to 18°C). The refrigerant in the heat pump coils absorbs the heat in the water and discharges it into the building through an air handling unit and ducts. The unit can reverse the process and remove heat from inside the building and discharge it to the water, which is disposed of in a disposal well (Fig. 46.24).

Water-to-water heat pumps can be used to heat water and at the same time cool water. In an operation in which both are required at the same time, the heat pump can remove heat from a liquid source, such as some fluid in a manufacturing operation that needs chilling, and move the heat to another liquid, such as water for cleaning operations.

Air-to-water heat pumps can be used to cool air in a room and transfer the heat to a water or other liquid that has to be heated. An apartment complex could cool air in the living space and use the heat to keep the water in the swimming pool at the desired temperature.

The *ground loop heat pump* is used in areas where groundwater is not readily available, too costly to acquire, or of poor quality for heat pump use. The system does avoid the cleaning requirements of groundwater systems due to minerals in the water. The system may use an earth coil in a horizontal loop (Fig. 46.25) or be installed vertically in drilled holes (Fig. 46.26). The earth coil system transfers heat from or to a water/antifreeze solution circulated through plastic pipes buried in the earth. The heat from the earth is transferred to the refrigerant in a heat exchanger and compressed and distributed by a blower through ducts. In the cooling mode the reverse occurs with indoor heat being transferred to the water/antifreeze solution, which disperses it to the cooler ground.

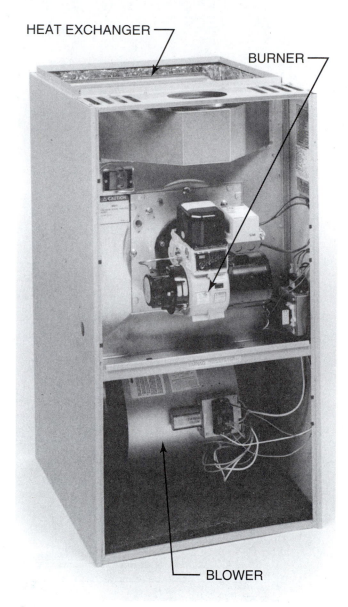

Figure 46.20 This oil-fired furnace is available in upflow, downflow, and horizontal units with a Btu/hr. heating capacity up to 150,000. *(Courtesy Rheem Manufacturing Company)*

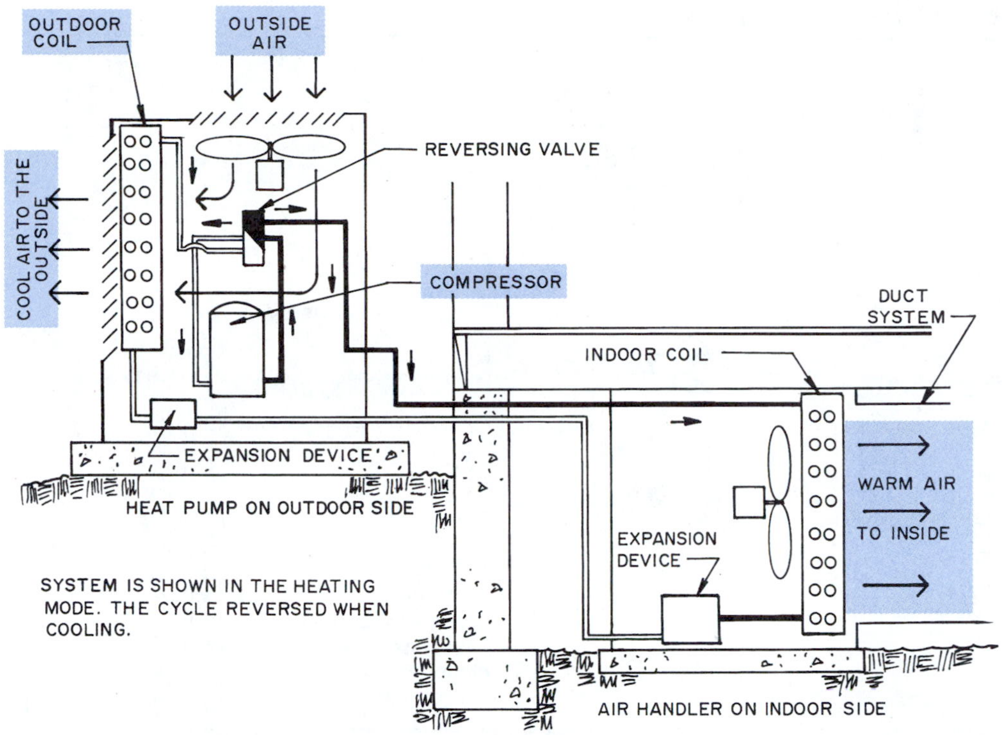

OUTDOOR
COIL

OUTSIDE
AIR

REVERSING VALVE

COOL AIR TO THE OUTSIDE

COMPRESSOR

DUCT SYSTEM

INDOOR COIL

EXPANSION DEVICE

HEAT PUMP ON OUTDOOR SIDE

EXPANSION DEVICE

WARM AIR

TO INSIDE

SYSTEM IS SHOWN IN THE HEATING MODE. THE CYCLE REVERSED WHEN COOLING.

AIR HANDLER ON INDOOR SIDE

Figure 46.21 The heating cycle of an air-to-air electric heat pump.

FAN MOVES AIR THROUGH HEAT EXCHANGER

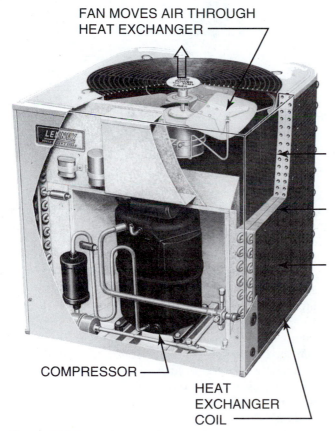

COMPRESSOR

HEAT EXCHANGER COIL

Figure 46.22 An air-to-air heat pump. *(Courtesy Lennox International, Inc.)*

Figure 46.23 The finished installation of the outdoor unit of an air-to-air heat pump.

A Well Water Heat Pump System

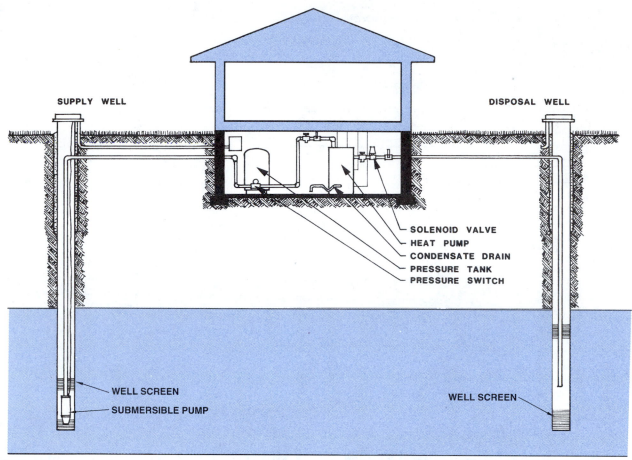

Figure 46.24 A water-to-air heat pump can be used wherever groundwater is available and wells can be drilled. The system also requires a water return system that returns the water to the earth. *(Courtesy Kansas Electric Utilities Research Program, research by Mark Hannifan and Joseph King, AIA)*

The choice of a horizontal or vertical ground loop depends on the results of a study of the surface conditions, including the amount of moisture in the soil, and subsurface conditions, including soil type, moisture content, rock strata, and ground temperatures. Horizontal loops require sufficient open space free from paved surfaces, underground utilities, and the number of trees and shrubs.

Efficient heat pumps designed to heat and cool one room with no ducts are manufactured as a single unit. The coil and fan are housed in a decorative interior case as shown in Fig. 46.27. The compressor and exterior coil extend through an opening in the wall and are exposed to the outside air (Fig. 46.28). Similar units are available with total electric heating and cooling.

A natural gas heating and cooling unit is shown in Fig. 46.29. This unit uses natural gas as the primary energy source for both heating and cooling, operating much like the all-electric heat pump. The compressor is driven by a quiet natural gas engine. The compressor function is the same as described for electric heat pumps. The unit has a recuperator that recovers engine heat, thus increasing the efficiency of the unit. An auxiliary gas heater is used to provide supplemental heat during extremely cold weather. The system is microprocessor-controlled to provide continuous comfort. A gas engine uses pressure formed by the combustion of the expanding gas to produce mechanical energy by supplying power to the motor shaft, which connects to the compressor.

A Horizontal GLHP System (Schematic)

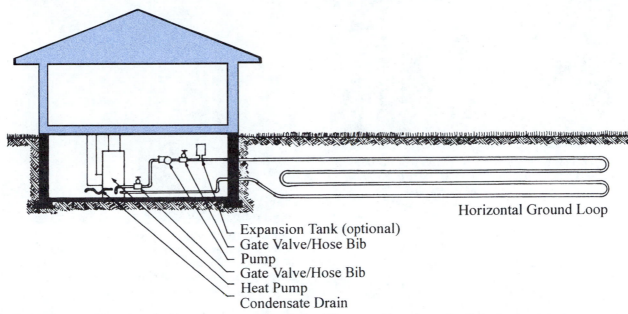

Horizontal Ground Loop

Expansion Tank (optional)
Gate Valve/Hose Bib
Pump
Gate Valve/Hose Bib
Heat Pump
Condensate Drain

Figure 46.25 A horizontal ground loop heat pump installation. The ground loop is made of plastic pipe carrying a water/antifreeze mixture that is circulated through the pipe to the water-to-refrigerant heat exchanger in the heat pump. *(Courtesy Kansas Electric Utilities Research Program, research by Mark Hannifan and Joseph King, AIA)*

A Vertical GLHP System (Schematic)

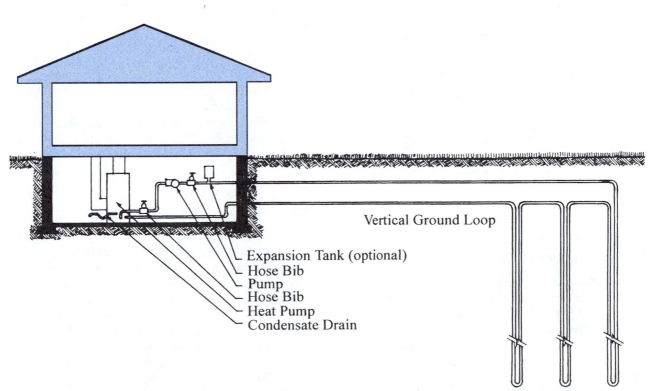

Vertical Ground Loop

Expansion Tank (optional)
Hose Bib
Pump
Hose Bib
Heat Pump
Condensate Drain

Figure 46.26 A vertical ground-loop heat pump installation. In areas with low soil moisture vertical loops are recommended if the soil permits drilling deep enough to get adequate lengths on the vertical loops. *(Courtesy Kansas Electric Utilities Research Program, research by Mark Hannifan and Joseph King, AIA)*

Figure 46.27 This is an efficient heat pump that fits through the wall and is used to heat one room. (*Used with permission of the copyright owner, General Electric Company*)

Figure 46.29 This gas-fired heating and cooling unit can be used for residential and small commercial buildings. It uses the basic heat pump cycle. However, it uses a high-efficiency natural gas engine to operate the compressor. (*Courtesy York International Corporation*)

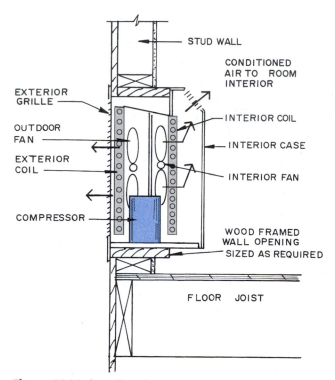

Figure 46.28 A section showing a typical through-the-wall heat pump installation.

AIR-CONDITIONING WITH DUCT SYSTEMS

Duct systems can also be used to air-condition a building. The heat pump previously discussed provides both heating and cooling modes. When oil, gas, or electric warm air furnaces are used, a cooling coil unit is placed on the furnace and connected to an outside air conditioner, as shown in Fig. 46.30. They are connected by pipes carrying the refrigerant from the compressor in the outdoor unit to the indoor coil. The furnace blower moves air over the cold coils and through the ducts into the building.

The external electric air-conditioning unit is shown in Fig. 46.31. The refrigerant absorbs heat from the air blown over the indoor coil. The refrigerant moves to the outdoor air-conditioning compressor, where the heat it has absorbed is transferred to the atmosphere. The refrigerant moves back to the interior coil to repeat the process.

If a building has hot water or steam heat, a similar system can be used to air-condition it. A duct system is installed for the cooling system, and the indoor unit is an air handler (Fig. 46.32) that has a blower and a coil. The outdoor air-conditioning unit is the same as just described. A typical system is shown in Fig. 46.33.

Figures 46.34 and 46.35 show roof-mounted heating/cooling units that are used on light commercial construction. When combined with a duct system as shown in Fig. 46.36 they can heat and cool zones within the building. Each zone has its own **thermostat** that operates a zone damper. It has automatic heating/cooling changeover.

ALL-AIR DISTRIBUTION SYSTEMS

Procedures for designing all-air duct distribution systems can be found in the 1989 ASHRAE Handbook-Fundamentals. All-air duct systems may be low or high velocity with the high-velocity design requiring smaller diameter pipes and high pressures. The designer must consider how the size of the ducts influences performance, outside air requirements, supply air temperatures, airflow, air changes, zoning humidity control, heat gain and heat loss of the space to be conditioned, control of noise in the system, possible use of heat recovery devices, and many special factors such as exist in hospitals, manufacturing plants, and laboratories.

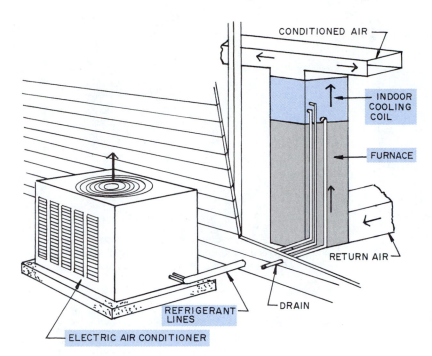

CONDITIONED AIR

INDOOR COOLING COIL

FURNACE

RETURN AIR

DRAIN

REFRIGERANT LINES

ELECTRIC AIR CONDITIONER

Figure 46.30 An electric air-conditioning unit is installed out of doors and supplies the chilled refrigerant to cooling coils on the furnace.

Figure 46.31 An electric air-conditioning unit that supplies cooled refrigerant to indoor cooling coils located on a furnace or in an air handler. *(Courtesy The Trane Company)*

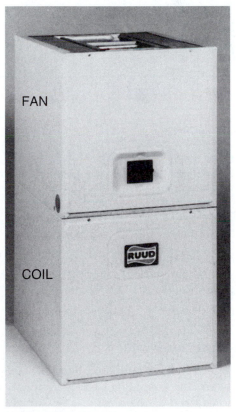

FAN

COIL

Figure 46.32 The air handler contains a set of coils and a fan to move the conditioned air through ducts. *(Courtesy Rudd Air Conditioning)*

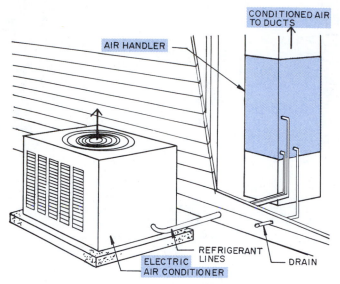

Figure 46.33 A unit air conditioner can supply cooled refrigerant to cooling coils in an air handler.

Figure 46.34 This roof-mounted unit is designed for commercial buildings. It is available as an electric cooling/gas heating unit or an electric cooling/electric heating unit. It could be installed with a duct system as shown in Fig. 46.36. *(Courtesy Rudd Air Conditioning)*

Figure 46.35 A heavy-duty rooftop heat pump for use on small commercial buildings. It could be installed with a duct system as shown in Fig. 46.36. *(Courtesy Rudd Air Conditioning)*

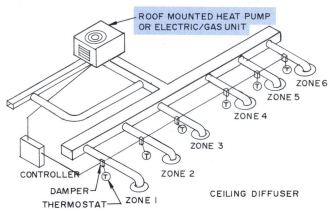

Figure 46.36 A typical duct system for a roof-mounted heat pump or electric/gas-fired unit. The controls permit it to condition the air in various zones within the building.

The systems in common use include single zone, multizone, reheat, variable air volume, and dual duct. Various combinations of these can be used to meet design requirements.

Single-Zone Systems

Single-zone systems use one air handling unit (AHU) to supply an entire building or a portion of a building that is considered a single zone. The furnace and air-handling unit can be installed outside of or within the space to be conditioned. A **return air** duct system is usually required. A typical example is a small residence. A large residence may have two or more heat and cooling sources using two or more single zone systems. A simple schematic is shown in Fig. 46.37. It should be

noted that this system supplies air at a constant rate. Therefore, room temperature is varied by changing the air temperature.

Multizone Systems

Multizone systems produce the heated and cooled air in the air-handling unit. The air for each zone is blended at the air-handling unit into a single temperature air and fed to the zone by a single duct (Fig. 46.38). This enables the zone requiring the lowest air temperature to have the cool air required or, if needed, the temperature can be raised to a higher level. Likewise, areas requiring higher temperatures can have it by blending little or no cool air with the heated air sent to the area.

The flow of air in the ducts is regulated by motorized dampers controlled by thermostats. Since this system uses a constant volume of air, it passes by the heating and cooling coils even if the air is not needed. Therefore, it is not a cost-efficient system. It is not widely used because more efficient zoning methods are available, as explained in the following reheat system.

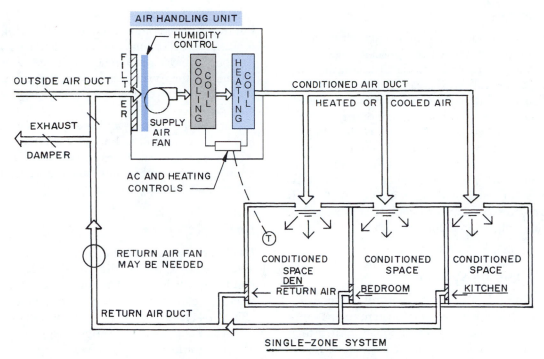

Figure 46.37 A single-zone heating/cooling duct system for a small building that has a single source of heating and cooling.

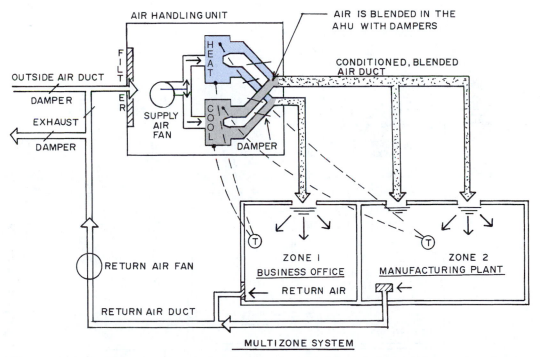

Figure 46.38 A multizone system designed to blend cool and heated air to produce various temperatures required for the various zones.

Reheat Systems

A reheat system is a variation of the single-duct single-zone and the multizone systems. It supplies a single source of preconditioned or recirculated air at a constant rate through ducts to several zoned areas. The air is processed through the air-handling unit, in which it can be filtered, humidified, and cooled. The temperature of the air is that required for the coolest zone. The cool air is then sent through ducts to each zone. The duct to each zone has a reheat coil that heats the incoming air to the temperature required for that zone. The reheat can be an electric resistance unit or a hot water or steam coil (Fig. 46.39). This system permits the simultaneous cooling and heating of areas with different requirements.

Variable Air Volume Systems

Variable air volume (VAV) systems control the temperature in a space by supplying air at a constant temperature and varying the *quantity of air* supplied rather than changing the temperature of the air as is done in the single duct, reheat, and multiple zone systems. The air from the air handling unit is moved through single ducts to each zone, where a variable air volume terminal is located. This terminal varies the air supply to the space, but the air temperature is held constant (Fig. 46.40).

Variable air volume systems may use a common or separate fan system. A reheat coil can be added at the entrance to the zone if necessary. If reheat is used, the volume of air can be reduced, producing a more cost-efficient operation.

Variable air volume systems are not useful in situations in which the control of humidity is important. They are well suited for use in offices and other situations in which temperature control for human comfort is important but precise humidity control is secondary.

Dual Duct Systems

Dual duct systems are similar to the multizone system. The difference is that the conditioned air in the air-handling unit is moved to the spaces by two parallel ducts. One duct carries cold air and the other warm air (Fig. 46.41). At each space, mixing valves combine the warm and cold air in the proportion needed to meet the air temperature requirements of the space. These may be constant volume or variable volume air systems. Constant volume systems may use a reheat. Variable air volume systems may use a single fan or dual supply fans.

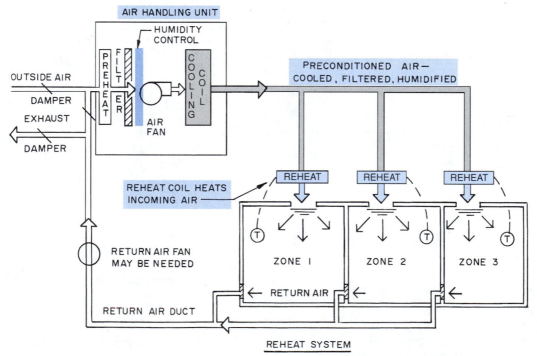

Figure 46.39 A reheat warm air system supplies conditioned air at the lowest required temperature and uses reheat units to raise the temperature at zones requiring higher air temperatures.

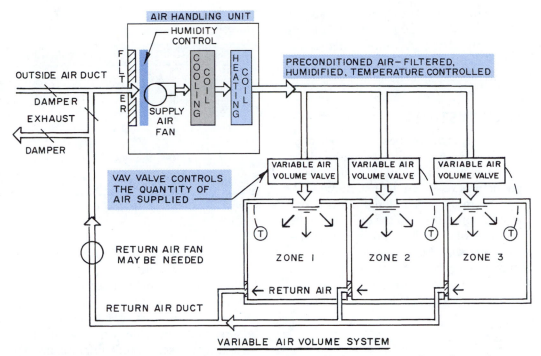

Figure 46.40 A variable air volume warm air system controls air temperature in each zone by varying the amount of air flowing into that zone.

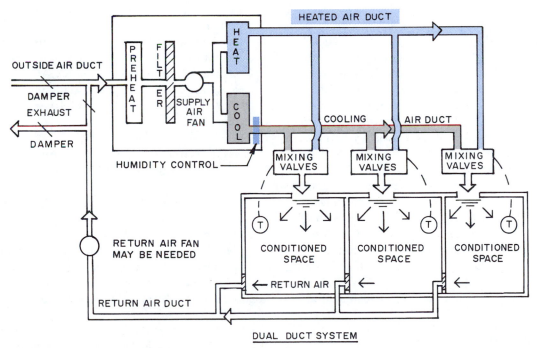

Figure 46.41 Dual duct warm air systems use separate warm and cool air ducts that meet at a mixing valve to produce air at the temperature required.

The dual duct system is not cost efficient because it uses energy to both heat and cool air and then mix them to get the desired temperature. Additional details on these systems can be found in the 1992 ASHRAE Handbook, HVAC Systems and Equipment.

TYPES OF DUCT SYSTEMS

The air distribution system receives heated or cooled air from the furnace. The air moves into the air-handling unit, where it may be filtered, humidified, or dehumidified. The air-handling unit has a fan, filters, humidifiers, coils, and dampers. From the air-handling unit the conditioned air is moved through ducts to the diffusers in the rooms that are to be heated or cooled, and a separate system of ducts moves air from these spaces back to the furnace for reconditioning, possibly exhausting some of the air and adding fresh air brought in from the outside. All of this is accomplished by a series of electrical controls.

Ducts

Warm air distribution systems use ducts to move the heated air from the furnace to the diffusers (outlets) in the various rooms. The design of the ducts and selection of materials are vital to a properly functioning system. The designer must carefully calculate the sizes required, ascertain the air velocity to use, and calculate the pressure.

Duct Pressure

The movement of air through a duct results in both velocity pressure and air pressure loads on the duct. The duct designer must calculate the actual static pressure on each section of the duct and specify the pressure classification. The pressure classifications for residential, commercial, and industrial ducts are shown in Table 46.5. The pressures are specified in inches of water. An inch of water in the inch-pound (I-P) system is a unit of head equal to a column of liquid water 1 inch high at 39.2°F. (Head is the energy per unit mass of fluid divided by gravitational acceleration.) The I-P system uses customary units (inches and pounds) rather than metric units.

A standard for duct design is published by the Air Conditioning Contractors of America (ACCA) Manual D, Duct Design for Residential Winter and Summer Air Conditioning. The Sheet Metal and Air Conditioning Contractors National Association (SMACNA) also has publications relating to duct design and installation.

Duct Classification

Duct systems are regulated by various laws, building codes, local ordinances, and standards. These must be considered by the engineer as the duct system is designed. Projects built for the federal government will have standards issued by various agencies such as

Table 46.5 Allowable Static Pressure Classifications for Ducts in Various Applications

Application	Allowable Static Pressure
Residences	±0.5 in. of water
	±1 in. of water
Commercial systems	±0.5 in. of water
	±1 in. of water
	±2 in. of water
	±3 in. of water
	+4 in. of water
	+6 in. of water
	+10 in. of water
Industrial systems	Any pressure

Courtesy American Society of Heating, Refrigerating and Air-Conditioning Engineers, Inc.

the General Services Administration and the Federal Construction Council.

Duct construction is classified in terms of the pressure and use. *Commercial duct systems* include HVAC systems for applications such as educational, business, general factory, and mercantile structures. *Industrial duct systems* include those used for industrial exhaust and air pollution control.

Model project specifications for the construction of ducts include *Masterspec*, produced by the American Institute of Architects (AIA), and *Spectext*, available from the Construction Specifications Institute (CSI).

Residential ducts are specified by local building codes. An often used source for multifamily dwellings is National Fire Protection Association (NFPA) Standard 90A. Supply ducts may be galvanized steel, aluminum, or other materials rated by Underwriters Laboratory (UL) Standard 181. Rigid and flexible fiberglass supply ducts must meet the standard, Fibrous Glass Duct Construction Standards of the SMACNA.

Commercial ducts are also usually regulated by NFPA Standard 90A and UL181. This classified ducts into two groups:

Class 0. Zero flame spread, zero smoke spread
Class 1. 25 flame spread, 50 smoke developed

Class 0 ducts are of iron, steel, aluminum, concrete, masonry, or clay tile. Class 1 ducts include many of the flexible and rigid fiberglass ducts manufactured.

Industrial ducts are specified by NFPA Standard 91. These are used for duct systems that might convey flammable vapors or air containing various particles. Particle-conveying ducts are available in four classifications:

Class 1. Nonparticulate applications (**makeup air,** general ventilation, and gaseous emission control)
Class 2. Moderately abrasive particles in the air (sanding or buffing)
Class 3. Highly abrasive material in low concentration (handling sand or abrasive cleaning)
Class 4. Highly abrasive particles in high concentration

Abrasive ratings are specified in Round Industrial Duct Construction Standards by the SMACNA. Industrial ducts are generally galvanized steel, uncoated carbon steel, or aluminum. Aluminum is not used if the air contains abrasive particles. Those carrying corrosive vapors must have appropriate protective coatings.

Duct standards include many other specifications, such as insulation, type and spacing of hangers, welding when necessary, seismic requirements, ducts exposed to outdoor atmospheric conditions, and ducts below ground.

Ducts are available in round, rectangular, and flat-oval shapes. Specifications for these vary depending on their use. They are often insulated to control heat loss or gain by the air being conducted. Insulation also controls condensation that occurs when moist air strikes a duct.

FORCED AIR DUCT SYSTEMS

The duct systems used in residential and small commercial buildings include the perimeter loop, perimeter radial, and extended plenum systems. The *perimeter loop system* is typically used with concrete slab floors and a downflow furnace. However, it could be used in a building with a basement or crawl space (Fig. 46.42). The perimeter duct is placed in the thickened edge of the slab, as shown in Fig. 46.43. It is essential that the edge and bottom of the slab be insulated. Registers are placed along the perimeter duct as needed. The return ducts in this system are in the attic.

The *perimeter radial system* is also used in concrete slab construction but can be used in basements and crawl spaces. It uses a downflow furnace (Fig. 46.44). A variation of this system uses an upflow furnace with the radial ducts in the attic (Fig. 46.45).

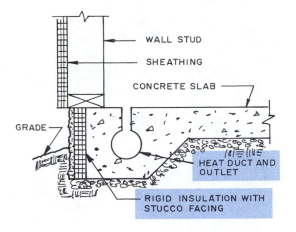

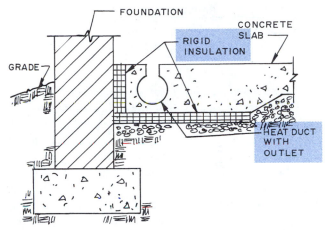

Figure 46.43 Concrete slab floors with ducts in the perimeter must be insulated around the outside of the slab.

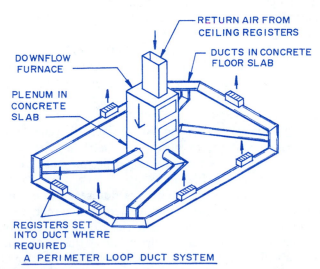

Figure 46.42 A perimeter loop warm air duct system circulates air through a continuous duct system fed by several horizontal ducts.

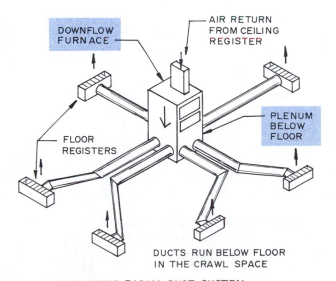

Figure 46.44 A perimeter radial duct system extends individual ducts to each space to be heated and cooled.

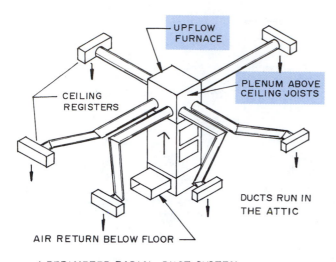

A PERIMETER RADIAL DUCT SYSTEM

Figure 46.45 A perimeter radial duct system can place the ducts in the attic and use ceiling diffusers.

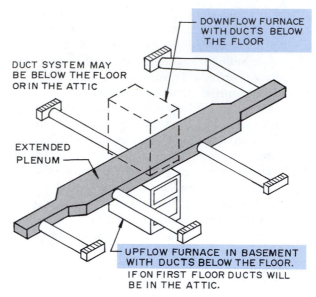

EXTENDED PLENUM DUCT SYSTEM

Figure 46.46 The extended plenum runs from the furnace and individual ducts are taken off it to the rooms to be heated and cooled.

The *extended plenum system* may have the furnace in the basement or on the first floor. It also can be used with horizontal furnaces in the crawl space or attic. The plenum is extended to provide the needed airflow to feed the ducts that run from it to outlets.

If the furnace is on the first floor the plenum could be in the attic and the ducts run over the ceiling joists with diffusers running through the ceiling into the room (Fig. 46.46). Horizontal warm air furnaces are commonly placed below the floor or in the attic (Fig. 46.47).

The ducts in residential and many small commercial systems are hidden from view, but in some large commercial and industrial buildings they are left exposed. Typically these systems are used where large open areas exist, such as in an airport or a sports complex (Fig. 46.48). Notice that this system uses slot diffusers.

SUPPLY AIR OUTLETS AND RETURN AIR INLETS FOR WARM AIR HEATING SYSTEMS

The design and selection of **supply air** outlets and return air inlets is critical to providing adequate conditioning of the air in a room. Poorly located or poorly chosen outlet and inlet grilles and diffusers can reduce the effectiveness of even the best heating and cooling units. The heating engineer must begin work on planning the system early in the planning stages of the building.

The engineer has a considerable number of factors to consider, and a decision is often one that is a compromise. Following are examples of some of the things to be considered.

Air velocity and temperature of the air in the supply duct must be greater than those permitted within the room. This air must therefore be emitted into the room

so air velocity is not offensive and people are not exposed to areas of high temperatures. This requires the system to mix the supply of air with the air in the room to control convection currents and uneven temperatures.

The air must be supplied to the room with a termination device (grille, diffuser, etc.) that provides the required air dispersion pattern needed for human comfort. The termination device must throw the air across the room far enough to enable it to blend with the air in the room, thus reducing velocity and temperature before it drops into the lower part of the room. The engineer recognizes that cool air is heavy and settles to the floor, but warm air is lighter and rises toward the ceiling. The windows in a room also cause design considerations because they introduce a source of cold air that produces discomfort for the occupants and cools the circulating warm air.

The engineer also considers **surface effect.** This is caused when the moving airstream from the duct outlet moves across a ceiling or has contact with a wall. This creates a low-pressure area along the surface, keeping the air in contact with it for most of the **throw.** Circular ceiling diffusers project an airstream across a portion of the ceiling and create surface effect if they are long enough to cover the ceiling area. Grilles may also cause some ceiling effect (Fig. 46.49).

Smudging occurs with both slot diffusers and ceiling diffusers. Smudging is a band of discoloration that eventually appears on the ceiling material around the edge of the diffuser. It is caused by dirt particles that are in the discharged airstream. They are held in the air turbulence around the diffuser and can cause smudging around the outlet. This can be combatted by placing antismudge rings around the outlet.

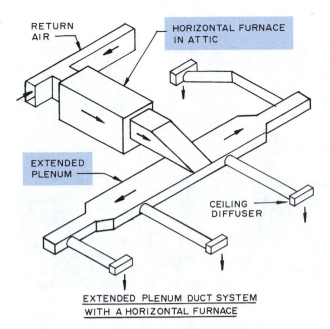

EXTENDED PLENUM DUCT SYSTEM
WITH A HORIZONTAL FURNACE

Figure 46.47 The extended plenum runs off the plenum on the end of the horizontal furnace. Notice the return air ducts at the rear of the furnace. *(Courtesy Bryant Air Conditioning)*

Another design consideration is the sound level of the outlet. The outlet will transmit noise caused by mechanical equipment and high air velocities through the duct. High-pitched noise is frequently caused by the air passing through the outlet, and a different design might be chosen. Recommended air velocities and decibel levels are found in publications of the American Society of Heating, Refrigerating and Air-Conditioning Engineers.

Types of Supply Outlets

The basic types of supply outlets are grilles, slot **diffusers,** air distributing ceilings, and ceiling diffusers.

A *grille* is a rectangular opening with vanes that direct the airstream as it leaves the outlet. *Adjustable* *grilles* have adjustable vertical or horizontal vanes that deflect the airstream in a horizontal or vertical direction. Some types have two sets of vanes at right angles to each other. *Fixed bar grilles* have vanes that cannot be moved. *Variable area grilles* (also called registers) have dampers that can vary the size of the outlet, thus varying the air that can pass through (Fig. 46.50). *Stamped grilles* are simply a flat sheet with openings that permit the air to pass through with no control of direction or volume (Fig. 46.51).

Slot diffusers are rectangular outlets with one or several slots often installed in long lengths (Fig. 46.52). *Perpendicular slot diffusers* discharge air perpendicular to or at a slight angle to the face of the diffuser. *Parallel slot diffusers* produce an airstream parallel with the face of the diffuser. A typical location is in the sill of large windows

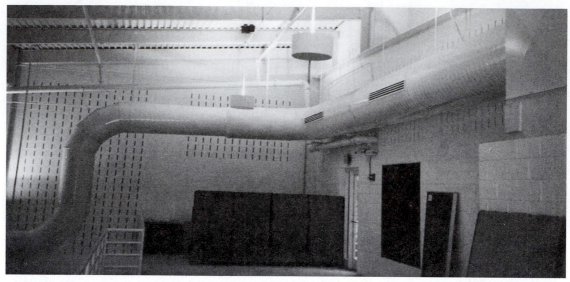

Figure 46.48 Exposed duct systems are used in buildings that have large open areas. Notice the use of a slot diffuser.

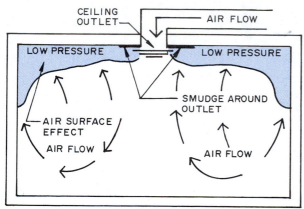

AIR SURFACE EFFECT WITH A CEILING DIFFUSER.

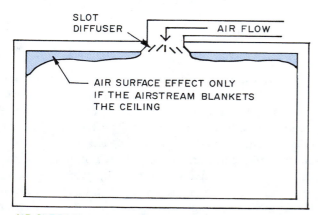

AIR SURFACE EFFECT WITH A SLOT DIFFUSER.

Figure 46.49 Air surface effect keeps the incoming air from a ceiling diffuser moving across the ceiling.

or in the floor below glass doors or floor-length glass windows. This creates an upflow of heated air over the cold glass, producing increased comfort for the occupants of the room.

Air distribution ceilings use the space between the bottom of the floor above and the top of a suspended ceiling as a plenum. The space is fed heated or cooled air by ducts located around the perimeter. The ceiling panels have perforations through which air enters the room (Fig. 46.53).

Ceiling diffusers may be round or square and contain a series of louvers that form air passages. They distribute the supply air in a uniform pattern around the diffuser (Fig. 46.54). *Flush ceiling diffusers* have concentric louvers that all extend the same distance from the core. *Stepped-down diffusers* have louvers that project beyond the shell of the unit, as shown in Fig. 46.54. *Perforated-face ceiling diffusers* have a flat perforated face that is flush with and blends in with the surface of the suspended ceiling. They may have deflection vanes behind the exposed face to direct the airstream discharge (Fig. 46.55). *Variable ceiling diffusers* have dampers that can be adjusted to control the amount of airflow from the diffuser.

Another type of ceiling air distribution diffuser is shown in Fig. 46.56. The supply duct runs above a suspended ceiling and supplies the conditioned air to the air diffuser. The diffuser is a rectangular unit that rests upon horizontal metal members that form the grid of the suspended ceiling (Fig. 46.57). Below this are installed parallel beams that have spaces between them used to distribute the air into the room. The direction of airflow is controlled in two directions by vanes (Fig. 46.58). Diffusers are available in several sizes, as shown in Fig. 46.59.

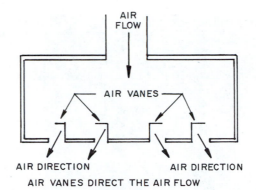

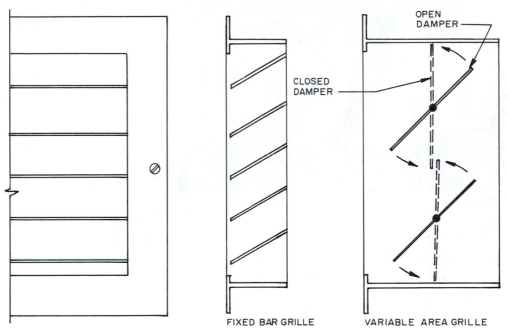

Figure 46.50 Grilles may have fixed bars or adjustable vanes that vary the direction and amount of air flow from the duct.

Figure 46.51 This is a decorative stamped grille typically used on floor outlets.

Figure 46.52 A slot diffuser used in a large-volume exposed air duct.

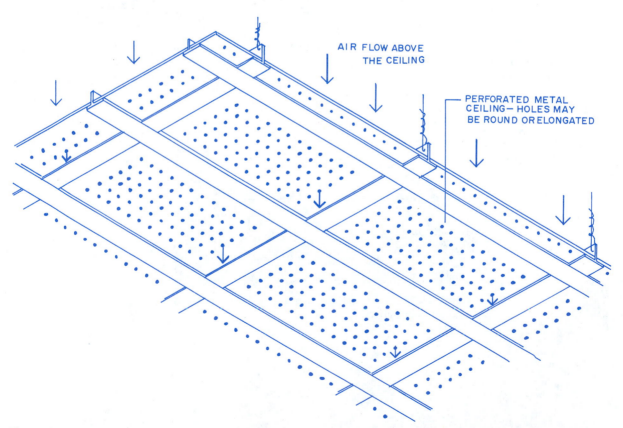

AIR FLOW ABOVE
THE CEILING

PERFORATED METAL
CEILING— HOLES MAY
BE ROUND OR ELONGATED

Figure 46.53 Perforated metal ceiling panels diffuse conditioned air flowing above them into the space below.

Figure 46.54 A round variable ceiling diffuser with dampers used to control the amount of air passing into the room.

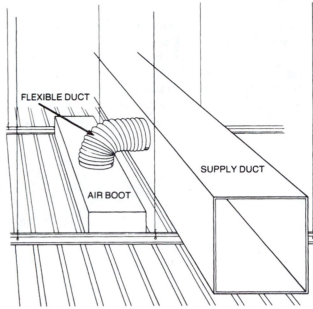

Figure 46.56 This diffuser fits into the grid of the suspended ceiling and is fed from a supply duct above the ceiling. *(Courtesy Chicago Metallic Corporation)*

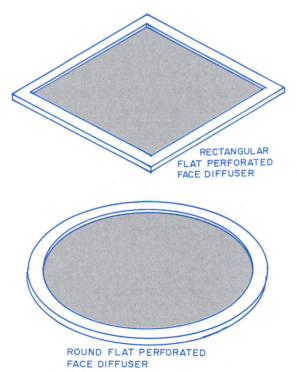

RECTANGULAR FLAT PERFORATED FACE DIFFUSER

ROUND FLAT PERFORATED FACE DIFFUSER

Figure 46.55 Perforated diffusers lie flat with the ceiling and are less obvious than ceiling grilles.

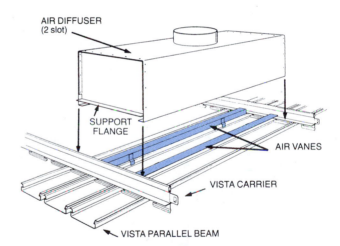

NOTE: Air diffusers should be seated on the Vista tees/carriers via the diffuser support flanges.

Figure 46.57 The diffuser is carried by the horizontal tees used in the suspended ceiling. *(Courtesy Chicago Metallic Corporation)*

Figure 46.60 shows a drop-in 2 ft. × 2 ft. air diffuser that fits into the grid of a suspended ceiling system manufactured by Chicago Metallic Corporation. It connects to a supply duct, and air reaches it through a flexible plastic duct (Fig. 46.61). The air is distributed into the room through a slot around the edge and is controlled by a vane (Fig. 46.62). The center of the diffuser is covered with a ceiling panel that matches those used in the overall ceiling.

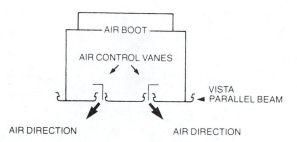

Air Control Vanes snap on the Vista Parallel Beam, one per slot. When the vanes are positioned as shown, a two-directional air pattern results. For a one-directional air pattern the appropriate vanes must be placed on the opposite side of the slot.

Figure 46.58 The air flow from the diffuser is through slots in the metal ceiling panels and can be directed with vanes. *(Courtesy Chicago Metallic Corporation)*

Figure 46.59 Diffusers are available to provide various amounts of air flow, and they can direct the flow in two directions. *(Courtesy Chicago Metallic Corporation)*

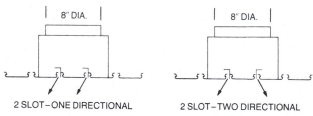

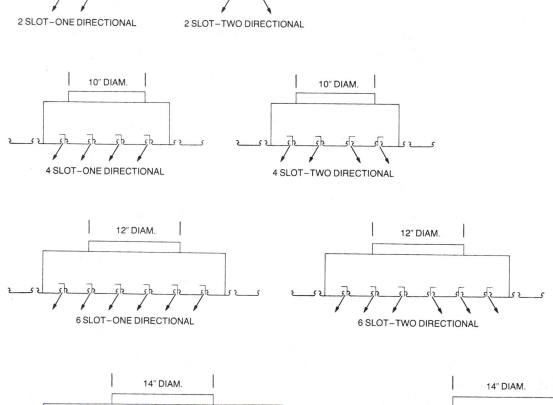

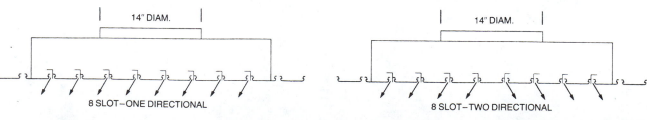

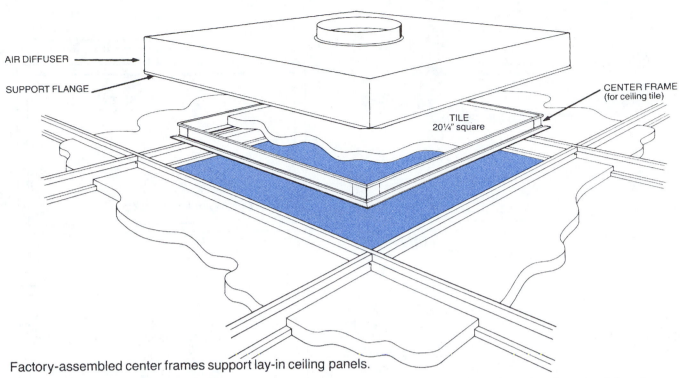

AIR DIFFUSER

SUPPORT FLANGE

CENTER FRAME
(for ceiling tile)

TILE
20¼" square

Factory-assembled center frames support lay-in ceiling panels.

Figure 46.60 This ceiling diffuser fits into the grid of a suspended ceiling and distributes air through a slot around the edges. *(Courtesy Chicago Metallic Corporation)*

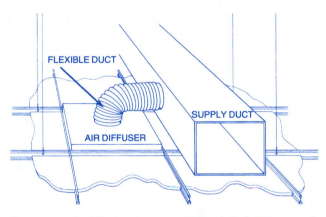

FLEXIBLE DUCT

SUPPLY DUCT

AIR DIFFUSER

The 2'x 2' air diffusers are manufactured of zinc coated sheet metal with acoustical and thermal insulation factory applied to the internal surfaces of the unit. They are equipped with top-entry inlet collars which vary in size based on required air delivery. The standard collar size is 8" in diameter with other sizes available upon request.

Figure 46.61 The square ceiling diffuser is fed from a supply duct run above the suspended ceiling. *(Courtesy Chicago Metallic Corporation)*

Some Applications

Following are some general applications for the location of outlet and inlet diffusers. Outlet diffusers bring air into a space, and inlet diffusers are used with the air returns or exhaust.

In general, in areas where the need for air cooling in a room is predominant, the outlets can be most effective in the ceiling and the air returns to the furnace on the wall at the floor or in the floor. This is possible because cool air sinks toward the floor as it is mixed with the air in the room. In areas where the need for heating in a room is predominant, the outlets can be in the floor or at the baseboard on the wall. The return air can be high on the wall (Fig. 46.63).

Outlet ducts are generally located on outside walls and under windows and return air ducts on inside walls. This permits a curtain of conditioned air to flow up the outside wall, warming or cooling its surface as needed to produce a more comfortable interior situation. A cold outside wall will pull body heat from an occupant even though the air temperature is adequate. Likewise a hot wall will radiate heat to the occupant (Fig. 46.64).

Return Air Inlets and Exhaust Air Inlets

Return air inlets connect to ducts that return the air from a room back to the furnace for reconditioning. Exhaust air inlets move air from inside the building through ducts and discharge it into the outside air. They typically do not require vanes to deflect or redirect the airstream. Some types do have dampers to control the amount of air flow. The *fixed bar grille* is commonly used and may have bars on a slight angle or perpendicular to the face of the grille (Fig. 46.65). A fixed bar louvered grille for

AIR CONTROL VANES

GRID

AIR DIRECTION   AIR DIRECTION

VERTICAL AIR FLOW

AIR DIRECTION   AIR DIRECTION

HORIZONTAL AIR FLOW

PERIMETER DETAIL

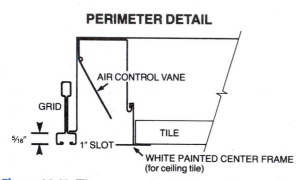

AIR CONTROL VANE

GRID

TILE

⁵⁄₁₆″   1″ SLOT

WHITE PAINTED CENTER FRAME
(for ceiling tile)

Figure 46.62 The square ceiling diffuser distributes air through a slot around the edge and can use a vane to control air direction.

exhaust air is shown in Fig. 46.66. Stamped grilles and ceiling and slot diffusers are also used.

HUMIDIFIERS AND DEHUMIDIFIERS

Control of the relative humidity in a space is an important factor in the overall conditioning of the environment. The requirements vary depending on the occupancy. **Humidity** is the water vapor within a space. **Relative humidity** is a ratio of the weight of water vapor actually in the air to the maximum possible weight of water vapor the air could contain when at the same temperature. It is expressed as a percentage. For example, if the relative humidity is 100 percent the air can hold no more water vapor and an increase in water vapor will cause moisture to condense and form water drops.

Human comfort depends a great deal on the relative humidity of the air. Typically, indoor relative humidity should

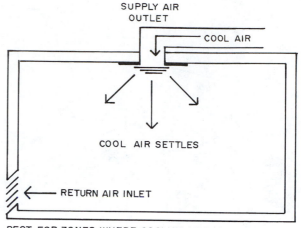

SUPPLY AIR OUTLET

COOL AIR

COOL AIR SETTLES

RETURN AIR INLET

BEST FOR ZONES WHERE COOLING NEEDS PREDOMINATE

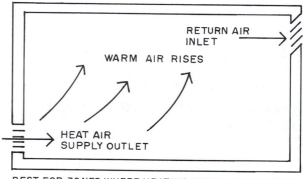

RETURN AIR INLET

WARM AIR RISES

HEAT AIR SUPPLY OUTLET

BEST FOR ZONES WHERE HEATING NEEDS PREDOMINATE

Figure 46.63 Frequently used locations for air outlets and inlets (returns).

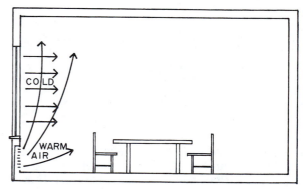

COLD

WARM AIR

Figure 46.64 Outlets under windows merge warm air with cold winter air or warm summer air with cool conditioned air to provide increased comfort to the occupants near the window.

be kept at 30 to 60 percent. Low humidity causes drying of the membranes of the nose, throat, skin, and hair. Furniture, cabinets, interior trim, and other wood products can shrink and check if the relative humidity is too low. Likewise high relative humidity can cause doors and drawers to swell and stick. Heating the air in the winter removes moisture, so a humidifier is used to increase the relative humidity. In the summer the air in many geographic areas has a

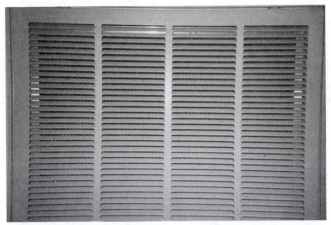

Figure 46.65 A fixed bar return air grille used on an interior wall.

high relative humidity. The air-conditioning system must dehumidify (remove water vapor from) the air.

Humidification Equipment

Humidifiers increase the amount of water vapor in the air. They operate in two different ways. One type adds heat, and as the air flows through the humidifier wet section it picks up moisture. The air is then distributed to the room. The other type takes heat from the air flow and the evaporation occurs in the room. Typical types of humidifiers in residential and small commercial buildings are shown in Fig. 46.67.

The *pan humidifier* has water automatically fed into the pan. Vertical water absorbent plates rest in the water and pass off moisture to the air as it passes over them, as shown in Fig. 46.67 part A. The water level is automatically controlled. The pan unit may have an electric heater that increases the rate of evaporation.

Wetted type humidifiers circulate air through or over a fibrous, porous material that is wetted by a spray, by gravity water flow over it, or by revolving through a water pan. One type uses a fan that moves air from the furnace plenum and blows it through the pad (Fig. 46.67 part B) or rotating drum (Fig. 46.67 part C) and back into the plenum.

A *bypass humidifier* is shown in Fig. 46.67 part D. The humidifier is mounted on the furnace air supply and uses air moved by the furnace blower. Small humidifiers are available that can be mounted in the duct.

A *spinning disc atomizing humidifier* is shown in Fig. 46.67 part E. Small particles of water are injected into the airflow and absorbed, raising the humidity. Another type uses a spray nozzle to inject a mist into the airflow.

A small *portable humidifier* is illustrated in Fig. 46.67 part F. It has a rotating belt that moves through a reservoir of water. A fan blows air through the belt, where the air picks up moisture, and distributes the air into the room.

Various industrial humidifiers are shown in Fig. 46.68. The *heated pan humidifier* shown can be heated by

Figure 46.66 A fixed bar louvered exhaust air grille.

hot water or steam coils or by an electric element. It may be installed on the duct as shown or be operated in a remote location and the hot water vapor carried to the main duct system by a connecting duct. Other types use an enclosed grid (Fig. 46.68 part B), or inject steam directly into the duct (Fig. 46.68 part C). The *jacketed steam humidifier* in Fig. 46.68 part D operates off a constant steam supply flowing into a steam trap in which any condensation is drained off. The flow into the dispersing tube in the duct is controlled by a steam valve.

A *self-contained steam humidifier* is shown in Fig. 46.68 part E. It uses electric heaters to convert water to steam and then injects it into the airflow in the duct. Various types of *atomizing humidifiers* (Fig. 46.68 part G) also are used in industrial applications.

Desiccant Dehumidification

Dehumidification involves the removal of water vapor from the air, gases, or other fluids by some mechanical or chemical means. A **dehumidifier** is a device used to remove moisture from the air, thus reducing the relative humidity.

Desiccation is the use of a desiccant for removing moisture from a material. A **desiccant** is any absorbent liquid or solid that is used to remove water or water vapor from a material. Dehumidification equipment uses both solid and liquid desiccant materials. **Absorbent** materials will extract substances from a liquid or gas medium with which they are in contact. **Adsorbent** materials have the ability to have molecules of gases, liquids, or solids to adhere to their surfaces without changing the adsorbent material chemically or physically.

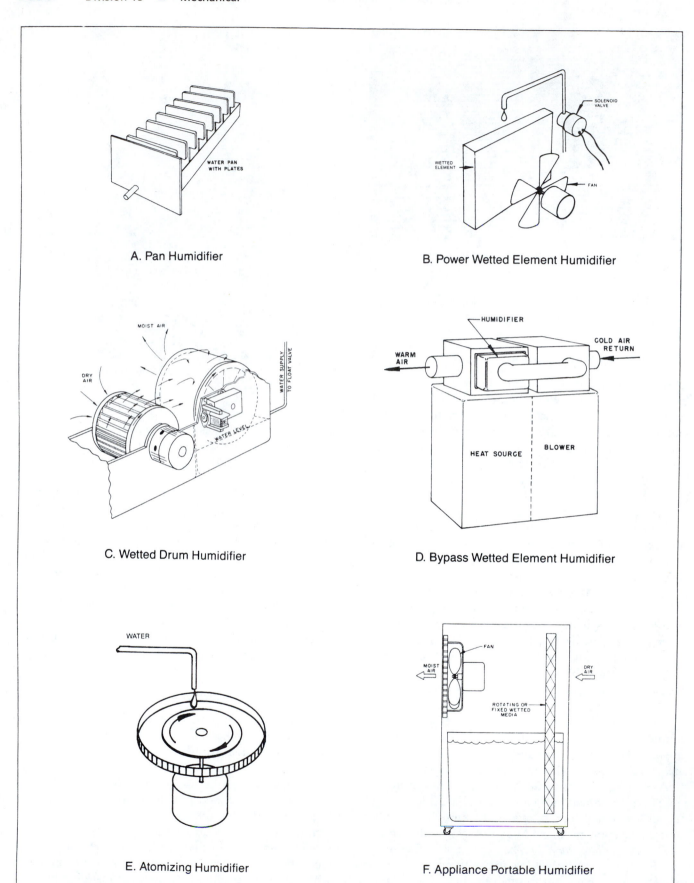

A. Pan Humidifier

B. Power Wetted Element Humidifier

C. Wetted Drum Humidifier

D. Bypass Wetted Element Humidifier

E. Atomizing Humidifier

F. Appliance Portable Humidifier

Residential Humidifiers

Figure 46.67 Types of humidification equipment commonly used in residences. *(Courtesy American Society of Heating, Refrigerating, and Air-Conditioning Engineers, Inc.)*

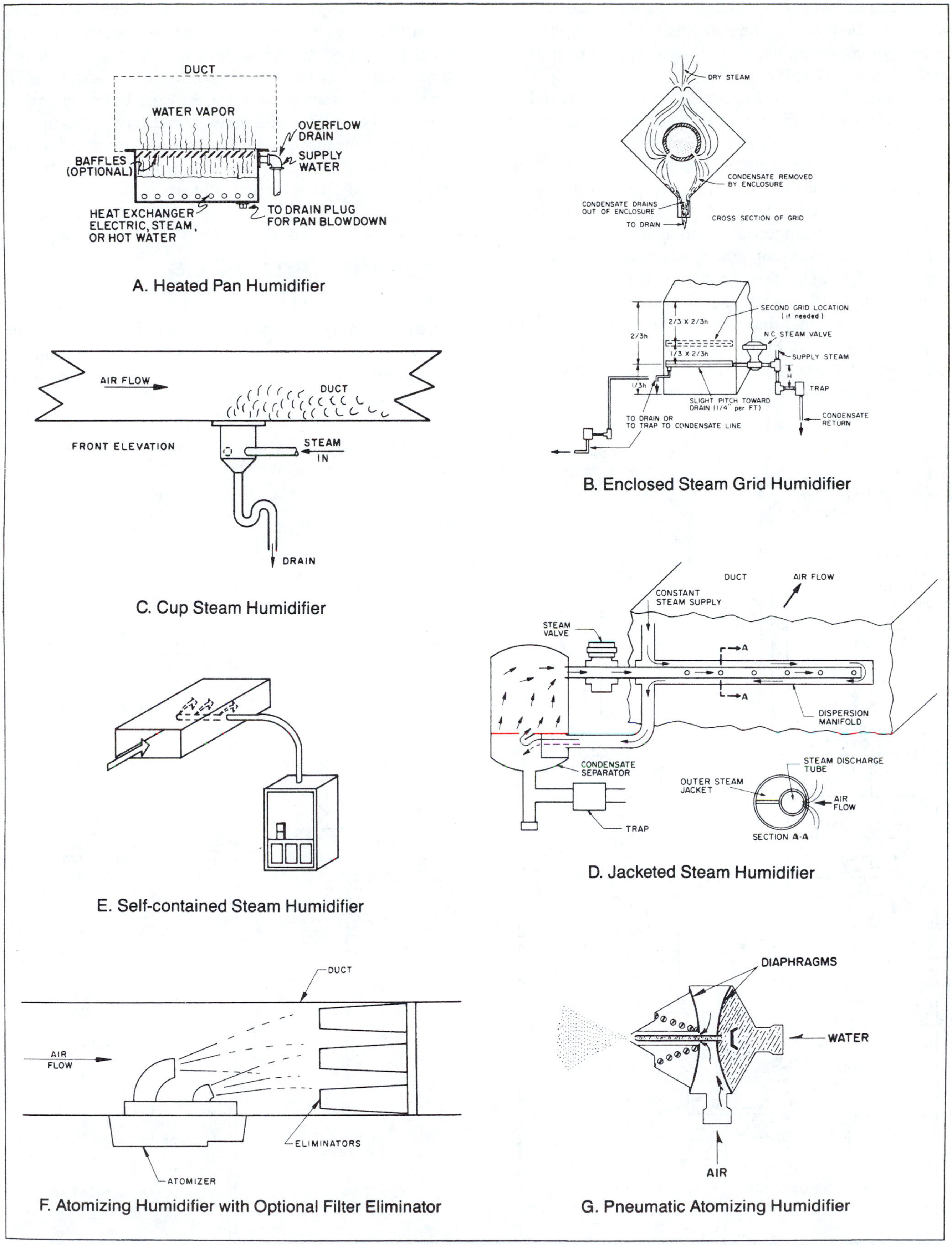

A. Heated Pan Humidifier

B. Enclosed Steam Grid Humidifier

C. Cup Steam Humidifier

D. Jacketed Steam Humidifier

E. Self-contained Steam Humidifier

F. Atomizing Humidifier with Optional Filter Eliminator

G. Pneumatic Atomizing Humidifier

Industrial Humidifiers

Figure 46.68 Types of humidification systems used in industrial buildings. *(Courtesy American Society of Heating, Refrigerating and Air-Conditioning Engineers, Inc.)*

Dehumidification is typically accomplished in residences by the cooling of air during the air-conditioning mode. For commercial and industrial applications other methods are available.

Dehumidification is required for many industrial applications. Among these are the need to maintain a dry atmosphere in a warehouse, to produce dry air when needed to aid in the drying of a material in an industrial process, to dry natural and liquified gas, and to lower the relative humidity in a plant manufacturing products using hygroscopic materials such as wood.

Liquid desiccant dehumidification systems remove moisture from the air by flowing the air through a mistlike spray of a liquid desiccant such as a glycol solution. The desiccant spray absorbs moisture from the air and is passed into a regeneration chamber where it is heated and gives off the moisture. The moisture is discharged by a flow of outside air (Fig. 46.69).

Solid sorption dehumidification systems remove moisture from the air by flowing it through a granular desiccant such as silica gel or hygroscopic salts. *Sorption* is a general term used to include both absorption and adsorption. The granular desiccant has a vapor pressure below that of the vapor in the air. This difference in pressure drives the water vapor in the air into the desiccant. When the desiccant becomes saturated, it is dried by heating it in a reactivation chamber and can be reused (Fig. 46.70).

HYDRONIC (HOT WATER) HEATING SYSTEMS

Hydronic heating systems are used in residential and commercial installations. A system consists of a boiler fueled by oil or natural gas, and a system of pipes, radiators, pumps, and controls. Three types of systems are typically used in residential and small commercial buildings. These are one-pipe, two-pipe direct-return, and two-pipe reverse-return systems.

One-Pipe Systems

One-pipe systems use a single loop of pipe to circulate water at 180°F (82.9°C) to radiators and to return the

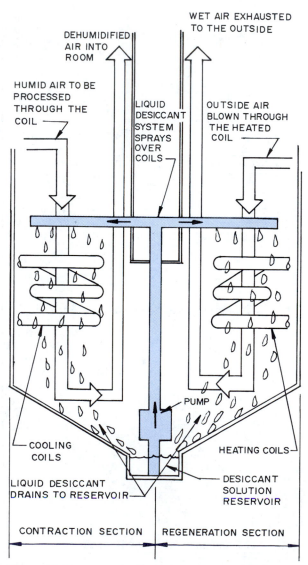

Figure 46.69 Liquid desiccant dehumidification uses a mistlike spray of the liquid desiccant to absorb moisture in the air.

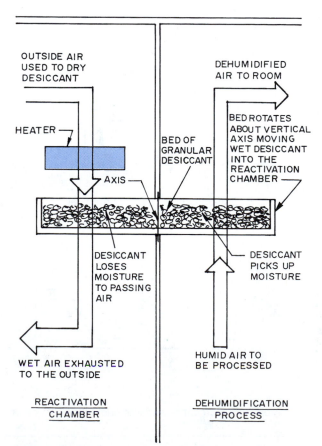

Figure 46.70 Solid sorption dehumidification flows the humid air through a granular desiccant.

cooler water to the boiler to be reheated and recirculated as shown in Fig. 46.71. Notice that the water flows directly from the boiler to radiator 1, from which it flows into radiator 2. The water temperature at radiator 2 will be lower than at radiator 1. This loss can be compensated for by having a larger No. 2 radiator, and the loss continues through the remainder of the radiators on the loop. This system is difficult to keep in balance.

Two-Pipe Direct-Return System

All the radiators in this system receive hot water directly from the boiler through the hot water supply line. The cool water from each radiator is returned directly to the boiler through a separate loop of pipe. The last radiator tends to receive less water because of higher pipe resistance to water flow. It has the longest hot water supply and cool water return lines. This is overcome by increasing the size of the hot water supply line and sizing the circulation pump to provide the needed flow at the radiators on the end of the circuit (Fig. 46.72). Balancing devices are used to regulate the flow of hot water through each radiator. In general, this system is not widely used because of the difficulty in getting a balanced distribution.

Two-Pipe Reverse-Return System

This system is much like the two-pipe direct-return except the return water flows in the same direction as the hot water. This provides a system having about the same pipe resistance on all radiators. For example, radiator 1 has the shortest hot water supply line but the longest return line. The reverse is true for radiator 3. This system is typically used in larger buildings with longer pipe runs (Fig. 46.73).

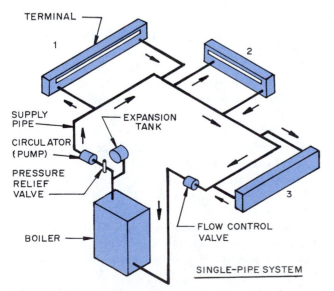

Figure 46.71 A single-pipe hydronic system carries the water from the boiler through each terminal and back to the boiler.

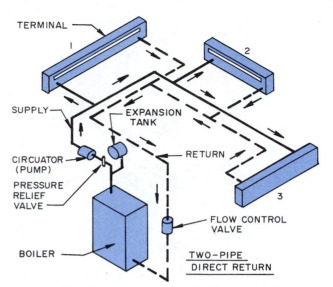

Figure 46.72 The two-pipe hydronic direct-return hot water system has a separate return line for the cooled water, which flows in a direction opposite the supply line.

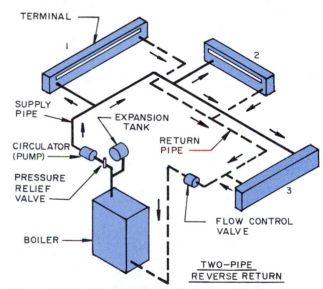

Figure 46.73 The two-pipe reverse-return hydronic hot water system has the return water flowing in the same direction as the hot supply water providing about the same pipe resistance to water flow for each terminal.

Multizone Two-Pipe Systems

A multizone system enables the temperatures in various parts of a building to be controlled separately. Each zone has a complete two-pipe system fed from a central boiler. The flow of hot water is individually controlled to each zone. This permits some zones to be kept at lower temperatures when not in use, resulting in a saving of energy costs. Large multistory and multiuse buildings will use multizone two-pipe hydronic systems (Fig. 46.74).

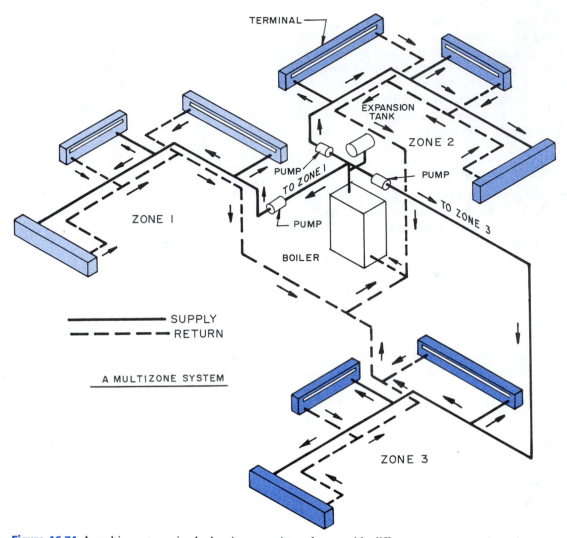

TERMINAL

EXPANSION TANK

ZONE 2

PUMP

TO ZONE 1

PUMP

TO ZONE 3

ZONE 1

PUMP

BOILER

——————— SUPPLY

— — — — — RETURN

A MULTIZONE SYSTEM

ZONE 3

Figure 46.74 A multizone two-pipe hydronic system is used to provide different temperatures in each zone.

Hydronic Controls

The water temperature in the boiler (180°F or 83°C) is controlled by a thermostat immersed in the water in the boiler. The water thermostat regulates the operation of the oil or gas burner. A room thermostat is used to start and stop the circulation pumps regulating the supply of water to the radiators. The boiler is often used to supply hot water for domestic use. In this case it must be sized to meet both heating and domestic water demands.

Expansion Chambers

Hydronic heating systems require some form of expansion chamber (tank). This serves as a space into which the water that is not compressible can flow as it expands within the system due to increase in the temperature. Three types of expansion systems are commonly used. These are open tank, closed tank, and a diaphragm tank (Fig. 46.75).

The *open expansion tank* is located above the highest radiator and is open to the atmosphere. It must be large enough to hold the maximum amount of water produced. These can cause poor performance and are not generally used.

The *closed expansion system* has a set volume of air that can be compressed and water that the tank will hold as the air is compressed. The air pressure changes as the volume of water increases or decreases. The only time water can flow into or out of the tank is when the water expands or shrinks. The tank is sized to handle expansion due to the maximum temperature of the water. The size is determined by equations that are available in the ASHRAE publication, HVAC Systems and Equipment.

The *diaphragm expansion system* has a flexible membrane in the tank to separate the air and water. As the expansive water enters the tank from the top the di-

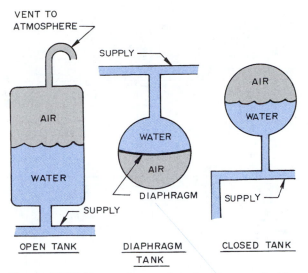

Figure 46.75 Types of expansion chambers used with hydronic heating systems.

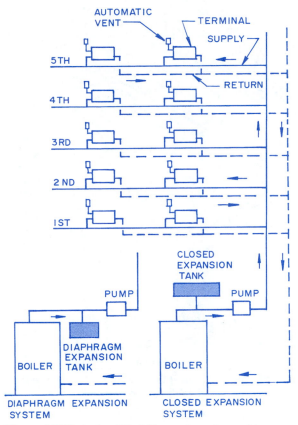

Figure 46.76 A simplified illustration of a multistory hot water heating system showing the diaphragm and closed expansion systems.

aphragm moves down, compressing the air. This system is widely used.

A simplified illustration of a multistory hot water heating system showing both closed and diaphragm type expansion systems is in Fig. 46.76. Each terminal unit has a vent that opens automatically to relieve excessive pressure.

PIPE SYSTEMS FOR WATER HEATING AND COOLING

Water distribution systems can be used for heating with hot water and cooling with chilled water. The systems are usually closed systems and use circulators (pumps) to move the water through the system and the terminal units. The hot water is supplied by a **boiler** and the cold water by a **chiller**. A chiller is a refrigerating machine used to remove heat from the water to be circulated for cooling air in a building. Three-pipe and four-pipe distributions are used.

Three-Pipe Systems

Three-pipe systems provide heating and cooling supply to the terminal units by running a heating supply pipe to each terminal. A third pipe is the return in which the hot and chilled water are mixed and returned to the boiler and chiller, which results in warmed chilled water going into the chiller and cooled warm water going into the boiler. This results in increased cost to reheat and rechill the water before it is recirculated. This is not economical and is not widely used.

Four-Pipe Systems

Four-pipe systems are used when the system provides both heating and cooling modes. The system provides each terminal with separate hot water and chilled water supply and return lines. This provides heating and cooling as required any time either is needed. Systems that use the same coil for heating and cooling control the flow with two valves on each of the heating and cooling supply at each terminal. Others put separate heating and cooling coils in the terminal unit (Fig. 46.77).

STEAM HEATING SYSTEMS

A steam heating system has a boiler or other steam generating device, a piping system, radiators or **convectors,** and controls. The boiler is usually oil or gas fired, but coal, wood, waste products, solar, nuclear, electrical energy, or cogeneration sources can also be used. The two types of steam heating systems are one-pipe and two-pipe.

Steam delivers considerably more heat per pound (0.45 k) than a pound (0.45 k) of water, but when it becomes vapor (steam) it expands much more than hot water. Therefore, a steam system requires larger diameter pipes than hot water systems. Steam produces high pressures that force it through the piping system without the use of pumps as are used in hot water systems. The pipes also must be sized to allow gravity flow of the condensed water back to the boiler without interfering with the flow of the steam. Engineers use the design data for pipe sizes

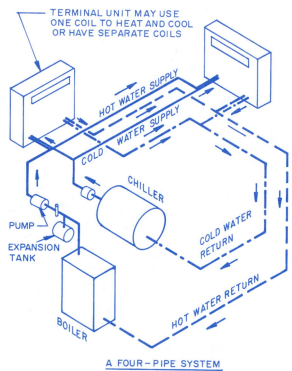

Figure 46.77 A four-pipe hydronic system providing both heating and cooling modes.

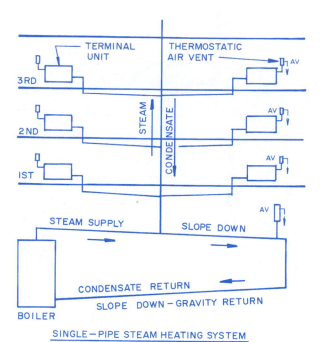

Figure 46.78 A single-pipe steam heating system uses the same pipe system for the steam supply and condensate return.

available in the ASHRAE Handbook. Steam space heating systems are usually classified as *low pressure*.

It is more difficult to control the temperature of steam than hot water, so hot water is more widely used for space heating systems. If a building requires steam for an industrial process the steam supply is generally used for space heating. This is usually *high-pressure steam* that requires the use of pressure-reducing valves in series to get the pressure to that acceptable for space heating.

One-Pipe Steam Heating System

The one-pipe system uses a single pipe that supplies steam to the radiators and also serves as the return line for the condensed water to flow back to the boiler (Fig. 46.78). Valves are used to control the steam flow to the radiators, and **condensates** flow back through the valve. Each radiator has thermostatically controlled air vents. The condensed water flows to the boiler by gravity, so the piping is sloped. Pressure in one-pipe steam systems is usually kept below 5 psi (3515 km^2). This system is used only in small buildings and is not widely used.

Two-Pipe Steam Heating Systems

A *two-pipe gravity return system* is shown in Fig. 46.79. This system has separate piping for the steam supply and the return of the condensate to the boiler. *Thermostatic traps* on the outlet line of each radiator or other terminal unit keep the steam within the unit until it has dis-

persed its latent heat. Then the trap opens, the condensate flows out the return line, and additional steam enters the terminal unit.

Two-pipe steam systems use either gravity or mechanical returns. Gravity flow two-pipe systems are used only in small systems. Most systems use higher steam pressures to supply steam to the terminal units and a condensate pump or vacuum pump to return the condensate to the boiler.

A *two-pipe vacuum system* is similar to the two-pipe mechanical return system but has a vacuum pump added to give an additional pressure difference. It circulates condensate, removing gas that is not condensable and discharging it to the atmosphere, which creates a vacuum on both the steam supply line and the condensate return line. The vacuum return system is used on larger buildings because it requires a lower steam pressure and can fill the system with steam rapidly (Fig. 46.80).

BOILERS

Boilers are used to transfer heat from a fuel source to a fluid, such as water. The liquid is contained in a cast iron, steel, or copper pressure vessel that transfers heat to the water to produce hot water or steam (Fig. 46.81). Boilers are constructed according to the American Society of Mechanical Engineers (ASME) Boiler and Pressure Vessel Code. The Hydronics Institute publishes the Testing and Rating Standard for Heating Boilers.

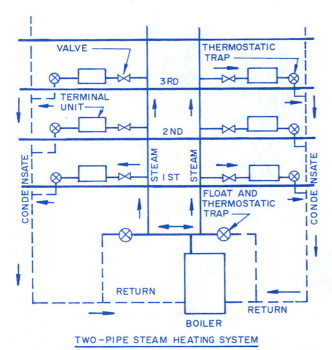

Figure 46.79 Two-pipe steam heating systems have separate piping lines for the steam supply and the return condensate.

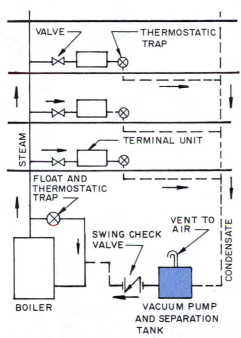

Figure 46.80 The two-pipe vacuum steam heating system uses a vacuum pump to increase the pressure difference and discharges gas that is not condensable to the atmosphere.

Boilers are classified by working pressure, temperature, fuel, size, and whether they are steam or water boilers. *Steam boilers* are used for space heating and for auxiliary uses such as in a commercial laundry or industrial processes in which steam is required. Steam typically is used in large commercial and industrial buildings. *Hot*

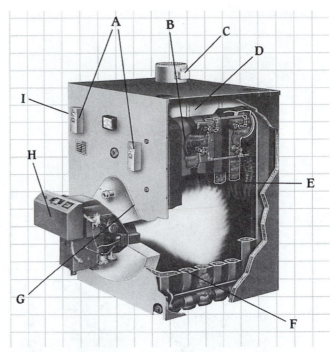

PF-5 Series Features
A. Front-mounted controls for easier adjustment and maintenance
B. Tankless heater for optimum domestic hot water
C. Adjustable lock-type damper for improved efficiency
D. Aluminized steel flue canopy for long life
E. Cast iron vertical flue design for maximum heat
F. Wet base thermal pump construction for improved circulation
G. Burner mounting plate with flame observation port
H. Four manufacturer burner options to best fit your needs
I. Left side cleanout for easy entrance to all flue surfaces

Figure 46.81 This is an oil-fired cast iron wet base boiler. The letter "F" points to the cast iron boiler sections that are assembled forming the boiler sections on all sides of the heat source. *(Courtesy Burnham Corporation, Hydronics Division, Lancaster, PA 17604)*

water boilers are used for space heating and domestic hot water supply. They are typically used in residential and small commercial buildings.

Boilers are classified as high-pressure or low-pressure. *High-pressure boilers* are referred to as *power boilers*. They produce steam pressures above 15 psi and hot water pressures and temperatures above 160 psi and 250°F (122°C). They are typically of steel construction and use firetube or watertube design.

Low-pressure boilers are referred to as *heating boilers* and are limited to a maximum steam pressure of 15 psi and a maximum hot water pressure of 160 psi and temperature of 250°F (122°C). They are made from cast iron, steel, or copper.

Firetube boilers have the hot combustion gases pass through tubes that are surrounded by water. They are used in both low- and high-pressure boilers. *Watertube boilers* have the water inside tubes, and the hot combustion gases

Figure 46.82 These are high-efficiency cast iron wet base boilers *(Courtesy H.B. Smith Company, Inc.)*

pass over the tubes. Some boilers have the tubes in horizontal or nearly horizontal position, and some are vertical. Small, low-pressure boilers used for residential units may have helical coils. Watertube boilers are available ranging from small low-pressure units to large high-pressure steam units.

Boilers are available as condensing and noncondensing units. For many years boilers were operated so they did not condense the flue gases in the boiler. This was done to prevent cast iron and steel parts from corroding. *Noncondensing boilers* are operated with a high flue gas return temperature to prevent moisture in the flue gas from condensing on the cast iron and steel parts.

Condensing boilers salvage latent heat from the products of combustion by condensing the flue gas, which increases the efficiency of the boiler. This means lower flue gas temperatures can be used and the water vapor in the flue gas can condense and drain. The condensing section of the boilers must be made of materials that will withstand the required temperatures and resist corrosion. Certain types of stainless steel are used.

A pair of high-efficiency cast iron wet base boilers are shown in Fig. 46.82. They operate on oil or natural/propane gas, and models are available for hot water and steam heating systems. Some types may have a combination gas/oil-fired burner/boiler unit typically used in schools, apartments, and commercial buildings, where a switch in fuels may be essential for continued heating service.

Cast iron boilers are made by assembling individual cast iron boiler sections called watertubes in which the water to be heated flows (Fig. 46.83). The size and rating of the boiler depends on the number of sections assembled (Fig. 46.84). The tubes may be vertical or horizontal. The boiler may be *dry base*, which means the firebox is beneath the fluid-backed sections, or *wet base*, which has the firebox surrounded by fluid-backed sections on all sides. *Wet*

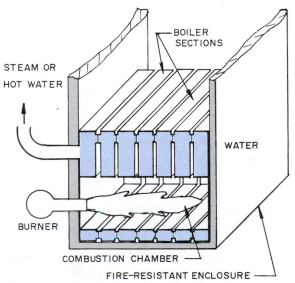

Figure 46.83 Cast iron boiler sections surround the heat source.

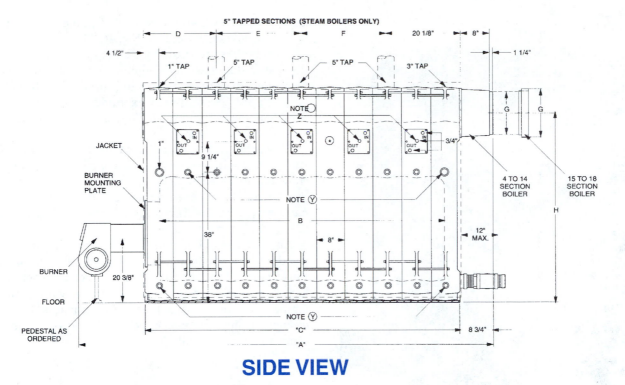

SIDE VIEW

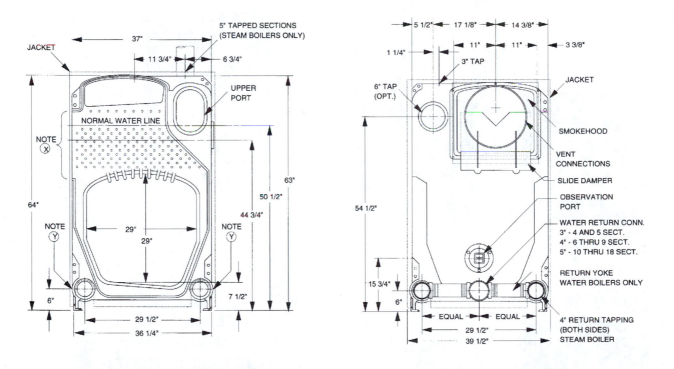

INTERMEDIATE SECTION

REAR VIEW

Figure 46.84 This shows the cast iron boiler sections of a high-performance wet base commercial boiler. They are used for hot water and steam boilers and are enclosed in an insulated metal jacket. (*Courtesy Smith Cast Iron Boilers*) (*continued on next page*)

Figure 46.84 *Continued*

leg boilers have the firebox surrounded on the top and sides by fluid-filled boiler sections (Fig. 46.85).

Steel boilers are made by welding the parts forming the water chambers. One type of heat-exchange surface may be slanted, horizontal, or vertical firetubes (Fig. 46.86). A simplified example of a steel water-tube boiler is in Fig. 46.87. They may be dry base, wet leg, or wet base design.

Copper boilers (Fig. 46.88) use finned copper tube coils run from headers or serpentine copper tubing coils as shown in Fig. 46.89. These are usually fired by natural gas. *Electric boilers* consume no fuel and produce no exhaust gas, so no flue is required. The electric electrodes are immersed in the water.

Another type of boiler used more in recent times is the gas-fired *pulse combustion boiler* (Figs. 46.90 and 46.91). Pulse combustion provides efficient combustion and high heat transfer rates. These boilers reach operating temperature much faster than conventional boilers, are cleaner burning than conventional gas boilers, and much cleaner than coal or oil.

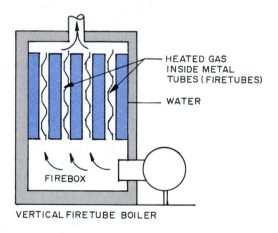

VERTICAL FIRETUBE BOILER

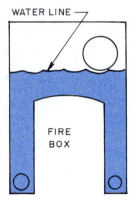

WET BASE BOILER SECTION   WET LEG BOILER SECTION

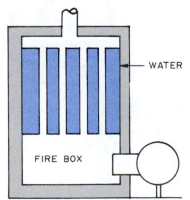

DRY BASE BOILER

Figure 46.85 The three types of cast iron boiler sections.

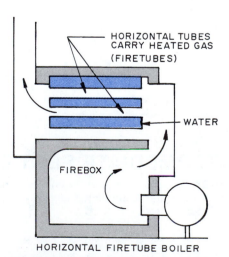

HORIZONTAL FIRETUBE BOILER

Figure 46.86 Some types of steel boilers use metal firetubes immersed in the supply water.

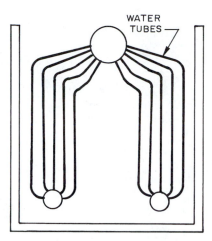

Figure 46.87 A simplified example of a commercial water-tube boiler core.

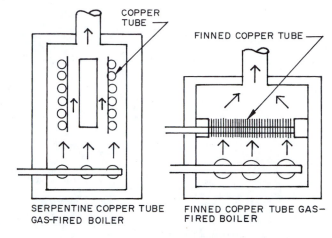

TYPICAL COPPER TUBE RESIDENTIAL BOILERS

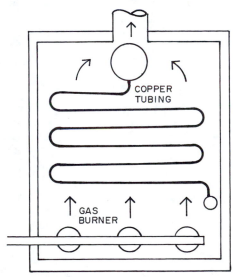

SIMPLIFIED ILLUSTRATION OF A COPPER TUBE COMMERCIAL BOILER

Figure 46.89 Simplified illustrations of commonly used copper tube residential and commercial boilers.

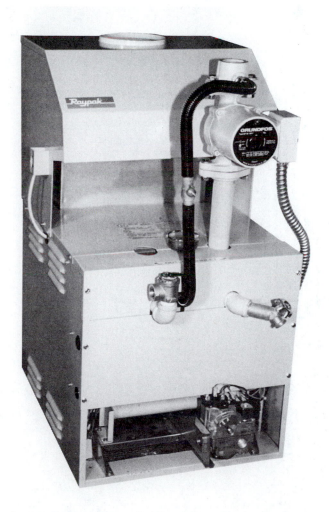

Figure 46.88 A gas-fired copper-finned tube residential boiler has a factory-installed circulator pump, temperature and pressure gauges, and automatic vent dampers. *(Courtesy Raypak, Inc.)*

Basic construction of pulse hydronic boilers

Photo shows the combustor in the center with air and gas intake pipes at the top and the radial design of the Thermaflex formed tailpipes.

Combustor and tailpipes surrounded by the pressure vessel ready for mounting to the exhaust decoupler at the bottom. Air metering valve is visible at the top.

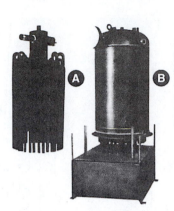

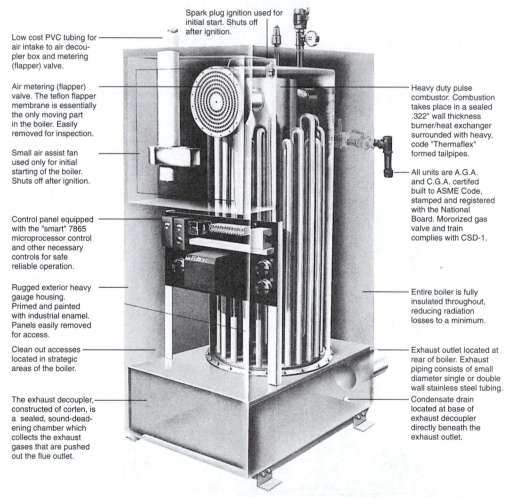

Spark plug ignition used for initial start. Shuts off after ignition.

Low cost PVC tubing for air intake to air decoupler box and metering (flapper) valve.

Air metering (flapper) valve. The teflon flapper membrane is essentially the only moving part in the boiler. Easily removed for inspection.

Small air assist fan used only for initial starting of the boiler. Shuts off after ignition.

Control panel equipped with the "smart" 7865 microprocessor control and other necessary controls for safe reliable operation.

Rugged exterior heavy gauge housing. Primed and painted with industrial enamel. Panels easily removed for access.

Clean out accesses located in strategic areas of the boiler.

The exhaust decoupler, constructed of corten, is a sealed, sound-deadening chamber which collects the exhaust gases that are pushed out the flue outlet.

Heavy duty pulse combustor. Combustion takes place in a sealed .322" wall thickness burner/heat exchanger surrounded with heavy, code "Thermaflex" formed tailpipes.

All units are A.G.A. and C.G.A. certifed built to ASME Code, stamped and registered with the National Board. Mororized gas valve and train complies with CSD-1.

Entire boiler is fully insulated throughout, reducing radiation losses to a minimum.

Exhaust outlet located at rear of boiler. Exhaust piping consists of small diameter single or double wall stainless steel tubing.

Condensate drain located at base of exhaust decoupler directly beneath the exhaust outlet.

Figure 46.90 An internal view of a vertical gas-fired pulse combustion commercial/industrial hydronic heating boiler. *(Courtesy Fulton Boiler Works)*

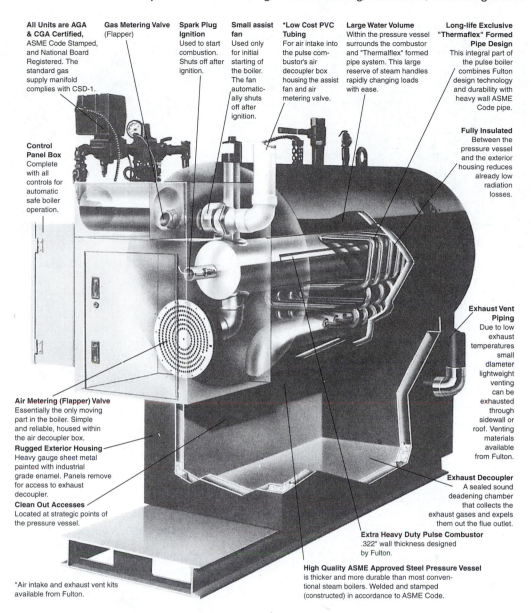

All Units are AGA & CGA Certified, ASME Code Stamped, and National Board Registered. The standard gas supply manifold complies with CSD-1.

Gas Metering Valve (Flapper)

Spark Plug Ignition Used to start combustion. Shuts off after ignition.

Small assist fan Used only for initial starting of the boiler. The fan automatically shuts off after ignition.

*Low Cost PVC Tubing For air intake into the pulse combustor's air decoupler box housing the assist fan and air metering valve.

Large Water Volume Within the pressure vessel surrounds the combustor and "Thermalflex" formed pipe system. This large reserve of steam handles rapidly changing loads with ease.

Long-life Exclusive "Thermaflex" Formed Pipe Design This integral part of the pulse boiler combines Fulton design technology and durability with heavy wall ASME Code pipe.

Control Panel Box Complete with all controls for automatic safe boiler operation.

Fully Insulated Between the pressure vessel and the exterior housing reduces already low radiation losses.

Exhaust Vent Piping Due to low exhaust temperatures small diameter lightweight venting can be exhausted through sidewall or roof. Venting materials available from Fulton.

Air Metering (Flapper) Valve Essentially the only moving part in the boiler. Simple and reliable, housed within the air decoupler box.

Rugged Exterior Housing Heavy gauge sheet metal painted with industrial grade enamel. Panels remove for access to exhaust decoupler.

Clean Out Accesses Located at strategic points of the pressure vessel.

Exhaust Decoupler A sealed sound deadening chamber that collects the exhaust gases and expels them out the flue outlet.

Extra Heavy Duty Pulse Combustor .322" wall thickness designed by Fulton.

High Quality ASME Approved Steel Pressure Vessel is thicker and more durable than most conventional steam boilers. Welded and stamped (constructed) in accordance to ASME Code.

*Air intake and exhaust vent kits available from Fulton.

Figure 46.91 An internal view of a horizontal gas-fired pulse combustion boiler used for low-pressure and high-pressure steam heating systems. *(Courtesy Fulton Boiler Works)*

The pulse combustion process has no power burner and no moving parts, and the combustion takes place in a sealed environment. The process begins with a charge of air and gas entering through metering valves into the burner/heat exchanger. The charge is initially ignited by a spark plug. Once combustion is started the spark plug shuts off. The positive pressure from combustion closes the metering valves and forces the hot gases created out the tailpipes into the exhaust decoupler. As this happens, the air and gas metering valves are sucked open, admitting a fresh charge of air and gas. This cycle repeats at a natural frequency of about thirty times a second (Fig. 46.92).

The combustion process is so clean that the boiler does not require a chimney but is vented through the roof or side wall with small-diameter tubing. Pulse-combustion boilers may be condensing or noncondensing.

Large commercial hot water and steam heating systems use large boilers often installed in series as shown in Fig. 46.93. Typical boiler connections for multiple boiler installations are controlled by local codes.

TERMINAL EQUIPMENT FOR HOT WATER AND STEAM HEATING SYSTEMS

The terminal units used on hot water and steam heating systems include natural convection units and forced convection units. Those used on high-pressure steam systems have heavier construction than those used for hot water and low-pressure steam. These include cast iron and steel radiators, convectors, finned tube units, and baseboards.

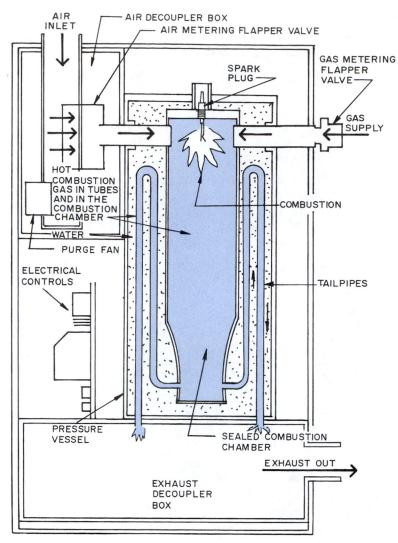

AIR INLET

AIR DECOUPLER BOX

AIR METERING FLAPPER VALVE

SPARK PLUG

GAS METERING FLAPPER VALVE

GAS SUPPLY

HOT COMBUSTION GAS IN TUBES AND IN THE COMBUSTION CHAMBER

COMBUSTION

WATER

PURGE FAN

ELECTRICAL CONTROLS

TAILPIPES

PRESSURE VESSEL

SEALED COMBUSTION CHAMBER

EXHAUST OUT

EXHAUST DECOUPLER BOX

NOTE: COMBUSTION CHAMBER SURROUNDED ON ALL SIDES BY TAILPIPES.

Figure 46.92 The gas-fired pulse combustion system mixes air and gas in a combustion chamber and ignites the mixture with a spark plug. *(Drawing based on materials of the Fulton Boiler Works)*

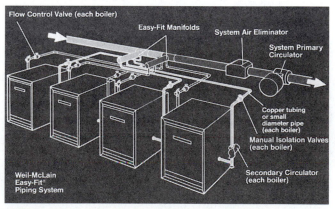

Flow Control Valve (each boiler)

Easy-Fit Manifolds

System Air Eliminator

System Primary Circulator

Copper tubing or small diameter pipe (each boiler)

Manual Isolation Valves (each boiler)

Secondary Circulator (each boiler)

Weil-McLain Easy-Fit® Piping System

Figure 46.93 Gas-fired boilers installed in series. *(Courtesy Weil-McLain)*

Natural Convection Terminal Units

Natural convection heating devices include various types of terminal units that may be wall hung, recessed, or in the form of a baseboard. These devices distribute heat by using natural air circulation. This occurs because heated air rises and cool air settles, producing a natural circulation. The cool air enters the unit below the finned heating tube and rises as it is heated, passing out through a grille or other opening at the top of the unit. These devices are available in various styles and sizes.

Finned tube units have metal fins secured to a metal tube and are a convection type heater. However, some radiant heat is produced. If placed so human contact may be possible, they have an enclosing cover; otherwise they need not be covered. The fins are usually aluminum or copper and are secured to copper tubing that conducts steam or hot water from the boiler. The fins become hot and the air flows by them and is heated. Several examples are in Fig. 46.94.

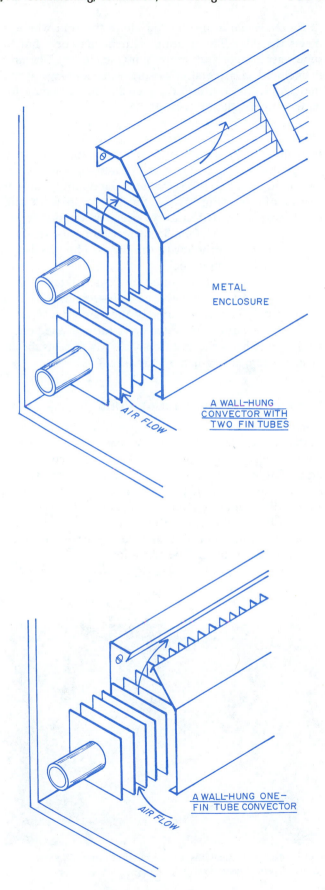

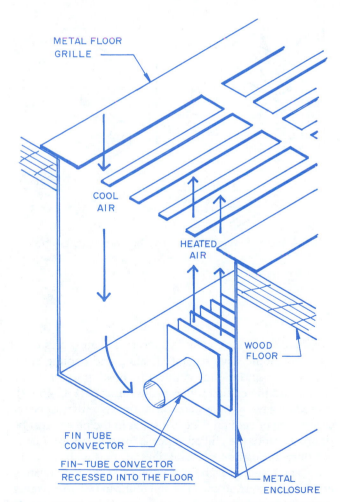

Figure 46.94 Typical finned tube convection terminal units.

Baseboard units are located along the wall where it meets the floor. The heating element may be a finned tube, cast iron, or aluminum unit (Fig. 46.95). The unit is enclosed in a metal enclosure with openings at the floor and near the top of the enclosure, providing for natural air circulation (Fig. 46.96).

Steel radiators are another device that heats by convection and radiation. These are in the form of panels typically mounted on a wall or ceiling, although they can be freestanding. They are heated with hot water circulating through tubes that feed flat hollow panels or a series of tubes (Fig. 46.97). Steel radiators work with low-temperature hydronic systems. Although they are not as hot as conventional hot water radiators, steel radiators are still fairly hot to the touch. The panels are made of heavy gauge steel.

Cast iron radiators are found in older buildings. They are assembled from hollow cast iron sections joined to form the unit. The size is varied by the number of sections used. The steam or hot water flows through the hollow sections and heats the cast iron shell, which heats the air by convection and radiation (Fig. 46.98).

Forced Convection Heating Units

Forced convection units are used for spaces that are to be heated and cooled. However, they can be used for heating or cooling only as is done with natural convection units. The units available include unit ventilators, unit heaters, induction units, fan coil units, and large central air handling units. Forced convection units use some form of fan or blower to produce air movement over a cooling or heating coil and to move the heated or cooled air through the space to be conditioned.

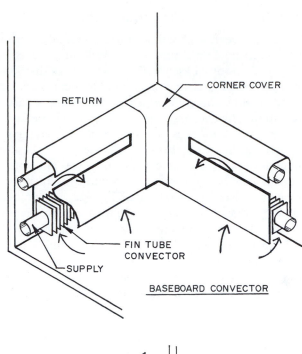

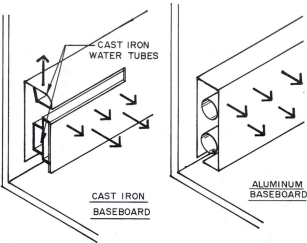

Figure 46.96 Several types of baseboard convectors.

Figure 46.95 Typical baseboard heating units. *(Courtesy Runtal North America, Inc.)*

Fan coil units have a fan, a filter, and heating and cooling coils. Some have separate heating and cooling coils, but others use the same coil for the heating water in the winter and the chilled water in the summer (Fig. 46.99). If used on a two-pipe system, the entire system must be set for heating or cooling. The hot water in the pipes must be drained before the chilled water is introduced. In a large building this can take several days. If used with three- or four-pipe systems, each fan coil unit has separate heating and cooling coils, so each fan coil unit can provide heating or cooling when required.

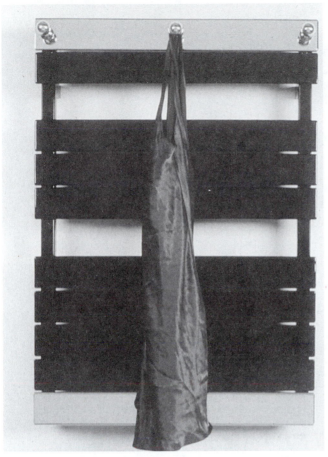

Figure 46.97 These steel hot water radiators are wall mounted, but baseboard units are available. They may be installed as free-standing units and serve as a room divider. *(Courtesy Runtal North America, Inc.)*

Fan coil units may be wall mounted as shown in Fig. 46.99, or they may be ceiling mounted or be high-rise vertical units as shown in Fig. 46.100. The ceiling-mounted units may be below or above the ceiling. If above the ceiling, short ducts are used for the air return and supply. High-rise units are usually placed in the corners of a room, but this is not mandatory.

Unit ventilators are much like fan coil units except they have an opening through the outside wall that allows them to bring in outside air. The opening is covered with a decorative louvered grille. The control system regulates dampers to control the influx of outside air. Unit ventilators typically are used in rooms to be occupied by large numbers of people where frequent air changes may be necessary. This helps meet code requirements for room ventilation.

Induction units are much like fan coil units, but the movement of the air in the room is through the use of high-pressure air that is piped through a nozzle behind the coil. This causes room air to circulate through the coil and out into the room. The nozzle and pressurized air replace the fan in the fan coil unit. They usually are mounted on an outside wall below a window.

Unit heaters have a fan that circulates air over some type of heat exchange surface, such as a hot water coil. They are enclosed in a metal case and usually are suspended from the ceiling. Most common uses are in large open industrial plants or in businesses such as auto repair shops. They can have electric heating elements, be fired by natural gas or propane, or be part of a steam or hot water heating system. A typical gas-fired unit is shown in Fig. 46.101.

Unit heaters provide a rapid flow of heated air. They also are used to temper cold outside air that is introduced into a building such as through doors. They fill the door area with warm air, protecting the occupants from sudden temperature changes. Unit heaters are controlled by thermostats and can be used to provide zoned heat by allowing one thermostat to control only the heaters in one part of the area to be heated.

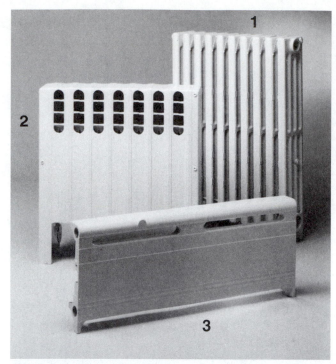

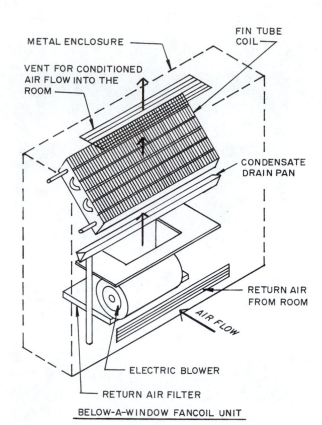

Figure 46.98 Several types of cast iron radiators. The slender model is made by assembling cast iron sections to form the size required for the desired heating capacity.
1. Slenderized cast iron radiator.
2. Cast iron radiant radiator.
3. Cast iron baseboard radiator.
(Courtesy Burnham Corporation)

BELOW-A-WINDOW FANCOIL UNIT

Figure 46.99 A typical wall-mounted fan coil. It is installed below a window so the conditioned air rises through a slot diffuser across the face of the window.

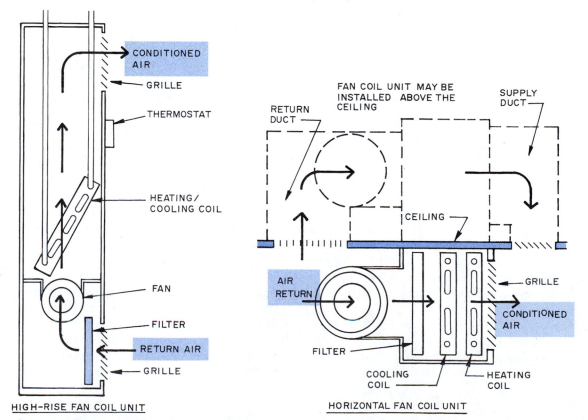

HIGH-RISE FAN COIL UNIT

HORIZONTAL FAN COIL UNIT

Figure 46.100 Simplified illustrations of horizontal and highrise fan coil units.

ITEM OF INTEREST

GAS METERS

A new gas meter is being developed to replace the old bulky meter. It is smaller than the old unit and may even reduce gas flow during periods of peak demand. It will have automatic shut-off valves that will reduce natural gas leaks occurring during a disaster.

Figure A A comparison of the new compact gas meter with the older meter design. *(Courtesy Gas Research Institute)*

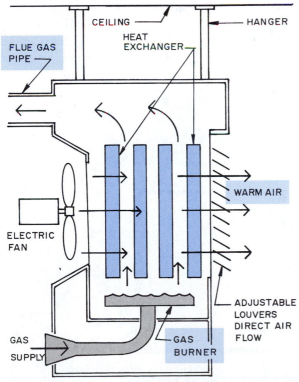

Figure 46.101 This illustrates a gas-fired heater designed to be hung from the ceiling. Heat also can be provided by electric coils, hot water, or steam.

Central air-handling units have a fan and a hot water or steam heating coil, which is used for cooling if a chilled water supply is used or a compressor if chilled water is not used. Heated and cooled air are distributed to the various rooms through a duct system. Each space has terminal equipment such as reheat coils, mixing boxes, and variable volume controls to regulate the air temperature. *Reheat coils* add additional heat to the air at the terminal device. *Mixing boxes* (also called blending boxes) are compartments in which two air supplies are mixed and then discharged into the room. *Variable volume controls* regulate the airflow to control the temperature of the air.

CHILLED WATER COOLING SYSTEMS FOR LARGE BUILDINGS

The most commonly used cooling system uses chilled water to remove heat from the air in a building. This involves the use of mechanical equipment, including a means of refrigeration, cooling coils, and heat exchangers. The chilled water system is a closed circuit of piping in which water is recirculated from a chiller where it is chilled and moved through other equipment used to remove heat from the water. It then circulates through terminal units in the space to be cooled and back to the chiller. The three types of liquid chillers are centrifugal, absorption, and positive displacement.

CENTRIFUGAL CHILLERS

Centrifugal chillers are driven by an electric motor, a steam or gas turbine, or an internal combustion engine, which may be diesel or natural gas fueled. They use a *vapor compression refrigeration system*. The major parts of a mechanical vapor-compression chilled water system include a compressor, a **condenser,** an evaporator, and an expansion valve. An air-cooled refrigerant *condenser* is a device in which heat removal from the refrigerant is accomplished entirely by heat absorption by air flowing over the condensing surfaces. A *compressor* is a machine that mechanically compresses the refrigerant. An *evaporator* is that part of a refrigerating system in which the refrigerant is evaporated to absorb heat from the contacting heat source. The *expansion valve* reduces the temperature and pressure of the refrigerant as it passes through the valve.

The *vapor compression refrigeration cycle* used by centrifugal chillers is illustrated in Fig. 46.102. This is a *direct expansion* (DX) cycle. The refrigerant in vapor form is compressed by the cylinder in the compressor. This causes it to become hot and at a high pressure. The refrigerant moves to the condenser where a fan blows outside air through the condenser coil, removing much of the heat in the refrigerant and causing it to condense into a warm liquid that is still at a high pressure. The pressure pushes the liquid refrigerant toward the evaporator. On the way it passes through a thermal expansion valve that imposes a pressure drop, causing it to expand, which reduces the temperature. The cool liquid refrigerant passes through the cooling coil of the evaporator where it absorbs heat from the air passing through the coil. When the refrigerant leaves the evaporator it is a vapor. The vapor moves to the compressor, and the cycle is repeated.

The centrifugal chiller system can also use a *shell-and-coil heat exchanger* and a cooling tower to dispose of the heat from the refrigerant in the shell-and-coil condenser. A shell-and-coil heat exchanger has a coil of pipes within a shell or container. The pipes carry the refrigerant through a second fluid held in the container. The shell-and-coil evaporator chills water in the shell or container and passes it through a cooling coil in the air handler. The air handler passes return air from the room and fresh outside air as required through the coil, conditioning the air supply to the rooms. The air flows through a system of ducts (Fig. 46.103).

The shell-and-coil condenser water absorbs heat from the refrigerant and moves it to the cooling tower on the roof. The hot condenser water drips or is sprayed in fine droplets. A fan pulls outside air through the spray, removing much of the heat. Other cooling tower designs are available. A centrifugal chiller is shown in Fig. 46.104.

THE ABSORPTION CHILLER REFRIGERATION SYSTEM

An absorption chiller can be described as a refrigerating machine that uses heat energy and absorption input to generate chilled water. Absorption cooling uses an evaporated refrigerant that is frequently water. It does not use the mechanical compression stage used in centrifugal chillers, but instead uses a process in which an absorbent such as lithium bromide solution is used. This solution pulls vapor off evaporator coils, creating a cooling reaction. A source of heat such as a gas burner, low-pressure steam, or hot water is used to regenerate the absorbent solution by separating it from the absorbed vapor. Since a source of heat is used to operate

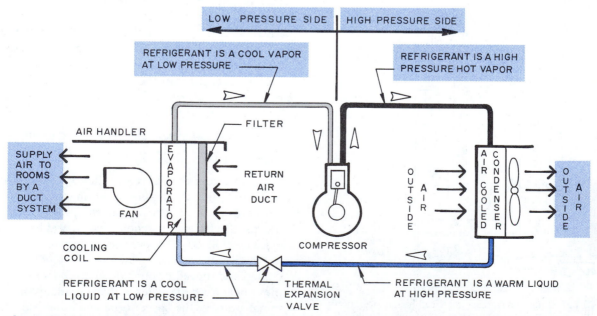

Figure 46.102 A simplified illustration of an air-cooled direct expansion (DX) vapor compression refrigeration cycle used in centrifugal chillers.

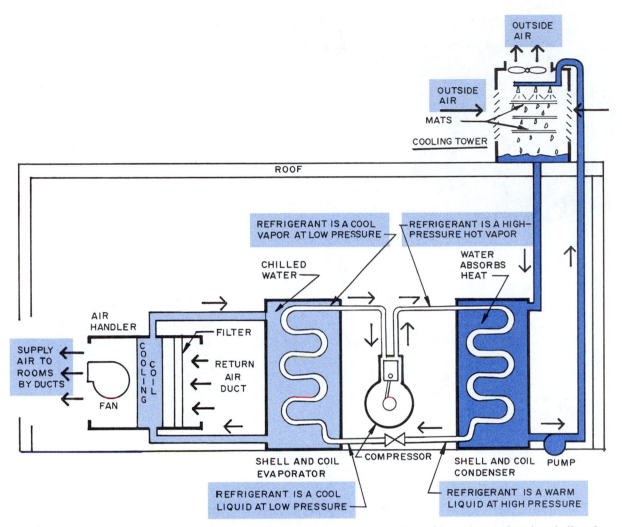

Figure 46.103 A simplified illustration of the chilled water vapor compression refrigeration cycle using shell-and-coil heat exchangers and cooling tower.

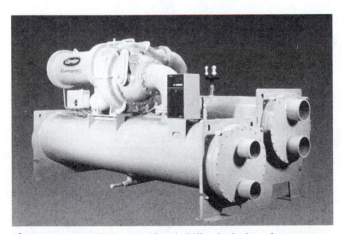

Figure 46.104 This centrifugal chiller is designed to accommodate non-ozone-depleting HFC-134a refrigerant. This is an environmentally responsible, nonrestricted refrigerant used in this high-efficiency chlorine-free chiller. The chiller is designed for use in buildings 300,000 sq. ft. (27,600 sq. m) or larger. (*Courtesy Carrier Corporation*)

an absorption chiller, it is especially economical to operate when a building has waste steam or other heat sources that can be used to provide a major amount of the heat required.

The absorption chiller is a quiet operating unit having few moving parts and generating little vibration, and it is lightweight compared with compression type chillers. An absorption chiller is shown in Fig. 46.105.

Single-Effect Absorption Chillers

The basic absorption chiller refrigeration process for a water-cooled chiller is shown in Fig. 46.106. Following is a description of this *single-effect* water-cooled absorption process.

The absorbent material, lithium bromide, has a great affinity for the refrigerant, water. The absorbent material, which contains some refrigerant, is heated in

Figure 46.105 A direct-fired double-effect absorption chiller/heater. (*Courtesy The Trane Company*)

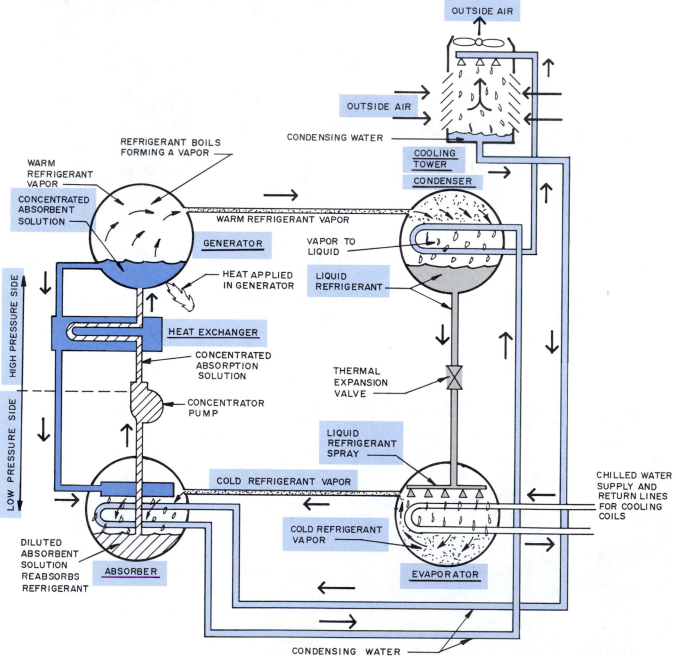

Figure 46.106 A simplified illustration showing the refrigeration process for a single-effect lithium bromide water cycle absorbent chiller.

the generator. The refrigerant vapor is then driven to the condenser where the process converts it to a liquid (water). The concentrated absorbent moves from the generator to the absorber.

The liquid refrigerant flows from the condenser through an expansion valve to a low-pressure evaporator where it flashes to a vapor, providing cooling for the chilled water system refrigerant that is circulated to cooling coils within the building. This is a refrigerant that is in a system separate from the refrigerant with the chiller system. The expansion valve reduces the refrigerant pressure as it enters the evaporator, which is on the low-pressure side of the cycle.

Next the chiller refrigerant moves to the absorber where it is reabsorbed by the concentrated absorbent solution. This produces reduced pressure in the system beyond the expansion valve. The absorber is cooled to increase the rate of absorption of refrigerant into the concentrated absorption solution. This absorption solution is pumped back into the generator, where the cycle repeats. The generator and condenser side operates at high pressure, and the evaporator and absorber operate at low pressure.

Double-Effect Absorption Chillers

A gas-fired double-effect absorption chiller-heater is shown in Fig. 46.107. Units of this type are used in commercial applications in which chilled water is used for cooling and hot water for heating, using a central air-conditioning system. The condenser is water cooled during the cooling period, and the heat is rejected through a cooling tower. The double-effect absorption cooling cycle has two generators. One is heated by natural gas and the other by the hot semiconcentrated refrigerant vapor.

Cooling Cycle

The cooling cycle for a double-effect chiller/heater is illustrated in Fig. 46.108. The gas burner heats a dilute absorbent solution (lithium bromide) in the *high-temperature generator*. The boiling process drives the refrigerant vapor and droplets of semiconcentrated solution to the primary separator. The semiconcentrated solution flows through a *heat exchanger* where it is precooled before flowing to the low-temperature generator.

In the *low-temperature generator* the hot refrigerant vapor flowing from the primary separator heats the semiconcentrated solution. This liberates refrigerant vapor, which flows to the *condenser*. The concentrated absorbent solution is precooled in the condenser and flows to the absorber.

In the condenser, refrigerant vapor is condensed on the surfaces of the cooling coils, and the latent heat is moved by the cooling water to a cooling tower where it is expelled to the outside air. Refrigerant liquid accumulates

Figure 46.107 A gas-fired double-effect absorption chiller/heater. (*Courtesy American Yazaki Corporation*)

in the condenser and passes through an orifice into the evaporator.

The pressure in the *evaporator* is lower than that in the condenser. As the refrigerant liquid flows into the evaporator, it boils on the surface of the chilled/hot water coil. Heat, equivalent to the latent heat of the refrigerant, is removed from the recirculating water to the cooling coils, which chills the water to about 44°F (6.7°C). The refrigerant vapor flows to the absorber.

In the *absorber* the concentrated lithium bromide solution (the absorbent) absorbs the refrigerant vapor as it flows across the absorber coil. The cooling water in the coil removes heat from the solution after which the dilute lithium bromide flows through the heat exchanger, is preheated, and moves to the high-temperature generator to start through the cycle again.

Heating Cycle

The heating cycle for a double-effect absorption chiller/heater is shown in Fig. 46.109. The lithium bromide absorbent solution is brought to a boil in the high-temperature *generator* and the vapor with concentrated lithium bromide solution is lifted to the *primary separator*. The hot refrigerant vapor and droplets of concentrated absorbent solution flow through the open changeover valve into the *evaporator* and the *absorber*. The heat is transferred to recirculating water coils in the evaporator. This heated water flows to the fan coil units within the building, providing the required space heating. The refrigerant (water) vapor mixes with the concentrated absorbent and returns to the generator, where the cycle is repeated.

Absorption chillers may be water cooled, as with a cooling tower, or air cooled. Air-cooled absorption chillers generally use ammonia as the refrigerant and water as the absorbent.

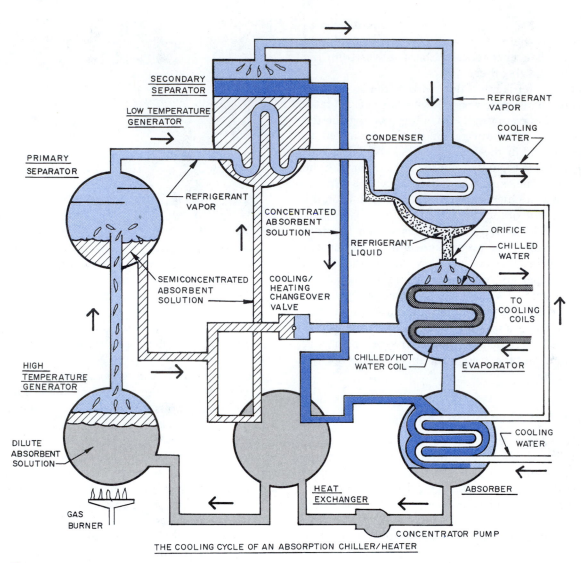

SECONDARY
SEPARATOR

LOW TEMPERATURE
GENERATOR

PRIMARY
SEPARATOR

REFRIGERANT
VAPOR

REFRIGERANT
VAPOR

CONDENSER

COOLING
WATER

CONCENTRATED
ABSORBENT
SOLUTION

REFRIGERANT
LIQUID

ORIFICE

CHILLED
WATER

SEMICONCENTRATED
ABSORBENT
SOLUTION

COOLING/
HEATING
CHANGEOVER
VALVE

TO
COOLING
COILS

HIGH
TEMPERATURE
GENERATOR

CHILLED/HOT
WATER COIL

EVAPORATOR

DILUTE
ABSORBENT
SOLUTION

COOLING
WATER

HEAT
EXCHANGER

ABSORBER

GAS
BURNER

CONCENTRATOR PUMP

THE COOLING CYCLE OF AN ABSORPTION CHILLER/HEATER

Figure 46.108 A simplified illustration of a gas-fired double-effect water-cooled absorption chiller/heater in the cooling mode. (*Developed from materials from American Yazaki Corporation*)

POSITIVE DISPLACEMENT CHILLERS

Positive displacement chillers use scroll, reciprocating, and rotary screw compressors, and they typically use electric motor drives. A *scroll compressor* is a positive displacement compressor in which the reduction of the internal volume of the compression chamber is produced by a rotating scroll within a stationary scroll. A scroll is an involute spiral. A *reciprocating compressor* is a positive-displacement compressor in which the change in the volume of the compression chamber is produced by the reciprocating movement of a piston. *Positive-displacement compressors* increase the pressure of the refrigerant vapor by reducing the volume of the compression chamber in which it has been injected using a mechanical means such as a piston. A *screw com-*

pressor uses a rotary motion to drive two intermeshing helical rotors to produce compression.

The capacity of the system is varied by having several compressors and turning them on and off as capacity requirements vary. Some reciprocating chiller compressors have an inlet suction valve. Reciprocating chillers use a compressive refrigeration cycle and are usually driven by electric motors. They are smaller than centrifugal chillers and usually dispose of the condenser heat into the air rather than through a cooling tower.

A water-cooled package chiller with a rotary screw compressor is shown in Fig. 46.110. Screw-type compressors adjust the capacity with a slide valve that changes the active area of the screw that provides the intake to the combustion process.

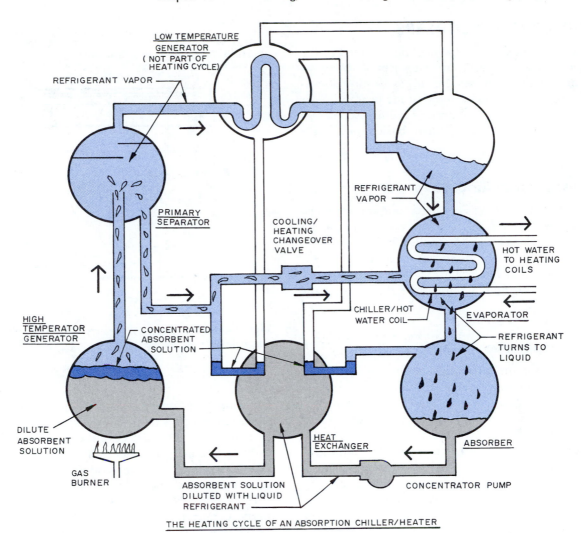

Figure 46.109 A simplified illustration of a gas-fired double-effect water-cooled absorption chiller/heater in the heating mode. *(Developed from materials from American Yazaki Corporation)*

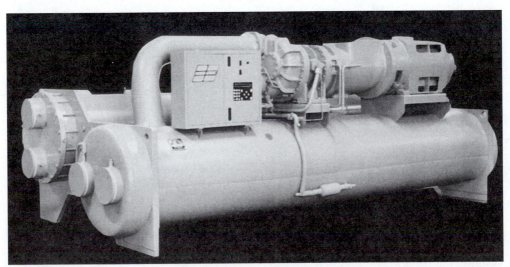

Figure 46.110 This is a water-cooled package chiller with a rotary screw compressor that uses highly efficient microcomputer controls. *(Courtesy Dunham-Bush, Inc.)*

EVAPORATIVE AIR COOLERS

Evaporative cooling is sensible cooling obtained by the exchange of heat produced by water jets or sprays injected into an airstream. As the air passes through water vapor, the vapor retains some of the heat in the air thus cooling it. This cooling process is effective in areas that are hot and dry. There are two basic types, direct and indirect.

Direct Evaporative Air Coolers

Direct evaporative air coolers have evaporative pads, usually of a wood fiber, a water-circulating pump, and a large fan (Fig. 46.111). The fan pulls the air through the pads, which cool the air, and moves the air to the spaces within the building. The water that does not evaporate collects in a pan, and a pump lifts this water back through the cycle.

Indirect Evaporative Air Coolers

A variety of systems in use fall under this type. The following describes a package indirect evaporative air cooler. The unit contains a heat exchanger, a system to provide water vapor, a secondary fan, and a secondary air inlet. The *heat exchanger* typically is constructed with tubes, allowing one airstream to flow inside them and another over their exterior surface. Air filters are used to keep the system from becoming clogged with dust. The circulated water usually is treated to remove minerals, thus reducing corrosion in the heat exchanger.

The system is much like the direct evaporative system except the conditioned air is kept separate from the secondary air (which causes the water evaporation). Therefore, the conditioned air does not pick up moisture during the cooling process. This is why it is called the *indirect system*.

RADIANT HEATING AND COOLING

Radiant panel systems use panels whose surface temperatures can be controlled. They may be on the walls or ceiling or in the floor. The panels usually are operated by electrical resistance units or by circulating water or air. Radiant energy is transmitted through the air in straight lines. It does not heat the air, but it does raise the temperature of solid objects upon which it falls. The objects obtain the heat by absorption. It can be reflected off a surface.

Piping in Ceilings

Several piping systems are used for radiant heating and cooling in ceilings. Hydronic ceiling panels can use a

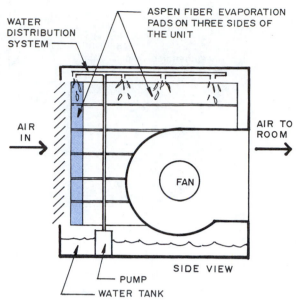

Figure 46.111 A direct evaporative air cooler pulls the air through wet evaporative pads.

two-pipe or a four-pipe distribution system similar to those discussed for hot water systems. The design of the system is critical to a successful end result and must be done by a qualified engineer. Typically, the panels are installed with the pipes embedded above or in a finished plaster ceiling, as shown in Fig. 46.112.

The suspended ceiling has the pipes tied to an overhead supporting member with metal lath and plaster below. When wood or metal joists are used, the pipes are secured to them with metal pipe hangers, and the metal lath and finished plaster are placed below. The coils may be embedded in the plaster coat by securing the metal lath above the pipes and plastering over them. Other types of finished ceilings can be used, but plaster is most common.

Another type of ceiling panel consists of copper tubing bonded to flat metal panels that will be the exposed finished ceiling. The individual panels are hung between the channels of a suspended ceiling as shown in Fig. 46.113. This system is used to heat and cool the interior air. Several panel designs are available, but most use copper tubing and aluminum panels.

Piping in Floors

Radiant heating piping may be embedded in concrete floors or placed above or below a wood subfloor. A detail for concrete slab heating is shown in Fig. 46.114. Plastic, ferrous, and nonferrous pipe and tube may be used. The piping may be arranged in continuous coils or have header coils (Fig. 46.115). Usually 1½ to 4 in. (38 to 101 mm) of concrete covers the pipe. The edges

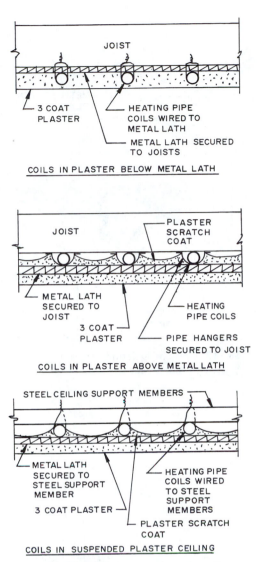

Figure 46.112 Typical radiant-heat ceiling panel piping installations.

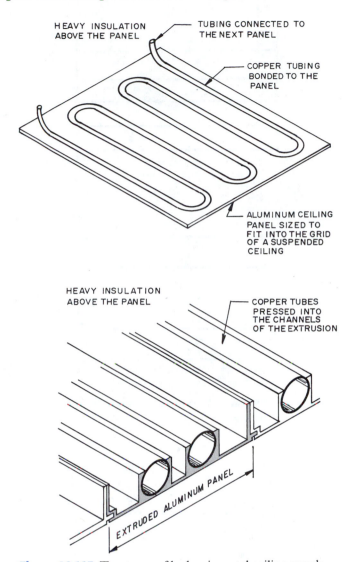

Figure 46.113 Two types of hydronic metal ceiling panels.

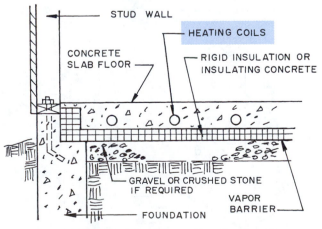

Figure 46.114 A typical detail for radiant heating pipes embedded in a concrete slab.

of the slab must be fully insulated with rigid insulation. Sometimes insulating concrete is used. Piping may be placed on top of a wood subfloor and covered with 1 to 2 in. (25 to 50 mm) of concrete or gypsum underlayment (Fig. 46.116). Gypsum products designed specifically for floor heating are available. Concrete should be of structural quality.

Piping under the subfloor is attached to the subfloor. Metal heat emission plates are used to improve heat transfer (Fig. 46.117).

Piping intertwined with the subfloor is shown in Fig. 46.118. The subfloor consists of spaced strips, which allows the piping to be between them and above the floor joists. Metal heat transfer plates are used. This is covered with wood subflooring and the finished floor covering.

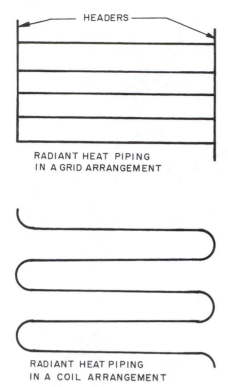

RADIANT HEAT PIPING
IN A GRID ARRANGEMENT

RADIANT HEAT PIPING
IN A COIL ARRANGEMENT

Figure 46.115 Two arrangements for
radiant heating pipes in a concrete slab.

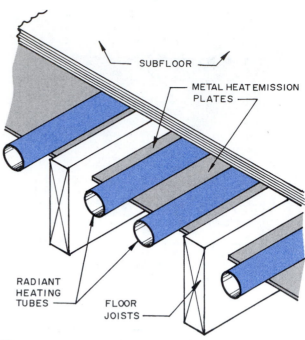

Figure 46.117 This construction places the radiant
heating pipes and metal heat emission plates below the
subfloor.

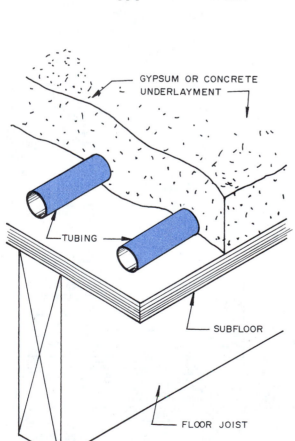

Figure 46.116 Radiant heating pipes can be installed
over a wood subfloor and then embedded in a gypsum
or concrete underlayment.

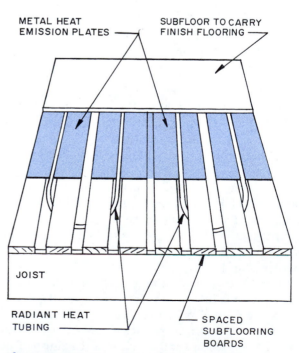

Figure 46.118 The radiant tubing is placed between
the spaced subfloor boards, and metal heat emission
plates are on the top of the spaced subfloor.

ELECTRIC HEATING SYSTEMS

The types of electrically heated systems include factory assembled panels that mount on walls or ceiling, fabrics and wall covering material containing resistance heating wires, and various types of electric resistance cables that may be embedded in concrete or laminated in drywall ceilings or in plaster.

One type of ceiling panel is made to fit into the grid formed by the structural members of suspended ceilings (Fig. 46.119). The panels are available with various constructions, such as conductors embedded in a panel (maybe a gypsum panel) or some form of laminated panel.

Another ceiling heating system uses electric heating cable stapled to a ceiling covering material such as gypsum board or plaster lath or other fire-resistant material. The wires are covered by the various coats of plaster. If metal lath is used it must first be covered with a brown coat of plaster so a nonelectrical conducting surface is available (Fig. 46.120).

Electrical heating cable can also be laid in concrete slabs. The floor is laid in two pours. The first pour is 3 in. (76 mm) or more of insulating concrete. The cable is laid on this slab and is fastened in place by stapling into the slab or using special nail anchors. These hold the cables so the spacing is maintained when the second layer of concrete is poured. The second layer is usually 1½ to 2 in. (38 to 51 mm) thick and must *not* be insulating concrete. A finish floor can be laid over this slab (Fig. 46.121).

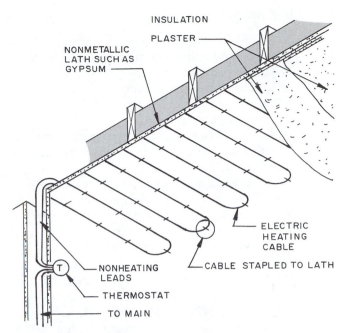

Figure 46.120 A ceiling electric heating system composed of heating cable secured to lath and covered with plaster.

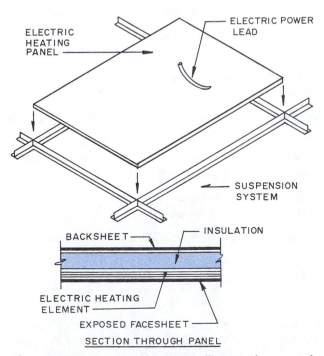

Figure 46.119 An electric heating ceiling panel supported by the structural grid of a suspended ceiling system.

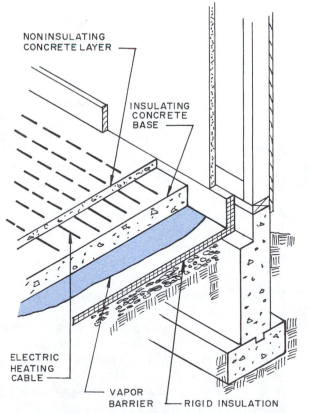

Figure 46.121 Electric heating cable can be installed in a two-layer concrete floor.

REVIEW QUESTIONS

1. What is the difference between thermal radiation, thermal conduction, and thermal convection?
2. What is a major source of information about thermal conditions necessary for human comfort?
3. What types of ventilation are used to assist with the condition of inside air?
4. What types of air cleaners are currently available?
5. What are the three basic directions of heat flow from warm air furnaces?
6. What fuels are used to fire warm air furnaces?
7. How does an air-to-air heat pump produce heat?
8. What is meant by a ground loop heat pump?
9. How can a gas-fired warm air furnace be used to cool a building?
10. What are the types of all-air distribution systems?
11. What factors must a designer consider as an all-air duct distribution system is being designed?
12. How does a reheat warm air distribution system differ from a variable air volume system?
13. How does a dual duct warm air system regulate air temperature?
14. What are the classes of commercial warm air ducts?
15. What is the difference between the perimeter loop and perimeter radial duct systems?
16. What is meant by the surface effect from a ceiling warm air diffuser?
17. What is a comfortable relative humidity for indoor air?
18. What types of residential and industrial humidifiers are available?
19. Why is the control of humidity in the air important?
20. How does a liquid desiccant dehumidification system remove moisture from the air?
21. What types of hydronic heating systems are in use?
22. What is the difference between a two-pipe direct-return and the two-pipe reverse-return hydronic system?
23. Why do hydronic heating systems need an expansion chamber?
24. What is the advantage of a four-pipe system for hydronic heating and cooling?
25. How does the condensate in a one-pipe steam heating system return to the boiler?
26. What purpose do thermostatic traps on each radiator of a steam heating serve?
27. Why is a two-pipe vacuum steam heating system used in large buildings?
28. What is the major difference in the way water is heated in firetube and watertube boilers?
29. What is the difference in the operation of condensing and noncondensing boilers?
30. What is the unique feature identifying a wet-base boiler?
31. What are the commonly used natural convection terminal units?
32. What are the major components of a fan coil unit?
33. What types of refrigeration units are used to chill water for air-conditioning purposes?
34. How does a shell-and-coil condenser remove heat from the refrigerant?
35. How does a single-effect water-cooled absorption chiller cool the water for the air-conditioning system?

KEY TERMS

absorbent A material that extracts a substance for which it has an affinity from a liquid or gas and that changes physically and chemically during the process.

adsorbent A material that allows molecules of gases, liquids, or solids to adhere to its surfaces without changing the adsorbent physically or chemically.

boiler A device in which a liquid is heated to form hot water or steam.

Btu The British thermal unit, which is the amount of heat required to raise the temperature of one pound of water by one degree Fahrenheit.

chiller A refrigerating machine composed of a compressor, a condenser, and an evaporator, used to transfer heat from one fluid to another.

compressor A device that mechanically increases the pressure of a gas.

condensate A liquid formed by the condensation of a vapor.

condenser A heat exchanger unit in which a vapor has some heat removed causing it to form a liquid.

convector A unit whose surface is used to transfer its heat to a surrounding fluid mainly by convection.

dehumidifier A device used to remove moisture from the air.

desiccant An absorbent or adsorbent that removes water vapor from a material.

desiccation The process for evaporating or removing water vapor from a material.

diffuser A unit with deflecting vanes used on heat duct outlets to discharge the air in various directions.

humidifier A device that adds water vapor to the air.

humidity The amount of water vapor within a given space.

hydronics The science of cooling and heating water.

latent heat Heat involved with the action of changing the state of a substance, such as changing water to steam.

makeup air Air brought into the building from the outside to replace air that has been exhausted.

radiant heat Heat transferred by radiation.

refrigerant The substance that transfers heat by gaining heat by evaporation at a low temperature and pressure and losing heat when it condenses at a higher temperature and pressure.

relative humidity The ratio of the density of water vapor in the air to the saturation density.

return air Air removed from a space and vented or reconditioned by a furnace, air conditioner, or other apparatus.

sensible heat Heat that causes a change in temperature.

steam Water in a vapor state.

supply air Conditioned air entering a space from an air-conditioning, heating, or ventilating unit.

surface effect The effect caused by the entrainment of secondary air against or parallel to a wall or ceiling when an outlet discharges air against or parallel to the wall or ceiling.

thermal conduction The process of heat transfer through a solid by transmitting kinetic energy from one molecule to the next.

thermal convection Heat transmission by the circulation of a liquid or a heated air or gas.

thermal radiation The transmission of heat from a hot surface to a cool one by means of electromagnetic waves.

thermostat A control device that automatically responds to temperature changes by opening and closing an electric circuit to regulate the temperature of the space in which it is located.

throw The horizontal or vertical distance an airstream travels after leaving the air outlet before it loses velocity.

SUGGESTED ACTIVITIES

1. Arrange visits to large multistory buildings and ask to see the equipment used for heating and cooling. Prepare a report describing what you saw. Cite positive and negative comments about the system made by the person conducting the tour.

2. Invite an engineer who designs multistory heating and cooling systems to address the class and review the steps of the design process and how decisions are reached.

3. Review the local building code and cite the general requirements relating to heating and air-conditioning systems for residential and commercial buildings. Report what the local building official checks as the building is under construction.

ADDITIONAL INFORMATION

Ambrose, J. E., *Building Construction: Service Systems*, Van Nostrand Reinhold, New York, 1990.

American Society of Heating, Refrigerating, and Air-Conditioning Engineers, Inc., *ASHRAE Handbook* and numerous other technical publications, Atlanta, Ga.

Bovay, Jr., H. E. *Handbook of Mechanical and Electrical Systems for Buildings*, McGraw-Hill, New York, 1990.

Bradshaw, V., *Building Control Systems*, John Wiley and Sons, New York, 1993.

Haines, R. W., and Hittle, D. C., *Control Systems for HVAC*, Van Nostrand Reinhold, New York.

Havrella, R. A., *Heating, Ventilating, and Air Conditioning Fundamentals*, McGraw-Hill, New York, 1981.

Levenhagen, J. I., and Spethmann, D. H., *HVAC Controls and Systems*, McGraw-Hill, New York, 1993.

National Electrical Code, National Fire Protection Association, P.O. Box 9101, Quincy, Mass. 02269.

Stein, B., and Reynolds, J. S., *Mechanical and Electrical Equipment for Buildings*, John Wiley and Sons, New York.

Other resources are

Numerous related publications from the organizations listed in Appendix B.

Publications from the Technical Information Program, National Renewable Energy Laboratory, 1617 Cloe Blvd., Golden, Colo. 80401–3393.

Publications of the Waste Material Management Division, U.S. Department of Energy, 1000 Independence Ave. SW, Washington, D.C. 20585.

DIVISION 16 ELECTRICAL

CSI MASTERFORMAT™

Courtesy Square D Company

CHAPTER 47

Electrical Equipment and Systems

This chapter will help you to:

1. Be aware of the various sources of electrical power and how it is distributed from the generating source.

2. Describe an electrical current and the difference between AC and DC current.

3. Specify the types of electrical conductors for various applications.

4. Identify and discuss the various units used to measure, control, and distribute electrical power within a building.

5. Specify lighting installations for meeting needs in a wide range of situations.

The modern building depends heavily on electricity to make it functional and habitable. Electricity provides power for the lighting; runs motors for heating, ventilating, and air-conditioning; provides power to operate the many devices brought into the building; powers elevators, escalators, and other conveying systems; supplies the power to operate the communications, fire, and security systems; and is used to operate a vast array of manufacturing machinery.

ELECTRICAL LOADS

The loads put on an electrical system vary widely, depending on the building's occupancy. All buildings have extensive lighting requirements. This is a major interior wiring system. Many functions, such as air-conditioning and heating, require the use of electric motors. Other systems, such as refrigeration and ventilation, have motors, compressors, and other electricity-consuming devices. The extensive range of appliances and other electrical devices produces varying loads as periods of demand occur.

Industrial plants have heavy electrical demands to operate the machinery used to perform manufacturing operations, such as melting, fusing, and otherwise processing materials. Internal transportation (elevators, escalators, moving walks, conveyors) requires electric power. Internal communications and controls are electrically operated, and many things would not function without them. Then there are hundreds of special equipment items such as those found in hospitals, computer centers, and radio and television studios. The determination of electrical loads and internal systems is a major part of an adequately designed building.

BASICS OF ELECTRICITY

Electric current can be defined as the flow of electrons along a conductor, such as a copper wire. It is produced by a generator or battery that forces electrons to follow the conductor to a consuming device, such as a light, and back to the producing source.

The flow of electric current in this continuing circuit resembles the flow of a fluid in a hydraulic circuit (Fig. 47.1). In a hydraulic system a pump puts the fluid

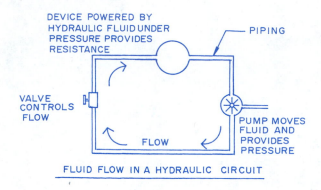

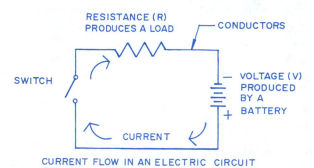

Figure 47.1 A comparison of electrical and hydraulic circuits reveals similarities. Electric switches and hydraulic valves control flow; electric current and hydraulic fluid flow in the circuit. Power is supplied by an electric device (a battery and a hydraulic pump), electric wire and hydraulic piping form the circuit, and resistance to flow occurs in both circuits.

under pressure (pounds per square inch). The fluid flow is measured in a quantity such as gallons. The fluid meets some resistance as it enters a fixture, and the flow is controlled by a valve. An electric circuit has a battery or generator to produce electricity and the electromotive force (volts) to move it along the conductor in quantities measured in amperes. The flow of electricity finds **resistance** (ohms) when it enters a fixture, and the flow is controlled by a switch.

The units used to identify the factors related to the flow of electric current are:

Ampere (A).   A unit of the rate of flow of electric current. An electromotive force of 1 ohm results in a current flow of 1 ampere.

Volt (V).   The unit of electromotive force (pressure) that causes electric current to flow along a conductor.

Ohm (V).   The unit of electrical resistance of a conductor. The symbol for ohm is the Greek capital letter omega (V).

Ohm's Law

Amperes, volts, and ohms are related to each other, and a variance in one will affect the others. This relationship is

identified as Ohm's Law. The relationship for each measure is shown by the following formula, in which I is the electric current or intensity of electron flow (measured in amperes), R is resistance (measured in ohms), and E is the electromotive force (measured in volts).

$$I = \frac{E}{R} \qquad\qquad amperes = \frac{volts}{ohms}$$

$$R = \frac{E}{I} \qquad\qquad ohms = \frac{volts}{amperes}$$

$$E = I \times R \qquad\qquad volts = amperes \times ohms$$

Conductors

Electric current flows along materials called **conductors.** The commonly used conductors for electric wiring are copper and aluminum. Copper is a better conductor than aluminum, so if aluminum wire is used it must have a larger diameter to carry the same amount of current. Good conductors have a low resistance to the flow of electricity. Materials that have a very high resistance to electrical flow get hot because of the friction generated by the flow of electrons. These materials, such as nichrome, are used in applications, such as electric heaters, in which the production of heat is wanted.

Other materials do not conduct electricity and are called **insulators.** Glass, ceramics, and plastics are examples. They are used on electrical devices to provide protection from the electricity. For example, the switch lever on a light switch is a nonconductor type material.

For electricity to do work, it must flow through a circuit. It flows from the positive connection to the negative connection, as shown in Fig. 47.2. The negative electrons move to the positive pole, through the circuit

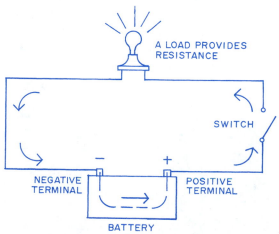

Figure 47.2 Electricity flows through a circuit from the positive connection to the negative connection.

to the consuming device, and back to the negative pole. A switch can be put in the circuit to interrupt the flow when desired. The switch when closed completes the circuit.

AC and DC Current

The two types of electric current are direct current (DC) and alternating current (AC). *Direct current* has a constant flow in one direction. *Alternating current* varies periodically in value and directions, first flowing in one direction in the circuit and then flowing in the opposite direction. Each complete repetition is called a *cycle*. The number of repetitions per second is called the *frequency*. The frequency at which this occurs is measured in *hertz* (Hz). In the United States the frequency for alternating current is 60 cycles per second or 60 hertz.

POWER AND ENERGY

Energy is the term used to express *work*. Energy is expressed in units of kilowatt-hours, footpounds, Btu, joules, or calories. **Power** is the *rate* at which energy is used. Power is expressed in terms of watts, kilowatts, and other units shown in Table 47.1. Since power is the rate at which energy is used, time is a factor. The relationship between power and energy is shown by the following equation:

$$Power = \frac{energy}{time}$$

$$Energy = power \times time$$

The unit of electric power in electric circuits is expressed in watts (W) or kilowatts (kW). One kilowatt = 1000 watts. One watt-hour of energy represents one watt of power used for one hour. A **watt** is one ampere flowing under an electromotive force of one volt. The power W (watts) flowing into an electrical device having a resistance of R (ohms) in which the current is I (amperes) is found by the equation

$$W = I^2 R$$

$$watts = amperes^2 \times ohms$$

It can be seen, therefore, that power is related to current (amperes), electromotive force (voltage), and resistance (ohms).

ELECTRICAL CODES

The design and installation of electrical systems are strictly regulated by electrical codes. The National Electrical Code (NEC) of the National Fire Protection As-

Table 47.1 Units of Power and Energy

English System	Metric System
Units of Power[a]	
Horsepower (hp)	Joule per second (J/sec.)
Btu per second (Btu/sec.)	Calorie per second (cal/sec.)
Watt (W)	Watt (W)
Kilowatt (kW)	Kilowatt (kW)
Units of Energy[b]	
Btu	Calorie (cal)
Foot-pound (ft. • lb.)	Joule (J)
Kilowatt-hour (kWh)	Kilowatt-hour (kWh)

[a]The rate at which work is done
[b]The amount of work done

sociation (NFPA) specifies the safety principles that must be followed. Local governments may have their own electrical codes. Local codes typically include all NEC requirements, but they may contain additional regulations. The Underwriters Laboratories, Inc. (UL) test and certify electrical devices of all types. Approved items carry the UL label.

ELECTRIC POWER SOURCES

Most electrical current is produced by some type of generator. These include hydroelectric, fossil fuel, and nuclear powered electrical generators.

Alternating current is produced by an AC generator, also called an alternator. The alternator is powered by any of the three sources of energy mentioned above.

Hydroelectric Power Generation

Electric power is generated by hydroelectric generation plants using the energy of falling water to turn a generator and produce electricity. The water turns the turbine, which drives the generator that produces the power. Transformers step up the voltage to the requirements needed to transmit it over electric lines to various destinations. A typical installation has many turbines, generators, and transformers (Fig. 47.3).

Fossil Fuel Powered Generators

Fossil fuels used to produce steam to produce electricity include coal, oil, and natural gas. Some experiments using the burning of waste materials is underway. The steam produced by burning the fuels is used to drive steam turbines that turn a generator and produce electricity.

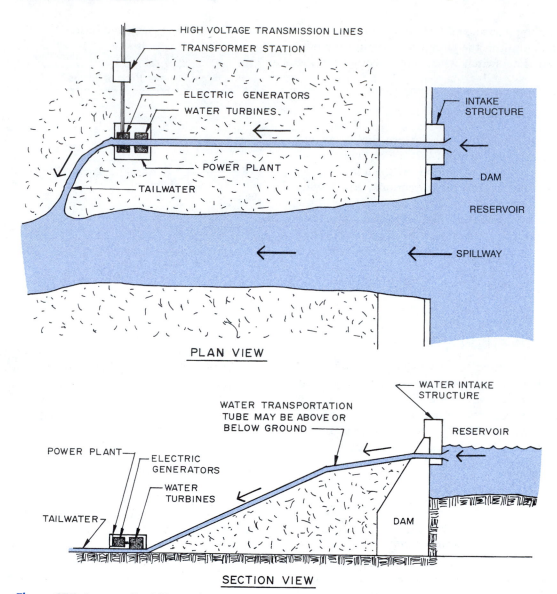

Figure 47.3 A generalized illustration showing the use of hydropower to generate electricity using water turbines to drive electric generators.

Since fossil fuels are nonrenewable resources and are becoming increasingly expensive, alternate sources such as nuclear fission are becoming commonly used. Some kinds of coal and oil fuels cause air pollution problems and require that exhaust fumes at the stack be monitored and cleaned if necessary (Fig. 47.4).

Nuclear Powered Generators

Figure 47.5 is an overall photo of the Harris nuclear power plant located at New Hill, North Carolina, and operated by Carolina Power and Light. The facility occupies a 10,723 acre site and has a 4,100 acre lake. It is capable of producing 900,000 kilowatts of electricity and is licensed to operate for forty years. A plan siting the facilities of the plant and related structures is in Fig. 47.6.

Nuclear powered electric generation plants are similar to fossil fueled plants in that they produce steam to run a steam turbine that drives the generator. The energy is produced in the nuclear reactor by fission. *Fission* is a process in which the centers or nucleuses of certain atoms are split when struck by subatomic particles called neutrons. The products of fission fly apart at high speed and generate heat as they collide with

Figure 47.4 The Mayo fossil fuel electric generating plant located at Roxboro, N.C. This coal-burning electric power generating plant was built in 1983 and can produce up to 750,000 kW. At full capacity it takes 73 rail cars to supply the 7,300 tons of coal per day required to run the plant. *(Courtesy Carolina Power and Light Company)*

Figure 47.5 The Harris nuclear electric generating plant located at New Hill, N.C., required 24 million pounds of reinforced steel, 500,000 cubic yards of concrete, and 2,000 miles of power and control cable. It is capable of producing 900,000 kW of electricity. *(Courtesy Carolina Power and Light Company)*

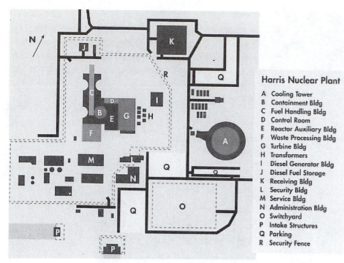

Harris Nuclear Plant

A Cooling Tower
B Containment Bldg
C Fuel Handling Bldg
D Control Room
E Reactor Auxiliary Bldg
F Waste Processing Bldg
G Turbine Bldg
H Transformers
I Diesel Generator Bldg
J Diesel Fuel Storage
K Receiving Bldg
L Security Bldg
M Service Bldg
N Administration Bldg
O Switchyard
P Intake Structures
Q Parking
R Security Fence

Figure 47.6 The siting plan for the Harris nuclear plant at New Hill, N.C. *(Courtesy Carolina Power and Light Company)*

surrounding matter (Fig. 47.7). The fission reaction is controlled in the nuclear core of the reactor, which consists of fuel rods in a chemical form of plutonium or uranium and thorium. Heat energy is produced by the fission reaction of the nuclear fuel. The heat is removed by a coolant and used to produce steam to drive a steam turbine, which drives the electrical generator.

Figure 47.8 shows how the Harris nuclear plant works. To ensure safety in the nuclear plant the concept of "defense in depth" is employed. There are several layers of protection, each of which is independent of the others, so if one should fail others will continue to protect the plant, the workers, and the general public.

Three fission product barriers are designed to keep the radioactive by-products of the fission process away from the environment during both normal and accident conditions:

▲ The fuel rod. Most fission products remain bound inside the ceramic uranium pellets. These pellets are stacked and sealed inside fuel rods made of zirconium alloy, which are joined together into fuel assemblies. The zirconium alloy cladding provides the first of three fission product barriers.

▲ The reactor coolant system. The heat generated by the fuel is removed by the reactor coolant system, which circulates through the fuel assemblies. The reactor coolant system is totally enclosed and provides the second of the three fission product barriers.

▲ The containment building. The reactor coolant system is totally enclosed in this building. This massive building is an airtight, steel-reinforced cylindrical concrete building with 4½ foot concrete walls reinforced with steel, a dome 2½ feet thick, and a base 12 feet thick. This is the third fission product barrier.

ON-SITE POWER GENERATION

On-site power generation is used when a utility system is not available or is unable to provide reliable service. Some systems are used to provide additional service during peak periods. Some facilities, such as hospitals, have on-site backup power generation systems that are used when the utility system fails.

The methods used to generate electricity on-site include wind turbines, solar photovoltaic cells, thermal sources, gas or steam-powered turbines, and cogeneration.

Wind turbines utilize prevailing winds to drive a propellor mounted on a generator that is mounted on a tower (Fig. 47.9). These are often built in a group and may be tied into the utility power system. Details are given in Chapter 42.

Solar photovoltaic cells convert some of the sunlight that falls on them into electricity. They are set in a large grouping and are connected together. Often they are connected to a battery. They charge the battery when the sun is shining and use the battery power at night or on cloudy days. Details are given in Chapter 42.

Thermal sources use an engine or turbine coupled to the shaft of a generator to produce electricity. *Internal combustion engines* run on fuels such as diesel fuel, methane gas, or natural gas (Fig. 47.10). These are typically used as emergency backup sources of electricity for a vast range of commercial facilities, such as hospitals, banks, computer centers, stores, schools, and waste water treatment plants. The backup system includes the generator set, transfer switches, and paralleling switchgear.

Turbines are either gas or steam powered. Gas turbines burn various types of gaseous and liquid fuels, such as natural gas or fuel oil. Steam turbines are driven by a source that produces a large quantity of high-pressure steam. The revolving turbine drives the generator, producing electricity. The steam is produced by a boiler that is fired with a fuel such as coal, solid waste, natural gas, or oil.

Cogeneration

Cogeneration involves the utilization of normally wasted heat energy produced by a reliable, steady source to generate electricity on-site or to use it to heat or cool the building (Fig. 47.11). For example, if an internal combustion engine or a turbine is used to produce on-site electricity, the heat produced by these power sources can be reclaimed and used to heat water or produce steam, which can be used to heat the building or to provide cooling using an absorption chiller. An auxiliary conventionally fired boiler is used in the system to provide extra hot water or steam if needed.

Figure 47.7 A view inside the reactor vessel during the refueling of the reactor. The large crane is used to load and unload fuel rods from the reactor. Each rod is placed in a precise location in the reactor vessel. *(Courtesy Carolina Power and Light Company)*

Salvaged heat may also be used to produce steam that is used to drive a turbine, producing a supplemental on-site supply of electricity (Fig. 47.12). This is especially effective in areas where high-pressure steam is required for an industrial process, such as food processing or pulp and paper manufacturing, which provides a steady, high-temperature source of wasted heat.

ELECTRICAL POWER CONDUCTORS

A major consideration in designing and installing an electric power system is to safely transmit the power to the end use. Electric power systems have the potential for causing fires, property damage, and human injury and death. Electrical codes are strict, and inspection during construction must be thorough. The key to the

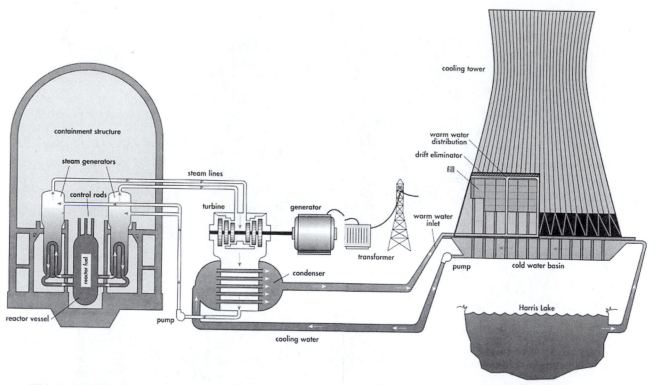

Figure 47.8 A simplified illustration showing how a nuclear electric generating plant operates. (*Courtesy Carolina Power and Light Company*)

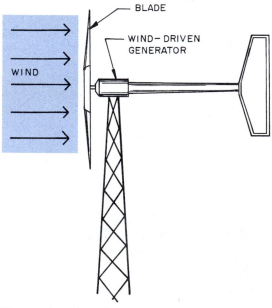

Figure 47.9 Wind-driven generators are one source of electricity.

Figure 47.10 Emergency standby generators may be powered by diesel or gas engines. They are part of a total system including the generator set, transfer switches, and paralleling switchgear. (*Courtesy Onan Corporation*)

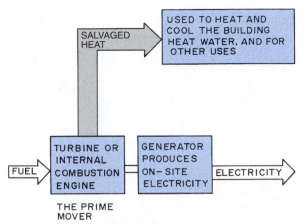

Figure 47.11 Cogeneration utilizes heat that normally would be wasted to produce on-site electricity or to heat or to cool a building.

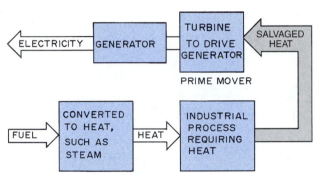

Figure 47.12 Cogeneration can utilize heat generated by processes within a building to produce steam that can drive a turbine producing on-site electricity.

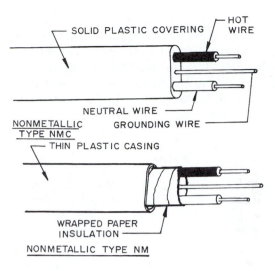

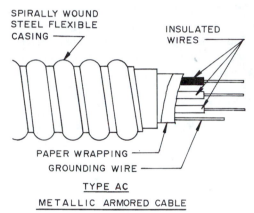

Figure 47.13 Several of the commonly used electrical conductors.

situation is to isolate the electrical conductors as they pass through the building until they reach the point of use, such as an electric light.

Cables

Electric cable design and use are strictly regulated by codes such as the National Electrical Code published by the National Fire Protection Association. The types of conductors (Fig. 47.13) rated by this code are:

▲ Type AC. Insulated conductors are wrapped in paper and enclosed in a flexible, spiral-wrapped, metal covering. An internal copper bonding strip in contact with the metal covering provides a means of grounding. It is used only in dry locations. This wire is referred to as **BX cable.**

▲ Type ACL. This type has the same insulation and covering as Type AC, but it has lead-covered conductors that make it useful in wet applications.

▲ Type ACT. The individual copper conductors have a moisture-resistant fibrous covering and run inside a spiral metal sheath.

▲ Type MC. Insulated copper conductors are sheathed in a flexible metal casing. If it has a lead sheath it can be used in wet locations. It is a heavy-duty industrial feeder cable.

▲ Type MI. The conductors are mineral insulated and sheathed in a gas-tight and watertight metal tube. This type can be used in hazardous locations and underground. It can be fire-rated.

▲ Type NM or NMC. A nonmetallic-sheathed cable used in protected areas. It is also called **Romex.** NM has a flame-retardant and moisture-resistant outer casing and is restricted to interior use. NMC is also fungus resistant and corrosion resistant and can be used on exterior applications.

▲ Type SE or USE. A moisture-resistant, fire-resistant insulated cable that has a braid of armor providing protection against atmospheric corrosion. Type USE has a lead covering, permitting it to be used underground. This is used as the underground service entrance cable. Type SE cable is used for service entrance wiring or general interior use.

▲ Type SNM. The conductors are in a core of moisture-resistant, flame-resistant, nonmetallic material. This assembly is covered with a metal tape and a wire shield and is sheathed in an extruded nonmetallic material impervious to oil, moisture, fire, sunlight, corrosion, and fungus. It can be used for hazardous applications.

▲ Type UF. The conductors are enclosed in a sheath resistant to corrosion, fire, fungus, and moisture and can be directly buried in the earth.

Flat Conductor Cables

Flat conductor cables consist of copper cables formed flat and embedded in a plastic sheathing. A typical cable is 0.030 in. (0.78 mm) thick and around $2\frac{1}{2}$ in. (63.5 mm) wide (Fig. 47.14). The conductor is placed on the subfloor, a shielding material is placed over it, and the carpet is installed. These cables are used for standard 120V electric power and in communication and data systems. Outlets are installed as required by making connections through the carpet (Fig. 47.15). Codes generally require that flat cable be covered with carpet squares so it is easily accessible.

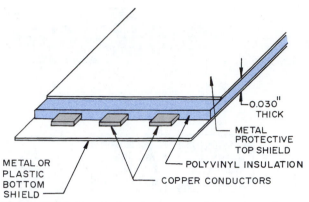

Figure 47.14 A typical under-carpet wiring system. It uses a flat power cable that runs under the carpet. Outlets are placed as needed along the cable.

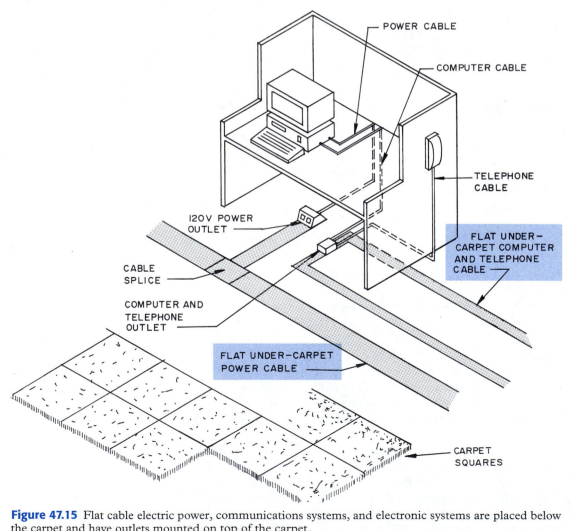

Figure 47.15 Flat cable electric power, communications systems, and electronic systems are placed below the carpet and have outlets mounted on top of the carpet.

 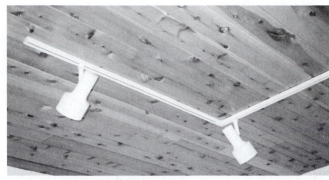

Figure 47.16 These lights are mounted on busways secured to the ceiling. The lights can be placed anywhere along the busway.

Cable Bus and Busways

A **bus** is a bare conductor run in a metal trough, called a **busway.** The conductors are mounted on insulators to keep them clear of the trough. Electrical connections are made to the conductors with various types of plug-ins (Fig. 47.16).

RACEWAYS

Raceways are used to support, enclose, and protect electrical wires.

Cable Trays

Open raceways or **cable trays** are open-faced metal channels used to provide support for electric wires that have adequate insulation and do not require extra protection. Cable trays only support the wires. Wires in this system are open to inspection and modifications (Fig. 47.17).

Conduit

Conduit is a form of *closed raceway.* It supports insulated electric wire and provides protection. Conduit does not have conductors inside when it is installed. The wires are run in later (Fig. 47.18). Conduit can be run inside walls, ceilings below and through floors, and in concrete slabs. Codes regulate the use and locations of the various types. Metal and nonmetal conduit can be left exposed to view when appearance is not important (Fig. 47.19).

One type of conduit is a steel pipe available in three thicknesses. The heavy-wall conduit is referred to as *rigid steel conduit* (RSC). The intermediate-wall type is called *intermediate metal conduit* (IMC), and the thin-wall type is called *electric metallic tubing* (EMT) (Fig. 47.20).

Flexible metal conduit called Greenfield is used for short runs, such as connecting a furnace to the power source. It is available as a watertight conduit (Fig. 47.21). Rigid nonmetallic conduit is also available in polyvinyl chloride (PVC) and high-density polyethylene (PE), as shown in Fig. 47.22.

The rigid metal conduit is available with inside diameters ranging from $\frac{1}{2}$ to 4 in. (12.7 to 101.6 mm). The thin-wall conduit can be bent to form curved corners. Junctions and sharp turns are made with metal fittings. The steel heavy-wall conduit can be used in concrete slabs.

Conduit made from aluminum is also available in the same sizes. It is lightweight and nonsparking, weathers well, and is easy to work. If embedded in concrete it may cause cracking. If buried in the earth it should be coated with asphalt or another type of protective coating.

Other raceways are made in rectangular shapes and are intended to be surface mounted and exposed to view. They are painted to match the wall and ceiling and can have electric outlets, telephone connections, and computer and communications wiring connections (Fig. 47.23).

Metal raceways are also part of cellular steel floor decking. The decking serves as a substrate to support the cast-in-place concrete floor. The cells in the decking are used to carry electrical and communications wires (Fig. 47.24 and 47.25). Perpendicular to the cells in the steel decking are metal ducts spaced as required having outlets through which wires may be pulled to provide power in that area. These are typically spaced on a grid, permitting electric power to be available in a number of places in the floor.

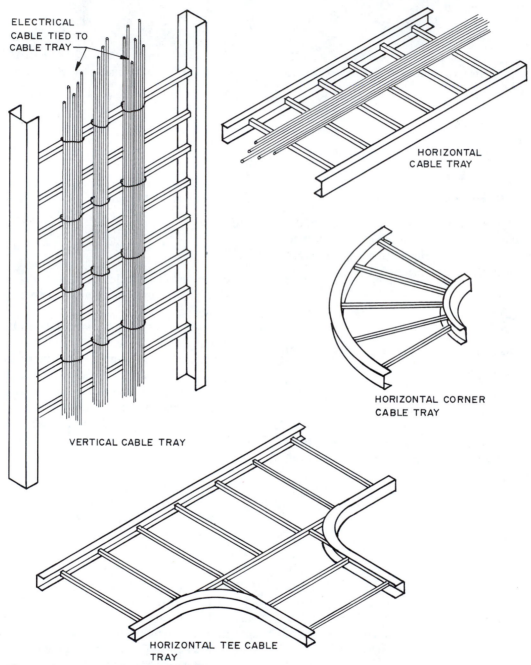

Figure 47.17 Some typical cable trays.

Figure 47.18 This plastic conduit protects the feeder cables into the meter pan.

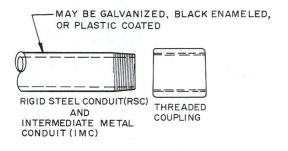

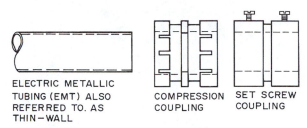

MAY BE GALVANIZED, BLACK ENAMELED, OR PLASTIC COATED

RIGID STEEL CONDUIT(RSC) AND INTERMEDIATE METAL CONDUIT (IMC)

THREADED COUPLING

ELECTRIC METALLIC TUBING (EMT) ALSO REFERRED TO. AS THIN—WALL

COMPRESSION COUPLING

SET SCREW COUPLING

EMT CONDUIT WALL THICKNESS TOO THIN TO PERMIT THREADING.

Figure 47.20 Types of rigid metal conduit.

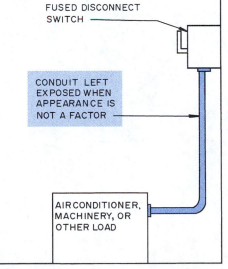

FUSED DISCONNECT SWITCH

CONDUIT LEFT EXPOSED WHEN APPEARANCE IS NOT A FACTOR

AIR CONDITIONER, MACHINERY, OR OTHER LOAD

Figure 47.19 Conduit is often left exposed to view. Conductors are not pulled into it until the raceway installation is complete.

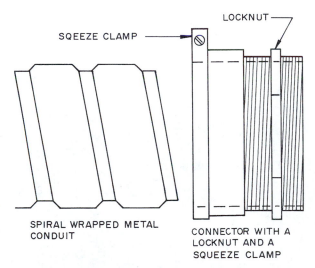

LOCKNUT

SQEEZE CLAMP

SPIRAL WRAPPED METAL CONDUIT

CONNECTOR WITH A LOCKNUT AND A SQUEEZE CLAMP

FLEXIBLE METALLIC CONDUIT (FMC) IS AVAILABLE UP TO 3"(76.2 mm) IN DIAMETER.

Figure 47.21 Flexible metal conduit is made with spiral-wrapped metal bands.

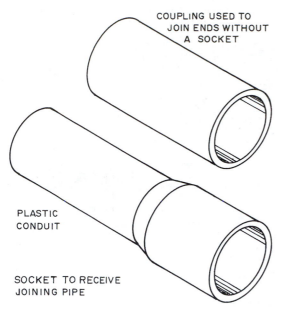

COUPLING USED TO
JOIN ENDS WITHOUT
A SOCKET

PLASTIC
CONDUIT

SOCKET TO RECEIVE
JOINING PIPE

Figure 47.22 Rigid nonmetallic conduit is made with a socket for joining sections. A coupling is available to join ends that may not have a socket.

METERS

The amount of electric power used is measured in watt-hours (Wh). Since the amount increases rapidly, it is reported in kilowatt-hours (kWh). One kWh is equal to 1000 Wh. The amount used is measured by a kilowatt-hour **meter** (Fig. 47.26). Three-wire meters are typically used for residential applications. Commercial and industrial applications having heavy demands and higher voltage requirements typically use four-wire meters.

The owner of the building provides the meter pans and any current transformers needed to step down the voltage within the building. The electric utility supplies the meter (Figs. 47.27). The meter usually is placed outside the building so the utility staff have ready access. Meters can be inside the building if easy access is provided.

Wall-mounted raceway.

Surface-mounted baseboard raceway.

Figure 47.23 Rectangular raceways are used for surface-mounted wiring installations. *(Courtesy Wiremold Company)*

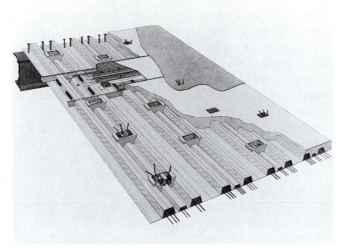

Figure 47.24 This cellular composite steel floor, Robertson Q-Floor, provides channels for running wiring and punchouts for the location of electrical outlets into the building. The outlet boxes are set in place before the concrete floor is poured. *(Courtesy H.H. Robertson)*

Figure 47.26 A meter is used to measure the kilowatt-hours of electricity used.

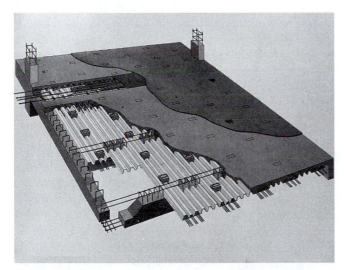

Figure 47.25 This is a structural cast-in-place monolithic-poured wire distribution system using the Robertson Cellcast Floor System. The cellcast units allow the floor to carry live and dead loads and provide access for electrical wiring with outlets as needed. *(Courtesy H.H. Robertson)*

MOTOR CONTROL CENTERS

Motor control centers are used to start and stop motors and to protect them from overloads. They have a disconnect switch that must be located within sight of the controller and the motor. The overload protection is typically a heat-operated relay that opens the circuit when line temperatures rise. The motor circuit is closed after the overload situation has been corrected by pressing a reset button (Figs. 47.28 and 47.29).

TRANSFORMERS

Transformers are used to change (transform) alternating current from one voltage to another. The voltage coming into the transformer is the *primary* voltage, and voltage leaving the transformer is the *secondary* voltage. For example, a transformer is used to step down a primary 4160V current received from the utility distribution system to a secondary voltage such as 480V as it enters the building. Another transformer in

Figure 47.27 The incoming power is run into the meter pan. Power enters the building electrical system after the utility company installs the meter.

the building's vault (or closet) could step this down to the 120V required for use within the building. The generally available primary voltages include 2400V, 4160V, 7200V, 12,470V and 13,200V. Secondary voltages include 480V, 277V, 240V, 208V, and 120V.

Transformers may be dry (air cooled) or liquid cooled. Transformers generate heat, which must be removed to prevent overheating. Dry transformers remove heat by circulating air through spaces in the transformer. Liquid-cooled transformers circulate an oil through coolers that absorb the heat and transfer it to the outside air. Mineral oil is the lowest cost liquid used, but it is flammable and cannot be used in all locations. A number of other liquid coolants, such as silicone liquids, have low flammability and are widely used. Liquid transformers are used on large installations (Fig. 47.30).

Transformers are specified by the voltages in and out, the amount of power they can handle in kilovolt-amps (kVA) (thousands of volts times the rated maximum am-

perage), also called insulation class, and sound rating. The insulation specification pertains to the electrical insulation temperature ratings in degrees centigrade. Sound ratings have to do with the sound vibrations produced by the different types and sizes of transformers. Transformers are available in single-phase and three-phase construction. The term "phase" refers to the number of circuits, voltages, and currents in an alternating current system. Single-phase means only one circuit or path for the flow of current; three-phase has three paths of flow.

Transformers may be located indoors or outdoors. Dry-type outdoor installations must have weatherproof enclosures, and larger sizes must be kept away from combustible materials. Liquid-type transformers on or adjacent to buildings in which combustion is possible must have a means for protecting against fires caused by excess heat or oil leaks. This can be provided with fire-resistant enclosures, various types of barriers, or a sprinkler system (Fig. 47.31).

Dry-type transformers installed indoors do not present the fire hazard of oil-type transformers. They need a fire-resistant barrier between them and combustible material, and they require adequate ventilation. Large sizes are usually placed in a fire-resistant vault.

Oil-type transformers placed indoors are generally required to be in a vault. The vault is fire resistant and must meet codes (Fig. 47.32). The vault walls, ceiling, and floor must have a fire rating to meet the code. They also must be ventilated, and the doors, frames, and sills also must meet fire ratings. No valves, piping, ducts, or other fittings not connected with the installation of the transformer are permitted in the vault.

Some of the smaller sizes of liquid-cooled transformers that use a nonflammable fluid may be installed indoors and outdoors in any location.

SWITCHES

Switches are used to open electrical circuits and interrupt the flow of current. A typical switch used on low-power applications, such as lighting, has some means of physically opening and closing the circuit—manually moving a lever or pushing a button using an electric coil, a motor, or a spring. This action moves metal contacts that separate to open the circuit or close to complete the circuit. Solid-state switches also interrupt the circuit, but they do so by electronically creating a conducting or nonconducting condition. There are no moving parts. They are classified by the National Electric Manufacturers Association (NEMA) and the Underwriters Laboratories (UL).

Switches have different actions that enable them to perform specific functions. These are illustrated by the simple knife switches shown in Fig. 47.33. A *single-pole, single-throw* (SPST) switch is moved to make or break a connection between two contact points. When the

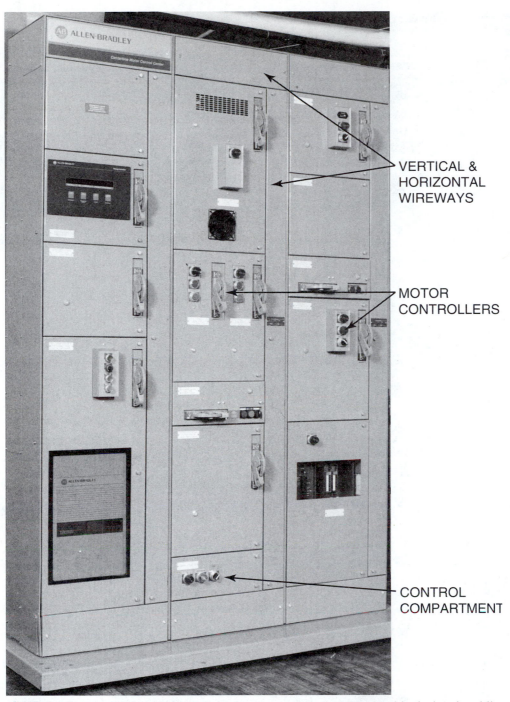

VERTICAL & HORIZONTAL WIREWAYS

MOTOR CONTROLLERS

CONTROL COMPARTMENT

Figure 47.28 A motor control center utilizing electronic and programmable devices in addition to electromechanical controls. Motor controls start and stop motors and provide overload protection. *(Courtesy Allen-Bradley Co., A Rockwell Automation Business)*

points are in contact the circuit is complete. When they separate or open, the circuit is broken.

A *single-pole, double-throw* (SPDT) action is used to control a unit such as a light from two locations. This requires a SPDT switch at both locations.

A *double-pole, single-throw* (DPST) action is similar to the SPST action. However, it opens both wires in the circuit. DPST switches are used when neither of the wires are grounded, as in switches used on 240V

motors and appliances. They can also be used to control two circuits at the same time. Both circuits may be open or closed simultaneously.

A *double-pole, double-throw* (DPDT) action can be used to control more than one circuit at a time. It can be used to reverse the direction of rotation of a DC motor by reversing the polarity.

Switches that control circuits with loads up to 30A, such as lighting, are in Fig. 47.34. They are rated to

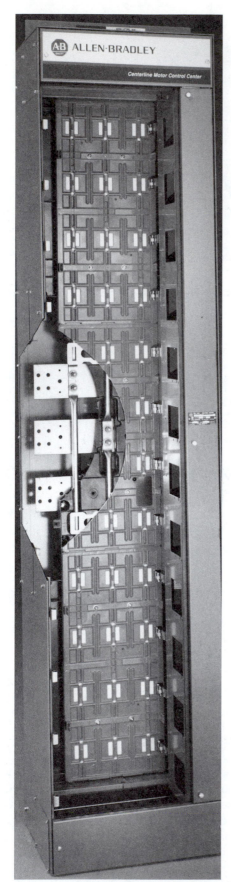

Figure 47.30 A liquid-cooled transformer used on a commercial building.

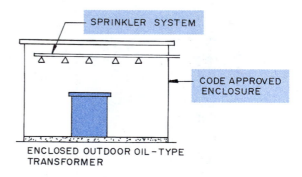

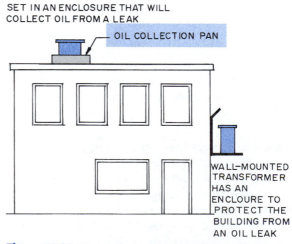

Figure 47.29 An internal view of the vertical motor control bus before electrical components are installed. It is made of molded polycarbonate plastic. *(Courtesy Allen-Bradley Co., A Rockwell Automation Business)*

Figure 47.31 Various ways codes permit a building to be protected from fire by oil leaks from oil-type transformers.

VENTS SIZED PER CODE
VENTS OPEN DIRECTLY TO OUTSIDE WITHOUT FLUES OR DUCTS
CONCRETE CEILINGS, WALLS AND FLOORS PER CODE
FIRE-RATED DOOR, FRAME AND SILL
CONCRETE FLOOR IN CONTACT WITH THE EARTH

Figure 47.32 A typical fire-resistant vault for oil-type transformers installed inside a building.

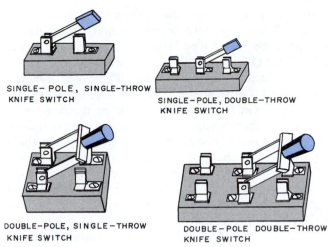

SINGLE-POLE, SINGLE-THROW KNIFE SWITCH
SINGLE-POLE, DOUBLE-THROW KNIFE SWITCH
DOUBLE-POLE, SINGLE-THROW KNIFE SWITCH
DOUBLE-POLE DOUBLE-THROW KNIFE SWITCH

Figure 47.33 Typical switch actions are illustrated by these knife switches.

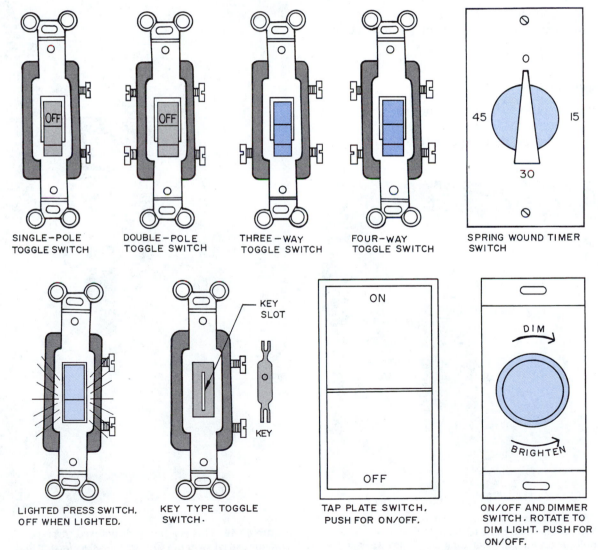

SINGLE-POLE TOGGLE SWITCH
DOUBLE-POLE TOGGLE SWITCH
THREE-WAY TOGGLE SWITCH
FOUR-WAY TOGGLE SWITCH
SPRING WOUND TIMER SWITCH

LIGHTED PRESS SWITCH. OFF WHEN LIGHTED.
KEY TYPE TOGGLE SWITCH.
KEY SLOT
KEY
TAP PLATE SWITCH, PUSH FOR ON/OFF.
ON/OFF AND DIMMER SWITCH. ROTATE TO DIM LIGHT. PUSH FOR ON/OFF.

Figure 47.34 These are examples of switches typically used to control circuits with loads up to 30A.

carry various loads that typically run 15A, 20A, or 30A at 120V. They may be toggle, push, keyed, rocker, touch, tap-plate, or rotary type. They are available as SPST, DPST, DPDT, and SPDT. They are also available as three-way and four-way switches. Three-way switches are used to control a light from two different locations. Four-way switches control a light from three different locations. Other switches can operate as timers that allow a unit, such as a fan, to operate for a set time after which it automatically shuts off. This type uses a spring-wound timer. A keyed switch provides security because a key is needed to operate it. Programmable switches are solid state switches that can be programmed to switch a circuit on and off at preset times (Fig. 47.35).

The switches described above do not contain a **fuse.** However, fusible switches are available (Fig. 47.36). The fuses and switches are enclosed in a metal box that can be closed and locked. The switch is manually operated with a handle on the outside of the box. Another type of switch using a solid state rectifier enables incandescent lights to have high, low, and off control. A rectifier is an apparatus in which electric current more readily flows in one direction than in the reverse direction for changing alternating current into a direct current.

Other Switches

There are many other switches designed to serve special purposes. A *service disconnect* disconnects all electric power to a building except for specific emergency equip-

ment requirements. Generally, these are located outside the building. They may be part of a switchboard.

A *contactor* is a form of switch that uses an electromagnet to close the contact blocks. It is operated from a remote location and can be activated by various devices, such as pushbutton or thermostat.

A *remote-control switch* is a form of contactor that remains latched until its electric circuit is energized. In this way the coil is energized only when the circuit is to be opened. It is used on applications where the circuit must be kept closed (as in a light installation) for long periods of time.

Time-controlled switches use some type of timer, such as an electronic timer or a miniaturized low-speed motor that rotates a disc that contacts the switch to open and close the circuit (Fig. 47.37).

Figure 47.36 This is a heavy-duty fused three-pole three-wire industrial switch. *(Courtesy Siemens Energy and Automation, Inc.)*

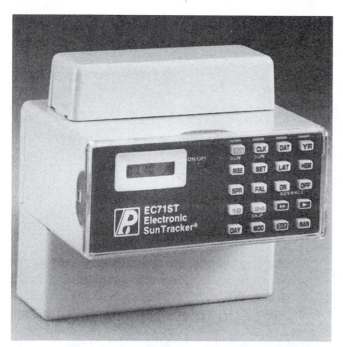

Figure 47.35 A programmable time switch that permits the independent programming of 8 to 20A, 240V circuits. It can provide multiple on-off daily switching on a 7 day week and special 365 day functions. *(Courtesy Paragon Electric Company)*

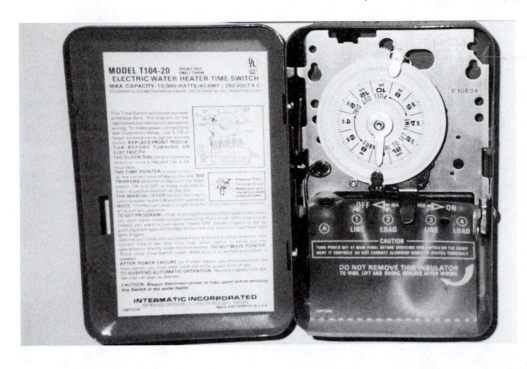

Figure 47.37 Time switches permit setting the clock so the circuit is open and closed as frequently as necessary. This switch operates on a 24 hour period and repeats the settings every 24 hours.

An *automatic transfer switch* is a double-throw switch that switches to a source of emergency power when the normal electrical service is interrupted (Fig. 47.38). A typical application would be in the power supply for a hospital operating room.

Isolation switches are opened only after the current flow in the circuit has been interrupted by another regular-use switch. They are not used to interrupt the flow of current in a circuit.

OVERCURRENT PROTECTION DEVICES

Circuit breakers are electromechanical units disconnecting a circuit automatically whenever the current attains an established value (Fig. 47.39) that would cause overheating and a possible fire in the circuit. Examples of occurrences that could cause a circuit breaker to open the circuit are excess current flow caused by overloading the circuit or a short in the circuit. Circuit breakers are available in a molded case that is inserted in the circuit at the panel (Fig. 47.40) or a large air type breaker that is not in a protective case. When a circuit breaker opens a circuit, the breaker is reset by moving the switch handle after the deficiency has been corrected (Fig. 47.41). Circuit breakers are mounted in a circuit breaker box or a panelboard (Fig. 47.42).

Fuses are also used to protect circuits from overloads and shorts in the circuits. They have an internal metal link that melts when overheated, breaking the circuit. The two types of fuses are cartridge and plug (Fig. 47.43).

The two types of cartridge fuses are *knife-blade* and *ferrule*. The blades on the knife-blade type slip into metal clips on the panelboard, and the ferrule type fits the copper rings on each end into clips. This connects the incoming power at the service entrance to the bus bars in the panelboard. Most cartridge fuses are discarded after they have blown, but some types have replaceable links.

Ferrule type cartridge fuses are rated from 10A to 60A and are generally used to protect currents for individual appliances, such as an electric stove. Knife-blade cartridge fuses are used for service over 60A and are used in the service entrance between the incoming power line and the circuits in the panelboard.

Figure 47.38 An 800A three-pole automatic transfer switch with a control panel. When the normal flow of electricity fails it will automatically transfer to the emergency service. *(Courtesy Automatic Switch Company)*

Figure 47.39 A circuit breaker disconnects a circuit automatically whenever there is excess current flow, a short circuit, a voltage surge, or overheating. This is a single-pole 20A plug-in circuit breaker. It also provides ground fault protection. *(Courtesy Square D Company)*

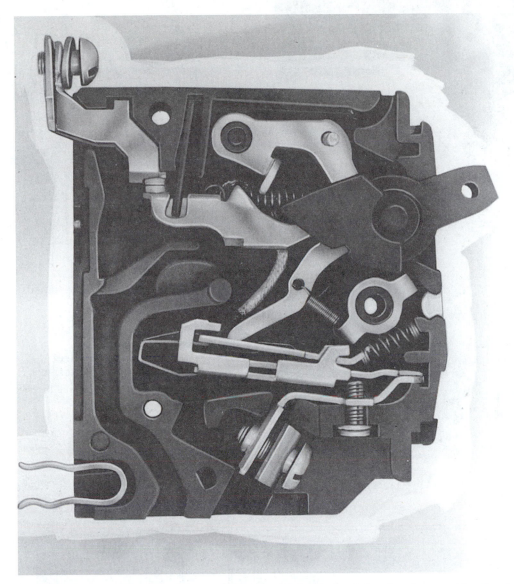

Figure 47.40 An interior view of a single-pole molded case circuit breaker that provides circuit and ground fault protection. *(Courtesy Square D Company)*

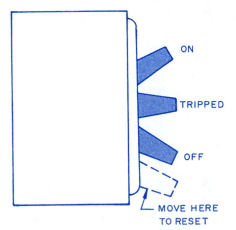

Figure 47.41 Positions of the handle on a circuit breaker. To restore power, move the lever to the "reset" position and back to the "on" position.

Plug fuses are available in 15A, 20A, 25A, and 30A sizes. They are used to protect individual circuits requiring small current requirements, such as a series of lights. They are installed in a fuse box or panelboard (Fig. 47.44) and serve the same function as a circuit breaker. However, when plug fuses are blown they are discarded then replaced with a new fuse when the deficiency has been corrected.

There are several varieties of plug fuses. The *standard plug fuse* will blow when the fusible link is overheated. The *time-delay plug fuse* has a fusible link that will melt immediately only when a short circuit occurs. If there is an overload the link softens but does not break. If the overload is quickly removed, such as a momentary load when starting a large electric motor, it will not break. However, if the overload continues the link will melt.

Figure 47.42 This circuit breaker box has one circuit breaker in place with blanks for six additional breakers.

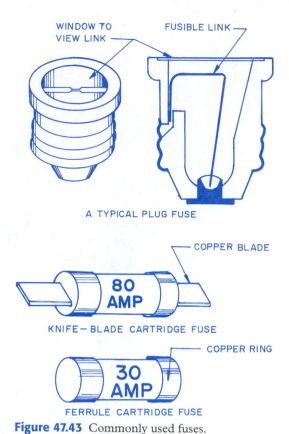

A TYPICAL PLUG FUSE

KNIFE—BLADE CARTRIDGE FUSE

FERRULE CARTRIDGE FUSE

Figure 47.43 Commonly used fuses.

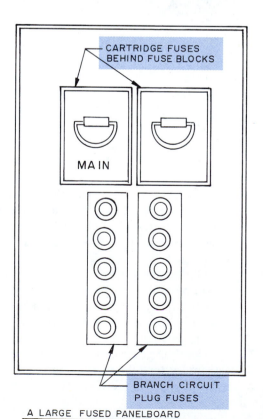

A LARGE FUSED PANELBOARD

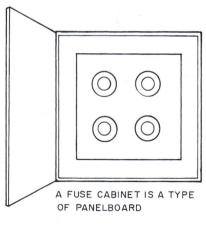

A FUSE CABINET IS A TYPE OF PANELBOARD

Figure 47.44 Plug type fuses are installed in a fuse box or a panelboard.

A *type S plug fuse* functions in the same manner as a time-delay fuse. However, the threaded base is too small to be screwed into the fuse panel. An adapter base is required. The adapters are threaded for fuses of different ampere capacities. The adapter is screwed into the threaded fuse socket on the panelboard. Since the adapter will only accept a plug fuse designed to fit it, a fuse of the wrong amperage cannot be screwed in place (Fig. 47.45).

SWITCHGEAR AND SWITCHBOARDS

Switchgear and **switchboards** are freestanding units made up of an assembly of fuses and/or circuit breakers, switches, and other line components. They provide switching and feeder protection to the circuits connected to the main power source. The switchboard distributes the incoming electrical power in smaller amounts within the building. Switchboards have all the components enclosed within a metal cabinet, and they are controlled by insulated handles and push buttons located on the front panel (Fig. 47.46).

Switchgear contain the same components and serve the same function as switchboards but handle higher voltages, usually above 600V. The term switchgear is also used to identify switching units that are individual switching units not assembled in a panel.

Switchboards and switchgear used in commercial and industrial buildings are commonly placed in a basement vault designed specifically to house the switchgear. It must be adequately ventilated and meet the requirements of the National Electrical Code. Typical space requirements are shown in Fig. 47.47 and in Table 47.2. The design of the vault should include adequate entrances and exits so equipment can be installed and removed and workers in the room can exit quickly in an emergency. Vaults must be located in permanently dry conditions. Exceptions must meet specific code requirements. Switchgear installed outdoors may be housed inside a small building or in a weatherproof metal case (Fig. 47.48).

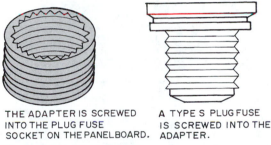

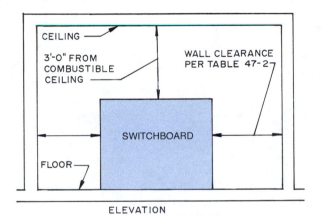

Figure 47.45 Threaded adapters are used to prevent someone from replacing a type S plug fuse of a specified amperage with a fuse of the wrong amperage.

Figure 47.46 A switchboard is a single electric control panel or assembly of panels on which are mounted switches, overcurrent devices, and instrumentation enclosed in an insulated metal structure. The devices are controlled by handles on the front of the panel. *(Courtesy Square D Company)*

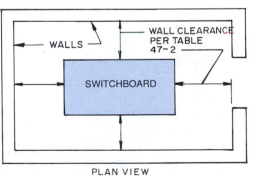

Figure 47.47 Typical minimum switchboard space requirements. *(Reprinted with permission from NFPA 70-1996, the National Electrical Code®, Copyright © 1996, National Fire Protection Association, Quincy, Mass. 02269. This reprinted material is not the complete and official position of the National Fire Protection Association, on the referenced subject, which is represented only by the standard in its entirety. National Electrical Code® and NEC® are registered trademarks of the National Fire Protection Association, Inc., Quincy, Mass. 02269.)*

Table 47.2 Switchboard Working Clearances

Nominal Voltage to Ground	Minimum Clear Distance		
	Condition 1[a]	Condition 2[b]	Condition 3[c]
0–150	3 ft. (914 mm)	3 ft. (914 mm)	3 ft. (914 mm)
151–600	3 ft. (914 mm)	3½ ft. (1066 mm)	4 ft. (1219 mm)

[a] Exposed live parts on one side and no live or grounded parts on the other side of the working space, or exposed live parts on both sides effectively guarded by suitable wood or other insulating materials. Insulated wire or insulated busbars operating at not over 300 volts shall not be considered live parts.
[b] Exposed live parts on one side and grounded parts on the other side.
[c] Exposed live parts on both sides of the work space (not guarded as provided in Condition 1) with the operator between.

Figure 47.48 Switchgear installed outdoors must be installed in a weatherproof enclosure.

Figure 47.49 This panelboard has the main breaker at the bottom and two vertical rows of circuit breakers.

PANELBOARDS

A **panelboard** receives a large amount of electrical power from the public utility and distributes it in smaller amounts through a number of circuits. In small buildings it serves the same basic purpose as a switchboard but on a smaller scale. A panelboard is typically used in residential and small commercial construction to distribute power to each of the circuits within the building after receiving input from the public utility line (Fig. 47.49).

A typical panelboard using circuit breakers is illustrated in Fig. 47.50. The incoming service entrance cable connects to the neutral bus bar and the main circuit breaker. The main controls the flow of power from the utility into the service panel. The white wire is the neutral and is connected to the neutral bus bar. The red and black wires

connect to the main and through it to the snap-on connections for the circuit breaker. The circuit breakers control the power to each circuit.

Panelboards have locked metal covers so no live parts are exposed unless the door is opened. Medium-duty panelboards are used for lighting and general purpose electrical outlets. Heavy-duty panels are used for distribution of power in industrial applications.

The panel may be surface mounted, semirecessed, or recessed into the wall. Each circuit is numbered, and an identifying description is given on a chart, such as No. 6–Kitchen wall outlets.

In large construction, panelboards are located on the various floors and are fed from the service switches and switchgear. The lights and outlets in an area of a floor are controlled from a panelboard on that floor.

ELECTRICAL SUPPLY

A typical system for supplying electricity to buildings and for other uses is shown in Fig. 47.51. The source is a public utility that uses some form of energy or fuel to operate the generators at the utility power plant to produce electricity for transmission to the consumer. The electricity passes to a step-up transformer station, which increases the voltage so the power can be transmitted with less current. This reduces the line power losses and permits the use of smaller conductors. The power is moved by high-voltage transmission lines to area transformer stations (Figs. 47.52 and 47.53).

Area transformer stations step down the voltage of the power (Figs. 47.54 and 47.55), and it is transmitted with overhead and underground lines to specified areas

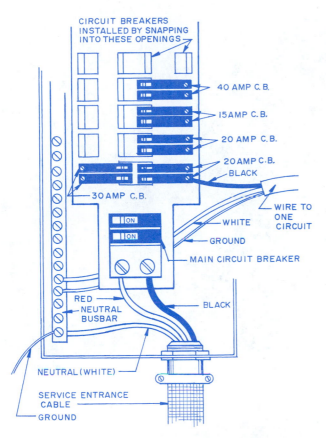

Figure 47.50 Assembly details for a typical panelboard using circuit breakers.

Figure 47.52 High-voltage power lines move the electrical power to areas requiring service.

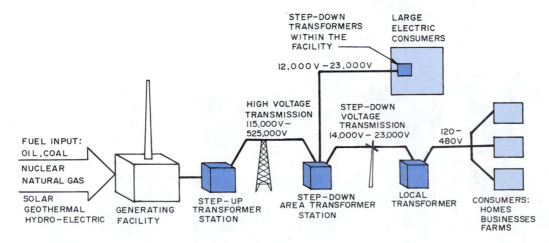

Figure 47.51 A general illustration showing the distribution of electricity from the generating facility to the consumer.

Figure 47.53 High-voltage line construction for long-distance power transmission. Cable is run from reels to temporary tower supports and on to the steel towers in the distance.

Figure 47.54 An area transformer station that receives power at 115,000V and steps it down to 23,000V three-phase for distribution to commercial and industrial uses and 13,200V single-phase for distribution to residential uses, where the on-site transformers step it down to 120/240V.

ELECTRICAL DISTRIBUTION SYSTEMS IN BUILDINGS

Electrical distribution systems inside a building supply the electrical power as needed to the various sections of the building and transmit information through an internal communications system. The basic standard for the design and installation is the National Electrical Code published by the National Fire Protection Association. The Underwriters Laboratory certifies electrical equipment and materials through the use of testing specifications and procedures.

The electrical power system of a commercial or industrial building includes the service entrance equipment, the interior distribution system, and the equipment that uses the electricity (Fig. 47.58). The service entrance includes equipment such as meters, switches, panelboards, switchgear, switchboards, fuses, and circuit breakers. The distribution system includes trays, wiring, raceways, wireways, conduit, and busways. The third part of the interior electrical system, the items to use the electricity, includes things such as lights, motors, heaters, industrial equipment, and communications equipment. These are the loads considered when the system is designed.

The following are generalized examples. The actual design of a system varies considerably depending on the requirements of the building and the equipment to be installed. System design requires the services of an experienced electrical engineer.

The Service Entrance

The service entrance is the system bringing the electrical power from the lines of the electrical utility into the building. A typical service entrance for a residential or small commercial building is shown in Fig. 47.59. The service from the public utility may be overhead or underground. The meter is generally on the exterior of the building. This enables the utility to read the meter and to install or remove it without entering the building. From the service entrance, panelboard branch circuits are run to various parts of the building for lights, outlets, appliances, furnace, and other required services. Each circuit has a circuit breaker overload device in the panelboard. Service can be run to a subpanel in an area some distance from the panelboard where branch circuits can be taken from it (Fig. 47.60). If the building has several occupants, as a duplex, individual meters may be set up for each apartment.

A service entrance for commercial buildings can take various forms depending on the requirements of the occupants. It requires a service switch or circuit breaker, which disconnects power to the entire building. This is typically located outside the building, but exceptions can be made. The power enters through a meter installation that includes a meter pan, a meter cabinet, and

Figure 47.55 A close-up of a transformer in the station in Fig. 47.54.

(Fig. 47.56). Industries using large amounts of electricity often receive this power into their own transformer stations for distribution and voltage regulation within their plants. For most commercial and private consumers the power is sent to a local transformer where it is stepped down to single-phase 120V or 120/240V power (Fig. 47.57). The 120/240V line can be used to operate appliances (electric dryer, range, water heater) and small industrial equipment requiring 240V. Larger commercial buildings may require either single-phase or three-phase at 120/208V, 120/240V, or 277/480V.

The power is connected to a commercial building through a **service entrance** consisting of a disconnect switch, a meter, and distribution switchgear. Residential and small commercial buildings service entrance consists of a meter and a panelboard.

Figure 47.56 Many transmission lines are run below ground. *(Courtesy International Brotherhood of Electrical Workers)*

a current transformer in a cabinet or vault. The service switch and metering equipment may be assembled in one unit. The metering transformer steps down the incoming electric current from the primary feeder to the building to a voltage required (as 120/240V) for use in the building. It may be outside the building on a concrete pad or inside or outside in a vault (Fig. 47.61). The power is fed into the switchboard, from which it is fed into a number of individual circuits as shown in Fig. 47.62.

LIGHTING

The use of natural lighting is a major consideration when a designer is working on the design of the illumination for a building interior. Most spaces have natural and artificial light integrated so they flow into the space and work together to provide the type and degree of illumination required. Natural lighting can be controlled by various shading devices and the type of glazing used. Review Chapter 29 for information on glazing.

Illumination Design Considerations

As the lighting system is designed, the activities in each area must be carefully analyzed. Certainly the purpose of the lighting system is to enable the occupants to have a pleasant atmosphere and to be visually able to perform required tasks. All of this must be accomplished using as little electrical power as possible.

When designing a lighting system the ceiling height, the reflectance values of the ceiling, walls, floors, and furniture, and the footcandle requirements all relate to each other. The choice of the type of **luminaire,** its spacing, how it is mounted, and its height above the area to be lighted must be considered.

Figure 47.57 Pole-mounted and pad-mounted transformers are connected to high-voltage transmission lines. They step down the voltage for use by the consumer.

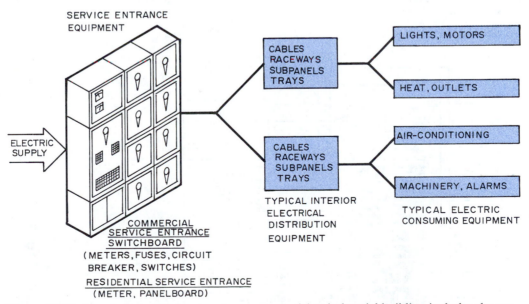

Figure 47.58 A typical electrical system for a commercial or industrial building includes the service entrance equipment, distribution equipment, and the equipment that consumes the electrical power.

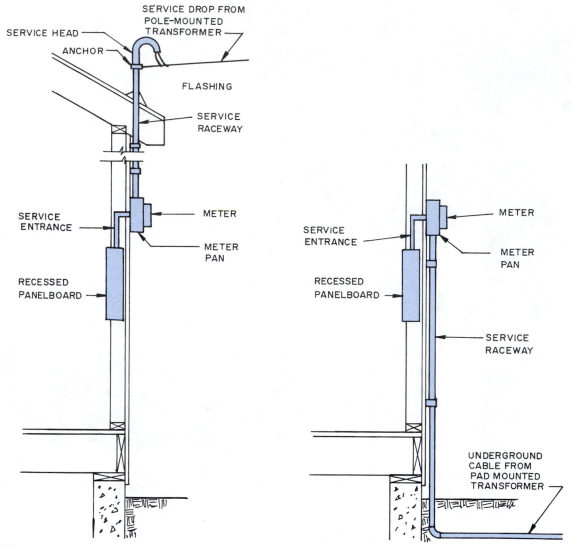

Figure 47.59 Typical above- and below-ground service entrances for residential and small commercial buildings.

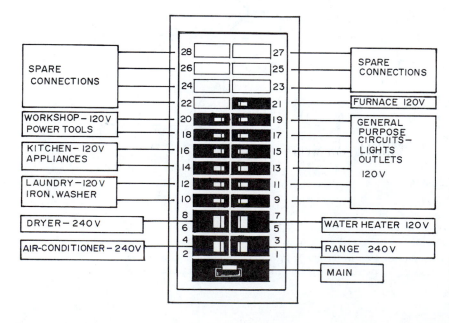

Figure 47.60 The service entrance panelboard is wired to distribute the electricity through individual circuits to various parts of the building and to equipment.

I T E M   O F   I N T E R E S T

CHOOSING LIGHT AND COLOR

Light quality is measured by the color temperature and the color rendering index. *Color temperature* gives a measurement of the visual warmth or coolness of the light source, expressed in degrees Kelvin (°K). This refers to the quality of the actual light from the lamp. The higher the color temperature the cooler the light. Likewise, the lower the color temperature the warmer the light. For example, typical inexpensive cool white fluorescent tube has a color temperature of about 4100°K. Typical incandescent lamps have a color temperature of about 2700°K. Light from an incandescent lamp that is warmer contains red and yellow, contributing to the tones of surrounding materials people find warm and pleasing. Likewise, if you replace the incandescent lamps with high-temperature fluorescent tubes you will see they cast a hot bluish to greenish light that give surrounding materials a gray and flat appearance.

The *color rendering index (CRI)* gives a measure of light quality that indicates how natural an object looks when under the light source. This index is given as a percentage. The closer the CRI is to 100 percent the more natural things appear. When a natural appearance is critical, as in an art museum, seldom is a lamp with less than 70 percent used. Incandescent lamps have CRI ratings close to 100 percent, but fluorescent tubes vary from below 50 to about 90 percent.

Following are some rules of thumb to help make color selection easier.*

1. All space colors (wall and floor coverings, furniture, drapes, accents, etc.) should be chosen under the lamp color specified for the installation. Experience suggests that warm sources should be used at low lighting levels, cool sources at high levels. However, the choice may be influenced by space colors and by degree of luminaire brightness control.
2. Warm color schemes may appear overpoweringly warm if lighted with a warm source to relatively high levels—use a cooler source.
3. Cool color schemes may need warm sources, particularly at low lighting levels.
4. Well-shielded lighting systems, such as those using wedge louvers, or low brightness lenses will "cool off" a room. Solution: Use a slightly warmer lamp to counteract the effect.

5. Where color rendition is highly critical, use high-CRI continuous-spectrum sources such as Chroma 50 (C50) or Chroma 75 (C75).
6. Where both color and lighting level are important, use either the SP30, SP35, or SP41 GE Specification Series Color lamps, with three-peak rare earth phosphors, for good color rendering and high efficiency.
7. The Specification Series Colors tend to make spaces look more colorful because the three-peak phosphors compress all colors into the blue, green, and red-orange bands. This increases the contrast between colors. Three-peak lamps do not, however, increase the contrast of black-on-white tasks. Claims that less light is needed for typical office and industrial tasks when three-peak lamps are used are scientifically unfounded.
8. The color of a light source does not affect the visual performance of people doing black-on-white seeing tasks. Vision and productivity studies, however, indicate that productivity may be affected by the color contrast and appearance of the visual environment and that color can contribute strongly to appearance.

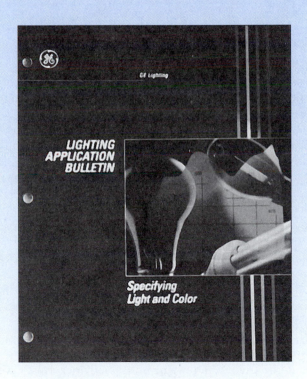

*Principles adapted with permission from the General Electric publication, *Specifying Light and Color.*

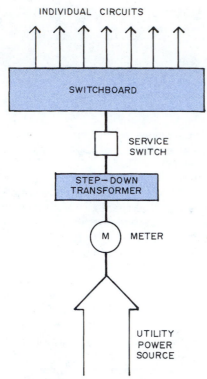

INDIVIDUAL CIRCUITS

SWITCHBOARD

SERVICE SWITCH

STEP–DOWN TRANSFORMER

M   METER

UTILITY POWER SOURCE

Figure 47.61 A typical service entrance for a commercial or industrial building.

Task Lighting

Task lighting provides an adequate level of light for a person to perform work of a specific nature, such as electronic assembly on a production line, without undue eyestrain. It should be possible to work at a specific task without having the same high levels of lighting in the entire room. The tasks performed, as in a manufacturing plant, need to be identified. Their location within the building and the area involved must be specified, as well as the number of people performing the task, their relationship (nearness) to each other, and any patterns of movement of the occupants performing the tasks. These and other activities are considered as the lighting is planned. In many types of businesses change in activities is normal, so the system must allow for easy revision, including relocating luminaires or redirecting light from existing luminaires. Since the level of light required may vary from time to time (as when natural light is plentiful), luminaires must be individually switched so only those needed can be turned on or dimmer switches should be used. Windows near task activities must have blinds, shutters, or operable louvers to regulate the natural light. Available daylight can vary considerably over a period of hours.

The surface upon which the task is to be performed also influences lighting decisions. The most common situation involves some form of horizontal surface, but inclined and vertical work surfaces are frequently found. The widespread use of computers presents multiple problems—a horizontal keyboard, a vertical rack to hold the copy being set, and a monitor that produces light directly at the operator. The reflective values of the surface also require consideration because highly reflective work surfaces can be very disturbing.

General Illumination

General illumination provides a degree of light over a large area, such as a room in which specific task lighting luminaires also are present. General illumination provides a way to ease visual discomfort caused when looking from a high-level lighted area (as a task area) out across a large area (a room). General illumination reduces the difference in brightness levels, thus maintaining visual comfort.

Since the level of lighting intensity is less, the general illumination luminaires can be widely spaced and use less electricity. An example would be in a library where the study carrels have individual lights with a level required for reading, yet the overall reading room has luminaires in the ceiling that provide a lower level of illumination and visual comfort when a person looks up from the book. General illumination also provides for a warm and more pleasant surrounding by reducing darkness and brightening a gloomy atmosphere. The designer attempts to get a balance between the brightness of a task-oriented area and the larger general area of the room.

Selective Lighting

Selective lighting is used to focus light on a specific object or area, such as light focused on a painting or on a display case containing jewelry (Fig. 47.63). The general illumination becomes the background and the specific lighting focuses visual attention on the object.

Glare

Glare refers to light that is intense and offending to the viewer. It may be from luminaires that shine *directly* into the eyes of the viewer. This can be reduced by lowering the brightness of the luminaires or relocating or shielding them if possible. *Reflected glare* is found more frequently. It occurs when the light is reflected off a surface. It can be controlled by selecting surfaces with low reflectance coefficients, reducing the brightness of the luminaire, or changing the angle of the reflecting surface (Fig. 47.64).

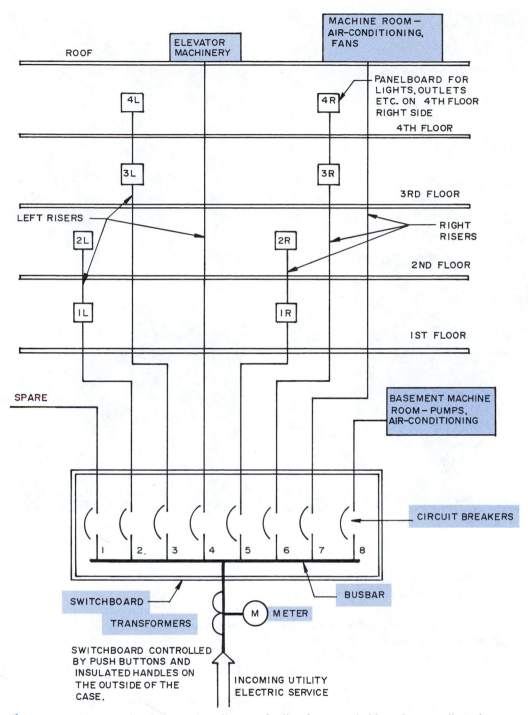

Figure 47.62 A generalized power riser diagram feeding from a switchboard to panelboards on the various floors and to equipment on the roof and in the basement.

Reflection

The **reflectance coefficient** of walls and ceiling is an important part of lighting planning. The **luminous intensity** of a luminaire can be reduced or enhanced by the reflective potential of the surrounding ceiling, walls, and floor. Dark colors absorb light, and light colors reflect light. Therefore, more light produced by the luminaire is used if light-colored ceilings and walls are available. Bright colors, such as red or yellow, are not used often because they tend to provide a glare.

Manufacturers of wall and ceiling materials and paints generally have had them tested and can give their reflectance coefficients. For example, if half the incident light on a surface is reflected, the reflectance coefficient is 50 percent or 0.50. The other 50 percent is absorbed by or transmitted through the material. The amount

Figure 47.63 Selective lighting is used to highlight an object or architectural feature.

Figure 47.64 Reflected glare causes visual discomfort and can obscure a view of an object.

transmitted is called **luminous transmittance.** The rate of flow of light energy through a surface is called **luminous flux.**

The reflectance value of a ceiling used with a direct lighting system has little influence on the light projected, because *direct lighting* projects from the luminaire directly down into the room. Ceiling reflectance does greatly influence the usable light if *indirect lighting* is used. Indirect lights project the light up to the ceiling where it is reflected down into the room. The reflectance value of walls also influences the reflected light. However, walls typically have doors, windows, cabinets, and other items that reduce the actual wall surface available. Floors have little influence on lighting; however, light-colored floors do reflect some light and provide visual comfort.

Room Cavity Ratio

The size of the space to be lighted affects the way light is distributed throughout the space. High ceilings reduce the efficiency of lighting. Open spaces can be more efficiently lighted than small rooms or partitioned space. The relation between the space proportions, room height, room perimeter, and the height of the work surface to be illuminated divided by the floor area is called the **room cavity ratio.**

LIGHTING-RELATED MEASUREMENTS

Levels of Illumination

Various activities require different levels of illumination. Adequate illumination depends on the **footcandles** (quantity) plus the brightness, distribution, and color of the light (quality). Drafting activities require about 200 footcandles, but a corridor only needs 10 to 20 footcandles. Recommended levels of illumination have been established by the Illuminating Engineering Society of North America. Some of these are shown in Table 47.3. The values in this table are given in footcandles, which have the same value as metric **candela.** They can be changed to **lux** by multiplying by 10.76. Lux is a unit of illumination equal to 1 **lumen** per square meter. Lighting unit conversion factors are shown in Table 47.4.

Table 47.3 Recommended Illumination Levels

Type of Space	Guideline (footcandles)[a]
Commercial and Institutional Interiors	
Art galleries	30–100
Auditoriums/assembly spaces	15–30
Banks	
Lobby	50
Customer areas	70
Teller stations and accounting areas	150
Hospitals	
Corridors, toilets, waiting rooms	20
Patient rooms	
General	20
Supplementary for reading	30
Supplementary for examination	100
Recovery rooms	30
Lab, exam, treatment rooms	
General lighting	50
Closework and examining table	100
Autopsy	
General lighting	100
Supplementary lighting	1,000
Emergency rooms	
General lighting	100
Supplementary lighting	2,000
Surgery	
General lighting	200
Supplementary on table	2,500
Hotels (rooms and lobbies)	10–50
Labs	50–100
Libraries	
Stacks	30
Reading rooms, carrels, book repair and binding, check-out	70
Catalogs, card files	100
Offices	
Corridors, stairways, washrooms	10–20[b]
Filing cabinets, bookshelves, conference tables	30
Secretarial desks (with task lighting as needed)	50–70
Routine work (reading, transcribing, filing, mail sorting, etc.)	100
Accounting, auditing, bookkeeping	150
Drafting	200
Post Offices	
Storage, corridors, stairways	20
Lobby	30
Sorting, mailing	100
Restaurants	50
Schools	
Auditoriums	15–30
Reading or writing, libraries	70
Lecture halls	70–150
Drafting labs, shops	100
Sewing rooms	150
Stores	
Circulation areas, stock rooms	30
Merchandise areas	
Serviced	100
Self-service areas	200
Showcases and wall cases	
Serviced	200
Self-service	500
Feature displays	
Serviced	500
Self-service	1,000

(Continued on next column)

Table 47.3 *(Continued)*

Type of Space	Guideline (footcandles)[a]
Industrial Interiors (Manufacturing Areas)	
Storage areas	
Inactive	5
Active (rough, bulky)	10
Active (medium)	20
Active (fine)	50
Loading, stairways, washrooms	20
Ordinary tasks (rough bench and machine work, light inspection, packing/wrapping)	50
Difficult tasks (medium bench and machine work, medium inspection)	100
Very difficult tasks (fine bench and machine work, difficult inspection)	200–500
Extremely difficult tasks	500–1,000
Garages	
Active traffic areas	20
Service and repair	100
Exteriors	
Building security	1–5
Parking	
Self-parking	1
Attendant parking	2
Shopping centers (to attract customers)	5
Floodlighting	5–50
Bulletins and poster panels	20–100

[a]To convert to lux, multiply value by 10.76.
[b]Must be at least 20 percent of the level of the adjacent work space.
Courtesy Illuminating Engineering Society of North America

Table 47.4 Lighting Unit Conversion Factors

Unit	To convert from ...	Multiply by the correction factor ...	To convert to ...
Luminous Intensity (I)	Candlepower	1.00	Candela
Illuminance (E)	Footcandle	10.76	Lux
	Lux	0.09	Footcandle
Luminance (L)	Candela/m^2	0.29	Footlambert
	Candela/in.^{2}	1550.00	Candela/m^2
	Candela/ft.^{2}	10.76	Candela/m^2
	Footlambert	3.43	Candela/m^2

Measuring Illuminance Levels

The measure of **illuminance** is a measure of the *quantity* of illumination. One lumen of luminous flux uniformly imposed on one square foot of area results in an illuminance of one footcandle.

The level of illuminance is commonly measured with a illuminance meter, often referred to as a light meter (Fig. 47.65). It has a photoelectric panel connected to a microammeter and an electronic control circuitry. It is calibrated to read in footcandles or lux. Some types have a remote sensor. The meter is held so the photoelectric panel is parallel to the plane of the area to be tested. For example, to measure ceiling illuminance the meter is held parallel to the ceiling and to measure wall illuminance it is held parallel to the wall. It is important to follow the manufacturer's instructions so accurate data are recorded.

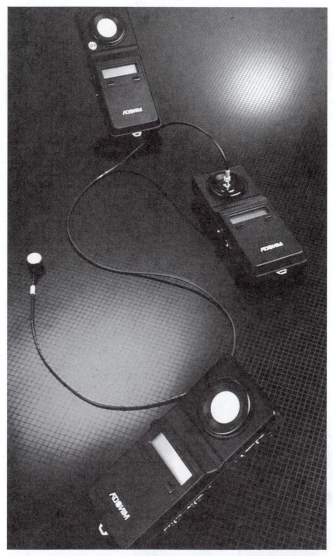

Figure 47.65 The level of illuminance is measured with an illuminance (light) meter. This meter measures in units of lux (lx) and footcandles (fc). *(Courtesy Minolta Corporation)*

Measuring Luminance

To totally evaluate the illumination available, it is necessary to get a measure of the **luminance** or *quality* of illumination. Luminance is a measure of the brightness and brightness contrasts caused by photometric luminance. It is what is seen rather than illuminance (the quantity in footcandles), and it is an important part of the evaluation of the available total illumination. Luminance is measured by a **luminance meter.**

LIGHTING SYSTEMS

The most frequently used lighting system is possibly a direct system, but indirect, semi-indirect, general-diffuse, and semi-direct find many applications (Fig. 47.66). The exact classification depends on the amount of light from the luminaire directed up and down (Table 47.5).

Direct lighting systems project most of the light down to the floor or work surface. They are the most efficient systems. A light color reflective floor will reflect some light toward the ceiling, reducing the darkness at the ceiling somewhat. *Indirect lighting systems* project most of the light up to the ceiling and some onto the walls. The light is then reflected to the floor and work surfaces. This system requires that ceilings and walls have high reflectance coefficients. It produces an illumination that is rather uniform, free of glare, and diffuse. Indirect lighting is rather inefficient and requires that luminaires have more powerful lamps than direct lighting.

Semi-indirect lighting systems allow more light to project down than indirect systems do. They do this by using a diffuser that is translucent, permitting some light to project down yet reflecting most of it up to the ceiling. This produces a diffuse low-glare illumination.

General-diffuse systems (also called direct-indirect) project about equal amounts of light up and down. They produce a light ceiling and upper wall area yet project light down to the floor and work surfaces. This system typically uses a globe diffuser, which allows light to project in all directions.

Semi-direct lighting systems project most of the light down with only a very small percentage upward to light the ceiling. This produces efficient direct lighting yet brightens the dark ceiling area that exists when direct lighting is used.

TYPES OF LAMPS

A variety of **lamps** are available, and each has particular advantages and disadvantages. Some are useful only for special applications.

Incandescent Lamps

An incandescent lamp is a glass bulb containing a filament joined to a metal base with copper lead-in wires (Fig. 47.67). The filament is a tungsten wire that resists

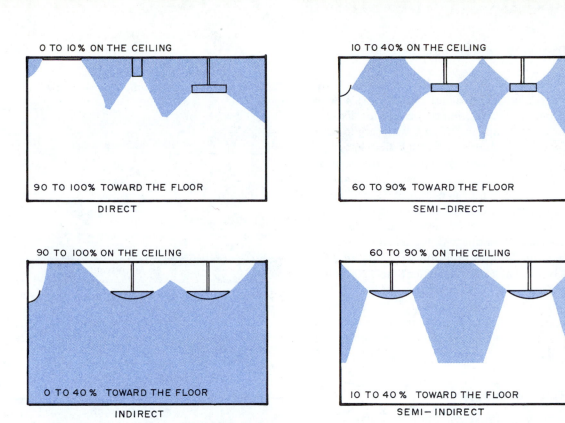

Figure 47.66 The distribution of light can be varied by using these types of lighting systems.

Table 47.5 Types of Lighting Systems

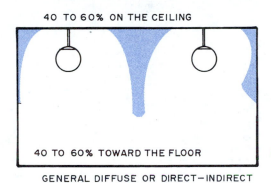

	Distribution of Light Emitted by Luminaire	
Type	Upward (%)	Downward (%)
Indirect	90–100	0–10
Semi-indirect	60–90	10–40
General diffuse	40–60	40–60
Semi-direct	10–40	60–90
Direct	0–10	90–100

Courtesy Illuminating Engineering Society of North America

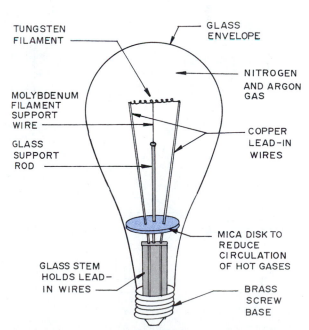

Figure 47.67 The major parts of an incandescent lamp.

the flow of electricity. In doing so, the wire gets hot and glows, producing light. The metal base screws into a socket that is part of the luminaire, or lighting fixture. Most incandescent lamps are filled with argon and nitrogen, which makes the use of a high filament temperature possible. The melting point of tungsten is 3655K (6170°F). Lamps filled with krypton gas have a longer life than argon and nitrogen lamps and cost more.

Incandescent lamps are rated in watts, as shown in Table 47.6. They are made with several size bases. The larger bases are used on lamps with higher wattages. The voltage at which the lamp is used influences its life. If a 120V lamp is burned at a voltage higher than 120, its life will be shortened, but it will produce more lumens per watt. If operated at a voltage below this, it will have a prolonged life but produce fewer lumens per watt. Some of the commonly available incandescent lamps are shown in Fig. 47.68.

Tungsten-Halogen Lamps

Tungsten-halogen lamps (quartz-iodine) are a type of incandescent lamp. They have a tungsten filament and are filled with a halogen, such as iodine or bromine, and an inert gas that reduces the evaporation of the filament. The bulb is made from quartz because it will withstand higher temperatures than glass. The halogen additive in the lamp reacts chemically with any tungsten deposited on the bulb and redeposits it on the tungsten filament, improving the efficiency of the lamp. Tungsten-halogen lamps are available in a wide range of wattages, as shown in Table 47.6.

Table 47.6 Summary of Lamp Characteristics[a]

Type	Light Production	Typical Wattage Range
High-Intensity Lamps		
Mercury	Argon gas and mercury vapor	50–1,000
Metal-halide	Argon gas, mercury, and halide salts	175–1,500
High-pressures sodium	Sodium, mercury amalgam, and xenon gas	100–1,000
Incandescent		
Standard incandescent	Tungsten filament in a gas	6–2000
	Tungsten filament in a vacuum	5,000 and 10,000
Tungsten-halogen	Tungsten filament with a halogen, such as iodine or bromine	12–2,000
Fluorescent		
Preheat	Mercury vapor in an inert gas, such as krypton, argon, or neon, with the tube interior coated with a fluorescent material	15–90
Instant heat		20–75
Rapid start, high output		30–100
Rapid start, very high output		100–215

[a]Check manufacturers' catalogs for specific light production and wattage data.

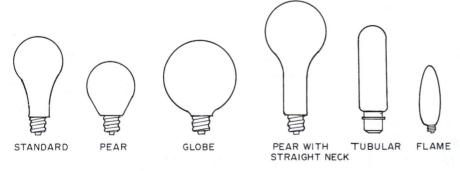

STANDARD   PEAR   GLOBE   PEAR WITH STRAIGHT NECK   TUBULAR   FLAME

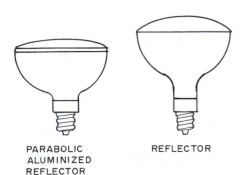

PARABOLIC ALUMINIZED REFLECTOR   REFLECTOR

Figure 47.68 Some of the types of incandescent lamps available.

I T E M   O F   I N T E R E S T

LOW-MERCURY FLUORESCENT LAMP

Standard fluorescent lamps contain mercury and are classified as hazardous waste by the Environmental Protection Agency. A new long-mercury fluorescent lamp is now available. It uses a chemical buffer to slow down the absorption of mercury by the phosphor crystals that coat the inside of the tube so less mercury is required in the lamp.

These lamps require a very precise amount of mercury. Each lamp receives a premeasured amount of mercury encapsulated in a tiny glass bubble. As the lamp leaves the factory, a radio-frequency beam heats a wire wrapped around the bubble, which breaks, releasing the mercury in the tube.

Low-mercury lamp meets the EPA standard. (*Courtesy Phillips Lighting Company*)

Examples of some typical lamps are in Fig. 47.69. Two of the lamps shown have quartz tubes. One has a screw base and is installed vertically. The double-end lamp is installed horizontally. Both require reflectors as shown in Fig. 47.70.

Tungsten-halogen lamps with parabolic reflectors are available from 15W to 1500W. They are available in a number of shapes. The slanted back side of the bulb is a reflector. Both flood and spot lamps are available.

Tungsten-halogen lamps operate at high temperatures, and even though the quartz envelope can withstand high temperatures the lamps could explode. Therefore, manufacturers recommend some type of shielding or protective screen be used to contain flying fragments from unprotected quartz lamps.

Fluorescent Lamps

Fluorescent lamps have a long glass tube that is sealed at each end. A mixture of an inert gas, such as argon, and low-pressure mercury vapor is in the tube. A cathode is in each end. The cathode produces electrons that start and maintain operation of a mercury arc, which produces an ultraviolet arc. The ultraviolet arc is absorbed by phosphors that coat the inside of the tube and cause it to fluoresce (radiate) light (Fig. 47.71).

Fluorescent lamps will not operate directly off 120V alternating current because it will not cause the arc dis-

charge. The luminaire has a *ballast* that provides the starting and operating voltages. *Standard lamps* have a starter that *preheats* the cathodes, producing the high-voltage arc needed to start the lamp. *Rapid-start* lamps have the same basic construction as standard lamps but a circuit keeps the lamp electrodes constantly preheated by means of low voltage windings that are a part of the ballast.

High-output (HO) lamps and *very-high-output* (VHO) lamps require special ballasts. They are used where high output is required in a limited area, such as a merchandise display or an outdoor sign. They generate considerable heat, and this must be considered when they are used. They will function in cold situations in which standard lamps will not light.

Instant-start fluorescent lamps use a high-voltage transformer to generate an arc, lighting the lamp without the preheating delay of standard lamps. They have a single pin at each end. The high-voltage start greatly reduces the life of the lamp, but it will start in temperatures below that of rapid-start lamps.

Fluorescent lamps are available in straight tubes, circles, and U-shapes. A small lamp is available that will fit in a table lamp socket (Fig. 47.72). Several different bases and lampholders are used depending on the type of lamp. Some of these are illustrated in Fig. 47.73.

Since fluorescent lamps are much more efficient than incandescent lamps, lower wattage lamps can be used. Typical wattage ratings are shown in Table 47.6.

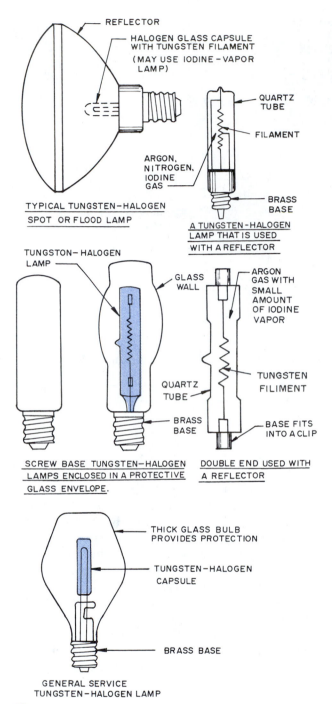

REFLECTOR

HALOGEN GLASS CAPSULE
WITH TUNGSTEN FILAMENT
(MAY USE IODINE-VAPOR
LAMP)

TYPICAL TUNGSTEN-HALOGEN
SPOT OR FLOOD LAMP

QUARTZ
TUBE

FILAMENT

ARGON,
NITROGEN,
IODINE
GAS

BRASS
BASE

A TUNGSTEN-HALOGEN
LAMP THAT IS USED
WITH A REFLECTOR

TUNGSTON-HALOGEN
LAMP

GLASS
WALL

QUARTZ
TUBE

BRASS
BASE

ARGON
GAS WITH
SMALL
AMOUNT
OF IODINE
VAPOR

TUNGSTEN
FILIMENT

BASE FITS
INTO A CLIP

SCREW BASE TUNGSTEN-HALOGEN
LAMPS ENCLOSED IN A PROTECTIVE
GLASS ENVELOPE.

DOUBLE END USED WITH
A REFLECTOR

THICK GLASS BULB
PROVIDES PROTECTION

TUNGSTEN-HALOGEN
CAPSULE

BRASS BASE

GENERAL SERVICE
TUNGSTEN-HALOGEN LAMP

Figure 47.69 Typical tungsten-halogen lamps.

Figure 47.70 A tungsten-halogen quartz-iodine lamp installed horizontally in a reflector to provide indirect lighting.

High-Intensity Discharge Lamps

High-intensity discharge lamps (HID) produce light by passing an electric arc through a metallic vapor that is confined in a sealed quartz or ceramic tube. The high-intensity discharge lamps available include mercury vapor, metal-halide, and high- and low-pressure sodium. With appropriate color correction they can be used in many indoor and outdoor situations.

Mercury Vapor Lamps

Mercury vapor lamps (M-V) are available in clear, white, white-delux, and color-corrected. The clear lamp produces a blue-green light that causes distortion of other colors. Color correction is accomplished by coating the outer bulb with phosphors to make the lamp useful for some indoor applications (Fig. 47.74).

Mercury vapor lamps use mercury and argon gas, which helps in starting since mercury has a low vapor

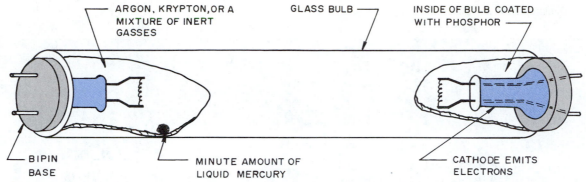

ARGON, KRYPTON, OR A MIXTURE OF INERT GASSES

GLASS BULB

INSIDE OF BULB COATED WITH PHOSPHOR

BIPIN BASE

MINUTE AMOUNT OF LIQUID MERCURY

CATHODE EMITS ELECTRONS

Figure 47.71 A generalized illustration showing the construction of a fluorescent lamp with bipin bases.

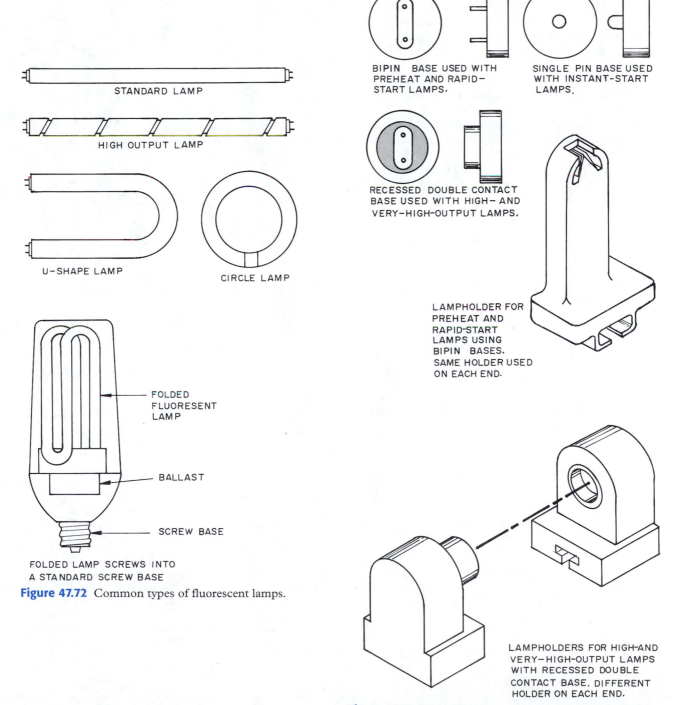

STANDARD LAMP

HIGH OUTPUT LAMP

U-SHAPE LAMP

CIRCLE LAMP

BIPIN BASE USED WITH PREHEAT AND RAPID-START LAMPS.

SINGLE PIN BASE USED WITH INSTANT-START LAMPS.

RECESSED DOUBLE CONTACT BASE USED WITH HIGH- AND VERY-HIGH-OUTPUT LAMPS.

LAMPHOLDER FOR PREHEAT AND RAPID-START LAMPS USING BIPIN BASES. SAME HOLDER USED ON EACH END.

FOLDED FLUORESENT LAMP

BALLAST

SCREW BASE

FOLDED LAMP SCREWS INTO A STANDARD SCREW BASE

Figure 47.72 Common types of fluorescent lamps.

LAMPHOLDERS FOR HIGH-AND VERY-HIGH-OUTPUT LAMPS WITH RECESSED DOUBLE CONTACT BASE. DIFFERENT HOLDER ON EACH END.

Figure 47.73 Typical fluorescent lamp bases and lampholders.

I T E M   O F   I N T E R E S T

PLUG-IN ELECTRICAL TESTERS

Inexpensive plug-in electrical testers are used to troubleshoot a receptacle to see if it has been miswired. The tester is plugged into the outlet and the results are shown by a series of lights on the front of the meter. Each combination of lighted lights tells what is wrong with the wiring of the outlet. If the tester indicates excess voltage drop under load it means there may be poor connections or damaged or undersized wires that could cause a fire. It can also indicate improper wiring, false grounds, inadequate grounds, and high-resistance grounds, which are all safety hazards. These testers only work with 120 volt receptacles (Fig. A).

Examples of three commonly found wiring situations are:

1. Wired correctly. White wire (neutral) to silver color screw, hot (black or red) to brass color screw, ground wire to green screw (Fig. B).
2. Reversed polarity. Hot and neutral wire reversed. The danger here is that even though a light plugged in works when the switch is off, the cylinder that holds the bulb is hot and a person could get a shock (even though the switch is off). (Fig. C).
3. Ground wire not connected. This outlet has the three holes for a three-prong grounded appliance, but actually the appliance is not grounded, and if it develops a hot-to-ground fault the user will get shocked and possibly killed. (Fig. D).

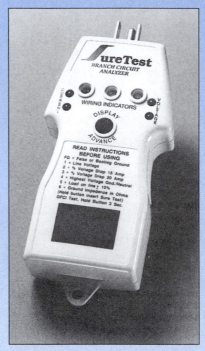

Figure A This is a microprocessor-based branch circuit analyzer that measures the effect on a circuit of a full 15 amp load. *(Courtesy Industrial Commercial Electronics, Inc.)*

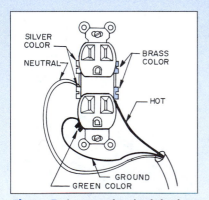

Figure B A correctly wired duplex receptacle.

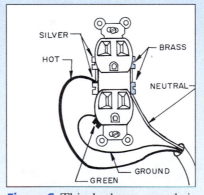

Figure C This duplex receptacle is wired with the polarity reversed. Although things plugged into it will work, the neutral wire is hot and even if the switch is off and the neutral wire is touched you will get a powerful electric shock.

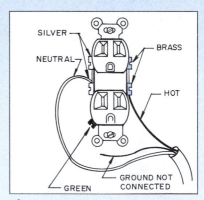

Figure D This duplex receptacle does not have the ground wire connected, so the expected protection does not actually exist. However, those using the receptacle assume they are protected.

pressure at room temperature. A starting voltage energizes the circuit and moves a voltage across the space between the starting electrode and the main electrode. This creates an argon arc that vaporizes the mercury (Fig. 47.75).

A mercury vapor lamp takes three to six minutes to reach its full warmup and maximum output. A ballast is used to start the lamp and control the arc. After the lamp is turned off it must cool and the pressure must be reduced before it can be relighted. This takes from three to eight minutes, depending on the construction of the lamp. Since this wait time would leave an area in total darkness if a power failure of even a few seconds would occur, some form of emergency backup lighting is usually required. This could be a system using incandescent lamps connected to the electric system or a battery-powered emergency lighting system.

Mercury vapor lamps are not as efficient as fluorescent lamps, but they are more efficient than incandescent lamps. They are used on applications in which they remain in use for long periods, such as overnight on a parking lot. They should not be used if frequent switching on and off is required.

Mercury vapor lamp spectrum is in the ultraviolet (UV) range, but it is not harmful to those exposed because the outer glass absorbs much of the UV. However, if the glass breaks, the lamp continues to function and UV exposure could be hazardous; therefore, a safety warning is on all mercury vapor lamps. They are available in various wattages, as shown in Table 47.6.

Metal-Halide Lamps

Metal-halide lamps are another type of high-intensity discharge lamp (Fig. 47.76). They are designed much like the mercury vapor lamp, but halides of metals, such as sodium, indium, or thallium, are added. These salts

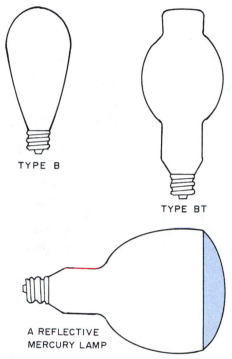

TYPE B

TYPE BT

A REFLECTIVE MERCURY LAMP

Figure 47.74 Typical mercury vapor lamps.

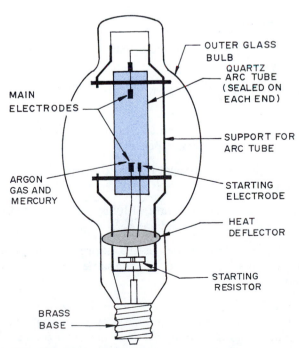

MAIN ELECTRODES

ARGON GAS AND MERCURY

OUTER GLASS BULB

QUARTZ ARC TUBE (SEALED ON EACH END)

SUPPORT FOR ARC TUBE

STARTING ELECTRODE

HEAT DEFLECTOR

STARTING RESISTOR

BRASS BASE

Figure 47.75 A simplified schematic of a typical mercury vapor lamp.

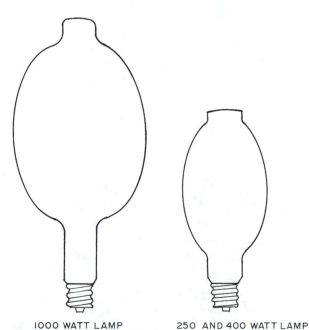

1000 WATT LAMP   250 AND 400 WATT LAMP

Figure 47.76 Typical metal-halide lamps.

produce a light that is radiated at frequencies other than the colors radiated by mercury, producing a lamp color that is better than mercury vapor lamps. It has about the same starting delay and restart times as the mercury vapor.

It should be noted that metal-halide lamps carry the same safety warning as mercury vapor and that they tend to explode and must be encased in an approved enclosing fixture. Some types have a plastic coating that makes them safe to use in an open fixture. They also are marked for their proper burning position. They are designed to be installed with their base up, base down, or horizontally. If installed in the wrong position their operating characteristics will be limited. Metal-halide lamps are available in various wattages, as shown in Table 47.6.

High- and Low-Pressure Sodium Lamps

High-pressure sodium (HPS) lamps are arc discharge lamps with sodium under high pressure in the glass arc tube. This produces a light with a yellow tint. They are highly efficient lamps, superior to all other types, and in areas where the color is not objectionable they can replace mercury vapor and metal-halide lamps (Fig. 47.77). Lamps are available in various wattages, as shown in Table 47.6.

Low-pressure sodium lamps, referred to as SOX, produce a deep yellow light that is usually not acceptable for general interior lighting. They are widely used for street, highway, and parking lighting. They have a long life and are the most economical type of lighting.

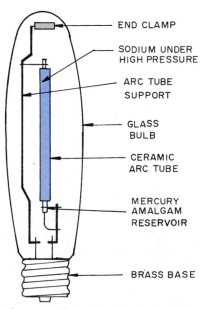

Figure 47.77 A simplified schematic of a high-pressure sodium lamp.

- END CLAMP
- SODIUM UNDER HIGH PRESSURE
- ARC TUBE SUPPORT
- GLASS BULB
- CERAMIC ARC TUBE
- MERCURY AMALGAM RESERVOIR
- BRASS BASE

REVIEW QUESTIONS

1. What is the rate of flow of electric current if it is available at 120V and has a resistance of 15 ohms?
2. What is the difference between alternating current and direct current?
3. What are the various ways electric power can be produced?
4. What fuels are used to power fossil fuel electric generating plants?
5. What power source is used to drive emergency generators?
6. How is electricity produced by utilizing normally wasted heat?
7. What are the types of electrical conductors recognized by codes?
8. What type of electrical conductor is used for underground applications?
9. What voltage current can be carried by flat conductor cables?
10. What types of conduit are in general use?
11. One kilowatt equals how many watt-hours?
12. What purpose do motor control centers serve?
13. What is the difference between the primary transformer voltage and the secondary voltage?
14. What are the commonly available switch actions?
15. What ways are various types of switches activated?
16. What devices are used to protect a circuit from overloads?
17. How does a type S plug fuse keep someone from putting the wrong size fuse in the panelboard?
18. What are the types of fuses in common use?

KEY TERMS

bus A rigid electric conductor enclosed in a protective busway.

busway A rigid conduit used to protect a bus running through it.

BX cable A cable sheathed with a spirally wrapped metal strip identified as type AC.

cable tray Open raceways used to support insulated electrical conductors.

candela (cd) A metric unit of luminous intensity that closely approximates candlepower.

candlepower (cp) A term used to express the luminous intensity of a light source. It is the same magnitude as a candela.

circuit breaker An electric device that automatically opens a circuit when an overload occurs and that can be reset and reused once the overload has been corrected.

cogeneration The utilization of normally wasted heat energy to produce electricity or to heat or cool a building.

conductivity, electric A measure of the ability of a material to conduct electric current.

conductor, electric Wires through which electric current flows.

conduit A pipe through which electric conductors are drawn.

current, electric The flow of electrons along a conductor.

footcandle (fc) A unit of illumination equal to 1 lumen per square foot.

footlambert A unit for measuring brightness or luminance. It is equal to 1 lumen per square foot when brightness is measured from the surface.

fuse A single-use overcurrent protection device.

illuminance (E) The density of luminous power in lumens per a specified area.

incandescence The emission of visible light produced by heating.

insulator, electric A material that is a poor conductor of electricity.

lamp A general term used to describe a source of artificial light.

light meter *See luminance meter.*

lighting fixture *See luminaire.*

lumen (lm) A unit for measuring the flow of light energy. *See luminous flux.*

luminaire A complete lighting unit consisting of one or more lamps plus elements needed to distribute light, hold and protect the lamps, and connect electric power to the lamps. Also called a **lighting fixture.**

luminance (L) The luminous intensity of a surface of a given area viewed from a given direction.

luminance meter A photoelectric instrument used to measure luminance. Also called a **light meter.**

luminescence The emission of light not directly caused by incandescence.

luminous flux The rate of flow of light energy through a surface, expressed in lumens (lm).

luminous intensity The force that generates visible light expressed by candela, lumens per steradian, or candlepower.

luminous transmittance A measure of the capacity of a material to transmit incident light in relation to the total incident light striking it.

lux A unit of illumination equal to 1 lumen per square meter.

meter, electric A device measuring and recording the amount of electricity passing through it in kilowatt-hours.

motor control center Controller used to start and stop electric motors and protect them from overloads.

panelboard A panel that includes fuses or circuit breakers used to protect the circuits in a building from overloads.

power The rate at which work is performed, expressed in watts or horsepower.

raceway A channel designed to enclose insulated electrical conductors.

reflectance coefficient A measure stated as a percentage of the amount of light reflected from a surface.

resistance, electric The physical property of a conductor or electric-consuming device to resist the flow of electricity, reducing power and generating heat.

romex A nonmetallic-sheathed cable of type NM or NMC.

room cavity ratio A relationship between the height of a room, its perimeter, and the height of the work surface above the floor divided by the floor area.

service entrance The electrical equipment at the point where electric service is introduced into the building.

switch An electrical device used to open or close an electric circuit or to change a connection within a circuit.

switchboard An assembly of electrical switches, overcurrent protection devices, buses, and other line components.

transformer An electrical device used to convert an incoming electric current at one voltage to another voltage.

watt A unit of power required to do work at the rate of 1 joule per second.

SUGGESTED ACTIVITIES

1. Invite the local building official to speak to the class on the electrical code requirements.

2. Collect various electrical control and metering devices. Prepare a display and label each, giving the manufacturer's specifications.

3. Erect a section of a stud wall and install boxes for fuses, outlets, switches, lights, etc., and wire these. Bring

power to the beginning end of the wiring and activate the system. Does it work?

4. Walk through various buildings and note the types of lighting used. Prepare a report citing the location for each type found.

5. Arrange a visit to the electrical equipment room of a major building and have the person in charge explain the function of each unit. Describe the flow of power from the service entrance to the various floors of the building. View distribution equipment used on one of these floors.

ADDITIONAL INFORMATION

Ambrose, J. E., *Building Construction: Service Systems,* Van Nostrand Reinhold, New York, 1990.

Bradshaw, V., *Building Control Systems,* John Wiley and Sons, New York, 1993.

Kaufman, J. E., *IES Lighting Handbook Application Volume,* Illuminating Engineering Society of America, New York, 1997.

Merritt, F. S., and Ricketts, J. T., *Building Design and Construction Handbook,* McGraw-Hill, New York, 1994.

Spence, W. P., *Architectural Working Drawings: Residential and Commercial Buildings,* John Wiley and Sons, New York, 1993.

Spence, W. P., *Architecture: Design, Engineering, Drawing,* Glencoe/McGraw-Hill Educational Division, Mission Hills, Calif., 1991.

Stein, B., and Reynolds, J. S., *Mechanical and Electrical Equipment for Buildings,* John Wiley and Sons, New York, 1992.

Other resources are

National Fire Protection Association, National Electrical Code, and other related publications, Quincy, Mass.

Numerous related publications from the organizations listed in Appendix B.

APPENDIX A

Division 1 General Requirements

The level two titles within Division 1, General Requirements, of the CSI MasterFormat™ can be found in Chapter 2. A detailed listing of the level three titles is available in the 1995 edition of MasterFormat, published jointly by the Construction Specifications Institute (CSI), Alexandria, Virginia, and the Construction Specifications Canada (CSC), Toronto, Ontario, Canada. Following is descriptive information about the contents under each level 2 title.

SUMMARY

The first level two title is Summary, which includes Summary of Work, Multiple Contract Summary, Work Restrictions, and Project Utility Sources.

▲ Summary of Work identifies the work to be covered in a single contract and describes provisions for future construction. It may include work by the owner, work covered in the contract documents, and specific things such as products ordered in advance, owner-furnished and installed products, and the use of salvaged material and products.
▲ Multiple Contract Summary describes the construction delivered under *more* than one construction contract, such as construction management contract and a multiple-prime contract. It includes the sequence of construction, contract interface, any construction by the owner, and a summary of the contracts.
▲ Work Restrictions pertains to restrictions that affect construction operations, such as the location of construction, occupancy requirements, and the use of the building premises and site during construction. If the construction involves a building already occupied, procedures for coordination with the occupants are detailed.
▲ Project Utility Sources includes the identity of utility companies that are to provide permanent services to the project.

PRICE AND PAYMENT PROCEDURES

The second level two title is Price and Payment Procedures. It details the allowance-adjusting procedures for cash and quantity allowances for products, installation, testing, and contingencies.

▲ Allowances includes the adjusting procedures for cash and quantity allowances for products, installation, inspection, and testing contingencies.
▲ Alternates provides for the submission and acceptance procedures for alternate bids including, when desired, a list and description of each alternate.
▲ Value Analysis includes the procedures and submittal requirements for value analysis, value engineering, application for consideration, and consideration of proposals.
▲ Contraction Modification Procedures details the procedures for making clarifications and proposals for change and for changing the contract.
▲ Unit Prices establishes the procedures associated with unit prices and measurement and payment. It may include a list and descriptions of the actual unit price items.
▲ Payment Procedures describes the procedures for submitting schedules of values and applications for payment.

ADMINISTRATIVE REQUIREMENTS

The third level two title, Administrative Requirements, encompasses the areas of Project Management and Coordination, Construction Progress Documentation, Submittal Procedures, and Special Procedures.

▲ Project Management and Coordination includes the administration of subcontractors and coordination with other contractors and the owner. This includes project meeting and on-site administration.

▲ Construction Progress Documentation covers the requirements for scheduling, recording, and reporting progress. This includes such things as construction photographs, progress reports, site observation, purchase order tracking, scheduling of construction, and survey and layout data.

▲ Submittal Procedures includes the general procedures and requirements for submittals during the course of the construction. This can include items such as certificates, design data, field test reports, shop drawings, product data and samples, and source quality control reports.

▲ Special Procedures includes any special procedures that any of the project situations require, such as historic restoration, renovation or alteration, preservation, or hazardous material abatement.

QUALITY REQUIREMENTS

The fourth level two title of Division 1 is Quality Requirements. This covers four areas: Regulatory Requirements, References, Quality Assurance, and Quality Control.

▲ Regulatory Requirements provides information required for conformance to requirements such as building codes, mechanical codes, electrical codes, and other regulations and to fee payments applicable to the project.

▲ References includes lists of reference standards cited in the contract documents and the organizations whose standards are cited. This includes items such as symbols used, abbreviations, and acronyms.

▲ Quality Assurance references are used to provide for procedures to ensure the quality of construction. This includes field observations and tests performed by manufacturers' representatives during installation. Quality assurance is provided by fabricators, installers, manufacturers, suppliers, and testing agencies.

▲ Quality Control establishes procedures to measure and report the quality and performance of construction. It may require field samples and mock-ups assembled at the site. The contractor provides quality control plans. Control can be provided by inspections, inspection services, and testing laboratories.

TEMPORARY FACILITIES AND CONTROLS

The fifth level two title, Temporary Facilities and Controls, provides requirements for installation, maintenance, and removal of temporary utilities, controls, facilities, and construction aids during construction. This includes Temporary Utilities, Construction Facilities, Temporary Construction, Construction Aids, Vehicular Access and Parking, Temporary Barriers and Enclosures, Temporary Controls, and Project Identification.

▲ Temporary Utilities includes all such services used during construction, such as electrical, water, lighting, gas, telephone, fire protection, fuel oil, gasoline, and diesel fuel.

▲ Construction Facilities includes any temporary facilities built on the site for use during construction. Typical examples include field offices, storage buildings, sanitary facilities, and first aid stations.

▲ Temporary Construction includes those facilities built to provide access to various parts of the site and project so as to facilitate the construction process or to accommodate the needs of the owner and occupants. Typical examples include temporary bridges, ramps, overpasses, decking, and turnarounds.

▲ Construction Aids include all requirements and procedures related to tools and equipment used during construction, such as scaffolding, cranes, hoists, and construction elevators.

▲ Vehicular Access and Parking provides requirements for and procedures related to access to the site and parking facilities to meet needs of the construction and owner operations. Typically this can include access roads, haul routes, parking areas, temporary roads, and control of traffic.

▲ Temporary Barriers and Enclosures includes all facilities and procedures for the protection of the occupants or existing spaces during construction. This can include air barriers, barricades, dust barriers, fences, noise barriers, pollution control, security, and the protection of trees and other vegetation.

▲ Temporary Controls includes site or environment controls required to allow construction to proceed, including such things as erosion and sediment control and pest control.

▲ Project Identification includes any signs used to identify the construction site and areas on the site.

PRODUCT REQUIREMENTS

The sixth level two title, Product Requirements, includes a comprehensive series of requirements.

▲ Basic Product Requirements includes the basic requirements for new, salvaged, and reused products used in construction.

▲ Product Options includes the basic requirements for options the contractor may have in selecting products and how the determination is made for equal products.

▲ Product Substitution Procedures includes the basic requirements and procedures when proposals for the substitution of a product are made.

▲ Owner-Furnished Products includes the basic requirements for products that are to be furnished by the owner. This also involves scheduling, coordinating, handling, and storing owner-furnished products.

▲ Product Delivery Requirements specifies the basic requirements for packing, shipping, delivery, and acceptance of products at the site.

▲ Product Storage and Handling Requirements sets forth the basic requirements for storing and handling products on the site.

EXECUTION REQUIREMENTS

The seventh level two title, Execution Requirements, has eight sections.

▲ Examination involves the acceptance of conditions, existing conditions, and the basic requirements for determining acceptable conditions for installation.

▲ Preparation details the requirements for preparing to install, erect, or apply products, including activities such as field engineering, protection of adjacent construction, surveying, and construction layout.

▲ Execution includes the basic requirements for installing, applying, or erecting products that are new, prepurchased, salvaged, or owner furnished.

▲ Cleaning sets the requirements for maintaining the site in a neat condition during construction and the final cleaning in preparation for turning the project over to the owner.

▲ Starting and Adjusting involves establishing the initial checkout and startup procedures and any adjustments needed to ensure safe operation during the acceptance testing and commissioning.

▲ Protecting Installed Construction provides the requirements and procedures for protecting installed construction.

▲ Closeout Procedures includes the administrative procedures for substantial completion and final completion of the work.

▲ Closeout Submittals sets the procedures for closeout submittals, revised project documents, and delivery and distribution of spare parts and maintenance materials.

FACILITY OPERATION

The eighth level two title, Facility Operation, has to do with the final requirements for preparing the facility for decommissioning.

▲ Commissioning requirements include commissioning summary, system performance evaluation, testing, adjusting, and balancing procedures.

▲ Demonstration and Training includes the requirements and procedures for the demonstration of the products and systems within the facility, including training the owner's operating and maintenance personnel.

▲ Operation and Maintenance sets forth the requirements and procedures for operating the facility after commissioning.

▲ Reconstruction involves any renovation or reconstruction of the existing facilities that may be required.

FACILITY DECOMMISSIONING

The last and ninth level two title, Facility Decommissioning, includes the basic requirements for deactivating a facility or a portion of it from operation. This includes activities such as facility demolition and removal; hazardous materials abatement, removal, and disposal; and protection of deactivated facilities.

I T E M   O F   I N T E R E S T

CHANGE ORDERS

A very common owner-contractor dispute is over change orders. "Get it done as soon as possible regardless of the cost" is an often heard order. Disputes result about what was said and how much a change was supposed to cost, resulting in the contractor losing money and the trust of the owner and the subcontractor who expected to get paid for the extra work or effort. Sometimes these disputes lead to litigation.

To avoid disputes over changes it is necessary to get written, signed change orders that clearly establish what is to be done and what it will cost. Verbal change orders can be easily challenged in court. Written, signed change orders almost always keep a claim from being filed against the contractor.

Do not do the work until the written change order is signed. Carefully written change orders will almost always eliminate arguments over any subcontractors' obligations and the owner's responsibility to pay. The pressure to put off writing change orders is great because the contractor is always very busy, but the writing takes a lot less time than going to court. Prepare and have printed a change order form that your attorney has approved. After the form has been filled out indicating the changes and cost in detail, the owner and the contractor should sign it and date it, and each should get a copy. Number each order (as Change Order No. 1), and arrange the office procedures so change orders are incorporated into the billing procedure.

The general contractor must work with the client and subcontractors and require that change orders be in writing.

APPENDIX B

Professional and Technical Organizations

Division 2. Site Work

American Association of State Highway and
Transportation Officials System
449 Capital St. NW, Suite 225
Washington, DC 20001

American Planning Association (APA)
1313 E. 60th St.
Chicago, IL 60637

American Society of Civil Engineers (ASCE)
345 E. 47th St.
New York, NY 10017

American Society of Landscape Architects (ASLA)
4401 Connecticut Ave. NW
Washington, DC 20008

American Water Works Association (AWWA)
6660 W. Quincy Ave.
Denver, CO 80235

ASFE: Professional Firms Practicing in the Geo-
sciences
8811 Colesville Road
Silver Spring, MD 20910

Asphalt Institute (AI)
P.O. Box 14052
Lexington, KY 40512–4052

Canadian Society of Landscape Architects (CSLA)
1339 Fifteenth Ave. SW, Apt. 310
Calgary, Alberta, Canada T3C 3V3

National Sanitation Foundation International
(NSFI)
P.O. Box 130140
Ann Arbor, MI 48113

Urban Land Institute (ULI)
624 Indiana Ave. NW, Suite 400
Washington, DC 20004

Division 3. Concrete

American Concrete Institute (ACI)
P.O. Box 9094
Farmington Hills, MI 48333–9094

American Society of Concrete Construction
(ASCC)
1902 Techny Ct.
Northbrook, IL 60062

Architectural Precast Association (APA)
1850 Lee Rd., Suite 230
Winter Park, FL 32789

Concrete Reinforcing Steel Institute (CRSI)
933 North Plum Grove Rd.
Schaumberg, IL 60173

National Concrete Masonry Association
(NCMA)
2302 Horsepen Rd.
Herndon, VA 22071

National Precast Concrete Association (NPCA)
10333 N. Meridian St., Suite 272
Indianapolis, IN 46290

Portland Cement Association (PCA)
5420 Old Orchard Rd.
Skokie, IL 60077–1083

Post-Tensioning Institute (PTI)
1717 W. Northern Ave., Suite 114
Phoenix, AZ 85021–5471

Precast/Prestressed Concrete Institute (PCI)
175 W. Jackson Blvd., Suite 1859
Chicago, IL 60604

Tilt-Up Concrete Association (TCA)
P.O. Box 204
Mt. Vernon, IA 52310

Wire Reinforcement Institute (WRI)
2911 Gleneagle Dr.
Findlay, OH 45840

Division 4. Masonry

Brick Institute of America (BIA)
11490 Commerce Park Dr.
Reston, VA 22091–1525

Building Stone Institute (BSI)
P.O. Box 507
Purdys, NY 10578

Cast Stone Institute (CastSI)
1850 Lee Rd., Suite 230
Winter Park, FL 32789

Indiana Limestone Institute of America (ILIA)
Stone City Bank Bldg., Suite 400
Bedford, IN 47421

International Masonry Institute (IMI)
823 Fifteenth St. NW, Suite 1001
Washington, DC 20005

Marble Institute of America (MIA)
33505 State St.
Farmington, MI 48335

The Masonry Council (MC)
2619 Spruce St.
Boulder, CO 80302

Masonry Institute of America (MasIA)
2550 Beverly Blvd.
Los Angeles, CA 90057

National Concrete Masonry Association (NCMA)
2302 HorsePen Rd.
Herndon, VA 22071

National Lime Association (NLA)
200 N. Glebe Rd., Suite 800
Arlington, VA 22203

Division 5. Metals

Aluminum Anodizers Council (AAC)
1000 N. Rand, Suite 214
Wauconda, IL 60084

Aluminum Association (AA)
900 Nineteenth St. NW, Suite 300
Washington, DC 20006

American Hot Dip Galvanizers Association
1133 Fifteenth St. NW
Washington, DC 20005

American Institute for Hollow Structural Sections (AIHSS)
929 McLaughin Run Road, Suite 8
Pittsburgh, PA 15017

American Institute of Steel Construction (AISC)
One Wacker Drive, Suite 3100
Chicago, IL 60601–2011

American Iron and Steel Institute (AISI)
1101 Seventeenth St. NW, Suite 1300
Washington, DC 20036

American Zinc Association
1112 Sixteenth St. NW, Suite 240
Washington, DC 20036

Copper Development Association (CDA)
260 Madison Ave.
New York, NY 10016

International Lead Zinc Research Organization, Inc.
2525 Meridian Parkway
Research Triangle Park, NC 27709–2306

Iron and Steel Society (ISS)
410 Commonwealth Dr.
Warrendale, PA 15086

Lead Industries Association, Inc.
295 Madison Ave.
New York, NY 10017

Metal Lath/Steel Framing Association
600 S. Federal St., Suite 400
Chicago, IL 60605

National Association of Architectural Metal Manufacturers (NAAMM)
600 S. Federal St., Suite 400
Chicago, IL 60605

National Ornamental and Miscellaneous Metals Association (NOMMA)
804-10 Main St., Suite E
Forest Park, GA 30050

Nickel Development Institute (NDI)
15 Toronto St., Suite 402
Toronto, Ontario, Canada M5C 2E3

Sheet Metal and Air-Conditioning Contractors National Association (SMACNA)
P.O. Box 221230
Chantilly, VA 22022–1230

Society of Automotive Engineers, Inc.
400 Commonwealth Drive
Warrendale, PA 15096–0001

Specialty Steel Industry of North America
3050 K St. NW
Washington, DC 20007

Steel Deck Institute (SDI)
P.O. Box 9506
Canton, Ohio 44711

Steel Joist Institute (SJI)
1205 Forty-Eighth Ave. N, Suite A
Myrtle Beach, SC 29577

Steel Structures Painting Council (SSPC)
4516 Henry St.
Pittsburgh, PA 15213–3728

Structural Insulated Panel Association
(SIPA)
1511 K Street NW, Suite 600
Washington, DC 20005

Wire Reinforcement Institute
2911 Gleneagle Drive
Findlay, OH 45840–2908

Division 6. Wood and Plastics

Wood Organizations

American Fiberboard Association
and
American Hardboard Association
1210 W. Northwest Highway
Palatine, IL 60067

American Forest and Paper Association
1111 Nineteenth St. NW, Suite 800
Washington, DC 20036

American Institute of Timber Construction
7012 South Revere Parkway, Suite 140
Englewood, CO 80112

American Lumber Standards Committee
P.O. Box 210
Germantown, MD 20875–0210

American Wood Council
1111 Nineteenth St. NW, Suite 800
Washington, DC 29936

American Wood Preservers Association
P.O. Box 286
Woodstock, MD 21163–0286

American Wood Preservers Institute
Tysons International Building,
1945 Old Gallows Road, Suite 550
Vienna, VA 22182

APA - The Engineered Wood Association
P.O. Box 11700
Tacoma, WA 98411

Architectural Woodwork Institute
13924 Braddock Rd., Suite 100
Centerville, VA 22020

California Redwood Association (CRA)
405 Enfrente Dr., Suite 200
Novato, CA 94949

Canadian Wood Council
1730 St. Laurent Blvd., Suite 350
Ottawa, Ontario, Canada K1G 5L1

Cedar Shake and Shingle Bureau
515 One Hundred Sixteenth Ave. NE, Suite 275
Bellevue, WA 98004–5294

Cultured Marble Institute (CMI)
1735 N. Lynn St., suite 950
Arlington, VA 22209

Fine Hardwood Veneer Association
260 S. First St., Suite 2
Zionsville, IN 46077

Forest Products Laboratory
One Gifford Pinchot Dr.
Madison, WI 53705

Forest Products Society
2801 Marshall Ct.
Madison, WI 53705

Hardwood Council
P.O. Box 525
Oakmont, PA 15139

Hardwood Manufacturers Association
400 Penn Center Blvd., Suite 530
Pittsburgh, PA 15235

Hardwood Plywood and Veneer Association
P.O. Box 2789
Reston, VA 22090

Maple Flooring Manufacturers Association
(MFMA)
60 Revere Drive, Suite 500
Northbrook, IL 60062

The Modular Building Systems Council and
The Panelized Building Systems Council Divisions
of The Building Systems Council

National Association of Home Builders
1201 Fifteenth St. NW
Washington, DC 20005

National Association of Home Builders
1201 Fifteenth St. NW
Washington, DC 20005

National Hardwood Lumber Association (NHLA)
P.O. Box 34518
Memphis, TN 38184–0518

National Oak Flooring Manufacturers Association
(NOFMA)
P.O. Box 3009
Memphis, TN 38173–0009

National Particleboard Association (NPA)
18928 Premiere Ct.
Gaithersburg, MD 20879

Oak Flooring Institute
P.O. Box 3009
Memphis, TN 38173–0009

Southern Forest Products Association (SFPA)
P.O. Box 641700
Kenner, LA 70064–1700

Southern Pine Association
P.O. Box 641700
Kenner, LA 70064–1700

Structural Board Association
45 Sheppard Ave. East, Suite 412
Willowdale, Ontario, Canada M2N 5W9

Structural Insulation Board Association
1511 K St. NW, Suite 600
Washington, DC 20005

Western Wood Products Association (WWPA)
Yeon Building
522 SW Fifth Ave.
Portland, OR 97204

Wood Protection Council (WPC)
1201 L Street NW, Suite 400
Washington, DC 20005

Plastics Organizations

Decorative Laminate Products Association
(DLPA)
600 S. Federal St., Suite 400
Chicago, IL 60605

Plastics Institute of America (PIA)
227 Fairfield Road
Fairfield, NJ 07004

Society of Plastics Engineers (SPE)
14 Fairfield Drive
Brookfield, CT 06804–0403

The Society of the Plastics Industry, Inc. (SPI)
1275 K Street NW, Suite 400
Washington, DC 20005

Vinyl Institute (VI)
65 Madison Ave.
Morristown, NJ 07960

Vinyl Siding Institute (VSI)
355 Lexington Ave.
New York, NY 10017

Division 7. Thermal and Moisture Protection

Adhesive and Sealant Council (ASC)
1627 K Street NW, Suite 1000
Washington, DC 20006–1707

American Society of Heating, Refrigerating and
Air-Conditioning Engineers (ASHRAE)
1791 Tullie Circle NE
Atlanta, GA 30329

Asphalt Institute (AI)
P.O. Box 14052
Lexington, KY 40512–4052

Asphalt Roofing Manufacturers Association (ARMA)
6288 Montrose Road
Rockville, MD 20852

Associated Foam Manufacturers (AFM)
P.O. Box 246
Excelsion, MN 55331

Cellulose Insulation Manufacturers Association
136 South Keowee St.
Dayton, OH 45402

Exterior Insulation Manufacturers Association (EIMA)
2759 State Road 580, Suite 112
Clearwater, FL 34621

Institute of Roofing and Waterproofing Consultants
(IRWC)
4242 Kirchoff Rd.
Rolling Meadows, IL 60008

Insulation Contractors Association of America
P.O. Box 26237
Alexandria, VA 22313

Mineral Insulation Manufacturers Association
(MIMA)
44 Canal Center Plaza, Suite 310
Alexandria, VA 22314

National Insulation and Abatement Contractors
Association (NIACA)
99 Canal Center Plaza
Alexandria, VA 22314

National Roofing Contractor Association
(NRCA)
10255 W. Higgins Rd., Suite 600
O'Hare International Center
Rosemont, IL 60018

National Tile Roofing Manufacturers Association
(NTRMA)
P.O. Box 947
Eugene, OR 97440

Noise Control Association (NCA)
680 Rainier Lane
Port Ludlow, WA 98365

North American Insulation Manufacturers Association
(NAIMA)
44 Canal Center Plaza
Alexandria, VA 22314

Perlite Institute (PI)
88 New Dorp Plaza
Staten Island, NY 10306–2994

Polyisocyanurate Insulation Manufacturers Association
(PIMA)
1001 Pennsylvania Ave. NW
Washington, DC 20004

Roof Coatings Manufacturers Association
(RCMA)
60 Revere Drive, Suite 500
Northbrook, IL 60062

Roof Consultants Institute (RCI)
7424 Chapel Hill Rd.
Raleigh, NC 27607

Roofing Industry Educational Institute (RIEI)
14 Inverness Dr. E., Bldg H, Suite 110
Englewood, CO 80112–5608

Roofing Products Division/Rubber Manufacturers Association (RPD/RMA)
1400 K Street NW
Washington, DC 20005

Sealant, Waterproofing and Restoration Institute (SWRI)
3101 Broadway, Suite 585
Kansas City, MO 64111

Single Ply Roofing Institute
20 Walnut St., Suite 8
Wellesley Hills, MA 02181

Society of the Plastics Industry (SPI)
1275 K Street NW, Suite 400
Washington, DC 20005

Specialty Steel Industry of North America
3050 K St. NW
Washington, DC 20007

Vermiculite Association (VA)
600 S. Federal St., Suite 400
Chicago, IL 60605

Division 8. Doors and Windows

Aluminum Extruders Council (AEC)
1000 N. Rand Rd., Suite 214
Wauconda, IL 60084

Aluminum Fenestration Products Association (AFPA)
1000 N. Rand Rd., Suite 214
Wauconda, IL 60084

American Architectural Manufacturers Association (AAMA)
1540 E. Dundee Rd., Suite 310
Palatine, IL 60067

Architectural Translucent Skylight and Curtain Wall Association (ATSCWA)
7120 Stewart Ave.
Wausau, WI 54401

Builders Hardware Manufacturers Association (BHMA)
355 Lexington Ave., Seventeenth Floor
New York, NY 10017

Flat Glass Marketing Association (FGMA)
Glass Tempering Association (GTA)
Laminators Safety Glass Association (LSGA)
3310 SW Harrison St.
Topeka, KS 66611

National Fenestration Rating Council (NFRC)
1300 Spring St., Suite 120
Silver Spring, MD 20910

National Glass Association (NGA)
8200 Greensboro Dr., Suite 302
McLean, VA 22102

National Wood Window and Door Association (NWWDA)
1400 E. Touhy Ave.
Des Plaines, IL 60018

Sealed Insulating Glass Manufacturers (SIGM)
401 N. Michigan Ave.
Chicago, IL 60611–4267

Steel Door Institute (SDI)
30200 Detroit Rd.
Cleveland, OH 44145

Steel Window Institute
1300 Summer Ave.
Cleveland, OH 44115

Vinyl Window and Door Institute (VWDI)
355 Lexington Ave.
New York, NY 10017

Division 9. Finishes

Acoustical Society of America (ASA)
500 Sunnyside Blvd.
Woodbury, NY 11797

American Fiber Manufacturers Association (AFMA)
1150 Seventeenth St. NW, Suite 310
Washington, DC 20036

Architectural Spray Coaters Association (ASCA)
230 W. Wells, Suite 311
Milwaukee, WI 53203

Audio Engineering Society (AES)
60 East 42nd St., Rm. 2520
New York, NY 10165

Carpet and Rug Institute (CRI)
P.O. Box 2048
Dalton, GA 30720

Carpet Cushion Council (CCC)
P.O. Box 546
Riverside, CT 06878

Ceilings and Interior Systems Construction (CISC)
579 W. North Ave., Suite 301
Elmhurst, IL 60126

Facing Tile Institute
P.O. Box 8880
Canton, OH 44711

Foundation of the Wall and Ceiling Industry (FWCI)
307 E. Annandale, Suite 200
Falls Church, VA 22042

Gypsum Association (GA)
810 First St. NE, Suite 510
Washington, DC 20002

International Institute of Lath and Plaster (IILP)
820 Transfer Road
St. Paul, MN 55114

Maple Flooring Manufacturers Association (MFMA)
60 Revere Dr., Suite 500
Northbrook, IL 60062

Materials and Methods Standards Association
(MMSA)
P.O. Box 350
Grand Haven, MI 49417

National Council of Acoustical Consultants
66 Morris Ave., Suite 1A
Springfield, NJ 07081–1409

National Oak Flooring Manufacturers Association
(NOFMA)
P.O. Box 3009
Memphis, TN 38173–0009

National Paint and Coatings Association (NPCA)
1500 Rhode Island Ave. NW
Washington, DC 20005

National Terrazzo and Mosaic Association (NTMA)
3166 Des Plaines Ave., Suite 132
Des Plaines, IL 60018

National Tile Contractors Association (NTCA)
P.O. Box 13629
Jackson, MS 39236

National Wood Flooring Association (NWFA)
233 Old Meramec Station Rd.
Manchester, MO 63021

Painting and Decorating Contractors of America
(PDCA)
3913 Old Lee Highway, Suite 33B
Fairfax, VA 22030

Resilient Floor Covering Institute (RFCI)
966 Hungerford Dr., Suite 12B
Rockville, MD 20850

Rubber Manufacturers Association (RMA)
1400 K Street NW
Washington, DC 20005

Tile Council of America (TCA)
P.O. Box 1787
Clemson, SC 29633

Wallcovering Manufacturers Association and Wallcovering Information Bureau (WMA and WIB)
401 N. Michigan Ave., Suite 2200
Chicago, IL 60611

Wood and Synthetic Flooring Institute (WSFI)
4415 W. Harrison St., Suite 242C
Hillside, IL 60162

Division 11. Equipment

American Society for Industrial Security (ASIS)
1655 N. Fort Meyer Dr., Suite 1200
Arlington, VA 22209

National Burglar and Fire Alarm Association
(NBFAA)
7101 Wisconsin Ave., Suite 1390
Bethesda, MD 20814

National Crime Prevention Council (NCPC)
1700 K Street NW, 2nd Floor
Washington, DC 20006–3817

National Crime Prevention Institute (NCPI)
University of Louisville, Belknap Campus
Burhans Hall
Louisville, KY 40292

Security Industry Association (SIA)
1801 K Street NW, Suite 1203L
Washington, DC 20006–1301

Division 12. Furnishings

American Society of Furniture Designers (ASFD)
521 S. Hamilton St.
P.O. Box 2688
High Point, NC 27261

Business and Institutional Furniture Manufacturers
Association (BIFMA)
2335 Burton St. SE
Grand Rapids, MI 49506

Contract Furnishings Council (CFC)
1190 Merchandise Mart
Chicago, IL 60654

Division 13. Special Construction

Acoustical Society of America (ASA)
500 Sunnyside Blvd.
Woodbury, NY 11797

American Solar Energy Society (ASES)
2400 Central Ave., Suite G-1
Boulder, CO 80301

Audio Engineering Society (AES)
60 East 42nd St., Room 2520
New York, NY 10165

Building and Fire Research Laboratory
National Institute of Standards and Technology
Building 226, Rm 216
Gaithersburg, MD 20899

Building Seismic Safety Council (BSSC)
National Institute of Building Sciences
1201 L Street NW, Suite 400
Washington, DC 20005

Fire Equipment Manufacturers Association
(FEMA)
1300 Summer Ave.
Cleveland, OH 44115

Florida Solar Energy Center (FSEC)
300 State Rd. 401
Cape Canaveral, FL 32920

Industrial Fabrics Association International (IFAI)
345 Cedar St., Suite 800
St. Paul, MN 55101–1088

International Fire Code Institute (IFCI)
5360 S. Workman Mill Rd.
Whittier, CA 90601

Metal Building Manufacturers Association (MBMA)
1300 Summer Ave.
Cleveland, OH 44115

National Council of Acoustical Consultants (NCAC)
66 Morris Ave., Suite 1A
Springfield, NJ 07081

National Fire Laboratory (NFL)
Institute for Research in Construction
Ottawa, Ontario, Canada K1A 0R6

National Fire Protection Association (NFPA)
1 Batterymarch Park
Quincy, MA 02269

National Fire Sprinkler Association, Inc. (NFSA)
P.O. Box 1000
Patterson, NY 12563

National Pool and Spa Institute (NPSI)
2111 Eisenhower Ave.
Alexandria, VA 22314

National Renewable Energy Laboratory (NREL)
1617 Cole Blvd.
Golden, CO 80401

Northeast Sustainable Energy Association (NSEA)
23 Ames St.
Greenfield, MA 01301

Society of Fire Protection Engineers (SFPE)
One Liberty Square
Boston, MA 02109

Solar Energy Industries Association (SEIA)
122 C St. NW
Washington, DC 20001

Division 14. Conveying Systems

Conveyor Equipment Manufacturing Association
1000 Vermont Ave. NW
Washington, DC 20005

National Association of Elevator Contractors (NAEC)
1298 Wellbrook Circle NE
Conyers, GA 30207

National Elevator Industry, Inc.
185 Bridge Plaza North
Fort Lee, NV 07024

Division 15. Mechanical

Air Conditioning and Refrigeration Institute
4301 N. Fairfax Dr., Suite 425
Arlington, VA 22203

Air Conditioning Contractors of America
1513 Sixteenth St. NW
Washington, DC 20036

Air Movement and Control Association (AMCA)
30 W. University Dr.
Arlington Heights, IL 60004

American Boiler Manufacturers Association
950 N. Glebe Rd., Suite 160
Arlington, VA 22203

American Gas Association (AGA)
1515 Wilson Blvd.
Arlington, VA 22209

American Society of Heating, Refrigerating and Air-Conditioning Engineers (ASHRAE)
1791 Tullie Circle NE
Atlanta, GA 30329

American Society of Mechanical Engineers (ASME)
345 E. 47th St.
New York, NY 10017

American Society of Plumbing Engineers (ASPE)
3617 Thousand Oaks Blvd., Suite 210
Westlake Village, CA 91362

American Society of Sanitary Engineers (ASSE)
P.O. Box 40362
Bay Village, OH 44140

American Water Works Association (AWWA)
6666 W. Quincy Ave.
Denver, CO 80235

Cast Iron Soil Pipe Institute (CISPI)
5959 Shallowford Rd., Suite 419
Chattanooga, TN 37412

Cooling Tower Institute (CTI)
P.O. Box 73383
Houston, TX 77273

Gas Research Institute
8600 West Bryn Mawr Ave.
Chicago, IL 60631

Heat Exchange Institute
1621 Euclid Ave., Suite 1230
Cleveland, OH 44113

The Hydronics Institute, Inc. (HI)
P.O. Box 218
Berkeley Heights, NJ 07922–0218

Institute of Heating and Air Conditioning Industries (IHACI)
606 N. Larchmont Blvd., Suite 4A
Los Angeles, CA 90004

International Association of Plumbing and Mechanical Officials (IAPMO)
5360 South Workman Mill Rd.
Whittier, CA 90601

International Society for Indoor Air Quality and Climate
Box 22038, Sub 32
Ottawa, Ontario, Canada K1V 0W2

National Association of Plumbing, Heating, and
Cooling Contractors (NAPHCC)
P.O. Box 6808
Falls Church, VA 22046

The National Burglar and Fire Alarm Association
7101 Wisconsin Ave.
Bethesda, MD 29814

National Fire Sprinkler Association, Inc.
Robin Hill Corporate Park, Route 22
P.O. Box 1000
Patterson, NY 12563

National Propane Gas Association (APGA)
1600 Eisenhower Lane, Suite 100
Lisle, IL 60532

National Sanitation Foundation
International (NSFI)
P.O. Box 130140
Ann Arbor, MI 48113

Plastic Pipe and Fittings Association (PPFA)
800 Roosvelt Rd., Bldg. C, Suite 20
Glen Ellyn, IL 60137

Plastic Pipe Institute
1275 K St. NW
Washington, DC 20005

Plumbing Manufacturers Institute (PMI)
800 Roosevelt Rd., Bldg. C, Suite 20
Glen Ellyn, IL 60137

Security Industries Association
1801 K St. NW, Suite 1203L
Washington, DC 20006

Sheet Metal and Air Conditioning Contractors
Association International (SMACNA)
4201 Lafayette Center Drive
Chantilly, VA 22022–1209

Solar Energy Industries Association (SEIA)
122 C St. NW
Washington, DC 20001

Waste Material Management Division
U.S. Department of Energy
1000 Independence Ave. SW
Washington, DC 20585

Water Quality Association (WQA)
4151 Naperville Rd.
Lisle, IL 60532

Division 16. Electrical

American Nuclear Society
555 North Kensington Ave.
LaGrange Park, IL 60525

American Public Power Association
2301 M St. NW, Suite 300
Washington, DC 20037

Edison Electric Institute (EEI)
701 Pennsylvania Ave. NW
Washington, DC 20004

Electric Power Research Institute
207 Coggins Drive
P.O. Box 23205
Pleasant Hill, CA 94523

Illuminating Engineering Society of
North America (IES)
120 Wall St., 17th Floor
New York, NY 10005

International Association of Lighting Designers (IALD)
18 E. 16th St., Suite 208
New York, NY 10003

Lighting Research Center (LRC)
Rensselaer Polytechnic Institute
Greens Building, Room 115
Troy, NY 12180

Lighting Research Institute (LRI)
120 Wall St., 17th Floor
New York, NY 10005

National Electrical Manufacturers Association
(NEMA)
2000 L St. NW
Washington, DC 20036

National Lighting Bureau (NLB)
2101 L St. NW
Washington, DC 20037

Underwriters Laboratories (UL)
333 Pfingsten Rd.
Northbrook, IL 60062

Information Sources for Recycled and Energy-Efficient Materials

Alliance to Save Energy
1725 K Street NW, Suite 914
Washington, DC 20006

American Council for an Energy-Efficient Economy
1001 Connecticut Ave. NW
Washington, DC 20036

American Solar Energy Society
2400 Central Ave., Suite G-1
Boulder, CO 80301

American Wind Energy Association
122 C St. NW, Suite 400
Washington, DC 20001

Association of Energy Engineers
4025 Pleasantdale Road, Suite 420
Atlanta, GA 30340

The Boston Society of Architects
52 Broad St.
Boston, MA 02109–4301

Center for Resourceful Building Technology
P.O. Box 3413
Missoula, MT 59806

Energy Efficiency and Renewable Energy Clearing
House
P.O. Box 3048
Merrifield, VA 22116

Environmental Resource Guide
John Wiley and Sons, Inc.
605 Third Ave.
New York, NY 10158–0012

EnviroSafe Products
81 Winent Place
Staten Island, NY 10309

National Energy Foundation
5160 Wiley Post Way, Suite 200
Salt Lake City, UT 84116

Passive Solar Industries Group
1090 Vermont Ave. NW, Suite 1200
Washington, DC 20005

Resource Conservation Technology, Inc.
2633 North Calvert St.
Baltimore, MD 21218

Technical Information Program
National Renewable Energy Laboratory
1617 Cole Blvd.
Golden, CO 80401–3393

Wastebusters, Inc.
1390 Richmond Terrace
Staten Island, NY 10310

Building Code Agencies

Building Officials and Code Administrators
International (BOCA)
4051 West Fossmore Rd.
Country Club Hills, IL 60478

Council of American Building Officials (CABO)
5203 Leesburg Pike, Suite 708
Falls Church, VA 22041

International Association of Plumbing
and Mechanical Officials
20001 Walnut Drive South
Walnut, CA 91789

International Conference of Building
Officials (ICBO)
5360 Workmen Mill Rd.
Whittier, CA 90601–2298

National Conference of States on Building Codes and
Standards (NCSBCS)
505 Huntmar Park Dr., Suite 210
Herndon, VA 22070

Southern Building Code Congress International
(SBCC)
900 Montclair Rd.
Birmingham, AL 35213–1206

Other Organizations

American Association of State and Highway
Transportation Officials
444 N. Capitol St. NW, Suite 249
Washington, DC 20001

American Institute of Architects
1735 New York Ave. NW
Washington, DC 20006

American National Metric Council
1735 N. Lynn St., Suite 950
Arlington, VA 22209–2022

American National Standards Institute
(ANSI)
11 West 42nd St., 13th Floor
New York, NY 10036

American Society for Testing and Materials
(ASTM)
1916 Race St.
Philadelphia, PA 19103

American Society of Civil Engineers
345 East 47th Street
New York, NY 10017–2398

Building and Fire Research Laboratory
National Institute of Standards and Technology
Bldg. 226, Room B216
Gaithersburg, MD 20899

Canada Mortgage and Housing Corporation
700 Montreal Road
Ottawa, Ontario, Canada K1A 0P7

Canada Standards Association (CSA)
178 Rexdale Blvd.
Rexdale, Ontario, Canada M9W 1R3

Canadian Construction Materials Centre
(CCMC)
Institute for Research in Construction
National Research Council Canada
Ottawa, Ontario, Canada K1A 0R6

Canadian Home Builders Association (CHBA)
150 Laurier Ave. West, Suite 200
Ottawa, Ontario, Canada K1P 5J4

Construction Specifications Institute (CSI)
601 Madison St.
Alexandria, VA 22314–1791

Department of Housing and Urban Development
(HUD)
451 Seventh Street SW
Washington, DC 20410

Environmental Hazards Management Institute
10 Newmarker Road
P.O. Box 932
Durham, NH 03824

Environmental Protection Agency
401 M Street NW
Washington, DC 20460

Federal Specifications
General Services Administrations, Specification
Activity Office, Bldg. 197
Washington Navy Yard
2nd and M Streets SE
Washington, DC 20407

Institute for Research in Construction and the
National Research Council of Canada
Montreal Road Building M-24
Ottawa, Ontario, Canada K1A 0R6

National Association of Home Builders (NAHB)
1201 Fifteenth Street NW
Washington, DC 20005–2800

National Fire Protection Association
1 Batterymarch Park
Quincy, MA 02269

National Institute of Building Science (NIBS)
1201 L Street NW
Washington, DC 20005

National Institute of Science and Technology
U.S. Department of Commerce
Bldg. 101, Room 813
Gaithersburg, MD 20899

National Institute of Standards and Technology
U.S. Department of Commerce
Bldg. 202, Rm. 204
Gaithersburg, MD 20899

National Research Center of the National Association
of Home Builders
400 Prince Georges Center Blvd.
Upper Marlboro, MD 20772–8731

National Standards of Canada
350 Sparks St.
Ottawa, Ontario, Canada K1R 7S8

Occupational Safety and Health Administration
(OSHA)
Francis Perkins Department of Labor Building
200 Constitution Ave. NW
Washington, DC 20210

Small Homes Council-Building Research Council
(SHC-BRC)
University of Illinois
1 East St. Mary's Road
Champaign, IL 61820

Underwriters Laboratories of Canada (ULC)
7 Crouse Rd.
Scarborough, Ontario, Canada M1R 3A9

Underwriters Laboratories, Inc.
333 Pfingsten Rd.
Northbrook, IL 60062–2096

U.S. Metric Association
10245 Andasol Ave.
Northridge, CA 91325

APPENDIX C

Metric Information

Tables giving metric equivalents for common fractions, two-place decimal inches, and millimeter-to-decimal-inches are printed inside the front and rear covers for ready reference.

Base SI Units

Quantity	Unit	Symbol
Length	Meter	m
Mass	Kilogram	kg
Time	Second	s
Electric current	Ampere	A
Thermodynamic temperature	Kelvin	K
Amount of substance	Mole	mo
Luminous intensity	Candela	cd

Supplementary SI Units

Quantity	Unit	Symbol
Plane angle	Radian	rad
Solid angle	Steradian	sr

Derived Metric Units with Compound Names

Physical Quantity	Unit	Symbol
Area	Square meter	m^2
Volume	Cubic meter	m^3
Density	Kilogram per cubic meter	kg/m^3
Velocity	Meter per second	m/s
Angular velocity	Radian per second	rad/s
Acceleration	Meter per second squared	m/s^2
Angular acceleration	Radian per second squared	rad/s^2
Volume rate of flow	Cubic meter per second	m^3/s
Moment of inertia	Kilogram meter squared	$kg{\cdot}m^2$
Moment of force	Newton meter	$N{\cdot}m$
Intensity of heat flow	Watt per square meter	W/m^2
Thermal conductivity	Watt per meter Kelvin	$W/m{\cdot}K$
Luminance	Candela per square meter	cd/m^2

SI Prefixes

Multiplication Factor			Prefix	Symbol
1 000 000 000 000 000 000	=	10^{18}	exa	E
1 000 000 000 000 000	=	10^{15}	peta	P
1 000 000 000 000	=	10^{12}	tera	T
1 000 000 000	=	10^{9}	giga	G
1 000 000	=	10^{6}	mega	M
1 000	=	10^{3}	kilo	k
100	=	10^{2}	hecto	h
10	=	10^{1}	deka	da
0.1	=	10^{-1}	deci	d
0.01	=	10^{-2}	centi	c
0.001	=	10^{-3}	milli	m
0.000 001	=	10^{-6}	micro	μ
0.000 000 001	=	10^{-9}	nano	n
0.000 000 000 001	=	10^{-12}	pico	p
0.000 000 000 000 001	=	10^{-15}	femto	f
0.000 000 000 000 000 001	=	10^{-18}	atto	a

Metric Unit to Imperial Unit Conversion Factors[a]

Metric Units		Imperial Equivalents
Length		
1 millimeter (mm)	=	0.0393701 inch
1 meter (m)	=	39.3701 inches
	=	3.28084 feet
	=	1.09361 yards
1 kilometer (km)	=	0.621371 mile
Length/Time		
1 meter per second (m/s)	=	3.28084 feet per second
1 kilometer per hour (km/h)	=	0.621371 mile per hour
Area		
1 square millimeter (mm^2)	=	0.001550 square inch
1 square meter (m^2)	=	10.7639 square feet
1 hectare (ha)	=	2.47105 acres
1 square kilometer (km^2)	=	0.386102 square mile

(Continued)

Metric Unit to Imperial Unit Conversion Factors[a] *(continued)*

Metric Units		Imperial Equivalents
Volume		
1 cubic millimeter (mm^3)	=	0.0000610237 cubic inch
1 cubic meter (m^3)	=	35.3147 cubic feet
	=	1.30795 cubic yards
1 milliliter (mL)	=	0.0351951 fluid ounce
1 liter (L)	=	0.219969 gallon
Mass		
1 gram (g)	=	0.0352740 ounce
1 kilogram (kg)	=	2.20462 pounds
1 tonne (t) (= 1,000 kg)	=	1.10231 tons (2,000 lb.)
	=	2204.62 pounds
Mass/Volume		
1 kilogram per cubic meter (kg/m^3)	=	0.0622480 pound per cubic foot
Force		
1 newton (N)	=	0.224809 pound-force
Stress		
1 megapascal (MPa) (=1 N/mm^2)	=	145.038 pounds-force per sq. in.
Loading		
1 kilonewton per sq. meter (kN/m^2)	=	20.8854 pounds-force per sq. ft.
1 kilonewton per meter (kN/m) (=1 N/mm)	=	68.5218 pounds-force per ft.
Moment		
1 kilonewtonmeter (kNm)	=	737.562 pound-force ft.
Miscellaneous		
1 joule (J)	=	0.00094781 Btu
1 joule (J)	=	1 watt-second
1 watt (W)	=	0.00134048 electric horsepower
1 degree Celsius (°C)	=	32 + 1.8 (°C) degrees Fahrenheit

Notes:
1. 1.0 newton = 1.0 kilogram × 9.80665 m/s^2 (International Standard Gravity Value)
2. 1.0 pascal = 1.0 newton per square meter
[a]Multiply the number of metric units by the imperial (English) equivalent to convert a measurement from metric units to imperial (English) units.

Imperial Unit to Metric Conversion Factors[b]

Imperial Units		Metric Equivalents
Length		
1 inch	=	25.4 mm
	=	0.0254 m
1 foot	=	0.3048 m
1 yard	=	0.9144 m
1 mile	=	1.60934 km
Length/Time		
1 foot per second	=	0.3048 m/s
1 mile per hour	=	1.60934 km/h
Area		
1 square inch	=	645.16 mm^2
1 square foot	=	0.0929030 m^2
1 acre	=	0.404686 ha
1 square mile	=	2.58999 km^2
Volume		
1 cubic inch	=	16387.1 mm^3
1 cubic foot	=	0.0283168 m^3
1 cubic yard	=	0.764555 m^3
1 fluid ounce	=	28.4131 mL
1 gallon	=	4.54609 L
Mass		
1 ounce	=	28.3495 g
1 pound	=	0.453592 kg
1 ton (2,000 lb.)	=	0.907185 t
1 pound	=	0.0004539 t
Mass/Volume		
1 pcf	=	16.1085 kg/m^3
Force		
1 pound	=	4.44822 N
Stress		
1 psi	=	0.00689476 MPa
Loading		
1 psf	=	0.0478803 kN/m^2
1 plf	=	0.0145939 kN/m
Moment		
1 pound-force ft.	=	0.00135582 kNm
Miscellaneous		
1 Btu	=	1055.06 J
1 watt-second	=	1 J
1 horsepower	=	746 W
1 degree Fahrenheit	=	(°F − 32)/1.8 °C

Notes:
1. 1.0 newton = 1.0 kilogram × 9.80665 m/s^2 (International Standard Gravity Value)
2. 1.0 pascal = 1.0 newton per square meter
[b]Multiply the number of imperial (English) units by the metric equivalent to convert a measurement from imperial (English) units to metric units.

Weights of Building Materials

Brick and Block Masonry	lb./ft.²	kg/m²
4″ brickwall	40	196
4″ concrete brick, stone or gravel	46	225
4″ concrete brick, lightweight	33	161
4″ concrete block, stone or gravel	34	167
4″ concrete block, lightweight	22	108
6″ concrete, stone or gravel	50	245
6″ concrete block, lightweight	31	152
8″ concrete block, stone or gravel	55	270
8″ concrete block, lightweight	35	172
12″ concrete block, stone or gravel	85	417
12″ concrete block, lightweight	55	270

Concrete	lb./ft.³	kg/m³
Plain, slag	132	2155
Plain, stone	144	2307
Reinforced, slag	138	2211
Reinforced, stone	150	2403

Lightweight Concrete	lb./ft.³	kg/m³
Concrete, perlite	35–50	561–801
Concrete, pumice	60–90	961–1442
Concrete, vermiculite	25–60	400–961

Wall, Ceiling, and Floor	lb./ft.²	kg/m²
Acoustical tile, ¹/₂″	0.8	3.9
Gypsum wallboard, ¹/₂″	2	9.8
Plaster, 2″ partition	20	98
Plaster, 4″ partition	32	157
Plaster, ¹/₂″	4.5	22
Plaster on lath	10	49
Tile, glazed, ³/₈″	3	14.7
Tile, quarry, ¹/₂″	5.8	28.4
Terrazzo, 1″	25	122.5
Vinyl composition floor tile	1.4	69
Hardwood flooring, ²⁵/₃₂″	4	19.6
Flexicore 6″, lightweight concrete	30	14.7

Wall, Ceiling, and Floor	lb./ft.²	kg/m²
Flexicore 6″, stone concrete	40	196
Plank, cinder concrete, 2″	15	73.5
Plank, gypsum, 2″	12	58.8
Concrete reinforced, stone, 1″	12.5	61.3
Concrete reinforced, lightweight, 1″	6–10	29.4–49
Concrete plain, stone, 1″	12	58.8
Concrete plain, lightweight, 1″	3–9	14.7–44.1

Partitions	lb./ft.²	kg/m²
2 × 4 wood studs, gypsum wallboard 2 sides	8	39.2
4″ metal stud, gypsum wallboard 2 sides	6	29.4
6″ concrete block, gypsum wallboard 2 sides	35	171.5

Roofing	lb./ft.²	kg/m²
Built-up	6.5	31.9
Concrete roof tile	9.5	46.6
Copper	1.5–2.5	7.4–12.3
Steel deck alone	2.5	12.3
Shingles, asphalt	1.7–2.8	8.3–13.7
Shingles, wood	2–3	9.8–14.7
Slate, ¹/₂″	14–18	68.6–88.2
Tile, clay	8–16	39.2–78.4

Stone Veneer	lb./ft.³	kg/m³
2″ granite, ¹/₂″ parging	30	481
4″ limestone, ¹/₂″ parging	36	577
4″ sandstone, ¹/₂″ parging	49	785
1″ marble	13	208

Structural Clay Tile	lb./ft.²	kg/m²
4″ hollow	23	368
6″ hollow	38	609
8″ hollow	45	721

(Continued)

Structural Facing Tile	lb./ft.²	kg/m³
2″ facing tile	14	68.6
4″ facing tile	24	118
6″ facing tile	34	167

Wood	lb./ft.²	kg/m²
Ash, white	40.5	198
Birch	44	202
Cedar	22	108
Cypress	33	162
Douglas fir	32	157
White pine	27	132
Pine, southern yellow	26	127
Redwood	26	127
Plywood, ½″	1.5	7.4

Residential Assemblies	lb./ft.²	kg/m²
Wood framed floor	10	49
Ceiling	10	49
Frame exterior wall, 4″ studs	10	49
Frame exterior wall, 6″ studs	13	64
Brick veneer of 4″ frame	50	245
Brick veneer over 4″ concrete block	74	363
Interior partitions with gypsum both sides (allowance per sq. ft. of floor area—not weight of material)	20	320

Suspended Ceilings	lb./ft.²	kg/m²
Acoustic plaster on gypsum lath	10–11	49–54
Mineral fiberboard	1.4	6.9

Metals	lb./ft.³	kg/m³
Aluminum	165	2643
Copper	556	8907
Iron, cast	450	7209
Steel	490	7850
Steel, stainless	490–510	7850–8170

Glass	lb./ft.²	kg/m²
¼″ (6.3 mm) plate or float	3.3	16.2
½″ (12.7 mm) plate or float	6.6	32.3
1/32″ (0.79 mm) sheet	2.8	13.7
¼″ (6.3 mm) sheet	3.5	17.2
⅛″ (3.2 mm) double strength	1.6	7.8
7/32″ (5.6 mm) sheet	2.85	14.0
¼″ (6.3 mm) laminated	3.30	16.2
½″ (12.7 mm) laminated	6.35	31.1
2″ (50.2 mm) bullet resistant	26.2	128.4
1″ (25.4 mm) insulating, ½″, (6.3 mm) air space	6.54	32.0
¼″ (6.3 mm) wired	3.5	17.1
⅜″ (9.5 mm) wired	5.0	24.4
3 7/8″ × 5 ¾″ square, (98.0 × 146 mm) glass block	16.0	78.4

APPENDIX

Names and Atomic Symbols of Selected Chemical Elements

Name	Atomic Symbol	Name	Atomic Symbol
Aluminum	Al	Gallium	Ga
Antimony	Sb	Germanium	Ge
Argon	Ar	Gold	Au
Arsenic	As	Hafnium	Hf
Barium	Ba	Helium	He
Beryllium	Be	Holmium	Ho
Bismuth	Bi	Hydrogen	H
Boron	B	Illinium	Il
Bromine	Br	Indium	In
Cadmium	Cd	Iodine	I
Calcium	Ca	Iridium	Ir
Carbon	C	Iron	Fe
Cerium	Ce	Krypton	Kr
Cesium	Cs	Lanthanum	La
Chlorine	Cl	Lead	Pb
Chromium	Cr	Lithium	Li
Cobalt	Co	Lutecium	Lu
Copper	Cu	Magnesium	Mg
Dysprosium	Dy	Manganese	Mn
Erbium	Er	Masurium	Ma
Europium	Eu	Mercury	Hg
Fluorine	F	Molybdenum	Mo
Gadolinium	Gd	Neodymium	Nd

Name	Atomic Symbol	Name	Atomic Symbol
Neon	Ne	Silicon	Si
Nickel	Ni	Silver	Ag
Niobium	Nb	Sodium	Na
Nitrogen	N	Strontium	Sr
Osmium	Os	Sulfur	S
Oxygen	O	Tantalum	Ta
Palladium	Pd	Tellurium	Te
Phosphorus	P	Terbium	Tb
Platinum	Pt	Thallium	Tl
Polonium	Po	Thorium	Th
Potassium	K	Thulium	Tm
Praseodymium	Pr	Tin	Sn
Protactinium	Pa	Titanium	Ti
Radium	Ra	Tungsten	W
Radon	Rn	Uranium	U
Rhenium	Re	Vanadium	V
Rhodium	Rh	Xenon	Xe
Rubidium	Rb	Ytterbium	Yb
Ruthenium	Ru	Yttrium	Y
Samarium	Sm	Zinc	Zn
Scandium	Sc	Zirconium	Zr
Selenium	Se		

APPENDIX F

Coefficients of Thermal Expansion[a] for Selected Construction Materials

Material	Multiply by 10^{-6} in./in./°F	Multiply by 10^{-6} mm/mm/°C
Concrete		
Normal weight concrete	5.5	9.8
Gypsum		
Gypsum panels	9.0	16.2
Gypsum plaster	7.0	12.6
Wood fiber plaster	8.0	14.4
Masonry		
Brick (varies some)	3.0	5.6
Concrete masonry units	5.2	9.4
Marble	7.3	13.1
Granite	4.7	8.5
Limestone	4.4	7.9

Material	Multiply by 10^{-6} in./in./°F	Multiply by 10^{-6} mm/mm/°C
Metal		
Iron, gray cast	5.7	10.5
Iron, malleable	5.6	10.5
Steel, carbon (ASTM A285)	5.6	10.5
Steel, high strength (ASTM A141)	6.4	11.7
Steel, stainless (type 201)	8.7	15.7
Steel, stainless (type 405)	6.0	10.8
Nickel (211)	7.4	13.3
Copper (CA110)	9.4	16.5
Bronze, commercial	10.2	19.3
Brass, red	10.4	19.9
Aluminum, wrought	12.8	23.0

Material	Multiply by 10^{-6} in./in.°/F	Multiply by 10^{-6} mm/mm/°C
Polymer, Thermosetting		
Phenolics	45.0	81.0
Urea-melamine	20.0	36.0
Polyesters	45.0	75.5
Epoxies	33.0	72.0

(Continued)

Material	Multiply by 10^{-6} in./in.°/F	Multiply by 10^{-6} mm/mm/°C
Polymer, Thermoplastic		
Polyethylene, high density	70.0	120.0
Polypropylene	50.0	90.0
Polystyrene	38.0	68.5
Polyvinyl chloride (PVC)	30.0	54.0
Acrylonitrile-butadiene-styrene (ABS)	50.0	90.0
Acrylics	40.0	72.0
Wood		

For most hardwoods and softwoods the *parallel-to-grain* thermal expansion values range from 1.7×10^{-6} to 2.5×10^{-6} in./in./°F or 3.1×10^{-6} to 4.5×10^{-6} mm/mm/°C. Linear expansion coefficients *across the grain* are proportional to wood density. They range from 5 to 10 times greater than the parallel-to-grain coefficients and, therefore, are of more concern.

Material	Multiply by 10^{-4} in./in.°/F	Multiply by 10^{-4} mm/mm/°C
Foam Insulation		
Polystyrene	3.5	6.3
Polyurethane	2.7	4.9

Material	Multiply by 10^{-7} in./in./°F	Multiply by 10^{-7} mm/mm/°C
Glass		
Glass, soda lime window sheet	47.0	85.0
Glass, soda lime plate	48.0	87.0

[a]The change in dimension of a material per unit of dimension per degree change in temperature.

Glossary

Additional definitions are available in construction dictionaries such as the following:

Harris, C. M., *Dictionary of Architecture and Construction*, McGraw-Hill, New York, 1987.

Kennedy, F., *The Wiley Dictionary of Civil Engineering and Construction*, John Wiley & Sons, Somerset, N.J., 1990.

Philbin, T., *The Illustrated Dictionary of Building Terms*, McGraw-Hill, New York, 1996.

Putnam, R., *Builders Comprehensive Dictionary*, Prentice-Hall, Englewood Cliffs, N.J., 1984.

abrasion resistance Resistance to being worn away by rubbing or friction.

absorption The process in which a liquid or mixture of liquid and gases is drawn into the pores of a porous solid material.

absorptivity The relative ability to absorb sound and light.

accelerator An admixture used to speed up the setting of concrete.

access floor A freestanding floor raised above the basic floor.

acoustical ceiling A ceiling of fibrous tiles, panels, or other sound-deadening material.

acoustical glass A glazing unit used to reduce the transmission of sound through the glazed opening by bonding a soft interlayer between the layers of glass.

acoustical plaster Calcined gypsum mixed with lightweight aggregates.

acoustics The science of sound generation, sound transmission, and the effects of sound waves.

acrylic A transparent thermoplastic made from esters of acrylic acid.

active solar system A solar heating system that uses mechanical means to move and store solar energy.

additive Materials mixed with the basic plastic resin to alter its properties.

adhesion The ability of a coating to stick to a surface.

adhesive A substance used to hold materials together by surface attachment.

admixture A material other than portland cement, aggregate, and water that is added to concrete to alter its properties.

adsorbent A material that has the ability to cause molecules of gases, liquids, or solids to adhere to its surfaces without changing the adsorbent physically or chemically.

agglomeration A process that bonds ground iron ore particles into pellets to facilitate handling.

aggregate Inert granules such as crushed stone, gravel, and expanded minerals mixed with portland cement and sand to form concrete.

air, combustion Air used to provide for the combustion of a fuel.

air conditioner A mechanical device used to provide air-conditioning.

air-conditioning The process of treating air to control simultaneously its humidity, cleanliness, and temperature and to provide distribution within a building.

air-dried lumber Wood dried by exposing it to the air.

air-entrained cement A portland cement with an admixture that causes a controlled quantity of stable, microscopic air bubbles to form in the concrete.

air-entrained concrete Concrete with an admixture added that produces millions of microscopic air bubbles in the concrete.

air gap An unobstructed vertical distance between the lowest opening of pipe that supplies a plumbing fixture and the level at which the fixture will overflow.

air-supported structures Structures with the enclosing envelope made from a flexible material that is pneumatically supported.

alkali A substance such as lye, soda, or lime that can be destructive to paint films.

alkyd Synthetic resin modified with oil for good adhesion, gloss, color retention, and flexibility.

alloy A metallic material composed of two or more chemical elements one of which is a metal.

alloying element Any substance added to a molten metal to change its mechanical or physical properties.

alternating current An electric current that varies periodically in value and direction by flowing first in one direction and then in the opposite direction.

alumina A hydrated form of aluminum oxide from which aluminum is made.

ambient temperature The temperature of the surrounding air.

ampere A basic SI unit that measures the rate of flow of electric current.

anaerobic bonding agents Bonding agents that set hard when not exposed to oxygen.

angle of repose The angle of the sloped surface of the sides of an excavation.

annealing Heating a metal to a high temperature followed by controlled cooling to relieve internal stresses.

anodizing An electrolytic process that forms a permanent, protective oxide coating on aluminum.

APA performance-rated panels Plywood manufactured to the structural specifications and standards of APA - The Engineered Wood Association.

arc resistance The total elapsed time in seconds an electric current must arc to cause a part to fail.

architectural terra-cotta Clay masonry units made with a textured or sculptured face.

asphalt Dark brown to black hydrocarbon solids or semi-solids having bituminous constituents that gradually liquify when heated.

autoclave A high-pressure steam room that rapidly cures green concrete units.

awning window A window that pivots near the top edge of the sash and projects toward the exterior.

Axminster construction Carpet formed by weaving on a loom that inserts each tuft of pile individually into the backing.

backfill Earth filled in around a foundation wall to replace earth removed for construction of the foundation.

bagasse Crushed sugar cane or beet refuse from sugar making.

ballast An electrical device to provide the starting voltage and operating current for fluorescent, mercury, and other electric discharge lamps.

balloon frame A system of framing a wood-framed building in which all vertical structural members (studs) of the exterior bearing walls and partitions extend the full height of the frame, from the bottom plate to the top plate, and support the floor joists and roof.

bank measure The volume of soil in situ in cubic yards.

batch The amount of concrete mixed at one time.

bauxite Ore containing high percentages of aluminum oxide.

beam A straight horizontal structural member whose main purpose is to carry transverse loads.

bearing pile A pile that carries a vertical load.

bearing plate A steel plate placed under a beam, column, or truss to distribute the end reaction from the beam to the supporting member.

bearing wall A wall that supports a roof, floor, or ceiling.

bedrock The hard, solid rock formation at or below the surface of the earth.

bending moment The moment that produces bending on a beam or other structural member.

bending stress A compressive or tensile stress developed by applying nonaxial force to a structural member.

beneficiation A process of grinding and concentration that removes unwanted elements from iron ore before the ore is used to produce steel.

Bentonite clay An absorptive clay that swells several times its dry volume when saturated with water.

binder Film-forming ingredient in paint that binds the suspended pigment particles together.

bitumen A generic term describing a material that is a mixture of predominantly hydrocarbons in solid or viscous form. It is derived from coal and petroleum.

bituminous coatings Coatings formulated by dissolving natural bitumens in an organic solvent.

bleeding Excess water that rises to the surface of concrete shortly after it has been poured.

blocking Wood pieces inserted between joists, studs, rafters, and other structural members to stabilize the frame, provide a nailing surface for finish materials, and block the passage of fire between the members.

board foot The measure of lumber having a volume of 144 in.^{3}

boards Lumber less than 2 in. (50.8 mm) thick and 1 in. (25.4 mm) or more wide.

boiler A closed vessel used to produce hot water or steam.

bond beam A continuous reinforced beam formed from horizontal masonry members bonded with reinforced concrete.

bond breaker A material used to prevent adjoining materials from adhering.

bonding agent A compound that holds materials together by bonding the surfaces to be joined.

box beam A structural member of metal or plywood whose cross section is a closed rectangular box shape.

box sill A type of sill used in frame construction in which the floor joists butt and are nailed to a header joist and rest on the sill.

branch A pipe in a plumbing system into which no other branch pipes discharge and that discharges into a main or submain.

branch circuit The electrical wiring between the overcurrent protection device and the connected outlets.

branch circuit, appliance An electric circuit supplying energy to outlets to which appliances are to be connected.

branch circuit, general purpose An electric circuit that supplies energy to a number of outlets for lighting and small appliances.

branch circuit, individual An electric circuit that supplies energy to only one piece of equipment.

branch interval A length of soil or waste stack 8 ft. or more in height (equal to one story) within which the horizontal branches from one floor or story of a building are connected to a stack.

branch, plumbing A horizontal run of waste piping that carries waste material to a vertical riser.

branch vent A vent connecting one or more individual vents into a vent stack or stack vent.

breaking strength The point at which a material actually begins to break.

Brinell hardness A measure of resistance of a material to indention.

Brinell hardness number A measure of Brinell hardness that is obtained by dividing the load in kilograms by the area of the indention given in square millimeters.

British thermal unit (Btu) The amount of heat required to raise the temperature of 1 lb. of water 1°F.

British thermal unit per hour (Btu/h) The rate of heat flow per hour.

brittleness The characteristic of a material that tends to crack or break without appreciable plastic deformation.

brown coat The second coat of plaster in a three-coat plaster finish.

buffer, elevator Energy-absorbing units placed in the elevator pit.

building code A set of legal regulations that ensure a minimum standard of health and safety in buildings.

building drain The lowest horizontal piping of a plumbing drainage system that receives the discharge from soil, waste, and other drainage pipes within the building and carries the waste to the building sewer.

building sewer Horizontal piping that carries the waste discharge from the building drain to the public sewer or septic tank.

built-up roof membrane A continuous, semiflexible roof membrane built up of plies of saturated felts, coated felts, fabrics, or mats that have surface coats of bitumens. The last ply is covered with mineral aggregates, bituminous materials, or a granular-surface roofing sheet.

bulletin board A surface used to display announcements and other material usually attached with tacks.

burning Curing bricks by placing them in a kiln and subjecting them to a high temperature.

bus A rigid electric conductor enclosed in a protective busway.

busway A rigid conduit used to protect a bus running through it.

BX cable A cable sheathed with spirally wrapped metal strip identified as Type AC.

cable tray A ladderlike metal frame open on the top used to support insulated electrical cables.

caisson A watertight structure within which work can be carried out below the surface of water.

calcareous clays Clays containing at least 15 percent calcium carbonate.

calcined gypsum Ground gypsum that has been heated to drive off the water content.

camber An arching in a structural member due to tension applied to the steel framing.

candela A metric unit of luminous intensity that closely approximates candlepower.

candlepower A term used to express the luminous intensity of a light source. It is the same magnitude as a candela.

cant strip A triangular molding secured over the joint between a wall and a roof deck.

capillary action The movement of a liquid through small openings of fibrous material by the adhesive force between the liquid and the material.

capillary break A groove in a member used to create an opening that is too wide to be bridged by a drop of water, thus eliminating the passage of water by capillary action.

car, elevator The load-carrying unit of an elevator, consisting of a platform, walls, ceiling, door, and a structural frame.

car safeties, elevator Devices used to stop a car and hold it in position should it travel at excessive speed or go into a free fall.

carbohydrates Organic compounds that form the supporting tissue of plants.

carbon steel Any steel for which no minimum content for alloying agents is specified but for which the carbon content is the element used to determine its properties.

cast-in-place concrete Concrete members formed and poured on the building site in the locations where they are needed.

cast-in-place piles Concrete piles cast in a hollow metal shell driven into the earth or an uncased hole.

casting A metal part produced by pouring a molten metal into a mold.

cast iron A hard, brittle metal made of iron that contains a high percentage of carbon.

caulking A resilient material used to seal cracks and prevent leakage of water.

cavity wall A masonry wall made up of two wythes of masonry units separated by an air space.

cellulose An inert carbohydrate that is the chief constituent of the cell walls of plants, wood, and paper.

Celsius temperature The temperature scale used with the SI system in which the boiling point of water is 100°C and freezing is 0°C.

cement 1. A powder with adhesive and cohesive properties that sets into a hard, solid mass when mixed with water. 2. Bonding agent made from synthetic rubber suspended in a liquid.

cement board A panel product made with an aggregate portland cement core reinforced with polymer-coated glass-fiber mesh on each side.

cementitious materials Materials that have cementing properties.

cement-lime mortar Mortar made with the addition of slaked lime to the cement.

central service core A fire-resistant vertical shaft through a multistory building used to route electrical, mechanical, and transportation systems.

ceramic A class of products made of clay fired at high temperatures.

ceramic glaze A compound of metallic oxides, chemicals, and clays fused to a material at high temperature, providing a hard, smooth surface.

chalkboard A surface that can be written on with chalk and from which the chalk can be easily removed.

chase A recessed area in a wall for holding pipes and conduit that passes vertically between floors.

chemical strengthening A process for strengthening glass that involves immersing the glass in a molten salt bath.

chiller A refrigerating machine composed of a compressor, a condenser, and an evaporator, used to transfer heat from one fluid to another.

chord, bottom A horizontal or inclined structural member forming the lower edge of a truss.

chord, top A horizontal or inclined structural member forming the top edge of a truss.

circuit breaker An electrical device used to open and close a circuit by nonautomatic means or to open a circuit by automatic means at a predetermined overcurrent without damage to itself.

cladding The external finish covering the base material on a wall.

Class A,B,C roofing Classification of roofing materials by their resistance to fire when tested in accordance with ASTM E108.

clay A very cohesive material made up of microscopic particles (less than 0.00008 in. or 0.002 mm).

clay tile A unit made from fired and sometimes glazed clay and used as a finish surface on floors and walls.

cleanouts Openings in the waste piping system that permit cleaning obstructions from the pipe.

clear coating A transparent protective and/or decorative film.

clear span The horizontal distance between the interior edges of supporting members.

coal tar Tar produced through the destructive distillation of coal during the conversion of coal to coke.

coal tar pitch A dark brown to almost black hydrocarbon solid or semisolid material derived by distilling coke-oven tar.

coating A paint, varnish, lacquer, or other finish used to create a protective and/or decorative layer.

coefficient of heat transmission The total amount of heat that passes through an assembly of materials, including any air spaces and surface air films. It is expressed in Btu per hr., per ft.^{2}, per °F temperature difference between the inside and outside air.

coefficient of thermal conductance The amount of heat, expressed in Btu, that can pass through a specified thickness of a material per hr., per ft.^{2}, per °F temperature difference between the surfaces.

coefficient of thermal expansion The total amount of heat, expressed in Btu, that passes by conduction through a 1 in. thickness of a homogeneous material per hr., per ft.^{2}, per °F, which is measured as the temperature difference between the two surfaces of the material.

cofferdam A temporary watertight enclosure around an area of water-bearing soil or an area of water from which water is pumped allowing construction to take place in the water-free area.

cogeneration systems Systems using fossil fuel, geothermal energy, wind, or solar energy to produce electricity and heat.

cohesion When referring to soils, it is the sticking together of soil particles whose forces of attraction exceed the forces that tend to separate them.

cohesion The molecular forces between particles within a body which acts to unite them.

cohesionless soil A soil that when unconfined has little or no cohesion when submerged and no significant strength when air dried.

cohesive soil A soil that when unconfined has considerable cohesion when submerged and considerable strength when air dried.

cold-rolled steel Steel rolled to the final desired shape at a temperature at which it is no longer plastic.

compaction Compressing soil to increase its density.

compartment A small area within a larger area enclosed by partitions.

composite materials Materials made by combining several layers of different materials.

composite panels Panels having a reconstituted wood core bonded between layers of solid veneer.

compression The condition of being shortened (compressed) by force.

compression test A test used to determine the behavior of materials under compression.

compressive strength The maximum stress a material can withstand before it is crushed.

compressive stresses Stresses created when forces push on a member and tend to shorten it.

compressor A mechanical device for increasing the pressure of a gas.

concentrated load Any load that acts on a very small area of a structure.

concrete A solid, hard material produced by combining portland cement, aggregates, sand, and water and sometimes admixtures.

concrete masonry Factory manufactured concrete units, such as concrete brick or block.

concrete pump A pump that moves concrete through hoses to the area where it is to be placed.

condensate A liquid formed by the condensation of a vapor.

condensation The process of changing from a gaseous to a liquid state.

condensation point The temperature at which a vapor liquefies if the latent heat is removed at standard or a stated pressure.

condenser A heat-exchanger unit in which a vapor has some heat removed, causing it to form a liquid.

conduction, thermal The process of heat transfer through a material to another part of that material or to a material touching it.

conductivity, electric A measure of the ability of a material to conduct electric current.

conductor, electric Wire through which electric current flows.

conduit A steel or plastic tube through which electrical wires are run.

consolidation The process of compacting freshly placed concrete in a form.

control joint A groove formed in concrete or masonry structures to allow a place where cracking can occur, thus reducing the development of high stresses.

controller An electric device or a group of devices used to govern the electric power delivered to the equipment to which it is connected.

convection The process of carrying heat from one spot to another by movement of a liquid or gas. The heated liquid or gas expands and becomes lighter, causing it to rise while the cooler, heavier dense liquid or air settles.

convector A unit designed to transfer heat from hot water or steam to the air by convection.

cooling tower A heat-transfer device in which the atmospheric air cools warm water flowing through the tower, usually by evaporation.

corrosion The deterioration of a metal or of concrete by chemical or electrochemical reaction caused by exposure to the weather.

covalent bonding A process in which small numbers of atoms are bonded into molecules.

creep Permanent dimensional deformation occurring over a period of time in a material subjected to constant stress at elevated temperatures.

creep strength The stress that produces a given size change when constantly applied for a given period of time at a specific temperature.

creep test A test to determine the creep behavior of materials subjected to constant stress at a constant temperature.

cricket A small false roof used to divert water from behind a projection above the roof, such as a chimney.

cubicle A very small enclosed space often large enough for just one person.

cupola A small roofed structure built on a roof, usually to vent the area below the roof.

curb A low wall of wood or masonry extending above the level of the roof and surrounding an opening in the roof.

cure To chemically cross-link polymer chains by heating and/or adding a chemical agent.

curing Protecting concrete after placing so that proper hydration occurs.

curing agent Part of a two-part compound that, when added to the second part, sets up the curing action. Also referred to as the *catalyst.*

current, electric The flow of electrons along a conductor.

curtain wall An exterior wall of lightweight construction that is non–load bearing, supporting only its own weight.

cushion A layer of resilient material applied to a floor over which a carpet is to be laid.

damper A movable vane used to vary the volume of air passing through a duct, inlet, or outlet.

damping capacity The ability of a material to absorb vibrational energy.

darby A tool used to level concrete in a form after it has been screeded.

dead load A permanent load that provides steady pressure on the building structure, such as roofing materials.

decibel A unit for measuring sound energy or power.

deflection Displacement of a member from its static position as a result of forces acting on that member.

deformation A change in shape of a member caused by a load or force acting on it without a breach of the continuity of its parts.

dehumidification The removal of water vapor from the air.

dehumidifier A cooling, absorption, or adsorption device used for removing moisture from the air.

dehydration The removal of water vapor from any substance.

delamination The separation of the plies in a laminate or plies from a base material.

desiccant Any absorbent, adsorbent, liquid, or solid that removes water or water vapor from a material.

desiccation The process of evaporating or removing water vapor from a material.

dew point The temperature at which air must be cooled to reach a given pressure and water content for it to reach saturation (100 percent relative humidity).

dewatering Pumping subsurface water from an excavation to maintain dry and stable working conditions.

diaphragm A horizontal roof or floor structural element designed to resist lateral loads and transmit them to shear walls (vertical resisting elements).

dielectric strength The maximum voltage a dielectric (nonconductor) can withstand without fracture.

diffuser A circular, square, or rectangular air distributing outlet, usually in the ceiling, that has members to discharge supply air in several directions, mixing the supply air with the secondary air in the room.

dimension lumber Lumber from 2 in. (50.8 mm) up to but not including 5 in. (127 mm) thick and 2 in. (50.8 mm) or more wide.

direct current Electricity that flows in one direction.

double glazing Two parallel sheets of glass with an air space between.

double-hung window A window having two vertically sliding sashes.

double tees T-shaped precast floor and roof units that span long distances unsupported.

dressed lumber Lumber having one or more sides planed smooth.

drypack A stiff granular grout.

dry-press process The process used to make bricks when the clay contains 10 percent or less moisture.

duct A hollow tube through which air is circulated.

ductile Capable of being stretched or deformed without fracturing (plastic deformation).

ductility A measure of the capability of a material to be stretched or deformed without breaking.

durability of a coating The ability of a coating to hold up against destructive agents, such as weathering and sunlight.

dynamic load Any load that is nonstatic.

E value The ratio of stress to strain.

eaves The lower part of a roof that projects over the exterior wall.

EER (Energy Efficiency Ratio) The Btu output divided by the input in watts. The higher the EER the more efficient the equipment.

efflorescence A white soluble salt deposit on the surface of concrete and masonry, usually caused by free alkalies leached from the mortar by moisture moving through it.

effluent In plumbing, the liquid discharge from a waste disposal system.

elastic deformation The ability of a material to return to its original position after a load has been removed.

elastic limit The greatest stress a material can withstand without permanent deformation upon the release of the stress.

elasticity The property of a material that causes it to return to its original shape upon removal of a deforming load.

elastomer A macromolecular material that returns to its approximate initial dimensions and shape after being subjected to substantial deformation.

elastomeric Having the properties of an elastomer.

electric conduction The ability of a material to conduct an electric current.

electric current The movement of electrons in an electric conductor.

electric current, alternating An electric current that reverses the direction of flow periodically.

electric current, direct An electric current that does not reverse its polarity.

electric power The rate of generating, transferring, or using electric energy. It is expressed in watts (W) and kilowatts (kW).

elevator A hoisting and lowering mechanism equipped with an enclosed car that moves between floors in a building.

elongation Drawing out to a greater length when under load or expansion due to temperature increases.

enamel A classification of paints that dry to a hard flat semi-gloss or gloss finish.

epoxy finish A clear finish having excellent adhesion qualities, abrasion and chemical resistance, and water resistance.

epoxy resin A class of synthetic thermosetting resins derived from certain special types of organic chemicals.

equilibrium The state of being equally balanced.

equilibrium moisture content The moisture content at which wood neither gains nor loses moisture when surrounded by air at a specified relative humidity and temperature.

erection plan An assembly drawing showing where each structural steel member is located on the building frame.

escalator A continuous moving stair used to move people up and down between floors.

evaporator That part of a refrigerating system in which the refrigerant is evaporated, allowing it to absorb heat from the contacting heat source.

expansion joint A joint used to separate two parts of a building to allow expansion and contraction movement of the parts.

exposed aggregate finish A finished concrete surface in which a coarse aggregate is exposed to view.

extruding A process in which a billet of material is shaped into a strip having a uniform cross section by forcing the material through a die.

fabric A cloth made by weaving, knitting, or felting fibers.

face brick Brick made or selected to produce an attractive exterior wall.

faceted glass window A window made by bonding 1 in. (25.4 mm) thick glass pieces with an epoxy resin matrix or reinforced concrete.

Fahrenheit temperature The temperature scale on which at standard atmospheric pressure the boiling point of water is 212°F, the freezing point is 32°F, and absolute zero is −459.69°F.

fan coil unit The fan and heat exchanger for cooling and heating that are assembled in a common cabinet.

fascia The finish board covering the edges of rafters at the eaves.

fatigue A condition that occurs in a material when it is subject to fluctuating or cyclic strains and stresses that lead to permanent deformation.

fatigue limit The number of cycles of loading of a specified type that a specified material can withstand before failure.

fatigue strength A measure of the ability of a material or structural member to carry a load without failure when the loading is applied a specified number of times.

fatigue test A test to determine the behavior of a material under fluctuating stresses.

felt A sheet material made using a fiber mat that has been saturated and topped with asphalt.

fenestration An area that allows light to pass into a building, commonly referring to glazed windows. Also, the arrangement of windows in an exterior wall.

ferrous Iron-based metallic materials.

fiber saturation point The moisture content of wood at which the cell walls are saturated but there is no water in the cell cavities.

fiberboard A panel made from vegetable fibers and binding agents.

fibered gypsum A neat gypsum with cattle hair or organic fillers added.

filler Inert material added to a plastic resin to alter the strength and working properties and to lower the cost.

finish coat The third or final coat of gypsum plaster.

finish floor The flooring that is left exposed to view.

finish lime A hydrated lime used in finish coats of plaster and in ornamental plasters.

finish plaster The topcoat of plaster on a wall or ceiling.

firebrick A brick made from special clays that will withstand high temperatures.

fireclays Deep mined clays that withstand heat.

fire endurance The time during which a material or an assembly of materials provides resistance against the passage of fire.

fireproofing Material used to protect various members from damage due to fire.

fire-rated partition A partition assembly that has been tested and given a rating indicating the length of time it will resist a fire in hours.

fire resistance The capacity of a material or assembly of materials to withstand fire or give protection from it.

fire-resistant gypsum A gypsum product that has increased fire-resistance properties due to the addition of fire-resistant materials in the gypsum core.

fire-stop A member used to close openings between studs, joists, and other members to retard the spread of fire through openings between them.

fire wall A construction of noncombustible materials that subdivides a building or separates adjoining buildings to retard the spread of fire.

flame spread Flaming combustion that occurs along the surface of a material.

flames spread rate The rate at which flames will spread across a surface of a material.

flame spread rating A numerical designation given to a material to indicate its comparative ability to restrict flaming combustion over its surface.

flammability The ability of a material to resist burning.

flash point The temperature at which a flammable material will suddenly break into flame.

flash set Very rapid setting of the cement in concrete.

flashing A thin impervious material used to prevent water from penetrating the joints between building elements.

float A flat hand tool used to smooth the surface of freshly placed concrete after it has been leveled with a darby.

float process A glass manufacturing process in which the molten glass ribbon flows through a furnace supported on a bed of molten metal.

flocked construction Carpet formed by electrostatically spraying short strands onto an adhesive-coated backing material.

flood coat A heavy coating of asphalt poured and spread over a surface.

fluorescence The emission of visible light from a substance as a result of the absorption of radiation of short wavelengths.

flux A mineral added to molten iron to cause impurities to separate into a layer of molten slag on top of the iron.

flying formwork Large sections of formwork for pouring concrete slabs that are lifted from story to story by a crane in an assembled condition.

footcandle The unit of illumination equal to 1 lumen per square foot.

footing The lowest, widest part of the foundation that distributes the load over a broad area of the soil.

footlambert A unit for measuring brightness or luminance. It is equal to 1 lumen per square foot when brightness is measured from the surface.

formwork Temporary construction used to contain and give shape and support to concrete as it cures.

foundation The lower part of a building, which transfers structural loads from the building to the soil.

framed connections Connections joining structural steel members with a metal, such as an angle, that is secured to the web of the beam.

framing plan A drawing showing the location of structural members.

freezing cycle day A day when the temperature of the air rises above or falls below 32°F or 0°C.

freezing point The temperature at which a given substance will solidify (freeze).

frequency The number of cycles per second of current or voltage in alternating current, of a sound wave, or of a vibrating solid expressed in hertz.

frost line The maximum depth in the earth to which the soil can be expected to freeze during a severe winter.

frost point The temperature at which frost forms on exposed, chilled surfaces.

fuel contributed A rating of the amount of combustible material in a coating.

furring Strips applied over a surface to increase thickness or to provide a base for the attachment of other material.

fuse An overcurrent protection device that opens an electric circuit when the fusible element is broken by heat due to overcurrent passing through it.

fusion bonded construction Carpet formed by bonding pile yarn between two sheets of backing material and cutting the pile yarn in the center, forming two pieces of carpet.

galling The wearing or abrading of one material against another under extreme pressure.

galvanic corrosion Corrosion that develops by galvanic action when two dissimilar metals are in contact in the atmosphere.

gearless traction elevator An elevator with the traction sheave connected to a spur gear that is driven by a worm gear connected to the shaft of the electric motor.

girder A major structural member used to support beams.

glass An amorphous, noncrystalline solid made by fusing silica with a basic oxide.

glazed structural clay tile Hollow clay tile products with glazed faces typically used to build interior walls.

glued laminated lumber (glulam) A structural wood member made by bonding together laminations of dimension lumber.

glues Bonding agents made from animal and vegetable products.

grade (1) Related to soil, the elevation or slope of the ground. (2) In relation to lumber, a means of classifying lumber or other wood products based on specified quality characteristics.

grade beam A ground-level reinforced structural member that supports the exterior wall of a structure and bears directly upon columns or piers.

grade level The elevation of the soil at a specific location.

grade mark A stamp on a product, such as wood, plywood, or steel, indicating the product's quality.

grading Adjusting the level of the ground on a site.

gravel Hard rock material in particles larger than $\frac{1}{4}$ in. (6.4 mm) in diameter but smaller than 3 in. (76 mm).

green lumber Lumber having a moisture content more than 19 percent.

grille An open grate used to cover or conceal, protect, or decorate an opening.

ground A conducting connection between an electrical circuit and the earth or a conducting body that serves in place of the earth.

ground-fault circuit interrupter A device providing protection from electric shock by de-energizing a circuit within an established period of time when the current to ground exceeds a predetermined value that is less than that needed to activate a standard overcurrent protective device.

groundwater Water that exists below the surface of the earth and passes through the subsoil.

grout A viscous mixture of portland cement, water, and aggregate used to fill cavities in concrete. Also refers to a specially formulated mortar used to fill under the baseplates of steel columns and in connections in precast concrete.

gypsum Hydrous calcium sulfate.

gypsum backerboard A gypsum panel used as the base on which to bond tile or gypsum wallboard.

gypsum board A gypsum panel used for interior wall and ceiling surfaces. It contains a gypsum core and surfaces covered with paper.

gypsum lath A panel having a gypsum core and a paper covering providing a bonding surface for plaster.

gypsum plaster Ground gypsum that has been calcined and mixed with additives to control setting time and working qualities.

gypsum sheathing A gypsum panel with a water-repellent core. Used for sheathing exterior walls.

hardboard A general term used to describe a panel made from interfelted ligno-cellulosic fibers consolidated under heat and pressure.

hardness A measure of the ability of a material to resist indention or surface scratching.

hardwood A botanical group of trees that have broad leaves that are shed in the winter. (It does *not* refer to the hardness of the wood.)

hardwood plywood Plywood with various species of hardwoods used on the outer veneers.

haunch A projection used to support a member, such as a beam.

header joist A structural member fastened between two parallel full-length framing members to support cut off members at openings.

heartwood The wood extending from the pith to the sapwood.

heat exchanger A device to transfer heat between two physically separated fluids.

heating value The amount of heat produced by the complete combustion of a unit quantity of fuel.

heat loss The energy needed to warm outside air leaking into a building through cracks around doors, windows, and other places.

heat pump A heating/refrigerating system in which heat is taken from a heat source, such as the air, and given up to the space to be heated. For cooling it takes heat from the air in the space and gives it up outdoors.

heat-strengthened glass Glass that has been strengthened by heat treatment.

heat-treatable alloys Aluminum alloys whose strength characteristics can be improved by heat treating.

heat treating Heating and cooling a solid metal to produce changes in physical and mechanical properties.

heavy timber construction A type of wood-frame construction using heavy timbers for the columns, beams, joists, and rafters.

hiding power The ability of a paint to hide the previous color or substrate.

hinge joint A joint that permits some action similar to a hinge and in which there is no appreciable separation of the joining members.

hip roof A roof consisting of four sloping planes that intersect forming a pyramidal shape.

hoistway, elevator A fire-resistant vertical shaft in which the elevator moves.

hollow clay masonry A unit whose core area is 25 to 40 percent of the gross cross-sectional area of the unit.

hollow concrete masonry Concrete masonry units that have open cores.

hollow-core door A door with face veneers on the outer surfaces, wood spacers around the edges, and a hollow interior supported with a honeycomb grid.

hollow-core slab A precast concrete structural slab that uses internal cavities to reduce its weight.

horizontal shear The tendency of the top wood fibers to move horizontally in relationship to the bottom fibers.

hot melt Adhesives that bond when they are heated to a liquid form.

humidifier A device used to add moisture to the air.

humidify Add water vapor to the air.

humidity The amount of water vapor within a given space.

hydrate The capacity of lime to soak up water several times its weight.

hydrated lime Calcium hydroxide made by burning calcium carbonate, which forms calcium oxide that can then chemically combine with water.

hydration A chemical reaction between water and cement that produces heat and causes the cement to cure or harden.

hydraulic elevator An elevator having the car mounted on top of a hydraulic piston that is moved by the action of hydraulic oil under pressure.

hydraulic mortar A mortar that is capable of setting and hardening under water.

hydronic heating system A system that circulates hot water through a system of pipes and convectors to heat a building.

hydronics The science of cooling and heating water.

hydrostatic pressure The pressure equivalent to that exerted on a surface by a column of water of a specified height.

hydroxide of lime The product produced by the chemical reaction during the slaking or hydrating of lime.

hygrometer An instrument used to measure humidity conditions of the air.

hygrometric expansion The expansion and contraction of materials in relation to their moisture content.

hygroscopic The ability to readily absorb and retain moisture from the air.

Hz The abbreviation for hertz, the unit of measurement of the frequency of electric current. It represents the number of cycles per second.

I joist A wood joist made of an assembly of laminated veneer wood top and bottom flanges and a web of plywood or oriented strandboard.

igneous rock Rock formed by the solidification of molten material to a solid state.

illuminance The density of luminous power in lumens per a specified area.

impact insulation class An index of the extent to which a floor assembly transmits impact noise from a room above to the room below.

impact noise Sound generated by impact on the floor or other parts of the building that is carried through the building.

impact strength The energy required to fracture a specimen when struck with a rapidly applied load.

impact test A test for determining the resistance of a specimen fracture from a high-velocity blow.

in situ Undisturbed soil.

incandescence The emission of visible light produced by heating.

independent footings Footings supporting a single structural element, such as a column.

ingot A mass of molten metal cast in a mold and solidified to be stored until used for forging or rolling into a finished product.

insulating glass A glazing unit used to reduce the transfer of heat through a glazed opening by leaving an air space between layers of glass.

insulation, electric A material that is a poor conductor of electricity.

interceptor A trapping device designed to collect materials that will not be able to be handled by a sewage treatment plant, such as grease, glass or metal chips, and hair.

intumescence The swelling of a fire-retardant coating when heated, which forms a low-density film that provides some resistance to the spread of flame on the surface.

iron A metallic element existing in the crust of the earth from which ferrous alloys, such as cast iron, are made.

jamb The vertical member forming the side of a door or window frame.

joint compound A plastic gypsum mixture used to cover the joints and fasteners in gypsum wallboard installations.

jointing Forming control joints in a concrete slab.

joist A horizontal structural member used to carry the floor and ceiling loads.

joule A meter-kilogram-second unit of work or energy.

jute A coarse fiber obtained from two East Indian tiliaceous plants.

Keene's cement A hard, high-strength, white, quick-setting finishing plaster made from burnt gypsum and alum.

kiln (1) A chamber with controlled humidity, temperature, and airflow in which lumber is dried. (2) A low-pressure steam room in which green concrete units are cured.

kiln dried Wood products dried in a kiln.

kinetic energy The energy of a body with respect to the motion of the body.

knitted construction Carpet formed by looping pile yarn, stitching, and backing together.

lacquer A fast-drying clear or pigmented coating that dries by solvent evaporation.

laminate A material made by bonding several layers of material.

laminate, wood A product made by bonding layers of wood or other material to a wood substrate.

laminated glass Glass panels that have outer layers of glass laminated to an inner layer of transparent plastic.

laminated veneer lumber A structural lumber manufactured from wood veneers so that the grain of all veneers runs parallel to the axis of the member.

lamp A general term used to describe the source of artificial light. Often called a *bulb* or *tube*.

landing zone, elevator The area 18 in. (5490 mm) above or below the landing floor.

latent heat Heat involved with the action of changing the state of a substance, such as changing water to steam.

lateral force A force acting generally in a horizontal direction, such as wind against an exterior wall or soil pressure against a foundation wall.

lateral loads Loads moving in a horizontal direction, such as the wind.

latex A water-based coating, such as styrene, butadiene, acrylic, and polyvinyl acetate.

lath The base material for the application of plaster.

leveling plate A steel plate set in grout on top of a concrete foundation to create a level bearing surface for the base of a steel column.

lift-slab construction A method of building site-cast concrete buildings by casting all the floor and roof slabs in a stack on the ground and lifting them up the columns with a series of jacks and welding them in place.

light A pane of glass.

light-gauge steel structural members Load-bearing members formed from light-gauge steel rolled into structural shapes.

lighting fixture See *luminaire.*

lighting outlet An electrical outlet to which a light fixture is connected.

light meter See *luminance meter.*

lightweight steel framing Structural steel framing members made from cold-rolled lightweight sheet steel.

lignin An amorphous substance that penetrates and surrounds the cellulose strands in wood, binding them together.

lime A white to gray powder produced by burning limestone, marble, coral, or shells.

limestone A sedimentary rock consisting of calcium and magnesium.

lintel A beam spanning an opening in a wall.

liquid limit Related to soils, the water content expressed as a percentage of dry weight at which the soil will start to flow when tested by the shaking method.

live load Nonpermanent moving or movable external loads on a structure, such as furniture or snow.

load-bearing Carrying an imposed load.

loomed construction Carpet formed by bonding the pile yarn to a rubber cushion.

louver A unit composed of sloping vanes used to restrict the entry of rain into openings in exterior walls yet permit the flow of air through the opening.

low-E glass Low emissivity glass that has a thin metallic coating that selectively reflects ultraviolet and infrared wavelengths of the energy spectrum.

low-emissivity coating A surface coating used on glass that permits the passage of most shortwave electromagnetic radiation (light and heat) but reflects longer-wave radiation (heat).

low-slope roofs Roofs that are nearly flat.

lumber A product produced by harvesting, sawing, drying, and processing wood.

lumber, boards Lumber nominally less than 2 in. thick and 2 in. or more wide.

lumber, dimension Lumber cut and dressed to standard sizes.

lumber, dressed size The size of lumber after it has been cut to size and the surfaces planed.

lumber, machine stress-rated Lumber that has been mechanically tested to determine its stiffness and bending strength.

lumber, matched Lumber that is edge dressed to make close tongue-and-groove edge joints.

lumber, nominal size The size of lumber after it has been sawn to size but has not been surfaced.

lumber, patterned Lumber that is shaped to a pattern or to a molded form.

lumber, rough Lumber that has not been surfaced but may be sawn, edged, and trimmed.

lumber, shiplapped Lumber that is edge dressed to make a lapped joint.

lumber, shop and factory Lumber intended to be cut up and used in some manufacturing process.

lumber, structural Lumber that is intended for use where allowable strength or stiffness of the piece is known.

lumber grader A person who inspects each piece of lumber and assigns a grade to it.

lumen A unit for measuring the flow of light energy. See *luminous flux*.

luminaire A complete lighting unit consisting of one or more lamps plus elements needed to distribute light, hold and protect the lamps, and connect power to the lamps. Also called a *lighting fixture*.

luminance The luminous intensity of a surface of a given area viewed from a given direction.

luminance meter A photoelectric instrument used to measure luminance. Also called a *light meter*.

luminescence The emission of light not directly caused by incandescence.

luminous flux The rate of flow of light energy through a surface, expressed in lumens.

luminous intensity The force that generates visible light expressed by candela, lumens per steradian, or candlepower.

luminous transmittance A measure of the capacity of a material to transmit incident light in relation to the total incident light striking it.

lux A unit of illumination equal to 1 lumen per square meter.

makeup air Air brought into the building from the outside to replace air that has been exhausted.

malleability The characteristic of a material that allows plastic deformation in compression without rupture.

malleability The property of a metal that permits it to be formed mechanically, such as by rolling or forging, without fracturing.

marble A metamorphic rock formed largely of calcite, dolomite, or dense limestone.

marker board A surface that can be written on with a water-based or semipermanent ink that can be removed.

masonry cement A hydraulic cement used in mortars to increase plasticity and water retention.

MasterFormat The trademarked title of a uniform system for indexing construction specifications published by the Construction Specifications Institute and Construction Specifications Canada.

mastic Trowelable bituminous adhesive compound.

mastic A doughlike compound available in many different formulations designed for use as sealants and adhesives.

mat foundation A large, single concrete footing equal in area to the area covered by the footprint of the building.

mat foundation A large reinforced concrete slab on the earth that covers the entire area of the building foundation.

mechanical action The bonding of materials by adhesives that enter the pores and harden, forming a mechanical link.

mechanical properties Properties exhibited by a material's reaction to applied forces, such as tensile strength and compressive strength.

melamine A white crystalline made from calcium cyanamide.

melting temperature The temperature at which a material turns from a solid to a liquid.

membrane A continuous, unbroken roof covering.

metal lath Perforated sheets of thin metal secured to studs that serve as the base for a finished plaster wall.

metamorphic rock Rock formed by the action of pressure and/or heat on sedimentary soil or rock.

meter, electric A device measuring and recording the amount of electricity passing through it in kilowatt-hours.

mildewcide An agent that helps prevent the growth of mold and mildew on painted surfaces.

mild steel Steel containing less than 0.3 percent carbon.

millwork Wood interior finish items manufactured in a factory, such as doors, windows, and cabinets.

modified bitumens A roofing membrane composed of a polyester or fiberglass mat saturated with a polymer-modified asphalt.

modular size A dimension that conforms to a given module, such as the 48 in. width of plywood panels.

module A repetitive dimension or other unit.

modulus of elasticity The property of a material that indicates its resistance to bending. It is the ratio of unit stress to unit strain.

modulus of rupture A measure of the ultimate load-carrying capacity of a structural member.

moisture barrier A membrane used to block the passage of water, and water vapor through an assembly of materials, such as a wall.

moisture content The amount of water contained in wood, expressed as a percentage of the weight of the wet wood to the weight of an oven-dry sample.

molding A wood strip material that has curved and shaped surfaces. It is used for decorative purposes.

moment A force that acts at a distance from a point and that tends to cause the body to rotate about that point.

moment of inertia The sum of the products of the mass and the square of the perpendicular distance to the axis of rotation of each particle in a body rotating about an axis.

monolithic concrete Concrete cast with no joints except construction joints; a continuous pour.

monomer An organic molecule that can be converted into a polymer by chemical reaction with similar molecules or organic molecules.

mortar A plastic mixture of cementitious materials, water, and a fine aggregate.

mortar flow A measure of the consistency of freshly mixed mortar related to the diameter of a molded truncated cone specimen after the sample has been vibrated a specified number of times.

mosaic A decoration made up of small pieces of inlaid stone, glass, or tile.

motor control A device that governs the electrical power delivered to one or more electric motors.

motor control center Controllers used to start and stop electric motors and protect them from overloads.

moving ramp A conveyor belt system used to move people or packages up or down an incline.

moving walk A conveyor belt system operating at floor level used to move people in a horizontal direction.

mullion Horizontal or vertical members between adjacent window or door units.

muntin Small horizontal and vertical bars between small lights of glass in windows and doors.

nail popping The loosening of nails holding gypsum board to a wall or ceiling. It produces a bulge in the surface of the gypsum panel.

natural fibers Fibers found in nature, such as wool and cotton.

neat plaster A gypsum plaster with no aggregates or fillers added. Sometimes called unfibered gypsum.

needle beam A steel or wood beam that is run through an opening in a bearing wall and used to support the wall and related loads as work on the foundation below the wall is performed.

noise reduction coefficient A single number indicated by the amount of airborne sound energy absorbed into a material.

nonbearing Refers to a structural part that does not carry a load.

noncalcareous clays Clays containing silicate of alumina, feldspar, and iron oxide.

nondestructive testing Methods of testing an item that do not destroy the item being tested.

nonferrous Metallic materials in which iron is not a principal element.

non-heat-treatable alloys Alloys that do not increase in strength when they are heat treated but that do gain strength by the addition of alloying elements.

nonprestressed units Concrete structural members in which the reinforcing steel is not subject to prestressing or posttensioning.

nylon A synthetic plastic made from coal, tar, and water.

oakum A caulking material made from hemp fibers treated with tar.

oil-based paint Paint composed of resins requiring solvent for reduction purposes.

opaque coatings Coatings that completely obscure the color and much of the texture of the substrate.

open-web joist A prefabricated steel truss made of welded members, used for floor and roof construction.

organic material A class of compounds comprising only those existing in plants and animals.

oriented strand board A panel made from wood strands that have their strand face oriented in the long direction of the panel.

outlet box A box that is part of the electrical wiring system that contains one or more receptacles.

overlaid plywood Plywood panels whose exterior surfaces are covered with a resin-impregnated fiber ply.

overload The operation of electrical equipment in excess of the normal full-loaded electrical rating or of a conductor carrying current in excess of its rated capacity.

oxidation A reaction between a material and oxygen in the atmosphere.

oxide layer In aluminum, a very thin protective layer formed naturally on aluminum due to its reaction to oxygen.

oxidize To convert an element into its oxide, such as rusting steel.

pan A metal form used to form the cavities between joists in cast-in-place concrete floors and roofs.

panelboard A panel that includes fuses or circuit breakers used to protect the circuits in a building from overloads.

panelized construction Construction that uses preassembled panels for walls, floors, and roof.

parallel strand lumber Lumber made from lengths of wood veneer bonded to produce a solid member.

parapet The top of an exterior wall that extends above the line of the roof.

parging The application of a portland cement plaster on masonry and concrete walls to make them less permeable to water.

particleboard A sheet product manufactured from wood particles and a synthetic resin or other binder.

partition A non-load-bearing interior wall.

party wall A wall that is common to two buildings and is on the boundary between them.

passive solar systems Solar systems that use natural means to store solar energy.

paver A thin brick used as the finished floor covering.

perlite A lightweight material made from volcanic rock.

perm The unit of vapor permeability.

phenol A class of acid organic compounds used in the manufacture of various resins, plastics, and wood preservatives.

phenolic A synthetic resin made by the reaction of a phenol with an aldehyde.

photovoltaic cells Thin, flat semiconductors that convert light energy into direct-current electricity.

physical properties The properties associated with the physical characteristics of a material, such as thermal expansion and density.

pier A column designed to support a load.

pig An ingot of cast iron.

pig iron A high-carbon-content iron produced by the blast furnace and used to produce cast iron and steel.

pigments Paint ingredients mainly used to provide color and hiding power.

pilaster A vertical projection from a masonry or concrete wall providing increased stiffening.

pile A wood, steel, or concrete column usually driven into the soil to be used to carry a vertical load.

pile cap A concrete slab or beam that covers the head of several piles, tying them together.

pile hammer A machine for delivering blows to the top of a pile, driving it into the earth.

pit, elevator The part of the hoistway that extends below the floor of the lowest landing to the floor at the bottom of the hoistway.

pitch The slope of a roof or other plane surface.

pitch Related to carpets, the number of tufts in a 27 in. width of carpet.

plaster A cementitious material, usually a mixture of portland cement, lime or gypsum, sand, and water. Used to finish interior walls and ceilings.

plaster base Any material suitable for the application of plaster.

plaster of Paris A calcined gypsum mixed with water to form a thick, pastelike mixture.

plastic An organic material that is solid in its finished state but is capable of being molded or of receiving form.

plastic behavior The ability of a material to become soft and formed into desired shapes.

plastic deformation The deformation of a material beyond the point at which it will recover its original shape.

plastic limit Related to soils, the percent moisture content at which the soil begins to crumble when it is rolled into a thread $1/8$ in. (3 mm) in diameter.

plasticity The ability of a material to be deformed into a different shape.

plasticizer Liquid material added to some plastics to reduce their hardness and increase pliability. Also, an additive to concrete and mortar to increase plasticity.

plate glass A high-quality glass sheet that has both surfaces ground flat and carefully polished.

platform frame A wood structural frame for light construction with the studs extending only one floor high upon which the second floor is constructed.

plenum The space above a suspended ceiling.

plenum chamber A chamber in a heating and air conditioning system that has air pressure higher than the surrounding air and that is connected to ducts supplying conditioned air to the rooms.

ply One of a number of layers in a layered construction.

plywood A glued wood panel made up of thin layers of wood veneer with the grain of adjacent layers at right angles to each other or of outer veneers glued to a core of solid wood or reconstituted wood.

plywood, cold-pressed Interior type plywood manufactured in a press without external applications of heat.

plywood, exterior Plywood bonded with a type of adhesive that is highly resistant to moisture and heat.

plywood, interior Plywood manufactured for indoor use or in locations in which it would be subject to moisture for only a brief time.

plywood, marine Plywood panels with the same glue as exterior plywood but with more restrictive veneer specifications.

plywood, molded Plywood that is glued to the desired shape either between curved forms or by fluid pressure applied with flexible bags or blankets.

plywood, postformed Panels formed when flat plywood sheets are reshaped into a curved configuration by steaming or the use of plasticizing agents.

plywood stressed-skin panel A structural panel constructed with outer skins of plywood applied over an internal frame of wood members forming a rigid panel.

pole construction Construction using large-diameter log poles in a vertical position to carry the loads of the floors and roof.

polycarbonate A polyester made by linking certain phenols through carbonate groups.

polyester A linear polymer made by linear linking of oxybenzoyl units.

polyethylene A thermoplastic resin made by polymerizing ethylene.

polyimide A polymer based on the combination of certain anhydrides with aromatic diamines.

polymer A chemical compound formed by the union of simple molecules to form more complex molecules.

polymerization A chemical reaction in which molecules of a monomer are linked together to form large molecules whose molecular weight is a multiple of that of the original substance.

polyolefin A polymer composed of open-chain hydrocarbons having double bonds.

polypropylene A polymer produced by the linking of repeated propylene monomers.

polystyrene A clear, colorless plastic resin made by polymerizing styrene.

polyurethane A thermoplastic or thermosetting resin derived by condensation reaction of a polyisocyanate and a hydroxyle.

polyvinyl chloride A thermoplastic resin derived by the polymerization of vinyl and acetate.

ponding The collection of water in shallow pools on the top surface of the roof.

porcelain A strong vitreous material bonded to metal at a high temperature.

porcelain enamel An inorganic metal oxide coating bonded to metal by fusion at a high temperature.

portland cement A cementitious binder used in concrete, mortar, and stucco, obtained from pulverizing clinker consisting of hydraulic calcium silicates.

post, plank, and beam framing A wood framing system using beams for horizontal structural members that rest on posts, forming the vertical members.

posttensioning A method used to place concrete under tension in which steel tenons are tensioned after the concrete has been poured and hardened.

potable water Water that is safe to drink and meets standards of the local health authority.

power The rate at which work is performed, expressed in watts or horsepower.

pozzolan A siliceous or siliceous and aluminus material blended with portland cement that chemically reacts with calcium hydroxide to form compounds possessing cementitious properties.

pozzolan cement A cement made from volcanic rock that contains considerable silica.

precast concrete Concrete cast in a form and cured before it is lifted into its intended position.

pressure-treated lumber Lumber that has chemicals forced into it under pressure to retard decay and provide resistance to fire.

prestressed concrete Concrete that has been pretensioned or posttensioned.

pretensioning A method used to place a concrete member under tension by pouring concrete over steel tendons that are under tension before the concrete is poured.

primer A base coat in a paint system. It is applied before the finish coats.

proportional limit The upper limit at which stress is proportional to strain.

purlin A horizontal structural member that spans beams, frames, or trusses and supports the roof deck or rafters or a joist supporting a roof deck.

quarry An open excavation in the earth from which building stone is removed.

quarry tile A large clay tile used for finished flooring.

quartersawing Sawing lumber so the hard annual rings are nearly perpendicular to the surface.

R-value The numerical value used to indicate the resistance to the flow of heat.

raceway An enclosed channel designed to carry wires and cables.

racking The distortion of a rectangular frame when subject to shear forces.

radiant heat Heat transferred by radiation.

radiation The transfer of heat through space by means of electromagnetic waves.

rafter Member of a roof structural frame that supports the sheathing and other roof loads.

rainscreen principle The principle that states that wall cladding can be made watertight by placing wind-pressurized air chambers behind the joints, which reduces the air pressure differentials between the inside and outside that could cause water to move through the joints.

rake The board along the sloping edge of a gable.

ready-mix concrete Concrete mixed in a central plant and delivered to the site by truck.

rebar Steel bar used to reinforce concrete.

receptacle A device installed in an electrical outlet box to receive a plug to supply electric current to portable equipment.

receptacle outlet An outlet box in which one or more receptacles are installed.

reduction (1) A process in which iron is separated from oxygen with which it is chemically mixed by smelting the ore in a blast furnace. (2) In regard to aluminum, the electrolytic process used to separate molten aluminum from the alumina.

reflectance The ratio of the intensity of light reflected by a material to the intensity of the incident light.

reflectance coefficient A measure stated as a percentage of the amount of light reflected off a surface.

reflective coated glass Glass having a thin layer of a metal or metal oxide deposited on the surface to reflect heat and light.

reflectivity The relative ability of a surface to reflect sound or light.

refractive index The ratio of the speed of light in a material to the speed of light in a vacuum.

refractory Nonmetallic ceramic material used where high temperatures (above 2700°F) are present, such as in furnace linings.

refrigerant The medium of heat transfer that absorbs heat by evaporating at low temperatures and pressure and giving up heat when it condenses at a higher temperature and pressure.

reinforced brick masonry Brick masonry construction that has steel reinforcing bars inserted to provide tensile strength.

relative humidity The ratio of the amount of water vapor present in the air to that which the air would hold if saturated at the same temperature.

remote-control circuit An electric circuit that controls any other circuit using a relay or other such device.

resilience The property of a material whereby it gives up its stored energy when the deforming force is removed.

resiliency The ability to regain the initial shape after being deformed.

resilient flooring Finished flooring made from a resilient material, such as polyvinyl chloride or rubber.

resin A natural or synthetic material that is the main ingredient of paint and binds the ingredients together.

resistance, electric The physical property of a conductor or electric-consuming device to resist the flow of electricity, reducing power and generating heat.

retaining wall A wall that bears against soil or other material and resists lateral and other forces from the material being held in place.

retarder An admixture used to slow the setting of concrete.

return air Air removed from a space and vented or reconditioned by a furnace, air conditioner, or other apparatus.

rigid frame A structural framework in which the beams and columns are rigidly connected (no hinged joints).

riser, plumbing A water supply running vertically through the building to supply water to the various branches and fixtures on each floor.

rivet A metal fastener used to hold metal plates by passing through holes in each and having a head formed on the protruding end.

rock A solid mineral material found naturally in large masses.

rock anchor A posttensioned cable or steel rod inserted into a hole drilled in rock and grouted in place.

romex A nonmetallic-sheathed electric cable of type NM or NMC.

roof drain A drain on the roof to carry away water to a downspout.

roof planks Precast gypsum concrete members used for decking roofs.

room cavity ratio A relationship between the height of a room, its perimeter, and the height of the work surface above the floor divided by the floor area.

rows Related to carpets, the number of tufts per inch.

saddle A ridge in a roof deck that divides two sloping parts, diverting water toward roof drains.

safing Fire-stopping material placed in spaces between the floor and curtain walls in high-rise construction.

sand Fine rock particles from 0.002 in. (0.05 mm) in diameter to less than 0.25 in. (6.4 mm) in diameter.

sandstone A sedimentary rock formed from sand.

sanitary sewage Waste material containing human excrement and other liquid wastes.

sanitary sewer A sewer to receive sanitary sewage without the infusion of other water such as rain, surface water, or other clear water drainage.

sapwood The wood near the outside of the log just under the bark.

scratch coat The first coat of gypsum plaster that is applied to the lath.

screed A tool used to strike off the surface of freshly poured concrete so it is flush with the top of the form.

screeding The process of striking off the surface of freshly poured concrete with a screed so it is flush with the top of the form.

sculpture A three-dimensional work of art.

scupper An outlet in a parapet wall for the drainage of overflow water from the roof to the outside of the building.

scuttle An opening through the ceiling and roof to provide access to the roof. It is covered with a waterproof cover. Also referred to as a *roof hatch*.

sealant A mastic used to seal joints and seams.

sealer A material used to seal the surface of a material against moisture.

seasoning Removing moisture from green wood.

seated connections Connections that join structural steel members with metal connectors, such as an angle, upon which one member, such as a beam, rests.

security glass Glass panels assembled with multiple layers of glass and plastic to produce a panel that will resist impact.

sedimentary rock Rock formed from the deposit of sedimentary materials on the bottom of a body of water or on the surface of the earth.

segregation The tendency of large aggregate to separate from the sand-cement mortar in the concrete mix.

seismic area A geographic area where earthquake activity may occur.

seismic load Forces produced on a structural mass by the movements caused by an earthquake.

semitransparent Coatings that allow some of the texture and color of the substrate to show through.

sensible heat Heat that causes a detectable change in temperature.

septic tank A watertight tank into which sewage is run and where it remains for a period of time to permit hydrolysis and gasification of the contents, which then flow from the tank and are absorbed into the soil.

service entrance The point at which power is supplied to a building and where electrical service equipment, such as the service switch, meter, overcurrent devices, and raceways, are located.

service equipment The equipment needed to control and cut off the power supply to the building, such as switches and circuit breakers.

service temperature The maximum temperature at which a plastic can be used without altering its properties.

shaft wall A fire-resistant wall that protects the elevator, stairwell, and vertical mechanical chases in high-rise construction.

shales Clays that have been subjected to high pressures, causing them to become relatively hard.

shear A deformation in which planes of a material slide so as to remain parallel.

shear panel A floor, wall, or roof designed to serve as a deep beam to assist in stabilizing a building against deformation by lateral forces.

shear plate connector A circular metal connector recessed into a wood member that is to be bolted to a steel member.

shear stress The result of forces acting parallel to an area but in opposite directions, causing one portion of the material to "slide" past another.

shear studs Metal studs welded to a steel frame that protrude up into the cast-in-place concrete deck.

sheathing A covering placed over exterior studs or rafters that serves as a base below the exterior cladding.

sheave A pulley over which the elevator wire hoisting rope runs.

sheeting Wood, metal, or concrete members used to hold up the face of an excavation.

shop drawings Related to steel construction, working drawings giving the information needed to fabricate structural steel members.

shoring Bracing used to temporarily hold a wall in position.

shrinkage limit Related to soils, the water content at which the soil volume is at its minimum.

siamese connection A connection outside a building to which firefighters connect an alternate water supply to boost the water used by the fire suppression system.

sill The wood member anchored to the top of a foundation to receive joists; the horizontal bottom part of a door or window frame.

silt Fine sand with particles smaller than 0.002 in. (0.05 mm) and larger than 0.00008 in. (0.002 mm).

single-ply roofing A roofing membrane composed of a sheet of waterproof material secured to the roof deck.

sintering A process that fuses iron ore dust with coke and fluxes into a clinker.

site-cast concrete Concrete poured and cured in its final position.

site investigation An investigation and testing of the surface and subsoil of the site to record information needed to design the foundation and the structure.

site plan A drawing of a construction site, showing the location of the building, contours of the land, and other features.

skylight A roof opening that is covered with a watertight transparent cover.

slab-on-grade A concrete slab poured and hardened directly on the surface of the earth.

slag A molten mass composed of fluxes and impurities removed from iron ore in the furnace.

slake The process of adding water to quicklime, hydrating it and forming lime putty.

slip form A form designed to move upward as concrete is poured in it.

slump A measure of the consistency of freshly mixed concrete, mortar, or stucco.

slump test A test to ascertain the slump of concrete samples.

slurry A liquid mixture of water, bentonite clay, or portland cement.

slurry wall A wall built of a slurry used to hold up the sides of an area to be excavated.

smelting A process in which iron ore is heated, separating the iron from the impurities.

smoke barriers Continuous membranes used to resist the passage of smoke.

smoke developed rating A relative numerical classification of the fumes developed by a burning material.

soffit The undersurface of an eave.

soft-mud process A process used to make bricks when the clay contains moisture in excess of 15 percent.

softwood A botanical group of trees that have needles and are evergreen.

soil anchors Metal shafts grouted into holes drilled into the sides of an excavation to stabilize it.

soil stack A vertical plumbing pipe into which waste flows through waste pipes from each fixture.

soil vent That portion of a soil stack above the highest fixture waste connection to it.

solar energy Radiant energy originating from the sun.

solar screen A device used to divert solar energy from windows.

solid clay masonry A unit whose core does not exceed 15 percent of the gross cross-sectional area of the unit.

solvents Liquids used in paint and other finishing materials that give the coating workability and that evaporate, permitting the finish material to harden.

sound A sensation produced by the stimulation of the organs of hearing by vibrations transmitted through the air. It is the movement of air molecules in a wavelike motion.

sound transmission class A single number rating of the resistance of a construction to the passage of airborne sound.

span The distance between two supporting members.

span rating Number indicating the distance a sheet of plywood or other material can span between supports.

spandrel beam A beam running from column to column in outside walls that carries the curtain wall above it.

special units Concrete masonry units that are designed and made for a special use.

specific adhesion Bonding dense materials using the attraction of unlike electrical charges.

specific gravity The ratio of the weight of one cubic foot of a material to the weight of one cubic foot of water.

specifications A written document in which the scope of the work, materials to be used, installation procedures, and quality of workmanship are detailed.

spire A tall pyramidal roof built upon a tower or steeple.

split ring connector A ring-shaped metal insert placed in circular recesses cut in joining wood members that are held together with a bolt or lag screw.

spoil bank An area where soil from the excavation is stored.

spread foundation A foundation that distributes the load over a large area.

spreading rate The area over which a paint can be spread expressed in square feet per gallon.

square In roofing, 100 square feet of roofing material.

stabilizers Additives used to stabilize plastic by helping it resist heat, loss of strength, and the effect of radiation on the bonds between the molecular chains.

stain A solution of coloring matter in a vehicle used to enhance the grain of wood during the finishing operation.

stained-glass window A window made of small colored glass pieces joined with leaded cames.

standing seam The vertical seam formed when two sheets of metal roofing are joined.

static loads Any load that does not change in magnitude or position with time.

steam Water in a vapor state.

steam separator A device used to remove moisture from steam after it flows from the boiler.

steam trap A device used to allow the passage of condensate while preventing the passage of steam.

steel A malleable alloy of iron and a small carefully controlled carbon content.

steeple A towerlike ornamental construction, usually square or hexagonal, placed on the roof of a building and topped with a spire.

steep-slope roofs Roofs with sufficient slope to permit rapid runoff of rain.

stiff-mud process A process used to make bricks from clay that has 12 to 15 percent moisture.

stiffness Resistance to bending or flexing.

stone Rock selected or processed by shaping to size for building or other use.

strain The deformation of a body or material when it is subjected to an external force.

strain hardening Increasing the strength of a metal by cold-rolling.

strand-casting machine A machine that casts molten steel into a continuous strand of metal that hardens and is cut into required lengths.

stress Force applied to a structural member, assembly of materials, or a component per unit of its area.

stressed-skin panels An assembly with high-strength facing panels separated by wood spacing strips and bonded firmly to them.

stress-rated lumber Lumber that has its modulus of elasticity determined by actual tests.

structural sandwich construction A wood construction consisting of a high-strength facing material bonded to and acting integrally with a low-density core material.

stucco A portland cement plaster used as the finish material on building exteriors.

stud The vertical framing member in frame wall construction.

subfloor The structural floor joined to the joists that supports the finish flooring.

substrate A subsurface to which another material is bonded.

suction The rate at which clay masonry units absorb moisture.

suction rate The weight of water absorbed when a brick is partially immersed in water for one minute expressed in grams per minute or ounces per minute.

superplasticizer An admixture used with concrete to make the wet concrete very fluid without adding additional water.

supply air Conditioned air entering a space from an air conditioning, heating, or ventilating unit.

surface burning rating The rating of interior and surface finish material providing indexes for flame spread and smoke developed.

surface clays Clays obtained by open-pit mining.

surface effect The effect caused by the entrainment of secondary air against or parallel to a wall or ceiling when an outlet discharges air against or parallel to the wall or ceiling.

suspended ceiling A finish ceiling hung from the overhead by a series of wires.

switch, bypass isolation A manually operated switch used in connection with a transfer switch to provide a means of directly connecting load conductors to a power source and for disconnecting the transfer switch.

switch, general-use A switch used for general electrical control.

switch, general-use snap A general-use switch built so it can be installed in a device box or on the box cover.

switch, isolating A switch used to isolate an electric circuit from the source of power.

switch, motor-control A switch used to interrupt the maximum operating overload current of a motor.

switch, transfer A device used to transfer one or more conductors from one power source to another.

switchboard A large single panel or an assembly of panels with switches, overcurrent protection devices, and buses that are mounted on the face or both sides of the panel.

switches Devices to open and close an electric circuit or to change the connection within a circuit.

synthetic fibers Fibers formed by chemical reactions.

synthetic materials Materials formed by the artificial building up of simple compounds.

T-sill A type of sill construction used in balloon framing in which the header joist is placed inside the studs and is butted by the floor joists.

tapestry A fabric upon which colored threads are woven by hand to produce a design.

tee Precast concrete or metal structural members in the shape of the letter T.

temper designation A specification of the temper or metallurgical condition of an aluminum alloy.

tempered glass Heat-treated glass that has great resistance to breakage and increased toughness.

tempering The reheating of hardened steel to decrease hardness and increase toughness.

tempering glass A process used to strengthen glass by raising the temperature of the glass to near the softening point and then blowing jets of cold air on both sides suddenly to chill it and create surface tension in the glass.

tendon A steel bar or cable in prestressed concrete used to impart stress in the concrete.

tensile bond strength The ability of a mortar to resist forces tending to pull the masonry apart.

tensile strength The maximum tensile stress that a material can sustain at the point of failure.

tensile stress The stress per unit area of the cross section of a material that resists elongation.

tension The condition of being pulled or stretched.

terra-cotta A hard unglazed clay tile used for ornamental work.

terrazzo A finish-floor material made up of concrete and an aggregate of marble chips which after curing is ground smooth and polished.

texture plaster A finish plaster used to produce rough, textured finished surfaces.

thermal break Material with a low thermal conductivity that is inserted between materials, such as metal with a high thermal conductivity, to retard the passage of cold or heat through the highly conductive material.

thermal bridge A thermal conducting material that conducts heat through an insulated assembly of materials.

thermal conductance (C) Thermal conductance is the same as thermal conductivity except it is based on a specified thickness of a material rather than on one inch as used for conductivity.

$$C = \frac{k}{thickness\ of\ the\ material}$$

thermal conduction The process of heat transfer through a solid by transmitting kinetic energy from one molecule to the next.

thermal conductivity (k) The rate of heat flow through one square foot of material one inch thick expressed in Btu per hour when a temperature difference of one degree Fahrenheit is maintained between the two surfaces.

k = Btu/hr./ft.^{2}/°F

thermal convection Heat transmission by the circulation of a liquid or a heated air or gas.

thermal insulation A material that has a high resistance to heat flow.

thermal properties The behavior of a material when subjected to a change in temperature.

thermal radiation The transmission of heat from a hot surface to a cool one by means of electromagnetic waves.

thermal resistance (R) An index of the rate of heat flow through a material or assembly of materials. It is the reciprocal of thermal conductance (C).

$$R = \frac{1}{C}$$

thermoforming A process in which heated plastic sheets are made to assume the contour of a mold by using the force of air pressure, vacuum, or mechanical stretching.

thermoplastics Plastics that soften by heating and reharden when cooled without changing the chemical composition.

thermosetting plastics Cured plastics that are chemically cross-linked and when heated will not soften but will be degraded.

thermostat A temperature-sensitive instrument that controls the flow of electricity to units used to heat and cool spaces in a building.

throw The horizontal or vertical distance an airstream travels after leaving the air outlet before it loses velocity.

tieback anchors Steel anchors grouted into holes drilled in the excavation wall to hold the sheeting, thus reducing the number of braces required.

timber Wood structural members having a minimum thickness of 6 in. (140 mm).

timber joinery The joining of structural wood members using wood joints, such as the mortise and tenon.

tolerance The permissible deviation from a given dimension or the acceptable variation in size from the given dimension.

topcoat The final coat of paint.

torque A twisting or rotating action.

torsion strength The maximum stress a material will withstand before fracturing under a twisting force.

torsion test A test used to ascertain the behavior of materials subject to torsion.

toughness A measure of the ability of a material to absorb energy from a blow or shock without fracturing.

transformer An electrical device used to convert an incoming electric current from one voltage to another voltage.

trap Device used to maintain a water seal against sewer gases that back up the waste pipe. Usually each fixture has a trap.

tread The horizontal part of a stair upon which the foot is placed.

troweling Producing a final smooth finish on freshly poured concrete with a steel-bladed tool after the concrete has been floated.

truss An assembly of structural members joined to form a rigid framework, usually connected to form triangles.

truss-framed system An assembled truss unit made up of a floor truss, wall stud, and a roof truss.

truss plate A steel plate used to strengthen the joints in truss assemblies.

tufted construction Carpet formed by stitching the pile yarn through the backing material.

twist Warping in which one or more corners of a piece of wood twist out of the plane of the piece.

two-way concrete joist system Floor and roof construction that has two perpendicular systems of parallel intersecting joists.

two-way flat plate Reinforced concrete construction in which the main reinforcement runs in two directions and both surfaces are flat planes and it is supported by columns.

two-way flat slab Reinforced concrete floor or roof construction in which a two-way flat plate is supported by columns with drop panels or column capitals.

U The overall coefficient of heat transmission; that is, the combined thermal value of all the materials in an assembly of building materials, such as a wall. U is expressed in Btuh per square foot of an area per degree Fahrenheit temperature difference.

U_o The average overall coefficient of heat transmission; that is, the average of the U-value of all the materials that make up the component assembly.

ultimate strength The maximum stress, such as tensile, compressive, or shear, that a material can withstand.

ultimate tensile strength The maximum tensile stress of a material up to the point of rupture.

ultrasonic testing A method of nondestructive testing of materials that uses high frequency sound vibrations to find defects in the material.

under-carpet wiring A flat, insulated electric wire that is run under the carpet.

underlayment Sheet material, such as hardboard, that is laid over the subfloor to provide a smooth, stiff surface for the finish flooring.

underpinning Placing a new foundation below the existing foundation.

unfibered gypsum A neat gypsum. It has no additives.

uniform load Any load that is spread out evenly over a large area.

unreinforced concrete Concrete placed without steel reinforcing bars or welded wire fabric.

valence The points on an atom to which valences of other elements can bond.

valley The intersection of two inclined surfaces.

vapor retarder A material having a high resistance to the passage of vapor through it.

varnish A transparent coating that dries on exposure to air, providing a protective coating.

vehicle The liquid portion of a paint composed mainly of solvents, resins, or oils.

velvet construction Carpet formed by joining the pile, stuffer, and weft yarns with double warp yarns.

veneer A thin sheet of material used to cover another surface.

veneer gypsum base A gypsum board product designed to serve as the base for the application of gypsum veneer plaster.

veneer plaster A thin layer of plaster applied over a special veneer gypsum base sheet.

vent stack That part of the soil stack above the highest vent branch.

vents, plumbing Pipes permitting the waste system to operate under atmospheric pressure. They allow air to enter and leave the system, preventing water in the traps from being siphoned off. If this occurs, sewer gases can enter the building.

vermiculite An insulation material or aggregate made of expanded mica.

vertical load A load acting in a direction perpendicular to the plane of the horizon.

vertical shear The tendency of one part of a member to move vertically in relationship to the adjacent part.

viscosity The resistance of a liquid to flow under an applied load or pressure.

vitrification A process of using high kiln temperatures to fuse the surface grains of clay products so they are impervious to the passage of water.

volatile organic compounds Compounds released to the atmosphere as a coating dries.

volt The unit of potential difference or electromotive force. One volt applied across a resistance of one ohm results in a current flow of one ampere.

voltage The force, pressure, or electromotive force that causes electric current to flow in an electric circuit.

waferboard A mat-formed panel made of wood wafers, randomly arranged and bonded with a waterproof binder.

waffle slab A concrete slab that has ribs running in two directions forming a wafflelike grid.

wainscot A protective or decorative finish wall covering applied to the lower part of an interior wall.

warp A variation in a board from a flat, plane condition.

waste pipe Horizontal plumbing pipes that connect a fixture to the soil pipe.

water-based coatings Coatings formulated with water as the solvent.

water-cement ratio In a concrete or mortar mixture, the ratio of the amount of water (minus that held by the aggregates) to the amount of cement used.

waterproofing Material used to make a surface impervious to the penetration of water.

water repellent Liquid that penetrates the pores of wood and prevents moisture from penetrating without altering the desirable qualities of the wood.

water retention The property of a mortar that prevents the rapid loss of water by absorption into the masonry units.

water-smoking A process used to drive off the remaining water from clay products before they are fired in the kiln.

water stop A rubber or plastic diaphragm placed across a joint in cast concrete to prevent the passage of water through the joint.

water-struck brick Brick made in a mold that was wetted before the clay was placed in the mold.

water table The level below the ground where the soil is saturated with water.

water-vapor permeability The rate of water-vapor transmission through a given area of flat material of a given thickness induced by a given vapor pressure difference between the two surfaces under specified temperature and humidity conditions.

water-vapor transmission rate The steady-state vapor flow in a given time through a given area of a body, normal to specified parallel surfaces, under specific conditions of temperature and humidity at each surface.

watt The unit of measurement of electrical power or rate of work. It is a pressure of one volt flowing at the rate of one ampere.

weatherability The ability of a plastic to resist deterioration due to moisture, ultraviolet light, heat, and chemicals found in the air.

weathered joint A mortar joint finished so the mortar slopes outward, allowing water to shed away from the joint.

weathering Changes in the strength, color, surface, or other properties of a material due to the action of the weather.

weathering index A value that reflects the ability of clay masonry units to resist the effects of weathering.

weathering steel A steel alloy that forms a natural self-protecting rust.

weep hole Small openings at the bottom of exterior cavity walls to allow moisture in the cavity to drain out.

welded-wire fabric A form of steel reinforcing made from wire strands welded where they cross, forming a mesh.

Wilton construction Carpet formed on a loom capable of feeding yarns of various colors.

wind load Any load on a building caused by pressure or suction developed by the wind.

wind uplift Upward forces on a building caused by negative air pressures produced under certain wind conditions.

winning A term used to describe the mining of clay.

wired glass Glass made with a wire grid embedded in it.

wood preservative Substance that is toxic to fungi, insects, borers, and other wood-destroying organisms.

workability (1) Describes the ease or difficulty with which concrete can be placed and worked into its final location. (2) In relation to mortar, the property of freshly mixed mortar which determines the ease and homogeneity with which it can be spread and finished.

working joints Joints in exterior walls that allow for expansion and contraction of materials in the wall.

wracking When a component, such as a wall, is forced out of plumb.

wrought products Products formed by any of the standard manufacturing processes, such as drawing, rolling, forging, or extruding.

yield point The point at which strains increase without a corresponding increase in stress.

yield strength The load at which a limited permanent deformation occurs.

zoning ordinances Local regulations that control the use and development of land.

Index